T0331101

The Navier–Stokes Problem in the 21st Century
Second Edition

The complete resolution of the Navier–Stokes equation—one of the Clay Millennium Prize Problems—remains an important open challenge in partial differential equations (PDEs) research despite substantial studies on turbulence and three-dimensional fluids. *The Navier–Stokes Problem in the 21st Century, Second Edition* continues to provide a self-contained guide to the role of harmonic analysis in the PDEs of fluid mechanics, now revised to include fresh examples, theorems, results, and references that have become relevant since the first edition published in 2016.

Pierre Gilles Lemarié-Rieusset is a professor at the University of Evry Val d'Essonne. Dr. Lemarié-Rieusset has constructed many widely used bases, such as the Meyer-Lemarié wavelet basis and the Battle-Lemarié spline wavelet basis. His current research focuses on the application of harmonic analysis to the study of nonlinear PDEs in fluid mechanics. He is the author or co-author of several books, including *Recent Developments in the Navier-Stokes Problem*.

The Navier–Stokes Problem
in the 21st Century

Second Edition

Pierre Gilles Lemarié-Rieusset
University of Evry Val d'Essonne

CRC Press
Taylor & Francis Group
Boca Raton London New York

CRC Press is an imprint of the
Taylor & Francis Group, an **informa** business

A CHAPMAN & HALL BOOK

Designed cover image: Bibliothèque nationale de France

Second edition published 2024
by CRC Press
6000 Broken Sound Parkway NW, Suite 300, Boca Raton, FL 33487-2742

and by CRC Press
4 Park Square, Milton Park, Abingdon, Oxon, OX14 4RN

CRC Press is an imprint of Taylor & Francis Group, LLC

© 2024 Taylor & Francis Group, LLC

First edition published by CRC Press 2016

ISBN: 978-0-367-48726-3 (hbk)
ISBN: 978-1-032-62373-3 (pbk)
ISBN: 978-1-003-04259-4 (ebk)

DOI: 10.1201/ 9781003042594

Typeset in CMR10 font
by KnowledgeWorks Global Ltd.

Publisher's note: This book has been prepared from camera-ready copy provided by the authors.

Contents

Preface to the First Edition

In a provocative paper published in 1974, Irigaray [240] muses on fluid mechanics with, according to Hayles [230], "elliptical prose and incendiary reasoning." Her thesis has been turned into a kind of "post-modern myth" by the French-Theory bashers Sokal and Bricmont, relayed by Dawkins [147] in his survey in *Nature* of Sokal and Bricmont's book *Fashionable Nonsense* [67].

Dawkins's irony focused on Hayles's 1992 paper where "Katherine Hayles made the mistake of re-expressing Irigaray's thoughts in (comparatively) clear language. For once, we get a reasonably unobstructed look at the emperor and, yes, he has no clothes." According to Hayles, the meaning of Irigaray's paper is the following one:

> *The privileging of solid over fluid mechanics, and indeed the inability of science to deal with turbulent flow at all, she attributes to the association of fluidity with femininity. Whereas men have sex organs that protrude and become rigid, women have openings that leak menstrual blood and vaginal fluids... From this perspective it is no wonder that science has not been able to arrive at a successful model for turbulence. The problem of turbulent flow cannot be solved because the conceptions of fluids (and of women) have been formulated so as necessarily to leave unarticulated remainders.*

And Dawkins adds:

> *It helps to have Sokal and Bricmont on hand to tell us the real reason why turbulent flow is a hard problem: the Navier–Stokes equations are difficult to solve.*

There is at least one point when one should disagree with Dawkins: whom can it help to have Sokal's libel on hand? Irigaray's essay is a seven-page paper in the journal *L'Arc*, in a special issue dedicated to the psychoanalyst Lacan; it is clearly not a treatise on the meaning of fluid mechanics but a variation on Lacanian themes. The stake of the paper is the confrontation between the symbolic order and reality: in Irigaray's views, rationality deals with universality and reality with singularity; she opposes the (masculine) "tout" [every] to the (feminine) "pas-toute" [not every] in light of the theses Lacan developed in the early seventies. Hayles's exposition of Irigaray's theses is purely provocative: she completely omits the psychoanalytic context of enunciation and tries to give the most dramatic presentation to an "outrageous" thesis that would uprise engineers and hydraulicians.

There is at least one point where one may agree with Dawkins: "the Navier–Stokes equations are difficult to solve." In a parody of Churchill's analysis of the Soviet policy in the thirties, Constantin writes [126]:

> *The Reynolds equations are still a riddle. They are based on the Navier–Stokes equations, which are still a mystery. The Navier–Stokes equations are a viscous regularization of the Euler equations, which are still an enigma. Turbulence is a riddle wrapped in a mistery inside an enigma.*

In this book, we are going to address the mystery part of the riddle-mistery-enigma trilogy. Obviously, with no great hopes of success: Tao expressed in 2007 his views on "Why global regularity for Navier-Stokes is hard" [460]. He insists that there is no hope to find an explicit formula to solve the equations, or even to re-write explicitly the equations in simpler ones. It may seem an obvious point in the light of centuries of studies in mathematical physics; however, one may find every now and then authors who try to find such a miraculous formula. Another point is that we lack controlled quantities that would impede any eventual blow-up; for instance, the vectorial structure of the equations does not allow arguments based on the control of the signs of associated quantities. The only control we have is Leray's energy inequality which provides the control of the L^2 norm of the velocity of the fluid, while we should control the vorticity of the fluid, i.e., the curl of the velocity. In dimension 2, the evolution equation of the vorticity allows such a control, and we know global existence of regular solutions; in dimension 3, however, we have no control on the vorticity (to the exception of the case of some symmetrical fluids, as the axisymmetrical fluids with no swirl studied by Ladyzhenskaya [295]) unless the data are very small.

The fact is that we have barely any control on the non-linear terms of the equations. The Navier–Stokes equations combine a diffusive term and a convective term. Diffusion will damp the high gradients, but convection may transfer energy from small gradients to high gradients at a rate faster than the damping induced by diffusion. We still do not know, eighty years after the seminal work of Leray on the Navier–Stokes equations [328], whether this mechanism of energy transfer between different scales may open the way to the blow-up of solutions or whether the damping effect of the diffusion will prevail and block the way to the blow-up.

When in the end of 2011, my editor Sunil Nair invited me to write another book on the Navier–Stokes equations, ten years after *Recent Developments in the Navier–Stokes Problem* [313], I felt that I had included in my first book all I knew about the Navier–Stokes equations; thus, I had the project of writing a book on everything I did not know about Navier–Stokes. Such a project turned out to be unrealistic, as the topic is immense. I then came back to my field of expertise, the sole study of the incompressible deterministic Navier–Stokes equations in the case of a fluid filling the whole space. Even within such a restricted frame, I discovered that I had a lot to learn, so the content of the book is much larger than foreseen, and the resulting book has very few parts in common with the old one. The references in the Bibliography are essentially either very old (historical references to the (pre)history of the Navier–Stokes equations) or extremely recent, testifying that those equations are a very active field in contemporary mathematical research.

It might be interesting to list what is not included in the book, so that a reader whose interest into fluid mechanics is piqued would be tempted to get further information elsewhere on further topics:

- the book is not a treatise on hydraulics, nor on whatever application of fluid mechanics. There will be no pipes nor vessels, as the fluid is assumed to fill the whole space; thus, no physically reasonable modelization of the real world is proposed hereby. In the same way, it is not a treatise on aerodynamics; no drag forces are investigated, and no body is immersed in the fluid at any page of the book.

- the book is not a treatise on turbulence. No modelization is commented, no statistical study is presented and no stochastical theory is introduced. Stochastic fluid mechanics is a very active research field, as there is some hope that the possible deterministic singular solutions are rare and unstable enough to be ignored by a random description of the equations.

- the book does not study general fluids that may have various behaviors (compressibility, inhomogeneity) and may be subject to various forces, whether external (as the Coriolis force) or internal (due to conductivity or thermal effects). We shall stick to the very simplified frame of a Newtonian, incompressible, homogeneous isotropic fluid subject only to the internal forces of friction.

- the book does not study fluids in a bounded domain, or an exterior domain. Thus, we do not have to consider many delicate problems in handling the pressure at the boundary, or to deal with the vorticity generated at the boundary.

- the book does not address issues from the computational fluid dynamics. No Galerkin bases are constructed in any place to provide the basis of algorithms. Some notions derived from the Large Eddy Simulation (as the α-models) will be studied but with no aim at practical computations.

Now that we know what this book is not, we may serenely comment on what this book is actually about, and in which way it differs from the 2002 book:

- One of the main differences between the two books is the systematic inclusion of forces in the theorems. Sometimes, it is a mere adaptation of the same theorems on equations without forcing terms. But very often, the choice of the hypotheses on the forces is not obvious. This is especially clear for the chapters on mild solutions: if you want to get a solution in $L_t^4 L^6$ for instance, then a natural choice for the space where to pick the initial value is the Besov space $\dot{B}_{6,4}^{-1/2}$; the choice of the space where to pick the force $\vec{f}$ is more complex.

- An obvious benefit of this inclusion of forcing terms in the equations is, of course, that one may consider problems that are linked to the behavior of the forces: if the force is stationary or time-periodic, will the solutions be asymptotically stationary or time-periodic?

- Another difference between the two books is the stress put in this new book on Morrey spaces, whereas the old book insisted on Besov spaces. In both cases, the idea is to deal with critical hypotheses on the data, where the criticality is intended with respect to the scaling properties of the equations. Littlewood–Paley analysis which was extensively used in the old book is very often replaced by Hedberg's inequality in the new book. In order to see the usefulness of such an approach, the reader may compare the proofs of Theorems 10.2 and 10.3 in the book with Gallagher's original proofs [195, 197], which used the Littlewood–Paley decomposition, or the proof of global existence of helical solutions (Theorem 10.7) with the same proof by Titi and co-workers in [347].

- The stress on Morrey spaces and on the related theory of singular multipliers has given rise to a wholly new chapter on parabolic Morrey spaces and capacitary theory applied to the existence of mild solutions (Chapter 5). (Let us remark, however, that parabolic Morrey spaces have already been used by several authors in the study of partial regularity results for the Navier–Stokes equations; see for instance, the papers by O'Leary [379], Ladyzhenskaya and Seregin [297] or Kukavica [287])

- Younger readers are often puzzled by the way the Navier–Stokes equations are stated and solved. I found interesting to include chapters that would introduce the equations in a simple context. Thus, a chapter is devoted to the physical meaning of each term involved in the equation (viscosity, density, velocity, vorticity, pressure, stress tensor, ...). Another chapter is devoted to the history of the equations up to Leray; I hope those historical indications will help the reader to better understand the inner logic of this theory. A chapter is devoted to a classical resolution of the equations, with tools pertaining to the nineteenth century, before the birth of functional analysis. This chapter thus introduces basic notions, as the Green function, the Leray projection operator, the heat kernel, the Oseen tensor and so on.

- In relation with Morrey spaces, special emphasis will be put on scaling properties of the Navier–Stokes equations. A related important result will be Jia and Šverák's theorem on the existence of self-similar solutions for large homogeneous initial values [245].

- Another important result related to scaling properties is the partial regularity result of Caffarelli, Kohn and Nirenberg for suitable solutions [74]. We shall discuss some recent variants of this theorem, and explore in a systematical way the various approximating processes of weak solutions, and show how they lead to those suitable solutions.

Summary of the Book

The book is divided into 19 chapters. We give here a brief presentation of those chapters.

1. **Presentation of the Clay Millenium Prizes.**
   This chapter gives a loose presentation of the Clay Millenium Prizes, based on the book published by the Clay Mathematics Institute [91]. We present more especially the formulation of the Clay Millennium Problem on Navier–Stokes equations, as given by Fefferman [171].

2. **The physical meaning of the Navier–Stokes equations.**
   This chapter gives a short presentation of each term in the Navier–Stokes equations in order to explain how and why they are introduced in fluid mechanics. In the case of an isotropic Newtonian fluid, and in the absence of other internal forces than the forces exerted by the hydrostatic pressure or the friction due to viscosity, we find the equations of hydrodynamics:

$$\frac{D}{Dt}\rho + \rho \operatorname{div} \vec{u} = 0 \tag{0.1}$$

and

$$\rho \frac{D}{Dt}\vec{u} = -\vec{\nabla}p + \mu \Delta \vec{u} + \lambda \vec{\nabla}(\operatorname{div}\vec{u}) + \vec{f}_{ext} \tag{0.2}$$

Equation (0.1) describes the mass conservation: ρ is the density of the fluid, $\vec{u} = \vec{u}(t,x)$ the velocity of the parcel of fluid that occupies at time t the position x and $\frac{D}{Dt}$ is the material derivative

$$\frac{D}{Dt}h = h + \sum_{i=1}^{3} u_i \partial_i h.$$

Equation (0.2) expresses Newton's second law on the momentum balance in the presence of forces. The forces are induced by the pressure p (the force density is given by $-\vec{\nabla}p$), by viscosity (the force density is given by the divergence of the viscous stress tensor $\operatorname{div} \mathbb{T}$; in the case of a Newtonian fluid, the tensor $\mathbb{T}$ depends linearly on the strain tensor ϵ described by Cauchy $- \epsilon_{i,j} = \frac{1}{2}(\partial_i u_j + \partial_j u_i) -$; more precisely,

$\mathbb{T} = 2\mu\epsilon + \eta\,\mathrm{tr}(\epsilon)\,\mathbb{I}_3$, where μ is the **dynamical viscosity** of the fluid and $\eta\,[= \lambda - \mu]$ is the **volume viscosity** of the fluid), and by external forces whose density $\vec{f}_{ext}$ is assumed to be independent of $\vec{u}$.

In the case of a Newtonian, isotropic, homogeneous and incompressible fluid, those equations of hydrodynamics are transformed into the **Navier–Stokes equations**. ρ is then constant (it does not depend neither on time t nor on position x); one then divides the equations by ρ, and replaces the force density $\vec{f}_{ext}$ with a reduced density $\vec{f}_r = \frac{1}{\rho}\vec{f}_{ext}$, the pressure p with a reduced pressure $p_r = \frac{1}{\rho}p$ (the **kinematic pressure**), and the dynamical viscosity μ by the **kinematic viscosity**. We then have:

The Navier–Stokes equations

$$\partial_t \vec{u} + (\vec{u}.\vec{\nabla})\vec{u} = -\vec{\nabla}p_r + \nu\Delta\vec{u} + \vec{f}_r \qquad (0.3)$$

$$\mathrm{div}\,\vec{u} = 0 \qquad (0.4)$$

---

3. **History of the equation.**

In this chapter, we give a short history of the Navier–Stokes equations, based on Darrigol's recent book *Worlds of flow* [145] and on the classical papers of Truesdell [479, 480, 481]. We recall the first works on hydrodynamics by Bernoulli [37], D'Alembert [137, 138, 139] and Euler [167]. Then we describe how the Navier–Stokes equations were introduced by Navier [373], Cauchy [96], Poisson [403], Saint-Venant [418] and Stokes [451].

Then we show how Lorentz computed the Green function for the steady Stokes problem [342], paving the way to modern Navier–Stokes theory: Oseen [384, 385] extended the work of Lorentz to the case of evolutionary Stokes equations and then to the Navier–Stokes equations, and proved local-in-time existence of classical solutions; Leray [328] extended Oseen's work to the existence of global-in-time weak solutions.

The formulas derived by Lorentz in 1896 and Oseen in 1911 for hydrodynamic potentials were explicitly known only for very simple domains and not available for more complex domains. Hopf [238] in 1951 and Ladyzhenskaya in 1957 [262] then used a Faedo–Galerkin method to deal with the case of a general domain, where no explicit formula could be used.

4. **Classical solutions.**

In this chapter, we solve the Navier–Stokes equations, using only classical tools of differential calculus, as they were used in the end of the 19th century or the beginning of the 20th century. More precisely, we stick to the spirit of Oseen's paper, which was published in 1911 [384] (a similar treatment can be found in a 1966 paper of Knightly [263]).

This chapter introduces the main equations and fundamental solutions used in the book: the heat equation and the heat kernel, the Poisson equation and the Green function, the Helmholtz decomposition and the Leray projection operator, the Stokes problem and the Oseen tensor.

5. **A capacitary approach of the Navier–Stokes integral equations.**
   In this chapter, we use a **new method** for solving the Navier–Stokes equations: we re-write the problem as a quadratic integral equation, and we solve it by the classical Picard iterative scheme. The novelty is the fact that we prove convergence by use of a dominating function that solves a quadratic integral problem with a positive symmetric kernel. We may then use a 1999 result of Kalton and Verbitsky [250] to describe those functions.

   This chapter introduces important tools for the study of parabolic equations: parabolic Morrey spaces, parabolic Riesz potentials and Hedberg's inequality. New functional spaces are introduced for the study of the Navier–Stokes equations, such as the space of pointwise multipliers between the parabolic Sobolev space $\dot{H}_{t,x}^{1/2,1} = L_t^2 \dot{H}_x^1 \cap L_x^2 \dot{H}_t^{1/2}$ and $L_{t,x}^2$, or the Triebel–Lizorkin–Morrey–Campanato spaces.

6. **The differential and the integral Navier–Stokes equations.**
   In this chapter, we discuss the relations between the differential version and the integral version of the Navier–Stokes equations and the way to get rid of the pressure through the Leray projection operator.

   Thus, we discuss various definitions of a solution of the Cauchy initial value problem for the Navier–Stokes equations:

   - **very weak solution:**
     - div $\vec{u} = 0$
     - $\vec{u}$ is locally square integrable on $(0, T) \times \mathbb{R}^3$
     - the map $t \in (0, T) \mapsto \vec{u}(t, .)$ is continuous from $(0, T)$ to $\mathcal{D}'(\mathbb{R}^3)$ and $\lim_{t \to 0^+} \vec{u}(t, .) = \vec{u}_0$
     - for all $\vec{\varphi} \in \mathcal{D}((0, T) \times \mathbb{R}^3)$ with div $\vec{\varphi} = 0$, we have

     $$\langle \partial_t \vec{u} - \nu \Delta \vec{u} + \mathrm{div}(\vec{u} \otimes \vec{u}) - \vec{f} | \vec{\varphi} \rangle_{\mathcal{D}', \mathcal{D}} = 0 \qquad (0.5)$$

     No other regularity is assumed on $\vec{u}$ than the continuity of $t \in [0, T) \mapsto \vec{u} \in \mathcal{D}'$, and no regularity is required on the distribution $\vec{f}$. The pressure p is only defined implicitly by the property (0.5).

   - **Oseen solution:**
     under appropriate assumptions on $\vec{f}$ and $\vec{u}$, we may get rid of the pressure with the help of the Leray projection operator $\mathbb{P}$ and write

     $$\vec{\nabla} p = (Id - \mathbb{P})(\vec{f} - \mathrm{div}(\vec{u} \otimes \vec{u})) = \vec{\nabla} \frac{1}{\Delta} \mathrm{div}(\vec{f} - \mathrm{div}(\vec{u} \otimes \vec{u})).$$

     An Oseen solution $\vec{u}$ of the Navier–Stokes equations on $(0, T) \times \mathbb{R}^3$, for initial value $\vec{u}_0$ and forcing term $\vec{f}$ is then a distribution vector field $\vec{u}(t, x) \in \mathcal{D}'((0, T) \times \mathbb{R}^3)$ is a very weak solution such that moreover:
     - $\vec{u} \in (L^2 L^2)_{\mathrm{uloc}}$
     - $\mathrm{div}(\vec{u} \otimes \vec{u}) + \vec{\nabla} p - \vec{f} = \mathbb{P}(\mathrm{div}(\vec{u} \otimes \vec{u}) - \vec{f})$

   - **mild solutions:**
     when the Oseen solution $\vec{u}$ may be computed by Picard's iteration method, we shall speak of mild solution.

   - **weak solutions:**
     when Picard's iterative scheme does not work, the existence of solutions is provided by energy estimates involving the (local) L^2 norm of the gradient of $\vec{u}$. Thus, one is led to consider weak solutions, i.e., Oseen solutions $\vec{u}$ such that

– $\vec{u} \in (L_t^\infty L_x^2)_{\text{uloc}}$

– $\vec{\nabla} \otimes \vec{u} \in (L_t^2 L_x^2)_{\text{uloc}}$

associated to an initial value $\vec{u}_0$ and to a forcing term $\vec{f}$ such that

– $\vec{u}_0 \in L_{\text{uloc}}^2$ with $\text{div } \vec{u}_0 = 0$

– $\vec{f} = \text{div } F$, where the tensor F is such that $F \in (L_t^2 L_x^2)_{\text{uloc}}$

- **suitable solutions:**
  a suitable solution is a weak solution that satisfies in $\mathcal{D}'$ the local energy inequality

$$\partial_t \left(\frac{|\vec{u}|^2}{2}\right) \leq \nu \Delta \left(\frac{|\vec{u}|^2}{2}\right) - \nu |\vec{\nabla} \otimes \vec{u}|^2 - \text{div}\left((p + \frac{|\vec{u}|^2}{2})\vec{u}\right) + \vec{u}.\vec{f} \qquad (0.6)$$

- **Leray weak solution:**
  a weak solution $\vec{u}$ of the Navier–Stokes equations is called a Leray weak solution if it satisfies the Leray energy inequality:

  – $\vec{u} \in L_t^\infty L_x^2 \cap L_t^2 \dot{H}_x^1$

  – $\vec{f} \in L_t^2 H_x^{-1}$

  – for every $t \in (0, T)$,

$$\|\vec{u}(t,.)\|_2^2 \leq \|\vec{u}_0\|_2^2 - 2\nu \int_0^t \|\vec{\nabla} \otimes \vec{u}\|_2^2 \, ds + 2 \int_0^t \langle \vec{u}|\vec{f}\rangle_{H^1, H^{-1}} \, ds.$$

7. **Mild solutions in Lebesgue or Sobolev spaces.**
   This chapter is devoted to the classical results of Kato on mild solutions: solutions in Sobolev spaces H^s for $s \geq 1/2$ (Fujita and Kato [185]) and in Lebesgue spaces L^p for $p \geq 3$ (Kato [255]).

   We also give the proof of uniqueness in $\mathcal{C}([0, T), (L^3)^3)$ (Furioli, Lemarié-Rieusset, and Terraneo [187]).

8. **Mild solutions in Besov or Morrey spaces.**
   This chapter is devoted to the study of mild solutions in Besov or Morrey spaces, in the spirit of the books of Cannone [81] and Lemarié-Rieusset [313].

   At the end of the chapter, one also considers the case of Fourier-Herz spaces (as in the results of Le Jan and Sznitman [305], of Lei and Lin [306] or of Cannone and Wu [86]).

9. **The space BMO^{-1} and the Koch and Tataru theorem.**
   This chapter deals with initial values in the largest critical spaces associated to the Navier–Stokes equations: the space BMO^{-1} or the Besov space $\dot{B}_{\infty,\infty}^{-1}$

   Koch and Tataru's theorem [266] is proved by following the strategy recently given by Auscher and Frey [11].

   Then we consider the Navier–Stokes problem with a null force ($\vec{f} = 0$), and we present many important results:

   - We prove ill-posedness in $\dot{B}_{\infty,\infty}^{-1}$ (Bourgain and Pavlović [52]).

   - We develop an example of global mild solutions associated to large initial value given by Chemin and Gallagher [108].

   - We prove the stability theorem of Auscher, Dubois and Tchamitchian [10] for global solutions in BMO^{-1}.

- We present the persistence theory of Furioli, Lemarié-Rieusset, Zahrouni and Zhioua in [188] for the propagation of initial regularity for mild solutions in BMO^{-1}.

- We give a simple proof of time and space analyticity, following Cannon and Knightly [80].

10. **Special examples of solutions.**

The symmetries for the Navier–Stokes equations were discussed one century ago by Wilczynski [502]. In this chapter, we study the solutions that are invariant with respect to those symmetries:

- **Two-and-a-half dimensional flows**: $\vec{u}$ is invariant under the action of space translations parallel to the x_3 axis. Global existence and regularity are similar to the case of the 2D Navier–Stokes equations, a case well understood since the works of Leray [327, 328, 329], and fully developed by Ladyzhenskaya, Lions and Prodi [293, 339].

- **Axisymmetrical solutions**: $\vec{u}$ is invariant under the action of rotations around the x_3 axis. In the case of axisymmetric flows with no swirl, Ladyzhenskaya [295], Uchovskii and Yudovich [486] proved global existence under regularity assumptions on $\vec{u}_0$ and $\vec{f}$ but without any size requirements on the data.

- **Helical solutions**: $\vec{u}$ is invariant under the action of a one-parameter group of screw motions

$$R_\theta(x_1, x_2, x_3) = (x_1 \cos\theta - x_2 \sin\theta, x_1 \sin\theta + x_2 \cos\theta, x_3 + \alpha\theta)$$

(where $\alpha \neq 0$ is fixed). Global existence of helical flows has been studied by Mahalov, Titi and Leibovich [347].

- **Brandolese's symmetrical solutions**: $\vec{u}$ is invariant under the action of a finite (non-trivial) group of isometries of $\mathbb{R}^3$ (Brandolese [60]).

- **Self–similar solutions**: $\vec{u}$ is invariant under the action of time-space rescalings, i.e., we consider self–similar solutions:

$$\text{for every } \lambda > 0, \quad \lambda \vec{u}(\lambda^2 t, \lambda x) = \vec{u}(t, x).$$

Backward self–similar solutions were first considered by Leray [328], but ruled out by Nečas, Růžička and Šverák [375] and by Tsai [482]. Forward self-similar mild solutions have been studied by many authors (see, for instance, Cannone, Meyer and Planchon [83, 84]) in the case of small data; the case of forward self-similar weak solutions associated with large data has recently been solved by Jia and Šverák [245].

- **Stationary solutions**: $\vec{u}$ is invariant under the action of time translations, i.e., we consider steady solutions. We present the results of Kozono and Yamazaki [280], Bjorland, Brandolese, Iftimie and Schonbek [44] and Phan and Phuc [395].

- **Landau's solutions**: those special solutions were described first (quite implicitly) by Slezkin [439], then independently by Landau [301] and Squire [447]. They are self-similar, axisymmetrical with no swirl and steady.

- **Time-periodic solutions**: $\vec{u}$ is invariant under the action of a discrete group of time translations, i.e., $\vec{u}$ is time-periodic. Such solutions have been considered by Maremonti [349], Kozono and Nakao [272], Yamazaki [509] and Kyed [290].

11. **Blow-up?**
    This chapter discusses various refinements of Serrin's criterion [435] for blow-up of the solutions, including the classical criterion of Beale, Kato and Majda [27]. We present the extension of the criterion to the setting of Besov spaces (Kozono and Shimada [274], Chen and Zhang [116], May [354] and Kozono, Ogawa and Taniuchi [273]).

12. **Leray's weak solutions.**
    Classical theory on existence and weak-strong uniqueness of Leray solutions are presented (Leray [328], Prodi [406], Serrin [435]). Extensions of the Prodi–Serrin criterion to larger classes of solutions (Kozono and Taniuchi [277], Kozono and Sohr [275], Lemarié-Rieusset [313, 316], Chen, Miao and Zhang [115]) are described. Uniqueness for "almost strong" solutions is proved (Chemin [106], Lemarié-Rieusset [317], May [355], Chen, Miao and Zhang [115]). Results on stability of mild solutions through L^2 perturbation will be discussed (Karch, Pilarczyk and Schonbek [251]).

13. **Partial regularity results for weak solutions.**
    This chapter is devoted to Serrin's theory of interior regularity (Serrin [434], Struwe [455] and Takahashi [459]), and to the celebrated theorem of Caffarelli, Kohn and Nirenberg [74]. For this theorem, new proofs are provided, including a new result where the pressure is submitted to (quite) no assumptions at all; this extension is based on the notion of dissipative solutions introduced by Duchon and Robert [159].

14. **A theory of uniformly locally L^2 solutions.**
    This chapter recalls the theory developed in *Recent Developments in the Navier–Stokes Problem* about suitable weak solutions with infinite energy.

15. **The L^3 theory of suitable solutions.**
    This chapter applies the theory of uniformly locally L^2 solutions to the case of solutions with values in L^3. We prove the recent results of Jia, Rusin and Šverák on minimal data for blowing-up solutions [417, 244] on the (potential) existence of a minimal-norm initial value for a blowing-up mild solution to the Navier–Stokes Cauchy problem. We prove as well the $L_t^\infty L_x^3$ regularity result of Escauriaza, Seregin and Šverák [163] for suitable solutions of the Navier–Stokes equations.

16. **Self-similarity and the Leray–Schauder principle.**
    The theory of self-similar solutions has known an impressive advance with the publication in 2014 of a paper of Jia and Šverák [245] establishing the existence of such solutions for any large homogeneous initial value. This chapter presents this result with some slight extensions.

17. **α-models.**
    α-models were developed (mainly by Holm) in recent years to provide efficient solvers for the Reynolds equations associated to turbulent flows. In this chapter, we discuss the existence of solutions of various α-models and their convergence to weak solutions of the Navier–Stokes equations when α goes to 0. Those α-models are:

    - the Leray–α model (discussed by Cheskidov, Holm, Olson and Titi [119])
    - the Navier–Stokes α-model, also known as viscous Camassa–Holm equations (studied by Chen, Foias, Holm, Olson, Titi and Wynne [117])
    - the Clark-α model (studied by Cao, Holm and Titi [87])
    - the simplified Bardina model (studied by Cao, Lunasin and Titi [89])

18. **Other approximations of the Navier–Stokes equations.**

In this chapter, we discuss various approximations of the Navier–Stokes equations, including frequency cut-off, hyperviscosity (Beirão da Vega [28], and damping (Cai and Jiu [75]). We present an important example of Ladyzhenskaya's model of a Stokesian fluid with a non-linear damping of the high frequencies through the friction tensor (Ladyzhenskaya [294]).

19. **Artificial compressibility.**

In order to simplify the estimation of the pressure in the Navier–Stokes equations, some authors have presented an approximation of the equations by introducing a small amount of compressibility on $\vec{u}$ in order to turn the Navier–Stokes equations, which contains a non-local term $\vec{\nabla} p$ (given by the Leray projection operator, thus by a singular integral), into a system of partial differential operators that contain no non-local terms. We present in this chapter two classical models (given by Temam [469, 470] and by Višik and Fursikov [189]) and Hachicha's recent model [225] of a hyperbolic approximation with finite speed of propagation.

Preface to the Second Edition

Navier–Stokes equations is a very difficult problem and a highly competitive field of research. This book is the second edition of a second book I wrote on the topic. Between the first book, released in 2002 [313], and the first edition of this book, released in 2016 [319], the number of references grew from 116 to 441 (including 180 references published after the first book was written) and the number of pages grew from approximately 400 pages to 740. This second edition contains more than 900 pages[1] and more than 500 references (including 50 references published after the first edition).

Beyond those quantitative data, we must underline that the field has known major breakthroughs that we could only allude to (the book would have been very much heavier if we detailed the lengthy and technical proofs involved). Concerning uniqueness of weak solutions, Buckmaster and Vicol [71] proved in 2019 non-uniqueness of very weak solutions in $\mathcal{C}([0,T],L^2)$ and in 2021, Albritton, Brué and Colombo [5] gave an example of non-uniqueness of suitable Leray solutions to the Navier–Stokes equations on $(0,T) \times \mathbb{R}^3$ with body force $\vec{f} \in L^1((0,T),L^2(\mathbb{R}^3))$ and initial condition $\vec{u}_0 = 0$. Another important result was Tao's result in 2019 [463] giving an explicit a priori bound for the L^3 norm of a blowing up mild solution, based on Fourier analysis of the solution, while Barker and Prange [19, 20, 21] developed an analysis of the blow up in spatially localized estimates.

The new (or not so new) results we chose to include in the second edition (besides correcting many typos and some serious mistakes, such as in pages 361 or 438) are the following ones:

- In Chapter 4, section 4.11, we added Swann's beautiful theorem [458] on the existence time for very regular solutions. We deal again with Swann's theorem in Chapter 12, where we added Section 12.8 on Kato's theorem on the inviscid limit of the Navier–Stokes equations [254].

- In Chapter 6, we modified the presentation of general weak solutions, replacing the role of uniform estimates by weighted estimates. In particular, we added a section (Section 6.3) on Leray's projection operator in order to take into account the recent results of by Bradshaw and Tsai [58] and Fernández-Dalgo and Lemarié-Rieusset [174].

- In Chapter 8, we added a small section (Section 8.9) on solutions expressed as a countable superposition of plane waves, as discussed by Dinaburg and Sinai [153].

- In Chapter 9, we added a variation on the Koch and Tataru theorem (Section 9.2) based on recent results of Lemarié-Rieusset [323]. We included a proof of Wang's result on norm inflation in the critical Besov space $\dot{B}_{\infty,2}^{-1}$ [495] .

- In Chapter 10, Section 10.3, we corrected the statement of Theorem 10.3 on Muckenhoupt weights and we enriched the section with 15 pages devoted to axisymmetric solutions in Morrey spaces (including the theorem of Gallay and Šverák [203]).

---

[1]The amount of pages given in this preface is estimated in the trim size of the first edition. This size has been changed for the second edition, so that the final result contains "only" 800 pages, and not 930 pages.

- In Chapter 11, we added a section (Section 11.8) devoted to the role of the second eigenvalue of the strain matrix, as discussed by Miller [361].

- In Chapter 12, we corrected the proof of Proposition 12.1 on the strong Leray energy inequality and we added two small sections, Section 12.7 on non-uniqueness of weak solutions and Section 12.8 on inviscid limits.

- In Chapter 13, we corrected the proof of Lemma 13.4.

- With 40 new pages, Chapter 14 has been largely extended, with new sections on weighted Leray solutions (Section 14.6, based on the results of Fernández-Dalgo and Lemarié-Rieusset [173] and of Bradshaw, Kukavica and Tsai [56]), global existence for local Leray solution (Section 14.5, which generalizes results presented in [313]), Barker's theorem on weak-strong uniqueness (Section 14.8, based on the papers of Barker [18], and Lemarié-Rieusset [324]) and a final section (Section 14.9) where we present a theorem (Theorem 19.2) loosely based on the theory of homogeneous statistical solutions (Višik and Fursikov [190], Basson [23]).

- In Chapter 15, we added a small section (Section 15.5) on the recent results on the L^3 norm of blowing up solutions (Tao [463], Barker and Prange [19, 20, 21])

- With 40 new pages, Chapter 16 has been largely extended. The section on existence of steady solution has been completed with a section (Section 16.3) on the Liouville problem for steady solutions (based mainly on Seregin's work [430]). New examples of application of the Leray–Schauder principle have been given: existence of discretely self-similar solutions for large data (Section 16.9, following Chae and Wolf [100], Bradshaw and Tsai [57], Fernández-Dalgo and Lemarié-Rieusset [173]), existence of time-periodic weak solutions for large data (Section 16.10, following Kyed [290]).

- In Chapter 17, we completed Theorem 17.2 on the Navier–Stokes-α model.

- In Chapter 19, we corrected the proof of Theorem 19.1.

- In conclusion (Chapter 20), we added a final section (Section 20.5) with a small list of open questions.

Chapter 1

Presentation of the Clay Millennium Prizes

1.1 Regularity of the Three-Dimensional Fluid Flows: A Mathematical Challenge for the 21st Century

Modern mathematical hydrodynamics was born in the 18th century. In 1750, Euler [166] expressed the conviction that the mechanics of continuous media could be reduced to the application of Newton's law to the infinitely small elements constituting the continuum. In 1755, Euler [167] presented a memoir (published in 1757) entitled *Principes généraux du mouvement des fluides* [General principles concerning the motion of fluids], where he could derive the equations for a general fluid, compressible or not, in the presence of arbitrary external forces. The **Euler equations** use Newton's law when the fluid element is submitted only to the external forces and to the pressure exerted by the other elements.

However successful Euler had been in applying his program, his results suffered from two severe limitations. The first one was underlined by Euler himself in his conclusion:

> *Cependant tout ce que la Théorie des Fluides renferme est contenu dans ces deux équations, de sorte que ce ne sont pas les principes de Méchanique qui nous manquent dans la poursuite de ces recherches, mais uniquement l'Analyse, qui n'est pas encore assés cultivée, pour ce dessein*[1]

As a matter of fact, the complete resolution of the Euler equations is still an open problem nowadays.

The second limitation is even more severe. In Euler's equations, the internal forces (i.e., the forces exerted on parts of the fluid by the other parts of the fluid) are described only in terms of the pressure. If we consider a fluid element as a little cube, the pressure exerts a force on the faces of the cube in the normal direction to the faces. But, due to the fact that the other elements of fluids have a different velocity, there is another force (the friction) exerted on the fluid element, in directions that are tangential to the faces. This shear stress has been described in the 19th century as the effect of **viscosity**. Viscous fluids behave drastically differently from the inviscid ones. Von Neumann coined the term "dry water"[2] to underline the inefficiency of modelization that would neglect viscosity forces, as commented by Feynman [175]:

> *When we drop the viscosity term, we will be making an approximation which describes some ideal stuff rather than real water. John von Neumann was well aware of the tremendous difference between what happens when you don't have*

---

[1]Everything that is held within the Theory of Fluids is contained in those two equations, so that it is not the principles of Mechanics that are lacking for the continuation of our research, but only the Analysis, which is still not developed enough for that purpose.

[2]Ironically enough, dry water exists. It was patented in 1968. Dry water, in this acceptance of the term, is a water-air emulsion in which tiny water droplets are surrounded by a sandy silica coating. The silica coating prevents the water droplets from combining and turning back into a bulk liquid.

DOI: 10.1201/9781003042594-1

the viscous terms and when you do; and he was also aware that, during most of the development of hydrodynamics until about 1900, almost the main interest was in solving beautiful mathematical problems with this approximation which had almost nothing to do with real fluids. He characterized the theorist who made such analyses as a man who studied "dry water." Such analyses leave out an essential property of the fluid.

Taking into account the viscosity led to the **Navier–Stokes equations**. Those equations were first introduced by Navier in 1822 [373]. Though they have been rediscovered by many authors, such as Cauchy, Poisson or de Saint-Venant, they remained quite controversial until they were settled on a firmer basis by Stokes in 1845 [451] (see the paper of Darrigol [144] on the "five births" of the Navier–Stokes equations). They still had to wait dozens of years before being definitely adopted by physicists, after that they were proven to be in accordance with Maxwell's kinetic theory of gases. Again, the complete resolution of the Navier–Stokes equations is still an open problem nowadays. Given some initial value that is smooth and well localized, we are not able to prove the existence of a global-in-time solution (except when the initial value is small enough). Local existence was rigorously established by Oseen [385] and his co-workers at the beginning of the 20th century. Then, in 1934, Leray [328] proved that those local-in-time solutions could be prolongated in global-in-time weak solutions that might be no longer smooth, so that the derivatives are to be taken in some weak sense. Thereafter, very few further results could be obtained for the 3D fluids, and the question of global existence of classical solutions remained an important challenge.

A major issue in the theory of fluid mechanics is the understanding of **turbulence**. Turbulence occurs when the motion of the fluid becomes disordered. The flow then turns out to be highly irregular and quite unpredictable. Reynolds [410] studied the instability of steady flows and gave experimental evidence of the transition from laminar flows (i.e., regular flows) to turbulence through the increase in the velocities. The Navier–Stokes equations are believed to be a good frame to establish transition to turbulence in a rigorous mathematical setting.

The study of three-dimensional fluids remains an important issue nowadays. It is considered as an important challenge for the 21st century. Such challenges have been presented by the International Mathematical Union (IMU) at the occasion of the World Mathematical Year 2000. More precisely, in 1992 in Rio de Janeiro, IMU, with support of the UNESCO, declared the year 2000 to be the World Mathematical Year. The purpose was to highlight mathematics for a larger audience, in an effort of world-wide promotion. The Declaration of Rio set three aims:

- The great challenges of the 21st century

- Mathematics, as a key for development

- The image of mathematics

In August 2000, the American Mathematical Society held an extraordinary meeting on the UCLA campus under the title "Mathematical Challenges of the 21st Century." In the editorial of the *Notices of the AMS* [69], Browder, the President of the AMS, explained the aims of the meeting:

1. *To exhibit the vitality of mathematical research and to indicate some of its potential major growing points: these include some of the major classical problems (the Riemann Hypothesis, the Poincaré conjecture, the regularity of three-dimensional fluid flows) as well as some of the recently developed major research programs like those associated with the names of Langlands and Thurston.*

2. *To point up the growing connections between the frontiers of research in the mathematical sciences and cutting-edge developments in such areas as physics, biology, computational science, and finance.*

Browder was not the only one to promote the issue of the regularity of three-dimensional fluid flows to such a prestigious neighborhood as the one of the Riemann Hypothesis and of the Poincaré conjecture. In 1997 in a conference at the Fields Institute at Toronto, the 1966 Fields medalist Smale [442] gave a list of problems he selected as *"likely to have great importance for mathematics and its development in the next century."* That list was an answer to an invitation of Arnold, on behalf of the International Mathematical Union, to describe some great problems for the 21st century, in a reminiscent way of Hilbert who described in the 1900 meeting of the IMU in Paris a list of twenty-three great problems for the 20th century. Smale listed eighteen problems, including what he considered as the three greatest open problems of mathematics: the Riemann Hypothesis, the Poincaré conjecture and the "Does P=NP?" problem. Smale's fifteenth problem is the question about global existence and regularity for the three-dimensional Navier–Stokes equations, *"perhaps the most celebrated problem in partial differential equations,"* whose solution *"might well be a fundamental step toward the very big problem of understanding turbulence."*

The most spectacular effort to promote mathematics, however, had been the establishment of the Millennium Prizes by the Clay Mathematics Institute. Seven $1 million prizes were established to reward the solution of seven classical mathematical problems that have resisted solutions for many years. Once more, the Navier–Stokes equations were selected, as well as other great problems such as the Riemann Hypothesis, the Poincaré conjecture and the "Does P=NP?" problem.

1.2   The Clay Millennium Prizes

The Clay Mathematics Institute is a non-profit foundation. It was established in 1998 by the American businessman Landon T. Clay. As indicated on its Web site (http://www.claymath.org/), the primary objectives and purposes of The Clay Mathematics Institute are:

- to increase and disseminate mathematical knowledge,

- to educate mathematicians and other scientists about new discoveries in the field of mathematics,

- to encourage gifted students to pursue mathematical careers,

- and to recognize extraordinary achievements and advances in mathematical research.

According to this mission, the Institute offers postdoctoral grants, funds summer schools and conferences, co-publishes with the American Mathematical Society monographs devoted to the *"exposition of recent developments, both in emerging areas and in older subjects transformed by new insights or unifying ideas,"* and has a program of providing to a large readership digital facsimiles of major mathematical works from the past.

The Institute is best known for establishing the Millennium Prize Problems in 2000. *"The Prizes were conceived to record some of the most difficult problems with which mathematicians were grappling at the turn of the second Millennium; to elevate in the consciousness of the general public the fact that in mathematics, the frontier is still open and abounds in important unsolved problems; to emphasize the importance of working towards a solution of the deepest, most difficult problems; and to recognize achievement in mathematics of historical magnitude."*

A committee of experts (Michael Atiyah, Andrew Wiles, John Tate, Arthur Jaffe, Alain Connes, Edward Witten) selected seven problems that were officially presented at a conference in the Collège de France (Paris) in May 2000. For each problem, the first person to solve it would be awarded $1,000,000 by the CMI. The problems to be solved are:

P versus NP

The Hodge Conjecture

The Poincaré Conjecture

The Riemann Hypothesis

Yang–Mills Existence and Mass Gap

Navier–Stokes Existence and Smoothness

The Birch and Swinnerton-Dyer Conjecture

The announcement of the prizes drew the attention of media to this area of science. As early as 2003, Perelman solved the Poincaré conjecture; he was awarded the Fields Medal in 2006 and the Clay Millennium Prize in 2010, and declined both. This had a rather large echo in the media. If most journals focused on Perelman's personality and his alleged eccentricity, many papers tried and explained the Poincaré conjecture to their laymen readers.

In 2006, *Science* [345] labeled Perelman's achievements *"Breakthrough of the year,"* a distinction that *Science* had never given to any mathematical result. In the same year, Smith posted a paper on arXiv which was supposed to have solved the question about the Navier–Stokes equations. Smith withdrew her paper within two weeks, after a serious flaw was found in her proof, but *Nature* [235] had already given a large echo to Smith's paper.

This buzz in the media has been severely criticized by mathematicians. Vershik [489] expressed his doubts about the utility of the Millennium Prizes.

> *Around the year 2000 /.../ I met my old friend Arthur Jaffe, who was then president of the Clay Mathematics Institute. I asked him: "What is this being done for?" At the time I felt that the assignment of huge (million dollar) prizes was more in keeping with the style of show business, aiming at drawing attention to something or somebody at any price, whereas scientific life should avoid cheap popularization.*

> */.../ Arthur answered me decidedly and professionally: 'You understand nothing about the American way of life. If a politician, a businessman, a housewife will see that one can earn a million by doing mathematics, they will not discourage their children from choosing that profession, will not insist on their doing medicine, law, or going in for some lucrative activity. And other rich philanthropists will be more likely to give money to mathematics, which is in such need of it'.*

One interesting remark of Vershik about the Prizes states:

> *I would also like to note that the stir created around the seven "Millennium prizes" creates the wrong impression in society about the work of mathematicians, supporting the hackneyed notion that it consists only in solving concrete problems. You don't have to be an expert to understand how misleading that notion is. The discovery of new domains and relationships between different branches of mathematics, the setting of new problems, the development and perfection of the mathematical apparatus, and so on, are no less important and difficult parts of our science, without which it cannot exist.*

In *The Millennium Prize Problems* [91], a collective book edited by the Clay Mathematics Institute, Gray [216] makes a presentation of the history of prizes and challenges in mathematics, with a special emphasis on the 18th and the 19th centuries and with a section devoted to the Hilbert problems. The aim of the chapter is, of course, to illustrate the *"tradition of stimulating problems that the Clay Mathematics Institute has also sought to promote,"* but it indicates as well that financial motivation could lead to some disaster, as in the case of the Wolfskehl prize, offered for a solution of Fermat's Last Theorem:

> *From some perspectives, such as generating enthusiasm for mathematics, the prize was a great success; from others, such as the advancement of knowledge, it was a complete disaster. In the first year [1907] no fewer than 621 solutions were submitted, and over the years more than 5,000 came in. These had to be read, the errors spotted, and the authors informed, who often replied with attempts to fix their 'proofs'.*

This *Millennium Prize Problems* book [91] presents the official description of each of the seven problems, given by eminent specialists of the field (while a book by Devlin [152] gives a loose description of the problems at stake, just sketching the general background of each problem in an effort of popularization toward the largest audience):

- The Birch and Swinnerton-Dyer Conjecture: the official description has been given by Wiles [503], the famous number theorist who proved Fermat's last theorem. This conjecture is a problem in Diophantine analysis. The question is to determine whether an algebraic equation $f(x,y) = 0$, where f is a polynomial with rational coefficients has rational solutions and how many. The answer depends on the genus of the curve defined by the equation $f(x,y) = 0$. When this genus is equal to 0, Hilbert and Hurwitz proved in 1890 that either there are no solutions or there are infinitely many. When this genus is no less than 2, then Faltings proved in 1980 that there are only finitely many solutions. The question is more complex when the genus is equal to 1. In that case, one can take a model of the curve of the form $y^2 = x^3 + ax + b$ with integer coefficients $a, b \in \mathbb{Z}$ and a non-null discriminant $\Delta = 4a^3 + 27b^2$. The Birch and Swinnerton-Dyer Conjecture relates the number of rational points to the numbers N_p of solutions of the same equation in the field $\mathbb{Z}/p$ for every prime p, which is not a divisor of 2Δ.

In 1965, Birch and Swinnerton-Dyer based their conjecture on computer simulations. If the curve $\mathcal{C}$ has a rational point, then there is a natural law group on the rational points of the curve that makes the set of rational points an Abelian group $\mathcal{C}(\mathbb{Q})$. In 1922, Mordell proved that the group is finitely generated, thus isomorphic to some $\mathbb{Z}^r \otimes F$, where F is finite; r is called the rank of the group $\mathcal{C}(\mathbb{Q})$. The Hasse L-function of the curve $\mathcal{C}$ is defined for $\Re s > 3/2$ as

$$L(\mathcal{C}, s) = \prod_{p \text{ prime } ; \ 2\Delta \notin p\mathbb{N}} \frac{1}{\left(1 - \frac{p - N_p}{p^s} + \frac{1}{p^{2s-1}}\right)}$$

A by-product of the proof by Wiles and Taylor of Fermat's Last Theorem gives that the Hasse function may be continued as a holomorphic function on the plane. The official Clay Millennium Problem is then to solve the following conjecture:

Conjecture (Birch and Swinnerton-Dyer)

The Taylor expansion of $L(C, s)$ at $s = 1$ has the form

$$L(C, s) = c(s - 1)^r + \text{ higher order terms}$$

with $c \neq 0$ and $r = \text{rank}\,(C(\mathbb{Q}))$.

Note that, if this conjecture is true, the equation has infinitely many rational solutions if and only if $L(\mathcal{C}, 1) = 0$.

- The Hodge Conjecture: the official description has been given by the 1978 Fields medalist Deligne [151]. The official Clay Millennium Problem is to solve the following conjecture:

Hodge Conjecture

On a projective non-singular algebraic variety over $\mathbf{C}$, any Hodge class is a rational linear combination of classes $cl(Z)$ of algebraic cycles.

The conjecture was presented by Hodge at the International Congress of Mathematicians in 1950. This conjecture concerns harmonic differential forms on a projective non-singular complex algebraic variety. It states that every rational harmonic (p, p)-form on the variety is (modulo exact forms) a rational linear combination of algebraic cycles, i.e., of classes induced by algebraic subvarieties of complex co-dimension p.

- Navier–Stokes Existence and Smoothness: the official description has been given by the 1978 Fields medalist Fefferman [171]. The official Clay Millennium Problem is then the following one:

Navier–Stokes existence and smoothness

We ask for a proof of one of the four following statements:

- A) Existence and smoothness of Navier–Stokes solutions on $\mathbb{R}^3$
- B) Existence and smoothness of Navier–Stokes solutions on $\mathbb{R}^3/\mathbb{Z}^3$
- C) Breakdown of Navier–Stokes solutions on $\mathbb{R}^3$
- D) Breakdown of Navier–Stokes solutions on $\mathbb{R}^3/\mathbb{Z}^3$

The problem concerns the initial value problem for a fluid that fills the whole space (so that there is no boundary problem) and which is viscous, homogeneous and incompressible. The question raised is whether, for a smooth initial value, the Navier–Stokes problem has a (unique) global smooth solution or whether one can exhibit an example of initial value for which the solution blows up in finite time. This question appears in the work of Leray who proved in 1934 global existence of weak solutions which may be non-unique and irregular. We shall discuss in Section 1.3 the terms of this problem to a greater extent.

- **P** versus **NP**: the official description has been given by the 1982 Turing Award winner Cook [130]. The official Clay Millennium Problem has a very simple statement:

Problem statement

Does **P**=**NP**?

The "Does **P**= **NP**?" problem appeared in 1971–1973 in the independent works of Cook, Karp and Levin in complexity theory. The class **P** is the class of decision problems solvable by some algorithm within a number of steps bounded by some fixed polynomial in the length of the input, while the class **NP** is the class of problems whose proposed solutions can be checked in polynomial time.

- The Poincaré Conjecture: the official description has been given by the 1962 Fields medalist and 2011 Abel Prize winner Milnor [362]. While the classification of all possible orientable compact two-dimensional surfaces has been well understood in the 19th century, the problem turned out to be much more complex in higher dimensions. In 1904, Poincaré formulated a conjecture that remained unsolved all along the 20th century. The Clay Millennium Problem was to prove the Poincaré Conjecture:

Question [the Poincaré Conjecture]

If a compact three-dimensional manifold M^3 has the property that every simple closed curve within the manifold can be deformed continuously to a point, does it follow that M^3 is homeomorphic to the sphere S^3? [The manifold M^3 is assumed to be connected and with no border.]

The analogue of this conjecture had been proved in higher dimension, by Smale in 1961 for dimensions greater than four, then by Freedman in 1982 for the four-dimensional case. The conjecture was finally proven by Perelman in 2002–2003. On March 18, 2010, Carlson, on behalf of the Clay Mathematics Institute, announced that the conjecture was proved and the prize awarded [90].

- The Riemann Hypothesis: the official description has been given by the 1974 Fields medalist Bombieri [50]. This is a famous problem in mathematics history. It deals with the Riemann zeta function $\zeta(s)$; this function is defined for $\Re(s) > 1$ as the series

$$\zeta(s) = \sum_{n=1}^{\infty} \frac{1}{n^s}$$

and then is prolongated by analytic continuation to a holomorphic function of $s \neq 1$ (with a simple pole at $s = 1$). It is easy to see that the negative even numbers $-2, -4, -6, \ldots$ are zeroes of the function ζ. They are called trivial zeroes. Riemann conjectured in 1859 that all the other zeroes should satisfy $\Re(s) = 1/2$. The Clay Millennium Problem is to prove the Riemann Hypothesis:

Riemann Hypothesis

The non-trivial zeroes of $\zeta(s)$ have real part equal to $\frac{1}{2}$.

This conjecture is important to our knowledge of prime numbers but has other far-reaching consequences as evoked by Bombieri in his presentation of the problem. To prove the Riemann Hypothesis was already one of the twenty-three problems Hilbert had listed for the 20th century.

- Quantum Yang–Mills theory: the official description has been given by the specialist of constructive quantum field theory and founding President of the Clay Mathematics Institute Jaffe and the 1990 Fields medalist Witten [243]. The problem concerns quantum field theory. In 1954, Yang and Mills introduced a non-Abelian gauge theory to modelize quantum electrodynamics and obtained a non-linear generalization of Maxwell's equations. The problem at stake now is to develop a gauge theory for the modelization of weak interactions and strong interactions. Those forces involve massive particles and require new tools, since the model of Yang and Mills dealt with long-range fields describing massless particles. Nowadays, we still are lacking a mathematically complete example of a quantum gauge theory in four-dimensional space-time. The official Clay Millennium Problem is thus the following one:

Yang–Mills Existence and Mass Gap

Prove that for any compact simple gauge group G, a non-trivial quantum Yang–Mills theory exists on $\mathbb{R}^4$ and has a positive mass gap $\Delta > 0$. Existence includes establishing axiomatic properties at least as strong as those cited in [387, 388, 453].

1.3 The Clay Millennium Prize for the Navier–Stokes Equations

We now turn to the precise formulation of the Clay Millennium Problem on Navier–Stokes equations. The **Navier–Stokes equations** considered in this formulation are the following partial differential equations:

$$\partial_t \vec{u}(t,x) = \nu \Delta \vec{u} - \sum_{i=1}^{3} u_i \partial_i \vec{u} - \vec{\nabla} p + \vec{f} \quad \text{for } t > 0 \text{ and } x \in \mathbb{R}^3 \tag{1.1}$$

$$\operatorname{div} \vec{u} = \sum_{i=1}^{3} \partial_i u_i = 0 \tag{1.2}$$

with initial condition

$$\vec{u}(0,x) = \vec{u}_0(x) \quad \text{for } x \in \mathbb{R}^3 \tag{1.3}$$

where ∂_i stands for $\frac{\partial}{\partial x_i}$, ∂_t for $\frac{\partial}{\partial t}$ and $\vec{\nabla}$ for the gradient operator $\vec{\nabla} = \begin{pmatrix} \partial_1 \\ \partial_2 \\ \partial_3 \end{pmatrix}$.

The unknown are $\vec{u}(t,x) = \begin{pmatrix} u_1(t,x) \\ u_2(t,x) \\ u_3(t,x) \end{pmatrix}$ and $p(t,x)$. The equations describe the motion of a fluid filling the whole space $\mathbb{R}^3$. The vector $\vec{u}$ is the velocity of the fluid element that, at time t, occupies the position x. The scalar quantity p measures the pressure exerted on the fluid element.

The fluid is assumed to be homogeneous and incompressible. Incompressibility is expressed by Equation (1.2). The constant density ρ is taken equal to 1. The fluid is assumed to be viscous and Newtonian, i.e., the friction of fluid elements of different velocities generates a force of the form $\nu \Delta \vec{u}$, where ν is a positive constant (the viscosity) and Δ is the Laplacian $\Delta = \sum_{i=1}^{3} \partial_i^2$. Finally, the fluid may be submitted to external forces; the force density is expressed by the vector $\vec{f}(t,x)$. The forces expressed by $\vec{f}$ are assumed to be independent from the velocity field $\vec{u}$ (for instance, the problem does not concern fluids in a rotating frame, submitted to the Coriolis force $\vec{\Omega} \wedge \vec{u}$).

Solving Equations (1.1), (1.2) and (1.3) is a Cauchy initial value problem. Given the initial state $\vec{u}_0$ at time $t = 0$ and the force $\vec{f}$ for $t > 0$, one wants to determine the evolution of the system for $t > 0$. For the Clay Millennium Problem, one assumes that the initial data $\vec{u}_0$ and the force $\vec{f}$ are given by smooth and well-localized functions: $\vec{u}_0$ is a C^{∞} divergence-free vector field on $\mathbb{R}^3$ such that, for all $\alpha \in \mathbb{N}^3$ and all $K > 0$,

$$|\partial^{\alpha} \vec{u}_0(x)| \leq C_{\alpha,K}(1+|x|)^{-K} \quad \text{on } \mathbb{R}^3 \tag{1.4}$$

(where $\left| \begin{pmatrix} x_1 \\ x_2 \\ x_3 \end{pmatrix} \right| = \sqrt{x_1^2 + x_2^2 + x_3^2}$); similarly, $\vec{f}$ is C^{∞} on $[0,+\infty) \times \mathbb{R}^3$ and satisfies for all $\alpha \in \mathbb{N}^3$, all $m \in \mathbb{N}$ and all $K > 0$

$$|\partial_x^{\alpha} \partial_t^m \vec{f}(t,x)| \leq C_{\alpha,m,K}(1+t+|x|)^{-K} \quad \text{on } [0,+\infty) \times \mathbb{R}^3 \tag{1.5}$$

Admissible solutions are smooth functions $\vec{u}$ and p with bounded energy:

$$p, \vec{u} \in C^{\infty}([0,+\infty) \times \mathbb{R}^3) \tag{1.6}$$

$$\int_{\mathbb{R}^3} |\vec{u}(t,x)|^2 \, dx < C \quad \text{for all } t > 0 \quad \text{(bounded energy)} \tag{1.7}$$

There is no need to specify the value of p at time $t = 0$. Indeed, we have:

$$\Delta \vec{u} = -\operatorname{curl}(\operatorname{curl} \vec{u}) + \vec{\nabla}(\operatorname{div} \vec{u}) = -\operatorname{curl}(\operatorname{curl} \vec{u})$$

and, since the size of $\vec{u}$ at $x = \infty$ is limited by the condition of integrability (1.7), $\vec{u}$ is uniquely determined through its curl. Let

$$\vec{\omega} = \operatorname{curl} \vec{u} = \vec{\nabla} \wedge \vec{u} = \begin{pmatrix} \partial_2 u_3 - \partial_3 u_2 \\ \partial_3 u_1 - \partial_1 u_3 \\ \partial_1 u_2 - \partial_2 u_1 \end{pmatrix};$$

ω is called the **vorticity** of the fluid.

Taking the curl of Equation (1.1) gives

$$\partial_t \vec{\omega}(t,x) = \nu \Delta \vec{\omega} - \operatorname{curl}(\sum_{i=1}^{3} u_i \partial_i \vec{u}) + \operatorname{curl} \vec{f} \quad \text{for } t > 0 \text{ and } x \in \mathbb{R}^3 \qquad (1.8)$$

with initial condition

$$\vec{\omega}(0,x) = \operatorname{curl} \vec{u}_0(x) \quad \text{for } x \in \mathbb{R}^3. \qquad (1.9)$$

Thus, we have a Cauchy initial value problem for $\vec{\omega}$ with no dependence on p. When $\vec{\omega}$, and thus $\vec{u}$, is known, p is determined by Equation (1.1).

Now, we can state precisely the Clay Millennium Problem for a viscous fluid ($\nu > 0$):

Navier–Stokes existence and smoothness (whole space)

We ask for a proof of one of the two following statements:

- A) Existence and smoothness of Navier–Stokes solutions on $\mathbb{R}^3$: Let $\vec{u}_0$ be any smooth, divergence-free vector field satisfying (1.4). Take $\vec{f}(t,x)$ to be identically zero. Then there exist smooth functions $p(t,x)$, $u_i(t,x)$ on $\mathbb{R}^3 \times [0,\infty)$ that satisfy (1.1), (1.2), (1.3), (1.6) and (1.7).

- C) Breakdown of Navier–Stokes solutions on $\mathbb{R}^3$: There exist a smooth, divergence-free vector field $\vec{u}_0$ on $\mathbb{R}^3$ and a smooth $\vec{f}$ on $[0,+\infty) \times \mathbb{R}^3$ satisfying (1.4) and (1.5) for which there exists no solution $(p,\vec{u})$ of (1.1), (1.2), (1.3), (1.6) and (1.7) on $[0,+\infty) \times \mathbb{R}^3$.

The Clay Millennium Problem may also be solved on a compact domain instead of the whole space. In order to avoid boundary terms, the domain is assumed to be the torus $\mathbb{R}^3/\mathbb{Z}^3$, i.e., one deals with periodical functions. Hypotheses (1.4) and (1.5) are replaced with:

- $\vec{u}_0$ and $\vec{f}$ are smooth and satisfy

$$\vec{u}_0(x+k) = \vec{u}_0(x) \text{ and } \vec{f}(t,x+k) = \vec{f}(t,x) \quad \text{for all } k \in \mathbb{R}^3 \qquad (1.10)$$

- for all $\alpha \in \mathbb{N}^3$, all $m \in \mathbb{N}$ and all $K > 0$

$$|\partial_x^\alpha \partial_t^m \vec{f}(t,x)| \leq C_{\alpha,m,K}(1+t)^{-K} \quad \text{on } [0,+\infty) \times \mathbb{R}^3 \qquad (1.11)$$

Admissible solutions are smooth functions $\vec{u}$ and p such that:

$$p, \vec{u} \in \mathcal{C}^\infty([0,+\infty) \times \mathbb{R}^3) \qquad (1.12)$$

$$\vec{u}(t,x+k) = \vec{u}(t,x) \quad \text{for all } k \in \mathbb{R}^3 \qquad (1.13)$$

The statement of the Clay Millennium Problem in the periodical case is then the following one:

Navier–Stokes existence and smoothness (torus)

We ask for a proof of one of the two following statements:

- B) Existence and smoothness of Navier–Stokes solutions on $\mathbb{R}^3/\mathbb{Z}^3$: Let $\vec{u}_0$ be any smooth, divergence-free vector field satisfying (1.10). Take $\vec{f}(t,x)$ to be identically zero. Then there exist smooth functions $p(t,x)$, $u_i(t,x)$ on $\mathbb{R}^3 \times [0,\infty)$ that satisfy (1.1), (1.2), (1.3), (1.12) and (1.13).

- D) Breakdown of Navier–Stokes solutions on $\mathbb{R}^3/\mathbb{Z}^3$: There exist a smooth, divergence-free vector field $\vec{u}_0$ on $\mathbb{R}^3$ and a smooth $\vec{f}$ on $[0,+\infty) \times \mathbb{R}^3$ satisfying (1.10) and (1.11) for which there exists no solution $(p, \vec{u})$ of (1.1), (1.2), (1.3), (1.12) and (1.13) on $[0,+\infty) \times \mathbb{R}^3$.

Remark: If we want to get rid of the pressure, we may take the equations (1.8) on the vorticity; but $\vec{u}$ has to be uniquely determined from $\vec{\omega}$. In the setting of the whole space, this is ensured by the spatial decay hypothesis on $\vec{u}$ at infinity. In the setting of periodic solutions, this is ensured by the hypothesis that $\int \vec{u}(t,x)\,dx = 0$ (or, equivalently when $\vec{f} = 0$, that p is periodical). Otherwise, we have the trivial example of non-uniqueness for $\vec{u}_0 = 0$ given by $\vec{u}(t,x) = \vec{v}(t)$ with no dependence on x, associated with the pressure $p(t,x) = -\vec{x}.\frac{d}{dt}\vec{v}(t)$, where $\vec{v}$ is any smooth function with $\vec{v}(0) = 0$. This trivial example can be turned into an example of blow-up by choosing a blowing up arbitrary function $\vec{v}(t)$.

Such examples were discussed by Giga, Inui and Matsui in 1999 when they considered non-decaying initial data for the Navier–Stokes problem [210], and by Koch, Nadirashvili, Seregin and Šverák in 2007 [265] when they considered a Liouville theorem for the Navier-Stokes problem and had to rule out those "parasitic solutions."

1.4   Boundaries and the Navier–Stokes Clay Millennium Problem

On the website of the Clay Mathematics Institute, the problem is presented in these words (http://www.claymath.org/Millennium/Navier-Stokes_Equations/):

> *Waves follow our boat as we meander across the lake, and turbulent air currents follow our flight in a modern jet. Mathematicians and physicists believe that an explanation for and the prediction of both the breeze and the turbulence can be found through an understanding of solutions to the Navier–Stokes equations. Although these equations were written down in the 19th Century, our understanding of them remains minimal. The challenge is to make substantial progress toward a mathematical theory which will unlock the secrets hidden in the Navier–Stokes equations.*

Thus, the aim is to eventually understand turbulence. But some severe doubts have been raised about the model case proposed for the Millennium Prize. For instance, Tartar writes in the preface of his book [464]:

> *Reading the text of the conjecture to be solved for winning that particular prize leaves the impression that the subject was not chosen by people interested in*

continuum mechanics, as the selected question has almost no physical content. Invariance by translation or scaling is mentioned, but why is invariance by rotations not pointed out and why is Galilean invariance omitted, as it is the essential fact which makes the equation introduced by Navier much better than that introduced by Stokes? If one used the word 'turbulence' to make the donator believe that he would be giving one million dollars away for an important realistic problem in continuum mechanics, why has attention been restricted to unrealistic domains without boundary (the whole space $\mathbb{R}^3$, or a torus for periodic solutions), as if one did not know that vorticity is created at the boundary of the domain? The problems seem to have been chosen in the hope that they will be solved by specialists of harmonic analysis.

However, the question of regularity of the solutions to the Navier–Stokes equations even when neglecting the influence of the boundary is generally considered as a major issue in hydrodynamics. For instance, Moffatt says about singularities in fluid dynamics [366]:

Singularities may be associated with the geometry of the fluid boundary or with some singular feature of the motion of the boundaries; they may arise spontaneously at a free surface as a result of viscous stresses and despite the smoothing effect of surface tension; or they may conceivably occur at interior points of a fluid due to unbounded vortex stretching at high (or infinite) Reynolds number. In the last case, we are up against the unsolved and extremely challenging 'finite-time-singularity' problem for the Euler and/or Navier–Stokes equations. The question of existence of finite-time singularities is still open. Solution of this problem would have far-reaching consequences for our understanding of the smallest-scale features of turbulent flow.

When dealing with models of fully developed turbulence, one often uses an asymptotic model, which has spatially homogeneous statistical properties; then, one neglects the boundary effects and works in the setting of the whole space (and in most of the cases in a space-periodical setting, as it is easier to define and compute the statistical quantities that are involved in the model).

In this book, we shall stick to the boundary-free[3] Navier–Stokes problem, i.e., when the domain is the whole space $\mathbb{R}^3$.

---

[3]I.e., on a domain without boundary (and not on a domain with free boundary).

Chapter 2

The Physical Meaning of the Navier–Stokes Equations

In this chapter, we try to give a short presentation of each term in the Navier–Stokes equations, to explain how and why they are introduced in fluid mechanics. A classical treaty on hydrodynamics from the physicists' point of view is the book by Landau and Lifschitz [302], and a rapid introduction can be found in Feynman's lecture notes [175]. The mathematicians' point of view can be found in the classical treaty of Batchelor [25], or for a modern point of view in the book by Childress [121].

2.1 Frames of References

Fluid theory is based on a continuum hypothesis [25] which states that the macroscopic behavior of a fluid is the same as if the fluid was perfectly continuous: density, pressure, temperature, and velocity are taken to be well-defined at infinitely small points and are assumed to vary continuously from one point to another.

Since the seminal memoir of Euler [167], one describes the laws of fluid mechanics as applied to **fluid parcels**, very small volumes δV of fluids that contain many molecules but whose size is "infinitesimal" with respect to the macroscopic scale. Then the physical properties of the parcels are defined as averages of the associated continuously varying quantities: for instance, the temperature $\theta_{\delta V}$ of the parcel is given by $\theta_{\delta V} = \frac{1}{|\delta V|} \int_{\delta V} \theta(t,x) \, dx$, where θ is the temperature defined at point x and at time t.

There are then two representations of the fluid motion and of the associated physical quantities. In the **Eulerian reference frame**, the reference frame is fixed while the fluid moves. Thus, the quantities are measured at a position x attached to the fixed frame (one often speaks of the "laboratory frame"). The velocity $\vec{u}(t,x)$ is the velocity at time t of the fluid parcel that occupies the position x at that very instant t. In the **Lagrangian reference frame**, the reference frame is the initial state of the fluid. The quantities are attached to the parcels as they move.

More precisely, if $X_{x_0}(t)$ is the position of the parcel at time t whose position at time 0 was x_0, and if Q is some quantity attached to the parcels, we have two descriptions of the distribution of the values taken by Q at time t: the value $Q(t,x)$ taken at time t for the parcel which is located at this time at position x, and $Q_{x_0}(t)$ the value taken at time t for the parcel which was located at time 0 at position x_0. In particular, the velocity field $\vec{u}(t,x)$ describes the velocities of the parcels as they move: $\frac{d}{dt}X_{x_0}(t) = \vec{u}(t, X_{x_0}(t))$. This gives us the link between the variations of $Q_{x_0}(t)$ and those of $Q(t,x)$: from the chain rule for differentiation, we get

$$\frac{d}{dt}Q_{x_0}(t) = \partial_t Q(x,t)_{|x=X_{x_0}(t)} + \sum_{i=1}^{3} \partial_i Q(x,t)_{|x=X_{x_0}(t)} \frac{d}{dt}X_{x_0,i}(t)$$

DOI: 10.1201/9781003042594-2

The quantity $\frac{d}{dt}Q_{x_0}(t)$ is called the **material derivative** of Q and is designed as $\frac{D}{Dt}Q$. We have thus obtained the following formula:

The material derivative

$$\frac{D}{Dt}Q = \partial_t Q(x,t) + \sum_{i=1}^{3} u_i(t,x)\partial_i Q(x,t) \qquad (2.1)$$

2.2   The Convection Theorem

If we consider a volume V_0 at time 0 filled of fluid parcels, and define V_t the volume filled by the parcels as they moved, we have

$$V_t = \{y \in \mathbb{R}^3 \ / \ y = X_x(t) \text{ for some } x \in V_0\}.$$

The volume element dy of V_t is given by $J(t,x)\,dx$, where J is the Jacobian of the transform $x \mapsto X_x(t)$. We have

$$J = \left| \det\left(\frac{\partial}{\partial x_i} y_j \right)_{1 \le i,j \le 3} \right|.$$

Let $\mathcal{J}(t,x) = \det\left(\frac{\partial}{\partial x_i} y_j \right)_{1 \le i,j \le 3}$; we have

$$\partial_t \frac{\partial}{\partial x_i} y_j = \frac{\partial}{\partial x_i} \partial_t y_j = \frac{\partial}{\partial x_i} u_j(t,y) = \sum_{k=1}^{3} \frac{\partial}{\partial y_k} u_j(t,y) \frac{\partial}{\partial x_i} y_k$$

and thus

$$\partial_t \mathcal{J} = \det(\partial_t \frac{\partial}{\partial x} y_1, \frac{\partial}{\partial x} y_2, \frac{\partial}{\partial x} y_3) + \det(\frac{\partial}{\partial x} y_1, \partial_t \frac{\partial}{\partial x} y_2, \frac{\partial}{\partial x} y_3)$$
$$+ \det(\frac{\partial}{\partial x} y_1, \frac{\partial}{\partial x} y_2, \partial_t \frac{\partial}{\partial x} y_3)$$
$$= \sum_{k=1}^{3} \frac{\partial}{\partial y_k} u_1(t,y) \det(\frac{\partial}{\partial x} y_k, \frac{\partial}{\partial x} y_2, \frac{\partial}{\partial x} y_3)$$
$$+ \sum_{k=1}^{3} \frac{\partial}{\partial y_k} u_2(t,y) \det(\frac{\partial}{\partial x} y_1, \frac{\partial}{\partial x} y_k, \frac{\partial}{\partial x} y_3)$$
$$+ \sum_{k=1}^{3} \frac{\partial}{\partial y_k} u_3(t,y) \det(\frac{\partial}{\partial x} y_1, \frac{\partial}{\partial x} y_2, \frac{\partial}{\partial x} y_k)$$
$$= \operatorname{div} \vec{u}(t,y) \, \mathcal{J}$$

so that, since $J(0,x) = 1$,

$$J(t,x) = e^{\int_0^t \operatorname{div} \vec{u}(s, X_x(s))\, ds} \qquad (2.2)$$

Thus, we have seen that the **divergence** of $\vec{u}$ is the quantity that governs the deflation or the inflation of the volume of V_t.

Now, if $f(t, x)$ is a time-dependent field over $\mathbb{R}^3$, we may define $F(t) = \int_{V_t} f(t, y)\, dy$. We have

$$F(t) = \int_{V_0} f(t, X_x(t)) J(t, x)\, dx$$

We use the fact that $\partial_t[f(t, X_x(t))] = \frac{D}{Dt}f(t, y)$ and $\partial_t J(t, x) = \operatorname{div} \vec{u}(t, y) J(t, x)$ and $J(t, x)\, dx = dy$ to get the convection theorem:

The convection theorem

$$\frac{d}{dt} \int_{V_t} f(t, y)\, dy = \int_{V_t} \frac{D}{Dt}f(t, y) + f(t, y) \operatorname{div} \vec{u}(t, y)\, dy \qquad (2.3)$$

Writing $\frac{D}{Dt}f + f \operatorname{div} \vec{u} = \partial_t f + \vec{u}.\vec{\nabla}f + f \operatorname{div} \vec{u} = \partial_t f + \operatorname{div}(f\vec{u})$, and using Ostrogradski's formula, we find, writing $d\sigma$ for the surface element of the boundary ∂V_t and $\vec{\nu}$ for the normal at $\partial_t V$ pointing outward:

$$\frac{d}{dt} \int_{V_t} f(t, y)\, dy = \int_{V_t} \partial_t f\, dy + \int_{\partial V_t} f\vec{u}.\vec{\nu}\, d\sigma \qquad (2.4)$$

This is a special case of Reynolds' transport theorem.

2.3 Conservation of Mass

We apply the convection theorem to the mass m of the parcels included in the volume V_t. If $\rho(t, y)$ is the density at time t and at position y, we have $m = \int_{V_t} \rho(t, y)\, dy$. When the parcels move, their mass is conserved, so we find that $\frac{d}{dt}m = 0$. For this identity to be valid for any initial volume V_0, this gives the equation of conservation of mass:

Conservation of mass

$$\frac{D}{Dt}\rho + \rho \operatorname{div} \vec{u} = 0 \qquad (2.5)$$

When the fluid is incompressible, the density of a given parcel cannot change, so that $\frac{D}{Dt}\rho = 0$, hence we find (in absence of vacuum or null-density areas)

Incompressibility

$$\operatorname{div} \vec{u} = 0 \qquad (2.6)$$

This is consistent with Equation (2.2): if div $\vec{u} = 0$, then the volume occupied by a parcel never varies.

For an incompressible fluid, we find that $\partial_t \rho = -\vec{u}.\vec{\nabla}\rho$. If the fluid is homogeneous, the density does not depend on the position, thus we find $\frac{d}{dt}\rho(t) = 0$; the density is constant in time and in space:

Incompressibility and homogeneity

$$\rho = Constant \tag{2.7}$$

2.4   Newton's Second Law

We apply Newton's second law to a moving parcel of fluid. The momentum of the parcel at time t is given by $M = \int_{V_t} \rho(t,y)\vec{u}(t,y)\ dy$. If $\vec{f}(t,y)$ is the force density at time t and position y, the force applied to the parcel is $\vec{F} = \int_{V_t} \vec{f}(t,y)\ dy$. Newton's second law of mechanics then gives that

$$\frac{d}{dt}M = \vec{F}.$$

The convection theorem gives then

$$\int_{V_t} \frac{D}{Dt}(\rho\vec{u}) + \rho\vec{u}\operatorname{div}\vec{u} - \vec{f}\ dy = 0$$

Equation (2.5) gives $\frac{D}{Dt}\rho + \rho\operatorname{div}\vec{u} = 0$, hence we have (taking infinitesimal volume V_0)

Newton's second law

$$\rho\frac{D}{Dt}\vec{u} = \vec{f} \tag{2.8}$$

This can be written as well as

$$\rho\left(\partial_t\vec{u} + (\vec{u}.\vec{\nabla})\vec{u}\right) = \vec{f} \tag{2.9}$$

where $\vec{u}.\vec{\nabla} = \sum_{i=1}^{3} u_i\partial_i$. Of course, there remains to describe the force density $\vec{f}$. This is the resultant of several forces: exterior forces (such as gravity) and internal forces. In the next sections, we consider two important types of internal forces: the force induced by pressure and the force induced by friction.

Remark: This balance of momentum is classical in fluid mechanics since the seminal memoir of Euler [167]. However, it has been recently disputed by H. Brenner [65] who argues that one must distinguish between the (Eulerian) mass transportation velocity $\vec{u}_m$ and the (Lagrangian) particle velocity $\vec{u}_v$. Thus, we would have instead of (2.1) the equation $\frac{D}{Dt}Q = \partial_t Q(x,t) + \sum_{i=1}^{3} u_{m,i}(t,x)\partial_i Q(x,t)$, the continuity Equation (2.5) would become

$\frac{D}{Dt}\rho + \rho \operatorname{div}\vec{u}_m = 0$ and the balance of momentum (2.8) would become $\rho\frac{D}{Dt}\vec{u}_v = \vec{f}$. One then needs a constitutive law to describe the difference $\vec{u}_v - \vec{u}_m$. Brenner proposed the law

$$\vec{u}_v - \vec{u}_m = K\vec{\nabla}\rho.$$

Thus, the equations should be modified in case of compressible fluids with high density gradients, while for uncompressible homogeneous fluids the classical equations of fluid mechanics would still be valid.

A study of the Brenner model has been performed by Feireisl and Vasseur [172] who showed that the weak solutions for this model are more regular than the weak solutions for the classical Navier–Stokes equations for highly compressible fluids.

The story does not stop with Brenner's model, which remains disputed. Various models of extended Navier–Stokes or Euler equations have been recently discussed, as for instance by Svärd in 2018 [457] or by Reddy, Dadzie, Ocone, Borg and Reese in 2019 [407].

2.5 Pressure

When a fluid is in contact with a body, it exerts on the surface of the body a force that is normal to the surface and called the pressure. The pressure is a scalar quantity, which does not depend on the direction of the normal. Positive pressure gives a compression force that points inward of the body, so that is opposed to the normal.

Internal pressure (or **static pressure**) is defined in an analogous way. The fluid parcel occupies a volume δV; the force exerted on the parcel induced by the pressure is then $\vec{F}_P = -\int_{\partial \delta V} p\vec{\nu}\, d\sigma$. This can be rewritten with Ostrogradski's formula into the following equation:

$$\vec{F}_P = -\int_V \vec{\nabla}p\, dx.$$

This gives us the density for the pressure force:

Force density for the pressure

$$\vec{f}_P = -\vec{\nabla}p \tag{2.10}$$

2.6 Strain

Fluids are not rigid bodies. Thus, their motion implies deformations. Those deformations may be illustrated through the strain tensor. If the velocities and their derivatives are small enough, we may estimate for two initial points x_0 and y_0 how the distance of the parcels will evolve. Indeed, if $x(t) = X_{x_0}(t)$ and $y(t) = X_{y_0}(t)$, we have

$$\|x - y\|^2 = \|x_0 - y_0\|^2 + 2\int_0^t (x(s) - y(s)).(\vec{u}(s, x(s)) - \vec{u}(s, y(s)))\, ds$$

and, neglecting terms of higher order, we get

$$\|x - y\|^2 \approx \|x_0 - y_0\|^2 + 2 \int_0^t (x(s) - y(s)).Du(s, x(s))(x(s) - y(s)) \, ds$$

where the matrix Du is the matrix

$$Du = (\partial_j u_i(s, x))_{1 \leq i,j \leq 3}. \tag{2.11}$$

Cauchy's strain tensor ϵ is defined as the symmetric part of Du:

$$\epsilon = \frac{1}{2}\left(Du + (Du)^T\right). \tag{2.12}$$

The antisymmetric part has a null contribution to the integral, and we find:

$$\|x - y\|^2 \approx \|x_0 - y_0\|^2 + 2 \int_0^t (x(s) - y(s)).\epsilon(s, x(s))(x(s) - y(s)) \, ds$$

Cauchy's strain tensor

The strain tensor at time t and position x is the matrix ϵ given by

$$\epsilon_{i,j} = \frac{1}{2}(\partial_i u_j + \partial_j u_i) \quad \text{for } 1 \leq i, j \leq 3 \tag{2.13}$$

If we look at the infinitesimal displacement of y, we have

$$\frac{D}{Dt}y = \vec{u}(t, y) = \vec{u}(t, x) + \epsilon(y - x) + \frac{1}{2}(Du - (Du)^T)(y - x) + O((y - x)^2).$$

$\vec{u}(t, x)$ does not depend on y: it corresponds to an (infinitesimal) translation; $\frac{1}{2}(Du - (Du)^T)$ does not contribute to the distortion of distances, it corresponds to an (infinitesimal) rotation. ϵ corresponds to the (infinitesimal) deformation.

2.7   Stress

When a fluid is viscous, it reacts like an elastic body that resists deformations. Applying the theory of elasticity to the fluid motion, one can see that the deformations induce forces. If δV is a small parcel, the deformation of the parcel induces a force exerted on the border of δV; this force $\vec{F}_{visc}$ is given by a tensor $\mathbb{T}$ (the **viscous stress tensor**)[1] and we have

$$\vec{F}_{visc} = \int_{\partial \delta V} \mathbb{T}\vec{\nu} \, d\sigma$$

or, equivalently,

$$F_{visc,i} = \int_{\partial \delta V} \sum_{j=1}^3 T_{i,j}\nu_j \, d\sigma.$$

---

[1]The stress tensor is the sum $\mathbb{T} - p\mathbb{I}_3$, where p is the hydrostatic pressure.

Ostrogradski's formula gives us the force density $\vec{f}_{visc}$ associated to the stress:

$$f_{visc,i} = \sum_{j=1}^{3} \partial_j T_{i,j} = \operatorname{div} T_{i,.} \tag{2.14}$$

When the fluid velocity and its derivatives are small enough, Stokes has shown that the relation between the stress tensor and the strain tensor is linear. In the case of an isotropic fluid (so that the linear relation is the same at all points) we find that $\vec{f}_{visc}$ is a sum of second derivatives of $\vec{u}$. But, due to the isotropy of the fluid, a change of referential through a rotation should not alter the relation between the force and the velocity. This gives that $\vec{f}_{visc}$ is determined only by two viscosity coefficients[2]:

Force density associated to the stress

In an isotropic fluid with small velocities, we have

$$\vec{f}_{visc} = \mu\Delta\vec{u} + \lambda\vec{\nabla}(\operatorname{div}\vec{u}) \tag{2.15}$$

Equation (2.15) corresponds to a very simple relationship between the tensor ϵ and the tensor $\mathbb{T}$:

$$\mathbb{T} = 2\mu\epsilon + \eta\operatorname{tr}(\epsilon)\,\mathbb{I}_3 \tag{2.16}$$

with $\operatorname{tr}(\epsilon) = \epsilon_{1,1} + \epsilon_{2,2} + \epsilon_{3,3}$ and $\lambda = \mu + \eta$. μ is called the **dynamical viscosity** of the fluid, and η the **volume viscosity** of the fluid. Fluids for which the relation (2.16) holds are called **Newtonian fluids**. All gases and most liquids which have simple molecular formula and low molecular weight such as water, benzene, ethyl alcohol, etc. are Newtonian fluids. In contrast, polymer solutions are non-Newtonian.

Stokes [451] has expressed the notion of internal pressure in a very general principle that allowed, a hundred years later, Reiner [408] and Rivlin [414] to describe a more general class of fluids. For a Stokesian fluid, the stress tensor $\mathbb{T}$ is still related to the strain tensor ϵ in a homogeneous and isotropic way, but the relationship is no longer linear. Following Serrin [431, 432] and Aris [6], a Stokesian fluid satisfies the following four assumptions:

- the stress tensor $\mathbb{T}$ is a continuous function of the strain tensor ϵ and the local thermodynamical state, but independent of other kinematical properties

- $\mathbb{T}$ does not depend explicitly on x (fluid homogeneity)

- the fluid is isotropic

- when there is no deformation ($\epsilon = 0$), the fluid is hydrostatic ($\mathbb{T} = 0$).

Then, using the symmetries induced by the principle of *material objectivity* or of *frame indifference* (see Noll and Truesdell [377]) which states that "the constitutive laws governing the internal conditions of a physical system and the interactions between its parts should not depend on whatever external frame of reference," Serrin showed that the viscous stress tensor can be expressed as

$$\mathbb{T} = \alpha\,\mathbb{I}_3 + \beta\,\epsilon + \gamma\epsilon^2 \tag{2.17}$$

---

[2]This is expressed by Feynman [175] in the following terms:

the most general form of second derivatives that can occur in a vector equation is a sum of a term in the Laplacian ($\nabla.\nabla\mathbf{v} = \nabla^2 v$), and a term in the gradient of the divergence ($\nabla(\nabla.\mathbf{v})$).

where $\alpha(0,0,0) = 0$ and $\alpha = \alpha(\Theta, \Phi, \Psi)$, $\beta = \beta(\Theta, \Phi, \Psi)$ and $\gamma = \gamma(\Theta, \Phi, \Psi)$ are functions of the three invariants of the symmetric matrix ϵ : if the eigenvalues of ϵ are λ_1, λ_2 and λ_3, then $\Theta = \lambda_1 + \lambda_2 + \lambda_3 = \text{tr}(\epsilon)$, $\Phi = \lambda_1\lambda_2 + \lambda_2\lambda_3 + \lambda_3\lambda_1$ and $\Psi = \lambda_1\lambda_2\lambda_3 = \det(\epsilon)$.

2.8   The Equations of Hydrodynamics

Let us consider a Newtonian isotropic fluid. We have seen that we have

$$\frac{D}{Dt}\rho + \rho\,\text{div}\,\vec{u} = 0$$

and

$$\rho\frac{D}{Dt}\vec{u} = \vec{f}.$$

The force density $\vec{f}$ is a superposition of external forces $\vec{f}_{ext}$ and internal forces $\vec{f}_{int}$. In the external forces, one may have the gravity, or the Coriolis force. In the internal forces, one has seen the force due to the pressure:

$$\vec{f}_P = -\vec{\nabla}p$$

and the force due to the viscosity:

$$\vec{f}_{visc} = \mu\Delta\vec{u} + \lambda\vec{\nabla}(\text{div}\,\vec{u})$$

In the absence of other internal forces, we obtain the equations of hydrodynamics:

The equations of hydrodynamics

For a Newtonian isotropic fluid, we have

$$\frac{D}{Dt}\rho + \rho\,\text{div}\,\vec{u} = 0 \tag{2.18}$$

and

$$\rho\frac{D}{Dt}\vec{u} = -\vec{\nabla}p + \mu\Delta\vec{u} + \lambda\vec{\nabla}(\text{div}\,\vec{u}) + \vec{f}_{ext} \tag{2.19}$$

Those equations are in number of four scalar equations with five unknown scalar quantities (u_1, u_2, u_3, ρ and p). The fifth equation depends on the nature of the fluid: it is a thermodynamical **equation of state** that links the pressure, the density and the temperature (one usually assumes that temperature is constant).

Remark:

1. In the case of an incompressible fluid, the equation of state is very simple:

$$\rho = Constant$$

2. When there is no viscosity, one speaks of **ideal fluids**: $\lambda = \mu = 0$.

3. Writing $\mu\Delta\vec{u} + \lambda\vec{\nabla}(\text{div } \vec{u})$ as the divergence of the symmetrical tensor

$$\mathbb{T} = \mu(\partial_i u_j + \partial_j u_i)_{1\leq i,j\leq 3} + \eta\,(\text{div } \vec{u})\,\mathbb{I}_3$$

with $\eta = \lambda - \mu$, we find that the trace of $\mathbb{T}$ is given by $(2\mu + 3\eta)\,\text{div } \vec{u}$; it leads to add to the gradient of the (thermodynamical) pressure another gradient of pressure; the total mechanical pressure is then $p - (2\mu + 3\eta)\,\text{div } \vec{u}$. The coefficient $2\mu + 3\eta$ is called the **bulk viscosity**. An important case is the **Stokes hypothesis** where the tensor $\mathbb{T}$ has no trace: $2\mu + 3\eta = 0$. This corresponds to $\lambda = 0$.

Sometimes, one considers other internal forces, such as those linked to electric or thermal conductivity of the fluid. One then has to add new internal forces to the equations that are dependent on the velocity and influence the velocity. One then quits the domain of hydrodynamics and enters the domain of magnetohydrodynamics (a discipline founded by the 1970 Nobel Prize winner Alfvén) or of the Boussinesq equations that link the velocity and the temperature.

2.9   The Navier–Stokes Equations

In this section we consider the case of a Newtonian, isotropic, homogeneous and incompressible fluid. The equations of hydrodynamics (2.18) and (2.19) then are transformed into the **Navier–Stokes equations**. Since ρ is constant, it is customary to divide the equations by ρ, and to replace the force density $\vec{f}_{ext}$ with a reduced density $\vec{f}_r = \frac{1}{\rho}\vec{f}_{ext}$, the pressure p with a reduced pressure $p_r = \frac{1}{\rho}p$ (which is called the **kinematic pressure**), and the dynamical viscosity μ by the **kinematic viscosity**[3] $\nu = \frac{1}{\rho}\mu$. We then have:

The Navier–Stokes equations

$$\partial_t \vec{u} + (\vec{u}.\vec{\nabla})\vec{u} = -\vec{\nabla}p_r + \nu\Delta\vec{u} + \vec{f}_r \qquad (2.20)$$

$$\text{div } \vec{u} = 0 \qquad (2.21)$$

ν is positive for a viscous fluid. In case of an ideal fluid, $(\nu = 0)$, we obtain the **Euler equations**:

The Euler equations

$$\partial_t \vec{u} + (\vec{u}.\vec{\nabla})\vec{u} = -\vec{\nabla}p_r + \vec{f}_r \qquad (2.22)$$

$$\text{div } \vec{u} = 0 \qquad (2.23)$$

---

[3]In the 19th century, the difference between *kinetics* and *kinematics* was a keystone in mechanics. This difference seems to be less understood in the 21st century: on the website www.answers.com, one can read

kinematic is the study of state of motion of a body i.e. includes both rest and moving bodies..
but kinetic is study of moving bodies only....

(https://www.answers.com/Q/Difference_between_kinetic_and_kinematic)

2.10   Vorticity

The Navier–Stokes equations may be rewritten to underline the role played by vorticity. We start from the identity

$$(\text{curl}\,\vec{u}) \wedge \vec{u} + \vec{\nabla}\frac{|\vec{u}|^2}{2} = (\vec{u}.\vec{\nabla})\vec{u}$$

We thus can write the Navier–Stokes equations as

Another formulation of the Navier–Stokes equations

$$\partial_t \vec{u} + \vec{\omega} \wedge \vec{u} = -\vec{\nabla}Q_r + \nu\Delta\vec{u} + \vec{f}_r \qquad (2.24)$$

$$\text{div}\,\vec{u} = 0 \qquad (2.25)$$

where $\vec{\omega} = \text{curl}\,\vec{u}$ is the **vorticity** of the flow and Q_r the (reduced) **total pressure**

The total pressure $Q = \rho Q_r$ is thus the sum of the **hydrostatic pressure** p and the **dynamic pressure** $q = \rho\frac{1}{2}|\vec{u}|^2$.

Taking the curl of the Navier–Stokes equations gives the following equations for $\vec{\omega}$:

$$\partial_t\vec{\omega} + (\vec{u}.\vec{\nabla})\vec{\omega} = \nu\Delta\vec{\omega} + (\vec{\omega}.\vec{\nabla})\vec{u} + \text{curl}\,\vec{f}_r. \qquad (2.26)$$

We find again the phenomenon of diffusion (induced by $\Delta\vec{\omega}$), the advection by the vector field $\vec{u}$ (described by the term $(\vec{u}.\vec{\nabla})\,\vec{\omega}$) and we have a third term $(\vec{\omega}.\vec{\nabla})\vec{u}$, which corresponds to **stretching forces**. This term is very important in 3D fluid mechanics. When the fluid is planar $\vec{u}(t,x_1,x_2,x_3) = (u_1(x_1,x_2), u_2(x_1,x_2), 0)$, the stretching force vanishes: $(\vec{\omega}.\vec{\nabla})\vec{u} = 0$.

2.11   Boundary Terms

To make the Navier–Stokes system complete, one must specify the conditions at the boundary of the domain of the fluid. In this book, all along, we will consider a problem with no boundary (the fluid fills the whole space). However, in this section, we shall give a few words on the boundary value problem.

When the fluid occupies only a domain Ω, the problem of the boundary conditions is raised. The domain may vary with time. A particular problem is the free-boundary problem: the boundary of Ω evolves through a partial differential equation which describes the evolution of the curvature of the boundary through the action of the deformation tensor of the fluid (see the paper by Solonnikov [445]).

For a rigid domain, one has to prescribe the behavior at the boundary and at infinity (when the domain is unbounded). The most used condition is the **no-slip condition** which says that, at a point of the border, the normal part of the velocity should vanish ($\vec{u}.\vec{\nu} = 0$) and the tangential part of the velocity should equal the velocity of the solid point of the boundary (if the boundary is moving). If the boundary points do not move, the no-slip condition is the homogeneous Dirichlet condition: $\vec{u}_{|\partial\Omega} = 0$.

For Euler equations on a fixed domain, the no-slip condition is replaced by an impermeability condition (that expresses that no fluid crosses over the boundary) $\vec{u}.\vec{\nu} = 0$ on $\partial\Omega$.

The no-slip condition was introduced by Stokes [451] in 1849 and has been in accordance with many experimental data. However, there are some cases where some slip is to be considered, as for instance in microfluidics (see the review paper [304]) that deals with very small quantities of fluids (between an attoliter [10^{-18} l.] and a nanoliter [10^{-9} l.]), where the macroscopic properties of fluids are no longer valid. For such fluids, the slip condition introduced by Navier in 1822 [373] has been experimentally validated. The Navier slip condition stipulates that the normal part of the fluid velocity at the boundary vanishes, but that the tangential part is governed by the stress tensor: if $Q_{\|}$ is the projection $Q_{\|}(\vec{g}) = \vec{g} - \langle \vec{g}|\vec{\nu}\rangle\vec{\nu}$ on the tangent plane to the boundary, $Q_{\|}\vec{u}$ is proportional to $Q_{\|}(\mathbb{T}.\vec{\nu})$.

For the Navier slip condition, one assumes more precisely that we have, for a constant $\sigma \geq 0$, the equality

$$Q_{\|}(\mathbb{T}\vec{\nu} + \alpha u) = 0.$$

α is called the friction coefficient. A popular choice is $\alpha = 0$, the *pure slip* condition. The pure slip condition may be rewritten in the following way. If τ is a tangent vector in the tangent plane of $\partial\Omega$, we have the identity

$$Q_{\|}(\mathbb{T}\vec{\nu}) = 2\mu\epsilon\vec{\nu} - 2\mu(\epsilon\vec{\nu}\cdot\vec{\nu})\vec{\nu} = 2\mu\epsilon\vec{\nu} - 2\mu(\vec{\nabla}\otimes\vec{u}\cdot\vec{\nu}\otimes\vec{\nu})\vec{\nu}.$$

Thus, $Q_{\|}(\mathbb{T}\vec{\nu}) = 0$ if and only for every tangent vector $\vec{\tau}$ of $\partial\Omega$, we have

$$\vec{\tau} \cdot \epsilon\vec{\nu} = 0. \tag{2.27}$$

The study of the Navier–Stokes equations with this pure slip boundary condition has been initiated by Solonnikov and Ščadilov in 1973 [446], while studying a model for flow with free boundary.

Recently, another type of boundary condition has been considered. Equation (2.27) may be rewritten, due to the identity

$$\vec{\tau} \cdot \epsilon\vec{\nu} = \frac{1}{2}\vec{\tau} \cdot (\vec{\omega} \wedge \vec{\nu}) + \vec{\tau} \cdot \left(\sum_{i=1}^{3} \nu_i \vec{\nabla}u_i\right) = \frac{1}{2}\vec{\tau} \cdot (\vec{\omega} \wedge \vec{\nu}) + \partial_\tau(\vec{u} \cdot \vec{\nu}) - \vec{u} \cdot \partial_\tau\vec{\nu}$$

and due to the fact that on the boundary $\vec{u} \cdot \vec{\nu} = 0$ so that the tangential derivative $\partial_\tau(\vec{u} \cdot \vec{\nu}) = 0$, as

$$\frac{1}{2}\vec{\tau} \cdot (\vec{\omega} \wedge \vec{\nu}) = \vec{u} \cdot \partial_\tau\vec{\nu}.$$

In the regions where the boundary of Ω is flat (so that the normal $\vec{\nu}$ is constant), we thus have (for all tangential directions)

$$\frac{1}{2}\vec{\tau} \cdot (\vec{\omega} \wedge \vec{\nu}) = 0$$

or equivalently

$$\vec{\omega} \wedge \vec{\nu} = 0. \tag{2.28}$$

The boundary conditions $\vec{u} \cdot \vec{\nu} = 0$ and $\vec{\omega} \wedge \vec{\nu} = 0$ on general (non-flat) domains were considered by Xiao and Xin [508] and Beirão da Veiga and Crispo [32] for the study of the inviscid limit of the equations. Those equations were coined as Hodge–Navier–Stokes equations by Mitrea and Monniaux [363], since those boundary conditions are natural for the Hodge-Laplacian operator.

2.12   Blow-up

Let us consider the Clay Millennium Problem for the Navier–Stokes equations in absence of external forces. As we shall see, a classical result on the Navier–Stokes equations shows that the Cauchy initial value problem will have a smooth solution as long as the velocity $\vec{u}$ remains bounded. Thus, in order to have a breakdown in regularity, the L^∞ norm must blow up. But this blow-up has no physical meaning; for various reasons, one has to drop the equations long before the blow-up can occur. For instance,

- the incompressibilty of the fluid is an approximation that is valid only if the velocity of the fluid is much smaller than the speed of sound

- the Newtonian character of the fluid was derived under the hypothesis of small velocities and small derivative of the velocities

- when velocities are too important, classical mechanics should be corrected into relativistic mechanics

Thus, the blow-up issue is essentially a mathematical problem, not a physical one. However, it is hoped that the understanding of the mechanism that leads to blow-up or blocks it would shed a good light on the mechanism that leads physical fluids to turbulent states.

2.13   Turbulence

Smooth flows are called *laminar*, whereas disordered flows are called *turbulent*. For turbulent flows, it is quite hopeless to try and find a description of all the fluid parcels, as the number of degrees of freedom is too important. Since the works of Reynolds (1894) and Taylor (1921), one tries only to describe the evolution of the flow on a large scale, and to discuss the behavior of the flow at small scales as a dissipative correction of the equations for the large scales.

This separation between the large-scale components and the small-scale ones relies on several physical observations. The large-scale components are sensitive to the geometry of the boundary and to the nature of external forces that are impressed on the fluid, whereas the small scale components can be analyzed in a more universal way.

To separate the large-scale component from the small-scale component, one uses an averaging process that gives a mean value $\bar{\mathbf{u}}$ of the velocity $\vec{u}$. The Navier–Stokes equations then give new equations for $\bar{\mathbf{u}}$:

$$\partial_t \bar{\mathbf{u}} + \bar{\mathbf{u}}.\vec{\nabla}\bar{\mathbf{u}} = \nu\Delta\bar{\mathbf{u}} - \vec{\nabla}\bar{p} + \bar{\mathbf{f}} + \operatorname{div}\mathcal{R} \qquad (2.29)$$

(together with $\operatorname{div}\bar{\mathbf{u}} = 0$ and $\bar{\mathbf{u}}_{|t=0} = \overline{\mathbf{u_0}}$) where the *Reynolds stress* $\mathcal{R}$ is given by

$$\mathcal{R} = \bar{\mathbf{u}} \otimes \bar{\mathbf{u}} - \overline{\vec{u} \otimes \vec{u}}. \qquad (2.30)$$

As the mean value $\overline{\vec{u} \otimes \vec{u}}$ does not depend on the mean value $\bar{\mathbf{u}}$, those equations are not closed. The problem is then to give a satisfying modelization of the Reynolds stress.

The theory of Kolmogorov (1941) gives a modelization of $\vec{u} - \bar{\mathbf{u}}$ as a random field obeying some universal laws due to the (local) homogeneity and isotropy of the fluctuations. Whereas this theory has been confirmed experimentally, it remains far from being completely understood and is the core of a very active research field (see the classical book of Monin and Yaglom [367]).

Chapter 3

History of the Equation

In this chapter, we sketch some points of the history of the Navier–Stokes equations. The reader will find a comprehensive study of the period 1750–1900 in Darrigol's book *Worlds of flow* [145], which studies the origin of the equations as well from the mathematical theoretical point of view as from the point of view of physical experiments and observation. Other stimulating references on the infancy of mathematical hydrodynamics are the papers of Truesdell [479, 480, 481].

3.1   Mechanics in the Scientific Revolution Era

Hydrodynamics appeared in 1738. The word *hydrodynamica* was coined by D. Bernoulli in his treatise *Hydrodynamica, sive De viribus et motibus fluidorum commentarii* [37], where he wanted to propose a unified theory of hydrostatics and hydraulics.

Hydraulics is a very old science. Irrigation has been known since the 6th millennium BCE in Ancient Persia. Managing water supply for human settlements and irrigation has been an important technique in human development. The technology of "qanats" has been developed by Iranians in the early 1st millennium BCE and then spread toward Asia, Africa and Europe. In the kingdom of Saba' (now, in Yemen), dams were constructed as soon as 2000 BCE in order to irrigate the crops; the great Dam of Ma'rib (built about the 8th century BCE) is counted as one of the most wonderful feats of engineering in the ancient world (its remains were severely damaged by a Saudi airstrike in 2015). Working machines using hydraulic power, such as the force pump, have been developed by Hellenistic scientists (as, for instance, Hero of Alexandria) and by Roman engineers for raising water. Modern hydraulics was initiated in Italy, in the 16th century as an experimental science, then in the 17th century in a more theoretical approach with the influential treatises of Castelli (1628) and Fontana (1696).

Hydrostatics has ancient roots as well. The phenomenon of buoyancy has been explained by Archimedes in the 3rd century BCE. Pressure has been explained in the 17th century: a fluid is a substance that continually deforms under an applied shear stress; thus, in hydrostatics (the science that studies fluids at rest), there cannot exist a shear stress; however, fluids can exert pressure normal to any contacting surface. Due to gravitation, liquids exert pressure on the sides of a container as well as on anything within the liquid itself. This pressure is transmitted in all directions and increases with depth, as established by Pascal. Atmospheric pressure had been revealed by Torricelli who invented barometers.

The 17th century is the century of the so-called *Scientific Revolution*. Mechanics was deeply refounded in that period, culminating with the work of Newton (1687). Basic concepts of physics emerged throughout the century. Kepler gave in 1609 the laws ruling planetary motion. The study of free fall by Galileo (1638) clearly put in light the notion of accelerated motion. Aristotle's notion of uniform velocity and of proportionality to describe motions had already been criticized by many medieval authors, including the philosopher Buridan and the Oxford Calculators (as Bradwardine), but the mathematical law of motion in free fall was stated and experimentally checked by Galileo. Huygens replaced Buridan's *impetus* with **momentum** (1673). Newton's second law expresses the variation of momentum through the action of forces. While free fall was caused by *gravity*, other forces were explored in this century: *hydrostatic pressure* (Pascal's law in 1648, extending previous work of Stevin (1586)), *tension* in elasticity (Hooke's law in 1660), resistance to motion, etc. Another important concept emerged in 1676: the *vis viva* introduced by Leibniz in the study of elastic shocks, which has been fiercely debated all along the 18th century and became the *kinetic energy* in the 19th century.

Physics had been reshaped through two main tools: experimentation, with the invention of new observation devices, and mathematization. The celebrated sentence of Galileo states:

> *La filosofia è scritta in questo grandissimo libro che continuamente ci sta aperto innanzi a gli occhi (io dico l'universo), ma non si può intendere se prima non s'impara a intender la lingua, e conoscer i caratteri, ne'quali è scritto. Egli è scritto in lingua matematica, e i caratteri son triangoli, cerchi, ed altre figure geometriche, senza i quali mezzi è impossibile a intenderne umanamente parola; senza questi è un aggirarsi vanamente per un oscuro laberinto[1].*

Mathematics in the 17th century was drastically reshaped as well, in a parallel dynamics. The new algebra introduced by Viète (1591) allowed symbolic computations. Algebraization of geometry was then proposed by Descartes (1637), through the use of numerical coordinates in a reference frame. Fermat's rule for the determination of *maxima* and *minima* and Barrow's duality principle between problems on tangents and problems of area [7] lead to the foundation of modern calculus by Newton and Leibniz (1684), with the notion of derivative and primitive functions.

3.2   Bernoulli's *Hydrodymica*

Mathematicians tried to apply those new tools to explain the empirical rules of physics. Understanding the laws of statics could be reduced to geometrical reasoning, as in the work of Stevin, who discovered the hydrostatic paradox [449] in 1586: the downward pressure of any given liquid is independent of the shape of the vessel, and depends only on its height and base. This was illustrated in Pascal's barrel experiment in 1646 [392]: Pascal inserted a 10-m long vertical tube into a barrel filled with water; when water was poured into the vertical tube, the increase in pressure caused the barrel to burst.

Understanding the laws of dynamics needed the invention of calculus. The model for mathematicians was then the derivation by Newton of Kepler's laws on planetary motion. In fluid motion, one of the first laws investigated was Torricelli's law on the efflux (1644) [474]: the speed of efflux of a fluid through a sharp-edged hole at the bottom of a tank filled to a depth h is the same as the speed that a body would acquire in falling freely from the same height h. As soon as 1695, Varignon, developing analytic dynamics by adapting Leibniz's calculus to the inertial mechanics of Newton's *Principia*, proposed a derivation

---

[1]Philosophy is written in this grand book – I mean the universe – which stands continually open to our gaze, but it cannot be understood unless one first learns to comprehend the language and interpret the characters in which it is written. It is written in the language of mathematics, and its characters are triangles, circles, and other geometrical figures, without which it is humanly impossible to understand a single word of it; without these, one is wandering around in a dark labyrinth.

of Torricelli's law based on the momentum principle. Varignon's derivation was based on the hypothesis that the force causing the outflow was given by the weight of the column of water over the opening. This assumption was proven to be false, but is very common in the attempts of mathematical derivations of Torricelli's law in the early 18th century (as, for instance, Hermann (1716) and J. Bernoulli (1716)). In the section X "Principes de l'hydrodynamique" of Part II of his famous treatise *Traité de méchanique analitique* [299], Lagrange comments on Varignon's proof and on Newton's attempt (in the second edition of the *Principia*, (1713)) of Torricelli's law. (For a discussion of those various attempts, see Mikhailov [360] or Blay [45]).

D. Bernoulli's approach in *Hydrodynamica* [37] is totally different. He does not rely on the momentum principle, but on the Leibnizian theory of *vis viva*. Leibniz's theory was contested both by the Newtonians in England and by the Cartesians in France, but was gaining stronger support: 's Gravesande in 1722 made an experiment in which brass balls were dropped with varying velocity onto a soft clay surface; the results of the experiment clearly proved that their penetration depth was proportional to the square of their impact speed. The French physicist and mathematician Émilie du Châtelet recognized the implications of the experiment and published an explanation in 1740 in her influential treatise on physics [104]. Bernoulli's treatise was a determining example of the interest of the principle of conservation of *vis viva*. In 1757, D'Alembert could write in the *Encyclopédie* [140]:

> *On peut voir par différens mémoires répandus dans les volumes des académies des Sciences de Paris, de Berlin, de Petersbourg, combien le principe de la conservation des forces vives facilite la solution d'un grand nombre de problemes de Dynamique; nous croyons même qu'il a été un tems où on auroit été fort embarrassé de résoudre plusieurs de ces problemes sans employer ce principe[2].*

Bernoulli's treatise has been considered as the first successful attempt of mathematical derivation of Torricelli's law. It contained other results on hydraulics, such as the prefiguration of *Bernoulli's law* that explains how the pressure exerted by a moving fluid is lesser than the pression of the fluid at rest.

3.3   D'Alembert

The solution proposed by D. Bernoulli for the derivation of Torricelli's law was felt insecure by many mathematicians as it relied on a controversial principle. There was at that time a strong discussion of what was the meaning of forces that put bodies in motion. Cartesians insisted on the *momentum*, a quantity that was clearly defined for a moving mass point (the mass times the velocity). Newtonians insisted on the variation of momentum, hence gave an important role to *acceleration*, according to Newton's second law:

> *II. The change of motion is proportional to the motive force impressed, and it takes place along the right line in which that force is impressed.*

But the nature of the motive force remained obscure, and coined as metaphysical by Cartesians who could not accept the principle of distant action and especially the theory of universal attraction to explain gravity. Leibnizians, following ideas from Huygens and Leibniz,

---

[2]One can see through various memoirs that can be found in the volumes of the science academies in Paris, Berlin or Petersburg, how the principle of conservation of living forces eases the solution of many problems in Dynamics; we even believe that there has been a time when one would have been most embarassed to solve many of those problems without using this principle.

used the notions of *vis viva* and *vis morta*, and expressed the laws of motion as a conversion of *vis morta* into *vis viva*. This conception was rejected by Cartesians as metaphysical as well, since the *vis viva* seemed an inherent property attached to the moving bodies, and not a measurable kinetic quantity.

Johann Bernoulli, Daniel's father, published *Hydraulica*, a treatise in 1742, with the aim of rewriting his son's results rather with help of Newtonian mechanics rather than of Leibnizian *vis viva*. Johann Bernoulli had to identify precisely the acceleration of the fluid, and he was thus led to describe the convective part of the acceleration and the internal pressure, i.e., the pressure that the moving parts of the fluid exerted on the other parts. Later, those two innovations would be crucial elements for Euler's derivation of the equations of hydrodynamics.

D'Alembert tried to avoid any use of the concept of force, as it seemed to be linked to metaphysical issues. In 1743, he founded his *Traité de dynamique* [136] on a principle that avoided the use of internal forces to describe the motion of a constrained system of bodies.

> *Tout ce que nous voyons bien distinctement dans le Mouvement d'un Corps, c'est qu'il parcourt un certain espace, & qu'il employe un certain tems à le parcourir. C'est donc de cette seule idée qu'on doit tirer tous les Principes de la Méchanique, quand on veut les démontrer d'une manière nette & précise; ainsi on ne sera point surpris qu'en conséquence de cette réflexion, j'aie, pour ainsi dire, détourné la vûe de dessus les causes motrices, pour n'envisager uniquement que le Mouvement qu'elles produisent; que j'aie entièrement proscrit les forces inhérentes au Corps en Mouvement, êtres obscurs & Métaphysiques, qui ne sont capables que de répandre les ténèbres sur une Science claire par elle-même.*[3]

He claimed that one did not need to use Newton's second law:

> *Pourquoi donc aurions-nous recours à ce principe dont tout le monde fait usage aujourd'hui, que la force accélératrice ou retardatrice est proportionnelle à l'élément de la vitesse? principe appuyé sur cet unique axiome vague & obscur, que l'effet est proportionnel à sa cause.*[4]

He based his theory of dynamics on three principles:

> *Le Principe de l'équilibre joint à ceux de la force d'inertie & du Mouvement composé, nous conduit donc à la solution de tous les Problèmes où l'on considère le Mouvement d'un Corps.*[5]

According to those three principles, he decomposed the motion of a constrained body into a natural one, described through the law of inertia, and the motion due to the presence of constraints; for this latter one, his principle of equilibrium asserts that the forces corresponding to the accelerations due to the presence of constraints form a system in static equilibrium.

---

[3]All that we can distinctly see in the Motion of a Body is the fact that it covers a certain space and that it takes a certain time to cover that space. One must draw all the Principles of Mechanics from that sole idea, when one wants to give a neat and precise demonstration of them. Thus, it won't be a surprise that, as a consequence of this reflection, I have turned my view away from the motive forces and considered but the Motion they produce; that I entirely banished the forces inherent to the Body in Motion, as obscure and Metaphysical beings that can only shed darkness on a Science that is clear by itself.

[4]Why should we appeal to that principle used by everybody nowadays, that the accelerating or retarding force is proportional to the element of velocity, a principle resting only on that vague and obscure axiom that the effect is proportional to the cause?

[5]The principle of equilibrium joined with the principles of the law of inertia and of the composition of motions leads us to the solution of all the problems where the Motion of a Body is considered.

With those simple principles, D'Alembert was able to prove the conservation of living forces. In 1744, right after the *Traité de dynamique*, he published the *Traité des fluides* [137] where he applied his dynamical theory to the proof of Daniel Bernoulli and Johann Bernoulli's results. In the *Traité des fluides* as well as in D. Bernoulli's *Hydrodynamica* or J. Bernoulli's *Hydraulica*, the fluid considered has only one degree of freedom: in their models, the fluid is decomposed into horizontal slices and the velocity is uniform on each slice.

In 1747, in his treatise *Réflexions sur la cause générale des Vents* [138], he developed the notion of a velocity field, with velocities that depended on the position. The differential equations were then turned into partial differential equations. D'Alembert is known as a pioneer of the use of partial derivatives in mathematical physics, with the famous example of the wave equation which he gave in 1749 for describing vibrating strings. While partial differential equations were already known in the setting of the prehistory of variational calculus, D'Alembert was the first to use them in a mechanical context. Later, D'Alembert worked on the resistance opposed to the motion of an immersed body, as in his 1752 treatise *Essai d'une nouvelle théorie de la résistance des fluides* [139].

In 1768, he noticed that his theory of (inviscid) incompressible fluids led to a paradox, the celebrated D'Alembert paradox [141]. He considered an axisymmetric body with a head-tail symmetry, immersed in an inviscid incompressible fluid and moving with constant velocity relative to the fluid, and proved that the drag force exerted on the body is then zero. This result was in direct contradiction to the observation of substantial drag on bodies moving relative to fluids:

> *Je ne vois donc pas, je l'avoue, comment on peut expliquer par la théorie, d'une manière satisfaisante, la résistance des fluides. Il me paroît au contraire que cette théorie, traitée & approfondie avec toute la rigueur possible, donne, au moins en plusieurs cas, la résistance absolument nulle; paradoxe singulier que je laisse à éclaircir aux Géomètres.*[6]

This paradox, and the fact that the equations derived for the description of fluid motions had in general no easily computed solutions, caused a deep gap between mathematicians dealing with fluid mechanics and engineers dealing with hydraulics. This situation lasted for decades, before eventually the mathematical theory evolved to a frame more adapted to the real-world situations, taking into account the viscosity effects.

3.4 Euler

Then Euler came...

Newtonian mechanics implied a new vision of geometry, as it has been underlined by Bochner [47]:

> *Several significant physical entities of the* Principia, *namely, velocities, moments, and forces are, by mathematical structure, vectors, that is, elements of vector fields, and vectorial composition and decomposition of these entities constitute an innermost scheme of the entire theory. This means that the mathematical*

---

[6]Thus, I do not see, I admit, how one can satisfactorily explain by theory the resistance of fluids. On the contrary, it seems to me that the theory, developed in all possible rigor, gives, at least in several cases, a strictly vanishing resistance, a singular paradox which I leave to future Geometers to elucidate.

space of the Principia, *in addition to being the Greek Euclidean substratum,
also carries a so-called affine structure, in the sense that with each point of the
space there is associated a three-dimensional vector space over real coefficients,
and that parallelism and equality between vectors which emanate from different
points are also envisaged.*

However, the celebrated Newton formula $\vec{f} = m\vec{a}$ (where $\vec{a}$ is the *acceleration*) was not
expressed in such a vectorial form in the *Principia* and was not well understood in the fifty
years following the release of the *Principia*. MacLaurin in 1742 [346] and Euler in 1747 [165]
were the first ones to express Newton's second law in its full 3D expression.

In 1750, Euler [166] applied Newton's second law to the mechanics of continuous media.
He expressed the opinion that no other mechanical principles were needed. Euler's mechanics
is an important turning point: Newton's mechanics was essentially a kinematic theory for
a mass point; Euler extended this theory to the case of a continuous medium.

In 1755, Euler [167] presented a memoir (published in 1757) entitled *Principes généraux
du mouvement des fluides*, where he applied his theory to the theory of fluid motions. While
his predecessors worked on incompressible flows with one degree of freedom (D. Bernoulli
and J. Bernoulli) or two degrees of freedom (D'Alembert), Euler could derive the equations
for a general fluid, compressible or not, in the presence of arbitrary external forces.

In his seminal memoir, Euler described the laws of fluid mechanics as applied to fluid
parcels, very small volumes of fluids that are fictitiously isolated. With this notion of parcels,
he could introduce the internal pressure (or static pressure), as the density of the force
exerted on the parcel by the other parcels of fluid. With those two ideas, he could derive
Euler's equation for an ideal fluid submitted to external forces (with force density $\vec{f}_{ext}$): the
equation expressing the conservation of mass

$$\frac{D}{Dt}\rho + \rho \operatorname{div} \vec{u} = 0 \qquad (3.1)$$

and the equation corresponding to Newton's second law

$$\rho \frac{D}{Dt}\vec{u} = -\vec{\nabla}p + \vec{f}_{ext} \qquad (3.2)$$

Lagrange underlined the importance of Euler's equations [299]:

*C'est à Euler qu'on doit les premières formules générales pour le mouvement
des fluides, fondées sur les lois de leur équilibre, et présentées avec la notation simple et lumineuse des différences partielles. Par cette découverte, toute la
Mécanique des fluides fut réduite à un seul point d'analyse, et si les équations
qui la renferment étaient intégrables, on pourrait, dans tous les cas, déterminer
complètement les circonstances du mouvement et de l'action d'un fluide mû par
des forces quelconques; malheureusement, elles sont si rebelles, qu'on n'a pu,
jusqu'à présent, en venir à bout que dans des cas très-limités.*[7]

Truesdell sketches the legacy of Euler in those words [481]:

*Judged from a positivist philosophy, Euler's hydrodynamic researches are mis-
conceived and unsuccessful: Their basic assumptions cannot be established experimentally, nor did Euler obtain from them numbers which can be read on a*

---

[7]Euler gave the first general formulas for the motion of fluids, based on the laws of their equilibrium, and
presented with the simple and bright notation of partial differences. By this discovery, the entire mechanics
of fluids was reduced to a single point of analysis, and if the equations which include it were integrable,
one could determine completely the circumstances of motion and of action of a fluid moved by any forces.
Unfortunately, they are so rebellious that up to the present time only a few very limited cases have been
worked out.

dial. Yet, after Euler's death, special solutions of his equations have given us the theories of the tides, the winds, the ship, and the airplane, and every year new practical as well as physical discoveries are found by their aid.

*Euler's success in this most difficult matter lay in his **analysis of concepts**. After years of trial, sometimes adopting some semi-empirical compromise with experimental data, Euler saw that experiments had to be set aside for a time. They concerned phenomena too complicated for treatment then; some remain not fully understood today. By creating a **simple** field model for fluids, defined by a set of partial differential equations, Euler opened to us a new range of vision in physical science. It is the range we all work in today. In this great insight, looking within the interior moving fluid, where neither eye nor experiment may reach, he called upon the "imagination, fancy, and invention" which Swift could find neither in music nor in mathematics.*

3.5 Laplacian Physics

The mathematical physics developed in the 18th century by Euler, D'Alembert and Lagrange rested on partial differential equations describing the regular behavior of continuous quantities. However powerful this theory turned out to be, it sufffered from many drawbacks for engineers as well as for physicists.

The equations obtained in this setting remained unsolved but in some very special cases. Moreover, they described idealized situations that were very different from the real life events. They could not explain the deformation of solid bodies, nor the creation of eddies in turbulent flows. Engineers went on applying empirical formulae that were not derived from those theories (and sometimes were in contradiction with those theories).

In opposition to Lagrange's analytical mechanics, Laplace tried to develop a molecular model of nature that could explain the laws of physics through the role of inter-molecular forces, in analogy to the Newtonian theory of celestial mechanics and the Laplacian theory of capillarity. This Laplacian physics was fiercely sustained by Poisson, Laplace's disciple [402]:

Il serait à désirer que les géomètres reprissent sous ce point de vue physique et conforme à la nature, les principales questions de la mécanique. Il a fallu les traiter d'une manière tout à fait abstraite, pour découvrir les lois générales de l'équilibre et du mouvement, et en ce genre de généralité et d'abstraction, Lagrange est allé aussi loin qu'on puisse le concevoir, lorsqu'il a remplacé les liens physiques des corps par des équations entre les coordonnées des différents points, c'est là ce qui constitue la mécanique analytique; *mais à côté de cette admirable conception, on pourrait maintenant élever la* mécanique physique *dont le principe unique serait de ramener tout aux actions moléculaires qui transmettent d'un point à un autre l'action des forces données et sont l'intermédiaire de leur équilibre.*[8]

---

[8]Translated in [145]: It would be desirable that geometers reconsider the main equations of mechanics under this physical point of view which better agrees with nature. In order to discover the general laws of equilibrium and motion, one had to treat these questions in a quite abstract manner; in this kind of generality and abstraction, Lagrange went as far as can be conceived when he replaced the physical connections of bodies with equations between the coordinates of their various points: this is what *analytical mechanics* is about; but next to this admirable conception, one could now erect a *physical mechanics*, whose unique

Molecular models or atomistic ones were as old as the antique science. History of atomism and molecular theories is well documented in the books by Whyte [500] or Kubbinga [283]. Atomism was proposed as a model by Leucippus and Democritus in the 5th century BCE. This model was revived in the 17th century by Basson, Beeckman, Gassendi (who coined the word *molecula*) and Boyle.

In 1745 Bošković [54, 55] published in *De Viribus Vivis* an explanation of elasticity and inelasticity of collisions through an atomistic theory of matter, that tried to find a middle way between Isaac Newton's gravitational theory and Gottfried Leibniz's metaphysical theory of monad-points. In this theory, however, atoms are no longer the ontological primitive of nature: forces become the primary property of the material world. This was underlined by Nietzsche [376]:

> *Während nämlich Kopernikus uns überredet hat zu glauben, wider alle Sinne, dass die Erde nicht fest steht, lehrte Boscovich dem Glauben an das Letzte, was von der Erde 'feststand', abschwören, dem Glauben an den 'Stoff', an die 'Materie', an das Erdenrest - und Klümpchen-Atom; es war der grösste Triumph über die Sinne, der bisher auf Erden errungen worden ist.*[9]

In the model of Bošković, molecules have no extension, they are just points that are center of forces. Those forces attract or repell the other molecules: when the distance r is large, the force is attractive (with a decrease in $1/r^2$ to fit Newton's theory of gravitation), while when the distance is small, the force is repelling (and becomes infinite for vanishing distance) in order to avoid direct contact between distinct molecules. Molecules thus remain at a positive distance from the other ones, and they are separated by vacuum. This force of interaction was expected to provide the explanation for all the properties of matter: gravitation, collision, cohesion, flexibility, sound propagation, crystalline states, phase transition, and so on.

The molecular model proposed in 1808 by Laplace [303] explained as well many physical phenomena such as optical refraction, elasticity, hardness and viscosity as the result of short-range forces between molecules. Laplace, together with his friend Berthollet, played a prominent role in the scientific field at the beginning of the 19th century [182] and many French physicists developed Laplace's model as a key to understand the physical phenomena on what they felt as a firm and non-hypothetical basis. However, the claim that the molecular model developed by Laplace could explain all the physical phenomena was rapidly discarded, as alternative methods were developed by Fourier (1822: theory of heat), Fresnel (1818: wave optics) and Germain (1821: elasticity theory).

Poisson fought the Lagrangian method of *virtual works* and promoted Laplace's discrete molecular distributions and inter-molecular forces instead of the forces of constraint of the continuous media used in analytical mechanics. In his treatise *L'évolution de la mécanique* (1905) [161], Duhem quotes de Saint-Venant and Boussinesq as Poisson's followers in the rejection of forces of constraint and in the privileged use of molecular forces. In the same treatise, Duhem shows how physical experiments on elasticity (such as Wertheim's experiments on metals), however, eventually disproved Poisson's hypotheses and confirmed the results that Cauchy, Green and Lamé obtained by means of analytical mechanics. Duhem's conclusion is as severe as Poisson's influence was still strong:

---

principle would be to reduce everything to molecular actions that transmit from one point to another the given action of forces and mediate their equilibrium.

[9]Translated in [499]: While Copernicus has persuaded us, against all senses, that the Earth does *not* stand still, Boscovich taught us to renounce belief in the last thing of earth to "stand fast," belief in "substance," in "matter," in the last remnant of Earth, the corpuscular atom: it was the greatest triumph over the senses achieved on Earth to this time.

Il est donc impossible de garder les principes sur lesquels Poisson voulait faire reposer la Mécanique physique, à moins d'avoir recours à des subtilités et à des faux-fuyants.[10]

Fifty years later, the conclusion of Truesdell [478] was less severe on the Poissonian approach, even if Truesdell prefered to employ a continuum analysis:

There are two methods of constructing a theory of elasticity or fluid dynamics. The first, used originally by Boscovich, Navier, Cauchy, and Poisson and after long discredit now again in favor among physicists, deduces macroscopic equations from special assumptions relative to the behavior of the supposed ultimate discrete entities comprising the medium. In the present article I employ only the continuum approach of Clairaut, D'Alembert, Euler, Lagrange, Fresnel, Cauchy, Green, St. Venant, and Stokes, in which molecular speculations are avoided, and gross phenomena are described in gross variables and gross hypotheses alone.

3.6   Navier, Cauchy, Poisson, Saint-Venant and Stokes

The discovery of the Navier–Stokes equations is linked to new formulations for elasticity theory. Elasticity was an important issue at the beginning of the 19th century. While engineers were facing the absence of a convincing theory for the problem of beam flexions, there had been a fierce debate around the prize proposed by the French Académie des Sciences on the problem of explaining Chladni's experiment of vibrating plates in 1808. The prize was eventually won by Germain in 1818: her work was based on and enriched by Lagrange's contributions and violently criticized by Poisson who derived in 1814 a molecular model based on the Laplacian system.

In this context, Navier contributed to the emergence of a new understanding of elasticity. In 1820, he proposed a Lagrangian approach of the problem of vibrating plates. He analyzed the continuous deformations of the plates as composed of isotropic stretching - as in Lagrange's computations - and anisotropic flexion.

In 1821, he gave two proofs of his results in elasticity, one was based this time on a Laplacian molecular model and the other one was based on the Lagrangian method which relied on the balance of virtual moments [372]. The equations he derived were valid for more general elastic bodies.

The idea developed by Navier was that the restoring forces appearing in elasticity could be modelized as a response to the change of distances between molecules. For small deformations, this intermolecular force would be proportional (and opposite) to the variation of the distance (with a proportionality coefficient depending on the distance).

In 1822, Navier [373] extended his theory to hydrodynamics. Once again, he introduced restoring forces generated by the opposition to the change of distances between molecules. Of course, those changes are due to the difference of velocities between molecules, and the computations led Navier to the introduction to a new internal force in Euler's equations of hydrodynamics: the internal forces included not only the pressure gradient $-\vec{\nabla}p$ (a force which was present in static fluids as well as in moving fluids), but a new force $\mu(\Delta\vec{u} + 2\vec{\nabla}(\text{div } \vec{u}))$ which was generated by the motion of the fluid (and more precisely by the non-uniformity of the motion of the fluid).

---

[10]It is thus impossible to keep the principles on which Poisson wanted to base physical Mechanics, unless resorting to subtleties and to evasions.

Truesdell [479] comments on Navier's approach:

> *In 1821 Navier, a French engineer, constructed imaginary models both for solid bodies and for fluids by regarding them as nearly static assemblages of 'molecules', mass-points obeying certain intermolecular force laws. Forces of cohesion were regarded as arising from summation of the multitudinous intermolecular actions. Such models were not new, having occurred in philosophical or qualitative speculations for millenia past. Navier's magnificent achievement was to put these notions into sufficiently concrete form that he could derive equations of motion from them.*

As soon as 1822, Cauchy [95] gave a new interpretation of Navier's results. From the theory of Navier, he could see that the internal force exerted on the surface of a fluid parcel was no longer perpendicular to the surface but contained a tangential part. Thus, he developed a theory of elastic bodies, introducing the notion of internal stress that would generate forces exerted on the surface of (imaginarily isolated) small elements of the body. In a modern language, the force exerted on the surface element with normal $\vec{\nu}$ would be given by a vector $\vec{f}_{\text{surf}}$ that depend linearly on $\vec{\nu}$ (but no longer directed in the direction of $\vec{\nu}$). He obtained a relationship $\vec{f}_{\text{surf}} = \sigma\vec{\nu}$, where σ is now defined as a 3×3 matrix. For physicists, σ is a second-rank tensor.

Bochner [47] underlined the importance of Cauchy's stress tensor:

> *Archimedes also accomplished basic work, perhaps his most famous one, in the mechanics of floating bodies. Here again he did not introduce the physical concept which is central to the subject matter, namely, the concept of hydrostatic pressure. But in this case Archimedes may be "excused". Modern mechanics had great difficulties in conceptualizing the notion of pressure, although Stevin immediately mapped out the task of doing so and everybody after him was pursuing it. Even Newton was not yet quite certain of it. In a sense the clear-cut mathematization of the concept of pressure was arrived at only in the course of the nineteenth century beginning perhaps with work by A. Cauchy on equations of motion for a continuous medium in general. In the nineteenth century the mathematical "image" of pressure became a tensor, albeit a very special one, and the actual formalization of the concept of a tensor and a full realization of its mathematical status took a long time to emerge.*

Cauchy showed that the tensor σ should be symmetrical. Then, he studied the quadratic form associated to this symmetric matrix, and compared it to the quadratic form induced by the strain tensor (or tensor of deformations: the quadratic form corresponds to the first-order development of the variation of distance between two close points). For an isotropic body, he argued that the stress tensor should have the same principal axes as the strain tensor; then, generalizing Hooke's law on elastic deformations, he assumed that the tensors were proportional, with a proportionality coefficient independent of the deformation.

Applying his theory to hydrodynamics, he then obtained Navier's equations, except that the new force was given $\mu(\Delta\vec{u} + \vec{\nabla}(\text{div }\vec{u}))$ (instead of $\mu(\Delta\vec{u} + 2\vec{\nabla}(\text{div }\vec{u}))$).

Poisson fought against Navier (who used Lagrangian methods of virtual works) and of Cauchy (who studied continuous media) in the name of a strict and rigorous Laplacian molecular approach. He proposed, as well as Cauchy, a strictly molecular theory; both [96, 403] rederived Navier equations (with a more general force $(\mu + \lambda)\Delta\vec{u} + 2\mu\vec{\nabla}(\text{div }\vec{u})$, i.e., the general form for a compressible Newtonian fluid). The arguments between Poisson, Cauchy and Navier are described in Darrigol's paper [144].

Saint-Venant tried to concile the experimental laws of engineers with a rigorous mathematical derivation of physical laws for elasticity. He applied his theory to fluid mechanics in an unpublished memoir to the Académie des Sciences in 1834 [418, 419]. The main idea in Saint-Venant's approach was the introduction of a varying viscosity and a non-linear dependency of the stress tensor on the strain tensor. A cause for the variations of the viscosity, he indicated in 1850, was to be found in the presence of eddies in the flow [420], an idea that turned out to be very influential in turbulence theory.

Stokes' first academic works were devoted to hydrodynamics. In his first paper, in 1842 [450], he studied steady flows and introduced the seminal notion of *stability*, underlining the fact that the mathematical possibility of a given motion did not imply its existence if this motion were unstable. In 1845, Stokes [451] derived his own modelization for elasticity and hydrodynamics. Studying the variation $d\vec{u}$ of the velocity, he decomposed

$$d\vec{u} = \sum_{i=1}^{3} \partial_i \vec{u}\, dx_i = (\partial_i u_j) d\vec{x}$$

into

$$d\vec{u} = \frac{1}{2}((\partial_i u_j) + (\partial_i u_j)^T) d\vec{x} + \frac{1}{2}((\partial_i u_j) - (\partial_i u_j)^T) d\vec{x}.$$

He identified the antisymmetric part to an infinitesimal rotation; nowadays, the matrix $\frac{1}{2}((\partial_i u_j) - (\partial_i u_j)^T)$ is identified with the vorticity: if $\vec{\omega} = \operatorname{curl} \vec{u}$, we have $\frac{1}{2}((\partial_i u_j) - (\partial_i u_j)^T) = \frac{1}{2}\begin{pmatrix} 0 & \omega_3 & -\omega_2 \\ -\omega_3 & 0 & \omega_1 \\ \omega_2 & -\omega_1 & 0 \end{pmatrix}$. The symmetrical part corresponded to a symmetric tensor, whose principal axes described the infinitesimal deformation axes. Thus, he found back Cauchy's tensor $\epsilon = \frac{1}{2}((\partial_i u_j) + (\partial_i u_j)^T)$ which gives the infinitesimal distortion $(d\vec{x})^T \epsilon\, d\vec{x}$ of the distances.

Stokes then required that the shear pressure be given by a tensor whose axes were superposed with the axes of the infinitesimal deformation, and whose coefficients were determined as functions of the tensor ϵ. Further, for small velocity gradients, he privileged a linear relation between the stress tensor and the strain tensor (based on a principle of superposition[11]). For symmetry (or frame-indifference) reasons, he obtained that the stress tensor should be a combination of the strain tensor ϵ and of $\operatorname{tr}(\epsilon)\, \mathbb{I}_3$. This gives, taking the divergence, an internal viscous force expressed as the sum $\mu \Delta \vec{u} + \lambda \vec{\nabla}(\operatorname{div} \vec{u})$. Thus, he found again the Navier–Stokes equations. Stokes' derivation is commented by Truesdell [479]:

> Stokes /.../ derived the same equations as had Poisson, but in doing so he put the theory on a sound and clear phenomenological basis. As far as the received theory of fluids with linear viscous response is concerned, this paper was final.

Stokes thoroughly investigated the case of *creeping flows*, where the velocities are so small that the advective term $\vec{u}.\vec{\nabla}\vec{u}$ can be neglected in a first-order approximation. The equations then become linear, and thus Stokes could give analytical formulas to express the solutions. Those equations are now labeled as the Stokes equations. He discussed at length the boundary conditions to impose on the velocity and privileged the no-slip condition, while Navier and Poisson used a tangential slip condition. Navier was influenced by former works of Girard on capillar vessels [213]. Stokes disagreed with the conclusions of Navier and, in 1850, in his memoir on the pendulum [452], where he computed the resistance to the motion of a sphere through a fluid (Stokes' law), he explained the physical reasons why the no-slip condition should rather be privileged. In his book *Recherches sur l'hydrodynamique* (1904) [160], Duhem explains the various hesitations of hydrodynamicians between the two

---

[11]This very principle that D'Alembert condemned as "a principle resting only on that vague and obscure axiom that the effect is proportional to the cause."

kinds of boundary conditions throughout the 19th century, and how the experiments of Poiseuille (1846) [401], Warburg (1870) [497] and Couette (1890) [132] eventually gave the advantage to Stokes' no-slip condition.

3.7   Reynolds

Following the works of Navier and Stokes, a theory had been established that enjoyed a good experimental validation in the case of laminar flows. However, the Navier–Stokes equations seemed inappropriate to describe turbulent flows, a major concern for practical applications.

Even nowadays, turbulence is difficult to define. Roughly speaking, laminar flows move peacefully and it is easy to follow their streamlines, while turbulent flows are constantly eddying, new vortices being generated from old vortices in a process that drastically increases energy dissipation, drag forces and heat transfers.

Whirling flows had interested scientists and artists for centuries. In the Renaissance era, Leonardo sketched many drawings of turbulent flows. Frisch [184] quotes a fragment of the *Codice Atlantico* where Leonardo uses the term "turbolenza" to describe the whirling flows:[12]

> *Doue la turbolenza dellacqua rigenera,*
> *doue la turbolenza dellacqua simantiene plugho,*
> *doue la turbolenza dellacqua siposa*

Frisch underlines that those lines of Leonardo point exactly to the characteristic features of turbulence that are at the basis of the modern scientific theories of Richardson and Kolmogorov.

Later, hydraulicians like Venturi (1797) [488] commented on the retardation effect of the creation of whirls in the streaming of rivers. Saint-Venant was deeply interested in this retardation effect of eddies. In 1850 [420], he vindicated that the presence of eddies in the flow generated an extra internal friction that modified the viscosity of the flow. He viewed the eddies as local variations around an average value, and thus distinguished two scales: the average value corresponded to a laminar flow obeying the Navier–Stokes equations, while smaller structures were oscillating and provoking extra viscosity. This theory of eddy viscosity was further extended by Boussinesq (1870) [53].

At the same time, the dynamics of vortices had been explored by Helmholtz. In 1858 [233], he studied fluid motion in the presence of dissipative forces. As the forces could not derive from a potential, the vorticity could not be equal to 0. Helmholtz identified vorticity with an infinitesimal rotation and wanted to exhibit the dynamics vorticity generated.

---

[12]Codice Atlantico, Biblioteca Ambrosiana di Milano, f. 74v. In modern Italian, this reads as

> dove la turbolenza dell'acqua si genera,
> dove la turbolenza dell'acqua si mantiene per lungo,
> dove la turbolenza dell'acqua si posa

and in English as

> Where the turbulence of water is generated
> Where the turbulence of water maintains for long
> Where the turbulence of water comes to rest.

For an ideal fluid in a potential field of forces, he defined *vortex lines* as lines that were everywhere tangent to the vorticity vector and *vortex filaments*[13] as the union of all vortex lines crossing a given surface element of the fluid, and he showed that those vortex filaments were stable structures of the fluid.

He introduced the *Helmholtz decomposition* of a velocity field into its irrotational part and its divergence-free part, and showed that the formula that reconstructs a divergence-free vector field from its curl was analogous to the *Biot-Savart law* in electromagnetism.

For fluid mechanicians, there were two regimes of flows. The laminar flows were very regular and obeyed the Navier–Stokes equations. Turbulent flows were very irregular, with vortices of all scales making impossible to describe the flow except for average values. The term "turbulent" was coined by Thomson (Lord Kelvin) in 1887 [472] in a paper entitled *On the propagation of laminar motion through a turbulently moving inviscid liquid*. Navier distinguished "linear" flows and "non-linear flows," Reynolds "direct" flows and "sinuous" flows [410]. Later, Oseen [385] would call turbulent the blowing up solutions of the Navier–Stokes equations, and he was followed in this by Leray [328] and Ladyzhenskaya [293].

Experimental investigation of turbulence was initiated by Reynolds in 1883 [410]. The sudden transition from laminar flows to turbulent flows in pipes has been first decribed by Hagen in 1839 [227] and 1854 [228]. However, Reynolds was the first to try and understand the dichotomy between the two regimes of flows:

> *The internal motion of water assumes one or other of two broadly distinguishable forms – either the elements of the fluid follow one another along lines of motion which lead in the most direct manner to their destination, or they eddy about in sinuous paths the most indirect possible.*

Reynolds made a decisive experiment on the visualization of the transition from laminarity to turbulence. Injecting dye in a moving fluid inside a glass tank, he could visualize the streak lines and show that when the velocity of the fluid at the entrance of the pipe was increased the flow began to develop eddies, and, varying the velocities and the pipes, he could show that the passage from laminar flows to turbulent flows was determined by the size of a dimensionless number, which is now called the Reynolds number $\mathbf{Re}$ and is given by $\mathbf{Re} = \frac{UL}{\nu}$, where ν is the kinematical viscosity, U is the characteristic velocity of the fluid and L the characteristic length of the device (such as the radius of the pipe, for instance).

Note that Reynolds' original apparatus is still used for experiments on turbulence transition in pipe flow (see Eckhardt in 2008 [162] or Mullin in 2011 [371]).

Another important contribution of Reynolds was his analytical study of turbulence, published in 1895 [411], with the introduction of the decomposition of the flow into mean and fluctuating parts. Averaging the velocity to get the mean part gives an equation (the Reynolds equations) on this mean velocity that is not closed, due to the non-linearity of the Navier–Stokes equations: the advective term $\vec{u}.\vec{\nabla}\vec{u}$ leads to a correction of the Navier–Stokes equations due to the interaction with the fluctuating part, that modifies the stress tensor with the Reynolds stress.

The theory of turbulence would then be developed by the modelization of this Reynolds stress and by the study of the transfer of kinetic energy from the mean flow to turbulent parts.

---

[13]Note that vortex filaments for mathematicians are highly more singular. In fluid mechanics, vortex filaments are a tube of vortex lines with cross sectional radius $\delta = O(\mathbf{Re}^{-\frac{1}{2}})$, where $\mathbf{Re}$ is the Reynolds number. In the vanishing viscosity limit, the tube is reduced to a line.

3.8   Oseen, Leray, Hopf and Ladyzhenskaya

Lorentz was awarded the 1902 Nobel prize for his works on the electron and paved the way to Einstein's relativity theory by discovering in 1904 the key role of the Lorentz group. But Lorentz gave as well some important contributions to hydrodynamics. In 1896 [342], he studied Stokes' steady creeping flows. Looking at the flow associated to the motion of a sphere of radius R and velocity c, and letting R go to 0, he obtained the Green function $\mathbb{J}_\nu$ for the steady Stokes equations

$$-\nu\Delta\vec{u} = \vec{F} - \vec{\nabla}p, \qquad \operatorname{div}\vec{u} = 0$$

$\mathbb{J}_\nu$ is a second-rank tensor and we have $\vec{u}(x) = \int \mathbb{J}_\nu(x-y)\vec{F}(y)\,dy$. This Green function is now called a *stokeslet,* as proposed in 1953 by Hancok [229] (though Kuiken states that it should rather be called *lorentzlet* [285] and it is sometimes called the *Oseen tensor*).

In 1911, Oseen [384, 385] extended the work of Lorentz to the case of evolutionary Stokes equations and then to the Navier–Stokes equations. He obtained an explicit tensor $\mathcal{O}_\nu$ (the *Oseen tensor*) such that the Stokes equations

$$\partial_t\vec{u} = \nu\Delta\vec{u} + \vec{F} - \vec{\nabla}p, \qquad \operatorname{div}\vec{u} = 0, \qquad \vec{u}(0,x) = 0$$

have the solution, for positive t,

$$\vec{u}(t,x) = \int_0^t \int_{\mathbb{R}^3} \mathcal{O}_\nu(t-s, x-y)\vec{F}(s,y)\,dy\,ds.$$

Then, he turned the Navier–Stokes equations

$$\partial_t\vec{u} = \nu\Delta\vec{u} - (\vec{u}.\vec{\nabla})\vec{u} + \vec{f} - \vec{\nabla}p$$
$$\operatorname{div}\vec{u} = 0 \tag{3.3}$$
$$\vec{u}_{|t=0} = \vec{u}_0$$

into an integro-differential equation

$$\vec{u}(t,x) = \int_{\mathbb{R}^3} W_\nu(t, x-y)\vec{u}_0(y)\,dy$$
$$+ \int_0^t \int_{\mathbb{R}^3} \mathcal{O}_\nu(t-s, x-y)\left(\vec{f}(s,y) - \sum_{i=1}^3 u_i(s,y)\partial_i\vec{u}(s,y)\right)dy\,ds \tag{3.4}$$

where W_ν is the heat kernel associated to the heat equation $\partial_t G = \nu\Delta G$. He was then able to get a solution for a small positive time interval $[0, T]$ when the initial data $\vec{u}_0$ and the force $\vec{f}(t,x)$ were regular and localized. More precisely, he considered the solution $\vec{u}_\epsilon$ associated to the initial value $\epsilon\vec{u}_0$ and the force $\epsilon\vec{f}$, and obtained a power series expansion of $\vec{u}_\epsilon$ with respect to the powers of ϵ (by identification of the coefficients of the expansion, which were computed inductively as solutions of linear Stokes equations) with a convergence radius greater than 1; taking $\epsilon = 1$ gives the solution of the Navier–Stokes equations.

In 1934, Leray [328] studied the problem of *turbulent solutions* that Oseen left open: when the estimates found by Oseen blow up, what can be said of the solutions? Leray found that one had an estimate that did not blow up: the energy $\int_{\mathbb{R}^3} |\vec{u}(t,x)|^2\,dx$. However, the control of the L^2 norm of $\vec{u}$ is not enough to ensure that the solution does not blow up. Leray introduced a new concept of solutions, that he called turbulent solutions and are now called weak solutions, for which the derivatives in the differential equations were no longer classical derivatives, but generalized derivatives (now called derivatives in the sense of distributions[14]). Two years before Sobolev [443], he thus introduced the space of functions that are Lebesgue measurable, square-integrable and that have a generalized square-integrable gradient: this space would later be called the Sobolev space H^1.

He was then able to prove that Oseen's classical solutions may be extended to global turbulent solutions, the loss of control on the size of $\vec{u}$ and its derivatives being compensated by a modification of the meaning of derivatives in the equations. The existence of those new solutions were proved by compactness arguments, due to the strong development of topological theory in the beginning of the 20th century. However, the prize to pay for using those methods (replacing a unique limit by a (possibly non-unique) limit point) was severe: uniqueness of solutions to the Cauchy initial value problem was no longer granted. The issue of uniqueness of weak solutions or of globalness of strong solutions has remained open since Leray's seminal thesis.

Leray went even further in the use of topological methods, by introducing with Schauder [331] index methods of algebraic topology to get solutions for functional equations, a method he could apply to the stationary Navier–Stokes problem in a domain. This theorem was a major turning point in the resolution of equations, as underlined by Leray [330]:[15]

> *Pour nous, résoudre une équation, c'est majorer les inconnues et préciser leur allure le plus possible; ce n'est pas en construire, par des développements compliqués, une solution dont l'emploi pratique sera presque toujours impossible.*

To construct his weak solutions, Leray used the formulas derived by Lorentz and Oseen for hydrodynamic potentials. Such formula were explicitly known only for very simple domains, and not available for more complex domains. Hopf [238] in 1951 and Ladyzhenskaya in 1957 [262] used another approach, by approximating the equations on finite-dimensional subspaces of L^2, which is now widely used in the numerical analysis of the equations [471]. This is the Faedo–Galerkin method, initially introduced by Galerkin for solving elliptic equations, then extended by Faedo for evolution problems; Hopf's work was one of the first applications of this method to non-linear equations.

Ladyzhenskaya developed a full mathematic theory for the use of weak solutions for partial differential equations, beginning in 1953 with her book on hyperbolic equations [291]. She described her theory in a review paper on the Clay millenium prize [296]. She comments on those weak solutions in those terms:

> *This ideology (program) was partially contained in the formulations of the 19th and the 20th Hilbert problems. Namely, these Hilbert problems contain the important idea of seeking solutions of variational problems in spaces dictated by the functional rather than in spaces of smooth functions. After those 'bad' solutions*

---

[14]Lützen underlines the influence of Leray on the birth of distribution theory [344] :

> *Leray used the test function generalization in a third way, namely to generalize the divergence operator. His consistent use of this generalization method occupies a central position in the prehistory of the theory of distribution as he taught it to his student L. Schwartz at the Ecole Normale.*

[15]See the translation in the review of Mawhin [352]: "For us, to solve an equation consists in bounding its unknowns and precise their shape as much as possible; it is not to construct, through complicated developments, a solution whose practical use will be almost always impossible."

In contrast, see the comments by Devlin [152] about the Clay millenium problem for the Navier–Stokes equations: "There is just one problem. No one has been able to find a formula that solves the Navier–Stokes equations. In fact, no one has been able to show in principle whether a solution even exists! (More precisely, we do not know whether there is a **mathematical** solution - a **formula** that satisfies the equations. Nature "solves" the equations every time a real fluid flows, of course.)"

are snared, one can then consider them in more detail, that is, study their actual smoothness in dependence on the smoothness of the data.

The use of weak solutions in mathematical physics was then unusual, as she recalls:

> *I still remember years (the 1940s and 50s) when the majority of maîtres (and first and foremost I.G. Petrovskii) regarded a problem as unsolved if on the chosen path of investigation the researcher did not guarantee the existence of a classical solution.*

3.9   Turbulence Models

All along the 20th century, engineers, physicists and mathematicians paid a great attention to the complexity of flows, focusing on instabilities and on turbulence. There are many references on the history of turbulence, including the historical introduction to the book of Monin and Yaglom [367], the book by Frisch [184], the review by Lumley and Yaglom [343] or the biographical book *A voyage through turbulence* [146].

Turbulence turned out to be a crucial issue with the development of aviation. Engineers were mostly interested in understanding the turbulence effects on large scales, as it had important consequences on drag forces. The description of the large scales involves the interaction with the boundaries, while in the small scales turbulence has a universal behavior which does not depend on the geometry.

The main tools for the large-scale description of turbulence were introduced in the twenties and the thirties by Prandtl and von Kármán with logarithmic velocity profile laws or logarithmic skin friction laws. The log-law [252] that expresses the profile of the mean velocity of a turbulent flow bounded by parallell walls in terms of the logarithm of the normal distance to the walls, was published in 1930 by von Kármán.

Another important contribution to the understanding of turbulence was in 1904 the boundary layer theory of Prandtl [405], which modelizes the behavior of the fluid when the viscosity vanishes.

The study of small scales was initiated by Taylor and Richardson. Richardson introduced in 1922 [413] his model of energy cascade in turbulent flows. To describe the energy transfer from large scales to small scales, he wrote a parody of a poem of Swift:

> *Big whirls have little whirls*
> *that feed on their velocity.*
> *Little whirls have lesser whirls*
> *and so on to viscosity*
> *... in the molecular sense.*[16]

Let us remark, however, that the model of energy cascade has been criticized recently: according to Tsinober [484], the non-local aspects of incompressible fluid mechanics (i.e.,

---

[16]Swift's poem reads as:

> *So, naturalists observe, a flea*
> *Hath smaller fleas that on him prey :*
> *And these have smaller fleas to bite 'em*
> *And so proceed ad infinitum.*
> *Thus every poet in his kind,*
> *Is hit by him that comes behind.*

the non-local dependence of the pressure on the velocity [Laplace equation] and of velocity on the vorticiity [Biot-Savard law])

contradict the idea of cascade in physical space, which is local by definition

so that Richardson's verse should be replaced by Betchov's [42]:

Big whirls lack smaller whirls
to feed on their velocity.
They crush and form the finest curls
permitted by viscosity.

Homogeneous and isotropic turbulence was introduced in 1935 by Taylor [465] as a statistical modelization of turbulence. Whereas such a turbulent model cannot be applied to an actual fluid (because of the presence of boundaries that prevent isotropy), this model can be applied to the asymptotical behavior of the fluid at small scales and the simplifications induced by the isotropy hypothesis allows an easier handling of the equations.

In 1941, Kolmogorov [268, 269] developed the analysis of fully developed turbulence based on a stochastic modelization, through the precise description of random fields he introduced in [267]. The basis of Kolmogorov's theory was a modelization of the universal equilibrium regime of small-scale components. Discrepancies between experimental results and his theory led Kolmogorov to modify his theory in 1962 in order to take into account intermittency in the distribution of dissipative structures in turbulent flows [270].

Nowadays, numerical simulation of turbulence provides a large number of quantitative data to support or disprove the theoric ideas that are formulated on the qualitative behavior of turbulent flows. The main technique is the *direct numerical simulation* (DNS), introduced in 1972 by Orszag and Patterson [382]. At the same time appeared the *large eddy simulation* (LES) where only the large scales are numerically resolved, the fine scales being parametrized (Deardorff (1970) [150], using a numerical model introduced in 1962 by Smagorinsky and Manabe [441]).

Chapter 4

Classical Solutions

In this chapter, we study classical solutions of the Cauchy initial value problem for the Navier–Stokes equations (with reduced (unknown) pressure p, reduced force density $\vec{f}$ and kinematic viscosity $\nu > 0$):

Navier–Stokes equations

Given a divergence-free vector field $\vec{u}_0$ on $\mathbb{R}^3$ and a force $\vec{f}$ on $(0, +\infty) \times \mathbb{R}^3$, find a positive T and regular functions $\vec{u}$ and p on $[0, T] \times \mathbb{R}^3$ solutions to

$$\partial_t \vec{u} = \nu \Delta \vec{u} - (\vec{u}.\vec{\nabla})\vec{u} + \vec{f} - \vec{\nabla}p$$
$$\text{div } \vec{u} = 0 \tag{4.1}$$
$$\vec{u}_{|t=0} = \vec{u}_0$$

We are going to solve Equations (4.1). We shall use only classical tools of differential calculus, as they were used in the end of the 19th century or the beginning of the 20th century. More precisely, we will stick to the spirit of Oseen's paper, which was published in 1911 [384, 385]. A similar treatment can be found in a 1966 paper of Knightly [263].

4.1   The Heat Kernel

The *heat kernel* W_t is the function

$$W_t(x) = \frac{1}{t^{3/2}} W\left(\frac{x}{\sqrt{t}}\right) = \frac{1}{(4\pi t)^{3/2}} e^{-\frac{x^2}{4t}} \tag{4.2}$$

This kernel is used to solve the *heat equation*:

Heat equation

Theorem 4.1.
Let u_0 be a bounded continuous function on $\mathbb{R}^3$. Then the function

$$u(t,x) = \int u_0(x - \sqrt{t}y)W(y)\,dy \tag{4.3}$$

DOI: 10.1201/9781003042594-4

is continuous on $[0, +\infty) \times \mathbb{R}^3$ and C^∞ on $(0, +\infty) \times \mathbb{R}^3$ and solution of the heat equation

$$\begin{aligned} \partial_t u &= \Delta u && \text{on } (0, +\infty) \times \mathbb{R}^3 \\ u(0, x) &= u_0(x) && \text{on } \mathbb{R}^3 \end{aligned} \qquad (4.4)$$

Proof. From the formula (4.3), we see that u is continuous on $[0, +\infty) \times \mathbb{R}^3$ and that $u(0,.) = u_0$. From the equality, for $t > 0$,

$$u(t, x) = W_t * u_0 = \int W_t(x - y) u(y) \, dy,$$

we see that u is C^∞ on $(0, +\infty) \times \mathbb{R}^3$, since all the derivatives of W_t have exponential decay in the space variable.

We have $(\partial_t - \Delta) u = u * (\partial_t - \Delta) W_t = 0$ since $\partial_i W(x) = -\frac{x_i}{2} W(x)$, hence

$$\partial_t \left(\frac{1}{t^{3/2}} W\left(\frac{x}{\sqrt{t}}\right) \right) = -\frac{1}{t^{5/2}} \left(\frac{3}{2} - \sum_{i=1}^{3} \frac{x_i^2}{4t} \right) W\left(\frac{x}{\sqrt{t}}\right)$$

and

$$\Delta \left(\frac{1}{t^{3/2}} W\left(\frac{x}{\sqrt{t}}\right) \right) = -\sum_{i=1}^{3} \partial_i \left(\frac{1}{2t^{5/2}} x_i W\left(\frac{x}{\sqrt{t}}\right) \right) = -\frac{1}{t^{5/2}} \left(\frac{3}{2} - \sum_{i=1}^{3} \frac{x_i^2}{4t} \right) W\left(\frac{x}{\sqrt{t}}\right).$$

The theorem is proved. □

We may consider as well classical solutions of the non-homogeneous heat equations, where we add a forcing term f:

Theorem 4.2.
Let f be a continuous function on $[0, +\infty) \times \mathbb{R}^3$, which is C^1 in the space variable on $(0, +\infty) \times \mathbb{R}^3$ and is uniformly bounded with uniformly bounded spatial derivatives. Then the function

$$F(t, x) = \int_0^t \int f(s, x - \sqrt{t - s}\, y) W(y) \, dy \, ds \qquad (4.5)$$

is continuous on $[0, +\infty) \times \mathbb{R}^3$, C^1 on $(0, +\infty) \times \mathbb{R}^3$ and is C^2 in the space variable on $(0, +\infty) \times \mathbb{R}^3$. Moreover, F is solution of the heat equation

$$\begin{aligned} \partial_t F &= \Delta F + f && \text{on } (0, +\infty) \times \mathbb{R}^3 \\ F(0, x) &= 0 && \text{on } \mathbb{R}^3 \end{aligned} \qquad (4.6)$$

Proof. Just write that

$$\partial_i \partial_j F(t, x) = \int_0^t \int \partial_i f(s, x - \sqrt{t - s}\, y) \partial_j W(y) \, dy \frac{ds}{\sqrt{t - s}}. \qquad \square$$

Remark: If $\vec{u}_0$ is C^2 with bounded derivatives, the solutions u and F given in Theorems 4.1 and 4.2 are clearly unique in the class of continuous functions on $[0, +\infty) \times \mathbb{R}^3$ with the property that, for all $T > 0$,

$$\sup_{0 < t < T, x \in \mathbb{R}^3, |\alpha| \leq 2} |\partial_x^\alpha u(t,x)| + |\partial_x^\alpha F(t,x)| < +\infty.$$

By linearity, we may assume that $u_0 = 0$ or $f = 0$ and prove that u or F is 0. This is obvious: if $\partial_t u = \Delta u$ and $u(0,.) = 0$, we write

$$\frac{d}{dt} \int |u(t,x)|^2 e^{-|x|}\, dx = 2 \int u\, \partial_t u\, e^{-|x|}\, dx = 2 \int u\, \Delta u\, e^{-|x|}\, dx$$

$$= -2 \int |\vec{\nabla}u|^2 e^{-|x|}\, dx + 2 \int u e^{-|x|} \sum_{j=1}^{3} \frac{x_j}{|x|} \partial_j u\, dx$$

$$\leq \frac{1}{2} \int |u(t,x)|^2 e^{-|x|}\, dx$$

so that $\int |u(t,x)|^2 e^{-|x|}\, dx \leq e^{\frac{t}{2}} \int |u(0,x)|^2 e^{-|x|}\, dx = 0.$[1]

If we consider now a heat equation with a viscosity $\nu > 0$, we have to modify very slightly the formula: the solution u of the equation

$$\begin{aligned} \partial_t u &= \nu \Delta u + f &&\text{on } (0, +\infty) \times \mathbb{R}^3 \\ u(0,x) &= u_0(x) &&\text{on } \mathbb{R}^3 \end{aligned} \qquad (4.7)$$

is given by

$$u = W_{\nu t} * u_0 + \int_0^t W_{\nu(t-s)} * f(s,.)\, ds. \qquad (4.8)$$

4.2   The Poisson Equation

We now solve the *Poisson equation* with help of the *Green function*

$$G(x) = \frac{1}{4\pi|x|}$$

where $|x| = \sqrt{x_1^2 + x_2^2 + x_3^2}$.

We begin with an easy lemma:

Lemma 4.1.

Let $0 < \gamma < 3$ and let

$$I_\gamma(x) = \int \frac{1}{(1+|x-y|)^4} \frac{1}{|y|^\gamma}\, dy.$$

We have

$$I_\gamma(x) \leq C_\gamma (1+|x|)^{-\gamma}.$$

---

[1] Actually, according to Tychonov [485], we even have uniqueness in the class of functions that grow no faster than $O(e^{C|x|^2})$, but uniqueness fails in the class of smooth functions with faster increase.

Proof. We first write:

$$I_\gamma(x) \le \int_{|y|\le 1} \frac{1}{|y|^\gamma}\,dy + \int_{|y|\ge 1} \frac{1}{(1+|x-y|)^4}\,dy$$

so that $I_\gamma \in L^\infty$. We then write

$$I_\gamma(x) \le \int_{|y|\ge \frac{|x|}{2}} \frac{2^\gamma}{|x|^\gamma}\frac{1}{(1+|x-y|)^4}\,dy + \int_{|x-y|\ge \frac{|x|}{2}} \frac{1}{|x-y|^3}\frac{1}{|y|^\gamma}\,dy$$

which gives $|x|^\gamma I_\gamma \in L^\infty$. $\qquad\square$

Corollary 4.1. *If $0 < \gamma < 3$, then, for $t > 0$, we have*

$$\int W_t(x-y)\frac{1}{|y|^\gamma}\,dy \le C_\gamma \frac{1}{(\sqrt{t}+|x|)^\gamma}$$

and

$$\int W_t(x-y)\frac{1}{(1+|y|)^\gamma}\,dy \le C_\gamma \frac{1}{(1+|x|)^\gamma}$$

where the constant C_γ does not depend on t nor x.
Similarly, we have

$$\int \frac{1}{(\sqrt{t}+|x-y|)^4}\frac{1}{(1+|y|)^\gamma}\,dy \le C\frac{1}{\sqrt{t}}\frac{1}{(1+|x|)^\gamma}$$

and

$$\int \frac{1}{(\sqrt{t}+|x-y|)^4}\frac{1}{(\sqrt{t}+|y|)^\gamma}\,dy \le C\frac{1}{\sqrt{t}}\frac{1}{(\sqrt{t}+|x|)^\gamma}$$

Proof. Let us notice that $W_t(x-y) \le C\frac{\sqrt{t}}{(\sqrt{t}+|x-y|)^4}$. We have

$$\int \frac{1}{(\sqrt{t}+|x-y|)^4}\frac{1}{(1+|y|)^\gamma}\,dy \le \int \frac{1}{(\sqrt{t}+|x-y|)^4}\,dy = \frac{1}{\sqrt{t}}\int \frac{dz}{(1+|z|)^4}.$$

On the other hand, letting $y = \sqrt{t}z$, we get

$$\int \frac{1}{(\sqrt{t}+|x-y|)^4}\frac{1}{|y|^\gamma}\,dy = \frac{1}{(\sqrt{t})^{1+\gamma}}\int \frac{1}{(1+|\frac{x}{\sqrt{t}}-z|)^4}\frac{1}{|z|^\gamma}\,dz$$

and we conclude by Lemma 4.1. $\qquad\square$

The Poisson equation

Theorem 4.3.
Let u_0 be a $\mathcal{C}^1$ function on $\mathbb{R}^3$ such that

$$\sup_{|\alpha|\le 1}\sup_{x\in\mathbb{R}^3}(1+|x|)^4|\partial^\alpha u_0(x)| < +\infty.$$

Then the function

$$U(x) = \int u_0(x-y)\,G(y)\,dy \qquad (4.9)$$

is C^2 on $\mathbb{R}^3$ and solution of the Poisson equation

$$\Delta U = -u_0 \qquad (4.10)$$

Moreover, we have

$$\sup_{x \in \mathbb{R}^3} (1 + |x|)|U(x)| < +\infty$$

and

$$\sup_{1 \leq |\alpha| \leq 2} \sup_{x \in \mathbb{R}^3} (1 + |x|)^2 |\partial^\alpha U(x)| < +\infty.$$

Proof. Let $G_j(x) = -\frac{x_j}{4\pi|x|^3}$. We have $\partial_j U = u_0 * G_j$ and $\partial_i \partial_j U = \partial_i u_0 * G_j$. Thus, we control U and its derivatives by

$$|U(x)| \leq C \int \frac{1}{(1 + |x - y|^4)} \frac{1}{|y|} \, dy$$

and, for $1 \leq |\alpha| \leq 2$,

$$|\partial^\alpha U(x)| \leq C \int \frac{1}{(1 + |x - y|^4)} \frac{1}{|y|^2} \, dy.$$

Lemma 4.1 then gives the control of the sizes of U and of its derivatives.

In particular, we have

$$\Delta U(x) = \lim_{\epsilon \to 0, R \to +\infty} \int_{\epsilon < |y| < R} \sum_{j=1}^3 \partial_j u_0(x - y) \partial_j G(y) \, dy$$

so that, by Ostrogradski's divergence theorem, we have

$$\Delta U(x) = \lim_{\epsilon \to 0, R \to +\infty} \Big(\epsilon^2 \int_{S^2} u_0(x - \epsilon\sigma) \sum_{j=1}^3 \sigma_j \partial_j G(\epsilon\sigma) \, d\sigma$$

$$- R^2 \int_{S^2} u_0(x - R\sigma) \sum_{j=1}^3 \sigma_j \partial_j G(R\sigma) \, d\sigma$$

$$+ \int_{\epsilon < |y| < R} u_0(x - y) \Delta G(y) \, dy \Big).$$

Since $\Delta G = 0$ on $\mathbb{R}^3 - \{0\}$, and $\sum_{j=1}^3 \sigma_j \partial_j G(R\sigma) = -\frac{1}{4\pi R^2}$, we get that

$$\Delta U(x) = \lim_{\epsilon \to 0, R \to +\infty} -\frac{1}{4\pi} \int_{S^2} u_0(x - \epsilon\sigma) - u_0(x - R\sigma) \, d\sigma$$

so that $\Delta U(x) = -u_0(x)$. $\qquad\square$

4.3   The Helmholtz Decomposition

A direct consequence of Theorem 4.3 is the following decomposition theorem of vector fields:

The Helmholtz decomposition

Theorem 4.4.
Let $\vec{F}_0$ be a C^1 vector field on $\mathbb{R}^3$ such that

$$\sup_{|\alpha| \leq 1} \sup_{x \in \mathbb{R}^3} (1 + |x|)^4 |\partial^\alpha \vec{F}_0(x)| < +\infty$$

and assume that the divergence of $\vec{F}_0$ is C^1 with

$$\sup_{|\alpha| = 1} \sup_{x \in \mathbb{R}^3} (1 + |x|)^4 |\partial^\alpha \operatorname{div} \vec{F}_0(x)| < +\infty.$$

Then there exists unique C^1 vector fields $\vec{F}$ and $\vec{H}$ such that:

- $\vec{F}_0 = \vec{F} + \vec{H}$

- $\vec{F}$ *is solenoidal (i.e., divergence free):* $\operatorname{div} \vec{F} = 0$

- $\vec{H}$ *is irrotational:* $\operatorname{curl} \vec{H} = 0$

- $\sup_{0 \leq |\alpha| \leq 1} \sup_{x \in \mathbb{R}^3} (1 + |x|)^2 |\partial^\alpha \vec{F}(x)| < +\infty.$

- $\sup_{0 \leq |\alpha| \leq 1} \sup_{x \in \mathbb{R}^3} (1 + |x|)^2 |\partial^\alpha \vec{H}(x)| < +\infty.$

Moreover, there exists a unique function p which is C^2 on $\mathbb{R}^3$ with $\sup_{x \in \mathbb{R}^3}(1 + |x|)|p(x)| < +\infty$ and

$$\vec{H} = \vec{\nabla} p.$$

Proof. We begin by proving the uniqueness. Let us assume that we have two solutions $(\vec{F}_1, \vec{H}_1)$ and $(\vec{F}_2, \vec{H}_2)$. Since $\vec{H}_2 - \vec{H}_1$ is irrotational and C^1, one can find a C^2 function q such that $\vec{H}_2 - \vec{H}_1 = \vec{\nabla} q$. We then have

$$\Delta q = \operatorname{div}(\vec{H}_2 - \vec{H}_1) = \operatorname{div}(\vec{F}_1 - \vec{F}_2) = 0.$$

Thus q is harmonic; its derivatives are harmonic functions that are $O(|x|^{-2})$ at infinity, hence its derivatives are equal to 0 by the maximum principle. Thus, $\vec{H}_2 = \vec{H}_1$, and $\vec{F}_2 = \vec{F}_1$.

For proving the existence, it is enough to use Theorem 4.3 and to define $\vec{H}$ as $\vec{H} = \vec{\nabla} p$, where p solves the Poisson equation $\Delta p = \operatorname{div} \vec{F}_0$. $\qquad\square$

Definition 4.1 (Leray projection operator).
For a regular vector field $\vec{F}_0$ and its Helmholtz decomposition $\vec{F}_0 = \vec{F} + \vec{H}$ into the sum of a solenoidal vector field $\vec{F}$ and an irrotational vector field $\vec{H}$, we shall write

$$\vec{F} = \mathbb{P}\vec{F}_0.$$

*The operator $\mathbb{P} : \vec{F}_0 \mapsto \vec{F}$ is called the **Leray projection operator**.*

We may define $\vec{F} = \mathbb{P}\vec{F}_0$ as the unique solution of the Poisson equation

$$-\Delta\vec{F} = \vec{\nabla} \wedge (\vec{\nabla} \wedge \vec{F}_0)$$

which is $o(1)$ at infinity.

4.4 The Stokes Equation

We have gathered enough results to be able to solve the Stokes equations, i.e., the Navier–Stokes equations when the convective bilinear term is neglected.

The Stokes problem

Given a divergence-free vector field $\vec{u}_0$ on $\mathbb{R}^3$ and a force $\vec{f}$ on $[0, +\infty) \times \mathbb{R}^3$, find regular functions $\vec{u}$ and p on $(0, +\infty) \times \mathbb{R}^3$ solutions to

$$\partial_t \vec{u} = \nu \Delta \vec{u} + \vec{f} - \vec{\nabla} p$$
$$\operatorname{div} \vec{u} = 0 \qquad\qquad (4.11)$$
$$\vec{u}_{|t=0} = \vec{u}_0$$

The solution of this problem is easy: we use the Helmholtz decomposition of $\vec{f}$ into $\vec{f} = \vec{F} + \vec{H}$, where $\vec{F}$ is divergence free and $\vec{H} = \vec{\nabla} q$ is irrotational. Then the Helmholtz decomposition of $\partial_t \vec{u}$ will give

$$\partial_t \vec{u} = \nu \Delta \vec{u} + \vec{F} \text{ and } 0 = \vec{\nabla} q - \vec{\nabla} p.$$

Thus, we know that p is determined through the Poisson equation $\Delta p = \operatorname{div} \vec{f}$, while $\vec{u}$ is a solution of the heat equation with initial value $\vec{u}_0$ and forcing term $\vec{F} = \mathbb{P}\vec{f}$. We thus use Theorem 4.1, Theorem 4.2 and Theorem 4.4 and get the following result:

The Stokes equation

Theorem 4.5.
Let $\vec{u}_0$ be a $\mathcal{C}^2$ divergence-free vector field on $\mathbb{R}^3$ and let $\vec{f}$ be a time-dependent vector field such that:

- $\sup_{|\alpha| \leq 2} \sup_{x \in \mathbb{R}^3} (1 + |x|)^2 |\partial^\alpha \vec{u}_0(x)| < +\infty$

- *for $|\alpha| \leq 1$, $\partial_x^\alpha \vec{f}$ is continuous on $[0, +\infty) \times \mathbb{R}^3$*

- $\sup_{|\alpha| \leq 1} \sup_{t \geq 0, x \in \mathbb{R}^3} (1 + |x|)^4 |\partial_x^\alpha \vec{f}(t,x)| < +\infty$

- *we have furthermore a control on the derivatives of $\operatorname{div} \vec{f}$:*

$$\sup_{|\alpha|=1} \sup_{t>0, x \in \mathbb{R}^3} (1 + |x|)^4 |\partial_x^\alpha \operatorname{div} \vec{f}(t,x)| < +\infty.$$

Then, there exists a unique solution $(\vec{u}, p)$ of the Stokes problem

$$\partial_t \vec{u} = \nu \Delta \vec{u} + \vec{f} - \vec{\nabla} p$$
$$\operatorname{div} \vec{u} = 0$$
$$\vec{u}_{|t=0} = \vec{u}_0$$

such that:

- *for $|\alpha| \leq 2$, $\partial_x^\alpha p$ is continuous on $[0, +\infty) \times \mathbb{R}^3$*

- $\sup_{t \geq 0, x \in \mathbb{R}^3} (1 + |x|) |p(t, x)| < +\infty$

- $\sup_{1 \leq |\alpha| \leq 2} \sup_{t \geq 0, x \in \mathbb{R}^3} (1 + |x|)^2 |\partial_x^\alpha p(t, x)| < +\infty$

- *for $|\alpha| \leq 2$, $\partial_x^\alpha \vec{u}$ is continuous on $[0, +\infty) \times \mathbb{R}^3$ and, for every $0 < T < +\infty$,*

$$\sup_{|\alpha| \leq 2} \sup_{0 < t < T, x \in \mathbb{R}^3} (1 + |x|)^2 |\partial_x^\alpha \vec{u}(t, x)| < +\infty$$

- $\partial_t \vec{u}$ *is continuous on $[0, +\infty) \times \mathbb{R}^3$.*

Proof. We write

$$\Delta p = \operatorname{div} \vec{f}$$

and use Theorem 4.3 to have a control on p and its derivatives:

$$\sup_{t \geq 0, x \in \mathbb{R}^3} (1 + |x|) |p(t, x)| + \sum_{1 \leq |\alpha| \leq 2} (1 + |x|)^2 |\partial_x^\alpha p(t, x)|$$

$$\leq C \sum_{|\alpha| \leq 1} \sup_{t > 0, x \in \mathbb{R}^3} (1 + |x|)^4 |\partial_x^\alpha \operatorname{div} \vec{f}(t, x)|.$$

Tnen, we let $\vec{F} = \vec{f} - \vec{\nabla} p$ and we write

$$\vec{u} = W_{\nu t} * \vec{u}_0 + \int_0^t W_{\nu(t-s)} * \vec{F}(s, .)\, ds = \vec{u}_1 + \vec{u}_2.$$

For $|\alpha \leq 2$, we have

$$|\partial_x^\alpha (W_{\nu t} * \vec{u}_0)| \leq \sup_{y \in \mathbb{R}^3} (1 + |y|)^2 |\partial_x^\alpha \vec{u}_0(y)| \int W_{\nu t}(x - y) \frac{1}{(1 + |y|)^2}\, dy$$

and Corollary 4.1 gives us the control of the size of $\partial_x^\alpha \vec{u}_1$. Similarly, we have, for $|\alpha| \leq 1$,

$$|\partial_x^\alpha (W_{\nu(t-s)} * \vec{F}(s, .))| \leq \sup_{y \in \mathbb{R}^3} (1 + |y|)^2 |\partial_x^\alpha \vec{F}(s, y)| \int W_{\nu(t-s)}(x - y) \frac{1}{(1 + |y|)^2}\, dy$$

and thus

$$|\partial_x^\alpha \vec{u}_1(t, x)| \leq C \frac{t}{(1 + |x|)^2} \sup_{0 < s < t, y \in \mathbb{R}^3} (1 + |y|)^2 |\partial_x^\alpha \vec{F}(s, y)|.$$

For $|\alpha| = 2$, we write $\partial_x^\alpha = \partial_i \partial_j$ and we remark that

$$|\partial_i (W_{\nu(t-s)}(x))| \leq C \frac{1}{\sqrt{\nu(t-s)}} W_{\nu \frac{(t-s)}{2}}(x)$$

and get

$$|\partial_i \partial_j (W_{\nu(t-s)} * \vec{F}(s, .))| \leq \frac{C}{\sqrt{\nu(t-s)}} \sup_{y \in \mathbb{R}^3} (1 + |y|)^2 |\partial_j \vec{F}(s, y)| \int W_{\nu \frac{t-s}{2}}(x - y) \frac{dy}{(1 + |y|)^2}$$

and finally

$$|\partial_i \partial_j \vec{u}_1(t, x)| \leq C \frac{\sqrt{t}}{\sqrt{\nu}(1 + |x|)^2} \sup_{0 < s < t, y \in \mathbb{R}^3} (1 + |y|)^2 |\partial_j \vec{F}(s, y)|. \qquad \square$$

4.5   The Oseen Tensor

In Theorem 4.5, the solution $(\vec{u}, p)$ is given by

$$p(t, x) = -\frac{1}{4\pi|x|} * \operatorname{div} \vec{f}(t, x) = -\sum_{j=1}^{3} f_j(t, x) * \partial_j G$$

and

$$\vec{u} = W_{\nu t} * \vec{u}_0 + \int_0^t W_{\nu(t-s)} * (\vec{f}(s, x) - \vec{\nabla} p(s, x))\, ds$$

Thus, the k-th component of $\vec{u}$ is given by

$$u_k = W_{\nu t} * u_{0,k} + \int_0^t W_{\nu(t-s)} * (f_k + \partial_k \sum_{j=1}^{3} \partial_j G * f_j)\, ds$$

We have

$$W_{\nu(t-s)} * (f_k + \partial_k \sum_{j=1}^{3} \partial_j G * f_j) = \sum_{j=1}^{3} f_j * (\delta_{j,k} W_{\nu(t-s)} + G * \partial_j \partial_k W_{\nu(t-s)}).$$

Definition 4.2 (Oseen tensor).
The **Oseen tensor** *is the tensor* $(O_{j,k}(\nu t, x))_{1 \leq j,k \leq 3}$ *given by*

$$O_{j,k}(\nu t, x) = \delta_{j,k} W_{\nu t} + G * \partial_j \partial_k W_{\nu t}$$

Let $O_{j,k}(x) = O_{j,k}(1, x)$; it is easy to see that

$$O_{j,k}(\nu t, x) = \frac{1}{(\nu t)^{3/2}} O_{j,k}\left(\frac{x}{\sqrt{\nu t}}\right).$$

The functions $O_{j,k}$ are easily determined through Oseen's formula[2]:

The Oseen tensor

Theorem 4.6.
We have

$$O_{j,k}(x) = \delta_{j,k} W(x) + 2\partial_j \partial_k \left(\frac{1}{(4\pi)^{3/2}|x|} \int_0^{|x|} e^{-\frac{s^2}{4}}\, ds\right). \qquad (4.12)$$

When x is close to 0, it is more convenient to write

$$O_{j,k}(x) = \delta_{j,k} W(x) + 2\partial_j \partial_k \left(\int_0^1 W(\theta x)\, d\theta\right) \qquad (4.13)$$

and when x is close to infinity, it is more convenient to write

$$O_{j,k}(x) = \partial_j \partial_k \left(\frac{1}{4\pi|x|}\right) + \delta_{j,k} W(x)$$
$$- 2\partial_j \partial_k \left(\frac{1}{(4\pi)^{3/2}|x|} \int_{|x|}^{\infty} e^{-\frac{s^2}{4}}\, ds\right). \qquad (4.14)$$

---

[2]Lerner has recently given an explicit expression of those kernels that involve the incomplete gamma function and the confluent hypergeometric functions of the first kind [332].

Proof. We may write
$$O_{j,k}(x) = \delta_{j,k} W(x) + \partial_j \partial_k (G * W).$$

The function $G * W = \Phi$ is radial: $\Phi(x) = F(|x|)$, and satisfies $-\Delta\Phi = W(x) = \frac{1}{(4\pi)^{3/2}} e^{-\frac{|x|^2}{4}} = H(|x|)$. Thus, we must have $F''(r) + \frac{2}{r} F'(r) = -H(r)$, hence $(rF)'' = -rH(r) = 2H'(r)$, and finally $F(r) = \frac{A}{r} + B + \frac{2}{r} \int_0^r H(s)\,ds$. Since Φ is bounded near 0 and vanishes at infinity, we find that $A = B = 0$. $\qquad\square$

Remark: This proof has been taken in Oseen's book [385]. Another simple proof is to write that

$$G * W = \lim_{t \to +\infty} G * W_1 - G * W_t = \lim_{t \to +\infty} -G * \int_1^t \Delta W_u \, du = \int_1^{+\infty} W_u \, du$$

and thus

$$G * W(x) = \frac{1}{(4\pi)^{3/2}} \int_1^{+\infty} e^{-\frac{|x|^2}{4u}} \frac{du}{u^{3/2}}.$$

Writing $u = \frac{|x|^2}{s^2}$ and thus $\frac{du}{u^{3/2}} = -2 \frac{s^3}{|x|^3} \frac{|x|^2}{s^3}\,ds$, one gets

$$G * W(x) = \frac{2}{(4\pi)^{3/2}} \int_0^{|x|} e^{-\frac{s^2}{4}} \frac{ds}{|x|}.$$

Theorem 4.6 allows precise estimates on the derivatives of $O_{j,k}$:

Corollary 4.2.
$O_{j,k}$ is $\mathcal{C}^\infty$ and satisfies:

- *for all $\alpha \in \mathbb{N}^3$, $|\partial_x^\alpha O_{j,k}(x)| \leq C_\alpha (1 + |x|)^{-3-|\alpha|}$*

- *for all $\alpha \in \mathbb{N}^3$ and $|x| > 1$, $|\partial_x^\alpha (O_{j,k}(x) - \partial_j \partial_k G(x))| \leq C_\alpha e^{-\frac{x^2}{8}}$*

where G is the Green function $G(x) = \frac{1}{4\pi|x|}$.

4.6 Classical Solutions for the Navier–Stokes Problem

Interpreting the Navier–Stokes equations with given force $\vec{f}$ as a Stokes equation with given force $\vec{f} - (\vec{u}.\vec{\nabla})\vec{u}$ allows one to turn the differential equations (4.1) into an integro-differential equation:

Integro-differential Navier–Stokes equations

Given a divergence-free vector field $\vec{u}_0$ on $\mathbb{R}^3$ and a force $\vec{f}$ on $(0, +\infty) \times \mathbb{R}^3$, find a positive T and regular functions $\vec{u}$ and p on $[0, T] \times \mathbb{R}^3$ solutions to

$$u_k = W_{\nu t} * u_{0,k} + \int_0^t \sum_{j=1}^3 O_{j,k}(\nu(t-s), .) * \left(f_j - (\vec{u}.\vec{\nabla})u_j \right) ds \qquad (4.15)$$

---

for $k = 1, \ldots, 3$, and

$$p(t, x) = -\sum_{j=1}^{3} \left(f_j - (\vec{u}.\vec{\nabla})u_j \right) * \partial_j G \tag{4.16}$$

Let us write $\mathcal{O}(\nu(t - s)) :: \vec{f}$ for the vector $\vec{g} = \mathcal{O}(\nu(t - s)) :: \vec{f}$ with components $g_k = \sum_{j=1}^{3} O_{j,k}(\nu(t - s)) * f_j$, we have to solve the quadratic equation

$$\vec{u} = \vec{v}_0 - \int_0^t \mathcal{O}(\nu(t - s)) :: \left((\vec{u}.\vec{\nabla})\vec{u} \right) ds \tag{4.17}$$

with

$$\vec{v}_0 = W_{\nu t} * \vec{u}_0 + \int_0^t \mathcal{O}(\nu(t - s)) :: \vec{f} \, ds. \tag{4.18}$$

Oseen's idea is to solve the same equation with an extra parameter ϵ:

$$\vec{u}_\epsilon = \vec{v}_0 - \epsilon \int_0^t \mathcal{O}(\nu(t - s)) :: \left((\vec{u}_\epsilon.\vec{\nabla})\vec{u}_\epsilon \right) ds \tag{4.19}$$

and to develop the solution $\vec{u}_\epsilon$ as a power series in ϵ:

$$\vec{u}_\epsilon = \sum_{n=0}^{\infty} \epsilon^n \vec{v}_n.$$

We get a cascade of equalities (which amounts to solve a cascade of Stokes equations)

$$\vec{v}_{n+1} = -\sum_{k=0}^{n} \int_0^t \mathcal{O}(\nu(t - s)) :: \left((\vec{v}_k.\vec{\nabla})\vec{v}_{n-k} \right) ds.$$

We have

$$\partial_t \vec{v}_0 = \nu \Delta \vec{v}_0 + \vec{f} - \vec{\nabla} q_0$$
$$\text{div } \vec{v}_0 = 0 \tag{4.20}$$
$$\vec{v}_0(0, .) = \vec{u}_0$$

and

$$\partial_t \vec{v}_{n+1} = \nu \Delta \vec{v}_{n+1} - \sum_{k=0}^{n} (\vec{v}_k.\vec{\nabla})\vec{v}_{n-k} - \vec{\nabla} q_{n+1}$$
$$\text{div } \vec{v}_{n+1} = 0 \tag{4.21}$$
$$\vec{v}_{n+1}(0, .) = 0$$

Thus, in order to find a classical solution for the Navier–Stokes equations (4.1), it will be enough to find a positive time T such that

$$\sum_{n=0}^{\infty} \sup_{0 \leq t \leq T, x \in \mathbb{R}^3, |\alpha| \leq 2} |\partial_x^\alpha \vec{v}_n(t, x)| < +\infty \text{ and } \sum_{n=0}^{\infty} \sup_{0 \leq t \leq T, x \in \mathbb{R}^3, |\alpha| \leq 1} |\partial_x^\alpha q_n(t, x)| < +\infty$$

– note that, as well, we will get

$$\sum_{n=0}^{\infty} \sup_{0 \le t \le T, x \in \mathbb{R}^3} |\partial_t \vec{v}_n(t,x)| < +\infty.$$

This can be easily done under the assumptions of the Millennium problem:

Navier–Stokes equations

Theorem 4.7.
Let $\vec{u}_0$ be a $\mathcal{C}^2$ divergence-free vector field on $\mathbb{R}^3$ and let $\vec{f}$ be a time-dependent vector field such that:

- $\sup_{|\alpha| \le 2} \sup_{x \in \mathbb{R}^3} (1 + |x|)^2 |\partial_x^\alpha \vec{u}_0(x)| < +\infty$

- *for $|\alpha| \le 2$, $\partial_x^\alpha \vec{f}$ is continuous on $[0, +\infty) \times \mathbb{R}^3$*

- $\sup_{|\alpha| \le 2} \sup_{t \ge 0, x \in \mathbb{R}^3} (1 + |x|)^4 |\partial_x^\alpha \vec{f}(t,x)| < +\infty.$

Then, there exists a positive time T and a unique solution $(\vec{u}, p)$ of the Navier–Stokes problem

$$\partial_t \vec{u} = \nu \Delta \vec{u} + \vec{f} - (\vec{u}.\vec{\nabla})\vec{u} - \vec{\nabla}p \text{ on } (0,T) \times \mathbb{R}^3$$
$$\operatorname{div} \vec{u} = 0$$
$$\vec{u}_{|t=0} = \vec{u}_0$$

---

such that:

- *for $|\alpha| \le 2$, $\partial_x^\alpha p$ and $\partial_x^\alpha \vec{u}$ are continuous on $[0,T] \times \mathbb{R}^3$*

- $\sup_{0 \le t \le T, x \in \mathbb{R}^3} (1 + |x|)|p(t,x)| < +\infty.$

- $\sup_{1 \le |\alpha| \le 2} \sup_{0 \le t \le T, x \in \mathbb{R}^3} (1 + |x|)^2 |\partial_x^\alpha p(t,x)| < +\infty.$

- $\sup_{|\alpha| \le 2} \sup_{0 \le t \le T, x \in \mathbb{R}^3} (1 + |x|)^2 |\partial_x^\alpha \vec{u}(t,x)| < +\infty.$

- $\partial_t \vec{u}$ *is continuous on $[0,T] \times \mathbb{R}^3$.*

Proof. We are going to estimate the size of the vector fields $\vec{v}_n$ given by the equations (4.20) and (4.21). First of all, we rewrite the "forces" in those equations in a divergence form: for $n \ge 0$, we have

$$\partial_t \vec{v}_n = \nu \Delta \vec{v}_n + \sum_{j=1}^{3} \partial_j \vec{g}_{j,n} - \vec{\nabla} q_n, \quad \operatorname{div} \vec{v}_n = 0, \quad \vec{v}_n(0,.) = \delta_{n,0} \vec{u}_0 \qquad (4.22)$$

with

$$\vec{g}_{j,0} = -\vec{f} * \partial_j G \qquad (4.23)$$

and

$$\vec{g}_{j,n+1} = -\sum_{k=0}^{n} v_{j,k} \, \vec{v}_{n-k}. \qquad (4.24)$$

This gives

$$\vec{v}_n = \delta_{n,0} W_{\nu t} * \vec{u}_0 + \sum_{j=1}^{3} \int_0^t \partial_j \mathcal{O}(\nu(t-s)) :: \vec{g}_{j,n} \, ds. \qquad (4.25)$$

We thus get, for $|\alpha| \le 2$ and $0 \le t \le T$,

$$|\partial_x^\alpha \vec{v}_n(t,x)|$$

$$\le \delta_{n,0} \sup_{y \in \mathbb{R}^3} (1+|y|)^2 |\partial_x^\alpha \vec{u}_0(y)| |W_{\nu t} * \frac{1}{(1+|x|)^2}|$$

$$+ \sum_{j=1}^{3} \left(\sup_{0 \le s \le T, y \in \mathbb{R}^3} |(1+|y|)^2 \partial_x^\alpha \vec{g}_{j,n}(s,y)| \right) \left(\iint_0^t |\partial_j \mathcal{O}(\nu(t-s)), x-y)| \frac{ds \, dy}{(1+|y|)^2} \right).$$

From Theorem 4.6 (and Corollary 4.2), we have the estimate

$$|\partial_j \mathcal{O}(\nu(t-s)), x-y)| \le C \frac{1}{(\sqrt{\nu(t-s)} + |x-y|)^4}$$

and thus, from Corollary 4.1,

$$(1+|x|)^2 |\partial_x^\alpha \vec{v}_n(t,x)|$$

$$\le C_0 \left(\delta_{n,0} \sup_{y \in \mathbb{R}^3} (1+|y|)^2 |\partial_x^\alpha \vec{u}_0(y)| + \sqrt{\frac{T}{\nu}} \sum_{j=1}^{3} \sup_{0 \le s \le T, y \in \mathbb{R}^3} |(1+|y|)^2 \partial_x^\alpha \vec{g}_{j,n}(s,y)| \right)$$

where the constant C_0 does not depend on T nor ν.

We are now going to estimate inductively the size of $\vec{v}_n$ and of its derivatives. Let us define

$$Z_n(T) = \sup_{|\alpha| \le 2} \sup_{0 < t < T, x \in \mathbb{R}^3} (1+|x|)^2 |\partial_x^\alpha \vec{v}_n(t,x)|.$$

- From Lemma 4.1, we know that

$$\sup_{t>0, x \in \mathbb{R}^3} (1+|x|)^2 |\partial_x^\alpha \vec{g}_{j,0}(t,x)| \le C_1 \sup_{t>0, x \in \mathbb{R}^3} (1+|x|)^4 |\partial_x^\alpha \vec{f}(t,x)|.$$

- Thus, we know that we have

$$Z_0(T) = \sup_{|\alpha| \le 2} \sup_{0 < t < T, x \in \mathbb{R}^3} (1+|x|)^2 |\partial_x^\alpha \vec{v}_0(t,x)| \le C_0(A_0 + C_1 \sqrt{\frac{T}{\nu}} B_0)$$

  with

$$A_0 = \sup_{|\alpha| \le 2} \sup_{x \in \mathbb{R}^3} (1+|x|)^2 |\partial_x^\alpha \vec{u}_0(x)| \text{ and } B_0 = \sup_{|\alpha| \le 2} \sup_{0 < t, x \in \mathbb{R}^3} (1+|x|)^4 |\partial_x^\alpha \vec{f}(t,x)|.$$

- From equality (4.24), we find (through Leinitz's rule on derivatives) that

$$\sup_{|\alpha| \le 2} \sup_{0 < t < T, x \in \mathbb{R}^3} (1+|x|)^4 |\partial_x^\alpha \vec{g}_{j,n+1}(t,x)| \le 4 \sum_{k=0}^{n} Z_k(T) Z_{n-k}(T).$$

Thus, we get that

$$Z_{n+1} \le 12 C_0 \sqrt{\frac{T}{\nu}} \sum_{k=0}^{n} Z_k(T) Z_{n-k}(T).$$

- We shall prove that there exists a constant C_2 (which does not depend on T nor ν) such that we have, for every $T > 0$ and every $n \in \mathbb{N}$,

$$Z_n(T) \leq \frac{\left(C_2\sqrt{\frac{T}{\nu}}C_0(A_0 + C_1\sqrt{\frac{T}{\nu}}B_0)\right)^n}{(1+n)^4} C_0(A_0 + C_1\sqrt{\frac{T}{\nu}}B_0)$$

This inequality is true when $n = 0$. Assume that it is true kor $n = 0, \ldots, N$; we are going to prove that it is true for $n = N + 1$. We have

$$Z_{N+1}(T) \leq 12C_0\sqrt{\frac{T}{\nu}}\sum_{k=0}^{n} Z_k(T)Z_{n-k}(T) \leq 12C_0\sqrt{\frac{T}{\nu}}D_{N+1}\left(C_0(A_0 + C_1\sqrt{\frac{T}{\nu}}B_0)\right)^2$$

with

$$D_{N+1} = \sum_{k=0}^{N} \frac{\left(C_2\sqrt{\frac{T}{\nu}}C_0(A_0 + C_1\sqrt{\frac{T}{\nu}}B_0)\right)^k}{(1+k)^4} \frac{\left(C_2\sqrt{\frac{T}{\nu}}C_0(A_0 + C_1\sqrt{\frac{T}{\nu}}B_0)\right)^{N-k}}{(1+(N-k))^4}$$

$$= \left(C_2\sqrt{\frac{T}{\nu}}C_0(A_0 + C_1\sqrt{\frac{T}{\nu}}B_0)\right)^N \sum_{k=0}^{N} \frac{1}{(1+k)^4}\frac{1}{(1+(N-k))^4}$$

$$\leq \left(C_2\sqrt{\frac{T}{\nu}}C_0(A_0 + C_1\sqrt{\frac{T}{\nu}}B_0)\right)^N \frac{32}{(N+2)^4}\sum_{k=1}^{+\infty}\frac{1}{k^4}.$$

Thus, writing $C_3 = 32\sum_{k=1}^{+\infty}\frac{1}{k^4}$, we get

$$Z_{N+1}(T) \leq 12C_0 C_3 C_2^N \frac{\left(\sqrt{\frac{T}{\nu}}C_0(A_0 + C_1\sqrt{\frac{T}{\nu}}B_0)\right)^{N+1}}{(N+2)^4} C_0(A_0 + C_1\sqrt{\frac{T}{\nu}}B_0)$$

The proof of the induction is over, if we choose $C_2 = 12C_0 C_3$.

- Size of $\partial_x^\alpha q_n$: in order to estimate q_n and its derivatives, we just write

$$\Delta q_0 = \operatorname{div}\vec{f} \quad \text{and} \quad \Delta q_{n+1} = \operatorname{div}\left(\sum_{j=1}^{3}\partial_j\vec{g}_{j,n+1}\right) = -\sum_{k=0}^{n}\sum_{j=1}^{3}\sum_{i=1}^{3}\partial_i v_{j,k}\partial_j v_{i,n-k}$$

and we use Theorem 4.3.

If we fix T small enough to have

$$C_2\sqrt{\frac{T}{\nu}}C_0(A_0 + C_1\sqrt{\frac{T}{\nu}}B_0) \leq 1, \tag{4.26}$$

we get the normal convergence of $\sum_{n=0}^{+\infty}\partial_x^\alpha\vec{v}_n$ and of $\sum_{n=0}^{+\infty}\partial_x^\alpha q_n$, which proves the existence of a solution on $(0, T)$.

Uniqueness follows the same lines: if $\vec{u}$ and $\vec{v}$ are two solutions of the Cauchy problem for the Navier–Stokes equations on $(0, T) \times \mathbb{R}^3$ (with associated pressures p and q) which fulfill the conclusions of Theorem 4.7, then we find that $\vec{w} = \vec{u} - \vec{w}$ is solution of

$$\partial_t\vec{w} = \nu\Delta\vec{w} - \sum_{j=1}^{3}\partial_j(u_j\vec{w} + w_j\vec{v}) - \vec{\nabla}(p - q), \operatorname{div}\vec{w} = 0, \vec{w}(0, .) = 0.$$

In particular, for every $S \in (0, T)$,

$$\sup_{0 \leq t \leq S, x \in \mathbb{R}^3} (1 + |x|)^2 |\vec{w}(t, x)|$$

$$\leq 3C_0 \sqrt{\frac{S}{\nu}} \left(\sup_{0 < s < T} \|\vec{u}(s, .)\|_\infty + \|\vec{v}(s, .)\|_\infty \right) \sup_{0 \leq t \leq S, x \in \mathbb{R}^3} (1 + |x|)^2 |\vec{w}(t, x)|.$$

If S is small enough (so that $3C_0 \sqrt{\frac{S}{\nu}} (\sup_{0 < s < T} \|\vec{u}(s, .)\|_\infty + \|\vec{v}(s, .)\|_\infty) < 1$), we find that $\vec{u} = \vec{v}$ on $[0, S]$. Then by reiteration from S to $2S$ and so on, we find that $\vec{u} = \vec{v}$ on $[0, T]$. □

We shall define regular data and classical solutions as the data that fulfill the assumptions of Theorem 4.7 and solutions that fulfill its conclusions:

Regular data

Definition 4.3.
Regular data for the initial-value problem for the Navier–Stokes equations on $(0, +\infty) \times \mathbb{R}^3$ are a C^2 divergence-free vector field $\vec{u}_0$ on $\mathbb{R}^3$ and a time-dependent vector field $\vec{f}(t, x)$ on $[0, +\infty) \times \mathbb{R}^3$ such that:

- $\sup_{|\alpha| \leq 2} \sup_{x \in \mathbb{R}^3} (1 + |x|)^2 |\partial^\alpha \vec{u}_0(x)| < +\infty$

- *for $|\alpha| \leq 2$, $\partial_x^\alpha \vec{f}$ is continuous on $[0, +\infty) \times \mathbb{R}^3$*

- *for every $0 < T$, $\sup_{|\alpha| \leq 2} \sup_{0 \leq t \leq T, x \in \mathbb{R}^3} (1 + |x|)^4 |\partial^\alpha \vec{f}(t, x)| < +\infty$.*

Classical solutions

Definition 4.4.
For $0 < T_1$, a classical solution of the Navier–Stokes problem

$$\partial_t \vec{u} = \nu \Delta \vec{u} + \vec{f} - (\vec{u}.\vec{\nabla})\vec{u} - \vec{\nabla}p \text{ on } (0, T_1) \times \mathbb{R}^3$$
$$\text{div } \vec{u} = 0$$
$$\vec{u}_{|t=0} = \vec{u}_0$$

associated to regular data $(\vec{u}_0, \vec{f})$ is a solution $(\vec{u}, p)$ such that, for every $0 < T < T_1$,

- *for $|\alpha| \leq 2$, $\partial_x^\alpha p$ and $\partial_x^\alpha \vec{u}$ are continuous on $[0, T] \times \mathbb{R}^3$*

- $\sup_{0 \leq t \leq T, x \in \mathbb{R}^3} (1 + |x|) |p(t, x)| < +\infty$.

- $\sup_{1 \leq |\alpha| \leq 2} \sup_{0 \leq t \leq T, x \in \mathbb{R}^3} (1 + |x|)^2 |\partial_x^\alpha p(t, x)| < +\infty$.

- $\sup_{|\alpha| \leq 2} \sup_{0 \leq t \leq T, x \in \mathbb{R}^3} (1 + |x|)^2 |\partial_x^\alpha \vec{u}(t, x)| < +\infty$.

4.7 Maximal Classical Solutions and Estimates in L^∞ Norms

Let $(\vec{u}_0, \vec{f})$ be regular data for the Cauchy problem for Navier–Stokes equations. Let $T_0 > 0$. We define $\vec{f}_{T_0}(t, x) = \vec{f}(\min(t, T_0), x)$, so that $\vec{f}_{T_0}$ fulfills the assumptions of Theorem 4.7 (uniform in time control on $[0, +\infty)$). The Cauchy problem on $(0, T_0) \times \mathbb{R}^3$ for data $(\vec{u}_0, \vec{f})$ or $(\vec{u}_0, \vec{f}_{T_0})$ coincide. In particular, from inequality (4.26), we know that we have existence of a solution on $(0, T)$ with $T \leq T_0$ as soon as T fulfills the conditions

$$\begin{cases} T \leq T_0 \\ T \leq \dfrac{\nu}{4C_2^2 C_0^2} \dfrac{1}{(\sup_{|\alpha| \leq 2} \sup_{x \in \mathbb{R}^3} (1 + |x|)^2 |\partial_x^\alpha \vec{u}_0(x)|)^2} \\ T \leq \dfrac{\nu}{2C_2 C_0 C_1} \dfrac{1}{\sup_{|\alpha| \leq 2} \sup_{0 < t < T_0, x \in \mathbb{R}^3} (1 + |x|)^4 |\partial_x^\alpha \vec{f}(t, x)|} \end{cases}.$$

Thus, we have local existence of a solution. Moreover, we know that we have uniqueness. We may then conclude that we have a unique maximal solution:

Maximal classical solution

Proposition 4.1.
Let $(\vec{u}_0, \vec{f})$ be regular data. Let T_{MAX} be the maximal time where one can find a classical solution $\vec{u}$ of the Cauchy problem for the Navier–Stokes equations on $(0, T_{\mathrm{MAX}}) \times \mathbb{R}^3$.
  If $T_{\mathrm{MAX}} < +\infty$, then

$$\lim_{t \to T_{\mathrm{MAX}}} \sup_{|\alpha| \leq 2} \sup_{x \in \mathbb{R}^3} (1 + |x|)^2 |\partial_x^\alpha \vec{u}(t, x)| = +\infty.$$

Proof. Assume that $T_{\mathrm{MAX}} < +\infty$. Let $0 < T_1 < T_{\mathrm{MAX}} < T_0 < +\infty$. We have a solution for the Cauchy problem at time T_1 which is defined on $[T_1, T_1 + T] \times \mathbb{R}^3$, for T satisfying

$$T \leq T_0 - T_1$$

$$T \leq T_2 = \dfrac{\nu}{2C_2 C_0 C_1} \dfrac{1}{\sup_{|\alpha| \leq 2} \sup_{0 < t < T_0, x \in \mathbb{R}^3} (1 + |x|)^4 |\partial_x^\alpha \vec{f}(t, x)|}$$

and

$$T \leq \dfrac{\nu}{4C_2^2 C_0^2} \dfrac{1}{(\sup_{|\alpha| \leq 2} \sup_{x \in \mathbb{R}^3} (1 + |x|)^2 |\partial_x^\alpha \vec{u}(T_1, x)|)^2}.$$

We must have

$$T + T_1 \leq T_{\mathrm{MAX}}$$

hence, if $T_1 > T_{\mathrm{MAX}} - T_2$, we find that

$$\dfrac{\nu}{4C_2^2 C_0^2} \dfrac{1}{(\sup_{|\alpha| \leq 2} \sup_{x \in \mathbb{R}^3} (1 + |x|)^2 |\partial_x^\alpha \vec{u}(T_1, x)|)^2} \leq T_{\mathrm{MAX}} - T_1. \qquad \square$$

Actually, it is enough to control the L^∞ norm of $\vec{u}$:

Theorem 4.8.
Let $(\vec{u}_0, \vec{f})$ be regular data. Let T_{MAX} be the maximal time where one can find a classical solution $\vec{u}$ of the Cauchy problem for the Navier–Stokes equations on $(0, T_{\mathrm{MAX}}) \times \mathbb{R}^3$.
   If $T_{\mathrm{MAX}} < +\infty$, then

$$\sup_{0 \leq t < T_{\mathrm{MAX}}, \, x \in \mathbb{R}^3} |\vec{u}(t, x)| = +\infty.$$

Proof. For $0 \leq t < T_{\mathrm{MAX}}$, let us define, for $0 \leq k \leq 2$,

$$A_k(t) = \sup_{|\alpha|=k} \sup_{x \in \mathbb{R}^3} (1 + |x|)^2 |\partial_x^\alpha \vec{u}(t, x)|$$

$$B_0(t) = \sup_{x \in \mathbb{R}^3} |\vec{u}(t, x)|$$

and

$$F_k(t) = \sup_{|\alpha|=k} \sup_{x \in \mathbb{R}^3} (1 + |x|)^4 |\partial_x^\alpha \vec{f}(t, x)|.$$

We write the Navier–Stokes equations as

$$\vec{u} = W_{\nu t} * \vec{u}_0 + \sum_{j=1}^{3} \int_0^t \partial_j \mathcal{O}(\nu(t - s)) :: \vec{g}_j \, ds. \tag{4.27}$$

with

$$\vec{g}_j = -\vec{f} * \partial_j G - u_j \, \vec{u}. \tag{4.28}$$

We proved (from Corollaries 4.1 and 4.2) that

$$(1 + |x|)^2 |\partial_x^\alpha \vec{u}(t, x)|$$

$$\leq C_0 \left(\sup_{y \in \mathbb{R}^3} (1 + |y|)^2 |\partial_x^\alpha \vec{u}_0(y)| + \int_0^t \frac{1}{\sqrt{\nu(t - s)}} \sum_{j=1}^{3} \sup_{0 \leq s \leq t, y \in \mathbb{R}^3} |(1 + |y|)^2 \partial_x^\alpha \vec{g}_j(s, y)| \, ds \right)$$

which gives

$$A_0(t) \leq C_0 A_0(0) + C_1 \sqrt{\frac{t}{\nu}} \sup_{0 < s < t} F_0(s) + C_1 \int_0^t \frac{1}{\sqrt{\nu(t - s)}} B_0(s) A_0(s) \, ds$$

$$A_1(t) \leq C_0 A_1(0) + C_1 \sqrt{\frac{t}{\nu}} \sup_{0 < s < t} F_1(s) + C_1 \int_0^t \frac{1}{\sqrt{\nu(t - s)}} B_0(s) A_1(s) \, ds$$

and

$$A_2(t) \leq C_0 A_2(0) + C_1 \sqrt{\frac{t}{\nu}} \sup_{0 < s < t} F_2(s) + C_1 \int_0^t \frac{1}{\sqrt{\nu(t - s)}} (B_0(s) A_2(s) + A_1(s)^2) \, ds.$$

We remark that a Gronwall-like inequality

$$\alpha(t) \leq A + B \int_0^t \alpha(s) \frac{ds}{\sqrt{t - s}}$$

can be reiterated into a Gronwall inequality, as

$$\alpha(t) \leq A + B \int_0^t A \frac{ds}{\sqrt{t-s}} + B^2 \iint_{0 \leq \tau \leq s \leq t} \alpha(\tau) \frac{d\tau\, ds}{\sqrt{t-s}\sqrt{s-\tau}} = A + \frac{AB\sqrt{t}}{2} + \pi B^2 \int_0^t \alpha(\tau)\, d\tau$$

and thus into

$$\alpha(t) \leq e^{\pi B^2 t}\left(A + \frac{AB\sqrt{t}}{2}\right).$$

Thus, we find that, if B_0 remains bounded, A_0 and A_1 remain bounded, and finally A_2 remains bounded. We then conclude with Proposition 4.1 that $T_{\text{MAX}} < +\infty$ implies that $\sup_{0<t<T_{\text{MAX}}} \|\vec{u}(t,.)\|_\infty = 0$. $\qquad\square$

4.8 Small Data

In case of small regular data, we prove easily that we have global solutions. In particular, we have the following result:

Global solutions

Theorem 4.9. *There exists a positive constant ϵ_0 such that, if*

- $\sup_{x \in \mathbb{R}^3} |x||\vec{u}_0(x)| < \epsilon_0 \nu$,

- $\sup_{t \geq 0, x \in \mathbb{R}^3} (\sqrt{\nu t} + |x|)^3 |\vec{f}(t,x)| < \epsilon_0 \nu^2$,

- $\sup_{t \geq 0, R > 0} |\int_{|x|<R} \vec{f}(t,x)\, dx| < \epsilon_0 \nu^2$,

then the classical solution $(\vec{u}, p)$ associated to the regular data $\vec{u}_0$ and $\vec{f}$ is defined for all times and satisfies

$$\sup_{0 \leq t, x \in \mathbb{R}^3} (\sqrt{\nu t} + |x|)\, |\vec{u}(t,x)| < +\infty.$$

Proof. We begin with a remark on the assumptions on $\vec{f}$. As we shall see in the proof, we only need to know that $\vec{f}$ may be written as $\vec{f} = \sum_{j=1}^3 \partial_j \vec{g}_j$, where

$$\sup_{t \geq 0, x \in \mathbb{R}^3} (\sqrt{\nu t} + |x|)^2 |\vec{g}_j(t,x)| < C_0 \epsilon_0 \nu^2. \qquad (4.29)$$

If (4.29) is satisfied, then, by Stokes' formula,

$$|\int_{|x|<R} \vec{f}(t,x)\, dx| \leq 4\pi \sum_{j=1}^3 \||x|^2 \vec{g}_j\|_\infty \leq 4\pi C_0 \epsilon_0 \nu^2.$$

Conversely, under the assumptions of Theorem 4.9, we write

$$\vec{f} = -\sum_{j=1}^3 \partial_j (\vec{f} * \partial_j G).$$

We have

$$|\vec{f} * \partial_j G| \le \epsilon_0 \nu^2 \int \frac{1}{(\sqrt{\nu t} + |x - y|)^3} \frac{1}{|y|^2} \, dy \le \frac{\epsilon_0 \nu^2}{\nu t} \|\frac{1}{(1 + |y|)^3}\|_{L^\infty \cap L^2} \|\frac{1}{|y|^2}\|_{L^1 + L^2}.$$

On the other hand, we have

$$\vec{f} * \partial_j G = \int_{|x-y| > \frac{|x|}{2}} \vec{f}(t, x - y) \partial_j G(y) \, dy$$

$$+ \int_{|x-y| < \frac{|x|}{2}} \vec{f}(t, x - y)(\partial_j G(y) - \partial_j G(x)) \, dy$$

$$+ \int_{|x-y| < \frac{|x|}{2}} \vec{f}(t, x - y) \partial_j G(x) \, dy$$

and we get

$$|\vec{f} * \partial_j G| \le \epsilon_0 \nu^2 \int_{|x-y| > \frac{|x|}{2}} \frac{1}{|x - y|^3} \frac{1}{4\pi |y|^2} \, dy$$

$$+ \epsilon_0 \nu^2 \int_{|x-y| < \frac{|x|}{2}} \frac{1}{|x - y|^3} \frac{|x - y|}{\pi |y|^3} \, dy + \epsilon_0 \nu^2 \frac{1}{4\pi |x|^2}$$

$$= C_1 \epsilon_0 \nu^2 \frac{1}{|x|^2}.$$

Finally, let us remark that $\vec{f}$ will satisfy the assumptions of Theorem 4.9 whenever

$$|\vec{f}(t, x)| < \frac{3}{2\pi^2} \epsilon_0 \nu^2 \frac{\sqrt{\nu t}}{(\sqrt{\nu t} + |x|)^4}.$$

The proof of Theorem 4.9 relies on the following inequality: there exists a constant $C_0 > 0$ such that, for all $t > 0$ and $x \in \mathbb{R}^3$, we have

$$I(t, x) = \int_0^t \int \frac{dy \, ds}{(\sqrt{\nu(t - s)} + |x - y|)^4 (\sqrt{\nu s} + |y|)^2} \le \frac{C_0}{\nu(\sqrt{\nu t} + |x|)}. \qquad (4.30)$$

This inequality is proved by Fubini's theorem and Hölder's inequality:

$$I(t, x) \le \int_0^t \|\frac{1}{(\sqrt{\nu(t - s)} + |y|)^4}\|_{L^{6/5}(dy)} \|\frac{1}{(\sqrt{\nu s} + |y|)^2}\|_{L^6(dy)} \, ds$$

$$= C \int_0^T \frac{1}{\nu^{3/2}} \frac{1}{(t - s)^{3/4}} \frac{1}{s^{3/4}} \, ds = C' \frac{1}{\nu \sqrt{t \nu}}$$

and

$$I(t, x) \le \int_{\mathbb{R}^3} \|\frac{1}{(\sqrt{\nu s} + |x - y|)^4}\|_{L^1(ds)} \|\frac{1}{(\sqrt{\nu s} + |y|)^2}\|_{L^\infty(ds)} \, dy$$

$$= C \int \frac{1}{\nu |x - y|^2} \frac{1}{|y|^2} \, dy = C' \frac{1}{\nu |x|}.$$

Now, let us consider the regular solution $(\vec{u}, p)$ on $[0, T_{\mathrm{MAX}})$. We write

$$f' = -\sum_{j=1}^3 \partial_j (\vec{f} * \partial_j G)$$

and we define

$$\alpha(t) = \sup_{x \in \mathbb{R}^3} (\sqrt{\nu t} + |x|)|\vec{u}(t,x)|$$

and

$$\beta(t) = \sup_{x \in \mathbb{R}^3} (\sqrt{\nu t} + |x|)^2 \sum_{j=1}^{3} |\vec{g}_j(t,x)|$$

where $g_j = -\vec{f} * \partial_j G$. We start from the equality

$$\vec{u} = W_{\nu t} * \vec{u}_0 + \sum_{j=1}^{3} \int_0^t \partial_j \mathcal{O}(\nu(t-s)) :: (\vec{g}_j - u_j \vec{u}) \, ds. \tag{4.31}$$

By Corollary 4.1, we know that

$$(\sqrt{t} + |x|)|W_{\nu t} * \vec{u}_0(x)| \leq C_1 \||x|\vec{u}_0\|_\infty = C_1 \alpha(0).$$

By Corollary 4.2, we know that

$$\left| \int_0^t \partial_j \mathcal{O}(\nu(t-s)) :: (\vec{g}_j - u_j \vec{u}) \, ds \right| \leq C_2 \int_0^t \int \left(|\vec{u}(s,y)|^2 + \sum_{j=1}^{3} |\vec{g}_j(s,y)| \right) \frac{dy \, ds}{\left(\sqrt{\nu(t-s)} + |x-y|\right)^4}.$$

Inequality (4.30) gives us that:

$$\alpha(t) \leq C_1 \alpha(0) + C_0 C_2 \frac{1}{\nu} \sup_{0<s<t} (\beta(s) + \alpha(s)^2).$$

Using the assumptions on $\vec{u}_0$ and $\vec{f}$ (and thus inequality (4.29) on $\vec{g}_j$), we find

$$\alpha(t) \leq C_3 \epsilon_0 \nu + C_3 \sup_{0<s<t} \frac{\alpha(s)^2}{\nu}.$$

Assuming that $\alpha(s) < 3C_3 \epsilon_0 \nu$ on $[0,t)$, we find that

$$\alpha(t) \leq C_3 \epsilon_0 \nu (1 + 9 C_3 \epsilon_0).$$

As α is a continuous function of t, we find that, if $\epsilon_0 < \frac{1}{9C_3}$, $\alpha(t)$ will remain bounded by $2C_3 \epsilon_0 \nu$ on the whole interval $[0, T_{\text{MAX}})$.

Finally, we find that

$$\|\vec{u}(t,.)\|_\infty \leq \frac{\alpha(t)}{\sqrt{\nu t}} \leq 2C_3 \epsilon_0 \sqrt{\frac{\nu}{t}}$$

and Theorem 4.8 gives us that $T_{\text{MAX}} = +\infty$. □

4.9 Spatial Asymptotics

In this section, we will show that, even if $\vec{u}_0$ has good decay properties, we (generically) cannot hope for a good decay for the solution $\vec{u}$. Dobrokhotov and Shafarevich [154] proved that, unless some algebraic conditions are satisfied by the initial value $\vec{u}_0$, there is an instantaneous spreading of the velocity that cannot decay faster than $O(|x|^{-4})$. This instantaneous spreading has been studied by Brandolese in his Ph.D. [59] and in several papers [14, 61, 62, 63]:

Spatial decay estimates

Theorem 4.10.
Let $(\vec{u}, p)$ be the classical solution of the Navier–Stokes problem on a strip $[0,T] \times \mathbb{R}^3$, associated to the regular data $(\vec{u}_0, \vec{f})$. Assume moreover that we have:

$$\begin{cases} \lim_{x\to\infty} |x|^4 |\vec{u}_0(x)| = 0 \\ \sup_{0\leq t\leq T, x\in\mathbb{R}^3} (1+|x|)^5 |\vec{f}(t,x)| < +\infty \end{cases} \tag{4.32}$$

then, for fixed $t \in (0,T]$, a necessary condition to ensure that

$$\lim_{x\to\infty} |x|^4 |\vec{u}(t,x)| = 0$$

is that $\vec{u}$ satisfies the Dobrokhotov and Shafarevich conditions

$$\begin{cases} \text{for } 1 \leq i \leq 3, \ \int_0^t \int f_i \, dx \, ds = 0 \\ \text{for } 1 \leq i < j \leq 3, \ \int_0^t \int 2u_i u_j + x_i f_j + x_j f_i \, dx \, ds = 0 \\ \text{for } 1 \leq i < j \leq 3, \ \int_0^t \int u_i^2 - u_j^2 + x_i f_i - x_j f_j \, dx \, ds = 0 \end{cases} \tag{4.33}$$

Following Brandolese's results, we are going to prove Theorem 4.10 by giving a precise asymptotic formula for the solution $\vec{u}$. This formula will prove that the Dobrokhotov and Shafarevich conditions are sufficient as well.

Spatial asymptotics

Theorem 4.11.
Let $\vec{u}_0$ be a C^2 vector field on $\mathbb{R}^3$ and let $\vec{f}$ satisfy the assumptions of Theorem 4.10, including the decay estimates (4.32). Then the classical solution $(\vec{u}, p)$ of the Navier–Stokes equations with initial data $\vec{u}_0$ and with forcing term $\vec{f}$ has, for any fixed $t \in (0,T]$, the following asymptotic development when x goes to ∞:

$$\vec{u} = \sum_{i,=1}^3 c_i(t) \vec{\nabla}\partial_i G(x) - \sum_{i=1}^3 \sum_{j=1}^3 d_{i,j}(t) \vec{\nabla}\partial_i\partial_j G(x) + o(|x|^{-4}) \tag{4.34}$$

where G is the Green function

$$G(x) = \frac{1}{4\pi|x|}$$

and where the coefficients c_i and $d_{i,j}$ are given by

$$\begin{cases} \text{for } 1 \leq i \leq 3, & c_i(t) = \int_0^t \int f_i \, dx \, ds \\ \text{for } 1 \leq i \leq 3, 1 \leq j \leq 3, & d_{i,j}(t) = \int_0^t \int u_i u_j + x_i f_j \, dx \, ds \end{cases}$$

Proof. We first write $\vec{u}$ as a solution of a Stokes system

$$\vec{u} = W_{\nu t} * \vec{u}_0 + \int_0^t \mathcal{O}(\nu(t-s)) :: \vec{g}(s) \, ds$$

with forcing term

$$\vec{g} = \vec{f} - (\vec{u}.\vec{\nabla})\vec{u}.$$

We already know that

$$|\vec{u}_0 * W_t(x)| \leq C(1+|x|)^{-2},$$

$$|\vec{u}(t,x)| \leq C(1+|x|)^{-2},$$

$$\text{for } 1 \leq i \leq 3, \ |\partial_i \vec{u}(t,x)| \leq C(1+|x|)^{-2},$$

and hence that

$$|\vec{g}(t,x)| \leq C(1+|x|)^{-4}.$$

We begin with checking that $W_{\nu t} * \vec{u}_0$ is small at infinity. We have

$$|W_{\nu t} * \vec{u}_0(x)| \leq \int_{|y|>\frac{1}{2}|x|} |\vec{u}_0(y)| |W_{\nu t}(x-y) \, dy + \|\vec{u}_0\|_\infty \int_{|y|<\frac{1}{2}|x|} W_{\nu t}(x-y) \, dy$$

$$\leq 16 \frac{\sup_{|y|\geq\frac{1}{2}|x|} |y|^4 |\vec{u}_0(y)|}{|x|^4} + C\|\vec{u}_0\|_\infty \int_{|x-y|>\frac{1}{2}|x|} \frac{\nu t^{5/2}}{|x-y|^8} \, dy$$

$$= o(|x|^{-4})$$

where the remainder $o(|x|^{-4})$ is small with respect to $|x|^{-4}$ (at fixed ν, uniformly on t in the compact interval $[0,T]$).

We then use the estimate, for any ϵ with $0 < \epsilon < 1$,

$$|\mathcal{O}(\nu(t-s), x-y)| \leq C_\epsilon \frac{1}{\nu^\epsilon (t-s)^\epsilon} \frac{1}{|x-y|^{3-2\epsilon}}.$$

This proves that if

$$|\vec{g}(t,x)| \leq C_\nu |x|^{-\gamma}$$

on $[0,T] \times \mathbb{R}^3$ with $0 < \gamma < 3$, then

$$|\vec{u}(t,x) - W_{\nu t} * \vec{u}_0(x)| \leq C_\nu' T^{1-\epsilon} |x|^{2\epsilon-\gamma}$$

From $\vec{g} = O(|x|^{-3+\epsilon})$, we get $\vec{u} = O(|x|^{-3+3\epsilon})$. As $\partial_j \vec{u} = O(|x|^{-2})$, we find that, for any $\gamma \in (0,1)$, $\vec{g} = O(|x|^{-4-\gamma})$.

Thus we are led to estimate

$$\vec{U} = \int_0^t \mathcal{O}(\nu(t-s)) :: \vec{g}(s) \, ds$$

when

$$\sup_{0 \leq t \leq T, x \in \mathbb{R}^3} (1+|x|)^{4+\gamma} |\vec{g}(t,x)| < +\infty.$$

We fix some $\beta \in (0,1)$ close enough to 1 and cut the integral

$$\int_0^t \int O_{j,k}(\nu(t-s), x-y) g_j(s,y) \, dy \, ds$$

into three domains of integration:

$$\Delta_1 = \{y \; / \; |y| > |x|^\beta \text{ and } |x - y| < |x|^\beta\},$$

$$\Delta_2 = \{y \; / \; |y| > |x|^\beta \text{ and } |x - y| > |x|^\beta\},$$

and

$$\Delta_3 = \{y \; / \; |y| < |x|^\beta\}.$$

For $1 \leq p \leq 3$, let

$$I_p = \int_0^t \int_{\Delta_p} O_{j,k}(\nu(t - s), x - y)g_j(s, y) \, dy \, ds.$$

We have, provided that $\epsilon < \gamma/2$,

$$|I_1| \leq C \int_0^t \int_{|x-y|<|x|^\beta} \frac{1}{|x|^{(4+\gamma)\beta}} \frac{dy \, ds}{(t - s)^\epsilon |x - y|^{3-2\epsilon}} = O(\frac{t^{1-\epsilon}}{|x|^{\beta(4+\gamma-2\epsilon)}})$$

and similarly

$$|I_2| \leq C \int_0^t \int_{|y|>|x|^\beta} \frac{1}{|y|^{4+\gamma}} \frac{dy \, ds}{(t - s)^\epsilon |x|^{\beta(3-2\epsilon)}} = O(\frac{t^{1-\epsilon}}{|x|^{\beta(4+\gamma-2\epsilon)}}).$$

If we choose β and ϵ such that $\beta(4 + \gamma - 2\epsilon) > 4$, we obtain

$$|I_1| + |I_2| = o(|x|^{-4}).$$

For I_3, we have

$$I_3 = A_3 + B_3 + C_3 + D_3$$

where

$$A_3 = \int_0^t \int_{|y|<|x|^\beta} (O_{j,k}(\nu(t - s), x - y) - \partial_j \partial_k G(x - y))g_j(s, y) \, dy \, ds,$$

$$B_3 = \int_0^t \int_{|y|<|x|^\beta} (\partial_j \partial_k G(x - y) - \partial_j \partial_k G(x) + \sum_{l=1}^{3} \partial_j \partial_k \partial_l G(x)y_l)g_j(s, y) \, dy \, ds,$$

$$C_3 = \int_0^t \int_{|y|>|x|^\beta} (-\partial_j \partial_k G(x) + \sum_{l=1}^{3} \partial_j \partial_k \partial_l G(x)y_l)g_j(s, y) \, dy \, ds,$$

and

$$D_3 = \int_0^t \int_{\mathbb{R}^3} (\partial_j \partial_k G(x) - \sum_{l=1}^{3} \partial_j \partial_k \partial_l G(x)y_l)g_j(s, y) \, dy \, ds.$$

We check that the three first terms are negligible. We have

$$|A_3| \leq C \int_0^t \int_{|y|\leq|x|^\beta} \frac{\nu^{5/2}(t - s)^{5/2}}{|x|^8} \|g_j\|_\infty \, dy \, ds = o(|x|^{-4})$$

$$|B_3| \leq C \int_0^t \int \frac{|y|^{1+\epsilon}}{|x|^{4+\epsilon}} \frac{1}{(1 + |y|)^{4+\gamma}} \, dy \, ds = o(|x|^{-4})$$

$$|C_3| \leq C \int_0^t \int_{|y|\geq|x|^\beta} (\frac{1}{|x|^3} + \frac{|y|}{|x|^4}) \frac{1}{|y|^{4+\gamma}} \, dy = o(|x|^{-4}).$$

Thus, only D_3 cannot be neglected. Hence, we have obtained

$$\vec{u} = \sum_{i,=1}^{3} c_i(t)\vec{\nabla}\partial_i G(x) - \sum_{i=1}^{3}\sum_{j=1}^{3} d_{i,j}(t)\vec{\nabla}\partial_i\partial_j G(x) + o(|x|^{-4}) \qquad (4.35)$$

where the coefficients c_i and $d_{i,j}$ are given by

$$\begin{cases} \text{for } 1 \leq i \leq 3, & c_i(t) = \int_0^t \int g_i \; dx \; ds \\ \text{for } 1 \leq i \leq 3, 1 \leq j \leq 3, & d_{i,j}(t) = \int_0^t \int x_j g_i \; dx \; ds \end{cases}$$

with $g_i = f_i - \sum_{k=1}^{3} \partial_k(u_k u_i)$. Since $|u_k u_i| \leq C(1+|x|)^{-5-\gamma}$ and $|\partial_k(u_k u_i)| \leq C(1+|x|)^{-4-\gamma}$, we have $\int \partial_k(u_k u_i) \; dx = 0$ and $\int x_j \partial_k(u_k u_j) \; dx = -\delta_{j,k} \int u_j u_i \; ds$. Thus, Theorem 4.11 is proved.

Thus, we have proved that $\vec{u} = \vec{\nabla}A + \vec{\nabla}B + o(|x|^{-4})$ with $A = \sum_{i,=1}^{3} c_i(t)\partial_i G(x)$ (hence $\vec{\nabla}A$ is homogeneous of degree -3) and $B = -\sum_{i=1}^{3}\sum_{j=1}^{3} d_{i,j}(t)\partial_i\partial_j G(x)$ (hence, $\vec{\nabla}B$ is homogeneous of degree -4). If $\vec{u} = o(|x|^{-4})$, we have that $A(t,x)$ and $B(t,x)$ must be constant on $x \neq 0$. Thus, by homogeneity of A and B, we have $A = B = 0$ and thus

$$\sum_{i,=1}^{3} c_i(t)x_i = A(t,x)|x|^3 = 0$$

and

$$\sum_{i=1}^{3}\sum_{j=1}^{3} d_{i,j}(t)(\delta_{i,j}|x|^2 - 3x_i x_j) = B(t,x)|x|^5 = 0.$$

Thus, we get $c_i(t) = 0$ for $1 \leq i \leq 3$, $d_{i,j}(t) + d_{j,i}(t) = 0$ for $1 \leq i < j \leq 3$ and $d_{1,1}(t) = d_{2,2}(t) = d_{3,3}(t)$. Theorem 4.10 is proved. $\quad\square$

This proves that we have (generically) instantaneous spreading:

Corollary 4.3.
Under the assumptions of Theorem 4.10:

- *if for some i, we have $\int f_i(0,x) \neq 0$, then there exists a positive time t_0 such that for all $t \in (0, t_0)$, $\limsup_{x\to\infty} |x|^3 |\vec{u}(t,x)| > 0$;*

- *if for some i and j with $i \neq j$, we have*

$$\int 2u_{0,i}(x)u_{0,j}(x) + x_i f_j(0,x) + x_j f_i(0,x) \; dx \neq 0$$

or

$$\int u_{0,i}(x)^2 + x_i f_i(0,x) \; dx \neq \int u_{0,j}(x)^2 + x_j f_i(0,x) \; dx$$

then there exists a positive time t_0 such that for all $t \in (0, t_0)$, $\limsup_{x\to\infty} |x|^4 |\vec{u}(t,x)| > 0$.

4.10   Spatial Asymptotics for the Vorticity

In contrast with the phenomenon of instantaneous spreading for the velocities, there is no such spreading for the vorticity:

Vorticity's decay

Theorem 4.12.

Let $(\vec{u}, p)$ be the classical solution of the Navier–Stokes problem on a strip $[0, T] \times \mathbb{R}^3$, associated to the regular data $(\vec{u}_0, \vec{f})$.

Then, we have the following property for the vorticity

$$\vec{\omega} = \vec{\nabla} \wedge \vec{u} :$$

if for some $N \in \mathbb{N}$, we have

$$\sup_{x \in \mathbb{R}^3} |x|^N |\vec{\omega}_0(x)| < +\infty,$$

and

$$\sup_{0 \leq t \leq T, x \in \mathbb{R}^3} |x|^N |\vec{f}(t, x)| < +\infty$$

then

$$\sup_{0 \leq t \leq T, x \in \mathbb{R}^3} |x|^N |\vec{\omega}(t, x)| < +\infty.$$

Proof. The proof is easy. It is enough to write $\vec{\omega}$ as a solution of a heat equation

$$\vec{\omega} = W_{\nu t} * \vec{\omega}_0 + \int_0^t W_{\nu(t-s)} * \vec{g}(s) \, ds$$

with forcing term

$$\vec{g} = \operatorname{curl} \vec{f} + \operatorname{div}(\vec{\omega} \otimes \vec{u} - \vec{u} \otimes \vec{\omega})$$

(where $\operatorname{div}(\vec{a} \otimes \vec{b}) = \sum_{j=1}^{3} \partial_j (a_j \vec{b})$). We already know that

$$|\vec{u}(t, x)| \leq C(1 + |x|)^{-2},$$

so that every information on $\vec{\omega} = O(|x|^{-\delta})$ will be converted into an estimate $\vec{\omega} = W_{\nu t} * \vec{\omega}_0 + \int_0^t \operatorname{curl}(W_{\nu(t-s)} * \vec{f}) \, ds + O(|x|^{-\delta - 2})$. □

We thus have precise information on the localization of the vorticity, and not on the velocity. However, there is a relationship between the velocity $\vec{u}$ and the vorticity $\vec{\omega}$: from $\vec{\omega} = \operatorname{curl} \vec{u}$, we have $\operatorname{curl} \vec{\omega} = -\Delta \vec{u}$ and, since $\vec{u}$ vanishes at infinity, we may determine $\vec{u}$ through *the Biot-Savart law*

$$\vec{u} = \vec{\nabla} G(\wedge *) \vec{\omega} \tag{4.36}$$

where the operation $\wedge *$ is defined with the Fourier transform $\mathcal{F}$ by

$$\vec{a}(\wedge *) \vec{b} = \mathcal{F}^{-1}(\mathcal{F}(\vec{a}) \wedge \mathcal{F}(\vec{b})).$$

If $\vec{\omega}$ is rapidly decaying ($|\vec{\omega}| \le C(1+|x|^{-5-\gamma})$), we find that

$$\vec{u} = \vec{\nabla}G(x) \wedge \vec{A}(t) + \sum_{i,=1}^{3} \vec{\nabla}\partial_i G(x) \wedge \vec{B}_i(t) + \sum_{i=1}^{3}\sum_{j=1}^{3} \vec{\nabla}\partial_i\partial_j G(x) \wedge \vec{C}_{i,j}(t) + o(|x|^{-4}) \quad (4.37)$$

with

$$\vec{A}(t) = \int \vec{\omega}(t,y)\,dy, \quad \vec{B}_i(t) = -\int y_i\vec{\omega}(t,y)\,dy, \quad \vec{C}_{i,j}(t) = \frac{1}{2}\int y_iy_j\vec{\omega}(t,y)\,dy.$$

From the decay of $\vec{\omega}$ in $O(|x|^{-5-\gamma})$ and the fact that $\operatorname{div}\vec{\omega} = 0$, we find that $\int \vec{\omega}\,dy = 0$, hence $\vec{A} = 0$ and $\vec{u} = O(|x|^{-3})$.

If we want $\vec{u}(t_0,.) = o(|x|^{-4})$ at some time t_0, we must have

$$\sum_{i,=1}^{3} \vec{\nabla}\partial_i G(x) \wedge \vec{B}_i(t_0) + \sum_{i=1}^{3}\sum_{j=1}^{3} \vec{\nabla}\partial_i\partial_j G(x) \wedge \vec{C}_{i,j}(t_0) = 0. \quad (4.38)$$

If $\vec{u}_0 = o(|x|^{-4})$, then (4.38) is satisfied at $t = 0$, and we find

$$\sum_{i,=1}^{3} \vec{\nabla}\partial_i G(x) \wedge \int_0^{t_0} \frac{d}{dt}\vec{B}_i(t)\,dt + \sum_{i=1}^{3}\sum_{j=1}^{3} \vec{\nabla}\partial_i\partial_j G(x) \wedge \int_0^{t_0} \frac{d}{dt}\vec{C}_{i,j}(t)\,dt = 0.$$

We now write

$$\partial_t\vec{\omega} = \nu\Delta\vec{\omega} + \operatorname{curl}\vec{f} + \operatorname{div}(\vec{\omega}\otimes\vec{u} - \vec{u}\otimes\vec{\omega}) = \nu\Delta\vec{\omega} + \operatorname{curl}\vec{f} - \operatorname{curl}(\operatorname{div}(\vec{u}\otimes\vec{u})).$$

From the decay of $\vec{f}$ in $|x|^{-5}$, of $\vec{u}$ in $|x|^{-3}$ and of $\vec{\omega}$ in $|x|^{-5-\gamma}$ we find that

$$\frac{d}{dt}\vec{B}_i(t) = -\int y_i\operatorname{curl}\vec{f}(t,y)\,dy = -\int \vec{f}(t,y) \wedge \vec{\nabla}(y_i)\,dy$$

and

$$\frac{d}{dt}\vec{C}_{i,j}(t) = \int \vec{f}(t,y) \wedge \vec{\nabla}\left(\frac{y_iy_j}{2}\right) + \sum_{k=1}^{3} u_k(t,y)\vec{u}(t,y) \wedge \vec{\nabla}\partial_k\left(\frac{y_iy_j}{2}\right)\,dy.$$

This gives

$$\vec{\nabla}\partial_i G(x) \wedge \frac{d}{dt}\vec{B}_i(t) = \sum_k \partial_i\partial_k G(x)\int f_k(t,y)\vec{\nabla}(y_i)\,dy - \partial_i^2 G(x)\int \vec{f}(t,y)\,dy$$

and thus (since $\Delta G = 0$ for $x \ne 0$)

$$\frac{d}{dt}\sum_{i=1}^{3} \vec{\nabla}\partial_i G(x) \wedge \vec{B}_i(t) = \sum_{k=1}^{3}\left(\int f_k(t,y)\,dy\right)\partial_k\vec{\nabla}G(x).$$

Similarly, we write

$$\vec{\nabla}\partial_i\partial_j G(x) \wedge \frac{d}{dt}\vec{C}_{i,j}(t) = \sum_{l=1}^{3} \partial_i\partial_j\partial_l G(x)\int \partial_l\left(\frac{y_iy_j}{2}\right)\vec{f}(t,y)\,dy$$

$$+ \sum_{l=1}^{3} \partial_i\partial_j\partial_l G(x)\int \sum_{k=1}^{3}\partial_k\partial_l\left(\frac{y_iy_j}{2}\right)u_k(t,y)\vec{u}(t,y)dy$$

$$- \sum_{l=1}^{3} \partial_i\partial_j\partial_l G(x)\int f_l(t,y)\vec{\nabla}\left(\frac{y_iy_j}{2}\right)\,dy$$

$$- \sum_{l=1}^{3} \partial_i\partial_j\partial_l G(x)\int \sum_{k=1}^{3} u_k(t,y)u_l(t,y)\partial_k\vec{\nabla}\left(\frac{y_iy_j}{2}\right)\,dy$$

and thus (using again $\Delta G(x) = 0$ for $x \neq 0$)

$$\sum_{i=1}^{3}\sum_{j=1}^{3} \vec{\nabla}\partial_i\partial_j G(x) \wedge \frac{d}{dt}\vec{C}_{i,j}(t) = -\sum_{i=1}^{3}\sum_{l=1}^{3}\partial_i\partial_l\vec{\nabla}G(x)\int f_l(t,y)y_i\,dy$$

$$-2\sum_{i=1}^{3}\sum_{l=1}^{3}\partial_i\partial_l\vec{\nabla}G(x)\int u_i(t,y)u_l(t,y)\partial_k(y_i)\,dy$$

We thus recover the Dobrokhotov and Shafarevich conditions. We can conclude, in the case of a null force $\vec{f} = 0$ and of a rapidly decaying vorticity $\vec{\omega}_0$, that we have:

- $\int \vec{\omega}(t,x)dx = 0$

- if $\int \vec{\omega}_0(x)x_i\,dx = 0$ for all $1 \leq i \leq 3$, then for all $t \in (0,T]$, $\int \vec{\omega}(t,x)x_i\,dx = 0$

- even if $\int \vec{\omega}_0(x)x_ix_j\,dx = 0$ for all $1 \leq i,j \leq 3$, we may have that for all $t \in (0,t_0]$, t_0 small enough, there exists i and j such that $\int \vec{\omega}(t,x)\,x_ix_jdx \neq 0$

- An easy example of such a $\vec{\omega}_0$ is $\vec{u}_0 = (-\partial_2\psi, \partial_1\psi, 0)$ and $\vec{\omega}_0 = (-\partial_1\partial_3\psi, -\partial_2\partial_3\psi, (\partial_1^2 + \partial_2^2)\psi)$ with $\psi \in \mathcal{D}$ and $\int \psi\,dx = 0$ (and $\psi \neq 0$). $\vec{u}$ does not satisfy the Dobrokhotov and Shafarevich conditions since $\int u_{0,1}^2\,dx \neq \int u_{0,3}^2\,dx$.

4.11 Maximal Classical Solutions and Estimates in L^2 Norms

Let $(\vec{u}_0, \vec{f})$ be regular data for the Cauchy problem for Navier–Stokes equations. We know that we have a unique maximal solution defined on an interval $[0, T_{\mathrm{MAX}})$. In order to prove that we have a global solution, i.e. that $T_{\mathrm{MAX}} = +\infty$, we have seen that it is enough to get an a priori control on the L^∞ norm of the solution (Theorem 4.8). However, we have no such control (except on the case of small data [Theorem 4.9]). Actually, the only control we have is a control on the L^2 norm:

Energy balance

Proposition 4.2.
Let $(\vec{u}, p)$ be the classical solution of the Navier–Stokes problem on the strip $[0, T_{\mathrm{MAX}}) \times \mathbb{R}^3$, associated to the regular data $(\vec{u}_0, \vec{f})$. Then we have

$$\partial_t(|\vec{u}|^2) = \nu\Delta(|\vec{u}|^2) - 2\nu|\vec{\nabla} \otimes \vec{u}|^2 - \mathrm{div}\left((|\vec{u}|^2 + 2p)\vec{u}\right) + 2\vec{u}\cdot\vec{f} \qquad (4.39)$$

(where $|\vec{\nabla} \otimes \vec{u}|^2 = \sum_{1\leq i,j\leq 3}|\partial_i u_j|^2$). In particular, writing $\vec{g}_j = \vec{f} * \partial_j G$ (so that $\vec{f} = -\sum_{j=1}^{3}\partial_j\vec{g}_j$), we have

$$\|\vec{u}(t,.)\|^2 + \nu\int_0^t \|\vec{\nabla}\otimes\vec{u}\|_2^2\,ds \leq \|\vec{u}_0\|_2^2 + \frac{1}{\nu}\sum_{j=1}^{3}\int_0^t\|\vec{g}_j\|_2^2\,ds \qquad (4.40)$$

and, for every finite T with $0 < T \leq T_{\mathrm{MAX}}$,

$$\vec{u} \in L_t^\infty L_x^2 \cap L_t^2 H_x^1((0,T) \times \mathbb{R}^3).$$

Proof. As $\vec{u}$, $\vec{f}$ and p are C^2, equation (4.39) is obtained easily by writing $\partial_t(|\vec{u}|^2) = 2\vec{u}\cdot\partial_t\vec{u}$ and using the fact that $\operatorname{div}\vec{u} = 0$. Then, due to the decay of $\vec{u}$, $\vec{f}$ and p and of their derivatives, we may integrate this equality on $(0,t)\times\mathbb{R}^3$ and obtain

$$\|\vec{u}(t,.)\|^2 - \|\vec{u}_0\|_2^2 = -2\nu\int_0^t \|\vec{\nabla}\otimes\vec{u}\|_2^2 + 2\int_0^t\int \vec{u}\cdot\vec{f}\,dx\,ds.$$

We then finish the proof by integration by parts:

$$\int \vec{u}\cdot\vec{f}\,dx = \sum_{j=1}^3 \int \partial_j\vec{u}\cdot\vec{g}_j\,dx \leq \sum_{j=1}^3 \nu\|\partial_j\vec{u}\|_2^2 + \frac{1}{\nu}\|\vec{g}_j\|_2^2. \qquad \square$$

Energy estimates can be useful for getting other accurate estimates. Here, we shall give an example of control on the L^2 norms of derivatives, i.e. on Sobolev norms, of the classical solutions of the Navier–Stokes equations. We begin with energy estimates for the heat kernel:

Energy estimates for the heat kernel

Proposition 4.3.
A) *if $u_0 \in L^2$, then $W_{\nu t} * u_0 \in L^\infty((0,+\infty), L^2)$ and $\vec{\nabla}(W_{\nu t} * u_0) \in L^2((0,+\infty), L^2)$. Moreover,*

$$\|W_{\nu t} * u_0\|_2^2 + 2\nu\int_0^t \|\vec{\nabla}(W_{\nu s} * u_0)\|_2^2\,ds = \|u_0\|_2^2.$$

B) *if $\vec{g} \in L^2((0,+\infty), L^2)$ and $U = \int_0^t W_{\nu(t-s)} * \operatorname{div}\vec{g}\,ds$, then $U \in L^\infty((0,+\infty), L^2)$ and $\vec{\nabla}U \in L^2((0,+\infty), L^2)$. Moreover,*

$$\|U(t,.)\|_2^2 + 2\nu\int_0^t \|\vec{\nabla}U(s,.)\|_2^2\,ds = -2\int_0^t\int \vec{\nabla}U(s,.)\cdot\vec{g}(s,.)\,dx\,ds$$

and thus

$$\|U(t,.)\|_2^2 + \nu\int_0^t \|\vec{\nabla}U(s,.)\|_2^2\,ds \leq \frac{1}{\nu}\int_0^t \|\vec{g}(s,.)\|_2^2\,ds.$$

C) *(maximal regularity) if $h \in L^2((0,+\infty), L^2)$ and $V = \int_0^t W_{\nu(t-s)} * \Delta h\,ds$, then $V \in L^2((0,+\infty), L^2)$ and*

$$\nu^2\int_0^{+\infty} \|V(s,.)\|_2^2\,ds \leq \int_0^{+\infty} \|h(s,.)\|_2^2\,ds$$

Proof.

A) By Plancherel and Fubini, we have

$$\int |\widehat{u_0}(\xi)|^2\,(e^{-2t|\xi|^2} - 1)\,d\xi = -2\int_0^t\int |\xi|^2\,e^{-2s|\xi|}|\widehat{u_0}(\xi)|^2\,d\xi\,ds$$

B) If $\vec{g} \in \mathcal{D}((0,+\infty)\times\mathbb{R}^3)$, then we have seen that U and its derivatives is smooth and rapidly decaying in space (uniformly in time). Moreover, we have

$$\partial_t(|U|^2) = 2U\partial_t U = 2\nu U\Delta U + 2U\operatorname{div}\vec{g}.$$

Intregrating this equatlity between 0 and t with respect to time and space variables gives (since $U(0, .) = 0$)

$$\|U(t, .)\|_2^2 + 2\nu \int_0^t \|\vec{\nabla} U(s, .)\|_2^2 \, ds = -2 \int_0^t \int \vec{\nabla} U(s, .) \cdot \vec{g}(s, .) \, dx \, ds.$$

The inequality

$$\int \vec{\nabla} U(s, .) \cdot \vec{g}(s, .) \, dx \leq \|\vec{\nabla} U(s, .)\|_2 \|\vec{g}(s, .)\|_2 \leq \nu \vec{\nabla} U(s, .) + \frac{1}{\nu} \|\vec{g}(s, .)\|_2^2$$

then gives that

$$\|U(t, .)\|_2^2 + \nu \int_0^t \|\vec{\nabla} U(s, .)\|_2^2 \, ds \leq \frac{1}{\nu} \int_0^t \|\vec{g}(s, .)\|_2^2 \, ds.$$

Thus, $\vec{g} \mapsto U$ is linear and continuous (in the $L_t^2 L_x^2$ norm) from the space $\mathcal{D}((0, +\infty) \times \mathbb{R}^3)$ to $\mathcal{C}_b([0, +\infty), L^2(\mathbb{R}^3)) \cap L^2((0, +\infty), \dot{H}_x^1(\mathbb{R}^3))$ (where the norm of the homogeneous Sobolev space $\dot{H}^1$ is given by $\|f\|_{\dot{H}^1} = \|\vec{\nabla} f\|_2$). We may then conclude, due to the density of the space $\mathcal{D}((0, +\infty) \times \mathbb{R}^3)$ in $L^2((0, +\infty) \times \mathbb{R}^3) \approx L^2((0, +\infty), L^2)$.

C) We define $\vec{g}_{j,\epsilon} = (g_{1,j,\epsilon}, g_{2,j,\epsilon}, g_{3,j,\epsilon})$ by the Fourier transforms

$$\widehat{g_{k,j,\epsilon}}(\xi) = -\frac{\xi_j \xi_k}{\epsilon + |\xi|^2} \hat{h}(\xi)$$

and

$$U_{j,\epsilon} = \int_0^t W_{\nu(t-s)} * \operatorname{div} \vec{g}_{j,\epsilon} \, ds.$$

We have

$$\nu \int_0^{+\infty} \|\vec{\nabla} U_{j,\epsilon}(s, .)\|_2^2 \, ds \leq \frac{1}{\nu} \int_0^{+\infty} \|\vec{g}_{j,\epsilon}(s, .)\|_2^2 \, ds.$$

or, equivalently, by the Plancherel formula,

$$\nu \int_0^\infty \int \frac{|\xi_j|^2 |\xi|^2}{(\epsilon + |\xi|^2)^2} |\hat{V}(s, \xi)|^2 \, d\xi \, ds \leq \frac{1}{\nu} \int_0^{+\infty} \int \frac{|\xi_j|^2 |\xi|^2}{(\epsilon + |\xi|^2)^2} |\hat{h}(s, \xi)|^2 \, d\xi \, ds.$$

Then, we sum on j and let ϵ go to 0; by monotonous convergence we find

$$\nu \int_0^\infty \int |\hat{V}(s, \xi)|^2 \, d\xi \, ds \leq \frac{1}{\nu} \int_0^{+\infty} \int |\hat{h}(s, \xi)|^2 \, d\xi \, ds. \qquad \square$$

Another interesting application of energy balances is the following result:

Energy equality

Proposition 4.4.
If $u \in L^2((0, T), H^1)$ and $\partial_t u \in L^2((0, T), H^{-1})$ (where H^{-1} is defined by: $f \in H^{-1} \Leftrightarrow (1 + |\xi|^2)^{-\frac{1}{2}} \hat{f} \in L^2$) then u belongs to $\mathcal{C}([0, T], L^2)$ and, for $0 \leq t_0 \leq t \leq T$,

$$\|u(t, .)\|_2^2 = \|u(t_0, .)\|_2^2 + 2 \int_{t_0}^t \langle u | \partial_t u \rangle_{H^1, H^{-1}} \, ds.$$

Proof. We prove the theorem for $u \in C_c^\infty([0,T] \times \mathbb{R}^3)$. We have

$$\partial_t |u|^2 = 2u\partial_t u$$

and integration in time and space variables between t_0 and t gives the result. In particular, if θ_1 and θ_2 are smooth functions on $[0,T]$ such that $0 \le \theta_1 \le 1$, $\theta_1(t) = 1$ for $t \le T/2$ and $= 0$ for $t \ge 3T/4$ while $0 \le \theta_2 \le 1$, $\theta_2(t) = 0$ for $t \le T/4$ and $= 0$ for $t \ge T/2$, we find that, for $t \le T/2$

$$\|u(t,.)\|_2^2 \le 2 \int_t^T \theta_1(s) |\langle u | \partial_t u \rangle_{H^1, H^{-1}}| + |\theta_1'(s)| \|u\|_2^2 \, ds$$

while, for $t \ge T/2$,

$$\|u(t,.)\|_2^2 \le 2 \int_0^t \theta_2(s) |\langle u | \partial_t u \rangle_{H^1, H^{-1}}| + |\theta_2'(s)| \|u\|_2^2 \, ds.$$

Thus, we find that the linear map $u \in C_c^\infty([0,T] \times \mathbb{R}^3) \mapsto u \in C([0,T], L^2)$ is bounded for the norm of $\|u\|_{L^2 H^1} + \|\partial_t u\|_{L^2 H^{-1}}$. We then finish the proof by a density argument. $\square$

An interesting application is the control on the size of $\vec{u}$ in the Sobolev space H^3 when $\vec{u}_0$ and $\vec{f}$ are one-derivative more regular. Let us recall some elementary results on Sobolev spaces H^k, $k \in \mathbb{N}$:

- definition: H^k is the space of functions in L^2 whose derivatives (in the sense of distributions) of order less or equal to k are square integrable, normed with

$$\|f\|_{H^k} = \sqrt{\sum_{|\alpha| \le k} \|\partial^\alpha f\|_2^2}.$$

- H^k is a Hilbert space. Its norm is equivalent with

$$\|(Id - \Delta)^{k/2} f\|_2 = \frac{1}{(2\pi)^{3/2}} \|(1 + |\xi|)^{k/2} \hat{f}\|_2$$

  (where $\hat{f}$ is the Fourier transform of f) and with

$$\sqrt{\|f\|_2^2 + \sum_{|\alpha|=k} \|\partial^\alpha f\|_2^2}$$

- for $k \ge 2$ and $f \in H^k$, $\hat{f} \in L^1$, so that H^k is continuously embedded in L^∞ :

$$\|f\|_\infty \le \frac{1}{(2\pi)^3} \|\hat{f}\|_1 \le \frac{1}{(2\pi)^3} \|(1 + |\xi|)^{-k/2}\|_2 \|(1 + |\xi|)^{k/2} \hat{f}\|_2$$

- for $f \in H^2$, we have

$$\int |\partial_i f|^4 \, dx = -3 \int f(\partial_i f)^2 \partial_i^2 f \, dx \le 3\|f\|_\infty \|\partial_i f\|_4^2 \|\partial_i^2 f\|_2 \le C\|f\|_{H^2}^2 \|\partial_i f\|_4^2$$

- for $k \ge 2$ and $f, g \in H^k$, $fg \in H^k$ and $\|fg\|_{H^k} \le C_k \|f\|_{H^k} \|g\|_{H^k}$ [obvious by the Leibniz rule]

We first collect the results that may be induced from a straightforward adaptation of Theorems 4.7 and 4.8:

Theorem 4.13 (Navier–Stokes equations and $\mathcal{C}^3$ solutions).
Let $\vec{u}_0$ be a $\mathcal{C}^3$ divergence-free vector field on $\mathbb{R}^3$ and let $\vec{f}$ be a time-dependent vector field such that:

- $\sup_{|\alpha| \leq 3} \sup_{x \in \mathbb{R}^3} (1 + |x|)^2 |\partial_x^\alpha \vec{u}_0(x)| < +\infty$

- *for $|\alpha| \leq 3$, $\partial_x^\alpha \vec{f}$ is continuous on $[0, +\infty) \times \mathbb{R}^3$*

- $\sup_{|\alpha| \leq 3} \sup_{t \geq 0, x \in \mathbb{R}^3} (1 + |x|)^4 |\partial_x^\alpha \vec{f}(t, x)| < +\infty.$

Then, there exists a positive time T and a unique solution $(\vec{u}, p)$ of the Navier–Stokes problem

$$\partial_t \vec{u} = \nu \Delta \vec{u} + \vec{f} - (\vec{u}.\vec{\nabla})\vec{u} - \vec{\nabla} p \text{ on } (0, T) \times \mathbb{R}^3$$
$$\text{div } \vec{u} = 0 \quad \vec{u}_{|t=0} = \vec{u}_0$$

such that:

- *for $|\alpha| \leq 3$, $\partial_x^\alpha p$ and $\partial_x^\alpha \vec{u}$ are continuous on $[0, T] \times \mathbb{R}^3$*

- $\sup_{0 \leq t \leq T, x \in \mathbb{R}^3} (1 + |x|)|p(t, x)| < +\infty.$

- $\sup_{1 \leq |\alpha| \leq 3} \sup_{0 \leq t \leq T, x \in \mathbb{R}^3} (1 + |x|)^2 |\partial_x^\alpha p(t, x)| < +\infty.$

- $\sup_{|\alpha| \leq 3} \sup_{0 \leq t \leq T, x \in \mathbb{R}^3} (1 + |x|)^2 |\partial_x^\alpha \vec{u}(t, x)| < +\infty.$

Let T_{MAX} be the maximal time where one can find a $\mathcal{C}^3$ classical solution $\vec{u}$ of the Cauchy problem for the Navier–Stokes equations on $(0, T_{\text{MAX}}) \times \mathbb{R}^3$. Then we have the a priori estimate:

$$T_{\text{MAX}} \geq \min(T_1, T_2)$$

with

$$\begin{cases} T_1 = \dfrac{\nu}{C_0} \dfrac{1}{(\sup_{|\alpha| \leq 3} \sup_{x \in \mathbb{R}^3} (1 + |x|)^2 |\partial_x^\alpha \vec{u}_0(x)|)^2} \\ T_2 = \dfrac{\nu}{C_0} \dfrac{1}{\sup_{|\alpha| \leq 3} \sup_{0 < t < T_0, x \in \mathbb{R}^3} (1 + |x|)^4 |\partial_x^\alpha \vec{f}(t, x)|} \end{cases}.$$

and the following blow up criterion: If $T_{\text{MAX}} < +\infty$, then

$$\sup_{0 \leq t < T_{\text{MAX}}, x \in \mathbb{R}^3} |\vec{u}(t, x)| = +\infty.$$

Thus, the lower bound on T_{MAX} seems to depend on ν, and to go to 0 as ν goes to 0. However, Swann [458] obtained a bound that depends only on the H^3 norm of $\vec{u}_0$ and on the size of $\vec{f}$:

Navier–Stokes equations and H^3 norms

Theorem 4.14.
Under the assumptions of Theorem 4.13, we have the following size estimates for

$$\|\vec{u}(t, .)\|_{H^3} = \sqrt{\|\vec{u}\|_2 + \sum_{|\alpha|=3} \|\partial_x^\alpha \vec{u}\|_2^2} :$$

---

for $0 < t < T_{\text{MAX}}$, we have

$$\|\vec{u}(t,.)\|_{H^3}^2 \leq \|\vec{u}_0\|_{H^3}^2 + \int_0^t \|\vec{f}\|_{H^3}^{3/2}\, ds + C_0 \int_0^t \|\vec{u}(s,.)\|_{H^1}^3\, ds$$

for a constant C_0 which does not depend on ν.

In particular, if $T_0 < +\infty$, and if

$$T_1 = \min(T_0, \frac{1}{8C_0(\|\vec{u}_0\|_{H^3}^2 + \int_0^{T_0} \|\vec{f}\|_{H^3}^{3/2}\, ds)^2})$$

then $T_{\text{MAX}} \geq T_1$.

Proof. We have

$$\partial_t \vec{u} = \nu \Delta \vec{u} - \mathbb{P}\operatorname{div}(\vec{u} \otimes \vec{u}) + \mathbb{P}\vec{f}$$

and

$$\vec{u} = W_{\nu t} * \vec{u}_0 + \int_0^t W_{\nu(t-s)} * (-\mathbb{P}\operatorname{div}(\vec{u} \otimes \vec{u}) + \mathbb{P}\vec{f})\, ds.$$

If $T < T_{\text{MAX}}$, we have that $\vec{u}$ belongs to $L^\infty((0,T), H^3)$; thus $\vec{f} - \operatorname{div}(\vec{u} \otimes \vec{u})$ belongs to $L^\infty((0,T), H^2)$. Recall that the Leray projection operator $\mathbb{P}$ is defined as

$$\mathbb{P}\vec{F} = \vec{F} + \vec{\nabla}(G * \operatorname{div} \vec{F})$$

and thus

$$\widehat{\mathbb{P}\vec{F}}(\xi) = \widehat{\vec{F}}(\xi) - \xi \frac{\xi \cdot \widehat{\vec{F}}(\xi)}{|\xi|^2};$$

thus, $\mathbb{P}$ is bounded on H^2 (as $|\widehat{\mathbb{P}\vec{F}}(\xi)| \leq |\widehat{\vec{F}}(\xi)|$). Thus, writing

$$\Delta^2 \vec{u} = \sum_{i=1}^3 \partial_i(W_{\nu t} * \partial_i \Delta \vec{u}_0) + \int_0^t W_{\nu(t-s)} * \Delta\left(\Delta \mathbb{P}(\vec{f} - \operatorname{div}(\vec{u} \otimes \vec{u}))\right)\, ds,$$

we find that $\vec{u} \in L^2((0,T), H^4)$: of course, $\|\vec{u}\|_{L^2 L^2} \leq \sqrt{T}\|\vec{u}\|_{L^\infty L^2}$; on the other hand, using the maximal regularity of the heat kernel (Proposition 4.3), we get

$$\|\Delta^2 \vec{u}\|_{L^2 L^2} \leq C\frac{1}{\sqrt{\nu}}\|\vec{u}_0\|_{H^3} + C\frac{1}{\nu}\|\Delta \mathbb{P}(\vec{f} - \operatorname{div}(\vec{u} \otimes \vec{u}))\|_{L^2 L^2}$$

$$\leq C\frac{1}{\sqrt{\nu}}\|\vec{u}_0\|_{H^3} + C'\frac{\sqrt{T}}{\nu}\|\vec{f}\|_{L^\infty H^2} + C'\frac{\sqrt{T}}{\nu}\|\vec{u}\|_{L^\infty H^3}^2.$$

Now, we write, for $|\alpha| = 3$,

$$\partial_t \partial_x^\alpha \vec{u} = \nu \partial_x^\alpha \Delta \vec{u} - \partial_x^\alpha \mathbb{P}\operatorname{div}(\vec{u} \otimes \vec{u}) + \partial_x^\alpha \mathbb{P}\vec{f};$$

as $\vec{u} \in L^2 H^4$, we find that $\partial_x^\alpha \vec{u} \in L^2 H^1$ and $\partial_t \partial_x^\alpha \vec{u} \in L^2 H^{-1}$, so that (Proposition 4.4) we have

$$\|\partial_x^\alpha \vec{u}(t,.)\|_2^2 = \|\partial_x^\alpha \vec{u}_0\|_2^2 + 2 \int_0^t \langle \partial_x^\alpha \vec{u} | \nu \partial_x^\alpha \Delta \vec{u} - \partial_x^\alpha \mathbb{P} \operatorname{div}(\vec{u} \otimes \vec{u}) + \partial_x^\alpha \mathbb{P} \vec{f} \rangle_{H^1, H^{-1}} \, ds$$

$$= \|\partial_x^\alpha \vec{u}_0\|_2^2 - 2\nu \|\vec{\nabla} \otimes \partial_x^\alpha \vec{u}\|_2^2 + 2 \int_0^t \int \partial_x^\alpha \vec{u} \cdot \partial_x^\alpha \vec{f} \, dx \, ds$$

$$\quad - 2 \int_0^t \langle \partial_x^\alpha \vec{u} | \partial_x^\alpha (\vec{u} \cdot \vec{\nabla} \vec{u}) \rangle_{H^1, H^{-1}} \, ds$$

$$= \|\partial_x^\alpha \vec{u}_0\|_2^2 - 2\nu \|\vec{\nabla} \otimes \partial_x^\alpha \vec{u}\|_2^2 + 2 \int_0^t \int \partial_x^\alpha \vec{u} \cdot \partial_x^\alpha \vec{f} \, dx \, ds$$

$$\quad - 2 \sum_{\gamma \leq \alpha, \gamma \neq \alpha} \frac{\alpha!}{\gamma!(\alpha - \gamma)!} \int_0^t \int \partial_x^\alpha \vec{u} \cdot \left((\partial_x^{\alpha-\gamma} \vec{u}) \cdot \partial_x^\gamma \vec{\nabla} \vec{u} \right) dx \, ds$$

$$\quad - 2 \int_0^t \langle \partial_x^\alpha \vec{u} | \vec{u} \cdot \vec{\nabla} \partial_x^\alpha \vec{u} \rangle_{H^1, H^{-1}} \, ds.$$

For $\vec{v} \in H^3$ and $w_1, w_2 \in H^1$, we have

$$\langle w_1 | \vec{v} \cdot \vec{\nabla} w_2 \rangle_{H^1, H^{-1}} = -\langle w_2 | \vec{v} \cdot \vec{\nabla} w_1 \rangle_{H^1, H^{-1}} - \int w_1 w_2 \operatorname{div} \vec{v} \, dx$$

so that

$$\langle \partial_x^\alpha \vec{u} | \vec{u} \cdot \vec{\nabla} \partial_x^\alpha \vec{u} \rangle_{H^1, H^{-1}} = 0.$$

On the other hand, we have:
* when $\gamma = 0$,

$$\|(\partial_x^{\alpha-\gamma} \vec{u}) \cdot \partial_x^\gamma \vec{\nabla} \vec{u}\|_2 \leq \|\partial_x^\alpha \vec{u}\|_2 \|\vec{\nabla} \otimes \vec{u}\|_\infty \leq C \|\vec{u}\|_{H^3}^2$$

* when $|\gamma| = 1$,

$$\|(\partial_x^{\alpha-\gamma} \vec{u}) \cdot \partial_x^\gamma \vec{\nabla} \vec{u}\|_2 \leq \|\partial_x^{\alpha-\gamma} \vec{u}\|_4 \|\partial_x^\gamma \vec{\nabla} \otimes \vec{u}\|_4 \leq C \|\vec{u}\|_{H^3}^2$$

* when $|\gamma| = 2$,

$$\|(\partial_x^{\alpha-\gamma} \vec{u}) \cdot \partial_x^\gamma \vec{\nabla} \vec{u}\|_2 \leq \|\partial_x^{\alpha-\gamma} \vec{u}\|_\infty \|\partial_x^\gamma \vec{\nabla} \otimes \vec{u}\|_2 \leq C \|\vec{u}\|_{H^3}^2.$$

Thus, if $\|\vec{u}\|_{H^3}$ is defined as

$$\|\vec{u}(t,.)\|_{H^3} = \sqrt{\|\vec{u}\|_2 + \sum_{|\alpha|=3} \|\partial_x^\alpha \vec{u}\|_2^2},$$

we find that

$$\|\vec{u}(t,.)\|_{H^3}^2 \leq \|\vec{u}_0\|_{H^3}^2 + 2 \int_0^t \|\vec{u}\|_{H^3} \|\vec{f}\|_{H^3} + C \int_0^t \|\vec{u}(s,.)\|_{H^3}^3 \, ds$$

and finally

$$\|\vec{u}(t,.)\|_{H^3}^2 \leq \|\vec{u}_0\|_{H^3}^2 + \int_0^t \|\vec{f}\|_{H^3}^{3/2} \, ds + C_0 \int_0^t \|\vec{u}(s,.)\|_{H^3}^3 \, ds.$$

We may now easily finish the proof. If $T_0 < +\infty$, and if

$$0 \leq t < T_{\text{MAX}}$$

and

$$t \leq T_1 = \min(T_0, \frac{1}{8C_0(\|\vec{u}_0\|_{H^3}^2 + \int_0^{T_0} \|\vec{f}\|_{H^3}^{3/2} ds)^2})$$

we get that

$$\|\vec{u}(t,.)\|_{H^3}^2 \leq 2(\|\vec{u}_0\|_{H^3}^2 + \int_0^{T_0} \|\vec{f}\|_{H^3}^{3/2} ds).$$

In particular,

$$\sup_{0<t<\min(T_{\text{MAX}},T_1)} \|\vec{u}(t,.)\|_\infty < +\infty$$

and $T_{\text{MAX}} \geq T_1$. $\qquad \square$

4.12   Intermediate Conclusion

What have we seen in this chapter? Given regular initial value $\vec{u}_0$ and forcing term $\vec{f}$, as in the setting of the Clay Millenium problem, we have been able to prove with elementary tools of calculus the following results:

- existence of a (classical) solution on $[0,T] \times \mathbb{R}^3$ for a positive time T

- global existence of the solution when the data $\vec{u}_0$ and $\vec{f}$ are small

- instantaneous spreading of the velocity (so that the assumptions of the Clay problem on the initial value cannot be kept)

- localization of the vorticity

In the following chapters, we are going to extend the class of solutions (weak solutions instead of classical ones) in order to grant global existence, and to use tools from functional analysis and real harmonic analysis to try and get better insight into the properties of those extended solutions. As a matter of fact, when the data are large, we do not know whether the solutions that we are able to construct are unique, nor whether they are regular.

Chapter 5

A Capacitary Approach of the Navier–Stokes Integral Equations

In Chapter 4, we have studied classical solutions of the Cauchy initial value problem for the Navier–Stokes equations (with reduced (unknown) pressure p, reduced force density $\vec{f}$ and kinematic viscosity $\nu > 0$): given a divergence-free vector field $\vec{u}_0$ on $\mathbb{R}^3$ and a force $\vec{f}$ on $(0, +\infty) \times \mathbb{R}^3$, find a positive T and regular functions $\vec{u}$ and p on $[0, T] \times \mathbb{R}^3$ solutions to

$$\partial_t \vec{u} = \nu \Delta \vec{u} - (\vec{u}.\vec{\nabla})\vec{u} + \vec{f} - \vec{\nabla}p$$
$$\operatorname{div} \vec{u} = 0 \tag{5.1}$$
$$\vec{u}_{|t=0} = \vec{u}_0$$

We have reformulated this problem into an integral equation: find $\vec{u}$ such that

$$\vec{u} = W_{\nu t} * \vec{u}_0 - \int_0^t \sum_{j=1}^3 \partial_j \mathcal{O}(\nu(t-s)) :: \left(\vec{f} * \partial_j G + u_j \vec{u}\right) \, ds \tag{5.2}$$

We can see in the formulation of Equation (5.2) that we do not need any regularity on $\vec{u}$ to compute the right-hand side of the equation, but just integrability properties. In this chapter, we shall discuss the existence of measurable solutions of the integral equation, and we shall see in Chapter 6 to what extent they are a solution of the differential equations.

5.1   The Integral Navier–Stokes Problem

Throughout the chapter, we are going to study generalized solutions of the Navier–Stokes equations. More precisely, starting with the initial data $\vec{u}_0$ and the force $\vec{f}$, we assume that

$$\vec{U}(t, x) = W_{\nu t} * \vec{u}_0 - \int_0^t \sum_{j=1}^3 \partial_j \mathcal{O}(\nu(t-s)) :: \left(\vec{f} * \partial_j G\right)$$

is a measurable function of t and x such that, for all $R > 0$, we have, for $1 \leq k \leq 3$,

$$\int_0^T \int_{B(0,R)} |U_k(t, x)| \, dx \, dt < \infty$$

and we study the measurable functions $\vec{u}(t, x)$ defined on $(0, T) \times \mathbb{R}^3$ such that, for all $R > 0$, we have, for $1 \leq j, k, l \leq 3$,

$$\int_0^T \int_{B(0,R)} \int_0^t \int |\partial_j \mathcal{O}_{k,l}(\nu(t-s), x-y)| \, |u_j(s,y)||u_l(s,y)| \, dy \, ds \, dx \, dt < \infty$$

DOI: 10.1201/9781003042594-5

and such that

$$\vec{u} = \vec{U} - \int_0^t \sum_{j=1}^3 \partial_j \mathcal{O}(\nu(t-s)) :: (u_j \vec{u}) \; ds. \tag{5.3}$$

5.2 Quadratic Equations in Banach Spaces

Solving the Navier–Stokes equations when written as integro-differential equations is solving a quadratic equation in the unknown $\vec{u}$ (Equation (5.3)). In this section, we show how to solve general quadratic equations with small data in a Banach space:

Quadratic equations

Theorem 5.1.
Let B be a bounded bilinear operator on a Banach space E:

$$\|B(u,v)\|_E \leq C_0 \|u\|_E \|v\|_E.$$

Then, when $\|u_0\|_E \leq \frac{1}{4C_0}$, the equation

$$u = u_0 + B(u,u)$$

has a unique solution in E such that $\|u\|_E \leq \frac{1}{2C_0}$. Moreover, $\|u\|_E \leq 2\|u_0\|_E$.

Proof. The method used by Oseen, which we developed in Chapter 4, Section 4.6, is very efficient. We introduce a development of the solution u_ϵ of the equation $u_\epsilon = u_0 + \epsilon B(u_\epsilon, u_\epsilon)$ as a power series in ϵ. We find (at least formally) $u_\epsilon = \sum_{n=0}^{\infty} \epsilon^n u_n$, where

$$u_{k+1} = \sum_{n=0}^k B(u_n, u_{k-n}).$$

Thus, we have

$$\|u_{k+1}\|_E \leq C_0 \sum_{n=0}^k \|u_n\|_E \|u_{k-n}\|_E.$$

The norm of u_n is thus dominated by α_n where $\alpha_0 = \|y_0\|_E$, and

$$\alpha_{k+1} = C_0 \sum_{n=0}^k \alpha_k \alpha_{k-n}.$$

The function

$$\alpha_\epsilon = \sum_{n=0}^{\infty} \epsilon^n \alpha_n \tag{5.4}$$

is solution of

$$\alpha_\epsilon = \alpha_0 + C_0 \epsilon \, \alpha_\epsilon^2.$$

We find that the series in Equation (5.4) converges for $1 - 4C_0\epsilon\alpha_0 \geq 0$ and that

$$\alpha_\epsilon = \frac{1 - \sqrt{1 - 4C_0\epsilon\,\alpha_0}}{2C_0\epsilon} = \frac{2\alpha_0}{1 + \sqrt{1 - 4C_0\epsilon\,\alpha_0}}.$$

This proves the existence of the solution u of $u = u_0 + B(u, u)$ when $4C_0\|u_0\|_E \leq 1$ and that $\|u\|_E \leq 2\|u_0\|_E$.

An alternative way to describe this series expansion of the solution is the following one. Consider B as an internal operation on E. We write $B(u, v) = u \circledast v$. This operation is not associative, hence we must use parentheses when defining the "product" of three (or more) terms: $(u \circledast v) \circledast w$ is not the same as $u \circledast (v \circledast w)$. Let A_n be the number of different ways to introduce parentheses for defining the product of n terms. Obviously, we have $A_1 = A_2 = 1$ (no need of parentheses) and $A_3 = 2$. For defining the product of n terms, $n \geq 2$, we must choose the order of priority in the computations of the $\circledast$ products: the product that will be computed at last will involve a product of k terms on the left-hand side and $n - k$ terms on the right-hand side, so that we see easily that

$$A_n = \sum_{k=1}^{n-1} A_k A_{n-k}.$$

Now, let us call a word an expression that is defined inductively in the following way:

- u_0 is a word

- if w_1 and w_2 are two words, then $w_1 \circledast w_2$ is a word.

Let W_n be the set of words where u_0 appears n times in the word, and $W = \cup_{n=1}^{+\infty} W_n$. The cardinal of the set W_n is A_n. Moreover, the norm of a word $w \in W_n$ is controlled by $C_0^{n-1}\|u_0\|_E^n$. If u_0 is such that $\sum_{n=1}^{+\infty} A_n C_0^{n-1}\|u_0\|_E^n < +\infty$, then the series $u = \sum_{w \in W} w$ is normally convergent. A word is either equal to u_0 or of the form $w_1 \circledast w_2$, so that

$$u = u_0 + \sum_{(w_1,w_2) \in W \times W} w_1 \circledast w_2 = u_0 + \left(\sum_{w_1 \in W} w_1\right) \circledast \left(\sum_{w_2 \in W} w_2\right) = u_0 + B(u, u).$$

Hence, we find a solution of $u = u_0 + B(u, u)$. In the case of $E = \mathbb{C}$ and $u \circledast v = uv$, we find that

$$\sum_{w \in W} w = \sum_{n=1}^{+\infty} A_n u_0^n$$

while $u = u_0 + u^2$; thus the generating series of the sequence A_n is given by

$$A(z) = \sum_{n=1}^{+\infty} A_n z^n = \frac{1 - \sqrt{1 - 4z}}{2}$$

(where the determination of the square root is given on $\Re(z) \geq 0$ by $(\sqrt{z})^2 = z$ and $\sqrt{1} = 1$). As the numbers A_n are positive and $A(1/4) < +\infty$, we find that the series is convergent for $|z| \leq 1/4$. Thus, the series $\sum_{w \in W} w$ will be normally convergent for $C_0\|u_0\|_E \leq 1/4$. The Oseen solution then consists in writing the solution as

$$u = \sum_{n=1}^{+\infty} \left(\sum_{w \in W_n} w\right).$$

As a final remark, let us recall that the numbers A_n are very well known: they are the Catalan numbers which are widely used in combinatorics.

When $4C_0\|u_0\|_E < 1$, we can yet use another proof, based on the Banach contraction principle. We consider the map $v \mapsto F(v) = u_0 + B(u, u)$ on the ball $B_0 = \{v \in E \;/\; 2C_0\|v\|_E \leq 1\}$. We have $\|F(v)\|_E \leq \|u_0\|_E + C_0\|v\|_E^2 \leq \|u_0\|_E + \frac{1}{4C_0} = \delta_0 < \frac{1}{2C_0}$. Moreover, on the ball $B_1 = \{v \in E \;/\; \|v\|_E \leq \delta_0\}$, we have

$$\|F(v) - F(w)\|_E = \|B(v, v - w) + B(v - w, w)\|_E \leq C_0\|v - w\|_E(\|v\|_E + \|w\|_E)$$
$$\leq 2\delta_0 C_0 \|v - w\|_E$$

with $2C_0\delta_0 < 1$. Thus, F is contractive on B_1 and has a unique fixed point in B_0.

In the case when $\|u_0\| = \frac{1}{4C_0}$, we already know that there exists a solution u in the ball B_0. If $2C_0\|u\|_E < 1$, then we see easily that this solution is unique in B_0, since, for $v \in B_0$, we have

$$\|u - (u_0 + B(v, v))\|_E \leq \left(\frac{1}{2} + C_0\|u\|_E\right)\|u - v\|_E.$$

In order to finish the proof of Theorem 5.1, it remains to deal with uniqueness in the case $4C_0\|u\|_E = 1$. We follow the proof of Auscher and Tchamitchian [12]. Let $v_0 \in B_0$ and let $v_{n+1} = F(v_n)$. Recall that the fixed-point u has been given as $u = \sum_{n=0}^{\infty} u_n$ with $\|u_n\|_E \leq \alpha_n$, $\sum_{n=0}^{\infty} \alpha_n \leq \frac{1}{2C_0}$; thus, if $\|u\|_E = \frac{1}{2C_0}$, we must have $\|u_n\|_E = \alpha_n$.

We have $\|v_0\|_E \leq \frac{1}{2C_0} = \sum_{n=0}^{\infty} \|u_n\|_E$. We are going to prove inductively that $\|v_k - \sum_{n=0}^{k-1} u_n\|_E \leq \sum_{n=k}^{\infty} \|u_n\|_E$. We write

$$v_{k+1} - \sum_{n=0}^{k} u_n = \left(F(v_k) - F\left(\sum_{n=0}^{k-1} u_n\right)\right) + \left(F\left(\sum_{n=0}^{k-1} u_n\right) - \sum_{n=0}^{k} u_n\right) = A_k + B_k.$$

We have

$$A_k = B\left(v_k - \sum_{n=0}^{k-1} u_n, v_k - \sum_{n=0}^{k-1} u_n\right) + B\left(v_k - \sum_{n=0}^{k-1} u_n, \sum_{n=0}^{k-1} u_n\right) + B\left(\sum_{n=0}^{k-1} u_n, v_k - \sum_{n=0}^{k-1} u_n\right).$$

Hence, we find

$$\|A_k\|_E \leq C_0\|v_k - \sum_{n=0}^{k-1} u_n\|_E\left(\|v_k - \sum_{n=0}^{k-1} u_n\|_E + 2\|\sum_{n=0}^{k-1} u_n\|_E\right)$$
$$\leq C_0 \sum_{n=k}^{\infty} \|u_n\|_E\left(\sum_{n=k}^{\infty} \|u_n\|_E + 2\sum_{n=0}^{k-1} \|u_n\|_E\right)$$

On the other hand, we have

$$B_k = u_0 + B\left(\sum_{n=0}^{k-1} u_n, \sum_{n=0}^{k-1} u_n\right) - u_0 - \sum_{p=1}^{k}\sum_{q=0}^{p-1} B(u_q, u_{p-1-q})$$
$$= \sum_{q=1}^{k-1}\sum_{p=k-q}^{k-1} B(u_p, u_q)$$

and

$$\|B_k\|_E \leq C_0 \sum_{q=1}^{k-1} \sum_{p=k-q}^{k-1} \|u_p\|_E \|u_q\|_E.$$

Thus, we get

$$\|v_{k+1} - \sum_{n=0}^{k} u_n\|_E \leq C_0 \left(\left(\sum_{n=0}^{\infty} \|u_n\|_E \right)^2 - \left(\sum_{n=0}^{k-1} \|u_n\|_E \right)^2 \right)$$

$$\leq 2C_0 \left(\sum_{n=0}^{\infty} \|u_n\|_E \right) \left(\sum_{n=k+1}^{\infty} \|u_n\|_E \right)$$

$$= \sum_{n=k+1}^{\infty} \|u_n\|_E$$

Thus, v_n converges to u and u is the unique fixed point of F in B_0. □

We may again interpret the proof through Picard's iterates in terms of the formal expansion in words of the form $w_1 \circledast w_2$. Indeed, the set W of words generated from u_0 through repeated combination of words may be described inductively: starting from the set $V_0 = \{u_0\}$, we describe inductively the set V_n by: $w \in W$ belongs to V_{n+1} if either $w = u_0$ or there exists two words w_1 and w_2 in V_n such that $w = w_1 \circledast w_2$. Then we easily check that $V_n \subset V_{n+1}$ and that $W = \cup_{n \in \mathbb{N}} V_n$. If $C_0 \|u_0\|_E \leq \frac{1}{4}$, we know that $\sum_{w \in W} \|w\|_E < +\infty$. Thus, we have, writing $v_n = \sum_{w \in V_n} w$ and $u = \sum_{w \in W} w$, $\lim_{n \to +\infty} \|u - v_n\|_E = 0$. As we have

$$v_{n+1} = u_0 + \sum_{(w_1,w_2) \in V_n \times V_n} w_1 \circledast w_2 = u_0 + \left(\sum_{w_1 \in V_n} w_1 \right) \circledast \left(\sum_{w_2 \in V_n} w_2 \right) = u_0 + B(v_n, v_n),$$

we find that v_n is the n-th iterate of Picard starting from $v_0 = u_0$ (for the function $v \mapsto F(v) = u_0 + v \circledast v$).

The method used by Oseen for solving a non-linear equation is based on a power series development, hence on analyticity. Combining with the method of analytic majorization introduced by Cauchy, we have obtained uniqueness in the limit case $\|u_0\| = \frac{1}{2C_0}$.

Oseen's method works in a more general setting than quadratic equations. In order to solve a non-linear equation $y = y_0 + N(y)$ in a (complete) vector space, Oseen's method solves more generally $y_\epsilon = y_0 + \epsilon N(y_\epsilon)$. If y_ϵ is analytical with respect to ϵ, we search for a Taylor development $y_\epsilon = \sum_{n=0}^{\infty} \epsilon^n y_n$, where $y_k = \frac{1}{k!} \frac{d^k}{d\epsilon^k} \left(\sum_{n=0}^{\infty} \epsilon^n y_n \right)_{|\epsilon=0}$. If N is analytical with respect to y, we will have a Taylor series for $N(\sum_{n=0}^{\infty} \epsilon^n y_n)$ given by $N(\sum_{n=0}^{\infty} \epsilon^n y_n) = \sum_{n=0}^{\infty} \epsilon^n A_n$ with $A_k = \frac{1}{k!} \frac{d^k}{d\epsilon^k} (N(\sum_{n=0}^{\infty} \epsilon^n y_n))_{|\epsilon=0} = \frac{1}{k!} \frac{d^k}{d\epsilon^k} \left(N(\sum_{n=0}^{k} \epsilon^n y_n) \right)_{|\epsilon=0}$. Thus, we find that

$$y_{k+1} = \frac{1}{k!} \frac{d^k}{d\epsilon^k} \left(N(\sum_{n=0}^{k} \epsilon^n y_n) \right)_{|\epsilon=0}. \tag{5.5}$$

Hence, we get a cascade of equations. The problem is then to solve the equations and to show that the radius of convergence is greater than 1[1].

---

[1] The old method used by Oseen has known modern developments in numerical analysis, where it is known as Adomian's decomposition method, a method introduced by Adomian in the eighties in George Adomian, *Solving Frontier problems of Physics: The decomposition method*, Kluwer Academic Publishers, 1994.

We have followed Oseen's method in Section 4.6. In modern texts on the Navier–Stokes equations, such as the book of Cannone for instance [81], one uses a less stringent approach: Picard's iteration method. This method was introduced by Picard in 1890 [398] for solving PDEs: one starts from $z_0 = y_0$ and one defines inductively z_{k+1} as $z_{k+1} = y_0 + N(z_k)$ and y_{k+1} as $y_{k+1} = z_{k+1} - z_k$; if N is a contraction, the series $\sum_{n=0}^{\infty} y_n$ converges to the solution y. This is known as the Banach contraction principle, as Banach stated the principle in abstract vector spaces in 1922 [16].

5.3 A Capacitary Approach of Quadratic Integral Equations

In this section, we discuss the general integral equation

$$f(x) = f_0(x) + \int_X K(x,y)f^2(y)\,d\mu(y) \tag{5.6}$$

where μ is a non-negative σ-finite measure on a space X ($X = \cup_{n\in\mathbb{N}} Y_n$ with $\mu(Y_n) < +\infty$), and K is a positive measurable function on $X \times X$: $K(x,y) > 0$ almost everywhere. We shall make a stronger assumption on K: there exists a sequence X_n of measurable subsets of X such that $X = \cup_{n\in\mathbb{N}} X_n$ and

$$\int_{X_n}\int_{X_n} \frac{d\mu(x)\,d\mu(y)}{K(x,y)} < +\infty. \tag{5.7}$$

We start with the following easy lemma:

Lemma 5.1.
Let f_0 be non-negative and measurable and let f_n be inductively defined as

$$f_{n+1}(x) = f_0(x) + \int_X K(x,y)f_n^2(y)\,d\mu(y) \tag{5.8}$$

Let $f = \sup_{n\in\mathbb{N}} f_n(x)$. Then either $f = +\infty$ almost everywhere or $f < +\infty$ almost everywhere. If $f < +\infty$, then f is a solution to Equation (5.6).

Proof. Due to the inequalities $f_0 \geq 0$ and $K \geq 0$, we find by induction that $0 \leq f_n$, so that f_{n+1} is well defined (with values in $[0, +\infty]$); we get moreover (by induction, as well) that $f_n \leq f_{n+1}$. We thus may apply the theorem of monotone convergence and get that $f(x) = f_0(x) + \int_X K(x,y)f^2(y)\,d\mu(y)$. If $f = +\infty$ on a set of positive measure, then $\int_X K(x,y)f^2(y)\,d\mu(y) = +\infty$ almost everywhere and $f = +\infty$ almost everywhere. □

We see that if f_0 is such that Equation (5.6) has a solution f which is finite almost everywhere, then we have $f_0 \leq f$ and $\int_X K(x,y)f^2(y)\,d\mu(y) \leq f(x)$. This is almost a characterization of such functions f_0:

Proposition 5.1.
Let C_K be the set of non-negative measurable functions Ω such that $\Omega < +\infty$ (almost everywhere) and $\int_X K(x,y)\Omega^2(y)\,d\mu(y) \leq \Omega(x)$. Then

A) *if $\Omega \in C_K$ and if f_0 is a non-negative measurable function such that $f_0 \leq \frac{1}{4}\Omega$, Equation (5.6) has a solution f which is finite almost everywhere. Moreover, we have $\int_X K(x,y)f_0^2(y)\,d\mu(y) \leq \frac{1}{2}\Omega(x)$.*

B) If $\Omega \in C_K$ and if f_0 is a non-negative measurable function such that

$$\int_X K(x,y)f_0^2(y) \, d\mu(y) \le \frac{1}{16}\Omega(x),$$

Equation (5.6) has a solution f which is finite almost everywhere.

Proof. A) Take the sequence of functions $(f_n)_{n\in\mathbb{N}}$ defined in Lemma 5.1. By induction, we see that $f_n \le \frac{1}{2}\Omega$, and thus $f = \sup_n f_n \le \frac{1}{2}\Omega$.

B) Take the sequence of functions $(f_n)_{n\in\mathbb{N}}$ defined in Lemma 5.1. By induction, we see that $f_n \le f_0 + \frac{1}{4}\Omega$, so that

$$\int K(x,y)f_n(y)^2 \, d\mu(y) \le 2\int K(x,y)(f_0(y)^2 + \frac{1}{16}\Omega(y)^2) \, d\mu(y) \le \frac{1}{4}\Omega(x).$$

Thus, $f = \sup_n f_n \le f_0 + \frac{1}{4}\Omega$.

$\square$

This remark leads us to define a Banach space of measurable functions in which it is natural to solve Equation (5.6):

Proposition 5.2.
Let $\mathcal{E}_K$ be the space of measurable functions f on X such that there exists $\lambda \ge 0$ and $\Omega \in C_K$ such that $|f(x)| \le \lambda\Omega$ almost everywhere. Then:

- *$\mathcal{E}_K$ is a linear space.*

- *The function $f \in \mathcal{E}_K \mapsto \|f\|_K = \inf\{\lambda \ / \ \exists\Omega \in C_k \ |f| \le \lambda\Omega\}$ is a semi-norm on $\mathcal{E}_K$.*

- *$\|f\|_K = 0 \Leftrightarrow f = 0$ almost everywhere.*

- *The normed linear space E_K (obtained from $\mathcal{E}_K$ by quotienting with the relationship $f \sim g \Leftrightarrow f = g$ a.e.) is a Banach space.*

- *If $f_0 \in \mathcal{E}_K$ is non-negative and satisfies $\|f_0\|_K < \frac{1}{4}$, then Equation (5.6) has a non-negative solution $f \in \mathcal{E}_K$.*

Proof. Since $t \mapsto t^2$ is a convex function, we find that C_K is a balanced convex set and thus that $\mathcal{E}_K$ is a linear space and $\| \ \|_K$ is a semi-norm on $\mathcal{E}_K$.
Next, we see that, for $\Omega \in C_K$, $p, q \in \mathbb{N}$, we have

$$\int_{Y_p \cap X_q} \Omega(x) \, d\mu(x) \le \frac{\int_{X_q} \int_{X_q} \frac{d\mu(x) \, d\mu(y)}{K(x,y)}}{(\mu(Y_p \cap X_q))^2} \tag{5.9}$$

This is easily checked by writing that

$$\int\int_{(Y_p \cap X_q)^2} \Omega(y) \, d\mu(x) \, d\mu(y) \le$$

$$\sqrt{\int_{X_q} \int_{X_q} \frac{d\mu(x) \, d\mu(y)}{K(x,y)}} \sqrt{\int_{Y_p \cap X_q} [\int K(x,y)\Omega^2(y) \, d\mu(y)] \, d\mu(x)} \tag{5.10}$$

Thus we find that, when $\|f\|_K = 0$, we have $\int \int_{Y_p \cap X_q} |f(x)| \, d\mu(x) = 0$ for all p and q, so that $f = 0$ almost everywhere.

Similarly, we find that if $\lambda_n \geq 0$, $\Omega_n \in C_K$ and $\sum_{n \in \mathbb{N}} \lambda_n = 1$, then, if $\Omega = \sum_{n \in \mathbb{N}} \lambda_n \Omega_n$, we have (by dominated convergence),

$$\int_{Y_p \cap X_q} \Omega(x) \, d\mu(x) \leq \frac{\int_{X_n} \int_{X_n} \frac{d\mu(x) \, d\mu(y)}{K(x,y)}}{(\mu(Y_p \cap X_q))^2} \tag{5.11}$$

so that $\Omega < +\infty$ almost everywhere. Moreover (by dominated convergence), we have $\Omega \in C_K$. From that, we easily get that E_K is complete.

Finally, existence of a solution of (5.6) when $\|f_0\|_K < \frac{1}{4}$ is a consequence of Proposition 5.1. □

Remark: We have obviously (by Proposition 5.1) that, for a measurable function f,

$$f \in E_K \Leftrightarrow |f| \in E_K \Leftrightarrow \int K(x,y) f(y)^2 \, d\mu(y) \in E_K.$$

An easy corollary of Proposition 5.2 is the following one:

Proposition 5.3.
If E is a Banach space of measurable functions such that:

- $f \in E \to |f| \in E$ *and* $\| \, |f| \, \|_E \leq C_E \|f\|_E$
- $\| \int_X K(x,y) f^2(y) \, d\mu(y) \|_E \leq C_E \|f\|_E^2$

then E is continuously embedded into E_K.

Proof. By Theorem 5.1, we know that the equation

$$\Omega = \Omega_0 + \int_X K(x,y) \Omega(y)^2 \, d\mu(y)$$

has a unique solution in E when $\|\Omega_0\|_E \leq \frac{1}{4C_E}$. Moreover, this solution Ω is non-negative if Ω_0 is non-negative (as it is obtained by the series method), and thus $\Omega \in C_K$. Thus, for $f \in E$, $\frac{1}{4C_E^2 \|f\|_E} |f| \in C_K$, and $\|f\|_{E_K} \leq 4C_E^2 \|f\|_E$. □

Now, we recall a result of Kalton and Verbitsky that characterizes the space E_K for a general class of kernels K [250].

Kalton and Verbitsky's theorem

Theorem 5.2.
Assume that the kernel K satisfies:

- $\rho(x,y) = \frac{1}{K(x,y)}$ *is a quasi-metric:*

  1. $\rho(x,y) = \rho(y,x) \geq 0$
  2. $\rho(x,y) = 0 \Leftrightarrow x = y$
  3. $\rho(x,y) \leq \kappa(\rho(x,z) + \rho(z,y))$

- *K satisfies the following inequality: there exists a constant $C > 0$ such that, for all $x \in X$ and all $R > 0$, we have*

$$\int_0^R \int_{\rho(x,y)<t} d\mu(y)\frac{dt}{t^2} \le CR \int_R^{+\infty} \int_{\rho(x,y)<t} d\mu(y)\frac{dt}{t^3} \tag{5.12}$$

Then the following assertions are equivalent for a measurable function f on X:

- *(A) $f \in E_K$*

- *(B) There exists a constant C such that, for all $g \in L^2$, we have*

$$\int_X |f(x)|^2 |\int_X K(x,y)g(y)\, d\mu(y)|^2\, d\mu(x) \le C\|g\|_2^2 \tag{5.13}$$

- *(C) There exists a constant C such that, for almost every x,*

$$\int_X K(x,y)(\int_X K(y,z)f^2(z)d\mu(z))^2\, d\mu(y)) \le C \int_X K(x,y)f^2(y)\, d\mu(y) \tag{5.14}$$

5.4 Generalized Riesz Potentials on Spaces of Homogeneous Type

A direct consequence of Theorem 5.2 concerns generalized Riesz potentials on spaces of homogeneous type.

Definition 5.1.
(X,δ,μ) is a space of homogeneous type if the quasi-metric δ and the measure μ satisfy:

- *for all $x,y \in X$, $\delta(x,y) \ge 0$*

- *$\delta(x,y) = \delta(y,x)$*

- *$\delta(x,y) = 0 \Leftrightarrow x = y$*

- *there is a positive constant κ such that:*

$$\text{for all } x,y,z \in X, \delta(x,y) \le \kappa(\delta(x,z) + \delta(z,y)) \tag{5.15}$$

- *there exists postive A, B and Q which satisfy:*

$$\text{for all } x \in X, \text{ for all } r > 0, Ar^Q \le \int_{\delta(x,y)<r} d\mu(y) \le Br^Q \tag{5.16}$$

Q is the homogeneous dimension of (X,δ,μ).

Riesz potentials

Theorem 5.3.
Let (X,δ,μ) be a space of homogeneous type, with homogeneous dimension Q. Let

$$K_\alpha(x,y) = \frac{1}{\delta(x,y)^{Q-\alpha}} \tag{5.17}$$

(where $0 < \alpha < Q/2$) and E_{K_α} the associated Banach space (defined in Proposition 5.2). Let $\mathcal{I}_\alpha$ be the Riesz operator associated K_α:

$$\mathcal{I}_\alpha f(x) = \int_X K_\alpha(x,y) f(y) \, d\mu(y). \tag{5.18}$$

We define two further linear spaces associated to K_α:

- *the Sobolev space W^α defined by*

$$g \in W^\alpha \Leftrightarrow \exists h \in L^2 \ \ g = \mathcal{I}_\alpha h \tag{5.19}$$

- *the multiplier space $\mathcal{V}^\alpha$ defined by*

$$f \in \mathcal{V}^\alpha \Leftrightarrow \|f\|_{\mathcal{V}^\alpha} = \Big(\sup_{\|h\|_2 \le 1} \int_X |f(x)|^2 |\mathcal{I}_\alpha h(x)|^2 \, d\mu(x) \Big)^{1/2} < +\infty \tag{5.20}$$

(so that pointwise multiplication by a function in $\mathcal{V}^\alpha$ maps boundedly W^α to L^2). Then, we have (with equivalence of norms) for $0 < \alpha < Q/2$:

$$E_{K_\alpha} = \mathcal{V}^\alpha. \tag{5.21}$$

Proof. It is enough to see that $At^{\frac{Q}{Q-\alpha}} \le \int_{\rho(x,y)<t} d\mu(y) \le Bt^{\frac{Q}{Q-\alpha}}$ (with $\rho(x,y) = \frac{1}{K(x,y)}$) and that $1 < \frac{Q}{Q-\alpha} < 2$, then use Theorem 5.2. $\square$

As $\mathcal{V}^\alpha$ is defined as the space of pointwise multipliers from W^α to L^2, we shall write $\mathcal{V}^\alpha = \mathcal{M}(W^\alpha \mapsto L^2)$. This space of multipliers is not easy to handle: it can be characterized through capacitary inequalities (for the case of Riesz potentials on $\mathbb{R}^n$, this is a theorem of Maz'ya [357]).

The space of multipliers however can be compared to easier spaces, the Morrey–Campanato spaces.

Definition 5.2.
The (homogeneous) Morrey–Campanato space $\dot{M}^{p,q}(X)$ $(1 < p \le q < +\infty)$ is the space of the functions that are locally L^p and satisfy

$$\|f\|_{\dot{M}^{p,q}} = \sup_{x \in X} \sup_{R>0} R^{Q(\frac{1}{q}-\frac{1}{p})} \Big(\int_{|\delta(x,y)|<R} |f(y)|^p \, d\mu(y) \Big)^{1/p} < +\infty. \tag{5.22}$$

Remark that $L^q \subset \dot{M}^{p,q}(X)$, as a direct consequence of Hölder inequality.

We shall need two technical lemmas on Morrey–Campanato spaces. The first lemma deals with the Hardy–Littlewood maximal function:

Lemma 5.2.
Let $\mathcal{M}_f$ be the Hardy–Littlewood maximal function of f:

$$\mathcal{M}_f(x) = \sup_{R>0} \frac{1}{\mu(B(x,R))} \int_{B(x,R)} |f(y)| \, d\mu(y) \tag{5.23}$$

where $B(x,R) = \{y \in X \ / \ \delta(x,y) < R\}$. Then there exist constants C_p and $C_{p,q}$ such that:

- *for every $f \in L^1$ and every $\lambda > 0$,*

$$\mu(\{x \in X \ / \ \mathcal{M}_f(x) > \lambda\}) \le C_1 \frac{\|f\|_1}{\lambda}$$

- *for $1 < p \le +\infty$ and for every $f \in L^p$*

$$\|\mathcal{M}_f\|_p \le C_p \|f\|_p$$

- *for every $1 < p \le q < +\infty$ and for every $f \in \dot{M}^{p,q}(X)$*

$$\|\mathcal{M}_f\|_{\dot{M}^{p,q}} \le C_{p,q} \|f\|_{\dot{M}^{p,q}}$$

Proof. The weak type (1,1) of the Hardy–Littlewood maximal function is a classical result (see Coifman and Weiss [125] for the spaces of homogeneous type). The boundedness of the maximal function on L^p for $1 < p \le +\infty$ is then a direct consequence of the Marcinkiewicz interpolation theorem [215].

Thus, we shall be interested in the proof for $\dot{M}^{p,q}(X)$. Let $f \in \dot{M}^{p,q}(X)$. For $x \in X$ and $R > 0$, we need to estimate $\int_{B(x,R)} |\mathcal{M}_f(y)|^p \, d\mu(y)$. We write $f = f_1 + f_2$, where $f_1(y) = f(y) 1_{B(x,2\kappa R)}(y)$. We have $\mathcal{M}_f \le \mathcal{M}_{f_1} + \mathcal{M}_{f_2}$. We have

$$\int_{B(x,R)} \mathcal{M}_{f_1}(y)^p \, d\mu(y) \le (C_p \|f_1\|_p)^p \le C_p^p \|f\|_{\dot{M}^{p,q}}^p (2\kappa R)^{Q(1-\frac{p}{q})}.$$

On the other hand, for $\delta(x,y) \le R$,

$$\mathcal{M}_{f_2}(y) = \sup_{\rho > R} \frac{1}{\mu(B(y,\rho))} \int_{B(y,\rho)} |f_2(z)| \, d\mu(z) \le \sup_{\rho > R} \frac{1}{A\rho^Q} \|f\|_{\dot{M}^{p,q}} \rho^{Q(1-\frac{1}{q})}$$

so that $1_{B(x,R)} \mathcal{M}_{f_2} \le \frac{\|f\|_{\dot{M}^{p,q}}}{A R^{\frac{1}{q}}}$ and

$$\int_{B(x,R)} \mathcal{M}_{f_2}(y)^p \, d\mu(y) \le \mu(B(x,R)) \|1_{B(x,R)} \mathcal{M}_{f_2}\|_\infty^p \le \frac{B}{A^p} \|f\|_{\dot{M}^{p,q}}^p R^{Q(1-\frac{p}{q})}.$$

□

The second lemma is a pointwise estimate for the Riesz potential, known as the *Hedberg inequality* [3, 232].

Lemma 5.3 (Adams–Hedberg inequality).
If $f \in \dot{M}^{p,q}(X)$ and if $0 < \alpha < \frac{Q}{q}$, then

$$\left| \int_X \frac{1}{\delta(x,y)^{Q-\alpha}} f(y) \, d\mu(y) \right| \le C_{p,q,\alpha} (\mathcal{M}_f(x))^{1-\frac{\alpha q}{Q}} \|f\|_{\dot{M}^{p,q}}^{\frac{\alpha q}{Q}}. \tag{5.24}$$

Proof. Let $R > 0$. We have

$$\left| \int_{\rho(x,y)<R} \frac{f(y)}{\delta(x,y)^{Q-\alpha}} d\mu(y) \right| \le \sum_{j=0}^{+\infty} \int_{\frac{R}{2^{j+1}} \le \rho(x,y) < \frac{R}{2^j}} \frac{|f(y)|}{\delta(x,y)^{Q-\alpha}} \, d\mu(y)$$

$$\le \sum_{j=0}^{+\infty} B 2^{-j\alpha} R^\alpha \frac{1}{\mu(B(x,2^{-j}R))} \int_{B(x,2^{-j}R)} |f(y)| \, d\mu(y)$$

$$\le B \frac{1}{1-2^{-\alpha}} R^\alpha \mathcal{M}_f(x)$$

and

$$|\int_{\rho(x,y)\geq R}\frac{f(y)}{\delta(x,y)^{Q-\alpha}}d\mu(y)|\leq\sum_{j=0}^{+\infty}\int_{2^jR\leq\rho(x,y)<2^{j+1}R}\frac{|f(y)|}{\delta(x,y)^{Q-\alpha}}\,d\mu(y)$$

$$\leq\sum_{j=0}^{+\infty}\frac{1}{(2^jR)^{Q-\alpha}}B^{1-\frac{1}{p}}(2^{j+1}R)^{Q(1-\frac{1}{p})}(2^{j+1}R)^{Q(\frac{1}{p}-\frac{1}{q})}\|f\|_{\dot{M}^{p,q}}$$

$$\leq B^{1-\frac{1}{p}}\frac{2^{Q(1-\frac{1}{q})}}{1-2^{\alpha-\frac{Q}{q}}}R^{\alpha-\frac{Q}{q}}\|f\|_{\dot{M}^{p,q}}$$

We then end the proof by taking $R^{\frac{Q}{q}}=\frac{\|f\|_{\dot{M}^{p,q}}}{M_f(x)}$. $\qquad\square$

As a direct corollary of Lemma 5.3, we get the following result of Adams [2] on Riesz potentials[2]:

Corollary 5.1.
For $0<\alpha<\frac{Q}{q}$, the Riesz potential $\mathcal{I}_\alpha$ is bounded from $\dot{M}^{p,q}(X)$ to $\dot{M}^{\frac{p}{\lambda},\frac{q}{\lambda}}(X)$, with $\lambda=1-\frac{\alpha q}{Q}$.

We may now state the comparison result between spaces of multipliers and Morrey–Campanato spaces, a result which is known as the *Fefferman–Phong inequality* [170]:

Proposition 5.4.
Let $0<\alpha<Q/2$ and $2<p\leq\frac{Q}{\alpha}$. Then we have:

$$\dot{M}^{p,\frac{Q}{\alpha}}(X)\subset\mathcal{V}^\alpha=\mathcal{M}(W^\alpha\mapsto L^2)\subset\dot{M}^{2,\frac{Q}{\alpha}}(X)\tag{5.25}$$

Proof. For $f\in\dot{M}^{p,\frac{Q}{\alpha}}(X)$ and $g\in\dot{M}^{p,\frac{Q}{\alpha}}(X)$, we have $fg\in\dot{M}^{\frac{p}{2},\frac{Q}{2\alpha}}(X)$. We have $p/2>1$ and $\alpha<Q/q$ with $q=\frac{Q}{2\alpha}$, hence, since $\lambda=1-\frac{\alpha q}{Q}=1/2$, $\mathcal{I}_\alpha(fg)\in\dot{M}^{p,\frac{Q}{\alpha}}(X)$. Thus, from Proposition 5.3, we see that $\dot{M}^{p,\frac{Q}{\alpha}}(X)\subset\mathcal{V}^\alpha$.

The embedding $\mathcal{V}^\alpha\subset\dot{M}^{2,\frac{Q}{\alpha}}(X)$ is easy to check. Indeed, if $F=1_{B(x,2\kappa R)}$, we have for $y\in B(x,R)$

$$\mathcal{I}_\alpha F(y)\geq\int_{\rho(z,y)<R}\frac{d\mu(z)}{\rho(z,y)^{Q-\alpha}}\geq\frac{\mu(B(y,R))}{R^{Q-\alpha}}\geq AR^\alpha$$

hence, for $f\in\mathcal{V}^\alpha$,

$$\int_{B(x,R)}|f(y)|^2\,d\mu(y)\leq\frac{\|F\|_2^2}{A^2R^{2\alpha}}\|f\|_{\mathcal{V}^\alpha}^2\leq\frac{B(2\kappa)^Q}{A^2}\|f\|_{\mathcal{V}^\alpha}^2R^{Q-2\alpha}.\qquad\square$$

Remark: The embeddings are strict. For a proof in the case of the Euclidean space, see for instance [318].

For the Navier–Stokes equations, we shall be interested in two examples of Riesz potentials: classical Riesz potentials on the usual Euclidean space $\mathbb{R}^n$ (with $\delta(x,y)=|x-y|=\sqrt{\sum_{i=1}^n|x_i-y_i|^2}$) and parabolic Riesz potentials on $\mathbb{R}\times\mathbb{R}^n$ (with the parabolic [quasi]-distance $\delta_2((t,x),(s,y))=|t-s|^{1/2}+|x-y|$).

---

[2]This is sometimes called the Olsen inequality; see the paper by Olsen on Schrödinger potentials [380].

Riesz potentials on $\mathbb{R}^n$

Proposition 5.5.
In the case of the usual Euclidean space $\mathbb{R}^n$ with $\delta(x,y) = |x-y| = \sqrt{\sum_{i=1}^n |x_i - y_i|^2}$, W^α is the homogeneous Sobolev space $\dot{H}^\alpha$, i.e., the Banach space of tempered distributions such that their Fourier transforms $\hat{f}$ are locally integrable and satisfy $\int |\xi|^{2\alpha} |\hat{f}(\xi)|^2 \, d\xi < +\infty$.
Thus, $\mathcal{V}^\alpha(\mathbb{R}^n) = \mathcal{M}(\dot{H}^\alpha \mapsto L^2)$.

Proof. Just check that the Fourier transform of $\frac{1}{|x|^{n-\alpha}}$ is equal to $c_{\alpha,n} \frac{1}{|\xi|^\alpha}$ for a positive constant $c_{\alpha,n}$. $\qquad\square$

Parabolic Riesz potential on $\mathbb{R} \times \mathbb{R}^n$

Proposition 5.6.
Let δ_α be the parabolic (quasi)-distance

$$\delta_\alpha((t,x),(s,y)) = |t-s|^{1/\alpha} + |x-y| \tag{5.26}$$

on $\mathbb{R} \times \mathbb{R}^n$, where $0 < \alpha$. The associated homogeneous dimension (for the Lebesgue measure) is $Q = n + \alpha$.
For $0 < \beta < \alpha$, we consider the kernel

$$K_{\alpha,\beta}(t-s, x-y) = \frac{1}{\delta_\alpha((t,x),(s,y))^{Q-(\alpha-\beta)}} \tag{5.27}$$

or equivalently

$$K_{\alpha,\beta}(t,x) = \frac{1}{(|t|^{1/\alpha} + |x|)^{n+\beta}} \tag{5.28}$$

For $0 < \alpha - \beta < Q/2$, we consider the associated Banach spaces $W^{\alpha,\beta} = K_{\alpha,\beta} * L^2$ and $\mathcal{V}^{\alpha,\beta} = \mathcal{M}(W^{\alpha,\beta} \mapsto L^2)$.

If, moreover, $\beta < 2$, we define the Banach space $\dot{H}_{t,x}^{1-\frac{\beta}{\alpha},\alpha-\beta}$ of tempered distributions such that their Fourier transforms $\hat{f}$ are locally integrable and satisfy

$$\iint (|\xi|^{\alpha-\beta} + |\tau|^{1-\frac{\beta}{\alpha}})^2 |\hat{f}(\tau,\xi)|^2 \, d\xi \, d\tau < +\infty \tag{5.29}$$

Then (for $\beta < 2$), $\mathcal{V}^{\alpha,\beta} = \mathcal{M}(\dot{H}_{t,x}^{1-\frac{\beta}{\alpha},\alpha-\beta} \mapsto L^2)$.

Proof. We shall use the Landau notation $\Omega(.)$: $F \approx \Omega(G)$ if there are two positive constants c_1 and c_2 such that $c_1 < F/G < c_2$. The proposition will be proved through the following lemma:

Lemma 5.4.
Let $W_{\beta,n}(x)$ be defined as

$$W_{\beta,n}(x) = \frac{1}{(2\pi)^n} \int_{\mathbb{R}^n} e^{-|\xi|^\beta} e^{i\,x \cdot \xi} \, d\xi \tag{5.30}$$

Let $\mathcal{K}_{\alpha,\beta}(t,x)$ be defined on $\mathbb{R} \times \mathbb{R}^n$ as

$$\mathcal{K}_{\alpha,\beta}(t,x) = \frac{1}{|t|^{\frac{n+\beta}{\alpha}}} W_{\beta,n}\left(\frac{x}{|t|^{\frac{1}{\alpha}}}\right) \tag{5.31}$$

Then, for $0 < \beta < 2,:$

$$\mathcal{K}_{\alpha,\beta}(t,x) \approx \Omega(K_{\alpha,\beta}(t,x)). \tag{5.32}$$

Let $M_{\alpha,\beta}(\tau,\xi)$ be the Fourier transform of $\mathcal{K}_{\alpha,\beta}(t,x)$. Then

$$M_{\alpha,\beta}(\tau,\xi) \approx \Omega\left(\frac{1}{|\xi|^{\alpha-\beta} + |\tau|^{1-\frac{\beta}{\alpha}}}\right). \tag{5.33}$$

The first step of the proof is the estimation of $W_{\beta,n}(x)$. When $\beta = 2$, we get the Gaussian function

$$W_{2,n}(x) = \frac{1}{(4\pi)^{n/2}} e^{-\frac{|x|^2}{4}}. \tag{5.34}$$

When $0 < \beta < 2$, we have a subordination of $W_{\beta,n}$ to $W_{2,n}$:

$$W_{\beta,n}(x) = \int_0^{+\infty} \frac{1}{\sigma^{n/2}} W_{2,n}\left(\frac{x}{\sqrt{\sigma}}\right) d\mu_\beta(\sigma) \tag{5.35}$$

where $d\mu_\beta$ is a probability measure on $(0, +\infty)$ [421].

We have the following important result of Blumenthal and Getoor [46]: for $0 < \beta < 2$, there exists a positive constant $c_{\beta,n}$ such that

$$\lim_{|x| \to +\infty} W_{\beta,n}(x)|x|^{n+\beta} = c_{\beta,n}. \tag{5.36}$$

Thus, we have

$$W_{\beta,n}(x) \approx \Omega\left(\frac{1}{(1+|x|)^{n+\beta}}\right). \tag{5.37}$$

Recall that

$$K_{\alpha,\beta}(x,y) = \frac{1}{(|t|^{1/\alpha} + |x|)^{n+\beta}}$$

which may be rewritten as

$$K_{\alpha,\beta}(x,y) = \frac{1}{|t|^{\frac{n+\beta}{\alpha}}} \frac{1}{(1+\frac{|x|}{|t|^{1/\alpha}})^{n+\beta}} \approx \Omega(\mathcal{K}_{\alpha,\beta}(t,x)).$$

We now compute the Fourier transform $M_{\alpha,\beta}(\tau,\xi)$ of $\mathcal{K}_{\alpha,\beta}$ as the Fourier transform in the time variable t of the Fourier transform $N(t,\xi)$ in the space variable x of $\mathcal{K}_{\alpha,\beta}$. We have

$$N(t,\xi) = \frac{1}{|t|^{\frac{\beta}{\alpha}}} e^{-|t|^{\frac{\beta}{\alpha}}|\xi|^\beta} \tag{5.38}$$

so that

$$M_{\alpha,\beta}(\tau,\xi) = C \int_{\mathbb{R}} \frac{1}{|\tau-\eta|^{1-\frac{\beta}{\alpha}}} \frac{1}{|\xi|^\alpha} W_{\frac{\beta}{\alpha},1}\left(\frac{\eta}{|\xi|^\alpha}\right) d\eta \tag{5.39}$$

Thus, we have

$$M_{\alpha,\beta}(\tau,\xi) \approx \Omega\left(\int_{\mathbb{R}} \frac{1}{|\tau-\eta|^{1-\frac{\beta}{\alpha}}} \frac{|\xi|^\beta}{(|\xi|^\alpha + |\eta|)^{1+\frac{\beta}{\alpha}}} d\eta\right). \tag{5.40}$$

We may rewrite that estimate as

$$M_{\alpha,\beta}(\tau,\xi) \approx \Omega\left(\frac{1}{|\xi|^{\alpha-\beta}} A_{\alpha,\beta}\left(\frac{\tau}{|\xi|^\alpha}\right)\right) \tag{5.41}$$

with

$$A_{\alpha,\beta}(\tau) = \int_{\mathbb{R}} \frac{1}{|\tau-\eta|^{1-\frac{\beta}{\alpha}}} \frac{1}{(1+|\eta|)^{1+\frac{\beta}{\alpha}}}\,d\eta. \tag{5.42}$$

Let $G(\tau) = \frac{1}{|\tau|^{1-\frac{\beta}{\alpha}}}$ and $H(\tau) = \frac{1}{(1+|\tau|)^{1+\frac{\beta}{\alpha}}}$, so that $A_{\alpha,\beta} = G * H$. Since $G \in L^1 + L^\infty(\mathbb{R})$ and $H \in L^1 \cap L^\infty(\mathbb{R})$, we have that $H * G$ is continuous, positive and bounded, so that we have, for $|\tau| \le 2$, $A_{\alpha,\beta}(\tau) \approx \Omega(1)$. For $|\tau| > 2$, we write:

- $H * G(\tau) \ge \left(\frac{2}{|\tau|}\right)^{1-\frac{\beta}{\alpha}} \int_{-1}^1 H(\eta)\,d\eta$

- $\int_{-|\tau|/2}^{|\tau|/2} G(\tau-\eta)H(\eta)\,d\eta \le \left(\frac{2}{|\tau|}\right)^{1-\frac{\beta}{\alpha}} \|H\|_1$

- $\int_{|\eta|>|\tau|/2} G(\tau-\eta)H(\eta)\,d\eta \le \int_{|\eta|>|\tau|/2} \frac{1}{|\tau-\eta|^{1-\frac{\beta}{\alpha}}} \frac{1}{|\eta|^{1+\frac{\beta}{\alpha}}}\,d\eta = C\frac{1}{|\tau|} \le C\left(\frac{1}{|\tau|}\right)^{1-\frac{\beta}{\alpha}}$

so that $A_{\alpha,\beta}(\tau) \approx \Omega\left(\frac{1}{|\tau|^{1-\frac{\beta}{\alpha}}}\right)$.

Now, the end of the proof is easy. Using Kalton and Verbitsky's theorem (Theorem Theorem 5.2), we begin with the inequality (5.13): from (5.32), we find that $E_{K_{\alpha,\beta}} = E_{K_{\alpha,\beta}}$ (with equivalence of norms). We now endow $\mathbb{R}\times\mathbb{R}^n$ with the quasi-metric $\tilde{\rho}_{\alpha,\beta}((t,x),(s,y)) = (K_{\alpha,\beta}(t-s,x-y))^{-\frac{1}{n+\beta}}$ and apply again Kalton and Verbitsky's theorem. We find that $\mathcal{V}^{\alpha,\beta} = \mathcal{M}(\tilde{W}^{\alpha,\beta} \mapsto L^2)$ whith $\tilde{W}^{\alpha,\beta} = K_{\alpha,\beta} * L^2$ Taking the Fourier transform in time and space variables, we see that

$$\tilde{W}^{\alpha,\beta} = \mathcal{F}_{t,x}^{-1}\left(\frac{1}{|\xi|^{\alpha-\beta}+|\tau|^{1-\frac{\beta}{\alpha}}}L^2\right) = \dot{H}_{t,x}^{1-\frac{\beta}{\alpha},\alpha-\beta}$$

and thus $\mathcal{V}^{\alpha,\beta} = \mathcal{M}(\dot{H}_{t,x}^{1-\frac{\beta}{\alpha},\alpha-\beta} \mapsto L^2)$. This ends the proof.[3]

5.5 Dominating Functions for the Navier–Stokes Integral Equations

In this section, we are going to solve Equation (5.3) through simple estimates on the associated Picard iterates: we start from $\vec{U}_0 = \vec{U}$ and we define inductively $\vec{U}_{n+1}$ as

$$\vec{U}_{n=1} = \vec{U} - \int_0^t \sum_{j=1}^3 \partial_j O(\nu(t-s)) :: (U_{n,j}\vec{U}_n)\,ds.$$

[3]In the first edition of this book [319], we concluded that $W^{\alpha,\beta} = \dot{H}_{t,x}^{1-\frac{\beta}{\alpha},\alpha-\beta}$, but this seems dubious. The correct statement has been given in [321].

Our starting point is the estimate

$$\left| \int_0^t \sum_{j=1}^3 \partial_j \mathcal{O}(\nu(t-s)) :: \left(V_j \vec{W} \right) ds \right|$$

$$\leq C_0 \int_0^t \int \frac{1}{\nu^2(t-s)^2 + |x-y|^4} |\vec{V}(s,y)|\, |\vec{W}(s,y)|\, ds\, dy.$$

Definition 5.3.
For $0 < T \leq +\infty$, a function $\Omega(t,x)$ belongs to the set $\Gamma_{\nu,T}$ of dominating functions for the Navier–Stokes equations on $(0,T)$ if, for all $0 < t < T$ and all $x \in \mathbb{R}^3$,

$$4C_0 \int_0^t \int \frac{1}{\nu^2(t-s)^2 + |x-y|^4} \Omega^2(s,y)\, ds\, dy \leq \Omega(t,x) \tag{5.43}$$

Similarly, a function $\Omega(t,x)$ belongs to the set Γ_ν of dominating functions for the Navier–Stokes equations if, for all $t \in \mathbb{R}$ and all $x \in \mathbb{R}^3$,

$$4C_0 \iint_{\mathbb{R} \times \mathbb{R}^n} \frac{1}{\nu^2(t-s)^2 + |x-y|^4} \Omega^2(s,y)\, ds\, dy \leq \Omega(t,x) \tag{5.44}$$

Of course, we have $\Gamma_\nu \subset \Gamma_{\nu,+\infty}$.

Dominating function may be used to establish the existence of solutions to the integral Navier–Stokes equations:

Navier–Stokes equations and dominating functions

Theorem 5.4.
If, for all $0 < t < T$ and $x \in \mathbb{R}^3$, $|\vec{U}(t,x)| \leq \Omega(t,x)$ with $\Omega \in \Gamma_{\nu,T}$, then the equation

$$\vec{u} = \vec{U} - \int_0^t \sum_{j=1}^3 \partial_j \mathcal{O}(\nu(t-s)) :: \left(u_j \vec{u} \right) ds$$

has a solution $\vec{u}$ on $(0,T) \times \mathbb{R}^3$ such that $|\vec{u}(t,x)| \leq 2\Omega(t,x)$.

Proof. We define $\Omega_0(t,x)$ as $\Omega_0(t,x) = |\vec{U}(t,x)|$ for $0 < t < T$, and

$$\Omega_{n+1}(t,x) = \Omega_0(t,x) + C_0 \int_0^t \int_{\mathbb{R}^3} \frac{1}{\nu^2(t-s)^2 + |x-y|^4} \Omega_n(s,y)^2\, dy\, ds.$$

By induction on n, we find that $\Omega_n(t,x) \leq 2\Omega(t,x)$. Thus, the non-decreasing sequence $\Omega_n(t,x)$ converge to $\Omega_\infty(t,x)$ which satisfies $\Omega_\infty \leq 2\Omega$ and

$$\Omega_\infty(t,x) = \Omega_0(t,x) + C_0 \int_0^t \int_{\mathbb{R}^3} \frac{1}{\nu^2(t-s)^2 + |x-y|^4} \Omega_\infty(s,y)^2\, dy\, ds.$$

For $\vec{W}_n = \vec{U}_{n+1} - \vec{U}_n$, we have

$$|\vec{W}_{n+1}| = \left| \int_0^t \sum_{j=1}^3 \partial_j \mathcal{O}(\nu(t-s)) :: \left(U_{n,j}\vec{W}_n + W_{n,j}\vec{U}_n + W_{n,j}\vec{W}_n \right) ds \right|$$

$$\leq C_0 \int_0^t \int \frac{1}{\nu^2(t-s)^2 + |x-y|^4} \left(|\vec{W}_n(s,y)|^2 + 2|\vec{U}_n(s,y)| \right) |\vec{W}_n(s,y)|\, ds\, dy$$

By induction on n, we find that $|\vec{W}_n(t,x)| \leq \Omega_{n+1}(t,x) - \Omega_n(t,x)$. Thus, $\vec{U}_n$ converges almost everywhere. $\qquad\qquad\qquad\qquad\qquad\qquad\qquad\qquad\qquad\qquad\qquad\qquad\square$

What we did was just applying Proposition 5.2 to the kernel

$$K_\nu(t-s, x-y) = C_0 1_{t-s>0} \frac{1}{\nu^2(t-s)^2 + |x-y|^4}$$

to solve

$$\Omega_\infty = \Omega_0 + \iint_{(0,T)\times\mathbb{R}^3} K_\nu(t-s, x-y)\Omega_\infty^2(y)\,dy.$$

This proposition associates a Banach space $E_{K_\nu,T}$ to K_ν, and the sufficient condition we find on $\vec{U}$ to get a solution to the Navier–Stokes integral equations on $(0,T)\times\mathbb{R}^3$ is $|\vec{U}| \in E_{K_\nu,T}$ and $\||\vec{U}|\|_{E_{K_\nu,T}} \leq 1/4$. Note that the space $E_{K_\nu,T}$ does not depend on ν, different values of ν give equivalent norms.

Recall that

$$\vec{U}(t,x) = W_{\nu t} * \vec{u}_0 - \int_0^t \sum_{j=1}^3 \partial_j \mathcal{O}(\nu(t-s)) :: \left(\vec{f} * \partial_j G\right) ds$$

so that

$$|\vec{U}(t,x)| \leq |W_{\nu t} * \vec{u}_0| + C_0 \int_0^t \int \frac{1}{\nu^2(t-s)^2 + |x-y|^4} |\vec{f} * \vec{\nabla} G|\,ds\,dy$$

with

$$|\vec{f} * \vec{\nabla} G| = \left(\sum_{i=1}^3 \sum_{j=1}^3 |f_i * \partial_j G|^2\right)^{1/2}.$$

Navier–Stokes equations and $E_{K_\nu,T}$ spaces

Corollary 5.2.

Let

$$\vec{U}(t,x) = W_{\nu t} * \vec{u}_0 - \int_0^t \sum_{j=1}^3 \partial_j \mathcal{O}(\nu(t-s)) :: \left(\vec{f} * \partial_j G\right) ds.$$

If

- $|W_{\nu t} * \vec{u}_0| \in E_{K_\nu,T}$ *and* $\||W_{\nu t} * \vec{u}_0|\|_{E_{K_\nu,T}} \leq \frac{1}{8}$

- $\sqrt{|\vec{f} * \vec{\nabla} G|} \in E_{K_\nu,T}$ *and* $\|\sqrt{|\vec{f} * \vec{\nabla} G|}\|_{E_{K_\nu,T}} \leq \frac{1}{2\sqrt{2}}$

then the equation $\vec{u} = \vec{U} - \int_0^t \sum_{j=1}^3 \partial_j \mathcal{O}(\nu(t-s)) :: (u_j\vec{u})\,ds$ has a solution $\vec{u}$ on $(0,T)\times\mathbb{R}^3$ such that $\||\vec{u}|\|_{E_{K_\nu,T}} \leq \frac{1}{2}$.

Similarly, the kernek $\mathcal{K}_\nu = C_0 \frac{1}{\nu^2(t-s)^2 + |x-y|^4}$ induces a norm $\|\ \|_{\mathcal{K}_\nu}$ on the space $\mathcal{V}^{2,1}(\mathbb{R}\times\mathbb{R}^3) = \mathcal{M}(\dot{H}_{t,x}^{\frac{1}{2},1} \mapsto L^2)$, and we have:

Navier–Stokes equations and the multiplier space

Corollary 5.3. *Let*

$$\vec{U}(t,x) = W_{\nu t} * \vec{u}_0 - \int_0^t \sum_{j=1}^3 \partial_j \mathcal{O}(\nu(t-s)) :: (\vec{f} * \partial_j G) \ ds.$$

Let $0 < T \le +\infty$. *If*

- $1_{0<t<T}|W_{\nu t} * \vec{u}_0| \in \mathcal{V}^{2,1}(\mathbb{R} \times \mathbb{R}^3)$ *and* $\||1_{0<t<T}|W_{\nu t} * \vec{u}_0|\||_{\mathcal{K}_\nu} \le \frac{1}{8}$

- $1_{0<t<T}\sqrt{|\vec{f} * \vec{\nabla} G|} \in \mathcal{V}^{2,1}(\mathbb{R} \times \mathbb{R}^3)$ *and* $\||1_{0<t<T}\sqrt{|\vec{f} * \vec{\nabla} G|}\||_{\mathcal{K}_\nu} \le \frac{1}{2\sqrt{2}}$

then the equation

$$\vec{u} = \vec{U} - \int_0^t \sum_{j=1}^3 \partial_j \mathcal{O}(\nu(t-s)) :: (u_j \vec{u}) \ ds$$

has a solution $\vec{u}$ *on* $(0, +T) \times \mathbb{R}^3$ *such that* $1_{0<t<T}|\vec{u}| \in \mathcal{V}^{2,1}(\mathbb{R} \times \mathbb{R}^3)$ *and* $\||1_{0<t<T}|\vec{u}|\||_{\mathcal{K}_\nu} \le \frac{1}{2}.$

Remark: note that $\vec{U}$ may be computed as

$$\vec{U} = W_{\nu t} * \vec{u}_0 + \int_0^t \sum_{j=1}^3 \partial_j \mathcal{O}(\nu(t-s)) :: \vec{F}_j \ ds \tag{5.45}$$

where $\vec{F}_j$ satisfies

$$\vec{f} = \sum_{j=1}^3 \partial_j \vec{F}_j \tag{5.46}$$

One then replaces estimates on $\sqrt{|\vec{f} * \vec{\nabla} G|}$ by similar estimates on $(\sum_{j=1}^3 |\vec{F}_j|^2)^{1/4}$. Conditions expressed on $\vec{F}_j$ are easier to deal with than for $\vec{f}$.

5.6 Oseen's Theorem and Dominating Functions

In this section, we partly reprove Theorem 4.9 in the light of Theorem 5.4.

Lemma 5.5. *There exists constants* ϵ_0, ϵ_1 *and* ϵ_2 *such that, for all* $t \in \mathbb{R}$, *all* $x \in \mathbb{R}^3$ *and all* $\nu > 0$, *we have*

$$\iint_{\mathbb{R} \times \mathbb{R}^3} \frac{1}{\nu^2(t-s)^2 + |x-y|^4} \frac{1}{(\sqrt{\nu|s|} + |y|)^2} \ dy \ ds \le \epsilon_0 \frac{1}{\nu} \frac{1}{\sqrt{\nu|t|} + |x|} \tag{5.47}$$

$$\int_{\mathbb{R}^3} W_{\nu t}(x-y)\frac{1}{|y|} \ dy \le \epsilon_1 \frac{1}{\sqrt{\nu|t|} + |x|} \tag{5.48}$$

and

$$\int_{\mathbb{R}^3} \frac{\sqrt{\nu|t|}}{(\nu t)^2 + |x - y|^4} \frac{1}{|y|^2} \, dy \leq \epsilon_2 \frac{1}{(\sqrt{\nu|t|} + |x|)^2} \tag{5.49}$$

Proof. See formula (4.30) and Corollary 4.1. □

Combining this lemma with Theorem 5.4, we then find:

Rough data (global existence)

Theorem 5.5. *There exists a constant $\eta_0 > 0$ (which does not depend on ν) such that the function*

$$G_\nu(t, x) = \eta_0 \frac{\nu}{\sqrt{\nu|t|} + |x|}$$

belongs to $\mathcal{V}^{2,1}(\mathbb{R} \times \mathbb{R}^3)$ with $\|G_\nu\|_{\mathcal{K}_\nu} \leq 1$. Moreover, there exists a positive constant η_1 (which does not depend on ν) such that, if $\vec{u}_0$ satisfies:

$$|\vec{u}_0(x)| \leq \eta_1 \frac{\nu}{|x|} \tag{5.50}$$

and $\vec{f} = \sum_{j=1}^3 \partial_j \vec{F}_j$ satisfies

$$|\vec{f}(t, x)| \leq \eta_1 \frac{\nu^2 \sqrt{\nu|t|}}{\nu t^2 + |x|^4} \tag{5.51}$$

or, for $j = 1, \ldots, 3$,

$$|\vec{F}_j| \leq \eta_1 \frac{\nu^2}{\nu|t| + |x|^2} \tag{5.52}$$

then there exists a unique solution $\vec{u}$ of

$$\vec{u} = W_{\nu t} * \vec{u}_0 + \int_0^t \sum_{j=1}^3 \partial_j \mathcal{O}(\nu(t-s)) :: \left(\vec{F}_j - u_j \vec{u}\right) \, ds \tag{5.53}$$

on $[0, +\infty) \times \mathbb{R}^3$ such that:

$$|\vec{u}(t, x)| \leq \frac{1}{2} G_\nu(t, x) \tag{5.54}$$

Remark: Inequality (5.50) may be viewed as a localization of the smallness condition on the Reynolds number of the fluid. If U is the characteristic velocity of the fluid, L the characteristic length and ν the kinematic viscosity, the Reynolds number is $\mathbf{Re} = \frac{UL}{\nu}$; here, we have a condition on the pointwise estimate of $\frac{|\vec{u}_0(x)||x|}{\nu}$.

5.7   Functional Spaces and Multipliers

We shall be interested in this section in the following functional spaces:

Definition 5.4.
$X_{\nu,T}$ *is the space of distributions $u_0 \in \mathcal{S}'(\mathbb{R}^3)$ such that*

$$1_{0<t<T} W_{\nu t} * u_0 \in E_{K_\nu,T}$$

and X_ν is the space of distributions u_0 such that

$$1_{0<t} W_{\nu t} * u_0 \in \mathcal{V}^{1,2}(\mathbb{R} \times \mathbb{R}^3).$$

Remark: The space X_ν has been introduced by Lemarié-Rieusset in 2013 in a conference organized by Warwick University in Venice [320] as a near optimal space for solving the Navier–Stokes equations; the same conclusion has been reached independently by Dao and Nguyen in 2017 [143].

We shall discuss some examples of subspaces E of X_ν, characterized by $u_0 \in E \Leftrightarrow 1_{0<t} W_{\nu t} * u_0 \in \mathcal{E}$, where $\mathcal{E}$ is a subspace of $\mathcal{V}^{1,2}(\mathbb{R} \times \mathbb{R}^3)$.

- **Example 1:** A classical example is the space $L_t^p L_x^q$ with $\frac{2}{p} + \frac{3}{q} = 1$ and $3 < q < +\infty$:

$$\mathcal{E} = L_t^p L_x^q = \{F \text{ Lebesgue measurable}/ \int_{\mathbb{R}} \left(\int_{\mathbb{R}^3} |F(t,x)|^q \, dx\right)^{\frac{p}{q}} dt < +\infty\}.$$

In order to check that $L_t^p L_x^q \subset \mathcal{V}^{1,2}(\mathbb{R} \times \mathbb{R}^3)$, we just write $L_t^p L_x^q \subset \dot{M}^{\min(p,q),5}(\mathbb{R} \times \mathbb{R}^3)$, with $2 < \min(p,q) \leq 5$ and then apply the Fefferman–Phong inequality. Indeed, we have, if $p \leq q$

$$\iint_{[t-R^2,t+R^2] \times B(x,R)} |u(s,y)|^p \, ds \, dy \leq \int_{t-R^2}^{t+R^2} \|u(s,.)\|_q^p |B(x,R)|^{1-\frac{p}{q}} \, ds$$

$$\leq C\|u\|_{L_t^p L_x^q}^p R^{3(1-\frac{p}{q})} = C\|u\|_{L_t^p L_x^q}^p R^{5-p}.$$

If $q \leq p$, we have

$$\iint_{[t-R^2,t+R^2] \times B(x,R)} |u(s,y)|^q \, ds \, dy \leq \int_{t-R^2}^{t+R^2} \|u(s,.)\|_q^q \, ds$$

$$\leq \|u\|_{L_t^p L_x^q}^q (2R^2)^{1-\frac{q}{p}} = C\|u\|_{L_t^p L_x^q}^q R^{5-q}.$$

Solutions in $L^p L^q$ were first described in 1972 by Fabes, Jones and Rivière [168]. The corresponding initial values belong to a homogeneous Besov space [36, 313, 475]:

$$1_{0<t} W_{\nu t} * u_0 \in L_t^p L_x^q \Leftrightarrow u_0 \in \dot{B}_{q,p}^{-\frac{2}{p}}$$

- **Example 2:** The same proof works when one changes the order of integration in t and in x and considers $L_x^q L_t^p$ with $\frac{2}{p} + \frac{3}{q} = 1$ and $3 < q < +\infty$:

$$\mathcal{E} = L_x^q L_t^p = \{F \text{ Lebesgue measurable}/ \int_{\mathbb{R}^3} \left(\int_{\mathbb{R}} |F(t,x)|^p \, dt\right)^{\frac{q}{p}} dx < +\infty\}.$$

The corresponding initial values belong to a homogeneous Triebel–Lizorkin space [36, 475]:

$$1_{0<t} W_{\nu t} * u_0 \in L_x^q L_t^p \Leftrightarrow u_0 \in \dot{F}_{q,p}^{-\frac{2}{p}}$$

- **Example 3:** When $q = 3$, the limiting cases of Example 1 and Example 2 corresponds to $L_t^\infty L_x^3$ and $L_x^3 L_t^\infty$, which we define [313] as

$$L_t^\infty L_x^3 = \{F(t,x) \text{ Lebesgue measurable}/ \ \| \|F(t,x)\|_{L^3(dx)} \|_{L^\infty(dt)} < +\infty\}$$

and

$$L_x^3 L_t^\infty = \{F(t,x) \text{ Lebesgue measurable}/ \ \| \|F(t,x)\|_{L^\infty(dt)} \|_{L^3(dx)} < +\infty\}.$$

We have

$$L_x^3 L_t^\infty \subset L_t^\infty L_x^3 \subset \dot{M}^{3,5}(\mathbb{R} \times \mathbb{R}^3)$$

The corresponding initial values then belong to L^3 :

$$1_{0<t} W_{\nu t} * u_0 \in L_x^3 L_t^\infty \Leftrightarrow 1_{0<t} W_{\nu t} * u_0 \in L_t^\infty L_x^3 \Leftrightarrow u_0 \in L^3$$

This is based on the inequality $|W_{\nu t} * u(x)| \leq \mathcal{M}_{u_0}(x)$, where $\mathcal{M}_{u_0}$ is the Hardy–Littlewood maximal function of u_0. The idea of using the maximal function in order to estimate the integrals in the Navier–Stokes problem goes back to Calderón in 1993 [78].

- **Example 4:** A variation on example 1 is the case of

$$\mathcal{E} = \{F(t,x) \text{ Lebesgue measurable}/ \ \sup_{t\in\mathbb{R}} |t|^{\frac{1}{p}} \|F(t,x)\|_{L^q(dx)} < +\infty\}$$

with $\frac{2}{p} + \frac{3}{q} = 1$ and $3 < q < +\infty$ (and where $\sup_{t\in\mathbb{R}}$ is taken as the essential supremum). Indeed, let $2 < r < \min(p,q)$; we shall see that $\mathcal{E} \subset \dot{M}^{r,5}(\mathbb{R} \times \mathbb{R}^3)$. We just write

$$\iint_{[t-R^2,t+R^2]\times B(x,R)} |u(s,y)|^r \ ds \ dy \leq \int_{t-R^2}^{t+R^2} \|u(s,.)\|_q^r |B(x,R)|^{1-\frac{r}{q}} \ ds$$

$$\leq C \| |s|^{-\frac{r}{p}} \|_{\dot{M}^{1,\frac{p}{r}}} R^{2(1-\frac{r}{p})} (\sup_{s\in\mathbb{R}} |s|^{\frac{1}{p}} \|u\|_q)^r R^{3(1-\frac{r}{q})}$$

$$= C \| |s|^{-\frac{r}{p}} \|_{\dot{M}^{1,\frac{p}{r}}} (\sup_{s\in\mathbb{R}} |s|^{\frac{1}{p}} \|u\|_q)^r R^{5-r}.$$

Solutions in this space $\mathcal{E}$ were first described in 1995 by Cannone [81]. The corresponding initial values belong to a homogeneous Besov space [36, 81, 313, 475]:

$$\sup_{0<t} t^{\frac{1}{p}} \|W_{\nu t} * u_0\|_q < +\infty \Leftrightarrow u_0 \in \dot{B}_{q,\infty}^{-\frac{2}{p}}$$

- **Example 5:** Of course, the same proof works when one changes the order of integration in t and in x and consider the space

$$\mathcal{E} = \{F(t,x) \text{ Lebesgue measurable}/ \ \| \sup_{t\in\mathbb{R}} |t|^{\frac{1}{p}} |F(t,x)| \|_{L^q(dx)} < +\infty\}$$

with $\frac{2}{p} + \frac{3}{q} = 1$ and $3 < q < +\infty$ (and where $\sup_{t\in\mathbb{R}}$ is taken as the essential supremum). Let us consider again $2 < r < \min(p,q)$; we shall see that $\mathcal{E} \subset \dot{M}^{r,5}$

$(\mathbb{R} \times \mathbb{R}^3)$. We just write

$$
\iint_{[t-R^2,t+R^2]\times B(x,R)} |u(s,y)|^r \, ds \, dy
$$

$$
\leq \int_{B(x,r)} \| |s|^{-\frac{r}{p}} \|_{\dot{M}^{1,\frac{p}{r}}} (2R)^{2(1-\frac{r}{p})} (\sup_{s\in\mathbb{R}} |s|^{\frac{1}{p}} |u(s,y)|)^r \, dy
$$

$$
\leq C \| |s|^{-\frac{r}{p}} \|_{\dot{M}^{1,\frac{p}{r}}} R^{2(1-\frac{r}{p})} \| \sup_{s\in\mathbb{R}} |s|^{\frac{1}{p}} |u(s,y)| \|_q^r R^{3(1-\frac{r}{q})}
$$

$$
= C \| |s|^{-\frac{r}{p}} \|_{\dot{M}^{1,\frac{p}{r}}} \| \sup_{s\in\mathbb{R}} |s|^{\frac{1}{p}} |u(s,y)| \|_q^r R^{5-r}.
$$

The corresponding initial values belong to a homogeneous Triebel–Lizorkin space [36, 475]:

$$
\sup_{0<t} t^{\frac{1}{p}} |W_{\nu t} * u_0| \in L^q \Leftrightarrow u_0 \in \dot{F}_{q,\infty}^{-\frac{2}{p}}
$$

- **Example 6:** In example 5, the elements of $\mathcal{E}$ satisfy $|F(t,x)| \leq t^{-\frac{1}{p}} G(x)$, with $G \in L^q$. More generally, one may look at the dominating functions which satisfy $|F(t,x)| = t^\alpha G(x)$ with $0 < \alpha < \frac{1}{2}$. We have

$$
\iint K_\nu(t-s, x-y) s^{-2\alpha} G^2(y) \, ds \, dy \leq
$$

$$
\sup_{t\in\mathbb{R}} \int \frac{C_0}{\nu^2(t-s)^2+1} s^{-2\alpha} \, ds \int \frac{1}{|x-y|^{2+2\alpha}} G^2(y) \, dy
$$

so that we look for a dominating function G for the kernel of the Riesz transform $\mathcal{I}_{1-2\alpha}$, hence $G \in \mathcal{V}^{1-2\alpha}(\mathbb{R}^3)$.

This leads us to consider

$$
\mathcal{E} = \{F(t,x) \text{ Lebesgue measurable}/ \| \sup_{t\in\mathbb{R}} |t|^{\frac{1}{p}} |F(t,x)| \|_{\mathcal{V}^{3/q}(dx)} < +\infty\}
$$

with $\frac{2}{p} + \frac{3}{q} = 1$ and $3 < q < +\infty$ (and where $\sup_{t\in\mathbb{R}}$ is taken as the essential supremum).

The corresponding initial values belong to a homogeneous Triebel–Lizorkin space $\dot{F}_{\mathcal{V}^{3/q},\infty}^{-2/p}$ (based on the multiplier space $\mathcal{V}^{3/q}$), which has not yet been defined in the literature. Besov spaces on multiplier spaces were introduced by Lemarié-Rieusset in 2002 [313]. Besov spaces and Triebel–Lizorkin spaces based on Morrey–Campanato spaces were defined by Kozono and Yamazaki in 1994 [279] and are extensively studied by Sickel, Yang and Yuan in [436].

- **Example 7:** If, in Example 6, we take $p = +\infty$, we are led to consider

$$
\mathcal{E} = \{F(t,x) \text{ Lebesgue measurable}/ \| \sup_{t\in\mathbb{R}} |F(t,x)| \|_{\mathcal{V}^1(dx)} < +\infty\}
$$

(where $\sup_{t\in\mathbb{R}}$ is taken as the essential supremum).
The corresponding initial values then belong to $\mathcal{V}^1(\mathbb{R}^3)$:

$$
\sup_{t>0} |W_{\nu t} * u_0| \in \mathcal{V}^1(\mathbb{R}^3) \Leftrightarrow u_0 \in \mathcal{V}^1(\mathbb{R}^3)
$$

This is based once again on the inequality $|W_{\nu t} * u(x)| \leq \mathcal{M}_{u_0}(x)$, where $\mathcal{M}_{u_0}$ is the Hardy–Littlewood maximal function of u_0. The boundedness of the maximal function on the multiplier space $\mathcal{V}^1$ has been proven by Maz'ya and Verbitsky [358].

- **Example 8:** If we take, in Example 6, $q = +\infty$, we meet a slight disappointment. Indeed, if we look for a dominating function $F(t,x) = H(t)$, we find that

$$\iint K_\nu(t-s, x-y)H^2(s)\, ds\, dy = \int \frac{C_0}{1+|y|^4} \int \frac{1}{\sqrt{\nu|t-s|}} H^2(s)\, ds$$

but the only dominating function H for the kernel $\frac{1}{\sqrt{\nu|t-s|}}$ is the null function.

Thus, we must work on a bounded time interval and look for a dominating function H_T such that:

$$\text{for } 0 < t < T, \int_0^t \frac{1}{\sqrt{\nu|t-s|}} H^2(s)\, ds \leq C_T H(t). \tag{5.55}$$

The inequality (5.55) has a solution

$$H_\nu(t) = \frac{\sqrt{\nu}}{\sqrt{t}\ln(\frac{eT}{t})}.$$

The condition $\sup_{0<t<T} t^{\frac{1}{2}} \ln(\frac{eT}{t})|W_{\nu t} * u_0(x)| < +\infty$ is equivalent to $u_0 \in B_{\infty,\infty}^{-1(\ln)}$, a space close to the Besov space $B_{\infty,\infty}^{-1}$. Such initial values for the Navier–Stokes equations have been studied by Yoneda [510].

- **Example 9:** From all the previous examples, one can see that it is worthwhile to consider the space

$$\mathcal{E} = \dot{M}^{p,5}(\mathbb{R} \times \mathbb{R}^3)$$

with $2 < p \leq 5$.

We thus must characterize the space E of distributions u_0 such that $1_{t>0} W_{\nu t} * u_0$ belong to $\dot{M}^{p,5}(\mathbb{R} \times \mathbb{R}^3)$.

If u_0 belongs to E, then obviously $u_0(x - x_0)$ $(x_0 \in \mathbb{R}^3)$ belongs to E with the same norm, and $\lambda u_0(\lambda x)$ $(\lambda > 0)$ belongs to E with the same norm. Thus, we find that $E \subset \dot{B}_{\infty,\infty}^{-1}$ [313] and $|W_{\nu t} * u_0(x)| \leq C \frac{1}{\sqrt{\nu t}} \|u_0\|_E$. Conversely, if $u_0 \in \dot{B}_{\infty,\infty}^{-1}$, if $R > 0$, if $x \in \mathbb{R}^3$ and if $|t| > 2R^2$, let $Q_R(t,x) = [t - R^2, t + R^2] \times B(x,R)$; we have

$$\iint_{Q_R(t,x)} |1_{s>0} W_{\nu s} * u_0(y)|^p \, ds\, dy \leq C\|u_0\|_{\dot{B}_{\infty,\infty}^{-1}}^p R^3 \int_{[t-R^2, t+R^2]} |1_{s>0} s^{-\frac{p}{2}}|\, ds$$

with $|s| \geq \frac{1}{2}|t| \geq R^2$, hence

$$\iint_{Q_R(t,x)} |1_{s>0} W_{\nu s} * u_0(y)|^p \, ds\, dy \leq C\|u_0\|_{\dot{B}_{\infty,\infty}^{-1}}^p R^{5-p}.$$

On the other hand, if $|t| < 2R^2$, we have $Q_R(t,x) \subset Q_{\sqrt{3}R}(0,x)$. Thus we see that u_0 belongs to E if and only if we have

$$\sup_{t>0, x\in\mathbb{R}^3} \sqrt{t}|W_{\nu t} * u_0(x)| < +\infty \tag{5.56}$$

and

$$\sup_{R>0, x\in\mathbb{R}^3} \frac{1}{R^{5-p}} \iint_{[0,R^2]\times B(x,R)} |W_{\nu s} * u_0(y)|^p \, ds\, dy < +\infty. \tag{5.57}$$

From (5.56) and $p > 2$, we have

$$\int_{R^2}^{+\infty} |W_{\nu s} * u_0(y)|^p \, ds \leq C\|u_0\|_{\dot{B}_{\infty,\infty}^{-1}}^p R^{2-p}.$$

Thus, we have that u_0 belongs to E if and only if we have

$$\sup_{R>0, x \in \mathbb{R}^3} \frac{1}{R^{5-p}} \int_{B(x,R)} \left(\int_0^{+\infty} |W_{\nu s} * u_0(y)|^p \, ds \right) dy < +\infty. \tag{5.58}$$

This is equivalent to the fact that $\left(\int_0^{+\infty} |W_{\nu s} * u_0(y)|^p \, ds \right)^{\frac{1}{p}}$ belongs to $\dot{M}^{p,q}(\mathbb{R}^3)$ with $\frac{2}{p} + \frac{3}{q} = 1$.

Thus, we have obtained: for $2 < p \leq 5$,

$$1_{0<t} W_{\nu t} * u_0 \in \dot{M}^{p,5}(\mathbb{R} \times \mathbb{R}^3) \Leftrightarrow u_0 \in \dot{F}_{\dot{M}^{p,q},p}^{-\frac{2}{p}}$$

with $\frac{2}{p} + \frac{3}{q} = 1$. The space $\dot{F}_{\dot{M}^{p,q},p}^{-\frac{2}{p}}$ is a Triebel–Lizorkin-type space based on Morrey spaces instead of Lebesgue spaces. In the notations of the book [436], this is the space $\dot{F}_{p,p}^{-\frac{2}{p}, \frac{1}{p} - \frac{1}{q}}$.

- **Example 10:** One may try to explore the space E_p of the distributions that satisfy (5.57) with $1 \leq p \leq 2$.

  For $p = 2$, one obtains the space BMO^{-1}. We shall see in Chapter 9 that the Navier–Stokes equations may be solved for a small data in BMO^{-1} (this is the theorem of Koch and Tataru [266]). However, one must use a new tool, using cancellation properties of the convolution kernels that occur in the Oseen tensor, and no longer deal only with absolute values.

  For $p < 2$, we obtain the Besov space $\dot{B}_{\infty,\infty}^{-1}$, for which the formalism of capacitary inequalities is clearly not working (see the cheap Navier–Stokes equation of Montgomery–Smith [369]).

- **Example 11:** If we want to take into account the results of Kozono and Yamazaki [279], we should look for an initial value in $\dot{B}_{\dot{M}^{1,q},\infty}^{-\frac{2}{p}}$ with $\frac{2}{p} + \frac{3}{q} = 1$. This is the larger space in the scale considered by Kozono and Yamazaki. In particular, we have the inclusions $\dot{B}_{q,p}^{-\frac{2}{p}} \subset \dot{B}_{\dot{M}^{1,q},\infty}^{-\frac{2}{p}}$ (Example 1), $\dot{F}_{q,p}^{-\frac{2}{p}} \subset \dot{B}_{\dot{M}^{1,q},\infty}^{-\frac{2}{p}}$ (Example 2), $\dot{B}_{q,\infty}^{-\frac{2}{p}} \subset \dot{B}_{\dot{M}^{1,q},\infty}^{-\frac{2}{p}}$ (Example 4), $\dot{F}_{q,\infty}^{-\frac{2}{p}} \subset \dot{B}_{\dot{M}^{1,q},\infty}^{-\frac{2}{p}}$ (Example 5), $\dot{F}_{V^{3/q},\infty}^{-2/p} \subset \dot{B}_{\dot{M}^{1,q},\infty}^{-\frac{2}{p}}$ (Example 6) and $\dot{F}_{\dot{M}^{p,q},p}^{-\frac{2}{p}} = \dot{F}_{p,p}^{-\frac{2}{p}, \frac{1}{p} - \frac{1}{q}} \subset \dot{B}_{\dot{M}^{1,q},\infty}^{-\frac{2}{p}}$ (Example 9).

The condition $u_0 \in \dot{B}_{\dot{M}^{1,q},\infty}^{-\frac{2}{p}}$ is equivalent to

$$\sup_{t>0} t^{\frac{1}{p}} \|W_{\nu t} * u_0(x)\|_{\dot{M}^{1,q}(dx)} < +\infty \quad \text{and} \quad \sup_{0<t} \sqrt{t} \|W_{\nu t} * u_0(x)\|_{L^\infty(dx)} < +\infty.$$

This gives, for all $0 < \theta < 1$,

$$\sup_{t>0} t^{\frac{\theta}{p} + \frac{1-\theta}{2}} \|W_{\nu t} * u_0(x)\|_{\dot{M}^{\frac{1}{\theta}, \frac{q}{\theta}}} < +\infty.$$

For $\theta < \frac{1}{2}$, we obtain that $1_{t>0} W_{\nu t} * u_0(x)$ belongs to $\dot{M}^{r,5}(\mathbb{R} \times \mathbb{R}^3)$ for $\max(\theta, \frac{\theta}{p} + \frac{1-\theta}{2}) < \frac{1}{r} < \frac{1}{2}$ (with the same proof as Example 4).

- **Example 12:** A variation on example 9 is the case of parabolic Morrey spaces in mixed norms considered by Krylov for the heat equation [282, 281, 322]: defining $Qr(t,x) = (t - r^2, t + r^2) \times B(x,r)$,

$$\mathcal{E} = \{F(t,x) \text{ measurable}/ \sup_{r>0,t>0,x\in\mathbb{R}^3} r^{\frac{2}{q}+3q-1}\|F(t,x)\|_{L_t^p L_x^q(Q_r(t,x))} < +\infty\}$$

or

$$\mathcal{E} = \{F(t,x) \text{ measurable}/ \sup_{r>0,t>0,x\in\mathbb{R}^3} r^{\frac{2}{q}+3q-1}\|F(t,x)\|_{L_x^q L_t^p(Q_r(t,x))} < +\infty\}$$

with $\frac{2}{p} + \frac{3}{q} > 1$ and $2 < p, q < +\infty$. [When $p = q$, we find again the Morrrey space $\dot{M}^{p,5}(\mathbb{R} \times \mathbb{R}^3)$ of example 9.]

Chapter 6

The Differential and the Integral Navier–Stokes Equations

In Chapter 4, we have seen classical solutions of the Navier–Stokes equations: the solution $\vec{u}$ was C^2 in space variable, and C^1 in time variable, and the pressure was C^1 in space variable, so that all the derivatives in the Navier–Stokes equations were classical derivatives. In Chapter 5, we considered measurable solutions of the integral equations derived from the Navier–Stokes equations, and we did not assume any differentiability on the solutions.

In the following chapters, we will study solutions in the sense of distributions of the differential equations, such as Kato's mild solutions or Leray's weak solutions.

In this chapter, we shall discuss the relations between the differential equations and the integral equations, where the pressure has been eliminated through the Leray projection operator. In the absence of external forces, such discussion has been developed by Furioli, Lemarié-Rieusset and Terraneo in [187] (see also Lemarié-Rieusset [313] and Dubois [158])[1]. The elimination of the pressure has also been discussed in 2011 by Tao in [461] and in 2020 by Bradshaw and Tsai [58] and Fernández-Dalgo and Lemarié-Rieusset [174]

6.1 Very Weak Solutions for the Navier–Stokes Equations

We now consider the Navier–Stokes equations:

$$\partial_t \vec{u} = \nu \Delta \vec{u} - (\vec{u}.\vec{\nabla})\vec{u} + \vec{f} - \vec{\nabla}p$$
$$\operatorname{div} \vec{u} = 0 \qquad (6.1)$$
$$\vec{u}_{|t=0} = \vec{u}_0$$

Although they were derived under the assumptions that $\vec{u}$ was a regular vector field with small derivatives, we shall consider solutions in the sense of distributions. In order to relax regularity assumptions on $\vec{u}$, it is better to write the term $(\vec{u}.\vec{\nabla})\vec{u}$ in the form $\operatorname{div}(\vec{u} \otimes \vec{u})$. In order to be able to define $\operatorname{div}(\vec{u} \otimes \vec{u})$ as a distribution on $(0,T) \times \mathbb{R}^3$, we shall assume that $\vec{u}$ is locally square integrable on $(0,T) \times \mathbb{R}^3$.

Another problem is to be able to define the initial value of $\vec{u}$. This will be usually done by an integrability assumption on $\partial_t \vec{u}$ up to time $t = 0$:

Definition 6.1.
Let $T > 0$, $1 \leq p \leq \infty$, and $\sigma \in \mathbb{R}$. The local spaces $(L_t^p H^\sigma)_{\mathrm{loc}}$ are the spaces of distributions $u \in \mathcal{D}'((0,T) \times \mathbb{R}^3)$ such that, for every $0 < T_0 < T$ and every $\varphi \in \mathcal{D}(\mathbb{R}^3)$,

$$\varphi u \in L_t^p((0,T_0), H^\sigma).$$

---

[1]This discussion goes back to the 70's, with, for instance, the paper of Fabes, Jones and Rivière [168] in the case of $L_t^p L_x^q$ mild solutions.

DOI: 10.1201/9781003042594-6

The following lemmas will allow us to define the value of a time-dependent distribution $\vec{u}$ at a given time t:

Lemma 6.1.
Let v be a distribution on $(0,T) \times \mathbb{R}^3$ such that, for some $\sigma \in \mathbb{R}$, $v \in (L_t^1 H^\sigma)_{\mathrm{loc}}$. We define $V = \int_0^t v(s,.)\,ds$. We have the following properties:

- $V \in (L_t^\infty H^\sigma)_{\mathrm{loc}}$.

- $t \in (0,T) \mapsto V(t,.)$ *is continuous from $(0,T)$ to $\mathcal{D}'(\mathbb{R}^3)$, $\lim_{t \to 0} V(t,.) = 0$ in $\mathcal{D}'(\mathbb{R}^3)$.*

- $\partial_t V = v$ *in $\mathcal{D}'((0,T) \times \mathbb{R}^3)$.*

Proof. To define V as a distribution can be done locally: we define φV for $\varphi \in \mathcal{D}(\mathbb{R}^3)$. Let $w = \varphi v$ on $(0,T_0) \times \mathbb{R}^3$. We extend w to $\mathbb{R} \times \mathbb{R}^3$ by defining $w = 0$ if $t < 0$ or $t > T_0$. Then, $w \in L^1(\mathbb{R}, H^\sigma)$. By standard arguments (truncation and regularization), one sees that $\mathcal{D}(\mathbb{R} \times \mathbb{R}^3)$ is dense in $L^1(\mathbb{R}, H^\sigma)$. The map

$$w \in \mathcal{D}(\mathbb{R} \times \mathbb{R}^3) \mapsto F(w) = \int_0^t w(s,.)\,ds \in \mathcal{C}_b(\mathbb{R}, H^1)$$

is bounded for the norm $\|w\|_{L^1 H^1}$. Thus, it can be extended to $L^1 H^1$ with values in $\mathcal{C}_b(\mathbb{R}, H^1)$. As $\mathcal{C}_b(\mathbb{R}, H^1)$ is continuously embedded in $\mathcal{D}'$, we find that $\partial_t F(w) = w$ in $\mathcal{D}'$, as it is obvious if $w \in \mathcal{D}$. $\qquad\square$

Corollary 6.1.
Let u be a distribution on $(0,T) \times \mathbb{R}^3$ such that, for some $\sigma_0, \sigma_1 \in \mathbb{R}$, $u \in (L_t^1 H^{\sigma_0})_{\mathrm{loc}}$ and, $\partial_t u \in (L_t^1 H^{\sigma_1})_{\mathrm{loc}}$. We define $U = \int_0^t \partial_t u(s,.)\,ds$. We have the following properties:

- $U \in (L_t^\infty H^\sigma)_{\mathrm{loc}}$.

- $t \in (0,T) \mapsto U(t,.)$ *is continuous from $(0,T)$ to $\mathcal{D}'(\mathbb{R}^3)$, $\lim_{t \to 0} U(t,.) = 0$ in $\mathcal{D}'(\mathbb{R}^3)$.*

- $\partial_t U = \partial_t u$ *in $\mathcal{D}'((0,T) \times \mathbb{R}^3)$.*

- *There exists a $u_0 \in H_{\mathrm{loc}}^{\min(\sigma_1, \sigma_2)}(\mathbb{R}^3)$ such that $u = U + \mathbb{1} \otimes u_0$*

- *Representing u as $u = U + \mathbb{1} \otimes u_0$, we find that $t \in (0,T) \mapsto u(t,.) = U(t,.) + u_0$ is continuous from $(0,T)$ to $\mathcal{D}'(\mathbb{R}^3)$, and $\lim_{t \to 0} u(t,.) = u_0$ in $\mathcal{D}'(\mathbb{R}^3)$.*

We shall often use a lemma on energy estimates:

Energy estimates

Lemma 6.2.
Let u be a distribution on $(0,T) \times \mathbb{R}^3$ such that $u \in L^2((0,T), H^1(\mathbb{R}^3))$ and $\partial_t u \in L^2((0,T), H^{-1}(\mathbb{R}^3))$. Then u has representant such that $u \in \mathcal{C}([0,T], L^2)$ and

$$\|u(t,.)\|_2^2 = \|u_0\|_2^2 + 2 \int_0^t \langle \partial_t u(s,.) | u(s,.) \rangle_{H^{-1}, H^1}\,ds.$$

Proof. We may extend u to $(-T, T)$ by defining $u(t, x) = u(-t, x)$ for $t < 0$. Then, $u \in L^2((-T, T), H^1(\mathbb{R}^3))$ and $\partial_t u \in L^2((-T, T), H^1(\mathbb{R}^3))$. If $T_0 < T$ and if θ is a smooth function on $\mathbb{R}$ which is compactly supported in $(-T, T)$ and is equal to 1 on a neighborhood of $[-T_0, T_0]$, then $\theta u \in L^2(\mathbb{R}, H^1(\mathbb{R}^3))$ and $\partial_t(\theta u) \in L^2(\mathbb{R}, H^1(\mathbb{R}^3))$. By standard arguments (truncation and regularization), one sees that $\mathcal{D}(\mathbb{R} \times \mathbb{R}^3)$ is dense in the space

$$E = \{u \in L^2(\mathbb{R}, H^1(\mathbb{R}^3)) \ / \ \partial_t(\theta u) \in L^2(\mathbb{R}, H^1(\mathbb{R}^3))\}.$$

We may then define the trace of $u \in E$ for time $t = t_0$ by extending to E the map $u \in \mathcal{D} \mapsto u(t_0, .) \in L^2$:

$$\|u(t_0, .)\|_2^2 = 2 \int_\infty^{t_0} \langle \partial_t u(s, .) | u(s, .) \rangle_{H^{-1}, H^1} \, ds.$$

Similarly, the bilinear form

$$(u, v) \mapsto B(u, v) = \int_{t_0}^{t_1} \langle \partial_t u(s, .) | v(s, .) \rangle_{H^{-1}, H^1} \, ds$$

is bounded on $E \times E$ and $B(u, u) = \|u(t_1, .)\|_2^2 - \|u(t_0, .)\|_2^2$ on $\mathcal{D}$. $\qquad \square$

We then have the concept of a **very weak solution**:

Very weak solution

Definition 6.2.
A very weak solution $\vec{u}$ of equations (6.1) on $(0, T) \times \mathbb{R}^3$, for data $\vec{u}_0 \in \mathcal{D}'(\mathbb{R}^3)$ with $\operatorname{div} \vec{u}_0 = 0$ and $\vec{f} \in \mathcal{D}'((0, T) \times \mathbb{R}^3)$ is a distribution vector field $\vec{u}(t, x) \in \mathcal{D}'((0, T) \times \mathbb{R}^3)$ such that:

- *$\operatorname{div} \vec{u} = 0$*

- *$\vec{u}$ is locally square integrable on $(0, T) \times \mathbb{R}^3$*

- *the map $t \in (0, T) \mapsto \vec{u}(t, .)$ is continuous from $(0, T)$ to $\mathcal{D}'(\mathbb{R}^3)$ and $\lim_{t \to 0^+} \vec{u}(t, .) = \vec{u}_0$*

- *for all $\vec{\varphi} \in \mathcal{D}((0, T) \times \mathbb{R}^3)$ with $\operatorname{div} \vec{\varphi} = 0$, we have*

$$\langle \partial_t \vec{u} - \nu \Delta \vec{u} + \operatorname{div}(\vec{u} \otimes \vec{u}) - \vec{f} | \vec{\varphi} \rangle_{\mathcal{D}', \mathcal{D}} = 0 \qquad (6.2)$$

The term $\vec{\nabla} p$ is implicitly defined by Equation (6.2): if $\vec{H} = -\partial_t \vec{u} + \nu \Delta \vec{u} - \operatorname{div}(\vec{u} \otimes \vec{u}) + \vec{f}$, then (6.2) implies that $\operatorname{curl} \vec{H} = 0$. We then conclude with the following classical lemma:

Lemma 6.3.
If $\vec{H}$ is a time-dependent distribution vector field on $(0, T) \times \mathbb{R}^3$ such that $\operatorname{curl} \vec{H} = 0$, then there exists a distribution $p \in \mathcal{D}'((0, T) \times \mathbb{R}^3)$ such that $\vec{H} = \vec{\nabla} p$.

Proof. Let $\omega \in \mathcal{D}(\mathbb{R})$ with $\int \omega(s) \, ds = 1$. Define h the distribution on $(0, T) \times \mathbb{R}^3$ defined by

$$\langle h | \varphi \rangle_{\mathcal{D}', \mathcal{D}} = \langle H_1 | \int_{x_1}^{+\infty} \left(\varphi(t, y_1, x_2, x_3) - \left(\int_{\mathbb{R}} \varphi(t, z_1, x_2, x_3) \, dz_1 \right) \omega(y_1) \right) dy_1 \rangle_{\mathcal{D}', \mathcal{D}}$$

We have $\partial_1 h = H_1$ Let $\vec{K} = \vec{H} - \vec{\nabla} h$. We have $\operatorname{curl} \vec{K} = 0$ and $K_1 = 0$. Thus $\partial_1 K_2 = \partial_2 K_1 = 0$, and similarly $\partial_1 K_3 = 0$. Thus $\vec{K}$ does not depend on x_1. We now write in a similar way $K_2 = \partial_2 k$, where k does not depend on x_1:

$$\langle k | \varphi \rangle_{\mathcal{D}'\mathcal{D}} = \langle K_2 | \int_{x_2}^{+\infty} \left(\varphi(t, x_1, y_2, x_3) - \left(\int_{\mathbb{R}} \varphi(t, x_1, z_2, x_3)\, dz_2 \right) \omega(y_2) \right) dy_2 \rangle_{\mathcal{D}',\mathcal{D}}$$

Let $\vec{L} = \vec{K} - \vec{\nabla} k$. We have $\operatorname{curl} \vec{L} = 0$ and $L_1 = L_2 = 0$. Thus L depends only on t and x_3. We write $L = \partial_3 l$, where the distribution l depends only on t and x_3, and we conclude by taking $p = h + k + l + q$, where q is any distribution on $(0, T) \times \mathbb{R}^3$ which depends only on t. □

6.2   Heat Equation

As for the case of classical solutions, the study of solutions for the Navier–Stokes equations will be dealt with by studying fisrt the Stokes equations and the heat equation. In this section, we begin with basic lemmas for the heat equation.

We first define general spaces where we shall study the heat equation:

Definition 6.3 (Distribution spaces for the heat equations).
The space $\mathcal{L}^1((0,T) \times \mathbb{R}^3)$ is the space of distributions F on $(0,T) \times \mathbb{R}^3$ that can be written for some $k \in \mathbb{N}_0$ and some $K \in \mathbb{N}_0$ as

$$F = \sum_{|\alpha| \leq k} \partial^\alpha F_\alpha \ \text{with}\ F_\alpha \in L^1((0,T), L^1(\frac{dx}{(1+|x|)^K})).$$

Similarly, the space $\Lambda^1(\mathbb{R}^3)$ is the space of distributions u on $\mathbb{R}^3$ that can be written for some $k \in \mathbb{N}_0$ and some $K \in \mathbb{N}_0$ as

$$u = \sum_{|\alpha| \leq k} \partial^\alpha u_\alpha \ \text{with}\ u_\alpha \in L^1(\frac{dx}{(1+|x|)^K}).$$

The space $\mathcal{L}^\infty((0,T) \times \mathbb{R}^3)$ is the space of distributions F on $(0,T) \times \mathbb{R}^3$ that can be written for some $k \in \mathbb{N}_0$ and some $K \in \mathbb{N}_0$ as

$$F = \sum_{|\alpha| \leq k} \partial^\alpha F_\alpha \ \text{with}\ F_\alpha \in L^\infty((0,T), L^1(\frac{dx}{(1+|x|)^K})).$$

Heat equation

Proposition 6.1.
Let $\nu > 0$, $0 < T < \infty$, $u_0 \in \Lambda^1(\mathbb{R}^3)$ and $f \in \mathcal{L}^1((0,T) \times \mathbb{R}^3)$. Then the equation

$$\begin{cases} \partial_t u = \nu \Delta u + f \\ u_{|t=0} = u_0 \end{cases} \tag{6.3}$$

has a unique solution $u \in \mathcal{L}^{\infty}((0,T) \times \mathbb{R}^3)$. Moreover,

$$u = W_{\nu t} * u_0 + \int_0^t W_{\nu(t-s)} * f(s,.) \, ds. \tag{6.4}$$

Proof. Let us first remark that if

$$u \in \mathcal{L}^{\infty}((0,T) \times \mathbb{R}^3), \ u = \sum_{|\alpha| \le k} \partial^{\alpha} u_{\alpha}, \ u_{\alpha} \in L^{\infty}((0,T), L^1(\frac{dx}{(1+|x|)^K})),$$

$$f \in \mathcal{L}^1((0,T) \times \mathbb{R}^3), \ f = \sum_{|\alpha| \le k} \partial^{\alpha} f_{\alpha}, \ f_{\alpha} \in L^1((0,T), L^1(\frac{dx}{(1+|x|)^K})),$$

and if $\partial_t u = \nu \Delta u + f$ and $T < +\infty$, then $\partial_t u \in (L_t^1 H^{-4-k})_{\mathrm{loc}}$ so that Lemma 6.1 applies and $t \mapsto u(t,.)$ is continuous on $(0,T)$ and has a limit when $t = 0$, and thus the initial value u_0 is well defined.

Existence: Due to the linearity of the equation, we discuss existence of the solution by considering the cases $f = 0$, $u_0 = \partial^{\alpha} u_{0,\alpha}$ and $u_0 = 0$, $f = \partial^{\alpha} f_{\alpha}$.[2]

First, we consider the cases $u_0 \in L^1(\frac{dx}{(1+|x|)^K})$ and $f \in L^1((0,T), L^1(\frac{dx}{(1+|x|)^K}))$. We may, of course, assume that $K \ge 4$. If $v \in L^1(\frac{dx}{(1+|x|)^K})$ and $0 < t < T$, we have

$$\int (\int W_{\nu t}(x-y)|v(y)| \, dy) \frac{dx}{(1+|x|)^K} = \int (\int W_{\nu t}(x-y) \frac{dx}{(1+|x|)^K}) f(y) \, dy$$

with

$$\int W_{\nu t}(x-y) \frac{dx}{(1+|x|)^K} \le C \int \frac{(\nu t)^{(K-3)/2}}{(\sqrt{\nu t} + |x-y|)^K} \frac{dx}{(1+|x|^K} \le C' \frac{1+(\nu t)^{(K-3)/2}}{(1+|y|)^K}.$$

This gives

$$\|W_{\nu t} * u_0\|_{L^1(\frac{dx}{(1+|x|)^K})} \le C(1 + (\nu T)^{(K-3)/2}) \|u_0\|_{L^1(\frac{dx}{(1+|x|)^K})}$$

and

$$\|\int_0^t W_{\nu(t-s)} * f(s,.) \, ds\|_{L^1(\frac{dx}{(1+|x|)^K})} \le C(1 + (\nu T)^{(K-3)/2}) \|f\|_{L^1((0,t),L^1(\frac{dx}{(1+|x|)^K}))}.$$

Thus, $U = W_{\nu t} * u_0$ and $V = \int_0^t W_{\nu(t-s)} * f(s,.) \, ds$ belong to $\mathcal{L}^{\infty}((0,T) \times \mathbb{R}^3)$. Moreover, $\partial_t U = \nu \Delta U$ in $\mathcal{D}'$, $\lim_{t \to 0} U(t,.) = u_0$ and $\partial_t V = \nu \Delta V + f$, $\lim_{t \to 0} V(t,.) = 0$ (check it when $u_0 \in \mathcal{D}(\mathbb{R}^3)$ and $f \in \mathcal{D}((0,T) \times \mathbb{R}^3)$ and conclude by a density argument).

The cases of $u_0 = \partial^{\alpha} u_{0,\alpha}$ and of $f = \partial^{\alpha} f_{\alpha}$ are then straightforward, as $W_{\nu t} * u_0 = \partial^{\alpha}(W_{\nu t} * u_{0,\alpha})$ and

$$\int_0^t W_{\nu(t-s)} * f(s,.) \, ds = \partial^{\alpha} (\int_0^t W_{\nu(t-s)} * f_{\alpha}(s,.) \, ds).$$

---

[2]In the first edition of this book [319], we studied the case $u_0 \in B_{\infty,\infty}^{\alpha}$ anf $f \in L^1 B_{\infty,\infty}^{\alpha}$. Remark that, if $v \in B_{\infty,\infty}^{\alpha}$, then $v = v_1 + \Delta^N v_2$, with $2N > \alpha$, $\int (|v_1| + |v_2|) \frac{dx}{(1+|x|)^4} \le C\|v\|_{B_{\infty,\infty}^{\alpha}}$.

Uniqueness: Let $u \in \mathcal{L}^{\infty}((0,T) \times \mathbb{R}^3)$ be such that $\partial_t u = \nu \Delta u$ and $u_{|t=0} = 0$. Let $\varphi \in \mathcal{D}(\mathbb{R}^3)$, and let $v = e^{-|x|}(\varphi * u)$. We have $v \in L^2((0,T), H^1)$ and $\partial_t v \in L^2((0,T), H^{-1})$, so that (by Lemma 6.2) we may write

$$\|v(t, .)\|_2^2 = \|v(0, .)\|_2^2 + 2 \int_0^t \int v \, \partial_t v \, dx \, ds.$$

We have $v(0, .) = 0$ by the assumption $u(0, .) = 0$. Moreover, we have

$$\int e^{-2|x|}(\varphi * u)\nu \Delta(\varphi * u) \, dx$$

$$= -\nu \int e^{-2|x|} |\vec{\nabla}(\varphi * u)|^2 \, dx + 2\nu \int e^{-2|x|} (\varphi * u) \frac{x}{|x|} \cdot \vec{\nabla}(\varphi * u)|^2 \, dx$$

$$\leq \nu \int e^{-2|x|} |\varphi * u|^2 \, dx.$$

We have

$$\|v(t, .)\|_2^2 \leq 2\nu \int_0^t \|v(s, .)\|_2^2 \, ds$$

and thus $v = 0$. As $\varphi * u = 0$ for all $\varphi \in \mathcal{D}$, we find that $u = 0$ (taking φ an approximation of identity). □

Corollary 6.2.
Let $u_0 \in \Lambda^1(\mathbb{R}^3)$ and $f \in \mathcal{L}^1((0,T) \times \mathbb{R}^3)$. Let $u \in \mathcal{L}^{\infty}((0,T) \times \mathbb{R}^3)$ be the solution of

$$\begin{cases} \partial_t u = \nu \Delta u + f \\ u_{|t=0} = u_0 \end{cases}$$

If $\operatorname{div} \vec{u}_0 = 0$ *and* $\operatorname{div} \vec{f} = 0$, *then* $\operatorname{div} \vec{u} = 0$.

Proof. If $v = \operatorname{div} \vec{u}$, we have $v \in \mathcal{L}^{\infty}((0,T) \times \mathbb{R}^3)$, $\partial_t v = \nu \Delta v$ and $v_{|t=0} = 0$. Thus, $v = 0$. □

6.3   The Leray Projection Operator

Assume that $\vec{u}$ is a very weak solution of

$$\partial_t \vec{u} = \nu \Delta \vec{u} - \operatorname{div}(\vec{u} \otimes \vec{u}) + \vec{f} - \vec{\nabla}p$$

$$\operatorname{div} \vec{u} = 0 \qquad\qquad (6.5)$$

$$\vec{u}_{|t=0} = \vec{u}_0$$

The vector field $\vec{f} - \operatorname{div}\vec{u} \otimes \vec{u}$ is decomposed into the sum of a divergence-free vector field $\partial_t \vec{u} - \Delta \vec{u}$ and a curl-free vector field $\vec{\nabla}p$.

The decomposition of a vector field $\vec{F}_0$ into $\vec{F} = \vec{F} + \vec{H}$, where $\vec{F}$ is solenoidal (i.e. divergence free) and $\vec{H}$ is irrotational (i.e. curl free) is not unique: if ψ is harmonic (i.e. $\Delta \psi = 0$), then we have another decomposition $\vec{F}_0 = (\vec{F} - \vec{\nabla}\psi) + (\vec{H} + \vec{\nabla}\psi)$. To exclude harmonic corrrections, we may require, if possible, that $\vec{H}$ be equal to 0 at infinity, in which case $\vec{F}$ will be called the Leray projection of $\vec{F}_0$:

Leray projection operator

Definition 6.4.
Let $\vec{F}_0$ be a distribution vector field on $\mathbb{R}^3$ such that $\vec{F}_0 \in \mathcal{S}'$. If there exists a vector field $\vec{H} \in \mathcal{S}'$ such that

- *$\vec{H}$ is curl free: $\vec{\nabla} \wedge \vec{H} = 0$*

- *$\vec{F}_0 - \vec{H}$ is divergence free: $\operatorname{div} \vec{H} = \operatorname{div} \vec{F}_0$*

- *$\lim_{t \to +\infty} e^{t\Delta} \vec{H} = 0$ in $\mathcal{S}'$*

*then $\vec{H}$ is unique and $\vec{F} = \vec{F}_0 - \vec{H}$ is called the **Leray projection** of $\vec{F}_0$. We shall write $\vec{F} = \mathbb{P}\vec{F}_0$.*

Uniqueness of $\vec{H}$ is easily checked: if $\vec{\nabla} \wedge \vec{H} = 0$ and $\operatorname{div} \vec{H} = 0$, then $\Delta \vec{H} = \vec{\nabla} \operatorname{div} \vec{H} - \vec{\nabla} \wedge (\vec{\nabla} \wedge \vec{H}) = 0$, so that $\vec{H}$ is harmonic; if moreover $\vec{H} \in \mathcal{S}'$, then the support of the Fourier transform of $\vec{H}$ is included in $\{0\}$ so that $\vec{H}$ is a polynomial; if moreover $\lim_{t \to +\infty} e^{t\Delta} \vec{H} = 0$ in $\mathcal{S}'$, then $H = 0$. We have straightforward examples of Leray projections:

- Obviously, if $\vec{F}_0 \in \mathcal{S}'$, then if $\operatorname{div} \vec{F}_0 = 0$, we have $\mathbb{P}(\vec{F}_0) = \vec{F}_0$.

- Let L^2_σ be the space of divergence-free square integrable vector fields. If $\vec{F}_0 \in L^2$, then $\mathbb{P}\vec{F}_0$ is the orthogonal projection of $\vec{F}_0$ on L^2_σ.

- in Theorem 4.4, we described the Leray projection of a regular and localized vector field $\vec{F}_0$

- if $\vec{F}_0$ is compactly supported, then we may use the Green function $G = \frac{1}{4\pi|x|}$ which is a fundamental solution of $-\Delta G = \delta$ and define $\mathbb{P}\vec{F}_0$ as

$$\mathbb{P}\vec{F}_0 = \vec{F}_0 + \vec{\nabla} G * \operatorname{div} \vec{F}_0.$$

Outside of the support of $\vec{F}_0$, $\vec{\nabla} G * \operatorname{div} \vec{F}_0$ is a smooth function that is $O(|x|^{-3})$ at infinity.

Since

$$\vec{\nabla} \wedge (\vec{\nabla} \wedge \vec{F}_0) = \vec{\nabla} \wedge (\vec{\nabla} \wedge \mathbb{P}\vec{F}_0) = -\Delta \mathbb{P}\vec{F}_0,$$

we find that, formally, we have

$$\mathbb{P}\vec{F}_0 = \frac{\vec{\nabla}}{\sqrt{-\Delta}} \wedge \left(\frac{\vec{\nabla}}{\sqrt{-\Delta}} \wedge \vec{F}_0\right).$$

This gives a direct way to compute $\mathbb{P}\vec{F}_0$:

Proposition 6.2.
Let $E \subset \mathcal{S}'$ be a Banach space of distributions such that:

- the Riesz transforms $\frac{\partial_j}{\sqrt{-\Delta}}$ operate boundedly on E

- the elements of E vanish at infinity: for $f \in E$, $\lim_{t \to +\infty} e^{t\Delta} f = 0$ in $\mathcal{S}'$. Then, if

$$\vec{\mathcal{R}} = \frac{\vec{\nabla}}{\sqrt{-\Delta}}$$

is the vectorial operator defined by the Riesz transforms, and if $\vec{F}_0$ is vector field with components in E, then the Leray projection of $\vec{F}_0$ is well-defined and we have

$$\mathbb{P}\vec{F}_0 = \vec{\mathcal{R}} \wedge (\vec{\mathcal{R}} \wedge \vec{F}_0).$$

An example of such Banach space E is the Besov-like space $\dot{B}^{0,\alpha}$ defined by the Littlewood-Paley decomposition[3] as

$$f \in \dot{B}^{1,\alpha} \Leftrightarrow f \in \mathcal{S}', f = \sum_{j \in \mathbb{Z}} \Delta_j f \text{ in } \mathcal{S}', \sum_{j \in \mathbb{Z}} \min(1, 2^{\alpha j}) \|\Delta_j f\|_\infty < +\infty.$$

In particular, the Leray projection of $\vec{F}_0$ is well defined if $\vec{F}_0 \in L^p$ $(1 \leq p < +\infty)$, or $\vec{F}_0 \in B^s_{p,q}$ $(s \in \mathbb{R}, 1 \leq p < +\infty, 1 \leq q \leq +\infty)$, or $\vec{F}_0 \in \dot{B}^s_{\infty,q}$ $(s < 0, 1 \leq q \leq +\infty)$ or $\vec{F}_0 \in \dot{B}^0_{\infty,1}$.

We now introduce another class of vector fields for which the Leray projection is well defined:

Proposition 6.3.
Let $k \in \mathbb{N}_0$. Let $\vec{F}_1$ be a distribution vector field on $\mathbb{R}^3$ such that $\vec{F}_1 \in L^1(\frac{1}{(1+|x|)^{3+k}} dx)$. If $\vec{F}_0 = \partial^\alpha \vec{F}_1$ with $|\alpha| = k$, then the Leray projection of $\vec{F}_0$ is well defined and may be computed as

$$\mathbb{P}\vec{F}_0 = \vec{F}_0 + \lim_{R \to +\infty} \vec{\nabla}(\theta_R G) * \text{div } \vec{F}_0$$

where $\theta \in \mathcal{D}$ is equal to 1 on a neighborhood of 0.

Proof. Let us remark that the function $\partial_i \partial_j \partial^\alpha ((1 - \theta_1)G)$ is controlled by

$$|\partial_i \partial_j \partial^\alpha ((1 - \theta_1)G)(x)| \leq C \frac{1}{(1 + |x|)^{3+k}}.$$

---

[3] Recall that the Littlewood–Paley decomposition of a tempered distribution if the equality

$$f = S_N f + \sum_{j=0}^{+\infty} \Delta_j f$$

where $S_N f = \mathcal{F}^{-1}(\varphi(\frac{\xi}{2^N})\hat{f})$ (with $\varphi \in \mathcal{D}$ equal to 1. for $|\xi| < \frac{1}{2}$ and to 0 fot $|\xi| > 1$) and $\Delta_j f = S_{j+1} f - S_j f$. If $S_N f \to 0$ in $\mathcal{S}'$ as $N \to -\infty$, we have the homogeneous Littlewood–Paley decomposition

$$f = \sum_{j \in \mathbb{Z}} \Delta_j f.$$

We have a similar control for $\partial_j \partial_k \partial^\alpha ((1-\theta_R)G)$, with $R > 1$,:

$$|\partial_i \partial_j \partial^\alpha ((1-\theta_R)G)(x)| \le C \mathbb{1}_{\{|x|>\gamma R\}} \frac{1}{(1+|x|)^{3+k}}$$

(where C and $\gamma > 0$ don't depend on R). Moreover, we have

$$\int \frac{1}{(1+|x|)^{3+k}} \frac{1}{(1+|x-y|)^{3+k}} \, dy \le C \frac{1}{(1+|y|)^{3+k}}$$

if $k \ge 1$, while, if $\epsilon > 0$,

$$\int \frac{1}{(1+|x|)^{3+\epsilon}} \frac{1}{(1+|x-y|)^3} \, dy \le C_\epsilon \frac{1}{(1+|y|)^3}.$$

Thus, $\lim_{R \to +\infty} \vec\nabla(\theta_R G) * \operatorname{div} \vec F_0$ is well defined as

$$\lim_{R \to +\infty} \vec\nabla(\theta_R G) * \operatorname{div} \vec F_0 = \vec\nabla(\theta_1 G) * \operatorname{div} \vec F_0 + \sum_{i=1}^{3} \partial_i \partial^\alpha \vec\nabla((1-\theta_1)G) * \vec F_{1,i}.$$

We write

$$\operatorname{div} \vec\nabla(\theta_R G) = -\delta + \Delta \theta_R \, G + 2\vec\nabla \theta_R \cdot \vec\nabla G$$

with

$$|\partial_i \partial^\alpha (\Delta \theta_R \, G + 2\vec\nabla \theta_R \cdot \vec\nabla G)(x)| \le \frac{C}{R} \frac{1}{(1+|x|)^{3+k}}.$$

This gives that

$$\lim_{R \to +\infty} (\Delta \theta_R \, G + 2\vec\nabla \theta_R \cdot \vec\nabla G) * \operatorname{div} \vec F_0 = 0$$

and thus

$$\operatorname{div} \lim_{R \to +\infty} \vec\nabla(\theta_R G) * \operatorname{div} \vec F_0 = -\operatorname{div} \vec F_0.$$

Finally, we write

$$e^{t\Delta}\left(\lim_{R \to +\infty} \vec\nabla(\theta_R G) * \operatorname{div} \vec F_0 \right) = \sum_{i=1}^{3} \partial_i \partial^\alpha \vec\nabla e^{t\Delta} G * F_{1,i}$$

with

$$|\partial_i \partial^\alpha \vec\nabla e^{t\Delta} G(x)| \le C \frac{1}{(\sqrt t + |x|)^{3+k}}.$$

By dominated convergence, we find that, for $\epsilon > 0$,

$$\lim_{t \to +\infty} \int \frac{1}{(1+|x|)^{3+k+\epsilon}} |e^{t\Delta}\left(\lim_{R \to +\infty} \vec\nabla(\theta_R G) * \operatorname{div} \vec F_0 \right)| \, dx = 0. \qquad \square$$

Let us remark that many usual spaces on which the Riesz transforms operate are included in $L^1(\frac{1}{(1+|x|)^3} \, dx)$:

- Lebesgue spaces ($E = L^p$ with $1 < p < +\infty$): let $q = \frac{p}{p-1}$; we have $\frac{1}{(1+|x|)^3} \in L^q$, so that $L^p \subset L^1(\frac{1}{(1+|x|)^3} \, dx)$.

- Morrey spaces ($E = \dot{M}^{p,q}$ with $1 < p \le q < +\infty$): we have $\int_{B(0,2^j)} |f(x)|\, dx \le C\|f\|_{\dot{M}^{p,q}} 2^{j(3-\frac{3}{q})}$ and thus

$$\int |f(x)|\frac{1}{(1+|x|)^3}\, dx \le \int_{B(0,1)} |f(x)|\, dx + \sum_{j=0}^{+\infty} 2^{-3j} \int_{B(0,2^{j+1})} |f(x)|\, dx < +\infty.$$

- weighted Lebesgue spaces ($E = L^p(w\, dx)$ with $1 < p < +\infty$ and $w \in \mathcal{A}_p$ (the Muckenhoupt class of weights)): let $q = \frac{p}{p-1}$; we have

$$\int_{B(0,2^j)} |f(x)|\, dx \le \|f\|_{L^p(w\, dx)}\left(\int_{B(0,2^j)} w^{-\frac{q}{p}}\, dx\right)^{1/q}.$$

$w^{-\frac{q}{p}} \in \mathcal{A}_q$, and thus there exists $r < q$ such that $w^{-\frac{q}{p}} \in \mathcal{A}_r$ [448]. Thus, we have

$$\left(\int_{B(0,2^j)} w^{-\frac{q}{p}}\, dx\right)^{\frac{1}{r}}\left(\int_{B(0,2^j)} w^{\frac{q}{p(r-1)}}\, dx\right)^{1-\frac{1}{r}} \le C2^{3j}$$

and we find

$$\int_{B(0,2^j)} |f(x)|\, dx \le \|f\|_{L^p(w\, dx)}\left(\frac{C}{(\int_{B(0,1} w^{\frac{q}{p(r-1)}}\, dx)^{1-\frac{1}{r}}}\right)^{\frac{r}{q}} 2^{3j\frac{r}{q}}$$

and thus

$$\int |f(x)|\frac{1}{(1+|x|)^3}\, dx \le \int_{B(0,1)} |f(x)|\, dx + \sum_{j=0}^{+\infty} 2^{-3j} \int_{B(0,2^{j+1})} |f(x)|\, dx < +\infty.$$

We shall need to express, if possible, the Leray projection of time-dependent distributions. We first define distributions that vanish at infinity:

Definition 6.5.
Let F be a distribution on $(0,T) \times \mathbb{R}^3$. We say that F vanishes at infinity if

- *F belongs to $(L^1 H^\sigma)_{\text{loc}}$ for some $\sigma \in \mathbb{R}$*

- *for every $\theta \in \mathcal{D}((0,T))$, $F_\theta = \int \theta(t)F(t,.)\, dt$ belongs to $\mathcal{S}'$ and $\lim_{\tau \to +\infty} e^{\tau\Delta}F_\theta = 0$ in $\mathcal{S}'$.*

Leray projection operator II

Definition 6.6.
Let $\vec{F}_0$ be a distribution vector field on $(0,T) \times \mathbb{R}^3$. We say that $\vec{F}_0$ admits a Leray projection if there exists a distribution vector field $\vec{F}$ such that,

- *$\vec{F}_0$ and $\vec{F}$ belong to $(L^1 H^\sigma)_{\text{loc}}$ for some $\sigma \in \mathbb{R}$*

- *$\vec{F}_0 - \vec{F}$ vanishes at infinity*

- *for almost every t, $\vec{F}(t,.) = \mathbb{P}(\vec{F}_0(t,.))$*

*Then, $\vec{F}$ is unique and is called the **Leray projection** of $\vec{F}_0$.*
We shall write $\vec{F} = \mathbb{P}\vec{F}_0$.

6.4 Stokes Equations

In this section, we are going to consider the Stokes equations as a preliminary step to the study of Navier–Stokes equations. The Cauchy problem for Stokes equations reads as

$$\begin{cases} \partial_t \vec{u} = & \nu \Delta \vec{u} - \vec{\nabla} p + \vec{f} \\ \operatorname{div} \vec{u} = 0 \\ \vec{u}_{|t=0} = \vec{u}_0 \end{cases} \tag{6.6}$$

We then have a decomposition of $\vec{f}$ into the sum of a divergence free vector field $\partial_t \vec{u} - \Delta \vec{u}$ and a curl free vector field $\vec{\nabla} p$. This suggests, if $\vec{f}$ admits a Leray projection $\mathbb{P}\vec{f}$, to rather study the Stokes equations defined as

$$\begin{cases} \partial_t \vec{u} = & \nu \Delta \vec{u} + \mathbb{P}\vec{f} \\ \operatorname{div} \vec{u} = 0 \\ \vec{u}_{|t=0} = \vec{u}_0 \end{cases} \tag{6.7}$$

Lemma 6.4.
Let $\vec{f}$ admit a Leray projection $\mathbb{P}f$.

(A) If $\vec{u}$ is solution of equations (6.7), then $\vec{u}$ is solution of equations (6.6) with $\vec{\nabla} p = \vec{f} - \mathbb{P}\vec{f}$.

(B) If $\vec{u}$ is solution of equations (6.6) and if $\vec{u}$ and $\vec{f}$ vanish at infinity, then $\vec{u}$ is solution of equations (6.7) and $\vec{\nabla} p = \vec{f} - \mathbb{P}\vec{f}$.

Proof. Let $\vec{u}$ be solution of equations (6.6) and assume that $\vec{u}$ and $\vec{f}$ vanish at infinity. We write $\vec{\nabla} p + \vec{f} = \mathbb{P}\vec{f} - \vec{\nabla} q$, and we want to prove that $\vec{\nabla} q = 0$. We take $\theta \in \mathcal{D}((-\epsilon, \epsilon))$ and consider, for $t \in (\epsilon, T - \epsilon)$,

$$Q_{\theta,t} = \int \theta(t-s)(\mathbb{P}\vec{f}(s,.) + \nu\Delta\vec{u}(s,.) - \partial_t\vec{u}(s,.)) \, ds$$

$$= \int \theta(t-s)\mathbb{P}\vec{f}(s,.) \, ds + \nu\Delta \int \theta(t-s)\vec{u}(s,.) \, ds + \int \theta'(t-s)\vec{u}(s,.) \, ds.$$

We have $\operatorname{div} Q_{\theta,t} = \operatorname{curl} Q_{\theta,t} = 0$, while $Q_{\theta,t} \in \mathcal{S}'$ and $\lim_{\tau \to +\infty} e^{\tau\Delta} Q_{\theta,t} = 0$. Thus, we have $Q_{\theta,t} = 0$. This gives $\theta * \vec{\nabla} q = 0$, and considering an approximation of identity, $\vec{\nabla} q = 0$. □

Thus, equations (6.6) and (6.7) are almost equivalent. We can even simplify equations (6.7) by dropping the requirement that $\vec{u}$ is divergence free, and thus reducing equations (6.7) to the heat equation:

$$\begin{cases} \partial_t \vec{u} = \nu \Delta \vec{u} + \mathbb{P}\vec{f} \\ \\ \vec{u}_{|t=0} = \vec{u}_0 \end{cases} \tag{6.8}$$

Lemma 6.5.
Let $\vec{f}$ admit a Leray projection $\mathbb{P}f \in \mathcal{L}^1((0,T) \times \mathbb{R}^3)$ and let $\vec{u}_0 \in \Lambda^1(\mathbb{R}^3)$ with $\operatorname{div} \vec{u}_0 = 0$. Let $\vec{u} \in \mathcal{L}^\infty((0,T))$ be the solution of the heat equation (6.8). Then $\vec{u}$ is solution of the Stokes equations (6.7).

Proof. This is a consequence of Corollary 6.2.　　　　□

Definition 6.7 (Distribution space for the Stokes equations).
The space $\mathcal{L}_0^1((0,T) \times \mathbb{R}^3)$ is the space of distributions F on $(0,T) \times \mathbb{R}^3$ that can be written for some $k \in \mathbb{N}_0$ and some $\sigma \in \mathbb{R}$ as

$$F = G_\sigma + \sum_{|\alpha| \le k} \partial^\alpha F_\alpha$$

with

$$G_\sigma \in L^1((0,T), \dot{B}^{0,\sigma}) \text{ and } F_\alpha \in L^1((0,T), L^1(\frac{dx}{(1+|x|)^{3+|\alpha|}})).$$

Stokes equations

Proposition 6.4.
Let $\nu > 0$, $0 < T < \infty$. Let

$$\vec{u}_0 \in \Lambda^1(\mathbb{R}^3) \text{ with } \operatorname{div} \vec{u}_0 = 0 \text{ and } \vec{f} \in \mathcal{L}_0^1((0,T) \times \mathbb{R}^3).$$

Then the vector field $\mathbb{P}\vec{f}$ is well defined, $\mathbb{P}\vec{f}$ belongs to $\mathcal{L}^1((0,T) \times \mathbb{R}^3)$ and the equation

$$\begin{cases} \partial_t \vec{u} = \nu \Delta \vec{u} + \mathbb{P}\vec{f} \\ \\ \vec{u}_{|t=0} = \vec{u}_0 \end{cases} \tag{6.9}$$

has a unique solution $\vec{u} \in \mathcal{L}^\infty((0,T) \times \mathbb{R}^3)$. This solution is given by the formula

$$\vec{u} = W_{\nu t} * \vec{u}_0 + \int_0^t W_{\nu(t-s)} * \mathbb{P}\vec{f}(s,.)\,ds. \tag{6.10}$$

Proof. We only need to check that $\mathbb{P}\vec{f}$ is well defined and belongs to $\mathcal{L}^1((0,T) \times \mathbb{R}^3)$. If $\vec{g}_\sigma \in L^1((0,T), \dot{B}^{0,\sigma})$, then $\mathbb{P}\vec{g}_\sigma \in L^1((0,T), \dot{B}^{0,\sigma})$. If $\vec{f}_\alpha \in L^1((0,T), L^1(\frac{dx}{(1+|x|)^{3+|\alpha|}}))$, then

$$\mathbb{P}(\partial^\alpha \vec{f}_\alpha) = \partial^\alpha \vec{f}_\alpha + \sum_{i=1}^3 (\partial_i \partial^\alpha \vec{\nabla}((1-\theta)G)) * \vec{f}_{\alpha,i} + \sum_{i=1}^3 \partial_i \partial^\alpha \vec{\nabla}((\theta G) * \vec{f}_{\alpha,i})$$

where $\theta \in \mathcal{D}(\mathbb{R}^3)$ is equal to 1 on a neighborhood of 0. We have $(\partial_i \partial^\alpha \vec{\nabla}((1-\theta)G)) * \vec{f}_{\alpha,i} \in L^1((0,T), L^1(\frac{dx}{(1+|x|)^{3+|\alpha|}}))$ if $|\alpha| > 0$, $(\partial_i \partial^\alpha \vec{\nabla}((1-\theta)G)) * \vec{f}_{\alpha,i} \in L^1((0,T), L^1(\frac{dx}{(1+|x|)^{4+|\alpha|}}))$ if $|\alpha| = 0$, and $(\theta G) * \vec{f}_{\alpha,i} \in L^1((0,T), L^1(\frac{dx}{(1+|x|)^{3+|\alpha|}}))$. Hence, $\mathbb{P}\vec{f}$ belongs to $\mathcal{L}^1((0,T) \times \mathbb{R}^3)$　　　　□

6.5   Oseen Equations

From Proposition 6.4, we see that we have (almost) equivalence between the differential and the integral formulation of the Navier–Stokes equations in the formalism described by Oseen's tensor, provided that the term $\vec{u} \otimes \vec{u}$ may be defined and integrated[4]:

Oseen's equations

Theorem 6.1.
Let $\nu > 0$, $0 < T < \infty$. Let

$$\vec{u}_0 \in \Lambda^1(\mathbb{R}^3) \text{ with } \operatorname{div} \vec{u}_0 = 0 \text{ and } \vec{f} \in L^1_0((0,T) \times \mathbb{R}^3).$$

Then for a vector field $\vec{u}$ such that $\vec{u} \in L^2((0,T), L^2(\frac{dx}{(1+|x|)^4}))$ the following assertions are equivalent:

- *(A) on $(0,T) \times \mathbb{R}^3$, $\vec{u}$ is a solution of the differential equation*

$$\begin{cases} \partial_t \vec{u} = \nu \Delta \vec{u} + \mathbb{P}(\vec{f} - \operatorname{div}(\vec{u} \otimes \vec{u})) \\ \vec{u}_{|t=0} = \vec{u}_0 \end{cases} \tag{6.11}$$

- *(B) on $(0,T) \times \mathbb{R}^3$, $\vec{u}$ is a solution of the integral equation*

$$\vec{u} = W_{\nu t} * \vec{u}_0 + \int_0^t W_{\nu(t-s)} * \mathbb{P}\vec{f}(s,.) \, ds$$

$$- \int_0^t \sum_{j=1}^3 \partial_j \mathcal{O}(\nu(t-s)) :: (u_j \vec{u}) \, ds \tag{6.12}$$

6.6   Mild Solutions for the Navier–Stokes Equations

Due to Lemma 6.3, we see that a very weak solution is solution in $\mathcal{D}'((0,T) \times \mathbb{R}^3)$ of the problem

$$\partial_t \vec{u} = \nu \Delta \vec{u} - \operatorname{div}(\vec{u} \otimes \vec{u}) + \vec{f} - \vec{\nabla} p$$
$$\operatorname{div} \vec{u} = 0 \tag{6.13}$$
$$\vec{u}_{|t=0} = \vec{u}_0$$

which is a slight modification of Problem (6.1).

Of course, we cannot in general rewrite $\operatorname{div}(\vec{u} \otimes \vec{u})$ as $(\vec{u}.\vec{\nabla})\vec{u}$ for a very weak solution, as the latter expression cannot be defined as a distribution when $\vec{u}$ is only assumed to be locally square integrable. This will be possible if we assume a little amount of differentiability on $\vec{u}$:

---

[4]In the first edition of this book [319], we studied the case $\vec{u} \in (L^2 L^2)_{\text{uloc}}$. Remark that $(L^2 L^2)_{\text{uloc}} \subset L^2((0,T), L^2(\frac{dx}{(1+|x|)^4}))$.

> ## Weakly regular very weak solution
>
> **Definition 6.8.**
> *A very weak solution $\vec{u}$ of equations (6.1) on $(0,T) \times \mathbb{R}^3$ is said to be weakly regular if there exists $p \in [2, \infty)$ and $r \in [2, \infty)$ such that, for every compact subset K of $(0,T) \times \mathbb{R}^3$, we have $1_K(t,x)\, \vec{u} \in L_t^r L_x^p$ and $1_K(t,x)\, \vec{\nabla} \otimes \vec{u} \in L_t^{\frac{r}{r-1}} L_x^{\frac{p}{p-1}}$.*
> *In that case, we have $\operatorname{div}(\vec{u} \otimes \vec{u}) = (\vec{u}.\vec{\nabla})\vec{u}$ in $\mathcal{D}'$.*

Due to Theorem 6.1, we may introduce the class of solutions such that moreover the pressure p is determined by the Leray projection operator:

> ## Oseen solution
>
> **Definition 6.9.**
> *Let $\nu > 0$, $0 < T < \infty$. Let*
>
> $$\vec{u}_0 \in \Lambda^1(\mathbb{R}^3) \text{ with } \operatorname{div} \vec{u}_0 = 0 \text{ and } \vec{f} \in \mathcal{L}_0^1((0,T) \times \mathbb{R}^3).$$
>
> *An Oseen solution $\vec{u}$ of Equations (6.1) on $(0,T) \times \mathbb{R}^3$, for initial value $\vec{u}_0$ and forcing term $\vec{f}$ is a very weak solution $\vec{u}(t,x) \in \mathcal{D}'((0,T) \times \mathbb{R}^3)$ such that moreover:*
>
> - $\vec{u} \in L^2((0,T), L^2(\frac{dx}{(1+|x|)^4}))$
>
> - $\operatorname{div}(\vec{u} \otimes \vec{u}) + \vec{\nabla} p - \vec{f} = \mathbb{P}(\operatorname{div}(\vec{u} \otimes \vec{u}) - \vec{f})$

In the paper by Furioli, Terraneo and Lemarié-Rieusset [186] and the book by Lemarié-Rieusset [313], a criterion was given on $\vec{u}$ and $\vec{f}$ to ensure that a very weak solution is indeed an Oseen solution. Their criterion was stated in terms of uniform local square integrability:

Definition 6.10 (The space $(L^2L^2)_{\text{uloc}}$).
A distribution u on $(0,T) \times \mathbb{R}^3$ is said to be uniformly locally square integrable if u is locally square integrable and if

$$\|u\|_{(L^2L^2)_{\text{uloc}}} = \sup_{x_0 \in \mathbb{R}^3} \sqrt{\int_0^T \int_{B(x_0,1)} |u(s,x)|^2 \, dx \, ds} < +\infty.$$

Proposition 6.5.
Let $\nu > 0$, $0 < T < \infty$. Let

$$\vec{u}_0 \in \Lambda^1(\mathbb{R}^3) \text{ with } \operatorname{div} \vec{u}_0 = 0 \text{ and } \vec{f} \in \mathcal{L}_0^1((0,T) \times \mathbb{R}^3).$$

Let $\vec{u}$ be a very weak solution of equations (6.1) on $(0,T) \times \mathbb{R}^3$. If $\vec{u}$ belongs to $L^2((0,T), L^2(\frac{dx}{(1+|x|)^3}))$ or if $\vec{u}$ belongs to the closure of $\mathcal{D}((0,T) \times \mathbb{R}^3)$ in $(L^2L^2)_{\text{uloc}}$, then the very weak solution $\vec{u}$ is an Oseen solution.

Proof. Just check that $\vec{u}$ vanishes at infinity and apply Lemma 6.4   □

A special case of Oseen solutions are solutions that may be determined by Picard's iteration method: the **mild solutions** that belong to adapted spaces.

Definition 6.11 (Adapted space).
A Banach space E of distribution vector fields on $(0, T) \times \mathbb{R}^3$ is called an adapted space for the Navier–Stokes equations if

- E *is continuously embedded into* $(L^2 L^2)_{\text{uloc}}$, *i.e.*

$$\sup_{x_0 \in \mathbb{R}^3} \int_0^t \int_{B(x_0, 1)} |\vec{f}(s, x)|^2 \, dx \, ds \leq C \|\vec{f}\|_E^2.$$

- *for every $\vec{v}$ and $\vec{w}$ in E, the solution $\vec{z}$ of the Stokes problem*

$$\begin{cases} \partial_t \vec{z} = \nu \Delta \vec{z} + \mathbb{P} \operatorname{div}(\vec{v} \otimes \vec{w}) \\ \\ \vec{z}_{|t=0} = 0 \end{cases}$$

still belongs to E and $\|\vec{z}\|_E \leq C_0 \|\vec{v}\|_E \|\vec{w}\|_E$ for a positive constant C_0 which depends only on E (and ν). Thus the bilinear operator $B_\nu : (\vec{v}, \vec{w}) \mapsto B_\nu(\vec{v}, \vec{w}) = \vec{z}$ is bounded on E.

The operator norm of B_ν will be called the Oseen constant of E (and will be denoted as $\Omega_{E,\nu}$):

$$\Omega_{E,\nu} = \sup_{\|\vec{v}\|_E \leq 1, \|\vec{w}\|/_E \leq 1} \|B_\nu(\vec{v}, \vec{w})\|_E. \tag{6.14}$$

If E is an adapted space, with Oseen constant $\Omega_{E,\nu}$, then a solution $\vec{u}$ of $\vec{u} = \vec{U}_0 - B_\nu(\vec{u}, \vec{u})$, where

$$\vec{U}_0 = W_{\nu t} * \vec{u}_0 + \int_0^t W_{\nu(t-s)} * \mathbb{P}\vec{f}(s, .) \, ds$$

$$B_\nu(\vec{u}, \vec{v}) = \int_0^t \sum_{j=1}^3 \partial_j \mathcal{O}(\nu(t-s)) :: (u_j \vec{v}) \, ds$$

may be found by Picard's iteration method as soon as $\|\vec{U}_0\|_E \leq \frac{1}{4\Omega_{E,\nu}}$ and will be unique in the closed ball $B = \{\vec{u} \in E \mid \|\vec{u}\|_E \leq \frac{1}{2\Omega_{E,\nu}}\}$.

Mild solution

Definition 6.12.
An Oseen solution $\vec{u}$ of Equations (6.1) on $(0, T) \times \mathbb{R}^3$, for initial value $\vec{u}_0$ and forcing term $\vec{f}$ is a mild solution if there exists an adapted Banach space E of distribution vector fields on $(0, T) \times \mathbb{R}^3$ such that:

- $W_{\nu t} * \vec{u}_0 \in E$ *and* $\int_0^t W_{\nu(t-s)} * \mathbb{P}\vec{f}(s, .) \, ds \in E$

- $\|W_{\nu t} * \vec{u}_0 + \int_0^t W_{\nu(t-s)} * \mathbb{P}\vec{f}(s, .) \, ds\|_E \leq \frac{1}{4\Omega_{E,\nu}}$

and if $\vec{u}$ is the unique solution of Equations (6.1) such that $\|\vec{u}\|_E \leq \frac{1}{2\Omega_{E,\nu}}$.

In the book [313], the Navier–Stokes problem with null external force is considered in a large collection of adapted spaces; those mild solutions are smooth on $(0, T) \times \mathbb{R}^3$, due to the regularization properties of the heat kernel. In presence of singular forces, mild solutions might fail to be weakly regular in the sense of Definition 6.8: if $\vec{u}$ is a Landau solution, i.e., a steady solution of $\Delta\vec{u} + \mathbb{P}(\vec{f} - \mathrm{div}(\vec{u} \otimes \vec{u})) = 0$ with $\vec{f} = c_0(0, 0, \delta(x))$ (where δ is the Dirac mass)[5] and if c_0 is small enough, then $\vec{u}$ is a mild solution with initial value $\vec{u}_0 = \vec{u}$ and forcing term $\vec{f}$ (in the adapted space $L_t^\infty L_x^{3,\infty}$) but the product $(\vec{u}.\vec{\nabla})\vec{u}$ is not locally integrable, as it is homogeneous of degree -3. [Let us remark however that $\vec{f} \notin \mathcal{L}_0^1((0, T) \times \mathbb{R}^3).$]

6.7 Suitable Solutions for the Navier–Stokes Equations

Another important class of weak solutions is based on quadratic energy estimates. In that case, we must assume not only that $\vec{u}$ is square integrable, but we must make a similar assumption on $\vec{\nabla} \otimes \vec{u}$.

Weak solution

Definition 6.13.
A weak solution $\vec{u}$ of Equations (6.1) on $(0, T) \times \mathbb{R}^3$ is an Oseen solution such that

- $\vec{u} \in L^\infty((0, T), L^2(\frac{dx}{(1+|x|)^4}))$

- $\vec{\nabla} \otimes \vec{u} \in L^2((0, T), L^2(\frac{dx}{(1+|x|)^4}))$

associated to an initial value $\vec{u}_0$ and to a forcing term $\vec{f}$ such that

- $\vec{u}_0 \in L^2(\frac{dx}{(1+|x|)^4})$ *with* $\mathrm{div}\, \vec{u}_0 = 0$

- $\vec{f} = \mathrm{div}\, \mathbb{F}$, *where the tensor $\mathbb{F}$ is such that $\mathbb{F} \in L^2((0, T), L^2(\frac{dx}{(1+|x|)^4}))$.*

Of course, a weak solution is weakly regular, so that $\mathrm{div}(\vec{u} \otimes \vec{u}) = (\vec{u}.\vec{\nabla})\vec{u}$, but it is not regular enough to grant that $\partial_t(\frac{|\vec{u}|^2}{2}) = \vec{u}.\partial_t\vec{u}$. While $\vec{u}.\Delta\vec{u}$ and $\vec{u}.\vec{f}$ are well defined, we cannot define $\vec{u}.(\vec{u}.\vec{\nabla})\vec{u}$ nor $\vec{u}.\vec{\nabla}p$ as distributions. If $\vec{u}$ and p were regular enough, we could write (since $\mathrm{div}\,\vec{u} = 0$)

$$\partial_t(\frac{|\vec{u}|^2}{2}) = \nu\vec{u}.\Delta\vec{u} - \vec{u}.(\vec{u}.\vec{\nabla})\vec{u} - \vec{u}.\vec{\nabla}p + \vec{u}.\vec{f}$$

$$= \nu\Delta(\frac{|\vec{u}|^2}{2}) - \nu|\vec{\nabla} \otimes \vec{u}|^2 - \mathrm{div}\left((p + \frac{|\vec{u}|^2}{2})\vec{u}\right) + \vec{u}.\vec{f} \tag{6.15}$$

We do not need much regularity to give meaning to the second line of equality (6.15):

Lemma 6.6.
If $\vec{u}$ is a weak solution of Equations (6.1) on $(0, T) \times \mathbb{R}^3$ (in the sense of Definition 6.13), then one can choose p (which is defined up to a time-dependent additive factor $q(t) \in \mathcal{D}'((0, T))$) so that $\vec{u}$ is locally $L_t^4 L_x^3$ and p locally $L_t^2 L_x^{3/2}$ on $(0, T) \times \mathbb{R}^3$.

---

[5]See Theorem 10.13.

Proof. $\vec{u}$ is locally $L_t^\infty L_x^2$ and $L_t^2 H_x^1 \subset L_t^2 L_x^6$, thus locally $L_t^4 L_x^3$. Moreover, we know (from Proposition 6.3) that

$$\vec{\nabla} p = \sum_i \sum_j \vec{\nabla} \partial_i \partial_j (G\theta) * (F_{i,j} - u_i u_j) + \sum_i \sum_j \vec{\nabla} \partial_i \partial_j (G(1-\theta)) * (F_{i,j} - u_i u_j).$$

[Recall that $\theta \in \mathcal{D}$ is equal to 1 on a neighborhood of 0.]

Let $K_{i,j} = \partial_i \partial_j (G(1-\theta))$. We define p as

$$p(t,x) = \sum_i \sum_j \partial_i \partial_j (G\theta) * (F_{i,j} - u_i u_j)$$

$$+ \sum_i \sum_j \int (K_{i,j}(x-y) - K_{i,j}(-y))(F_{i,j}(t,y) - u_i(t,y)u_j(t,y)) \, dy$$

$$= A(t,x) + B(t,x).$$

Assume that $\theta = 1$ on $B(0,1)$ and $\theta = 0$ outside from ball $B(0,2)$. On $(0,T) \times B(0,R)$, we have

$$A(t,x) = \sum_i \sum_j \partial_i \partial_j (G\theta) * (\theta_{4R}(F_{i,j} - u_i u_j))$$

where $\theta_{4R}(x) = \theta(\frac{x}{4R})$. We have, for $x \neq 0$,

$$|\partial_i \partial_j (G\theta)(x)| \leq C \frac{1}{|x|^3}$$

and

$$|\nabla \partial_i \partial_j (G\theta)(x)| \leq C \frac{1}{|x|^4}.$$

Moreover, the Fourier transform of $\partial_i \partial_j (G\theta)$ is bounded, as $\partial_i \partial_j (G\theta) = \theta \partial_i \partial_j G + (\partial_i \theta) \partial_j G + (\partial_j \theta) \partial_i G + (\partial_i \partial_j \theta)G$; the second, the third and the fourth terms are integrable, while the Fourier transform of $\partial_i \partial_j G$ is bounded and the Fourier transform of θ is integrable. Thus, convolution with $\partial_i \partial_j (G\theta)$ is Calderón–Zygmund operator and is bounded on every L^p, $1 < p < +\infty$ [215]. As $\vec{u}$ is locallly $L^4 L3$ and $\mathbb{F}$ is locally $L^2 L^2$, we find that A is $L^2 L^{3/2}$ on $(0,T) \times B(0,R)$.

On the other hand, if $|x| < R$, $R \geq 1$, we have for every y

$$|K_{i,j}(x-y) - K_{i,j}(-y)| \leq 2\|K_{i,j}\|_\infty < +\infty$$

and, when $|y| > 2R$,

$$|K_{i,j}(x-y) - K_{i,j}(-y)| \leq 2 \frac{R}{|y|^4}.$$

As $F_{i,j}$ and $u_i u_j$ belong to $L^2((0,T), L^1(\frac{dx}{(1+|x|)^4}))$, we find that B is in $L_t^2 L_x^\infty$ (hence is $L^2 L^{3/2}$) on $(0,T) \times B(0,R)$. $\qquad\square$

Thus, for a weak solution, we may define the distribution

$$\mu_{\vec{u}} = \nu \Delta\left(\frac{|\vec{u}|^2}{2}\right) - \nu |\vec{\nabla} \otimes \vec{u}|^2 - \text{div}\left((p + \frac{|\vec{u}|^2}{2})\vec{u}\right) + \vec{u}.\vec{f} - \partial_t\left(\frac{|\vec{u}|^2}{2}\right). \tag{6.16}$$

We have a semi-continuity result for the map $\vec{u} \mapsto \mu_{\vec{u}}$:

Convergence of weak solutions

Theorem 6.2.

Let $(\vec{u}_n)_{n\in\mathbb{N}}$ be a sequence of weak solutions of Equations (6.1) on $(0,T)\times\mathbb{R}^3$ (with initial value $\vec{u}_{0,n}$ and forcing term $\vec{f}_n = \operatorname{div} F_n$) such that

- $\sup_{n\in\mathbb{N}}\|\vec{u}_n\|_{L^\infty((0,T),L^2(\frac{dx}{(1+|x|)^4}))} < \infty$

- $\sup_{n\in\mathbb{N}}\|\vec{\nabla}\otimes\vec{u}_n\|_{L^2((0,T),L^2(\frac{dx}{(1+|x|)^4}))} < \infty$

- $\sup_{n\in\mathbb{N}}\|F_n\|_{L^2((0,T),L^2(\frac{dx}{(1+|x|)^4}))} < +\infty$

and assume that $\vec{u}_n$ converges to $\vec{u}$ in $\mathcal{D}'((0,T)\times\mathbb{R}^3)$ and that F_n converges strongly to F in $L^2_{\mathrm{loc}}((0,T)\times\mathbb{R}^3)$. Then:

- $\vec{u}$ *is a weak solution of the Navier–Stokes equations*

- *for every $\phi \in \mathcal{D}((0,T)\times\mathbb{R}^3)$ such that $\phi \geq 0$, we have*

$$\langle\mu_{\vec{u}}|\phi\rangle_{\mathcal{D}',\mathcal{D}} \geq \limsup_{n\to+\infty}\langle\mu_{\vec{u}_n}|\phi\rangle_{\mathcal{D}',\mathcal{D}} \tag{6.17}$$

Proof. If $\Phi \in \mathcal{D}((0,T)\times\mathbb{R}^3)$, we find that $\Phi\,\vec{u}_n$ converges in $\mathcal{D}'$ to $\Phi\,\vec{u}$; moreover, the sequence $\Phi\,\vec{u}_n$ is bounded in $L^2_t H^1_x$ and the sequence $\partial_t(\Phi\,\vec{u}_n)$ is bounded in $L^2 H^{-3/2}_x$ (by Lemma 6.6). Thus, the sequence $\Phi\,\vec{u}_n$ is bounded in the Sobolev space $H^{2/7}(\mathbb{R}\times\mathbb{R}^3)$: just write

$$(1+\tau^2+\xi^2)^{2/7} \leq \left((1+\tau^2)(1+\xi^2)^{-3/2}\right)^{2/7}(1+\xi^2)^{5/7}.$$

Since the functions $\Phi\,\vec{u}_n$ are all supported in the same compact set (the support of Φ), Rellich's theorem gives that $\Phi\vec{u}_n$ converges strongly to $\Phi\vec{u}$ in $L^2_t L^2_x$. Since Φu_n converges weakly in $L^2_t H^1_x$, hence in $L^2_t L^6_x$, we see that it converges strongly in $L^4_t L^3_x$.

Using again Lemma 6.6, we find that p_n converges *-weakly to p. Thus, we find that $\vec{u}$ is solution to the Navier–Stokes equations. Moreover, in the terms defining $\mu_{\vec{u}_n}$, one has the following convergences in $\mathcal{D}'$: $\Delta(|\vec{u}_n|^2) \to \Delta(|\vec{u}|^2)$, $\operatorname{div}\left((p_n + \frac{|\vec{u}_n|^2}{2})\vec{u}_n\right) \to \operatorname{div}\left((p + \frac{|\vec{u}|^2}{2})\vec{u}\right)$, $\vec{u}_n.\vec{f}_n \to \vec{u}.\vec{f}$ and $\partial_t(|\vec{u}_n|^2) \to \partial_t(|\vec{u}|^2)$. The only term for which we do not have convergence is $|\vec{\nabla}\otimes\vec{u}_n|^2$. But if $\Phi \in \mathcal{D}$ is a non-negative function, we have that $\sqrt{\Phi}(\vec{\nabla}\otimes\vec{u}_n)$ converges weakly to $\sqrt{\Phi}(\vec{\nabla}\otimes\vec{u})$ in $L^2_t L^2_x$, hence by the Banach–Steinhaus theorem we get that

$$\iint \Phi|\vec{\nabla}\otimes\vec{u}|^2\,dt\,dx = \|\sqrt{\Phi}(\vec{\nabla}\otimes\vec{u})\|_2^2 \leq \liminf \iint \Phi|\vec{\nabla}\otimes\vec{u}_n|^2\,dt\,dx.$$

The theorem is proved. $\qquad\qquad\qquad\qquad\qquad\qquad\qquad\qquad\qquad\qquad\qquad\qquad\square$

Thus, if we assume that the $\vec{u}_n$ and p_n are regular enough to satisfy equality (6.15) (so that $\mu_{\vec{u}_n} = 0$), we find that the limit $\vec{u}$ satisfies $\mu_{\vec{u}} \geq 0$, i.e., that the distribution $\mu_{\vec{u}}$ is associated to a non-negative locally finite measure $m_{\vec{u}}$:

$$\langle\mu_{\vec{u}}|\phi\rangle_{\mathcal{D}',\mathcal{D}} = \int_{(0,T)\times\mathbb{R}^3}\phi(t,x)\,dm_{\vec{u}}.$$

Such a solution is called a **suitable solution**:

Suitable solution

Definition 6.14.
A weak solution $\vec{u}$ of Equations (6.1) on $(0,T) \times \mathbb{R}^3$ is called a suitable solution if $\mu_{\vec{u}} \geq 0$, i.e., if it satisfies in $\mathcal{D}'$ the local energy inequality

$$\partial_t\left(\frac{|\vec{u}|^2}{2}\right) \leq \nu\Delta\left(\frac{|\vec{u}|^2}{2}\right) - \nu|\vec{\nabla} \otimes \vec{u}|^2 - \operatorname{div}\left((p + \frac{|\vec{u}|^2}{2})\vec{u}\right) + \vec{u}.\vec{f} \qquad (6.18)$$

An interesting case is when we have global estimates: assume that $\vec{u} \in L^\infty_t L^2_x \cap L^2_t H^1_x$ and $\mathbb{F} \in L^2_t L^2_x$ (in which case $p \in L^2_t L^2_x + L^2_t L^{3/2}_x$), and that $\lim_{t\to 0} \|\vec{u}(,t.) - \vec{u}_0\|_2 = 0$. Integrating inequality (6.18) against $\Phi(t,x) = \theta_\epsilon(t)\varphi^2(x/R)$, where $\varphi \in \mathcal{D}(\mathbb{R}^3)$ satisfies $\varphi(x) = 1$ on $B(0,1)$, and where $\theta_\epsilon(t) = \alpha(\frac{t-\epsilon}{\epsilon}) - \alpha(\frac{t-t_0+2\epsilon}{\epsilon})$ with α a smooth non-decreasing function on $\mathbb{R}$ such that $\alpha(s) = 0$ when $s \leq 1$ and $\alpha(s) = 1$ for $s \geq 2$, $R > 0$, $0 < \epsilon < t_0/3$, we find that:

$$-\frac{1}{2}\int_0^T \theta'_\epsilon(t)\|\vec{u}(t,.)\varphi(\frac{.}{R})\|_2^2\,dt \leq \frac{\nu}{2R^2}\int_0^T \theta_\epsilon(t)(\int |\vec{u}(t,x)|^2\Delta(\varphi^2)(\frac{x}{R})\,dx)\,dt$$

$$-\nu\int_0^T \theta_\epsilon(t)\|\varphi(\frac{.}{R})(\vec{\nabla} \otimes \vec{u})\|_2^2\,dt$$

$$+\frac{2}{R}\int_0^T \theta_\epsilon(t)(\int \varphi(\frac{x}{R})(p + \frac{1}{2}|\vec{u}|^2)\vec{u} \cdot \vec{\nabla}\varphi(\frac{x}{R})\,dx)\,dt$$

$$-\int_0^T \theta_\epsilon(t)(\int \varphi^2(\frac{x}{R})(\vec{\nabla} \otimes \vec{u}) \cdot \mathbb{F}\,dx)\,dt$$

$$-\frac{2}{R}\int_0^T \theta_\epsilon(t)(\int \varphi(\frac{x}{R})(\vec{\nabla}\varphi(\frac{x}{R}) \otimes \vec{u}) \cdot \mathbb{F}\,dx)\,dt$$

Letting R go to ∞, we get

$$-\frac{1}{2}\int_0^T \theta'_\epsilon(t)\|\vec{u}(t,.)\|_2^2\,dt \leq -\nu\int_0^T \theta_\epsilon(t)\|\vec{\nabla} \otimes \vec{u}\|_2^2\,dt + \int_0^T \theta_\epsilon(t)\langle\vec{u}|\vec{f}\rangle_{H^1,H^{-1}}\,dt$$

If t_0 is a Lebesgue point of $t \mapsto \|\vec{u}(t,.)\|_2^2$, we find that

$$\|\vec{u}(t_0,.)\|_2^2 \leq \|\vec{u}_0\|_2^2 - 2\nu\int_0^{t_0} \|\vec{\nabla} \otimes \vec{u}\|_2^2\,dt + 2\int_0^{t_0} \langle\vec{u}|\vec{f}\rangle_{H^1,H^{-1}}\,dt. \qquad (6.19)$$

This inequality is thus satisfied for almost every t_0, and even for every t_0 as $t \mapsto \vec{u}(t,.)$ is weakly continuous from $[0,T)$ to L^2, so that $t \mapsto \|\vec{u}(t,.)\|_2^2$ is semi-continuous. This inequality is called the **Leray energy inequality**[6].

---

[6] A similar proof gives the **strong Leray inequality**: for every Lebesgue point t_0 of $t \mapsto \|\vec{u}(t,.)\|_2^2$ and for every $t \in [t_0, T)$, we have:

$$\|\vec{u}(t,.)\|_2 \leq \|\vec{u}(t_0,.)\|_2^2 - 2\nu\int_{t_0}^t \|\vec{\nabla} \otimes \vec{u}\|_2^2\,ds + 2\int_{t_0}^t \langle\vec{u}|\vec{f}\rangle_{H^1,H^{-1}}\,ds.$$

This global inequality might be satisfied even if the local inequality is not fulfilled. For instance, solutions constructed by a Galerkin method are known to satisfy (6.19), but we do not know whether they are suitable (see the discussion by Biryuk, Craig and Ibrahim in [43]).

Leray weak solution

Definition 6.15.
A weak solution $\vec{u}$ of Equations (6.1) on $(0,T) \times \mathbb{R}^3$ is called a Leray weak solution if it satisfies

- $\vec{u} \in L_t^\infty L_x^2 \cap L_t^2 \dot{H}_x^1$

- $\vec{f} \in L_t^2 H_x^{-1}$

- *for every $t \in (0,T)$,*

$$\|\vec{u}(t,.)\|_2 \leq \|\vec{u}_0\|_2^2 - 2\nu \int_0^t \|\vec{\nabla} \otimes \vec{u}\|_2^2 \, ds + 2 \int_0^t \langle \vec{u}|\vec{f}\rangle_{H^1,H^{-1}} \, ds.$$

Chapter 7

Mild Solutions in Lebesgue or Sobolev Spaces

7.1 Kato's Mild Solutions

The search for solutions to the Navier–Stokes equations has known three eras. The first one was based on explicit formulas for hydrodynamic potentials, given by Lorentz (1896 [342]) and Oseen (1911 [384]), and further used by Leray in his seminal work introducing weak solutions (1934 [328]). Then, in the fifties, a second approach was developed by Hopf [238] and Ladyzhenskaya [262], based on the Faedo–Galerkin approximation method who turned the partial differential equations into the study of an ordinary differential equation in a finite-dimensional space.

The third period began in the mid-sixties, when the theory of accretive operators was developed, leading to the theory of semi-groups of operators. The problem was to find solution for non-linear equations of the type

$$\begin{cases} \frac{d}{dt}u = \mathcal{L}u + f(t, u) \\ \\ u(0) = u_0 \end{cases} \tag{7.1}$$

where $\mathcal{L}$ was an unbounded operator on a Banach space E. The solution began by studying the properties of the semi-group $U(t) = e^{t\mathcal{L}}$. Then the equation (7.1) was turned into an integral equation (due to Duhamel's formula)

$$u(t) = U(t)u_0 + \int_0^t U(t-s)f(s, u(s)) \, ds \tag{7.2}$$

The study of the properties of the integral term could then allow the use of Banach's contraction principle. The solutions obtained by this formalism were called *mild solutions* by Browder (1964 [68]) and Kato (1965 [253]).

Recall that we consider the Cauchy initial value problem for the Navier–Stokes equations (with reduced (unknown) pressure p, reduced force density $\vec{f}$ and kinematic viscosity $\nu > 0$): given a divergence-free vector field $\vec{u}_0$ on $\mathbb{R}^3$ and a force $\vec{f}$ on $(0, +\infty) \times \mathbb{R}^3$, find a positive T and regular functions $\vec{u}$ and p on $[0, T] \times \mathbb{R}^3$ solutions to

$$\begin{aligned} \partial_t \vec{u} = \nu \Delta \vec{u} - (\vec{u}.\vec{\nabla})\vec{u} + \vec{f} - \vec{\nabla}p \\ \operatorname{div} \vec{u} = 0 \\ \vec{u}_{|t=0} = \vec{u}_0 \end{aligned} \tag{7.3}$$

We have reformulated this problem into an integral equation: find $\vec{u}$ such that

$$\vec{u} = W_{\nu t} * \vec{u}_0 - \int_0^t \sum_{j=1}^3 \partial_j \mathcal{O}(\nu(t-s)) :: \left(\vec{f} * \partial_j G + u_j \vec{u} \right) \, ds \tag{7.4}$$

DOI: 10.1201/9781003042594-7

or, equivalently,

$$\vec{u} = W_{\nu t} * \vec{u}_0 - \int_0^t W_{\nu(t-s)} * \mathbb{P}(-\vec{f} + \operatorname{div} \vec{u} \otimes \vec{u})\, ds \tag{7.5}$$

Throughout this chapter and the following one, we are going to study mild solutions of the Navier–Stokes equations. More precisely, we shall seek to exhibit Banach spaces Y_T of Lebesgue measurable functions $F(t, x)$ defined on $[0, T] \times \mathbb{R}^3$ so that the operator

$$B(\vec{F}, \vec{G}) = \int_0^t W_{\nu(t-s)} * \mathbb{P}\operatorname{div}(\vec{F} \otimes \vec{G})\, ds \tag{7.6}$$

is a bounded bilinear operator on $Y_T^3 \times Y_T^3$:

$$\|B(\vec{F}, \vec{G})\|_{Y_T} \leq C_{Y_T} \|\vec{F}\|_{Y_T} \|\vec{G}\|_{Y_T}. \tag{7.7}$$

Then, starting with the initial data $\vec{u}_0$ and the force $\vec{f}$, if

$$\vec{U}(t, x) = W_{\nu t} * \vec{u}_0 - \int_0^t W_{\nu(t-s)} * \mathbb{P}\vec{f}\, ds$$

is a measurable function of t and x such that

$$\|\vec{U}\|_{Y_T} < \frac{1}{4C_{Y_T}},$$

we will find a solution $\vec{u}(t, x)$ of the equation $\vec{u} = \vec{U} - B(\vec{u}, \vec{u})$, defined on $(0, T) \times \mathbb{R}^3$, such that $\|\vec{u}\|_{Y_T} < \frac{1}{2C_{Y_T}}$.

The next step will then be to identify Banach spaces X_T of distributions on $\mathbb{R}^3$ and Z_T on $(0, T) \times \mathbb{R}^3$ such that $\vec{u}_0 \in (X_T)^3 \Leftrightarrow ($ or $\Rightarrow)1_{0<t<T} W_{\nu t} * \vec{u}_0 \in (Y_T)^3$ and $\vec{f} \in (Z_T)^3 \Rightarrow \int_0^t W_{\nu(t-s)} * \mathbb{P}\vec{f}\, ds \in (Y_T)^3$.

7.2   Local Solutions in the Hilbertian Setting

The simplest framework where to look for mild solutions to the Navier–Stokes equations is for initial values in the Sobolev space $H^1(\mathbb{R}^3)$ and forces in $L^2((0, T), L^2)$. This has been done in 1964 by Fujita and Kato [185]. Before stating their results, we begin with three easy lemmas on Sobolev spaces.

We recall that the Sobolev space H^s, $s \in \mathbb{R}$, is defined as the space of tempered distributions f such that the Fourier transform $\hat{f}$ of f is locally integrable and satisfies:

$$\|f\|_{H^s} = \sqrt{\frac{1}{(2\pi)^3} \int (1 + |\xi|^2)^s |\hat{f}(\xi)|^2\, d\xi} < +\infty.$$

Similarly, when $s < 3/2$, the *homogeneous Sobolev space* $\dot{H}^s$, $s \in \mathbb{R}$, is defined as the space of tempered distributions f such that the Fourier transform $\hat{f}$ of f is locally integrable and satisfies:

$$\|f\|_{\dot{H}^s} = \sqrt{\frac{1}{(2\pi)^3} \int |\xi|^{2s} |\hat{f}(\xi)|^2\, d\xi} < +\infty.$$

If $0 \leq s < 3/2$, we have the embedding $\dot{H}^s(\mathbb{R}^3) \subset L^p$ with $\frac{1}{p} = \frac{1}{2} - \frac{s}{3}$.

For $s \geq 3/2$, we will not use the space $\dot{H}^s$ (which is no longer a space of distrributions), but we will use the notation $\| \ \|_{\dot{H}^s}$, when dealing with tempered distributions f such that the Fourier transform $\hat{f}$ of f is locally integrable and satisfies:

$$\|f\|_{\dot{H}^s} = \sqrt{\frac{1}{(2\pi)^3} \int |\xi|^{2s}|\hat{f}(\xi)|^2 \, d\xi} < +\infty.$$

In particular, we will never work in a space $\dot{H}^s$, $s \geq 3/2$, but we may work in spaces $\dot{H}^{\sigma_1} \cap \dot{H}^{\sigma_2}$, where $\sigma_1 < 3/2$. For instance, for $s \geq 0$, $H^s = L^2 \cap \dot{H}^s$.

Lemma 7.1.
If $u_0 \in L^2$, then $W_{\nu t} * u_0 \in \mathcal{C}([0,+\infty), L^2)$ with

$$\sup_{t>0} \|W_{\nu t} * u_0\|_2 = \|u_0\|_2. \tag{7.8}$$

Moreover, $W_{\nu t} * u_0 \in L^2((0,+\infty), \dot{H}^1)$ with

$$\|W_{\nu t} * u_0\|_{L^2 \dot{H}^1} = \frac{1}{\sqrt{2\nu}}\|u_0\|_2 \tag{7.9}$$

Proof. To check that $W_{\nu t} * u_0 \in \mathcal{C}([0,+\infty), L^2)$, just use the spatial Fourier transform

$$\mathcal{F}_x(W_{\nu t} * u_0)(\xi) = e^{-\nu t|\xi|^2}\hat{u}_0(\xi).$$

To check that it belongs to $L^2((0,+\infty), \dot{H}^1)$ (where $\dot{H}^1$ is the homogeneous Sobolev space), just write:

$$\int_0^{+\infty} \|\vec{\nabla}(W_{\nu t} * u_0)\|_2^2 \, dt = \frac{1}{(2\pi)^3} \int_0^{+\infty} \int |\xi|^2 e^{-2\nu t|\xi|^2}|\hat{u}_0(\xi)|^2 \, d\xi \, dt = \frac{\|\hat{u}_0\|_2^2}{(2\pi)^3 2\nu} \qquad \square$$

Lemma 7.2.
If $f \in L^2(0,+\infty), L^2)$ and $F(t,x) = \int_0^t W_{\nu(t-s)} * f(s,.) \, ds$, then F belongs to $\mathcal{C}([0,+\infty), \dot{H}^1)$ and we have

$$\|\vec{\nabla}F(t,.)\|_2 \leq \frac{1}{\sqrt{2\nu}}\|f\|_{L^2 L^2}. \tag{7.10}$$

Moreover, $F \in L^2((0,+\infty), \dot{H}^2)$ and we have

$$\|\Delta F\|_{L^2 L^2} \leq \frac{1}{\nu}\|f\|_{L^2 L^2} \tag{7.11}$$

Proof. Just write:

$$\|\vec{\nabla}F(t,.)\|_2 = \|\sqrt{-\Delta}F(t,.)\|_2$$
$$= \sup_{\|u_0\|_2=1} |\int (\int_0^t \sqrt{-\Delta}(W_{\nu(t-s)} * f(s,.) \, ds)u_0(x) \, dx|$$
$$= \sup_{\|u_0\|_2=1} |\int \int_0^t f(s,x)\sqrt{-\Delta}\left(W_{\nu(t-s)} * u_0(x)\right) \, ds \, dx|$$
$$\leq \sup_{\|u_0\|_2=1} \|f\|_{L^2 L^2}\|W_{\nu(t-s)} * u_0\|_{L^2 \dot{H}^1}$$
$$\leq \frac{1}{\sqrt{2\nu}}\|f\|_{L^2 L^2}$$

From this inequality, and from the density of $\mathcal{D}((0,+\infty)\times\mathbb{R}^3)$, we find that F belongs to $\mathcal{C}([0,+\infty), L^2)$ with $F(0,.) = 0$. We may extend f and F to $t < 0$ by taking $f = F = 0$. We then have in the distributional sense that

$$\partial_t F = \nu\Delta F + f \tag{7.12}$$

If $G = \Delta F$, we find, taking the time-space Fourier transform on $\mathbb{R}\times\mathbb{R}^3$, that

$$\hat{G}(\tau,\xi) = -\frac{|\xi|^2}{i\tau + \nu|\xi|^2}\hat{f}(\tau,\xi)$$

and finally (by Plancherel inequality)

$$\|\Delta F\|_{L^2 L^2} \leq \frac{1}{\nu}\|f\|_{L^2 L^2}. \qquad \square$$

Lemma 7.3.
Let $0 < \delta < 3/2$ and $s \geq 0$. Then, if u and v belong to $H^s(\mathbb{R}^3) \cap \dot{H}^\delta$, we have that $uv \in H^{s+\delta-\frac{3}{2}}$ and

$$\|uv\|_{H^{s+\delta-\frac{3}{2}}} \leq C_{s,\delta}(\|u\|_{\dot{H}^\delta}\|v\|_{H^s} + \|v\|_{\dot{H}^\delta}\|u\|_{H^s}) \tag{7.13}$$

The same estimate holds for homogeneous norms:

$$\|uv\|_{\dot{H}^{s+\delta-\frac{3}{2}}} \leq C_{s,\delta}(\|u\|_{\dot{H}^\delta}\|v\|_{\dot{H}^s} + \|v\|_{\dot{H}^\delta}\|u\|_{\dot{H}^s}) \tag{7.14}$$

Proof. If $s + \delta - \frac{3}{2} \leq 0$, we use thrice the Sobolev embedding inequality: $0 \leq s < 3/2$, hence $H^s \subset L^{\frac{6}{3-2s}}$; $0 < \delta < 3/2$, hence $\dot{H}^\delta \subset L^{\frac{6}{3-2\delta}}$; $0 \leq 3/2 - s - \delta$, hence $H^{\frac{3}{2}-s-\delta} \subset L^{\frac{6}{2(s+\delta)}}$. The conclusion follows from the Hölder inequality:

$$\left|\int uvw\, dx\right| \leq \|u\|_{\frac{6}{3-2s}}\|v\|_{\frac{6}{3-2\delta}}\|w\|_{\frac{6}{2(s+\delta)}} \leq C\|u\|_{H^s}\|v\|_{\dot{H}^\delta}\|w\|_{H^{\frac{3}{2}-s-\delta}}.$$

If $s + \delta - \frac{3}{2} > 0$, we use the Plancherel formula and compute the norms of the Fourier transforms. Let

$$I = \int (1+|\xi|^2)^{s+\delta-\frac{3}{2}}\left|\int \hat{u}(\xi-\eta)\hat{v}(\eta)\, d\eta\right|^2 d\xi \leq 2(I_1 + I_2)$$

where

$$I_1 = \int (1+|\xi|^2)^{s+\delta-\frac{3}{2}}\left|\int_{|\eta|<|\xi-\eta|} \hat{u}(\xi-\eta)\hat{v}(\eta)\, d\eta\right|^2 d\xi$$

and

$$I_2 = \int (1+|\xi|^2)^{s+\delta-\frac{3}{2}}\left|\int_{|\gamma|<|\xi-\gamma|} \hat{u}(\gamma)\hat{v}(\xi-\gamma)\, d\gamma\right|^2 d\xi.$$

We have

$$I_1 \leq \|v\|_{\dot{H}^\delta}^2 \iint_{|\eta|<|\xi-\eta|} (1+|\xi|^2)^{s+\delta-\frac{3}{2}}|\eta|^{-2\delta}|\hat{u}(\xi-\eta)|^2 \, d\eta\, d\xi$$

$$\leq \|v\|_{\dot{H}^\delta}^2 \iint_{|\eta|<|\xi-\eta|} (1+2|\xi-\eta|^2)^{s+\delta-\frac{3}{2}}|\eta|^{-2\delta}|\hat{u}(\xi-\eta)|^2 \, d\eta\, d\xi$$

$$\leq C_{s,\delta}\|v\|_{\dot{H}^\delta}^2\|u\|_{H^s}^2$$

I_2 provides a symmetric control by $C_{s,\delta}\|u\|_{\dot{H}^\delta}^2\|v\|_{H^s}^2$. The case of homogeneous norms is similar. The lemma is proved. $\qquad\square$

We may now state and prove the result of Fujita and Kato:

Navier–Stokes equations and Sobolev spaces: local solutions

Theorem 7.1.
If $\vec{u}_0 \in (H^1(\mathbb{R}^3))^3$ and $\vec{f} \in L^2((0,T),(L^2(\mathbb{R}^3)^3)$, then there exists a $T_0 \in (0,T)$ and a mild solution $\vec{u}$ of Equation (7.4) on $(0,T_0) \times \mathbb{R}^3$ such that $\vec{u} \in \mathcal{C}([0,T_0],(H^1)^3) \cap L^2((0,T_0),(H^2)^3)$.

Proof. Let $\vec{U}$ be defined as

$$\vec{U}(t,x) = W_{\nu t} * \vec{u}_0 - \int_0^t W_{\nu(t-s)} * \mathbb{P}\vec{f}(s,.)\, ds.$$

We have:

$$\|W_{\nu t} * \vec{u}_0\|_{L^\infty H^1} \leq \|\vec{u}_0\|_{H^1} \text{ and } \|W_{\nu t} * \vec{u}_0\|_{L^2 \dot{H}^2} \leq \frac{1}{\sqrt{2\nu}}\|\vec{u}_0\|_{H^1}.$$

Similarly, we may extend $\vec{f}$ beyond T by $\vec{f} = 0$ when $t > T$. We find, for $\vec{F} = \int_0^t W_{\nu(t-s)} * \mathbb{P}\vec{f}(s,.)\, ds$,

$$\|\vec{F}\|_{L^\infty \dot{H}^1} \leq \frac{1}{\sqrt{2\nu}}\|\vec{f}\|_{L^2 L^2}, \quad \|\vec{F}\|_{L^2 \dot{H}^2} \leq \frac{1}{\nu}\|\vec{f}\|_{L^2 L^2},$$

while we have on $(0,T_0) \times \mathbb{R}^3$,

$$\|\vec{F}\|_{L^\infty L^2} \leq C T_0^{\frac{1}{2}}\|\vec{f}\|_{L^2 L^2}.$$

Thus, we find that, for any positive $T_0 \leq T$, $\vec{U}$ belongs to $\mathcal{C}([0,T_0],(H^1)^3) \cap L^2((0,T_0),(H^2)^3)$.

We now take $Y_{T_0} = \mathcal{C}([0,T_0],H^1) \cap L^2((0,T_0),H^2)$. We have $\vec{U} \in (Y_{T_0})^3$. Moreover, B is bounded on $(Y_{T_0})^3 \times (Y_{T_0})^3$. Indeed, we have, for $\vec{u}$ in $(Y_{T_0})^3$, the inclusion $\vec{u} \in L^4 H^{3/2} \cap L^4 \dot{H}^1$ with the inequality $\|\vec{u}\|_{L^4 \dot{H}^1} \leq T_0^{1/4}\|\vec{u}\|_{Y_{T_0}}$; thus we get from Lemma 7.3 that, for $\vec{u}$ and $\vec{v}$ in $(Y_{T_0})^3$,

$$\|\vec{u} \otimes \vec{v}\|_{L^2((0,T_0),H^1)} \leq C T_0^{1/4}\|\vec{u}\|_{Y_{T_0}}\|\vec{v}\|_{Y_{T_0}}.$$

We thus get $\mathbb{P}\operatorname{div}(\vec{u} \otimes \vec{v}) \in L^2 L^2 \cap L^2 \dot{H}^{-1}$. Using Lemma 7.2, we find that

$$\sqrt{\nu}\|B(\vec{u},\vec{v})\|_{L^\infty H^1} + \nu\|B(\vec{u},\vec{v})\|_{L^2 \dot{H}^2} \leq C T_0^{1/4}\|\vec{u}\|_{Y_{T_0}}\|\vec{v}\|_{Y_{T_0}} \tag{7.15}$$

so that

$$\|B(\vec{u},\vec{v})\|_{Y_{T_0}} \leq C\|\vec{u}\|_{Y_{T_0}}\|\vec{v}\|_{Y_{T_0}} T_0^{1/4}\left(\frac{1}{\sqrt{\nu}}(1 + T_0^{1/2}) + \frac{1}{\nu}\right) \tag{7.16}$$

Thus, for T_0 small enough

$$T_0 \leq \min\left(1, C_\nu \frac{1}{(\|\vec{u}_0\|_{H^1} + \|\vec{f}\|_{L^2 L^2})^4}\right), \tag{7.17}$$

we may find a solution for the Navier–Stokes equations $\vec{u} \in (Y_{T_0})^3$. $\qquad \square$

The proof of the theorem gives, as a corollary, uniqueness of mild solutions:

Proposition 7.1.
If $\vec{u}_0 \in (H^1(\mathbb{R}^3))^3$ and $\vec{f} \in L^2((0,T),(L^2(\mathbb{R}^3)^3)$, if $\vec{u}$ and $\vec{v}$ are two mild solutions of Equation (7.4) on $(0,T) \times \mathbb{R}^3$ such that $\vec{u}$ and $\vec{v}$ belong to $\mathcal{C}([0,T),(H^1)^3) \cap L^2((0,T),(H^2)^3)$, then $\vec{u} = \vec{v}$.

Proof. If $\vec{w} = \vec{u} - \vec{v}$, then $\vec{w} = B(\vec{v}, \vec{v}) - B(\vec{u}, \vec{u}) = -B(\vec{w}, \vec{v}) - B(\vec{u}, \vec{w})$. We have $\vec{w}(0,.) = 0$. Let $T^* \in [0,T]$ the maximal time such that $\vec{w} = 0$ on $[0,T^*) \times \mathbb{R}^3$. If $T^* < T$, and if $T_0 \in (T^*, T)$, we find that

$$\|\vec{w} \otimes \vec{v} + \vec{u} \otimes \vec{w}\|_{L^2((0,T_0),H^1)} \leq C(T_0 - T^*)^{1/4} \|\vec{w}\|_{Y_{T_0}} (\|\vec{u}\|_{Y_{T_0}} + \|\vec{v}\|_{Y_{T_0}}).$$

Thus, we get

$$\|\vec{w}\|_{Y_{T_0}} \leq C(\|\vec{u}\|_{Y_{T_0}} + \|\vec{v}\|_{Y_{T_0}})(T_0 - T^*)^{1/4} \left(\frac{1}{\sqrt{\nu}}(1 + T_0^{1/2}) + \frac{1}{\nu} \right) \|\vec{w}\|_{Y_{T_0}}$$

For $T_0 - T^*$ small enough, we get $\|\vec{w}\|_{Y_{T_0}} = 0$, hence $\vec{w} = 0$ on $[0,T_0) \times \mathbb{R}^3$, which is absurd since T^* is maximal. Thus, $T^* = T$.   □

7.3   Global Solutions in the Hilbertian Setting

Kato and Fujita moreover gave a criterion for the existence of global solutions:

Navier–Stokes equations and Sobolev spaces: global solutions

Theorem 7.2.
Let $T_\infty \in (0, +\infty]$. Let $\vec{u}_0 \in (H^1(\mathbb{R}^3))^3$ with $\operatorname{div} \vec{u}_0 = 0$ and $\vec{f}$ be defined on $(0, T_\infty)$ be such that $\vec{f} \in L^2((0,T),(L^2(\mathbb{R}^3)^3)$ for all $T < T_\infty$. Let T_{MAX} be the maximal time where one can find a mild solution $\vec{u}$ of Equation (7.4) on $(0, T_{\mathrm{MAX}}) \times \mathbb{R}^3$ such that, for all $T < T_{\mathrm{MAX}}$, $\vec{u}$ belongs to $\mathcal{C}([0,T],(H^1)^3) \cap L^2((0,T),(H^2)^3)$.

- *If $T_{\mathrm{MAX}} < T_\infty$, then $\sup_{0 < t < T_{\mathrm{MAX}}} \|\vec{u}(t,.)\|_{H^1} = +\infty$.*

- *If $T_{\mathrm{MAX}} < T_\infty$, then $\int_0^{T_{\mathrm{MAX}}} \|\vec{u}(s,.)\|_{\dot{H}^{3/2}}^2 \, ds = +\infty$.*

- *There exists a positive constant ϵ_0 (independent of T_∞, ν, $\vec{u}_0$ and $\vec{f}$), such that, if $\|\vec{u}_0\|_{\dot{H}^{1/2}} < \epsilon_0 \nu$ and $\int_0^{T_\infty} \|\vec{f}(s,.)\|_{\dot{H}^{-\frac{1}{2}}}^2 \, ds < \epsilon_0^2 \nu^3$, then $T_{\mathrm{MAX}} = T_\infty$.*

Proof. If $T < T_{\mathrm{MAX}} < T_1 < T_\infty$, we have a solution $\vec{u}$ on $[0,T]$ of

$$\vec{u} = W_{\nu t} * \vec{u}_0 + \int_0^t W_{\nu(t-s)} \mathbb{P} \vec{f}(s,.) \, ds - \int_0^t W_{\nu(t-s)} * \mathbb{P} \operatorname{div}(\vec{u}(s,.) \otimes \vec{u}(s,.)) \, ds$$

and a solution $\vec{u}$ on $[T, T+\delta]$ of

$$\vec{u} = W_{\nu(t-T)} * \vec{u}(T,.) + \int_T^t W_{\nu(t-s)} \mathbb{P} \vec{f}(s,.) \, ds - \int_T^t W_{\nu(t-s)} * \mathbb{P} \operatorname{div}(\vec{u}(s,.) \otimes \vec{u}(s,.)) \, ds$$

with $\delta = \min(T_1 - T, 1, C_\nu \frac{1}{(\|\vec{u}(T,.)\|_{H^1} + \|\vec{f}\|_{L^2((T,T_1),L^2)})^4})$. This gives a mild solution on $(0, T + \delta)$, and, due to the maximality of T_{MAX}, we must have $T + \delta \leq T_{\text{MAX}}$. This is possible only if $\liminf_{T \to T_{\text{MAX}}^-} \|\vec{u}(T,.)\|_{H^1} = +\infty$. This proves the first point of the theorem.

If $\vec{u}$ is a mild solution on $(0, T_{\text{MAX}})$, we have as well that

$$\partial_t \vec{u} = \nu \Delta \vec{u} - \mathbb{P} \operatorname{div}(\vec{u} \otimes \vec{u}) + \mathbb{P}\vec{f}$$

so that, for all $T < T_{\text{MAX}}$, $\vec{u} \in L^2((0,T), H^2)$ and $\partial_t \vec{u} \in L^2((0,T), L^2)$. Thus, we find:

$$\frac{d}{dt} \int |\vec{u}(t,x)|^2 \, dx = 2 \int \vec{u}.\partial_t \vec{u} dx = -2\nu \int |\vec{\nabla} \otimes \vec{u}|^2 \, dx + 2 \int \vec{u}.\vec{f} \, dx \qquad (7.18)$$

since $\vec{u}$ is divergence-free so that

$$\int \vec{u}.\mathbb{P} \operatorname{div}(\vec{u} \otimes \vec{u}) \, dx = \int \vec{u}.(\vec{u}.\vec{\nabla})\vec{u} \, dx$$

and

$$I = \int \vec{u}.(\vec{u}.\vec{\nabla})\vec{u} \, dx = -\int \vec{u}.(\vec{u}.\vec{\nabla})\vec{u} + |\vec{u}|^2 \operatorname{div} \vec{u} \, dx = -I = 0.$$

Thus, the L^2 norm of $\vec{u}$ remains bounded:

$$\|\vec{u}(t,.)\|_2 \leq \|\vec{u}_0\|_2 + \int_0^t \|\vec{f}(s,.)\|_2 \, ds. \qquad (7.19)$$

Similarly, we have:

$$\frac{d}{dt} \int |\vec{\nabla} \otimes \vec{u}(t,x)|^2 \, dx = 2 \sum_{i=1}^3 \int \partial_i \vec{u}.\partial_i \partial_t \vec{u} dx$$

$$= -2\nu \int |\Delta \vec{u}|^2 \, dx - 2 \sum_{i=1}^3 \int \partial_i \vec{u}.\partial_i((\vec{u}.\vec{\nabla})\vec{u}) \, dx - 2 \int \Delta \vec{u}.\vec{f} dx \qquad (7.20)$$

$$= -2\nu \int |\Delta \vec{u}|^2 \, dx - 2 \sum_{i=1}^3 \int \partial_i \vec{u}.((\partial_i \vec{u}).\vec{\nabla})\vec{u} \, dx - 2 \int \Delta \vec{u}.\vec{f} dx$$

since

$$J = \int \partial_i \vec{u}.(\vec{u}.\vec{\nabla})\partial_i \vec{u} \, dx = -\int \partial_i \vec{u}.(\vec{u}.\vec{\nabla})\partial_i \vec{u} + |\partial_i \vec{u}|^2 \operatorname{div} \vec{u} \, dx = -J = 0.$$

We then get

$$\frac{d}{dt} \int |\vec{\nabla} \otimes \vec{u}(t,x)|^2 \, dx \leq -2\nu \|\Delta \vec{u}\|_2^2 + 2 \sum_{i=1}^3 \|\vec{\nabla} \otimes \vec{u}\|_2 \|\partial_i \vec{u}\|_3 \|\partial_i \vec{u}\|_6 + 2\|\Delta \vec{u}\|_2 \|\vec{f}\|_2$$

$$\leq -\nu \|\Delta \vec{u}\|_2^2 + C\|\vec{\nabla} \otimes \vec{u}\|_2 \|\Delta \vec{u}\|_2 \|\vec{u}\|_{\dot{H}^{3/2}} + \frac{1}{\nu} \|\vec{f}\|_2^2 \qquad (7.21)$$

$$\leq \frac{C}{4\nu} \|\vec{\nabla} \otimes \vec{u}\|_2^2 \|\vec{u}\|_{\dot{H}^{3/2}}^2 + \frac{1}{\nu} \|\vec{f}\|_2^2.$$

Thus, $\|\vec{u}\|_{\dot{H}^1}^2 \leq (\|\vec{u}_0\|_{\dot{H}^1}^2 + \frac{1}{\nu} \int_0^t \|\vec{f}\|_2^2) e^{\frac{C}{4\nu} \int_0^t \|\vec{u}(s,.)\|_{\dot{H}^{3/2}}^2 \, ds}$. This proves the second point of the theorem.

Finally, we may write

$$\frac{d}{dt}\int |(-\Delta)^{1/4}\vec{u}(t,x)|^2 \, dx = 2\int ((-\Delta)^{1/2}\vec{u}).\partial_t\vec{u}dx$$

$$= -2\nu\int |(-\Delta)^{3/4}\vec{u}|^2 \, dx - 2\int ((-\Delta)^{1/2}\vec{u}).\operatorname{div}(\vec{u}.\otimes\vec{u}) \, dx$$

$$+ 2\int ((-\Delta)^{1/2})\vec{u}.\vec{f}dx \tag{7.22}$$

$$\leq -\nu\|\vec{u}\|^2_{\dot{H}^{3/2}} + C\|\vec{u}\|_{\dot{H}^{3/2}}\|\vec{u}\otimes\vec{u}\|_{\dot{H}^{1/2}} + \frac{1}{\nu}\|\vec{f}\|^2_{\dot{H}^{-\frac{1}{2}}}$$

Using Lemma 7.3, we get

$$\frac{d}{dt}\int |(-\Delta)^{1/4}\vec{u}(t,x)|^2 \, dx \leq (C_0\|\vec{u}\|_{\dot{H}^{1/2}} - \nu)\|\vec{u}\|^2_{\dot{H}^{3/2}} + \frac{1}{\nu}\|\vec{f}\|^2_{\dot{H}^{-\frac{1}{2}}} \tag{7.23}$$

Thus, if

$$\|\vec{u}_0\|_{\dot{H}^{1/2}} \leq \frac{\nu}{2C_0} \quad \text{and} \quad \int_0^{T_\infty}\|\vec{f}\|^2_{\dot{H}^{-\frac{1}{2}}} \, ds \leq \frac{\nu^3}{4C_0^2}, \tag{7.24}$$

we find that $\|\vec{u}\|_{\dot{H}^{1/2}} \leq \frac{\nu}{\sqrt{2}C_0}$, and finally

$$(1 - \frac{1}{\sqrt{2}})\nu\int_0^{T_{\text{MAX}}}\|\vec{u}\|^2_{\dot{H}^{3/2}} \, ds \leq \|\vec{u}_0\|^2_{\dot{H}^{1/2}} + \frac{1}{\nu}\int_0^{T_{\text{MAX}}}\|\vec{f}\|^2_{\dot{H}^{-\frac{1}{2}}} \, ds. \tag{7.25}$$

Thus, the third point is proved.   □

7.4   Sobolev Spaces

Fujita and Kato's theorems may be extended to the case of Sobolev spaces H^s with $s > 1/2$ (see Chemin [105]).

Navier–Stokes equations and Sobolev spaces

Theorem 7.3.

(A) *If $\vec{u}_0 \in (H^s(\mathbb{R}^3))^3$ and $\vec{f} \in L^2((0,T),(H^{s-1}(\mathbb{R}^3)^3)$ with $s > 1/2$, then there exists a $T_0 \in (0,T)$ and a unique mild solution $\vec{u}$ of Equation (7.4) on $(0,T_0) \times \mathbb{R}^3$ such that $\vec{u} \in \mathcal{C}([0,T_0],(H^s)^3) \cap L^2((0,T_0),(H^{s+1})^3)$.*

(B) *Let $s > 1/2$. Let $T_\infty \in (0,+\infty]$. Let $\vec{u}_0 \in (H^s(\mathbb{R}^3)^3$ with $\operatorname{div}\vec{u}_0 = 0$ and $\vec{f}$ be defined on $(0,T_\infty)$ be such that $\vec{f} \in L^2((0,T),(H^{s-1}(\mathbb{R}^3)^3)$ for all $T < T_\infty$. Let T_{MAX} be the maximal time where one can find a mild solution $\vec{u}$ of Equation (7.4) on $(0,T_{\text{MAX}}) \times \mathbb{R}^3$ such that, for all $T < T_{\text{MAX}}$, $\vec{u}$ belongs to $\mathcal{C}([0,T],(H^s)^3) \cap L^2((0,T),(H^{s+1})^3)$.*

- *If $T_{\text{MAX}} < T_\infty$, then $\sup_{0<t<T_{\text{MAX}}}\|\vec{u}(t,.)\|_{H^s} = +\infty$.*

- *If $T_{\text{MAX}} < T_\infty$, then $\int_0^{T_{\text{MAX}}}\|\vec{u}(s,.)\|^2_{\dot{H}^{3/2}} \, ds = +\infty$.*

- *There exists a positive constant ϵ_0 (independent of T_∞, ν, $\vec{u}_0$ and $\vec{f}$), such that, if $\|\vec{u}_0\|_{\dot{H}^{1/2}} < \epsilon_0 \nu$ and $\int_0^{T_\infty} \|\vec{f}(s,.)\|_{\dot{H}^{-\frac{1}{2}}}^2 \, ds < \epsilon_0^2 \nu^3$, then $T_{\text{MAX}} = T_\infty$.*

Proof. The proof of (A) is similar to the proof of Theorem 7.1. We check easily that, for any positive $T_0 \leq T$, $\vec{U}$ belongs to $\mathcal{C}([0,T_0], (H^s)^3) \cap L^2((0,T_0), (H^{s+1})^3)$, since

$$\|W_{\nu t} * \vec{u}_0\|_{L^\infty H^s} \leq \|\vec{u}_0\|_{H^s}, \quad \|W_{\nu t} * \vec{u}_0\|_{L^2 \dot{H}^{s+1}} \leq \frac{1}{\sqrt{2\nu}}\|\vec{u}_0\|_{H^s}$$

$$\|\vec{\nabla} \vec{F}\|_{L^\infty H^{s-1}} \leq \frac{1}{\sqrt{2\nu}}\|\vec{f}\|_{L^2 H^{s-1}}, \quad \|\Delta \vec{F}\|_{L^2 \dot{H}^{s-1}} \leq \frac{1}{\nu}\|\vec{f}\|_{L^2 H^{s-1}},$$

$$\|\vec{F}\|_{L^\infty L^2} \leq C T_0^{\frac{1}{2}}\|\vec{f}\|_{L^2 L^2}.$$

We take $Y_{T_0} = \mathcal{C}([0,T_0], H^s) \cap L^2((0,T_0), H^{s+1})$. We find again that $Y_{T_0} \subset L^4((0,T_0), H^{s+\frac{1}{2}}) \cap L^4((0,T_0), \dot{H}^1)$: if $s \geq 1$, $H^s \subset \dot{H}^1$ and $\|u\|_{L^4 \dot{H}^1} \leq T_0^{1/4}\|u\|_{Y_{T_0}}$; if $1/2 < s < 1$, we have $Y_{T_0} \subset L^{\frac{2}{1-s}} \dot{H}^1$, hence $\|u\|_{L^4 \dot{H}^1} \leq T_0^{(2s-1)/4}\|u\|_{Y_{T_0}}$. This gives (if $T_0 \leq 1$)

$$\|\vec{u} \otimes \vec{v}\|_{L^2((0,T_0), H^s)} \leq C T_0^{\min(1,2s-1)/4}\|\vec{u}\|_{Y_{T_0}}\|\vec{v}\|_{Y_{T_0}}$$

so that

$$\|B(\vec{u},\vec{v})\|_{Y_{T_0}} \leq C\|\vec{u}\|_{Y_{T_0}}\|\vec{v}\|_{Y_{T_0}} T_0^{\min(2s-1,1)/4}\left(\frac{1}{\sqrt{\nu}}(1 + T_0^{1/2}) + \frac{1}{\nu}\right) \quad (7.26)$$

Thus, for T_0 small enough, we may find a solution for the Navier–Stokes equations $\vec{u} \in (Y_{T_0})^3$.

The proof of (B) is similar to the proof of Theorem 7.2. The only difference is on the study of the role of the norm of $\vec{u}$ in $L^2 \dot{H}^{3/2}$, as we may no longer use Leibnitz's rule on derivatives.

First, we have $H^{s-1} \subset H^{-3/2}$, hence

$$\frac{d}{dt}\|\vec{u}\|_2^2 = -2\nu\|\vec{u}\|_{\dot{H}^1}^2 + 2\int \vec{u}.\vec{f}\,dx$$

$$\leq 2\|\vec{u}\|_{H^{3/2}}\|\vec{f}\|_{H^{s-1}}$$

$$\leq \|\vec{u}\|_2^2 + \|\vec{u}\|_{\dot{H}^{3/2}}^2 + \|\vec{f}\|_{H^{s-1}}^2$$

and

$$\|\vec{u}\|_2^2 \leq \|\vec{u}_0\|_2^2 e^t + \int_0^t e^{t-\tau}(\|\vec{u}\|_{\dot{H}^{3/2}}^2 + \|\vec{f}\|_{H^{s-1}}^2)\,d\tau.$$

We have moreover:

$$\frac{d}{dt}\int |(-\Delta)^{s/2}\vec{u}(t,x)|^2\,dx = 2\int (-\Delta)^{s/2}\vec{u}.(-\Delta)^{s/2}\partial_t \vec{u}\,dx$$

$$= -2\nu \int |(-\Delta)^{(s+1)/2}\vec{u}|^2\,dx$$

$$\qquad - 2\int (-\Delta)^{s/2}\vec{u}.(-\Delta)^{s/2}((\vec{u}.\vec{\nabla})\vec{u})\,dx \qquad (7.27)$$

$$\qquad - 2\int (-\Delta)^{s/2}\vec{u}.(-\Delta)^{s/2}\vec{f}\,dx$$

We then get

$$\frac{d}{dt}\|\vec{u}\|^2_{\dot{H}^s} \leq -2\nu\|\vec{u}\|^2_{\dot{H}^{s+1}} + \|\vec{u}\|_{H^{s+1}}\|\vec{f}\|_{H^{s-1}}$$

$$+ 2|\int (-\Delta)^{s/2}\vec{u}.(-\Delta)^{s/2}((\vec{u}.\vec{\nabla})\vec{u})\,dx|$$

$$\leq -\nu\|\vec{u}\|^2_{\dot{H}^{s+1}} + \frac{1}{\nu}\|\vec{f}\|^2_{H^{s-1}} + 2\|\vec{u}\|_2\|\vec{f}\|_{H^{s-1}}$$

$$+ 2|\int (-\Delta)^{s/2}\vec{u}.(-\Delta)^{s/2}((\vec{u}.\vec{\nabla})\vec{u})\,dx|.$$

We will prove in the next section (Section 7.5) that

$$|\int (-\Delta)^{s/2}\vec{u}.(-\Delta)^{s/2}((\vec{u}.\vec{\nabla})\vec{u})\,dx| \leq C\|\vec{u}\|_{\dot{H}^s}\|\vec{u}\|_{\dot{H}^{s+1}}\|\vec{u}\|_{\dot{H}^{3/2}}. \tag{7.28}$$

This gives

$$\frac{d}{dt}\|\vec{u}\|^2_{\dot{H}^s} \leq \frac{C^2}{4\nu}\|\vec{u}\|^2_{\dot{H}^s}\|\vec{u}\|^2_{\dot{H}^{3/2}} + \frac{1}{\nu}\|\vec{f}\|^2_{H^{s-1}} + 2\|\vec{u}\|_2\|\vec{f}\|_{H^{s-1}}$$

and

$$\|\vec{u}\|^2_{\dot{H}^s} \leq \|\vec{u}_0\|^2_{\dot{H}^s}e^{\frac{C^2}{4\nu}\int_0^t \|\vec{u}\|^2_{\dot{H}^{3/2}}\,d\tau}$$

$$+ \int_0^t e^{\frac{C^2}{4\nu}\int_\tau^t \|\vec{u}\|^2_{\dot{H}^{3/2}}\,d\sigma}(\frac{1}{\nu}\|\vec{f}\|^2_{H^{s-1}} + 2\|\vec{u}\|_2\|\vec{f}\|_{H^{s-1}})\,d\tau$$

Thus, the theorem is proved.   □

The case $s = 1/2$ is similar, except that the existence time is no more controlled by a power of the norm of the initial value:

Navier–Stokes equations and the critical Sobolev space

Theorem 7.4.

(A) *If $\vec{u}_0 \in (H^{1/2}(\mathbb{R}^3))^3$ and $\vec{f} \in L^2((0,T),(H^{-1/2}(\mathbb{R}^3)^3)$, then there exists a $T_0 \in (0,T)$ and a unique mild solution $\vec{u}$ of Equation (7.4) on $(0,T_0) \times \mathbb{R}^3$ such that $\vec{u} \in \mathcal{C}([0,T_0],(H^{1/2})^3) \cap L^2((0,T_0),(H^{3/2})^3)$.*

(B) *Let $\vec{u}_0 \in (H^{1/2}(\mathbb{R}^3))^3$ with $\operatorname{div}\vec{u}_0 = 0$ and $\vec{f}$ be defined on $(0,+\infty)$ be such that $\vec{f} \in L^2((0,+\infty),(\dot{H}^{-1/2}(\mathbb{R}^3)^3)$. There exists a positive constant ϵ_0 (independent of ν, $\vec{u}_0$ and $\vec{f}$), such that, if $\|\vec{u}_0\|_{\dot{H}^{1/2}} < \epsilon_0\nu$ and $\int_0^{+\infty}\|\vec{f}(s,.)\|^2_{\dot{H}^{-\frac{1}{2}}}\,ds < \epsilon_0^2\nu^3$, then the mild solution is defined on $(0,+\infty)$.*

Proof. We check easily that, for any finite T, $\vec{U}$ belongs to $\mathcal{C}([0,T],(H^{1/2})^3) \cap L^2((0,T),(H^{3/2})^3)$. If $T = +\infty$, and $\vec{f} \in L^2\dot{H}^{-1/2}$, then $\vec{U}$ belongs to $\mathcal{C}_b([0,+\infty),(\dot{H}^{1/2})^3) \cap L^2((0,+\infty),(\dot{H}^{3/2})^3)$. Indeed, if we split $\vec{f}$ in $\vec{f} = \vec{f}_L + \vec{f}_H$ with $\vec{f}_L \in L^2L^2$ and $\vec{f}_H \in L^2\dot{H}^{-1/2}$, we may use the following estimates for the high-frequency $\vec{f}_H$:

$$\|\int_0^t W_{\nu(t-s)} * \mathbb{P}\vec{f}_H\,ds\|_{L^\infty((0,T_0),\dot{H}^{1/2})} \leq C\frac{1}{\sqrt{\nu}}\|\vec{f}_H\|_{L^2((0,T_0),\dot{H}^{-1/2})}$$

$$\|\int_0^t W_{\nu(t-s)} * \mathbb{P}\vec{f}_H \, ds\|_{L^2((0,T_0),\dot{H}^{1/2})} \leq C\frac{1}{\nu}\|\vec{f}_H\|_{L^2((0,T_0),\dot{H}^{-1/2})}$$

$$\|\int_0^t W_{\nu(t-s)} * \mathbb{P}\vec{f}_H \, ds\|_{L^2(\mathbb{R}^3)} \leq \int_0^t \|W_{\nu(t-s)} * \mathbb{P}\vec{f}_H\|_2 \, ds$$

$$\leq C\int_0^t \frac{1}{(\nu(t-s))^{1/4}}\|\vec{f}_H\|_{\dot{H}^{-1/2}} \, ds$$

$$\leq \frac{4C}{3}\frac{t^{3/4}}{\sqrt{\nu}}\|\vec{f}_H\|_{L^2((0,T_0),\dot{H}^{-1/2})}$$

and the following estimates for the low-frequency $\vec{f}_L$:

$$\|\int_0^t W_{\nu(t-s)} * \mathbb{P}\vec{f}_L \, ds\|_{L^2(\mathbb{R}^3)} \leq \int_0^t \|W_{\nu(t-s)} * \mathbb{P}\vec{f}_L\|_2 \, ds$$

$$\leq \int_0^t \|\vec{f}_L\|_2 \, ds$$

$$\leq \sqrt{t}\|\vec{f}_H\|_{L^2((0,T_0),L^2)}$$

$$\|\int_0^t W_{\nu(t-s)} * \mathbb{P}\vec{f}_L \, ds\|_{\dot{H}^{1/2}(\mathbb{R}^3)} \leq \int_0^t \|W_{\nu(t-s)} * \mathbb{P}\vec{f}_L\|_{\dot{H}^{1/2}} \, ds$$

$$\leq C\int_0^t \frac{1}{(\nu(t-s))^{1/4}}\|\vec{f}_L\|_2 \, ds$$

$$\leq \frac{4C}{3}\frac{t^{3/4}}{\sqrt{\nu}}\|\vec{f}_H\|_{L^2((0,T_0),L^2)}$$

$$\|\int_0^t W_{\nu(t-s)} * \mathbb{P}\vec{f}_L \, ds\|_{L^2((0,T_0),\dot{H}^{3/2})} \leq \|\int_0^t \|W_{\nu(t-s)} * \mathbb{P}\vec{f}_L\|_{\dot{H}^{3/2}} \, ds\|_{L^2((0,T_0))}$$

$$\leq C\|\int_0^t \frac{1}{(\nu(t-s))^{3/4}}\|\vec{f}_L\|_2 \, ds\|_{L^2((0,T_0))}$$

$$\leq 4C\frac{T_0^{1/4}}{\nu^{3/4}}\|\vec{f}_H\|_{L^2((0,T_0),L^2)}$$

We take $Y_{T_0} = L^4((0,T_0),\dot{H}^1)$. We find

$$\|\vec{u} \otimes \vec{v}\|_{L^2((0,T_0),\dot{H}^{1/2})} \leq C\|\vec{u}\|_{Y_{T_0}}\|\vec{v}\|_{Y_{T_0}}$$

so that

$$\|B(\vec{u},\vec{v})\|_{Y_{T_0}} \leq C\|B(\vec{u},\vec{v})\|_{L^\infty((0,T_0),\dot{H}^{1/2})}^{1/2}\|B(\vec{u},\vec{v})\|_{L^2((0,T_0),\dot{H}^{3/2})}^{1/2}$$

$$\leq C'\|\vec{u}\|_{Y_{T_0}}\|\vec{v}\|_{Y_{T_0}}\frac{1}{\nu^{3/4}}. \tag{7.29}$$

Thus, for $\|\vec{U}\|_{Y_{T_0}}$ small enough (i.e., $\vec{u}_0$ small enough in $\dot{H}^{1/2}$ and $\vec{f}$ small enough in $L^2\dot{H}^{-1/2}$ with $T_0 = +\infty$, or T_0 small enough depending on $\vec{u}_0$ and $\vec{f}$), we may find a solution for the Navier–Stokes equations $\vec{u} \in (Y_{T_0})^3$.

For this solution $\vec{u}$, $B(\vec{u},\vec{u})$ belongs to $\mathcal{C}_b([0,T_0),(\dot{H}^{1/2})^3) \cap L^2((0,T_0),(\dot{H}^{3/2})^3)$ and, for any finite $T_1 < T_0$, to $\mathcal{C}([0,T_1],(L^2)^3)$. $\qquad\square$

7.5   A Commutator Estimate

In this section, we prove that, for $\vec{u} \in (\dot{H}^{1/2} \cap \dot{H}^{s+1})^3$ with div $\vec{u} = 0$, we have the inequality (7.28). Indeed, we have

$$\int (-\Delta)^{s/2} \vec{u} . \left(\vec{u}.\vec{\nabla} (-\Delta)^{s/2} \vec{u} \right) \, dx = 0$$

Hence, we have

$$\int (-\Delta)^{s/2} \vec{u} . (-\Delta)^{s/2} ((\vec{u}.\vec{\nabla}) \vec{u}) \, dx$$

$$= \int (-\Delta)^{s/2} \vec{u} . \sum_{i=1}^{3} \left((-\Delta)^{s/2} (u_i \partial_i \vec{u}) - u_i (-\Delta)^{s/2} \partial_i \vec{u} \right) \, dx$$

and thus

$$\left| \int (-\Delta)^{s/2} \vec{u} . (-\Delta)^{s/2} ((\vec{u}.\vec{\nabla}) \vec{u}) \, dx \right| \leq \| \vec{u} \|_{\dot{H}^s} \sum_{i=1}^{3} \| (-\Delta)^{s/2} (u_i \partial_i \vec{u}) - u_i (-\Delta)^{s/2} \partial_i \vec{u} \|_2$$

We then end the proof of (7.28) with the following commutator estimate [105]:

Proposition 7.2.
Let $s > 1/2$. Then we have

$$\| (-\Delta)^{s/2} (uv) - u (-\Delta)^{s/2} v \|_2 \leq C (\| u \|_{\dot{H}^{3/2}} \| v \|_{\dot{H}^s} + \| u \|_{\dot{H}^{s+1}} \| v \|_{\dot{H}^{1/2}}) \tag{7.30}$$

Proof. We compute the norm of the Fourier transform; let

$$I = \int \left| \int \hat{u}(\xi - \eta) \hat{v}(\eta) \, (|\xi|^s - |\eta|^s) \, d\eta \right|^2 \, d\xi \leq 2(I_1 + I_2)$$

where

$$I_1 = \int \left| \int_{|\eta| < 2|\xi - \eta|} \hat{u}(\xi - \eta) \hat{v}(\eta) \, (|\xi|^s - |\eta|^s) d\eta \right|^2 \, d\xi$$

and

$$I_2 = \int \left| \int_{|\gamma| < \frac{1}{2}|\xi - \gamma|} \hat{u}(\gamma) \hat{v}(\xi - \gamma) \, (|\xi|^s - |\xi - \gamma|^s) d\gamma \right|^2 \, d\xi.$$

We have

$$I_1 \leq \| v \|_{\dot{H}^{1/2}}^2 \iint_{|\eta| < 2|\xi - \eta|} \big| |\xi|^s - |\eta|^s \big|^2 |\eta|^{-1} |\hat{u}(\xi - \eta)|^2 \, d\eta \, d\xi$$

$$\leq 3^{2s} \| v \|_{\dot{H}^{1/2}}^2 \iint_{|\eta| < 2|\xi - \eta|} |\xi - \eta|^{2s} |\eta|^{-1} |\hat{u}(\xi - \eta)|^2 \, d\eta \, d\xi$$

$$= C 3^{2s} \| v \|_{\dot{H}^{1/2}}^2 \| u \|_{\dot{H}^{s+1}}^2$$

and

$$I_2 \leq \| u \|_{\dot{H}^{3/2}}^2 \iint_{|\gamma| < \frac{1}{2}|\xi - \gamma|} \frac{\big| |\xi|^s - |\xi - \gamma|^s \big|^2}{|\gamma|^3} |\hat{v}(\xi - \gamma)|^2 \, d\gamma \, d\xi$$

$$\leq C_s \| u \|_{\dot{H}^{3/2}}^2 \int_{|\gamma| < \frac{1}{2}|\xi - \gamma|} |\xi - \gamma|^{2s-2} |\gamma|^{-1} |\hat{v}(\xi - \gamma)|^2 \, d\eta \, d\xi$$

$$= C C_s \| u \|_{\dot{H}^{3/2}}^2 \| v \|_{\dot{H}^s}^2. \qquad \square$$

7.6 Lebesgue Spaces

Another simple framework where to look for mild solutions to the Navier–Stokes equations is for initial values in the Sobolev space $L^p(\mathbb{R}^3)$ and forces in $L^r((0,T),L^q)$ with $\frac{2}{r}+\frac{3}{q}<2+\frac{3}{p}$. This has been done in 1984 by Kato [255]. (see also Cannone and Planchon for the discussion on external forces [85]).

The case $p>3$ is very easy.

Proposition 7.3.
Let $3<p<+\infty$. Then:

- *The bilinear operator B defined as*

$$B(\vec{F},\vec{G})=\int_0^t W_{\nu(t-s)}*\mathbb{P}\,\mathrm{div}(\vec{F}\otimes\vec{G})\,ds$$

is continuous on $Y_T=\mathcal{C}([0,T],(L^p)^3)$ for every finite T:

$$\sup_{0<t<T}\|B(\vec{F},\vec{G})\|_p\leq C_p(\nu T)^{\frac{1}{2}-\frac{3}{2p}}\frac{1}{\nu}\sup_{0<t<T}\|\vec{F}(t,.)\|_p\sup_{0<t<T}\|\vec{G}(t,.)\|_p \qquad (7.31)$$

where the constant C_p depends only on p (and not on T, ν, $\vec{F}$ nor $\vec{G}$).

- *If $\vec{u}_0\in(L^p(\mathbb{R}^3))^3$ and $\int_0^t W_{\nu(t-s)}*\mathbb{P}\vec{f}(s,.)\,ds\in\mathcal{C}([0,T],(L^p(\mathbb{R}^3)^3)$, then there exists a $T_0\in(0,T)$ and a mild solution $\vec{u}$ of Equation (7.4) on $(0,T_0)\times\mathbb{R}^3$ such that $\vec{u}\in\mathcal{C}([0,T_0],(L^p)^3)$.*

- *If $\vec{f}\in L^r((0,T),(L^q)^3)$ with $1<q<p$ and $\frac{2}{r}+\frac{3}{q}<2+\frac{3}{p}$, or $q=p$ and $r\geq 1$, or if $\sup_{0<t<T}t^\beta\|\vec{f}\|_q<+\infty$ and $\lim_{t\to 0}t^\beta\|\vec{f}\|_q=0$ with $\frac{3p}{2p+3}<q<p$ and $\beta=1-\frac{3}{2q}+\frac{3}{2p}$, then $\int_0^t W_{\nu(t-s)}*\mathbb{P}\vec{f}(s,.)\,ds\in\mathcal{C}([0,T],(L^p(\mathbb{R}^3)^3)$*

Proof. Indeed, we use the estimate on the size of $\partial_j\mathcal{O}(\nu t)$ that is derived from Theorem 4.6 and Corollary 4.2 and write the inequality:

$$|B(\vec{F},\vec{G})|\leq C_0\int_0^t\int\frac{1}{\nu^2(t-s)^2+|x-y|^4}|\vec{F}(s,y)|\,|\vec{G}(s,y)|\,ds\,dy \qquad (7.32)$$

We get the inequality

$$\|B(\vec{F},\vec{G})\|_p\leq C_0\int_0^t\|\int\frac{1}{\nu^2(t-s)^2+|x-y|^4}|\vec{F}(s,y)|\,|\vec{G}(s,y)|\,dy\|_p\,ds$$

$$\leq C_0\int_0^t\|\frac{1}{\nu^2(t-s)^2+|x|^4}\|_{\frac{p}{p-1}}\|\vec{F}(s,.)\|_p\|\vec{G}(s,.)\|_p\,ds \qquad (7.33)$$

$$= C_p\int_0^t\frac{1}{(\nu|t-s|)^{\frac{1}{2}+\frac{3}{2p}}}\|\vec{F}(s,.)\|_p\|\vec{G}(s,.)\|_p\,ds$$

so that (for $p>3$)

$$\sup_{0<t<T}\|B(\vec{F},\vec{G})\|_p\leq C_pT^{\frac{1}{2}-\frac{3}{2p}}\nu^{-\frac{1}{2}-\frac{3}{2p}}\sup_{0<t<T}\|\vec{F}(t,.)\|_p\sup_{0<t<T}\|\vec{G}(t,.)\|_p \qquad (7.34)$$

where the constant C_p depends only on p (and not on T, ν, $\vec{F}$ nor $\vec{G}$).

Thus, we may find a mild solution in $\mathcal{C}([0,T_0],(L^p)^3)$, as soon as $\vec{U}$ belongs to $\mathcal{C}([0,T_0],(L^p)^3)$ and T_0 is small enough.

In order to check that $\vec{F} = \int_0^t W_{\nu(t-s)} * \mathbb{P}\vec{f}\,ds$ belongs to $\mathcal{C}([0,T_0],(L^p)^3)$, we may write, if $1 < q \leq p$,

$$\|W_{\nu(t-s)} * \mathbb{P}\vec{f}\|_p \leq C\|\vec{f}\|_q \frac{1}{(\nu(t-s))^{\frac{3}{2}(\frac{1}{q}-\frac{1}{p})}}$$

Thus, if $\vec{f} \in L_t^r L_x^q$ with

$$\frac{3}{2}\left(\frac{1}{q}-\frac{1}{p}\right) < 1 - \frac{1}{r}$$

we get

$$\|\vec{F}(t,.)\|_p \leq C_\nu t^{1-\frac{1}{r}-\frac{3}{2}(\frac{1}{q}-\frac{1}{p})}\|\vec{f}\|_{L^r L^q}. \tag{7.35}$$

The same estimate is valid in the case $q = p$ and $r = 1$. We may assume $r < +\infty$; from (7.35) and the density of test functions in $L^r L^q$, we find that $\vec{F}$ belongs to $\mathcal{C}([0,T_0],(L^p)^3)$.

Similarly, we have (for $1 < q \leq p$)

$$\|W_{\nu(t-s)} * \mathbb{P}\vec{f}\|_p \leq C(s^\beta\|\vec{f}\|_q)\frac{1}{(\nu(t-s))^{\frac{3}{2}(\frac{1}{q}-\frac{1}{p})}s^\beta}$$

Thus, if $\beta + \frac{3}{2}(\frac{1}{q}-\frac{1}{p}) = 1$, we find that

$$\|\vec{F}(t,.)\|_p \leq C_\nu \sup_{0<s<t} s^\beta\|\vec{f}(s,.)\|_q. \tag{7.36}$$

Regularity of the heat kernel shows that $\vec{F}$ belongs to $\mathcal{C}((0,T_0],(L^p)^3)$. The continuity at $t = 0$ is then ensured by (7.36) and the assumption that $\lim_{s\to 0} s^\beta\|\vec{f}(s,.)\|_q = 0$. $\quad\square$

The critical case $p = 3$ is not as simple, as the bilinear operator B is no longer bounded on $\mathcal{C}([0,T],(L^3)^3)$. The cancellation properties of solenoidal vector fields do not provide enough compensation, as it has been proved by Oru [383] that B is not bounded on the smaller space $\{\vec{u} \in \mathcal{C}([0,T],(L^3)^3) \ / \ \operatorname{div}\vec{u} = 0\}$. The solution proposed by Weissler [498] is then to use the smoothing properties of the heat kernel: if $u_0 \in L^3$, then for any positive σ we have

$$\|(-\Delta)^\sigma W_{\nu t} * u_0\|_3 \leq C_\sigma(\nu t)^{-\sigma}\|u_0\|_3. \tag{7.37}$$

Moreover, since for a regular function u_0 (such that $(-\Delta)^\sigma u_0 \in L^3$), we have $\|(-\Delta)^\sigma W_{\nu t} * u_0\|_3 = O(1)$, we find that, for $u_0 \in L^3$,

$$\lim_{t\to 0+} t^\sigma\|(-\Delta)^\sigma W_{\nu t} * u_0\|_3 = 0. \tag{7.38}$$

Kato's solution [255] was even simpler: his proof uses only direct estimations on the absolute values of the integrands[1], beginning with the estimate, for any $q > 3$,

$$\|W_{\nu t} * u_0\|_q \leq C_q(\nu t)^{\frac{3}{2q}-\frac{1}{2}}\|u_0\|_3.$$

---

[1]This is this approach that we have followed in a systematic way in Chapter 5.

Navier–Stokes equations and the critical Lebesgue space

Theorem 7.5.

- Let $\sigma \in (0, 1/2]$. The bilinear operator B defined as

$$B(\vec{F}, \vec{G}) = \int_0^t W_{\nu(t-s)} * \mathbb{P} \operatorname{div}(\vec{F} \otimes \vec{G}) \, ds$$

is continuous on

$$Y_{T,\sigma} = \{\vec{u} \ / \ \sup_{0<t<T} t^{\sigma/2} \|\vec{u}\|_{\dot{H}_3^\sigma} < +\infty \ \text{and} \ \lim_{t \to 0} t^{\sigma/2} \|\vec{u}\|_{\dot{H}_3^\sigma} = 0\}$$

(with $\|\vec{u}\|_{\dot{H}_3^\sigma} = \|(-\Delta)^{\sigma/2}\vec{u}\|_3$) for every finite or infinite $T \in (0, +\infty]$:

$$\sup_{0<t<T} t^{\frac{\sigma}{2}} \|B(\vec{F}, \vec{G})\|_{\dot{H}_3^\sigma} \leq C_{\sigma,\nu} \sup_{0<t<T} t^{\frac{\sigma}{2}} \|\vec{F}(t,.)\|_{\dot{H}_3^\sigma} \sup_{0<t<T} t^{\frac{\sigma}{2}} \|\vec{G}(t,.)\|_{\dot{H}_3^\sigma} \quad (7.39)$$

where the constant $C_{\sigma,\nu}$ does not depend on T, $\vec{F}$ nor $\vec{G}$.

- Similarly, for $q \in (3, +\infty)$ and $\beta = \frac{1}{2} - \frac{3}{2q}$, the bilinear operator B is continuous on

$$\tilde{Y}_{T,q} = \{\vec{u} \ / \ \sup_{0<t<T} t^\beta \|\vec{u}\|_q < +\infty \ \text{and} \ \lim_{t \to 0} t^\beta \|\vec{u}\|_q = 0\}$$

for every finite or infinite $T \in (0, +\infty]$:

$$\sup_{0<t<T} t^\beta \|B(\vec{F}, \vec{G})\|_q \leq C_{\nu,q} \sup_{0<t<T} t^\beta \|\vec{F}(t,.)\|_q \sup_{0<t<T} t^\beta \|\vec{G}(t,.)\|_q \quad (7.40)$$

- If $\vec{u}_0 \in (L^3(\mathbb{R}^3)^3$ and $\int_0^t W_{\nu(t-s)} * \mathbb{P}\vec{f}(s,.) \, ds \in \mathcal{C}([0,T], (L^3(\mathbb{R}^3)^3) \cap Y_{T,\sigma}$, then there exists a $T_0 \in (0,T)$ and a mild solution $\vec{u}$ of Equation (7.4) on $(0, T_0) \times \mathbb{R}^3$ such that $\vec{u} \in \mathcal{C}([0, T_0], (L^3)^3) \cap Y_{T_0,\sigma}$. (A similar result holds for $\mathcal{C}([0, T], (L^3(\mathbb{R}^3)^3) \cap \tilde{Y}_{T,q}$).

- If $\vec{u}_0 \in (L^3(\mathbb{R}^3)^3$ is small enough and $\int_0^t W_{\nu(t-s)} * \mathbb{P}\vec{f}(s,.) \, ds$ is small enough in $Y_{+\infty,\sigma}$, then this mild solution is defined on $(0, +\infty)$: we have a global solution in $\mathcal{C}([0, +\infty)(L^3)^3) \cap Y_{+\infty,\sigma}$. (A similar result holds for $\mathcal{C}([0, +\infty), (L^3(\mathbb{R}^3)^3) \cap \tilde{Y}_{+\infty,q}$).

- If $\sup_{0<t<T} t^\gamma \|\vec{f}\|_p < +\infty$ and $\lim_{t \to 0} t^\gamma \|\vec{f}\|_p = 0$ with $1 < p < 3$ and $\gamma = \frac{3}{2} - \frac{3}{2p}$, then $\int_0^t W_{\nu(t-s)} * \mathbb{P}\vec{f}(s,.) \, ds \in \mathcal{C}([0, T], (L^3(\mathbb{R}^3)^3) \cap Y_{T,\sigma} \cap \tilde{Y}_{T,q}$ for $\sigma \in (0, 1 - \gamma)$ and $q \in (3, \frac{3}{\gamma})$.

Remark: The restriction $\sigma \leq 1/2$ in the first point of the theorem may be changed into $\sigma < 1$. But in that case the proof would be more technical, as we shoud use, instead of the Sobolev embedding inequalities, the Littlewood–Paley decomposition [15, 313] and Bony's paraproduct operators [51].

Proof. We find from the Sobolev embedding inequalities that, for $0 < \sigma < 1$, $\|u\|_q \leq C_\sigma \|u\|_{\dot{H}_3^\sigma}$, with $\frac{1}{q} = \frac{1-\sigma}{3}$. If $\sigma \leq 1/2$, we have $3 < q \leq 6$ and thus $L^{q/2} \subset \dot{H}_3^{3(\frac{1}{3} - \frac{2}{q})}$, hence, for u and v in $\dot{H}_3^\sigma$, we have

$$\|uv\|_{\dot{H}_3^{2\sigma-1}} \leq C \|uv\|_{q/2} \leq C' \|u\|_{\dot{H}_3^\sigma} \|v\|_{\dot{H}_3^\sigma}.$$

We thus find

$$\|B(\vec{F}, \vec{G})\|_{\dot{H}_3^\sigma} \leq C \int_0^t \frac{1}{(t-s)^{\frac{1}{2} + \frac{1-\sigma}{2}}} \frac{ds}{s^\sigma} \|\vec{F}\|_{Y_{T,\sigma}} \|\vec{G}\|_{Y_{T,\sigma}} = C' t^{-\frac{\sigma}{2}} \|\vec{F}\|_{Y_{T,\sigma}} \|\vec{G}\|_{Y_{T,\sigma}}.$$

Thus, B is bounded on $Y_{T,\sigma}$.

In the case of $Y_{T,\sigma}$, we used the regularizing property of the heat kernel. If we use only the size of the heat kernel (or more precisely of the kernel of $W_{\nu(t-s)} * \mathbb{P} \operatorname{div}$), we shall work with Lebesgue norms and use the Young inequality on convolution between $L^{q/2}$ and L^r with $\frac{1}{r} + \frac{2}{q} - 1 = \frac{1}{q}$, to get

$$\|B(\vec{F}, \vec{G})\|_q \leq C \int_0^t \frac{1}{(t-s)^{\frac{1}{2} + \frac{3(r-1)}{2r}}} \frac{ds}{s^{2\beta}} \|\vec{F}\|_{\tilde{Y}_{T,q}} \|\vec{G}\|_{Y_{T,q}} = C' t^{-\beta} \|\vec{F}\|_{\tilde{Y}_{T,q}} \|\vec{G}\|_{\tilde{Y}_{T,q}}.$$

Since $\frac{1}{2} + \frac{3(r-1)}{2r} = 1 - (\frac{3}{2r} - 1) = 1 - \beta$ with $0 < \beta < 1$, we find that B is bounded on $\tilde{Y}_{T,q}$.

If $\vec{U} = W_{\nu t} * \vec{u}_0 + \int_0^t W_{\nu(t-s)} * \mathbb{P}\vec{f}(s,.)\,ds$ belongs to the space $\mathcal{C}([0,T], (L^3(\mathbb{R}^3)^3) \cap Y_{T,\sigma}$, we have $\lim_{T_0 \to 0} \|\vec{U}\|_{Y_{T_0,\sigma}} = 0$, thus, for T_0 small enough, we may find a solution $\vec{u}$ in $Y_{T_0,\sigma}$ of $\vec{u} = \vec{U} - B(\vec{u}, \vec{u})$. We must check that $B(\vec{u}, \vec{u})$ belongs to $\mathcal{C}([0,T], (L^3(\mathbb{R}^3)^3)$. We discuss only the case of $\vec{u} \in \tilde{Y}_{T,q}$ with $3 < q \leq 6$ (since, for $0 < \sigma < 1$, $\frac{1}{r} = \frac{1-\sigma}{3}$ and $q = \min(r,6)$, we have $\vec{U} \in \mathcal{C}([0,T], (L^3(\mathbb{R}^3)^3) \cap Y_{T,\sigma} \subset \mathcal{C}([0,T], (L^3(\mathbb{R}^3)^3) \cap \tilde{Y}_{T,r} \subset \mathcal{C}([0,T], (L^3(\mathbb{R}^3)^3) \cap \tilde{Y}_{T,q})$. Of course, we know that, for $\vec{F}$ and $\vec{G}$ in $\tilde{Y}_{T,q}$ with $q \in (3,6]$) the L^3 norm of $B(\vec{F}, \vec{G})$ is bounded:

$$\|B(\vec{F}, \vec{G})\|_3 \leq C \int_0^t \frac{1}{(t-s)^{1-2\beta}} \frac{1}{s^{2\beta}}\,ds \|\vec{F}\|_{\tilde{Y}_{T,q}} \|\vec{G}\|_{\tilde{Y}_{T,q}} = C' \|\vec{F}\|_{\tilde{Y}_{T,q}} \|\vec{G}\|_{\tilde{Y}_{T,q}}.$$

Moreover, we see that $\lim_{t \to 0^+} \|B(\vec{F}, \vec{G})\|_3 = 0$, so that continuity at $t = 0^+$ is obvious.

For proving continuity at time $t > 0$, we consider θ close to t: $|t - \theta| < \frac{1}{3}t$; let $\eta = |t - \theta|$. For $\vec{H} = B(\vec{F}, \vec{G})$, we write, for $s < \min(t, \theta)$,

$$W_{\nu(t-s)} - W_{\nu(\theta-s)} = \int_t^\theta \nu \Delta W_{\nu(\tau-s)}\,d\tau$$

so that

$$\vec{H}(t,x) - \vec{H}(\theta, x) = \int_t^\theta \left(\int_0^{t-2\eta} \Delta W_{\nu(\tau-s)} * \mathbb{P}\operatorname{div}(\vec{F} \otimes \vec{G})\,ds \right) d\tau$$

$$+ \int_{t-2\eta}^t W_{\nu(t-s)} * \mathbb{P}\operatorname{div}(\vec{F} \otimes \vec{G})\,ds$$

$$- \int_{t-2\eta}^\theta W_{\nu(\theta-s)} * \mathbb{P}\operatorname{div}(\vec{F} \otimes \vec{G})\,ds$$

and

$$\|\vec{H}(t,.) - \vec{H}(\theta,.)\|_3 \le C_{\nu,q} \int_t^\theta \left(\int_0^{t/3} \frac{1}{(\tau-s)^{2-2\beta}} \frac{ds}{s^{2\beta}} \right) d\tau \, \|\vec{F}\|_{\tilde{Y}_{T,q}} \|\vec{G}\|_{\tilde{Y}_{T,q}}$$

$$+ C_{\nu,q} \int_t^\theta \left(\int_{t/3}^{t-2\eta} \frac{1}{(\tau-s)^{2-2\beta}} \frac{ds}{s^{2\beta}} \right) d\tau \, \|\vec{F}\|_{\tilde{Y}_{T,q}} \|\vec{G}\|_{\tilde{Y}_{T,q}}$$

$$+ C_{\nu,q} \int_{t-2\eta}^t \frac{1}{(t-s)^{1-2\beta}} \frac{ds}{t^{2\beta}} \|\vec{F}\|_{\tilde{Y}_{T,q}} \|\vec{G}\|_{\tilde{Y}_{T,q}}$$

$$+ C_{\nu,q} \int_{t-2\eta}^\theta \frac{1}{(\theta-s)^{1-2\beta}} \frac{ds}{t^{2\beta}} \|\vec{F}\|_{\tilde{Y}_{T,q}} \|\vec{G}\|_{\tilde{Y}_{T,q}}$$

$$\le C'_{\nu,q} \|\vec{F}\|_{\tilde{Y}_{T,q}} \|\vec{G}\|_{\tilde{Y}_{T,q}} \left(|t-\theta|\frac{1}{t} + |t-\theta|\frac{1}{\eta^{1-2\beta}t^{2\beta}} + \eta^{2\beta}\frac{1}{t^{2\beta}} \right)$$

$$\le 3 C'_{\nu,q} \|\vec{F}\|_{\tilde{Y}_{T,q}} \|\vec{G}\|_{\tilde{Y}_{T,q}} \frac{|t-\theta|^{2\beta}}{t^{2\beta}}.$$

In order to finish the proof, we consider the case of a force density $\vec{f}$ which satisfies $\sup_{0<t<T} t^\gamma \|\vec{f}\|_p < +\infty$ and $\lim_{t\to 0} t^\gamma \|\vec{f}\|_p = 0$ with $1 < p < 3$ and $\gamma = \frac{3}{2} - \frac{3}{2p}$, and we define $\vec{F} = \int_0^t W_{\nu(t-s)} * \mathbb{P}\vec{f}(s,.) \, ds$. We have

$$\|(-\Delta)^{\sigma/2}\vec{F}(t,.)\|_3 \le C_{\nu,\gamma,\sigma} \int_0^t \frac{1}{(t-s)^{1-\gamma-\sigma}} \frac{1}{s^\gamma} ds \sup_{0<t<T} t^\gamma \|\vec{f}\|_p$$

Thus, for $0 < \sigma < 1-\gamma$, and $q = \frac{3}{1-\sigma}$, we have $\vec{F} \in Y_{T,\sigma} \cap \tilde{Y}_{T,q}$. Moreover, $\vec{F}(t,.)$ is bounded in L^3 and $\lim_{t\to 0^+} \|\vec{F}(t,.)\|_3 = 0$, whereas, for $t > 0$ and $|t - \theta| < \frac{1}{3}t$, we have

$$\|\vec{F}(t,.) - \vec{F}(\theta,.)\|_3 \le C_{\nu,\gamma} \sup_{0<s<T} s^\gamma \|\vec{f}\|_p \frac{|t-\theta|^\gamma}{t^\gamma}.$$

Thus, the theorem is proved. $\qquad\square$

7.7 Maximal Functions

For small data, existence of global solutions in L^3 may be proved in a simpler way. As a matter of fact, while the bilinear operator B is not bounded on $\mathcal{C}([0,T],(L^3)^3)$ (for the norm $L_t^\infty L_x^3$: $\|u\|_{L_t^\infty L_x^3} = \sup_{0<t<T} \|u(t,.)\|_3$), Calderón [78] and Cannone [81] showed that B is bounded on the smaller space $\{\vec{u} \in \mathcal{C}([0,T],(L^3)^3) \ / \ \text{ess sup}_{0<t<T} |\vec{u}(t,x)| \in L^3\}$ (for the norm $L_x^3 L_t^\infty$: $\|u\|_{L_x^3 L_t^\infty} = \|\text{ess sup}_{0<t<T} |u(t,x)|\|_3$).

> **Navier–Stokes equations and maximal functions**
>
> **Theorem 7.6.**
>
> - *Let $0 < T \le +\infty$. The bilinear operator B is continuous on*
>
> $$Y_T = \{\vec{u} \in \mathcal{C}([0,T),(L^3)^3) \ / \ \sup_{0<t<T} |\vec{u}(t,x)| \in L^3\}$$

- For $\vec{u}_0 \in (L^3)^3$, we have $W_{\nu t} * \vec{u}_0 \in Y_T$ with

$$\sup_{0 < t < T} |W_{\nu t} * \vec{u}_0(x)| \leq \mathcal{M}_{\vec{u}_0}(x)$$

(where $\mathcal{M}_{\vec{u}_0}$ is the Hardy–Littlewood maximal function of $\vec{u}_0$).

- If $\vec{f} \in L^1((0,T),(L^3)^3)$, then $\int_0^t W_{\nu(t-s)} * \mathbb{P}\vec{f}(s,.)\,ds \in Y_T$.

- Let $\vec{u}_0 \in (L^3(\mathbb{R}^3))^3$ and $\vec{f} \in L^1((0,T),(L^3)^3)$. There exists a positive constant ϵ_0 (independent of ν, T, $\vec{u}_0$ and $\vec{f}$), such that, if $\|\vec{u}_0\|_3 < \epsilon_0 \nu$ and $\int_0^T \|\vec{f}(s,.)\|_3\,ds < \epsilon_0 \nu$, then there exists a mild solution $\vec{u}$ of equation (7.4) on $(0,T) \times \mathbb{R}^3$ such that $\vec{u} \in Y_T$.

Remark: We have as well the following properties:

- If $\sup_{0 < t < T} t^\gamma |\mathbb{P}\vec{f}(t,x)| \in L^p$ with $1 < p < 3$ and $\gamma = \frac{3}{2} - \frac{3}{2p}$, then $\int_0^t W_{\nu(t-s)} * \mathbb{P}\vec{f}(s,.)\,ds \in Y_T$.

- Let G be the Green function $G(x) = \frac{1}{4\pi|x|}$. If, for $j = 1, \ldots, 3$, $\sup_{0 < t < T} |\vec{f}(t,x) * \partial_j G| \in L^{3/2}$, then $\int_0^t W_{\nu(t-s)} * \mathbb{P}\vec{f}(s,.)\,ds \in Y_T$.

Proof. If $\vec{u}$ and $\vec{v}$ belong to $(L_x^3 L_t^\infty)^3$, $|\vec{u}(t,x)| \leq U(x) \in L^3$, $|\vec{v}(t,x)| \leq V(x) \in L^3$, we have

$$|B(\vec{u},\vec{v})| \leq C \int \left(\int_0^t \frac{1}{\nu^2|t-s|^2 + |x-y|^4}\,ds\right) U(y)V(y)\,dy$$

$$\leq \frac{\pi C}{2\nu} \int \frac{1}{|x-y|^2} U(y)V(y)\,dy$$

Since the Riesz potential $\mathcal{I}_1$ maps $L^{3/2}$ to L^3, we find that

$$\|B(\vec{u},\vec{v})\|_{L_x^3 L_t^\infty} \leq \frac{C}{\nu} \|\vec{u}\|_{L_x^3 L_t^\infty} \|\vec{v}\|_{L_x^3 L_t^\infty}. \tag{7.41}$$

Moreover, if $\vec{u}_R = \vec{u}(t,x)1_{U(x)<R}$, $\vec{v}_R = \vec{v}(t,x)1_{V(x)<R}$, and if $T_0 \leq T$ is finite, we have: $\lim_{R \to +\infty} \|\vec{u} - \vec{u}_R\|_{L_x^3 L_t^\infty} = \lim_{R \to +\infty} \|\vec{v} - \vec{v}_R\|_{L_x^3 L_t^\infty} = 0$, while $\vec{u}_R$ and $\vec{v}_R$ belong to the space $\tilde{Y}_{T_0,6}$ described in Theorem 7.5, so that $B(\vec{u}_R, \vec{v}_R) \in \mathcal{C}([0,T_0],(L^3)^3)$. By uniform convergence, we find that $B(\vec{u},\vec{v}) \in \mathcal{C}([0,T_0],(L^3)^3)$. Thus, B is bounded on Y_T.

Now, we recall a classical lemma (see [215] for instance):

Lemma 7.4.

If ω is a radially decreasing function on $\mathbb{R}^3$ and f a locally integrable function, then

$$\left|\int_{\mathbb{R}^3} \omega(x-y)f(y)\,dy\right| \leq \|\omega\|_1 \sup_{r>0} \frac{1}{|B(0,r)|} \int_{|y|<r} |f(x-y)|\,dy \tag{7.42}$$

or equivalently

$$|\omega * f| \leq \|\omega\|_1 M_f \tag{7.43}$$

where M_f is the Hardy–Littlewood maximal function of f.

Using this lemma, it is obvious that, for $\vec{u}_0 \in (L^3)^3$, we have $W_{\nu t} * \vec{u}_0 \in Y_T$:

$$|W_{\nu t} * \vec{u}_0(x)| \le \mathcal{M}_{\vec{u}_0}(x) \in L^3 \tag{7.44}$$

Similarly, if $\vec{f} \in L^1_t L^3_x$, then $\mathcal{M}_{\mathbb{P}\vec{f}(t,.)}(x) = F(t,x) \in L^1_t L^3_x$ and $F(x) = \int_0^T F(t,x)\,dt \in L^3(\mathbb{R}^3)$; we have

$$\left| \int_0^t W_{\nu(t-s)} * \mathbb{P}\vec{f}(s,.)\,ds \right| \le \int_0^t F(s,x)\,ds \le F(x) \tag{7.45}$$

so that

$$\left\| \int_0^t W_{\nu(t-s)} * \mathbb{P}\vec{f}(s,.)\,ds \right\|_{L^3_x L^\infty_t} \le C \|\vec{f}\|_{L^1_t L^3_x} \tag{7.46}$$

Moreover, if $\vec{f}_\epsilon = \vec{f}(t,x)1_{\epsilon < |\vec{f}(t,x)| < 1/\epsilon}$, and if $T_0 \le T$ is finite, we have: $\lim_{\epsilon \to 0^+} \|\vec{f} - \vec{f}_\epsilon\|_{L^1_t L^3_x} = 0$, while $\vec{f}_\epsilon$ satisfies the assumptions of Theorem 7.5, so that $\int_0^t W_{\nu(t-s)} * \mathbb{P}\vec{f}_\epsilon(s,.)\,ds \in \mathcal{C}([0,T_0],(L^3)^3)$. By uniform convergence, we find that $\int_0^t W_{\nu(t-s)} * \mathbb{P}\vec{f}(s,.)\,ds \in \mathcal{C}([0,T_0],(L^3)^3)$. Thus, the theorem is proved. □

7.8 Basic Lemmas on Real Interpolation Spaces

Real interpolation spaces $[A_0, A_1]_{\theta,q}$ have been introduced by Lions and Peetre [338], but we shall use in the next sections only basic properties of those spaces (as described in [36, 313]), mainly for the values $q = 1$ and $q = \infty$. We shall give here a definition of those interpolation spaces and of their norms which is slightly different (but equivalent) to the classical ones:

Definition 7.1.
Let A_0, A_1 be two Banach spaces. The real interpolation spaces $[A_0, A_1]_{\theta,1}$ and $[A_0, A_1]_{\theta,\infty}$ ($0 < \theta < 1$) can be characterized through the following properties:

- $f \in [A_0, A_1]_{\theta,1}$ *if and only if it can be written in $A_0 + A_1$ as a sum $\sum_{j \in \mathbb{N}} \lambda_j f_j$ with $f_j \in A_0 \cap A_1$, $\|f\|_{A_0}^{1-\theta} \|f\|_{A_1}^{\theta} \le 1$ and $\sum_{j \in \mathbb{N}} |\lambda_j| < +\infty$. We define its norm in the following way:*

$$\|f\|_{[A_0,A_1]_{\theta,1}} = \inf_{f = \sum_{j \in \mathbb{N}} \lambda_j f_j} \sum_{j \in \mathbb{N}} |\lambda_j| \|f_j\|_{A_0}^{1-\theta} \|f_j\|_{A_1}^{\theta} \tag{7.47}$$

- $f \in [A_0, A_1]_{\theta,\infty}$ *if and only if $f \in A_0 + A_1$ and there exists a constant C such that for every $\lambda > 0$ we may decompose f as $f = f_\lambda + g_\lambda$ with $f_\lambda \in A_0$, $g_\lambda \in A_1$, $\|f_\lambda\|_{A_0} \le C\lambda^\theta$ and $\|g_\lambda\|_{A_1} \le C\lambda^{\theta-1}$. We define its norm in the following way:*

$$\|f\|_{[A_0,A_1]_{\theta,\infty}} = \sup_{\lambda > 0} \inf_{f = f_\lambda + g_\lambda} \lambda^{-\theta} \|f_\lambda\|_{A_0} + \lambda^{1-\theta} \|g_\lambda\|_{A_1} \tag{7.48}$$

Of course, for a Banach space E, we have $[E,E]_{\theta,1} = [E,E]_{\theta,\infty} = E$.
We shall need the classical lemma [36]:

Lemma 7.5.
If $A_0 \cap A_1$ is dense in A_0 and in A_1, then $[A_0, A_1]'_{\theta,1} = [A'_0, A'_1]_{\theta,\infty}$.

We shall use in the following sections the following easy interpolation results:

Lemma 7.6.

For $\eta = 1$ or $\eta = +\infty$, we write $[A_0, A_1]_{0,\eta} = A_0$ and $[A_0, A_1]_{1,\eta} = A_1$. For $\gamma \in \{1, +\infty\}$, $0 \le \tau_0 < \tau_1 \le 1$, $0 < \theta < 1$ and $\tau = (1-\theta)\tau_0 + \theta\tau_1$, if $[A_0, A_1]_{\tau_0, 1} \subset B_0 \subset [A_0, A_1]_{\tau_0, \infty}$ and $[A_0, A_1]_{\tau_1, 1} \subset B_1 \subset [A_0, A_1]_{\tau_1, \infty}$, we have

$$[B_0, B_1]_{\theta,\gamma} = [A_0, A_1]_{\tau,\gamma} \tag{7.49}$$

Proof. Case $\gamma = \infty$: If $u \in [A_0, A_1]_{\tau,\infty}$, with norm C_0, we may write $u = u_j + v_j$ with $\|u_j\|_{A_0} \le C_0 2^{-j\tau}$ and $\|v_j\|_{A_1} \le C_0 2^{j(1-\tau)}$. If $w_j = u_j - u_{j+1} = v_{j+1} - v_j$, we have $u_0 = \sum_{j \ge 0} w_j$ in A_0 and $v_0 = \sum_{j < 0} w_j \in A_1$. Thus, $u = \sum_{j \in \mathbb{Z}} w_j$. We have $w_j \in A_0 \cap A_1 \subset [A_0, A_1]_{\tau_0, 1} \subset B_0$ with

$$\|w_j\|_{B_0} \le C\|w_j\|_{A_0}^{1-\tau_0}\|w_j\|_{A_1}^{\tau_0} \le CC_0 2^{j(\tau_0(1-\tau) - \tau(1-\tau_0))} = CC_0 2^{j(\tau_0 - \tau_1)\theta}.$$

Similarly, we have

$$\|w_j\|_{B_1} \le CC_0 2^{j(\tau_0 - \tau_1)(\theta - 1)}.$$

This gives $[A_0, A_1]_{\tau,\infty} \subset [B_0, B_1]_{\theta,\infty}$.

Conversely, let $u \in [B_0, B_1]_{\theta,\infty}$, with norm C_0. For $\lambda > 0$, we may write $u = u_\lambda + v_\lambda$ with $\|u_\lambda\|_{B_0} \le C_0 \lambda^{-\theta}$ and $\|v_\lambda\|_{B_1} \le C_0 \lambda^{1-\theta}$. Since $B_0 \subset [A_0, A_1]_{\tau_0,\infty}$, we may write $u_\lambda = u_{0,\lambda} + u_{1,\lambda}$ with $\mu = \lambda^{\frac{1}{\tau_0 - \tau_1}}$ and

$$\|u_{0,\lambda}\|_{A_0} \le \|u_\lambda\|_{B_0} \mu^{\tau_0} \le C_0 \lambda^{-\theta + \frac{\tau_0}{\tau_0 - \tau_1}} = C_0 \lambda^{-\tau \frac{1}{\tau_1 - \tau_0}}$$

and similarly

$$\|u_{1,\lambda}\|_{A_1} \le \|u_\lambda\|_{B_0} \mu^{(\tau_0 - 1)} \le C_0 \lambda^{-\theta + \frac{\tau_0 - 1}{\tau_0 - \tau_1}} = C_0 \lambda^{(1-\tau)\frac{1}{\tau_1 - \tau_0}},$$

while $v_\lambda = v_{0,\lambda} + v_{1,\lambda}$ with $\nu = \lambda^{\frac{1}{\tau_0 - \tau_1}}$ and

$$\|v_{0,\lambda}\|_{A_0} \le \|v_\lambda\|_{B_1} \nu^{\tau_1} \le C_0 \lambda^{1-\theta + \frac{\tau_1}{\tau_0 - \tau_1}} = C_0 \lambda^{-\tau \frac{1}{\tau_1 - \tau_0}}$$

and similarly

$$\|v_{1,\lambda}\|_{A_1} \le \|v_\lambda\|_{B_1} \lambda^{\gamma(\tau_1 - 1)} \le C_0 \lambda^{1-\theta + \frac{\tau_1 - 1}{\tau_0 - \tau_1}} = C_0 \lambda^{(1-\tau)\frac{1}{\tau_1 - \tau_0}}.$$

This gives $[B_0, B_1]_{\theta,\infty} \subset [A_0, A_1]_{\tau,\infty}$.

Case $\gamma = 1$: Let us consider $u \in A_0 \cap A_1$. We have $\|u\|_{B_0} \le C\|u\|_{A_0}^{1-\tau_0}\|u\|_{A_1}^{\tau_0}$, $\|u\|_{B_1} \le C\|u\|_{A_0}^{1-\tau_1}\|u\|_{A_1}^{\tau_1}$, hence

$$\|u\|_{B_0}^{1-\theta}\|u\|_{B_1}^{\theta} \le C\|u\|_{A_0}^{1-\tau}\|u\|_{A_1}^{\tau}$$

and thus $[A_0, A_1]_{\tau,1} \subset [B_0, B_1]_{\theta,1}$.

Conversely, let $u \in B_0 \cap B_1$. We may decompose u into $u = u_j + v_j$ with $\|u_j\|_{A_0} \le 2^{-j}$ and $\|v_j\|_{A_1} \le C \min(\|u\|_{B_0}^{\frac{1}{\tau_0}} 2^{j\frac{1-\tau_0}{\tau_0}}, \|u\|_{B_1}^{\frac{1}{\tau_1}} 2^{j\frac{1-\tau_1}{\tau_1}})$. Hence, we have $u = \sum_{j \in \mathbb{Z}} w_j$ with

$$\|w_j\|_{A_0}^{1-\tau}\|w_j\|_{A_1}^{\tau} \le C \min(\|u\|_{B_0}^{\frac{\tau}{\tau_0}} 2^{j\frac{\tau - \tau_0}{\tau_0}}, \|u\|_{B_1}^{\frac{\tau}{\tau_1}} 2^{j\frac{\tau - \tau_1}{\tau_1}})$$

so that

$$\|u\|_{[A_0, A_1]_{\tau,1}} \le C \sum_{j \in \mathbb{Z}} \min(\|u\|_{B_0}^{\frac{\tau}{\tau_0}} 2^{j\frac{\tau - \tau_0}{\tau_0}}, \|u\|_{B_1}^{\frac{\tau}{\tau_1}} 2^{j\frac{\tau - \tau_1}{\tau_1}}) \le C'\|u\|_{B_1}^{\frac{\tau - \tau_0}{\tau_1 - \tau_0}} \|u\|_{B_0}^{\frac{\tau_1 - \tau}{\tau_1 - \tau_0}}$$

Since $\theta = \frac{\tau - \tau_0}{\tau_1 - \tau_0}$, this gives $[B_0, B_1]_{\theta,1} \subset [A_0, A_1]_{\tau,1}$. $\qquad \square$

Lemma 7.7.

Let $0 < \theta < 1$, $0 < \eta < 1$, $\theta + \eta \leq 1$.

(a) *If T is a linear operator which is bounded from A_0 to B_0 and bounded from A_1 to B_1, then it is bounded from $[A_0, A_1]_{\theta,1}$ to $[B_0, B_1]_{\theta,1}$ with operator norm*

$$\|T\|_{\mathcal{L}([A_0,A_1]_{\theta,1} \to [B_0,B_1]_{\theta,1})} \leq \|T\|_{\mathcal{L}(A_0 \to B_0)}^{1-\theta} \|T\|_{\mathcal{L}(A_1 \to B_1)}^{\theta}$$

and similarly it is bounded from $[A_0, A_1]_{\theta,\infty}$ to $[B_0, B_1]_{\theta,\infty}$ with operator norm

$$\|T\|_{\mathcal{L}([A_0,A_1]_{\theta,\infty} \to [B_0,B_1]_{\theta,\infty})} \leq \|T\|_{\mathcal{L}(A_0 \to B_0)}^{1-\theta} \|T\|_{\mathcal{L}(A_1 \to B_1)}^{\theta}$$

(b) *If T is a bilinear operator which is bounded from $A_0 \times B_0$ to C_0 and from $A_1 \times B_1$ to C_1, then it is bounded from $[A_0, A_1]_{\theta,1} \times [B_0, B_1]_{\theta,1}$ to $[C_0, C_1]_{\theta,1}$.*

(c) *If T is a bilinear operator which is bounded from $A_0 \times B_0$ to C_0, from $A_1 \times B_0$ to C_1 and from $A_0 \times B_1$ to C_1, then it is bounded from $[A_0, A_1]_{\theta,\infty} \times [B_0, B_1]_{\eta,\infty}$ to $[C_0, C_1]_{\theta+\eta,\infty}$, from $[A_0, A_1]_{\theta,1} \times [B_0, B_1]_{\eta,\infty}$ to $[C_0, C_1]_{\theta+\eta,1}$ and from $[A_0, A_1]_{\theta,\infty} \times [B_0, B_1]_{\eta,1}$ to $[C_0, C_1]_{\theta+\eta,1}$.*

Proof. (a) Assume that we have $\|T(u)\|_{B_0} \leq M_0 \|u\|_{A_0}$ and $\|T(u)\|_{B_1} \leq M_1 \|u\|_{A_1}$. For $u \in A_0 \cap A_1$, we have

$$\|T(u)\|_{B_0}^{1-\theta} \|T(u)\|_{B_1}^{1-\theta} \leq M_0^{1-\theta} M_1^{\theta} \|u\|_{A_0}^{1-\theta} \|u\|_{A_1}^{\theta}$$

and we find that T maps $[A_0, A_1]_{\theta,1}$ to $[B_0, B_1]_{\theta,1}$ with operator norm $M_\theta \leq M_0^{1-\theta} M_1^{\theta}$. Similarly, if $u \in [A_0, A_1]_{\theta,\infty}$ with norm N_θ, we may decompose u for every $\lambda > 0$ (and $\mu = \lambda \frac{M_0}{M_1}$) into $u = u_\lambda + v_\lambda$ with $\|u_\lambda\|_{A_0} \leq N_\theta \mu^{-\theta}$ and $\|v_\lambda\|_{A_1} \leq N^\theta \mu^{1-\theta}$ and we find that $T(u) = T(u_\lambda) + T(v_\lambda)$ with

$$\|T(u_\lambda)\|_{B_0} \leq M_0 N_\theta \mu^{-\theta} = M_0^{1-\theta} M_1^{\theta} N_\theta \lambda^{-\theta}$$

and

$$\|T(v_\lambda)\|_{B_1} \leq M_1 N_\theta \mu^{1-\theta} = M_0^{1-\theta} M_1^{\theta} N_\theta \lambda^{1-\theta}$$

Thus, we find that T maps $[A_0, A_1]_{\theta,\infty}$ to $[B_0, B_1]_{\theta,\infty}$ with operator norm $M_\theta \leq M_0^{1-\theta} M_1^{\theta}$.

(b) Assume that we have $\|T(u,v)\|_{C_0} \leq M_0 \|u\|_{A_0} \|v\|_{B_0}$ and $\|T(u,v)\|_{C_1} \leq M_1 \|u\|_{A_1} \|v\|_{B_1}$. If $u \in A_0 \cap A_1$ and $v \in B_0 \cap B_1$, we have

$$\|T(u,v)\|_{C_0}^{1-\theta} \|T(u,v)\|_{C_1}^{\theta} \leq M_0^{1-\theta} M_1^{\theta} \|u\|_{A_0}^{1-\theta} \|u\|_{A_1}^{\theta} \|v\|_{B_0}^{1-\theta} \|v\|_{B_1}^{\theta}$$

and we find that T maps $[A_0, A_1]_{\theta,1} \times [B_0, B_1]_{\theta,1}$ to $[C_0, C_1]_{\theta,1}$ with operator norm $M_\theta \leq M_0^{1-\theta} M_1^{\theta}$.

(c) Assume that we have $\|T(u,v)\|_{C_0} \leq M_0 \|u\|_{A_0} \|v\|_{B_0}$, $\|T(u,v)\|_{C_1} \leq M_1 \|u\|_{A_0} \|v\|_{B_1}$ and $\|T(u,v)\|_{C_1} \leq M_2 \|u\|_{A_1} \|v\|_{B_0}$. By interpolation, we find that T is bounded from $[A_0, A_1]_{\theta,1} \times B_0$ to $[C_0, C_1]_{\theta,1}$ with norm less or equal to $M_0^{1-\theta} M_2^{\theta}$ and from $[A_0, A_1]_{\theta,1} \times [B_0, B_1]_{1-\theta,1}$ to C_1 with norm less or equal to $M_0^{1-\theta} M_1^{\theta}$. By interpolation, we find that, for $0 < \eta < 1 - \theta$, T is bounded from $[A_0, A_1]_{\theta,1} \times [B_0, B_1]_{\eta,1}$ to $[C_0, C_1]_{\theta+\eta,1}$ with norm less or equal to $C(M_0^{1-\theta} M_2^{\theta})^{1-\frac{\eta}{1-\theta}} (M_0^{1-\theta} M_1^{\theta})^{\frac{\eta}{1-\theta}}$. Interpolating between two values of η gives that T is bounded from $[A_0, A_1]_{\theta,1} \times [B_0, B_1]_{\eta,\infty}$ to $[C_0, C_1]_{\theta+\eta,\infty}$; interpolating between two values of θ gives that T is bounded from $[A_0, A_1]_{\theta,\infty} \times [B_0, B_1]_{\eta,\infty}$ to $[C_0, C_1]_{\theta+\eta,\infty}$ and from $[A_0, A_1]_{\theta,1} \times [B_0, B_1]_{\eta,\infty}$ to $[C_0, C_1]_{\theta+\eta,1}$. $\quad\square$

Example: Lorentz spaces.

In this example, we are interested in $[L^1, L^\infty]_{1-\frac{1}{p},1} = L^{p,1}$ and $[L^1, L^\infty]_{1-\frac{1}{p},\infty} = L^{p,\infty}$ for $1 < p < +\infty$. Those interpolation spaces are Lorentz spaces [36] but we shall not use the measure-theoretical definition of Lorentz spaces.

Another important identification of interpolation spaces for Lebesgue spaces is the equality $[L^1, L^\infty]_{1-\frac{1}{p},\infty} = L^{p,*}$, where $L^{p,*}$ is the weak Lebesgue space introduced by Marcinkiewicz. More precisely, we have:

Lemma 7.8.

Let $1 \leq p_0 < p_1 \leq +\infty$ and $0 < \theta < 1$. Then $[L^{p_0}, L^{p_1}]_{\theta,\infty} = L^{p,*}$ with $\frac{1}{p} = \frac{1-\theta}{p_0} + \frac{\theta}{p_1}$.

Proof. If $f \in [L^{p_0}, L^{p_1}]_{\theta,\infty}$ with norm M and $\lambda > 0$, we write $f = f_\lambda + g_\lambda$ with $\|f_\lambda\|_{p_0} \leq M\mu^{-\theta}$, $\|g_\lambda\|_{p_1} \leq M\mu^{1-\theta}$ and $\mu = \left(\frac{M}{\lambda}\right)^{p\left(\frac{1}{p_1} - \frac{1}{p_0}\right)}$. We have

$$|\{x \ / \ |f(x)| > \lambda\}| \leq |\{x \ / \ |f_\lambda(x)| > \lambda/2\}| + |\{x \ / \ |g_\lambda(x)| > \lambda/2\}|$$

$$\leq M^{p_0}\mu^{-\theta p_0}\left(\frac{2}{\lambda}\right)^{p_0} + M^{p_1}\mu^{(1-\theta)p_1}\left(\frac{2}{\lambda}\right)^{p_1}$$

$$= CM^p\frac{1}{\lambda^p}.$$

Conversely, if $f \in L^{p,*}$, ($\|f\|_{p,*} = \sup_{\lambda > 0} \lambda\left(|\{x \ / \ |f(x)| > \lambda\}|\right)^{1/p} < +\infty$) we write $f_\lambda = f1_{|f(x)| > \lambda}$ and $g_\lambda = f - f_\lambda$. We have

$$\|f_\lambda\|_{p_0} \leq \sum_{j \in \mathbb{N}} \|f1_{2^j\lambda < |f(x)| \leq 2^{j+1}\lambda}\|_{p_0} \leq \sum_{j \in \mathbb{N}} 2^{(j+1)}\lambda\|f\|_{p,*}(2^j\lambda)^{-\frac{p}{p_0}}$$

$$= C\|f\|_{p,*}\lambda^{\theta\frac{p}{p_1-p_0}}$$

and

$$\|g_\lambda\|_{p_0} \leq \sum_{j \in \mathbb{N}} \|f1_{2^{-(j+1)}\lambda < |f(x)| \leq 2^{-j}\lambda}\|_{p_0} \leq \sum_{j \in \mathbb{N}} 2^{-j}\lambda\|f\|_{p,*}(2^{-(j+1)}\lambda)^{-\frac{p}{p_1}}$$

$$= C\|f\|_{p,*}\lambda^{(\theta-1)\frac{p}{p_1-p_0}}$$

Thus, the lemma is proved.   □

Consequences of those lemmas are the following ones:

- Let $1 \leq p_0 < p_1 \leq +\infty$ and $0 < \theta < 1$. Then $[L^{p_0}, L^{p_1}]_{\theta,1} = L^{p,1}$ with $\frac{1}{p} = \frac{1-\theta}{p_0} + \frac{\theta}{p_1}$.

- For $1 < p < \infty$, the dual of $L^{p,1}$ is $L^{\frac{p}{p-1},\infty}$ and the pointwise product maps $L^{p,1} \times L^{\frac{p}{p-1},\infty}$ to L^1.

- For $1 < p < \infty$, and the pointwise product maps $L^{p,1} \times L^\infty$ to $L^{p,1}$ and $L^{p,\infty} \times L^\infty$ to $L^{p,\infty}$.

- For $1 < p < \infty$ and $\frac{p}{p-1} < q < \infty$, the pointwise product maps $L^{p,1} \times L^{q,\infty}$ to $L^{r,1}$ and $L^{p,\infty} \times L^{q,\infty}$ to $L^{r,\infty}$ with $\frac{1}{r} = \frac{1}{p} + \frac{1}{q}$.

Example: Besov spaces.

Let E be a Banach space such that:

- $E \subset \mathcal{S}'(\mathbb{R}^3)$

- E is stable under convolution with L^1: $\|f * g\|_E \leq \|f\|_1 \|g\|_E$.

We define the potential space H_E^s as the space of distributions f such that $(Id-\Delta)^{s/2}f \in E$ (with norm $\|f\|_{H_E^s} = \|(Id-\Delta)^{s/2}f\|_E$). Then, the Besov spaces $B_{E,\infty}^s$ and $B_{E,1}^s$ are defined in the following way: let $s_0 < s < s_1$ and $\theta \in (0,1)$ such that $s = (1-\theta)s_0 + \theta s_1$, then $[H_E^{s_0}, H_E^{s_1}]_{\theta,\infty} = B_{E,\infty}^s$ and $[H_E^{s_0}, H_E^{s_1}]_{\theta,1} = B_{E,1}^s$. Of course, we must check that $B_{E,\infty}^s$ and $B_{E,1}^s$ depend only on s and not on s_0 and s_1.

Since $(Id-\Delta)^{\sigma/2}$ is an isomorphism between H^s and $H^{s-\sigma}$, we may assume with no loss of generality that $s_1 < 0$:

Lemma 7.9.
If $s_0 < s < s_1 < 0$, $\theta \in (0,1)$ and $s = (1-\theta)s_0 + \theta s_1$, then $f \in [H_E^{s_0}, H_E^{s_1}]_{\theta,\infty}$ if and only if $\sup_{0<t<1} t^{|s|/2}\|W_t * f\|_E < +\infty$. Moreover, the norm of f in $[H_E^{s_0}, H_E^{s_1}]_{\theta,\infty}$ is equivalent to $\sup_{0<t<1} t^{|s|/2}\|W_t * f\|_E < +\infty$.

Proof. The proof is based on the following representation of $(Id-\Delta)^{-\tau/2}$ when τ is a positive real number:

$$(Id-\Delta)^{-\tau/2}f = \frac{1}{\Gamma(\tau/2)} \int_0^{+\infty} e^{-t} W_t * f \; t^{\tau/2} \frac{dt}{t} \tag{7.50}$$

(easily checked through the Fourier transform) and on the inequality

$$\text{for } \sigma > 0, \; \sup_{0<t} \min(1,t)^{\sigma/2}\|(Id-\Delta)^{\sigma/2}W_t\|_1 < +\infty. \tag{7.51}$$

(7.51) is obvious for σ an even integer $\sigma = 2N$. For σ a general positive number, we write $\sigma = 2N + (\sigma - 2N) = 2N - \tau$ with $2N > \sigma$ and we use (7.50) to get

$$\|(Id-\Delta)^{\sigma/2}W_t\|_1 \leq \frac{1}{\Gamma(\frac{2N-\sigma}{2})} \int_0^{+\infty} e^{-s}\|(Id-\Delta)^{2N}W_{t+s}\|_1 \; s^{\frac{2N-\sigma}{2}} \frac{ds}{s}$$

$$\leq C \int_0^{+\infty} e^{-s} \max(1,(t+s)^{-N}) \; s^{\frac{2N-\sigma}{2}} \frac{ds}{s}.$$

If $t > 1$, we find

$$\|(Id-\Delta)^{\sigma/2}W_t\|_1 \leq C \int_0^{+\infty} e^{-s} \; s^{\frac{2N-\sigma}{2}} \frac{ds}{s} = C\,\Gamma(N - \frac{s}{2}).$$

If $t \leq 1$, we have

$$\|(Id-\Delta)^{\sigma/2}W_t\|_1 \leq Ct^{-N} \int_0^t s^{\frac{2N-\sigma}{2}} \frac{ds}{s}$$

$$+ C \int_t^1 \frac{ds}{s^{1+\frac{\sigma}{2}}} + C \int_1^{+\infty} e^{-s} \; s^{\frac{2N-\sigma}{2}} \frac{ds}{s}$$

$$\leq C't^{-\sigma/2}.$$

Thus, (7.51) is proven.

Now, if $f \in [H_E^{s_0}, H_E^{s_1}]_{\theta,\infty}$ with norm M, we decompose f into $f = f_\lambda + g_\lambda$ with $\|(Id-\Delta)^{s_0/2}f_\lambda\|_E \leq M\lambda^{-\theta}$ and $\|(Id-\Delta)^{s_1/2}g_\lambda\|_E \leq M\lambda^{1-\theta}$ for $\lambda = t^{\frac{|s_1|-|s_0|}{2}}$, and we get

$$\|W_t * f\|_E \leq C(M\lambda^{-\theta}t^{-|s_0|/2} + M\lambda^{1-\theta}t^{-|s_1|/2}) = 2CMt^{-\frac{|s|}{2}}.$$

Conversely, if $\sup_{0<t<1} t^{|s|/2}\|W_t * f\|_E = M < +\infty$, we use the identity $(Id-\Delta)^{s_0/2}f = \frac{1}{\Gamma(|s_0|/2)} \int_0^{+\infty} e^{-t} W_t * f \; t^{|s_0|/2} \frac{dt}{t}$, hence

$$\|(Id-\Delta)^{s_0/2}f\|_E \leq \frac{1}{\Gamma(|s_0|/2)} M \int_0^{+\infty} e^{-t}t^{|s_0|/2} \max(1,t^{-|s|/2})\frac{dt}{t} < +\infty.$$

Thus, $f \in H_E^{s_0}$. In order to prove that $f \in [H_E^{s_0}, H_E^{s_1}]_{\theta,\infty}$, we must check that for all $\lambda > 1$, we have $f = f_\lambda + g_\lambda$ with $\|f_\lambda\|_{H_E^{s_0}} \leq CM\lambda^{-\theta}$ and $\|g_\lambda\|_{H_E^{s_1}} \leq CM\lambda^{1-\theta}$. Let $A \in (0,1)$. We write $f = u_A + v_A$ with

$$u_A = \frac{1}{\Gamma(|s_0|/2)}(Id - \Delta)^{|s_0|/2} \int_0^A e^{-t} W_t * f \, t^{|s_0|/2} \frac{dt}{t}$$

and

$$v_A = \frac{1}{\Gamma(|s_0|/2)}(Id - \Delta)^{|s_0|/2} \int_A^{+\infty} e^{-t} W_{t/2} * W_{t/2}t * f \, t^{|s_0|/2} \frac{dt}{t}$$

We have

$$\|u_A\|_{H_E^{s_0}} \leq \frac{1}{\Gamma(|s_0|/2)} M \int_0^A t^{\frac{|s_0|-|s|}{2}} \frac{dt}{t} = \frac{2M}{\Gamma(|s_0|/2)(|s_0| - |s|)} A^{\frac{s-s_0}{2}}$$

and

$$\|v_A\|_{H_E^{s_1}} \leq CM \int_A^{+\infty} e^{-t} \max(1, (\tfrac{t}{2})^{(s_0-s_1)/2}) \max(1, (\tfrac{t}{2})^{s/2}) t^{-s_0/2} \frac{dt}{t}$$
$$\leq C'M A^{\frac{s-s_1}{2}}$$

We conclude the proof by taking $A = \lambda^{-\frac{2}{s_1-s_0}}$. $\qquad\qquad\square$

With Lemma 7.9, we have now a non-ambiguous definition of Besov spaces $B_{E,1}^s$ and $B_{E,\infty}^s$. We have the following properties for Besov spaces:

- for $E = L^p$, we find the usual Besov spaces $B_{L^p,1}^s = B_{p,1}^s$ and $B_{L^p,\infty}^s = B_{p,\infty}^s$ [36, 475].

- If $\mathcal{S}$ is dense in E (so that $E' \subset \mathcal{S}'$), we have $(B_{E,1}^s)' = B_{E',\infty}^{-s}$.

Example: homogeneous Besov spaces

Homogeneous Besov spaces may be defined in a similar way, at least for negative indexes, replacing $(Id - \Delta)^{\sigma/2}$ with the fractional Laplacian operators $(-\Delta)^{\sigma/2}$.

Let again E be a Banach space such that:

- $E \subset \mathcal{S}'(\mathbb{R}^3)$

- E is stable under convolution with L^1: $\|f * g\|_E \leq \|f\|_1 \|g\|_E$.

Then, for a positive σ, $(-\Delta)^{\sigma/2}$ may be defined on E in the following way: if $\omega \in \mathcal{S}$ is such that its Fourier transform $\hat{\omega}$ is compactly supported and is identically equal to 1 on a neighborhood of 0, then $|\xi|^\sigma (1 - \hat{\omega}(\xi))$ is a pointwise multiplier of $\mathcal{S}'$ (so that $(-\Delta)^{\sigma/2}(\delta - \omega)$ is a convolutor of $\mathcal{S}'$, where δ is the Dirac mass at the origin) while $|\xi|^\sigma \hat{\omega}(\xi)$ is the Fourier transform of a function $(-\Delta)^{-\sigma/2}\omega \in L^1$. Then, $(-\Delta)^{\sigma/2}$ is defined on E by:

$$(-\Delta)^{\sigma/2} f = ((-\Delta)^{\sigma/2}\omega) * f + ((-\Delta)^{\sigma/2}(\delta - \omega)) * f \qquad (7.52)$$

This definition does not depend on the choice of ω, and we have, for positive σ and τ, $(-\Delta)^{\sigma/2}((-\Delta)^{\tau/2}f) = (-\Delta)^{(\sigma+\tau)/2}f$.

Definition 7.2.
Let E be a Banach space of tempered distributions that is stable under convolution with L^1. We define, for positive σ, the Banach space $\dot{H}_E^{-\sigma}$ in the following way:

$$f \in \dot{H}_E^{-\sigma} \Leftrightarrow \exists g \in E \ f = (-\Delta)^{\sigma/2} g$$

and

$$\|f\|_{\dot{H}_E^{-\sigma}} = \inf_{f=(-\Delta)^{\sigma/2}g} \|g\|_E.$$

If $g \in E$ satisfies $(-\Delta)^{\sigma/2}g = 0$, then $\hat{g}$ is supported in $\{0\}$, hence g is a polynomial function; moreover, we have

$$|W * g(x)| = |\langle W_{1/2}(-z)| \int W_{1/2}(x+z-y)g(y)\,dy\rangle| \leq C\|W_{1/2}(x+.)*g\|_E \leq C'\|g\|_E$$

Since $W * g$ is a polynomial with the same degree as g, we find that g must be constant. Thus, $\dot{H}_E^{-\sigma}$ is isomorphic to the Banach space $E/(E \cap \mathbb{R}\,1)$.

The definition of $\dot{H}_E^{\sigma}$ for positive σ is not so direct. If there exists another Banach space F of tempered distributions that is stable under convolution with L^1 and a positive β_E such that

$$\sup_{t\geq 1} t^{\beta_E/2}\|W_t * f\|_F \leq C\|f\|_E$$

then we may define $(-\Delta)^{-\sigma/2}$ on E (with values in $E+F$) for $0 < \sigma < \beta$ as:

$$(-\Delta)^{-\sigma/2}f = \frac{1}{\Gamma(\sigma/2)} \int_0^{+\infty} W_t * f \ t^{\sigma/2}\frac{dt}{t} \tag{7.53}$$

We then define $\dot{H}_E^{\sigma}$ in the following way:

$$f \in \dot{H}_E^{\sigma} \Leftrightarrow \exists g \in E \ f = (-\Delta)^{-\sigma/2}g$$

and

$$\|(-\Delta)^{-\sigma/2}g\|_{\dot{H}_E^{\sigma}} = \|g\|_E.$$

Similarly, we define $\dot{H}_E^0 = E$. If $-\infty < \sigma < \beta_E$, $-\infty < \tau < \beta_E$ then $(-\Delta)^{(\sigma-\tau)/2}$ is an isomorphism from $\dot{H}_E^{\sigma}$ to $\dot{H}_E^{\tau}$.

We define the homogeneous Besov spaces $\dot{B}_{E,\infty}^s$ and $\dot{B}_{E,1}^s$ in the following way: let $s_0 < s < s_1 < \beta_E$ and $\theta \in (0,1)$ such that $s = (1-\theta)s_0 + \theta s_1$, then $[\dot{H}_E^{s_0}, \dot{H}_E^{s_1}]_{\theta,\infty} = \dot{B}_{E,\infty}^s$ and $[\dot{H}_E^{s_0}, \dot{H}_E^{s_1}]_{\theta,1} = \dot{B}_{E,1}^s$. Of course, $\dot{B}_{E,\infty}^s$ and $\dot{B}_{E,1}^s$ depend only on s and not on s_0 and s_1. In particular, we have

Lemma 7.10.
If $s < 0$, then

$$f \in \dot{B}_{E,\infty}^s \text{ if and only if } \sup_{0<t} t^{|s|/2}\|W_t * f\|_E < +\infty.$$

Moreover, the norm of f in $\dot{B}_{E,\infty}^s$ is equivalent to $\sup_{0<t} t^{|s|/2}\|W_t * f\|_E$.

For $E = L^p$, we find the usual homogeneous Besov spaces $\dot{B}_{L^p,1}^s = \dot{B}_{p,1}^s$ and $\dot{B}_{L^p,\infty}^s = \dot{B}_{p,\infty}^s$ with $\beta_E = \frac{3}{p}$ [36, 475].

Lemma 7.11.
If S is dense in E (so that $E' \subset S'$), we have $(\dot{B}_{E,1}^s)' = \dot{B}_{E',\infty}^{-s}$ for $-\beta_{E'} < s < \beta_E$.

Proof. Let $\omega \in \mathcal{S}$ be such that its Fourier transform $\hat{\omega}$ is compactly supported and is identically equal to 1 on a neighborhood of 0. Let Δ_j be the convolution operator with $2^{3(j+1)}\omega(2^{j+1}x) - 2^{3j}\omega(2^j x)$. If $f \in \dot{H}^{s_0}_E \cap \dot{H}^{s_1}_E$, we have $\|\Delta_j f\|_E \leq C \min(2^{-js_0}\|f\|_{\dot{H}^{s_0}_E}, 2^{-js_1}\|f\|_{\dot{H}^{s_1}_E})$. From this, we get that $\sum_{j\in\mathbb{Z}} 2^{js}\|\Delta_j f\|_E \leq C\|f\|_{\dot{H}^{s_0}_E}^{\frac{s_1-s}{s_1-s_0}}\|f\|_{\dot{H}^{s_1}_E}^{\frac{s-s_0}{s_1-s_0}}$. Finally, we get[2]

$$f \in \dot{B}^s_{E,1} \Leftrightarrow f = \sum_{j\in\mathbb{Z}} \Delta_j f \quad \text{with} \quad \sum_{j\in\mathbb{Z}} 2^{js}\|\Delta_j f\|_E < +\infty.$$

Hence, we get that $T \in (\dot{B}^s_{E,1})'$ if and only if it can be written as a *-weak convergent series $T = \sum_{j\in\mathbb{Z}} \Delta_j T_j$ with $\sup_{j\in\mathbb{Z}}\|T_j\|_{E'} 2^{-js} < +\infty$. We have $\|\Delta_j T_j\|_{\dot{H}^{-s_0}_{E'}} \leq C 2^{-js_0}\|T\|_{E'}$ and $\|\Delta_j T_j\|_{\dot{H}^{-s_1}_{E'}} \leq C 2^{-js_1}\|T\|_{E'}$, so that $T \in \dot{B}^{-s}_{E',\infty}$. $\qquad\square$

7.9   Uniqueness of L^3 Solutions

The results of Kato [255] left an open question: uniqueness for mild solutions in $\mathcal{C}([0,T),(L^3)^3)$. Given $\vec{u}_0 \in (L^3)^3$, we may not directly construct a solution in $\mathcal{C}([0,T),(L^3)^3)$, since B is not continuous on $\mathcal{C}([0,T),(L^3)^3)$ (Oru [383]). Solutions are always constructed in a smaller space (see Kato [255], Giga [209], Cannone [81], or Planchon [399]) and, thus, uniqueness was first granted only in the subspaces of $\mathcal{C}([0,T),(L^3)^3)$ where the iteration algorithm was convergent. In 1997, Furioli, Lemarié-Rieusset and Terraneo [186, 187] proved uniqueness in $\mathcal{C}([0,T),(L^3)^3)$:

Uniqueness of L^3 solutions

Theorem 7.7.
If $\vec{u}$ and $\vec{v}$ are two solutions of the Equation (7.4) on $(0,T) \times \mathbb{R}^3$ with $\vec{u}_0 \in (L^3)^3$ and $\int_0^t W_{\nu(t-s)} * \mathbb{P}\vec{f}(s,.)\,ds \in \mathcal{C}([0,T),(L^3(\mathbb{R}^3))^3)$ so that $\vec{u}$ and $\vec{v}$ belong to $\mathcal{C}([0,T),(L^3(\mathbb{R}^3))^3)$, then $\vec{u} = \vec{v}$.

Proof. In the two years following its proof by Furioli, Lemarié-Rieusset and Terraneo [186], this theorem was reproved by many authors through various methods (Meyer [359], Monniaux [368], Lions and Masmoudi [340]) and was extended to the case of Morrey-Campanato spaces by Furioli, Lemarié-Rieusset and Terraneo [187] and May [325, 353], as we shall see in Section 8.3.

Here, we will sketch the proof of Furioli, Lemarié-Rieusset and Terraneo [186] and its adaptation by Meyer [359] and Monniaux [368].

The first step of the proof is the reduction to prove local uniqueness. If we prove that under the assumptions of Theorem 7.7 there exists some positive ϵ so that $\vec{u} = \vec{v}$ on $[0, \epsilon]$, we can end the proof in the following way:

- let $E = \{\tau \in [0,T) \ / \ \|\vec{u}(.,.) - \vec{v}(t,.)\|_3 = 0 \text{ on } [0, \tau]\}$; we have $0 \in E$: $\vec{u}(0,.) = \vec{v}(0,.)$;

---

[2]This is the homogeneous Littlewood–Paley decomposition of f.

- let $\tau^* = \sup_{\tau \in E} \tau$; if $\tau^* < T$, then, by continuity, we have $\vec{u}(\tau^*, .) = \vec{v}(\tau^*, .)$ in $(L^3)^3$, so that $\tau^* \in E$; moreover, we have, on $[0, T - \tau^*)$,

$$\vec{u}(t + \tau^*, x) = W_{\nu t} * \vec{u}(\tau^*, x) + \int_0^t W_{\nu(t-s)} * \mathbb{P}\vec{f}(s + \tau^*, .) \, ds$$

$$- \int_0^t W_{\nu(t-s)} * \mathbb{P} \operatorname{div}(\vec{u}(s + \tau^*, .) \otimes \vec{u}(s + \tau^*, ;)) \, ds$$

and the same equation for $\vec{v}$. Remark that $\int_0^t W_{\nu(t-s)} * \mathbb{P}\vec{f}(s+\tau^*, .) \, ds$ may be written as

$$\int_0^{t+\tau^*} W_{\nu(t-s)} * \mathbb{P}\vec{f}(s, .) \, ds - W_{\nu t} * \int_0^{\tau^*} W_{\nu(\tau^*-s)} * \mathbb{P}\vec{f}(s, .) \, ds$$

and thus fulfills hypotheses of Theorem 7.7. Thus, if we have local uniqueness, there exists a positive ϵ such that $\vec{u}(s+\tau^*, .) = \vec{v}(s+\tau^*, .)$ for $0 \le s \le \epsilon$, so that $\tau^* + \epsilon \in E$. This is a contradiction with the definition of τ^*. Thus $\tau^* = T$ and $\vec{u} = \vec{v}$ for all $t \in [0, T)$.

The second step is to write the equation $\vec{u} = \vec{v}$ as a fixed-point problem in $\vec{w} = \vec{u} - \vec{v}$: we have $\vec{u} = \vec{U} - B(\vec{u}, \vec{u})$ and $\vec{v} = \vec{U} - B(\vec{v}, \vec{v})$, so that, writing $\vec{u} = W_{\nu t} * \vec{u}_0 - \vec{u}_1$, $\vec{v} = W_{\nu t} * u_0 - \vec{v}_1$, we get

$$\vec{w} = B(\vec{u}_1, \vec{w}) + B(\vec{w}, \vec{v}_1) - B(W_{\nu t} * \vec{u}_0, \vec{w}) - B(\vec{w}, W_{\nu t} * \vec{u}_0). \tag{7.54}$$

The idea is that $\vec{u}_1$ and $\vec{v}_1$ are small in L^3 norm: $\lim_{t \to 0} \|\vec{u}_1(t, .)\|_3 = \lim_{t \to 0} \|\vec{v}_1(t, .)\|_3 = 0$, while $W_{\nu t} * \vec{u}_0$ is small in other norms (for instance, $\lim_{t \to 0} \sqrt{t} \|W_{\nu t} * \vec{u}_0\|_\infty = 0$), so that we may hope to find a contractive estimate to prove that $\vec{w} = 0$.

This contractive estimate is not to be hoped in terms of the norm of $\vec{w}$ in $\mathcal{C}([0, \epsilon), (L^3(\mathbb{R}^3))^3$, as we know that the bilinear operator B is not bounded on $\mathcal{C}([0, \epsilon), (L^3(\mathbb{R}^3))^3$. Thus, the third step is to identify a norm on $\vec{w}$ for which one has a contractive estimate.

- **Besov norms:** The proof of Furioli, Lemarié–Rieusset and Terraneo is based on basic inequalities on Besov norms. In the following inequalities, the constants C_i, $i = 1, \ldots,$ depend on ν and p but not on τ:

1. for $3/2 < p < 3$ and $0 < \eta < 1$ and $\varphi \in \mathcal{D}(\mathbb{R}^3)$, we have the inequality $\int_0^{+\infty} \|W_{\nu t} * \sqrt{-\Delta}\varphi\|_{\dot{B}^{1-\eta}_{\frac{p}{p-1},1}} \, ds \le C_1 \|\varphi\|_{\dot{B}^{-\eta}_{\frac{p}{p-1},1}}$: it is enough to check that

$$\int_0^{+\infty} \|W_{\nu t} * \sqrt{-\Delta}\varphi\|_{\dot{B}^{1-\eta}_{\frac{p}{p-1},1}} \, ds \le C_0 \int_0^{+\infty} \min\left(\frac{\|\varphi\|_{\dot{B}^{-\eta-1}_{\frac{p}{p-1},1}}}{s^{3/2}}, \frac{\|\varphi\|_{\dot{B}^{-\eta+1}_{\frac{p}{p-1},1}}}{s^{1/2}}\right) \, ds$$

$$= C_1 \sqrt{\|\varphi\|_{\dot{B}^{-\eta-1}_{\frac{p}{p-1},1}} \|\varphi\|_{\dot{B}^{-\eta+1}_{\frac{p}{p-1},1}}}$$

2. writing

$$\left| \int (\vec{u}(t, x) - \vec{U}(t, x)) . \vec{\varphi}(x) \, dx \right|$$

$$\le C_2 \int_0^t \|\vec{u}(s, .) \otimes \vec{u}(s, .)\|_{\dot{B}^{\eta-1}_{p,\infty}} \|W_{\nu t} * \sqrt{-\Delta}\varphi\|_{\dot{B}^{1-\eta}_{\frac{p}{p-1},1}} \, ds$$

for $3/2 < p < 3$ and $\eta = \frac{3}{p} - 1$, we get

$$\sup_{0<t<\tau} \|\vec{u}(t,.) - \vec{U}(t,.)\|_{\dot{B}_{p,\infty}^{\frac{3}{p}-1}} \leq C_3 \sup_{0<t<\tau} \|\vec{u}(t,.) \otimes \vec{u}(t,.)\|_{\dot{B}_{p,\infty}^{\frac{3}{p}-2}}$$

$$\leq C_4 \sup_{0<t<\tau} \|\vec{u}(t,.) \otimes \vec{u}(t,.)\|_{3/2}$$

$$\leq C_5 \sup_{0<t<\tau} \|\vec{u}(t,.)\|_3^2$$

3. similarly, $\sup_{0<t<\tau} \|\vec{v}(t,.) - \vec{U}(t,.)\|_{\dot{B}_{p,\infty}^{\frac{3}{p}-1}} \leq C_5 \sup_{0<t<\tau} \|\vec{v}(t,.)\|_3^2$, so that $\sup_{0<t<\tau} \|\vec{w}(t,.)\|_{\dot{B}_{p,\infty}^{\frac{3}{p}-1}} < +\infty$

4. for $3/2 < p < 3$, $\eta = \frac{3}{p} - 1$ and $0 < \epsilon < \min(\eta, 1 - \eta)$, we have $\dot{H}_p^{\eta+\epsilon} \subset L^r$ with $\frac{1}{r} = \frac{1-\epsilon}{3}$ and $\dot{H}_p^{\eta-\epsilon} \subset L^\rho$ with $\frac{1}{\rho} = \frac{1+\epsilon}{3}$ (Sobolev embeddings) and, by duality of the Sobolev embeddings, we have $L^s \subset \dot{H}_p^{\eta+\epsilon-1}$ with $\frac{1}{s} = \frac{2-\epsilon}{3}$ and $L^\sigma \subset \dot{H}_p^{\eta-\epsilon-1}$ with $\frac{1}{\sigma} = \frac{2+\epsilon}{3}$. thus, pointwise multiplication with a function in L^3 maps $\dot{H}_p^{\eta+\epsilon}$ to $\dot{H}_p^{\eta+\epsilon-1}$ and $\dot{H}_p^{\eta-\epsilon}$ to $\dot{H}_p^{\eta-\epsilon-1}$, and by interpolation $\dot{B}_{p,\infty}^\eta$ to $\dot{B}_{p,\infty}^{\eta-1}$.

5. for $3/2 < p < 3$,

$$\sup_{0<t<\tau} \|B(\vec{u}_1, \vec{w})\|_{\dot{B}_{p,\infty}^{\frac{3}{p}-1}} \leq C_3 \sup_{0<t<\tau} \|\vec{u}_1(t,.) \otimes \vec{w}(t,.)\|_{\dot{B}_{p,\infty}^{\frac{3}{p}-2}}$$

$$\leq C_6 \sup_{0<t<\tau} \|\vec{u}_1(t,.)\|_3 \|\vec{w}(t,.)\|_{\dot{B}_{p,\infty}^{\frac{3}{p}-1}}$$

$$\leq C_6 \sup_{0<t<\tau} \|\vec{u}_1(t,.)\|_3 \sup_{0<t<\tau} \|\vec{w}(t,.)\|_{\dot{B}_{p,\infty}^{\frac{3}{p}-1}}$$

and

$$\sup_{0<t<\tau} \|B(\vec{w}, \vec{v}_1)\|_{\dot{B}_{p,\infty}^{\frac{3}{p}-1}} \leq C_6 \sup_{0<t<\tau} \|\vec{v}_1(t,.)\|_3 \sup_{0<t<\tau} \|\vec{w}(t,.)\|_{\dot{B}_{p,\infty}^{\frac{3}{p}-1}}$$

6. similarly, if $3 < r < \frac{3}{\eta}$, we find (by interpolating Sobolev embedding inequalities) that pointwise multiplication with a function in L^r maps $\dot{B}_{p,\infty}^\eta$ to $\dot{B}_{p,\infty}^{\eta-\frac{3}{r}}$.

7. for $3/2 < p < 3$ and $3 < r < \frac{3p}{3-p}$, we find

$$\|B(W_{\nu t} * \vec{u}_0, \vec{w})(t,.)\|_{\dot{B}_{p,\infty}^{\frac{3}{p}-1}}$$

$$\leq C_7 \int_0^t \frac{1}{(t-s)^{\frac{1}{2}+\frac{3}{2r}}} \|(W_{\nu s} * \vec{u}_0) \otimes \vec{w}(s,.)\|_{\dot{B}_{p,\infty}^{\frac{3}{p}-1-\frac{3}{r}}} \, ds$$

$$\leq C_8 \sup_{0<s<t} s^{\frac{1}{2}-\frac{3}{2r}} \|W_{\nu s} * \vec{u}_0\|_r \|\vec{w}(s,.)\|_{\dot{B}_{p,\infty}^{\frac{3}{p}-1}}$$

8. Thus, we find

$$\sup_{0<t<\tau} \|\vec{w}(t,.)\|_{\dot{B}_{p,\infty}^{\frac{3}{p}-1}} \leq C_9 A(\tau) \sup_{0<t<\tau} \|\vec{w}(t,.)\|_{\dot{B}_{p,\infty}^{\frac{3}{p}-1}}$$

with

$$A(\tau) = \sup_{0<t<\tau} \|\vec{u}_1(t,.)\|_3 + \sup_{0<t<\tau} \|\vec{v}_1(t,.)\|_3 + \sup_{0<t<\tau} t^{\frac{1}{2}-\frac{3}{2r}} \|W_{\nu t} * \vec{u}_0\|_r.$$

As we have $\lim_{\tau \to 0^+} A(\tau) = 0$, the theorem is proved.

- **Lorentz norms**: Meyer's proof follows the lines of the proof of Furioli, Lemarié-Rieusset and Terraneo, but deals with the Lorentz space $L^{p,\infty}$ instead of the Besov space $\dot{B}_{p,\infty}^{\frac{3}{p}-1}$:

  1. for $\varphi \in \mathcal{D}(\mathbb{R}^3)$, we have the inequality $\int_0^{+\infty} \|W_{\nu t} * \sqrt{-\Delta}\varphi\|_{L^{3,1}} ds \leq C_1 \|\varphi\|_{L^{3/2,1}}$: it is enough to check that

$$\int_0^{+\infty} \|W_{\nu t} * \sqrt{-\Delta}\varphi\|_{L^{3,1}} ds \leq C_0 \int_0^{+\infty} \min\left(\frac{\|\varphi\|_{L^{\frac{3}{2+3\epsilon}}}}{s^{1+\frac{3\epsilon}{2}}}, \frac{\|\varphi\|_{L^{\frac{3}{2-3\epsilon}}}}{s^{1-\frac{3\epsilon}{2}}}\right) ds$$

$$= C_1 \sqrt{\|\varphi\|_{L^{\frac{3}{2-3\epsilon}}} \|\varphi\|_{L^{\frac{3}{2+3\epsilon}}}}$$

  2. from this inequality, we get

$$\sup_{0<t<\tau} \|B(\vec{u}_1, \vec{w})\|_{L^{3,\infty}} \leq C_2 \sup_{0<t<\tau} \|\vec{u}_1(t,.) \otimes \vec{w}(t,.)\|_{L^{3/2,\infty}}$$

$$\leq C_3 \sup_{0<t<\tau} \|\vec{u}_1(t,.)\|_3 \sup_{0<t<\tau} \|\vec{w}(t,.)\|_{L^{3,\infty}}$$

and

$$\sup_{0<t<\tau} \|B(\vec{w}, \vec{v}_1)\|_{L^{3,\infty}} \leq C_3 \sup_{0<t<\tau} \|\vec{v}_1(t,.)\|_3 \sup_{0<t<\tau} \|\vec{w}(t,.)\|_{L^{3,\infty}}$$

  3. moreover, we have

$$\|B(W_{\nu t} * \vec{u}_0, \vec{w})(t,.)\|_{L^{3,\infty}}$$

$$\leq C_4 \int_0^t \frac{1}{(t-s)^{\frac{3}{4}}} \|(W_{\nu s} * \vec{u}_0) \otimes \vec{w}(t,.)\|_{L^{2,\infty}} ds$$

$$\leq C_5 \sup_{0<s<t} s^{\frac{1}{4}} \|W_{\nu s} * \vec{u}_0\|_6 \|\vec{w}(s,.)\|_{L^{3,\infty}}$$

  4. thus, we find

$$\sup_{0<t<\tau} \|\vec{w}(t,.)\|_{L^{3,\infty}} \leq C_6 A(\tau) \sup_{0<t<\tau} \|\vec{w}(t,.)\|_{L^{3,\infty}}$$

with

$$A(\tau) = \sup_{0<t<\tau} \|\vec{u}_1(t,.)\|_3 + \sup_{0<t<\tau} \|\vec{v}_1(t,.)\|_3 + \sup_{0<t<\tau} t^{\frac{1}{4}} \|W_{\nu t} * \vec{u}_0\|_6.$$

As we have $\lim_{\tau \to 0^+} A(\tau) = 0$, the theorem is proved.

- **$L^p L^q$ maximal regularity**: Monniaux's proof replaces the role of real interpolation by the $L^p L^q$ maximal regularity for the heat kernel [313] and shows that it is easier to estimate $\int_0^\tau \|\vec{w}(t,.)\|_3^p dt$ (for $2 < p < +\infty$) than $\sup_{0<t<\tau} \|\vec{w}(t,.)\|_3$. Monniaux's proof is then the following one:

  1. $L^p L^q$ maximal regularity for the heat kernel states that, for $1 < p < +\infty$ and $1 < q < +\infty$, there exists a constant C_1 (which depends on p, q and ν, but not on τ) such that

$$\left\| \int_0^t W_{\nu(t-s)} * \Delta F(s,.) ds \right\|_{L^p((0,\tau), L^q)} \leq C_1 \|F\|_{L^p((0,\tau), L^q)}$$

2. we have

$$\|\frac{1}{\Delta}\mathbb{P}\operatorname{div}(\vec{u}_1(s,.)\otimes\vec{w}(s,.))\|_3 \leq C_2\|\mathcal{I}_1(\vec{u}_1(s,.)\otimes\vec{w}(s,.))\|_3$$

$$\leq C_3\|\vec{u}_1(s,.)\|_3\|\vec{w}(s,.)\|_3$$

3. from this inequality and maximal regularity, we get, for $1 < p < +\infty$,

$$\|B(\vec{u}_1,\vec{w})\|_{L^p((0,\tau),L^3)} \leq C_4 \sup_{0<t<\tau} \|\vec{u}_1(t,.)\|_3\|\vec{w}(t,.)\|_{L^p((0,\tau),L^3)}$$

4. moreover, we have

$$\|B(W_{\nu t}*\vec{u}_0,\vec{w})(t,.)\|_3 \leq C_5 \int_0^t \frac{1}{(t-s)^{\frac{1}{2}}}\|(W_{\nu s}*\vec{u}_0)\otimes\vec{w}(t,.)\|_3\,ds$$

$$\leq C_6 \sup_{0<s<t} s^{\frac{1}{2}}\|W_{\nu s}*\vec{u}_0\|_\infty \int_0^t \frac{1}{s^{\frac{1}{2}}}\frac{1}{(t-s)^{\frac{1}{2}}}\|\vec{w}(s,.)\|_3\,ds$$

If $2 < p < +\infty$, we find

$$\|B(W_{\nu t}*\vec{u}_0,\vec{w})(t,.)\|_{L^p((0,\tau),L^3)}$$

$$\leq C_7 \sup_{0<s<t} s^{\frac{1}{2}}\|W_{\nu s}*\vec{u}_0\|_\infty \left\|\int_0^t \frac{1}{s^{\frac{1}{2}}}\frac{1}{(t-s)^{\frac{1}{2}}}\|\vec{w}(s,.)\|_3\,ds\right\|_{L^p((0,\tau))})$$

$$\leq C_8 \sup_{0<s<t} s^{\frac{1}{2}}\|W_{\nu s}*\vec{u}_0\|_\infty \left\|\frac{1}{s^{\frac{1}{2}}}\|\vec{w}(s,.)\|_3\,ds\right\|_{L^{\frac{2p}{2+p},p}((0,\tau))}$$

$$\leq C_9 \sup_{0<s<t} s^{\frac{1}{2}}\|W_{\nu s}*\vec{u}_0\|_\infty \|\|\vec{w}(s,.)\|_3\,ds\|_{L^p((0,\tau))}$$

5. thus, we find

$$\|\vec{w}(t,.)\|_{L^p((0,\tau),L^3)} \leq C_{10} A(\tau)\|\vec{w}(t,.)\|_{L^p((0,\tau),L^3)}$$

with

$$A(\tau) = \sup_{0<t<\tau} \|\vec{u}_1(t,.)\|_3 + \sup_{0<t<\tau} \|\vec{v}_1(t,.)\|_3 + \sup_{0<t<\tau} t^{\frac{1}{2}}\|W_{\nu t}*\vec{u}_0\|_\infty.$$

As we have $\lim_{\tau\to 0+} A(\tau) = 0$, the theorem is proved.

$\square$

Chapter 8

Mild Solutions in Besov or Morrey Spaces

Soon after the release of Kato's paper on strong solutions in Lebesgue spaces, there has been a flourishing of papers on solutions in Morrey spaces: Giga and Miyakawa [212] in 1989, Kato [256] and Taylor [467] in 1992, Federbush [169] in 1993. Those examples have been extended in the books of Cannone [81] in 1995 and Lemarié-Rieusset [313] in 2002, and recently in 2013 in the book of Triebel [476].

8.1 Morrey Spaces

The simplest generalization one can introduce uses only basic properties of those spaces: scaling, shift invariance and stability under bounded pointwise multiplication. We shall use real interpolation spaces as auxiliary spaces where to look for solutions.

We begin with an easy lemma on shift invariance:

Shift invariant estimates

Lemma 8.1. *Let E be a Banach space such that:*

- $E \subset \mathcal{S}'(\mathbb{R}^3)$ *(continuous embedding)*

- E *is stable under convolution with* L^1: $\|f * g\|_E \leq \|f\|_1 \|g\|_E$

Then

- *for every* $\varphi \in \mathcal{S}$ *and every* $x_0 \in \mathbb{R}^3$, $\|\varphi(x - x_0)\|_{E'} = \|\varphi\|_{E'}$.

- *for every* $\varphi \in \mathcal{S}$, *the map* $f \mapsto f * \varphi$ *is bounded from* E *to* L^∞.

Proof. If $\theta \in \mathcal{D}$ be an even function with $\int \theta dx = 1$, then $\varphi(x - x_0)$ is the limit (for $\epsilon \to 0$) in $\mathcal{S}$ of $\varphi * \theta_{\epsilon, x_0}$, where $\theta_{\epsilon, x_0}(x) = \frac{1}{\epsilon^3}\theta(\frac{x - x_0}{\epsilon})$. Thus, we have

$$\langle \varphi(x - x_0) | f \rangle_{E', E} = \langle \varphi(x - x_0) | f \rangle_{\mathcal{S}, \mathcal{S}'}$$
$$= \lim_{\epsilon \to 0} \langle \varphi | \theta_{\epsilon, -x_0} * f \rangle_{\mathcal{S}, \mathcal{S}'}$$

and thus

$$\|\varphi(x - x_0)\|_{E'} \leq \|\varphi\|_{E'}.$$

In particular, we have

$$|f * \varphi(x)| = |\langle \varphi(x - y) | f(y) \rangle_{\mathcal{S}, \mathcal{S}'}| \leq \|f\|_E \|\varphi(-x)\|_{E'}. \qquad \square$$

DOI: 10.1201/9781003042594-8

Navier–Stokes equations and local measures: the bilinear operator

Theorem 8.1.

Let E be a Banach space such that:

- $E \subset \mathcal{S}'(\mathbb{R}^3)$ *(continuous embedding)*

- *E is stable under convolution with L^1: $\|f * g\|_E \leq \|f\|_1 \|g\|_E$*

- *E is stable under bounded pointwise multiplication: $\|fg\|_E \leq \|f\|_\infty \|g\|_E$*

- *E is stable under dilations with a (sub)critical scaling: for $\lambda \leq 1$, $\|f(\lambda x)\|_E \leq C\lambda^{-\alpha}\|f\|_E$ for two constants $C > 0$ and $0 \leq \alpha \leq 1$.*

The bilinear operator B defined as

$$B(\vec{F}, \vec{G}) = \int_0^t W_{\nu(t-s)} * \mathbb{P}\operatorname{div}(\vec{F} \otimes \vec{G})\, ds$$

is continuous on the space Y_T of Lebesgue measurable vector fields on $(0,T) \times \mathbb{R}^3$ such that $t \mapsto \vec{u}(t,.)$ is continuous from $(0,T)$ to $([E, L^\infty]_{1/2,1})^3$ and

$$\|\vec{u}\|_{Y_T} = \sup_{0<t<T} t^{\alpha/4}\|\vec{u}(t,.)\|_{[E,L^\infty]_{1/2,1}} < +\infty.$$

More precisely, we have:

$$\sup_{0<t<T} t^{\frac{\alpha}{4}}\|B(\vec{F},\vec{G})\|_{[E,L^\infty]_{1/2,1}} \leq C_\nu T^{\frac{1-\alpha}{2}}(1 + (\nu T)^{\alpha/4})\|\vec{F}\|_{Y_T}\|\vec{G}\|_{Y_T} \qquad (8.1)$$

where the constant C_ν does not depend on T, $\vec{F}$ nor $\vec{G}$.

Moreover $t \mapsto B(\vec{F}, \vec{G})(t,.)$ is continuous from $[0,T]$ to E^3 if $\alpha < 1$. If $\alpha = 1$, $t \mapsto B(\vec{F}, \vec{G})(t,.)$ is continuous and bounded from $(0,T]$ to E^3; if moreover $\vec{F}$ or $\vec{G}$ belong to the space $Y_{T,0} = \{\vec{f} \in Y_T \mid \lim_{t\to 0} t^{1/4}\|\vec{f}(t,.)\|_{[E,L^\infty]_{1/2,1}} = 0\}$, then $B(\vec{F}, \vec{G}) \in Y_{T,0} \cap \mathcal{C}([0,T], E^3)$.

Proof. Pointwise product maps $E \times L^\infty$ and $L^\infty \times E$ to E, hence maps $[E, L^\infty]_{1/2,1} \times [E, L^\infty]_{1/2,1}$ to E. Moreover convolution with W_t maps E to E with a bounded operator norm while it maps E to L^∞ with an operator norm which is $O(\max(1, t^{-\alpha/2}))$. Thus, we find, writing

$$W_{\nu(t-s)} * \mathbb{P}\operatorname{div}(\vec{F} \otimes \vec{G}) = W_{\nu(t-s)/2} * \mathbb{P}\operatorname{div}(W_{\nu(t-s)/2} * (\vec{F} \otimes \vec{G})),$$

that (using the inequality $\max(1, (\nu(t-s))^{-\alpha/4}) \leq (\nu(t-s))^{-\alpha/4}\max(1, (\nu T)^{\alpha/4})$),

$$\|W_{\nu(t-s)} * \mathbb{P}\operatorname{div}(\vec{F} \otimes \vec{G})\|_{[E,L^\infty]_{1/2,1}} \leq C\frac{1}{\sqrt{\nu(t-s)}}\frac{\max(1, (\nu T)^{\alpha/4})}{(\nu(t-s))^{\alpha/4}}\frac{1}{s^{\frac{\alpha}{2}}}\|\vec{F}\|_{Y_T}\|\vec{G}\|_{Y_T}$$

so that

$$t^{\alpha/4}\|B(\vec{F},\vec{G})(t,.)\|_{[E,L^\infty]_{1/2,1}} \leq C'\frac{1 + (\nu T)^{\alpha/4}}{\nu^{\frac{2+\alpha}{4}}}t^{\frac{1-\alpha}{2}}\|\vec{F}\|_{Y_T}\|\vec{G}\|_{Y_T}.$$

Similarly, we have

$$\|W_{\nu(t-s)} * \mathbb{P}\operatorname{div}(\vec{F} \otimes \vec{G})\|_E \leq C \frac{1}{\sqrt{\nu(t-s)}} \frac{1}{s^{\frac{\alpha}{2}}} \|\vec{F}\|_{Y_T} \|\vec{G}\|_{Y_T}$$

and

$$\|B(\vec{F}, \vec{G})(t, .)\|_E \leq C t^{\frac{1-\alpha}{2}} \frac{1}{\sqrt{\nu}} \|\vec{F}\|_{Y_T} \|\vec{G}\|_{Y_T}.$$

The proof of continuity follows the same line as for Theorem 7.5. Given a time $t > 0$, we consider θ close to t: $|t - \theta| < \frac{1}{3}t$; let $\eta = |t - \theta|$. For $\vec{H} = B(\vec{F}, \vec{G})$, we write

$$\vec{H}(t, x) - \vec{H}(\theta, x) = \int_t^\theta \left(\int_0^{t-2\eta} \Delta W_{\nu(\tau-s)} * \mathbb{P}\operatorname{div}(\vec{F} \otimes \vec{G}) \, ds \right) d\tau$$

$$+ \int_{t-2\eta}^t W_{\nu(t-s)} * \mathbb{P}\operatorname{div}(\vec{F} \otimes \vec{G}) \, ds - \int_{t-2\eta}^\theta W_{\nu(\theta-s)} * \mathbb{P}\operatorname{div}(\vec{F} \otimes \vec{G}) \, ds$$

so that

$$\|\vec{H}(t, .) - \vec{H}(\theta, .)\|_{[E, L^\infty]_{1/2, 1}} \leq$$

$$C_{\nu, T} \int_{[t, \theta]} \left(\int_0^{t/3} \frac{1}{(\tau - s)^{3/2 + \alpha/4}} \frac{ds}{s^{\alpha/2}} \right) d\tau \, \|\vec{F}\|_{Y_T} \|\vec{G}\|_{Y_T}$$

$$+ C_{\nu, T} \int_{[t, \theta]} \left(\int_{t/3}^{t-2\eta} \frac{1}{(\tau - s)^{3/2 + \alpha/4}} \frac{ds}{s^{\alpha/2}} \right) d\tau \, \|\vec{F}\|_{Y_T} \|\vec{G}\|_{Y_T}$$

$$+ C_{\nu, T} \int_{t-2\eta}^t \frac{1}{(t-s)^{\frac{1}{2} + \frac{\alpha}{4}}} \frac{ds}{t^{\alpha/2}} \|\vec{F}\|_{Y_T} \|\vec{G}\|_{Y_T}$$

$$+ C_{\nu, T} \int_{t-2\eta}^\theta \frac{1}{(\theta-s)^{\frac{1}{2} + \frac{\alpha}{4}}} \frac{ds}{t^{\alpha/2}} \|\vec{F}\|_{Y_T} \|\vec{G}\|_{Y_T}$$

$$\leq C'_{\nu, T} \|\vec{F}\|_{Y_T} \|\vec{G}\|_{Y_T} \left(|t - \theta| \frac{1}{t^{\frac{1}{2} + \frac{3\alpha}{4}}} + |t - \theta| \frac{1}{\eta^{\frac{1}{2} + \frac{\alpha}{4}} t^{\alpha/2}} + \eta^{\frac{1}{2} - \frac{\alpha}{4}} \frac{1}{t^{\alpha/2}} \right)$$

$$\leq 3 C'_{\nu, T} \|\vec{F}\|_{Y_T} \|\vec{G}\|_{Y_T} \frac{|t - \theta|^{\frac{1}{2} - \frac{\alpha}{4}}}{t^{\alpha/2}}$$

and

$$\|\vec{H}(t, .) - \vec{H}(\theta, .)\|_E \leq C_{\nu, T} \int_{[t, \theta]} \left(\int_0^{t/3} \frac{1}{(\tau - s)^{3/2}} \frac{ds}{s^{\alpha/2}} \right) d\tau \, \|\vec{F}\|_{Y_T} \|\vec{G}\|_{Y_T}$$

$$+ C_{\nu, T} \int_{[t, \theta]} \left(\int_{t/3}^{t-2\eta} \frac{1}{(\tau - s)^{3/2}} \frac{ds}{s^{\alpha/2}} \right) d\tau \, \|\vec{F}\|_{Y_T} \|\vec{G}\|_{Y_T}$$

$$+ C_{\nu, T} \int_{t-2\eta}^t \frac{1}{(t-s)^{\frac{1}{2}}} \frac{ds}{t^{\alpha/2}} \|\vec{F}\|_{Y_T} \|\vec{G}\|_{Y_T}$$

$$+ C_{\nu, T} \int_{t-2\eta}^\theta \frac{1}{(\theta-s)^{\frac{1}{2}}} \frac{ds}{t^{\alpha/2}} \|\vec{F}\|_{Y_T} \|\vec{G}\|_{Y_T}$$

$$\leq C'_{\nu, T} \|\vec{F}\|_{Y_T} \|\vec{G}\|_{Y_T} \left(|t - \theta| \frac{1}{t^{\frac{1}{2} + \frac{\alpha}{2}}} + |t - \theta| \frac{1}{\eta^{\frac{1}{2}} t^{\alpha/2}} + \eta^{\frac{1}{2}} \frac{1}{t^{\alpha/2}} \right)$$

$$\leq 3 C'_{\nu, T} \|\vec{F}\|_{Y_T} \|\vec{G}\|_{Y_T} \frac{|t - \theta|^{\frac{1}{2}}}{t^{\alpha/2}}. \qquad \square$$

Navier–Stokes equations and local measures: mild solutions

Theorem 8.2.

Let E be a Banach space such that:

- $E \subset \mathcal{S}'(\mathbb{R}^3)$

- *E is stable under convolution with L^1: $\|f * g\|_E \leq \|f\|_1 \|g\|_E$*

- *E is stable under bounded pointwise multiplication: $\|fg\|_E \leq \|f\|_\infty \|g\|_E$*

- *E is stable under dilations with a (sub)critical scaling: for $\lambda \leq 1$, $\|f(\lambda x)\|_E \leq C\lambda^{-\alpha}\|f\|_E$ for two constants $C > 0$ and $0 \leq \alpha \leq 1$.*

Let Y_T be the space of Lebesgue measurable vector fields on $(0,T) \times \mathbb{R}^3$ such that $t \mapsto \vec{u}(t,.)$ is continuous from $(0,T)$ to $([E, L^\infty]_{1/2,1})^3$ and

$$\|\vec{u}\|_{Y_T} = \sup_{0 < t < T} t^{\alpha/4}\|\vec{u}(t,.)\|_{[E,L^\infty]_{1/2,1}} < +\infty.$$

Then

- *If $\vec{u}_0 \in E^3$, then $W_{\nu t} * \vec{u}_0 \in Y_T$.*

- *If $\vec{f}$ is defined on $(0,T) \times \mathbb{R}^3$ and satisfies $\mathbb{P}\vec{f} \in L^1((0,T), E^3)$ and $\sup_{0 < t < T} t\|\mathbb{P}\vec{f}(t,.)\|_E < +\infty$, then $\int_0^t W_{\nu(t-s)} * \mathbb{P}\vec{f}(s,.)\,ds \in \mathcal{C}((0,T], E^3) \cap Y_T$ and $\lim_{t\to 0} t^{\alpha/4}\|\mathbb{P}\vec{f}(t,.)\|_{[E,L^\infty]_{1/2,1}} = 0$.*

- *Case $\alpha < 1$: If $\vec{u}_0 \in E$ and $\int_0^t W_{\nu(t-s)} * \mathbb{P}\vec{f}(s,.)\,ds \in Y_T$, then there exists a $T_0 \in (0,T)$ and a mild solution*

$$\vec{u} = W_{\nu t} * \vec{u}_0 + \int_0^t W_{\nu(t-s)} * \mathbb{P}\vec{f}(s,.)\,ds - B(\vec{u}, \vec{u})$$

of Equation (7.4) on $(0,T_0) \times \mathbb{R}^3$ such that $\vec{u} \in \mathcal{C}([0,T_0], E^3) \cap Y_{T_0}$.

- *Case $\alpha = 1$: If $\vec{u}_0 \in E$ and $\int_0^t W_{\nu(t-s)} * \mathbb{P}\vec{f}(s,.)\,ds \in Y_T$ and if $W_{\nu t} * \vec{u}_0$ and $\int_0^t W_{\nu(t-s)} * \mathbb{P}\vec{f}(s,.)\,ds$ are small enough in Y_T, then there exists a mild solution*

$$\vec{u} = W_{\nu t} * \vec{u}_0 + \int_0^t W_{\nu(t-s)} * \mathbb{P}\vec{f}(s,.)\,ds - B(\vec{u}, \vec{u})$$

of Equation (7.4) on $(0,T) \times \mathbb{R}^3$ such that $\vec{u} \in \mathcal{C}((0,T], E^3) \cap Y_T$.

- *Case $\alpha = 1$: if $\mathcal{S} \subset E$, if $\vec{u}_0$ belongs to the closure of $\mathcal{S}^3$ in E^3, if $\vec{f}$ is defined on $(0,T) \times \mathbb{R}^3$ and satisfies $\mathbb{P}\vec{f} \in L^1((0,T), E^3)$ and $\sup_{0 < t < T} t\|\mathbb{P}\vec{f}(t,.)\|_E < +\infty$, then there exists a $T_0 \in (0,T)$ and a mild solution $\vec{u}$ of Equation (7.4) in $\mathcal{C}([0,T_0], E^3) \cap Y_{T_0}$.*

Remark: Under those assumptions on E, in the case $\alpha = 1$, we have $\lim_{T\to 0}\|W_{\nu t} * \vec{u}_0\|_{Y_T}$ as soon as $\lim_{t\to 0}\sup_{0 < t < T}\sqrt{t}\|W_{\nu t} * \vec{u}_0\|_\infty = 0$. In particular, when $\vec{u}_0$ belongs to the closure of $\mathcal{S}^3$ in E^3.

Proof. If $\vec{u}_0 \in E^3$, we know that $\|W_{\nu t} * \vec{u}_0\|_{[E,L^\infty]_{1/2,1}} \leq C \max(1, \nu t)^{-\alpha/2}\|\vec{u}_0\|_E$; moreover, we have $\|\partial_t W_{\nu t} * \vec{u}_0\|_{[E,L^\infty]_{1/2,1}} \leq C\frac{1}{\nu t} \max(1, \nu t)^{-\alpha/2}\|\vec{u}_0\|_E$. Thus, $W_{\nu t} * \vec{u}_0 \in Y_T$. Similarly, we have $\|W_{\nu t} * \vec{u}_0\|_E \leq \|\vec{u}_0\|_E$ and $\|\partial_t W_{\nu t} * \vec{u}_0\|_E \leq C\frac{1}{\nu t}\|\vec{u}_0\|_E$; thus, $W_{\nu t} * \vec{u}_0 \in \mathcal{C}((0,T], E^3)$. If $\vec{u}_0$ belongs to $\mathcal{S}^3 \subset E^3$, then we have $\|W_{\nu t} * \vec{u}_0 - \vec{u}_0\|_E \leq t\|\Delta\vec{u}_0\|_E$ and $\|W_{\nu t} * \vec{u}_0\|_\infty \leq \|\vec{u}_0\|_\infty$; thus, if $\vec{u}_0$ belongs to the closure of $\mathcal{S}^3$ in E^3, we have $\lim_{t\to 0} t^{\alpha/4}\|W_{\nu t} * \vec{u}_0\|_{[E,L^\infty]_{1/2,1}} = 0$ and $\lim_{t\to 0}\|W_{\nu t} * \vec{u}_0 - \vec{u}_0\|_E = 0$, so that $W_{\nu t} * \vec{u}_0 \in \mathcal{C}([0,T], E^3)$.

If $\mathbb{P}\vec{f} \in L^1((0,T), E^3)$ and $\sup_{0<t<T} t\|\mathbb{P}\vec{f}(t,.)\|_E < +\infty$, we find that

$$\|\int_0^t W_{\nu(t-s)} * \mathbb{P}\vec{f}(s,.) \, ds\|_{[E,L^\infty]_{1/2,1}} \leq C_{\nu,T} \int_0^{t/2} t^{-\alpha/4}\|\mathbb{P}\vec{f}\|_E \, ds$$

$$+ C_{\nu,T}\int_{t/2}^t (t-s)^{-\alpha/4}\frac{ds}{t} \sup_{0<s<t} s\|\mathbb{P}\vec{f}(s,.)\|_E$$

$$\leq C'_{\nu,T} t^{-\alpha/4}(\|\mathbb{P}\vec{f}\|_{L_t^1 E} + \sup_{0<s<T} s\|\mathbb{P}\vec{f}(s,.)\|_E)$$

and

$$\|\int_0^t W_{\nu(t-s)} * \mathbb{P}\vec{f}(s,.) \, ds\|_E \leq \|\mathbb{P}\vec{f}\|_{L_t^1 E}.$$

Moreover, given a time $t > 0$, we consider θ close to t: $|t - \theta| < \frac{1}{3}t$; let $\eta = |t - \theta|$. For $\vec{F} = \int_0^t W_{\nu(t-s)} * \mathbb{P}\vec{f}(s,.) \, ds$, we write

$$\vec{F}(t,x) - \vec{F}(\theta,x) = \int_t^\theta \left(\int_0^{t-2\eta} \Delta W_{\nu(\tau-s)} * \mathbb{P}\vec{f}(s,.) \, ds\right) d\tau$$

$$+ \int_{t-2\eta}^t W_{\nu(t-s)} * \mathbb{P}\vec{f} \, ds - \int_{t-2\eta}^\theta W_{\nu(\theta-s)} * \mathbb{P}\vec{f} \, ds$$

so that

$$\|\vec{F}(t,.) - \vec{F}(\theta,.)\|_{[E,L^\infty]_{1/2,1}} \leq C_{\nu,T} \int_{[t,\theta]} \left(\int_0^{t/3} \frac{1}{(\tau-s)^{1+\frac{\alpha}{4}}}\|\mathbb{P}\vec{f}(s,.)\|_E \, ds\right) d\tau$$

$$+ C_{\nu,T}\int_{[t,\theta]} \left(\int_{t/3}^{t-2\eta} \frac{1}{(\tau-s)^{1+\frac{\alpha}{4}}}\frac{ds}{s}\right) d\tau \sup_{0<s<T} s\|\mathbb{P}\vec{f}(s,.)\|_E$$

$$+ C_{\nu,T}\int_{t-2\eta}^t \frac{1}{(t-s)^{\frac{\alpha}{4}}}\frac{ds}{s} \sup_{0<s<T} s\|\mathbb{P}\vec{f}(s,.)\|_E$$

$$+ C_{\nu,T}\int_{t-2\eta}^\theta \frac{1}{(\theta-s)^{\frac{\alpha}{4}}}\frac{ds}{s} \sup_{0<s<T} s\|\mathbb{P}\vec{f}(s,.)\|_E$$

$$\leq C'_{\nu,T}\frac{\eta}{t^{1+\frac{\alpha}{4}}} \int_0^{t/3} \|\mathbb{P}\vec{f}\|_E \, ds + C'_{\nu,T}\frac{\eta^{1-\frac{\alpha}{4}}}{t} \sup_{0<s<T} s\|\mathbb{P}\vec{f}(s,.)\|_E$$

$$\leq C''_{\nu,T}\frac{\eta^{1-\frac{\alpha}{4}}}{t}(\int_0^T \|\mathbb{P}\vec{f}\|_E \, ds + \sup_{0<s<T} s\|\mathbb{P}\vec{f}(s,.)\|_E)$$

and similarly

$$\|\vec{F}(t,.) - \vec{F}(\theta,.)\|_E \leq C_{\nu,T} \int_{[t,\theta]} \left(\int_0^{t/3} \frac{1}{(\tau - s)} \|\mathbb{P}\vec{f}(s,.)\|_E \, ds \right) d\tau$$

$$+ C_{\nu,T} \int_{[t,\theta]} \left(\int_{t/3}^{t-2\eta} \frac{1}{(\tau - s)} \frac{ds}{s} \right) d\tau \sup_{0<s<T} s\|\mathbb{P}\vec{f}(s,.)\|_E$$

$$+ C_{\nu,T} \int_{t-2\eta}^t \frac{ds}{s} \sup_{0<s<T} s\|\mathbb{P}\vec{f}(s,.)\|_E$$

$$+ C_{\nu,T} \int_{t-2\eta}^\theta \frac{ds}{s} \sup_{0<s<T} s\|\mathbb{P}\vec{f}(s,.)\|_E$$

$$\leq C'_{\nu,T} \frac{\eta}{t} \int_0^{t/3} \|\mathbb{P}\vec{f}\|_E \, ds + C'_{\nu,T} \frac{\eta}{t} \sup_{0<s<T} s\|\mathbb{P}\vec{f}(s,.)\|_E$$

$$+ C'_{\nu,T} \frac{1}{t} \int_{t/3}^{t-2\eta} |\int_{[t,\theta]} \frac{d\tau}{\tau - s}| \, ds \sup_{0<s<T} s\|\mathbb{P}\vec{f}(s,.)\|_E$$

$$\leq C''_{\nu,T} \frac{\eta}{t} \left(\int_0^T \|\mathbb{P}\vec{f}\|_E \, ds + \sup_{0<s<T} s\|\mathbb{P}\vec{f}(s,.)\|_E \right)$$

We now discuss the behavior near $t = 0$. We have, under the sole assumption that $\mathbb{P}\vec{f} \in L^1((0,T), E^3)$ that $\lim_{t\to 0} \|\vec{F}\|_E = 0$: for every $M > 0$ and $t \in (0,T)$, we have

$$\|\vec{F}(t,.)\|_E \leq Mt + |\{s \in (0,T) \,/\, \|\mathbb{P}\vec{f}(s,.)\|_E > M\}|.$$

Moreover, we have

$$\|\vec{F}(t,.)\|_{[E,L^\infty]_{3/4,1}} \leq C_{\nu,T} t^{-3\alpha/8} (\|\mathbb{P}\vec{f}\|_{L^1_t E} + \sup_{0<s<T} s\|\mathbb{P}\vec{f}(s,.)\|_E),$$

hence

$$t^{\alpha/2} \|\vec{F}(t,.)\|_{[E,L^\infty]_{1/2,1}} \leq C_{\nu,T} (\|\mathbb{P}\vec{f}\|_{L^1_t E} + \sup_{0<s<T} s\|\mathbb{P}\vec{f}(s,.)\|_E)^{2/3} \|\vec{F}(t,.)\|_E^{1/3}$$

$$\to_{t\to 0} 0.$$

The rest of the proof (on existence of mild solutions) is then a direct application of Theorem 8.1.   □

Examples: There are many examples of spaces satisfying the hypotheses of Theorem 8.2. We may quote for instance

- Lebesgue space L^p with $p \geq 3$

- uniform local Lebesgue space L^p_{uloc} with $p \geq 3$: $f \in L^p_{uloc}$ if $\sup_{x_0 \in \mathbb{R}^3} (\int_{B(x_0,1)} |f(y)| \, dy)^{1/p} < +\infty$

- weak Lebesgue space $L^{p,*}$ with $p \geq 3$

- more generally, Lorentz spaces $L^{p,q}$ with $p \geq 3$ and $1 \leq q \leq +\infty$ [remark: $L^{p,p} = L^p$ and $L^{p,\infty} = L^{p,*}$]

- Morrey spaces $M^{p,q}$ with $q \geq 3$ and $1 \leq p \leq q$: if $p > 1$, this is the space of locally p-integrable functions such that

$$\|f\|_{M^{p,q}} = \sup_{0 < R \leq 1, x_0 \in \mathbb{R}^3} \left(\frac{1}{R^{3(1-\frac{p}{q})}} \int_{|x-x_0|<R} |f(x)|^p \, dx\right)^{1/p} < +\infty. \qquad (8.2)$$

For $p = 1$, this is the space of locally finite Borel measures $d\mu$ (a larger space than the spaces of locally integrable functions f, i.e., of absolutely continuous measures $f(x)dx$) such that

$$\sup_{0 < R \leq 1, x_0 \in \mathbb{R}^3} \left(\frac{1}{R^{3(1-\frac{1}{q})}} \int_{|x-x_0|<R} d|\mu|(x)\right)^{1/p} < +\infty$$

- homogeneous Morrey spaces $\dot{M}^{p,q}$ with $q \geq 3$ and $1 \leq p \leq q$, defined (with the usual modification when $p = 1$) by

$$\|f\|_{\dot{M}^{p,q}} = \sup_{0 < R, x_0 \in \mathbb{R}^3} \left(\frac{1}{R^{3(1-\frac{p}{q})}} \int_{|x-x_0|<R} |f(x)|^p \, dx\right)^{1/p} < +\infty. \qquad (8.3)$$

- multiplier spaces $\mathcal{M}(H^\alpha \to L^2)$ and $\mathcal{V}^\alpha = \mathcal{M}(\dot{H}^\alpha \to L^2)$ with $0 \leq \alpha \leq 1$.

When there is no forcing term, another way to state the results in Theorem 8.2 could be the following one:

Navier–Stokes equations and local measures: mild solutions

Theorem 8.3.
Let E be a Banach space such that:

- $E \subset \mathcal{S}'(\mathbb{R}^3)$

- E *is stable under convolution with L^1:* $\|f * g\|_E \leq \|f\|_1 \|g\|_E$

- E *is stable under bounded pointwise multiplication:* $\|fg\|_E \leq \|f\|_\infty \|g\|_E$

- E *is stable under dilations with a (sub)critical scaling: for $\lambda \leq 1$, $\|f(\lambda x)\|_E \leq C\lambda^{-\alpha}\|f\|_E$ for two constants $C > 0$ and $0 \leq \alpha \leq 1$.*

For $T > 0$, there exists a constant $\epsilon_{\nu,T,E} > 0$ such that, if $\vec{u}_0 \in E^3$ with div $\vec{u}_0 = 0$ and

$$\|\vec{u}_0\|_E \sup_{0<t<T} \sqrt{t}\|W_{\nu t} * \vec{u}_0\|_\infty < \epsilon_{\nu,T,E},$$

then there exists a mild solution of Equation

$$\vec{u} = W_{\nu t} * \vec{u}_0 - B(\vec{u}, \vec{u})$$

on $(0,T) \times \mathbb{R}^3$ such that $\vec{u} \in L^\infty((0,T], E^3)$ with $\sup_{0<t<T} \sqrt{t}\|\vec{u}(t,.)\|_\infty < +\infty$.

Proof. First, we remark that $W_{\nu t} * \vec{u}_0$ belongs to Y_T (the space introduced in Theorem 8.2), with

$$\|W_{\nu t} * \vec{u}_0\|_{Y_T} \leq C\|\vec{u}_0\|_E \sup_{0<t<T} \sqrt{t}\|W_{\nu t} * \vec{u}_0\|_\infty$$

(since $\|v\|_{[E,L^\infty]_{1/2,1}} \leq \|v\|_E^{1/2}\|v\|_\infty^{1/2}$). We know that, if $\|W_{\nu t} * \vec{u}_0\|_{Y_T}$ is small enough, the Picard iterates $\vec{U}_0 = W_{\nu t} * \vec{u}_0$ and $\vec{U}_{n+1} = \vec{U}_0 - B(\vec{U}_n, \vec{U}_n)$ converge to a solution $\vec{u}$ in Y_T. In particular, if $\epsilon_n = \|\vec{U}_n - \vec{U}_{n-1}\|_{Y_T}$ (with $\vec{U}_{-1} = 0$), we have $\sum_{n=0}^{+\infty} \epsilon_n < +\infty$.

If $\beta_n = \sup_{0<t<T} \sqrt{t}\|\vec{U}_n(t,.)\|_\infty$, we have $\beta_0 \leq C_T\|\vec{u}_0\|_E$ and

$$\|\vec{U}_{n+1} - \vec{U}_n\|_\infty \leq C_{\nu,T} \int_0^t \frac{1}{(t-s)^{\frac{1}{2}+\frac{\alpha}{4}}} \|\vec{U}_{n+1} - \vec{U}_n\|_{[E,L^\infty]_{\frac{1}{2},1}} (\|\vec{U}_n\|_\infty + \|\vec{U}_{n-1}\|_\infty)\, ds$$

$$\leq C_{\nu,T}\epsilon_n(\beta_n + \beta_{n-1}) \int_0^t \frac{1}{(t-s)^{\frac{1}{2}+\frac{\alpha}{4}}} \frac{ds}{s^{\frac{1}{2}+\frac{\alpha}{4}}}$$

$$\leq C'_{\nu,T} \frac{1}{\sqrt{t}}\epsilon_n(\beta_n + \beta_{n-1}).$$

Thus, if $B_N = \sum_{n=0}^N \beta_n$, we have $B_{N+1} \leq B_N(1 + C\epsilon_N) \leq B_0 \prod_{n\geq 0}(1 + C\epsilon_n)$ and

$$\sup_{0<t<T} \sqrt{t}\|\vec{U}_{n+1} - \vec{U}_n\|_\infty \leq \epsilon_N B_0 \prod_{n\geq 0}(1 + C\epsilon_n).$$

If $\gamma_n = \sup_{0<t<T} \|\vec{U}_n\|_E$, we have

$$\gamma_0 \leq C_T\|\vec{u}_0\|_E$$

and

$$\|\vec{U}_{n+1} - \vec{U}_n\|_E \leq C_{\nu,T} \int_0^t \frac{1}{(t-s)^{\frac{1}{2}}} \|\vec{U}_{n+1} - \vec{U}_n\|_\infty (\|\vec{U}_n\|_E + \|\vec{U}_{n-1}\|_E)\, ds$$

$$\leq C_{\nu,T}\beta_n(\gamma_n + \gamma_{n-1}) \int_0^t \frac{1}{(t-s)^{\frac{1}{2}}} \frac{ds}{s^{\frac{1}{2}}}$$

$$\leq C'_{\nu,T}\beta_n(\gamma_n + \gamma_{n-1}).$$

Thus, if $C_N = \sum_{n=0}^N \gamma_n$, we have $C_{N+1} \leq C_N(1 + C\beta_N) \leq B_0 \prod_{n\geq 0}(1 + C\beta_n)$ and

$$\sup_{0<t<T} \|\vec{U}_{n+1} - \vec{U}_n\|_E \leq \beta_N B_0 \prod_{n\geq 0}(1 + C\beta_n). \qquad \square$$

8.2   Morrey Spaces and Maximal Functions

An alternative proof for existence of mild solutions can be given as a generalization of Theorem 7.6. More precisely, we will say that a Banach space of distributions E satisfy hypothesis (H_α) for some $\alpha \in (0,1]$ when

- $E \subset \mathcal{S}'(\mathbb{R}^3)$

- E is stable under convolution with L^1: $\|f * g\|_E \leq \|f\|_1\|g\|_E$

- E is stable under bounded pointwise multiplication: $\|fg\|_E \leq \|f\|_\infty\|g\|_E$

- The Hardy–Littlewood maximal function operator is bounded on E.

- The operator $(u,v) \mapsto I_\alpha(uv)$ is continuous from $E \times E$ to E, where I_α is the Riesz potential $I_\alpha(g) = c_\alpha \int_{\mathbb{R}^3} \frac{1}{|x-y|^{3-\alpha}} g(y)\, dy$.

The last point implies that $E \subset \mathcal{V}^\alpha = \mathcal{M}(\dot{H}^\alpha \to L^2) \subset \dot{B}_{\infty,\infty}^{-\alpha}$. In particular, when $\vec{f} \in E^3$, $\mathbb{P}\vec{f}$ is well defined in $(\mathcal{V}^\alpha)^3$. Another important point is that we have $\sup_{t>0} t^{\alpha/2} \|W_t * u\|_\infty \leq C\|u\|_E$.

We shall be interested in functions $f(t,x)$ such that $f \in L_t^1 E$ or $f \in L_t^\infty E$. We do not need vector integrals or measurability, as we are dealing with Lebesgue measurable functions; instead of using vector integration, we shall use the Fubini theorem combined with norm estimates and duality to give a meaning to integrals in E [1]. More precisely, as $E \subset L_{loc}^1$, the notation $f \in L_t^1 E$ or $f \in L_t^\infty E$ will mean that $f \in L_{loc}^1((0,T) \times \mathbb{R}^3$ so that for almost every $t \in (0,T)$ the locally integrable $f(t,x)$ is well defined and belongs to E, with $\int_0^T \|f(t,.)\|_E < +\infty$ or ess $\sup_{0<t<T} \|f(t,.)\|_E < +\infty$. We shall as well assume that E is a dual space and that $\mathcal{S}$ is dense in the pre-dual of E. In that case, we have $f \in L_t^1 E \Rightarrow \int_0^T f(s,.)\,ds \in E$ and $\| \int_0^T f(s,.)\,ds\|_E \leq \int_0^T \|f(s,.)\|_E\,ds$.

Navier–Stokes equations, Morrey spaces and maximal functions

Theorem 8.4.
Let E be a Banach space that satisfies hypothesis (H_α) for some $\alpha \in (0,1]$. We define the following Banach space Y_T:

$$Y_T = \{\vec{u} \in L^\infty((0,T), E^3) \ / \ \operatorname*{ess\,sup}_{0<t<T} |\vec{u}(t,x)| \in E\}.$$

We then have the following results:

- *The bilinear operator B is continuous on Y_T.*

- *For $\vec{u}_0 \in E^3$, we have $W_{\nu t} * \vec{u}_0 \in Y_T$ with*

$$\sup_{0<t<T} |W_{\nu t} * \vec{u}_0(x)| \leq \mathcal{M}_{\vec{u}_0}(x)$$

  (where $\mathcal{M}_{\vec{u}_0}$ is the Hardy–Littlewood maximal function of $\vec{u}_0$).

- *If $\mathbb{P}\vec{f} \in L^1((0,T), E^3)$, then $\int_0^t W_{\nu(t-s)} * \mathbb{P}\vec{f}(s,.)\,ds \in Y_T$.*

- *Let $\vec{u}_0 \in E^3$ and $\mathbb{P}\vec{f} \in L^1((0,T), E^3)$. There exists a positive constant ϵ_0 (independent of ν, T, $\vec{u}_0$ and $\vec{f}$), such that, if $\|\vec{u}_0\|_E < \epsilon_0 \nu^{\frac{1+\alpha}{2}} T^{-\frac{1-\alpha}{2}}$ and $\int_0^T \|\mathbb{P}\vec{f}(s,.)\|_E\,ds < \epsilon_0 \nu^{\frac{1+\alpha}{2}} T^{-\frac{1-\alpha}{2}}$, then there exists a mild solution $\vec{u}$ of Equation (7.4) on $(0,T) \times \mathbb{R}^3$ such that $\vec{u} \in Y_T$.*

Proof. Let $\vec{u}$ and $\vec{v}$ in Y_T: $|\vec{u}(t,x)| \leq U(x) \in E$, $|\vec{v}(t,x)| \leq V(x) \in E$. We have

$$|B(\vec{u}, \vec{v})| \leq C \int \left(\int_0^t \frac{1}{\nu^2 |t-s|^2 + |x-y|^4}\,ds \right) U(y)V(y)\,dy \leq C_\alpha \frac{(\nu T)^{\frac{1-\alpha}{2}}}{\nu} I_\alpha(UV)(x)$$

Thus, the bilinear operator B is continuous on Y_T.

The inequality $\sup_{0<t<T} |W_{\nu t} * \vec{u}_0(x)| \leq \mathcal{M}_{\vec{u}_0}(x))$ is given by Lemma 7.4. Thus, for $\vec{u}_0 \in E^3$, $W_{\nu t} * \vec{u}_0 \in Y_T$.

---

[1] See the discussion in [313].

Now, we consider $\vec{F} = \int_0^t W_{\nu(t-s)} * \mathbb{P}\vec{f}(s,.)\,ds$ with $\mathbb{P}\vec{f} \in L^1((0,T),E^3)$. We have

$$|\vec{F}(t,x)| \le \int_0^T \mathcal{M}\mathbb{P}\vec{f}(s,.)(x)\,ds$$

hence

$$\|\sup_{0<t<T} |\vec{F}(t,x)|\|_E \le \int_0^T \|\mathcal{M}\mathbb{P}\vec{f}(s,.)(x)\|_E\,ds \le C\int_0^T \|\mathbb{P}\vec{f}(s,.)\|_E\,ds.$$

The end of the proof is now easy. □

Examples: There are many examples of spaces E satisfying hypothesis (H_α) for some $\alpha \in (0,1]$. We may quote for instance

- Lebesgue space L^p with $3 \le p < \infty$
- weak Lebesgue space $L^{p,*}$ with $p \ge 3$
- more generally, Lorentz spaces $L^{p,q}$ with $p \ge 3$ and $1 \le q \le +\infty$
- homogeneous Morrey spaces $\dot{M}^{p,q}$ with $3 \le q < \infty$ and $2 < p \le q$
- multiplier spaces $\mathcal{V}^\alpha = \mathcal{M}(\dot{H}^\alpha \to L^2)$ with $0 < \alpha \le 1$

In all those examples, the Leray projection operator $\mathbb{P}$ is bounded on E^3.

The role played by the maximal function may be underlined in another way. We know that the Navier–Stokes equations can be written as

$$\vec{u} = W_{\nu t} * \vec{u}_0 - B(\vec{u},\vec{u}) - \int_0^t W_{\nu(t-s)}\mathbb{P}\vec{f}\,ds$$

where

$$B(\vec{u},\vec{v}) = \int_0^t \sum_{j=1}^3 \partial_j \mathcal{O}(\nu(t-s)) :: (u_j\vec{v})\,ds.$$

As we have

$$|\partial_j \mathcal{O}(\nu(t-s))(x-y)| \le C\frac{1}{(\sqrt{t-s}+|x-y|)^4},$$

we write

$$\partial_j \mathcal{O}(\nu(t-s)) :: (u_j\vec{v}) = W_{\nu(t-s)/2} * (\partial_j\mathcal{O}(\nu(t-s)/2) :: (u_j\vec{v}))$$

and get

$$|B(\vec{u},\vec{v})| \le C\int_0^t \frac{1}{\sqrt{t-s}}W_{\nu(t-s)/2} * \mathcal{M}_{|\vec{u}(s,.)||\vec{v}(s,.)|}\,ds.$$

This gives us the following version of Theorem 8.3:

Theorem 8.5.
Let E be a Banach space such that:

- $E \subset \mathcal{S}'(\mathbb{R}^3)$
- *E is stable under bounded pointwise multiplication: $\|fg\|_E \le \|f\|_\infty\|g\|_E$*

- *The Hardy–Littlewood maximal function operator is bounded on E.*

- *E is embedded in $B_{\infty,\infty}^{-1}$*

For $T > 0$, there exists a constant $\epsilon_{\nu,T,E} > 0$ such that, if $\vec{u}_0 \in E^3$ with $\operatorname{div} \vec{u}_0 = 0$ and

$$\|\vec{u}_0\|_E \sup_{0<t<T} \sqrt{t}\|W_{\nu t} * \vec{u}_0\|_\infty < \epsilon_{\nu,T,E},$$

then there exists a mild solution of Equation

$$\vec{u} = W_{\nu t} * \vec{u}_0 - B(\vec{u}, \vec{u})$$

on $(0,T) \times \mathbb{R}^3$ such that $\vec{u} \in L^\infty((0,T], E^3)$ with $\sup_{0<t<T} \sqrt{t}\|\vec{u}(t,.)\|_\infty < +\infty$.

Proof. Once again, the proof is given by a Picard iteration in the space Y_T defined as the space of Lebesgue measurable vector fields on $(0,T) \times \mathbb{R}^3$ such that $t \mapsto \vec{u}(t,.)$ is continuous from $(0,T)$ to $([E, L^\infty]_{1/2,1})^3$ and

$$\|\vec{u}\|_{Y_T} = \sup_{0<t<T} t^{1/4}\|\vec{u}(t,.)\|_{[E,L^\infty]_{1/2,1}} < +\infty.$$

Pointwise multiplications maps $[E, L^\infty]_{1/2,1} \times [E, L^\infty]_{1/2,1}$ to E, so that

$$\left\|W_{\nu(t-s)/2} * \mathcal{M}_{|\vec{u}(s,.)||\vec{v}(s,.)|}\right\|_E \leq C\|\mathcal{M}_{\mathcal{M}_{|\vec{u}||\vec{v}|}}\|_E \leq C'\||\vec{u}||\vec{v}|\|_E$$
$$\leq C''\|\vec{u}\|_{[E,L^\infty]_{1/2,1}}\|\vec{v}\|_{[E,L^\infty]_{1/2,1}}$$

and

$$\left\|W_{\nu(t-s)/2} * \mathcal{M}_{|\vec{u}(s,.)||\vec{v}(s,.)|}\right\|_\infty \leq C \max\left(1, \frac{1}{\sqrt{\nu(t-s)}}\right)\|\mathcal{M}_{|\vec{u}||\vec{v}|}\|_E$$
$$\leq C' \max\left(1, \frac{1}{\sqrt{\nu(t-s)}}\right)\|\vec{u}\|_{[E,L^\infty]_{1/2,1}}\|\vec{v}\|_{[E,L^\infty]_{1/2,1}},$$

so that

$$\left\|W_{\nu(t-s)/2} * \mathcal{M}_{|\vec{u}(s,.)||\vec{v}(s,.)|}\right\|_{[E,L^\infty]_{1/2,1}}$$
$$\leq C' \max\left(1, \frac{1}{(\nu(t-s))^{1/4}}\right)\|\vec{u}\|_{[E,L^\infty]_{1/2,1}}\|\vec{v}\|_{[E,L^\infty]_{1/2,1}}.$$

Now, the solution $\vec{u}$ we get in Y_T satisfies

$$\|B(\vec{u},\vec{u})\|_E \leq C_\nu \int_0^t \frac{1}{\sqrt{t-s}} \frac{1}{\sqrt{s}} (s^{1/4}\|\vec{u}(s,.)\|_{[E,L^\infty]_{1/2,1}})^2 \, ds,$$

$$\|B(\vec{u},\vec{u})\|_{[E,L^\infty]_{3/4,1}}$$
$$\leq C_\nu \int_0^t \frac{1}{\sqrt{t-s}} \max\left(1, \frac{1}{(t-s)^{3/8}}\right) \frac{1}{\sqrt{s}} (s^{1/4}\|\vec{u}(s,.)\|_{[E,L^\infty]_{1/2,1}})^2 \, ds,$$

so that $\vec{u} \in L^\infty((0,T,E))$ and $t^{3/8}\vec{u} \in L^\infty((0,T), [E, L^\infty]_{3/4,1})$. Moreover, pointwise multiplication maps $[E, L^\infty]_{1/2,1} \times [E, L^\infty]_{1/2,1}$ to E, and maps $[E, L^\infty]_{1/2,1} \times L^\infty$ to $[E, L^\infty]_{1/2,1}$, thus maps $[E, L^\infty]_{1/2,1} \times [E, L^\infty]_{3/4,1}$ to $[E, L^\infty]_{1/4,1} \subset B_{\infty,\infty}^{-3/4}$; thus, we find

$$\|B(\vec{u}, \vec{v})\|_\infty$$

$$\leq \int_0^t \frac{C_\nu}{\sqrt{t-s}} \max(1, \frac{1}{(t-s)^{3/8}}) \frac{1}{s^{5/8}} (s^{1/4}\|\vec{u}\|_{[E,L^\infty]_{1/2,1}})(s^{3/8}\|\vec{u}\|_{[E,L^\infty]_{3/4,1}}) \, ds$$

and $\sqrt{t}\vec{u} \in L^\infty((0,T), L^\infty)$.   □

Example: An easy example of a space E that satisfies the assumptions of Theorem 8.5 is the variable exponent Lebesgue space $L^{p()}$ under some conditions on $p()$. Recall that, if p is a continuous function on $\mathbb{R}^3$ such that $1 \leq p_- \leq p(x) \leq p_+ < +\infty$, $f \in L^{p()}$ means that f is measurable and

$$\int_{\mathbb{R}^3} |f(x)|^{p(x)} dx < +\infty.$$

The space $L^{p()}$ is normed by

$$\|f\|_{L^{p()}} = \inf\{\lambda > 0 \; / \int_{\mathbb{R}^3} |\frac{f(x)}{\lambda}|^{p(x)} dx \leq 1\}.$$

Let us remark that, if p is not constant, then $L^{p()}$ is stable neither under convolution with L^1 nor under dilations. Thus, two assumptions in Theorem 8.3 are not fulfilled. Let us consider whether $L^{p()}$ satisfies the assumptions of Theorem 8.5. Stability under bounded pointwise multiplication is obvious. As we have the embedding $L^{p()} \subset L^{p_-} + L^{p_+}$, we find that $L^{p()} \subset B_{\infty,\infty}^{-1}$ if $p_- \geq 3$. Finally, Cruz-Uribe and Fiorenza [133] have shown that the Hardy–Littelwood maximal function defines a bounded operator on $L^{p()}$ if $p_- > 1$, $|p(x) - p(y)| \leq C\frac{1}{\ln(\frac{1}{|x-y|})}$ when $|x - y| < 1/2$ and, for some constant p_∞, $|\frac{1}{p(x)} - \frac{1}{p_\infty}| \leq C\frac{1}{\ln(e+|x|)}$.

8.3   Uniqueness of Morrey Solutions

Analogous to the case of uniqueness for mild solutions in Lebesgue spaces, we shall study uniqueness for mild solutions in $\mathcal{C}([0,T), E^3)$, where E is a space of local measures. We consider a space E that satisfies the hypotheses of Theorem 8.2:

- $E \subset \mathcal{S}'(\mathbb{R}^3)$

- E is stable under convolution with L^1: $\|f * g\|_E \leq \|f\|_1 \|g\|_E$

- E is stable under bounded pointwise multiplication: $\|fg\|_E \leq \|f\|_\infty \|g\|_E$

- E is stable under dilations with a (sub)critical scaling: for $\lambda \leq 1$, $\|f(\lambda x)\|_E \leq C\lambda^{-\alpha}\|f\|_E$ for two constants $C > 0$ and $0 \leq \alpha \leq 1$.

To give meaning to the Equation (7.4) with the sole assumption that $\vec{u} \in \mathcal{C}([0,T), E^3)$, we need to define the term $\vec{u} \otimes \vec{u}$, hence to assume that $E \subset L_{loc}^2$. In order to adapt the proof of Theorem 7.7, we need that $W_{\nu t} * \vec{u}_0$ is small enough when t is close to 0; this will be granted if $\mathcal{S}$ is dense in E.

From those assumptions, we get that $E \subset \tilde{M}^{2,\frac{3}{\alpha}}$, the closure of $\mathcal{S}$ in the Morrey space $M^{2,\frac{3}{\alpha}}$:

$$\|f\|_{M^{2,\frac{3}{\alpha}}} = \sup_{x_0 \in \mathbb{R}^3, 0 < R \leq 1} \left(\frac{1}{R^{3-2\alpha}} \int_{|y-x_0|<R} |f(y)|^2 \, dy \right)^{1/2}.$$

In the case $\alpha = 1$, however, uniqueness in $\mathcal{C}([0,T),(\tilde{M}^{2,3})^3)$ is still an open problem. Uniqueness has been proved for some subspaces close to $\tilde{M}^{2,3}$: $\tilde{M}^{p,3}$ with $2 < p \leq 3$ by Furioli, Lemarié-Rieusset and Terraneo [187], $\tilde{\mathcal{M}}(H^1 \to L^2)$ (the closure of $\mathcal{S}$ in $\mathcal{M}(H^1 \to L^2)$) by May [325] and $\tilde{M}^{L^{2,1},3}$ (the closure of $\mathcal{S}$ in $M^{L^{2,1},3}$) by Lemarié-Rieusset [316], where $L^{2,1}$ is a Lorentz space and

$$\|f\|_{M^{L^{2,1},3}} = \sup_{x_0 \in \mathbb{R}^3, 0 < R \leq 1} \left(\frac{1}{R} \|1_{|y-x_0|<R} f(y)\|_{L^{2,1}} \right)^{1/2}.$$

Uniqueness of Morrey solutions

Theorem 8.6.
Let $E = \tilde{M}^{2,\frac{3}{\alpha}}$ with $0 < \alpha < 1$, $\tilde{M}^{p,3}$ with $2 < p \leq 3$, $\tilde{\mathcal{M}}(H^1 \to L^2)$ or $\tilde{M}^{L^{2,1},3}$. Let $\vec{u}_0 \in E^3$ and $\int_0^t W_{\nu(t-s)} * \mathbb{P}\vec{f}(s,.) \, ds \in \mathcal{C}([0,T),E^3)$. If $\vec{u}$ and $\vec{v}$ are two solutions of the Equation (7.4) on $(0,T) \times \mathbb{R}^3$ that belong to $\mathcal{C}([0,T),E^3)$, then $\vec{u} = \vec{v}$.

Proof. As for the proof of Theorem 7.7, we reduce the problem to local uniqueness and we look for a contractive estimate for the fixed-point problem

$$\vec{w} = B(\vec{u}_1, \vec{w}) + B(\vec{w}, \vec{v}_1) - B(W_{\nu t} * \vec{u}_0, \vec{w}) - B(\vec{w}, W_{\nu t} * \vec{u}_0). \tag{8.4}$$

with $\vec{w} = \vec{u} - \vec{v}$, $\vec{u}_1 = W_{\nu t} * \vec{u}_0 - \vec{u}$ and $\vec{v}_1 = W_{\nu t} * u_0 - \vec{v}$.

The case $\alpha < 1$ is easy: if $\vec{F}$ and $\vec{G}$ belong to $\mathcal{C}([0,\epsilon],(\tilde{M}^{2,\frac{3}{\alpha}})^3)$ (with $0 < \epsilon < 1$), then $\vec{F}(s,.) \otimes \vec{G}(s,.) \in (\tilde{M}^{1,\frac{3}{2\alpha}})^9$ and we have, for $0 < s < t < \epsilon$,

$$\|W_{\nu(t-s)} * \mathbb{P} \operatorname{div}(\vec{F}(s,.) \otimes \vec{G}(s,.))\|_{M^{1,\frac{3}{2\alpha}}} \leq C_\nu \frac{1}{\sqrt{t-s}} \|\vec{F}(s,.)\|_{M^{2,\frac{3}{2\alpha}}} \|\vec{F}(G,.)\|_{M^{2,\frac{3}{2\alpha}}}$$

and

$$\|W_{\nu(t-s)} * \mathbb{P} \operatorname{div}(\vec{F}(s,.) \otimes \vec{G}(s,.))\|_\infty \leq C_\nu \frac{1}{(t-s)^{\frac{1}{2}+\alpha}} \|\vec{F}(s,.)\|_{M^{2,\frac{3}{2\alpha}}} \|\vec{F}(G,.)\|_{M^{2,\frac{3}{2\alpha}}}$$

so that

$$\sup_{0<t<\epsilon} \|\vec{w}\|_{M^{2,\frac{3}{2\alpha}}} \leq C \epsilon^{\frac{1-\alpha}{2}} \sup_{0<t<\epsilon} \|\vec{w}\|_{M^{2,\frac{3}{2\alpha}}} \left(\sup_{0<t<\epsilon} \|\vec{u}\|_{M^{2,\frac{3}{2\alpha}}} + \sup_{0<t<\epsilon} \|\vec{v}\|_{M^{2,\frac{3}{2\alpha}}} \right).$$

(Remark: in the case $\alpha < 1$, this proof (and the result) is still valid for $\vec{u}$ and $\vec{v}$ *-weakly continuous from $[0,T)$ with values in the (non-separable) space $(M^{2,\frac{3}{\alpha}})^3$).

The case $\alpha = 1$ is more difficult. As $M^{p,3} \subset M^{L^{2,1},3}$ for $p > 2$ we consider only the cases $E = M^{L^{2,1},3}$ and $E = \mathcal{M}(H^1 \to L^2)$. The proof for $M^{p,3}$ given by Furioli, Lemarié-Rieusset and Terraneo [187] used Morrey-Besov spaces, following their use of Besov spaces for the proof of uniqueness in the case $E = L^3$. For the case $M^{L^{2,1},3}$, we shall adapt the proof by Lemarié-Rieusset [316], which is a variation of the proof by Meyer [359] who used the Lorentz space $L^{3,\infty}$ to prove uniqueness in the case $E = L^3$. Similarly, for the case

$\mathcal{M}(H^1 \to L^2)$, we shall follow the proof by May [325], which is a variation of the proof by Monniaux [368] who used the maximal regularity of the heat kernel to prove uniqueness in the case $E = L^3$.

Case $E = M^{L^{2,1},3}$: We introduce the space $M^{L^{2,\infty},3}$, where

$$\|f\|_{M^{L^{2,\infty},3}} = \sup_{x_0 \in \mathbb{R}^3, 0 < R \leq 1} \left(\frac{1}{R} \|1_{|y-x_0|<R} f(y)\|_{L^{2,\infty}} \right)^{1/2}.$$

We have $M^{L^{2,1},3} \subset M^{L^{2,\infty},3}$. We are going to estimate

$$\sup_{0<t<\epsilon} \|\vec{w}(t,.)\|_{M^{L^{2,\infty},3}} = \sup_{0<t<\epsilon} \sup_{x_0 \in \mathbb{R}^3, 0 < R \leq 1} \left(\frac{1}{R} \|1_{|y-x_0|<R} \vec{w}(t,y)\|_{L^{2,\infty}} \right)^{1/2}.$$

We have

$$\|B(W_{\nu t} * \vec{u}_0, \vec{w})\|_{M^{L^{2,\infty},3}} \leq C \sup_{0<s<t} \sqrt{s} \|W_{\nu s} * \vec{u}_0\|_{\infty} \sup_{0<s<t} \|\vec{w}(s,.)\|_{M^{L^{2,\infty},3}}$$

and

$$\|B(\vec{w}, W_{\nu t} * \vec{u}_0)\|_{M^{L^{2,\infty},3}} \leq C \sup_{0<s<t} \sqrt{s} \|W_{\nu s} * \vec{u}_0\|_{\infty} \sup_{0<s<t} \|\vec{w}(s,.)\|_{M^{L^{2,\infty},3}}.$$

In order to estimate $B(\vec{u}_1, \vec{w})$, we write $\vec{u}_1(s,.) \otimes \vec{w}(s,.) \in (M^{1,\frac{3}{2}})^9$. For $0 < t < \epsilon < 1$, we have

$$\|W_{\nu(t-s)} * \mathbb{P} \operatorname{div} \vec{u}_1(s,.) \otimes \vec{w}(s,.)\|_{M^{1,\frac{3}{2}}} \leq C_\nu \frac{1}{\sqrt{t-s}} \|\vec{u}_1(s,.)\|_{M^{L^{2,1},3}} \|\vec{w}(s,.)\|_{M^{L^{2,\infty},3}}$$

and

$$\|W_{\nu(t-s)} * \mathbb{P} \operatorname{div} \vec{u}_1(s,.) \otimes \vec{w}(s,.)\|_{\infty} \leq C_\nu \frac{1}{(t-s)^{3/2}} \|\vec{u}_1(s,.)\|_{M^{L^{2,1},3}} \|\vec{w}(s,.)\|_{M^{L^{2,\infty},3}}$$

On the ball $B(x_0, R)$ with $0 < R \leq 1$, for $-\infty < A < t$, we write $B(\vec{u}_1, \vec{w})(y) = \vec{W}_\alpha + \vec{Z}_\alpha$ with

$$W_\alpha(t,y) = \int_0^{\max(0,A)} W_{\nu(t-s)} * \mathbb{P} \operatorname{div} \vec{u}_1(s,.) \otimes \vec{w}(s,.) \, ds.$$

We have

$$\|W_\alpha\|_{\infty} \leq C_\nu \int_0^{\max(0,A)} \frac{1}{(t-s)^{3/2}} \|\vec{u}_1(s,.)\|_{M^{L^{2,1},3}} \|\vec{w}(s,.)\|_{M^{L^{2,\infty},3}} \, ds$$

hence

$$\|1_{B(x_0,R)} W_\alpha\|_{\infty} \leq C'_\nu \frac{1}{\sqrt{t-A}} \sup_{0<s<t} \|\vec{u}_1(s,.)\|_{M^{L^{2,1},3}} \sup_{0<s<t} \|\vec{w}(s,.)\|_{M^{L^{2,\infty},3}} \, ds$$

Similarly, we have

$$\|Z_\alpha\|_{M^{1,\frac{3}{2}}} \leq C_\nu \int_{\max(0,A)}^{t} \frac{1}{(t-s)^{1/2}} \|\vec{u}_1(s,.)\|_{M^{L^{2,1},3}} \|\vec{w}(s,.)\|_{M^{L^{2,\infty},3}} \, ds$$

hence

$$\|1_{B(x_0,R)} Z_\alpha\|_{1} \leq C'_\nu \sqrt{t-A} \sup_{0<s<t} \|\vec{u}_1(s,.)\|_{M^{L^{2,1},3}} \sup_{0<s<t} \|\vec{w}(s,.)\|_{M^{L^{2,\infty},3}} \, ds \, R$$

From this, we find that

$$\|1_{B(x_0,R)}B(\vec{u}_1,\vec{w})\|_{L^{2,\infty}} \leq C_\nu \sup_{0<s<t} \|\vec{u}_1(s,.)\|_{ML^{2,1},3} \sup_{0<s<t} \|\vec{w}(s,.)\|_{ML^{2,\infty},3} \, ds \, \sqrt{R}$$

hence we get the control of $B(\vec{u}_1,\vec{w})(t,.)\|_{ML^{2,\infty},3}$. Finally, we get

$$\sup_{0<t<\epsilon} \|\vec{w}(t,.)\|_{ML^{2,\infty},3} \leq C_\nu A(\epsilon)\|\vec{w}(t,.)\|_{ML^{2,\infty},3}$$

with

$$A(\epsilon) = \sup_{0<s<\epsilon} \|\vec{u}_1(s,.)\|_{ML^{2,1},3} + \sup_{0<s<\epsilon} \|\vec{v}_1(s,.)\|_{ML^{2,1},3} + \sup_{0<s<\epsilon} \sqrt{s}\|W_{\nu s} * \vec{u}_0\|_\infty$$

and

$$\lim_{\epsilon \to 0} A(\epsilon) = 0.$$

Case $E = \mathcal{M}(H^1 \to L^2)$: We are going to estimate

$$I_\epsilon(\vec{w}) = \sup_{0<t<\epsilon} \sup_{x_0 \in \mathbb{R}^3} \left(\int_0^\epsilon \left(\int_{B(x_0,1)} |\vec{w}(t,.)|^2 \, dx \right)^{p/2} dt \right)^{1/p}$$

for $2 < p < +\infty$ and $0 < \epsilon < 1$.

Fix $x_0 \in \mathbb{R}^3$. Let $\vec{W} = 1_{B(x_0,4)}\vec{w}$ and $\vec{Z} = \vec{w} - \vec{W}$. We write

$$\vec{w} = B(\vec{u}_1, \vec{W}) + B(\vec{W}, \vec{v}_1) - B(W_{\nu t} * \vec{u}_0, \vec{W}) - B(\vec{W}, W_{\nu t} * \vec{u}_0) - B(\vec{u}, \vec{Z}) - B(\vec{Z}, \vec{v}).$$

We want to estimate $\|1_{B(x_0,1)}\vec{w}\|_{L_t^p L_x^2((0,\epsilon)\times\mathbb{R}^3)}$. Estimating the terms involving $\vec{Z}$ is easy: we have, for $|x - x_0| < 1$,

$$|B(\vec{u}, \vec{Z})(t,x)| \leq C \int_0^t \int_{|x_0-y|>4} \frac{1}{|x-y|^4}|\vec{Z}(s,y)||\vec{u}(s,y)| \, ds \, dy$$

$$\leq C' \sum_{k\in\mathbb{Z}^3} \int_0^t \int_{x_0+k+[0,1]^3} \frac{1}{1+|k|^4}|\vec{Z}(s,y)||\vec{u}(s,y)| \, ds \, dy$$

$$\leq C_p'' \sup_{0<s<t} \|\vec{u}(s,.)\|_{M^{2,3}} \sum_{k\in\mathbb{Z}^3} \frac{1}{1+|k|^4} I_\epsilon(\vec{w}) t^{\frac{p-1}{p}}$$

and we find finally

$$\|1_{B(x_0,1)}B(\vec{u}, \vec{Z})\|_{L_t^p L_x^2((0,\epsilon)\times\mathbb{R}^3)} \leq C_p \epsilon I_\epsilon(\vec{w}) \sup_{0<s<\epsilon} \|\vec{u}(s,.)\|_{\mathcal{M}(H^1\to L^2)}.$$

A similar estimate holds for $B(\vec{Z}, \vec{v})$.

For the terms involving $W_{\nu t} * \vec{u}_0$, we just write that the operator $h \mapsto H$ with $H(t) = \int_0^t \frac{1}{\sqrt{t-s}} \frac{1}{\sqrt{s}} H(s), ds$ is bounded on $L^p((0,+\infty),dt)$ for $p > 2$ (as it can be checked with the estimates for convolution and products in Lorentz spaces [313]: $1_{t>0}\frac{1}{\sqrt{t}} \in L^{2,\infty}$, the pointwise product maps $L^p \times L^{2,\infty}$ to $L^{\frac{2p}{p+2},p}$ and the convolution maps $L^{\frac{2p}{p+2},p} \times L^{2,\infty}$ to L^p). Writing

$$\|B(W_{\nu s} * \vec{u}_0, \vec{W})(t,.)\|_2$$

$$\leq C \int_0^t \frac{1}{\sqrt{\nu(t-s)}} \frac{1}{\sqrt{s}} \|\vec{W}(s,.)\|_2 \, ds \sup_{0<s<t} \sqrt{s}\|W_{\nu s} * \vec{u}_0\|_{L^\infty(dx)}$$

we get

$$\|1_{B(x_0,1)}B(W_{\nu s}*\vec{u}_0,\vec{W})\|_{L_t^pL_x^2((0,\epsilon)\times\mathbb{R}^3)} \le C_p\frac{1}{\sqrt{\nu}}I_\epsilon(\vec{w})\sup_{0<s<\epsilon}\sqrt{s}\|W_{\nu s}*\vec{u}_0\|_{L^\infty(dx)}.$$

A similar estimate holds for $B(\vec{W},W_{\nu s}*\vec{u}_0)$.

For the terms involving $\vec{u}_1$ and $\vec{v}_1$, we use the maximal L^pL^2 regularity for the heat kernel [313]: for $1 < p < +\infty$

$$\|\int_0^t W_{\nu(t-s)}*\Delta f(s,.)\,ds\|_{L^p((0,+\infty),L^2(\mathbb{R}^3))} \le C_{\nu,p}\|f\|_{L^pL^2}.$$

On the other hand, we have

$$\|\int_0^t W_{\nu(t-s)}*f(s,.)\,ds\|_{L^p((0,\epsilon),L^2(\mathbb{R}^3))} \le \epsilon\|f\|_{L^pL^2}.$$

Thus, for $\epsilon < 1$, we have

$$\|\int_0^t W_{\nu(t-s)}*f(s,.)\,ds\|_{L^p((0,\epsilon),L^2(\mathbb{R}^3))} \le C_{\nu,p}\|f\|_{L^pH^{-2}}.$$

Moreover, $\mathcal{M}(H^1 \to L^2) = \mathcal{M}(L^2 \to H^{-1})$ (by duality), so that

$$\|\mathbb{P}\operatorname{div}(\vec{u}_1 \otimes \vec{W})\|_{H^{-2}} \le C\|\vec{u}_1(s,.)\|_{\mathcal{M}(H^1\to L^2)}\|\vec{W}(s,.)\|_2.$$

We thus get

$$\|1_{B(x_0,1)}B(\vec{u}_1,\vec{W})\|_{L_t^pL_x^2((0,\epsilon)\times\mathbb{R}^3)} \le C_{\nu,p}I_\epsilon(\vec{w})\sup_{0<s<\epsilon}\|\vec{u}_1(s,.)\|_{\mathcal{M}(H^1\to L^2)}.$$

A similar estimate holds for $B(\vec{W},\vec{v}_1)$.

Finally, we get

$$I_\epsilon(\vec{w}) = \sup_{0<t<\epsilon}\sup_{x_0\in\mathbb{R}^3}\left(\int_0^\epsilon\left(\int_{B(x_0,1)}|\vec{w}(t,.)|^2\,dx\right)^{p/2}dt\right)^{1/p} \le C_{\nu,p}A(\epsilon)I_\epsilon(\vec{w})$$

with

$$A(\epsilon) = \sup_{0<s<\epsilon}\|\vec{u}_1(s,.)\|_{\mathcal{M}(H^1\to L^2)} + \sup_{0<s<\epsilon}\|\vec{v}_1(s,.)\|_{\mathcal{M}(H^1\to L^2)}$$
$$+ \sup_{0<s<\epsilon}\sqrt{s}\|W_{\nu s}*\vec{u}_0\|_\infty + \epsilon\sup_{0<s<\epsilon}\|\vec{u}(s,.)\|_{\mathcal{M}(H^1\to L^2)}$$

and

$$\lim_{\epsilon\to 0}A(\epsilon) = 0.$$

The theorem is proved. □

8.4　Besov Spaces

Besov spaces play an important role in the analysis of the Navier–Stokes equations, as the regularization properties of the heat kernel may often be expressed in terms of Besov norms. We have seen in the previous sections estimates of the type:

$$\sup_{0<t<1} t^{\alpha/2}\|W_{\nu t}*u_0\|_\infty \le C\|u_0\|_E$$

for some $\alpha > 0$. Such an estimate is equivalent to the continuous embedding $E \subset B_{\infty,\infty}^{-\alpha}$.

Similarly, in Theorem 7.4, global existence of a solution $\vec{u} \in C_b([0,+\infty),(H^{1/2})^3) \cap L^2((0,T_0),(H^{3/2})^3)$ was granted under the hypothesis that $\vec{u}_0 \in (H^{1/2}(\mathbb{R}^3))^3$, $\vec{f} \in L^2((0,+\infty),(\dot{H}^{-1/2}(\mathbb{R}^3))^3)$ with $\|W_{\nu t} * \vec{u}_0\|_{L^4 \dot{H}^1}$ and $\|\int_0^t W_{\nu(t-s)} * \mathbb{P}\vec{f}(s,.)\,ds\|_{L^4 \dot{H}^1}$ small enough. The assumption on $\vec{u}_0$ can, again, be expressed in terms of Besov spaces, as we have the equivalence

$$\|W_{\nu t} * u_0\|_{L^4((0,+\infty),\dot{H}^1)} \approx \|u_0\|_{\dot{B}_{\dot{H}^1,4}^{-1/2}} \approx \|u_0\|_{\dot{B}_{2,4}^{1/2}} \tag{8.5}$$

(together with the embedding $H^{1/2} \subset \dot{H}^{1/2} = \dot{B}_{2,2}^{1/2} \subset \dot{B}_{2,4}^{1/2}$).

Similarly, in Theorem 7.5, global existence of a solution $\vec{u} \in C_b([0,+\infty),(L^3)^3)$ was proved under the assumptions that $\vec{u}_0$ and $\vec{f}$ were small enough, the assumptions on $\vec{u}_0$ being that $\vec{u}_0 \in (L^3)^3$ and $\sup_{0<t} t^\beta \|W_{\nu t} * \vec{u}_0\|_q$ small enough for some $q \in (3,+\infty)$ and $\beta = \frac{1}{2} - \frac{3}{2q}$. This is again an assumption on a Besov norm, as underlined by Cannone [81]:

$$\sup_{0<t} t^\beta \|W_{\nu t} * \vec{u}_0\|_q \approx \|\vec{u}_0\|_{\dot{B}_{q,\infty}^{-2\beta}}. \tag{8.6}$$

In the same way, the control norm on $\vec{u}_0$ we used in Theorem 8.1 was $\|W_{\nu t} * \vec{u}_0\|_{Y_T} = \sup_{0<t<T} t^{\alpha/4} \|\vec{u}(t,.)\|_{[E,L^\infty]_{1/2,1}}$, which is equivalent to the norm $\|\vec{u}_0\|_{B_{[E,L^\infty]_{1/2,1},\infty}^{-\alpha/2}}$.

All those examples suggest to investigate the Navier–Stokes equations with an initial value $\vec{u}_0$ in a Besov space $B_{E,q}^{-\alpha}$ with $\alpha > 0$ and with solutions in the space $Y_{\alpha,E,q} = \{\vec{u} \ / \ t^{\alpha/2}\|\vec{u}(t,.)\|_E \in L^q((0,T),\frac{dt}{t})\}$. For scaling arguments, we restrict to $E \subset B_{\infty,\infty}^{-1+\alpha}$. If B is not bounded on $Y_{\alpha,E}$, we may consider as well the space $Y_{\alpha,E,q} \cap Y_{1,L^\infty,\infty}$ and use the following lemma:

Lemma 8.2.
Let $0 \le \alpha < 1$. The operator $f \mapsto F$ defined by

$$F(t) = \int_0^t \frac{1}{\sqrt{t-s}\sqrt{s}} \left(\frac{t}{s}\right)^{\alpha/2} f(s)\,ds$$

is bounded on $L^p((0,+\infty),\frac{dt}{t})$ for $1 \le p \le +\infty$.

Proof. The case $p = +\infty$ is obvious, since

$$\int_0^t \frac{1}{\sqrt{t-s}\sqrt{s}} \left(\frac{t}{s}\right)^{\alpha/2} ds = \int_0^1 \frac{1}{\sqrt{1-s}\sqrt{s}} \left(\frac{1}{s}\right)^{\alpha/2} ds < +\infty.$$

For $p = 1$, we write

$$\int_0^{+\infty} |F(t)| \frac{dt}{t} \le \int_0^{+\infty} s \left(\int_s^{+\infty} \frac{1}{\sqrt{t-s}\sqrt{s}} \left(\frac{t}{s}\right)^{\alpha/2} \frac{dt}{t}\right) |f(s)| \frac{ds}{s}$$

and

$$s \left(\int_s^{+\infty} \frac{1}{\sqrt{t-s}\sqrt{s}} \left(\frac{t}{s}\right)^{\alpha/2} \frac{dt}{t}\right) = \int_1^{+\infty} \frac{1}{\sqrt{t-1}} t^{\alpha/2} \frac{dt}{t} < +\infty.$$

By interpolation, we get the result for all values of $p \in [1,+\infty]$. $\qquad\square$

The Navier–Stokes bilinear operator and Besov spaces

Theorem 8.7. *Let $E \subset \mathcal{S}'(\mathbb{R}^3)$ be a Banach space such that:*

- *E is stable under convolution with L^1: $\|f * g\|_E \leq \|f\|_1 \|g\|_E$*

- *E is stable under bounded pointwise multiplication: $\|fg\|_E \leq \|f\|_\infty \|g\|_E$*

- *$E \subset B_{\infty,\infty}^{-1+\alpha}$ for some $\alpha \in (0,1)$.*

The bilinear operator B defined as

$$B(\vec{F}, \vec{G}) = \int_0^t W_{\nu(t-s)} * \mathbb{P} \operatorname{div}(\vec{F} \otimes \vec{G}) \, ds$$

is continuous on the space Y_T of Lebesgue measurable vector fields on $(0, T) \times \mathbb{R}^3$ such that

$$\|\vec{u}\|_{Y_T} = \|t^{\alpha/2}\|\vec{u}(t,.)\|_E\|_{L^q((0,T),\frac{dt}{t})} + \|t^{1/2}\|\vec{u}(t,.)\|_\infty\|_{L^\infty((0,T),\frac{dt}{t})} < +\infty$$

where $1 \leq q \leq +\infty$.
Moreover, if $E \subset \dot{B}_{\infty,\infty}^{-1+\alpha}$, the same result holds with $T = +\infty$.

Proof. The proof is based on the inequality, for $0 < t < T$,

$$\|W_{\nu t} * f\|_\infty \leq C_{\nu,T} t^{(\alpha-1)/2} \|f\|_E$$

(which is valid as well for $T = +\infty$ when $E \subset \dot{B}_{\infty,\infty}^{-1+\alpha}$).
Let $\vec{H} = B(\vec{F}, \vec{G})$. We have

$$\|W_{\nu(t-s)} * \mathbb{P} \operatorname{div}(\vec{F}(s,.) \otimes \vec{G}(s,.))\|_E \leq C \frac{1}{\sqrt{\nu(t-s)}} \|\vec{F}(s,.)\|_E \|\vec{G}(s,.)\|_\infty$$

so that, by Lemma 8.2, we have

$$\|t^{\alpha/2}\|\vec{H}(t,.)\|_E\|_{L^q((0,T),\frac{dt}{t})}$$
$$\leq C_\nu \|t^{\alpha/2}\|\vec{F}(t,.)\|_E\|_{L^q((0,T),\frac{dt}{t})} \|t^{1/2}\|\vec{G}(t,.)\|_\infty\|_{L^\infty((0,T),\frac{dt}{t})}.$$

Moreover, we have

$$\|W_{\nu(t-s)} * \mathbb{P} \operatorname{div}(\vec{F}(s,.) \otimes \vec{G}(s,.))\|_\infty \leq C_{\nu,T} \frac{1}{(t-s)^{1/2}} \|\vec{F}(s,.)\|_\infty \|\vec{G}(s,.)\|_\infty$$

and

$$\|W_{\nu(t-s)} * \mathbb{P} \operatorname{div}(\vec{F}(s,.) \otimes \vec{G}(s,.))\|_\infty \leq C_{\nu,T} \frac{1}{(t-s)^{1-\frac{\alpha}{2}}} \|\vec{F}(s,.)\|_E \|\vec{G}(s,.)\|_\infty$$

so that

$$\|\vec{H}(t,.)\|_\infty \le C_{\nu,T} \|\vec{F}\|_{Y_T} \|\vec{G}\|_{Y_T} \times$$

$$\times \left(\left(\int_0^{t/2} \left[\frac{s^{\frac{1-\alpha}{2}}}{(t-s)^{1-\frac{\alpha}{2}}} \right]^{\frac{q}{q-1}} \frac{ds}{s} \right)^{1-\frac{1}{q}} + \int_{t/2}^t \frac{1}{\sqrt{t-s}} \frac{ds}{s} \right)$$

$$\le C'_{\nu,T} \|\vec{F}\|_{Y_T} \|\vec{G}\|_{Y_T} \frac{1}{\sqrt{t}}.$$

The theorem is proved. □

We shall be interested in a smaller class of Besov spaces for which we do not need the estimates on the size of $\vec{u}$ in L^∞:

The Navier–Stokes bilinear operator and Besov spaces II

Theorem 8.8.
Let $E, F \subset \mathcal{S}'(\mathbb{R}^3)$ be two Banach spaces such that:

- *$E \subset L^2_{loc}$ and $F \subset L^1_{loc}$*

- *E is stable under convolution with L^1: $\|f * g\|_E \le \|f\|_1 \|g\|_E$*

- *pointwise product is bounded from $E \times E$ to F*

- *For every finite positive T, we have $\sup_{0<t<T} t^{\frac{1-\alpha}{2}} \|W_{\nu t} * f\|_E \le C_T \|f\|_F$ for some $\alpha \in (0,1)$.*

The bilinear operator B defined as

$$B(\vec{F}, \vec{G}) = \int_0^t W_{\nu(t-s)} * \mathbb{P} \operatorname{div}(\vec{F} \otimes \vec{G}) \, ds$$

is continuous on the space Y_T of Lebesgue measurable vector fields on $(0,T) \times \mathbb{R}^3$ such that

$$\|\vec{u}\|_{Y_T} = \|t^{\alpha/2} \|\vec{u}(t,.)\|_E\|_{L^q((0,T), \frac{dt}{t})} < +\infty$$

where $\frac{2}{\alpha} \le q \le +\infty$.

Proof. Let $\vec{H} = B(\vec{F}, \vec{G})$. Writing

$$W_{\nu(t-s)} * \mathbb{P} \operatorname{div}(\vec{F} \otimes \vec{G}) = W_{\nu(t-s)/2} * \mathbb{P} \operatorname{div}(W_{\nu(t-s)/2} * [\vec{F} \otimes \vec{G}]),$$

we get

$$\|W_{\nu(t-s)} * \mathbb{P} \operatorname{div}(\vec{F}(s,.) \otimes \vec{G}(s,.))\|_E \le C_{\nu,T} \frac{1}{(t-s)^{1-\frac{\alpha}{2}}} \|\vec{F}(s,.)\|_E \|\vec{G}(s,.)\|_E$$

Hence, we need to estimate the $L^q(\frac{dt}{t})$ norm of

$$H(t) = t^{\alpha/2} \int_0^t \frac{1}{(t-s)^{1-\frac{\alpha}{2}}} s^{-\alpha} F(s) \, ds$$

with $F \in L^{q/2}(\frac{dt}{t})$. When $q = +\infty$, we just write:

$$t^{\alpha/2} \int_0^t \frac{1}{(t-s)^{1-\frac{\alpha}{2}}} s^{-\alpha} \, ds = \int_0^1 \frac{1}{(1-s)^{1-\frac{\alpha}{2}}} s^{-\alpha} \, ds < +\infty.$$

When $q = \frac{2}{\alpha}$, we need to estimate the $L^q(dt)$ norm of

$$K(t) = \int_0^t \frac{1}{(t-s)^{1-\frac{1}{q}}} G(s) \, ds$$

with $G \in L^{q/2}(dt)$ and $q > 2$. This is easy with the Hardy–Littlewood–Sobolev inequality:

$$\| \int_0^t \frac{1}{(t-s)^r} f(s) \, ds \|_q \leq C_{r,p} \|f\|_p$$

for $1 < p < +\infty$, $1 - \frac{1}{p} < r < 1$ and $\frac{1}{q} = r + \frac{1}{p} - 1$. For the other values of q, we use interpolation. Thus the theorem is proved. $\qquad\square$

From Theorems 8.7 and 8.8, it is easy to deduce some conditions to ensure existence of solutions to the Navier–Stokes equations with initial values in Besov spaces or to deduce some regularity estimates for solutions in more regular spaces. For example, Giga [208] described the $L_t^p L_x^q$ properties of the solutions associated to an initial value in L^3; the case of L^3 has been later fully commented by Cannone and Planchon [85].

Solutions in $L^p L^q$ were first described in 1972 by Fabes, Jones and Rivière [168]. The corresponding initial values belong to a homogeneous Besov space:

$$W_{\nu t} * u_0 \in L_t^p L_x^q \Leftrightarrow u_0 \in \dot{B}_{q,p}^{-\frac{2}{p}}$$

Solutions such that

$$\sup_{t \in \mathbb{R}} |t|^{\frac{1}{p}} \|\vec{u}(t,x)\|_{L^q(dx)} < +\infty$$

with $\frac{2}{p} + \frac{3}{q} = 1$ and $3 < q < +\infty$ were described in 1995 by Cannone [81]. The corresponding initial values belong to a homogeneous Besov space:

$$\sup_{0 < t} t^{\frac{1}{p}} \|W_{\nu t} * u_0\|_q < +\infty \Leftrightarrow u_0 \in \dot{B}_{q,\infty}^{-\frac{2}{p}}$$

Similarly, Besov spaces based on Morrey–Campanato spaces were defined by Kozono and Yamazaki in 1994 [279] and led to the existence of solutions such that $\sup_{t>0} t^{\frac{1}{p}} \|\vec{u}(t,x)\|_{\dot{M}^{r,q}(dx)} < +\infty$ and $\sup_{0<t} \sqrt{t} \|\vec{u}(t,x)\|_{L^\infty(dx)} < +\infty$ (with $1 \leq r \leq q$ and $\frac{2}{p} + \frac{3}{q} = 1$).

Here, we give an example of such an existence theorem:

Navier–Stokes equations and Besov spaces: mild solutions

Theorem 8.9.
Let $E, F, G \subset \mathcal{S}'(\mathbb{R}^3)$ be three Banach spaces such that:

- $E \subset L^2_{loc}$

- E is stable under convolution with L^1: $\|f * g\|_E \leq \|f\|_1 \|g\|_E$

- pointwise product is bounded from $E \times E$ to F

- For every finite positive T, we have $\sup_{0<t<T} t^{\frac{1-\alpha}{2}} \|W_{\nu t} * f\|_E \leq C_T \|f\|_F$ for some $\alpha \in (0,1)$

- For every finite positive T, we have $\sup_{0<t<T} t^{\frac{1-\beta}{2}} \|W_{\nu t} * f\|_E \leq C_T \|f\|_G$ for some $\beta \in (0,1)$, with $\alpha + \beta < 1$.

Then

- If $\vec{u}_0 \in (B_{E,\frac{2}{\alpha}}^{-\alpha})^3$, then $W_{\nu t} * \vec{u}_0 \in L_t^p E_x$ on $(0,T) \times \mathbb{R}^3$ for all (finite) positive T, with $p = \frac{2}{\alpha}$.

- If $\vec{f}$ is defined on $(0,T) \times \mathbb{R}^3$ and satisfies $\mathbb{P}\vec{f} \in L^q((0,T),G^3)$ with $\frac{1}{q} = \frac{1+\alpha+\beta}{2}$, then $\int_0^t W_{\nu(t-s)} * \mathbb{P}\vec{f}(s,.) \, ds \in L_t^p E_x$ on $(0,T) \times \mathbb{R}^3$ for all (finite) positive T, with $p = \frac{2}{\alpha}$.

- If $\vec{u}_0 \in (B_{E,\frac{2}{\alpha}}^{-\alpha})^3$ and $\mathbb{P}\vec{f} \in L^q((0,T),G^3)$, then there exists a $T_0 \in (0,T)$ and a mild solution $\vec{u}$ of Equation (7.4) on $(0,T_0) \times \mathbb{R}^3$ such that $\vec{u} \in L^p((0,T_0),E^3)$.

Assume moreover that we have the global inequalities

$$\sup_{0<t} t^{\frac{1-\alpha}{2}} \|W_{\nu t} * v\|_E \leq C_\infty \|v\|_F \quad \text{and} \quad \sup_{0<t} t^{\frac{1-\beta}{2}} \|W_{\nu t} * v\|_E \leq C_\infty \|v\|_G.$$

Then, if $\vec{u}_0$ is small enough in $(\dot{B}_{E,\frac{2}{\alpha}}^{-\alpha})^3$ and $\mathbb{P}\vec{f}$ is small enough in $L^q((0,+\infty),G^3)$, then there exists a mild solution $\vec{u}$ of Equation (7.4) on $(0,+\infty) \times \mathbb{R}^3$ such that $\vec{u} \in L^p((0,+\infty),E^3)$.

Proof. This is a direct consequence of Theorem 8.8; the only thing we need to check is that $\int_0^t W_{\nu(t-s)} * \mathbb{P}\vec{f}(s,.) \, ds \in L_t^p E_x$. We have (for $0 < t < T$)

$$\left\| \int_0^t W_{\nu(t-s)} * \mathbb{P}\vec{f}(s,.) \, ds \right\|_E \leq C_T \int_0^t \frac{1}{(t-s)^{\frac{1-\beta}{2}}} \|\mathbb{P}\vec{f}(s,.)\|_G \, ds$$

thus we have only to use the Hardy–Littlewood–Sobolev inequality, since $\frac{1}{q} + \frac{1-\beta}{2} = 1 + \frac{\alpha}{2}$. $\qquad\square$

Examples:

- Lebesgue spaces: $E = L^q$, $3 < q < +\infty$: in that case, we have global solutions in $L_t^p L_x^q$ with $\frac{2}{p} + \frac{3}{q} = 1$ when $\vec{u}_0$ is small enough in $(\dot{B}_{q,p}^{-1+\frac{3}{q}})^3$ and when $\vec{f}$ is small enough in $(L_t^r L_x^s)^3$ with $\frac{2}{r} + \frac{3}{s} = 3$ and $\frac{3q}{3+q} < s < 3$. The latter condition may be relaxed to $(L_t^r L_x^{s,*})^3$ or even to $(L_t^r \dot{B}_{q,\infty}^{3(\frac{1}{q}-\frac{1}{s})})^3$.
  For the case $q = 3$, see Theorem 15.12.

- Morrey spaces: $E = \dot{M}^{\rho,q}$ with $\max(2,\frac{q}{3}) < \rho \le q$ and $3 < q < +\infty$: in that case, we have global solutions in $L_t^p \dot{M}^{\rho,q}$ with $\frac{2}{p} + \frac{3}{q} = 1$ when $\vec{u}_0$ is small enough in $(\dot{B}_{\dot{M}^{\rho,q},p}^{-1+\frac{3}{q}})^3$ and when $\vec{f}$ is small enough in $(L_t^r \dot{M}^{\rho^{\frac{s}{q}},s})^3$ with $\frac{2}{r} + \frac{3}{s} = 3$ and $\max(\frac{q}{\rho}, \frac{3q}{3+q}) < s < 3$. The latter condition may be relaxed to $(L_t^r \dot{B}_{\dot{M}^{\rho,q},\infty}^{3(\frac{1}{q}-\frac{1}{s})})^3$.

8.5 Regular Besov Spaces

In the preceding section, we considered Besov spaces with negative regularity indexes. The case of Besov spaces with positive regularity indexes is easier to deal with, in a complete analogy to the case of Morrey spaces and Theorem 8.1.

Navier–Stokes equations and regular Besov spaces

Theorem 8.10.
Let $E \subset \mathcal{S}'(\mathbb{R}^3)$ be a Banach space such that:

- *E is stable under convolution with L^1: $\|f * g\|_E \le \|f\|_1 \|g\|_E$*

- *E is stable under bounded pointwise multiplication: $\|fg\|_E \le \|f\|_\infty \|g\|_E$*

- *$E \subset B_{\infty,\infty}^{-\alpha}$ for some $\alpha > 0$.*

Let $\beta > 0$ such that $\alpha - \beta \le 1$. Let Y_T be the space of Lebesgue measurable vector fields on $(0,T) \times \mathbb{R}^3$ such that

$$t^{(\alpha-\beta)/4} \|\vec{u}(t,.)\|_{B_{[E,L^\infty]_{1/2,1}}^{\frac{\beta}{2}}} \in L^\infty((0,T))$$

normed with

$$\|\vec{u}\|_{Y_T} = \sup_{0<t<T} t^{(\alpha-\beta)/4} \|\vec{u}(t,.)\|_{B_{[E,L^\infty]_{1/2,1}}^{\frac{\beta}{2}}} < +\infty.$$

Then

- *The bilinear operator B is continuous on the space Y_T:*

$$\sup_{0<t<T} t^{(\alpha-\beta)/4} \|B(\vec{F},\vec{G})\|_{[E,L^\infty]_{1/2,1}} \le C_\nu T^{\frac{1-\alpha}{2}} (1 + (\nu T)^{\alpha/4}) \|\vec{F}\|_{Y_T} \|\vec{G}\|_{Y_T} \quad (8.7)$$

  where the constant C_ν does not depend on T, $\vec{F}$ nor $\vec{G}$.
  Moreover $t \mapsto B(\vec{F},\vec{G})(t,.)$ is bounded from $(0,T]$ to $(B_{E,\infty}^\beta)^3$.

- *If $\vec{u}_0 \in (B_{E,\infty}^\beta)^3$, then $W_{\nu t} * \vec{u}_0 \in Y_T$.*

- If $\vec{f}$ is defined on $(0, T) \times \mathbb{R}^3$ and satisfies $\mathbb{P}\vec{f} \in L^1((0, T), E^3)$ and $\sup_{0 < t < T} t \|\mathbb{P}\vec{f}(t, .)\|_E < +\infty$, then $\int_0^t W_{\nu(t-s)} * \mathbb{P}\vec{f}(s, .) \, ds \in \mathcal{C}((0, T], E^3) \cap Y_T$ and $\lim_{t \to 0} t^{\alpha/4} \|\mathbb{P}\vec{f}(t, .)\|_{[E, L^\infty]_{1/2, 1}} = 0$.

- Case $\alpha < 1$: If $\vec{u}_0 \in E^3$ and $\int_0^t W_{\nu(t-s)} * \mathbb{P}\vec{f}(s, .) \, ds \in Y_T$, then there exists a $T_0 \in (0, T)$ and a mild solution $\vec{u}$ of Equation (7.4) on $(0, T_0) \times \mathbb{R}^3$ such that $\vec{u} - W_{\nu t * u_0} - \int_0^t W_{\nu(t-s)} * \mathbb{P}\vec{f}(s, .) \, ds \in \mathcal{C}([0, T_0], E^3) \cap Y_{T_0}$.

- Case $\alpha = 1$: If $\vec{u}_0 \in E$ and $\int_0^t W_{\nu(t-s)} * \mathbb{P}\vec{f}(s, .) \, ds \in Y_T$ are small enough, then there exists a mild solution $\vec{u}$ of Equation (7.4) on $(0, T) \times \mathbb{R}^3$ such that $\vec{u} - W_{\nu t * u_0} - \int_0^t W_{\nu(t-s)} * \mathbb{P}\vec{f}(s, .) \, ds \in \mathcal{C}((0, T], E^3) \cap Y_T$.

- Case $\alpha = 1$: if $\mathcal{S} \subset E^3$, if $\vec{u}_0$ belongs to the closure of $\mathcal{S}^3$ in E^3, if $\vec{f}$ is defined on $(0, T) \times \mathbb{R}^3$ and satisfies $\mathbb{P}\vec{f} \in L^1((0, T), E^3)$ and $\sup_{0 < t < T} t \|\mathbb{P}\vec{f}(t, .)\|_E < +\infty$, then there exists a $T_0 \in (0, T)$ and a mild solution $\vec{u}$ of Equation (7.4) in $\mathcal{C}([0, T_0], E^3) \cap Y_{T_0}$.

8.6 Triebel–Lizorkin Spaces

Our study of the Cauchy problem for the Navier–Stokes problem with initial value in a Besov space $B_{E,q}^{-\alpha}$ (with $\alpha > 0$) relied on the thermic characterization of the Besov space:

$$f \in \dot{B}_{q,E}^{-\alpha} \Leftrightarrow t^{\alpha/2} W_{\nu t} * f \in L_t^q E_x.$$

Similar results hold when we change the order of integration in t and in x:

The Navier–Stokes bilinear operator and Triebel–Lizorkin spaces

Theorem 8.11.
Let $E, F \subset \mathcal{S}'(\mathbb{R}^3)$ be two Banach spaces such that:

- $E \subset L_{loc}^2$ and $F \subset L_{loc}^1$.

- E is stable under convolution with L^1: $\|f * g\|_E \leq \|f\|_1 \|g\|_E$.

- If $h \in E$ and if $g \in L_{loc}^2$ with $|g| \leq |h|$, then $g \in E$ and $\|g\|_E \leq \|h\|_E$.

- Pointwise product is bounded from $E \times E$ to F.

- The Riesz potential $\mathcal{I}_{1-\alpha}$ is bounded from F to E for some $\alpha \in (0, 1)$.

The bilinear operator B defined as

$$B(\vec{F}, \vec{G}) = \int_0^t W_{\nu(t-s)} * \mathbb{P} \operatorname{div}(\vec{F} \otimes \vec{G}) \, ds$$

> *is continuous on the space Y_T of measurable vector fields on $(0,T) \times \mathbb{R}^3$ such that*
>
> $$\|\vec{u}\|_{Y_T} = \| \, \|t^{\alpha/2}|\vec{u}(t,x)| \, \|_{L^q((0,T), \frac{dt}{t})}\|_E < +\infty$$
>
> *where $\frac{2}{\alpha} \leq q \leq +\infty$.*

Proof. Let $\vec{H} = B(\vec{F}, \vec{G})$. Writing

$$W_{\nu(t-s)} * \mathbb{P} \operatorname{div}(\vec{F} \otimes \vec{G}) = W_{\nu(t-s)/2} * \mathbb{P} \operatorname{div}(W_{\nu(t-s)/2} * [\vec{F} \otimes \vec{G}]),$$

we get

$$|W_{\nu(t-s)} * \mathbb{P} \operatorname{div}(\vec{F}(s,.) \otimes \vec{G}(s,.))| \leq \frac{C_{\nu,\alpha}}{(t-s)^{1-\frac{\alpha}{2}}} \int \frac{1}{|x-y|^{2+\alpha}} |\vec{F}(s,y)||G(s,y)| \, dy$$

Thus,

$$\|t^{\alpha/2}|\vec{H}(t,x)| \, \|_{L^q(\frac{dt}{t})} \leq$$
$$\int \frac{C_{\nu,\alpha}}{|x-y|^{2+\alpha}} \|t^{\alpha/2} \int_0^t \frac{1}{(t-s)^{1-\frac{\alpha}{2}}} s^{-\alpha} s^{\alpha/2} |\vec{F}(s,y)| s^{\alpha/2} |\vec{G}(s,y)| \, ds\|_{L^q(\frac{dt}{t})} \, dy$$

We alreaby proved (for Theorem 8.8) that

$$\|t^{\alpha/2} \int_0^t \frac{1}{(t-s)^{1-\frac{\alpha}{2}}} s^{-\alpha} F(s) \, ds\|_{L^q(\frac{dt}{t})} \leq C\|F\|_{L^{q/2}(\frac{dt}{t})}$$

for $\frac{2}{\alpha} \leq q \leq +\infty$. Thus, we find

$$\| \, \|t^{\alpha/2}|\vec{H}(t,x)| \, \|_{L^q(\frac{dt}{t})}\|_E \leq$$
$$C\|\mathcal{I}_{1-\alpha}(\|t^{\alpha/2}|\vec{F}(s,.)|\|_{L^q(\frac{dt}{t})} \|t^{\alpha/2}|\vec{G}(s,.)|\|_{L^q(\frac{dt}{t})})\|_E$$

Thus the theorem is proved.   □

This gives readily the following theorem:

The Navier–Stokes bilinear operator and Triebel–Lizorkin spaces: mild solutions

Theorem 8.12.
Let $E, F, G \subset \mathcal{S}'(\mathbb{R}^3)$ be three Banach spaces such that:

- $E \subset L^2_{loc}$.

- *E is stable under convolution with L^1: $\|f * g\|_E \leq \|f\|_1\|g\|_E$.*

- *If $h \in E$ and if $g \in L^2_{loc}$ with $|g| \leq |h|$, then $g \in E$ and $\|g\|_E \leq \|h\|_E$.*

- *Pointwise product is bounded from $E \times E$ to F.*

- *The Riesz potential $\mathcal{I}_{1-\alpha}$ is bounded from F to E for some $\alpha \in (0,1)$.*

- *The Riesz potential $\mathcal{I}_{1-\beta}$ is bounded from G to E for some $\beta \in (0,1)$ with $\alpha + \beta < 1$.*

Then let $\dot{F}_{E,p}^{-\alpha}$ be defined by

$$h \in \dot{F}_{E,p}^{-\alpha} \Leftrightarrow \||t^{\alpha/2}|W_{\nu t} * h|\|_{L^p(\frac{dt}{t})}\|_E < +\infty.$$

If $\vec{u}_0$ is small enough in $(\dot{F}_{E,p}^{-\alpha})^3$ with $\frac{2}{\alpha} = p$ and $\|\mathbb{P}\vec{f}\|_{L_t^q}$ is small enough in G with $\frac{1}{q} = \frac{1+\alpha+\beta}{2}$, then there exists a mild solution $\vec{u}$ of Equation (7.4) on $(0, +\infty) \times \mathbb{R}^3$ such that $(\int_0^{+\infty} |\vec{u}|^p \, dt)^{1/p} \in E$.

Proof. This is a direct consequence of Theorem 8.11, the only thing we need to check is that $\| \int_0^t W_{\nu(t-s)} * \mathbb{P}\vec{f}(s,.) \, ds\|_{L_t^p} \in E$. We have

$$|W_{\nu(t-s)} * \mathbb{P}\vec{f}(s,.)| \leq \frac{C_{\nu,\beta}}{(t-s)^{\frac{1}{2}-\frac{\beta}{2}}} \int \frac{1}{|x-y|^{2+\beta}} |\mathbb{P}\vec{f}(s,y)| \, dy$$

Thus,

$$\| \int_0^t W_{\nu(t-s)} * \mathbb{P}\vec{f}(s,.) \, ds \|_{L_t^p} \leq$$

$$\int \frac{C_{\nu,\beta}}{|x-y|^{2+\beta}} \| \int_0^t \frac{1}{(t-s)^{\frac{1}{2}-\frac{\beta}{2}}} |\mathbb{P}\vec{f}(s,y)| \, ds\|_{L_t^p} \, dy$$

Thus, as $\frac{1}{p} = \frac{\alpha}{2} = \frac{1}{q} + \frac{1-\beta}{2} - 1$, we find (by the Hardy–Littlewood–Sobolev inequality) that

$$\| \int_0^t W_{\nu(t-s)} * \mathbb{P}\vec{f}(s,.) \, ds \|_{L_t^p} \leq$$

$$\int \frac{C_{\nu,\beta}}{|x-y|^{2+\beta}} \| |\mathbb{P}\vec{f}(t,y)| \|_{L_t^p} \, dy$$

Thus the theorem is proved. $\qquad \square$

In particular, we may easily find a solution in $L_x^q L_t^p$ (with $\frac{2}{p} + \frac{3}{q} = 1$) when the initial value belongs to a (classical) homogeneous Triebel–Lizorkin space [36, 475]:

$$W_{\nu t} * u_0 \in L_x^q L_t^p \Leftrightarrow u_0 \in \dot{F}_{q,p}^{-\frac{2}{p}}.$$

Another interesting example is the case of homogeneous Triebel–Lizorkin–Morrey spaces (studied by Sickel, Yang and Yuan in [436]): if $E = \dot{M}^{p,q}$ with $2 < p \leq q$ and $F = \dot{M}^{p/2,q/2}$ then pointwise product maps $E \times E$ to F and the Riesz potential $\mathcal{I}_{1-\alpha}$ maps F to E, provided that $1 - \alpha = \frac{3}{q}$ (and $3 < q < +\infty$, to ensure that $0 < \alpha < 1$). If $\vec{u}_0 \in \dot{F}_{\dot{M}^{p,q},r}^{-1+\frac{3}{q}}$ with $\frac{2}{r} + \frac{3}{q} = 1$, we may find a solution such that $(\int_0^{+\infty} |\vec{u}|^r \, dt)^{1/r} \in \dot{M}^{p,q}$. In the case $p = r$ and $2 < p \leq 5$, we have seen on page 99 that

$$1_{t>0} W_{\nu t} * u_0 \in \dot{M}^{p,5}(\mathbb{R} \times \mathbb{R}^3) \Leftrightarrow u_0 \in \dot{F}_{\dot{M}^{p,q},p}^{-\frac{2}{p}}.$$

An interesting subspace of $\dot{F}_{\dot{M}^{p,q},r}^{-1+\frac{3}{q}}$ with $2 < p \leq q$ and $\frac{2}{r} + \frac{3}{q} = 1$ is $\dot{F}_{\dot{M}^{p,q},2}^{-1+\frac{3}{q}} = \sqrt{-\Delta}^{1-\frac{3}{q}} \dot{M}^{p,q}$. The limit case $p = 2$ has been considered by Xiao in [507] within the theory of Q-spaces, as indeed we have

$$\dot{F}_{\dot{M}^{2,q},2}^{-1+\frac{3}{q}} = Q^{-1,\frac{3}{q}}.$$

8.7   Fourier Transform and Navier–Stokes Equations

The Navier–Stokes equations have a simple structure: they have constant coefficients and the non-linearity is quadratic. Thus, the Fourier transform turns out to be an efficient tool to describe some classes of solutions. This is well-known for periodic solutions [471], especially in the setting of Sobolev spaces.

But the Fourier transform is useful as well for the problem in the whole space. We shall give examples in the general setting of Fourier-Lebesgue spaces or Fourier–Herz spaces: existence of mild solutions (as in the results of Le Jan and Sznitman [305], of Lei and Lin [306] or of Cannone and Wu [86]); analyticity of the solutions (when $\vec{f} = 0$), following the formalism of Foias and Temam [181] or of Lemarié-Rieusset [312].[2]

Let us consider a mild solution of the Navier–Stokes equations

$$\vec{u} = W_{\nu t} * \vec{u}_0 + \int_0^t W_{\nu(t-s)} * \mathbb{P}(\vec{f} - \mathrm{div}(\vec{u} \otimes \vec{u})) \, ds.$$

Let $\vec{U}$ be the (spatial) Fourier transform of $\vec{u}$ and $\vec{F}$ be the (spatial) Fourier transform of $\vec{f}$. We find

$$
\begin{aligned}
\vec{U} = & e^{-\nu t |\xi|^2} \vec{U}_0(\xi) + \int_0^t e^{-\nu(t-s)|\xi|^2} (Id - \frac{\xi \otimes \xi}{|\xi|^2}) \vec{F}(s, \xi) \, ds \\
& - \int_0^t e^{-\nu(t-s)|\xi|^2} (Id - \frac{\xi \otimes \xi}{|\xi|^2})(i\xi) \cdot (\int \vec{U}(s, \xi - \eta) \vec{U}(s, \eta) \frac{d\eta}{(2\pi)^3}) \, ds.
\end{aligned}
\tag{8.8}
$$

Thus, we may look for Fourier transforms of mild solutions in some space $\mathbb{Y}$, with Fourier transform of the initial value in some space $\mathbb{X}$ and Fourier transform of the force in some space $\mathbb{Z}$, where X is a lattice Banach space of measurable functions on $\mathbb{R}^3$ and $\mathbb{Y}$ and $\mathbb{Z}$ are lattice Banach spaces of measurable functions on $(0, +\infty) \times \mathbb{R}^3$, provided we have

$$\|e^{-\nu t |\xi|^2} U_0\|_{\mathbb{Y}} \leq C_0 \|U_0\|_{\mathbb{X}}$$

$$\|\int_0^t e^{-\nu(t-s)|\xi|^2} F(s, \xi) \, ds\|_{\mathbb{Y}} \leq C_1 \|F\|_{\mathbb{Z}}$$

and

$$\|\int_0^t \int e^{-\nu(t-s)|\xi|^2} |\xi| |V(s, \eta)| |W(s, \xi - \eta)| \, ds \, d\eta\|_{\mathbb{Y}} \leq C_2 \|V\|_{\mathbb{Y}} \|W\|_{\mathbb{Y}}$$

for some constants that depend on ν. Then we know that we have a mild solution on $(0, +\infty) \times \mathbb{R}^3$ provided that

$$C_0 \|\vec{U}_0\|_{\mathbb{X}} + C_1 \|\vec{F}\|_{\mathbb{Z}} \leq \frac{(2\pi)^3}{4C_2}.$$

---

[2]A more general result on analyticity will be proved in Section 9.9.

First example:

Let $\alpha > 0$ and $1 < q < +\infty$ be such that moreover $2 < \frac{3}{q} + \alpha < 3$. We choose

$$\mathbb{Y} = \{U \ / \ |\xi|^\alpha U \in L_t^p L_\xi^q\}, \quad \text{where} \quad \frac{2}{p} = \frac{3}{q} + \alpha - 2.$$

We then have $2 < p < +\infty$. We write

$$e^{-(t-s)|\xi|^2}|\xi| \le C|\xi|^{\frac{2}{p}-1}\frac{1}{(t-s)^{1-\frac{1}{p}}}.$$

Thus, if

$$Z = \int_0^t \int e^{-\nu(t-s)|\xi|^2}|\xi||V(s,\eta)||W(s,\xi-\eta)|\, ds\, d\eta,$$

we have

$$|\xi|^\alpha Z \le C \int_0^t \frac{1}{(\nu(t-s))^{1-\frac{1}{p}}}|\xi|^{\alpha+\frac{2}{p}-1}\int |V(s,\eta)||W(s,\xi-\eta)|\, d\eta\, ds$$

and thus

$$\|Z\|_{\mathbb{Y}} \le C\left\|\int_0^t \frac{1}{\nu((t-s))^{1-\frac{1}{p}}}\||\xi|^{2\alpha+\frac{3}{q}-3}\int |V(s,\eta)||W(s,\xi-\eta)|\, d\eta\|_{L^q(d\xi)}\, ds\right\|_{L^p(dt)}$$

We now write, for $0 < \delta < \alpha$, $|\xi|^\delta \le C(|\xi-\eta|^\delta + |\eta|^\delta)$ and thus

$$|\xi|^\delta \int |V(s,\eta)||W(s,\xi-\eta)|\, d\eta \le C(W * (|\eta|^\delta V) + V * (|\eta|^\delta W)).$$

We have $|\eta|^\delta = |\eta|^{\delta-\alpha}|\eta|^\alpha V \in L^{r,q}$ and $W = |\eta|^{-\alpha}|\eta|^\alpha W \in L^{\rho,q}$ with $\frac{1}{r} = \frac{1}{q} + \frac{\alpha-\delta}{3}$ and $\frac{1}{\rho} = \frac{1}{q} + \frac{\alpha}{3}$. Thus, by convolution inequalities, we find that

$$|\xi|^\delta \int |V(s,\eta)||W(s,\xi-\eta)|\, d\eta \in L^{\tau,q}$$

where $\frac{1}{\tau} = \frac{2}{q} + \frac{2\alpha}{3} - \frac{\delta}{3} - 1$ (provided that $0 < \delta < 2(\frac{3}{q}+\alpha-\frac{3}{2})$). As we have $2\alpha+\frac{3}{q}-3-\delta < 0$, we find that

$$|\xi|^{2\alpha+\frac{3}{q}-3-\delta}|\xi|^\delta \int |V(s,\eta)||W(s,\xi-\eta)|\, d\eta \in L^{v,q}$$

with

$$\frac{1}{v} = -\frac{1}{3}(2\alpha+\frac{3}{q}-3-\delta) + \frac{2}{q} + \frac{2\alpha}{3} - \frac{\delta}{3} - 1 = \frac{1}{q}.$$

Thus, we find

$$\|Z\|_{\mathbb{Y}} \le C\left\|\int_0^t \frac{1}{(\nu(t-s))^{1-\frac{1}{p}}}\||\xi|^\alpha V(s,;)\|_q\||\xi|^\alpha W(s,.)\|_q\, ds\right\|_p \le \frac{C_2}{\nu^{1-\frac{1}{p}}}\|V\|_{\mathbb{Y}}\|W\|_{\mathbb{Y}}.$$

Of course, we have now to identify $\mathbb{X}$. We could define $\mathbb{X}$ just as the space of U_0 such that

$$\||\xi|^\alpha e^{-\nu t|\xi|^2}U_0\|_{L^p L^q} < +\infty.$$

If we perform a dyadic partition of unity on $\mathbb{R}^3$:

$$1 = \sum_{j \in \mathbb{Z}} \psi(\frac{\xi}{2^j})$$

for some smooth ψ supported in $\{\xi \; \frac{1}{2} \leq \xi \leq 4\}$ and similarly we have a partition of unity on $(0, +\infty)$

$$1 = \sum_{k \in \mathbb{Z}} \omega(\frac{t}{4^k}),$$

we find

$$\||\xi|^\alpha e^{-\nu t|\xi|^2} U_0\|_q^q \approx \sum_{j \in \mathbb{Z}} 2^{q\alpha j} \|e^{-\nu t|\xi|^2} \psi(\frac{\xi}{2^j}) U_0(\xi)\|_q^q$$

and thus

$$\|U_0\|_X \approx \left(\sum_{k \in \mathbb{Z}} \|\omega(\frac{t}{4^k}) \left(\sum_{j \in \mathbb{Z}} 2^{q\alpha j} \|e^{-\nu t|\xi|^2} \psi(\frac{\xi}{2^j}) U_0(\xi)\|_q^q \right)^{1/q} \|_p^p \right)^{1/p}$$

Thus, for two positive constants A and B, we have

$$A \left(\sum_{k \in \mathbb{Z}} 4^k \left(\sum_{j \in \mathbb{Z}} 2^{q\alpha j} e^{-256 \, \nu q 4^{k+j}} \|\psi(\frac{\xi}{2^j}) U_0(\xi)\|_q^q \right)^{p/q} \right)^{1/p}$$

$$\leq \|U_0\|_X$$

$$\leq B \left(\sum_{k \in \mathbb{Z}} 4^k \left(\sum_{j \in \mathbb{Z}} 2^{q\alpha j} e^{-\frac{1}{16} \nu q 4^{k+j}} \|\psi(\frac{\xi}{2^j}) U_0(\xi)\|_q^q \right)^{p/q} \right)^{1/p}$$

Restricting in the first term the sum over j to the sole value $j = -k$, we find

$$A e^{-256\nu} \left(\sum_{j \in \mathbb{Z}} 2^{-2j} 2^{j\alpha p} \|\psi(\frac{\xi}{2^j}) U_0(\xi)\|_q^p \right)^{1/p} \leq \|U_0\|_X \qquad (8.9)$$

The *Herz spaces* $\mathcal{B}_{q,p}^s$ are defined by

$$\|U_0\|_{\mathcal{B}_{q,p}^s} = \left(\sum_{j \in \mathbb{Z}} 2^{sjp} \|\psi(\frac{\xi}{2^j}) U_0(\xi)\|_q^p \right)^{1/p} < +\infty.$$

They have been introduced by Herz [234], and the Fourier-Herz spaces $\mathcal{FB}_{q,p}^s$ (i.e., the image of Herz spaces through the Fourier transform) have been recently used in the context of parabolic equations, first by Iwabuchi [242] for the Keller–Segel equation then by Cannone and Wu [86] for the Navier–Stokes equations. Let $1 < r < +\infty$ and $\frac{1}{q} + \frac{1}{r} = 1$. Then, if $r \leq 2$, we have $\dot{B}_{r,p}^{-1+\frac{3}{r}} \subset \mathcal{FB}_{q,p}^{2-\frac{3}{q}}$ while, if $r \geq 2$, we have $\mathcal{FB}_{q,p}^{2-\frac{3}{q}} \subset \dot{B}_{r,p}^{-1+\frac{3}{r}}$.

From (8.9), we find that $\mathbb{X} \subset \mathcal{B}_{q,p}^{\alpha - \frac{2}{p}} = \mathcal{B}_{q,p}^{2 - \frac{3}{q}}$. On the other hand, we have, by the Minkowski and Young inequalities,

$$\left(\sum_{k \in \mathbb{Z}} 4^k \left(\sum_{j \in \mathbb{Z}} 2^{q\alpha j} e^{-\frac{1}{16} \nu q 4^{k+j}} \| \psi(\frac{\xi}{2^j}) U_0(\xi) \|_q^q \right)^{p/q} \right)^{1/p}$$

$$\leq \left(\sum_{k \in \mathbb{Z}} 4^k \left(\sum_{j \in \mathbb{Z}} 2^{\alpha j} e^{-\frac{1}{16} \nu 4^{k+j}} \| \psi(\frac{\xi}{2^j}) U_0(\xi) \|_q \right)^p \right)^{1/p}$$

$$= \left(\sum_{k \in \mathbb{Z}} \left(\sum_{j \in \mathbb{Z}} 2^{(k+j)\frac{2}{p}} e^{-\frac{1}{16} \nu 4^{k+j}} 2^{(\alpha - \frac{2}{p})j} \| \psi(\frac{\xi}{2^j}) U_0(\xi) \|_q \right)^p \right)^{1/p}$$

$$\leq (\sum_{k \in \mathbb{Z}} 2^{k\frac{2}{p}} e^{-\frac{1}{16} \nu 4^k})(\sum_{j \in \mathbb{Z}} 2^{j(\alpha - \frac{2}{p})p} \| \psi(\frac{\xi}{2^j}) U_0(\xi) \|_q^p)^{1/p}$$

so that

$$\| U_0 \|_{\mathbb{X}} \leq B(\sum_{k \in \mathbb{Z}} 2^{k\frac{2}{p}} e^{-\frac{1}{16} \nu 4^k}) \| U_0 \|_{\mathcal{B}_{q,p}^{\alpha - \frac{2}{p}}}.$$

Thus, $\mathbb{X} = \mathcal{B}_{q,p}^{\alpha - \frac{2}{p}}$.

For the choice of the space $\mathbb{Z}$, we may take $\mathbb{Z} = \{ F \ / \ |\xi|^{\alpha - \delta} F \in L^r L^q \}$, where $\frac{2}{p} < \delta < 2$ and $\frac{1}{r} = 1 + \frac{1}{p} - \frac{\delta}{2}$ (so that $1 < r < p$): indeed, we have

$$|\xi|^\alpha \int_0^t e^{-\nu(t-s)|\xi|^2} |F(s,\xi)| \, ds \leq C \int_0^t \frac{1}{(\nu(t-s))^{\delta/2}} |\xi|^{\alpha - \delta} |F(s,\xi)| \, ds$$

so that

$$\| |\xi|^\alpha \int_0^t e^{-\nu(t-s)|\xi|^2} |F(s,\xi)| \, ds \|_q \leq C \int_0^t \| |\xi|^{\alpha - \delta} F(s,\xi) \|_q \frac{ds}{(\nu(t-s))^{\delta/2}}.$$

As $0 < \delta/2 < 1$, we use the Hardy–Littlewood–Sobolev inequality and find that $\| |\xi|^\alpha \int_0^t e^{-\nu(t-s)|\xi|^2} |F(s,\xi)| \, ds \|_q$ belongs to L^σ with $\frac{1}{\sigma} = \frac{1}{r} + \frac{\delta}{2} - 1 = \frac{1}{p}$.

Recollecting all those results, we find the theorem:

Theorem 8.13.
Let $\alpha > 0$ and $1 < q < +\infty$ be such that moreover $2 < \frac{3}{q} + \alpha < 3$. Let δ be such that $\frac{3}{q} + \alpha - 2 < \delta < 2$. Let $\frac{2}{p} = \frac{3}{q} + \alpha - 2$ and $\frac{1}{r} = 1 + \frac{1}{p} - \frac{\delta}{2}$. Then there exists a positive constant ϵ_0 (depending on ν, α, q and δ) such that, if $\vec{u}_0 \in \mathcal{F}\mathcal{B}_{q,p}^{2 - \frac{3}{q}}$ with $\operatorname{div} \vec{u}_0 = 0$ and $(-\Delta)^{(\alpha - \delta)/2} \vec{f} \in L^r \mathcal{F} L^q$, and if moreover

$$\| \mathcal{F} \vec{u}_0 \|_{\mathcal{B}_{q,p}^{2 - \frac{3}{q}}} + \| |\xi|^{\alpha - \delta} \mathcal{F} \vec{f} \|_{L^r L^q} < \epsilon_0,$$

then the Cauchy problem for the Navier–Stokes equations with initial value $\vec{u}_0$ and forcing term $\vec{f}$ has a global mild solution $\vec{u}$ such that $(-\Delta)^{\alpha/2} \vec{u} \in L^p \mathcal{F} L^q$.

Second example:
We now consider the case $p = +\infty$. Here, we interchange the order of integration between t and ξ. Thus, we choose, for $0 \leq \alpha < 2$,

$$\mathbb{Y} = \{U \ / \ |\xi|^\alpha \sup_{t>0} |U(t,\xi)| \in L^q_\xi\}, \text{ where } 0 = \frac{3}{q} + \alpha - 2.$$

Thus, if

$$Z = \int_0^t \int e^{-\nu(t-s)|\xi|^2} |\xi| |V(s,\eta)| |W(s,\xi-\eta)| \, ds \, d\eta,$$

with $|V(s,\xi)| \leq V(\xi)|\xi|^{-\alpha}$ and $|W(s,\xi)| \leq W(\xi)|\xi|^{-\alpha}$, we have

$$|\xi|^\alpha Z \leq \frac{1}{\nu} |\xi|^{\alpha-1} \int \frac{V(\xi-\eta)}{|\xi-\eta|^\alpha} \frac{W(\eta)}{|\eta|^\alpha} \, d\eta$$

$$\leq C \frac{1}{\nu|\xi|} \int V(\xi-\eta) \frac{W(\eta)}{|\eta|^\alpha} + \frac{V(\xi-\eta)}{|\xi-\eta|^\alpha} W(\eta) \, d\eta.$$

V belongs to L^q and $\frac{W}{|\eta|^\alpha}$ belongs to $L^{r,q}$ with $\frac{1}{r} = \frac{1}{q} + \frac{\alpha}{3}$. Thus, $\sup_{t>0} |\xi|^\alpha Z(t,\xi)$ belongs to $L^{\rho,q}$ with $\frac{1}{\rho} = \frac{2}{q} + \frac{\alpha}{3} - 1 + \frac{1}{3} = \frac{1}{q}$ and we find $\|Z\|_\mathbb{Y} \leq C_2 \|V\|_\mathbb{Y} \|W\|_\mathbb{Y}$.
 The associated space $\mathbb{X}$ is easily identified, as

$$\|e_0^{-\nu t|\xi|^2}\|_\mathbb{Y} = \||\xi|^\alpha U_0\|_q.$$

The associated space $\mathbb{Z}$ can be chosen as $L^1((0,+\infty), \mathbb{X})$. Indeed, we have

$$\left| \int_0^t e^{-\nu(t-s)|\xi|^2} F(s,\xi) \, ds \right| \leq \int_0^{+\infty} |F(s,\xi)| \, ds$$

so that

$$\||\xi|^\alpha \int_0^t e^{-\nu(t-s)|\xi|^2} F(s,\xi) \, ds\|_q \leq \int_0^{+\infty} \||\xi|^\alpha F(s,\xi)\|_q \, ds.$$

We thus find the theorem:

Theorem 8.14.
Let $\alpha \geq 0$ and $\frac{3}{2} \leq q < +\infty$ with $2 = \frac{3}{q} + \alpha$. Then there exists a positive constant ϵ_0 (depending on ν, and q) such that, if $(-\Delta)^{\alpha/2} \vec{u}_0 \in \mathcal{F}L^q$ with $\text{div } \vec{u}_0 = 0$ and $(-\Delta)^{\alpha/2} \vec{f} \in L^1 \mathcal{F}L^q$, and if moreover

$$\| \||\xi|^\alpha \mathcal{F}\vec{u}_0\|_q + \||\xi|^\alpha \mathcal{F}\vec{f}\|_{L^1 L^q} < \epsilon_0,$$

then the Cauchy problem for the Navier–Stokes equations with initial value $\vec{u}_0$ and forcing term $\vec{f}$ has a global mild solution $\vec{u}$ such that $\sup_{t>0} |\xi|^\alpha |\mathcal{F}\vec{u}(t,\xi)| \in L^q$.

In particular, we recover the results of Theorem 7.4 on global existence of solutions in the Sobolev space $\dot{H}^{1/2}$ (except that we replaced the condition $\vec{f}$ small in $L^2 \dot{H}^{-1/2}$ by $\vec{f}$ small in $L^1 \dot{H}^{1/2}$).

Third example:
We consider the case $q = +\infty$. This is a simple case. Let $\alpha > 0$ be such that moreover $2 \leq \alpha < 3$. We choose

$$\mathbb{Y} = \{U \ / \ |\xi|^\alpha U \in L^p_t L^\infty_\xi\}, \text{ where } \frac{2}{p} = \alpha - 2.$$

We then have $2 < p \leq +\infty$.

If

$$Z = \int_0^t \int e^{-\nu(t-s)|\xi|^2} |\xi| |V(s,\eta)| |W(s,\xi-\eta)| \, ds \, d\eta,$$

we write $|V(s,\eta)| \leq \frac{V(s)}{|\eta|^\alpha}$ with $V(s) \in L^p$ and similarly $|W(s,\xi-\eta)| \leq \frac{W(s)}{|\xi-\eta|^\alpha}$, so that

$$\int |V(s,\eta)| |W(s,\xi-\eta)| \, d\eta \leq V(s)W(s) \int \frac{d\eta}{|\eta|^\alpha |\xi-\eta|^\alpha} \leq C|\xi|^{3-2\alpha} V(s)W(s)$$

and

$$|\xi|^\alpha Z(t,\xi) \leq C \int_0^t e^{-\nu(t-s)|\xi|^2} |\xi|^{4-\alpha} V(s)W(s) \, ds.$$

If $\alpha = 2$, we obtain

$$|\xi|^\alpha Z(t,\xi) \leq C \|V(s)\|_\infty \|W(s)\|_\infty \int_0^t e^{-\nu(t-s)|\xi|^2} |\xi|^2 \, ds$$

$$\leq C \frac{1}{\nu} \|V(s)\|_\infty \|W(s)\|_\infty.$$

If $\alpha > 2$, we write $e^{-\nu(t-s)|\xi|^2} \leq C \left(\frac{1}{\nu(t-s)|\xi|^2} \right)^{2-\alpha/2}$ and obtain

$$|\xi|^\alpha Z(t,\xi) \leq C \frac{1}{\nu^{2-\alpha/2}} \int \frac{1}{(t-s)^{1-\frac{1}{p}}} V(s)W(s) \, ds.$$

In every case, we have $\|Z\|_{\mathbb{Y}} \leq C_2 \|V\|_{\mathbb{Y}} \|W\|_{\mathbb{Y}}$.

The space $\mathbb{X}$ will then again be a Herz space $\mathcal{B}_{\infty,p}^2$. The associated Fourier-Herz space is then a Besov space based on pseudo-measures[3], as studied by Cannone and Karch [82]: $\mathcal{F}\mathcal{B}_{\infty,p}^2 = \dot{B}_{\mathrm{PM},p}^2$. The case $p = +\infty$ and $\alpha = 2$ is the case considered by Le Jan and Sznitman [305].

For the choice of the space $\mathbb{Z}$, we may again take, if $p < +\infty$, $\mathbb{Z} = \{F \ / \ |\xi|^{\alpha-\delta} F \in L^r L^q\}$, where $\frac{2}{p} < \delta < 2$ and $\frac{1}{r} = 1 + \frac{1}{p} - \frac{\delta}{2}$ (so that $1 < r < p$). For $p = +\infty$, we want to get

$$|\xi|^2 \int_0^t e^{-\nu(t-s)|\xi|^2} |F(s,\xi)| \, ds \in L_t^\infty L_\xi^\infty.$$

As $\|e^{-\nu t |\xi|^2}\|_{L^r((0,+\infty))} = C_r(\nu|\xi|)^{-2/r}$, we may take $\mathbb{Z} = \{F \ / \ |\xi|^{2-\frac{2}{r}} F \in L_t^{\frac{r}{r-1}} L_\xi^\infty\}$. We thus get (in the case $p = +\infty$) the theorem:

Theorem 8.15.
Let $r \in [1, +\infty]$. Then there exists a positive constant ϵ_0 (depending on ν, and r) such that, if $\Delta \vec{u}_0 \in \mathcal{F}L^\infty$ with $\mathrm{div}\, \vec{u}_0 = 0$ and $(-\Delta)^{\frac{1}{r}} \vec{f} \in L^r \mathcal{F}L^\infty$, and if moreover

$$\||\xi|^2 \mathcal{F}\vec{u}_0\|_\infty + \||\xi|^{\frac{2}{r}} \mathcal{F}\vec{f}\|_{L^r L^\infty} < \epsilon_0,$$

then the Cauchy problem for the Navier–Stokes equations with initial value $\vec{u}_0$ and forcing term $\vec{f}$ has a global mild solution $\vec{u}$ such that $\sup_{t>0} |\xi|^2 |\mathcal{F}\vec{u}(t,\xi)| \in L^\infty$.

---

[3] A pseudo-measure is a tempered distribution whose Fourier transform belongs to L^∞.

Let us remark that the case $\vec{f} = \beta\delta\vec{e}_3$ (where $\vec{e}_3$ is the unit vector in the x_3 axis, β is a positive constant and δ is the Dirac mass at $x = 0$) has been discussed by Cannone and Karch [82], in relation with the Landau self-similar solutions [301, 439, 447] (see Section 10.8).

Fourth example:
We consider the case $q = 1$. Of course, the conditions $\alpha > 0$ and $\frac{3}{q} + \alpha < 3$ become incompatible, and we shall deal with the limit case $\alpha = 0$. Thus, we choose

$$\mathbb{Y} = L_t^2 L_\xi^1.$$

If

$$Z = \int_0^t \int e^{-\nu(t-s)|\xi|^2} |\xi| |V(s,\eta)| |W(s,\xi-\eta)| \, ds \, d\eta,$$

we write

$$\int Z(t,\xi) \frac{d\xi}{|\xi|} \leq \int_0^t \iint |V(s,\eta)| |W(s,\xi-\eta)| \, d\eta \, d\xi \, ds$$

$$= \int_0^t \|V(s,.)\|_1 \|W(s,.)\|_1 \, ds$$

$$\leq \|V\|_{L^2 L^1} \|W\|_{L^2 L^1}$$

so that $Z \in L_t^\infty(L^1(\frac{d\xi}{|\xi|})$. Moreover, we have

$$\int_0^{+\infty} \int Z(t,\xi) |\xi| \, d\xi \, dt$$

$$= \int_0^{+\infty} \iint \left(\int_s^{+\infty} e^{-\nu(t-s)|\xi|^2} |\xi|^2 \, dt \right) |V(s,\eta)| |W(s,\xi-\eta)| \, d\eta \, d\xi \, ds$$

$$= \frac{1}{\nu} \int_0^{+\infty} \iint |V(s,\eta)| |W(s,\xi-\eta)| \, d\eta \, d\xi \, ds$$

$$= \frac{1}{\nu} \int_0^{+\infty} \|V(s,.)\|_1 \|W(s,.)\|_1 \, ds$$

$$\leq \frac{1}{\nu} \|V\|_{L^2 L^1} \|W\|_{L^2 L^1}$$

so that $Z \in L^1(L^1(|\xi| \, d\xi))$. Thus, we find

$$\int_0^{+\infty} \left(\int Z(t,\xi) \, d\xi \right)^2 dt \leq \int_0^{+\infty} \left(\int Z \frac{d\xi}{|\xi|} \right) \left(\int Z|\xi| \, d\xi \right) dt \leq \frac{1}{\nu} \|V\|_{L^2 L^1}^2 \|W\|_{L^2 L^1}^2.$$

A similar proof gives that, if $F \in \mathbb{Z} = L_t^1(L^1(\frac{d\xi}{|\xi|}))$, then

$$\int \left(\int_0^t e^{-\nu(t-s)|\xi|^2} |F(s,\xi) \, ds| \right) \frac{d\xi}{|\xi|} \leq \|F\|_{L_t^1(L^1(\frac{d\xi}{|\xi|}))}$$

and

$$\int_0^{+\infty} \int \left(\int_0^t e^{-\nu(t-s)|\xi|^2} |F(s,\xi) \, ds| \right) |\xi| \, d\xi \, dt \leq \frac{1}{\nu} \|F\|_{L_t^1(L^1(\frac{d\xi}{|\xi|}))}$$

so that

$$\| \int_0^t e^{-\nu(t-s)|\xi|^2} |F(s,\xi) \, ds| \|_{L^2 L^1} \leq \frac{1}{\sqrt{\nu}} \|F\|_{L_t^1(L^1(\frac{d\xi}{|\xi|}))} \tag{8.10}$$

Moreover, we see that the associated space $\mathbb{X}$ for the initial value U_0 is the Herz space $\dot{B}_{1,2}^{-1}$. We thus find the theorem:

Theorem 8.16.
There exists a positive constant ϵ_0 (depending on ν) such that, if $\vec{u}_0 \in \mathcal{F}\mathcal{B}_{1,2}^{-1}$ with div $\vec{u}_0 = 0$ and $(-\Delta)^{(-1/2}\vec{f} \in L^1\mathcal{F}L^1$, and if moreover

$$\|\mathcal{F}\vec{u}_0\|_{\mathcal{B}_{1,2}^{-1}} + \|\frac{1}{|\xi|}\mathcal{F}\vec{f}\|_{L^1L^1} < \epsilon_0,$$

then the Cauchy problem for the Navier–Stokes equations with initial value $\vec{u}_0$ and forcing term $\vec{f}$ has a global mild solution $\vec{u}$ such that $\vec{u} \in L^2\mathcal{F}L^1$.

Cannone and Wu [86] studied the more general case of $\vec{u}_0 \in \mathcal{F}\mathcal{B}_{1,q}^{-1}$ with $1 \leq q \leq 2$. The case $q = 1$ corresponds to $\mathcal{F}\vec{u}_0 \in L^1(\frac{d\xi}{|\xi|})$, i.e., to the theorem of Lei and Lin [306]:

Corollary 8.1.
There exists a positive constant ϵ_0 (depending on ν) such that, if

$$\int |\mathcal{F}\vec{u}_0(\xi)|\frac{d\xi}{|\xi|} + \int_0^{+\infty}\int |\mathcal{F}\vec{f}(t,\xi)|\frac{d\xi}{|\xi|}\,dt < \epsilon_0,$$

and if div $\vec{u}_0 = 0$ then the Cauchy problem for the Navier–Stokes equations with initial value $\vec{u}_0$ and forcing term $\vec{f}$ has a global mild solution $\vec{u}$ such that $\mathcal{F}\vec{u} \in L_t^\infty(L^1(\frac{d\xi}{|\xi|})) \cap L^1(L^1(|\xi|\,d\xi))$.

Proof. It is enough to check that $e^{-\nu t|\xi|^2}\mathcal{F}\vec{u}_0(\xi) \in L_t^\infty(L^1(\frac{d\xi}{|\xi|}) \cap L^1(L^1(|\xi|\,d\xi))$. Then, we have that $\vec{u}_0$ belongs to $\mathcal{F}\mathcal{B}_{1,2}^{-1}$ and we may apply Theorem 8.16 to get the existence of the mild solution $\vec{u} \in L^2\mathcal{F}L^1$. But the proof of Theorem 8.16 shows that $\mathcal{F}(\vec{u} - W_{\nu t} * \vec{u}_0)$ belongs to $L_t^\infty(L^1(\frac{d\xi}{|\xi|}) \cap L^1(L^1(|\xi|\,d\xi))$. $\square$

It is very easy to slightly modify the proofs of Theorems 8.13 to 8.16 to get Gevrey-type estimates for our solutions. Indeed, we have

$$e^{-t\nu|\xi|^2} \leq Ce^{-\frac{1}{2}t\nu|\xi|^2}e^{-\sqrt{\nu t}|\xi|}.$$

Thus, if $U_0 \in \mathcal{B}_{q,p}^{\alpha-\frac{2}{p}}$, then

$$|\xi|^\alpha e^{\sqrt{\nu t}|\xi|}e^{-t\nu|\xi|^2}U_0 \in L^pL^q.$$

Moreover, if

$$Z = \int_0^t \int e^{-\nu(t-s)|\xi|^2}|\xi|e^{-\sqrt{\nu s}|\eta|}|V(s,\eta)|e^{-\sqrt{\nu s}|\xi-\eta|}|W(s,\xi-\eta)|\,ds\,d\eta,$$

we use the inequalities

$$|\xi - \eta| + |\eta| \geq |\xi|$$

and

$$\sqrt{s} + \sqrt{t-s} \geq \sqrt{t}$$

to get that

$$Z \leq Ce^{-\sqrt{\nu t}|\xi|}\int_0^t \int e^{-\frac{1}{2}\nu(t-s)|\xi|^2}|\xi||V(s,\eta)||W(s,\xi-\eta)|\,ds\,d\eta.$$

Thus, we find

Theorem 8.17.
For the following spaces $\mathbb{X}$, $\mathbb{Y}$ *and* $\mathbb{Z}$, *there exists a positive constant* ϵ_1 *(depending on* ν, $\mathbb{X}$, $\mathbb{Y}$ *and* $\mathbb{Z}$*) such that, if* $\vec{u}_0 \in \mathcal{F}\mathbb{X}$ *with* $\operatorname{div} \vec{u}_0 = 0$ *and* $e^{\sqrt{-\nu t \Delta}} \vec{f} \in \mathcal{F}\mathbb{Z}$, *and if moreover*

$$\|\mathcal{F}\vec{u}_0\|_{\mathbb{X}} + \|e^{\sqrt{\nu t}\,|\xi|}\mathcal{F}\vec{f}\|_{\mathbb{Z}} < \epsilon_1,$$

then the Cauchy problem for the Navier–Stokes equations with initial value $\vec{u}_0$ *and forcing term* $\vec{f}$ *has a global mild solution* $\vec{u}$ *such that* $e^{\sqrt{-\nu t \Delta}}\vec{u} \in \mathcal{F}\mathbb{Y}$:

- $\mathbb{X} = \mathcal{B}_{q,p}^{2-\frac{3}{q}}$, $\mathbb{Y} = \frac{1}{|\xi|^{\alpha}}L^p L^q$, $\mathbb{Z} = \frac{1}{|\xi|^{\alpha-\delta}}L^r L^q$, *with* $\alpha > 0$, $1 < q < +\infty$, $2 < \frac{3}{q} + \alpha < 3$, $\frac{3}{q} + \alpha - 2 < \delta < 2$, $\frac{2}{p} = \frac{3}{q} + \alpha - 2$ *and* $\frac{1}{r} = 1 + \frac{1}{p} - \frac{\delta}{2}$.

- $\mathbb{X} = \frac{1}{|\xi|^{\alpha}}L^q$, $\mathbb{Y} = \frac{1}{|\xi|^{\alpha}}L_\xi^q L_t^\infty$, $\mathbb{Z} = \frac{1}{|\xi|^{\alpha}}L_t^1 L_\xi^q$, *with* $\alpha \geq 0$ *and* $\frac{3}{2} \leq q < +\infty$.

- $\mathbb{X} = \frac{1}{|\xi|^{2}}L^q$, $\mathbb{Y} = \frac{1}{|\xi|^{2}}L_{t,\xi}^\infty$, $\mathbb{Z} = \frac{1}{|\xi|^{2/r}}L_t^r L_\xi^\infty$, *with* $r \in [1, +\infty]$.

- $\mathbb{X} = \mathcal{B}_{1,2}^{-1}$, $\mathbb{Y} = L^2 L^1$, $\mathbb{Z} = L^1(L^1(\frac{d\xi}{|\xi|}))$.

- $\mathbb{X} = L^1(\frac{d\xi}{|\xi|})$, $\mathbb{Y} = L_t^\infty(L^1(\frac{d\xi}{|\xi|}) \cap L^1(L^1(|\xi|\,d\xi))$, $\mathbb{Z} = L^1(L^1(\frac{d\xi}{|\xi|}))$.

This result can be extended to the case of more general spaces, where the absolute value does not operate on the Fourier transforms, so that the proofs given here do not apply and must be replaced by more delicate estimates on singular integrals. For instance, Lemarié-Rieusset [312, 313, 314] proved the following theorem:

Theorem 8.18.
There exists a positive constant ϵ_1 *(depending on* ν*) such that, if* $\vec{u}_0 \in L^3$ *with* $\operatorname{div} \vec{u}_0 = 0$ *and* $\|\vec{u}_0\|_3 < \epsilon_1$, *then the Cauchy problem for the Navier–Stokes equations with initial value* $\vec{u}_0$ *(and forcing term* $\vec{f} = 0$*) has a global mild solution* $\vec{u}$ *such that*

$$t^{1/8}\mathcal{F}^{-1}\left(e^{\sqrt{\nu t}(|\xi_1|+|\xi_2|+|\xi_3|)}\mathcal{F}\vec{u}\right) \in L^\infty L^4.$$

Theorem 8.17 may easily be adapted to Gevrey regularity of the form $e^{(\sqrt{-\nu t \Delta})^\beta}\vec{u} \in \mathbb{Y}$, where $0 < \beta \leq 1$:

Theorem 8.19.
Let $0 < \beta \leq 1$ *and let* $\mathbb{X}$, $\mathbb{Y}$ *and* $\mathbb{Z}$ *be the same spaces as in Theorem 8.17. There exists a positive constant* ϵ_1 *(depending on* ν, $\mathbb{X}$, $\mathbb{Y}$ *and* $\mathbb{Z}$*) such that, if* $\vec{u}_0 \in \mathcal{F}\mathbb{X}$ *with* $\operatorname{div} \vec{u}_0 = 0$ *and* $e^{(\sqrt{-\nu t \Delta})^\beta}\vec{f} \in \mathcal{F}\mathbb{Z}$, *and if moreover*

$$\|\mathcal{F}\vec{u}_0\|_{\mathbb{X}} + \|e^{(\sqrt{\nu t}\,|\xi|)^\beta}\mathcal{F}\vec{f}\|_{\mathbb{Z}} < \epsilon_1,$$

then the Cauchy problem for the Navier–Stokes equations with initial value $\vec{u}_0$ *and forcing term* $\vec{f}$ *has a global mild solution* $\vec{u}$ *such that* $e^{(\sqrt{-\nu t \Delta})^\beta}\vec{u} \in \mathcal{F}\mathbb{Y}$.

Proof. Same proof as for Theorem 8.17, replacing the inequality

$$e^{-\sqrt{\nu(t-s)}|\xi|}e^{-\sqrt{\nu s}|\eta|}e^{-\sqrt{\nu s}|\xi-\eta|} \leq e^{-\sqrt{\nu t}|\xi|}$$

with

$$e^{-(\sqrt{\nu(t-s)}|\xi|)^\beta}e^{-(\sqrt{\nu s}|\eta|)^\beta}e^{-(\sqrt{\nu s}|\xi-\eta|)^\beta} \leq e^{-(\sqrt{\nu t}|\xi|)^\beta}. \qquad \square$$

8.8 The Cheap Navier–Stokes Equation

In chapter 5, we replaced the integral vector equation

$$\vec{u} = W_{\nu t} * \vec{u}_0 - \int_0^t \sum_{j=1}^3 \partial_j \mathcal{O}(\nu(t-s)) :: (\vec{f} * \partial_j G + u_j \vec{u}) \, ds$$

with the study of the integral scalar equation

$$\Omega(t,x) = \Omega_0(t,x) + C_0 \int_0^t \int_{\mathbb{R}^3} \frac{1}{\nu^2(t-s)^2 + |x-y|^4} \Omega(s,y)^2 \, dy \, ds.$$

Of course, we can do the same for the Fourier transform of the Navier–Stokes equations (8.8)

$$\vec{U} = e^{-\nu t |\xi|^2} \vec{U}_0(\xi) + \int_0^t e^{-\nu(t-s)|\xi|^2} (Id - \frac{\xi \otimes \xi}{|\xi|^2}) \vec{F}(s,\xi) \, ds$$

$$- \int_0^t e^{-\nu(t-s)|\xi|^2} (Id - \frac{\xi \otimes \xi}{|\xi|^2})(i\xi) \cdot (\int \vec{U}(s, \xi - \eta) \otimes \vec{U}(s, \eta) \frac{d\eta}{(2\pi)^3}) \, ds$$

and study the scalar equation

$$W(t,\xi) = e^{-\nu t |\xi|^2} W_0(\xi) + \int_0^t e^{-\nu(t-s)|\xi|^2} F(s,\xi) \, ds$$

$$+ \frac{1}{(2\pi)^3} \int_0^t e^{-\nu(t-s)|\xi|^2} |\xi| \int_{\mathbb{R}^3} W(s,\eta) W(s, \xi - \eta) \, d\eta \, ds. \tag{8.11}$$

Taking the inverse Fourier transforms $w = \mathcal{F}^{-1} W$ of W and $f = \mathcal{F}^{-1} F$ of F, equation (8.11 becomes the equation

$$\begin{cases} \partial_t w = \Delta w + f + \sqrt{-\Delta}(w^2) \\ \qquad\qquad w(0,.) = w_0. \end{cases} \tag{8.12}$$

Equation (8.12) is known as the *cheap Navier–Stokes equation*. It has been introduced in 2001 by S. Montgomery-Smith [369] as a toy model for the Navier–Stokes equations. He gave an example of an initial value w_0 in the Schwartz class ($w_0 \in \mathcal{S}(\mathbb{R}^3)$) with a non-negative Fourier transform W_0 such that the solution w for the equation with a null force ($\vec{f} = 0$) blows up in finite time (see section 11.2).

This cheap equation allows very simple computations for the search of solutions. Indeed, let us assume that $\vec{U}_0$ is controlled by a function W_0 and $\vec{F}$ by a function F:

$$|\vec{U}_0(\xi)| \le \frac{1}{18} W_0(\xi) \text{ and } |\vec{F}(t,\xi)| \le \frac{1}{18} F(t,\xi)$$

and that $W(t,\xi)$ is measurable, almost everywhere finite and is a non-negative solution of the integral inequation for every $t \in [0,T]$ and every $\xi \in \mathbb{R}^3$

$$e^{-\nu t |\xi|^2} W_0(\xi) + \int_0^t e^{-\nu(t-s)|\xi|^2} F(s,\xi) \, d\xi + B_0(W,W)(t,\xi) \le W(t,\xi)$$

with

$$B_0(W,V)(t,\xi) = \frac{1}{(2\pi)^3} \int_0^t e^{-\nu(t-s)|\xi|^2} |\xi| \int_{\mathbb{R}^3} W(s,\eta) V(s, \xi - \eta) \, d\eta \, ds.$$

Define

- $W^{[0]}(t,\xi) := e^{-\nu t|\xi|^2} W_0(\xi) + \int_0^t e^{-\nu(t-s)|\xi|^2} F(s,\xi)\, d\xi$

- $W^{[n+1]}(t,\xi) := W^{[0]}(t,\xi) + B_0(W^{[n]}, W^{[n]})(t,\xi)$

- $\vec{U}^{[0]} := e^{-\nu t|\xi|^2}\vec{U}_0(\xi) + \int_0^t e^{-\nu(t-s)|\xi|^2}(Id - \frac{\xi \otimes \xi}{|\xi|^2})\vec{F}(s,\xi)\, ds$

- $\vec{U}^{[n+1]}(t,\xi) := \vec{U}^{[0]}(t,\xi) - B(\vec{U}^{[n]}, \vec{U}^{[n]})(t,\xi)$

where

$$B(\vec{U}, \vec{V}) = \int_0^t e^{-\nu(t-s)|\xi|^2}(Id - \frac{\xi \otimes \xi}{|\xi|^2})(i\xi) \cdot \left(\int \vec{U}(s,\xi - \eta) \otimes \vec{V}(s,\eta)\, \frac{d\eta}{(2\pi)^3}\right) ds.$$

By induction on n, we find that we have the pointwise inequalities

- $0 \le W^{[n]}(t,\xi) \le W^{[n+1]}(t,\xi) \le W(t,\xi)$

- $|\vec{U}^{[n]}(t,\xi)| \le \frac{1}{18}W^{[n]}(t,\xi)$

- $|\vec{U}^{[n+1]}(t,\xi) - \vec{U}^{[n]}(t,\xi)| \le \frac{1}{18}(W^{[n+1]}(t,\xi) - W^{[n]}(t,\xi)).$

We find that $W^{[n]}$ is pointwisely convergent to a function $W^{[\infty]} \le W$. By monotonous convergence, we have

$$W^{[\infty]} = W^{[0]} + B_0(W^{[\infty]}, W^{[\infty]}).$$

Then, by dominated convergence, we find that $\vec{U}^{[n]}$ converges to a limit $\vec{U}^{[\infty]}$ such that

$$\vec{U}^{[\infty]} = \vec{U}^{[0]} - B(\vec{U}^{[\infty]}, \vec{U}^{[\infty]}).$$

$\vec{U}^{[\infty]}$ is then the Fourier transform of a solution to the Navier–Stokes problem with initial value $\vec{u}_0$ and forcing term $\vec{f}$.

The same formalism allows one to get easily Gevrey-type analyticity estimates. Let us assume more precisely that $\vec{U}_0$ is controlled by a function X_0 and $\vec{F}$ by a function G:

$$|\vec{U}_0(\xi)| \le \frac{1}{18e}X_0(\xi) \quad \text{and} \quad |\vec{F}(t,\xi)| \le \frac{1}{18e}e^{-\sqrt{\nu t}|\xi|}G(t,\xi)$$

and that $X(t,\xi)$ is measurable, almost everywhere finite and is a non-negative solution of the integral inequation for every $t \in [0,T]$ and every $\xi \in \mathbb{R}^3$

$$e^{-\frac{\nu}{2}t|\xi|^2}X_0(\xi) + \int_0^t e^{-\frac{\nu}{2}(t-s)|\xi|^2}G(s,\xi)\, d\xi + B_1(X,X)(t,\xi) \le X(t,\xi)$$

with

$$B_1(X,Y)(t,\xi) = \frac{1}{(2\pi)^3}\int_0^t e^{-\frac{\nu}{2}(t-s)|\xi|^2}|\xi|\int_{\mathbb{R}^3} W(s,\eta)V(s,\xi-\eta)\, d\eta\, ds.$$

Define

- $Z^{[0]}(t,\xi) := e^{-\frac{\nu}{2}t|\xi|^2}X_0(\xi) + \int_0^t e^{-\frac{\nu}{2}(t-s)|\xi|^2}G(s,\xi)\, d\xi$

- $Z^{[n+1]}(t,\xi) := Z^{[0]}(t,\xi) + B_0(Z^{[n]}, Z^{[n]})(t,\xi)$

- $\vec{U}^{[0]} := e^{-\nu t|\xi|^2}\vec{U}_0(\xi) + \int_0^t e^{-\nu(t-s)|\xi|^2}(Id - \frac{\xi \otimes \xi}{|\xi|^2})\vec{F}(s,\xi)\, ds$

- $\vec{U}^{[n+1]}(t,\xi) := \vec{U}^{[0]}(t,\xi) - B(\vec{U}^{[n]}, \vec{U}^{[n]})(t,\xi)$.

Again, we find that $0 \leq Z^{[n]}(t,\xi) \leq Z^{[n+1]}(t,\xi) \leq X(t,\xi)$, so that $Z^{[n]}$ is pointwisely convergent to a function $Z^{[\infty]} \leq X$. By monotonous convergence, we have

$$Z^{[\infty]} = Z^{[0]} + B_0(Z^{[\infty]}, Z^{[\infty]}).$$

We have

$$\sup_{z \geq 0} e^{z - \frac{1}{2}z^2} = \sqrt{e}$$

and, for $0 \leq s \leq t$

$$e^{\sqrt{\nu t}|\xi|} e^{-\sqrt{\nu s}|\xi - \eta|} e^{-\sqrt{\nu s}|\eta|} \leq e^{(\sqrt{\nu t} - \sqrt{\nu s})|\xi|} \leq e^{\sqrt{\nu(t-s)}|\xi|}.$$

By induction on n, we then find that we have the pointwise inequalities

- $|\vec{U}^{[n]}(t,\xi)| \leq \frac{1}{18\sqrt{e}} e^{-\sqrt{\nu t}|\xi|} Z^{[n]}(t,\xi)$

- $|\vec{U}^{[n+1]}(t,\xi) - \vec{U}^{[n]}(t,\xi)| \leq \frac{1}{18\sqrt{e}} e^{-\sqrt{\nu t}|\xi|} (Z^{[n+1]}(t,\xi) - Z^{[n]}(t,\xi)).$

This gives the Gevrey estimate

$$|\vec{U}(t,\xi)| \leq \frac{1}{18\sqrt{e}} e^{-\sqrt{\nu t}|\xi|} Z^{[\infty]}(t,\xi). \tag{8.13}$$

The study of the cheap equations

$$W = W^{[0]} + B_0(W,W) \tag{8.14}$$

or

$$Z = Z^{[0]} + B_1(Z,Z) \tag{8.15}$$

thus provides simple classes of solutions to the Navier–Stokes equations.

For instance, if $M_0(\xi)$ and $M_1(\xi)$ are non-negative measurable functions that satisfy the following inequation

$$M_0(\xi) + \frac{1}{(2\pi)^3|\xi|} \int_{\mathbb{R}^3} M_1(\xi - \eta) M_1(\eta)\, d\eta \leq M_1(\xi),$$

and if $W^{[0]}(t,\xi) \leq M_0(\xi)$ for $0 \leq t \leq T$, we get, by induction on n, that $W^{[n]}(t,\xi) \leq M_1(\xi)$. This means that, if M_0 belongs to a lattice Banach space of functions E such that the operator $(Z,V) \mapsto \frac{1}{|\xi|}(Z * V)$ is bounded on E and if $\|M_0\|_E$ is small enough, then the Navier–Stokes equations with initial value $\vec{u}_0$ and force $\vec{f}$ such that their Fourier transforms satisfy

$$|\vec{U}_0| + \int_0^T |\vec{F}(s,\xi)|\, dt \leq \frac{1}{18} M_0(\xi)$$

have a global solution $\vec{u}$ with $\sup_{0<t<+T} |\mathcal{F}\vec{u}| \in E$. Two simple instances can be found in the litterature:

- the case where $E = L^2(|\xi|\, d\xi)$: if $Z \in E$, it means that $Z = \frac{1}{|\xi|^{1/2}} Z_0$ with $Z_0 \in L^2$; thus Z belongs to the Lorentz space $L^{3/2,2}$ (as a product of a function in $L^{6,\infty}$ by a function in L^2), thus $Z * Z$ belongs to $L^{3,1} \subset L^{3,2}$ and $\frac{1}{|\xi|^{1/2}} Z * Z \in L^{2,2} = L^2$. As $E = \mathcal{F}(\dot{H}^{\frac{1}{2}})$, we find the theorem of Fujita and Kato for the Sobolev space $\dot{H}^{1/2}$ (Theorem 7.4) .

- the equality

$$\int \frac{1}{|\xi - \eta|^2} \frac{1}{|\eta|^2} \, d\eta = C_0 \frac{1}{|\xi|}$$

allowed Le Jan and Sznitman [305] to consider the space $E = \mathcal{F}(\dot{B}_{\mathrm{PM},\infty}^{-2})$ defined by

$$Z \in E \Leftrightarrow Z \in L_{\mathrm{loc}}^1 \text{ and } |\xi|^2 Z \in L^\infty.$$

If we look for local-in-time solutions, we must include the time variable in our estimations. For instance, since

$$e^{-\nu(t-s)|\xi|^2} \le e^{\frac{3}{4}} \left(\frac{3}{4} \right)^{3/2} \frac{1}{(\nu(t-s))^{3/4}} \frac{1}{|\xi|^{3/2}},$$

then, if $M_0(\xi)$, $M_1(\xi)$ and $\alpha(t)$ are non-negative measurable functions that satisfy on $(0, T_0)$ the following inequation

$$\alpha(t) M_0(\xi) + \left(\frac{3}{4} \right)^{\frac{3}{2}} \frac{e^{\frac{3}{4}}}{(2\pi)^3} \int_0^t \frac{\alpha(s)^2}{(\nu(t-s))^{\frac{3}{4}}} \, ds \frac{1}{|\xi|^{\frac{1}{2}}} \int_{\mathbb{R}^3} M_1(\xi - \eta) M_1(\eta) \, d\eta \le \alpha(t) M_1(\xi),$$

and if $W^{[0]}(t, \xi) \le \alpha(t) M_0(t)$, we get, by induction on n, that $W^{[n]}(t, \xi) \le \alpha(t) M_1(\xi)$. Thus, if F is a lattice Banach space of functions such that the operator $(Z, V) \mapsto \frac{1}{|\xi|^{1/2}}(Z * V)$ is bounded on F, if $\sup_{0<t<T} t^{1/4} e^{-\nu t |\xi|^2} |\vec{U}_0(\xi)| \in F$ and $\sup_{0<t<T} t^{1/4} \int_0^t e^{-\nu(t-s)|\xi|^2} |\vec{F}(s, \xi)| \, ds \in F$ and if

$$\lim_{T_0 \to 0} \| \sup_{0<t<T_0} t^{1/4} e^{-\nu t |\xi|^2} |\vec{U}_0(\xi)| \|_F + \| \sup_{0<t<T_0} t^{1/4} \int_0^t e^{-\nu(t-s)|\xi|^2} |\vec{F}(s, \xi)| \, ds \|_F = 0$$

$\| \sup_{0<t<T} t^{1/4} e^{-t|\xi|^2} W^0(\xi) \|_F$ is small enough, then the Navier–Stokes equations with initial value $\vec{u}_0$ and force $\vec{f}$ have a solution $\vec{u}$ on $(0, T_0)$ for T_0 small enough, with $\sup_{0<t<T_0} t^{1/4} |\vec{U}(t, \xi)| \in F$. Let us look at our two simple instances:

Cheap Navier–Stokes equation.

Let us look at our two simple instances:

- the case where $E = L^2(|\xi| \, d\xi)$ and $F = L^2(|\xi|^2 \, d\xi)$: if $Z \in F$, it means that $Z = \frac{1}{|\xi|} Z_0$ with $Z_0 \in L^2$; thus Z belongs to the Lorentz space $L^{6/5,2}$ (as a product of a function in $L^{3,\infty}$ by a function in L^2) and $V = |\xi|^{1/2} Z \in L^{3/2,2}$, thus, writing

$$\frac{1}{|\xi|^{1/2}} |Z * Z| \le \frac{2}{|\xi|} (|Z| * |V|),$$

we get that $|Z| * |V|$ belongs to $L^{6/5,2} * L^{3/2,2} \subset L^{2,1} \subset L^{2,2} = L^2$, so that we have $\frac{1}{|\xi|^{1/2}}(Z * Z) \in F$. Moreover, if $A > 0$ and $W_0 \in E$, we find that, for $t > 0$,

$$|(\nu t)^{1/4} e^{-\nu t |\xi|^2} W_0(\xi)| \le 1_{|\xi| \le A} (\nu t)^{\frac{1}{4}} W_0(\xi) + C 1_{|\xi| > A} \frac{1}{|\xi|^{1/2}} W_0(\xi)$$

so that

$$\| \sup_{0<t<T} (\nu t)^{1/4} e^{-\nu t |\xi|^2} |W_0(\xi)| \|_F \le (\nu T)^{1/4} A^{1/2} \|W_0\|_E + C \|1_{|\xi|>A} W_0\|_E$$

and

$$\lim_{T \to 0^+} \| \sup_{0 < t < T} t^{1/4} e^{-\nu t |\xi|^2} |W_0(\xi)| \|_F = 0.$$

Similarly, if $F_0 \in L^1((0,T), E)$, $t^{1/4} F_0 \in L^1((0,T), F)$, $0 < \epsilon < 1$ and $A > 0$, we have

$$(\nu t)^{1/4} \int_0^t e^{-\nu(t-s)|\xi|^2} |F_0(s,\xi)| \, ds$$

$$\leq \left(\frac{\nu}{1-\epsilon} \right)^{\frac{1}{4}} \int_{(1-\epsilon)t}^t s^{\frac{1}{4}} |F_0(s,\xi)| \, ds$$

$$+ (\nu t)^{1/4} \int_0^{(1-\epsilon)t} 1_{|\xi| \leq A} |F_0(s,\xi)| \, d\xi + \frac{C}{\epsilon^{1/4}} \int_0^t 1_{|\xi|>A} \frac{1}{|\xi|^{1/2}} |F_0(s,\xi)| \, ds.$$

We find that

$$\sup_{0<t<T_0} (\nu t)^{1/4} \| \int_0^t e^{-\nu(t-s)|\xi|^2} |F_0(s,\xi)| \, ds \|_F$$

$$\leq \left(\frac{\nu}{1-\epsilon} \right)^{\frac{1}{4}} \sup_{|I| \leq \epsilon T} \int_I s^{\frac{1}{4}} \|F_0(s,\xi)\|_F \, ds$$

$$+ (\nu T_0)^{1/4} A^{1/2} \int_0^T \|F_0(s,\xi)\|_E \, ds + \frac{C}{\epsilon^{1/4}} \int_0^T \|1_{|\xi|>A} F_0(s,\xi)\|_E \, ds.$$

and thus

$$\lim_{T_0 \to 0} \sup_{0<t<T_0} (\nu t)^{1/4} \| \int_0^t e^{-\nu(t-s)|\xi|^2} |F_0(s,\xi)| \, ds \|_F = 0.$$

Thus, we find that if the initial value $\vec{u}_0$ belongs to the homogeneous Sobolev space $\dot{H}^{1/2}$ and the force $\vec{f}$ belongs to $L^1((0,T), \dot{H}^{1/2})$ and $\sqrt{t} \vec{f}$ belongs to $L^1((0,T), \dot{H}^1)$, then the Navier–Stokes problem has a local-in-time solution such that $t^{1/4} \vec{u} \in L^\infty((0,T_0), \dot{H}^1)$. This is the result of Fujita and Kato [185]. A similar proof willl give us Gevrey regularity estimates for a data in the Sobolev space, a result first given by Foias and Temam [181]:

Theorem 8.20.
If the initial value $\vec{u}_0$ belongs to the homogeneous Sobolev space $\dot{H}^{1/2}$ and the force $\vec{f}$ is such that $e^{\sqrt{-\nu t \Delta}} \vec{f}$ belongs to $L^1((0,T), \dot{H}^{1/2})$ and $t^{1/4} e^{\sqrt{-\nu t \Delta}} \vec{f}$ belongs to $L^1((0,T), \dot{H}^1)$, then the Navier–Stokes problem has a local-in-time solution such that $t^{1/4} e^{\sqrt{-\nu t \Delta}} \vec{u} \in L^\infty((0,T_0), \dot{H}^1)$.

- the case where $E = \frac{1}{|\xi|^2} L^\infty(d\xi)$ and $F = \frac{1}{|\xi|^{5/2}} L^\infty(d\xi)$: the equality

$$\int \frac{1}{|\xi - \eta|^{5/2}} \frac{1}{|\eta|^{5/2}} \, d\eta = C_0 \frac{1}{|\xi|^2}$$

shows that $\frac{1}{|\xi|^{1/2}} (F * F) \subset F$. Moreover, if $A > 0$ and $W_0 \in E$, we write again that, for $t > 0$,

$$|(\nu t)^{1/4} e^{-t|\xi|^2} W_0(\xi)| \leq 1_{|\xi| \leq A} (\nu t)^{\frac{1}{4}} W_0(\xi) + 1_{|\xi|>A} \frac{1}{|\xi|^{1/2}} W_0(\xi)$$

so that

$$\| \sup_{0<t<T} (\nu t)^{1/4} e^{-t|\xi|^2} W_0(\xi) \|_F \leq (\nu T)^{1/4} A^{1/2} \|W^0\|_E + \|1_{|\xi|>A} W^0\|_E$$

and

$$\limsup_{T\to 0^+} \| \sup_{0<t<T} (\nu t)^{1/4} e^{-t|\xi|^2} W^0(\xi)\|_F \le \limsup_{A\to +\infty} \sup_{|\xi|>A} |\xi|^2 |W_0(\xi)|.$$

Similarly, if $F_0 \in L^1((0,T), E)$ and $t^{1/4} F_0 \in L^1((0,T), F)$, we have, for every $0 < \epsilon < 1$ and every $A > 0$,

$$\limsup_{T\to 0^+} \sup_{0<t<T_0} (\nu t)^{1/4} \| \int_0^t e^{-\nu(t-s)|\xi|^2} |F_0(s,\xi)| \, ds \|_F$$

$$\le \left(\frac{\nu}{1-\epsilon} \right)^{\frac{1}{4}} \sup_{|I|\le \epsilon T} \int_I s^{\frac{1}{4}} \|F_0(s,\xi)\|_F \, ds + \frac{C}{\epsilon^{1/4}} \int_0^T \|1_{|\xi|>A} F_0(s,\xi)\|_E \, ds.$$

Hence, we get again the result of Le Jan and Sznitman [305]:

Theorem 8.21.
Let $\tilde{B}^2_{PM,\infty}$ be the closure of S in $\dot{B}^2_{PM,\infty}$. If the initial value $\vec{u}_0$ belongs to $\tilde{B}^2_{PM,\infty}$ and the force $\vec{f}$ is such that $\vec{f}$ belongs to $L^1((0,T), \tilde{B}^2_{PM,\infty})$ and $t^{1/4}\vec{f}$ belongs to $L^1((0,T), \dot{B}^{5/2}_{PM,\infty})$, then the Navier–Stokes problem has a local-in-time solution such that $t^{1/4}\vec{u}$ belongs to $L^1((0,T_0), \dot{B}^{5/2}_{PM,\infty})$.

8.9 Plane Waves

When dealing with Fourier transforms, we stated some results in terms of the L^1 norm of the Fourier transform. The results can be easily extended to the case of Fourier transforms that are finite Borel measures. We shall write $\mu(d\xi)$ for a locally finite measure and $|\mu|(d\xi)$ for its total variation. Let $\mathcal{M}(\mathbb{R}^3)$ be the space of finite Borel measure normed with $\|\mu\| = \int |\mu|(d\xi)$. For μ_1, μ_2 two finite Borel measures and σ a continuous bounded function on $\mathbb{R}^3$, we have the easy estimates

$$\|\mu_1 * \mu_2\|_{\mathcal{M}} \le \|\mu_1\|_{\mathcal{M}} \|\mu_2\|_{\mathcal{M}}, \quad \|\sigma \mu_1\|_{\mathcal{M}} \le \|\mu_1\|_{\mathcal{M}}.$$

Thus, if $\vec{u}$ and $\vec{v}$ have their Fourier transform in $(\mathcal{M})^3$, which we write $\vec{u}, \vec{v} \in \mathcal{FM}$, we find that

$$\|e^{\nu(t-s)\Delta} \mathbb{P}\, \text{div}(\vec{u} \otimes \vec{v})\|_{\mathcal{FM}} \le C \frac{1}{\sqrt{\nu(t-s)}} \|\vec{u}\|_{\mathcal{FM}} \|\vec{v}\|_{\mathcal{FM}}.$$

With these estimates one easily deals with the Navier–Stokes problem

$$\begin{cases} \partial_t \vec{u} = \Delta \vec{u} - \mathbb{P}\, \text{div}(\vec{u} \otimes \vec{u}) \\ \vec{u}(0,.) = \vec{u}_0 \text{ with } \text{div}\, \vec{u}_0 = 0 \end{cases} \tag{8.16}$$

when $\vec{u}_0 \in \mathcal{FM}$ and find a solution $\vec{u} \in L^\infty((0,T), \mathcal{FM})$ with $T = O(\frac{\nu}{\|\vec{u}_0\|^2_{\mathcal{FM}}})$ (where, to avoid measurability issues, $L^\infty \mathcal{FM}$ is defined as the dual space of $L^1 \mathcal{FC}_0$).

When looking for global solutions, one may extend Lei-Lin's theorem [306] (see Corollary 8.1):

Proposition 8.1.
There exists a positive constant ϵ_0 (depending on ν) such that, if

$$\int \frac{1}{|\xi|} |\mathcal{F}\vec{u}_0|(d\xi) < \epsilon_0,$$

and if div $\vec{u}_0 = 0$ then the Cauchy problem for the Navier–Stokes equations (8.16) with initial value $\vec{u}_0$ has a global mild solution $\vec{u}$ such that $\frac{1}{|\xi|}\mathcal{F}\vec{u} \in L_t^\infty \mathcal{M}$ and $|\xi|\mathcal{F}\vec{u} \in L^1\mathcal{M}$.

In 2008, Dinaburg and Sinai [153] discussed the special case of an initial value given by a finite combination of plane waves:

$$\vec{u}_0 = \sum_{j=1}^N e^{i\omega_j \cdot x} \vec{a}_j \text{ with } \omega_j \cdot \vec{a}_j = 0 \text{ and } \omega_j \neq 0.$$

One easily checks that the solution of (8.16) can be written as a sum

$$\vec{u}(t,x) = \sum_{\xi \in \Xi} e^{i\xi \cdot x} \vec{a}_\xi(t)$$

where Ξ is the set of finite sums $\xi = \sum_{j=1}^N k_j \omega_j$ with $N \in \mathbb{N}^*$, $k_j \in \mathbb{N}$ and $\xi \neq 0$. The equations on $\vec{a}_\xi(t)$ can be written as

$$\vec{a}_\xi(t) = \sum_{j=1}^N \delta_{\xi,\omega_j} e^{-\nu t |\omega_j|^2} \vec{a}_{\xi_j} - i \sum_{\eta \in \Xi, \xi - \eta \in \Xi} \int_0^t e^{-\nu(t-s)|\xi|^2} \xi \cdot \vec{a}_\eta(s) \vec{a}_{\xi - \eta}(s)\, ds$$

$$+i \sum_{\eta \in \Xi, \xi - \eta \in \Xi} \sum_{1 \leq p,q \leq 3} \int_0^t e^{-\nu(t-s)|\xi|^2} \frac{\xi_p \xi_q \xi}{|\xi|^2} a_{\eta,p}(s) a_{\xi-\eta,q}(s)\, ds.$$

Existence of the solutions can thus be proved directly by a fixed-point estimate on the set of coefficients $(\vec{a}_\xi)_{\xi \in \Xi}$. Local existence is proved using the norm $\sup_{0<t<T} \sum_{\xi \in \Xi} |\vec{a}_\xi(t)|$ and global existence is proved for small data using the norm $\sup_{0<t} \sum_{\xi \in \Xi} \frac{1}{|\xi|} |\vec{a}_\xi(t)| + \int_0^{+\infty} \sum_{\xi \in \Xi} |\xi| \|\vec{a}_\xi(t)\|$. This can be done by Picard's iteration and Banach contraction principle. Dinaburg and Sinai use the series method of Oseen (as discussed in Section 5.2), a method emphasized by Sinai in his approach of the Navier–Stokes equations in the frequency variable [438].

A very interesting feature of the series method is that it gives directly Gevrey regularity estimates on Dinaburg and Sinai's solutions. Indeed, let us use again the notations of Section 5.2: the solution $\vec{u}$ is written as a sum of basic words $\vec{w} \in W$ as $\vec{u} = \sum_{n=1}^{+\infty} \sum_{\vec{w} \in W_n} \vec{w}$ in the series method and as $\vec{u} = \vec{w}_1 + \sum_{n=1}^{+\infty} \sum_{\vec{w} \in V_n \setminus V_{n-1}} \vec{w}$ in the Picard method. In both methods, we have a control of norm of the n-th term in the sum (the norm is the norm of the Banach space $\mathbb{X}$ where we solve the quadratic equation):

$$\left\| \sum_{\vec{w} \in W_n} \vec{w} \right\|_{\mathbb{X}} \leq C_0 \epsilon^n \text{ and } \left\| \sum_{\vec{w} \in V_n \setminus V_{n-1}} \vec{w} \right\|_{\mathbb{X}} \leq C_0 \epsilon^n$$

where $\epsilon < 1$. Moreover, when considering the problem with plane waves as initial data, we have that $\mathcal{F}u_0$ is supported in the ball $B(0,R)$ with $R = \max_{1 \leq j \leq N}(|\omega_j|)$. If $\vec{w} \in W_n$, then $\mathcal{F}(\vec{w})$ is supported in $B(0,nR)$; if $\vec{w} \in V_n$, then $\mathcal{F}(\vec{w})$ is supported in $B(0,2^n R)$. It means that the Picard estimates seem to give at best polynomially decaying spectral estimates (in $O((\frac{|\xi|}{R})^{-\frac{\ln(1/\epsilon)}{\ln 2}})$)) while the Oseen estimates give exponentially decaying spectral estimates (in $O(e^{-\ln(1/\epsilon)\frac{|\xi|}{R}})^4$.

---

[4] One may, of course, recover Gevrey estimates through the Picard method by dealing with the corresponding Gevrey norm when studying the fixed-point problem.

Chapter 9

The Space BMO^{-1} and the Koch and Tataru Theorem

9.1   The Koch and Tataru Theorem

The Koch and Tataru theorem [266] deals with the largest space where to search for mild solutions, which is well fitted to the symmetries of the Navier–Stokes equations.

We recall that we have rewritten the Navier–Stokes equations

$$
\begin{cases}
\partial_t \vec{u} = \nu \Delta \vec{u} - (\vec{u}.\vec{\nabla})\vec{u} + \vec{f} - \vec{\nabla}p \\
\quad\quad \text{div } \vec{u} = 0 \\
\quad\quad \vec{u}_{|t=0} = \vec{u}_0
\end{cases}
\tag{9.1}
$$

into

$$
\vec{u} = W_{\nu t} * \vec{u}_0 - \int_0^t \sum_{j=1}^3 \partial_j \mathcal{O}(\nu(t-s)) :: \left(\vec{f} * \partial_j G + u_j \vec{u} \right) ds
\tag{9.2}
$$

In order to give some meaning to the integral in Equation (9.2), we shall suppose that $\vec{u}$ is locally square integrable on $[0,T) \times \mathbb{R}^3$. Moreover, we shall suppose that the estimates are uniform in x (in order to use translation invariance of the equations) and invariant under the scaling $\vec{u} \to \lambda \vec{u}(\lambda^2 t, \lambda x)$ (in order to use the scaling invariance of the equations). This gives:

$$
\sup_{x \in \mathbb{R}^3, 0 < t < T} \frac{1}{t^{3/2}} \iint_{(0,t) \times B(x,\sqrt{t})} |\vec{u}(s,y)|^2 \, ds \, dy < +\infty.
$$

Koch and Tataru characterized then the associated space of initial values for the Navier–Stokes equations [266, 313] as derivatives of functions in the BMO space of John and Nirenberg [247] or in the local bmo space of Goldberg [214]:

Proposition 9.1. *For a measurable function F on $(0,T) \times \mathbb{R}^3$ (with $T \in (0,+\infty]$), define*

$$
\|F\|_{X_T} = \sup_{x \in \mathbb{R}^3, \, 0 < t < T} \sqrt{\frac{1}{t^{3/2}} \iint_{(0,t) \times B(x,\sqrt{t})} |F(s,y)|^2 \, ds \, dy}.
$$

Then, for $T < +\infty$ and $u_0 \in \mathcal{S}'(\mathbb{R}^3)$, we have

$$
\|W_{\nu t} * u_0\|_{X_T} < +\infty \Leftrightarrow \exists (f_0, \dots, f_3) \in (bmo(\mathbb{R}^3))^4 \quad u_0 = f_0 + \sum_{j=1}^3 \partial_j f_j
$$

and the norm $\|W_{\nu t} * u_0\|_{X_T}$ is equivalent to the infimum of $\sum_{j=0}^3 \|f_j\|_{bmo}$ over all possible decompositions $u_0 = f_0 + \sum_{j=1}^3 \partial_j f_j$.

DOI: 10.1201/9781003042594-9

Simarly, we have

$$\|W_{\nu t} * u_0\|_{X_\infty} < +\infty \Leftrightarrow \exists (f_1, \ldots, f_3) \in (BMO(\mathbb{R}^3))^3 \quad u_0 = \sum_{j=1}^{3} \partial_j f_j$$

and the norm $\|W_{\nu t} * u_0\|_{X_\infty}$ is equivalent to the infimum of $\sum_{j=1}^{3} \|f_j\|_{BMO}$ over all possible decompositions $u_0 = \sum_{j=1}^{3} \partial_j f_j$.

Koch and Tataru coined bmo^{-1} the space of distributions u_0 such that $\|W_{\nu t} * u_0\|_{X_T} < +\infty$ for finite T, and BMO^{-1} the space of distributions u_0 such that $\|W_{\nu t} * u_0\|_{X_\infty} < +\infty$.

We define the norm of u_0 in BMO^{-1} as the infimum of $\sum_{j=1}^{3} \|f_j\|_{BMO}$ over all possible decompositions $u_0 = \sum_{j=1}^{3} \partial_j f_j$. In particular, we have

$$\|W_{\nu t} * u_0\|_{X_\infty} \leq C \sqrt{\frac{\ln(e + \frac{1}{\nu})}{\nu}} \|u_0\|_{BMO^{-1}}.$$

Koch and Tataru's theorem is then the following one:

Theorem 9.1.
The bilinear operator

$$B(\vec{F}, \vec{G}) = \int_0^t \sum_{j=1}^{3} \partial_j \mathcal{O}(\nu(t-s)) :: (F_j \vec{G}) \, ds = \int_0^t W_{\nu(t-s)} * \mathbb{P} \operatorname{div}(\vec{F} \otimes \vec{G}) \, ds$$

is bounded on the space

$$E_T = \{\vec{F} \ / \ \|\vec{F}\|_{X_T} < +\infty \text{ and } \sup_{0 < t < T} \sqrt{t} \|\vec{F}(t, .)\|_\infty < +\infty\}$$

for every $T \in (0, +\infty]$.

In order to prove Theorem 9.1, we follow the strategy of Auscher and Frey [11]. We begin with the following lemma:

Lemma 9.1.
If $\vec{F}$ and $\vec{G}$ belong to E_∞, then $\mathcal{A}(\vec{F}, \vec{G}) = \int_0^{+\infty} W_{\nu s} * (\vec{F} \otimes \vec{G}) \, ds$ belongs to $(BMO)^9$ and

$$\|\mathcal{A}(\vec{F}, \vec{G})\|_{BMO} \leq C_\nu \|\vec{F}\|_{E_\infty} \|\vec{G}\|_{E_\infty}.$$

Proof. We want to estimate the BMO norm of $H(x) = \int_0^{+\infty} W_{\nu s} * h(s, .) \, ds$, where h satisfies

$$\|h\|_{(1)} = \sup_{x \in \mathbb{R}^3, 0 < t} \frac{1}{t^{3/2}} \iint_{(0,t) \times B(x, \sqrt{t})} |h(s, y)| \, ds \, dy < +\infty$$

and

$$\sup_{t > 0} t \|h(t, .)\|_\infty < +\infty.$$

Using the fact that BMO is the dual space of the Hardy space $\mathcal{H}^1$ [313, 448], we must prove that, if $\mathbb{A}_\infty$ is the set of atoms for $\mathcal{H}^1$ (i.e., $a \in \mathbb{A}_\infty$ if for some $r > 0$ and $x_0 \in \mathbb{R}^3$, a is supported in $B(x,r)$, $\|a\|_\infty \leq \frac{1}{|B(x_0,r)|}$, and $\int a\, dx = 0$), then

$$\sup_{a \in \mathbb{A}_\infty} \left| \int a(x) H(x)\, dx \right| \leq C_\nu (\|h\|_{(1)} + \sup_{t>0} t \|h(t,.)\|_\infty).$$

If a is an atom (associated to a ball $B(x_0, r)$), then it can be written as $a = \sum_{i=1}^3 \partial_i \alpha_i$, with α_i supported in $x_0 + [-r, r]^3$ and $\|\alpha_j\|_\infty \leq Cr^{-2}$ (and thus $\|\alpha_j\|_1 \leq 8Cr$). This gives

$$\left| \int a(x) \int_{r^2}^{+\infty} W_{\nu s} * h(s,.)\, ds\, dx \right| \leq 24C \int_{r^2}^{+\infty} r \|\vec{\nabla} W_{\nu s} * h(s,.)\|_\infty\, ds$$

$$\leq C' r (\sup_{t>0} t \|h(t,.)\|_\infty) \int_{r^2}^{+\infty} \frac{dt}{t\sqrt{\nu t}}$$

$$= \frac{C'}{\sqrt{\nu}} \sup_{t>0} t \|h(t,.)\|_\infty$$

On the other hand, we write $h = h_1 + h_2$, where $h_1(s,y) = 1_{B(x_0,3r)}(y) h(s,y)$; we have

$$\left| \int a(x) \int_0^{r^2} W_{\nu s} * h_1(s,.)\, ds\, dx \right| \leq \|a\|_\infty \int_0^{r^2} \|h_1(s,.)\|_1\, ds$$

$$\leq C \|a\|_\infty r^3 \|h\|_{(1)}$$

$$\leq C' \|h\|_{(1)}$$

while

$$\left| \int a(x) \int_0^{r^2} W_{\nu s} * h_2(s,.)\, ds\, dx \right|$$

$$\leq C \int_{B(x_0,r)} \int_0^{r^2} \int_{|x_0-y|>3r} |a(x)| \frac{\nu s}{|x-y|^5} |h(s,y)|\, dy\, ds\, dx$$

$$\leq C' \nu \|a\|_1 r^2 \int_0^{r^2} \int_{|x_0-y|>3r} \frac{1}{|x_0-y|^5} |h(s,y)|\, ds\, dy$$

$$\leq C'' \nu r^{-3} \sum_{k \in \mathbb{Z}^3, k \neq 0} \frac{1}{|k|^5} \int_0^{r^2} \int_{x_0+kr+[-r/2,r/2]^3} |h(s,y)|\, ds\, dy$$

$$\leq C''' \nu \|h\|_{(1)}.$$

The lemma is proved. □

Proof of Theorem 9.1:
A first remark is that when F belongs to E_T and when G is defined on $(0, +\infty) \times \mathbb{R}^3$ by $G(t,x) = 1_{(0,T)}(t) F(t,x)$, then G belongs to E_∞. Indeed, for $t < T$, we have $\iint_{(0,t)\times B(x,\sqrt{t})} |G(s,y)|^2\, ds\, dy \leq \|F\|_{X_T}^2 t^{3/2}$; for $t \geq T$, we cover the ball $B(x, \sqrt{t})$ by a finite number N_t of balls $B(x_i, \sqrt{T})$ with $N_t = O((\frac{t}{T})^{3/2})$ and we write

$$\iint_{(0,t)\times B(x,\sqrt{t})} |G(s,y)|^2\, ds\, dy \leq \sum_{1 \leq i \leq N_t} \iint_{(0,T)\times B(x_i,\sqrt{T})} |F(s,y)|^2\, ds\, dy$$

$$\leq N_t \|F\|_{X_T}^2 T^{3/2}$$

$$\leq C \|F\|_{X_T}^2 t^{3/2}.$$

As

$$B(\vec{F}, \vec{G}) = \int_0^t W_{\nu(t-s)} * \mathbb{P} \operatorname{div}(\vec{F} \otimes \vec{G}) \, ds$$

only involves, for $t < T$, the values of $\vec{F}$ and $\vec{G}$ for $s < T$, we find that it is enough to prove the theorem for $T = +\infty$.

Now, we take $\vec{F}$ and $\vec{G}$ in E_∞ and we fix $T > 0$ and $x_0 \in \mathbb{R}^3$ and we want to prove that, for a constant C which does not depend on T nor on x_0, we have

$$|B(\vec{F}, \vec{G})(T, x_0)| \leq C \|\vec{F}\|_{E_\infty} \|\vec{G}\|_{E_\infty} T^{-1/2}$$

and

$$\int_0^T \int_{B(x_0, \sqrt{T})} |B(\vec{F}, \vec{G})(t, x)| \, dt \, dx \leq \|\vec{F}\|_{E_\infty} \|\vec{G}\|_{E_\infty} T^{3/2}.$$

Let $\chi_{x_0, T} = 1_{(0,T)}(t) 1_{B(x_0, 5\sqrt{T})}(x)$ and $\psi_{x_0, T} = 1_{(0,T)}(t)(1 - 1_{B(x_0, 5\sqrt{T})}(x))$. For $0 < t \leq T$, we have

$$B(\vec{F}, \vec{G})(t, x) = B(\chi_{x_0, T} \vec{F}, \vec{G})(t, x) + B(\psi_{x_0, T} \vec{F}, \vec{G})(t, x).$$

If moreover $x \in B(x_0, \sqrt{T})$, we have

$$|B(\psi_{x_0, T} \vec{F}, \vec{G})(t, x)| \leq C \int_0^t \int \psi_{x_0, T}(s, y) |\vec{F}(s, y)| \, |\vec{G}(s, y)| \, \frac{ds \, dy}{(\sqrt{\nu(t-s)} + |x - y|)^4}$$

$$\leq C' \int_0^T \int_{|y - x_0| > 4\sqrt{T}} \frac{1}{|x - y|^4} |\vec{F}(s, y)| \, |\vec{G}(s, y)| \, ds \, dy$$

$$\leq C' \sum_{k \in \mathbb{Z}^3, k \neq 0} \frac{1}{|k|^4 T^2} \int_0^T \int_{y \in x_0 + k\sqrt{T} + [-\sqrt{T}/2, \sqrt{T}/2]^3} |\vec{F}(s, y)| \, |\vec{G}(s, y)| \, ds \, dy$$

$$\leq C'' T^{-1//2} \sum_{k \in \mathbb{Z}^3, k \neq 0} \frac{1}{|k|^4} \|\vec{F}\|_{X_\infty} \|\vec{G}\|_{X_\infty}$$

and thus

$$|B(\psi_{x_0, T} \vec{F}, \vec{G})(T, x_0)| \leq C \|\vec{F}\|_{E_\infty} \|\vec{G}\|_{E_\infty} T^{-1/2}$$

and

$$\int_0^T \int_{B(x_0, \sqrt{T})} |B(\psi_{x_0, T} \vec{F}, \vec{G})(t, x)|^2 \, dt \, dx \leq \|\vec{F}\|_{E_\infty} \|\vec{G}\|_{E_\infty} T^{3/2}.$$

In order to estimate $B(\chi_{x_0, T} \vec{F}, \vec{G})$, we follow the strategy of Auscher and Frey [11] and decompose $B(\chi_{x_0, T} \vec{F}, \vec{G})$ into

$$B(\chi_{x_0, T} \vec{F}, \vec{G}) = \mathcal{A}_1(\chi_{x_0, T} \vec{F}, \vec{G}) + \mathcal{A}_2(\chi_{x_0, T} \vec{F}, \vec{G}) + \mathcal{A}_3(\chi_{x_0, T} \vec{F}, \vec{G})$$

with

$$\mathcal{A}_1(\chi_{x_0, T} \vec{F}, \vec{G}) = \int_0^t (W_{\nu(t-s)} - W_{\nu(t+s)}) * \mathbb{P} \operatorname{div}(\chi_{x_0, T} \vec{F} \otimes \vec{G}) \, ds$$

$$\mathcal{A}_2(\chi_{x_0, T} \vec{F}, \vec{G}) = \int_0^{+\infty} W_{\nu(t+s)} * \mathbb{P} \operatorname{div}(\chi_{x_0, T} \vec{F} \otimes \vec{G}) \, ds$$

$$\mathcal{A}_3(\chi_{x_0, T} \vec{F}, \vec{G}) = -\int_t^{+\infty} W_{\nu(t+s)} * \mathbb{P} \operatorname{div}(\chi_{x_0, T} \vec{F} \otimes \vec{G}) \, ds.$$

We further rewrite $\mathcal{A}_1(\chi_{x_0,T}\vec{F},\vec{G})$ as

$$\mathcal{A}_1(\chi_{x_0,T}\vec{F},\vec{G}) = -2\nu \int_0^1 \left(\int_0^t \Delta W_{\nu(t-s)} * W_{2\nu\theta s} * \mathbb{P}\operatorname{div}(\chi_{x_0,T}\vec{F}\otimes\vec{G})\, s\, ds \right) d\theta$$

or as

$$\mathcal{A}_1(\chi_{x_0,T}\vec{F},\vec{G}) = -2\nu \int_0^1 \mathcal{A}_4(\mathcal{A}_{5,\theta}(\chi_{x_0,T}\vec{F},\vec{G}))\, d\theta$$

with

$$\mathcal{A}_4(\vec{H}) = \int_0^t \Delta W_{\nu(t-s)} * \vec{H}\, ds$$

and

$$\mathcal{A}_{5,\theta}(\chi_{x_0,T}\vec{F},\vec{G})(s,.) = sW_{2\nu\theta s} * \mathbb{P}\operatorname{div}(\chi_{x_0,T}\vec{F}\otimes\vec{G}).$$

We rewrite $\mathcal{A}_2(\chi_{x_0,T}\vec{F},\vec{G})$ as

$$\mathcal{A}_2(\chi_{x_0,T}\vec{F},\vec{G}) = W_{\nu t} * \mathbb{P}\operatorname{div}\mathcal{A}_6(\chi_{x_0,T}\vec{F},\vec{G})$$

with

$$\mathcal{A}_6(\chi_{x_0,T}\vec{F},\vec{G}) = \int_0^{+\infty} W_{\nu s} * (\chi_{x_0,T}\vec{F}\otimes\vec{G})\, ds.$$

We thus get the decomposition

$$B(\chi_{x_0,T}\vec{F},\vec{G}) = -2\nu \int_0^1 \mathcal{A}_4(\mathcal{A}_{5,\theta}(\chi_{x_0,T}\vec{F},\vec{G}))\, d\theta + W_{\nu t} * \mathbb{P}\operatorname{div}\mathcal{A}_6(\chi_{x_0,T}\vec{F},\vec{G})$$
$$+ \mathcal{A}_3(\chi_{x_0,T}\vec{F},\vec{G}).$$

The last two terms are easily estimated:

- $\mathcal{A}_3(\chi_{x_0,T}\vec{F},\vec{G})(T,x_0) = 0$

- we have

$$\|W_{\nu T} * \mathbb{P}\operatorname{div}\mathcal{A}_6(\chi_{x_0,T}\vec{F},\vec{G})\|_\infty \leq CT^{-1/2}\|\mathcal{A}_6(\chi_{x_0,T}\vec{F},\vec{G})\|_{BMO}$$
$$\leq C_\nu T^{-1/2}\|\vec{F}\|_{E_\infty}\|\vec{G}\|_{E_\infty}.$$

- we have similarly

$$\int_0^T \int_{B(x_0,\sqrt{T})} |W_{\nu T} * \mathbb{P}\operatorname{div}\mathcal{A}_6(\chi_{x_0,T}\vec{F},\vec{G})|^2\, dt\, dx$$
$$\leq CT^{3/2}\|W_{\nu T} * \mathbb{P}\operatorname{div}\mathcal{A}_6(\chi_{x_0,T}\vec{F},\vec{G})\|_{X_\infty}^2$$
$$\leq C'T^{3/2}\|\mathcal{A}_6(\chi_{x_0,T}\vec{F},\vec{G})\|_{BMO}^2$$
$$\leq C''\|\vec{F}\|_{E_\infty}^2\|\vec{G}\|_{E_\infty}^2.$$

- Let $\alpha(t) = \|\chi_{x_0,T}(t,.)\vec{F}(t,.)\|_2$ and $\beta = \sup_{t>0}\sqrt{t}\|\vec{G}(t,.)\|_\infty$. We have

$$\int_0^T \int_{B(x_0,\sqrt{T})} |\mathcal{A}_3(\chi_{x_0,T}\vec{F},\vec{G})(t,x)|^2\, dt\, dx$$
$$\leq \int_0^{+\infty} \int |\mathcal{A}_3(\chi_{x_0,T}\vec{F},\vec{G})(t,x)|^2\, dt\, dx$$
$$\leq \int_0^{+\infty} \left(\int_t^{+\infty} \|W_{\nu(t+s)} * \mathbb{P}\operatorname{div}(\chi_{x_0,T}\vec{F}\otimes\vec{G})(s,.)\|_2\, ds \right)^2 dt$$
$$\leq C \int_0^{+\infty} \left(\int_t^{+\infty} \frac{1}{\sqrt{\nu(t+s)}}\|\chi_{x_0,T}\vec{F}\otimes\vec{G}(s,.)\|_2\, ds \right)^2 dt.$$

$$\leq \frac{C}{\nu} \int_0^{+\infty} \left(\int_t^{+\infty} \frac{1}{\sqrt{t+s}\sqrt{s}} \beta\alpha(s)\,ds \right)^2 dt$$

$$\leq \frac{C}{\nu} \int_0^{+\infty} \left(\int_t^{+\infty} \frac{1}{\sqrt{s}} \beta^2\alpha^2(s)\,ds \right) \left(\int_t^{+\infty} \frac{1}{(t+\tau)\sqrt{\tau}}\,d\tau \right) dt$$

$$= \frac{C'}{\nu} \int_0^{+\infty} \frac{1}{\sqrt{t}} \left(\int_t^{+\infty} \frac{1}{\sqrt{s}} \beta^2\alpha^2(s)\,ds \right) dt$$

$$= 2\frac{C'}{\nu} \int_0^{+\infty} \beta^2\alpha^2(s)\,ds$$

$$\leq \frac{C''}{\nu} T^{3/2} \|\vec{F}\|_{X_\infty}^2 (\sup_{t>0} \sqrt{t}\|\vec{G}\|_\infty)^2.$$

Thus, we are left with the estimation of the first term, i.e., $\mathcal{A}_1(\chi_{x_0,T}\vec{F},\vec{G}) = \int_0^t (W_{\nu(t-s)} - W_{\nu(t+s)}) * \mathbb{P}\operatorname{div}(\chi_{x_0,T}\vec{F}\otimes\vec{G})\,ds = -2\nu \int_0^1 \mathcal{A}_4(\mathcal{A}_{5,\theta}(\chi_{x_0,T}\vec{F},\vec{G}))\,d\theta.$

Let again $\alpha(t) = \|\chi_{x_0,T}(t,.)\vec{F}(t,.)\|_2$ and $\beta = \sup_{t>0}\sqrt{t}\|\vec{G}(t,.)\|_\infty$. We use the maximal regularity of the heat kernel [313] to get

$$\|\mathcal{A}_4(\vec{H})\|_{L^2L^2((0,+\infty)\times\mathbb{R}^3)} \leq C\|\vec{H}\|_{L^2L^2((0,+\infty)\times\mathbb{R}^3)}$$

and thus

$$\left\| \int_0^1 \mathcal{A}_4(\mathcal{A}_{5,\theta}(\chi_{x_0,T}\vec{F},\vec{G}))\,d\theta \right\|_{L^2L^2} \leq \int_0^1 \|\mathcal{A}_4(\mathcal{A}_{5,\theta}(\chi_{x_0,T}\vec{F},\vec{G}))\|_{L^2L^2}\,d\theta$$

$$\leq C \int_0^1 \|\mathcal{A}_{5,\theta}(\chi_{x_0,T}\vec{F},\vec{G})\|_{L^2L^2}\,d\theta$$

$$\leq C' \int_0^1 \left(\int_0^{+\infty} \frac{s}{\nu\theta}\alpha^2(s)\frac{\beta^2}{s}\,ds \right)^{1/2} d\theta$$

$$= 2C' \|\chi_{x_0,T}\vec{F}\|_{L^2L^2}\beta$$

$$\leq C'' T^{3/4} \|\vec{F}\|_{X_\infty} \sup_{t>0} \sqrt{t}\|\vec{G}\|_\infty.$$

Finally, we write

$$|\mathcal{A}_1(\chi_{x_0,T}\vec{F},\vec{G})(T,x_0)| \leq C \int_0^T \int \frac{\chi_{x_0,T}(s,y)}{(\sqrt{\nu(T-s)}+|x_0-y|)^4} |\vec{F}(s,y)||\vec{G}(s,y)|\,ds\,dy$$

$$\leq \frac{C'}{T^2} \int_0^{T/2} \int \chi_{x_0,T}(s,y)|\vec{F}(s,y)||\vec{G}(s,y)|\,ds\,dy$$

$$+ C \int_{T/2}^T \int \|\vec{F}(s,.)\|_\infty \|\vec{G}(s,.)\|_\infty \frac{ds\,dy}{(\sqrt{\nu(T-s)}+|x_0-y|)^4}$$

$$\leq C'' T^{-1/2} (\|\vec{F}\|_{X_\infty}\|\vec{G}\|_{X_\infty} + \frac{1}{\sqrt{\nu}}(\sup_{t>0}\sqrt{t}\|\vec{F}(t,.)\|_\infty)(\sup_{t>0}\sqrt{t}\|\vec{G}(t,.)\|_\infty))$$

The theorem is proved.   □

Theorem 9.1 gives the following theorem on Navier–Stokes equations:

Koch and Tataru theorem

Theorem 9.2.
Let, for $T \in (0, +\infty]$,

$$\|h\|_{E_T} = \sup_{0<t<T} \sqrt{t}\|h(t,.)\|_\infty + \sup_{0<t<T} \sup_{x_0 \in \mathbb{R}^3} \frac{1}{t^{3/4}} \sqrt{\int_0^t \int_{B(x_0,\sqrt{t})} |h(s,y)|^2 \, dy \, ds}$$

and

$$\|h\|_{F_T} = \sup_{0<t<T} t\|h(t,.)\|_\infty + \sup_{0<t<T} \sup_{x_0 \in \mathbb{R}^3} \frac{1}{t^{3/2}} \int_0^t \int_{B(x_0,\sqrt{t})} |h(s,y)|, dy \, ds.$$

There exist two constants ϵ_0 and C_0 which do not depend on T (but depend on ν) such that if

- *$\vec{f} = \operatorname{div} F$ with $F \in F_T$ and $\|F\|_{F_T} < \epsilon_0$*

- *$\vec{u}_0 \in bmo^{-1}$ and $\operatorname{div} \vec{u}_0 = 0$*

- *$\|1_{0<t<T} W_{\nu t} * \vec{u}_0\|_{E_T} < \epsilon_0$*

then the Navier–Stokes equations

$$\partial_t \vec{u} = \nu \Delta \vec{u} - \vec{u}.\vec{\nabla}\vec{u} - \vec{\nabla}p + \vec{f}$$

with $\operatorname{div} \vec{u} = 0$ and $\vec{u}(0,.) = \vec{u}_0$ have a unique solution $\vec{u}$ on $(0,T)$ such that $\vec{u} \in E_T$ and $\|\vec{u}\|_{E_T} \leq C_0 \epsilon_0$. This solution satisfies

$$\|\vec{u}\|_{E_T} \leq C_0(\|1_{0<t<T} W_{\nu t} * \vec{u}_0\|_{E_T} + \|F\|_{F_T}).$$

Proof. As usual, by Picard's iterative scheme, using the estimate given by Theorem 9.1. ☐

The Koch and Tataru theorem gives criteria for local or global existence:

- if $\vec{u}_0 \in bmo^{-1}$ and $W_{\nu t} * \vec{u}_0 \in E_\infty$, then $\vec{u}_0$ belongs to the smaller space BMO^{-1} and we have $\|W_{\nu t} * \vec{u}_0\|_{E_\infty} \approx \|\vec{u}_0\|_{BMO^{-1}}$.

- Hence, if $\|F\|_{F_\infty} < \epsilon_0$, then we have global existence of the solution of the Navier–Stokes equations provided that $\|\vec{u}_0\|_{BMO^{-1}}$ is small enough.

- If $\vec{u}_0$ belongs to the closure of test functions (or more generally of bounded functions) in bmo^{-1}, then we have

$$\lim_{T \to 0} \|1_{0<t<T} W_{\nu t} * \vec{u}_0\|_{E_T} = 0.$$

- Hence, if $\|F\|_{F_T} < \epsilon_0$, we have local existence (for some $T_0 \in (0,T]$) of a solution of the Navier–Stokes equations provided that $\vec{u}_0$ is regular enough (i.e., belongs to the closure of bounded functions in bmo^{-1}).

9.2   A Variation on the Koch and Tataru Theorem

From Theorem 9.2, we find existence of a global mild solution to the Cauchy problem for the Navier–Stokes equations if the initial value $\vec{u}_0$ is small enough in BMO^{-1} and if, for the forcing term $\vec{f} = \operatorname{div} F$, F is small enough in F_∞. However, one easily sees that we may allow more general forces. For instance, we have:

Theorem 9.3.

Let $2 < p \leq 5$.

A) The bilinear operator

$$B(\vec{F}, \vec{G}) = \int_0^t \sum_{j=1}^3 \partial_j \mathcal{O}(\nu(t-s)) :: (F_j \vec{G})\, ds = \int_0^t W_{\nu(t-s)} * \mathbb{P} \operatorname{div}(\vec{F} \otimes \vec{G})\, ds$$

is bounded on the space $E_\infty + \mathcal{M}_2^{p,5}$, where

$$\|\vec{F}\|_{E_\infty} = \sup_{x \in \mathbb{R}^3, 0 < t} \sqrt{\frac{1}{t^{3/2}} \iint_{(0,t) \times B(x,\sqrt{t})} |\vec{F}(s,y)|^2\, ds\, dy} + \sup_{0 < t} \sqrt{t}\|\vec{F}(t,.)\|_\infty$$

and

$$\|\vec{F}\|_{\mathcal{M}_2^{p,5}} = \sup_{x \in \mathbb{R}^3, t \in \mathbb{R}, r > 0} r^{1-\frac{5}{p}} \left(\iint_{s>0,\, (s,y) \in (t-r^2, t+r^2) \times B(x,r)} |\vec{F}(s,y)|^p\, ds\, dy \right)^{1/p}$$

B) If $\vec{u}_0$ is small enough in $BMO^{(-1)}$ (and is divergence free) and if $\int_0^t e^{(t-s)\Delta} \mathbb{P} \operatorname{div} F\, ds$ is small enough in $E_\infty + \mathcal{M}_2^{p,5}$ (in particular, if $F = F_1 + F_2$, where F_1 is small in F_∞ and F_2 is small in the Morrey space $\mathcal{M}_2^{p/2,5/2}$), then the Navier–Stokes equations

$$\partial_t \vec{u} = \nu \Delta \vec{u} - \vec{u}.\vec{\nabla}\vec{u} - \vec{\nabla}p + \operatorname{div} F$$

with $\operatorname{div} \vec{u} = 0$ and $\vec{u}(0,.) = \vec{u}_0$ have a global mild solution in $E_\infty + \mathcal{M}_2^{p,5}$.

Proof. A) We already know that B is bounded from $E_\infty \times E_\infty$ to E_∞ (Theorem 9.1). Moreover, we know that, by Hedberg's inequality (Lemma 5.3), that the parabolic Riesz potential $f \mapsto K_{2,1} *_{t,x} f$ is bounded from $\mathcal{M}_2^{\rho,r}$ to $\mathcal{M}_2^{\sigma,s}$ for $1 < \rho \leq r < 5$, $\frac{1}{s} = \frac{1}{r} - \frac{1}{5}$ and $\sigma = \frac{s}{r}\rho$. In particular, it is bounded from $\mathcal{M}_2^{p/2,5/2}$ to $\mathcal{M}_2^{p,5}$; as

$$|B(\vec{F}, \vec{G})| \leq C K_{2,1} *_{t,x} (|\vec{F}|\,|\vec{G}|),$$

B is bounded from $\mathcal{M}_2^{p,5} \times \mathcal{M}_2^{p,5}$ to $\mathcal{M}_2^{p,5}$. Finally, we check that $(F,) \mapsto K_{2,1} *_{t,x} (FG)$ is bounded from $L^{2,\infty}L^\infty \times \mathcal{M}_2^{p,5}$ to $\mathcal{M}_2^{p,5}$. Let $(t_0, x_0) \in \mathbb{R} \times \mathbb{R}^3$ and $r < 0$. We want to estimate the $L^p L^p$ norm of $K_{2,1} *_{t,x} (FG)$ on $(t_0 - r^2, t_0 + r^2) \times B(x_0, r)$; we write $G = G_1 + G_2$ with $G_1 = \mathbb{1}_{(t_0 - 16r^2, t_0 + 16r^2) \times B(x_0, 4r)} G$. We have

$$\|K_{2,1} *_{t,x} (FG_1)\|_{L^p(dx)} \leq \int \|K_{2,1}(t-s,.)\|_{L^1(dx)} \|F(s,.)\|_{L^\infty(dx)} \|G_1(s,.)\|_{L^p(dx)}\, ds.$$

As $\|A * (BC)\|_p \leq C_p \|A\|_{L^{2,\infty}} \|B\|_{L^{2,\infty}} \|C\|_p$, we find that

$$\|K_{2,1} *_{t,x} (FG_1)\|_{L^p L^p} \leq C\|F\|_{L^{2,\infty}L^\infty} \|G_1\|_{L^p L^p} \leq C' r^{\frac{5}{p}-1} \|F\|_{L^{2,\infty}L^\infty} \|G\|_{\mathcal{M}_2^{p,5}}.$$

On the other hand, $L^{2,\infty}L^\infty \subset \mathcal{M}_2^{\frac{p}{p-1},5}$ so that $\|FG\|_{\mathcal{M}_2^{1,5/2}} \leq C\|F\|_{L^{2,\infty}L^\infty}\|G\|_{\mathcal{M}_2^{p,5}}$. We then write, for $(t,x) \in (t_0 - r^2, t_0 + r^2) \times B(x_0, r)$,

$$|K_{2,1} *_{t,x} (FG_2)(t,x)|$$

$$\leq C \sum_{j=0}^{+\infty} \iint_{4^j r \leq \sqrt{|t-0-s|} + |x_0-y| < 4^{j+1}} \frac{1}{(\sqrt{|t-s|} + |x-y|)^4}|FG(s,y)|\,ds\,dy$$

$$\leq C' \sum_{j=0}^{+\infty} \frac{1}{(4^j r)^4}\|FG\|_{\mathcal{M}_2^{1,5/2}}(4^j r)^3$$

$$\leq C''\|F\|_{L^{2,\infty}L^\infty}\|G\|_{\mathcal{M}_2^{p,5}}\frac{1}{r};$$

we et that

$$\|\mathbb{1}_{(t_0-r^2,t_0+r^2)\times B(x_0,r)}K_{2,1} *_{t,x} (FG_2)\|_{L^p L^p} \leq C\|F\|_{L^{2,\infty}L^\infty}\|G\|_{\mathcal{M}_2^{p,5}}\frac{1}{r}r^{5/p}.$$

Thus, B is bounded from $E_\infty \times \mathcal{M}_2^{p,5}$ to $\mathcal{M}_2^{p,5}$ and from $\mathcal{M}_2^{p,5} \times E_\infty$ to $\mathcal{M}_2^{p,5}$. B) is then a direct consequence of A). $\qquad\square$

Remark: the assumptions of Theorem 9.3 allows one to consider singular tensors F for the Navier–Stokes problem; for instance, if F belongs to $L^p\dot{W}^{-1,q} = (\sqrt{-\Delta})(L^p L^q)$ with $2 < p < +\infty$ and $\frac{2}{p} + \frac{3}{q} = 1$, then $\int_0^t e^{(t-s)\Delta}\mathbb{P}\operatorname{div} F\,ds \in L^p L^q \subset \mathcal{M}_2^{\min(p,q),5}$.

We remark that, for $2 < p \leq 5$, $E_\infty + \mathcal{M}_2^{p,5} \subset \mathcal{M}_2^{2,5}$. However, $\mathcal{M}_2^{2,5}$ does not play the role of maximal space where to find mild solutions for the Navier–tokes equations, as B is not bounded from $\mathcal{M}_2^{2,5} \times \mathcal{M}_2^{2,5}$ to $\mathcal{M}_2^{2,5}$ [323]. On the other hand, some mild solutions don't satisfy the assumptions of Theorem 9.3 and the proof of the theorem cannot be applied to them. Consider for instance the space $L^2\mathcal{F}L^1$; we discussed mild solutions in this space in Theorem 8.16 and the Corolllary 8.1, corresponding to theorems of Cannone and Wu [86] and of Lei and Lin [306]. The bilinear operator B is not bounded from $E_\infty \times L^2\mathcal{F}L^1$ to $L^2\mathcal{F}L^1$ nor to E_∞. Thus, we need a new space where to work.

We have a similar problem with the space $\mathcal{V}^{2,1}(\mathbb{R} \times \mathbb{R}^3)$ we considered in Corollary 5.3. We have, for $2 < p$, $\mathcal{M}_2^{p,5} \subset \mathcal{V}^{2,1}(\mathbb{R} \times \mathbb{R}^3) \subset \mathcal{M}_2^{2,5}$. The idea developed by Lemarié-Rieusset in [323] is to consider a broader space than E_∞, namely the space $E_{\infty,q}$, for $5 < q < +\infty$, defined as the space of vector fields $\vec{u}$ on $(0, +\infty) \times \mathbb{R}^3$ such that

$$\sup_{T>0, x\in\mathbb{R}^3} T^{-3/4}\|\mathbb{1}_{(0,T)\times B(x,\sqrt{T})}\vec{u}\|_{L^2_{t,x}} < +\infty$$

and

$$\sup_{T>0, x\in\mathbb{R}^3} T^{-\frac{5}{2q}+\frac{1}{2}}\|\mathbb{1}_{(T/2,T)\times B(x,\sqrt{T})}\vec{u}\|_{\dot{\mathcal{M}}_2^{\frac{2q}{5},q}} < +\infty.$$

We have $E_\infty \subset E_{\infty,q} \subset \mathcal{M}_2^{2,5}$. Lemarié-Rieusset's theorem is then:

Theorem 9.4. *Let $5 < q < +\infty$.*

A) The bilinear operator

$$B(\vec{F}, \vec{G}) = \int_0^t \sum_{j=1}^3 \partial_j \mathcal{O}(\nu(t-s)) :: (F_j\vec{G})\,ds = \int_0^t W_{\nu(t-s)} * \mathbb{P}\operatorname{div}(\vec{F} \otimes \vec{G})\,ds$$

is bounded from $E_{\infty,q} \times \mathcal{M}_2^{2,5}$ to $E_{\infty,q}$ and from $\mathcal{M}_2^{2,5} \times E_{\infty,q}$ to $E_{\infty,q}$.

B) If $\vec{u}_0$ is small enough in $BMO^{(-1)}$ (and is divergence free) and if $\int_0^t e^{(t-s)\Delta}\mathbb{P}\operatorname{div}F\,ds$ is small enough in $E_{\infty,q}+\mathcal{V}^{2,1}(\mathbb{R}\times\mathbb{R}^3)$, then the Navier–Stokes equations

$$\partial_t\vec{u} = \nu\Delta\vec{u} - \vec{u}.\vec{\nabla}\vec{u} - \vec{\nabla}p + \operatorname{div}F, \quad \operatorname{div}\vec{u} = 0, \quad , \vec{u}(0,.) = \vec{u}_0$$

have a global mild solution in $E_{\infty,q}+\mathcal{V}^{2,1}(\mathbb{R}\times\mathbb{R}^3)$.

C) Let $2 < p \le 5$. If $\vec{u}_0$ is small enough in $BMO^{(-1)}$ (and is divergence free) and if $\int_0^t e^{(t-s)\Delta}\mathbb{P}\operatorname{div}F\,ds$ is small enough in $E_{\infty,q}+\mathcal{M}_2^{p,5}+L^2\mathcal{F}L^1$, then the Navier–Stokes equations

$$\partial_t\vec{u} = \nu\Delta\vec{u} - \vec{u}.\vec{\nabla}\vec{u} - \vec{\nabla}p + \operatorname{div}F, \quad \operatorname{div}\vec{u} = 0, \quad , \vec{u}(0,.) = \vec{u}_0$$

have a global mild solution in $E_{\infty,q}+\mathcal{M}_2^{p,5}+L^2\mathcal{F}L^1$.

9.3 Q-spaces

The origin of Q-spaces lies in the study of certain classes of holomorphic functions on the disk that are invariant under Möbius transforms, i.e., under bi-holomorphic automorphisms of the disk (see Xiao [506] for a survey). Those classes are connected to the notion of Carleson measures and, as such, appear as generalizations of the class $BMOA$.

Then the notion was exported from the setting of holomorphic functions in complex analysis to the setting of real variable harmonic analysis on $\mathbb{R}^d$. For $0 < \alpha < 1$, the space $Q_\alpha(\mathbb{R}^3)$ is defined as the space of measurable functions such that

$$\|f\|_{Q_\alpha} = \sup_{r>0,x\in\mathbb{R}^3}\left(\frac{1}{r^{3-2\alpha}}\iint_{B(x,r)\times B(x,r)}\frac{|f(y)-f(z)|^2}{|y-z|^{3+2\alpha}}\,dy\,dz\right)^{1/2} < +\infty.$$

This definition is reminiscent both of the characterization of homogeneous Sobolev spaces[1]

$$f \in \dot{H}^\alpha \Leftrightarrow \iint\frac{|f(y)-f(z)|^2}{|y-z|^{3+2\alpha}}\,dy\,dz < +\infty$$

and of Morrey–Campanato spaces

$$f \in \mathcal{L}^{2,-\alpha} \Leftrightarrow \sup_{x_0\in\mathbb{R}^3,r>0}\frac{1}{r^{3-2\alpha}}\int_{B(x_0,r)}|f(y)-m_{B(x_0,r)}f|^2\,dy < +\infty.$$

In the limit case $\alpha = 0$, we have $\mathcal{L}^{2,0} = BMO$; for $0 < \alpha < 1$, we find that $f \in \mathcal{L}^{2,-\alpha}$ if and only if $f = g + C$, where C is a constant $(C = \lim_{r\to+\infty}\frac{1}{|B(0,r)|}\int_{B(0,r)}f(x)\,dx)$ and g belongs to the Morrey space $\dot{M}^{2,3/\alpha}$.

As a matter of fact, it turns out that, for $0 < \alpha < 1$,

$$f \in Q_\alpha \Leftrightarrow (-\Delta)^{\alpha/2}f \in \dot{M}^{2,3/\alpha}.$$

---

[1]When the definition of $\dot{H}^\alpha$ is performed modulo constant functions.

This was proved by Xiao [507] (a similar result was proved earlier by May [353], following an idea of Meyer). In particular, we have $Q_\alpha \subset BMO$. In order to adapt Koch and Tataru's theorem to the setting of Q-spaces, one considers the derivatives of functions in Q_α:

$$Q^{-1,\alpha} = (-\Delta)^{1/2} Q_\alpha = (-\Delta)^{\frac{1-\alpha}{2}} \dot{M}^{2,3/\alpha}.$$

Proposition 9.1 then becomes:

Proposition 9.2.
For a measurable function F on $(0, +\infty) \times \mathbb{R}^3$, define

$$\|F\|_{X_\alpha} = \sup_{x \in \mathbb{R}^3, 0 < t} \sqrt{\frac{1}{t^{3/2}} \iint_{(0,t) \times B(x,\sqrt{t})} |F(s,y)|^2 \left(\frac{t}{s}\right)^\alpha ds\, dy}.$$

Then, we have the equivalence:

$$\|W_{\nu t} * u_0\|_{X_\alpha} < +\infty \Leftrightarrow \exists (f_1, \ldots, f_3) \in (Q_\alpha(\mathbb{R}^3))^3 \ u_0 = \sum_{j=1}^{3} \partial_j f_j$$

and the norm $\|W_{\nu t} * u_0\|_{X_\alpha}$ is equivalent to the infimum of $\sum_{j=1}^{3} \|f_j\|_{Q_\alpha}$ over all possible decompositions $u_0 = \sum_{j=1}^{3} \partial_j f_j$.

Theorem 9.1 was then adapted by May [353] and Xiao [507] to the setting of $Q^{-1,\alpha}$ spaces:

Theorem 9.5.
The bilinear operator

$$B(\vec{F}, \vec{G}) = \int_0^t \sum_{j=1}^{3} \partial_j \mathcal{O}(\nu(t-s)) :: (F_j \vec{G})\, ds = \int_0^t W_{\nu(t-s)} * \mathbb{P} \operatorname{div}(\vec{F} \otimes \vec{G})\, ds$$

is bounded on the space

$$E_\alpha = \{\vec{F} \ / \ \|\vec{F}\|_{X_\alpha} < +\infty \text{ and } \sup_{0 < t} \sqrt{t}\|\vec{F}(t,.)\|_\infty < +\infty\}.$$

Proof. As the space E_α is clearly embedded in the space E_∞ of Theorem 9.1, we may use the results of the proof of Theorem 9.1. For $T > 0$, $x_0 \in \mathbb{R}^3$, we decompose again $B(\vec{F}, \vec{G})$ on $(0, T) \times B(x_0, \sqrt{T})$ into

$$B(\vec{F}, \vec{G})(t, x) = B(\chi_{x_0, T} \vec{F}, \vec{G})(t, x) + B(\psi_{x_0, T} \vec{F}, \vec{G})(t, x)$$

and we already know that, on $(0, T) \times B(x_0, \sqrt{T})$,

$$|B(\psi_{x_0, T} \vec{F}, \vec{G})(t, x)| \leq CT^{-1/2} \|\vec{F}\|_{X_\infty} \|\vec{G}\|_{X_\infty} \leq CT^{-1/2} \|\vec{F}\|_{X_\alpha} \|\vec{G}\|_{X_\alpha}$$

so that

$$\int_0^T \int_{B(x_0, \sqrt{T})} |B(\psi_{x_0, T} \vec{F}, \vec{G})(t, x)(t, x)|^2 \left(\frac{T}{t}\right)^\alpha dt\, dx \leq CT^{3/2} \|\vec{F}\|_{X_\alpha} \|\vec{G}\|_{X_\alpha}.$$

Thus, we are left with the estimation of $I(t, x) = |B(\chi_{x_0, T} \vec{F}, \vec{G})(t, x)(t, x)|$. We write

$$|I(t, x)| \leq C \int_0^t \int \frac{1}{(\sqrt{\nu(t-s)} + |x-y|)^4} \chi_{x_0, T}(s, y) |\vec{F}(s, y)| \, |\vec{G}(s, y)|\, ds\, dy. \qquad (9.3)$$

In particular,

$$I(T, x_0) \leq C \frac{1}{(\nu T)^2} \int_0^{T/2} \int_{B(x_0, 5\sqrt{T})} |\vec{F}(s, y)| |\vec{G}(s, y)| \left(\frac{T}{t}\right)^\alpha ds\, dy$$

$$+ C \frac{1}{T} (\sup_{t>0} \sqrt{t} \|\vec{F}(t, .)\|_\infty)(\sup_{t>0} \sqrt{t} \|\vec{G}(t, .)\|_\infty) \int_{T/2}^T \int \frac{ds\, dy}{(\sqrt{\nu(T-s)} + |x_0 - y|)^4}$$

$$\leq C_\nu T^{-1/2} (\|\vec{F}\|_{X_\alpha} \|\vec{G}\|_{X_\alpha} + (\sup_{t>0} \sqrt{t} \|\vec{F}(t, .)\|_\infty)(\sup_{t>0} \sqrt{t} \|\vec{G}(t, .)\|_\infty))$$

Moreover,

$$H(t, x) = \chi_{x_0, T}(t, x) \frac{|\vec{F}(t, x)|}{t^{\frac{1}{2}}} \sqrt{t} |\vec{G}(t, x)|$$

belongs to $L^2 L^2$ and

$$\iint H(t, x)^2\, dt\, dx \leq C T^{\frac{3}{2} - \alpha} \|\vec{F}\|_{X_\alpha}^2 (\sup_{t>0} \sqrt{t} \|\vec{G}(t, .)\|_\infty)^2.$$

We then write, defining $h(s) = \|H(s, .)\|_2$,

$$\int_0^T \int \left(\frac{T}{t}\right)^\alpha I(t, x)^2\, dt\, dx$$

$$\leq C \int_0^{+\infty} \left(\frac{T}{t}\right)^\alpha \left(\int (\int_0^t \int \frac{1}{(\sqrt{\nu(t-s)} + |x-y|)^4} s^{\frac{\alpha-1}{2}} H(s, y)\, ds\, dy)^2\, dx\right) dt$$

$$\leq C \int_0^{+\infty} \left(\frac{T}{t}\right)^\alpha \left(\int_0^t \left(\int (\int H(s, y) \frac{s^{\frac{\alpha-1}{2}}\, dy}{(\sqrt{\nu(t-s)} + |x-y|)^4})^2\, dx\right)^{1/2} ds\right)^2 dt$$

$$\leq C' \int_0^{+\infty} \left(\frac{T}{t}\right)^\alpha (\int_0^t \frac{1}{\sqrt{\nu(t-s)}} s^{\frac{\alpha-1}{2}} h(s)\, ds)^2\, dt$$

We then get, fixing β such that $1 - \alpha < \beta < 1$,

$$\int_0^T \int \left(\frac{T}{t}\right)^\alpha I(t, x)^2\, dt\, dx$$

$$\leq C' \frac{1}{\nu} \int_0^\infty \left(\frac{T}{t}\right)^\alpha (\int_0^t \frac{1}{\sqrt{t-s}} s^{\beta+\alpha-1} h(s)^2\, ds)(\int_0^t \frac{1}{\sqrt{t-\tau}} \frac{d\tau}{\tau^\beta})\, dt$$

$$= C_{\beta, \nu} T^\alpha \int_0^{+\infty} \int_0^t \frac{1}{\sqrt{t-s}} \frac{1}{t^{\alpha+\beta-\frac{1}{2}}} s^{\alpha+\beta-1} h(s)^2\, ds\, dt$$

$$= C_{\alpha, \beta, \nu} T^\alpha \int_0^{+\infty} h(s)^2\, ds$$

$$\leq C'_{\alpha, \beta, \nu} T^{\frac{3}{2}} \|\vec{F}\|_{X_\alpha}^2 (\sup_{t>0} \sqrt{t} \|\vec{G}(t, .)\|_\infty)^2.$$

The theorem is proved. □

Remark: As we did not use the oscillations of the kernel but directly estimated integrals involving absolute values of the integrands (see inequality (9.3)), we may suspect that the space $Q^{-1,\alpha}$ is such that $\vec{u}_0 \in Q^{-1,\alpha} \Rightarrow 1_{t>0} W_{\nu t} * \vec{u}_0$ belongs to the space $\mathcal{V}^{1,2}(\mathbb{R} \times \mathbb{R}^3)$ we discussed in Chapter 5. This is indeed the case: as a matter of fact we have more precisely

$$\|1_{t>0} W_{\nu t} * \vec{u}_0\|_{\mathcal{M}_2^{2(1+\alpha),5}} \leq C \|\vec{u}_0\|_{Q^{-1,\alpha}}$$

where $\mathcal{M}_2^{2(1+\alpha),5}$ is a parabolic Morrey space on $\mathbb{R} \times \mathbb{R}^3$. The Fefferman–Phong inequality gives the required embedding.

9.4 A Special Subclass of BMO^{-1}

In this section, we offer a few words to 2D space–periodical problems. It is well known that, in the 2D-case, the L^2 theory works very well (see for existence the classical theory in Ladyzhenskaya's book [293]). In particular, the Cauchy problem for $\vec{u} = (\vec{u}_1, \vec{u}_2)$ in $\mathcal{C}([0,+\infty), L^2(\mathbb{R}^2/2\pi\mathbb{Z}^2)) \cap L^2(\dot{H}^1(\mathbb{R}^2/2\pi\mathbb{Z}^2))$ is well posed.

Here, we shall consider three-dimensional vector fields that depend only on the first two variables:

$$\vec{u} = (u_1(t, x_1, x_2), u_2(t, x_1, x_2), u_3(t, x_1, x_2)).$$

Bertozzi and Majda labeled those vector fields as "two-and-a-half dimensional flows" [40]. We consider the Navier–Stokes equations with a null force:

$$\begin{cases} \partial_t \vec{u} = \nu\Delta\vec{u} - (\vec{u}.\vec{\nabla})\vec{u} - \vec{\nabla}p \\ \operatorname{div} \vec{u} = 0 \\ \vec{u}_{|t=0} = \vec{u}_0 \end{cases} \tag{9.4}$$

with $\vec{u}_0 \in L^2(\mathbb{R}^2/2\pi\mathbb{Z}^2)$, $\operatorname{div} \vec{u}_0 = 0$ and $\iint_{[0,2\pi]^2} \vec{u}_0(t, x_1, x_2)\, dx_1\, dx_2 = 0$. As a matter of fact, such an initial value belongs to $BMO^{-1}(\mathbb{R}^3)$:

Proposition 9.3.
Let $\mathcal{E}_2$ be the space of measurable functions $u : \mathbb{R}^3 \mapsto \mathbb{R}$ such that

- *$\partial_3 u = 0$: $u(x_1, x_2, x_3)$ does not depend on x_3 (we shall write $u(x) = u(x_1, x_2)$ or $u(x) = u(x_1, x_2[, x_3])$)*

- *u is $2\pi\mathbb{Z}^2$ periodical:*

$$u(x_1, x_2[, x_3]) = u(x_1 + 2\pi, x_2[, x_3]) = u(x_1, x_2 + 2\pi[, x_3])$$

- *$\iint_{[0,2\pi]^2} |u(x_1, x_2[, x_3])|^2\, dx_1\, dx_2 < +\infty$*

- *$\iint_{[0,2\pi]^2} u(x_1, x_2[, x_3])\, dx_1\, dx_2 = 0$*

endowed with the norm $\|u\|_{\mathcal{E}_2} = (\iint_{[0,2\pi]^2} |u(x_1, x_2[, x_3])|^2\, dx_1\, dx_2)^{1/2}$.
Then we have the embeddings:

$$\mathcal{E}_2 \subset \dot{B}_{\infty,2}^{-1}(\mathbb{R}^3) \subset BMO^{-1}(\mathbb{R}^3).$$

Proof. For the proof of $\mathcal{E}_2 \subset \dot{B}_{\infty,2}^{-1}$, the simplest way is to use the characterization of $\dot{B}_{\infty,2}^{-1}$ through the Littlewood–Paley decomposition $u = \sum_{j\in\mathbb{Z}} \Delta_j u$ with $\Delta_j u = \psi(\frac{D}{2^j})u$, where the Fourier multiplier is an even smooth function ψ supported in the annulus $\{\xi \in \mathbb{R}^3 \ / \ 1 \le |\xi| \le 4\}$ such that, for $\xi \ne 0$, $\sum_{j \in \mathbb{Z}} \psi(2^{-j}\xi) = 1$.

For $u \in \mathcal{E}_2$, we have a Fourier decomposition

$$u = \sum_{k \in \mathbb{Z}^2, k \ne 0} a_k \cos(k_1 x_1 + k_2 x_2) + b_k \sin(k_1 x_1 + k_2 x_2)$$

with

$$\|u\|_{\mathcal{E}_2}^2 = 2\pi^2 \sum_{k \in \mathbb{Z}^2, k \ne 0} a_k^2 + b_k^2$$

We have

$$\Delta_j u = \sum_{2^j < |k| < 4 \ 2^j} \psi(2^{-j}(k,0))(a_k \cos(k_1 x_1 + k_2 x_2) + b_k \sin(k_1 x_1 + k_2 x_2))$$

so that $\Delta_j u = 0$ for $j \le -2$, while, for $j \ge -2$,

$$\|\Delta_j u\|_\infty \le \|\psi\|_\infty \left(\sum_{2^j < |k| < 4 \ 2^j} a_k^2 + b_k^2 \right)^{1/2} \left(\sum_{2^j < |k| < 4 \ 2^j} 2 \right)^{1/2}$$

$$\le C 2^j \left(\sum_{2^j < |k| < 4 \ 2^j} a_k^2 + b_k^2 \right)^{1/2}.$$

This gives the embedding $\mathcal{E}_2 \subset \dot{B}_{\infty,2}^{-1}(\mathbb{R}^3)$: $u = \sum_{j \in \mathbb{Z}} \Delta_j u$ (convergence in $\mathcal{S}'$) and

$$\sum_{j \in \mathbb{Z}} (2^{-j} \|\Delta_j u\|_\infty)^2 \le C \|u\|_{\mathcal{E}_2}.$$

For the proof of $\dot{B}_{\infty,2}^{-1} \subset BMO^{-1}$, the simplest way is to use the thermic characterization of $\dot{B}_{\infty,2}^{-1}$: if $u \in \dot{B}_{\infty,2}^{-1}$, then $W_t * u \in L^2((0, +\infty), L^\infty)$. We then write

$$\int_0^t \int_{B(x_0, \sqrt{t})} |W_s * u(s, y)|^2 \, ds \, dy \le C t^{3/2} \int_0^t \|W_s * u\|_\infty^2 \, ds \le C t^{3/2} \|W_s * u\|_{L^2 L^\infty}^2.$$

The proposition is proved. $\qquad\square$

As $L^\infty \cap \mathcal{E}_2$ is dense in $\mathcal{E}_2$ (proof: just truncate the Fourier series), we see that $\mathcal{E}_2$ is more precisely embedded into the closure of $L^\infty \cap BMO^{-1}$ in BMO^{-1}; thus, Theorem 9.2 ensures the local existence of a solution to Equations (9.4) when $\vec{u}_0$ belongs to $\mathcal{E}_2$. It turns out that this solution has global existence:

Two-and-a-half dimensional Navier–Stokes equations

Theorem 9.6.
If $\vec{u}_0 \in \mathcal{E}_2$ with div $\vec{u}_0 = 0$, then there exists a unique global mild solution $\vec{u}$ of equations

$$\begin{cases} \partial_t \vec{u} + (\vec{u}.\vec{\nabla})\vec{u} = \nu \Delta \vec{u} - \vec{\nabla} p \\ \vec{u}(0,.) = \vec{u}_0 \\ \text{div } \vec{u} = 0 \end{cases} \tag{9.5}$$

such that $\vec{u} \in L_t^\infty \mathcal{E}_2$ and $\vec{\nabla} \otimes \vec{u} \in L^2 \mathcal{E}_2$.

Proof. We know that there is a small time T for which the Picard iterates converge to a solution in the E_T norm (Theorem 9.2). Here, we shall use another norm that ensures the convergence.

First, we remark that the solution we construct does not depend on x_3 and is $2\pi\mathbb{Z}^2$ periodical (by the translation invariance of the Navier–Stokes equations), and the same holds for the associated pressure. More precisely, writing $u_1(t, x) = v_1(t, x_1, x_2)$, $u_2(t, x) = v_2(t, x_1, x_2)$, $u_3(t, x) = w(t, w_1, x_2)$, and $p(t, x) = q(t, x_1, x_2)$, we find that $(\vec{v}, w)$ solve the following equations

- $\vec{v}$ satisfies a $2D$ Navier–Stokes equation:

$$\begin{cases} \partial_t \vec{v} + (\vec{v}.\vec{\nabla})\vec{v} = \nu\Delta\vec{v} - \vec{\nabla}p \\ \vec{v}(0, x_1, x_2) = (u_{0,1}(x), u_{0,2}(x)) \\ \operatorname{div} \vec{v} = 0 \end{cases} \tag{9.6}$$

- w satisfies a linear advection-diffusion scalar equation:

$$\begin{cases} \partial_t w + (\vec{v}.\vec{\nabla})w = \nu\Delta w \\ w(0, x_1, x_2) = u_{0,3}(x) \end{cases} \tag{9.7}$$

Step 1: Local existence for the 2D Navier–Stokes equations.
The study of 2D Navier–Stokes equations with initial value $\vec{v}_0 \in L^2(\mathbb{R}^2/2\pi\mathbb{Z}^2)$ was initiated by the works of Leray [327, 328, 329], and fully developed by Ladyzhenskaya, Lions and Prodi [293, 339].

We rewrite (9.6) into

$$\vec{v} = W_{\nu t}^{(2)} * \vec{v}_0 - \int_0^t W_{\nu(t-s)}^{(2)} * \mathbb{P}^{(2)} \operatorname{div}(\vec{v} \otimes \vec{v}) \, ds \tag{9.8}$$

where $W_t^{(2)}(x_1, x_2)$ is the 2D heat kernel and $\mathbb{P}^{(2)}$ is the 2D Leray projection operator. For a periodical distribution vector field

$$\vec{V}(x_1, x_2) = \sum_{k \in \mathbb{Z}^2} \cos(k \cdot x)\vec{A}_k + \sin(k \cdot x)\vec{B}_k,$$

we have

$$W_{\nu t}^{(2)} * \vec{V} = \sum_{k \in \mathbb{Z}^2} e^{-\nu t|k|^2}(\cos(k \cdot x)\vec{A}_k + \sin(k \cdot x)\vec{B}_k)$$

and for a periodical distribution tensor

$$\mathbb{T}(x_1, x_2) = \sum_{k \in \mathbb{Z}^2} \cos(k \cdot x)\mathbb{A}_k + \sin(k \cdot x)\mathbb{B}_k,$$

we have

$$\mathbb{P}^{(2)} \operatorname{div} \mathbb{T} = \sum_{k \in \mathbb{Z}^2, k \neq 0} \cos(k \cdot x)(\vec{k}.\mathbb{B}_k - \frac{\vec{k} \otimes \vec{k} \cdot \mathbb{B}_k}{|k|^2}\,\vec{k}) - \sin(k \cdot x)(\vec{k} \cdot \mathbb{A}_k - \frac{\vec{k} \otimes \vec{k}.\mathbb{A}_k}{|k|^2}\,\vec{k})$$

We are going to look for a solution $\vec{v} \in L^4((0, T_0), L^4(\mathbb{R}^2/2\pi\mathbb{Z}^2))$ for T_0 small enough. Indeed, we have the following estimates:

$$\|W_{\nu t}^{(2)} * \vec{v}_0\|_{L^\infty L^2(\mathbb{R}^2/2\pi\mathbb{Z}^2)} = \|\vec{v}_0\|_{L^2(\mathbb{R}^2/2\pi\mathbb{Z}^2)}$$

$$\|W^{(2)}_{\nu t} * \vec{v}_0\|_{L^2 \dot{H}^1(\mathbb{R}^2/2\pi\mathbb{Z}^2)} = \frac{1}{\sqrt{2\nu}}\|\vec{v}_0\|_{L^2(\mathbb{R}^2/2\pi\mathbb{Z}^2)}$$

$$\|\int_0^t W^{(2)}_{\nu(t-s)} * \mathbb{P}^{(2)}\operatorname{div}\mathbb{T}\,ds\|_{L^\infty((0,T_0),L^2(\mathbb{R}^2/2\pi\mathbb{Z}^2))} \le \frac{1}{\sqrt{2\nu}}\|\mathbb{T}\|_{L^2 L^2(\mathbb{R}^2/2\pi\mathbb{Z}^2)}$$

$$\|\int_0^t W^{(2)}_{\nu(t-s)} * \mathbb{P}^{(2)}\operatorname{div}\mathbb{T}\,ds\|_{L^2((0,T_0),\dot{H}^1(\mathbb{R}^2/2\pi\mathbb{Z}^2))} \le \frac{1}{\nu}\|\mathbb{T}\|_{L^2 L^2(\mathbb{R}^2/2\pi\mathbb{Z}^2)}$$

Moreover, we have the Sobolev embedding

$$f \in \dot{H}^{1/2}(\mathbb{R}^2/2\pi\mathbb{Z}^2) \text{ and } \iint_{\mathbb{R}^2/2\pi\mathbb{Z}^2} f\,dx = 0 \Rightarrow f \in L^4(\mathbb{R}^2/2\pi\mathbb{Z}^2).$$

Thus, we find that:

- $\vec{V}_0 = W^{(2)}_{\nu t} * \vec{v}_0$ belongs to $L^4 L^4$. Moreover, $\lim_{T_0 \to 0}\|\vec{V}_0\|_{L^4((0,T_0),L^4)} = 0$.
- If $\vec{u}$ and $\vec{v}$ belong to $L^4 L^4(\mathbb{R}^2/2\pi\mathbb{Z}^2)$, we have that

$$B(\vec{u}, \vec{v}) = \int_0^t W^{(2)}_{\nu(t-s)} * \mathbb{P}\operatorname{div}(\vec{u} \otimes \vec{v})\,ds$$

  belongs to $L^\infty L^2 \cap L^2 \dot{H}^1 \cap L^4 L^4$ with

$$\|B(\vec{u}, \vec{v})\|_{L^4((0,T_0),L^4)} \le C_\nu \|\vec{u}\|_{L^4((0,T_0),L^4)}\|\vec{v}\|_{L^4((0,T_0),L^4)}.$$

- If T_0 is small enough, so that $\|\vec{V}_0\|_{L^4((0,T_0),L^4)} < \frac{1}{4C_\nu}$, we shall find a solution $\vec{v}$ through Picard's iterative process. The process will converge in $L^\infty L^2 \cap L^2 \dot{H}^1 \cap L^4 L^4$.

Step 2: Global existence.

If the solution $\vec{v}$ belongs to $L^\infty((0,T_0), L^2) \cap L^2((0,T_0), \dot{H}^1)$ then we find an estimate slightly better than just $\vec{v} \in L^\infty L^2$. As a matter of fact, we have

$$\vec{v} = W^{(2)}_{\nu t} * \vec{v}_0 - \int_0^t W^{(2)}_{\nu(t-s)} * \mathbb{P}^{(2)}\operatorname{div}(\vec{v} \otimes \vec{v}))\,ds = W^{(2)}_{\nu t} * \vec{v}_0 + \mathcal{L}(\mathbb{V})$$

where $\mathbb{V} = \vec{v} \otimes \vec{v}$. The operator $\mathcal{L}$ maps $L^2 L^2$ to $L^\infty L^2$ and $L^2 H^2$ to $\operatorname{Lip} L^2$; as $L^2 H^2$ is dense in $L^2 L^2$, we find that $\mathcal{L}$ actually maps $L^2 L^2$ to $\mathcal{C}([0,T_0], L^2)$.

If T^* is the maximal time of existence of the solution $\vec{v}$, so that $\vec{v}$ belongs to $L^\infty((0,T_0), L^2) \cap L^2((0,T_0), \dot{H}^1)$ for every $T_0 < T^*$, we find that $T^* = +\infty$ unless that $T^* < +\infty$ and $\vec{v}$ does not belong to $L^\infty((0,T^*), L^2) \cap L^2((0,T^*), \dot{H}^1)$: if $\vec{v}$ belonged to $L^\infty((0,T^*), L^2) \cap L^2((0,T^*), \dot{H}^1)$ with $T^* < T$, then it would belong to $\mathcal{C}([0,T^*], L^2)$ and we could solve the Cauchy problem for the Navier–Stokes equations on some interval $[T^*, T^* + T_0]$ with initial value $\vec{v}(T^*,.)$.

Thus, in order to show that we have a global solution, we only need to control the sizes of $\vec{v}$ and $\vec{\nabla} \otimes \vec{v}$ on $[0,T^*)$. As $\vec{v}$ belongs to $L^2 H^1$ and $\partial_t \vec{v}$ belongs to $L^2 H^{-1}$ on every compact interval of $[0,T^*)$, we may write

$$\frac{d}{dt}\|\vec{v}(t,.)\|^2_{L^2(\mathbb{R}^2/2\pi\mathbb{Z}^2)} = 2\langle \vec{v}(t,.)|\partial_t \vec{v}(t,.)\rangle_{H^1,H^{-1}} = -2\nu\|\vec{\nabla} \otimes \vec{v}(t,.)\|^2_2$$

so that

$$\|\vec{v}(t,.)\|^2_2 + 2\nu\int_0^t \|\vec{\nabla} \otimes \vec{v}(s,.)\|^2_2\,ds = \|\vec{v}_0\|^2_2.$$

Hence, $T^* = +\infty$: we have a global solution.

Step 3: Global existence of w.

For the existence of w, we write w as a fixed point of the transform

$$\omega \mapsto W^{(2)}_{\nu t} * w_0 - \int_0^t W^{(2)}_{\nu(t-s)} * (\text{div}(\omega \vec{v})) \, ds = W^{(2)}_{\nu t} * w_0 + \mathcal{L}(w).$$

$\mathcal{L}$ is a bounded linear operator on $L^\infty((0,T_0), L^2) \cap L^2((0,T_0), \dot{H}^1)$ and satisfies (uniformly in T_0)

$$\|\mathcal{L}(w)\|_{L^\infty((0,T_0),L^2) \cap L^2((0,T_0),\dot{H}^1)} \leq C_0 \|\vec{v}\|_{L^4((0,T_0),L^4)} \|w\|_{L^\infty((0,T_0),L^2) \cap L^2((0,T_0),\dot{H}^1)}.$$

Thus, $\mathcal{L}$ is a contraction as soon as T_0 is small enough to grant that $C_0 \|\vec{v}\|_{L^4((0,T_0),L^4)} < 1$.

Global existence of w is then proved by splitting any given interval $[0, T]$ into a finite union of intervals $[T_j, T_{j+1}]$ with $C_0 \|\vec{v}\|_{L^4((T_j,T_{j+1}),L^4)} < 1$: once w is constructed on $[0, T_{j+1}]$, one constructs w on $[T_{j+1}, T_{j+2}]$ by considering the Cauchy problem with initial value $w(T_{j+1}, .)$ at $t = T_{j+1}$. Thus, w exists up to the given arbitrary time T.

$\square$

9.5 Ill-posedness

Thus far, the largest space of initial values that are well fitted for the Cauchy problem for the Navier–Stokes equations is the space bmo^{-1} and its homogeneous counterpart BMO^{-1}. For scaling properties of the equations, any such space that respects the shift invariance of the equations and their scaling properties should be embedded into $B^{-1}_{\infty,\infty}$ (for local existence results) or $\dot{B}^{-1}_{\infty,\infty}$ (for global existence results).

Bourgain and Pavlović [52] proved that the problem was ill-posed in $\dot{B}^{-1}_{\infty,\infty}$. More precisely, they proved a phenomenon of *norm inflation*:

Theorem 9.7.
For every $\delta > 0$, there exists a smooth divergence-free $\vec{u}_0 \in \mathcal{E}_2$ with a small norm in $\dot{B}^{-1}_{\infty,\infty}$ (i.e., $\|\vec{u}_0\|_{\dot{B}^{-1}_{\infty,\infty}} < \delta$) which generates a solution $\vec{u}$ of the Navier–Stokes equations

$$\begin{cases} \partial_t \vec{u} = \nu \Delta \vec{u} - (\vec{u}.\vec{\nabla})\vec{u} - \vec{\nabla}p \\ \text{div } \vec{u} = 0 \\ \vec{u}_{|t=0} = \vec{u}_0 \end{cases} \tag{9.9}$$

which becomes very large in a very small time: for some $\tau \in (0, \delta)$,

$$\|\vec{u}(\tau, .)\|_{\dot{B}^{-1}_{\infty,\infty}} \geq \frac{1}{\delta}.$$

Of course, the norm of $\vec{u}_0$ must be large in BMO^{-1}: if $\|\vec{u}_0\|_{BMO^{-1}}$ is small enough, then the Koch and Tataru theorem (Theorem 9.2) implies that there exists a global solution $\vec{u}$ to (9.9) with

$$\|\vec{u}(t, .)\|_{\dot{B}^{-1}_{\infty,\infty}} \leq C \|\vec{u}\|_{BMO^{-1}} \leq C' \|\vec{u}_0\|_{BMO^{-1}}.$$

Proof. The mild solution satisfies:

$$\vec{u} = W_{\nu t} * \vec{u}_0 - B(\vec{u}, \vec{u})$$

with

$$B(\vec{u}, \vec{v}) = \int_0^t W_{\nu(t-s)} * \mathbb{P} \operatorname{div}(\vec{u} \otimes \vec{v}) \, ds.$$

The discussion will focus on the decomposition

$$\vec{u} = \vec{U}_0 - \vec{U}_1 + \vec{U}_2$$

with

$$\vec{U}_0 = W_{\nu t} * \vec{u}_0 \text{ and } \vec{U}_1 = B(\vec{U}_0, \vec{U}_0).$$

Bourgain and Pavlović's choice for $\vec{u}_0$ (discussed as well by Sawada [422]) is given by a lacunary sums of cosines

$$
\begin{aligned}
\vec{u}_0 &= \frac{Q}{\sqrt{N}} \sum_{j \in \Lambda} \vec{w}_j \\
&= \frac{Q}{\sqrt{N}} \sum_{j \in \Lambda} 2^j \left(\cos(2^j x_1) \begin{pmatrix} 0 \\ 1 \\ 1 \end{pmatrix} + \cos(2^j x_1 + x_2) \begin{pmatrix} -2^{-j} \\ 1 \\ 1 \end{pmatrix} \right)
\end{aligned}
\tag{9.10}
$$

with Q a (large) integer and Λ a lacunary finite subset of $\mathbb{N}$ of the type $\{j_0 < j_1 < \cdots < j_{N-1}\}$ with $j_0 \geq 5$ and $j_{q+1} > 4 j_q$. We assume that $Q^3 < N$. (Thus $\frac{Q}{\sqrt{N}} < Q^{-1/2}$ is small.)

In the following computations, $C_0, C_1, \ldots$ will denote positive constants that may depend on ν but depend neither on Q, N, nor Λ.

Estimates on $\vec{u}_0$:

We have obviously $\operatorname{div} \vec{u}_0 = 0$ and (using the Littlewood–Paley characterization of $\dot{B}^{-1}_{\infty,p}$)

$$\|\vec{u}_0\|_{\dot{B}^{-1}_{\infty,2}} \approx \frac{Q}{\sqrt{N}} \left(\sum_{j \in \Lambda} 2^{-2j} \|\vec{w}_j\|_\infty^2 \right)^{1/2} \leq C_0 Q$$

and

$$\|\vec{u}_0\|_{\dot{B}^{-1}_{\infty,\infty}} \approx \frac{Q}{\sqrt{N}} \sup_{j \in \Lambda} 2^{-j} \|\vec{w}_j\|_\infty \leq C_0 \frac{Q}{\sqrt{N}}.$$

Of course, we shall take $\frac{Q}{\sqrt{N}}$ small enough to ensure that

$$\|\vec{u}_0\|_{\dot{B}^{-1}_{\infty,\infty}} \leq C_0 \frac{Q}{\sqrt{N}} \leq \delta$$

while Q will be large with respect to δ^{-1}. (We shall see that $\|\vec{u}_0\|_{BMO^{-1}} \approx Q$).

Estimate on $\vec{U}_0$:

The initial assumption is that $\vec{u}_0 \in B^{-1}_{\infty,\infty}$ with $\|\vec{u}_0\|_{\dot{B}^{-1}_{\infty,\infty}} \leq \delta$. This gives, in particular, the following estimate on $\vec{U}_0 = W_{\nu t} * \vec{u}_0$:

$$\|\vec{U}_0(t, .)\|_{\dot{B}^{-1}_{\infty,\infty}} \leq \delta.$$

Thus, $\vec{U}_0$ remains small in $\dot{B}^{-1}_{\infty,\infty}$.

Estimates on $\vec{U}_1$:

Now, we compute $\vec{U}_1 = B(W_{\nu t} * \vec{u}_0, W_{\nu t} * \vec{u}_0)$. For $j \in \Lambda$, let $k_j = (2^j, 0, 0)$, $l_j = (2^j, 1, 0)$,

$$\vec{\alpha}_j = \begin{pmatrix} 0 \\ 1 \\ 1 \end{pmatrix} \text{ and } \vec{\beta}_j = \begin{pmatrix} -2^{-j} \\ 1 \\ 1 \end{pmatrix} \text{ so that}$$

$$\vec{u}_0 = \frac{Q}{\sqrt{N}} \sum_{j \in \Lambda} 2^j \left(\cos(k_j \cdot x)\vec{\alpha}_j + \cos(l_j \cdot x)\vec{\beta}_j \right) \tag{9.11}$$

and

$$W_{\nu t} * \vec{u}_0 = \frac{Q}{\sqrt{N}} \sum_{j \in \Lambda} 2^j e^{-\nu t 2^{2j}} \left(\cos(k_j \cdot x)\vec{\alpha}_j + e^{-\nu t}\cos(l_j \cdot x)\vec{\beta}_j \right) = \sum_{j \in \Lambda} \vec{\gamma}_j(t, x).$$

We may compute $\vec{U}_0 \otimes \vec{U}_0$ with the decomposition in paraproducts [313]: if $j < l$, then the frequency localization of $\vec{\gamma}_j \otimes \vec{\gamma}_l$ and of $\vec{\gamma}_l \otimes \vec{\gamma}_j$ will be for frequencies of order 2^l so that

$$\left\| \sum \sum_{j \neq l} \vec{\gamma}_j \otimes \vec{\gamma}_l \right\|_{L_t^\infty \dot{B}_{\infty,\infty}^{-2}} \leq C \sup_l \sum_{j < l} \frac{Q^2}{N} 2^j 2^l 2^{-2l} \sim C\frac{Q^2}{N}$$

and we find finally

$$\sup_{t > 0} \left\| \sum \sum_{j \neq l} B(\vec{\gamma}_j, \vec{\gamma}_l)(t, .) \right\|_{\dot{B}_{\infty,\infty}^{-1}} \leq C_1 \delta^2.$$

(We used the maximal regularity of the heat kernel for Besov spaces

$$\left\| \int_0^t W_{\nu(t-s)\Delta} * f(s, .)\, ds \right\|_{L_t^\infty \dot{B}_{\infty,\infty}^{-1}} \leq C\frac{1}{\nu} \|f\|_{L_t^\infty \dot{B}_{\infty,\infty}^{-3}}$$

– see Lemarié-Rieusset [313]).

If we now look at the square term $\vec{\gamma}_j \otimes \vec{\gamma}_j$, then we find low frequencies 0, $k_j - l_j$ and $l_j - k_j$ (which are all such that $|\xi| \leq 1$) and high frequencies $k_j + l_j$ and $-k_j - l_j$ which are of order 2^j. Let P_0 be the projection on frequencies less than 2. Then, again, we have

$$\sup_{t > 0} \left\| (Id - P_0) \sum_j B(\vec{\gamma}_j, \vec{\gamma}_j)(t, .) \right\|_{\dot{B}_{\infty,\infty}^{-1}} \leq C_1 \delta^2.$$

Moreover, the frequency 0 may be forgotten, as the constant terms will be killed by applying the divergence operator (equivalently, as the convolution kernel of $W_{\nu(t-s)} * \mathbb{P}\,\mathrm{div}$ has integral equal to 0). Thus, we are left with estimating

$$\vec{U}_3(t, x) = \frac{Q^2}{N} \sum_{j \in \Lambda} 2^{2j} P_0(B(W_{\nu s} * (\cos(k_j \cdot x)\vec{\alpha}_j), W_{\nu s} * (\cos l_j \cdot x)\vec{\beta}_j))$$

$$+ \frac{Q^2}{N} \sum_{j \in \Lambda} 2^{2j} P_0(B(W_{\nu s} * (\cos(l_j \cdot x)\vec{\beta}_j), W_{\nu s} * (\cos k_j \cdot x)\vec{\alpha}_j))$$

$$= \frac{Q^2}{2N} \sum_{j \in \Lambda} 2^{2j} \left(\int_0^t e^{-\nu(t-s)} e^{-\nu s(2^{2j+1}+1)}\, ds \right) \times$$

$$\times \mathbb{P}\,\mathrm{div}(\cos((k_j - l_j) \cdot x)(\vec{\alpha}_j \otimes \vec{\beta}_j + \vec{\beta}_j \otimes \vec{\alpha}_j))).$$

We have

$$\cos((k_j - l_j) \cdot x)(\vec{\alpha}_j \otimes \vec{\beta}_j + \vec{\beta}_j \otimes \vec{\alpha}_j) = \cos(x_2) \begin{pmatrix} 0 & -2^{-j} & -2^{-j} \\ -2^{-j} & 2 & 2 \\ -2^{-j} & 2 & 2 \end{pmatrix}$$

and thus

$$\mathbb{P}\operatorname{div}((\cos((k_j - l_j) \cdot x)(\vec{\alpha}_j \otimes \vec{\beta}_j + \vec{\beta}_j \otimes \vec{\alpha}_j))) = \begin{pmatrix} 2^{-j}\sin(x_2) \\ 0 \\ -2\sin x_2 \end{pmatrix}.$$

Writing

$$\vec{U}_3 = \begin{pmatrix} V_3 \\ 0 \\ W_3 \end{pmatrix},$$

we find that V_3 is small and W_3 large:

$$\|V_3(t,x)\|_{\dot{B}_{\infty,\infty}^{-1}} \le C\frac{1}{\nu}\frac{Q^2}{N}\sum_{j\in\Lambda}2^{-j} \le C_1\delta^2$$

and

$$\|W_3(t,x)\|_{\dot{B}_{\infty,\infty}^{-1}} \le C\frac{1}{\nu}\frac{Q^2}{N}\sum_{j\in\Lambda}1 \le C_1 Q^2.$$

This latter estimate is quite sharp:

$$\|W_3(t,x)\|_{\dot{B}_{\infty,\infty}^{-1}} \approx \frac{Q^2}{N}\sum_{j\in\Lambda}2^{2j}\int_0^t e^{-\nu(t-s)}e^{-\nu s(1+2^{2j+1})}\,ds$$

$$= \frac{Q^2}{N}\frac{1}{2\nu}e^{-\nu t}\sum_{j\in\Lambda}(1 - e^{-\nu t 2^{2j+1}})$$

so that, defining $j_0 = \min_{j\in\Lambda} j$, we find that

$$C_2 Q^2 \le \|W_3(t,x)\|_{\dot{B}_{\infty,\infty}^{-1}} \le C_3 Q^2$$

with positive constants C_2 and C_3 independent from Λ, Q, and N, as far as

$$\frac{1}{2^{2j_0}} \le \nu t \le 1.$$

Thus, $\vec{U}_1$ becomes very large in the $\dot{B}_{\infty,\infty}^{-1}$ norm in a very short time ($t \approx \nu^{-1}2^{-2j_0}$) and remains large on a rather long interval (up to $t \approx \nu^{-1}$).

Remark: In Lemma 9.1, we saw that if $\vec{F}$ and $\vec{G}$ belong to E_∞, then $\mathcal{A}(\vec{F}, \vec{G}) = \int_0^{+\infty} W_{\nu s} * (\vec{F} \otimes \vec{G})\,ds$ belongs to $(BMO)^9$ and

$$\|\mathcal{A}(\vec{F},\vec{G})\|_{BMO} \le C_\nu \|\vec{F}\|_{E_\infty}\|\vec{G}\|_{E_\infty}.$$

In particular, we find that, since $B(\vec{F},\vec{G})(t,.) = \mathbb{P}\operatorname{div}\mathcal{A}(1_{0<s<t}\vec{F}, 1_{0<s<t}\vec{G})$, that

$$\|B(W_{\nu s} * \vec{u}_0, W_{\nu s} * \vec{u}_0)\|_{BMO^{-1}} \le C_\nu\|\vec{u}_0\|_{BMO^{-1}}^2.$$

Thus, we find, for $t_0 = \frac{1}{\nu}$,

$$Q^2 \approx \|\vec{U}_1(t_0,.)\|_{\dot{B}_{\infty,\infty}^{-1}} \le C\|\vec{U}_1(t_0,.)\|_{BMO^{-1}} \le C'\|\vec{u}_0\|_{BMO^{-1}}^2$$

while we saw that

$$\|\vec{u}_0\|_{BMO^{-1}} \le C\|\vec{u}_0\|_{\dot{B}_{\infty,2}^{-1}} \approx Q.$$

We have thus clearly established that $\|\vec{u}_0\|_{BMO^{-1}} \approx Q$.

Estimates on $\vec{U}_2$:

The core idea in the Bourgain and Pavlović proof is to show that $\vec{U}_2$ remains small while $\vec{U}_1$ becomes very large.

$\vec{U}_2$ is the solution $\vec{U}_2 = \vec{z}$ of the equation

$$\vec{z} = \vec{Z}_0 + L(\vec{z}) - B(\vec{z}, \vec{z})$$

with

$$\vec{Z}_0 = B(\vec{U}_0, \vec{U}_1) + B(\vec{U}_1, \vec{U}_0) - B(\vec{U}_1, \vec{U}_1)$$

and

$$L(\vec{z}) = -B(\vec{U}_0, \vec{z}) - B(\vec{z}, \vec{U}_0) + B(\vec{U}_1, \vec{z}) + B(\vec{z}, \vec{U}_1).$$

Equivalently, $\vec{U}_2$ is a mild solution $\vec{U}_2 = \vec{z}$ of

$$\begin{cases} \partial_t \vec{z} = & \nu \Delta \vec{z} - \operatorname{div}(\vec{U}_1 \otimes \vec{U}_1 - \vec{U}_0 \otimes \vec{U}_1 - \vec{U}_1 \otimes \vec{U}_0 + \vec{z} \otimes \vec{z}) \\ & - \operatorname{div}(\vec{U}_0 \otimes \vec{z} + \vec{z} \otimes \vec{U}_0 - \vec{U}_1 \otimes \vec{z} - \vec{z} \otimes \vec{U}_1) - \vec{\nabla} q \\ \operatorname{div} \vec{z} = 0 \\ \vec{z}_{|t=0} = 0 \end{cases} \quad (9.12)$$

We are going to estimate the evolution of $\|\vec{U}_2(t,.)\|_{BMO^{-1}}$. Let $0 \le T_1 \le T_2$ with $T_1 = 0$ or $\frac{1}{\nu 2^{2J_1}}$ for some $J_1 \in \Lambda$, and $T_2 = \frac{1}{\nu 2^{2J_2}}$ for some $J_2 \in \Lambda$.

If we define, for a function h on $(T_1, T_2) \times \mathbb{R}^3$,

$$\|h\|_E = \sup_{T_1 < t < T_2} \sqrt{t - T_1} \|h(t,.)\|_\infty$$

$$+ \sup_{T_1 < t < T_2} \sup_{x_0 \in \mathbb{R}^3} \frac{1}{(t-T_1)^{3/4}} \sqrt{\int_{T_1}^t \int_{B(x_0, \sqrt{t-T_1})} |h(s,y)|^2 \, dy \, ds}$$

we find, from Lemma 9.1, that

$$\|\vec{U}_2(T_2,.)\|_{BMO^{-1}} \le \|\vec{U}_2(T_1,.)\|_{BMO^{-1}} + \|\vec{Z}_0(T_2,.) - \vec{Z}_0(T_1,.)\|_{BMO^{-1}} \\ + C_\nu \|\vec{U}_2\|_E (\|\vec{U}_2\|_E + \|\vec{U}_1\|_E + \|\vec{U}_0\|_E) \quad (9.13)$$

while, from Theorem 9.1, we know that

$$\|\vec{U}_2\|_E \le \|\vec{U}_2(T_1,.)\|_{BMO^{-1}} + \|\vec{Z}_0 - \vec{Z}_0(T_1,.)\|_E \\ + C_\nu \|\vec{U}_2\|_E (\|\vec{U}_2\|_E + \|\vec{U}_1\|_E + \|\vec{U}_0\|_E) \quad (9.14)$$

where

$$\vec{Z}_0(t,.) - \vec{Z}_0(T_1,.) = \int_{T_1}^t W_{\nu(t-s)} * \mathbb{P} \operatorname{div}(\vec{U}_0 \otimes \vec{U}_1 + \vec{U}_1 \otimes \vec{U}_0 - \vec{U}_1 \otimes \vec{U}_1) \, ds.$$

Again, by Lemma 9.1 and Theorem 9.1, we may estimate $\vec{Z}_0(t,.) - \vec{Z}_0(T_1,.)$, and find

$$\|\vec{U}_2(T_2,.)\|_{BMO^{-1}} \le \|\vec{U}_2(T_1,.)\|_{BMO^{-1}} \\ + C_\nu \|\vec{U}_1\|_E (\|\vec{U}_0\|_E + \|\vec{U}_1\|_E) \\ + C_\nu \|\vec{U}_2\|_E (\|\vec{U}_2\|_E + \|\vec{U}_1\|_E + \|\vec{U}_0\|_E) \quad (9.15)$$

and

$$
\begin{aligned}
\|\vec{U}_2\|_E \leq &\|\vec{U}_2(T_1,.)\|_{BMO^{-1}} \\
&+ C_\nu \|\vec{U}_1\|_E (\|\vec{U}_0\|_E + \|\vec{U}_1\|_E) \\
&+ C_\nu \|\vec{U}_2\|_E (\|\vec{U}_2\|_E + \|\vec{U}_1\|_E + \|\vec{U}_0\|_E).
\end{aligned}
\tag{9.16}
$$

Thus, we need to estimate $\|\vec{U}_0\|_E$ and $\|\vec{U}_1\|_E$. Recall that

$$
\vec{U}_0 = \frac{Q}{\sqrt{N}} \sum_{j \in \Lambda} 2^j e^{-\nu t 2^{2j}} \left(\cos(k_j \cdot x) \vec{\alpha}_j + e^{-\nu t} \cos(l_j \cdot x) \vec{\beta}_j \right) = \sum_{j \in \Lambda} \vec{\gamma}_j(t, x).
$$

We split the sum between the indexes j such that $j \leq J_2$, those such that $J_2 < j \leq J_1$ and those such that $j > J_1$:

$$
\vec{A} = \sum_{j \leq J_2} \vec{\gamma}_j(t, x), \quad \vec{B} = \sum_{J_2 < j \leq J_1} \vec{\gamma}_j(t, x) \text{ and } \vec{C} = \sum_{j > J_1} \vec{\gamma}_j(t, x).
$$

As, on (T_1, T_2), we have $\vec{A}(t,.) = W_{\nu(t-T_1)} * \vec{A}(T_1,.)$, $\vec{B}(t,.) = W_{\nu(t-T_1)} * \vec{B}(T_1,.)$, and $\vec{C}(t,.) = W_{\nu(t-T_1)} * \vec{C}(T_1,.)$, we find that

$$
\|\vec{U}_0\|_E \leq C_\nu (\sqrt{T_2 - T_1} \|\vec{A}(T_1,.)\|_\infty + \|\vec{B}(T_1,.)\|_{BMO^{-1}} + \|\vec{C}(T_1,.)\|_{BMO^{-1}})
$$

with

$$
\begin{aligned}
\sqrt{T_2 - T_1} \|\vec{A}(T_1,.)\|_\infty &\leq \sqrt{T_2} \|\vec{A}(0,.)\|_\infty \\
&\leq 2 \frac{Q}{\sqrt{N}} \sqrt{T_2} \sum_{j \in \Lambda, j \leq J_2} 2^j \\
&\leq 4 \frac{Q}{\sqrt{N}} \sqrt{T_2} 2^{J_2} \\
&\leq C_\nu \frac{Q}{\sqrt{N}}
\end{aligned}
$$

and

$$
\begin{aligned}
\|\vec{C}(T_1,.)\|_{BMO^{-1}} &\leq C \|\vec{C}(T_1,.)\|_{\dot{B}^{-1}_{\infty,2}} \\
&\leq C' \left(\sum_{j \in \Lambda, j > J_1} 2^{-2j} \|\vec{\gamma}_j(T_1,.)\|_\infty^2 \right)^{1/2} \\
&\leq C'' \frac{Q}{\sqrt{N}} \left(\sum_{j \in \Lambda, j > J_1} e^{-2\nu T_1 2^{2j}} \right)^{1/2} \\
&\leq C'' \frac{Q}{\sqrt{N}} \left(\sum_{p \in \mathbb{N}} e^{-2(16)^p} \right)^{1/2} \\
&= C''' \frac{Q}{\sqrt{N}}.
\end{aligned}
$$

The difficult term is, of course, $\vec{B}$, for which we have

$$\|\vec{B}(T_1,.)\|_{BMO^{-1}} \leq C\|\vec{B}(T_1,.)\|_{\dot{B}^{-1}_{\infty,2}}$$

$$\leq C' \left(\sum_{j\in\Lambda, J_2 < j \leq J_1} 2^{-2j}\|\vec{\gamma}_j(T_1,.)\|^2_\infty \right)^{1/2}$$

$$\leq C'' \frac{Q}{\sqrt{N}} \sqrt{\#(\{j \in \Lambda \ / \ J_2 < j \leq J_1\})}$$

If we want this quantity to be small, we need that the ratio $\frac{\#(\{j\in\Lambda \ / \ J_2 < j \leq J_1\})}{\#(\Lambda)}$ be small. Bourgain and Pavlović's choice is the ratio

$$\frac{\#(\{j \in \Lambda \ / \ J_2 < j \leq J_1\})}{\#(\Lambda)} = \frac{1}{Q^3}. \tag{9.17}$$

For that choice, we find that $\|\vec{B}(T_1,.)\|_{BMO^{-1}} \leq CQ^{-1/2}$. As $\frac{Q}{\sqrt{N}} \leq Q^{-1/2}$, we find that

$$\|\vec{U}_0\|_E \leq C_4 Q^{-1/2}.$$

Now, we estimate the norms of $\vec{U}_1$. Recall that we have split $\vec{U}_1$ into $\vec{U}_3$ and $\vec{U}_1 - \vec{U}_3$, and $\vec{U}_3$ into $P_0\vec{U}_3$ and $(Id - P_0)\vec{U}_3$. We have

$$\vec{U}_1 - \vec{U}_3 = \sum_{j\neq p}\sum B(\vec{\gamma}_j, \vec{\gamma}_p).$$

Let $\epsilon_j \in \{k_j, -k_j, l_l, -l_j\}$ and $\epsilon_p \in \{k_p, -k_p, l_p, -l_p\}$. Then

$$\int_0^t e^{-\nu(t-s)|\epsilon_j+\epsilon_p|^2}|\epsilon_j + \epsilon_p|2^j e^{-\nu s|\epsilon_j|^2} 2^p e^{-\nu s|\epsilon_p|^2} ds$$

$$= 2^j 2^p |\epsilon_j + \epsilon_p| e^{-\nu t|\epsilon_j+\epsilon_p|^2} \int_0^t e^{2\nu s\epsilon_j \cdot \epsilon_p} ds$$

Let us notice that, if $p < j$, and due to the lacunarity of Λ, we have $\frac{1}{2}2^j \leq |\epsilon_j + \epsilon_p| \leq \frac{3}{2}2^j$. If $\epsilon_j \cdot \epsilon_p \leq 0$, we find

$$\int_0^t e^{-\nu(t-s)|\epsilon_j+\epsilon_p|^2}|\epsilon_j + \epsilon_p|2^j e^{-\nu s|\epsilon_j|^2} 2^p e^{-\nu s|\epsilon_p|^2} ds$$

$$\leq 2^j 2^p \frac{3}{2}2^j e^{-\frac{1}{4}\nu t 2^{2j}} t,$$

while, if $p < j$ and $\epsilon_j.\epsilon_p \geq 0$, we find

$$\int_0^t e^{-\nu(t-s)|\epsilon_j+\epsilon_p|^2}|\epsilon_j + \epsilon_p|2^j e^{-\nu s|\epsilon_j|^2} 2^p e^{-\nu s|\epsilon_p|^2} ds$$

$$= 2^j 2^p |\epsilon_j + \epsilon_p| e^{-\nu t|\epsilon_j+\epsilon_p|^2} \frac{e^{2\nu t \epsilon_j.\epsilon_p} - 1}{2\nu\epsilon_j.\epsilon_p}$$

$$= 2^j 2^p |\epsilon_j + \epsilon_p| e^{-\nu t(|\epsilon_j|^2+|\epsilon_p|^2)} \frac{1 - e^{-2\nu t \epsilon_j.\epsilon_p}}{2\nu\epsilon_j.\epsilon_p}$$

$$\leq 2^j 2^p \frac{3}{2}2^j e^{-\frac{1}{4}\nu t 2^{2j}} t.$$

This gives

$$\|\vec{U}_1(t,.) - \vec{U}_3(t,.)\|_\infty \leq C \frac{Q^2}{N} \sum_{j\in\Lambda} \sum_{p\in\Lambda, p<j} 2^p t 2^{2j} e^{-\frac{1}{4}\nu t 2^{2j}}$$

$$\leq C' \frac{1}{\nu} \frac{Q^2}{N} \sum_{j\in\Lambda} \sum_{p\in\Lambda, p<j} 2^p e^{-\frac{1}{8}\nu t 2^{2j}}.$$

If $\sigma(j) = \sup\{p \in \Lambda \ / \ p < j\}$, we find

$$\|\vec{U}_1(t,.) - \vec{U}_3(t,.)\|_\infty \leq C \frac{1}{\nu} \frac{Q^2}{N} \sum_{j\in\Lambda, j>j_0} 2^{\sigma(j)} e^{-\frac{1}{8}\nu t 2^{2j}}$$

We then find, for $T_1 < t (< T_2)$,

$$\|\vec{U}_1(t,.) - \vec{U}_3(t,.)\|_\infty \leq C \frac{1}{\nu} \frac{Q^2}{N} \sum_{j\in\Lambda, j>j_0} 2^{\sigma(j)} e^{-\frac{1}{8}\nu(t-T_1) 2^{2j}}$$

$$\leq C' \frac{1}{\nu} \frac{Q^2}{N} \sum_{j\in\Lambda, j>j_0} \frac{2^{\sigma(j)}}{\sqrt{\nu(t-T_1)} 2^j}$$

so that, as $2^{\sigma(j)} \leq 2^{j/2}$, we have

$$\|\vec{U}_1(t,.) - \vec{U}_3(t,.)\|_\infty \leq C_\nu \frac{Q^2}{N} \frac{1}{\sqrt{t-T_i}}.$$

Similarly, we have

$$\|\vec{U}_1 - \vec{U}_3\|_{L^2((T_1,T_2),L^\infty)} \leq C \frac{1}{\nu} \frac{Q^2}{N} \sum_{j\in\Lambda, j>j_0} 2^{\sigma(j)} \left(\int_0^{+\infty} e^{-\frac{1}{4}\nu t 2^{2j}} \, dt \right)^{1/2}$$

so that

$$\|\vec{U}_1 - \vec{U}_3\|_{L^2((T_1,T_2),L^\infty)} \leq C \frac{1}{\nu} \frac{Q^2}{N} \sum_{j\in\Lambda, j>j_0} \frac{2^{\sigma(j)}}{\sqrt{\nu} 2^j}$$

Finally, we get

$$\|\vec{U}_1(t,.) - \vec{U}_3(t,.)\|_E \leq C_5 \frac{Q^2}{N}.$$

A similar estimate holds for $(Id - P_0)\vec{U}_3$:

$$\|(Id - P_0)\vec{U}_3(t,.)\|_\infty \leq C \frac{1}{\nu} \frac{Q^2}{N} \sum_{j\in\Lambda} 2^j e^{-2\nu t 2^{2j}}.$$

In order to estimate $\|(Id - P_0)\vec{U}_3\|_E$, we split again the sum between the indexes j such that $j \leq J_2$, those such that $J_2 < j \leq J_1$ and those such that $j > J_1$:

$$\|(Id - P_0)\vec{U}_3(t,.)\|_\infty \leq C \frac{1}{\nu} (D(t) + E(t) + F(t))$$

with

$$D(t) = \frac{Q^2}{N} \sum_{j\in\Lambda, j\leq J_2} 2^j e^{-2\nu t 2^{2j}}, \quad E(t) = \frac{Q^2}{N} \sum_{j\in\Lambda, J_2<j\leq J_1} 2^j e^{-2\nu t 2^{2j}}$$

and

$$F(t) = \frac{Q^2}{N} \sum_{j \in \Lambda, j > J_1} 2^j e^{-2\nu t 2^{2j}}.$$

Let us write

$$\|A(t)\|_{\mathcal{E}} = \sup_{T_1 < t < T_2} \sqrt{t - T_1} A(t) + \left(\int_{T_1}^{T_2} A(t)^2 \, dt \right)^{1/2}.$$

We have

$$\|(Id - P_0)\vec{U}_3(t, .)\|_E \leq C \frac{1}{\nu} (\|D(t)\|_{\mathcal{E}} + \|E(t)\|_{\mathcal{E}} + \|\|F(t)\|_{\mathcal{E}}.)$$

We have $\|D\|_\infty \leq 2 \frac{Q^2}{N} 2^{J_2}$ and thus

$$\|D\|_{\mathcal{E}} \leq C \sqrt{T_2 - T_1} \sup_{T_2 < t < T_1} D(t) \leq 2C \frac{Q^2}{N} 2^{J_2} \sqrt{T_2} \leq C_\nu \frac{Q^2}{N}.$$

To estimate $F(t)$, for $T_1 < t < T_2$, we write

$$F(t) \leq \frac{Q^2}{N} \sum_{j \in \Lambda, j > J_1} 2^j e^{-\frac{1}{8}\nu T_1 2^{2j}} e^{-\frac{1}{8}\nu(t - T_1)2^{2j}}$$

$$\leq C \frac{Q^2}{N} \sum_{j \in \Lambda, j > J_1} \frac{2^j}{\sqrt{\nu(t - T_1)2^j}} e^{-\frac{1}{8}\nu T_1 2^{2j}};$$

on the other hand, we have

$$\|F\|_{L^2(T_1, T_2)} \leq \frac{Q^2}{N} \sum_{j \in \Lambda, j > J_1} 2^j e^{-\frac{1}{8}\nu T_1 2^{2j}} \left(\int_0^{+\infty} e^{-\frac{1}{4}\nu t 2^{2j}} \, dt \right)^{1/2}$$

so that

$$\|F\|_{L^2(T_1, T_2)} \leq C \frac{Q^2}{N} \sum_{j \in \Lambda, j > J_1} \frac{2^j}{\sqrt{\nu} 2^j} e^{-\frac{1}{8}\nu T_1 2^{2j}}.$$

Finally, we get

$$\|F\|_{\mathcal{E}} \leq C_\nu \frac{Q^2}{N} \sum_{j \in \Lambda, j > J_1} e^{-\frac{1}{8}\frac{2^{2j}}{2^{2J_1}}} \leq C' \frac{Q^2}{N} \left(\sum_{p \in \mathbb{N}} e^{-\frac{1}{8}(16)^p} \right)^{1/2} = C'' \frac{Q^2}{N}.$$

For $E(t)$, we write (for $t > T_1$)

$$E(t) \leq \frac{Q^2}{N} \sqrt{\sum_{j \in \Lambda, J_2 < j \leq J_1} \sum_{p \in \Lambda, p \leq j} 2^p 2^j e^{-2\nu(t - T_1)2^{2j}}}$$

$$\leq \frac{Q^2}{N} \sqrt{2 \sum_{j \in \Lambda, J_2 < j \leq J_1} 2^{2j} e^{-2\nu(t - T_1)2^{2j}}}$$

From this, we find

$$\|E\|_{\mathcal{E}} \leq C_\nu \frac{Q^2}{N} \sqrt{\#(\{j \in \Lambda \ / \ J_2 < j \leq J_1\})}$$

so that, using inequality (9.17), we find

$$\|E\|_{\mathcal{E}} \leq C' \frac{Q^{1/2}}{\sqrt{N}}.$$

Collecting those three estimates on D, E and F, we find that

$$\|(Id - P_0)\vec{U}_3\|_E \leq C_6 \frac{Q^{1/2}}{\sqrt{N}}$$

For the low frequencies $P_0\vec{U}_3$, we have $\|P_0\vec{U}_3(t,.)\|_\infty \leq CQ^2$ so that

$$\|P_0\vec{U}_3\|_E \leq C_\nu \sqrt{T_2 - T_1} \sup_{t>0} \|P_0\vec{U}_3(t.)\|_\infty \leq C_7 \sqrt{T_2}Q^2$$

Collecting all those estimates we find

$$\begin{aligned}
\|\vec{U}_2(T_2,.)\|_{BMO^{-1}} &\leq \|\vec{U}_2(T_1,.)\|_{BMO^{-1}} \\
&+ C_8(Q^2\sqrt{T_2} + \frac{Q^{1/2}}{\sqrt{N}})(Q^{-1/2} + Q^2\sqrt{T_2} + \frac{Q^{1/2}}{\sqrt{N}}) \\
&+ C_8\|\vec{U}_2\|_E(\|\vec{U}_2\|_E + Q^{-1/2} + Q^2\sqrt{T_2} + \frac{Q^{1/2}}{\sqrt{N}})
\end{aligned} \tag{9.18}$$

and

$$\begin{aligned}
\|\vec{U}_2\|_E &\leq \|\vec{U}_2(T_1,.)\|_{BMO^{-1}} \\
&+ C_8(Q^2\sqrt{T_2} + \frac{Q^{1/2}}{\sqrt{N}})(Q^{-1/2} + Q^2\sqrt{T_2} + \frac{Q^{1/2}}{\sqrt{N}}) \\
&+ C_8\|\vec{U}_2\|_E(\|\vec{U}_2\|_E + Q^{-1/2} + Q^2\sqrt{T_2} + \frac{Q^{1/2}}{\sqrt{N}})
\end{aligned} \tag{9.19}$$

If we have the bounds

$$C_8(Q^{-1/2} + Q^2\sqrt{T_2} + \frac{Q^{1/2}}{\sqrt{N}}) \leq \frac{1}{4} \tag{9.20}$$

and

$$C_8\|\vec{U}_2(T_1,.)\|_{BMO^{-1}} \leq \frac{1}{16} \tag{9.21}$$

we will find

$$\|\vec{U}_2\|_E \leq \frac{1}{2}(4\|\vec{U}_2(T_1,.)\|_{BMO^{-1}} + Q^2\sqrt{T_2} + \frac{Q^{1/2}}{\sqrt{N}})$$

and

$$C_8\|\vec{U}_2\|_E \leq \frac{1}{4}.$$

Moreover, we will have

$$\|\vec{U}_2(T_2,.)\|_{BMO^{-1}} \leq 2\|\vec{U}_2(T_1,.)\|_{BMO^{-1}} + \frac{1}{2}(Q^2\sqrt{T_2} + \frac{Q^{1/2}}{\sqrt{N}})$$

As $\vec{U}_2(0,.) = 0$, we find that if we split the time interval $[0, \frac{1}{\nu 2^{2j_0}}]$ into Q^3 intervals $[T_i, T_{i+1}]$ with $T_0 = 0$ and, for $i \geq 1$,

$$T_i = \frac{1}{\nu 2^{2J_i}} \text{ with } J_i = j_{N-i\frac{N}{Q^3}}$$

(recall that $\Lambda = \{j_0 < j_1 < \cdots < j_{N-1}\}$) so that (9.17) is satisfied, we shall have, provided that

$$C_8(Q^{-1/2} + Q^2\frac{1}{\sqrt{\nu 2^{j_0}}} + \frac{Q^{1/2}}{\sqrt{N}}) \leq \frac{1}{4},$$

the inequality

$$\|\vec{U}_2(T_i,.)\|_{BMO^{-1}} \leq (2^i - 1)\frac{1}{2}(Q^2\frac{1}{\sqrt{\nu}2^{j_0}} + \frac{Q^{1/2}}{\sqrt{N}})$$

as long as

$$C_8 2^i (Q^2\frac{1}{\sqrt{\nu}2^{j_0}} + \frac{Q^{1/2}}{\sqrt{N}}) \leq \frac{1}{4}.$$

We end the proof by fixing Q large enough to ensure that $C_8 Q^{-1/2} \leq \frac{1}{8}$, then j_0 and N large enough to ensure that $N > Q^3$ and that

$$C_8 2^{Q^3}(Q^2\frac{1}{\sqrt{\nu}2^{j_0}} + \frac{Q^{1/2}}{\sqrt{N}}) \leq \frac{1}{4}.$$

We then get, for $T = \frac{1}{\nu 2^{2j_0}}$ (which we may assume less than δ, by fixing j_0 large enough)

$$\|\vec{U}_2(T,.)\|_{\dot{B}^{-1}_{\infty,\infty}} \leq C\|\vec{U}_2(T,.)\|_{BMO^{-1}} \leq \frac{1}{16C_8} \leq \frac{1}{2\delta}$$

(for small enough δ) while $\|\vec{U}_0\|_{\dot{B}^{-1}_{\infty,\infty}} \leq \delta \leq \frac{1}{2\delta}$ and $\|\vec{U}_1(T,.)\|_{\dot{B}^{-1}_{\infty,\infty}} \geq \frac{2}{\delta}$.
Theorem 9.7 is thus proved. $\qquad\square$

9.6 Further Results on Ill-posedness

Bourgain and Pavlović [52] proved that the Cauchy problem for the Navier–Stokes equations was ill-posed in $\dot{B}^{-1}_{\infty,\infty}$. Ill-posedness in the smaller spaces $\dot{B}^{-1}_{\infty,q}$ (for $2 < q < +\infty$) has been also discussed by Germain [205] and Yoneda [510]. The proof of Bourgain and Pavlović is still valid in this case: in their example, their solutions (which belong to $L^\infty(0,+\infty), \mathcal{E}2$) as we have seen it in section 9.4) belong to $L^\infty(\dot{B}^{-1}_{\infty,q})$ for $2 \leq q \leq +\infty$; we can take N arbitrarily large; as we have $\|\vec{u}_0\|_{\dot{B}^{-1}_{\infty,q}} \approx \frac{Q}{\sqrt{N}}N^{\frac{1}{q}}$ and $\|\vec{u}(T,.)\|_{\dot{B}^{-1}_{\infty,q}} \geq \frac{1}{\delta}$ for some $T < \frac{1}{\delta}$, we conclude easily that we have norm inflation in $\dot{B}^{-1}_{\infty,q}$ for $q > 2$ since $\|\vec{u}(T,.)\|_{\dot{B}^{-1}_{\infty,\infty}} \leq \|\vec{u}(T,.)\|_{\dot{B}^{-1}_{\infty,q}}$.

A common assumption was that the problem was well posed in $\dot{B}^{-1}_{\infty,q}$ when $1 \leq q \leq 2$, as in that case $\dot{B}^{-1}_{\infty,q} \subset BMO^{-1}$. However, recently, Wang [495] proved that the Cauchy problem for the Navier–Stokes equations is ill-posed on all Besov spaces $\dot{B}^{-1}_{\infty,q}, 1 \leq q \leq +\infty$. The norm inflation described by Theorem 9.7 occurs in all those spaces, while existence of mild solutions is granted for small data in $\dot{B}^{-1}_{\infty,q}$ with $1 \leq q \leq 2$. For such data, a global solution will be given by the Koch and Tataru theorem (Theorem 9.2) (as $\dot{B}^{-1}_{\infty,q} \subset BMO^{-1}$ when $1 \leq q \leq 2$), and we know by Lemma 9.1 that we will have control on the BMO^{-1} norm; but we will have no control at all on the Besov norm.

Another approach of ill-posedness can be found in a paper of Bejenaru and Tao [34]. Following their idea, we shall define ill-posedness or well-posedness in the following way:

Definition 9.1.
If $\vec{u}_0 \in \mathrm{bmo}^{-1}$ with $\mathrm{div}\,\vec{u}_0 = 0$, let $A_n(\vec{u}_0)$ be the n-th term which appears in the solution

$$\vec{u}_\epsilon = \sum_{n=1}^{+\infty} \epsilon^n A_n(\vec{u}_0)$$

of the Navier–Stokes problem

$$\partial_t \vec{u}_\epsilon = \nu \Delta \vec{u}_\epsilon - \mathbb{P}\,\mathrm{div}(\vec{u}_\epsilon, \vec{u}_\epsilon), \quad \mathrm{div}\,\vec{u}_\epsilon = 0, \quad \vec{u}_\epsilon(0,.) = \epsilon \vec{u}_0$$

for ϵ small enough.

A Banach space Y is adapted to the Navier–Stokes problem if $Y \subset \mathrm{bmo}^{-1}$ (continuous embedding) and if, for every $T > 0$, there exists a constant $C_{\nu,T}$ such that the operators A_n satisfy, for every $n \geq 1$,

$$\|A_n(\vec{u}_0)\|_{L^\infty((0,T),Y)} \leq C_{\nu,T}^n \|\vec{u}_0\|_Y^n.$$

Thus, if $\|\vec{u}_0\|_Y < \frac{1}{C_{\nu,T}}$, the Navier–Stokes problem with initial value $\vec{u}_0$ has a solution in $L^\infty((0,T),Y)$.

Definition 9.2.
The Navier–Stokes equations are well-posed in a Banach space X if $X \cap \mathrm{bmo}^{-1}$ is an adapted Banach space and for every $T > 0$, there exists $\epsilon_T > 0$ such that if $\vec{u}_{0,n} \in \mathrm{bmo}^{-1}$ with $\|\vec{u}_{0,n}\|_{\mathrm{bmo}^{-1}} + \|\vec{u}_{0,n}\|_X < \epsilon_T$ and $\lim_{n\to+\infty} \|\vec{u}_{0,n}\|_X = 0$ then the Navier–Stokes problem with initial value $\vec{u}_{0,n}$ has a solution $\vec{u}_n$ on $(0,T)$ and $\lim_{n\to+\infty} \|\vec{u}_n\|_{L^\infty((0,T),X)} = 0$.

Proposition 9.4.
Let $s > -1$, $1 \leq p \leq \infty$, $1 \leq q \leq \infty$. Then the Navier–Stokes equations are well-posed in $B_{p,q}^s$.

Proof. Recall that the operator $B(\vec{u}, \vec{v}) = \int_0^t W_{\nu(t-s)} * \mathbb{P}\,\mathrm{div}(\vec{u} \otimes \vec{v})\,ds$ is bounded on the space E_T described in Theorem 9.1:

$$\|B(u,v)\|_{E_T} \leq C_0 \|u\|_{E_T} \|v\|_{E_T}.$$

Thus, by Theorem 5.1, if $\|\vec{u}_0\|_{\mathrm{bmo}^{-1}} < \epsilon_T$ (so that $\|W_{\nu t} * \vec{u}_0\|_{E_T} < \frac{1}{4C_0}$) and $\mathrm{div}\,\vec{u}_0 = 0$, the Navier–Stokes problem with initial value $\vec{u}_0$ has a solution $\vec{u}$ in E_T with

$$\vec{u} = \sum_{k=1}^{+\infty} \vec{U}_k$$

where $\vec{U}_1 = W_{\nu t} * \vec{u}_0$ and $\vec{U}_{k+1} = -\sum_{j=1}^{k} B(\vec{U}_j C, \vec{U}_{k+1-j})$, and we have

$$\|\vec{U}_k\|_{E_T} \leq A_k C_0^{k-1} \|\vec{U}_1\|_{E_T}^k \leq C\left(\frac{\|\vec{u}_0\|_{\mathrm{bmo}^{-1}}}{\epsilon_T}\right)^k$$

(where A_k is the k-th Catalan number).
By Lemma 9.1, since the Riesz tranfoms are bounded on BMO, we have moreover

$$\|\vec{U}_{k+1}(t,.)\|_{\mathrm{BMO}^{-1}} \leq C_1 \sum_{j=1}^{k} \|\vec{U}_j\|_{E_T} \|\vec{U}_{k+1-j}\|_{E_T} \leq \frac{C_1}{C_0} A_{k+1} C_0^k \|\vec{U}_1\|_{E_T}^{k+1}.$$

Thus, we find, as $BMO^{-1} \subset bmo^{-1}$,

$$\|\vec{U}_k\|_{L^\infty((0,T),bmo^{-1})} \leq C \left(\frac{\|\vec{u}_0\|_{bmo^{-1}}}{\epsilon_T} \right)^k.$$

We now prove that $\vec{U}_k$ belongs to $\{\vec{U} \; / \; t^{\alpha/2}\vec{U} \in L^\infty((0,T), B_{p,q}^{\alpha+s})\}$ for $0 < \alpha < 1$ and $s + \alpha > 0$. This is based on the following classical estimates:

- if $u, v \in B_{p,q}^r \cap L^\infty$ with $r > 0$, then

$$\|uv\|_{B_{p,q}^r} \leq C(\|u\|_\infty \|v\|_{B_{p,q}^r} + \|v\|_\infty \|u\|_{B_{pq}^r})$$

  (where C depends only on r, p and q)

- if $u \in B_{p,q}^r \cap L^\infty$ with $r \in \mathbb{R}$ and if $\rho > r$, then

$$\|W_t * u\|_{B_{p,q}^\rho} \leq C(1 + t^{-\frac{\rho-r}{2}})\|u\|_{B_{p,q}^r}$$

  (where C depends only on $r - \rho$).

Thus, we have, for $0 < t < T(< +\infty)$,

$$t^{\alpha/2}\|\vec{U}_1\|_{B_{p,q}^{s+\alpha}} \leq C_{\nu,T}\|\vec{u}_0\|_{B_{p,q}^s}$$

and

$$\|\vec{U}_{k+1}\|_{B_{p,q}^{s+\alpha}} \leq C_{\nu,T} \int_0^t \frac{1}{(t-s)^{\frac{1}{2}}} \frac{ds}{s^{\frac{1+\alpha}{2}}} \sum_{j=1}^k \sup_{0<s<t} t^\alpha \|\vec{U}_j\|_{B_{p,q}^{\alpha+s}} \sup_{0<s<t} \sqrt{s}\|\vec{U}_{k+1-j}\|_\infty.$$

If $B_k = \sup_{0<t<T} t^{\alpha/2}\|\vec{U}_k\|_{B_{p,q}^{s+\alpha}}$, we find that

$$B_{k+1} \leq C_2 \sum_{j=1}^k B_j A_{k+1-j} C_0^{k-j} \|\vec{U}_1\|_{E_T}^{k+1-j}$$

and by induction (taking C_2 greater than 1), we find that

$$B_k \leq A_k(C_2C_0)^{k-1}\|\vec{U}_1\|_{E_T}^{k-1} B_1.$$

Moreover, we have

$$\|\vec{U}_{k+1}\|_{B_{p,q}^s} \leq C_{\nu,T} \int_0^t \frac{1}{(t-s)^{\frac{1-\alpha}{2}}} \frac{ds}{s^{\frac{1+\alpha}{2}}} \sum_{j=1}^k \sup_{0<s<t} t^\alpha \|\vec{U}_j\|_{B_{p,q}^{\alpha+s}} \sup_{0<s<t} \sqrt{s}\|\vec{U}_{k+1-j}\|_\infty$$

so that

$$\sup_{0<s<t} \|\vec{U}_{k+1}\|_{B_{p,q}^s} \leq C_3 \sum_{j=1}^k B_j A_{k+1-j} C_0^{k-j}\|\vec{U}_1\|_{E_T}^{k+1-j} \leq C_3 C_4^k \left(\frac{\|\vec{u}_0\|_{bmo^{-1}}}{\epsilon_T} \right)^k \|\vec{u}_0\|_{B_{p,q}^s}.$$

Thus, the Navier–Stokes equations are well-posed in $B_{p,q}^s$. $\square$

In order to disprove well-posedness, we shall use the following lemma, of Bejenaru and Tao [34]:

Lemma 9.2. *If the Navier–Stokes equations are well-posed in a Banach space X, then the bilinear operator*

$$A_2 : \vec{u}_0 \mapsto \int_0^t W_{\nu(t-s)} * \mathbb{P} \operatorname{div}((W_{\nu s} * \vec{u}_0) \otimes (W_{\nu s} * \vec{u}_0)) \, ds$$

maps the unit ball B_0 of bmo$^{-1} \cap X$ *to $L^\infty((0,T), \text{bmo}^{-1} \cap X)$ and is continuous at $\vec{u}_0 = 0$ in the X norm in the following sense: if $\vec{u}_n \in B_0$ and $\|\vec{u}_n\|_X \to 0$, then $\|A_2(\vec{u}_n)\|_{L^\infty((0,T),X)} \to 0$.*

Proof. As bmo$^{-1} \cap X$ is adapted, we know that

$$\|A_2(\vec{u}_0)\|_{L^\infty(\text{bmo}^{-1} \cap X)} \le C\|\vec{u}_0\|^2_{\text{bmo}^{-1} \cap X}.$$

Moreover, there exists a positive η_T such that, if $\|\vec{u}_0\|_{\text{bmo}^{-1} \cap X} < \eta_T$, then the Navier–Stokes problem with initial value $\vec{u}_0$ has a solution in $L^\infty((0,T), \text{bmo}^{-1} \cap X)$ given by

$$\vec{u} = \sum_{k=1}^{+\infty} A_k(\vec{u}_0)$$

where $A_k(\lambda \vec{u}_0) = \lambda^k A_k(\vec{u}_0)$ and

$$\|A_k(\vec{u}_0)\|_{L^\infty((0,T),\text{bmo}^{-1} \cap X)} \le C\eta_T^{-k} \|\vec{u}_0\|^k_{\text{bmo}^{-1} \cap X}.$$

Now, let us consider $\vec{u}_{0,n} \in B_0$ and $\|\vec{u}_{0,n}\|_X \to 0$. For $0 \le \delta < 1$, let $\vec{u}_{n,\delta}$ be the solution to the Navier–Stokes problem with initial value $\delta \frac{\eta_T}{2} \vec{u}_{0,n}$. We have

$$\vec{u}_{n,\delta} = \sum_{k=1}^{+\infty} \delta^k \left(\frac{\eta_T}{2}\right)^k A_k(\vec{u}_{0,n}).$$

Since we assume that the Navier–Stokes equations are well-posed in X, we have $\lim_{n \to +\infty} \|\vec{u}_{n,\delta}\|_{L^\infty((0,T),X)} = 0$. Dividing with $\delta \frac{\eta_T}{2}$, we find that

$$\lim_{n \to \infty} \|A_1(\vec{u}_{0,n}) + \sum_{k \ge 2} \delta^{k-1} \left(\frac{\eta_T}{2}\right)^{k-1} A_k(\vec{u}_{0,n})\|_{L^\infty((0,T),X)} = 0.$$

As we have

$$\sum_{k \ge 2} \delta^{k-1} \left(\frac{\eta_T}{2}\right)^{k-1} \|A_k(\vec{u}_{0,n})\|_{L^\infty((0,T),X)} \le \sum_{k \ge 2} \delta \left(\frac{1}{2}\right)^{k-1} \|\vec{u}_{0,n}\|^k_{\text{bmo}^{-1} \cap X} \le C\delta$$

we find that

$$\lim_{n \to \infty} \|A_1(\vec{u}_{0,n})\|_{L^\infty((0,T),X)} = 0.$$

In particular, we have

$$\lim_{n \to \infty} \|\sum_{k \ge 2} \delta^{k-1} \left(\frac{\eta_T}{2}\right)^{k-1} A_k(\vec{u}_{0,n})\|_{L^\infty((0,T),X)} = 0.$$

Dividing with $\left(\delta \frac{\eta_T}{2}\right)^2$, we find that

$$\lim_{n \to \infty} \|A_2(\vec{u}_{0,n}) + \sum_{k \ge 3} \delta^{k-2} \left(\frac{\eta_T}{2}\right)^{k-2} A_k(\vec{u}_{0,n})\|_{L^\infty((0,T),X)} = 0$$

with

$$\sum_{k\geq 3}\delta^{k-2}\left(\frac{\eta_T}{2}\right)^{k-2}\|A_k(\vec{u}_{0,n})\|_{L^\infty((0,T),X)}\leq\sum_{k\geq 3}\delta\left(\frac{1}{2}\right)^{k-2}\|\vec{u}_{0,n}\|^k_{\mathrm{bmo}^{-1}\cap X}\leq C\delta.$$

This gives

$$\lim_{n\to\infty}\|A_2(\vec{u}_{0,n})\|_{L^\infty((0,T),X)}=0. \qquad\qquad \square$$

Wang's result [495] is then the following one:

Theorem 9.8.
Let $1\leq q\leq\infty$. Then the Navier–Stokes equations are ill-posed in $B^{-1}_{\infty,q}$.

Proof. The case $q>2$ can be dealt with the example of Bourgain and Pavlović [52]: if we label $\vec{u}_{0,Q,N}$ their example given in equation (9.10), we have

$$\|\vec{u}_{0,Q,N}\|_{\mathrm{BMO}^{-1}}\leq CQ, \|\vec{u}_{0,Q,N}\|_{B^{-1}_{\infty,q}}\leq CQN^{\frac{1}{q}-\frac{1}{2}}$$

and

$$\left\|A_2(\vec{u}_{0,Q,N})(\frac{1}{4^{j_0}\nu},.)\right\|_{B^{-1}_{\infty,q}}\geq c\left\|A_2(\vec{u}_{0,Q,N})(\frac{1}{4^{j_0}\nu},.)\right\|_{B^{-1}_{\infty,\infty}}\geq c'Q^2.$$

If we fix $\delta>0$ small enough to get that $\frac{\delta}{Q}(\|\vec{u}_{0,Q,N}\|_{\mathrm{BMO}^{-1}}l+\|\vec{u}_{0,Q,N}\|_{B^{-1}_{\infty,q}})\leq 2C\delta<1$ and j_0 large enough to have so that $\frac{1}{4^{j_0}\nu}<T$, we find that

$$\|\frac{\delta}{Q}\vec{u}_{0,Q,N}\|_{\mathrm{bmo}^{-1}\cap B^{-1}_{\infty,q}}<1, \lim_{N\to+\infty}\|\frac{\delta}{Q}\vec{u}_{0,Q,N}\|_{B^{-1}_{\infty,q}}=0$$

while

$$\|A_2(\frac{\delta}{Q}\vec{u}_{0,Q,N})\|_{L^\infty((0,T),B^{-1}_{\infty,q})}\geq c'\delta^2.$$

We conclude with Lemma 9.2.

Wang's idea for dealing with the case $q\leq 2$ is to study the operator A_2 restricted to $X_k=\{u\in\dot{B}^{-1}_{\infty,1}\ /\ \hat{u}(\xi)=0\text{ for }|\xi|<\frac{1}{4}2^k\text{ or }|\xi|>4\ 2^k\}$. If $u\in X_k$, then we have, for $1\leq q\leq+\infty$, $\|u\|_{B^{-1}_{\infty,q}}\approx\|u\|_{\mathrm{bmo}^{-1}}\approx 2^{-k}\|u\|_\infty$. In particular, we have, for $\vec{u}_0\in X_k$,

$$\|A_2(\vec{u}_0)\|_{L^\infty((0,T)),B^{-1}_{\infty,\infty})}\leq C\|A_2(\vec{u}_0)\|_{L^\infty((0,T)),\mathrm{bmo}^{-1})}\leq C'\|\vec{u}_0\|^2_{\mathrm{bmo}^{-1}}\leq C''\|\vec{u}_0\|^2_{B^{-1}_{\infty,q}}.$$

As the spectrum of $A_2(\vec{u}_0)$ is contained in $\{\xi\ /\ |\xi|<2^{k+3}\}$, we have

$$\|A_2(\vec{u}_0)\|_{L^\infty((0,T)),B^{-1}_{\infty,q})}\leq Ck^{1/q}\|A_2(\vec{u}_0)\|_{L^\infty((0,T)),B^{-1}_{\infty,\infty})}\leq C'k^{1/q}\|\vec{u}_0\|^2_{B^{-1}_{\infty,q}}.$$

Wang has proved that the estimate is sharp:

$$\sup_{\vec{u}_0\in X_k,\mathrm{div}\,\vec{u}_0=0,\|\vec{u}_0\|_{B^{-1}_{\infty,q}}\leq 1}\|A_2(\vec{u}_0)\|_{L^\infty((0,T),B^{-1}_{\infty,q})}\geq c_Tk^{1/q} \qquad (9.22)$$

where $c_T>0$. Due to Lemma 9.2, such an estimate proves that the Navier–Stokes equations are ill-posed in $B^{-1}_{\infty,q}$ when $q<+\infty$.

In order to prove (9.22), Wang makes a first simplification: he will prove that

$$\sup_{\vec{u}_0 \in X_k, \operatorname{div} \vec{u}_0 = 0, \|\vec{u}_0\|_{B^{-1}_{\infty,q}} \leq 1} \|\mathbb{P} \operatorname{div}(\vec{u}_0 \otimes \vec{u}_0)\|_{B^{-1}_{\infty,q}} \geq \gamma \, k^{1/q} 2^{2k} \tag{9.23}$$

where $\gamma > 0$. If $\vec{u}_0 \in X_k$ with $\|\vec{u}_0\|_{B^{-1}_{\infty,q}} \leq 1$, then $\|\vec{u}_0\|_\infty \leq 2^k$, $\|\Delta \vec{u}_0\|_\infty \leq C 2^{3k}$ and

$$\|W_{\nu t} * \vec{u}_0 - \vec{u}_0\|_\infty = \nu \Big\| \int_0^t W_{\nu s} * \Delta \vec{u}_0 \, ds \Big\|_\infty \leq C \nu t 2^{3k};$$

this gives the following estimate

$$\Big\| A_2(\vec{u}_0) - \int_0^t W_{\nu(t-s)} * \mathbb{P} \operatorname{div}(\vec{u}_0 \otimes \vec{u}_0) \, ds \Big\|_{B^{-1}_{\infty,q}} \leq C t k^{1/q} 2^k \nu t 2^{3k} = C \nu t^2 2^{4k} k^{1/q}.$$

Similarly, we have

$$\|W_{\nu t} * (\vec{u}_0 \otimes \vec{u}_0) - \vec{u}_0 \otimes \vec{u}_0\|_\infty = \nu \Big\| \int_0^t W_{\nu s} * \Delta(\vec{u}_0 \otimes \vec{u}_0) \, ds \Big\|_\infty \leq C \nu t 2^{4k};$$

this gives the following estimate

$$\Big\| t \mathbb{P} \operatorname{div}(\vec{u}_0 \otimes \vec{u}_0) - \int_0^t W_{\nu(t-s)} * \mathbb{P} \operatorname{div}(\vec{u}_0 \otimes \vec{u}_0) \, ds \Big\|_{B^{-1}_{\infty,q}} \leq C t k^{1/q} \nu t 2^{4k} = C \nu t^2 2^{4k} k^{1/q}.$$

For $t_0 = \eta 2^{-2k} < T$, we find, assuming that (9.23) is true,

$$\sup_{\vec{u}_0 \in X_k, \operatorname{div} \vec{u}_0 = 0, \|\vec{u}_0\|_{B^{-1}_{\infty,q}} \leq 1} \|A_2(\vec{u}_0)\|_{L^\infty((0,T), B^{-1}_{\infty,q})}$$

$$\geq \sup_{\vec{u}_0 \in X_k, \operatorname{div} \vec{u}_0 = 0, \|\vec{u}_0\|_{B^{-1}_{\infty,q}} \leq 1} \|A_2(\vec{u}_0)(t_0, .)\|_{B^{-1}_{\infty,q}}$$

$$\geq t_0 \sup_{\vec{u}_0 \in X_k, \operatorname{div} \vec{u}_0 = 0, \|\vec{u}_0\|_{B^{-1}_{\infty,q}} \leq 1} \|\mathbb{P} \operatorname{div}(\vec{u}_0 \otimes \vec{u}_0)\|_{B^{-1}_{\infty,q}} - 2C \nu t_0^2 2^{4k} k^{1/q}$$

$$\geq (\gamma \eta - 2C \nu \eta^2) k^{1/q}$$

and thus, taking η small enough, we get (9.22).

We may now describe Wang's example which leads to estimate (9.23). We take $\Phi \in \mathcal{S}(\mathbb{R}^3)$, $\Phi(0) \neq 0$ and such that its Fourier transform $\hat{\Phi}$ is supported in $\{\xi \ / \ |\xi| \leq 1\}$. For $k \geq 10$, we define $N_k = \{l \in 4\mathbb{N} \ / \ \frac{k}{100} \leq l \leq \frac{k}{10}\}$. Let $\alpha_k = 2^k (\frac{1}{3}, \frac{2}{3}, \frac{2}{3})$, $\beta_l = 2^l (\epsilon, \epsilon, \sqrt{1 - 2\epsilon^2})$ and $x_l = (0, 0, 2^l)$. We define Ψ as

$$\Psi = 4\lambda \sum_{l \in N_k} \Phi(x - x_l) \cos(\alpha_k \cdot (x - x_l)) \cos(\beta_l \cdot (x - x_l))$$

and

$$\vec{u}_0 = (-\partial_2 \Psi, \partial_1 \Psi, 0)$$

where $\lambda > 0$ does not depend on k. We have

$$\hat{\Psi} = \lambda \sum_{l \in N_k} e^{-i 2^l \xi_3} (\hat{\Phi}(\xi - \alpha_k - \beta_l) + \hat{\Phi}(\xi - \alpha_k + \beta_l) + \hat{\Phi}(\xi + \alpha_k + \beta_l) + \hat{\Phi}(\xi + \alpha_k - \beta_l)).$$

The function $e^{-i 2^l \xi_3} \hat{\Phi}(\xi - \alpha_k - \beta_l)$ is supported in

$$\{\xi \in \mathbb{R}^3 \ / \ |\xi - \alpha_k - \beta_l| \leq 1\}$$

hence in the area where

$$|\xi| \geq |\alpha_k| - |\beta_l| - 1 = 2^k - 2^l - 1 \geq \frac{1}{2}2^k$$

and

$$|\xi| \leq |\alpha_k| + |\beta_l| + 1 = 2^k + 2^l + 1 \leq \frac{3}{2}2^k.$$

Moreover,

$$\|\Psi\|_\infty \leq 4\lambda\| \sum_{l \in 4\mathbb{N}} |\Phi(x - (0,0,2^l))|\|_\infty \leq C\lambda$$

so that

$$\|\vec{u}_0\|_{\dot{B}_{\infty,q}^{-1}} \leq C'\lambda.$$

For λ fixed small enough, we have $\vec{u}_0 \in X_k$.

Let $\Phi_{l,\epsilon_1,\epsilon_2} = \mathcal{F}^{-1}\left(e^{-i2^l\xi_3}\hat{\Phi}(\xi - \epsilon_1\alpha_k - \epsilon_2\beta_l)\right)$. If we look at the support of the Fourier transform of

$$\partial_p\mathcal{F}^{-1}\left(e^{-i2^l\xi_3}\hat{\Phi}(\xi - \epsilon_1\alpha_k - \epsilon_2\beta_l)\right)\partial_{q,r}^2\mathcal{F}^{-1}\left(e^{-i2^m\xi_3}\hat{\Phi}(\xi - \epsilon_3\alpha_k - \epsilon_4\beta_m)\right)$$

(with $\epsilon_i \in \{-1,1\}$), it is contained in

$$\{\xi \in \mathbb{R}^3 \ / \ |\xi - (\epsilon_1 + \epsilon_3)\alpha_k - \epsilon_2\beta_l - \epsilon_4\beta_m| \leq 2\}.$$

- if $\epsilon_1 = \epsilon_3$, we find $\frac{1}{2}2^{k+1} \leq |\xi| \leq \frac{3}{2}2^{k+1}$

- if $\epsilon_1 \neq \epsilon_3$ and $l \neq m$ (hence $\sup(l,m) \geq \inf(l,m) + 4$), we find that $\frac{3}{4}2^{\sup(l,m)} \leq |\xi| \leq \frac{5}{4}2^{\sup(l,m)}$

- if $\epsilon_1 \neq \epsilon_3$, $l = m$ and $\epsilon_2 \neq \epsilon_4$, we find $|\xi| \leq 2$

- if $\epsilon_1 \neq \epsilon_3$, $l = m$ and $\epsilon_2 = \epsilon_4$, we find $\frac{7}{4}2^l \leq |\xi| \leq \frac{9}{4}2^l$.

The spectral domains are well separated and we find that

$$\|\mathbb{P}\,\mathrm{div}(\vec{u}_0 \otimes \vec{u}_0)\|_{B_{\infty,q}^{-1}} \geq \frac{1}{C}\lambda\left(\sum_{l \in N_k} 2^{-lq}\|\mathbb{P}\vec{v}_l\|_\infty^q\right)^{1/q}$$

where

$$\vec{v}_l = \begin{pmatrix} \sum_{(\epsilon_1,\epsilon_2)\in\{-1,1\}^2} \partial_1\left(\partial_2\Phi_{l,\epsilon_1,\epsilon_2}\partial_2\Phi_{l,-\epsilon_1,\epsilon_2}\right) - \partial_2\left(\partial_1\Phi_{l,\epsilon_1,\epsilon_2}\partial_2\Phi_{l,-\epsilon_1,\epsilon_2}\right) \\ \sum_{(\epsilon_1,\epsilon_2)\in\{-1,1\}^2} \partial_2\left(\partial_1\Phi_{l,\epsilon_1,\epsilon_2}\partial_1\Phi_{l,-\epsilon_1,\epsilon_2}\right) - \partial_1\left(\partial_2\Phi_{l,\epsilon_1,\epsilon_2}\partial_1\Phi_{l,-\epsilon_1,\epsilon_2}\right) \\ 0 \end{pmatrix}.$$

The Fourier transform of $\vec{v}_l$ is supported in

$$\{\xi \in \mathbb{R}^3 \ / \ |\xi - 2\beta_l| \leq 2\} \cup \{\xi \in \mathbb{R}^3 \ / \ |\xi + 2\beta_l| \leq 2\}.$$

In particular, we have $\frac{7}{4}2^l \leq |\xi| \leq \frac{9}{4}2^l$, $\frac{7}{4}2^l \leq |\xi_3| \leq \frac{9}{4}2^l$ and $\frac{7}{4}\epsilon 2^l \leq |\xi_1|, |\xi_2| \leq \frac{9}{4}\epsilon 2^l$ (k large enough for a fixed ϵ). If $\theta \in \mathcal{D}(\mathbb{R})$ is equal to 1 on $7/4 \leq |t| \leq 9/4$ and is supported in $1 \leq t \leq 3$ and if $\theta_0(t) = t\theta(t)$ and $\theta_1(t) = \frac{1}{t}\theta(t)$, we control easily the Riesz ransforms R_j ($1 \leq j \leq 3$) of the components $v_{l,1}$, $v_{l,2}$ of $\vec{v}_l$:

$$R_3v_{l,p} = \mathcal{F}^{-1}\left(i\theta_0(\frac{\xi_3}{2^l})\theta_1(\frac{|\xi|}{2^l})\hat{v}_{l,p}\right)$$

and, for $j = 1$ or $j = 2$,

$$R_j v_{l,p} = \mathcal{F}^{-1}\left(i\epsilon\theta_0(\frac{\xi_j}{2^l\epsilon})\theta_1(\frac{|\xi|}{2^l})\hat{v}_{l,p}\right).$$

As $\theta_0 \in \mathcal{F}(L^1(\mathbb{R}))$ and $\theta_1(|\xi|) \in \mathcal{F}(L^1(\mathbb{R}^3))$, we find that

$$\|\mathbb{P}\vec{v}_l - \vec{v}_l\|_\infty \leq C\epsilon\|\vec{v}_l\|_\infty$$

and thus (for ϵ small enough)

$$\|\mathbb{P}\operatorname{div}(\vec{u}_0 \otimes \vec{u}_0)\|_{B_{\infty,q}^{-1}} \geq \frac{\lambda}{2C}\left(\sum_{l \in N_k} 2^{-lq}\|\vec{v}_l\|_\infty^q\right)^{1/q}.$$

Similarly, for $1 \leq p, q \leq 2$, the function $\partial_p\Phi_{l,\epsilon_1,\epsilon_2}\partial_q\Phi_{l,-\epsilon_1,\epsilon_2}$ is supported in $|\xi_1 - 2\epsilon_2 2^l|, |\xi_2 - 2\epsilon_2 2^l| \leq 2$, hence in $\frac{7}{4}\epsilon 2^l \leq |\frac{\xi_1 + \xi_2}{2}| \leq \frac{9}{4}\epsilon 2^l$, $|\frac{\xi_1 - \xi_2}{2}| \leq 2$.
We write

$$\vec{v}_l = \frac{\partial_1 + \partial_2}{2}\begin{pmatrix} \sum_{(\epsilon_1,\epsilon_2)\in\{-1,1\}^2}\partial_2\Phi_{l,\epsilon_1,\epsilon_2}\partial_2\Phi_{l,-\epsilon_1,\epsilon_2} - \partial_1\Phi_{l,\epsilon_1,\epsilon_2}\partial_2\Phi_{l,-\epsilon_1,\epsilon_2} \\ \sum_{(\epsilon_1,\epsilon_2)\in\{-1,1\}^2}\partial_1\Phi_{l,\epsilon_1,\epsilon_2}\partial_1\Phi_{l,-\epsilon_1,\epsilon_2} - \partial_2\Phi_{l,\epsilon_1,\epsilon_2}\partial_1\Phi_{l,-\epsilon_1,\epsilon_2} \\ 0 \end{pmatrix}$$

$$+ \frac{\partial_1 - \partial_2}{2}\begin{pmatrix} \sum_{(\epsilon_1,\epsilon_2)\in\{-1,1\}^2}\partial_2\Phi_{l,\epsilon_1,\epsilon_2}\partial_2\Phi_{l,-\epsilon_1,\epsilon_2} + \partial_1\Phi_{l,\epsilon_1,\epsilon_2}\partial_2\Phi_{l,-\epsilon_1,\epsilon_2} \\ \sum_{(\epsilon_1,\epsilon_2)\in\{-1,1\}^2}-\partial_1\Phi_{l,\epsilon_1,\epsilon_2}\partial_1\Phi_{l,-\epsilon_1,\epsilon_2} - \partial_2\Phi_{l,\epsilon_1,\epsilon_2}\partial_1\Phi_{l,-\epsilon_1,\epsilon_2} \\ 0 \end{pmatrix}$$

$$= \frac{\partial_1 + \partial_2}{2}\vec{V}_l + \frac{\partial_1 - \partial_2}{2}\vec{W}_l$$

with

$$\|\frac{\partial_1 + \partial_2}{2}\vec{V}_l\|_\infty \geq \frac{\epsilon}{C}2^l\|\vec{V}_l\|_\infty \text{ and } \|\frac{\partial_1 - \partial_2}{2}\vec{W}_l\|_\infty \leq C\|\vec{W}_l\|_\infty \leq C'2^{2k}.$$

We have

$$\Phi_{l,\epsilon_1,\epsilon_2} = \Phi(x - x_l)e^{i\epsilon_1\alpha_k\cdot(x-x_l)}e^{i\epsilon_2\beta_l\cdot(x-x_l)}$$

so that, for $p = 1$ or $p = 2$,

$$|\partial_p\Phi_{l,\epsilon_1,\epsilon_2} - i\epsilon_1\frac{2^k p}{3}\Phi_{l,\epsilon_1,\epsilon_2}| \leq C\epsilon 2^l$$

hence we have, for

$$V_{l,1} = \sum_{(\epsilon_1,\epsilon_2)\in\{-1,1\}^2}\partial_2\Phi_{l,\epsilon_1,\epsilon_2}\partial_2\Phi_{l,-\epsilon_1,\epsilon_2} - \partial_1\Phi_{l,\epsilon_1,\epsilon_2}\partial_2\Phi_{l,-\epsilon_1,\epsilon_2},$$

the estimate

$$|V_{l,1} - 2\frac{2^{2k}}{9}\sum_{(\epsilon_1,\epsilon_2)\in\{-1,1\}^2}\Phi_{l,\epsilon_1,\epsilon_2}\Phi_{l,-\epsilon_1,\epsilon_2}| = |V_{l,1} - 4\frac{2^{2k}}{9}\Phi(x-x_l)^2\cos\beta_l\cdot(x-x_l))| \leq C\epsilon 2^{2k}.$$

Thus,

$$\|V_{l,1}\|_\infty \geq |V_{l,1}(x_l)| \geq 4\frac{2^{2k}}{9}\Phi(0)^2 - C\epsilon 2^{2k}.$$

We may conclude that

$$\|\mathbb{P}\operatorname{div}(\vec{u}_0 \otimes \vec{u}_0)\|_{B_{\infty,q}^{-1}} \geq \frac{\lambda}{2C}\left(\sum_{l \in N_k} 2^{-lq}\|\vec{v}_l\|_\infty^q\right)^{1/q}$$

$$\geq \frac{\lambda}{C'}\left(\sum_{l \in N_k}\|\vec{V}_l\|_\infty^q\right)^{1/q} - C'\lambda\left(\sum_{l \in N_k} 2^{-lq}\|\vec{W}_l\|_\infty^q\right)^{1/q}$$

$$\geq \frac{\lambda}{C'}\#(N_k)^{1/q}(4\frac{2^{2k}}{9}\Phi(0)^2 - C\epsilon 2^{2k}) - C''\lambda 2^{2k}$$

$$\geq C_1\lambda k^{1/q}2^{2k}(1 - C_2\epsilon) - C_3\lambda 2^{2k}.$$

Thus, we get (9.23) and we prove the theorem. $\square$

Further, this result has been generalized by Cui [135] to the setting of logarithmically improved Besov spaces. Recall that the Besov spaces $\dot{B}_{\infty,q}^{-1}$ $(1 \leq q \leq +\infty)$ are characterized by the homogeneous Littlewood–Paley decomposition as

$$f \in \dot{B}_{\infty,q}^{-1} \Leftrightarrow f = \sum_{j \in \mathbb{Z}} \Delta_j f \text{ in } \mathbb{S}' \text{ and } (2^{-j}\|\Delta_j f\|_\infty)_{j \in \mathbb{Z}} \in l^q.$$

Homogeneity is important mainly for existence of global solutions (compare the Koch and Tataru theorem for bmo^{-1} [local solutions] and BMO^{-1} [global solutions]). The (non-homogeneous) Besov space $B_{\infty,q}^{-1}$ $(1 \leq q \leq +\infty)$ is characterized by the Littlewood–Paley decomposition as

$$f \in B_{\infty,q}^{-1} \Leftrightarrow S_0 f \in L^\infty \text{ and } (2^{-j}\|\Delta_j f\|_\infty)_{j \in \mathbb{N}} \in l^q.$$

We have $B_{\infty,q}^{-1} \subset bmo^{-1}$ when $1 \leq q \leq 2$.

Definition 9.3.
For $\sigma \geq 0$, the logarithmically improved Besov space $B_{\infty,q}^{-1,\sigma}$ is defined by

$$f \in B_{\infty,q}^{-1,\sigma} \Leftrightarrow S_0 f \in L^\infty \text{ and } (2^{-j}(j+1)^\sigma\|\Delta_j f\|_\infty)_{j \in \mathbb{N}} \in l^q.$$

Proposition 9.5.
Let $\sigma \geq 0$, $1 \leq q \leq \infty$ and $0 < T < +\infty$. Then the following assertions are equivalent

(A) $f \in B_{\infty,q}^{-1,\sigma}$

(B) $t^{1/2}(\ln(\frac{eT}{t}))^\sigma\|W_{\nu t} * f\|_\infty \in L^q((0,T), \frac{dt}{t})$ and the norms $\|f\|_{B_{\infty,q}^{-1,\sigma}}$,

$$\left\|t^{1/2}(\ln(\frac{eT}{t}))^\sigma\|W_{\nu t} * f\|_\infty\right\|_{L^q((0,T),\frac{dt}{t})}$$

and

$$\sup_{0<t<T} t^{1/2}(\ln(\frac{eT}{t}))^\sigma\|W_{\nu t} * f\|_\infty + \left\|t^{1/2}(\ln(\frac{eT}{t}))^\sigma\|W_{\nu t} * f\|_\infty\right\|_{L^q((0,T),\frac{dt}{t})}$$

are equivalent.

Proof.

(A) $\implies$ (B): We write $f = S_0 f + \sum_{j>0} \Delta_j f$ with $S_0 f \in L^\infty$ and $\|\Delta_j f\|_\infty = 2^j (1+j)^{-\sigma} \epsilon_j$ with $(\epsilon_j)_{j \in \mathbb{N}} \in l^q$. We have $\|W_{\nu t} * S_0 f\|_\infty \le \|S_0 f\|_\infty$; on the other hand, we have

$$\|W_{\nu t} * \Delta_j f\|_\infty = \|\Delta W_{\nu t} * \frac{1}{\Delta} \Delta_j f\|_\infty \le C \min(1, (\nu t)^{-1} 2^{-2j}) \|\Delta_j f\|_\infty.$$

Now, if $t \le T$, we choose j_0 so that $1 \le 4^{j_0} \frac{t}{T} < 4$ and we write

$$t^{1/2} (\ln(\frac{eT}{t}))^\sigma \|W_{\nu t} * f\|_\infty$$

$$\le C_{\nu,T} (1+j_0)^\sigma (2^{-j_0} \|S_0 f\|_\infty + \sum_{j \le j_0} 2^{(j-j_0)} \frac{\epsilon_j}{(1+j)^\sigma} + \sum_{j > j_0} 2^{(j_0-j)} \frac{\epsilon_j}{(1+j)^\sigma})$$

$$\le C_{\nu,T,\sigma} (2^{-\frac{j_0}{2}} \|S_0 f\|_\infty + \sum_{j \le j_0} 2^{\frac{(j-j_0)}{2}} \epsilon_j + \sum_{j > j_0} 2^{(j_0-j)} \epsilon_j).$$

(B) $\implies$ (A): Using the integrability of the kernel of the convolution operator $e^{-\nu T \Delta} S_0$, we write, for $T/2 \le t \le T$,

$$\|S_0 f\|_\infty \le C_{\nu,T} \|W_{\nu T} f\|_\infty \le C'_{\nu,T} (2t)^{1/2} (\ln(\frac{eT}{t}))^\sigma \|W_{\nu t} f\|_\infty$$

and get $S_0 f \in L^\infty$.

Similarly, we write, for $j \ge 0$ and $1 \le 4^j \frac{t}{T} \le 4$, using the integrability of the kernel of $e^{-4\nu T \Delta} \Delta_0$,

$$2^{-j} (1+j)^\sigma \|\Delta_j f\|_\infty \le C_{\nu,T} t^{1/2} (\ln(\frac{eT}{t}))^\sigma \|W_{4\nu \frac{T}{4j}} * f\|_\infty \le C_{\nu,T} t^{1/2} (\ln(\frac{eT}{t}))^\sigma \|W_{\nu t} * f\|_\infty.$$

Thus, we have the equivalence of (A) and (B). The control of $\sup_{0 < t < T} t^{1/2} (\ln(\frac{eT}{t}))^\sigma \|W_{\nu t} * f\|_\infty$ is then a consequence of the embeddding $B_{\infty,q}^{-1,\sigma} \subset B_{\infty,\infty}^{-1,\sigma}$. $\qquad\square$

If the case $q = 2$, we have $B_{\infty,2}^{-1,\sigma} \subset B_{\infty,2}^{-1} \subset \mathrm{bmo}^{-1}$. Thus, we know that we can solve the Navier–Stokes problem on $(0,T) \times \mathbb{R}^3$ with initial value $\vec{u}_0$ if the norm of $\vec{u}_0$ in $B_{\infty,2}^{-1,\sigma}$ is small enough. Cui's result reads as:

Theorem 9.9.

- for $\sigma \ge \frac{1}{2}$, the Cauchy problem is locally well posed for small data in $B_{\infty,2}^{-1,\sigma}$: for every $T > 0$,

$$\lim_{\delta \to 0} \sup_{\|\vec{u}_0\|_{B_{\infty,2}^{-1,\sigma}} < \delta} \sup_{0 < t < T} \|\vec{u}(t,.)\|_{B_{\infty,2}^{-1,\sigma}} = 0.$$

- for $0 \le \sigma < \frac{1}{2}$, the Cauchy problem is ill-posed in $B_{\infty,q}^{-1,\sigma}$.

Proof. The proof follows Yoneda [510] for well-posedness and Wang for ill-posedness [495].

Case $\sigma \ge 1/2$:

As $f \in B_{\infty,2}^{-1,\sigma} \Leftrightarrow W_{\nu t} * f \in \mathbb{X}_T$, where

$$\|F\|_{\mathbb{X}_T} = \sup_{0 < t < T} t^{1/2} (\ln(\frac{eT}{t}))^\sigma \|F(t,.)\|_\infty + \left\| t^{1/2} (\ln(\frac{eT}{t}))^\sigma \|F(t,.)\|_\infty \right\|_{L^2((0,T), \frac{dt}{t})},$$

we just need to prove that

$$B(\vec{u}, \vec{v}) = \int_0^t W_{\nu(t-s)} * \mathbb{P} \operatorname{div}(\vec{u} \otimes \vec{v}) \, ds$$

maps boundedly $\mathbb{X}_T \times \mathbb{X}_T$ to $\mathbb{X}_T \cap L^\infty((0,T), B_{\infty,2}^{-1,\sigma})$.

We start from the inequalities

$$\|B(\vec{u}, \vec{v})(t, .)\|_\infty \leq C \int_0^t \frac{1}{\sqrt{\nu(t-s)}} \|\vec{u}(s, .)\|_\infty \|\vec{v}(s, .)\|_\infty \, ds$$

and

$$\|W_{\nu\theta} * B(\vec{u}, \vec{v})(t, .)\|_\infty \leq C \int_0^t \frac{1}{\sqrt{\nu(t+\theta-s)}} \|\vec{u}(s, .)\|_\infty \|\vec{v}(s, .)\|_\infty \, ds.$$

Thus, defining

$$\omega(t) = \ln(\frac{eT}{t}),$$

$$\|f\|_{\mathbb{Y}_T} = \sup_{0<t<T} t^{1/2}(\omega(t))^\sigma |f(t)| + \|\omega^\sigma f\|_{L^2((0,T),dt)},$$

$$J(f,g)(t) = \int_0^t \frac{1}{\sqrt{t-s}} f(s)g(s) \, ds,$$

and

$$K_t(f,g)(\theta) = \int_0^t \frac{1}{\sqrt{t+\theta-s}} f(s)g(s) \, ds,$$

we are going to prove

$$\|J(f,g)\|_{\mathbb{Y}_T} \leq C_T \|f\|_{\mathbb{Y}_T} \|g\|_{\mathbb{Y}_T} \tag{9.24}$$

and

$$\sup_{0<t<T} \|\omega^\sigma(\theta) K_t(f,g)(\theta)\|_{L^2((0,T),d\theta)} \leq C_T \|f\|_{\mathbb{Y}_T} \|g\|_{\mathbb{Y}_T}. \tag{9.25}$$

For $0 < s < t < T$, we have

$$1 \leq \omega^\sigma(t) \leq \omega^\sigma(s) \leq \omega^{2\sigma}(s)$$

so that

$$\omega^\sigma(t)|J(f,g)(t)| \leq \int_0^t \frac{1}{\sqrt{t-s}} \omega^\sigma(s)|f(s)|\omega^\sigma(s)|g(s)| \, ds$$

$$\leq \sqrt{\frac{2}{t}}(\int_0^{t/2} \omega^{2\sigma} f^2 \, ds)^{1/2}(\int_0^{t/2} \omega^{2\sigma} g^2 \, ds)^{1/2}$$

$$+ (\sup_{0<s<t} s^{1/2}\omega^\sigma(s)|f(s)|)(\sup_{0<s<t} s^{1/2}\omega^\sigma(s)|g(s)|) \int_{t/2}^t \frac{1}{\sqrt{t-s}} \frac{ds}{s}$$

$$\leq C \frac{1}{\sqrt{t}} \|f\|_{\mathbb{Y}_T} \|g\|_{\mathbb{Y}_T}.$$

On the other hand, we have

$$\omega^\sigma(t)|J(f,g)(t)| \leq \sqrt{\frac{2}{t}}\omega^\sigma(t) \int_0^{t/2} |f(s)||g(s)| \, ds$$

$$+ (\sup_{0<s<t} s^{1/2}\omega^\sigma(s)|f(s)|) \int_{t/2}^t \frac{1}{\sqrt{t-s}\sqrt{s}} \omega^\sigma(s)|g(s)| \, ds. \tag{9.26}$$

The operator $h \mapsto \int_{t/2}^{t} \frac{1}{\sqrt{t-s}\sqrt{s}} h(s)\,ds$ maps boundedly $L^1((0,T),\,dt)$ to $L^1((0,T),\,dt)$ and $L^\infty((0,T),\,dt)$ to $L^\infty((0,T),\,dt)$, hence $L^2((0,T),\,dt)$ to $L^2((0,T),\,dt)$. Thus, the second term in the right-hand side of (9.26) is well controlled in $L^2((0,T),dt)$.

For the first term, we define

$$A(t,s) = \mathbb{1}_{s<t}\sqrt{\frac{1}{t}\omega^\sigma(t)}|f(s)||g(s)|$$

and write

$$\sqrt{\frac{2}{t}}\omega^\sigma(t)\int_0^{t/2}|f(s)||g(s)|\,ds \le \sqrt{2}\int_0^T A(t,s)\,ds.$$

The Minkowski inequality then gives

$$\left(\int_0^T \left(\int_0^T A(t,s)\,ds\right)^2 dt\right)^{1/2} \le \int_0^T \left(\int_0^T A(t,s)^2\,dt\right)^{1/2} ds$$

$$= \int_0^T \left(\int_s^T \omega^{2\sigma}(t)\frac{dt}{t}\right)^{1/2}|f(s)||g(s)|\,ds$$

$$= \frac{1}{\sqrt{2\sigma+1}}\int_0^T \omega^{\sigma+\frac{1}{2}}(s)|f(s)||g(s)|\,ds$$

$$\le \frac{1}{\sqrt{2\sigma+1}}\int_0^T \omega^{2\sigma}(s)|f(s)||g(s)|\,ds$$

(since $1 \le \omega$ and $1/2 \le \sigma$). Thus, inequality (9.24) is proved.

In order to estimate $K_t(f,g)(\theta)$, we write

$$|K_t(f,g)(\theta)| \le \int_0^\theta \frac{1}{\sqrt{\theta-s}}|f(s)|g(s)|\,ds + \mathbb{1}_{t>\theta}\int_\theta^t \frac{1}{\sqrt{t-s}}|f(s)|g(s)|\,ds.$$

We know that we can control $\omega^\sigma(\theta)\int_0^\theta \frac{1}{\sqrt{\theta-s}}|f(s)|g(s)|\,ds$ in $L^2((0,T),d\theta)$. For the second term, we define

$$A_t(\theta,s) = \mathbb{1}_{\theta<s}\frac{1}{\sqrt{t-s}}\omega^\sigma(\theta)|f(s)||g(s)|$$

and write

$$\omega^\sigma(\theta)\int_\theta^t \frac{1}{\sqrt{t-s}}|f(s)|g(s)|\,ds = \int_0^t A_t(\theta,s)\,ds.$$

The Minkowski inequality then gives

$$\left(\int_0^T \left(\int_0^t A_t(\theta,s)\,ds\right)^2 d\theta\right)^{1/2} \le \int_0^t \left(\int_0^T A_t(\theta,s)^2\,d\theta\right)^{1/2} ds$$

$$= \int_0^t \left(\int_0^s \omega^{2\sigma}(\theta)d\theta\right)^{1/2}\frac{1}{\sqrt{t-s}}|f(s)||g(s)|\,ds$$

$$\le C\int_0^t \omega^\sigma(s)\sqrt{s}\frac{1}{\sqrt{t-s}}|f(s)||g(s)|\,ds$$

$$\le C\int_0^t \frac{1}{\sqrt{t-s}}\omega^\sigma(s)\sqrt{s}|f(s)|\omega^\sigma(s)\sqrt{s}||g(s)|\frac{ds}{\sqrt{s}}$$

$$\le \pi C(\sup_{0<s<t} s^{1/2}\omega^\sigma(s)|f(s)|)(\sup_{0<s<t} s^{1/2}\omega^\sigma(s)|g(s)|).$$

Thus, inequality (9.25) is proved.

Case $\sigma < 1/2$:

If we look at the example of Wang, we have $\vec{u}_0 \in X_k$ with $\|\vec{u}_0\|_\infty \approx 2^k$. In particular, $\|\vec{u}_0\|_{B_{\infty,2}^{-1,\sigma}} \approx k^\sigma$.

We begin with getting rid of the time, as in Wang's proof: we write

$$\left\| A_2(\vec{u}_0) - \int_0^t W_{\nu(t-s)} * \mathbb{P}\operatorname{div}(\vec{u}_0 \otimes \vec{u}_0)\, ds \right\|_{B_{\infty,2}^{-1,\sigma}} \leq Ctk^{\sigma+1/2} 2^k \nu t 2^{3k} = C\nu t^2 2^{4k} k^{\sigma+1/2}.$$

and

$$\left\| t\mathbb{P}\operatorname{div}(\vec{u}_0 \otimes \vec{u}_0) - \int_0^t W_{\nu(t-s)} * \mathbb{P}\operatorname{div}(\vec{u}_0 \otimes \vec{u}_0)\, ds \right\|_{B_{\infty,2}^{-1,\sigma}} \leq Ctk^{\sigma+1/2} \nu t 2^{4k} = C\nu t^2 2^{4k} k^{\sigma+1/2}.$$

For $t_0 = \eta 2^{-2k} < T$, we find

$$\|A_2(\vec{u}_0)(t_0, .)\|_{B_{\infty,2}^{-1,\sigma}} \geq \eta 2^{-2k} \|\mathbb{P}\operatorname{div}(\vec{u}_0 \otimes \vec{u}_0)\|_{B_{\infty,2}^{-1,\sigma}} - 2C\nu \eta^2 k^{\sigma+1/2}.$$

Moreover, we have

$$\|\mathbb{P}\operatorname{div}(\vec{u}_0 \otimes \vec{u}_0)\|_{B_{\infty,2}^{-1,\sigma}} \geq \frac{1}{C} \left(\sum_{l \in N_k} l^{2\sigma} 2^{-2l} \|\mathbb{P}\vec{v}_l\|_\infty^q \right)^{1/2} \geq \frac{1}{C'} k^\sigma \left(\sum_{l \in N_k} 2^{-2l} \|\mathbb{P}\vec{v}_l\|_\infty^q \right)^{1/2}$$

and we have seen that

$$\left(\sum_{l \in N_k} 2^{-2l} \|\mathbb{P}\vec{v}_l\|_\infty^2 \right)^{1/2} \geq C_1 k^{1/2} 2^{2k} (1 - C_2 \epsilon) - C_3 2^{2k}.$$

Thus, we have $\|\vec{u}_0\|_{B_{\infty,2}^{-1,\sigma}} \approx k^\sigma$ and

$$\|A_2(\vec{u}_0)\|_{L^\infty((0,T), B_{\infty,2}^{-1,\sigma})} \geq \gamma k^{2\sigma} k^{1/2-\sigma}.$$

As $\sigma < 1/2$, we conclude that the Navier–Stokes equations are ill-posed in $B_{\infty,2}^{-1,\sigma}$. $\square$

9.7   Large Data for Mild Solutions

We saw by Theorem 9.2 that we have a global mild solution to

$$\partial_t \vec{u} = \nu \Delta \vec{u} - \operatorname{div}(\vec{u} \otimes \vec{u}) - \vec{\nabla}p + \operatorname{div} \mathbb{F}$$

with $\operatorname{div} \vec{u} = 0$ and $\vec{u}_{|t=0} = \vec{u}_0$, provided that $\vec{u}_0$ is small enough in BMO^{-1} (and $\mathbb{F}$ small enough in F_∞). Theorem 9.7 shows that there is little hope that the sole smallness of $\|\vec{u}_0\|_{\dot{B}_{\infty,\infty}^{-1}}$ (instead of $\|\vec{u}_0\|_{BMO^{-1}}$) should be sufficient for such a result.

On the other hand, we might wonder if global mild solutions exist for some large initial values in BMO^{-1} (and even in $\dot{B}_{\infty,\infty}^{-1}$). The answer is obviously positive, as can be seen by the example of $\vec{u}_0 \in \mathcal{E}_2$ (Theorem 9.6): for such initial value, there is no restriction at all on the size of $\vec{u}_0$.

Other examples of "two-dimensional" vector fields that lead to global mild solutions with no restriction on their size will be given in Chapter 10:

- Majda and Bertozzi's two-and-a-half dimensional flows [40] (Proposition 10.1); those initial values may be perturbated by a small enough vector field that belongs to $\dot{H}^{1/2}(\mathbb{R}^3)$, as shown by Gallagher [195] (Theorem 10.2)

- regular enough axisymmetric flows with no swirl, as shown by Ladyzhenskaya [295], Uchovskii and Yudovich [486] and Leonardi, Malek, Nečas, and Pokorný [326] (Theorem 10.4)

- vector fields with helical symmetry, as shown by Mahalov, Titi, and Leibovich in [347] (Theorem 10.7)

- strong Beltrami flows [40, 129], also known as Trkalian flows [300] as they were first studied by Trkal [477] (Theorem 10.16)

In a series of papers, Chemin and Gallagher presented more genuinely three-dimensional examples of vector fields that are large in $\dot{B}^{-1}_{\infty,\infty}$ and lead to global mild solutions. The series began with [107], followed by [108], [109], [111] (joint paper with Paicu) and [112] (joint paper with Zhang). This is currently an active field of research, as it can be seen in the recent papers of Kukavica, Rusin and Ziane [288], Paicu and Zhang [391] or Wong [505].

We present here (a slightly simplified version of) Chemin and Gallagher's result [108]:

Theorem 9.10.
Let $\omega \in \mathcal{D}(\mathbb{R})$ a non-negative even function supported in $[-1,1]$ (with $\omega \neq 0$), and let $\Omega \in \mathcal{S}(\mathbb{R})$ the inverse Fourier transform of ω: $\mathcal{F}\Omega = \omega$. For $\epsilon \in (0,1)$, let Φ_ϵ be the function

$$\Phi_\epsilon(x) = |\ln(\epsilon)|^{1/5}\Omega(x_1)\Omega(\frac{x_2}{\epsilon^{\frac{1}{2}}})\cos(\frac{x_3}{\epsilon})\Omega(x_3)$$

and let

$$\vec{u}_{0,\epsilon} = \vec{\nabla} \wedge \begin{pmatrix} 0 \\ 0 \\ -\partial_2\Phi_\epsilon \end{pmatrix} = \begin{pmatrix} -\partial_2^2\Phi_\epsilon \\ \partial_1\partial_2\Phi_\epsilon \\ 0 \end{pmatrix} = \begin{pmatrix} \alpha_\epsilon \\ \beta_\epsilon \\ 0 \end{pmatrix}.$$

Then div $\vec{u}_{0,\epsilon} = 0$, $\vec{u}_{0,\epsilon}$ is large in $\dot{B}^{-1}_{\infty,\infty}$:

$$\|\vec{u}_{0,\epsilon}\|_{\dot{B}^{-1}_{\infty,\infty}} \approx |\ln(\epsilon)|^{1/5}$$

and, for ϵ small enough, the Cauchy problem for the Navier–Stokes equations with initial value $\vec{u}_{0,\epsilon}$ (and forcing term $\vec{f} = 0$) has a global mild solution $\vec{u}$ such that $\vec{u} \in L^2L^\infty$.

Proof. In light of Theorem 8.16 and its Corollary 8.1, we will work mainly in the setting of the frequency variable ξ. Recall that we have the following embeddings between Fourier-Herz spaces and the Besov spaces

$$\mathcal{F}\mathcal{B}^{-1}_{1,1} \subset \mathcal{F}\mathcal{B}^{-1}_{1,2} \subset \dot{B}^{-1}_{\infty,2} \subset BMO^{-1} \subset \dot{B}^{-1}_{\infty,\infty}.$$

If $\mathbb{A}$ is the Wiener space $\mathbb{A} = \mathcal{F}L^1$, we have, for the Littlewood–Paley decomposition of a function h,

$$\|h\|_{\mathcal{F}\mathcal{B}^{-1}_{1,1}} \approx \sum_{j\in\mathbb{Z}} 2^{-j}\|\Delta_j h\|_{\mathbb{A}} \approx \int |\hat{h}(\xi)|\frac{d\xi}{|\xi|}$$

while

$$\|h\|_{\dot{B}^{-1}_{\infty,\infty}} \approx \sum_{j\in\mathbb{Z}} 2^{-j}\|\Delta_j h\|_\infty.$$

The function Φ_ϵ has a precise frequency localization:

$$\hat{\Phi}_\epsilon(\xi) = |\ln(\epsilon)|^{1/5}\epsilon^{1/2}\omega(\xi_1)\omega(\epsilon^{\frac{1}{2}}\xi_2)(\omega(\xi_3 + \frac{5}{\epsilon}) + \omega(\xi_3 - \frac{5}{\epsilon}))$$

so that

$$\hat{\Phi}_\epsilon(\xi) \neq 0 \Rightarrow \xi \in [-1,1] \times [-\frac{1}{\epsilon^{\frac{1}{2}}}, \frac{1}{\epsilon^{\frac{1}{2}}}] \times ([-1 - \frac{5}{\epsilon}, 1 - \frac{5}{\epsilon}] \cup [-1 + \frac{5}{\epsilon}, 1 + \frac{5}{\epsilon}]),$$

and thus (for ϵ small enough)

$$\mid |\xi| - \frac{5}{\epsilon}\mid \leq 2 + \frac{1}{\epsilon^{\frac{1}{2}}} \leq \frac{1}{\epsilon}.$$

Thus, we have

$$\|\beta_\epsilon\|_{\mathcal{F}\mathcal{B}_{1,1}^{-1}} = \|\partial_1\partial_2\Phi_\epsilon\|_{\mathcal{F}\mathcal{B}_{1,1}^{-1}} \approx \epsilon\int |\xi_1\xi_2|\,|\hat{\Phi}_\epsilon(\xi)|\,d\xi = 2\epsilon^{\frac{1}{2}}|\ln\epsilon|^{1/5}\|\omega\|_1\|s\omega\|_1^2 \qquad (9.27)$$

and

$$\|\alpha_\epsilon\|_{\mathcal{F}\mathcal{B}_{1,1}^{-1}} = \|\partial_2^2\Phi_\epsilon\|_{\mathcal{F}\mathcal{B}_{1,1}^{-1}} \approx \epsilon\int |\xi_2^2|\,|\hat{\Phi}_\epsilon(\xi)|\,d\xi = 2|\ln\epsilon|^{1/5}\|\omega\|_1^2\|s^2\omega\|_1. \qquad (9.28)$$

Similarly, we have

$$\|\beta_\epsilon\|_{\dot{B}_{\infty,\infty}^{-1}} \approx \epsilon\|\beta_\epsilon\|_\infty \leq \frac{\epsilon}{(2\pi)^3}\int |\hat{\beta}_\epsilon|\,d\xi \leq C\epsilon^{\frac{1}{2}}|\ln\epsilon|^{1/5}$$

and, since $\hat{\alpha}_\epsilon$ is non-negative,

$$\|\alpha_\epsilon\|_{\dot{B}_{\infty,\infty}^{-1}} \approx \epsilon\|\alpha_\epsilon\|_\infty = \frac{\epsilon}{(2\pi)^3}\int |\hat{\alpha}_\epsilon|\,d\xi \approx |\ln\epsilon|^{1/5}.$$

Thus, we get that

$$\|\vec{u}_{0,\epsilon}\|_{\dot{B}_{\infty,\infty}^{-1}} \approx \|\vec{u}_{0,\epsilon}\|_{\mathcal{B}_{1,\infty}^{-1}} \approx |\ln\epsilon|^{1/5}. \qquad (9.29)$$

Another useful estimate will be

$$\|\vec{u}_{0,\epsilon}\|_{\dot{H}^{-1}} \approx \epsilon\|\vec{u}_{0,\epsilon}\|_2 \approx \epsilon^{\frac{1}{4}}|\ln\epsilon|^{1/5}. \qquad (9.30)$$

We shall now look for the mild solution $\vec{u}$ in $L^2\mathbb{A} = L^2\mathcal{F}L^1$. We define the bilinear operator B as

$$B(\vec{v},\vec{w}) = \int_0^t W_{\nu(t-s)} * \mathbb{P}\,\text{div}(\vec{v}\otimes\vec{w})\,ds$$

and we look for a solution of

$$\vec{u} = W_{\nu t} * \vec{u}_{0,\epsilon} - B(\vec{u},\vec{u}).$$

We know by Theorem 8.16 that

$$\|B(\vec{v},\vec{w})\|_{L^2\mathbb{A}} \leq C_0\|\vec{v}\|_{L^2\mathbb{A}}\|\vec{w}\|_{L^2\mathbb{A}}$$

but we cannot proceed directly with the Banach contraction principle as

$$\|W_{\nu t} * \vec{u}_{0,\epsilon}\|_{L^2\mathbb{A}} \approx \|\vec{u}_{0,\epsilon}\|_{\mathcal{F}\mathcal{B}_{1,2}^{-1}} \approx |\ln\epsilon|^{1/5}.$$

Let $\vec{U}_0 = W_{\nu t} * \vec{u}_{0,\epsilon}$, $\vec{U}_1 = B(\vec{U}_0, \vec{U}_0)$ and let $\vec{V} = \vec{u} - \vec{U}_0$. $\vec{V}$ must be solution to the equation

$$\vec{V} = \vec{U}_1 - B(\vec{U}_0, \vec{V}) - B(\vec{V}, \vec{U}_0) - B(\vec{V}, \vec{V}). \tag{9.31}$$

We begin by checking that $\|\vec{U}_1\|_{L^2 \mathbb{A}}$ is much smaller than $C_0 \|\vec{U}_0\|^2_{L^2 \mathbb{A}} \approx |\ln \epsilon|^{2/5}$. We have seen (in Theorem 8.16) that

$$\|\vec{U}_1\|_{L^2 \mathbb{A}} \leq C_1 \|\frac{1}{\sqrt{-\Delta}} \operatorname{div}(\vec{U}_0 \otimes \vec{U}_0)\|_{L^1 \mathbb{A}}$$

We have

$$\vec{U}_0 = \begin{pmatrix} \gamma_1 \\ \gamma_2 \\ 0 \end{pmatrix} = \begin{pmatrix} W_{\nu t} * \alpha_\epsilon \\ W_{\nu t} * \beta_\epsilon \\ 0 \end{pmatrix} \text{ and } \operatorname{div}(\vec{U}_0 \otimes \vec{U}_0) = \begin{pmatrix} \partial_1(\gamma_1^2) + \partial_2(\gamma_1\gamma_2) \\ \partial_1(\gamma_1\gamma_2) + \partial_2(\gamma_2^2) \\ 0 \end{pmatrix}.$$

We have

$$\|\frac{1}{\sqrt{-\Delta}} \partial_j(\gamma_1\gamma_2)\|_{L^1 \mathbb{A}} \leq \int_0^{+\infty} \|\gamma_1\|_{\mathbb{A}} \|\gamma_2\|_{\mathbb{A}} \, ds \leq \|\gamma_1\|_{L^2 \mathbb{A}} \|\gamma_2\|_{L^2 \mathbb{A}}$$

and thus

$$\|\frac{1}{\sqrt{-\Delta}} \partial_j(\gamma_1\gamma_2)\|_{L^1 \mathbb{A}} \leq C \|\alpha_\epsilon\|_{\mathcal{F}\mathcal{B}_{1,2}^{-1}} \|\beta_\epsilon\|_{\mathcal{F}\mathcal{B}_{1,2}^{-1}} \leq C' \epsilon^{\frac{1}{2}} |\ln \epsilon|^{2/5}.$$

Similarly, we have

$$\|\frac{1}{\sqrt{-\Delta}} \partial_2(\gamma_2^2)\|_{L^1 \mathbb{A}} \leq C \|\beta_\epsilon\|^2_{\mathcal{F}\mathcal{B}_{1,2}^{-1}} \leq C' \epsilon |\ln \epsilon|^{2/5}.$$

For the last term $\frac{1}{\sqrt{-\Delta}} \partial_1(\gamma_1^2)$, we notice that the spectral support of γ_1^2 is contained in

$$E_\epsilon = [-2,2] \times [-\frac{2}{\epsilon^{\frac{1}{2}}}, \frac{2}{\epsilon^{\frac{1}{2}}}] \times ([-2 - \frac{5}{\epsilon}, 2 - \frac{5}{\epsilon}] \cup [-2,2] \cup [-2 + \frac{5}{\epsilon}, 2 + \frac{5}{\epsilon}]).$$

Thus, we have (splitting the integral on ξ between $|\xi| > R$ and $|\xi| \leq R$ with $R << \epsilon^{-\frac{1}{2}}$),

$$\|\frac{1}{\sqrt{-\Delta}} \partial_1(\gamma_1^2)\|_{L^1 \mathbb{A}} = (2\pi)^3 \int_0^{+\infty} \int_{E_\epsilon} \frac{|\xi_1|}{|\xi|} |\hat{\gamma}_1 * \hat{\gamma}_1| \, d\xi$$

$$\leq (2\pi)^3 \int_0^{+\infty} \int_{\xi \in E_\epsilon, |\xi| < R} \|\hat{\gamma}_1 * \hat{\gamma}_1\|_\infty \, d\xi$$

$$+ (2\pi)^3 \int_0^{+\infty} \int_{\xi \in E_\epsilon, |\xi| > R} \frac{2}{R} |\hat{\gamma}_1 * \hat{\gamma}_1| \, d\xi$$

$$\leq (2\pi)^3 (|E_\epsilon \cap B(0, R)| \|\hat{\gamma}_1\|^2_{L^2 L^2} + \frac{2}{R} \|\hat{\gamma}_1\|^2_{L^2 L^1})$$

with $|E_\epsilon \cap B(0, R)| \leq 32R$, $\|\hat{\gamma}_1\|^2_{L^2 L^2} \leq C \|\alpha_\epsilon\|^2_{\dot{H}^{-1}} \leq C' \epsilon^{\frac{1}{2}} |\ln \epsilon|^{2/5}$ and $\|\hat{\gamma}_1\|^2_{L^2 L^1} \leq C \|\alpha_\epsilon\|^2_{\mathcal{F}\mathcal{B}_{1,2}^{-1}} \leq C' |\ln \epsilon|^{2/5}$. Taking $R = \epsilon^{-\frac{1}{4}}$, we find that

$$\|\frac{1}{\sqrt{-\Delta}} \partial_1(\gamma_1^2)\|_{L^1 \mathbb{A}} \leq C \epsilon^{\frac{1}{4}} |\ln \epsilon|^{2/5}.$$

Thus, we have

$$\|\vec{U}_1\|_{L^2 \mathbb{A}} \leq C_1 \epsilon^{\frac{1}{4}} |\ln \epsilon|^{2/5}. \tag{9.32}$$

We may now proceed to the construction of the solution $\vec{V}$ of Equation (9.31). As $\|\vec{U}_0\|_{L^2\mathbb{A}}$ is large, we must change the norm on $L^2\mathbb{A}$. Let $\lambda > 0$ and

$$\|U\|_\lambda = \|\mu_\lambda(t)U\|_{L^2\mathbb{A}}, \text{ where } \mu_\lambda(t) = e^{-\lambda \int_0^t \|\vec{U}_0(s,.)\|_\mathbb{A}^2 \, ds}.$$

We have obviously, for $A_\lambda = e^{\lambda \int_0^{+\infty} \|\vec{U}_0(s,.)\|_\mathbb{A}^2 \, ds}$,

$$\|U\|_\lambda \le \|U\|_{L^2\mathbb{A}} \le A_\lambda \|U\|_\lambda.$$

Thus,

$$\|B(\vec{V},\vec{W})\|_\lambda \le \|B(\vec{V},\vec{W})\|_{L^2\mathbb{A}} \le C_0 A_\lambda^2 \|\vec{V}\|_\lambda \|\vec{W}\|_\lambda$$

and

$$\|\vec{U}_1\|_\lambda \le C_1 \epsilon^{1/4} |\ln \epsilon|^{2/5}.$$

In order to estimate $B(\vec{U}_0,\vec{V})$ and $B(\vec{V},\vec{U}_0)$, we define

$$Z(V,W) = \int_0^t \int e^{-\nu(t-s)|\xi|^2} |\xi| \, |V(s,\eta)| \, |W(s,\xi-\eta)| \, d\eta \, ds.$$

We write $U_0(t,\xi) = |\mathcal{F}\vec{U}_0(t,\xi)|$ and $V(t,\xi) = |\mathcal{F}\vec{V}(t,\xi)|$, so that

$$|\mathcal{F}(B(\vec{U}_0,\vec{V}) + B(\vec{V},\vec{U}_0))| \le 2(2\pi)^3 Z(U_0,V)(t,\xi).$$

Thus,

$$\|B(\vec{U}_0,\vec{V}) + B(\vec{V},\vec{U}_0)\|_\lambda \le 2(2\pi)^3 \|\mu_\lambda(t)Z(U_0,V)(t,\xi)\|_{L^2L^1}.$$

We write

$$\mu_\lambda(t)Z(U_0,V)(t,\xi) = \int_0^t \int e^{-\nu(t-s)|\xi|^2} |\xi| \frac{\mu_\lambda(t)}{\mu_\lambda(s)} U_0(s,\eta)\mu_\lambda(s)V(s,\xi-\eta) \, d\eta \, ds.$$

As $\mu_\lambda(t) \le \mu_\lambda(s)$, we find

$$\|\mu_\lambda(t)Z(U_0,V)(t,\xi)\|_{L^1L^1(|\xi|\,d\xi)} \le Z(U_0,\mu_\lambda V) \le C_0\|U_0\|_{L^2L^1}\|\mu_\lambda V\|_{L^2L^1}.$$

As $e^{-\nu(t-s)|\xi|^2} \le 1$, we find

$$\|\mu_\lambda(t)Z(U_0,V)(t,\xi)\|_{L^1(\frac{d\xi}{|\xi|})} \le \left(\int_0^t \left(\frac{\mu_\lambda(t)}{\mu_\lambda(s)}\right)^2 \|U_0(s,.)\|_1^2 \, ds\right)^{1/2}\|\mu_\lambda V\|_{L^2L^1}$$

with

$$\int_0^t \left(\frac{\mu_\lambda(t)}{\mu_\lambda(s)}\right)^2 \|U_0(s,.)\|_1^2 \, ds = \int_0^t e^{-2\lambda \int_s^t \|U_0(\sigma,.)\|_1^2 \, d\sigma}\|U_0(s,.)\|_1^2 \, ds$$

$$= \left[\frac{e^{-2\lambda \int_s^t \|U_0(\sigma,.)\|_1^2 \, d\sigma}}{-2\lambda}\right]_{s=0}^{s=t}$$

$$\le \frac{1}{2\lambda}.$$

This gives

$$\|\mu_\lambda(t)Z(U_0,V)(t,\xi)\|_{L^\infty L^1(\frac{d\xi}{|\xi|})} \le \sqrt{\frac{1}{2\lambda}}\|\mu_\lambda V\|_{L^2L^1}$$

and finally

$$\|\mu_\lambda(t)Z(U_0,V)(t,\xi)\|_{L^2L^1} \le \sqrt{\frac{C_0\|U_0\|_{L^2L^1}}{\sqrt{2\lambda}}}\|\mu_\lambda V\|_{L^2L^1}.$$

For $\lambda = C_4\|U_0\|_{L^2L^1}^2$, we find

- $\|B(\vec{U}_0, \vec{V}) + B(\vec{V}, \vec{U}_0)\|_\lambda \leq \frac{1}{2}\|\vec{V}\|_\lambda$

- $\|\vec{U}_1\|_\lambda \leq C_1\epsilon^{1/4}|\ln \epsilon|^{2/5}$

- $\|B(\vec{V}, \vec{W})\|_\lambda \leq \|B(\vec{V}, \vec{W})\|_{L^2\mathbb{A}} \leq C_0 A_\lambda^2 \|\vec{V}\|_\lambda \|\vec{W}\|_\lambda$

Thus, the Picard iterative scheme will work in $(L^2\mathbb{A}, \|\ \|_\lambda)$, provided that

$$C_1\epsilon^{1/4}|\ln \epsilon|^{2/5} \leq \frac{1}{16 C_0 A_\lambda^2}.$$

We have

$$\frac{1}{A_\lambda^2} = e^{-2\lambda\|U_0\|^2_{L^2 L^1}} = e^{-2C_4\|U_0\|^4_{L^2 L^1}}$$

with

$$\|U_0\|_{L^1 L^2} \approx \|\vec{u}_{0,\epsilon}\|_{\mathcal{F}\mathcal{B}^{-1}_{1,2}} \approx |\ln \epsilon|^{1/5}.$$

As

$$\epsilon^{1/4}|\ln \epsilon|^{2/5} = o(e^{-C_5|\ln \epsilon|^{4/5}})$$

as $\epsilon \to 0$, we have proven the theorem. $\qquad\square$

As proven by Chemin and Gallagher, we have found a non-linear smallness criterion for global existence: there exists a constant C_0 such that, if $\vec{u}_0 \in \mathcal{F}\mathcal{B}^{-1}_{1,2}$ and

$$\|B(W_{\nu t} * \vec{u}_0, W_{\nu t} * \vec{u}_0)\|_{L^2\mathbb{A}} \leq \frac{1}{C_0}e^{-C_0\|\vec{u}_0\|^4_{\mathcal{F}\mathcal{B}^{-1}_{1,2}}},$$

then we have global existence of a mild solution.

Chemin and Gallagher's example relies on the frequency anisotropy of their initial data. The role of this anisotropy has been commented by Chemin, Gallagher and Mullaert [110].

9.8 Stability of Global Solutions

When $\vec{u}_0$ is an (regular enough) initial value that generates a global mild solution for the Navier–Stokes problem (with no forcing term), then small (in BMO^{-1} norm) perturbations of $\vec{u}_0$ still lead to initial values of global mild solutions. This stability of global soutions has been studied by many authors including Kawanago [257], Gallagher, Iftimie and Planchon [198] or Auscher, Dubois and Tchamitchian [10]. In particular, we have:

Theorem 9.11.
Let $\mathcal{C}$ be the space of (smooth) vector fields $\vec{u}$ on $(0, +\infty) \times \mathbb{R}^3$ such that

- *for all $0 < T_1 < T_2 < +\infty$, $\sup_{T_1 < t < T_2} \|\vec{u}(t, .)\|_\infty < +\infty$*

- *$\operatorname{div} \vec{u} = 0$*

- *$\partial_t \vec{u} = \nu \Delta \vec{u} - \mathbb{P}\operatorname{div}(\vec{u} \otimes \vec{u})$*

- *$\vec{u}$ is *-weakly continuous from $[0, +\infty)$ to BMO^{-1}*

Let $\vec{u} \in \mathcal{C}$ such that moreover

- $\vec{u}_0 = \vec{u}(0, .)$ *belongs to the closure of $L^2 \cap BMO^{-1}$ in BMO^{-1}*

- $\lim_{t \to 0^+} \|\vec{u}(t, .) - \vec{u}_0\|_{BMO^{-1}} = 0$.

Then $\vec{u}$ satisfies the following properties:

- $\vec{u}$ *is strongly continuous from $[0, +\infty)$ to BMO^{-1}*

- $\lim_{t \to +\infty} \|\vec{u}(t, .)\|_{BMO^{-1}} = 0$

- *there exists a positive ϵ_0 (which depends on $\vec{u}$) such that, for all divergence-free vector field $\vec{v}_0$ with $\|\vec{v}_0 - \vec{u}_0\|_{BMO^{-1}} < \epsilon_0$, there exists $\vec{v} \in \mathcal{C}$ such that $\vec{v}(0, .) = \vec{v}_0$.*

Proof. **Continuity in BMO^{-1}-norm:**
The continuity of $t \mapsto \vec{u}$ is quite obvious: the continuity at initial point $t = 0$ is given by the assumption $\lim_{t \to 0^+} \|\vec{u}(t, .) - \vec{u}_0\|_{BMO^{-1}} = 0$; for $t > 0$, we have that $\vec{u}$ is bounded with all its derivatives on every compact subset $[T_0, T_1]$ of $(0, +\infty)$ (with values in L^∞), so that $\partial_t \vec{u}$ is bounded on $[T_0, T_1]$ with values in BMO^{-1}, and hence that $\vec{u}$ is locally Lipschitzian from $(0, +\infty)$ to BMO^{-1}.

Decay at $t = +\infty$:
We now prove that $\lim_{t \to +\infty} \|\vec{u}(t, .)\|_{BMO^{-1}} = 0$. Recall that we defined the space E_T, for $0 < T \leq +\infty$, as the space of measurable functions h on $(0, T) \times \mathbb{R}^3$ such that

$$\sup_{0 < t < T} \sqrt{t} \, \|h(t, .)\|_\infty < +\infty$$

and

$$\sup_{x \in \mathbb{R}^3, \, 0 < t < T} \sqrt{\frac{1}{t^{3/2}} \iint_{(0,t) \times B(x, \sqrt{t})} |h(s, y)|^2 \, ds \, dy} < +\infty.$$

By Theorem 9.1, we know that the bilinear operator

$$B(\vec{F}, \vec{G}) = \int_0^t W_{\nu(t-s)} * \mathbb{P} \operatorname{div}(\vec{F} \otimes \vec{G}) \, ds$$

is bounded on the space E_T for every $T \in (0, +\infty]$:

$$\|B(\vec{F}, \vec{G})\|_{E_T} \leq C_0 \|\vec{F}\|_{E_T} \|\vec{G}\|_{E_T} \tag{9.33}$$

for a constant C_0 which depends on ν but not on T.
  Moreover, we know that

$$\|W_{\nu t} * \vec{F}\|_{E_T} \leq C_1 \|\vec{F}\|_{BMO^{-1}} \tag{9.34}$$

(for a constant C_1 which depends on ν but not on T).
  As $\vec{u}_0$ is the limit of $\vec{u}(t, .)$ in BMO^{-1} and as $\vec{u}(t, .)$ is in L^∞ for $t > 0$, we have

$$\lim_{T \to 0^+} \|W_{\nu t} * \vec{u}_0\|_{E_T} = 0$$

so that, for some $T_0 > 0$, we have

$$\|W_{\nu t} * \vec{u}_0\|_{E_{T_0}} < \frac{1}{8C_0}.$$

By strong continuity of $t \mapsto \vec{u}$, we find that, for some $T_1 > 0$, we have

$$\sup_{0 \leq \tau \leq T_1} \|W_{\nu t} * \vec{u}(\tau, .)\|_{E_{T_0}} \leq \frac{1}{8C_0}.$$

Thus, we may construct, by Picard's iterative scheme, a solution $\vec{v}_\tau(t, x)$ of the Navier–Stokes equations on $(0, T_0) \times \mathbb{R}^3$ with initial value $\vec{v}_\tau(0, .) = \vec{u}(\tau, .)$: $\vec{v}_\tau = \lim_{n \to +\infty} \vec{v}_{\tau,n}$ with

$$\begin{cases} \vec{v}_{\tau,0} = & W_{\nu t} * \vec{u}(\tau, .) \\ \vec{v}_{\tau,n+1} = & \vec{v}_{\tau,0} - B(\vec{v}_{\tau,n}, \vec{v}_{\tau,n}) \end{cases}$$

and

$$\|\vec{v}_{\tau,n+1} - \vec{v}_{\tau,n}\|_{E_{T_0}} \leq \frac{1}{2^n} \frac{1}{4C_0}.$$

Moreover, for $\tau > 0$, we have $\vec{v}_{\tau,0} \in L_t^\infty L^\infty$ and

$$\sup_{0 < t < T_0} \|\vec{v}_{\tau,n+1}(t, .) - \vec{v}_{\tau,n}(t, .)\|_\infty$$

$$\leq C \sup_{0 < t < T_0} \int_0^t \frac{1}{\sqrt{\nu(t-s)}} \frac{1}{\sqrt{s}} \sqrt{s} \|\vec{v}_{\tau,n}(s, .) - \vec{v}_{\tau,n-1}(s, .)\|_\infty \times$$

$$\times (\|\vec{v}_{\tau,n}(s, .)\|_\infty + \|\vec{v}_{\tau,n-1}(s, .)\|_\infty) \, ds$$

$$\leq C' \frac{1}{\sqrt{\nu}} \frac{1}{2^n C_0} \sup_{0 < t < T_0} (\|\vec{v}_{\tau,n}(t, .)\|_\infty + \|\vec{v}_{\tau,n-1}(t, .)\|_\infty)$$

so that

$$\sup_{0 < t < T_0} \|\vec{v}_{\tau,n+1}(t, .) - \vec{v}_{\tau,n}(t, .)\|_\infty \leq C \frac{1}{2^n} \|\vec{u}(\tau, .)\|_\infty.$$

(where C depends on ν and C_0). Thus, $\vec{v}_\tau \in L^\infty((0, T_0), L^\infty)$. By uniqueness of mild solutions in $L^\infty L^\infty$, we have

$$\vec{v}_\tau(t, s) = \vec{u}(\tau + t, x).$$

Letting τ go to 0, we find that

$$\|\vec{u}\|_{E_{T_0}} \leq \liminf_{\tau \to 0} \|\vec{v}_\tau\|_{E_{T_0}} \leq \frac{1}{4C_0}.$$

As $\|\vec{v}_0\|_{E_{T_0}} \leq \frac{1}{4C_0}$, we find that $\vec{u} = \vec{v}_0$ on $(0, T_0)$, by uniqueness of small mild solutions[2].

We now use the fact that $\vec{u}_0$ belongs to the closure of $L^2 \cap BMO^{-1}$ in BMO^{-1}: for $0 < \delta \leq \frac{1}{32} \frac{1}{C_0 C_1}$ (where $\delta > 0$ may be taken as small as we like), we write $\vec{u}_0 = \vec{\alpha}_0 + \vec{\beta}_0$ with $\vec{\beta}_0 \in L^2 \cap BMO^{-1}$ and $\|\vec{\alpha}_0\|_{BMO^{-1}} \leq \delta$. (We may assume that div $\vec{\alpha}_0 = 0$, as the Leray projection operator $\mathbb{P}$ is bounded on both L^2 and BMO^{-1}). As $\|W_{\nu t} * \vec{\alpha}_0\|_{E_{+\infty}} \leq C_1 \delta \leq \frac{1}{32 C_0}$, we may construct a global mild solution $\vec{\alpha}$ of the Navier–Stokes equations on $(0, +\infty) \times \mathbb{R}^3$ with initial value $\vec{\alpha}(0, .) = \vec{\alpha}_0$. This solution satisfies $\|\vec{\alpha}\|_{E_{+\infty}} \leq 2C_1 \delta$, so that, by Lemma 9.1, $\sup_{0 < t} \|\vec{\alpha}(t, .)\|_{BMO^{-1}} \leq C_2 \delta$. We shall assume that δ is small enough to ensure that $C_2 \delta \leq \frac{1}{8 C_1 C_0}$.

Let $\vec{\beta} = \vec{u} - \vec{\alpha}$. It is a mild solution of

$$\vec{\beta} = W_{\nu t} * \vec{\beta}_0 - B(\vec{\alpha}, \vec{\beta}) - B(\vec{\beta}, \vec{\alpha}) - B(\vec{\beta}, \vec{\beta}).$$

---

[2] This uniqueness argument goes back to an old paper of Brezis [66]; see Miura for the use of this argument in the context of bmo^{-1} [364].

It may be constructed on a small interval $(0, T_1)$ as the limit of the Picard iterative scheme $\vec{\beta} = \lim_{n \to +\infty} \vec{\gamma}_n$ with

$$\begin{cases} \vec{\gamma}_0 = & W_{\nu t} * \vec{\beta}_0 \\ \vec{\gamma}_{n+1} = & \vec{\gamma}_0 - B(\vec{\alpha}, \vec{\gamma}_n) \quad -B(\vec{\gamma}_n, \vec{\alpha}) - B(\vec{\gamma}_n, \vec{\gamma}_n) \end{cases}$$

If we take δ small enough to get $24C_1^2 C_0 \delta < 1$ and T_1 small enough to have $\|\vec{u}\|_{E_{T_1}} < C_1 \delta$, we find that $\|\vec{\gamma}_0\|_{E_{T_1}} \le 3C_1\delta$, and $\|\vec{\gamma}_n\|_{E_{T_1}} \le 4C_1\delta$.

Moreover, we have

$$\sup_{0 < t < T_1} \|\vec{\gamma}_{n+1}(t, .) - \vec{\gamma}_n(t, .)\|_2$$

$$\le C \sup_{0 < t < T_1} \int_0^t \frac{1}{\sqrt{\nu(t-s)}} \frac{1}{\sqrt{s}} \|\vec{\gamma}_n(s, .) - \vec{\gamma}_{n-1}(s, .)\|_2 \times$$

$$\times \sqrt{s}(\|\vec{\gamma}_n(s, .)\|_\infty + \vec{\gamma}_{n-1}(s, .)\|_\infty + \|\vec{\alpha}(s, .)\|_\infty)\, ds$$

$$\le C_3 C_1 \delta \sup_{0 < t < T_1} \|\vec{\gamma}_n(t, .) - \vec{\gamma}_{t-1}(s, .)\|_2$$

If $C_3 C_1 \delta < 1$, we find convergence of $\vec{\gamma}_n$ in $L^\infty L^2$ as well.

This proves that $\vec{\beta} \in L^\infty((0, T_1), L^2)$ for T_1 small enough. As $\vec{\alpha}$ and $\vec{\beta}$ belong to $L^\infty((T_1, T_2), L^\infty)$ for every $T_2 > T_1$, we find (reiterating the contraction argument on small enough intervals with fixed length) that $\vec{\beta}$ belongs to $L^\infty((T_1, T_2), L^2)$. Moreover, we have $\text{div}(\vec{\alpha} \otimes \vec{\beta} + \vec{\beta} \otimes \vec{\alpha} + \vec{\beta} \otimes \vec{\beta}) \in L^\infty((0, T_1), \dot{H}^{-1})$ so that $\vec{\beta} - W_{\nu(t-T_1)} * \vec{\beta}(T_1, .) \in L^2((T_1, T_2), H^1)$. Finally, we conclude that $\vec{\beta}$ belongs to $L^\infty((T_1, T_2), L^2) \cap L^2((T_1, T_2), H^1)$ while $\partial_t \vec{\beta} \in L^2((T_1, T_2), H^{-1})$. We may write

$$\partial_t \|\vec{\beta}(t, .)\|_2^2 = 2 \int \vec{\beta}.\partial_t \vec{\beta}\, dx.$$

As $\text{div}\, \vec{u} = 0$, we have

$$\int \vec{\beta}.((\vec{\alpha} + \vec{\beta}).\vec{\nabla}\vec{\beta})\, dx = 0$$

so that

$$\partial_t \|\vec{\beta}(t, .)\|_2^2 = -2\nu\|\vec{\nabla} \otimes \vec{\beta}\|_2^2 + 2 \int \vec{\alpha}.(\vec{\beta}.\vec{\nabla}\vec{\beta})\, dx \le -\nu\|\vec{\nabla} \otimes \vec{\beta}\|_2^2 + \frac{1}{\nu}\|\vec{\alpha}\|_\infty^2 \|\vec{\beta}\|_2^2.$$

This gives

$$\|\vec{\beta}(t, .)\|_2^2 \le \|\vec{\beta}(T_1, .)\|_2^2\, e^{\frac{1}{\nu} \int_{T_1}^t \frac{C_1^2\delta^2}{s}\, ds} = \|\vec{\beta}(T_1, .)\|_2^2 \left(\frac{t}{T_1}\right)^{\frac{C_1^2\delta^2}{\nu}}$$

and

$$\int_{T_1}^t \|\vec{\nabla} \otimes \vec{\beta}\|_2^2\, ds \le \frac{1}{\nu}\|\vec{\beta}(T_1, .)\|_2^2(1 + \frac{1}{\nu}\frac{C_1^2\delta^2}{T_1^{\frac{C_1^2\delta^2}{\nu}}} \int_{T_1}^t s^{\frac{C_1^2\delta^2}{\nu}} \frac{ds}{s}).$$

By interpolation, we find that

$$\|\vec{\beta}\|_{L^4((T_1, T_2), L^3)} \le C_4 \|\vec{\beta}(T_1, .)\|_2 \left(\frac{T_2}{T_1}\right)^{\frac{C_1^2\delta^2}{2\nu}}.$$

If δ is small enough to have $\frac{C_1^2\delta^2}{2\nu} < \frac{1}{4}$, this gives

$$\liminf_{t \to +\infty} \|\vec{\beta}(t, .)\|_3 = 0$$

and thus

$$\liminf_{t \to +\infty} \|\vec{\beta}(t,.)\|_{BMO^{-1}} = 0$$

Thus, we may find a time T_3 such that $\|\vec{\beta}(T_3,.)\|_{BMO^{-1}} < C_2\delta$. This gives that $\|\vec{u}(T_3,.)\|_{BMO^{-1}} < 2C_2\delta$. Hence, we have $\|W_{\nu(t-T_3)} * \vec{u}(T_3,.)\|_{E_{+\infty}} \leq 2C_1C_2\delta \leq \frac{1}{4C_0}$; it means that we may control $\vec{u}$ on $(T_3, +\infty)$ by

$$\|\vec{u}(T_3 + t,.)\|_{E_{+\infty}} \leq 4C_1C_2\delta$$

and thus

$$\sup_{t > T_3} \|\vec{u}(t,.)\|_{BMO^{-1}} \leq 2C_2\delta + C_4(4C_1C_2\delta)^2.$$

As δ may be taken arbitrarily small, we find that $\lim_{t \to +\infty} \|\vec{u}(t,.)\|_{BMO^{-1}} = 0$.

Stability:
We know that we have

$$\|B(\vec{F}, \vec{G})\|_{E_T} \leq C_0 \|\vec{F}\|_{E_T} \|\vec{G}\|_{E_T} \tag{9.35}$$

and

$$\|B(\vec{F}, \vec{G})\|_{L^\infty BMO^{-1}} \leq C_0 \|\vec{F}\|_{E_T} \|\vec{G}\|_{E_T} \tag{9.36}$$

for a constant C_0 which depends on ν but not on T. Moreover, for a constant C_1, we have

$$\|\vec{W}_{\nu t} * \vec{w}_0\|_{E_T} \leq C_1 \|\vec{w}_0\|_{BMO^{-1}}$$

Similarly, we have

$$\|\vec{W}_{\nu t} * \vec{w}_0\|_{E_T} \leq C_2\sqrt{T}\|\vec{w}_0\|_\infty.$$

One may conclude that, if $\vec{\alpha}_0$, $\vec{\beta}_0$ and $\vec{\gamma}_0$ satisfy $\text{div}(\vec{\alpha}_0 + \vec{\beta}_0) = \text{div}\,\vec{\gamma}_0 = 0$, $\|\vec{\alpha}_0\|_{BMO^{-1}} \leq \frac{1}{16C_0C_1}$, $\|\vec{\beta}_0\|_\infty \leq M$ and $\|\vec{\gamma}_0\|_{BMO^{-1}} \leq \frac{1}{16C_0C_1}$, then

- for $T = \frac{1}{(16C_2C_0M)^2}$, $\|W_{\nu t} * (\vec{\alpha}_0 + \vec{\beta}_0)\|_{E_T} \leq \frac{1}{8C_0}$, so that the Navier–Stokes equations with initial value $\vec{\alpha}_0 + \vec{\beta}_0$ have a mild solution $\vec{w}$ on $(0, T) \times \mathbb{R}^3$, with $\|\vec{w}\|_{E_T} \leq \frac{1}{4C_0}$

- as $\|W_{\nu t} * \vec{\gamma}_0\|_{E_T} \leq \frac{1}{16C_0}$, the Navier–Stokes equations with initial value $\vec{\alpha}_0 + \vec{\beta}_0 + \vec{\gamma}_0$ then have a mild solution $\vec{w} + \vec{\gamma}$ on $(0, T) \times \mathbb{R}^3$, with $\|\vec{\gamma}\|_{E_T} \leq 2C_1\|\vec{\gamma}_0\|_{BMO^{-1}}$

- moreover, we have $\|\vec{\gamma}(T,.)\|_{BMO^{-1}} \leq (1 + \frac{5}{4}C_1)\|\vec{\gamma}_0\|_{BMO^{-1}}$.

Now, recall that $\vec{u}$ is strongly continuous from $[0, +\infty)$ to the closure of $L^\infty \cap BMO^{-1}$ in BMO^{-1} and that $\lim_{t \to +\infty} \|\vec{u}(t,.)\|_{BMO^{-1}} = 0$. Thus, we can find a time $T_0 > 0$ such that $\|\vec{u}(T_0,.)\|_{BMO^{-1}} \leq \frac{1}{8C_0C_1}$. By compactness of $\vec{u}([0, T_0])$ in BMO^{-1}, we can find a $M < +\infty$ such that, for every $t \in [0, T_0]$, we may decompose $\vec{u}(t,.)$ into $\vec{\alpha}_t + \vec{\beta}_t$ with $\|\vec{\beta}_t\|_\infty \leq M$ and $\|\vec{\alpha}_t\|_{BMO^{-1}} < \frac{1}{16C_0C_1}$. Let $N \in \mathbb{N}$ such that $\frac{1}{N}T_0 < \frac{1}{(16C_2C_0M)^2}$. If

$$\|\vec{v}_0 - \vec{u}_0\|_{BMO^{-1}} < (1 + \frac{5}{4}C_1)^{-N} \frac{1}{16C_0C_1}$$

then we may construct the mild solution $\vec{v}$ inductively on $[\frac{k}{N}T_0, \frac{k+1}{N}T_0]$ for $0 \leq k < N$: indeed, we will have $\vec{u}([\frac{k}{N}T_0,.) = \vec{\alpha}_{\frac{k}{N}T_0} + \vec{\beta}_{\frac{k}{N}T_0}$ with $\|\vec{\beta}_{\frac{k}{N}T_0}\|_\infty \leq M$ and $\|\vec{\alpha}_{\frac{k}{N}T_0}\|_{BMO^{-1}} < \frac{1}{16C_0C_1}$ and

$$\|\vec{u}(\tfrac{k}{N}T_0,.) - \vec{v}(\tfrac{k}{N}T_0,.)\|_{BMO^{-1}} \leq (1 + \frac{5}{4}C_1)^k \|\vec{u}_0 - \vec{v}_0\|_{BMO^{-1}}$$

$$\leq (1 + \frac{5}{4}C_1)^{k-N} \frac{1}{16C_0C_1}.$$

Thus, we may construct $\vec{v}$ on $[0, T_0]$ and we have

$$\|\vec{v}(T_0, .)\|_{BMO^{-1}} \leq \|\vec{u}(T_0, .)\|_{BMO^{-1}} + \|\vec{v}(T_0, .) - \vec{u}(T_0, .)\|_{BMO^{-1}} \leq \frac{3}{16C_0C_1}.$$

Thus, $\vec{v}(T_0, .)$ is small enough to grant existence of a mild solution $\vec{v}$ on $[T_0, +\infty)$ as well. $\square$

9.9 Analyticity

Let $\vec{u}$ be a mild solution of the Navier–Stokes problem generated from an initial value $\vec{u}_0$ in the closure of L^∞ in BMO^{-1} (or bmo^{-1}) (in absence of a force $\vec{f}$). We have seen that such a solution exists on a small interval $(0, T)$ and is locally bounded; more precisely,

$$\sup_{0 < t < T} \sqrt{t} \|\vec{u}(t, .)\|_\infty < +\infty.$$

This solution may be prolongated as long as it does not blow up, as existence time for $t > t_0$ is bounded by below by $\frac{1}{C\|\vec{u}(t_0,.)\|_\infty^2}$. If T^* is the maximal existence time of the mild solution $\vec{u}$, then $\vec{u}$ is actually analytical in the time and space variables on $(0, T^*) \times \mathbb{R}^3$. Analyticity was first proven by Kahane [249] and Masuda [351]. A simple proof (in case of solutions in Sobolev spaces) was given by Foias and Temam [181] (see Section 8.7 for a short presentation).

Spatial analyticity in the context of Lebesgue spaces has been studied by Grujić and Kukavica [218] and by Lemarié-Rieusset [312, 314]. Spatial analyticity has been more recently studied in the context of BMO^{-1} (Germain, Pavlović and Staffilani [206], Miura and Sawada [365], Guberović [219]), Besov spaces (Bae, Biswas and Tadmor [13]) or modulation spaces (Guo, Wang and Zhao [224]).

We follow in this section the proof of Cannon and Knightly [80] for time and space analyticity:

Analyticity

Theorem 9.12.
Let $\vec{u}$ be a mild solution of the Navier–Stokes equations on $(T_0, T_1) \times \mathbb{R}^3$:

$$\partial_t \vec{u} = \nu \Delta \vec{u} - \mathbb{P} \operatorname{div}(\vec{u} \otimes \vec{u}) \qquad (9.37)$$

that is bounded on $(T_0, T_1) \times \mathbb{R}^3$:

$$\vec{u} \in L^\infty((T_0, T_1) \times \mathbb{R}^3).$$

Then $\vec{u}$ is analytical in the time and space variables.

Proof. If C_0 is a constant such that, for bounded vector fields $\vec{v}$ and $\vec{w}$ and for $T > 0$, we have

$$\|B(\vec{v}, \vec{w})(t, .)\|_\infty \leq C_0 \sqrt{T} \sup_{0 < t < T} \|\vec{v}(t, .)\|_\infty \sup_{0 < t < T} \|\vec{w}(t, , .)\|_\infty$$

where

$$B(\vec{v}, \vec{w}) = \int_0^t W_{\nu(t-s)} * \mathbb{P} \operatorname{div}(\vec{v} \otimes \vec{w}) \, ds,$$

we know by the proof of Theorem 5.1 (Oseen's method) that for $t_0 \in (T_0, T_1)$ and for $t \in (t_0, t_0 + \frac{1}{(4C_0 \|\vec{u}(t_0,.)\|_\infty)^2})$, $\vec{u}(t,.)$ is given by

$$\vec{u}(t,.) = \sum_{k=0}^{+\infty} \vec{V}_k(t,.) \tag{9.38}$$

with

$$\vec{V}_0 = W_{\nu(t-t_0)} * \vec{u}(t_0,.)$$

and

$$\vec{V}_{k+1} = (-1)^{k+1} \int_{t_0}^t W_{\nu(t-s)} * \mathbb{P} \operatorname{div}\Big(\sum_{j=0}^k \vec{V}_j \otimes \vec{V}_{k-j}\Big) ds.$$

The idea of the proof is then to show that $\vec{V}_k$ is analytical in time and space variables and is defined as a holomorphic function on a neighborhood Ω of $(t_0, t_1) \times \mathbb{R}^3$ in $\mathbb{C} \times \mathbb{C}^3$ (with $t_1 = t_0 + \frac{1}{(4C_1 \|\vec{u}(t_0,.)\|_\infty)^2}$ for some positive constant C_1) and that the expansion (9.38) converges uniformly on every compact subset of Ω, so that $\vec{u}$ is still holomorphic on Ω.

Let us recall the results of Theorem 4.6. The heat kernel $W_{\nu t} = \frac{1}{(\nu t)^{3/2}} W(\frac{x}{\sqrt{\nu t}})$ is given by

$$W(x) = \frac{1}{(4\pi)^{3/2}} e^{-\frac{|x|^2}{4}}$$

and the Oseen tensor which is the matrix convolution kernel of the operator $W_{\nu t} * \mathbb{P}$ is the tensor $(O_{j,k}(\nu t, x))_{1 \le j,k \le 3}$ given by

$$O_{j,k}(\nu t, x) = \delta_{j,k} W_{\nu t} + G * \partial_j \partial_k W_{\nu t} = \frac{1}{(\nu t)^{3/2}} O_{j,k}(\frac{x}{\sqrt{\nu t}})$$

where the functions $O_{j,k}$ are determined through Oseen's formula:

$$O_{j,k}(x) = \delta_{j,k} W(x) + 2\partial_j \partial_k \left(\frac{1}{(4\pi)^{3/2}|x|} \int_0^{|x|} e^{-\frac{s^2}{4}} ds \right), \tag{9.39}$$

which may be rewritten as

$$O_{j,k}(x) = \delta_{j,k} W(x) + 2\partial_j \partial_k \left(\int_0^1 W(\theta x) d\theta \right) \tag{9.40}$$

or as

$$O_{j,k}(x) = \partial_j \partial_k \left(\frac{1}{4\pi|x|} \right) + \delta_{j,k} W(x)$$

$$- 2\partial_j \partial_k \left(\frac{1}{(4\pi)^{3/2}|x|} \int_{|x|}^\infty e^{-\frac{s^2}{4}} ds \right). \tag{9.41}$$

Let us now fix $\gamma > 0$ and let

$$\Omega_\gamma = \{(\tau, z) \in \mathbb{C} \times \mathbb{C}^3 \,/\, 0 < \Re(\tau), \quad |\Im(\tau)| < \gamma \Re(\tau) \text{ and } |\Im(z)| < \gamma(\nu|\tau|)^{1/2}\}.$$

On Ω_γ, the function $\sqrt{\tau}$ is well defined and holomorphic. Thus, the functions $W_{\nu t}(x)$ and $O_{j,k}(\nu t, x)$ have holomorphic extensions $W_{\nu \tau}(z)$ and $O_{j,k}(\nu \tau, z)$ to Ω_γ. We now define

$$\vec{V}_0(\tau, z) = \int W_{\nu(\tau - t_0)}(z - y)\vec{u}(t_0, y)\, dy$$

and

$$\vec{V}_{k+1}(\tau, z) = (-1)^{k+1} \int_{t_0}^{\tau} \int \mathbb{O}(\nu(\tau - t_0 - \sigma), z - y) \operatorname{div}(\sum_{j=0}^{k} \vec{V}_j(\sigma, y) \otimes \vec{V}_{k-j}(\sigma, y)) \, d\sigma \, dy.$$

On $t_0 + \Omega_\gamma$, we have, writing $\tau - t_0 = t + i\sigma$ and $z = \alpha + i\beta$ (real and imaginary parts),

$$|W_{\nu(\tau-t_0)}(z-y)| = \frac{1}{(4\pi\nu)^{3/2}} \frac{1}{|t+i\sigma|^{3/2}} e^{-\frac{t|\alpha-y|^2}{4\nu|t+i\sigma|^2}} e^{\frac{t|\beta|^2}{4\nu|t+i\sigma|^2}} e^{-\frac{2\sigma\beta.(\alpha-y)}{4\nu|t+i\sigma|^2}}$$

As $t \le |t+i\sigma| \le t\sqrt{1+\gamma^2}$, we find

$$|W_{\nu(\tau-t_0)}(z-y)| \le \frac{1}{(4\pi\nu t)^{3/2}} e^{-\frac{1}{4(1+\gamma^2)}\frac{|\alpha-y|^2}{\nu t}} e^{\frac{\gamma^2}{4}} e^{\frac{\gamma}{2}\frac{|\alpha-y|}{\sqrt{\nu t}}}.$$

Thus, we may conclude that $\vec{V}_0$ is holomorphic on $t_0 + \Omega_\gamma$ and that

$$\|\vec{V}_0\|_{L^\infty(t_0+\Omega_\gamma)} \le C_\gamma \|\vec{u}(t_0, .)\|_\infty.$$

Assuming now that $\vec{V}_j$ is holomorphic and (locally) bounded on $t_0 + \Omega_\gamma$ for $0 \le j \le k$, we may estimate $\vec{V}_{k+1}$. Indeed, we have, for $(\tau, z) \in \Omega_\gamma$,

$$|\vec{\nabla} \otimes \mathbb{O}(\nu\tau, z)| = \frac{1}{(\nu|\tau|)^2} |\vec{\nabla} \otimes \mathbb{O}|(\frac{z}{\sqrt{\nu\tau}}).$$

Let $Z = \frac{z}{\sqrt{\nu\tau}}$. We have, for $Z^2 = z_1^2 + z_2^2 + z_3^2$,

$$\Re(Z^2) \ge -\gamma^2 + \Re(\frac{(\Re(z)^2}{\nu\tau}) = -\gamma^2 + (\Re(z))^2 \frac{\Re(\tau)}{\nu|\tau|^2} \ge -\gamma^2 + \frac{1}{1+\gamma^2}\frac{(\Re(z))^2}{\nu\Re(\tau)}$$

and

$$|Z| \le \gamma + \frac{|\Re(z)|}{\sqrt{\nu\Re(\tau)}}.$$

From

$$\partial_l O_{j,k}(Z) = \delta_{j,k}\partial_l W(Z) + 2\partial_l\partial_j\partial_k \left(\int_0^1 W(\theta Z) \, d\theta \right) \tag{9.42}$$

we find that

$$|\partial_l O_{j,k}(Z)| \le C(|Z|e^{-\frac{\Re(Z^2)}{4}} + \int_0^1 (1+\theta^3|Z|^3)e^{-\theta^2\frac{\Re(Z^2)}{4}})$$

and thus

$$|\partial_l O_{j,k}(Z)| \le C_\gamma.$$

Moreover, if $|Z| > \gamma(1+\sqrt{1+\gamma^2})$, we have $\Re(Z^2) > 0$ and $(Z^2)^{1/2}$ is a holomorphic function of Z. We have

$$\partial_l O_{j,k}(Z) = \partial_l\partial_j\partial_k \left(\frac{1}{4\pi(Z^2)^{1/2}} \right) + \delta_{j,k}\partial_l W(Z)$$
$$- 2\partial_l\partial_j\partial_k \left(\frac{1}{(4\pi)^{3/2}} \int_1^\infty e^{-\theta^2\frac{Z^2}{4}} \, d\theta \right). \tag{9.43}$$

Thus, we get that

$$|Z|^4|\partial_l O_{j,k}(Z)| \le C_\gamma.$$

Writing again, on $t_0 + \Omega_\gamma$, $\tau - t_0 = t + i\sigma$ and $z = \alpha + i\beta$ (real and imaginary parts), we find that, for two vector fields $\vec{V}(\tau, z)$ and $\vec{W}(\tau, z)$ that are holomorphic on $t_0 + \Omega_\gamma$ and bounded on each $t_0 + \Omega_{\gamma, M} = \{(\tau, z) \in t_0 + \Omega_\gamma \ / \ |\Re(\tau) - t_0| < M\}$, we have

$$\left| \int_{t_0}^{\tau} \int (\vec{\nabla} \otimes \mathbb{O}(\nu(\tau - t_0 - \sigma), z - y).(\vec{V}(\sigma, y) \otimes \vec{W}(\sigma, y)) \, d\sigma \, dy \right|$$

$$\leq \int_{t_0}^{\tau} \int |\vec{\nabla} \otimes \mathbb{O}(\nu(\tau - t_0 - \sigma), z - y| \, |\vec{V}(\sigma, y) \otimes \vec{W}(\sigma, y))| \, d\sigma \, dy$$

$$\leq C_\gamma \sup_{|\sigma - t_0| < |\tau - t_0|} \|\vec{V}(\sigma, .)\|_\infty \sup_{|\sigma - t_0| < |\tau - t_0|} \|\vec{W}(\sigma, .)\|_\infty \int_{t_0}^{\Re\tau} \int \frac{dt \, d\alpha}{t^2 + |\alpha|^4}$$

$$= C'_\gamma \sqrt{\Re(\tau) - t_0} \sup_{|\sigma - t_0| < |\tau - t_0|} \|\vec{V}(\sigma, .)\|_\infty \sup_{|\sigma - t_0| < |\tau - t_0|} \|\vec{W}(\sigma, .)\|_\infty$$

Thus, we find by induction on k, that $\vec{V}_k$ is holomorphic on $t_0 + \Omega_\gamma$ and satisfies

$$|\vec{V}_k(\tau, z)| \leq \alpha_k |\Re(\tau) - t_0|^{k/2} \|\vec{u}(t_0, .)\|_\infty^{k+1}$$

where the constant α_k does not depend on τ, z, or $\vec{u}(t_0, .)$. We have

$$\alpha_0 \leq C_\gamma$$

and

$$\alpha_{k+1} \leq C_\gamma \sum_{j=0}^{k} \alpha_j \alpha_{k-j}.$$

If $\beta_0 = 1$ and $\beta_{k+1} = \sum_{j=0}^{k} \beta_j \beta_{k-j}$ and F is the formal series $F(z) = \sum_{k \in \mathbb{N}} \beta_k z^{k+1}$, we find that

$$F^2(z) = F(z) - z$$

and thus

$$F(z) = \frac{1}{2}(1 - \sqrt{1 - 4z}).$$

The radius of convergence of the Taylor expansion of F is $\frac{1}{4}$ and its coefficients are non-negative, thus we find

$$\sum_{k=0}^{+\infty} \left(\frac{1}{4}\right)^k \beta_k = \frac{1}{2}.$$

We thus find

$$\alpha_k \leq C_\gamma^{1+2k} \beta_k$$

and thus

$$|\vec{V}_k(\tau, z)| \leq C_\gamma^{1+2k} \beta_k |\Re(\tau) - t_0|^{k/2} \|\vec{u}(t_0, .)\|_\infty^{k+1}$$

The series $\sum_{k \in \mathbb{N}} \vec{V}_k(\tau, z)$ will converge normally on $t_0 + \Omega_{\gamma, M}$ if we choose M such that

$$4 C_\gamma^2 \|\vec{u}(t_0, .)\|_\infty \sqrt{M} < 1.$$

If

$$M_0 = \left(\frac{1}{8 C_\gamma^2 \|\vec{u}\|_{L^\infty((T_0, T_1) \times \mathbb{R}^3)}} \right)^2$$

we find that $\vec{u}$ has a holomorphic extension to $\cup_{T_0 < t_0 < T_1} t_0 + \Omega_{\gamma, M_0}$, thus $\vec{u}$ is analytic on $(T_0, T_1) \times \mathbb{R}^3$ with respect to the time and space variables. $\quad\square$

9.10   Small Data

If $\vec{u}_0$ is small enough in BMO^{-1}, we know that the Cauchy problem for the Navier–Stokes equations with initial value $\vec{u}_0$ and forcing term $\vec{f} = 0$ will have a global mild solution $\vec{u}$ that can be constructed by the Picard iterative scheme. If $\vec{u}_0$ belongs moreover to some Besov space $\dot{B}^s_{p,q}$ with $s > -1$, the sequence of the Picard iterates will converge in $L^\infty((0,+\infty), \dot{B}^s_{p,q})$ as well, without any restriction on the size of $\vec{u}$ in $\dot{B}^s_{p,q}$ nor on p, q, s. It means that not only $\vec{u}$ keeps its regularity at all times but also the Picard iterates behave well at all times. This property has been described as *persistency* by Furioli, Lemarié-Rieusset, Zahrouni, and Zhioua [188] (see a generalization to multilinear equations [315]).

Theorem 9.13.
Let $\vec{u}_0 \in BMO^{-1}$ with $\operatorname{div} \vec{u}_0 = 0$. Let B be the bilinear operator

$$B(\vec{V}, \vec{W}) = \int_0^t W_{\nu(t-s)} * \mathbb{P}\operatorname{div}(\vec{V}(s,.) \otimes \vec{W}(s,.))\, ds.$$

Let $\vec{U}_n$ be the sequence of Picard iterates defined by $\vec{U}_0 = W_{\nu t} * \vec{u}_0$ and $\vec{U}_{n+1} = \vec{U}_0 - B(\vec{U}_n, \vec{U}_n)$. Assume that we have

- $\sum_{n \in \mathbb{N}} \sup_{t>0} \sqrt{t}\|\vec{U}_{n+1}(t,.) - \vec{U}_n(t,.)\|_\infty < +\infty$

(for instance, assume that $\|\vec{u}_0\|_{BMO^{-1}}$ is small enough). Then the limit $\vec{u} = \lim_{n \to +\infty} \vec{U}_n$ satisfies the Navier–Stokes equations

$$\partial_t \vec{u} = \nu \Delta \vec{u} - \mathbb{P}\operatorname{div}(\vec{u} \otimes \vec{u})$$

and we have

$$\sup_{t>0} \sqrt{t}\|\vec{u}(t,.)\|_\infty < +\infty.$$

If moreover $\vec{u}_0$ belongs to some Besov space $\dot{B}^\sigma_{p,q}$ with $1 \le p, q \le +\infty$ and $-1 < \sigma \le 0$, then

$$\sup_{t>0} \|\vec{u}(t,.)\|_{\dot{B}^\sigma_{p,q}} < +\infty.$$

We have more precisely

$$\sum_{n \in \mathbb{N}} \sup_{t>0} \|\vec{U}_{n+1}(t,.) - \vec{U}_n(t,.)\|_{\dot{B}^\sigma_{p,1}} < +\infty.$$

Proof. **Case $-1 < \sigma < 0$:**
We introduce the path space

$$X = \{\vec{U} \ / \ t^{-\sigma/2}\|\vec{U}(t,.)\|_p \in L^q(\frac{dt}{t})\}.$$

We easily check that, if $\vec{V} \in X$ and $\sup_{t>0} \sqrt{t}\|\vec{W}(t,.)\|_\infty < +\infty$, then $B(\vec{V}, \vec{W})$ and $B(\vec{W}, \vec{V})$ still belong to X. Indeed, we have

$$\left\| \int_0^t W_{\nu(t-s)} * \mathbb{P}\operatorname{div}(\vec{v} \otimes \vec{W})\, ds \right\|_p \le C_p \int_0^t \frac{1}{\sqrt{t-s}\sqrt{s}} \|\vec{V}(s,.)\|_p\, ds \sup_{s>0} \sqrt{s}\|\vec{W}(s,.)\|_\infty.$$

Thus, we must only check that, for $0 < \alpha < 1/2$,

$$F \mapsto \int_0^t \frac{1}{\sqrt{t-s}\sqrt{s}} \frac{t^\alpha}{s^\alpha} F(s)\, ds$$

is bounded on $L^q(\frac{dt}{t})$ for $1 \leq q \leq +\infty$; this is obvious for $q = +\infty$ (direct estimation) and for $q = 1$ (first integrate on t, by Fubini's theorem), and thus for all q by interpolation.

Thus, by induction, we have that $\vec{U}_n$ belongs to X for every n (since $\vec{U}_0 \in X$ because $\vec{u}_0 \in \dot{B}_{p,q}^\sigma$). Moreover, writing

$$\vec{U}_{n+1} - \vec{U}_n = -B(\vec{U}_n, \vec{U}_n - \vec{U}_{n-1}) - B(\vec{U}_n - \vec{U}_{n-1}, \vec{U}_{n-1})$$

for all $n \in \mathbb{N}$ (with the convention that $\vec{U}_{-1} = 0$), we find, for

$$\alpha_n = \|\vec{U}_n - \vec{U}_{n-1}\|_X \text{ and } \epsilon_n = \sup_{t>0} \sqrt{t}\|\vec{U}_n(t,.) - \vec{U}_{n-1}(t,.)\|_\infty,$$

that

$$\alpha_{n+1} \leq C_0\epsilon_n \sum_{k=0}^n \alpha_k$$

where the constant C_0 depends only on ν, s and p. Thus, if

$$M_n = \sum_{k=0}^n \alpha_k,$$

we have

$$M_{n+1} \leq (1 + C_0\epsilon_n)M_n \leq \|\vec{U}_0\|_X \prod_{k=0}^n (1 + C_0\epsilon_k).$$

As $\sum_{k\in\mathbb{N}} \epsilon_k < +\infty$, we find that $\prod_{k\in\mathbb{N}}(1 + C_0\epsilon_k) < +\infty$, so that $\sum_{k\in\mathbb{N}} \alpha_k < +\infty$.

Now, if $f \in L^p$, we have $\vec{\nabla} f \in \dot{B}_{p,\infty}^{-1}$ and thus (as $s > -1$) $W_{\nu t} * \vec{\nabla} f \in \dot{B}_{p,1}^\sigma$ with $\|W_{\nu t} * \vec{\nabla} f\|_{\dot{B}_{p,1}^\sigma} \leq C \frac{1}{(t-s)^{(1+\sigma)/2}} \|f\|_p$. Thus, since $\dot{B}_{p,q}^\sigma \subset \dot{B}_{p,\infty}^\sigma$, and thus

$$\sup_{t>0} t^{-\frac{\sigma}{2}}\|\vec{U}_0(t,.)\|_p + \sum_{k=0}^{+\infty} \sup_{t>0} t^{-\frac{\sigma}{2}}\|\vec{U}_{k+1}(t,.) - \vec{U}_k(t,.)\|_p = M < +\infty,$$

we find

$$\|\vec{U}_{n+1}(t,.) - \vec{U}_n(t,.)\|_{\dot{B}_{p,1}^\sigma} \leq C \int_0^t \frac{1}{(t-s)^{(1+\sigma)/2} s^{(1-\sigma)/2}}\, ds M\epsilon_n = C'M\epsilon_n$$

and finally

$$\sum_{n\in\mathbb{N}} \sup_{t>0} \|\vec{U}_{n+1}(t,.) - \vec{U}_n(t,.)\|_{\dot{B}_{p,1}^\sigma} < +\infty.$$

Remark: this proves that, in contrast with the case $p = +\infty$, the Cauchy problem is well posed in $\dot{B}_{p,q}^\sigma$ for $3 < p < +\infty$ and $\sigma = -1 + \frac{3}{p}$ (for small data).

Case $\sigma = 0$:
We have $\dot{B}_{p,q}^0 \cap \dot{B}_{\infty,\infty}^{-1} \subset \dot{B}_{2p,\infty}^{-1/2}$. Thus, our analysis of the case $\vec{u}_0 \in \dot{B}_{2p,\infty}^{-1/2}$ shows that $\vec{U}_n$ will converge in the path space

$$X = \{\vec{U} \ / \ t^{1/4}\|\vec{U}(t,.)\|_{2p} \in L^\infty(\frac{dt}{t})\}.$$

More precisely, we proved that

$$\sum_{n\in\mathbb{N}} \|\vec{U}_n - \vec{U}_{n-1}\|_X < +\infty.$$

We write again

$$\vec{U}_{n+1} - \vec{U}_n = -B(\vec{U}_n, \vec{U}_n - \vec{U}_{n-1}) - B(\vec{U}_n - \vec{U}_{n-1}, \vec{U}_{n-1})$$

and consider the action of B on $X \times X$. We have

$$\|W_{\nu(t-s)} * \mathbb{P}\operatorname{div}(\vec{V} \otimes \vec{W})\|_{\dot{B}^0_{p,1}} \leq C\|\vec{V}\|_p \|\vec{W}\|_p \frac{1}{(\nu(t-s))^{1/2}}$$

so that

$$\|B(\vec{V}, \vec{W})(t, .)\|_{\dot{B}^0_{p,1}} \leq C\|\vec{V}\|_X \|\vec{W}\|_X.$$

Thus,

$$\sum_{n\in\mathbb{N}} \sup_{t>0} \|\vec{U}_{n+1}(t, .) - \vec{U}_n(t, .)\|_{\dot{B}^0_{p,1}} \leq C \sup_{n\in\mathbb{N}} \|\vec{U}_n\|_X \sum_{n\in\mathbb{N}} \|\vec{U}_n - \vec{U}_{n-1}\|_X < +\infty.$$

Remark: As $\dot{B}^0_{p,1} \subset L^p \subset \dot{B}^0_{p,\infty}$, this proves that we have persistency for the L^p norm as well.

$\square$

Theorem 9.14.
Let $\vec{u}_0 \in BMO^{-1}$ with $\operatorname{div} \vec{u}_0 = 0$. Let B be the bilinear operator

$$B(\vec{V}, \vec{W}) = \int_0^t W_{\nu(t-s)} * \mathbb{P}\operatorname{div}(\vec{V}(s, .) \otimes \vec{W}(s, .)) \, ds.$$

Let $\vec{U}_n$ be the sequence of Picard iterates defined by $\vec{U}_0 = W_{\nu t} * \vec{u}_0$ and $\vec{U}_{n+1} = \vec{U}_0 - B(\vec{U}_n, \vec{U}_n)$. Assume that we have

- $\sum_{n\in\mathbb{N}} \sup_{t>0} \sqrt{t} \|\vec{U}_{n+1}(t, .) - \vec{U}_n(t, .)\|_\infty < +\infty$

- $\sup_{n\in\mathbb{N}} \sup_{t>0} \|\vec{U}_n(t, .)\|_{\dot{B}^{-1}_{\infty,\infty}} < +\infty$

(for instance, assume that $\|\vec{u}_0\|_{BMO^{-1}}$ is small enough). Then the limit $\vec{u} = \lim_{n\to+\infty} \vec{U}_n$ satisfies the Navier-Stokes equations

$$\partial_t \vec{u} = \nu\Delta\vec{u} - \mathbb{P}\operatorname{div}(\vec{u} \otimes \vec{u})$$

and we have

$$\sup_{t>0} \sqrt{t} \|\vec{u}(t, .)\|_\infty < +\infty.$$

If moreover $\vec{u}_0$ belongs to some Besov space $\dot{B}^\sigma_{p,q}$ with $1 \leq p, q \leq +\infty$ and $\sigma > 0$, then

$$\sup_{t>0} \|\vec{u}(t, .)\|_{\dot{B}^\sigma_{p,q}} < +\infty.$$

We have more precisely

$$\sum_{n\in\mathbb{N}} \sup_{t>0} \|\vec{U}_{n+1}(t, .) - \vec{U}_n(t, .)\|_{\dot{B}^\sigma_{p,1}} < +\infty.$$

Proof. We would like to control a product fg when $f \in \dot{B}^\sigma_{p,q} \cap \dot{B}^{-1}_{\infty,\infty}$ and $g \in \dot{B}^\sigma_{p,q} \cap L^\infty$. Using the Littlewood–Paley decomposition, we write

$$fg = \pi(f,g) + \rho(f,g)$$

where

$$\pi(f,g) = \sum_{j\in\mathbb{Z}} S_{j+3}g\Delta_j f \text{ and } \rho(f,g) = \sum_{j\in\mathbb{Z}} S_{j-2}f\Delta_j g$$

We have

$$\|\pi(fg)\|_{\dot{B}^p_{\infty,q}} \leq C\|g\|_\infty\|f\|_{\dot{B}^\sigma_{p,q}}$$

and

$$\|W_{\nu(t-s)} * \vec{\nabla}(\pi(f,g))\|_{\dot{B}^\sigma_{p,1}} \leq C\frac{1}{\sqrt{\nu(t-s)}}\|g\|_\infty\|f\|_{\dot{B}^\sigma_{p,q}}.$$

If $\lambda \in (\frac{\sigma}{1+\sigma}, 1)$, we have

$$\|f\|_{\dot{B}^{\sigma(1-\lambda)-\lambda}_{\frac{p}{1-\lambda}},\infty} \leq C\|f\|^{1-\lambda}_{\dot{B}^\sigma_{p,q}}\|f\|^\lambda_{\dot{B}^{-1}_{\infty,\infty}}$$

and

$$\|g\|_{\dot{B}^{\lambda\sigma}_{\frac{p}{\lambda}},\infty} \leq C\|g\|^\lambda_{\dot{B}^\sigma_{p,q}}\|g\|^{1-\lambda}_\infty;$$

as $\sigma(1-\lambda) - \lambda < 0$, we find that

$$\|S_{j+2}f\|_{\frac{p}{1-\lambda}} \leq C\|f\|_{\dot{B}^{\sigma(1-\lambda)-\lambda}_{\frac{p}{1-\lambda}},\infty}2^{-j(\sigma(1-\lambda)-\lambda)}$$

and thus

$$\|\rho(f,g)\|_{\dot{B}^{\sigma-\lambda}_{p,\infty}} \leq C\|f\|_{\dot{B}^{\sigma(1-\lambda)-\lambda}_{\frac{p}{1-\lambda}},\infty}\|g\|_{\dot{B}^{\lambda\sigma}_{\frac{p}{\lambda}},\infty}$$

and

$$\|W_{\nu(t-s)} * \vec{\nabla}(\rho(f,g))\|_{\dot{B}^\sigma_{p,1}} \leq C\frac{1}{\nu(t-s))^{(1+\lambda)/2}}\|g\|^{1-\lambda}_\infty\|g\|^\lambda_{\dot{B}^\sigma_{p,q}}\|f\|^\lambda_{\dot{B}^{-1}_{\infty,\infty}}\|f\|^{1-\lambda}_{\dot{B}^\sigma_{p,q}}.$$

We write again

$$\vec{U}_{n+1} - \vec{U}_n = -B(\vec{U}_n, \vec{U}_n - \vec{U}_{n-1}) - B(\vec{U}_n - \vec{U}_{n-1}, \vec{U}_{n-1})$$

for all $n \in \mathbb{N}$ (with the convention that $\vec{U}_{-1} = 0$), we find, for

$$\alpha_0 = \|\vec{U}_0\|_{L^\infty\dot{B}^s_{p,q}}, \quad \alpha_{n+1} = \|\vec{U}_{n+1} - \vec{U}_n\|_{L^\infty\dot{B}^\sigma_{p,1}}, M_n = \sum_{k=0}^n \alpha_k,$$

$$\beta_n = \sup_{k\leq n}\|\vec{U}_k\|_{L^\infty\dot{B}^{-1}_{\infty,\infty}} \text{ and } \epsilon_n = \sup_{t>0}\sqrt{t}\|\vec{U}_n(t,.) - \vec{U}_{n-1}(t,.)\|_\infty,$$

that

$$\alpha_{n+1} \leq C_0(\epsilon_n M_n + \epsilon_n^{1-\lambda}\alpha_n^\lambda M_n^{1-\lambda}\beta_n^\lambda)$$

where the constant C_0 depends only on ν, σ, λ, and p. Young's inequality gives

$$\alpha_{n+1} \leq \frac{1}{2}\alpha_n + C_1\beta_\infty^{\frac{\lambda}{1-\lambda}}\epsilon_n M_n.$$

This gives (as $M_{n-1} \leq 2M_n - M_{n-1}$)

$$M_{n+1} - \frac{1}{2}M_n \leq (1 + C_1 \beta_\infty^{\frac{\lambda}{1-\lambda}} \epsilon_n) M_n - \frac{1}{2}M_{n-1}$$

$$\leq (1 + C_1 \beta_\infty^{\frac{\lambda}{1-\lambda}} \epsilon_n)(M_n - \frac{1}{2}M_{n-1}) + \frac{1}{2}C_1 \beta_\infty^{\frac{\lambda}{1-\lambda}} \epsilon_n M_{n-1}$$

$$\leq (1 + 2C_1 \beta_\infty^{\frac{\lambda}{1-\lambda}} \epsilon_n)(M_n - \frac{1}{2}M_{n-1})$$

We get finally

$$M_n \leq M_{n+1} - \frac{1}{2}M_n = (M_1 - \frac{1}{2}M_0) \prod_{k=1}^{+\infty}(1 + 2C_1 \beta_\infty^{\frac{\lambda}{1-\lambda}} \epsilon_k) < +\infty.$$

Thus, the theorem is proved.   □

Remark:

(i) The homogeneous Besov spaces $\dot{B}^\sigma_{p,q}$ are not well defined as Banach spaces of distributions, as we have not the convergence of $\sum_{j \in \mathbb{Z}} \Delta_j f$ to f in $\mathcal{S}'$ if $\sigma > 3/p$ (or $\sigma = 3/p$ and $q > 1$). However, we work here with $BMO^{-1} \cap \dot{B}^\sigma_{p,q}$ so that the infrared divergence is avoided.

(ii) As the homogeneous Sobolev spaces $\dot{W}^{-1+\frac{3}{p},p}$ for $1 < p < 3$ satisfies $-1 + \frac{3}{p} > 0$, and $\dot{B}^{-1+\frac{3}{p}}_{p,1} \subset \dot{W}^{-1+\frac{3}{p},p} \subset \dot{B}^{-1+\frac{3}{p}}_{p,\infty} \subset BMO^{-1}$, the Cauchy problem is well posed in $\dot{W}^{-1+\frac{3}{p},p}$ for $1 < p < 3$ (for small data).

Chapter 10

Special Examples of Solutions

10.1 Symmetries for the Navier–Stokes Equations

In this section, we consider the Navier–Stokes equations

$$\partial_t \vec{u} = \nu \Delta \vec{u} - \vec{u}.\vec{\nabla}\vec{u} - \vec{\nabla}p + \vec{f} \tag{10.1}$$

with div $\vec{u} = 0$, defined on $(T_0, T_1) \times \mathbb{R}^3$. We assume that p is vanishing at infinity, in the sense that, if S_j is the Littlewood–Paley operator $S_j f = 2^{3j}\phi(2^j x) * p(x,t)$, then

$$\lim_{j \to -\infty} \|S_j p\|_\infty = 0. \tag{10.2}$$

We consider especially the case $\vec{f} = 0$ and say that in that case we have $\vec{u} \in \mathcal{NS}_0$. We list there some known symmetries for the Navier–Stokes equations and the associated transforms on $\mathcal{NS}_0$. Such transforms have been discussed one century ago by Wilczynski [502].

- time translation: assume we change the origin of time (new coordinates $T = t - t_0, X = x$). Then we obtain for $\vec{u} \in \mathcal{NS}_0$ another solution $\vec{v} \in \mathcal{NS}_0$ (defined on $(T_0 - t_0, T_1 - t_0) \times \mathbb{R}^3$) given by

$$\vec{v}(T,X) = \vec{u}(T + t_0, X)$$

associated to the pressure

$$q(T,X) = p(T + t_0, X)$$

The fact that $\vec{v}$ obeys the same equation as $\vec{u}$ comes from the fact that the coefficients in Equation (10.1) are constant.

Of course, if we assume the force $\vec{f} \neq 0$, then we must modify the force $\vec{f}$ in (10.1) into $\vec{g}$ with

$$\vec{g}(T,X) = \vec{f}(T + t_0, X).$$

- space translation: assume we change the origin of space coordinates (new coordinates $T = t, X = x - x_0$). Then we obtain for $\vec{u} \in \mathcal{NS}_0$ another solution $\vec{v} \in \mathcal{NS}_0$ given by

$$\vec{v}(T,X) = \vec{u}(T, X + x_0)$$

associated to the pressure

$$q(T,X) = p(T, X + x_0)$$

The fact that $\vec{v}$ obeys the same equation as $\vec{u}$ comes again from the fact that the coefficients in Equation (10.1) are constant.

If we assume the force $\vec{f} \neq 0$, then we must modify the force $\vec{f}$ in (10.1) into $\vec{g}$ with

$$\vec{g}(T,X) = \vec{f}(T, X + x_0).$$

DOI: 10.1201/9781003042594-10

- space rotation: assume we change the reference axes by a rotation: $\vec{I} = R\vec{i}, \vec{J} = R\vec{j}, \vec{K} = R\vec{k}$. The new coordinates are then $T = t, X = R^{-1}x$. Then we obtain for $\vec{u} \in \mathcal{NS}_0$ another solution $\vec{v} \in \mathcal{NS}_0$ given by

$$\vec{v}(T, X) = R^{-1}\vec{u}(T, RX)$$

associated to the pressure

$$q(T, X) = p(T, RX)$$

The fact that $\vec{v}$ obeys the same equation as $\vec{u}$ comes from the fact that we have modelized the evolution of an isotropic fluid.

If we assume the force $\vec{f} \neq 0$, then we must modify the force $\vec{f}$ in (10.1) into $\vec{g}$ with

$$\vec{g}(T, X) = R^{-1}\vec{f}(T, RX).$$

- change of Galilean frame: if we change the reference frame into another one moving with uniform velocity $\vec{U}$, the laws of Newtonian physics do not change. We get the new coordinates $T = t, X = x - t\vec{U}$. Then we obtain for $\vec{u} \in \mathcal{NS}_0$ another solution $\vec{v} \in \mathcal{NS}_0$ given by

$$\vec{v}(T, X) = \vec{u}(T, X + T\vec{U}) - \vec{U}$$

associated to the pressure

$$q(T, X) = p(T, X + T\vec{U}).$$

If we assume the force $\vec{f} \neq 0$, then we must modify the force $\vec{f}$ in (10.1) into $\vec{g}$ with

$$\vec{g}(T, X) = \vec{f}(T, X + T\vec{U}).$$

- change of scale: if we want to change the scales of times and of space, and keep the viscosity coefficient constant (remember that the kinematic viscosity ν has the same dimension as UL, where U is a velocity and L a length), we use the new coordinates $T = t/\lambda^2, X = x/\lambda$ for some positive λ. Then we obtain for $\vec{u} \in \mathcal{NS}_0$ another solution $\vec{v} \in \mathcal{NS}_0$ given by

$$\vec{v}(T, X) = \lambda\vec{u}(\lambda^2 T, \lambda X)$$

associated to the pressure

$$q(T, X) = \lambda^2 p(\lambda^2 T, \lambda X).$$

If we assume the force $\vec{f} \neq 0$, then we must modify the force $\vec{f}$ in (10.1) into $\vec{g}$ with

$$\vec{g}(T, X) = \lambda^3 \vec{f}(\lambda^2 T, \lambda X).$$

- change of orientation: if we work with the new coordinates $T = t, X = -x$, we still find solutions in $\mathcal{NS}_0$: we obtain for $\vec{u} \in \mathcal{NS}_0$ another solution $\vec{v} \in \mathcal{NS}_0$ given by

$$\vec{v}(T, X) = -\vec{u}(T, -X)$$

associated to the pressure

$$q(T, X) = p(T, -X).$$

If we assume the force $\vec{f} \neq 0$, then we must modify the force $\vec{f}$ in (10.1) into $\vec{g}$ with

$$\vec{g}(T, X) = -\vec{f}(T, -X).$$

Those are the only symmetries for the Navier–Stokes equations (see Bytev [73] and Lloyd [341]). We did not consider the generalized symmetries described by Boisvert [49] that we obtain by dropping the request on the control (10.2) of p at large and larger scales. The tranforms we shall not consider in the following are then the following ones:

- uniform change of pressure: with the same coordinates t, x, the same velocity $\vec{u}$ and the same force $\vec{f}$, just change p into $q(t, x) = p(t, x) + \varpi(t)$, where ϖ is arbitrary. The new pressure q satisfies $S_j(q(t, x)) = S_j p(t, x) + \varpi(t)$, and thus satisfies (10.2) if and only if $\varpi = 0$.

- motion of the observer with no rotation of the axes: in the new frame, we have coordinates $T = t$ and $X = x - m(t)$, where $m(t)$ is an arbitrary $(\mathcal{C}^2)$ function of t. We then change $\vec{u}, p, \vec{f}$ into $\vec{v}, q, \vec{g}$ with $\vec{v}(T, X) = \vec{u}(T, X + m(t)) - \dot{m}(t)$, $q(T, X) = p(T, X + m(t)) + \ddot{m}(t).X$ and $\vec{g}(T, X) = \vec{f}(T, X + m(t))$. The new pressure q satisfies $S_j(q(T, X)) = (S_j p)(T, X + m(t)) + \ddot{m}(t).X = 0$, and thus satisfies (10.2) if and only if $\ddot{m}(t) = 0$ (hence the change of coordinates amounts to a mere space translation followed by a change of Galilean reference frame).

10.2 Two-and-a-Half Dimensional Flows

In this section, we consider the Navier–Stokes problem with the following symmetry property: $\vec{u}$ is invariant under the action of space translations parallel to the x_3 axis. Thus, $\vec{u}$ does not depend on x_3 and may be seen as a (time-dependent) bivariate function: $\vec{u}(t, x_1, x_2, x_3) = \vec{v}(t, x_1, x_2)$. As this is a three-dimensional bivariate vector field, Bertozzi and Majda label those vector fields as "two-and-a-half dimensional flows" [40].

Thus, if we consider the Cauchy problem, we start with a data $\vec{u}_0(x_1, x_2, x_3) = \vec{v}_0(x_1, x_2)$ and a force $\vec{f}(t, x_1, x_2, x_3) = \vec{g}(t, x_1, x_2)$.

If we may construct the solution by the Picard iterative scheme, then the symmetry of the Navier–Stokes equations gives us that the solution $(\vec{u}, \vec{p})$ will satisfy the same symmetry as $\vec{u}_0$ and $\vec{f}$: $\vec{u}(t, x_1, x_2, x_3) = \vec{v}(t, x_1, x_2)$ and $p(t, x_1, x_2, x_3) = q(t, x_1, x_2)$.

For the time being, let us assume that $\vec{f} = 0$. We know (by Theorem 8.3) that the problem

$$\partial_t \vec{u} = \nu \Delta \vec{u} - \vec{u}.\vec{\nabla}\vec{u} - \vec{\nabla}p \tag{10.3}$$

with div $\vec{u} = 0$ and $\vec{u}(0, .) = \vec{u}_0$ has a mild solution when two conditions are satisfied:

- $\vec{u}_0$ belongs to the Morrey space $\dot{M}^{2,3}$:

$$\sup_{R>0, x_0 \in \mathbb{R}^3} \frac{1}{R} \int_{B(x_0, R)} |\vec{u}_0(x)|^2 \, dx < +\infty$$

- $\vec{u}_0$ satisfies:

$$\lim_{t \to 0} t^{1/2} \|W_{\nu t} * \vec{u}_0\|_\infty = 0$$

In that case, we know that we have a solution in the space $\{\vec{u} \in L^\infty((0, T_0), \dot{M}^{2,3}) \, / \, \sup_{0 < t < T_0} \sqrt{t}\|\vec{u}(t, .)\|_\infty < +\infty\}$ for some positive T_0.

If we assume moreover that $\vec{u}_0(x_1, x_2, x_3) = \vec{v}_0(x_1, x_2)$, then we have

$$\vec{u}_0 \in \dot{M}^{2,3} \Leftrightarrow \vec{v}_0 \in L^2(\mathbb{R}^2)$$

and the condition $\lim_{t\to 0} t^{1/2}\|W_{\nu t} * \vec{u}_0\|_\infty = 0$ is automatically fulfilled, as $L^\infty \cap L^2$ is dense in L^2. Thus, a natural assumption on $\vec{v}_0$ is the fact that $\vec{v}_0 \in L^2$.

If we assume that $\vec{u}_0$ and $\vec{f}$ do not depend on x_3 and if we search for a solution $\vec{u}$ and a pressure p which do not depend on x_3, we find two equations. More precisely, writing $u_1(t,x) = v_1(t,x_1,x_2)$, $u_2(t,x) = v_2(t,x_1,x_2)$, $u_3(t,x) = w(t,w_1,x_2)$, $p(t,x) = q(t,x_1,x_2)$, $f_1(t,x) = g_1(t,x_1,x_2)$, $f_2(t,x) = g_2(t,x_1,x_2)$ and $f_3(t,x) = h(t,x_1,x_2)$, we have to solve the following equations

- $\vec{v}$ satisfies a $2D$ Navier–Stokes equation:

$$\begin{cases} \partial_t \vec{v} + (\vec{v}.\vec{\nabla})\vec{v} = \nu \Delta \vec{v} + \vec{g} - \vec{\nabla}p \\ \vec{v}(0,x_1,x_2) = (u_{0,1}(x), u_{0,2}(x)) \\ \qquad \operatorname{div} \vec{v} = 0 \end{cases} \qquad (10.4)$$

- w satisfies a linear advection-diffusion scalar equation:

$$\begin{cases} \partial_t w + (\vec{v}.\vec{\nabla})w = \nu \Delta w + h \\ \quad w(0,x_1,x_2) = u_{0,3}(x) \end{cases} \qquad (10.5)$$

The study of 2D Navier–Stokes equations with initial value $\vec{v}_0 \in L^2(\mathbb{R}^2)$ and $\vec{g} \in L^2 H^{-1}$ was initiated by the works of Leray [327, 328, 329], and fully developed by Ladyzhenskaya, Lions and Prodi [293, 339].

2D Navier–Stokes equations

Theorem 10.1.
If $\vec{v}_0 \in (L^2(\mathbb{R}^2))^2$ with $\operatorname{div} \vec{v}_0 = 0$ and let $\vec{g} \in L^2((0,T),(H^{-1}(\mathbb{R}^2)^2)$, then there exists a unique solution $\vec{v}$ of equation

$$\begin{cases} \partial_t \vec{v} + (\vec{v}.\vec{\nabla})\vec{v} = \nu \Delta \vec{v} + \vec{g} - \vec{\nabla}p \\ \qquad \vec{v}(0,.) = \vec{v}_0 \\ \qquad \operatorname{div} \vec{v} = 0 \end{cases} \qquad (10.6)$$

such that $\vec{v} \in L_t^\infty L_x^2 \cap L_t^2 H_x^1$ (if $T < +\infty$).

Proof. **Local existence:**
We rewrite (10.6) into

$$\vec{v} = W_{\nu t}^{(2)} * \vec{v}_0 + \int_0^t W_{\nu(t-s)}^{(2)} * \mathbb{P}^{(2)}(\vec{g} - \operatorname{div}(\vec{v} \otimes \vec{v}))\, ds \qquad (10.7)$$

where $W_t^{(2)}(x_1,x_2)$ is the 2D heat kernel and $\mathbb{P}^{(2)}$ is the 2D Leray projection operator. We are going to look for a solution $\vec{v} \in L^4((0,T_0), L^4(\mathbb{R}^2))$ for T_0 small enough.

Indeed, we have the following estimates:

$$\|W_{\nu t}^{(2)} * \vec{v}_0\|_{L^\infty L^2} = \|\vec{v}_0\|_2$$

$$\|W_{\nu t}^{(2)} * \vec{v}_0\|_{L^2 \dot{H}^1} = \frac{1}{\sqrt{2\nu}}\|\vec{v}_0\|_2$$

$$\left\| \int_0^t W^{(2)}_{\nu(t-s)} * \mathbb{P}^{(2)} \vec{g} \, ds \right\|_{L^\infty((0,T_0),L^2)} \le \left(\sqrt{T_0} + \frac{1}{\sqrt{2\nu}}\right) \|\vec{g}\|_{L^2 H^{-1}}$$

$$\left\| \int_0^t W^{(2)}_{\nu(t-s)} * \mathbb{P}^{(2)} \vec{g} \, ds \right\|_{L^2((0,T_0),\dot{H}^1)} \le \left(CT_0 + \frac{1}{\nu}\right) \|\vec{g}\|_{L^2 H^{-1}}$$

Moreover, we have the Sobolev embedding

$$H^{1/2}(\mathbb{R}^2) \subset L^4.$$

Thus, we find that $\vec{V}_0 = W^{(2)}_{\nu t} * \vec{v}_0 + \int_0^t W^{(2)}_{\nu(t-s)} * \mathbb{P}^{(2)} \vec{g} \, ds$ belongs to $L^4 L^4$. Moreover, $\lim_{t \to 0} \|\vec{V}_0\|_{L^4((0,T_0),L^4)} = 0$.

Now, if $\vec{u}$ and $\vec{v}$ belong to $L^4 L^4$, we have that $\operatorname{div}(\vec{u} \otimes \vec{v})$ belong to $L^2 \dot{H}^{-1}$, so that $B(\vec{u}, \vec{v}) = \int_0^t W^{(2)}_{\nu(t-s)} * \mathbb{P} \operatorname{div}(\vec{u} \otimes \vec{v}) \, ds$ belongs to $L^\infty L^2 \cap L^2 \dot{H}^1 \cap L^4 L^4$ with

$$\|B(\vec{u}, \vec{v})\|_{L^4((0,T_0),L^4} \le C_\nu \|\vec{u}\|_{L^4((0,T_0),L^4} \|\vec{v}\|_{L^4((0,T_0),L^4}.$$

If T_0 is small enough, so that $\|\vec{V}_0\|_{L^4((0,T_0),L^4} < \frac{1}{4C_\nu}$, we shall find a solution $\vec{v}$ through Picard's iterative process. The process will converge in $L^\infty L^2 \cap L^2 \dot{H}^1 \cap L^4 L^4$.

Global existence:
If the solution $\vec{v}$ belongs to $L^\infty((0,T_0),L^2) \cap L^2((0,T_0),\dot{H}^1)$ then we find an estimate slightly better than just $\vec{v} \in L^\infty L^2$. As a matter of fact, we have

$$\vec{v} = W^{(2)}_{\nu t} * \vec{v}_0 + \int_0^t W^{(2)}_{\nu(t-s)} * \vec{V} \, ds = W^{(2)}_{\nu t} * \vec{v}_0 + \mathcal{L}(\vec{V})$$

where $\vec{V} = \mathbb{P}^{(2)}(\vec{g} - \operatorname{div}(\vec{v} \otimes \vec{v}))$ belongs to $L^2 H^{-1}$; the operator $\mathcal{L}$ maps $L^2 H^{-1}$ to $L^\infty L^2$ and $L^2 H^2$ to $\operatorname{Lip} L^2$; as $L^2 H^2$ is dense in $L^2 H^{-1}$, we find that $\mathcal{L}$ actually maps $L^2 H^{-1}$ to $\mathcal{C}([0,T_0],L^2)$.

If T^* is the maximal time of existence of the solution $\vec{v}$, so that $\vec{v}$ belongs to $L^\infty((0,T_0),L^2) \cap L^2((0,T_0),\dot{H}^1)$ for every $T_0 < T^*$, we find that $T^* = T$ unless that $\vec{v}$ does not belong to $L^\infty((0,T^*),L^2) \cap L^2((0,T^*),\dot{H}^1)$: if $\vec{v}$ belonged to $L^\infty((0,T^*),L^2) \cap L^2((0,T^*),\dot{H}^1)$ with $T^* < T$, then it would belong to $\mathcal{C}([0,T^*],L^2)$ and we could solve the Cauchy problem for the Navier–Stokes equations on some interval $[T^*, T^* + T_0]$ with initial value $\vec{v}(T^*, .)$.

Thus, in order to show that we have a global solution, we only need to control the sizes of $\vec{v}$ and $\vec{\nabla} \otimes \vec{v}$ on $[0,T^*)$. As $\vec{v}$ belongs to $L^2 H^1$ and $\partial_t \vec{v}$ belongs to $L^2 H^{-1}$ on every compact interval of $[0,T^*)$, we may write

$$\frac{d}{dt} \|\vec{v}(t,.)\|_2^2 = 2\langle \vec{v}(t,.) | \partial_t \vec{v}(t,.) \rangle_{H^1,H^{-1}} = -2\nu \|\vec{\nabla} \otimes \vec{v}(t,.)\|_2^2 + 2\langle \vec{v}(t,.) | \vec{g}(t,.) \rangle_{H^1,H^{-1}}$$

so that

$$\frac{d}{dt} \|\vec{v}(t,.)\|_2^2 \le \nu \|\vec{\nabla} \otimes \vec{v}(t,.)\|_2^2 + \nu \|\vec{v}(t,.)\|_2^2 + \frac{1}{\nu} \|\vec{g}(t,.)\|_{\dot{H}^{-1}}^2.$$

By Grönwall's lemma, we get

$$\|\vec{v}(t,.)\|_2^2 + \nu \int_0^t \|\|\vec{\nabla} \otimes \vec{v}(s,.)\|_2^2 \, ds \le e^{\nu t} (\|\vec{v}_0\|_2^2 + \frac{1}{\nu} \int_0^t \|\vec{g}(s,.)\|_{\dot{H}^{-1}}^2 \, ds).$$

Hence, $T^* = T$: we have a global solution. $\qquad\square$

We may now solve the Navier–Stokes equations with initial value $\vec{u}_0 = (\vec{v}_0, w_0)$ and forcing term $\vec{f} = (\vec{g}, h)$:

Proposition 10.1.
If $(\vec{v}_0, w_0) \in (L^2(\mathbb{R}^2))^3$ with div $\vec{v}_0 = 0$ and $(\vec{g}, h) \in (L^2((0,T), H^{-1}(\mathbb{R}^2)))^3$ then the Equations (10.4) and (10.5) have global solutions $\vec{v}$ and w in $L^\infty L^2 \cap L^2 H^1$ (if $T < +\infty$).

Proof. The existence of $\vec{v}$ has been proved in Theorem 10.1. For the existence of w, we write w as a fixed point of the transform

$$ w \mapsto W^{(2)}_{\nu t} * w_0 + \int_0^t W^{(2)}_{\nu(t-s)} * (h - \mathrm{div}(w\vec{v}))\, ds = W^{(2)}_{\nu t} * w_0 + \mathcal{L}(w). $$

$\mathcal{L}$ is a bounded linear operator on $L^\infty((0,T_0), L^2) \cap L^2((0,T_0), L^2)$ and satisfies (uniformly in T_0)

$$ \|\mathcal{L}(w)\|_{L^\infty((0,T_0),L^2)\cap L^2((0,T_0),L^2)} \leq C_0 \|\vec{v}\|_{L^4((0,T_0),L^4)} \|w\|_{L^\infty((0,T_0),L^2)\cap L^2((0,T_0),L^2)}. $$

Thus, $\mathcal{L}$ is a contraction as soon as T_0 is small enough to grant that $C_0 \|\vec{v}\|_{L^4((0,T_0),L^4)} < 1$.

Global existence of w is then proved by splitting $[0,T]$ into a finite union of intervals $[T_j, T_{j+1}]$ with $C_0 \|\vec{v}\|_{L^4((T_j,T_{j+1}),L^4)} < 1$: once w is constructed on $[0, T_{j+1}]$, one constructs w on $[T_{j+1}, T_{j+2}]$ by considering the Cauchy problem with initial value $w(T_{j+1}, .)$ at $t = T_{j+1}$. □

Thus, we have global existence of unique solutions to the Navier–Stokes problem when the initial value $\vec{u}_0$ depends only on the first two variables (x_1, x_2) and when the force $\vec{f}$ depends only on t and (x_1, x_2) and when $\vec{u}_0 \in L^2(\mathbb{R}^2)$ and $\vec{f} \in L^2((0,T), H^{-1}(\mathbb{R}^2))$ (whatever their sizes). The case of $\vec{u}_0 \in L^2_{uloc}(\mathbb{R}^2)$ has been discussed by Basson [24] (when $\vec{f} = 0$). He found global existence as well in this case.

The stability of the solutions decribed in Theorem 10.1 and in Proposition 10.1 under small enough 3D perturbations has been discussed by Gallagher [195] and Iftimie [239]. We are going to present Gallagher's result in the following theorem:

3D perturbation of the 2D Navier–Stokes equations

Theorem 10.2.
Let $\vec{v}_0 \in (L^2(\mathbb{R}^2))^3$ with $\partial_1 v_1 + \partial_2 v_2 = 0$ and let $\vec{g} \in L^2((0,T), (H^{-1}(\mathbb{R}^2)^3)$ (where $T < +\infty$), and let $\vec{v}$ be the associated solution of the Navier–Stokes equations such that $\vec{v} \in (L^\infty((0,T), L^2(\mathbb{R}^2) \cap L^2((0,T), H^1(\mathbb{R}^2))^3$. Then there exists a positive ϵ which depends on T, $\vec{v}_0$ and $\vec{g}$ such that the Navier–Stokes equations

$$ \begin{cases} \partial_t \vec{u} + \vec{u}.\vec{\nabla}\vec{u} = \nu\Delta\vec{u} + \vec{f} - \vec{\nabla}p \\ \vec{u}(0,.) = \vec{v}_0 + \vec{w}_0 \\ \vec{f} = \vec{g} + \vec{h} \\ \mathrm{div}\,\vec{u} = 0 \end{cases} \qquad (10.8) $$

with $\|\vec{w}_0\|_{\dot{H}^{1/2}} < \epsilon$ and $\int_0^T \|\vec{h}\|^2_{\dot{H}^{-1/2}}\, ds < \epsilon^2$, have a global solution $\vec{u} = \vec{v} + \vec{w}$ with $\vec{w} \in L^\infty_t((0,T), \dot{H}^{1/2}_x) \cap L^2_t \dot{H}^{3/2}_x$.

The same result holds for $T = +\infty$ provided that we have $\vec{g} \in L^2((0,+\infty), (\dot{H}^{-1}(\mathbb{R}^2)^3)$ or $\vec{g} \in L^1((0,+\infty), (L^2(\mathbb{R}^2)^3)$.

Proof. We first construct our solution $\vec{w}$ on $(0, T_0)$ with T_0 small enough.

$\vec{w}$ is a solution of the fixed-point problem

$$\vec{w} = W_{\nu t} * \vec{w}_0 + \int_0^t W_{\nu(t-s)} * \mathbb{P}(\vec{h} - \text{div}(\vec{v} \otimes \vec{w} + \vec{w} \otimes \vec{v} + \vec{w} \otimes \vec{w}))\, ds.$$

We write

$$\vec{w} = \vec{W}_0 - \mathcal{L}(\vec{w}) - \mathcal{B}(\vec{w}, \vec{w})$$

with $\vec{W}_0 = W_{\nu t} * \vec{w}_0 + \int_0^t W_{\nu(t-s)} * \mathbb{P}\vec{h}\, ds$,

$$\mathcal{L}(\vec{w}) = \int_0^t W_{\nu(t-s)} * \mathbb{P}\,\text{div}(\vec{v} \otimes \vec{w} + \vec{w} \otimes \vec{v})\, ds$$

and

$$\mathcal{B}(\vec{w}_1, \vec{w}_2) = \int_0^t W_{\nu(t-s)} * \mathbb{P}\,\text{div}(\vec{w}_1 \otimes \vec{w}_2)\, ds$$

Let $Y_{T_0} = L_t^\infty((0,T), \dot{H}_x^{1/2}) \cap L_t^2 \dot{H}_x^{3/2}$. We have, for $\alpha \in \dot{H}^{1/2}(\mathbb{R}^3)$ and $\beta \in L^2((0,T_0), \dot{H}^{-1/2}(\mathbb{R}^3))$

$$\|W_{\nu t} * \alpha\|_{Y_{T_0}} \leq C_0 \|\alpha\|_{\dot{H}^{1/2}} \tag{10.9}$$

and

$$\left\| \int_0^t W_{\nu(-s)t} * \beta(s.)\, ds \right\|_{Y_{T_0}} \leq C_0 \|\beta\|_{L^2 \dot{H}^{-1/2}} \tag{10.10}$$

where C_0 does not depend on T_0.

We thus get

$$\|\vec{W}_0\|_{Y_{T_0}} \leq C_1 \left(\|\vec{w}_0\|_{\dot{H}^{1/2}} + \left(\int_0^{T_0} \|\vec{h}(s,.)\|_{\dot{H}^{-1/2}}^2\, ds \right)^{1/2} \right)$$

and, for $\vec{w}_1, \vec{w}_2 \in Y_{T_0}$,

$$\|\mathcal{B}(\vec{w}_1, \vec{w}_2)\|_{Y_{T_0}} \leq C_2 \|\vec{w}_1\|_{L_t^4 \dot{H}^1} \|\vec{w}_2\|_{L_t^4 \dot{H}^1} \leq C_2 \|\vec{w}_1\|_{Y_{T_0}} \|\vec{w}_2\|_{Y_{T_0}}$$

Moreover, we may see from inequality (10.9) that we have, for $\beta \in L^1((0,T_0), \dot{H}^{1/2}(\mathbb{R}^3))$

$$\left\| \int_0^t W_{\nu(-s)t} * \beta(s.)\, ds \right\|_{Y_{T_0}} \leq C_3 \|\beta\|_{L^1 \dot{H}^{1/2}} \tag{10.11}$$

Complex interpolation between (10.10) and (10.11) gives then

$$\left\| \int_0^t W_{\nu(-s)t} * \beta(s.)\, ds \right\|_{Y_{T_0}} \leq C_4 \|\beta\|_{L^{4/3} L^2} \tag{10.12}$$

We have

$$\|\mathbb{P}(\vec{v}.\vec{\nabla}\vec{w})\|_{L_t^{4/3} L_x^2} \leq \|\vec{v}\|_{L_t^4 L_{x_1,x_2}^4} \|\vec{\nabla} \otimes \vec{w}\|_{L_t^2 L_{x_3}^2 L_{x_1,x_2}^4}$$

with

$$\|\vec{\nabla} \otimes \vec{w}\|_{L_t^2 L_{x_3}^2 L_{x_1,x_2}^4} \leq C \|\vec{\nabla} \otimes \vec{w}\|_{L_t^2 L_{x_3}^2 \dot{H}_{x_1,x_2}^{1/2}} \leq C \|\vec{w}\|_{Y_{T_0}}$$

while

$$\|\mathbb{P}(\vec{w}.\vec{\nabla}\vec{v})\|_{L_t^{4/3} L_x^2} \leq \|\vec{\nabla} \otimes \vec{v}\|_{L_t^2 L_{x_1,x_2}^2} \|\vec{w}\|_{L_t^4 L_{x_3}^2 L_{x_1,x_2}^\infty}$$

with

$$\|\vec{w}\|_{L_t^4 L_{x_3}^2 L_{x_1,x_2}^\infty} \leq C \|\vec{w}\|_{L_t^4 L_{x_3}^2 (\dot{B}_{2,1}^1)_{x_1,x_2}} \leq C' \sqrt{\|\vec{w}\|_{L_t^\infty L_{x_3}^2 \dot{H}_{x_1,x_2}^{1/2}} \|\vec{w}\|_{L_t^2 L_{x_3}^2 \dot{H}_{x_1,x_2}^{3/2}}}$$

so that

$$\|\vec{w}\|_{L_t^4 L_{x_3}^2 L_{x_1,x_2}^\infty} \leq C' \|\vec{w}\|_{Y_{T_0}}.$$

Hence, we have

$$\|\mathcal{L}(\vec{w})\|_{Y_{T_0}} \leq C_5 (\|\vec{v}\|_{L_t^4 L_{x_1,x_2}^4} + \|\vec{\nabla} \otimes \vec{v}\|_{L_t^2 L_{x_1,x_2}^2}) \|\vec{w}\|_{Y_{T_0}}.$$

Thus, if T_0 is small enough to grant that

$$C_5 (\|\vec{v}\|_{L_t^4 L_{x_1,x_2}^4} + \|\vec{\nabla} \otimes \vec{v}\|_{L_t^2 L_{x_1,x_2}^2}) \leq \frac{1}{2}$$

and if $\vec{w}_0$ and T_0 are small enough to grant that

$$\|\vec{w}_0\|_{\dot{H}^{1/2}} + \left(\int_0^{T_0} \|\vec{h}(s,.)\|_{\dot{H}^{-1/2}}^2 \, ds \right)^{1/2} < \frac{1}{16 C_1 C_2}$$

then we shall find a fixed-point $\vec{w}$ with

$$\|\vec{w}\|_{Y_{T_0}} < 4 C_1 \left(\|\vec{w}_0\|_{\dot{H}^{1/2}} + \left(\int_0^{T_0} \|\vec{h}(s,.)\|_{\dot{H}^{-1/2}}^2 \, ds \right)^{1/2} \right).$$

We now turn to the global existence. If $\vec{w}$ is defined on $(0, T_1)$ with $T_1 < T$ and if $\sup_{0 < t < T_1} \|\vec{w}(t,.)\|_{\dot{H}^{1/2}} < \frac{1}{16 C_1 C_2}$, we find that the behavior of $\vec{w}$ in $\dot{H}^{1/2}$ is controlled by the behaviors of $\vec{W}_0$, $\mathcal{L}(\vec{w})$ and $\mathcal{B}(\vec{w}, \vec{w})$; due to the fact that smooth functions are dense in $L^2 \dot{H}^{-1/2}$ and in $L^{3/2} L^2$, we find that $\vec{w} \in \mathcal{C}([0, T_1], \dot{H}^{1/2})$; we can then reiterate the construction of $\vec{w}$ from the departure time $t = T_1$ and see that $\vec{w}$ may be defined on a larger interval. Thus, in order to check the existence of a global solution, we just have to check that the $\dot{H}^{1/2}$ norm of $\vec{w}$ is controlled.

We may write

$$\frac{d}{dt} \|\vec{w}\|_{\dot{H}^{1/2}}^2 = 2 \langle \sqrt{-\Delta} \vec{w} | \partial_t \vec{w} \rangle = 2 \langle \sqrt{-\Delta} \vec{w} | \nu \Delta \vec{w} + \vec{h} - \operatorname{div}(\vec{v} \otimes \vec{w} + \vec{w} \otimes \vec{v} + \vec{w} \otimes \vec{w}) \rangle$$

so that

$$\frac{d}{dt} \|\vec{w}\|_{\dot{H}^{1/2}}^2 = -2\nu \|(-\Delta)^{3/4} \vec{w}\|_2^2 + 2 \langle \sqrt{-\Delta} \vec{w} | \vec{h} \rangle_{H^{1/2}, H^{-1/2}}$$

$$-2 \langle \sqrt{-\Delta} \vec{w} | \vec{v} . \vec{\nabla} \vec{w} + \vec{w} . \vec{\nabla} \vec{v} \rangle_{L^2, L^2} - 2 \langle \sqrt{-\Delta} \vec{w} | \operatorname{div}(\vec{w} \otimes \vec{w}) \rangle_{H^{1/2}, H^{-1/2}}$$

$$\leq -2\nu \|\vec{w}\|_{H^{3/2}}^2 + 2 \|\vec{w}\|_{H^{3/2}} \|\vec{h}\|_{H^{-1/2}}$$

$$+C \|\vec{w}\|_{\dot{H}^1} (\|\vec{v}\|_{L^4(\mathbb{R}^2)} \|\vec{w}\|_{\dot{H}^{3/2}} + \|\vec{\nabla} \otimes \vec{v}\|_{L^2(\mathbb{R}^2)} \|\vec{w}\|_{\dot{H}^1}) + C \|\vec{w}\|_{H^{3/2}} \|\vec{w}\|_{\dot{H}^1}^2$$

$$\leq -2\nu \|\vec{w}\|_{H^{3/2}}^2 + 2 \|\vec{w}\|_{H^{3/2}} \|\vec{h}\|_{H^{-1/2}} + C \|\vec{w}\|_{\dot{H}^{1/2}}^{1/2} \|\vec{v}\|_{L^4(\mathbb{R}^2)} \|\vec{w}\|_{\dot{H}^{3/2}}^{3/2}$$

$$+C \|\vec{\nabla} \otimes \vec{v}\|_{L^2(\mathbb{R}^2)} \|\vec{w}\|_{\dot{H}^{1/2}} \|\vec{w}\|_{\dot{H}^{3/2}} + C \|\vec{w}\|_{H^{3/2}}^2 \|\vec{w}\|_{\dot{H}^{1/2}}$$

$$\leq -\frac{\nu}{2} \|\vec{w}\|_{H^{3/2}}^2 + \frac{1}{\nu} \|\vec{h}\|_{H^{-1/2}}^2$$

$$+C_6 \|\vec{w}\|_{\dot{H}^{1/2}}^2 \left(\frac{1}{\nu^4} \|\vec{v}\|_{L^4(\mathbb{R}^2)} + \frac{1}{\nu^2} \|\vec{\nabla} \otimes \vec{v}\|_{L^2(\mathbb{R}^2)}^2 \right) + C_7 \|\vec{w}\|_{H^{3/2}}^2 \|\vec{w}\|_{\dot{H}^{1/2}}$$

As long as $2C_7\|\vec{w}\|_{\dot{H}^{1/2}} < \nu$, we find that

$$\|\vec{w}\|_{\dot{H}^{1/2}} \leq (\|\vec{w}_0\|_{\dot{H}^{1/2}} + \frac{1}{\nu}\int_0^T \|\vec{h}\|_{\dot{H}^{-1/2}}^2\, ds)\, e^{C_6 \int_0^t \frac{1}{\nu^4}\|\vec{v}\|_{L^4(\mathbb{R}^2)} + \frac{1}{\nu^2}\|\vec{\nabla}\otimes\vec{v}\|_{L^2(\mathbb{R}^2)}^2\, ds}$$

Thus, we have global existence on $(0,T)$, provided that

$$\|\vec{w}_0\|_{\dot{H}^{1/2}} + \frac{1}{\nu}\int_0^T \|\vec{h}\|_{\dot{H}^{-1/2}}^2\, ds <$$
$$\min(\frac{1}{16 C_1 C_2}, \frac{\nu}{2 C_7})e^{-C_6 \int_0^T \frac{1}{\nu^4}\|\vec{v}\|_{L^4(\mathbb{R}^2)} + \frac{1}{\nu^2}\|\vec{\nabla}\otimes\vec{v}\|_{L^2(\mathbb{R}^2)}^2\, ds}.$$

If we want to get a criterion for existence on $(0,+\infty)$, we need that

$$\int_0^{+\infty} \frac{1}{\nu^4}\|\vec{v}\|_{L^4(\mathbb{R}^2)} + \frac{1}{\nu^2}\|\vec{\nabla}\otimes\vec{v}\|_{L^2(\mathbb{R}^2)}^2\, ds < +\infty.$$

This is the case if $\vec{g} \in L^2\dot{H}^{-1}$ or $\vec{g} \in L^1 L^2$: we start from the energy balance

$$\frac{d}{dt}\|\vec{v}\|_2^2 = -2\nu\|\vec{\nabla}\otimes\vec{v}\|_2^2 + 2\langle\vec{v}|\vec{g}\rangle.$$

- if $\vec{g} \in L^2\dot{H}^{-1}$, we get

$$\frac{d}{dt}\|\vec{v}\|_2^2 \leq -\nu\|\vec{\nabla}\otimes\vec{v}\|_2^2 + \frac{1}{\nu}\|\vec{g}\|_{\dot{H}^{-1}}^2$$

so that

$$\|\vec{v}(t,.)\|_2^2 + \nu\int_0^t \|\vec{\nabla}\otimes\vec{v}\|_2^2\, ds \leq \|\vec{v}_0\|_2^2 + \frac{1}{\nu}\int_0^t \|\vec{g}\|_{\dot{H}^{-1}}^2\, ds$$

- if $\vec{g} \in L^1 L^2$, we get

$$\frac{d}{dt}\|\vec{v}\|_2^2 \leq 2\|\vec{g}\|_2\|\vec{v}\|_2$$

so that

$$\|\vec{v}(t,.)\|_2 \leq \|\vec{v}_0\|_2 + \int_0^t \|\vec{g}\|_2\, ds$$

and

$$2\nu\int_0^t \|\vec{\nabla}\otimes\vec{v}\|_2^2\, ds \leq (\|\vec{v}_0\|_2^2 + \int_0^t \|\vec{g}\|_2\, ds)^2. \qquad \square$$

A special example of two-and-a-half dimensional flow is the *parallel flow*: assume that $\vec{u}$ is a solution of $\partial_t \vec{u} + \vec{u}.\vec{\nabla}\vec{u} = \nu\Delta\vec{u} - \vec{\nabla}p$, div $\vec{u} = 0$ that depends only on t, x_1, x_3 and that moreover $u_3 = 0$. Then the condition div $\vec{u} = 0$ gives $\partial_1 u_1 = 0$, so that $u_1 = u_1(t,x_3)$, while $u_2 = u_2(t,x_1,x_3)$. Moreover, $\vec{u}\cdot\vec{\nabla}\vec{u} = u_1\partial_1\vec{u} = (0, u_1\partial_1 u_2, 0)$ and div$(\vec{u}\cdot\vec{\nabla}\vec{u}) = 0$, so that the pressure is equal to 0. Finally, the Navier—Stokes system is transformed into a linear heat equation $\partial_t u_1 = \nu\partial_3^2 u_1$ and a linear advection-diffusion equation $\partial_t u_2 = \nu(\partial_1^2 + \partial_3^2)u_2 - u_1(t,x_3)\partial_1 u_2$, which are easily solved.

10.3   Axisymmetrical Solutions

In this section, we consider the Navier–Stokes problem with the following symmetry property: $\vec{u}$ is invariant under the action of rotations around the x_3 axis (i.e., $\vec{u}$ is axisymmetric). In order to describe those solutions, we shall use the cylindrical coordinates: $r > 0$, $\theta \in (-\pi, \pi)$, $z \in \mathbb{R}$, with $x_1 = r\cos\theta$, $x_2 = r\sin\theta$ and $x_3 = z$. We then write $\vec{u}(x_1, x_2, x_3) = U_r \vec{e}_r + U_\theta \vec{e}_\theta + U_z \vec{e}_z$, with $\vec{e}_r = (\cos\theta, \sin\theta, 0)$, $\vec{e}_\theta = (-\sin\theta, \cos\theta, 0)$ and $\vec{e}_z = (0, 0, 1)$. Let us remark however that this change of coordinates is degenerated on the axis $r = 0$.

Our hypothesis is that $\vec{u}$ is axisymmetric: it means that U_r, U_θ and U_z do not depend on θ, or equivalently:

$$\partial_\theta \vec{u} = \vec{e}_z \wedge \vec{u}. \tag{10.13}$$

If we want for a locally square integrable axisymmetric vector field $\vec{u} = U_r \vec{e}_r + U_\theta \vec{e}_\theta + U_z \vec{e}_z$ to belong to $\dot{M}^{2,3}$, we may suppose that

$$\iint_{(0,+\infty)\times\mathbb{R}} |U_r(r, z)|^2 + |U_\theta(r, z)|^2 + |U_z(r, z)|^2 \, dr \, dz < +\infty \tag{10.14}$$

i.e., $\vec{U} \in (L^2((0, +\infty) \times \mathbb{R}))^3$, or equivalently that $\frac{1}{(x_1^2 + x_2^2)^{1/4}} \vec{u} \in L^2(\mathbb{R}^3)$. Indeed, if we integrate $|\vec{u}|^2$ on $B(x_0, R)$, with $x_0 = (r_0 \cos\theta_0, r_0 \sin\theta_0, z_0)$, we find:

- if $r_0 < 9R$,

$$\int_{B(x_0, R)} |\vec{u}|^2 \, dx \leq \iint_{0 < r < 10R, |z - z_0| < R} |\vec{U}(r, z)|^2 \, r \, dr \, dz \leq 10R \|\vec{U}\|_2^2$$

- if $r_0 > 9R$ and $|x - x_0| < R$, then $\frac{8r_0}{9} < r < \frac{10r_0}{9}$, $|z - z_0| < R$ and $|\theta - \theta_0| < \frac{\pi}{2} \frac{9R}{8r_0}$ so that

$$\int_{B(x_0, R)} |\vec{u}|^2 \, dx \leq \pi \frac{9R}{8r_0} \iint_{r < \frac{10r_0}{9}, |z - z_0| < R} |\vec{U}(r, z)|^2 \, r \, dr \, dz \leq \frac{5\pi}{4} R \|\vec{U}\|_2^2$$

Thus, if $\vec{u}_0 \in L^2(\frac{1}{\sqrt{x_1^2 + x_2^2}} \, dx)$ and is axisymmetric (with $\vec{e}_z$ as symmetry axis), then $\vec{u}_0 \in \dot{M}^{2,3}$.

Conversely, if $\vec{u}$ is a regular (axisymmetric) field ($\vec{u}_0 \in \dot{H}^{1/2}(\mathbb{R}^3)$), we use the embedding $\dot{H}^{1/2} \subset L^2_{x_3} \dot{H}^{1/2}_{x_1, x_2}$ and the Hardy inequality $\dot{H}^{1/2}_{x_1, x_2} \subset L^2(\frac{1}{r} \, dx_1 \, dx_2)$, to get that $\vec{u}_0 \in L^2(\frac{1}{\sqrt{x_1^2 + x_2^2}} \, dx)$.

Thus, the assumption $\vec{u}_0 \in L^2(\frac{1}{\sqrt{x_1^2 + x_2^2}} \, dx)$ is quite natural for the study of axisymmetric fields. Since the smooth function that are compactly supported in $(0, +\infty) \times \mathbb{R}$ are dense in $L^2((0, +\infty) \times \mathbb{R})$, we see that smooth compactly supported axisymmetric fields are dense in the spaces of axisymmetric fields that belong to $L^2(\frac{1}{\sqrt{x_1^2 + x_2^2}} \, dx)$; for such vector fields $\vec{u}_0$, we have $\vec{u}_0 \in \dot{M}^{2,3}$ and $\lim_{t \to 0^+} \sqrt{t} \|W_{\nu t} * \vec{u}_0\|_\infty = 0$. When $\vec{f} = 0$, this ensures the local existence of a mild solution of the Navier–Stokes equations (and global existence if the $\dot{M}^{2,3}$ norm of $\vec{u}_0$ is small enough), by Theorem 8.3.

As underlined by Gallagher, Ibrahim and Majdoub [197], one may prove directly that the Picard algorithm works in the frame of weighted Lebesgue spaces, using the theory of Muckenhoupt weights [248, 448]:

Definition 10.1.
Let (X, ρ, μ) be a space of homogeneous type (see Definition 5.1) and $1 < p < +\infty$. A positive function w on X belongs to the Muckenhoupt class $\mathcal{A}_p$ *if it satisfies the reverse Hölder inequality:*

$$\sup_{x_0 \in X, r > 0} \frac{1}{\mu(B(x_0, r))} \Big(\int_{B(x_0,r)} w(x)\, d\mu \Big)^{1/p} \Big(\int_{B(x_0,r)} w(x)^{-\frac{1}{p-1}}\, d\mu \Big)^{1-1/p} < +\infty.$$

Then the theory of singular integrals on spaces of homogeneous type [125, 313] allows one to prove the following facts:

- if $w \in \mathcal{A}_p(X)$, then the Hardy–Littlewood maximal operator is bounded on $L^p(w\, d\mu)$:

$$\|\mathcal{M}f\|_{L^p(w\, d\mu)} \leq C \|f\|_{L^p(w\, d\mu)}.$$

- if T is a bounded Calderón–Zygmund operator on $L^2(X, d\mu)$, then it can be extended as a bounded operator on $L^p(X, w\, d\mu)$

- if T is a bounded Calderón–Zygmund operator from $L^q(X, d\mu, L^{p_0}(X_0, d\mu_0))$ to $L^q(X, d\mu, L^{p_1}(X_1, d\mu_1))$ for some $1 < q < +\infty$ (where X_0 and X_1 are locally compact σ-compact metric spaces and μ_0 and μ_1 are regular Borel measures on X_0 and X_1), it can be extended as a bounded operator on $L^p(X, w\, d\mu, L^{p_0}(X_0, d\mu_0))$ to $L^p(X, w\, d\mu, L^{p_1}(X_1, d\mu_1))$

We shall use as well a variant of Hedberg's inequality (see Lemma 5.3):

Lemma 10.1.
If $\sigma > 0$, $f \in \dot{B}^{-\sigma}_{\infty,\infty}(\mathbb{R}^3)$ and if $\vec\nabla f \in L^1_{\text{loc}}$, then we have the pointwise inequality

$$|f(x)| \leq C(\mathcal{M}_{\vec\nabla f}(x))^{\frac{\sigma}{1+\sigma}} \|f\|_{\dot{B}^{-\sigma}_{\infty,\infty}}^{\frac{1}{1+\sigma}}.$$

Proof. We write

$$f = -\int_0^{+\infty} \Delta W_t * f\, dt = \int_0^R \Delta W_t * f\, dt + \int_R^{+\infty} \Delta W_t * f\, dt = A_R(x) + B_R(x).$$

We have

$$|A_R(x)| \leq \int_0^R |\text{div}(W_t * \vec\nabla f)|\, dt \leq C \int_0^R \frac{1}{\sqrt{t}} \mathcal{M}_{\vec\nabla f}(x)\, dt = 2C\sqrt{R}\, \mathcal{M}_{\vec\nabla f}(x)$$

and

$$|B_R(x)| \leq \int_R^{+\infty} |(\Delta W_t) * f|\, dt \leq C \int_R^{+\infty} \frac{1}{t^{1+\frac{\sigma}{2}}} \|f\|_{\dot{B}^{-\sigma}_{\infty,\infty}}\, dt = \frac{2C}{\sigma} \frac{1}{R^{\frac{\sigma}{2}}} \|f\|_{\dot{B}^{-\sigma}_{\infty,\infty}}.$$

We end the proof by taking

$$R = \left(\frac{\|f\|_{\dot{B}^{-\sigma}_{\infty,\infty}}}{\mathcal{M}_{\vec\nabla f}(x)} \right)^{\frac{2}{1+\sigma}}. \qquad \square$$

Then the study of axisymmetric solutions $\vec u \in L^\infty((0,T), L^2(\frac{1}{\sqrt{x_1^2 + x_2^2}}\, dx))$ can be done by noticing that $\frac{1}{\sqrt{x_1^2 + x_2^2}}$ belongs to $\mathcal{A}_2(\mathbb{R}^3)$ and that

$$\|\vec u\|_{\dot{B}^{-1}_{\infty,\infty}} \leq C \|\vec u\|_{\dot{M}^{2,3}} \leq C' \|\vec u\|_{L^2(\frac{1}{\sqrt{x_1^2 + x_2^2}}\, dx)}.$$

Proposition 10.2.
Let w be a weight on $\mathbb{R}^3$ such that $w \in \mathcal{A}_2(\mathbb{R}^3, dx)$. Then:

- $|W_{\nu t} * f(x)| \leq \mathcal{M}_f(x)$ *for* $f \in L^2(w \, dx)$

- *if* $f \in L^2(w \, dx)$, *then* $W_{\nu t} * f$ *belongs to* $L_t^\infty L^2(w \, dx)$, *and* $\vec{\nabla} W_{\nu t} * f$ *belongs to* $L_t^2 L^2(w \, dx)$:

$$\sup_{t>0} \|W_{\nu t} * f(t, .)\|^2_{L^2(w \, dx)} + \nu \int_0^{+\infty} \|\vec{\nabla} W_{\nu t} * f\|^2_{L^2(w \, dx)} \, dt \leq C \|f\|^2_{L^2(w \, dx)} \quad (10.15)$$

- *if* $g \in L^2((0, +\infty), L^2(w \, dx)$ *and* $\vec{G} = \int_0^t \vec{\nabla} W_{\nu(t-s)} * g(s, .) \, ds$, *then* $\vec{G}$ *belongs to* $L_t^\infty L^2(w \, dx)$ *and* $\vec{\nabla} \otimes \vec{G}$ *belongs to* $L_t^2 L^2(w \, dx)$:

$$\nu \sup_{t>0} \|\vec{G}(t, .)\|^2_{L^2(w \, dx)} + \nu^2 \int_0^{+\infty} \|\vec{\nabla} \otimes \vec{G}(t, .)\|^2_{L^2(w \, dx)} \, dt \leq C \|g\|^2_{L^2(w \, dx)} \quad (10.16)$$

- *if g belongs to* $L_t^\infty \dot{B}_{\infty,\infty}^{-1}$ *and* $\vec{\nabla} g$ *belongs to* $L_t^2 L^2(w \, dx)$, *then g belongs to* $L_t^4 L^4(w \, dx)$ *and*

$$\|g\|_{L^4 L^4(w \, dx)} \leq C \sqrt{\|g\|_{L^\infty \dot{B}_{\infty,\infty}^{-1}} \|\vec{\nabla} g\|_{L^2 L^2(w \, dx)}} \quad (10.17)$$

Proof. From the inequality $|W_{\nu t} * f(x)| \leq \mathcal{M}_f(x)$ (Lemma 7.4), we get

$$\sup_{t>0} \|W_{\nu t} * f(t, .)\|_{L^2(w \, dx)} \leq \|\mathcal{M}_f\|_{L^2(w \, dx)} \leq C \|f\|_{L^2(w \, dx)}.$$

To estimate $\sqrt{\nu} \|\vec{\nabla} W_{\nu t} * f\|_{L^2 L^2(w \, dx)}$, we first remark that with no loss of generality we may assume that $\nu = 1$ (changing t into $\tau = \nu t$). We then write $L_t^2 L^2(w \, dx) = L^2(w \, dx) L_t^2$.

The mapping $f \mapsto (\vec{\nabla} W_t * f)_{t>0}$ is bounded from $L^2(\mathbb{R}^3)$ to $L^2(dx, L^2(dt))$. Moreover, we have, for $h \in L^2(\mathbb{R})$, $\| \int_{-\infty}^{+\infty} \vec{\nabla} W_{\nu(t-s)}(x) h(s) \, ds \|_{L^2(dt)} \leq \int_0^{+\infty} |\vec{\nabla} W_{\nu t}(x)| \, dt \|h\|_2 = C |x|^{-3} \|h\|_2$ and, for $i = 1, \ldots, 3$,

$$\| \int_{-\infty}^t \partial_i \vec{\nabla} W_{\nu(t-s)}(x) h(s) \, ds \|_{L^2(dt)} \leq \int_0^{+\infty} |\partial_i \vec{\nabla} W_{\nu t}(x)| \, dt \|h\|_2 = C |x|^{-4} \|h\|_2$$

Thus, we may apply the theory of singular integrals with values in $L^2(dt)$ and we find that $f \mapsto (\vec{\nabla} W_t * f)_{t>0}$ is bounded from $L^2(w \, dx)$ to $L^2(w \, dx, L^2(dt))$. Inequality (10.15) is proved.

Now, to estimate $\|\vec{G}(t, .)\|_{L^2(w \, dx)}$, we use the fact that $L^2(w \, dx)$ is the dual of $L^2(w^{-1} \, dx)$ and that $w^{-1} \in \mathcal{A}_2$ as well. If $f \in L^2(w^{-1} \, dx)$, we have

$$\int \vec{G}(t, x) f(x) \, dx = - \int_0^t g(s, x) \vec{\nabla} W_{(\nu(t-s))} * f(x) \, dx \, ds$$

so that

$$\left| \int \vec{G}(t, x) f(x) \, dx \right| \leq \|g\|_{L_t^2 L^2(w \, dx)} \|\vec{\nabla} W_{(\nu t} * f\|_{L_t^2 L^2(w^{-1} \, dx)}$$

$$\leq C \frac{1}{\sqrt{\nu}} \|g\|_{L_t^2 L^2(w \, dx)} \|f\|_{L^2(w^{-1} \, dx)}$$

Now, to estimate $\|\vec{\nabla} \otimes \vec{G}\|_{L_t^2 L^2(w\,dx)}$, we shall use the theory of the maximal regularity for the heat kernel [313]. As the Riesz transforms are bounded on $L^2(w\,dx)$, we just have to estimate $\|\int_0^t \Delta W_{\nu(t-s)} * g(s,.)\,ds\|_{L_t^2 L^2(w\,dx)}$. The operator $h \mapsto \int_{-\infty}^t \Delta W_{\nu(t-s)} * h(s,.)\,ds$ may be viewed as a Calderón–Zygmund operator on the parabolic space $\mathbb{R} \times \mathbb{R}^3$, which is a space of homogeneous type with quasi-metric $\rho((t,x),(s,y)) = ((t-s)^2 + |x-y|)^{1/4}$ and measure $d\mu = dt\,dx$. The boundedness of this operator on $L^2(\mathbb{R} \times \mathbb{R}^3)$ is obvious, as this is a convolution in $\mathbb{R}^4$ with a kernel $K(t,x)$ whose Fourier transform is $\hat{K}(\tau,\xi) = \frac{-|\xi|^2}{i\tau - |\xi|^2}$. As a function of (t,x) (independent of t), w still is a Muckenhoupt weight: $w(x) \in \mathcal{A}_2(\mathbb{R} \times \mathbb{R}^3)$. Thus, we have

$$\|\int_{-\infty}^t \Delta W_{\nu(t-s)} * h(s,.)\,ds\|_{L^2(w(x)\,dt\,dx)} \leq C\|h\|_{L^2(w(x)\,dt\,dx)}.$$

We conclude by taking $h = 1_{t>0} g$, since $L^2(w(x)\,dt\,dx) = L_t^2 L^2(w\,dx)$.

Let us now consider $g \in L_t^\infty \dot{B}_{\infty,\infty}^{-1}$ with $\vec{\nabla} g \in L_t^2 L^2(w\,dx)$. Hedberg's inequality (Lemma 10.1) gives

$$|g(t,x)| \leq C\|g(t,.)\|_{\dot{B}_{\infty,\infty}^{-1}}^{1/2} (\mathcal{M}_{\vec{\nabla} g(t,.)}(x))^{1/2}$$

with

$$\int ((\mathcal{M}_{\vec{\nabla} g(t,.)}(x))^{1/2})^4 \, w(x)\,dx \leq C \int |\vec{\nabla} g(t,x)|^2 \, w(x)\,dx. \qquad \square$$

Navier–Stokes equations and Muckenhoupt weights

Theorem 10.3.
Let $w \in \mathcal{A}_2(\mathbb{R}^3)$ be such that, for axisymmetric vector fields $\vec{u}$ in $L^2(w\,dx)$, we have the inequality

$$\|\vec{u}\|_{\dot{B}_{\infty,\infty}^{-1}} \leq C\|\vec{u}\|_{L^2(w\,dx)}.$$

Let $\vec{u}_0 \in L^2(w\,dx)$ be an axisymmetric vector field with $\operatorname{div} \vec{u}_0 = 0$ and let $\vec{f}$ be an axisymmetric forcing term that can be written as $\vec{f} = \sqrt{-\Delta} \vec{F}$, with $\vec{F} \in L^2(w\,dx)$. Then there exists a time $T > 0$ such that the problem

$$\begin{cases} \partial_t \vec{u} + \vec{u}.\vec{\nabla}\vec{u} = \nu \Delta \vec{u} + \vec{f} - \vec{\nabla} p \\ \vec{u}(0,.) = \vec{u}_0 \\ \operatorname{div} \vec{u} = 0 \end{cases} \qquad (10.18)$$

has a unique axisymmetric solution $\vec{u}$ on $(0,T) \times \mathbb{R}^3$ with $\vec{u} \in L_t^\infty L^2(w\,dx)$ and $\vec{\nabla} \otimes \vec{u} \in L_t^2 L^2(w\,dx)$.

Moreover, there exists a positive ϵ (which does not depend on ν, nor on $\vec{u}_0$ nor $\vec{f}$), such that, when $\|\vec{u}_0\|_{L^2(w\,dx)} < \epsilon\nu$ and $\|\vec{F}\|_{L_t^2 L^2(w\,dx)} < \epsilon\nu^{3/2}$, then the solution is global.

Proof. We are going to solve the problem in the space $L_t^4 L^4(w\,dx)$. Indeed, using (10.15) and (10.17), we find

$$\|W_{\nu t} * \vec{u}_0\|_{L^4 L^4(w\,dx)} \leq C\nu^{-1/4}\|\vec{u}_0\|_{L^2(w\,dx)}$$

and, using (10.16) and (10.17), we find

$$\|\int_0^t \sqrt{-\Delta} W_{\nu(t-s)} * \vec{F}\,ds\|_{L^4 L^4(w\,dx)} \leq C\nu^{-3/4}\|\vec{F}\|_{L^2 L^2(w\,dx)}$$

Moreover, if B is the bilinear operator

$$B(\vec{u}, \vec{v}) = \int_0^t W_{\nu(t-s)} * \operatorname{div}(\vec{u} \otimes \vec{v}) \, ds$$

we find

$$\|B(\vec{u}, \vec{v})\|_{L^4 L^4(w\,dx)} \leq \frac{C}{\nu^{3/4}} \|\vec{u} \otimes \vec{v}\|_{L^2 L^2(w\,dx)} \leq \frac{C_0}{\nu^{3/4}} \|\vec{u}\|_{L^4 L^4(w\,dx)} \|\vec{v}\|_{L^4 L^4(w\,dx)}$$

Thus, we can see that we get a solution of the fixed point problem on $(0, T_0)$ provided that, defining $\vec{U}_0 = W_{\nu t} * \vec{u}_0 + \int_0^t W_{\nu(t-s)} * \mathbb{P}\sqrt{-\Delta}\vec{F} \, ds$, we have

$$\int_0^{T_0} \int |\vec{U}_0(t, x)|^4 \, dx \, dt < \frac{\nu^3}{256 \, C_0^4}.$$

As

$$\int_0^{T_0} \int |\vec{U}_0(t, x)|^4 \, dx \, dt \leq \frac{C_1^4}{2} \left(\frac{1}{\nu} \|\vec{u}_0\|_{L^4(w\,dx)}^4 + \frac{1}{\nu^3} \|\vec{F}\|_{L^2 L^2(w\,dx)}^4 \right)$$

we get global existence for

$$\|\vec{u}_0\|_{L^2(w\,dx)} < \frac{\nu}{4 C_1 C_0} \quad \text{and} \quad \|\vec{F}\|_{L^2 L^2(w\,dx)} < \frac{\nu^{3/2}}{4 C_1 C_0}. \qquad \square$$

Theorem 10.3 gives then the result of [197]:

Locally square integrable axisymmetric solutions

Corollary 10.1.
Let $\vec{u}_0 \in L^2(\frac{1}{\sqrt{x_1^2 + x_2^2}} \, dx)$ with $\operatorname{div} \vec{u}_0 = 0$ be an axisymmetric vector field. If $\|\vec{u}_0\|_{L^2(w\,dx)} < \epsilon \nu$ (where the constant $\epsilon > 0$ does not depend on ν, nor on $\vec{u}_0$), then the problem

$$\begin{cases} \partial_t \vec{u} + \vec{u}.\vec{\nabla}\vec{u} = \nu \Delta \vec{u} - \vec{\nabla}p \\ \vec{u}(0, .) = \vec{u}_0 \\ \operatorname{div} \vec{u} = 0 \end{cases} \qquad (10.19)$$

has a unique axisymmetric solution $\vec{u}$ on $(0, +\infty) \times \mathbb{R}^3$ with $\vec{u} \in L_t^\infty L^2(\frac{1}{\sqrt{x_1^2+x_2^2}} \, dx)$ and $\vec{\nabla} \otimes \vec{u} \in L_t^2 L^2(\frac{1}{\sqrt{x_1^2+x_2^2}} \, dx)$.

In order to describe axisymmetric flows, it is convenient to use cylindrical coordinates: $x_1 = r \cos\theta$, $x_2 = r \sin\theta$ and $x_3 = z$. A scalar function $A(x)$ is axisymmetrical if $\partial_\theta A = 0$. A vector field $\vec{V}(x)$ is axisymmetrical if $\partial_\theta \vec{V} = \vec{e}_z \wedge \vec{V}$. The *swirl* of the vector field is the component V_θ, where $\vec{V} = V_r \vec{e}_r + V_\theta \vec{e}_\theta + V_z \vec{e}_z$.

Vector calculus in cylindrical coordinates

In the open set $\Omega = \{x \in \mathbb{R}^3 \ / \ (x_1, x_2) \neq (0,0)\}$, we have the following formula for scalar functions A and vector fields $\vec{V}$, $\vec{W}$:

- $\partial_1 = \cos\theta\,\partial_r - \sin\theta\,\frac{1}{r}\partial_\theta,\ \partial_2 = \sin\theta\,\partial_r + \cos\theta\,\frac{1}{r}\partial_\theta,\ \partial_3 = \partial_z$ so that, formally,
$\vec{\nabla} = \vec{e}_r\partial_r + \frac{1}{r}\vec{e}_\theta\partial_\theta + \vec{e}_z\partial_z$

- $\vec{\nabla}A = \partial_r A\,\vec{e}_r + \frac{1}{r}\partial_\theta A\,\vec{e}_\theta + \partial_z A\,\vec{e}_z$

- $\Delta A = \partial_r^2 A + \frac{1}{r}\partial_r A + \frac{1}{r^2}\partial_\theta^2 A + \partial_z^2 A$

- $\operatorname{div}\vec{V} = \partial_r V_r + \frac{1}{r}V_r + \frac{1}{r}\partial_\theta V_\theta + \partial_z V_z$

- $|\vec{\nabla}\otimes\vec{V}|^2 = |\partial_r V_r|^2 + |\partial_r V_\theta|^2 + |\partial_r V_z|^2 + |\partial_z V_r|^2 + |\partial_z V_\theta|^2 + |\partial_z V_z|^2 + \frac{1}{r^2}(|\partial_\theta V_r - V_\theta|^2 + |\partial_\theta V_\theta + V_r|^2 + |\partial_\theta V_z|^2)$

- $\vec{\nabla}\wedge\vec{V} = (\frac{1}{r}\partial_\theta V_z - \partial_z V_\theta)\vec{e}_r + (\partial_z V_r - \partial_r V_z)\vec{e}_\theta + (\partial_r V_\theta - \frac{1}{r}\partial_\theta V_r + \frac{1}{r}V_\theta)\vec{e}_z$

- $(\vec{V}.\vec{\nabla})A = V_r\partial_r A + \frac{1}{r}V_\theta\partial_\theta A + V_z\partial_z A$

- $(\vec{V}.\vec{\nabla})\vec{W} = (V_r\partial_r W_r + \frac{1}{r}V_\theta\partial_\theta W_r + V_z\partial_z W_r - \frac{1}{r}V_\theta W_\theta)\vec{e}_r + (V_r\partial_r W_\theta + \frac{1}{r}V_\theta\partial_\theta W_\theta + V_z\partial_z W_\theta + \frac{1}{r}V_\theta W_r)\vec{e}_\theta + (V_r\partial_r W_z + \frac{1}{r}V_\theta\partial_\theta W_z + V_z\partial_z W_z)\vec{e}_z$

- $\Delta\vec{V} = (\partial_r^2 V_r + \frac{1}{r}\partial_r V_r + \frac{1}{r^2}\partial_\theta^2 V_r + \partial_z^2 V_r - \frac{1}{r^2}V_r - 2\frac{1}{r^2}\partial_\theta V_\theta)\vec{e}_r + (\partial_r^2 V_\theta + \frac{1}{r}\partial_r V_\theta + \frac{1}{r^2}\partial_\theta^2 V_\theta + \partial_z^2 V_\theta - \frac{1}{r^2}V_\theta)\vec{e}_\theta + (\partial_r^2 V_z + \frac{1}{r}\partial_r V_z + \frac{1}{r^2}\partial_\theta^2 V_z + \partial_z^2 V_z)\vec{e}_z$

For axisymmetric functions or vector fields, the ∂_θ terms vanish and we find:

- $\vec{\nabla}A = \partial_r A\,\vec{e}_r + \partial_z A\,\vec{e}_z$

- $\Delta A = \partial_r^2 A + \frac{1}{r}\partial_r A + \partial_z^2 A$

- $\operatorname{div}\vec{V} = \partial_r V_r + \frac{1}{r}V_r + \partial_z V_z$

- $|\vec{\nabla}\otimes\vec{V}|^2 = |\partial_r V_r|^2 + |\partial_r V_\theta|^2 + |\partial_r V_z|^2 + |\partial_z V_r|^2 + |\partial_z V_\theta|^2 + |\partial_z V_z|^2 + \frac{1}{r^2}(|V_r|^2 + |V_\theta|^2)$

- $\vec{\nabla}\wedge\vec{V} = -\partial_z V_\theta\vec{e}_r + (\partial_z V_r - \partial_r V_z)\vec{e}_\theta + (\partial_r V_\theta + \frac{1}{r}V_\theta)\vec{e}_z$

- $(\vec{V}.\vec{\nabla})A = V_r\partial_r A + V_z\partial_z A$

- $(\vec{V}.\vec{\nabla})\vec{W} = (V_r\partial_r W_r + V_z\partial_z W_r - \frac{1}{r}V_\theta W_\theta)\vec{e}_r + (V_r\partial_r W_\theta + V_z\partial_z W_\theta + \frac{1}{r}V_\theta W_r)\vec{e}_\theta + (V_r\partial_r W_z + V_z\partial_z W_z)\vec{e}_z$

- $\Delta\vec{V} = (\partial_r^2 V_r + \frac{1}{r}\partial_r V_r + \partial_z^2 V_r - \frac{1}{r^2}V_r)\vec{e}_r + (\partial_r^2 V_\theta + \frac{1}{r}\partial_r V_\theta + \partial_z^2 V_\theta - \frac{1}{r^2}V_\theta)\vec{e}_\theta + (\partial_r^2 V_z + \frac{1}{r}\partial_r V_z + \partial_z^2 V_z)\vec{e}_z$

Thus, if we consider the axisymmetric solution $\vec{u}$ of the Navier–Stokes problem (with axisymmetric forcing term $\vec{f}$ and axisymmetric pressure p), we find the following evolution equation for the swirl u_θ of $\vec{u}$:

$$\partial_t u_\theta = \nu(\partial_r^2 u_\theta + \frac{1}{r}\partial_r u_\theta + \partial_z^2 u_\theta - \frac{1}{r^2}u_\theta) + f_\theta$$
$$- (u_r\partial_r u_\theta + u_z\partial_z u_\theta + \frac{1}{r}u_\theta u_r) \tag{10.20}$$

Thus, if the force $\vec{f}$ has no swirl ($f_\theta = 0$) and the initial value $\vec{u}(0,.)$ has no swirl ($u_\theta(0,.) = 0$), then the solution $\vec{u}$ will still have no swirl:

$$u_\theta = 0 \tag{10.21}$$

and in this case the vorticity $\vec{\omega}$ is very simple:

$$\vec{\omega} = \omega_\theta \vec{e}_\theta = (\partial_z u_r - \partial_r u_z)\vec{e}_\theta. \tag{10.22}$$

We are going to consider a solution $\vec{u}$ that is locally in time $L_t^\infty H^2 \cap L_t^2 H^3$. Under the condition that $\vec{u}$ is divergence free and axisymmetrical without swirl, we find that:

- the norm $\|\vec{u}\|_{\dot{H}^1}$ is equivalent to $\|\vec{\omega}\|_2 = \|\omega_\theta\|_2$

- the norm $\|\vec{u}\|_{\dot{H}^2}$ is equivalent to $\|\vec{\nabla} \otimes \vec{\omega}\|_2 = \sqrt{\|\partial_r \omega_\theta\|_2^2 + \|\partial_z \omega_\theta\|_2^2 + \|\frac{1}{r}\omega_\theta\|_2^2}$

- the norm $\|\vec{u}\|_{\dot{H}^3}$ is equivalent to $\|\Delta\vec{\omega}\|_2 = \|\partial_r^2 \omega_\theta + \partial_z^2 \omega_\theta + \partial_r(\frac{\omega_\theta}{r})\|_2$

Moreover, we have:

- $(\partial_z^2 \omega_\theta)\vec{e}_\theta = \partial_3^2\vec{\omega}$, hence $\|\partial_z^2 \omega_\theta\|_2 \leq \|\vec{\omega}\|_{\dot{H}^2}$

- $(\partial_r^2 \omega_\theta)\vec{e}_\theta = (\cos^2\theta\partial_1^2 + \sin^2\theta\partial_2^2 + 2\cos\theta\sin\theta\partial_1\partial_2)\vec{\omega}$ hence $\|\partial_r^2 \omega_\theta\|_2 \leq \|\vec{\omega}\|_{\dot{H}^2}$

- in particular, we get that $\|\partial_r(\frac{\omega_\theta}{r})\|_2 \leq 3\|\vec{\omega}\|_{\dot{H}^2}$

- We find that $\frac{\omega_\theta}{r}$ belongs to $L_{x_3}^2 H_{x_1,x_2}^1$, hence for every $\gamma \in (0,1)$, we have

$$\left\|\frac{\omega_\theta}{r^{1+\gamma}}\right\|_2 \leq C_\gamma \left\|\frac{\omega_\theta}{r}\right\|_{L_{x_3}^2 L_{x_1,x_2}^2}^{1-\gamma} \left\|\frac{\omega_\theta}{r}\right\|_{L_{x_3}^2 \dot{H}_{x_1,x_2}^1}^{\gamma} \leq C_\gamma \|\vec{\omega}\|_{\dot{H}^1}^{1-\gamma}\|\vec{\omega}\|_{\dot{H}^2}^{\gamma}. \tag{10.23}$$

In the case of axisymmetric flows with no swirl, Ladyzhenskaya [295], Uchovskii and Yudovich [486] proved global existence under regularity assumptions on $\vec{u}_0$ and $\vec{f}$ but without any size requirements on the data. We shall follow the very simple proof proposed by Leonardi, Malek, Nečas, and Pokorný [326]. (Abidi [1] proved global existence with very weak (close to optimality) regularity requirements: $\vec{u}_0 \in H^{1/2}$ and $\vec{f} \in L^2 H^{\frac{1}{4}+\epsilon}$ in Abidi's paper, while we assume $\vec{u}_0 \in H^2$ and $\vec{f} \in L^2 H^1$. The regularity on $\vec{u}_0$ is unessential: if $\vec{u}_0$ belongs to $H^{1/2}$ and $\vec{f} \in L^2 H^s$ with $s \geq 1/2$, then the mild solution $\vec{u}$ that belongs to $L^\infty((0,T), H^{1/2})$ for every $T < T^*$ will actually belong to $L^\infty((T_0,T), H^{s+1})$ for every $0 < T_0 < T < T^*$).

Global existence of axisymmetrical solutions without swirl

Theorem 10.4.
Let $\vec{u}_0 \in (H^2(\mathbb{R}^3))^3$ with div $\vec{u}_0 = 0$ be an axisymmetric vector field without swirl, and let $\vec{f} \in L^2((0,T), (H^1(\mathbb{R}^3)^3)$ be axisymmetric without swirl. Then the problem

$$\begin{cases} \partial_t \vec{u} + \vec{u}.\vec{\nabla}\vec{u} = \nu\Delta\vec{u} + \vec{f} - \vec{\nabla}p \\ \vec{u}(0,.) = \vec{u}_0 \\ \text{div } \vec{u} = 0 \end{cases} \tag{10.24}$$

has a unique global axisymmetric solution $\vec{u}$ on $(0,T) \times \mathbb{R}^3$ with $\vec{u} \in L_t^\infty H^2 \cap L^2 H^3$ (if $T < +\infty$).

Proof. Due to Theorem 7.3, we know that there exists a time $T^* < T$ and a unique solution $\vec{u}$ that belongs to $\cap_{0 < T_0 < T^*} L^\infty((0,T_0), H^2) \cap L^2((0,T_0), H^3)$. This solution will be axisymmetric with no swirl. Moreover, if T^* is the maximal existence time, then, if $T < T^*$, we have $\sup_{0 < t < T^*} \|\vec{u}\|_{H^1} = +\infty$.

Due to div $\vec{u} = 0$, we have

$$\int \vec{u}.(\vec{u}.\vec{\nabla})\vec{u}\,dx = 0$$

and

$$\int \vec{u}.\vec{\nabla}p = 0$$

so that the nonlinearity disappears in the energy balance and we find

$$\partial_t \|\vec{u}(t,.)\|_2^2 = 2\int (\nu\Delta\vec{u} + \vec{f}).\vec{u}\,dx \le -2\nu\|\vec{\nabla}\otimes\vec{u}\|_2^2 + \|\vec{f}\|_2^2 + \|\vec{u}\|_2^2$$

and

$$\|\vec{u}(t,.)\|_2^2 + 2\nu\int_0^t \|\vec{u}(s,.)\|_{H^1}^2\,ds \le e^t(\|\vec{u}_0\|_2^2 + \int_0^t \|\vec{f}(s,.)\|_2^2\,ds).$$

In the case of axisymmetric flows without swirl, we have another energy inequality. Let $\vec{\omega} = \operatorname{curl}\vec{u}$. We know that $\vec{\omega}$ is solution of:

$$\partial_t\vec{\omega} = \nu\Delta\vec{\omega} + (\vec{\omega}.\vec{\nabla})\vec{u} - (\vec{u}.\vec{\nabla})\vec{\omega} + \operatorname{curl}\vec{f}.$$

$\vec{\omega}$ belongs, for any $T_0 < T^*$, to $L^\infty((0,T_0), H^1) \cap L^2((0,T_0), H^2)$. Thus, $\vec{\nabla}\otimes\vec{\omega}$ belongs to $L^\infty((0,T_0), L^2)$. We have $\vec{\omega} = \omega_\theta \vec{e}_\theta = (\partial_z u_r - \partial_r u_z)\vec{e}_\theta$ and

$$|\vec{\nabla}\otimes\vec{\omega}|^2 = |\partial_r\omega_\theta|^2 + |\partial_z\omega_\theta|^2 + \frac{|\omega_\theta|^2}{r^2} = |\partial_r\omega_\theta|^2 + |\partial_z\omega_\theta|^2 + \frac{|\vec{\omega}|^2}{r^2}.$$

Ladyzhenskaya's key observation is that we have a uniform control of $\int \frac{|\vec{\omega}|^2}{r^2}\,dx$. This is based on the identity:

$$\int ((\vec{\omega}.\vec{\nabla})\vec{u} - (\vec{u}.\vec{\nabla})\vec{\omega}).\vec{\omega}\frac{dx}{r^2} = 0. \tag{10.25}$$

Before proving this identity, we notice that the integral is well defined: we have $\vec{\nabla}\otimes\vec{\omega} \in L^\infty((0,T_0), L^2) \cap L^2((0,T_0), H^1)$, hence $\frac{1}{r}\vec{\omega}$ belongs to $L^\infty L^2 \cap L^2 L^6 \subset L^4 L^3$, while $\vec{\nabla}\otimes\vec{u}$ belongs to $L^2((0,T_0), H^2) \subset L^2 L^3$, so that we may integrate $|\vec{\omega}|\|\vec{\nabla}\otimes\vec{u}|\frac{1}{r^2}$; similarly, we have $\vec{\nabla}\otimes\vec{\omega} \in L^\infty L^2$, $\frac{1}{r}\vec{\omega} \in L^2 L^6$ and $\frac{1}{r}\vec{u} \in L^2 L^3$.

In order to prove (10.25), we introduce a function α which is smooth on $(0,+\infty)$, equal to 0 on $(0,1)$ and to 1 on $(2,+\infty)$, and $\alpha_\epsilon(r) = \alpha(r/\epsilon)$. We have

$$\int ((\vec{\omega}.\vec{\nabla})\vec{u} - (\vec{u}.\vec{\nabla})\vec{\omega}).\vec{\omega}\alpha_\epsilon(r)\frac{dx}{r^2} = \int ((\vec{\omega}.\vec{\nabla})\vec{u}).\vec{\omega}\frac{\alpha_\epsilon(r)}{r^2} + \frac{1}{2}|\vec{\omega}|^2(\vec{u}.\vec{\nabla})\left(\frac{\alpha_\epsilon(r)}{r^2}\right)dx$$

$$= \int \omega_\theta^2 u_r \frac{\alpha_\epsilon(r)}{r^3} + \frac{1}{2}\omega_\theta^2 u_r \partial_r\left(\frac{\alpha_\epsilon(r)}{r^2}\right)dx$$

$$= \frac{1}{2}\int \omega_\theta^2 u_r \frac{1}{\epsilon}\alpha'(r/\epsilon)\frac{dx}{r^2}$$

with

$$|\frac{1}{2}\int \omega_\theta^2 u_r \frac{1}{\epsilon}\alpha'(r/\epsilon)\frac{dx}{r^2}| \le \|\alpha'\|_\infty \int_{\epsilon < r < 2\epsilon} |\omega_\theta|^2 |u_r| \frac{dx}{r^3}.$$

As $\frac{\omega_\theta}{r} \in L^4 L^3$ and $\frac{u_r}{r} \in L^2 L^3$, we have $\lim_{\epsilon \to 0} \int_{\epsilon < r < 2\epsilon} |\omega_\theta|^2 |u_r| \frac{dx}{r^3} = 0$, and (10.25) is proved.

We may now estimate $\int \frac{|\vec{\omega}|^2}{r^2} dx$. A direct proof would need $\frac{\omega_\theta}{r^2} \in L^2 L^2$, or at least

$$\lim_{\epsilon \to 0} \int_0^{T_0} \int_{\epsilon < r < 2\epsilon} \frac{\omega_\theta^2}{r^4} dx \, ds = 0$$

but we do not have such an estimate on $\vec{\omega}$. A weaker property is that, for $0 < \eta < 1$,

$$\lim_{\epsilon \to 0} \int_0^{T_0} \int_{\epsilon < r < 2\epsilon} r^\eta \frac{\omega_\theta^2}{r^4} dx \, ds = 0. \tag{10.26}$$

This is a consequence of $r^{\eta/2} \frac{\omega_\theta}{r^2} \in L^2 L^2$ for $0 < \eta < 2$ by inequality (10.23).

We thus follow [326] and replace in our computations the function α_ϵ by $r^\eta \alpha_\epsilon$, with $0 < \eta < 1$. We write

$$\partial_t \int \frac{|\vec{\omega}|^2}{r^2} r^\eta \alpha_\epsilon(r) dx = 2 \int \partial_t \vec{\omega} . \vec{\omega} r^\eta \frac{\alpha_\epsilon(r)}{r^2} dx$$

$$= 2 \int (\nu \Delta \vec{\omega} + \operatorname{curl} \vec{f}) . \vec{\omega} r^\eta \frac{\alpha_\epsilon(r)}{r^2} dx + \int \omega_\theta^2 u_r \partial_r (r^\eta \alpha(r/\epsilon)) \frac{dx}{r^2}$$

with

$$\int \operatorname{curl} \vec{f} . \vec{\omega} \frac{r^\eta \alpha_\epsilon(r)}{r^2} dx = \int \vec{f} . \operatorname{curl}(\vec{\omega} \frac{r^\eta \alpha_\epsilon(r)}{r^2}) dx$$

$$= \int (-f_r \partial_z \omega_\theta \frac{r^\eta \alpha_\epsilon(r)}{r^2} + f_r \partial_r (\omega_\theta \frac{r^\eta \alpha_\epsilon(r)}{r^2}) + f_r \omega_\theta \frac{r^\eta \alpha_\epsilon(r)}{r^3}) dx$$

$$= \int (-\frac{f_z}{r} \partial_z (\frac{\omega_\theta}{r}) + \frac{f_r}{r} \partial_r (\frac{\omega_\theta}{r})) r^\eta \alpha_\epsilon(r) + f_r \omega_\theta \frac{1}{r^2} \partial_r (r^\eta \alpha(r/\epsilon)) dx$$

and

$$\int \Delta \vec{\omega} . \vec{\omega} \frac{r^\eta \alpha_\epsilon(r)}{r^2} dx = \int (\frac{\omega_\theta}{r} \partial_z^2 (\frac{\omega_\theta}{r}) + \frac{1}{r} \partial_r (\frac{\omega_\theta}{r} \partial_r \omega_\theta) - (\partial_r (\frac{\omega_\theta}{r})^2) r^\eta \alpha_\epsilon(r) dx$$

$$= - \int ((\partial_r (\frac{\omega_\theta}{r})^2 + (\partial_z (\frac{\omega_\theta}{r})^2)) r^\eta \alpha_\epsilon(r) dx$$

$$+ 2\pi \int \int \partial_r (\frac{\omega_\theta}{r} \partial_r \omega_\theta) r^\eta \alpha_\epsilon(r) dz \, dr$$

$$= - \int ((\partial_r (\frac{\omega_\theta}{r})^2 + (\partial_z (\frac{\omega_\theta}{r})^2)) r^\eta \alpha_\epsilon(r) dx$$

$$+ \int \frac{\omega_\theta^2}{2} \partial_r (\frac{1}{r} \partial_r (r^\eta \alpha_\epsilon(r))) \frac{dx}{r}$$

We find that:

$$\int \frac{|\vec{\omega}|^2}{r^2} r^\eta \alpha_\epsilon(r) dx \leq \int \frac{|\vec{\omega}_0|^2}{r^2} r^\eta \alpha_\epsilon(r) dx + \frac{1}{\nu} \int_0^t \int \frac{|\vec{f}|^2}{r^2} r^\eta \alpha_\epsilon(r) dx \, ds$$

$$- \nu \int_0^t \int ((\partial_r (\frac{\omega_\theta}{r})^2 + (\partial_z (\frac{\omega_\theta}{r})^2)) r^\eta \alpha_\epsilon(r) dx \, ds$$

$$- \eta \int_0^t \int (\omega_\theta^2 u_r + 2 f_r \omega_\theta + \nu(2 - \eta) \frac{1}{r} \omega_\theta^2) \frac{r^\eta}{r^3} \alpha(r/\epsilon) dx \, ds$$

$$+ C \int_0^t \int_{\epsilon < r < 2\epsilon} (\omega_\theta^2 |u_r| + |f_r \omega_\theta|) \frac{r^\eta}{r^3} + \omega_\theta^2 \frac{r^\eta}{r^4} dx \, ds$$

Letting ϵ go to 0, we find

$$\int \frac{|\vec{\omega}|^2}{r^2} r^\eta dx \leq \int \frac{|\vec{\omega}_0|^2}{r^2} r^\eta dx + \frac{1}{\nu} \int_0^t \int \frac{|\vec{f}|^2}{r^2} r^\eta \, dx \, ds$$

$$- \nu \int_0^t \int ((\partial_r(\frac{\omega_\theta}{r}))^2 + (\partial_z(\frac{\omega_\theta}{r}))^2))r^\eta \, dx \, ds$$

$$- \eta \int_0^t \int (\omega_\theta^2 u_r + 2 f_r \omega_\theta + \nu(2 - \eta)\frac{1}{r}\omega_\theta^2)\frac{r^\eta}{r^3} \, dx \, ds$$

$$\leq \int \frac{|\vec{\omega}_0|^2}{r^2} r^\eta dx + +\frac{1}{\nu} \int_0^t \int \frac{|\vec{f}|^2}{r^2} r^\eta \, dx \, ds$$

$$- \eta \int_0^t \int \omega_\theta^2 u_r \frac{r^\eta}{r^3} \, dx \, ds + \frac{\eta}{\nu} \int_0^t \int \frac{|\vec{f}|^2}{r^2} r^\eta \, dx \, ds$$

Letting η go to 0, we find

$$\int \frac{|\vec{\omega}|^2}{r^2} dx \leq \|\vec{\omega}_0\|_{\dot{H}^1}^2 + \frac{1}{\nu} \int_0^t \|\vec{f}\|_{\dot{H}^1}^2 \, ds \qquad (10.27)$$

The end of the proof is now easy. We have

$$\|u_r\|_\infty \leq C\|\vec{u}\|_{\dot{H}^1}^{1/2}\|\vec{u}\|_{\dot{H}^2}^{1/2} \leq C'\|\vec{\omega}\|_2^{1/2}\|\vec{\omega}\|_{\dot{H}^1}^{1/2}$$

so that

$$\partial_t(\|\vec{\omega}\|_2^2) = 2\int \partial_t\vec{\omega}.\vec{\omega} \, dx$$

$$= 2\int (\nu\Delta\vec{\omega} + \text{curl}\,\vec{f} + \vec{\omega}.\vec{\nabla}\vec{u} - \vec{u}.\vec{\nabla}\vec{\omega}).\vec{\omega} \, dx$$

$$= - 2\nu\|\vec{\omega}\|_{\dot{H}^1}^2 + 2\int \vec{f}.\,\text{curl}\,\vec{\omega} \, dx + 2\int u_r\omega_\theta^2\frac{1}{r} \, dx$$

$$\leq - \nu\|\vec{\omega}\|_{\dot{H}^1}^2 + \frac{1}{\nu}\|\vec{f}\|_2^2 + \|u_r\|_\infty\|\vec{\omega}\|_2\|\frac{\vec{\omega}}{r}\|_2$$

$$\leq \frac{1}{\nu}\|\vec{f}\|_2^2 + \frac{C}{\nu^{1/3}}\|\vec{\omega}\|_2^2\|\frac{\vec{\omega}}{r}\|_2^{4/3}$$

We then conclude by Grönwall's lemma and (10.27) that $\|\vec{\omega}\|_2$ remains bounded. □

Gallay and Šverák [203] considered another class of axisymmetric vector fields which lead to global solutions. Their motivation was to investigate global existence in a space which corresponds to the scale invariance of the Navier–Stokes equations (i.e. in a space E such that, for $\vec{u}_0 \in E$, $\|\lambda\vec{u}_0(\lambda x)\|_E = \|\vec{u}_0\|_E$, such as the space $\dot{H}^{1/2}$ considered by Abidi [1] or the space $L^2(\frac{1}{\sqrt{x_1^2+x_2^2}} dx)$ considered by Gallagher, Ibrahim and Majdoub [197]). The case they considered concerns the vorticity and is $\vec{\omega}_0 \in L^1(\frac{1}{\sqrt{x_1^2+x_2^2}} dx)$; their choice of measures as initial vorticities aimed to help to understand the problem of vortex filaments.

Let us remark that, if $\vec{\omega}_0 \in L^1(\frac{1}{\sqrt{x_1^2+x_2^2}} dx)$ and is axisymmetric, then it belongs to the Morrey space $\dot{M}^{1,\frac{3}{2}}$: indeed, $|\vec{\omega}_0|^{1/2}$ is axisymmetric and belongs to $L^2(\frac{1}{\sqrt{x_1^2+x_2^2}} dx)$, and we saw that this implies that $|\vec{\omega}_0|^{1/2} \in \dot{M}^{2,3}$, and thus $\vec{\omega}_0 \in \dot{M}^{1,\frac{3}{2}}$.

For $\vec{\omega}_0 = \omega_{\theta,0}(r,z)\vec{e}_\theta$, the fact that $\vec{\omega}_0 \in L^1(\frac{1}{\sqrt{x_1^2+x_2^2}} dx)$ is equivalent to $\omega_{\theta,0} \in L^1((0,+\infty) \times \mathbb{R}, dr\,dz)$. More generally, if $\omega_{\theta,0}$ is a finite Borel measure $f d\mu$ on $(0,+\infty) \times \mathbb{R}$

(with $|f(r,z)| = 1$ and μ a non-negative finite measure), then $\vec{\omega}_0 \in \dot{M}^{1,3/2}$: if $\vec{\psi} \in (\mathcal{D}(\mathbb{R}^3))^3$ is supported in $B(x_0, R)$ with $x_0 = (r_0 \cos\theta_0, r_0 \sin\theta_0, z_0)$, then we may estimate

$$\langle \vec{\omega}(x) \mid \vec{\psi} \rangle_{\mathcal{D}',\mathcal{D}} = \int_{(0,+\infty)\times\mathbb{R}} f(r,z) r \left(\int_0^{2\pi} \vec{\psi}(r\cos\theta, r\sin\theta, z) \cdot \vec{e}_\theta \, d\theta \right) d\mu(r,z)$$

in the following way:

- if $r_0 < 9R$,

$$\left| \langle \vec{\omega}(x) \mid \vec{\psi} \rangle_{\mathcal{D}',\mathcal{D}} \right| \leq (R + r_0) 2\pi \|\vec{\psi}\|_\infty \mu((0,+\infty)\times\mathbb{R})$$

$$\leq 20\pi R \|\vec{\psi}\|_\infty \mu((0,+\infty)\times\mathbb{R})$$

- if $r_0 > 9R$ and $|x - x_0| < R$, then $\frac{8r_0}{9} < r < \frac{10r_0}{9}$, $|z - z_0| < R$ and $|\theta - \theta_0| < \frac{\pi}{2}\frac{9R}{8r_0}$ so that

$$\left| \langle \vec{\omega}(x) \mid \vec{\psi} \rangle_{\mathcal{D}',\mathcal{D}} \right| \leq \frac{10\, r_0}{9} \frac{9\pi R}{8r_0} \|\vec{\psi}\|_\infty \mu((0,+\infty)\times\mathbb{R})$$

$$\leq \frac{5\pi}{4} R \|\vec{\psi}\|_\infty \mu((0,+\infty)\times\mathbb{R}).$$

The Navier–Stokes equations in Morrey spaces have been studied for many years, with contributions by Giga and Miyakawa [212], Kato [256] and Taylor [467]. In particular, we have the following result (which is a variant of Theorem 8.3):

Proposition 10.3.
Let $\vec{\omega}_0 \in \dot{M}^{1,3/2}$ with $\operatorname{div} \vec{\omega}_0 = 0$. Let $\vec{u}_0$ be the solution of

$$\vec{\nabla} \wedge \vec{u}_0 = \vec{\omega}_0, \operatorname{div} \vec{u}_0 = 0, \vec{u}_0 \in \dot{B}^{-1}_{\infty,\infty}.$$

There exists a constant $\epsilon_\nu > 0$ such that, if, for $0 < T \leq +\infty$, we have

$$\|\vec{\omega}_0\|^3_{\dot{M}^{1,3/2}} \sup_{0<t<T} t \|W_{\nu t} * \vec{\omega}_0\|_\infty < \epsilon_\nu,$$

then the Navier–Stokes equations

$$\vec{u} = W_{\nu t} * \vec{u}_0 - \int_0^t W_{\nu(t-s)} * \mathbb{P} \operatorname{div}(\vec{u} \otimes \vec{u}) \, ds$$

have a solution such that

- $\sup_{0<t<T} \sqrt{t} \|\vec{u}(t,.)\|_\infty < +\infty$

- $\sup_{0<t<T} \|\vec{\omega}(t,.)\|_{\dot{M}^{1,3/2}} < +\infty$, where $\vec{\omega} = \vec{\nabla} \wedge \vec{u}$

- $\sup_{0<t<T} t \|\vec{\omega}(t,.)\|_\infty < +\infty$.

Proof. First, we remark that $\dot{M}^{1,3/2}$ is contained in $\dot{B}_{\infty,\infty}^{-2}$:

$$|\int W_t(x_0 - y)f(y)\,dy| = |\int W(y)f(x - \sqrt{t}y)\,dy|$$

$$\leq \|f(x_0 - \sqrt{t}x)\|_{M^{1,3/2}}\left(1 + \sum_{n=0}^{+\infty} 2^{n+1}\sup_{2^n < |y| \leq 2^{n+1}} W(y)\right)$$

$$\leq C\|f\|_{\dot{M}^{1,3/2}}t^{-1}.$$

If $\vec{\omega}_0$ is a divergence-free vector field in $\dot{M}^{1,3/2}$, we may define

$$\vec{u}_0 = \lim_{t \to +\infty}\int_0^t W_t * (\vec{\nabla} \wedge \vec{\omega}_0)\,ds$$

since

$$\int_1^{+\infty}\|W_t * (\vec{\nabla} \wedge \vec{\omega}_0)\|_\infty\,dt \leq C\|\vec{\omega}_0\|_{\dot{M}^{1,3/2}}\int_1^{+\infty}\frac{dt}{t^{3/2}}.$$

We have

$$\|W_\theta * \vec{u}_0\|_\infty \leq \int_0^{+\infty}\|W_{(t+\theta)/2} * \vec{\nabla} \wedge (W_{(t+\theta)/2} * \vec{\omega}_0)\|_\infty$$

$$\leq C\int_0^{+\infty}\frac{1}{\sqrt{t+\theta}}\frac{\|\vec{\omega}_0\|_{\dot{M}^{1,3/2}}}{t+\theta}\,dt = 2C\frac{\|\vec{\omega}_0\|_{\dot{M}^{1,3/2}}}{\sqrt{\theta}}.$$

Thus, $\vec{u}_0 \in \dot{B}_{\infty,\infty}^{-1}$. We have div $\vec{u}_0 = 0$ and

$$\vec{\nabla} \wedge \vec{u}_0 = -\int_0^{+\infty} W_t * \Delta\vec{\omega}_0\,dt = \vec{\omega}_0$$

since

$$\vec{\nabla} \wedge (\vec{\nabla}\vec{\omega}_0) = -\Delta\vec{\omega}_0 + \vec{\nabla}(\text{div } \vec{\omega}_0) = -\Delta\vec{\omega}_0.$$

We now solve the Navier–Stokes equations

$$\partial_t \vec{u} = \nu\Delta\vec{u} - \mathbb{P}(\vec{\omega} \wedge \vec{u}), \vec{u}(0,.) = \vec{u}_0, \text{div } \vec{u} = 0$$

through Picard iterations: we define $\vec{U}_N$ (and $\vec{\Omega}_N = \vec{\nabla} \wedge \vec{U}_N$) as

$$\vec{U}_0 = W_{\nu t} * \vec{u}_0 \text{ and } \vec{U}_{N+1} = \vec{U}_0 - \int_0^t W_{\nu(t-s)} * \mathbb{P}(\vec{\Omega}_N \wedge \vec{U}_N)\,ds.$$

Taking the curl of

$$\partial_t \vec{U}_{N+1} = \nu\Delta\vec{U}_{N+1} - \mathbb{P}(\vec{\Omega}_N \wedge \vec{U}_N),$$

we find

$$\partial_t \vec{\Omega}_{N+1} = \nu\Delta\Omega_{N+1} + \text{div}(\vec{\Omega}_N \otimes \vec{U}_N - \vec{U}_N \otimes \vec{\Omega}_N)$$

and thus

$$\vec{\Omega}_{N+1} = \vec{\Omega}_0 + \int_0^t W_{\nu(t-s)} * \text{div}(\vec{\Omega}_N \otimes \vec{U}_N - \vec{U}_N \otimes \vec{\Omega}_N)\,ds.$$

Thus, we study the bilinear operator

$$\mathcal{B}((\vec{U},\vec{\Omega}),(\vec{V},\vec{O})) = (-\int_0^t W_{\nu(t-s)} * \mathbb{P}(\vec{\Omega} \wedge \vec{V})\,ds, \int_0^t W_{\nu(t-s)} * \text{div}(\vec{\Omega} \otimes \vec{V} - \vec{U} \otimes \vec{O})\,ds).$$

We will work with the norm

$$\|(\vec{U}, \vec{\Omega})\|_{E_T} = \sup_{0 < t < T} \sqrt{t}\|\vec{U}(t,.)\|_\infty + t^{1/4}\|\vec{\Omega}(t,.)\|_{\dot{M}^{4/3,2}} + t^{3/4}\|\vec{\Omega}(t,.)\|_{\dot{M}^{4,6}}.$$

We have

$$\|\int_0^t W_{\nu(t-s)} * \mathbb{P}(\vec{\Omega} \wedge \vec{V})\, ds\|_\infty \leq \int_0^t \|W_{\nu(t-s)} * \mathbb{P}(\vec{\Omega} \wedge \vec{V})\|_\infty\, ds$$

$$\leq C \int_0^t \frac{1}{(\nu(t-s))^{3/4}}\|\mathbb{P}(\vec{\Omega} \wedge \vec{V})\|_{\dot{M}^{4/3,2}}\, ds$$

$$\leq C \int_0^t \frac{1}{(\nu(t-s))^{3/4}} \frac{ds}{s^{3/4}} \sup_{0<s<t} \sqrt{s}\|\vec{V}\|_\infty \sup_{0<s<t} s^{1/4}\|\vec{\Omega}\|_{M^{4/3,2}}$$

$$\leq C'\nu^{-3/4}\|(\vec{U}, \vec{\Omega})\|_{E_T}\|(\vec{V}, \vec{O})\|_{E_T}\frac{1}{\sqrt{t}}.$$

Similarly, we have

$$\|\int_0^t W_{\nu(t-s)} * \operatorname{div}(\vec{\Omega} \otimes \vec{V} - \vec{U} \otimes \vec{O})\, ds\|_{\dot{M}^{4/3,2}}$$

$$\leq \int_0^t \|W_{\nu(t-s)} * \operatorname{div}(\vec{\Omega} \otimes \vec{V} - \vec{U} \otimes \vec{O})\|_{\dot{M}^{4/3,2}}\, ds$$

$$\leq C \int_0^t \frac{1}{(\nu(t-s))^{1/2}}(\|\vec{\Omega} \otimes \vec{V}\|_{\dot{M}^{4/3,2}} + \|\vec{U} \otimes \vec{O}\|_{\dot{M}^{4/3,2}})\, ds$$

$$\leq C \int_0^t \frac{1}{(\nu(t-s))^{\frac{1}{2}}} \frac{ds}{s^{\frac{3}{4}}} \sup_{0<s<t} (\sqrt{s}\|\vec{V}\|_\infty s^{\frac{1}{4}}\|\vec{\Omega}\|_{M^{\frac{4}{3},2}} + \sqrt{s}\|\vec{U}\|_\infty s^{\frac{1}{4}}\|\vec{O}\|_{M^{\frac{4}{3},2}})$$

$$\leq C'\nu^{-1/2}\|(\vec{U}, \vec{\Omega})\|_{E_T}\|(\vec{V}, \vec{O})\|_{E_T}\frac{1}{t^{1/4}}.$$

Finally, we have

$$\|\int_0^t W_{\nu(t-s)} * \operatorname{div}(\vec{\Omega} \otimes \vec{V} - \vec{U} \otimes \vec{O})\, ds\|_{\dot{M}^{4,6}}$$

$$\leq \int_0^t \|W_{\nu(t-s)} * \operatorname{div}(\vec{\Omega} \otimes \vec{V} - \vec{U} \otimes \vec{O})\|_{\dot{M}^{4,6}}\, ds$$

$$\leq C \int_0^{t/2} \frac{1}{\nu(t-s)}(\|\vec{\Omega} \otimes \vec{V}\|_{\dot{M}^{4/3,2}} + \|\vec{U} \otimes \vec{O}\|_{\dot{M}^{4/3,2}})\, ds$$

$$+ C \int_{t/2}^t \frac{1}{(\nu(t-s))^{1/2}}(\|\vec{\Omega} \otimes \vec{V}\|_{\dot{M}^{4,6}} + \|\vec{U} \otimes \vec{O}\|_{\dot{M}^{4,6}})\, ds$$

$$\leq C \int_0^{t/2} \frac{1}{\nu(t-s)} \frac{ds}{s^{\frac{3}{4}}} \sup_{0<s<t} (\sqrt{s}\|\vec{V}\|_\infty s^{\frac{1}{4}}\|\vec{\Omega}\|_{M^{\frac{4}{3},2}} + \sqrt{s}\|\vec{U}\|_\infty s^{\frac{1}{4}}\|\vec{O}\|_{M^{\frac{4}{3},2}})$$

$$+ C \int_{t/2}^t \frac{1}{(\nu(t-s))^{\frac{1}{2}}} \frac{ds}{s^{\frac{5}{4}}} \sup_{0<s<t} (\sqrt{s}\|\vec{V}\|_\infty s^{\frac{3}{4}}\|\vec{\Omega}\|_{M^{4,6}} + \sqrt{s}\|\vec{U}\|_\infty s^{\frac{3}{4}}\|\vec{O}\|_{M^{4,6}})$$

$$\leq C'(\nu^{-1/2} + \nu^{-1})\|(\vec{U}, \vec{\Omega})\|_{E_T}\|(\vec{V}, \vec{O})\|_{E_T}\frac{1}{t^{3/4}}.$$

Thus, we have

$$\|\mathcal{B}((\vec{U}, \vec{\Omega}), (\vec{V}, \vec{O}))\|_{E_T} \leq C_\nu\|(\vec{U}, \vec{\Omega})\|_{E_T}\|(\vec{V}, \vec{O})\|_{E_T}$$

where C_ν does not depend on T. Moreover, we have

$$
\begin{aligned}
\|W_{\nu t} * \vec\omega_0\|_{\dot M^{4/3,2}} &\leq \|W_{\nu t} * \vec\omega_0\|_{\dot M^{1,3/2}}^{3/4} \|W_{\nu t} * \vec\omega_0\|_\infty^{1/4} \\
&\leq t^{-\frac14} \left(\|\vec\omega_0\|_{\dot M^{1,3/2}}^3 t \|W_{\nu t} * \vec\omega_0\|_\infty \right)^{1/4},
\end{aligned}
$$

$$
\begin{aligned}
\|W_{\nu t} * \vec\omega_0\|_{\dot M^{4,6}} &= \|W_{\nu t/2} * W_{\nu t/2} * \vec\omega_0\|_{\dot M^{4,6}} \\
&\leq \|W_{\nu t/2} * \vec\omega_0\|_{\dot M^{1,3/2}}^{3/4} \|W_{\nu t} * \vec\omega_0\|_\infty^{1/4} \\
&\leq C(\nu t)^{-\frac12} \|W_{\nu t/2} * \vec\omega_0\|_{\dot M^{4/3,2}} \\
&\leq C \frac{1}{\sqrt\nu} t^{-\frac34} \left(\|\vec\omega_0\|_{\dot M^{1,3/2}}^3 t \|W_{\nu t} * \vec\omega_0\|_\infty \right)^{1/4},
\end{aligned}
$$

and

$$
\begin{aligned}
\|W_{\nu t} * \vec u_0\|_\infty &= \| \int_0^{+\infty} W_\tau * \vec\nabla \wedge (W_{\nu t} * \vec\omega_0) \, d\tau \|_\infty \\
&\leq C \int_0^{\nu t} \tau^{-\frac34} \|W_{\nu t} * \vec\omega_0\|_{\dot M^{4,6}} \, d\tau \\
&\quad + C \int_{\nu t}^{+\infty} \tau^{-\frac54} \|W_{\nu t} * \vec\omega_0\|_{\dot M^{4/3,2}} \, d\tau \\
&\leq C \frac{1}{\nu^{1/4}} t^{-\frac34} \left(\|\vec\omega_0\|_{\dot M^{1,3/2}}^3 t \|W_{\nu t} * \vec\omega_0\|_\infty \right)^{1/4}.
\end{aligned}
$$

Thus,

$$
\|(W_{\nu t} * \vec u_0, W_{\nu t} * \vec\omega_0)\|_{E_T} \leq C_\nu \sup_{0<t<T} \left(\|\vec\omega_0\|_{\dot M^{1,3/2}}^3 t \|W_{\nu t} * \vec\omega_0\|_\infty \right)^{1/4}
$$

and the Picard iterates converge if $\|\vec\omega_0\|_{\dot M^{1,3/2}}^3 \sup_{0<t<T} t \|W_{\nu t} * \vec\omega_0\|_\infty$ is small enough.

It is easy to control the norm of the solution $\vec\omega$ in L^∞:

$$
\begin{aligned}
\| \int_0^t W_{\nu(t-s)} * \operatorname{div}(\vec\omega \otimes \vec u - \vec u \otimes \vec\omega) \, ds \|_\infty & \\
&\hspace{-6em} \leq C \int_0^{t/2} \frac{1}{(\nu(t-s))^{5/4}} \|\vec\omega\|_{\dot M^{4/3,2}} \|\vec u)\|_\infty \, ds \\
&\hspace{-6em} \quad + C \int_{t/2}^t \frac{1}{(\nu(t-s))^{3/4}} \|\vec\omega\|_{\dot M^{4,6}} \|\vec u\|_\infty \, ds \\
&\hspace{-6em} \leq C \frac1t \sup_{0<s<t} \sqrt s \|\vec u(s,\cdot)\|_\infty \left(\frac{\sup_{0<s<t} s^{1/4} \|\vec\omega\|_{\dot M^{4/3,2}}}{\nu^{5/4}} + \frac{\sup_{0<s<t} s^{5/4} \|\vec\omega\|_{\dot M^{4,6}}}{\nu^{3/4}} \right).
\end{aligned}
$$

For estimating the norm of $\vec\omega$ in $\dot M^{1,3/2}$, we define the following quantities: $\beta_N = \sup_{0<t<T} \|\vec\Omega_N - \vec\Omega_{N-1}\|_{\dot M^{1,3/2}}$, $B_N = \sum_{n=0}^N \beta_n$, $\epsilon_N = \sup_{0<t<T} \sqrt t \|\vec U_N - \vec U_{N-1}\|_\infty$ and $\eta = \sup_{N\geq0} \sup_{0<t<T} \sqrt t \|\vec U_N\|_\infty$. We have

$$
\beta_{N+1} \leq C_\nu (\eta\beta_N + (B_N + B_{N-1})\epsilon_N).
$$

This gives

$$
B_\infty \leq B_0 + C_\nu \eta B_\infty + 2C_\nu B_\infty \sum_{N\geq0} \epsilon_N.
$$

If if $\|\vec{\omega}_0\|^3_{\dot{M}^{1,3/2}} \sup_{0<t<T} t\|W_{\nu t} * \vec{\omega}_0\|_\infty$ is small enough, then we have

$$C_\nu \eta \leq C_\nu \sum_{N \geq 0} \epsilon_N \leq \frac{1}{4}$$

and thus $B_\infty \leq \frac{1}{4} B_0$. Thus,

$$\sup_{0<t<T} \|\vec{\omega}(t,.)\|_{\dot{M}^{1,3/2}} \leq B_\infty \leq 4B_0 \leq 4\|\vec{\omega}_0\|_{\dot{M}^{1,3/2}}. \qquad \square$$

Corollary 10.2.
Let $\vec{\omega}_0 \in \dot{M}^{1,3/2} \cap \dot{M}^{2,3}$ with $\text{div}\,\vec{\omega}_0 = 0$. Let $\vec{u}_0$ be the solution of

$$\vec{\nabla} \wedge \vec{u}_0 = \vec{\omega}_0, \text{div}\,\vec{u}_0 = 0, \vec{u}_0 \in \dot{B}^{-1}_{\infty,\infty}.$$

Then the Navier–Stokes equations

$$\vec{u} = W_{\nu t} * \vec{u}_0 - \int_0^t W_{\nu(t-s)} * \mathbb{P}\,\text{div}(\vec{u} \otimes \vec{u})\,ds$$

have a solution on $(0, T) \times \mathbb{R}^3$ with $T \geq C_0 \frac{1}{\|\vec{\omega}_0\|^3_{\dot{M}^{1,3/2}} \|\vec{\omega}_0\|_{\dot{M}^{2,3}}}$.

We may now state the result of Gallay and Šverák [203]:

Global existence of axisymmetrical solutions without swirl II

Theorem 10.5.
Let $\vec{u}_0$ with $\text{div}\,\vec{u}_0 = 0$ be an axisymmetric vector field without swirl, such that the vorticity $\vec{\omega}_0 = \omega_{\theta,0}\vec{e}_\theta$ with $\omega_{\theta,0}$ a finite Borel measure $f d\mu$ on $(0, +\infty) \times \mathbb{R}$ (with $|f(r, z)| = 1$ and μ a non-negative finite measure on $(0, +\infty) \times \mathbb{R}$.
We assume that

$$\lim_{T \to 0} \|\vec{\omega}_0\|^3_{\dot{M}^{1,3/2}} \sup_{0<t<T} t\|W_{\nu t} * \vec{\omega}_0\|_\infty < \epsilon_\nu,$$

where ϵ_ν is the constant in Theorem 10.3. (This is the case when $d\mu$ is absolutely continuous with respect to the Lebesgue measure).
Then the problem

$$\begin{cases} \partial_t \vec{u} + \vec{u}.\vec{\nabla}\vec{u} = \nu\Delta\vec{u} - \vec{\nabla}p \\ \vec{u}(0,.) = \vec{u}_0 \\ \text{div}\,\vec{u} = 0 \end{cases} \qquad (10.28)$$

has a global axisymmetric solution $\vec{u}$ on $(0, +\infty) \times \mathbb{R}^3$ with

$$\sup_{0<t} \iint |\omega_\theta(t, r, z)|\,dr\,dz < +\infty.$$

Proof.

Strategy of proof

Proposition 10.3 gives us a solution $\vec{u}$ on a small interval $(0, T_0)$ with $\sup_{0<t<T_0} \|\vec{\omega}(t,.)\|_{\dot{M}^{1,3/2}} +\infty$ and $\sup_{0<t<T_0} t\|\vec{\omega}(t,.)\|_\infty < +\infty$. In particular, $\sup_{0<t<T_0} \sqrt{t}\|\vec{\omega}(t,.)\|_{\dot{M}^{2,3}} < +\infty$. By

Corollary 10.2, if T^* is the maximal time of existence and if $T^* < +\infty$, we must have $\limsup_{t \to T^*} \|\vec{\omega}(t,.)\|_{\dot{M}^{1,3/2}} = +\infty$ or $\limsup_{t \to T^*} \|\vec{\omega}(t,.)\|_{\dot{M}^{2,3}} = +\infty$. Thus, we shall prove global existence by proving that, for every $T < +\infty$, we have[1]

$$\sup_{T_0/2 < t < \inf(T,T^*)} \|\vec{\omega}(t,.)\|_{\dot{M}^{1,3/2}} < \infty \tag{10.29}$$

and

$$\sup_{T_0/2 < t < \inf(T,T^*)} \|\vec{\omega}(t,.)\|_{\dot{M}^{2,3}} < \infty. \tag{10.30}$$

Small times

First, we show that the local solution $\vec{u}$ given by Proposition 10.3 is axisymmetric without swirl (it is enough to check that the iterates $\vec{U}_N$ are axisymmetric without swirl) and that $\vec{\omega}$ is given by $\vec{\omega} = \omega_\theta(t, \theta, r, z)\vec{e}_\theta$ with $\omega_\theta \in L^\infty((0,T), L^1(\,dr\,dz))$. As $\vec{\omega}$ is the pointwise limit of $\vec{\Omega}_N$, we shall prove that we have a uniform control of $I_N = \int |\vec{\Omega}_N| \frac{1}{\sqrt{x_1^2 + x_2^2}}\,dx$. For $\omega_{\theta,0} = f\,d\mu$, we write

$$|W_{\nu t} * \vec{\omega}_0| \le \int_0^{2\pi} \int_{(0,+\infty)\times\mathbb{R}} W_{\nu t}(x - (r\cos\theta, r\sin\theta, z)))|f(r,z)|\, r\,d\theta\,d\mu(r,z)$$

so that

$$
\begin{aligned}
I_0(t) &\le C \int_0^{2\pi} \int_{(0,+\infty)\times\mathbb{R}} \left(\int W_{\nu t}(x - (r\cos\theta, r\sin\theta, z)) \frac{dx}{\sqrt{x_1^2 + x_2^2}} \right) |f(r,z)|\, r\,d\theta\,d\mu \\
&\le C' \int_0^{2\pi} \int_{(0,+\infty)\times\mathbb{R}} \frac{1}{\sqrt{\nu t} + r} |f(r,z)|\, r\,d\theta\,d\mu(r,z) \\
&\le 2\pi C' \int_{(0,+\infty)\times\mathbb{R}} |f(r,z)|\, d\mu(r,z).
\end{aligned}
$$

Moreover, we have (writing again $\eta = \sup_{N \ge 0} \sup_{0 < t < T} \sqrt{t}\|\vec{U}_N\|_\infty$)

$$
\begin{aligned}
I_{N+1}(t) &\le I_0(t) + \int_0^t \int |W_{\nu(t-s)} * \mathrm{div}(\vec{\Omega}_N \otimes \vec{U}_N - \vec{U}_N \otimes \vec{\Omega}_N)|\, ds \frac{dx}{\sqrt{x_1^2 + x_2^2}} \\
&\le I_0(t) + 2\eta \int_0^t \int |\vec{\Omega}_N(s,y)| \left(\int |\vec{\nabla} W_{\nu(t-s)}(x-y)| \frac{dx}{\sqrt{x_1^2 + x_2^2}} \right) dy \frac{ds}{\sqrt{s}} \\
&\le I_0(t) + C\eta \int_0^t \int |\vec{\Omega}_N(s,y)| \frac{1}{\sqrt{\nu(t-s)}} \frac{1}{\sqrt{\nu(t-s)} + \sqrt{y_1^2 + y_2^2}}\, dy \frac{ds}{\sqrt{s}} \\
&\le I_0(t) + \frac{C'}{\sqrt{\nu}} \eta \sup_{0 < s < t} I_N(s).
\end{aligned}
$$

For ϵ_ν small enough, we have $\frac{C'}{\sqrt{\nu}}\eta < \frac{1}{2}$, so that

$$\sup_{N \ge 0} \sup_{0 < t < T} I_N(t) \le 2 \sup_{0 < t < T} I_0(t) \le C \int_{(0,+\infty)\times\mathbb{R}} |f(r,z)|\, d\mu(r,z).$$

---

[1] Actually, Gallay and Šverák proved a global control with $\sup_{t>0} \|\vec{\omega}(t,.)\|_{\dot{M}^{1,3/2}} < +\infty$ and $\sup_{t>0} \sqrt{t}\|\vec{\omega}(t,.)\|_{\dot{M}^{2,3}} < +\infty$.

Decay of $\int |\vec{\omega}| \frac{1}{\sqrt{x_1^2 + x_2^2}} \, dx$.

Now, we consider an interval of time (t_0, t_1) with $T_0/2 \leq t_0 \leq t_1 < T^*$. $\vec{u}$ is bounded on $(t_0, t_1) \times \mathbb{R}^3$ and

$$\sup_{t_0 < t < t_1} \int |\vec{\omega}| \frac{1}{\sqrt{x_1^2 + x_2^2}} \, dx < +\infty.$$

The next step in Gallay and Šverák's proof is to prove that the function $t \mapsto \int |\vec{\omega}(t, x)| \frac{1}{\sqrt{x_1^2 + x_2^2}} \, dx$ is non-increasing.

We know that $\vec{u}$ is smooth on $(t_0, t_1) \times \mathbb{R}^3$ (and even analytic, see Theorem 9.12). We consider a non-negative smooth function α which is compactly supported in $(0, +\infty)$ and a non-negative smooth function β which is compactly supported in $\mathbb{R}$, and we want to estimate

$$I_{\alpha,\beta}(t) = \int |\vec{\omega}(t, x)| \alpha(r) \beta(z) \frac{1}{r} \, dx = \iint_{(0,+\infty) \times \mathbb{R}} |\omega_\theta(t, r, z)| \alpha(r) \beta(z) \, dr \, dz.$$

For $t_0 < \tau_0 < \tau_1 < t_1$, we have

$$I_{\alpha,\beta}(\tau_1) - I_{\alpha,\beta}(\tau_0)$$

$$= \lim_{\epsilon \to 0^+} \int \left(\sqrt{\omega_\theta(\tau_1, x)^2 + \epsilon} - \sqrt{\omega_\theta(\tau_0, x)^2 + \epsilon}\right) \alpha(r) \beta(z) \frac{1}{r} \, dx$$

$$= \lim_{\epsilon \to 0^+} \int \int_{\tau_0}^{\tau_1} \partial_t \left(\sqrt{\omega_\theta(s, x)^2 + \epsilon}\right) \alpha(r) \beta(z) \frac{1}{r} \, dx \, ds.$$

We have

$$\partial_t \vec{\omega} = (\partial_t \omega_\theta) \vec{e}_\theta, \quad \text{hence} \quad \partial_t \omega_\theta = \Delta \omega_\theta - \frac{1}{r^2} \omega_\theta - \partial_r(u_r \omega_\theta) - \partial_z(u_z \omega_\theta)$$

and thus

$$\partial_t\left(\sqrt{\omega_\theta^2 + \epsilon}\right) = \frac{\omega_\theta}{\sqrt{\omega_\theta^2 + \epsilon}} \partial_t \omega_\theta$$

$$= \Delta\left(\sqrt{\omega_\theta^2 + \epsilon}\right) - \partial_r\left(u_r \sqrt{\omega_\theta^2 + \epsilon}\right) - \partial_z\left(u_z \sqrt{\omega_\theta^2 + \epsilon}\right) - \frac{1}{r^2}\sqrt{\omega_\theta^2 + \epsilon}$$

$$+ \sqrt{\epsilon} \frac{1}{\sqrt{1 + \frac{\omega_\theta^2}{\epsilon}}} \frac{1}{r^2} + \sqrt{\epsilon} \frac{1}{\sqrt{1 + \frac{\omega_\theta^2}{\epsilon}}} (\partial_r u_r + \partial_z u_z) - \epsilon \frac{|\vec{\nabla}\omega_\theta|^2}{(\epsilon + \omega_\theta^2)^{3/2}}$$

$$\leq \Delta\left(\sqrt{\omega_\theta^2 + \epsilon}\right) - \frac{1}{r^2}\sqrt{\omega_\theta^2 + \epsilon} - \partial_r\left(u_r\sqrt{\omega_\theta^2 + \epsilon}\right) - \partial_z\left(u_z\sqrt{\omega_\theta^2 + \epsilon}\right) + \sqrt{\epsilon}\left(\frac{1}{r^2} + |\partial_r u_r + \partial_z u_z|\right).$$

We get

$$\int \partial_t\left(\sqrt{\omega_\theta(s, x)^2 + \epsilon}\right) \alpha(r) \beta(z) \frac{1}{r} \, dx$$

$$\leq \int \Delta\left(\sqrt{\omega_\theta^2 + \epsilon}\right) \alpha(r) \beta(z) \frac{1}{r} \, dx - 2\pi \iint_{(0,+\infty) \times \mathbb{R}} \sqrt{\omega_\theta(s, r, z)^2 + \epsilon} \frac{1}{r^2} \alpha(r) \beta(z) \, dr \, dz$$

$$+ 2\pi \iint_{(0,+\infty) \times \mathbb{R}} \sqrt{\omega_\theta(s, r, z)^2 + \epsilon} (u_r \partial_r \alpha(r) \beta(z) + \alpha(r) u_z \partial_z \beta(z)) \, dr \, dz$$

$$+ 2\pi\sqrt{2\epsilon}(1 + \|\vec{\nabla} \otimes \vec{u}(s, .)\|_\infty) \iint_{(0,+\infty) \times \mathbb{R}} \left(1 + \frac{1}{r^2}\right) \alpha(r) \beta(z) \, dr \, dz.$$

Letting ϵ go to 0, we get

$$I_{\alpha,\beta}(\tau_1) - I_{\alpha,\beta}(\tau_0)$$
$$\leq \int_{\tau_0}^{\tau_1} \int \Delta |\omega_\theta| \, \alpha(r)\beta(z) \frac{1}{r} \, dx \, ds - 2\pi \iint_{(0,+\infty)\times\mathbb{R}} |\omega_\theta| \frac{1}{r^2} \alpha(r)\beta(z) \, dr \, dz \, ds$$
$$+ 2\pi \int_{\tau_0}^{\tau_1} \iint_{(0,+\infty)\times\mathbb{R}} |\omega_\theta| (u_r \partial_r \alpha(r)\beta(z) + \alpha(r)u_z \partial_z \beta(z)) \, dr \, dz \, ds$$
$$= 2\pi \int_{\tau_0}^{\tau_1} \iint_{(0,+\infty)\times\mathbb{R}} (\Delta |\omega_\theta| - |\omega_\theta| \frac{1}{r^2}) \alpha(r)\beta(z) \, dr \, dz \, ds$$
$$+ 2\pi \int_{\tau_0}^{\tau_1} \iint_{(0,+\infty)\times\mathbb{R}} |\omega_\theta| (u_r \partial_r \alpha(r)\beta(z) + \alpha(r)u_z \partial_z \beta(z)) \, dr \, dz \, ds$$

and thus

$$\frac{1}{2\pi} (I_{\alpha,\beta}(\tau_1) - I_{\alpha,\beta}(\tau_0)) \leq \int_{\tau_0}^{\tau_1} \iint_{(0,+\infty)\times\mathbb{R}} (\partial_r^2 + \partial_z^2 + \frac{1}{r}\partial_r - \frac{1}{r^2}) |\omega_\theta| \, \alpha(r)\beta(z) \, dr \, dz \, ds$$
$$+ \int_{\tau_0}^{\tau_1} \iint_{(0,+\infty)\times\mathbb{R}} |\omega_\theta| (u_r \partial_r \alpha(r)\beta(z) + \alpha(r)u_z \partial_z \beta(z)) \, dr \, dz \, ds$$
$$= - \int_{\tau_0}^{\tau_1} \iint_{(0,+\infty)\times\mathbb{R}} \partial_r |\omega_\theta| \, \partial_r \alpha(r)\beta(z) \, dr \, dz \, ds$$
$$- \int_{\tau_0}^{\tau_1} \iint_{(0,+\infty)\times\mathbb{R}} \frac{1}{r} |\omega_\theta| \, \partial_r \alpha(r)\beta(z) \, dr \, dz \, ds$$
$$+ \int_{\tau_0}^{\tau_1} \iint_{(0,+\infty)\times\mathbb{R}} |\omega_\theta| \, \alpha(r)\partial_z^2 \beta(z) \, dr \, dz \, ds$$
$$+ \int_{\tau_0}^{\tau_1} \iint_{(0,+\infty)\times\mathbb{R}} |\omega_\theta| u_r \partial_r \alpha(r)\beta(z) \, dr \, dz \, ds$$
$$+ \int_{\tau_0}^{\tau_1} \iint_{(0,+\infty)\times\mathbb{R}} |\omega_\theta| \alpha(r)u_z \partial_z \beta(z) \, dr \, dz \, ds$$
$$= A_{\alpha,\beta} + B_{\alpha,\beta} + C_{\alpha,\beta} + D_{\alpha,\beta} + E_{\alpha,\beta}.$$

Let $\gamma \in \mathcal{D}(\mathbb{R})$ be an even function, radially non-increasing and equal to 1 on $[-1,1]$ and to 0 on $(2,+\infty)$. For $R > 1$, we define the function $\alpha_R(r) = \gamma(\frac{r}{R}) - \gamma(Rr)$. We shall write $\lim_{\alpha\to1}$ meaning $\lim_{R\to+\infty}$. We have obviously

$$\lim_{\alpha\to1} I_{\alpha,\beta}(\tau) = I_\beta(\tau) = \int |\vec{\omega}(\tau,x)|\beta(z)\frac{1}{r} \, dx$$

$$\lim_{\alpha\to1} C_{\alpha,\beta} = C_\beta = \int_{\tau_0}^{\tau_1} \iint_{(0,+\infty)\times\mathbb{R}} |\omega_\theta| \, \partial_z^2 \beta(z) \, dr \, dz \, ds$$

$$\lim_{\alpha\to1} E_{\alpha,\beta} = E_\beta = \int_{\tau_0}^{\tau_1} \iint_{(0,+\infty)\times\mathbb{R}} |\omega_\theta| \, u_z \partial_z \beta(z) \, dr \, dz \, ds$$

Next, we remark that $\gamma' \leq 0$ so that $\partial_r \alpha \geq \frac{1}{R}\gamma'(\frac{r}{R}) \geq -\frac{1}{R}\|\gamma'\|_\infty \mathbb{1}_{r>R}$ while

$$|\partial_r \alpha| \leq \|\gamma'\|_\infty (\frac{2}{r} \mathbb{1}_{r<\frac{2}{R}} + \frac{1}{R} \mathbb{1}_{r>R}).$$

This gives

$$\limsup_{\alpha\to1} B_{\alpha,\beta} \leq 0$$

and

$$\limsup_{\alpha \to 1} D_{\alpha,\beta} \leq 2\|\gamma'\|_\infty \limsup_{\alpha \to 1} \int_{\tau_0}^{\tau_1} \iint_{(0,\frac{2}{R})\times\mathbb{R}} |\omega_\theta| \frac{|u_r|}{r} \beta(z) \, dr \, dz \, ds.$$

As $\frac{|u_r|}{r} \leq \|\vec{\nabla} \otimes \vec{u}\|_\infty$, we find that $\limsup_{\alpha \to 1} D_{\alpha,\beta} \leq 0$.
We now control $A_{\alpha,\beta}$. We have

$$A_{\alpha,\beta} = F_{\alpha,\beta} + G_{\alpha,\beta}$$

with

$$F_{\alpha,\beta} = \int_{\tau_0}^{\tau_1} \iint_{(0,+\infty)\times\mathbb{R}} |\omega_\theta| \frac{1}{R^2} \gamma''(\frac{r}{R}) \beta(z) \, dr \, dz \, ds$$

and

$$G_{\alpha,\beta} = -\int_{\tau_0}^{\tau_1} \iint_{(0,+\infty)\times\mathbb{R}} |\omega_\theta| \, R^2 \gamma''(Rr) \beta(z) \, dr \, dz \, ds.$$

We have

$$\lim_{\alpha \to 1} F_{\alpha,\beta} = 0.$$

In order to estimate $G_{\alpha,\beta}$, we introduce for $\eta > 0$

$$G_{\eta,\alpha,\beta} = -\int_{\tau_0}^{\tau_1} \iint_{(0,+\infty)\times\mathbb{R}} \sqrt{|\omega_\theta|^2 + \eta r^2} \, R^2 \gamma''(Rr) \beta(z) \, dr \, dz \, ds.$$

We have

$$|G_{\alpha,\beta} - G_{\eta,\alpha,\beta}| \leq \int_{\tau_0}^{\tau_1} \iint_{(0,+\infty)\times\mathbb{R}} \sqrt{\eta} r \, R^2 |\gamma''(Rr)| \beta(z) \, dr \, dz \, ds$$

and thus

$$\limsup_{\alpha \to 1} G_{\alpha,\beta} \leq \limsup_{\alpha \to 1} G_{\eta,\alpha,\beta} + \sqrt{\eta}(\tau_1 - \tau_0)\left(\int r|\gamma''(r)| \, dr\right)\|\beta\|_1.$$

We write

$$G_{\eta,\alpha,\beta} = \int_{\tau_0}^{\tau_1} \iint_{(0,+\infty)\times\mathbb{R}} \partial_r(\sqrt{|\omega_\theta|^2 + \eta r^2}) \, R\gamma'(Rr)\beta(z) \, dr \, dz \, ds$$

$$= -\frac{1}{2} \int_{\tau_0}^{\tau_1} \iint_{(0,+\infty)\times\mathbb{R}} \frac{r\partial_r|\omega_\theta^2| + 2\eta r^2}{r\sqrt{|\omega_\theta|^2 + \eta r^2}} \, R|\gamma'(Rr)|\beta(z) \, dr \, dz \, ds.$$

We have

$$r\partial_r|\omega_\theta^2| + 2\epsilon r^2 = x_1\partial_1|\vec{\omega}|^2 + x_2\partial_2|\vec{\omega}|^2 + 2\epsilon(x_1^2 + x_2^2).$$

The function $\vec{\omega}$ is smooth on $(\tau_0,\tau_1) \times \mathbb{R}^3$ and its derivatives are bounded. Moreover, it vanishes for $x_1 = x_2 = 0$, as $|\vec{\omega}| \leq r|\vec{\nabla} \otimes \vec{\omega}|$. Thus, $|\vec{\omega}|^2$ has a minimum at x when $x_1 = x_2 = 0$, so that the derivatives $\partial_1|\omega_\theta|^2(\tau,0,0,x_3)$ and $\partial_2|\omega_\theta|^2(\tau,0,0,x_3)$ are equal to 0 and the quadratic form on $\mathbb{R}^2$

$$Q(u,v) = u_1v_1\partial_1^2|\omega_\theta|^2(\tau,0,0,x_3) + u_2v_2\partial_2^2|\omega_\theta|^2(\tau,0,0,x_3)$$
$$+ (u_1v_2 + u_2v_1)\partial_1\partial_2|\omega_\theta|^2(\tau,0,0,x_3)$$

is non-negative: $Q(u, u) \geq 0$. We then write

$$x_1 \partial_1 |\vec{\omega}|^2 + x_2 \partial_2 |\vec{\omega}|^2$$

$$= x_1^2 \partial_1^2 |\omega_\theta|^2 (\tau, 0, 0, x_3) + x_2^2 \partial_2^2 |\omega_\theta|^2 (\tau, 0, 0, x_3) + 2 x_1 x_2 \partial_1 \partial_2 |\omega_\theta|^2 (\tau, 0, 0, x_3)$$

$$+ x_1 \sum_{1 \leq i,j \leq 2} \int_0^1 \partial_1 \partial_i \partial_j |\omega_\theta|^2 (\tau, \lambda x_1, \lambda x_2, x_3)(1 - \lambda) x_i x_j \, d\lambda$$

$$+ x_2 \sum_{1 \leq i,j \leq 2} \int_0^1 \partial_2 \partial_i \partial_j |\omega_\theta|^2 (\tau, \lambda x_1, \lambda x_2, x_3)(1 - \lambda) x_i x_j \, d\lambda$$

$$\geq - 4 \|\vec{\omega}\|_{W^{3,\infty}} r^3.$$

For $R > \frac{4}{\eta} \sup_{\tau_0 < t < \tau_1} \|\vec{\omega}\|_{W^{3,\infty}}$, we find $G_{\eta,\alpha,\beta} \leq 0$ and thus $\limsup_{\alpha \to 1} G_{\eta,\alpha,\beta} \leq 0$. We then conclude by letting β go to 1 ($\beta = \gamma(z/R)$ and $R \to +\infty$). Thus, we proved that

$$\sup_{T_0/2 < t < T^*} \iint |\vec{\omega}(t, r, z)| \, dr \, dz \leq \iint |\vec{\omega}(t_0, r, z)| \, dr \, dz.$$

As $\|\vec{\omega}\|_{\dot{M}^{1,3/2}} \leq C \iint |\vec{\omega}(t, r, z)| \, dr \, dz$, we proved inequality (10.29).

Control of $\frac{\omega_\theta}{r}$.

Let $\eta = \frac{\omega_\theta}{r}$. We have just proved that, on the maximal time interval of existence $(0, T^*)$ we have

$$\sup_{0 < t < T^*} \|\eta(t, .)\|_{L^1(\mathbb{R}^3)} = \liminf_{t \to 0} \|\eta(T, .)\|_{L^1(\mathbb{R}^3)};$$

if $\omega_{\theta,0} \in L^1((0, +\infty) \times \mathbb{R}, dr \, dz)$, we have

$$\liminf_{t \to 0} \|\eta(t, .)\|_{L^1(\mathbb{R}^3), dx)} = \|\eta_0\|_{L^1(\mathbb{R}^3), dx)} = \|\omega_{\theta,0}\|_{L^1(dr \, dz)}.$$

Recall that the proof of global existence for axisymmetric vector fields in H^1 relied on inequality (10.27), i.e. on the control of $\eta(t, .)$ in $L^2(dx)$. Here, we shall prove a similar control on $\|\eta(t, .)\|_{L^2(dx)}$, when $T_0/2 < t < T^*$.

We want to estimate, for $T_0/2 \leq \tau_0 \leq t \leq \tau_1 < T^*$,

$$I(t) = \int |\vec{\omega}(t, x)|^2 \frac{1}{r^2} \, dx.$$

A first remark is that $|\eta(t, x)| \leq |\vec{\nabla} \otimes \vec{\omega}(t, x)|$, so that η is bounded on $[\tau_0, \tau_1] \times \mathbb{R}^3$, and, since we control the L^1 norm of η, $I(t) < +\infty$. A second remark is that $|\vec{\omega}|^2$ is smooth and axisymmetrix; in particular, looking at the Taylor polynomial of order 4 for (x, y) close to $(0, 0)$ and writing

$$|\vec{\omega}(x, y, z)|^2 = |\vec{\omega}(x, -y, z)|^2 = |\vec{\omega}(-x, y, z)|^2 = |\vec{\omega}(y, x, z)|^2 = |\vec{\omega}(\frac{x + y}{\sqrt{2}}, \frac{x - y}{\sqrt{2}}, z)|^2,$$

one gets (since $\vec{\omega}(t, 0, 0, z) = 0$)

$$|\vec{\omega}(t, x, y, z)|^2 = \frac{r^2}{2!} \partial_1^2 (|\vec{\omega}|^2)(t, 0, 0, z) + \frac{r^4}{4!} \partial_1^4 (|\vec{\omega}|^2)(t, 0, 0, z) + r^6 \epsilon(t, x, y, z)$$

where $\epsilon(t, x, y, z)$ is bounded on $[\tau_0, \tau_1] \times \mathbb{R}^3$.

A third remark is that we have the identity

$$\frac{1}{r^2}\vec{\omega}\cdot\partial_t\vec{\omega} = \frac{1}{r^2}\nu\vec{\omega}\cdot\Delta\vec{\omega} - \frac{1}{r^2}(\vec{u}\cdot\vec{\nabla}\vec{\omega} - \vec{\omega}\cdot\vec{\nabla}\vec{u})\cdot\vec{\omega}$$

$$= \frac{1}{r^2}\nu\omega_\theta\Delta\omega_\theta - \frac{1}{r^4}\nu\omega_\theta^2 - \frac{1}{r^2}\vec{u}\cdot\vec{\nabla}(\frac{\omega_\theta^2}{2}) + \frac{u_r}{r^3}\omega_\theta^2$$

$$= \nu\frac{\omega_\theta}{r}\Delta(\frac{\omega_\theta}{r}) + 2\nu\frac{\omega_\theta}{r^2}\partial_r(\frac{\omega_\theta}{r}) - \vec{u}\cdot\vec{\nabla}(\frac{\omega_\theta^2}{2r^2})$$

or equivalently

$$\partial_t(\eta^2) = 2\nu\eta\Delta\eta + 2\frac{\nu}{r}\partial_r(\eta^2) - \vec{u}\cdot\vec{\nabla}(\eta^2). \tag{10.31}$$

As $\eta(t, .) \in L^1 \cap L^\infty$, we have, for $\phi_R = \phi(x/R)$, where $\phi \in \mathcal{D}$ is radial, non-negative and is equal to 1 on the ball $B(0, 1)$,

$$\int \eta\,\Delta\eta\,\phi_R\,dx = \int \frac{\eta^2}{2}\Delta(\phi_R)\,dx - \int |\vec{\nabla}\eta|^2\,dx;$$

the first term is $O(\frac{1}{R^2})$ and the second one is non-positive. Moreover, we have

$$\int \frac{1}{r}\partial_r(\eta^2)\phi_R\,dx = \iint_{(0,+\infty)\times\mathbb{R}} \partial_r(\eta^2)\phi_R(r, 0, z)\,dr\,dz$$

$$= -\int \eta^2(t, 0, 0, z)\phi_R(0, 0, z)\,dz - \int \eta^2 r\partial_r\phi_R\,dx.$$

The first term is non-positive and the second one is $O(\int_{|x|>R}|\eta|\,dx) = o(1)$. On the other hand, as $\vec{u}$ is bounded on $[\tau_0, \tau_1] \times \mathbb{R}^3$ and divergence-free, we find that

$$\int \phi_R\vec{u}\cdot\vec{\nabla}(\eta^2)\,dx = -\int \eta^2\vec{u}\cdot\vec{\nabla}\phi_R\,dx = O(\frac{1}{R}).$$

Thus, we easily check that I is non-increasing, as $\frac{d}{dt}(\int \eta^2(t, x)\phi_R(x)\,dx) \leq o(1)$. We finally get

$$\sup_{T_0/2\leq t<T^*} \|\eta(t, .)\|_2 \leq \|\eta(T_0, .)\|_2.$$

We control $\|\eta\|_4$ in a similar way. We write

$$\frac{d}{dt}\int \eta(t, x)^4\,\phi_R\,dx = 2\int \eta^2\partial_t(\eta^2)\,\phi_R\,dx.$$

Using identity (10.31), we get

$$\eta^2\partial_t(\eta^2) = 2\nu\eta^3\Delta\eta + 2\frac{\nu}{r}\eta^2\partial_r(\eta^2) - \eta^2\vec{u}\cdot\vec{\nabla}(\eta^2)$$

$$= \nu\eta^2\Delta(\eta^2) - 2\nu\eta^2|\vec{\nabla}\eta|^2 + \frac{\nu}{r}\partial_r(\eta^4) - \frac{1}{2}\vec{u}\cdot\vec{\nabla}(\eta^4).$$

this gives that $\frac{d}{dt}(\int \eta^4(t, x)\phi_R(x)\,dx) \leq o(1)$, and thus

$$\sup_{T_0/2\leq t<T^*} \|\eta(t, .)\|_4 \leq \|\eta(T_0, .)\|_4.$$

Control of $u_r(t, x)$.

Recall that
$$\vec{u} = G * \vec{\nabla} \wedge \vec{\omega} \text{ with } G = \frac{1}{4\pi |x|}.$$

We have
$$\vec{\nabla} \wedge \vec{\omega} = -\partial_z \omega_\theta \vec{e}_r + (\partial_r \omega_\theta + \frac{1}{r} \omega_\theta) \vec{e}_z.$$

Thus, we have, for $y = (\rho \cos \sigma, \rho \sin \sigma, w)$ and $x = (r \cos \theta, r \sin \theta, z)$,
$$u_r(t, r, z) = \vec{u} \cdot \vec{e}_r = \int \partial_z G(r\vec{e}_r - y, z - w) \omega_\theta(t, \rho, w)\, \vec{e}_r \cdot \vec{e}_\rho \, dy$$

We split the domain of integration in $(\rho, w) \in \Delta_1$ and $(\rho, w) \in \Delta_2$, where
$$\Delta_1 = \{(\rho, w) \in (0, +\infty) \times \mathbb{R}/ \ \rho \leq 2r\}.$$

We thus have
$$u_r = A(t, x) + B(t, x)$$

with
$$A(t, x) = \frac{1}{4\pi} \int_{(\rho, w) \in \Delta_1} \frac{(w - z)}{|y - x|^3} \omega_\theta(t, y)\, dy$$

and
$$B(t, x) = \iint_{(\rho, w) \in \Delta_2} \left(\frac{w - z}{4\pi} \int_0^{2\pi} \frac{\vec{e}_r \cdot \vec{e}_\rho}{(|r\vec{e}_r - \rho\vec{e}_\rho|^2 + (z - w)^2)^{3/2}} d\sigma \right) \omega_\theta(\rho, w)\rho \, d\rho \, dw$$
$$= \iint_{(\rho, w) \in \Delta_2} K(r, z, \rho, w) \omega_\theta(\rho, w)\, d\rho \, dw$$

with
$$K(r, z, \rho, w) = \frac{\rho(w - z)}{4\pi} \int_0^{2\pi} \frac{\cos \gamma}{(r^2 + \rho^2 - 2r\rho \cos \gamma + (z - w)^2)^{3/2}} d\gamma$$
$$= \frac{3r\rho^2(w - z)}{4\pi} \int_0^{2\pi} \frac{\sin^2 \gamma}{(r^2 + \rho^2 - 2r\rho \cos \gamma + (z - w)^2)^{5/2}} d\gamma.$$

We have
$$|A(t, x)| \leq \frac{1}{4\pi} \| \frac{1}{|y|^2} \|_{L^{3/2, \infty}} \| \mathbb{1}_{(\rho, w) \in \Delta_1} \omega_\theta \|_{L^{3,1}}$$
$$\leq C \| \mathbb{1}_{(\rho, w) \in \Delta_1} \omega_\theta \|_{L^1(\mathbb{R}^3)}^{1/9} \| \mathbb{1}_{(\rho, w) \in \Delta_1} \omega_\theta \|_{L^4(\mathbb{R}^3)}^{8/9}$$
$$\leq 2Cr \| \eta(t, .) \|_1^{1/9} \| \eta(t, .) \|_4^{8/9}.$$

On the other hand, we have, if $(\rho, w) \in \Delta_2$, $|r\vec{e}_r - \rho\vec{e}_\rho| \geq \rho/2$ and thus
$$K(r, z, \rho, w) \leq 24\, r \frac{1}{\rho^2 + (z - w)^2}$$

and
$$|B(t, x)| \leq 24\, r \int_{(\rho, w) \in \Delta_2} \frac{1}{\rho^2 + (z - w)^2} |\omega_\theta(\rho, w)| \, d\rho \, dw$$
$$\leq 24\, r \| \mathbb{1}_{\Delta_2}(\rho, w) \frac{\rho^{1/3}}{\rho^2 + (z - w)^2} \|_{L^3(d\rho\, dw)} \| \rho^{-1/3} \omega_\theta(t, \rho, w) \|_{L^{3/2}(d\rho\, dw)}$$
$$= C \| \eta(t, .) \|_{L^{3/2}(dx)}$$

We thus get
$$|u_r(t, x)| \leq C(\| \eta(t, .) \|_1 + \| \eta(t, .) \|_4)(r + 1).$$

Control of $\|\vec{\omega}(t, .)\|_{\dot{M}^{2,3}}$

We know that $\|\vec{\omega}(t, .)\|_{\dot{M}^{2,3}} \leq C\|\omega(t, .)\|_{L^2(\frac{dx}{r})}$, thus we will try and control $J(t) = \int |\vec{\omega}|^2 \frac{dx}{r}$. For $T_0/2 \leq \tau_0 \leq \tau_1 < T^*$, $\vec{\omega}$ is bounded on $(\tau_0, \tau_1) \times \mathbb{R}^3$ so that

$$J(t) \leq \|\vec{\omega}(t, .)\|_\infty \|\eta(t, .)\|_1.$$

We have

$$\frac{d}{dt} \int |\vec{\omega}|^2 \phi_R \frac{dx}{r} = \int \partial_t(\eta(t, x)^2) \phi_R r \, dx.$$

with

$$r\partial_t(\eta(t, x)^2) = r(2\nu\eta\Delta\eta + 2\frac{\nu}{r}\partial_r(\eta^2) - \vec{u} \cdot \vec{\nabla}(\eta^2))$$

$$= \nu r\Delta(\eta^2) - 2\nu r|\vec{\nabla}\eta|^2 + 2\nu\partial_r(\eta^2) - \vec{u} \cdot \vec{\nabla}(r\eta^2) + u_r\eta^2$$

$$= \nu r\partial_r^2(\eta^2) + \nu r\partial_z^2(\eta^2) + 3\nu\partial_r(\eta^2) - 2\nu r|\vec{\nabla}\eta|^2 + 2\nu\partial_r(\eta^2) - \vec{u} \cdot \vec{\nabla}(r\eta^2) + u_r\eta^2$$

$$= \frac{\nu}{r}(\partial_r^2(\omega_\theta^2) + \partial_z^2(\omega_\theta^2)) - \nu\partial_r(\eta^2) - \frac{2\nu}{r}\eta^2 - 2\nu r|\vec{\nabla}\eta|^2 - \vec{u} \cdot \vec{\nabla}(r\eta^2) + u_r\eta^2$$

$$= \frac{\nu}{r}(\partial_r^2(\omega_\theta^2) + \partial_z^2(\omega_\theta^2)) - \frac{\nu}{r}\partial_r(\frac{\omega_\theta^2}{r}) - \nu\frac{\omega_\theta^2}{r^3} - 2\nu r|\vec{\nabla}(\frac{\omega_\theta}{r})|^2 - \vec{u} \cdot \vec{\nabla}(\frac{\omega_\theta^2}{r}) + u_r\frac{\omega_\theta^2}{r^2}.$$

This gives

$$\frac{d}{dt} \int |\vec{\omega}|^2 \phi_R \frac{dx}{r} = 2\pi\nu \iint_{r>0} \omega_\theta^2(\partial_r^2\phi_R + \partial_z^2\phi_R + \frac{1}{r}\partial_r\phi_R) \, dr \, dz$$

$$- \int \nu(\frac{\omega_\theta^2}{r^2} + 2r^2|\vec{\nabla}(\frac{\omega_\theta}{r})|^2)\frac{dx}{r} + \int \frac{\omega_\theta^2}{r}u_r\partial_r\phi_R \, dx + \int u_r\frac{\omega_\theta^2}{r^2}\phi_R \, dx.$$

As $|u_r| \leq C_0(1 + r)$ on $(T_0/2, T^*)$, we obtain

$$J(t) \leq J(T_0/2) + \int_{T_0/2}^t \int u_r \frac{\omega_\theta(s, x)^2}{r^2} \, dx \, ds$$

$$\leq J(T_0/2) + C_0 \int_{T_0/2}^t J(s) \, ds + C_0 \int_{T_0/2}^t \|\eta(s, .)\|_2^2 \, ds.$$

We then conclude with the Grönwall lemma.   □

The case of axisymmetric flows with swirls has been studied by many authors. Regularity criteria have been given by Chen, Fang and T. Zhang in 2017 [114]; those criteria were used by Lei and Q. Zhang [307] to prove existence when the swirl component is small enough:

Theorem 10.6.
Let $\vec{u}_0 \in H^{1/2}$ with $\operatorname{div} \vec{u}_0 = 0$ be an axisymmetric vector field, with vorticity $\vec{\omega}_0 = \vec{\nabla} \wedge \vec{u}_0$, such that $\frac{\omega_{\theta,0}}{r} \in L^2$, $\frac{u_{\theta,0}^2}{r} \in L^2$ and $ru_{\theta,0} \in L^2 \cap L^\infty$. Then there is a constant ϵ_ν (which does not depend on $\vec{u}_0$) such that, if

$$(\|\frac{\omega_{\theta,0}}{r}\|_2 + \|\frac{u_{\theta,0}^2}{r}\|_2)\|ru_{\theta,0}\|_2\|ru_{\theta,0}\|_\infty < \epsilon_0,$$

then the problem

$$\begin{cases} \partial_t \vec{u} + \vec{u}.\vec{\nabla}\vec{u} = \nu\Delta\vec{u} - \vec{\nabla}p \\ \vec{u}(0,.) = \vec{u}_0 \\ \operatorname{div}\vec{u} = 0 \end{cases} \tag{10.32}$$

has a global regular axisymmetric solution $\vec{u}$ on $(0,+\infty) \times \mathbb{R}^3$.

10.4 Helical Solutions

In this section, we consider the Navier–Stokes problem with the following symmetry property: $\vec{u}$ is invariant under the action of a one-parameter group of screw motions $R_\theta(x_1, x_2, x_3) = (x_1\cos\theta - x_2\sin\theta, x_1\sin\theta + x_2\cos\theta, x_3 + \alpha\theta)$ (where $\alpha \neq 0$ is fixed): this is the case of helical symmetry.

In cylindrical cordinates, we find that $\vec{u} = u_r\vec{e}_r + u_\theta\vec{e}_\theta + u_z\vec{e}_z$, where u_r, u_θ and u_z depend only on r and $\eta = \theta\cos\gamma + z\sin\gamma$ with $\tan\gamma = -1/\alpha$. The case $\gamma = \pi/2$ would correspond to axisymmetrical solutions, the case $\gamma = 0$ to two-and-a-half dimensional flows. For helical symmetry, we consider $\gamma \in (-\pi/2, 0) \cup (0, \pi/2)$.

It will be more convenient to define $\xi = \theta - z/\alpha = \eta/\cos\gamma$. A scalar function $A(x)$ with helical symmetry may be written as $A(x) = B(r, \xi)$, where B is 2π-periodical in ξ. If Γ is a cylindrical domain of $\mathbb{R}^3$ of the form $\Gamma = \{x \in \mathbb{R}^3 \mid z \in I, r \in J\}$ for an interval $I \subset \mathbb{R}$ and an interval $J \subset (0, +\infty)$, we find

$$\int_\Gamma |A(x)|^2\,dx = |I| \iint_{r\in J} |A(x_1,x_2,0)|^2\,dx_1\,dx_2 = |I| \int_J \int_0^{2\pi} |B(r,\theta)|^2\,r\,dr\,d\theta.$$

Thus, we can see that we have the equivalence for a flow $\vec{u}_0$ with helical symmetry:

$$\vec{u}_0 \in \dot{M}^{2,3} \Leftrightarrow \vec{u}_0(x_1, x_2, 0) \in L^2(\mathbb{R}^2)$$

and the condition $\lim_{t\to 0} t^{1/2}\|W_{\nu t} * \vec{u}_0\|_\infty = 0$ is automatically fulfilled, as $L^\infty \cap L^2$ is dense in L^2. This situation is very similar to the case of two-and-a-half dimensional flows.

Helical flows have been studied by Mahalov, Titi, and Leibovich [347]. We have the following result:

Global existence of helical symmetrical solutions

Theorem 10.7.
Let $\vec{u}_0 \in \dot{M}^{2,3}$ with $\operatorname{div}\vec{u}_0 = 0$ be a vector field with helical symmetry, and let $\vec{f} \in L^1((0,T),, \dot{M}^{2,3})$ with helical symmetry. Then the problem

$$\begin{cases} \partial_t \vec{u} + \vec{u}.\vec{\nabla}\vec{u} = \nu\Delta\vec{u} + \vec{f} - \vec{\nabla}p \\ \vec{u}(0,.) = \vec{u}_0 \\ \operatorname{div}\vec{u} = 0 \end{cases} \tag{10.33}$$

has a unique global helical solution $\vec{u}$ on $(0,T) \times \mathbb{R}^3$ with $\vec{u} \in L_t^\infty \dot{M}^{2,3}$ and $\vec{\nabla} \otimes \vec{u} \in L_t^2 \dot{M}^{2,3}$.

Proof. First, we consider $\vec{u}_1 = W_{\nu t} * \vec{u}_0$. If $\vec{u}_0$ is helical, so is $\vec{u}_1$. If $\vec{u}_0$ belongs to $\dot{M}^{2,3}$, then $\vec{u}_1$ belongs to $L_t^\infty \dot{M}^{2,3}$. Moreover, $\vec{u}_0 \in L_{x_1,x_2}^2 L_{\mathrm{per},x_3}^2$ and thus $\vec{u}_1$ will satisfy

$$\partial_t \int_{0<x_3<2\pi\alpha} |\vec{u}_1(t,x)|^2 \, dx = 2\nu \int_{0<x_3<2\pi\alpha} \vec{u}_1.\Delta\vec{u}_1 \, dx = -2\nu \int_{0<x_3<2\pi\alpha} |\vec{\nabla}\otimes\vec{u}_1|^2 \, dx$$

Thus, $\vec{\nabla}\otimes\vec{u}_1$ belongs to $L^2 L_{x_1,x_2}^2 L_{\mathrm{per},x_3}^2$; as $|\vec{\nabla}\otimes\vec{u}_1|$ is helical, we find that $\vec{\nabla}\otimes\vec{u}_1$ belongs to $L_t^2 \dot{M}^{2,3}$.

We consider now $\vec{u}_2 = \int_0^t W_{\nu(t-s)} * \mathbb{P}\vec{f}(s,.) \, ds$. As $\vec{f}$ is helical, $\mathbb{P}\vec{f}$ is helical. Thus, writing

$$\|\vec{u}_2(t,.)\|_{\dot{M}^{2,3}} \le \int_0^t \|W_{\nu(t-s)} * \mathbb{P}\vec{f}(s,.)\|_{\dot{M}^{2,3}} \, ds,$$

we find that $\vec{u}_2$ belongs to $L_t^\infty \dot{M}^{2,3}$. Moreover,

$$\|\vec{\nabla}\otimes\vec{u}_2\|_{L^2\dot{M}^{2,3}} \le \int_0^T \|1_{t>s} W_{\nu(t-s)} * \mathbb{P}\vec{f}(s,.)\|_{L^2\dot{M}^{2,3}} \, ds$$

so that $\vec{\nabla}\otimes\vec{u}_2 \in L^2\dot{M}^{2,3}$.

Now, if v is a function such that $v \in \dot{M}^{2,3}$ and $\vec{\nabla}v \in \dot{M}^{2,3}$, we find that v belongs to $\dot{M}^{4,6}$: this is easily seen with Hedberg's inequality. Indeed, we write

$$|v(x)| = |\int_0^{+\infty} \Delta W_{t\Delta} * v(x) \, dt| \le C \int_0^\infty \min(\frac{\mathcal{M}_{\vec{\nabla}v(x)}}{t^{1/2}}, \frac{\|v\|_{\dot{M}^{2,3}}}{t^{3/2}}) \, dt$$

$$= C' \sqrt{\|v\|_{\dot{M}^{2,3}}} \sqrt{\mathcal{M}_{\vec{\nabla}v(x)}}.$$

In particular, we obtain, for $\vec{u}$ and $\vec{v}$ in $L^\infty \dot{M}^{2,3}$ with $\vec{\nabla}\otimes\vec{u}$ and $\vec{\nabla}\otimes\vec{v}$ in $L^2\dot{M}^{2,3}$ that $\vec{u}.\vec{\nabla}\vec{v}$ belongs to $L^{4/3}\dot{M}^{4/3,2}$. Moreover, if $\vec{u}$ is divergence free, we have that $\frac{1}{\sqrt{-\Delta}}\vec{u}\cdot\vec{\nabla}\vec{v} = \frac{1}{\sqrt{-\Delta}}\mathrm{div}(\vec{u}\otimes\vec{v})$ belongs to $L^2\dot{M}^{2,3}$.

The next step is to consider $\vec{u}_3 = \int_0^t W_{\nu(t-s)} * \mathbb{P}\vec{g}(s,.) \, ds$ for a helical $\vec{g} \in L^{4/3}\dot{M}^{4/3,2}$ such that $\frac{1}{\sqrt{-\Delta}}\vec{g}$ belongs to $L^2\dot{M}^{2,3}$. We have

$$\|\vec{\nabla}\otimes\vec{u}_3\|_{\dot{M}^{2,3}} \le \int_0^t \|\vec{\nabla}\otimes W_{\nu(t-s)} * \mathbb{P}\vec{g}(s,.)\|_{\dot{M}^{2,3}} \, ds$$

$$\le C \int_0^t \frac{1}{(t-s)^{3/4}} \|\vec{g}(s,.)\|_{\dot{M}^{4/3,2}} \, ds.$$

so that $\vec{\nabla}\otimes\vec{u}_3$ is controlled in $L^2\dot{M}^{2,3}$:

$$\|\vec{\nabla}\otimes\vec{u}_3\|_{L^2\dot{M}^{2,3}} \le C\|\vec{g}\|_{L^{4/3}\dot{M}^{4/3,2}}.$$

In order to estimate $\vec{u}_3$ in $\dot{M}^{2,3}$, we shall use the helicity of $\vec{u}_3$, and thus just try and estimate $\vec{u}_3(x_1, x_2, 0)$ in $(L^2(\mathbb{R}^2))^3$. Thus, we consider $\vec{v}_0$ in $(L^2(\mathbb{R}^2))^3$ and compute $I = \iint \vec{u}_3(t, x_1, x_2, 0)).\vec{v}_0(x_1, x_2) \, dx_1 \, dx_2$. We write $\vec{v}_0 = v_r'(r, \theta)\vec{e}_r + v_\theta(r, \theta)\vec{e}_\theta + v_z(r, \theta)\vec{e}_z$, and consider the helical extension of $\vec{v}_0$: $\vec{v}(r, \theta, z) = v_r'(r, \theta - z/\alpha)\vec{e}_r + v_\theta(r, \theta - z/\alpha)\vec{e}_\theta + v_z(r, \theta - z/\alpha)\vec{e}_z$. We have

$$I = \frac{1}{2\pi\alpha} \int_{0<x_3<2\pi\alpha} \vec{u}_3 \cdot \vec{v} \, dx$$

$$= \frac{1}{2\pi\alpha} \int_0^t \int_{0<x_3<2\pi\alpha} (\sqrt{-\Delta} W_{\nu(t-s)} * \vec{v}) \cdot \frac{1}{\sqrt{-\Delta}} \mathbb{P}\vec{g}(s,.) \, dx \, ds$$

so that

$$|I| \leq C\|\vec{\nabla} \otimes W_{\nu(t-s)} * \vec{v}\|_{L^2 \dot{M}^{2,3}}\|\frac{1}{\sqrt{-\Delta}}\vec{g}\|_{L^2 \dot{M}^{2,3}}$$

$$\leq C'\|\vec{v}\|_{\dot{M}^{2,3}}\|\frac{1}{\sqrt{-\Delta}}\vec{g}\|_{L^2 \dot{M}^{2,3}}$$

$$= C'\|\vec{v}_0\|_{L^2(\mathbb{R}^2)}\|\frac{1}{\sqrt{-\Delta}}\vec{g}\|_{L^2 \dot{M}^{2,3}}.$$

and we get

$$\|\vec{u}_3\|_{L^\infty \dot{M}^{2,3}} \leq C\|\frac{1}{\sqrt{-\Delta}}\vec{g}\|_{L^2 \dot{M}^{2,3}}.$$

Thus, if we look at the Picard iterates in the space

$$Y_{T_0} = \{\vec{u} \text{ is helical } / \text{ div } \vec{u} = 0, \vec{u} \in L^4((0, T_0), \dot{M}^{4,6}), \vec{\nabla} \otimes \vec{u} \in L^2((0, T_0), \dot{M}^{2,3})\}$$

we shall find convergence to a mild solution as soon as T_0 is small enough to grant that $\vec{U}_0 = W_{\nu t}\vec{u}_0 + \int_0^t W_{\nu(t-s)} * \mathbb{P}\vec{f}(s, .)\, ds$ satisfies for some constant ϵ_ν (which depends only on ν):

$$\|\vec{U}_0\|_{L^4((0,T_0),\dot{M}^{4,6})} + \|\vec{\nabla} \otimes \vec{U}_0\|_{L^2((0,T_0),\dot{M}^{2,3})} < \epsilon_\nu.$$

Let T^* be the maximal time of existence of our solution. If $\vec{u}$ is bounded in $L^4((0, T^*), \dot{M}^{4,6})$ and $\vec{\nabla} \otimes \vec{u}$ is bounded in $L^2((0, T^*), \dot{M}^{2,3})$, we find that $\vec{u} \in L^\infty((0, T^*), \dot{M}^{2,3})$. But we have a more precise statement: due to helicity, the norm of $\vec{u}$ in the non-separable space $L^4((0, T^*), \dot{M}^{4,6})$ is equal to the norm of $\vec{u}(t, x_1, x_2, 0)$ in the separable space $L^4((0, T^*), L^4(\mathbb{R}^2))$, and similarly the norm of $\vec{\nabla} \otimes \vec{u}$ in the non-separable space $L^2((0, T^*), \dot{M}^{2,3})$ is equal to the norm of $(\vec{\nabla} \otimes \vec{u})(t, x_1, x_2, 0)$ in the separable space $L^2((0, T^*), L^2(\mathbb{R}^2))$, Thus, we may approximate $\vec{u}$ by smooth functions and we get in return that $\vec{u}$ belongs actually to $\mathcal{C}([0, T^*], \dot{M}^{2,3})$. This would give that $\vec{u}(T^*, .) \in \dot{M}^{2,3}$ and if $T^* < T$ we might reiterate the construction of the solution by considering the Cauchy problem with initial time $t = T^*$.

Thus, in order to prove that the solution is global, we just need to control the size of $\vec{u}$ and of $\vec{\nabla} \otimes \vec{u}$. We have proven enough regularity on $\vec{u}$ to be allowed to write:

$$\partial_t \|\vec{u}(t, .)\|_{\dot{M}^{2,3}}^2 = \frac{1}{2\pi\alpha}\partial_t \int_{0 < x_3 < 2\alpha\pi} |\vec{u}|^2\, dx$$

$$= \frac{2}{2\pi\alpha} \int_{0 < x_3 < 2\alpha\pi} \vec{u}.\partial_t \vec{u}\, dx$$

$$= -\frac{2\nu}{2\pi\alpha} \int_{0 < x_3 < 2\alpha\pi} |\vec{\nabla} \otimes \vec{u}|^2\, dx + \frac{2}{2\pi\alpha} \int_{0 < x_3 < 2\alpha\pi} \vec{u}.\vec{f}\, dx$$

$$= -2\nu \iint |\vec{\nabla} \otimes \vec{u}(t, x_1, x_2, 0)|^2\, dx_1\, dx_2$$

$$+ 2 \iint \vec{u}(t, x_1, x_2, 0).\vec{f}(t, x_1, x_2, 0)\, dx_1\, dx_2$$

$$\leq -2\nu\|\vec{\nabla} \otimes \vec{u}(t, x_1, x_2, 0)\|_{L^2(\mathbb{R}^2)}^2$$

$$+ 2\|\vec{u}(t, x_1, x_2, 0)\|_{L^2(\mathbb{R}^2)}\|\vec{f}(t, x_1, x_2, 0)\|_{L^2(\mathbb{R}^2)}$$

$$= -2\nu\|\vec{\nabla} \otimes \vec{u}\|_{\dot{M}^{2,3}}^2 + 2\|\vec{u}\|_{\dot{M}^{2,3}}\|\vec{f}\|_{\dot{M}^{2,3}}.$$

This gives $\partial_t \|\vec{u}\|_{\dot{M}^{2,3}} \leq \|\vec{f}\|_{\dot{M}^{2,3}}$, so that

$$\|\vec{u}\|_{\dot{M}^{2,3}} \leq \|\vec{u}_0\|_{\dot{M}^{2,3}} + \|\vec{f}\|_{L^1 \dot{M}^{2,3}}.$$

Moreover, we have

$$2\nu \||| \vec{\nabla} \otimes \vec{u} \||^2_{L^2 M^{2,3}} \leq \|\vec{u}_0\|_{\dot{M}^{2,3}} + \int_0^{T^*} \|\vec{u}\|_{\dot{M}^{2,3}} \|\vec{f}\|_{\dot{M}^{2,3}} \, dt \leq (\|\vec{u}_0\|_{\dot{M}^{2,3}} + \|\vec{f}\|_{L^1 \dot{M}^{2,3}})^2$$

We thus have global existence.   □

10.5   Brandolese's Symmetrical Solutions

In this section, we consider the Navier–Stokes problem with the following symmetry property: $\vec{u}$ is invariant under the action of the discrete group generated by the isometries $W_1 : (x_1, x_2, x_3) \mapsto (x_2, x_3, x_1)$ and $W_2 : (x_1, x_2, x_3) \mapsto (-x_1, x_2, x_3)$. In that case, we get the symmetrical solutions of Brandolese, that satisfy $u_1(t, x_1, x_2, x_3) = u_2(t, x_3, x_1, x_2) = u_3(t, x_2, x_3, x_1)$ and $u_1(t, -x_1, x_2, x_3) = -u_1(t, x_1, x_2, x_3)$ [59, 62].

Thus, if we consider the solutions for the Cauchy problem described in Theorem 4.10, and if we start with a data $\vec{u}_0$ and a force $\vec{f}$ which are invariant under the action of the isometries W_1 and W_2, we obtain a solution that is still invariant. But such a solution and such a force clearly satisfy the Dobrokhotov and Shafarevich conditions

$$\begin{cases} \text{for } 1 \leq i \leq 3, \ \int_0^t \int f_i \, dx \, ds = 0 \\ \\ \text{for } 1 \leq i < j \leq 3, \ \int_0^t \int 2u_i u_j + x_i f_j + x_j f_i \, dx \, ds = 0 \\ \\ \text{for } 1 \leq i < j \leq 3, \ \int_0^t \int u_i^2 - u_j^2 + x_i f_i - x_j f_j \, dx \, ds = 0 \end{cases} \qquad (10.34)$$

Thus, we get a better decay at infinity ($\vec{u} = o(|x|^{-4})$) than for the generic solutions of the Navier–Stokes equations.

As a final remark, let us recall that Brandolese studied more generally the finite groups of isometry of $\mathbb{R}^3$ and the solutions that are invariant under the action of such a group, in order to determine which decay estimate was obtainable in that case [60].

10.6   Self-similar Solutions

In this section, we consider the Navier–Stokes problem with the following symmetry property: $\vec{u}$ is invariant under the action of time-space rescalings, i.e., we consider self-similar solutions:

$$\text{for every } \lambda > 0, \quad \lambda \vec{u}(\lambda^2 t, \lambda x) = \vec{u}(t, x). \qquad (10.35)$$

Those solutions are associated to homogeneous initial values

$$\lambda \vec{u}_0(\lambda x) = \vec{u}_0(x) \qquad (10.36)$$

and self-similar forcing terms

$$\lambda^3 \vec{f}(\lambda^2 t, \lambda x) = \vec{f}(t, x). \qquad (10.37)$$

It is easy to check that the only homogeneous $\vec{u}_0$ (satisfying (10.36)) belonging to a Lebesgue space L^p or a Sobolev space H^s or $\dot{H}^s$ is the null function $\vec{u}_0 = 0$. Thus, the study of self-similar solutions was an argument for the study of mild solutions in more general spaces as Morrey spaces (Giga and Miyakawa [212]), homogeneous Besov spaces (Cannone [81]) or Lorentz spaces (Barraza [22]).

Recall that the homogeneous distributions T (of homogeneity degree -1) may be written as $T(x) = \frac{\omega(\frac{x}{|x|})}{|x|}$, where ω is a distribution on the sphere S^2, in the sense that

$$\langle T|\varphi\rangle_{\mathcal{S}',\mathcal{S}} = \langle \omega(\sigma)| \int_0^{+\infty} \varphi(r\sigma)\,dr\rangle_{\mathcal{D}'(S^2),\mathcal{D}(S^2)}$$

(see Lemarié-Rieusset [313], chapter 23, for instance). In particular, we have $T \in L^{3,\infty}$ if and only if $\omega \in L^3(S^2)$, and $T \in \dot{M}^{p,3}$ with $2 \leq p < 3$ if and only if $\omega \in L^p(S^2)$.

Existence of self-similar solutions is then a direct consequence of the theory of mild solutions for small data in Besov spaces [81, 313].

Self-similar mild solutions

Theorem 10.8.
Let $\mathbb{X}$ be a Banach space such that

- *for $\lambda > 0$, $\|\lambda^\alpha u(\lambda x)\|_{\mathbb{X}} = \|u\|_{\mathbb{X}}$, where α is a positive constant*

- *$\|\varphi * u\|_{\mathbb{X}} \leq \|\varphi\|_1 \|u\|_{\mathbb{X}}$.*

Assume moreover that, for some $\beta \in (\max(0, \alpha - 1), \alpha)$, pointwise multiplication maps $\mathbb{X} \times \dot{B}_{\mathbb{X},1}^\beta$ to $\dot{B}_{\mathbb{X},\infty}^{\beta-\alpha}$:

$$\|uv\|_{\dot{B}_{\mathbb{X},\infty}^{\beta-\alpha}} \leq C\|u\|_{\mathbb{X}}\|v\|_{\dot{B}_{\mathbb{X},1}^\beta}.$$

(This is the case for instance if we have the inequality for Riesz potentials

$$\|\mathcal{I}_{\alpha-\beta}(u\,\mathcal{I}_\beta v)\|_{\mathbb{X}} \leq C\|u\|_{\mathbb{X}}\|u\|_{\mathbb{X}}.)$$

Finally, let γ such that $\max(0, 1 - \alpha) < \gamma < 2 - \beta$. Then, there exists an $\epsilon_0 > 0$ (depending on $\mathbb{X}$, on γ and on ν) such that, if $\vec{u}_0$ and $\vec{f}$ satisfy

- *$\vec{u}_0 \in \mathbb{Y}$, div $\vec{u}_0 = 0$ and $\vec{u}_0$ is homogeneous of degree -1 ($\lambda \vec{u}_0(\lambda x) = \vec{u}_0(x)$), where $\mathbb{Y} = \mathbb{X}$ if $\alpha = 1$, $\mathbb{Y} = \dot{B}_{\mathbb{X},\infty}^{-1+\alpha}$ if $\alpha < 1$, $\mathbb{Y} = \{0\}$ if $\alpha > 1$*

- *$\vec{f}(t,x) = \frac{1}{t^{3/2}}\vec{F}(\frac{x}{\sqrt{t}})$ with $\vec{F} \in \dot{B}_{\mathbb{X},\infty}^{-\gamma}$*

- *$\|\vec{u}_0\|_{\mathbb{Y}} + \|\vec{F}\|_{\dot{B}_{\mathbb{X},\infty}^{-\gamma}} < \epsilon_0$*

then the Navier–Stokes problem

$$\begin{cases} \partial_t \vec{u} + \mathrm{div}(\vec{u} \otimes \vec{u}) = \nu\Delta\vec{u} + \vec{f} - \vec{\nabla}p \\ \mathrm{div}\,\vec{u} = 0 \\ \vec{u}(0,.) = \vec{u}_0 \end{cases}$$

has a self-similar solution $\vec{u}(t,x) = \frac{1}{\sqrt{t}}\vec{U}(\frac{x}{\sqrt{t}})$, with $\vec{U} \in \mathbb{X} \cap \dot{B}_{\mathbb{X},1}^\beta$.

Proof. Let us write $\vec{U}_0 = W_\nu * \vec{u}_0$, $\vec{U}_1 = \int_0^1 W_{\nu(1-s)} * \mathbb{P}\vec{f}(s,.)\,ds$ and $\vec{V}_0 = W_{\nu t} * \vec{u}_0 + \int_0^t W_{\nu(t-s)} * \mathbb{P}\vec{f}(s,.)\,ds$. Then we have

$$\vec{V}_0(t,x) = \frac{1}{\sqrt{t}}(\vec{U}_0(\frac{x}{\sqrt{t}}) + \vec{U}_1(\frac{x}{\sqrt{t}})).$$

Moreover, we have

$$\|\vec{U}_0\|_{\mathbb{X}} \leq \|\vec{u}_0\|_{\mathbb{X}}$$

and, for all $\beta > 0$,

$$\|\vec{U}_0\|_{\dot{B}^{\beta}_{\mathbb{X},1}} \leq C_{\nu,\beta}\|\vec{u}_0\|_{\mathbb{X}}.$$

If $\alpha < 1$, we write for all positive δ,

$$\|\vec{U}_0\|_{\dot{B}^{-1+\alpha+\delta}_{\mathbb{X},1}} \leq C_{\nu,\alpha,\delta}\|\vec{u}_0\|_{\dot{B}^{-1+\alpha}_{\mathbb{X},\infty}};$$

we shall use it for $\delta = 1 - \alpha + \beta$ and for $\delta = 1 - \alpha$ (since we have $\dot{B}^0_{\mathbb{X},1} \subset \mathbb{X}$). Similarly, we have, for all $\delta > 0$,

$$\|W_{\nu(1-s)} * \vec{f}(s,.)\|_{\dot{B}^{-\gamma+\delta}_{\mathbb{X},1}} \leq C\|\vec{F}\|_{\dot{B}^{-\gamma}_{\mathbb{X},\infty}} s^{-\frac{(3-\alpha-\gamma)}{2}}(1-s)^{-\delta/2};$$

This will give a control on $\vec{U}_1$ in the $\dot{B}^{-\gamma+\delta}_{\mathbb{X},1}$ norm, provided that $\alpha + \gamma > 1$ and $\delta < 2$. Taking $\delta = \gamma$, then $\delta = \beta + \gamma$ gives the control on $\vec{U}_1$ in $\mathbb{X} \cap \dot{B}^{\beta}_{\mathbb{X},1}$ (provided $\gamma < 2 - \beta$).

Now, if $\vec{U}$ belongs to $\mathbb{X}$ and $\vec{V}$ belongs to $\dot{B}^{\beta}_{\mathbb{X},1}$ and if $\vec{W} = \mathrm{div}(\vec{U} \otimes \vec{V})$, we know that $\vec{W}$ belongs to $\dot{B}^{\beta-\alpha-1}_{\mathbb{X},\infty}$. On the other hand, when $\vec{U}$ and $\vec{V}$ belong to $\dot{B}^{\beta}_{\mathbb{X},1}$, we may use the paradifferrential calculus and write the Littlewood–Paley decomposition [313] of $\vec{W}$ as

$$\Delta_j\vec{W} = \Delta_j(\sum_{|k-j|\leq 2}(\mathrm{div}(S_k\vec{U} \otimes \Delta_k\vec{V} + \Delta_k\vec{U} \otimes S_k\vec{V})$$

$$+ \Delta_j(\sum_{k\geq j-3}\sum_{|k-l|\leq 1}\mathrm{div}(\Delta_k\vec{U} \otimes \Delta_l\vec{V}))$$

and thus

$$2^{j(\beta-\alpha)}\|\Delta_j\vec{W}\|_{\mathbb{X}} \leq C2^j(\sum_{|k-j|\leq 2}\|S_k\vec{U}\|_{\dot{B}^{\beta}_{\mathbb{X},1}}\|\Delta_k\vec{V}\|_{\mathbb{X}} + \|S_k\vec{V}\|_{\dot{B}^{\beta}_{\mathbb{X},1}}\|\Delta_k\vec{U}\|_{\mathbb{X}})$$

$$+ C2^j\sum_{k\geq j-3}\sum_{|k-l|\leq 1}\|\Delta_k\vec{U}\|_{\dot{B}^{\beta}_{\mathbb{X},1}}\|\Delta_l\vec{V}\|_{\mathbb{X}}$$

$$\leq C2^{j(1-\beta)}\|\vec{U}\|_{\dot{B}^{\beta}_{\mathbb{X},1}}\|\vec{V}\|_{\dot{B}^{\beta}_{\mathbb{X},1}}$$

so that $\vec{W}$ belongs to $\dot{B}^{2\beta-\alpha-1}_{\mathbb{X},\infty}$.

Thus, we may find $\delta \in (\max(0, 1-\alpha), 1+\alpha-\beta) \cap (1+\alpha-2\beta, 2-\beta)$ with $\vec{W} \in \dot{B}^{\delta}_{\mathbb{X},\infty}$. This gives

$$\|\int_0^1 W_{\nu(t-s)} * \mathbb{P}\,\mathrm{div}(\frac{1}{\sqrt{s}}\vec{U}(\frac{\cdot}{\sqrt{s}}) \otimes \frac{1}{\sqrt{s}}\vec{V}(\frac{\cdot}{\sqrt{s}}))\,ds\|_{\mathbb{X}\cap\dot{B}^{\beta}_{\mathbb{X},1}} \leq C\|\vec{U}\|_{\mathbb{X}\cap\dot{B}^{\beta}_{\mathbb{X},1}}\|\vec{V}\|_{\mathbb{X}\cap\dot{B}^{\beta}_{\mathbb{X},1}}.$$

This gives the existence of self-similar solutions for small data. $\square$

Theorem 10.8 may be applied to a lot of spaces $\mathbb{X}$. For instance, in the case $\mathbb{X} = L^p$, $p > 3$, we find self-similar solutions $\vec{u} = \frac{1}{\sqrt{t}}\vec{U}(\frac{x}{\sqrt{t}})$ with a profile $\vec{U} \in L^p \cap L^q$, where $p < q < \infty$, for a small enough homogeneous initial value $\vec{u} \in \dot{B}_{p,\infty}^{-1+\frac{3}{p}}$ and for a small enough forcing term $\vec{f} = \frac{1}{t^{3/2}}\vec{F}(\frac{x}{\sqrt{t}})$ with $\vec{F} \in L^r$ for some $r \in (r^*, p)$, where $\frac{1}{r^*} = \frac{1}{q} + \frac{2}{3}$ (just use the embeddings $\dot{B}_{p,1}^{\frac{3}{p}-\frac{3}{q}} \subset L^q$ and $L^r \subset \dot{B}_{p,\infty}^{\frac{3}{p}-\frac{3}{r}}$).

In Theorem 10.8, the force $\vec{F}$ is quite regular: it belongs to $\dot{B}_{\mathbb{X},\infty}^{-\gamma}$ with $\gamma < 2$. In some cases, one may even consider a force $\vec{F}$ in $\dot{B}_{\mathbb{X},\infty}^{-2}$; of course, we shall not have the extra-regularity $\vec{U} \in \dot{B}_{\mathbb{X},1}^{\beta}$, and may only hope that $\vec{U} \in \mathbb{X}$. We give here an easy lemma that allows as well to deal with discretely self-similar solutions (or DSS solutions), a class of solutions which has been considered by Tsai, when the initial data is not homogeneous, but only discretely homogeneous: the equality $\lambda\vec{u}_0(\lambda x) = \vec{u}_0(x)$ holds only for $\lambda \in \{\lambda = \lambda_0^k \ / \ k \in \mathbb{Z}\}$ (with $\lambda_0 > 1$), a discrete subgroup of $\mathbb{R}_+^*$.

Lemma 10.2.

Let $\mathbb{X}$ be a Banach space such that

- *for $\lambda > 0$, $\|\lambda u(\lambda x)\|_\mathbb{X} = \|u\|_\mathbb{X}$*

- *$\|\varphi * u\|_\mathbb{X} \leq \|\varphi\|_1 \|u\|_\mathbb{X}$*

Assume moreover that the pointwise product maps $\mathbb{X} \times \mathbb{X}$ to a shift-invariant space $\mathbb{Y}$ such that

- *for $\lambda > 0$, $\|\lambda^2 u(\lambda x)\|_\mathbb{Y} = \|u\|_\mathbb{Y}$*

- *$\|\varphi * u\|_\mathbb{Y} \leq \|\varphi\|_1 \|u\|_\mathbb{Y}$*

- *$[\mathbb{Y}, \dot{B}_{\mathbb{Y},1}^2]_{1/2,\infty} \subset \mathbb{X}$ (so that in particular we have $\mathbb{Y} \subset \dot{B}_{\mathbb{X},\infty}^{-1}$).*

If $\mathbb{F} \in L^\infty((0,+\infty), \mathbb{Y})$, then $\vec{v} = \int_0^t W_{\nu(t-s)} * \mathbb{P}\operatorname{div}\mathbb{F}\,ds$ belongs to $L^\infty((0,+\infty), \mathbb{X})$ and

$$\sup_{t>0} \|\vec{v}(t,.)\|_\mathbb{X} \leq C_0 \frac{1}{\nu} \sup_{t>0} \|\mathbb{F}(t,.)\|_\mathbb{Y}$$

where the constant C_0 does not depend on ν.

Proof. Recall that the norm of f in $[\mathbb{Y}, L^\infty]_{1/2,\infty}$ is equivalent to

$$\sup_{A>0} \inf_{f=g+h} A\|g\|_\mathbb{Y} + A^{-1}\|h\|_\infty.$$

We write

$$\|W_{\nu(t-s)} * \operatorname{div}\mathbb{F}\|_\mathbb{Y} \leq C \frac{1}{\sqrt{\nu(t-s)}} \|\mathbb{F}\|_\mathbb{Y}$$

and

$$\|W_{\nu(t-s)} * \operatorname{div}\mathbb{F}\|_\infty \leq C \frac{1}{(\nu(t-s))^{3/2}} \|\mathbb{F}\|_{L^{3/2,\infty}},$$

so that, for $A < t$,

$$\left\| \int_0^{\max(A,0)} W_{\nu(t-s)} * \operatorname{div}\mathbb{F}\,ds \right\|_\infty \leq 2C \frac{1}{\nu^{3/2}\sqrt{t-A}} \|\mathbb{F}\|_{L^\infty\mathbb{Y}}$$

and

$$\left\| \int_{\max(A,0)}^t W_{\nu(t-s)} * \operatorname{div}\mathbb{F}\,ds \right\|_\mathbb{Y} \leq 2C \frac{\sqrt{t-A}}{\sqrt{\nu}} \|\mathbb{F}\|_{L^\infty\mathbb{Y}}.$$

This gives the control of $\vec{v}(t,.)$ in $[\mathbb{Y}, L^\infty]_{1/2,\infty}$, hence in $\mathbb{X}$. $\qquad\square$

Examples of Banach spaces $\mathbb{X}$ that satisfy assumptions of Lemma 10.2 are the Lorentz space $\mathbb{X} = L^{3,\infty}$ (with $\mathbb{Y} = L^{3/2,\infty}$), the Besov spaces $\mathbb{X} = \dot{B}_{p,\infty}^{-1+\frac{3}{p}}$ (with $\mathbb{Y} = \dot{B}_{p,\infty}^{-2+\frac{3}{p}}$), where $1 \leq p < 3$ or the Besov space based on pseudo-measures $\mathbb{X} = \dot{B}_{\mathrm{PM},\infty}^{2}$ (with $\mathbb{Y} = \dot{B}_{\mathrm{PM},\infty}^{1}$).

Theorem 10.9.
Let $\mathbb{X}$ be a Banach space such that

- *for $\lambda > 0$, $\|\lambda u(\lambda x)\|_{\mathbb{X}} = \|u\|_{\mathbb{X}}$*

- *$\|\varphi * u\|_{\mathbb{X}} \leq \|\varphi\|_1 \|u\|_{\mathbb{X}}$*

Assume moreover that the pointwise product maps $\mathbb{X} \times \mathbb{X}$ to a shift-invariant space $\mathbb{Y}$ such that

- *for $\lambda > 0$, $\|\lambda^2 u(\lambda x)\|_{\mathbb{Y}} = \|u\|_{\mathbb{Y}}$*

- *$\|\varphi * u\|_{\mathbb{Y}} \leq \|\varphi\|_1 \|u\|_{\mathbb{Y}}$*

- *$[\mathbb{Y}, \dot{B}_{\mathbb{Y},1}^{2}]_{1/2,\infty} \subset \mathbb{X}$ (so that in particular we have $\mathbb{Y} \subset \dot{B}_{\mathbb{X},\infty}^{-1}$).*

Then, there exists an $\epsilon_0 > 0$ (depending on $\mathbb{X}$) and a constant $C_1 > 0$ such that, if $\vec{u}_0$ and $\vec{f}$ satisfy

- *$\vec{u}_0 \in \mathbb{X}$, $\operatorname{div} \vec{u}_0 = 0$*

- *$\mathbb{F} \in L^\infty((0,+\infty), \mathbb{Y})$*

- *$\frac{\|\vec{u}_0\|_{\mathbb{X}}}{\nu} + \frac{\|\mathbb{F}\|_{L^\infty \mathbb{Y}}}{\nu^2} < \epsilon_0$*

then the Navier–Stokes problem

$$\begin{cases} \partial_t \vec{u} = \nu \Delta \vec{u} + \mathbb{P} \operatorname{div}(\mathbb{F} - \vec{u} \otimes \vec{u}) \\ \operatorname{div} \vec{u} = 0 \\ \vec{u}(0,.) = \vec{u}_0 \end{cases}$$

has a unique solution $\vec{u}(t,x)$ such that $\sup_{t>0} \|\vec{u}(t,.)\|_{\mathbb{X}} \leq C_1$.
If there exists $\lambda_0 > 1$ such that $\lambda_0 \vec{u}_0(\lambda_0 x) = \vec{u}_0(x)$ and $\lambda_0^2 \mathbb{F}(\lambda_0^2 t, \lambda_0 x) = \mathbb{F}(t,x)$, then this solution $\vec{u}$ is discretely self-similar:

$$\lambda_0 \vec{u}(\lambda_0^2 t, \lambda_0 x) = \vec{u}(t,x).$$

Proof. As usual, the proof is performed by using Picard's iterates. By Lemma 10.2, we know that the bilinear operator $B(\vec{u}, \vec{v}) = \int_0^t W_{\nu(t-s)} * \mathbb{P} \operatorname{div}(\vec{u} \otimes \vec{v})\, ds$ is bounded on $L^\infty((0,+\infty), \mathbb{X})$, with an operator norm which is bounded by $C_0 \frac{1}{\nu}$. Thus the Picard iterates defined by

$$\vec{U}_0 = W_{\nu t} * \vec{u}_0 + \int_0^t W_{\nu(t-s)} * \mathbb{P} \operatorname{div} \mathbb{F}\, ds$$

and

$$\vec{U}_{N+1} = \vec{U}_0 - \int_0^t W_{\nu(t-s)} * \mathbb{P} \operatorname{div}(\vec{U}_N \otimes \vec{U}_N)\, ds$$

will converge to a solution $\vec{u} \in L^\infty \mathbb{X}$ if we have $\|\vec{U}_0\|_{L^\infty \mathbb{X}} < \frac{\nu}{4C_0}$. We conclude by the estimates $\|W_{\nu t} * \vec{u}_0\|_{L^\infty \mathbb{X}} \le \|\vec{u}_0\|_{\mathbb{X}}$ and, again by Lemma 10.2,

$$\left\| \int_0^t W_{\nu(t-s)} * \mathbb{P} \operatorname{div} \mathbb{F} \, ds \right\|_{L^\infty \mathbb{X}} \le C_0 \frac{1}{\nu} \|\mathbb{F}\|_{L^\infty \mathbb{Y}}.$$

Self-similarity is then obvious, due to the symmetry of the equations under scaling and the invariance of the norms and due to uniqueness of the solution in the ball. □

10.7 Stationary Solutions

In this section, we consider the Navier-Stokes problem with the following symmetry property: $\vec{u}$ is invariant under the action of time translations, i.e., we consider steady solutions.

When the force $\vec{f}$ is stationary (i.e., does not depend on time), one may ask whether one might find a steady-state solution $\vec{u}$ of the Navier–Stokes equations. The problem to be solved is then the following one:

Stationary Navier–Stokes equations

Given the force $\vec{f}(x)$, find the velocity $\vec{u}(x)$ and the pressure $p(x)$ such that

$$-\nu \Delta \vec{u} = -\vec{u}.\vec{\nabla}\vec{u} - \vec{\nabla}p + \vec{f}, \quad \operatorname{div} \vec{u} = 0 \tag{10.38}$$

This is an old and difficult problem, especially in bounded domains (Galdi's *Introduction* [194] to the steady state problem runs over more than 1000 pages). In the whole space, however, the problem is quite easy.

We are going to consider a special formulation of the problem, turning the differential equation into an integral one and considering only small data.

Stationary Navier–Stokes equations

Given the force $\vec{f}(x)$, find the velocity $\vec{u}(x)$ such that

$$\vec{u} = -\frac{1}{\nu \Delta} \mathbb{P} \vec{f} + \frac{1}{\nu} \mathbb{P} \frac{\vec{\nabla}}{\Delta}.(\vec{u} \otimes \vec{u}) \tag{10.39}$$

This problem has been studied for solutions in the Lorentz space $L^{3,\infty}$ or in the Morrey space $\dot{M}^{p,3}$ with $2 < p < 3$ (remark: $L^{3,\infty} \subset \dot{M}^{p,3}$ when $p < 3$) by Kozono ad Yamazaki [280]. In that case, one starts from $\vec{f} = \Delta \vec{F}$, with $\vec{F}$ small enough in $L^{3,\infty}$ or in $\dot{M}^{p,3}$. The case of $\vec{F} \in L^{3,\infty} \cap L^q$ with $\vec{F}$ small in $L^{3,\infty}$ ($3/2 < q < \infty$) has been discussed by Bjorland, Brandolese, Iftimie, and Schonbek [44].

Recently, Phan and Phuc [395] discussed the problem in the largest space where one can look for steady solutions: as in Section 5.3, one looks at a dominating quadratic equation with non-negative kernel

$$U = \frac{1}{\nu}F + C_0 \frac{1}{\nu\sqrt{-\Delta}}U^2,$$

for which the space where to look for solutions is the multiplier space $\mathcal{V}^1(\mathbb{R}^3) = \mathcal{M}(\dot{H}^1 \mapsto L^2)$.

Existence of steady solutions

Theorem 10.10.
Let $\mathbb{X}$ be a space such that

- *the Riesz transforms are bounded on $\mathbb{X}$*

- *the bilinear operator $(u,v) \mapsto \frac{1}{\sqrt{-\Delta}}(uv)$ is bounded on $\mathbb{X}$*

Then, there exists $\epsilon_0 > 0$ and $C_0 > 0$ (depending on $\mathbb{X}$) such that, if $\vec{f}(x)$ (independent from t) satisfies

- *$\vec{f} = \Delta \vec{F}$ with $\vec{F} \in \mathbb{X}$*

- *$\|\vec{F}\|_{\mathbb{X}} < \epsilon_0 \nu^2$*

then there is a unique solution $\vec{u} \in \mathbb{X}$ of the equation

$$\vec{u} = -\frac{1}{\nu\Delta}\mathbb{P}\vec{f} + \frac{1}{\nu}\mathbb{P}\frac{\vec{\nabla}}{\Delta}.(\vec{u} \otimes \vec{u}) \tag{10.40}$$

with $\|\vec{u}\|_{\mathbb{X}} < C_0\nu$.

Proof. This is obvious by Picard's fixed-point theorem: we have

$$\|\frac{1}{\nu}\mathbb{P}\frac{\vec{\nabla}}{\Delta}.(\vec{u} \otimes \vec{v})\|_{\mathbb{X}} \leq C_1 \frac{1}{\nu}\|\vec{u}\|_{\mathbb{X}}\|\vec{v}\|_{\mathbb{X}}$$

and we find that Equation (10.40) has a unique solution $\vec{u}$ with $\|\vec{u}\|_{\mathbb{X}} < \frac{\nu}{2C_1}$ as soon as $\|\frac{1}{\nu\Delta}\mathbb{P}\vec{f}\|_{\mathbb{X}} < \frac{\nu}{4C_1}$. ☐

Classical examples of such spaces $\mathbb{X}$ are the homogeneous Sobolev space $\dot{H}^{1/2}$, the Lebesgue space L^3, the Lorentz space $L^{3,\infty}$, the Morrey spaces $\dot{M}^{p,3}$ with $2 < p \leq 3$ and the mulitplier space $\mathcal{V}^1 = \mathcal{M}(\dot{H}^1 \mapsto L^2)$.

Notice that those examples, except $\dot{H}^{1/2}$ are lattice spaces of Lebesgue measurable functions: if $u \in \mathbb{X}$ and if v is a Lebesgue measurable function such that $|v| \leq |u|$, then $v \in \mathbb{X}$ and $\|v\|_{\mathbb{X}} \leq \|u\|_{\mathbb{X}}$.

Moreover, we have that the bilinear operator $(u,v) \mapsto \frac{1}{(-\Delta)^{1/4}}(u\frac{1}{(-\Delta)^{1/4}}v)$ is bounded on $\mathbb{X}$. For L^3, $L^{3,\infty}$ or $\dot{M}^{p,3}$, this is obvious from Hedberg's inequality:

$$|\frac{1}{(-\Delta)^{1/4}}(u\frac{1}{(-\Delta)^{1/4}}v)| \leq C(\mathcal{M}_{u\sqrt{\mathcal{M}_v}}(x))^{2/3}\|u\|_{\mathbb{X}}^{1/3}\|v\|_{\mathbb{X}}^{2/3}$$

For $\mathbb{X} = \mathcal{V}^1$, we may use the results of Gala and Lemarié-Rieusset [191]: if $v \in \mathcal{V}^1 = \mathcal{M}(\dot{H}^1 \mapsto L^2)$, then

$$\frac{1}{(-\Delta)^{1/4}} v \in \mathcal{M}(\dot{H}^1 \mapsto \dot{H}^{1/2}) = \mathcal{M}(\dot{H}^{-1/2} \mapsto \dot{H}^{-1}) \subset \mathcal{M}(L^2 \mapsto H^{-1/2});$$

thus, if we first multiply with u, we map $\dot{H}^1$ to L^2, then, multiplying the result by $\frac{1}{(-\Delta)^{1/4}} v$ (that maps L^2 to $H^{-1/2}$); we find that the product $u \frac{1}{(-\Delta)^{1/4}} v$ maps $\dot{H}^1$ to $\dot{H}^{-1/2}$, and finally (using again [191]) we get $\frac{1}{(-\Delta)^{1/4}}(u \frac{1}{(-\Delta)^{1/4}} v) \in \mathcal{M}(\dot{H}^1 \mapsto L^2) = \mathbb{X}$.

Following Kozono and Yamazaki [280], and Phan and Phuc [395], we may then give a simple proof of the stability of steady solutions under small perturbations:

• Stability of steady solutions

Theorem 10.11.
Let $\mathbb{X}$ be a Banach space of Lebesque measurable functions such that

- *$\mathbb{X}$ is a lattice*

- *the Hardy–Littlewood maximal function is a bounded operator on $\mathbb{X}$*

- *the Riesz transforms are bounded on $\mathbb{X}$*

- *the bilinear operator $(u, v) \mapsto \frac{1}{\sqrt{-\Delta}}(uv)$ is bounded on $\mathbb{X}$*

- *the bilinear operator $(u, v) \mapsto \frac{1}{(-\Delta)^{1/4}}(u \frac{1}{(-\Delta)^{1/4}} v)$ is bounded on $\mathbb{X}$*

Then, there exists $\epsilon_0 > 0$ and $C_0 > 0$ (depending on $\mathbb{X}$) such that, if $\vec{u}_0$ and $\vec{f}(x)$ (independent from t) satisfy

- *$\vec{u}_0 \in \mathbb{X}$ and $\operatorname{div} \vec{u}_0 = 0$*

- *$\vec{f} = \Delta \vec{F}$ with $\vec{F} \in \mathbb{X}$*

- *$\|\vec{u}_0\|_{\mathbb{X}} < \epsilon_0 \nu$ and $\|\vec{F}\|_{\mathbb{X}} < \epsilon_0 \nu^2$*

then

- *there is a unique solution $\vec{U} \in \mathbb{X}$ of the equation*

$$\vec{U} = -\frac{1}{\nu \Delta} \mathbb{P} \vec{f} + \frac{1}{\nu} \mathbb{P} \frac{\vec{\nabla}}{\Delta} \cdot (\vec{U} \otimes \vec{U})$$

with $\|\vec{U}\|_{\mathbb{X}} < C_0 \nu$

- *there exists a unique solution $\vec{u}$ on $(0, +\infty) \times \mathbb{R}^3$ of the problem*

$$\begin{cases} \partial_t \vec{u} - \operatorname{div}(\vec{u} \otimes \vec{u}) = \nu \Delta \vec{u} + \vec{f} - \vec{\nabla} p \\ \operatorname{div} \vec{u} = 0 \\ \vec{u}(0, .) = \vec{u}_0 \end{cases} \qquad (10.41)$$

such that $\sup_{t>0} |\vec{u}(t, .)| \in \mathbb{X}$ and $\|\sup_{t>0} |\vec{u}(t, .)|\|_{\mathbb{X}} < C_0 \nu$.

- *moreover, we have*

$$\sup_{t>0} t^{1/4} \|(-\Delta)^{1/4}(\vec{u}(t, x) - \vec{U}(x))\|_{\mathbb{X}} \leq C_0 \nu^{3/4} \qquad (10.42)$$

so that $\vec{u}(t, .)$ converges to $\vec{U}$ in $\mathcal{S}'$ (and in L^2_{loc}) as t goes to $+\infty$.

Proof. We have already proved the existence of the steady solution $\vec{U}$ (Theorem 10.10). For the existence of $\vec{u}$, we define $\vec{w} = \vec{u} - \vec{U}$. $\vec{w}$ is a solution of the problem

$$\begin{cases} \partial_t \vec{w} - \operatorname{div}(\vec{w} \otimes \vec{w} + \vec{U} \otimes \vec{w} + \vec{w} \otimes \vec{U}) = \nu \Delta \vec{w} - \vec{\nabla} q \\ \operatorname{div} \vec{w} = 0 \\ \vec{w}(0, .) = \vec{u}_0 - \vec{U} \end{cases}$$

We rewrite the problem as

$$\vec{w} = W_{\nu t} * \vec{w}_0 - B(\vec{w}, \vec{w}) - B(\vec{U}, \vec{w}) - B(\vec{w}, \vec{U})$$

with

$$B(\vec{v}, \vec{w}) = \int_0^t W_{\nu(t-s)} * \mathbb{P} \operatorname{div}(\vec{v} \otimes \vec{w}) \, ds.$$

We know that

$$|W_{\nu t} * \vec{w}_0(x)| \leq \mathcal{M}_{|\vec{w}_0|}$$

so that

$$\sup_{t>0} |W_{\nu t} * \vec{w}_0(x)| \in \mathbb{X}.$$

Moreover, if $|\vec{v}(t, x)| \leq V(x)$ and $|\vec{w}(t, x)| \leq W(x)$, we find that

$$|B(\vec{v}, \vec{w})(t, x)| \leq C \int_0^t \int \frac{1}{(\nu(t-s))^2 + |x-y|^4} W(y) V(y) \, dy \, ds \leq \frac{C'}{\nu} \mathcal{I}_1(VW)(x)$$

(where $\mathcal{I}_1 = \frac{1}{\sqrt{-\Delta}}$). This grants the existence of $\vec{w}$ (hence of $\vec{u}$), if $\vec{u}_0$ and $\vec{f}$ are small enough.

Moreover, we have

$$|(\nu t)^{1/4}(-\Delta)^{1/4} W_{\nu t} * \vec{w}_0(x)| \leq \mathcal{M}_{|\vec{w}_0|}$$

and thus

$$|W_{\nu t} * \vec{w}_0(x)| \leq C \frac{1}{(\nu t)^{1/4}} \mathcal{I}_{1/2} \mathcal{M}_{|\vec{w}_0|}(x)$$

(where $\mathcal{I}_{1/2} = \frac{1}{(-\Delta)^{1/4}}$). If $|\vec{v}(t, x)| \leq V(x)$ and $t^{1/4}|\vec{w}(t, x)| \leq \mathcal{I}_{1/2}W(x)$, we find that

$$t^{1/4}|(-\Delta)^{1/4} B(\vec{v}, \vec{w})(t, x)| \leq C \int_0^t \int \frac{(t-s)^{1/4} + s^{1/4}}{(\nu(t-s) + |x-y|^2)^{9/4}} V(y) \mathcal{I}_{1/2}W(y) \, dy \, \frac{ds}{s^{1/4}}$$

$$\leq C' \frac{1}{\nu} \mathcal{I}_{1/2}(V \mathcal{I}_{1/2}W)(x)$$

and a similar estimate holds for $|(-\Delta)^{1/4} B(\vec{w}, \vec{v})(t, x)|$. This will give the regularity estimate for our solution $\vec{w}$. $\square$

Remark:

- An interesting point is that, while the perturbation $\vec{w}$ is regular ($(-\Delta)^{1/4}\vec{w}$ is a locally square integrable function), the steady solution may be irregular (this proves that in presence of a singular forcing term, the mild solutions of the Navier–Stokes equations may be singular): for example, let $\theta \in L^3(\mathbb{R})$ be a compactly supported function with $\int \theta(s) \, ds = 0$ and let $\Theta(s) = \int_{-\infty}^s \theta(\sigma) \, d\sigma$; if $\vec{F} \in L^3$ is defined by

$$\vec{F} = (\Theta(x_1)\theta(x_2)\theta(x_3), \theta(x_1)\Theta(x_2)\theta(x_3), -2\theta(x_1)\theta(x_2)\Theta(x_3)),$$

  then the steady solution $\vec{U}$ of $\vec{U} = -\frac{1}{\nu}\mathbb{P}\vec{F} + \frac{1}{\nu}\mathbb{P}\frac{\vec{\nabla}}{\Delta}.(\vec{U} \otimes \vec{U})$ is the sum of a term $\vec{U}_1 = \frac{1}{\nu}\mathbb{P}\frac{\vec{\nabla}}{\Delta}.(\vec{U} \otimes \vec{U})$ that satisfies $\vec{U}_1 \in \dot{H}^{1/2}$ and of a term $\vec{U}_2 = -\frac{1}{\nu}\vec{F}$ that is very irregular if θ is irregular.

- One may consider more singular initial values. For instance, if $\vec{U}$ is the steady solution in $\mathcal{V}^1$ associated to the small force $\vec{f} = \Delta\vec{F}$, and if the initial value $\vec{u}_0$ for the Cauchy problem (10.41) is small enough in $\dot{M}^{2,3}$, then the Cauchy problem has a solution $\vec{u}$ such that:

$$\sup_{t>0} |\vec{u}(t,.)| \in \dot{M}^{2,3}$$

and

$$\sup_{t>0} t^{1/4} |(-\Delta)^{1/4}(\vec{u}(t,.) - \vec{U})| \in \dot{M}^{2,3}.$$

10.8 Landau's Solutions of the Navier–Stokes Equations

Let $\vec{f} \in L^1(\mathbb{R}^3)$. Then $\frac{1}{\Delta}\vec{f} \in L^{3,\infty}$; thus, we may apply Theorem 10.11 if $\|\vec{f}\|_1$ is small enough and find a steady solution $\vec{u} \in L^{3,\infty}$ of the equation

$$\vec{u} = -\frac{1}{\nu\Delta}\mathbb{P}\vec{f} + \frac{1}{\nu}\mathbb{P}\frac{\vec{\nabla}}{\Delta}.(\vec{u} \otimes \vec{u})$$

This solution is quite regular: as $L^{3/2,\infty} \subset \dot{B}_{2,\infty}^{-1/2}$, we find that $\vec{u} \in \dot{B}_{2,\infty}^{1/2}$.

Theorem 10.12.
There exist $\epsilon_0 > 0$ and $C_0 > 0$ such that, if $\vec{f}(x)$ (independent from t) satisfies

$$\|\vec{f}\|_1 < \epsilon_0\nu^2$$

then

- *there is a unique solution $\vec{u} \in L^{3,\infty}$ of the equation*

$$\vec{u} = -\frac{1}{\nu\Delta}\mathbb{P}\vec{f} + \frac{1}{\nu}\mathbb{P}\frac{\vec{\nabla}}{\Delta}.(\vec{u} \otimes \vec{u})$$

  with $\|\vec{u}\|_{L^{3,\infty}} < C_0\nu$

- *the functions $\lambda\vec{u}(\lambda x)$ is *-weakly convergent in $L^{3,\infty}$ (as $\lambda \to +\infty$) to the unique solution $\vec{u}_\infty$ on $(0,+\infty) \times \mathbb{R}^3$ of the problem*

$$\vec{u}_\infty = -\frac{1}{\nu\Delta}\mathbb{P}\vec{f}_\infty + \frac{1}{\nu}\mathbb{P}\frac{\vec{\nabla}}{\Delta}.(\vec{u}_\infty \otimes \vec{u}_\infty)$$

  with $\vec{f}_\infty = \delta(x-0)\int \vec{f}\,dy$ and $\|\vec{u}_\infty\|_{L^{3,\infty}} < C_0\nu$.

Proof. The existence of $\vec{u}$ and of $\vec{u}_\infty$ are proved by Theorem 10.11. Moreover, the functions $\lambda\vec{u}(\lambda x)$ are all contained in the closed ball $B = \{\vec{v} \in L^{3,\infty} / \|\vec{v}\|_{L^{3,\infty}} \leq \|\vec{u}\|_{L^{3,\infty}}\}$, which is a compact metrizable space for the *-weak convergence. Thus, in order to prove the theorem, we have only to prove that $\vec{u}_\infty$ is the only limit point of the family $\lambda\vec{u}(\lambda x)$ when $\lambda \to +\infty$.

Let us consider a limit point $\vec{v}$ of $\lambda\vec{u}(\lambda x)$. To $\vec{u}$, one may associate $p \in L^{3/2,\infty}$ such that

$$\partial_t\vec{u} = \nu\Delta\vec{u} - \text{div}(\vec{u} \otimes \vec{u}) - \vec{\nabla}p + \vec{f}$$

We have $\vec{v} = * - \lim \lambda_k\vec{u}(\lambda_k x)$; we may assume that $\lambda_k^2 p(\lambda_k x)$ is *-weakly convergent as well (to some $q \in L^{3/2,\infty}$). The, we get in the distribution sense that

$$\lim \text{div}(\lambda_k\vec{u}(\lambda_k x) \otimes \lambda_k\vec{u}(\lambda_k x)) = \nu\Delta\vec{v} + (\int \vec{f}\,dx)\delta(x-0) - \vec{\nabla}q.$$

Thus, the problem is just to study the convergence of the non-linear term. But this convergence is easy, due to the Rellich lemma: the functions $\lambda_k \vec{u}(\lambda_k x)$ are bounded in $\dot{B}_{2,\infty}^{1/2}$, hence in H_{loc}^s for $0 < s < 1/2$. Thus, applying Rellich theorem, we find that a subsequence will converge strongly in L_{loc}^2 to $\vec{v}$ and the convergence of the non-linear term is obtained. $\qquad\square$

Thus, we can see that the steady solutions associated to a Dirac mass

$$\vec{f} = \beta\delta(x - 0)\vec{e}$$

($\beta \in \mathbb{R}$, $\vec{e}$ unit vector) play a special role into the asymptotic behavior of steady solutions when x goes to ∞. Up to a rotation, we may assume that $\vec{e} = \vec{e}_3$ ($= \vec{e}_z$ in cylindrical coordinates); the solution will then be axisymmetrical with no swirl. Those solutions are known as Landau's (self-similar) solutions. Surprisingly enough, those solutions exist for all β (even large ones) and have been described first (quite implicitly) by Slezkin [439][2], then independently by Landau [301] and Squire [447]; recently, Tian and Xin provided another derivation of those solutions [473]. The role of Landau solutions in asymptotics of steady solution has been discussed by Šverák [492].

Landau solutions

Theorem 10.13.
For $\beta \in \mathbb{R}$, there exists one and only one solution of the problem

$$\nu\Delta\vec{u} - \text{div}(\vec{u} \otimes \vec{u}) - \vec{\nabla}p + \beta\delta(x - 0)\vec{e}_3, \quad \text{div}\,\vec{u} = 0 \qquad (10.43)$$

such that $\vec{u}$ is axisymmetric with no swirl, homogeneous of homogeneity degree -1 and C^2 on $|x| \neq 0$.
For $\beta \neq 0$, this solution is given by the formula

$$\begin{cases} u_1 = & 2\nu\dfrac{x_1(Ax_3 - |x|)}{|x|(A|x| - x_3)^2} \\[2ex] u_2 = & 2\nu\dfrac{x_2(Ax_3 - |x|)}{|x|(A|x| - x_3)^2} \\[2ex] u_3 = & 2\nu\dfrac{A|x|^2 + Ax_3^2 - 2x_3|x|}{|x|(A|x| - x_3)^2} \end{cases} \qquad (10.44)$$

where $A = A(\beta)$ is a constant with $|A| > 1$. The pressure p is given by

$$p = 4\nu^2 \frac{Ax_3 - |x|}{|x|(A|x| - x_3)^2}$$

Remark: If $\vec{u}$ is a steady solution of the Navier-Stokes equations on $\mathbb{R}^3 \setminus \{0\}$ (with null forcing term) which is regular and homogeneous of degree -1, then it is a Landau solution (see Šverák [492]).

Proof. In order to describe axisymmetric flows, recall that we found it convenient to use cylindrical coordinates: $x_1 = r\cos\theta$, $x_2 = r\sin\theta$ and $x_3 = z$. Another interesting system

---

[2]Galaktionov [192] provides an English translation of Slezkin's paper.

of coordinates is the system of spherical coordinates (which amounts to write $(r, z) = \rho(\sin\varphi, \cos\varphi)$).

We thus write our velocity as $\vec{u} = u_r(\rho, \varphi)\vec{e}_r + u_z(\rho, \varphi)\vec{e}_z$ and the vorticity as $\vec{\omega} = \omega_\theta \vec{e}_\theta$, with $\omega_\theta = \partial_z u_r - \partial_r u_z$.

The equation div $\vec{u} = 0$ gives

$$\frac{1}{r}(\partial_r(ru_r) + \partial_z(ru_z)) = 0.$$

Thus, we have (on the open set $r > 0$), $ru_r = -\partial_z\gamma$ and $ru_z = \partial_r\gamma$ for some function $\gamma(r, z)$, or equivalently

$$\vec{u} = \vec{\nabla} \wedge (\psi_\theta \vec{e}_\theta) \text{ with } \psi_\theta = \frac{1}{r}\gamma.$$

We then obtain an equation on γ: as $\operatorname{div}(\psi_\theta \vec{e}_\theta) = 0$, we have $\vec{\omega} = -\Delta(\psi_\theta \vec{e}_\theta)$; let D be the differential operator

$$Dh = \partial_r^2 h + \frac{1}{r}\partial_r h + \partial_z^2 h - \frac{1}{r^2}h;$$

we have $\Delta(\psi_\theta \vec{e}_\theta) = (D\psi_\theta)\vec{e}_\theta$ and $\Delta\vec{\omega} = (D\omega_\theta)\vec{e}_\theta$, so that, taking the curl of the Navier–Stokes equations, we have the equation on $\vec{\omega}$:

$$\nu\Delta\vec{\omega} + (\vec{\omega}.\vec{\nabla})\vec{u} - (\vec{u}.\vec{\nabla})\vec{\omega} + \operatorname{curl}\vec{f} = 0$$

and thus (on $r > 0$)

$$\nu D^2\psi_\theta = -\frac{1}{r}u_r D\psi_\theta + u_r\partial_r D\psi_\theta + u_z\partial_z D\psi_\theta$$

$$= \frac{1}{r^2}\partial_z\psi_\theta\, D\psi_\theta - \partial_z\psi_\theta\,\partial_r D\psi_\theta + \frac{1}{r}\partial_r(r\psi_\theta)\,\partial_z D\psi_\theta$$

Let

$$D_0 h = \partial_r^2 h - \frac{1}{r}\partial_r h + \partial_z^2 h.$$

We have

$$D(\frac{h}{r}) = \frac{D_0 h}{r} \text{ and } D^2(\frac{h}{r}) = \frac{D_0^2 h}{r}.$$

This gives

$$\nu D_0^2\gamma = r\left(\partial_z\gamma\frac{D_0\gamma}{r^3} - \partial_z\gamma\frac{1}{r}\partial_r(\frac{D_0\gamma}{r}) + \frac{1}{r}\partial_r\gamma\partial_z(\frac{D_0\gamma}{r})\right)$$

$$= 2\partial_z\gamma\frac{D_0\gamma}{r^2} - \frac{1}{r}\,\partial_z\gamma\,\partial_r D_0\gamma + \frac{1}{r}\partial_r\gamma\,\partial_z D_0\gamma$$

As ru_r and ru_z are homogeneous of degree 0, Sleznik's idea was to look for an axisymmetric function γ that would be homogeneous of degree 1, thus to write γ as

$$\gamma = \rho\, G(\cos\varphi).$$

Indeed, we have $\partial_\rho\gamma = \cos\varphi\partial_z\gamma + \sin\varphi\partial_r\gamma = -\cos\varphi\, ru_r + \sin\varphi\, ru_z$; thus, $\partial_\rho\gamma$ is homogeneous of order 0: $\partial_\rho\gamma = G(\cos\varphi)$, and thus $\gamma = \rho G(\cos\varphi) + H(\cos\varphi)$; moreover, $\vec{\nabla} \wedge (\frac{\gamma}{r}\vec{e}_\theta) = \vec{u}$ is homogeneous of degree -1: by homogeneity, we must have $\vec{\nabla}(\frac{H}{r}\vec{e}_\theta) = 0$; since $\vec{\nabla}(\frac{H}{r}\vec{e}_\theta) = -\frac{1}{r}\partial_z H\vec{e}_r$, we find $\partial_z H = 0$, so that H is constant (and may be taken equal to 0).

A further change of variable $\tau = \cos\varphi$ then gives:

$$r = \rho\sin(\varphi) = \rho\sqrt{1 - \tau^2} \text{ and } z = \rho\cos(\varphi) = \rho\tau$$

so that

$$\partial_z = \cos(\varphi)\partial_\rho - \frac{\sin(\varphi)}{\rho}\partial_\varphi = \tau\partial_\rho + \frac{1 - \tau^2}{\rho}\partial_\tau$$

and

$$\partial_r = \sin(\varphi)\partial_\rho + \frac{\cos(\varphi)}{\rho}\partial_\varphi = \frac{r}{\rho}\partial_\rho - \frac{r\tau}{\rho^2}\partial_\tau$$

This gives

$$-\frac{1}{r}\partial_z\gamma\,\partial_r D_0\gamma + \frac{1}{r}\partial_r\gamma\,\partial_z D_0\gamma = \frac{1}{\rho^2}(\partial_\rho\gamma\,\partial_\tau D_0\gamma - \partial_\tau\gamma\,\partial_\rho D_0\gamma)$$

and

$$2\partial_z\gamma\frac{D_0\gamma}{r^2} = \frac{2}{\rho^2}D_0\gamma\left(\frac{\tau}{1 - \tau^2}\partial_\rho\gamma + \frac{1}{\rho}\partial_\tau\gamma\right).$$

We then write

$$D_0 = \partial_\rho^2 + \frac{1}{\rho}\partial_\rho + \frac{1}{\rho^2}\partial_\phi^2 - \frac{1}{r}\partial_r = \partial_\rho^2 + \frac{1 - \tau^2}{\rho^2}\partial_\tau^2.$$

We have

- $\partial_\rho\gamma = G(\tau)$ and $\partial_\tau\gamma = \rho\frac{d}{d\tau}G(\tau)$

- $D_0\gamma = \frac{1 - \tau^2}{\rho}\frac{d^2}{d\tau^2}G(\tau)$

- $\partial_\rho D_0\gamma = -\frac{1 - \tau^2}{\rho^2}\frac{d^2}{d\tau^2}G(\tau)$ and $\partial_\tau D_0\gamma = \frac{1}{\rho}\frac{d}{d\tau}\left((1 - \tau)^2\frac{d^2}{d\tau^2}G(\tau)\right)$

- $D_0^2\gamma = 2\frac{1 - \tau^2}{\rho^3}\frac{d^2}{d\tau^2}G(\tau) + \frac{1 - \tau^2}{\rho^3}\frac{d^2}{d\tau^2}\left((1 - \tau^2)\frac{d^2}{d\tau^2}G(\tau)\right)$ or equivalently $D_0^2\gamma = \frac{1 - \tau^2}{\rho^3}((1 - \tau^2)\frac{d^4}{d\tau^4}G(\tau) - 4\tau\frac{d^3}{d\tau^3}G(\tau))$

We get an equation on G:

$$\nu\left((1 - \tau^2)\frac{d^4}{d\tau^4}G(\tau) - 4\tau\frac{d^3}{d\tau^3}G(\tau)\right) = G(\tau)\frac{d^3}{d\tau^3}G(\tau) + 3\frac{d}{d\tau}G(\tau)\frac{d^2}{d\tau^2}G(\tau)$$

which can be rewritten as

$$\nu\frac{d^3}{d\tau^3}\left((1 - \tau^2)\frac{d}{d\tau}G(\tau) + 2\tau G(\tau)\right) = \frac{1}{2}\frac{d^3}{d\tau^3}(G(\tau)^2)$$

and finally, for three constants of integration, we obtain Slezkin's equation

$$\nu\left((1 - \tau^2)\frac{d}{d\tau}G(\tau) + 2\tau G(\tau)\right) = \frac{1}{2}G(\tau)^2 + C_2\tau^2 + C_1\tau + C_0. \qquad (10.45)$$

General solutions of Slezkin's equation have been discussed by many authors (as Sedov [427] or Vyskrebtsov [493]). Landau's solutions correspond to the simple case

$$\nu\left((1 - \tau^2)\frac{d}{d\tau}G(\tau) + 2\tau G(\tau)\right) = \frac{1}{2}G(\tau)^2. \qquad (10.46)$$

Indeed, we have

$$\partial_\tau G(\tau) = \frac{1}{\rho}\partial_\tau\gamma = -\frac{\tau}{\sqrt{1 - \tau^2}}\partial_r\gamma + \partial_z\gamma = -\rho\tau u_z - \rho\sqrt{1 - \tau^2}u_r = -zu_z - ru_r$$

thus, as $\vec{u}$ is continuous on $|x| \neq 0$, $\partial_\tau G$ is continuous on $[-1, 1]$, and G is $\mathcal{C}^1$ on $[-1, 1]$. Moreover, $\vec{u}_r = -\frac{1}{r}\partial_z\gamma = -\frac{z}{\rho r}G(\tau) - \frac{r}{\rho^2}\partial_\tau G(\tau)$; as ρu_r is bounded, we find that $G(1) = G(-1) = 0$ and $\frac{G}{1-\tau^2} = \frac{\rho^2 G}{r^2}$ is bounded. If we define $H(\tau) = \frac{G}{1-\tau^2}$, we have that H is continuous on $[-1, 1]$, with $H(1) = -\frac{1}{2}\partial_\tau G(1)$ and $H(-1) = \frac{1}{2}\partial_\tau G(-1)$; this gives that, near 1 and -1, we have

$$C_2\tau^2 + C_1\tau + C_0 = \nu((1-\tau^2)\frac{d}{d\tau}G(\tau) + 2\tau G(\tau)) - \frac{1}{2}G(\tau)^2 = o(1-\tau^2)$$

which is possible only if $C_0 = C_1 = C_2 = 0$.

In the case $C_0 = C_1 = C_2 = 0$, we have

$$\nu\partial_\tau\left(\frac{G(\tau)}{1-\tau^2}\right) = \frac{1}{2}\left(\frac{G(\tau)}{1-\tau^2}\right)^2$$

hence, if G is not the null function, for some constant A

$$G(\tau) = \frac{2\nu(1-\tau^2)}{A-\tau}. \tag{10.47}$$

For G to be $\mathcal{C}^1$ with $G(1) = G(-1) = 0$, we must have $|A| > 1$.

Now, if $\vec{u} = \vec{\nabla} \wedge (\frac{\rho G(\tau)}{r}\vec{e}_\theta)$, with $G(\tau) = \frac{2\nu(1-\tau^2)}{A-\tau}$, we know that $\vec{\omega} = \vec{\nabla} \wedge \vec{u}$ satisfies on $|x| \neq 0$

$$\nu\Delta\vec{\omega} + (\vec{\omega}.\vec{\nabla})\vec{u} - (\vec{u}.\vec{\nabla})\vec{\omega} = 0$$

so that $\vec{u}$ satisfies

$$\vec{\nabla} \wedge (\nu\Delta\vec{u} - \vec{\omega} \wedge \vec{u}) = \vec{\nabla} \wedge (\nu\Delta\vec{u} - \vec{u}.\vec{\nabla}\vec{u}) = 0.$$

Thus, the support of $\vec{w} = -\vec{\nabla} \wedge (\nu\Delta\vec{u} - \operatorname{div}(\vec{u} \otimes \vec{u}))$ is reduced to $\{0\}$, and $\vec{w}$ is a sum of derivatives of Dirac masses. But $\vec{w}$ is homogeneous of homogeneous degree -4, so the derivatives are derivatives of order 1: $\vec{w} = \partial_1\delta\vec{E}_1 + \partial_2\delta\vec{E}_2 + \partial_3\delta\vec{E}_3$ for three constant vectors $\vec{E}_1, \vec{E}_2, \vec{E}_3$. As $\vec{w}$ is divergence free, we find that

$$\vec{w} = \alpha\begin{pmatrix} 0 \\ -\partial_z\delta \\ \partial_y\delta \end{pmatrix} + \beta\begin{pmatrix} -\partial_y\delta \\ \partial_x\delta \\ 0 \end{pmatrix} + \gamma\begin{pmatrix} \partial_z\delta \\ 0 \\ -\partial_x\delta \end{pmatrix}$$

for three constants α, β, γ. Moreover, $\vec{w}$ is axisymmetrical; rotating the axes in x_1 and x_2 should let the component on $\vec{e}_z$ invariant: this gives $\alpha = \gamma = 0$. Thus, $\vec{w} = \beta\begin{pmatrix} -\partial_y\delta \\ \partial_x\delta \\ 0 \end{pmatrix} = \vec{\nabla} \wedge (\beta\delta\vec{e}_3)$. We find that, on $\mathbb{R}^3$ we have

$$\nu\Delta\vec{u} - \operatorname{div}(\vec{u} \otimes \vec{u}) + \beta\delta\vec{e}_3 - \vec{\nabla}p = 0$$

for some distribution p. Thus, $\vec{u}$ satisfies the Navier–Stokes equations with forcing term $\beta\delta\vec{e}_3$.

It remains to state the exact range where β can be taken in. First, we have to compute β as a function of the constant A in Equation (10.47). This value of β is given in Batchelor's book [25] and in the paper of Cannone and Karch [82]:

$$\beta = \nu^2\frac{8\pi A}{3(A^2-1)}(2 + 6A^2 - 3A(A^2-1)\ln(\frac{A+1}{A-1})) \tag{10.48}$$

We have that β is an odd function and satisfies

$$\frac{d}{dA}\beta = -\nu^2 \left(\frac{6}{A^2-1} + \frac{4}{(A-1)^2} + \frac{4}{(A+1)^2} + 6A\ln(\frac{A+1}{A-1}) \right) < 0$$

with $\beta(1) = +\infty$ and $\beta(+\infty) = 0$. Thus, the mapping $A \in (1, +\infty) \mapsto \beta \in (0, +\infty)$ is a bijection. $\qquad\square$

10.9   Time-Periodic Solutions

In this section, we consider the Navier–Stokes problem with the following symmetry property: $\vec{u}$ is invariant under the action of a discrete group of time translations, i.e., $\vec{u}$ is time-periodic.

When the force $\vec{f}$ is time-periodic ($\vec{f}(t+T, x) = \vec{f}(t, x)$), one may ask whether one might find a time-periodic solution $\vec{u}$ of the Navier–Stokes equations. The problem to be solved is then the following one:

Time-periodic Navier–Stokes equations

Given a time-periodic force $\vec{f}(t, x)$, find a time-periodic velocity $\vec{u}(t, x)$ and a time-periodic pressure $p(t, x)$ such that

$$\partial_t \vec{u} = \nu\Delta\vec{u} - \vec{u}\cdot\vec{\nabla}\vec{u} - \vec{\nabla}p + \vec{f}, \quad \text{div }\vec{u} = 0 \qquad (10.49)$$

The study of the time-periodic Navier–Stokes problem is now ancient. In the fifties, Serrin studied the problem on bounded domains [433]. Then, in the nineties, there were several works on the whole space. Maremonti [349] constructed periodic solutions $\vec{U}_{\text{per}}(t, .) = \vec{U}_{\text{per}}(t + NT, .)$ as the asymptotic limit of $\vec{u}(t + NT, .)$ of the Cauchy initial value problem for a small arbitrary initial value $\vec{u}_0$. Then, Yamazaki [509], generalizing a previous work of Kozono and Nakao [272], proved that the formalism of mild solutions developed by Fujita and Kato [185] could be adapted to find solutions in the Lorentz space $L_t^\infty L^{3,\infty}$.

In order to solve the time-periodic problem, we begin by showing a simple inequality:

Lemma 10.3.
The operator $f \mapsto \int_{-\infty}^t W_{\nu(t-s)} * f \, ds$ is bounded from $L^1((-\infty, +\infty), \dot{B}_{\infty,\infty}^{-1}(\mathbb{R}^3))$, $L^\infty \dot{B}_{\infty,\infty}^{-3}$ or $L^{p,\infty}\dot{B}_{\infty,\infty}^{-3+\frac{2}{p}}$ ($1 < p < \infty$) to $L^\infty \dot{B}_{\infty,\infty}^{-1}$.

Proof. Let $u = \int_{-\infty}^t W_{\nu(t-s)} * f \, ds$. For $1 \le p \le +\infty$, and $\tau > 0$, we have

$$\|W_{\nu\tau\Delta} * u\|_\infty \le C_p \int_{-\infty}^t \frac{1}{(\nu(\tau+t-s))^{\frac{3}{2}-\frac{1}{p}}} \|f(s,.)\|_{\dot{B}^{-3+\frac{2}{p}}} \, ds.$$

Let $k_{p,\tau}(t) = 1_{t>0} (t+\tau)^{-\frac{3}{2}+\frac{1}{p}}$. Then $\|k_{1,\tau}\|_\infty = \tau^{-1/2}$ and $\|k_{\infty,\tau}\|_1 = 2\tau^{-1/2}$. If $1 < p < +\infty$, we remark that $(\frac{3}{2} - \frac{1}{p})\frac{p}{p-1} > 1$, so that $k_{p,\tau} \in L^{\frac{p}{p-1},1}$ and $\|k_{p,\tau}\|_{L^{\frac{p}{p-1},1}} = C_p\tau^{-1/2}$. Since convolution maps $L^\infty \times L^1$, $L^1 \times L^\infty$ and $L^{\frac{p}{p-1},1} \times L^{p,\infty}$ to L^∞, we find that u belongs to $L^\infty \dot{B}_{\infty,\infty}^{-1}$. $\qquad\square$

With this lemma, we may provide a simple exposition of the results of Kyed [290] on time-periodic solutions which belong to $L_t^\infty \dot{H}^{1/2} \cap L_{\text{per}}^2 \dot{H}^{3/2}$:

Time-periodic Navier–Stokes equations in Sobolev spaces

Theorem 10.14.
There exists a positive constant η such that: if $\vec{f}_{\text{per}}$ is a time-periodic vector field on $\mathbb{R} \times \mathbb{R}^3$ (with period T) such that

- *the mean value $\vec{f}_0 = \frac{1}{T} \int_0^T \vec{f}_{\text{per}}(s, .) \, ds$ belongs to $\dot{H}^{-3/2}$ and satisfies*

$$\|\mathbb{P}\vec{f}_0\|_{\dot{H}^{-3/2}} < \eta\nu$$

- *$\vec{f}_{\text{per}}$ belongs to $L_{\text{per}}^2 \dot{H}^{-1/2}$ with*

$$\|\vec{f}_{\text{per}}\|_{L_{\text{per}}^2 \dot{H}^{-1/2}} < \eta\sqrt{\nu}$$

then there exists a time-periodic solution $\vec{u}_{\text{per}}$ of the Navier–Stokes problem (10.49) such that $\vec{u}_{\text{per}} \in L_t^\infty \dot{H}^{1/2} \cap L_{\text{per}}^2 \dot{H}^{3/2}$.

Proof. We first study $\vec{U}_0 = \int_{-\infty}^t e^{\nu(t-s)\Delta} \mathbb{P}\vec{f}_{\text{per}} \, ds$. We expand $\mathbb{P}\vec{f}_{\text{per}}$ as a time-Fourier series

$$\mathbb{P}\vec{f}_{\text{per}} = \sum_{k \in \mathbb{Z}} \vec{g}_k(x) e^{\frac{2\pi}{T} ikt}.$$

We have

$$\int_0^T \|\mathbb{P}\vec{f}_{\text{per}}\|_{\dot{H}^{-1/2}}^2 \, dt = T \sum_{k \in \mathbb{Z}} \|\vec{g}_k\|_{\dot{H}^{-1/2}}^2.$$

The Fourier expansion of $\vec{U}_0$ is

$$\vec{U}_0 = \sum_{k \in \mathbb{Z}} \vec{W}_k(x) e^{\frac{2\pi}{T} ikt}, \quad \text{with} \quad \vec{W}_k = \frac{1}{ik\frac{2\pi}{T} - \nu\Delta} \mathbb{P}\vec{g}_k.$$

We have $\|\vec{W}_k\|_{\dot{H}^{3/2}} \le \frac{1}{\nu}\|\vec{g}_k\|_{\dot{H}^{-1/2}}$, and thus $\vec{U}_0 \in L_{\text{per}}^2 \dot{H}^{3/2}$. Moreover, $\vec{g}_0 = \mathbb{P}\vec{f}_0 \in \dot{H}^{-3/2}$ so that $\vec{W}_0 \in \dot{H}^{1/2}$. Let $\vec{\Omega}_k$ be the Fourier transform of $\vec{g}_k$. We have:

$$(2\pi)^3 \|\vec{U}_0(t, .) - \vec{W}_0\|_{\dot{H}^{1/2}}^2 = \int |\xi| \left| \sum_{k \neq 0} \frac{1}{ik\frac{2\pi}{T} + \nu|\xi|^2} \vec{\Omega}_k(\xi) e^{\frac{2\pi}{T} ikt} \right|^2 d\xi$$

$$\le \int |\xi| \left(\sum_{k \neq 0} \frac{1}{k^2 \frac{4\pi^2}{T^2} + \nu^2|\xi|^4} \right) \left(\sum_{k \neq 0} |\vec{\Omega}_k(\xi)|^2 \right) d\xi$$

If $\nu T|\xi|^2 \le 1$, we write

$$\sum_{k \neq 0} \frac{1}{k^2 \frac{4\pi^2}{T^2} + \nu^2|\xi|^4} \le \frac{T^2}{4\pi^2} \sum_{k \neq 0} \frac{1}{k^2} = \frac{T^2}{12} \le \frac{T}{12\nu|\xi|^2}.$$

If $\nu T |\xi|^2 > 1$, we write

$$\sum_{k \neq 0} \frac{1}{k^2 \frac{4\pi^2}{T^2} + \nu^2 |\xi|^4} \leq 2 \Big(\sum_{1 \leq k \leq 2\nu T |\xi|^2} \frac{1}{\nu^2 |\xi|^4} + \sum_{k > 2\nu T |\xi|^2} \frac{1}{k^2 \frac{4\pi^2}{T^2}} \Big) \leq \Big(4 + \frac{1}{2\pi^2}\Big) \frac{T}{\nu |\xi|^2}$$

Thus, we find that $\vec{U}_0 \in L_t^\infty \dot{H}^{1/2}$. We found more precisely:

$$\|\vec{U}_0\|_{L^\infty \dot{H}^{1/2}} \leq C \frac{1}{\nu} \|\vec{f}_0\|_{\dot{H}^{-3/2}} + C \frac{1}{\sqrt{\nu}} \|\vec{f}\|_{L^2_{\mathrm{per}} \dot{H}^{-1/2}}$$

and

$$\|\vec{U}_0\|_{L^2 \dot{H}^{3/2}} \leq C \frac{1}{\nu} \|\vec{f}\|_{L^2_{\mathrm{per}} \dot{H}^{-1/2}}.$$

It is now easy to check that the bilinear operator B

$$B(\vec{U}, \vec{V}) = \int_{-\infty}^t W_{\nu(t-s)} * \mathbb{P} \operatorname{div}(\vec{U} \otimes \vec{V}) \, ds$$

is bounded on $E = L_t^\infty \dot{H}^{1/2} \cap L^2_{\mathrm{per}} \dot{H}^{3/2}$. Indeed, we have, for $\vec{U}$ and $\vec{V}$ in E,

$$\|\vec{U}(t,.) \otimes \vec{V}(t,.)\|_{\dot{H}^{1/2}} \leq C \big(\|\vec{U}(t,.)\|_{\dot{H}^{1/2}} \|\vec{V}(t,.)\|_{\dot{H}^{3/2}} + \|\vec{V}(t,.)\|_{\dot{H}^{1/2}} \|\vec{U}(t,.)\|_{\dot{H}^{3/2}} \big)$$

and

$$\|\vec{U}(t,.) \otimes \vec{V}(t,.)\|_{\dot{H}^{-1/2}} \leq C \|\vec{U}(t,.)\|_{\dot{H}^{1/2}} \|\vec{V}(t,.)\|_{\dot{H}^{1/2}}$$

so that $\vec{F} = \operatorname{div}(\vec{U} \otimes \vec{V})$ satisfies $\vec{F} \in L^2_{per} \dot{H}^{-1/2}$ and $\int_0^T \vec{F}(s,.) \, ds \in \dot{H}^{-3/2}$. The proof we gave on $\vec{U}_0$ gives us as well that $B(\vec{U}, \vec{V}) \in E$: we have

$$\|B(\vec{U}, \vec{V})\|_{L^\infty \dot{H}^{1/2}} \leq C \frac{1}{\nu} \|\frac{1}{T} \int_0^T \vec{F}(s,.) \, ds\|_{\dot{H}^{-3/2}} + C \frac{1}{\sqrt{\nu}} \|\vec{F}\|_{L^2_{\mathrm{per}} \dot{H}^{-1/2}}$$

$$\leq C' \frac{1}{\nu} \|\vec{U}\|_{L^\infty \dot{H}^{1/2}} \|\vec{V}\|_{L^\infty \dot{H}^{1/2}}$$

$$+ C' \frac{1}{\sqrt{\nu}} \big(\|\vec{U}\|_{L^\infty \dot{H}^{1/2}} \|\vec{V}\|_{L^2_{\mathrm{per}} \dot{H}^{3/2}} + \|\vec{V}\|_{L^\infty \dot{H}^{1/2}} \|\vec{U}\|_{L^2_{\mathrm{per}} \dot{H}^{3/2}} \big)$$

and

$$\|B(\vec{U}, \vec{V})\|_{L^2 \dot{H}^{3/2}} \leq C \frac{1}{\nu} \|\vec{F}\|_{L^2_{\mathrm{per}} \dot{H}^{-1/2}}$$

$$\leq C' \frac{1}{\nu} \big(\|\vec{U}\|_{L^\infty \dot{H}^{1/2}} \|\vec{V}\|_{L^2_{\mathrm{per}} \dot{H}^{3/2}} + \|\vec{V}\|_{L^\infty \dot{H}^{1/2}} \|\vec{U}\|_{L^2_{\mathrm{per}} \dot{H}^{3/2}} \big)$$

The proof of the theorem is now reduced to the fixed–point theorem of Picard, by takin ion E the norm $\|\vec{U}\|_E = \|\vec{U}\|_{L^\infty \dot{H}^{1/2}} + \sqrt{\nu} \|\vec{U}\|_{L^2_{\mathrm{per}} \dot{H}^{3/2}}$. $\qquad \square$

We now give another theorem on the existence of time-periodic solutions, based on Kozono and Nakao's approach:

Time-periodic Navier–Stokes equations in Morrey spaces

Theorem 10.15.

We consider a Banach space of distributions on $\mathbb{R}^3$, $\mathcal{X} \subset L^2_{\mathrm{loc}}$, such that:

- (A1) *The pointwise product is bounded from* $L^\infty \times \mathcal{X}$ *to* $\mathcal{X}$.

- (A2) *The Riesz transforms are bounded on* $\mathcal{X}$.

- (A3) *The Hardy–Littlewood maximal function is bounded on* $\mathcal{X}$.

- (A4) *The operator* $(u,v) \mapsto \frac{1}{\sqrt{-\Delta}}(uv)$ *is bounded from* $\mathcal{X} \times \mathcal{X}$ *to* $\mathcal{X}$.

- (A5) *The operator* $(u,v) \mapsto \frac{1}{(-\Delta)^{1/4}}(u\frac{1}{(-\Delta)^{1/4}}v)$ *is bounded from* $\mathcal{X} \times \mathcal{X}$ *to* $\mathcal{X}$.

Then there exist positive constants $\epsilon_\mathcal{X}$ and $C_\mathcal{X}$ such that: if $\vec{f}_{\mathrm{per}}$ is a time-periodic vector field (with period T) on $\mathbb{R} \times \mathbb{R}^3$ such that

- $\vec{f}_{\mathrm{per}}$ *belongs to* $L^1_{\mathrm{per}}\mathcal{X}$ *with*

$$\int_0^T \|\vec{f}_{\mathrm{per}}\|_\mathcal{X}\, dt < \epsilon_\mathcal{X} \nu$$

- *the mean value $\vec{f}_0 = \frac{1}{T}\int_0^T \vec{f}_{\mathrm{per}}(s,.)\,ds$ belongs to $\dot{B}^{-3}_{\infty,\infty}$ and satisfies $\frac{1}{\Delta}\vec{f}_0 \in \mathcal{X}$ with*

$$\left\|\frac{1}{\Delta}\vec{f}_0\right\|_\mathcal{X} < \epsilon_\mathcal{X}\nu^2$$

then

- *there exists a time-periodic solution $\vec{u}_{\mathrm{per}}$ of the Navier–Stokes problem (10.49) such that $\sup_{t\in\mathbb{R}} |\vec{u}(t,x)| \in \mathcal{X}$ and $\|\sup_{t\in\mathbb{R}} |\vec{u}(t,x)|\,\|_\mathcal{X} < C_\mathcal{X}\nu$*

- $\vec{\nabla} \otimes \vec{u}_{\mathrm{per}} \in L^{2,\infty}_{\mathrm{per}}\mathcal{X}$

- *if $\|\vec{u}_0\|_\mathcal{X} < \epsilon_\mathcal{X}\nu$, there exists a unique solution $\vec{u}$ on $(0,+\infty)\times\mathbb{R}^3$ of the problem*

$$\begin{cases} \partial_t \vec{u} + \mathrm{div}(\vec{u}\otimes\vec{u}) = \nu\Delta\vec{u} + \vec{f}_{\mathrm{per}} - \vec{\nabla}p \\ \mathrm{div}\,\vec{u} = 0 \\ \vec{u}(0,.) = \vec{u}_0 \end{cases} \tag{10.50}$$

such that $\sup_{t>0} |\vec{u}(t,.)| \in \mathcal{X}$ and $\|\sup_{t>0} |\vec{u}(t,.)|\|_\mathcal{X} < C_\mathcal{X}\nu$.

- *moreover, we have*

$$\sup_{t>0} t^{1/4}\|(-\Delta)^{1/4}(\vec{u}(t,x) - \vec{u}_{\mathrm{per}}(x))\|_\mathcal{X} \leq C_\mathcal{X}\nu^{3/4} \tag{10.51}$$

so that $\vec{u}(t,.)$ converges to $\vec{u}_{\mathrm{per}}$ in $\mathcal{S}'$ (and in L^2_{loc}) as t goes to $+\infty$.

Examples of such Banach spaces $\mathcal{X}$ are the Lebesgue space L^3, the Lorentz spaces $L^{3,p}$ ($1 \leq p \leq +\infty$), the Morrey spaces $\dot{M}^{p,3}$ ($2 < p < 3$) and the multiplier space $\mathcal{V}^1 = \mathcal{M}(\dot{H}^1 \mapsto L^2)$. Recall that $L^{3,1} \subset L^3 \subset L^{3,\infty} \subset \dot{M}^{p,3} \subset \mathcal{V}^1$.

Let us further remark that if $\vec{f}$ is divergence-free and belongs to $L^1_{\mathrm{per}}\mathcal{V}^1$, and if moreover the support of $\vec{f}$ is bounded: $\vec{f}(t,x) = 0$ if $|x| \geq R$, then its mean value belongs to $\dot{B}^{-3}_{\infty,\infty}$ and satisfies $\frac{1}{\Delta}\vec{f}_0 \in L^{3,1}$. Indeed, $\vec{f}_0 \in \mathcal{V}^1 \subset L^2_{\mathrm{loc}}$ and the support of $\vec{f}_0$ is contained in the ball $\bar{B}(0, R)$. Thus, $\vec{f}_0 \in L^1$. Since $\vec{f}_0$ is divergence-free, we find that $\int \vec{f}_0 \, dx = 0$. Moreover, $\vec{f}_0 \in L^2$ and has compact support. Hence, $\vec{f}_0$ belongs to the Hardy space $\mathcal{H}^1$, hence $\frac{1}{\Delta} \vec{f}_0 \in L^{3,1}$.

We may now prove the theorem:

Proof. The solutions $\vec{u}_{\text{per}}$ and $\vec{u} = \vec{u}_{\text{per}} + \vec{w}$ are solutions of two-fixed point problems: $\vec{u}_{\text{per}}$ is solution of

$$\vec{U} = \int_{-\infty}^{t} W_{\nu(t-s)} * \mathbb{P}(\vec{f}_{\text{per}} - \operatorname{div}(\vec{U} \otimes \vec{U})) \, ds$$

and $\vec{w}$ is solution of

$$\vec{W} = W_{\nu t} * \vec{w}_0 - \int_0^t W_{\nu(t-s)} * \mathbb{P}\operatorname{div}(\vec{W} \otimes \vec{W} + \vec{W} \otimes \vec{u}_{\text{per}} + \vec{u}_{\text{per}} \otimes \vec{W}) \, ds$$

with $\vec{w}_0 = \vec{u}_0 - \vec{u}_{\text{per}}(0, .)$.

Let $\mathcal{E}$ be the space of divergence free vector fields $\vec{u}$ such that $\sup_{t \in \mathbb{R}} |\vec{u}(t, x)| \in X$. We are going to prove the existence of $\vec{u}_{\text{per}}$ by Picard's iterations in the space

$$\mathcal{F} = \{\vec{U} \in \mathcal{E} \ / \ \vec{U}(t + T, x) = \vec{U}(t, x), \vec{\nabla} \otimes \vec{U} \in L^{2,\infty}_{\text{per}} X \ \text{and} \ \operatorname{div} \vec{U} = 0\}$$

and the existence of $\vec{w}$ by Picard's iterations in the space

$$\mathcal{G} = \{\vec{W} \ / \ \sup_{t>0} |\vec{W}(t, x)| \in X \ \text{and} \ \sup_{t>0} t^{1/4}|(-\Delta)^{1/4}\vec{W}(t, x)| \in X\}$$

We thus define inductively $\vec{U}_n$ and $\vec{W}_n$ as

$$\vec{U}_0 = \int_{-\infty}^{t} W_{\nu(t-s)} * \mathbb{P}\vec{f}_{\text{per}} \, ds \quad \text{and} \quad \vec{U}_{n+1} = \vec{U}_0 - B(\vec{U}_n, U_n)$$

where

$$B(\vec{U}, \vec{V}) = \int_{-\infty}^{t} W_{\nu(t-s)} * \mathbb{P}\operatorname{div}(\vec{U} \otimes \vec{V}) \, ds$$

and

$$\vec{W}_0 = W_{\nu t} * \vec{w}_0 \quad \text{and} \quad \vec{W}_{n+1} = \vec{W}_0 - B_0(\vec{W}_n, \vec{W}_n) - B_0(\vec{W}_n, \vec{u}_{\text{per}}) - B_0(\vec{u}_{\text{per}}, \vec{W}_n)$$

where

$$B_0(\vec{U}, \vec{V}) = \int_0^t W_{\nu(t-s)} * \mathbb{P}\operatorname{div}(\vec{U} \otimes \vec{V}) \, ds.$$

We first study the existence of $\vec{u}_{\text{per}}$. First, we must check that $\vec{U}_0$ belongs to $\mathcal{F}$. Let us remark that, when $g \in X$ and $\vec{\nabla} g \in X$, we have for every $A > 0$

$$|(-\Delta)^{1/4}g| \le \int_0^A |(-\Delta)^{3/4}W_t * (-\Delta)^{1/2}g| \, dt + \int_A^{+\infty} |(-\Delta)^{5/4}W_t * g| \, dt$$

$$\le C(A^{1/4}\mathcal{M}_{(-\Delta)^{1/2}g} + A^{-1/4}\mathcal{M}_g)$$

so that

$$|(-\Delta)^{1/4}g(x)| \le C\sqrt{\mathcal{M}_{(-\Delta)^{1/2}g}(x)\mathcal{M}_g)(x)}$$

and

$$\|(-\Delta)^{1/4}g\|_X \le C\|g\|_X^{1/2}\|\vec{\nabla}g\|_X^{1/2}. \tag{10.52}$$

Similarly, when $g \in \mathcal{X}$ and $\Delta g \in \mathcal{X}$, we have for every $A > 0$

$$|(-\Delta)^{1/2}g| \leq \int_0^A |(-\Delta)^{1/2}W_t * (-\Delta)g| \, dt + \int_A^{+\infty} |(-\Delta)^{3/2}W_t * g| \, dt$$

$$\leq C(A^{1/2}\mathcal{M}_{\Delta g} + A^{-1/2}\mathcal{M}_g)$$

so that

$$|(-\Delta)^{1/2}g(x)| \leq C\sqrt{\mathcal{M}_{\Delta g}(x)\mathcal{M}_g)(x)}$$

and

$$\|(-\Delta)^{1/2}g\|_{\mathcal{X}} \leq C\|g\|_{\mathcal{X}}^{1/2}\|\Delta g\|_{\mathcal{X}}^{1/2}. \tag{10.53}$$

We write $\vec{U}_0 = \mathbb{P}(\vec{V}_0 + \vec{V}_1)$, with $\vec{V}_j = \int_{-\infty}^t W_{\nu(t-s)} * \vec{f}_j(s,.) \, ds$ and $\vec{f}_1 = \vec{f}_{\text{per}} - \vec{f}_0$. We are going to show that $\mathbb{P}\vec{V}_0$ and $\mathbb{P}\vec{V}_1$ belong to $\mathcal{E}$ and that $\vec{\nabla} \otimes \vec{V}_0$ and $\vec{\nabla} \otimes \vec{V}_1$ belong to $L_{\text{per}}^{2,\infty}\mathcal{X}$. First, we have $\vec{V}_0 = -\frac{1}{\nu\Delta}\vec{f}_0$ so that $\|\mathbb{P}\vec{V}_0\|_{\mathcal{E}} = \frac{1}{\nu}\|\frac{1}{\Delta}\mathbb{P}\vec{f}_0\|_{\mathcal{X}}$. Moreover, by inequality (10.53), we have

$$\|\vec{\nabla} \otimes \vec{V}_0\|_{\mathcal{X}} \leq C\frac{1}{\nu}\sqrt{\|\vec{f}_0\|_{\mathcal{X}}\|\frac{1}{\Delta}\vec{f}_0\|_{\mathcal{X}}} \leq C\frac{\sqrt{\nu}}{\sqrt{T}}\left(\frac{\|\vec{f}_0\|_{\mathcal{X}}}{\nu} + \frac{\|\frac{1}{\Delta}\vec{f}_0\|_{\mathcal{X}}}{\nu^2}\right)$$

Thus, $\mathbb{P}\vec{V}_0 \in \mathcal{F}$.

We now study $\vec{V}_1$ on the period interval $(0,T)$. We write $\vec{V}_1 = \vec{V}_2 + \vec{V}_3$ with

$$\vec{V}_2(t,x) = \int_{-\infty}^t W_{\nu(t-s)} * 1_{[-T,T]}(s)\vec{f}_1(s,.) \, ds$$

and

$$\vec{V}_3(t,x) = \int_{-\infty}^{-T} W_{\nu(t-s)} * \vec{f}_1(s,.) \, ds.$$

We have

$$|\mathbb{P}\vec{V}_2| \leq C \int_{-T}^T \mathcal{M}_{\mathbb{P}\vec{f}_1}(s,x) \, ds$$

so that

$$\|\mathbb{P}\vec{V}_2\|_{\mathcal{E}} \leq C\|\vec{f}_1\|_{L_{\text{par}}^1\mathcal{X}}.$$

We have also

$$|\vec{\nabla} \otimes \vec{V}_2| \leq C \int_{-T}^T \frac{1}{\sqrt{\nu(t-s)}}\mathcal{M}_{\vec{f}_1}(s,x) \, ds$$

so that

$$\|\vec{\nabla} \otimes \vec{V}_2\|_{L^{2,\infty}((0,T),\mathcal{X})} \leq C\frac{1}{\sqrt{\nu}}\|\vec{f}_1\|_{L_{\text{per}}^1\mathcal{X}}.$$

For $\vec{V}_3$, we integrate by parts, writing that $\vec{f}_1 = \partial_t \vec{f}_2$, where $\vec{f}_2 = \int_0^t \vec{f}_1(s,.) \, ds$ is T-periodic with $\vec{f}_2(0,.) = 0$ and $\|\vec{f}_2\|_{L^\infty\mathcal{X}} \leq \|\vec{f}_1\|_{L_{\text{per}}^1\mathcal{X}}$. Thus, we find that

$$\vec{V}_3(t,x) = \int_{-\infty}^{-T} \nu\Delta W_{\nu(t-s)} * \vec{f}_2(s,.) \, ds.$$

This gives

$$|\vec{\nabla} \otimes \vec{V}_3| \leq C \int_{-\infty}^{-T} \frac{\nu}{(\nu(t-s))^{3/2}}\mathcal{M}_{\vec{f}_2}(s,x) \, ds$$

so that

$$\|\vec{\nabla} \otimes \vec{V}_3\|_{L^{2,\infty}((0,T),\mathcal{X})} \le C \frac{1}{\sqrt{\nu}} \|\vec{f}_1\|_{L^1_{\mathrm{per}}\mathcal{X}}.$$

For the control of $\|\mathbb{P}\vec{V}_3\|_{\mathcal{E}}$, we need to perform one more integration by parts. We define the mean value $\vec{f}_3 = \frac{1}{T}\int_0^T \vec{f}_2(s,.)\,ds$, the fluctuation $\vec{f}_4 = \vec{f}_2 - \vec{f}_3$ and finally $\vec{f}_5 = \int_0^t \vec{f}_4(s,.)\,ds$. We write $\vec{V}_3 = \vec{V}_4 + \vec{V}_5$, with

$$\vec{V}_4 = \int_{-\infty}^{-T} \nu\Delta W_{\nu(t-s)} * \vec{f}_3 \, ds = -W_{\nu(T+t)} * \vec{f}_3.$$

Thus, $|\mathbb{P}\vec{V}_4(t,x)| \le \mathcal{M}_{\mathbb{P}\vec{f}_3}(x)$ and $\mathbb{P}\vec{V}_4 \in \mathcal{E}$. On the other hand, we have

$$\mathbb{P}\vec{V}_5 = \int_{-\infty}^{-T} (\nu\Delta)^2 W_{\nu(t-s)} * \mathbb{P}\vec{f}_5(s,.)\,ds$$

so that

$$|\mathbb{P}\vec{V}_5(t,x)| \le C \int_{-\infty}^{-T} \frac{1}{(t-s)^2} \mathcal{M}_{\mathbb{P}\vec{f}_5}(s,.)\,ds$$

with $|\mathbb{P}\vec{f}_5(t,x)| \le T\frac{1}{T}\int_0^T |\mathbb{P}\vec{f}_4(s,x)|\,ds = \frac{1}{T}f_6(x)$; thus, $|\mathbb{P}\vec{V}_5(t,x)| \le C\mathcal{M}_{f_6}(x)$ and

$$\|\mathbb{P}\vec{V}_5\|_{\mathcal{E}} \le C\|f_6\|_{\mathcal{X}} \le C'\|\vec{f}\|_{L^1_{\mathrm{per}}\mathcal{X}}.$$

Thus, we found that

$$\|\vec{U}_0\|_{\mathcal{E}} + \sqrt{\nu}\|\vec{\nabla} \otimes \vec{U}_0\|_{L^{2,\infty}_{\mathrm{per}}\mathcal{X}} \le C(\|\vec{f}\|_{L^1_{\mathrm{per}}\mathcal{X}} + \frac{\|\frac{1}{\Delta}\vec{f}_0\|_{\mathcal{X}}}{\nu}) \tag{10.54}$$

Let $\vec{U}$ and $\vec{V}$ in $\mathcal{F}$. The control of $B(\vec{U},\vec{V})$ in $\mathcal{E}$ is easy: the proof follows the proof of Calderón [78]. We write $U_{\max}(x) = \sup_{t\in\mathbb{R}} |\vec{U}(t,x)|$ and $V_{\max}(x) = \sup_{t\in\mathbb{R}} |\vec{V}(t,x)|$, and

$$|B(\vec{U},\vec{V})(t,x)| \le C \int_{-\infty}^t \int_{\mathbb{R}^3} \frac{1}{\nu^2(t-s)^2 + |x-y|^4} |\vec{U}(t,y)|\,|\vec{V}(t,y)|\,dy\,ds$$

$$\le C \int_{\mathbb{R}^3} \left(\int_{-\infty}^t \frac{1}{\nu^2(t-s)^2 + |x-y|^4}\,ds \right) U_{\max}(y) V_{\max}(y)\,dy$$

$$= \frac{\pi}{2\nu}C \int \frac{1}{|x-y|^2} U_{\max}(y) V_{\max}(y)\,dy$$

and thus

$$\sup_{t\in\mathbb{R}} |B(\vec{U},\vec{V})(t,x)| \le C\frac{1}{\nu\sqrt{-\Delta}}(U_{\max} V_{\max})(x)$$

and

$$\|\sup_{t\in\mathbb{R}} |B(\vec{U},\vec{V})(t,x)|\|_{\mathcal{X}} \le C\|\frac{1}{\nu\sqrt{-\Delta}}(U_{\max} V_{\max})\|_{\mathcal{X}} \le C'\frac{1}{\nu}\|U_{\max}\|_{\mathcal{X}}\|V_{\max}\|_{\mathcal{X}}$$

The control of $\vec{\nabla} \otimes B(\vec{U},\vec{V})$ is a little more delicate. We write

$$\vec{Z} = \mathrm{div}(\vec{U} \otimes \vec{V}).$$

Then we have

$$\|\vec{\nabla} \otimes B(\vec{U}, \vec{V})\|_{\mathcal{X}} \leq C \sum_{j=1}^{3} \| \int_{-\infty}^{t} \partial_j W_{\nu(t-s)} * \vec{Z}\, ds\|_{\mathcal{X}}.$$

First, we notice that $L^{\infty}_{\text{per}}\mathcal{X} \subset (L^3_t L^2_x)_{\text{loc}}$ and $L^{2,\infty}_{\text{per}}\mathcal{X} \subset (L^{3/2}_t L^2_x)_{\text{loc}}$, so that we may write $\vec{Z} = \vec{U}.\vec{\nabla}\vec{V}$. We thus have $\|\frac{1}{\sqrt{-\Delta}}\vec{Z}\|_{\mathcal{X}} \leq C\|\vec{U}\|_{\mathcal{X}}\|\vec{\nabla} \otimes \vec{V}\|_{\mathcal{X}}$. Let $\vec{Z}_0 = \frac{1}{T}\int_0^T \vec{Z}(s,.)\, ds$ be the mean value of $\vec{Z}$. We have

$$\|\frac{1}{\sqrt{-\Delta}}\vec{Z}_0\|_{\mathcal{X}} \leq \frac{1}{T}\int_0^T \|\frac{1}{\sqrt{-\Delta}}\vec{Z}\|_{\mathcal{X}}\, ds \leq C\frac{1}{\sqrt{T}}\|\vec{U}\|_{L^{\infty}\mathcal{X}}\|\vec{\nabla} \otimes \vec{V}\|_{L^{2,\infty}\mathcal{X}}$$

and thus

$$\|\int_{-\infty}^{t} \partial_j W_{\nu(t-s)} * \vec{Z}_0\, ds\|_{\mathcal{X}} = \|\frac{\partial_j}{\nu\Delta}\vec{Z}_0\|_{\mathcal{X}} \leq C\frac{1}{\nu\sqrt{T}}\|\vec{U}\|_{L^{\infty}\mathcal{X}}\|\vec{\nabla} \otimes \vec{V}\|_{L^{2,\infty}\mathcal{X}}$$

We now look at the contribution of the fluctuation $\vec{Z}_1 = \vec{Z} - \vec{Z}_0$. We write $\vec{Z}_2 = \int_0^t \vec{Z}_1(s,.)\, ds$, then $\vec{Z}_2$ is periodical and satisfies $\vec{Z}_2(kT) = 0$ for every $k \in \mathbb{Z}$. Thus, for $0 \leq t < T$, we may write

$$\int_{-\infty}^{t} \partial_j W_{\nu(t-s)} * \vec{Z}_1\, ds = \int_{-\infty}^{t} \partial_j W_{\nu(t-s)} * (1_{[-T,T]}(s)\vec{Z}_1)\, ds$$

$$- \sum_{k=1}^{+\infty} \int_{-(k+1)T}^{-kT} \nu\Delta\partial_j W_{\nu(t-s)} * \vec{Z}_2\, ds.$$

We then write

$$\frac{1}{(-\Delta)^{1/4}}\vec{Z} = \frac{1}{(-\Delta)^{1/4}}((\frac{1}{(-\Delta)^{1/4}}(-\Delta)^{1/4}\vec{U}).\vec{\nabla}\vec{V})$$

and, using inequality (10.52),

$$\|\frac{1}{(-\Delta)^{1/4}}\vec{Z}\|_{L^{4/3,\infty}_{\text{per}}\mathcal{X}} \leq C\|\vec{U}\|^{1/2}_{L^{\infty}_{\text{per}}\mathcal{X}}\|\vec{\nabla} \otimes \vec{U}\|^{1/2}_{L^{2,\infty}_{\text{per}}\mathcal{X}}\|\vec{\nabla} \otimes \vec{V}\|_{L^{2,\infty}_{\text{per}}\mathcal{X}}.$$

The same estimate holds for $\vec{Z}_1$. We then write

$$\|\int_{-\infty}^{t} \partial_j W_{\nu(t-s)} * (1_{[-T,T]}(s)\vec{Z}_1)\, ds\|_{\mathcal{X}} \leq$$

$$C\int_{-\infty}^{t} \frac{1}{(\nu(t-s))^{3/4}}1_{[-T,T]}(s)\|\frac{1}{(-\Delta)^{1/4}}\vec{Z}_1\|_{\mathcal{X}}\, ds$$

and, since $L^{4/3,\infty} * L^{4/3,\infty} \subset L^{2,\infty}$, we find that

$$\|\int_{-\infty}^{t} \partial_j W_{\nu(t-s)} * (1_{[-T,T]}(s)\vec{Z}_1)\, ds\|_{L^{2,\infty}\mathcal{X}} \leq$$

$$\frac{C}{\nu^{3/4}}\|\vec{U}\|^{1/2}_{L^{\infty}_{\text{per}}\mathcal{X}}\|\vec{\nabla} \otimes \vec{U}\|^{1/2}_{L^{2,\infty}_{\text{per}}\mathcal{X}}\|\vec{\nabla} \otimes \vec{V}\|_{L^{2,\infty}_{\text{per}}\mathcal{X}}.$$

Finally, we have $\|\frac{1}{(-\Delta)^{1/4}}\vec{Z}_2\|_{L^{\infty}\mathcal{X}} \leq T^{1/4}\|\frac{1}{(-\Delta)^{1/4}}\vec{Z}_1\|_{L^{4/3,\infty}_{\text{per}}\mathcal{X}}$ and thus, for $0 \leq t < T$,

$$\|\int_{-\infty}^{-T} \nu\Delta\partial_j W_{\nu(t-s)} * \vec{Z}_2\, ds\|_{\mathcal{X}} \leq C\frac{1}{\nu^{3/4}T^{3/4}}T^{1/4}\|\frac{1}{(-\Delta)^{1/4}}\vec{Z}_1\|_{L^{4/3,\infty}_{\text{per}}\mathcal{X}}$$

We thus have found (for $\|g\|_{\mathcal{E}} = \|\sup_{t>0} |g't, .)|\|_{\mathcal{X}}$) that

$$\|B(\vec{U}, \vec{V})\|_{\mathcal{E}} \leq C\frac{1}{\nu}\|\vec{U}\|_{\mathcal{E}}\|\vec{V}\|_{\mathcal{E}}$$

and

$$\|\vec{\nabla} \otimes B(\vec{U}, \vec{V})\|_{L^{2,\infty}_{\text{per}}\mathcal{X}} \leq C\frac{1}{\nu}\|\vec{U}\|_{\mathcal{E}}\|\vec{\nabla} \otimes \vec{V}\|_{L^{2,\infty}_{\text{per}}\mathcal{X}} + C\frac{1}{\nu^{1/2}}\|\vec{\nabla} \otimes \vec{U}\|_{L^{2,\infty}_{\text{per}}\mathcal{X}}\|\vec{\nabla} \otimes \vec{V}\|_{L^{2,\infty}_{\text{per}}\mathcal{X}}$$

Picard's iterative algorithm will then provide a solution $\vec{u}_{\text{per}}$ as soon as $\|\vec{U}_0\|_{\mathcal{E}} + \nu^{1/2}\|\vec{\nabla} \otimes \vec{U}_0\|_{L^{2,\infty}_{\text{per}}\mathcal{X}}$ will be less than $C_0\nu$ for a constant C_0 which does not depend on ν nor on T.
Existence of $\vec{w}$ is now easy: just follow the proof of the end of Theorem 10.11. □

10.10   Beltrami Flows

In this final section, we pay a few words on Beltrami flows. Beltrami flows have thoroughly been used as examples of incompressible fluid flows for Euler or Navier–Stokes equations [40, 129].

Recall that we may write the Navier–Stokes equations as

$$\partial_t \vec{u} + \vec{\omega} \wedge \vec{u} = -\vec{\nabla}Q + \nu\Delta\vec{u} + \vec{f}, \quad \text{div } \vec{u} = 0.$$

Beltrami flows are defined as flows for which vorticity and velocity are parallel:

$$\vec{\omega} \wedge \vec{u} = 0.$$

The Navier–Stokes equations then reduce to linear equations:

$$\begin{cases} \partial_t \vec{u} = -\vec{\nabla}Q + \nu\Delta\vec{u} + \vec{f} \\ \vec{\nabla} \wedge \vec{u} = \lambda(t,x)\vec{u} \end{cases} \tag{10.55}$$

or

$$\begin{cases} \partial_t \vec{u} = \nu\Delta\vec{u} + \mathbb{P}\vec{f} \\ \vec{\nabla} \wedge \vec{u} = \lambda(t,x)\vec{u} \end{cases} \tag{10.56}$$

The case $\vec{f} = 0$ and λ constant was first discussed by Trkal [300, 477]; the solutions are labeled as Strong Beltrami flows in [40].

Trkalian flows

Theorem 10.16.
Let $\vec{u}$ be a solution to

$$\begin{cases} \partial_t \vec{u} = \nu\Delta\vec{u} \\ \vec{\nabla} \wedge \vec{u} = \lambda\vec{u} \end{cases} \tag{10.57}$$

where $\lambda \neq 0$. Then

- $\Delta\vec{u} = -\lambda^2\vec{u}$

- $\vec{u}(t,x) - e^{-\nu\lambda^2 t}\vec{u}_0$ *with $\vec{u}_0 \in \mathcal{D}'(\mathbb{R}^3)$ and $\vec{\nabla} \wedge \vec{u}_0 = \lambda\vec{u}_0$*

If $\vec{u}_0 \in \mathcal{S}'$, then the equation $\nabla \wedge \vec{u}_0 = \lambda \vec{u}_0$ is equivalent to the existence of a distribution $\vec{A} \in \mathcal{D}'(\mathbb{S}^2)$ with

$$\sigma.\vec{A}(\sigma) = 0$$

and

$$\vec{u}_0 = \int_{\mathbb{S}^2} \cos(\lambda x.\sigma) \vec{A}(\sigma) - \sin(\lambda x.\sigma) \sigma \wedge \vec{A}(\sigma) \, d\sigma$$

The latter equality means that

$$\langle \vec{u}_0 | \vec{\varphi} \rangle_{\mathcal{S}',\mathcal{S}} = \langle \vec{A} | \int \vec{\varphi}(x) \cos(\lambda x.\sigma) \, dx \rangle_{\mathcal{D}'(\mathbb{S}^2), \mathcal{D}(\mathbb{S}^2)}$$

$$- \langle \sigma \wedge \vec{A} | \int \vec{\varphi}(x) \sin(\lambda x.\sigma) \, dx \rangle_{\mathcal{D}'(\mathbb{S}^2), \mathcal{D}(\mathbb{S}^2)}$$

Proof. From $\vec{\nabla} \wedge \vec{u} = \lambda \vec{u}$, we get that div $\vec{u} = 0$. Then, we have

$$\Delta \vec{u} = -\vec{\nabla} \wedge (\vec{\nabla} \wedge \vec{u}) = -\lambda^2 \vec{u}.$$

Thus, $\vec{u}(t,x) = e^{-\nu \lambda^2 t} \vec{u}_0$.

Now, we have $-\Delta \vec{u}_0 = \lambda^2 \vec{u}_0$. If $\vec{u}_0 \in \mathcal{S}'(\mathbb{R}^3)$, we find that the Fourier transform $\vec{U}_0$ of $\vec{u}_0$ is supported on the sphere $|\xi| = |\lambda|$ and satisfies

$$(|\lambda| - |\xi|) \vec{U}_0(\xi) = 0.$$

Thus, in spherical coordinates $\xi = \rho \sigma$, we find that $\vec{U}_0(\xi) = \vec{B}(\sigma) \otimes \delta(\rho - |\lambda|)$:

$$\langle \vec{U}_0(\xi) | \vec{\theta}(\xi) \rangle = \langle \vec{B}(\sigma) | \lambda^2 \vec{\theta}(|\lambda|\sigma) \rangle$$

or equivalently

$$\langle \vec{u}_0(x) | \vec{\varphi}(x) \rangle = \frac{1}{(2\pi)^3} \langle \vec{U}_0(\xi) | \int \vec{\varphi}(x) e^{-ix.\xi} \, dx \rangle = \frac{\lambda^2}{(2\pi)^3} \langle \vec{B}(\sigma) | \int \vec{\varphi}(x) e^{-i|\lambda| x.\sigma} \, dx \rangle$$

Moreover, we want $\vec{\nabla} \wedge \vec{u}_0 = \lambda \vec{u}_0$, so that $i\xi \wedge \vec{U}_0 = \lambda \vec{U}_0$ and

$$\vec{U}_0 = (i\frac{\lambda}{|\lambda|} \sigma \wedge \vec{B}(\sigma)) \otimes \delta(\rho - |\lambda|) = \vec{C}(\sigma) \otimes \delta(\rho - |\lambda|)$$

where $\sigma.\vec{C}(\sigma) = 0$. We want as well $\vec{u}_0 = \frac{1}{\lambda} \vec{\nabla} \wedge \vec{u}_0 = \frac{1}{2}(\vec{u}_0 + \frac{1}{\lambda} \vec{\nabla} \wedge \vec{u}_0)$, so that

$$\vec{U}_0 = \frac{1}{2}(\vec{C}(\sigma) + i\frac{\lambda}{|\lambda|} \sigma \wedge \vec{C}(\sigma)) \otimes \delta(\rho - |\lambda|).$$

Writing $\vec{C}(\sigma) = \vec{E}(\sigma) + i\vec{F}(\sigma)$ with $\vec{E}$ even and $\vec{F}$ odd, we have $\sigma.\vec{E} = \sigma.\vec{F} = 0$ and we get, with $\vec{D} = \frac{1}{2}(\vec{E} - \frac{\lambda}{|\lambda|} \sigma \wedge \vec{F})$,

$$\vec{U}_0 = (\vec{D} + i\frac{\lambda}{|\lambda|} \sigma \wedge \vec{D}) \otimes \delta(\rho - |\lambda|)$$

We then find the decomposition $\vec{u}_0 = \int_{\mathbb{S}^2} \cos(\lambda x.\sigma) \vec{A}(\sigma) - \sin(\lambda x.\sigma) \sigma \wedge \vec{A}(\sigma) \, d\sigma$ with

$$\vec{A} = \frac{2\lambda^2}{(2\pi)^3} \vec{D}(\sigma). \qquad \square$$

A classical example of Trkalian flow is the flow associated to

$$\vec{u}_0 = \begin{pmatrix} C\cos(\lambda x_3) - B\sin(\lambda x_2) \\ A\cos(\lambda x_1) - C\sin(\lambda x_3) \\ B\cos(\lambda x_2) - A\sin(\lambda x_1) \end{pmatrix}$$

for three constants A, B, C. $\vec{u}_0$ is known as the Arnold-Beltrami-Childress flow [155].

We may easily construct other Trkalian flows. For instance, starting from the axisymmetric flow $\vec{v} = \frac{\sin(\lambda|x|)}{|x|}\vec{e}_3$, we obtain the axisymmetric Trkalian flow associated to $\vec{u}_0 = \vec{v} + \frac{1}{\lambda^2}\vec{\nabla}\operatorname{div}\vec{v} + \frac{1}{\lambda}\vec{\nabla}\wedge\vec{v}$. This field is smooth and belongs to $L^{3,\infty} \cap L^\infty$.

Chapter 11

Blow-up?

11.1   First Criteria

Throughout this chapter, we shall consider the Navier–Stokes problem

$$\partial_t \vec{u} = \nu \Delta \vec{u} - (\vec{u} \cdot \vec{\nabla})\vec{u} + \vec{f} - \vec{\nabla}p$$
$$\mathrm{div}\, \vec{u} = 0 \qquad\qquad (11.1)$$
$$\vec{u}_{|t=0} = \vec{u}_0$$

where $\vec{u}_0 \in (H^1(\mathbb{R}^3))^3$ with $\mathrm{div}\, \vec{u}_0 = 0$ and $\vec{f} \in L^2((0, +\infty), (L^2(\mathbb{R}^3)^3)$.
Recall the results of Theorem 7.2: there exists a (positive) maximal time $T_{\mathrm{MAX}} \in (0, +\infty]$ for which one can find a mild solution $\vec{u}$ of Equation (11.1) on $(0, T_{\mathrm{MAX}}) \times \mathbb{R}^3$ which satisfies, for all $T < T_{\mathrm{MAX}}$, $\vec{u}$ belongs to $\mathcal{C}([0, T], (H^1)^3) \cap L^2((0, T), (H^2)^3)$.

Definition 11.1 (Blow-up).
If T_{MAX} is finite, we shall say that the solution $\vec{u}$ blows up in finite time and that T_{MAX} is the blow-up time of $\vec{u}$.

Theorem 7.2 gave us some criteria on the possibility of blow-up:

- If $T_{\mathrm{MAX}} < +\infty$, then $\sup_{0<t<T_{\mathrm{MAX}}} \|\vec{u}(t,.)\|_{H^1} = +\infty$.

- If $T_{\mathrm{MAX}} < +\infty$, then $\int_0^{T_{\mathrm{MAX}}} \|\vec{u}(s,.)\|_{\dot{H}^{3/2}}^2 \, ds = +\infty$

- There exists a positive constant ϵ_0 (independent of ν, $\vec{u}_0$ and $\vec{f}$), such that, if $\|\vec{u}_0\|_{\dot{H}^{1/2}} < \epsilon_0 \nu$ and $\int_0^{+\infty} \|\vec{f}(s,.)\|_{\dot{H}^{-\frac{1}{2}}}^2 \, ds < \epsilon_0^2 \nu^3$, then $T_{\mathrm{MAX}} = +\infty$.

The Clay Millennium problem is essentially to answer the following question:

Clay Millennium problem for the Navier–Stokes equations

Do we have global existence (i.e., $T_{\mathrm{MAX}} = +\infty$) when $\vec{f} = 0$?

11.2   Blow-up for the Cheap Navier–Stokes Equation

Let us recall that the proof of Theorem 7.2 was based on energy estimates:

DOI: 10.1201/9781003042594-11

- the L^2 norm of $\vec{u}(t,.)$ is estimated by

$$\frac{d}{dt}\int |\vec{u}(t,x)|^2\,dx = 2\int \vec{u}\cdot\partial_t\vec{u}dx = -2\nu\int |\vec{\nabla}\otimes\vec{u}|^2\,dx + 2\int \vec{u}\cdot\vec{f}\,dx \qquad (11.2)$$

so that

$$\|\vec{u}(t,.)\|_2 \le \|\vec{u}_0\|_2 + \int_0^t \|\vec{f}(s,.)\|_2\,ds. \qquad (11.3)$$

- similarly, the $\dot{H}^1$ norm of $\vec{u}(t,.)$ is estimated by

$$\frac{d}{dt}\int |\vec{\nabla}\otimes\vec{u}(t,x)|^2\,dx = -2\nu\int |\Delta\vec{u}|^2\,dx - 2\sum_{i=1}^{3}\int \partial_i\vec{u}\cdot((\partial_i\vec{u})\cdot\vec{\nabla})\vec{u}\,dx$$

$$-2\int \Delta\vec{u}\cdot\vec{f}dx$$

$$\le -\nu\|\Delta\vec{u}\|_2^2 + C\|\vec{\nabla}\otimes\vec{u}\|_2\|\|\Delta\vec{u}\|_2\|\vec{u}\|_{\dot{H}^{3/2}} + \frac{1}{\nu}\|\vec{f}\|_2^2$$

$$\le \frac{C}{4\nu}\|\vec{\nabla}\otimes\vec{u}\|_2^2\|\vec{u}\|_{\dot{H}^{3/2}}^2 + \frac{1}{\nu}\|\vec{f}\|_2^2 \qquad (11.4)$$

so that

$$\|\vec{u}\|_{\dot{H}^1}^2 \le \left(\|\vec{u}_0\|_{\dot{H}^1}^2 + \frac{1}{\nu}\int_0^t \|\vec{f}\|_2^2\right) e^{\frac{C}{4\nu}\int_0^t \|\vec{u}(s,.)\|_{\dot{H}^{3/2}}^2\,ds} \qquad (11.5)$$

In order to underline the role of those energy estimates, Montgomery–Smith studied a general form of (pseudo)-differential equation

$$\partial_t\vec{u} = \nu\Delta\vec{u} + \sigma(D)(\vec{u}\otimes\vec{u}) \qquad (11.6)$$

(generalizing the Navier–Stokes problem $\partial_t\vec{u} = \nu\Delta\vec{u} - \mathbb{P}\,\text{div}(\vec{u}\otimes\vec{u})$), where

- $\vec{u}(t,x)$ is defined on $(0,T)\times\mathbb{R}^3$ with values in $\mathbb{R}^d$

- $\sigma(D)$ is a matrix of Fourier multipliers $\sigma(\xi) = (\sigma_{j,(k,l)}(\xi))$ with d rows and d^2 columns, such that the coefficients $\sigma_{i,j}$ are smooth functions on $\mathbb{R}^3$ which are positively homogeneous of order 1: for $\lambda > 0$, $\sigma_{j,(k,l)}(\lambda\xi) = \lambda\sigma_{j,(k,l)}(\xi)$.

It is easy to check that, in the case of equation (11.6), the proofs of Fujita and Kato's theorem (Theorem 7.1) or of Koch and Tataru's theorem (Theorem 9.2) still work.

If we consider only the Hilbertian setting, one may even deal with a more general class of equations:

Proposition 11.1.
Let $\sigma_p(D)$ $(p = 0, 1, 2)$ be matrices of Fourier multipliers $\sigma_p(\xi) = (\sigma_{p,\alpha,\beta}(\xi))$ with respectively d rows and d^2 columns $(p = 0)$ or d rows and d columns $(p = 1$ or $p = 2)$. Assume that the coefficients $\sigma_{p,\alpha,\beta}$ are locally bounded functions on $\mathbb{R}^3\setminus\{0\}$ which are positively homogeneous of order λ_p with $0 \le \lambda_p$ and $\lambda_0 + \lambda_1 + \lambda_2 = 1$. Let $\nu > 0$. If $\vec{u}_0$ is a function on $\mathbb{R}^3$ with values in $\mathbb{R}^d$ and if $\vec{u}_0$ belongs to H^1 then:

- *there exists $T > 0$ and a function $\vec{u}$ defined on $[0,T]\times\mathbb{R}^3$, with values in $\mathbb{R}^d$ such that $\vec{u}\in\mathcal{C}([0,T],H^1)\cap L^2((0,T),H^2)$ and such that*

$$\partial_t\vec{u} = \nu\Delta + \sigma_0(D)(\sigma_1(D)\vec{u}\otimes\sigma_2(D)\vec{u}) \qquad (11.7)$$

- *The existence time T satisfies*

$$T \geq C_\nu \frac{1}{\|\vec{u}_0\|_{\dot{H}^1}^4}.$$

  In particular, let T_{MAX} be the maximal existence time. We have blow-up if and only if $\sup_{0<t<T_{\mathrm{MAX}}} \|\vec{u}(t,.)\|_{H^1} = +\infty$.

- *if the maximal time of existence T_{MAX} is finite, then*

$$\int_0^{T_{\mathrm{MAX}}} \|\vec{u}(t,.)\|_{\dot{H}^{3/2}}^2 \, ds = +\infty.$$

- *there exists $\epsilon_0 > 0$ such that if $\|\vec{u}_0\|_{\dot{H}^{1/2}} < \epsilon_0 \nu$ then $T_{\mathrm{MAX}} = +\infty$.*

Proof. We first solve the problem in $L^4((0,T), \dot{H}^1)$. By the product laws in Sobolev spaces, we have

$$\begin{aligned}
\|\sigma_0(D)(\sigma_1(D)\vec{u} \otimes \sigma_2(D)\vec{v})\|_{\dot{H}^{-1/2}} &\leq C\|\sigma_1(D)\vec{u} \otimes \sigma_2(D)\vec{v}\|_{\dot{H}^{\lambda_0 - \frac{1}{2}}} \\
&\leq C'\|\sigma(D)\vec{u}\|_{\dot{H}^{1-\lambda_1}}\|\sigma_2(D)\vec{v}\|_{\dot{H}^{\lambda_0+\lambda_1}} \\
&\leq C''\|\vec{u}\|_{\dot{H}^1}\|\vec{v}\|_{\dot{H}^1}.
\end{aligned}$$

Thus, if $\vec{u}$ and $\vec{v}$ belong to $L^4((0,T), \dot{H}^1)$, we find, for

$$B(\vec{u}, \vec{v}) = \int_0^t W_{\nu(t-s)} * \sigma_0(D)(\sigma_1(D)\vec{u} \otimes \sigma_2(D)\vec{v}) \, ds,$$

$$\begin{aligned}
\|B(\vec{u}, \vec{v})\|_{L^4((0,T),\dot{H}^1)} &\leq \sqrt{\|B(\vec{u}, \vec{v})\|_{L^\infty((0,T),\dot{H}^{1/2})}}\sqrt{\|B(\vec{u}, \vec{v})\|_{L^2((0,T),\dot{H}^{3/2})}} \\
&\leq C\nu^{-3/4}\|\sigma_0(D)(\sigma_1(D)\vec{u} \otimes \sigma_2(D)\vec{v})\|_{L^2((0,T),\dot{H}^{-1/2})} \\
&\leq C'\nu^{-3/4}\|\vec{u}\|_{L^4((0,T),\dot{H}^1)}\|\vec{v}\|_{L^4((0,T),\dot{H}^1)}.
\end{aligned}$$

Thus, the Picard iterates $\vec{U}_0 = W_{\nu t} * \vec{u}_0$ and $\vec{U}_{n+1} = \vec{U}_0 + B(\vec{U}_n, \vec{U}_n)$ will converge to a solution in $L^4((0,T), \dot{H}^1)$ as long as $\|\vec{U}_0\|_{L^4((0,T),\dot{H}^1)} \leq \frac{\nu^{3/4}}{4C'}$, hence if

$$T \leq \frac{\nu^3}{\|\vec{u}_0\|_{\dot{H}^1}^4 (4C')^4}.$$

Similarly, we have

$$\begin{aligned}
\|B(\vec{u}, \vec{u})\|_{L^\infty H^1} &+ \sqrt{\nu}\|B(\vec{u}, \vec{u})\|_{L^2 \dot{H}^1} + \sqrt{\nu}\|B(\vec{u}, \vec{u})\|_{L^2 \dot{H}^2} \\
&\leq C\nu^{-1/2}\|\sigma_0(D)(\sigma_1(D)\vec{u} \otimes \sigma_2(D)\vec{u})\|_{L^2((0,T),\dot{H}^{-1})} \\
&\quad + C\nu^{-1/2}\|\sigma_0(D)(\sigma_1(D)\vec{u} \otimes \sigma_2(D)\vec{u})\|_{L^2((0,T),L^2)} \\
&\leq C\nu^{-1/2}\|\vec{u}\|_{L^4 \dot{H}^1}(\|\vec{u}\|_{L^4((0,T),\dot{H}^{1/2})} + \|\vec{u}\|_{L^4((0,T),\dot{H}^{3/2})}) \\
&\leq C'\nu^{-3/4}\|\vec{u}\|_{L^4 \dot{H}^1}(\|\vec{u}\|_{L^\infty H^1} + \sqrt{\nu}\|\vec{u}\|_{L^2 \dot{H}^1} + \sqrt{\nu}\|\vec{u}\|_{L^2 \dot{H}^2}).
\end{aligned}$$

Thus, we find that the solution will belong to $L^\infty((0,T), H^1) \cap L^2((0,T), \dot{H}^2)$.

Moreover, we have, for $0 < T_0 < t < T_{\mathrm{MAX}}^*$,

$$\|\vec{u}(t,.)\|_{\dot{H}^1} \leq \|\vec{u}(T_0,.)\|_{\dot{H}^1} + C\nu^{-3/4} \sup_{T_0<s<t} \|\vec{u}(s,.)\|_{\dot{H}^1}\|\vec{u}\|_{L^2((T_0,t),\dot{H}^{3/2})}.$$

This gives that, if $\int_0^{T_{MAX}} \|\vec{u}(s,.)\|_{\dot{H}^{3/2}}^2 \, ds < +\infty$, then $\|\vec{u}\|_{\dot{H}^1}$ remains bounded on $(0, T_{MAX})$, so that $T_{MAX} = +\infty$.

Finally, we write

$$\|B(\vec{u}, \vec{u})\|_{L^\infty \dot{H}^{1/2}} + \sqrt{\nu}\|B(\vec{u}, \vec{u})\|_{L^2 \dot{H}^{3/2}}$$
$$\leq C\nu^{-1/2}\|\sigma_0(D)(\sigma_1(D)\vec{u} \otimes \sigma_2(D)\vec{u})\|_{L^2((0,T),\dot{H}^{-1/2})}$$
$$\leq C\nu^{-1} \sup_{0<s<T} \|\vec{u}\|_{\dot{H}^{1/2}} \sqrt{\nu}\|\vec{u}\|_{L^2((0,T),\dot{H}^{3/2})}.$$

Thus, if $\|\vec{U}_0\|_{L^\infty((0,T),\dot{H}^{1/2})} + \sqrt{\nu}\|\vec{U}_0\|_{L^2((0,T),\dot{H}^{3/2})} < \frac{\nu}{4C'}$, we have

$$\|\vec{u}\|_{L^\infty((0,T),\dot{H}^{1/2})} + \sqrt{\nu}\|\vec{u}\|_{L^2((0,T),\dot{H}^{3/2})} < \frac{\nu}{2C'}.$$

In particular, if $\vec{u}_0$ is small enough in $\dot{H}^{1/2}$, we find that $T_{MAX} = +\infty$.   □

The Navier–Stokes equations may be writen in the form of equations (11.7) in two ways. Let $\vec{\mathcal{R}} = \frac{1}{\sqrt{-\Delta}}\vec{\nabla}$. The Leray projection operator $\mathbb{P}$ may be written as $\mathbb{P}\vec{f} = \vec{\mathcal{R}} \wedge (\vec{\mathcal{R}} \wedge \vec{f})$. From the equations

$$\partial_t \vec{u} = \nu\Delta\vec{u} - \vec{u} \cdot \vec{\nabla}\vec{u} - \vec{\nabla}p,$$

we get

$$\partial_t \vec{u} = \nu\Delta\vec{u} - \vec{\mathcal{R}} \wedge (\vec{\mathcal{R}} \wedge \operatorname{div}(\vec{u} \otimes \vec{u})) = \nu\Delta\vec{u} + \sigma_0(D)(\vec{u} \otimes \vec{u})$$

where $\sigma_0(D)$ is the matrix of Fourier multipliers associated with $-\vec{\mathcal{R}} \wedge (\vec{\mathcal{R}} \wedge \operatorname{div})$:

$$\sigma_{0,j,(k,l)}(\xi) = -i\delta_{j,l}\xi_k - i\xi_j \frac{\xi_k\xi_l}{|\xi|^2}.$$

On the other hand, from the equations

$$\partial_t \vec{u} = \nu\Delta\vec{u} - \vec{\omega} \wedge \vec{u} - \vec{\nabla}q,$$

we get

$$\partial_t \vec{u} = \nu\Delta\vec{u} - \vec{\mathcal{R}} \wedge (\vec{\mathcal{R}} \wedge (\vec{\omega} \wedge \vec{u})) = \nu\Delta\vec{u} + \sigma_0(D)((\sigma_1(D)\vec{u} \otimes \vec{u})$$

where $\sigma_0(D)$ is the matrix of Fourier multipliers described by the cycle $\gamma : 1 \to 2 \to 3 \to 1$ as:

$$\sigma_{0,j,(k,l)}(\xi)$$
$$= -\delta_{k,\gamma(j)}\delta_{l,\gamma^2(j)} + \delta_{k,\gamma^2(j)}\delta_{l,\gamma(j)} + \sum_{q=1}^3 \frac{\xi_j\xi_q}{|\xi|^2}(-\delta_{k,\gamma(q)}\delta_{l,\gamma^2(q)} + \delta_{k,\gamma^2(q)}\delta_{l,\gamma(q)})$$

and $\sigma_1(D) = \vec{\nabla}\wedge$:

$$\sigma_1(\xi) = i \begin{pmatrix} 0 & -\xi_3 & \xi_2 \\ \xi_3 & 0 & -\xi_1 \\ -\xi_2 & \xi_1 & 0 \end{pmatrix}.$$

Many problems of the form (11.7) have been studied as models for blow ups (or no blow up):

- Montgomery–Smith proved blow-up in the case of the *cheap Navier–Stokes equation* where $d = 1$ and $\sigma(D) = \sqrt{-\Delta}$ [369]:

$$\partial_t u = \nu\Delta + \sqrt{-\Delta}(u^2).$$

We will describe below the result of Montgomery–Smith.

- The cheap equation has been recently adapted by Gallagher and Paicu [200] into a vector equation ($d = 3$) which preserves the divergence-free condition:

$$\partial_t \vec{u} = \nu \Delta \vec{u} + \mathbb{P} Q(u, u) = \nu \Delta \vec{u} + \sigma(D)(\vec{u} \otimes \vec{u})$$

with

$$\sigma_{j,(k,l)}(\xi) = \mathbb{1}_E(\xi) \frac{1}{|\xi|}(|\xi|^2 - \xi_k \xi_l \delta_{j,l})$$

and

$$E = \{\xi \ / \ \xi_1 \xi_2 < 0, \xi_1 \xi_3 < 0, |\xi_2| < \min(|\xi_1|, |\xi_2|)\}.$$

The key point is the fact that, similarly to the case of the cheap equation, when the components of the Fourier transform of $\vec{u}_0$ are non-negative then the components of the Fourier transform of the solution $\vec{u}$ remain non-negative.

- If we look for a complex–valued solution of the Navier–Stokes problem

$$\partial_t \vec{u} = \nu \Delta \vec{u} - \mathbb{P} \operatorname{div}(\vec{u} \otimes \vec{u}),$$

it is equivalent to find real-valued solutions $(\vec{v}, \vec{w})$ of the system

$$\begin{cases} \partial_t \vec{v} = \nu \Delta \vec{v} - \mathbb{P} \operatorname{div}(\vec{v} \otimes \vec{v} - \vec{w} \otimes \vec{w}) \\ \partial_t \vec{w} = \nu \Delta \vec{w} - \mathbb{P} \operatorname{div}(\vec{v} \otimes \vec{w} + \vec{w} \otimes \vec{v}) \end{cases}$$

which is of the form (11.7). Blow-up for this equation has been proved by Li and Sinai [334] in a difficult paper based on tools in renormalization group theory and on the theory of linear hydrodynamic instability.

- On the other hand, Wang [496] gave an example where no blow up occurs, namely the equations

$$\partial_t \vec{u} = \nu \Delta \vec{u} - \vec{\mathcal{R}} \wedge (\vec{\omega} \wedge (\vec{\mathcal{R}} \wedge \vec{u})) = \nu \Delta u + \sigma_0(D)(\sigma_1(D)\vec{u} \otimes \sigma_2(D)\vec{u})$$

with

$$\sigma_{0,j,(k,l)}(\xi) = i(\frac{\xi_l}{|\xi|}\delta_{j,k} - \frac{\xi_k}{|\xi|}\delta_{j,l})$$

and

$$\sigma_1(\xi) = i \begin{pmatrix} 0 & -\xi_3 & \xi_2 \\ \xi_3 & 0 & -\xi_1 \\ -\xi_2 & \xi_1 & 0 \end{pmatrix}, \sigma_2(\xi) = \frac{1}{|\xi|}\sigma_1(\xi).$$

We now present the result of Montgomery–Smith (blow up), then the result of Wang (no blow up):

Cheap Navier–Stokes equation

Theorem 11.1.
There exists a positive constant A_ν such that if u is a solution of

$$\partial_t u = \nu \Delta u + \sqrt{-\Delta}(u^2)$$

with $u \in \mathcal{C}([0, T], H^1) \cap L^2((0, T), H^2)$ for all $T < T_{\text{MAX}}$ and $u(0, .) = u_0$ satisfies

- $u_0 \in H^1$

- *the Fourier transform $\hat{u}_0$ is non-negative*

- *for some $\xi_0 \in \mathbb{R}^3$ with $|\xi_0| = 1$, $\int_{|\xi - \xi_0| < 1/3} |\hat{u}_0(\xi)| \, d\xi > A_\nu$*

then we have

$$T_{\text{MAX}} \leq 1.$$

Proof. First, we check that $\hat{u} \geq 0$. Indeed, this is true at time $t = 0$. Let $T_0 = \sup\{T \geq 0 \ / \ \hat{u} \geq 0 \text{ on } [0, T] \times \mathbb{R}^3\}$. If $T_0 < T_{\text{MAX}}$, then, by continuity, we find that $\hat{u}(T_0, .) \geq 0$. Moreover, there exists a small time T_1 such that on $[T_0, T_0 + T_1]$ u may be constructed by Picard's iterative scheme. It is easy to check that every Picard iterate has its Fourier transform non-negative, and so does their limit u; thus, we find $\hat{u} \geq 0$ on $[0, T_0 + T_1]$, in contradiction with the definition of T_0. Thus, $T_0 = T_{\text{MAX}}$.

We now start from Duhamel's formula

$$\hat{u}(t, \xi) = e^{-\nu t |\xi|^2} \hat{u}_0(\xi) + \int_0^t e^{-\nu(t-s)|\xi|^2} |\xi| \, (\hat{u}(s, .) * \hat{u}(s..))(\xi) \, ds$$

Let $w_0(\xi) = 1_{B(\xi_0, 1/3)} \hat{u}_0(\xi)$. Define $w_n(\xi)$ by induction as $w_{n+1} = w_n * w_n$. w_n is then supported in $B(2^n \xi_0, \frac{1}{3} 2^n) \subset \{\xi \ / \ 2^{n+1} < 3|\xi| < 2^{n+2}\}$. Thus, if $\hat{u}(t, \xi) \geq \alpha_n(t) w_n(\xi)$, we have

$$\int |\xi|^2 (\hat{u}(t, \xi))^2 \, d\xi \geq \sum_{n=0}^{+\infty} 4^{n+1} \|w_n(\xi)\|_2^2 \alpha_n(t)^2$$

On the other hand, we have

$$A_\nu^{2^n} \leq \int w_n(\xi) \, d\xi \leq C_0 2^{3n/2} \|w_n\|_2$$

and thus

$$\int |\xi|^2 (\hat{u}(t, \xi))^2 \, d\xi \geq \frac{4}{C_0^2} \sum_{n=0}^{+\infty} 2^{-n} \, A_\nu^{2^{n+1}} \alpha_n(t)^2.$$

We now turn to the estimation of $\alpha_n(t)$. As $\hat{u}(t, \xi) \geq e^{-\nu t |\xi|^2} \hat{u}_0(\xi)$, we find that

$$\alpha_0(t) \geq e^{-\frac{16}{9} \nu t}.$$

Further, we have

$$\alpha_{n+1}(t) \geq \min_{2^{n+1} < 3|\xi| < 2^{n+2}} \int_0^t e^{-\nu(t-s)|\xi|^2} |\xi| \alpha_n^2(s) \, ds \geq \frac{2}{3} 2^n \int_0^t e^{-\frac{16}{9}\nu(t-s)4^n} \alpha_n^2(s) \, ds$$

Now, we define

$$\beta_n = \min_{1 - 4^{-n} \leq t \leq 1} \alpha_n(t).$$

For $1 - 4^{-n-1} \leq t \leq 1$, we have

$$\alpha_{n+1}(t) \geq \frac{2}{3} 2^n \int_{t-4^{-n-1}}^t e^{-\frac{16}{9}\nu(t-s)4^n} \alpha_n^2(s) \, ds$$

which gives

$$\beta_{n+1} \geq \frac{1}{6} e^{-\frac{4}{9}\nu} \beta_n^2.$$

Recall that

$$\int |\xi|^2 (\hat{u}(1,\xi))^2 \, d\xi \geq \frac{4}{C_0^2} \sum_{n=0}^{+\infty} 2^{-n} A_\nu^{2^{n+1}} \beta_n^2.$$

Assume that $A_\nu > 2$; then $2^{-n} A_\nu^{2^n} \geq 1$, so that $2^{-n} A_\nu^{2^{n+1}} \beta_n^2 \geq A_\nu^{2^n} \beta_n^2$; moreover

$$A_\nu^{2^{n+1}} \beta_{n+1}^2 \geq \frac{1}{36} e^{-\frac{8}{9}\nu} \left(A_\nu^{2^n} \beta_n^2 \right)^2$$

Hence, if $A_\nu > 36 \, e^{\frac{40}{9}\nu}$, we find by induction on n that $A_\nu^{2^n} \beta_n^2 > 36 \, e^{\frac{8}{9}\nu}$ and finally that $\|u(1,.)\|_{H^1} = +\infty$. Thus $T_{\text{MAX}} \leq 1$. □

Proposition 11.2.
Let $\vec{u}$ be the solution in $C([0, T_{\text{MAX}}, H^1) \cap \cap_{T < T_{\text{MAX}}} L^2([0,T], \dot{H}^2)$ of the Cauchy problem for the equations

$$\partial_t \vec{u} = \nu \Delta \vec{u} - \vec{\mathcal{R}} \wedge (\vec{\omega} \wedge (\vec{\mathcal{R}} \wedge \vec{u})))$$

(where $\vec{\omega} = \text{curl}\, \vec{u}$) with initial value $\vec{u}_0 \in H^1$. Then $T_{\text{MAX}} = +\infty$.

Proof. We have

$$\frac{d}{dt} \|\vec{u}(t,.)\|_{\dot{H}^{1/2}}^2 = 2 \int \partial_t \vec{u} \cdot \sqrt{-\Delta} \vec{u} \, dx$$
$$= -2\nu \|\vec{u}\|_{\dot{H}^{3/2}}^2,$$

since

$$\int \sqrt{-\Delta} \vec{u} \cdot \vec{\mathcal{R}} \wedge (\vec{\omega} \wedge (\vec{\mathcal{R}} \wedge \vec{u})) \, dx = \int \vec{\omega} \cdot (\vec{\omega} \wedge (\vec{\mathcal{R}} \wedge \vec{u})) \, dx = 0.$$

Thus, we have

$$\int_0^{T_{\text{MAX}}} \|\vec{u}(s,.)\|_{\dot{H}^{3/2}}^2 \, ds \leq \frac{1}{2\nu} \|\vec{u}_0\|_{\dot{H}^{1/2}}. \qquad \square$$

In Wang's example, we have as well energy conservation:

$$\frac{d}{dt} \|\vec{u}\|_2^2 = -2\nu \|\vec{u}\|_{\dot{H}^1}^2.$$

This is not necessary: we could have dealt with the equation

$$\partial_t \vec{u} = \nu \Delta \vec{u} - \vec{\mathcal{R}} \wedge (\vec{\omega} \wedge \vec{u})).$$

On the other hand, Tao [462] considered the problem of blow-up in presence of energy conservation. More precisely, he considered the abstract problem $\partial_t \vec{u} = \nu \Delta \vec{u} - B(\vec{u}, \vec{u})$, where B would mimic the operator $B_{NS}(\vec{u}, \vec{v}) = \mathbb{P} \text{div}(\vec{u} \otimes \vec{v})$ on three points:

- $\text{div}\, B(\vec{u}, \vec{v}) = 0$

- $\|B(\vec{u}, \vec{v})\|_{L^2(\mathbb{R}^3)} \leq C(\|\vec{u}\|_4 \|\vec{\nabla} \otimes \vec{v}\|_4 + \|\vec{v}\|_4 \|\vec{\nabla} \otimes \vec{u}\|_4)$

- $\int B(\vec{u}, \vec{u}).\vec{u}\, dx = 0$ for $\vec{u} \in H^2$ with $\text{div}\, \vec{u} = 0$

He constructed an example of such an operator B for which blow-up occurs, thus invalidating the abstract Hilbertian approach of Otelbaev [389].

11.3   Serrin's Criterion

Recall that the proof of Theorem 7.2 relied on the differential equalities

$$\frac{d}{dt}\int |\vec{u}(t,x)|^2\,dx = -2\nu\int |\vec{\nabla}\otimes\vec{u}|^2\,dx + 2\int \vec{u}\cdot\vec{f}\,dx \tag{11.8}$$

and

$$\frac{d}{dt}\int |\vec{\nabla}\otimes\vec{u}(t,x)|^2\,dx = -2\nu\int |\Delta\vec{u}|^2\,dx + 2\int \Delta\vec{u}\cdot(\vec{u}\cdot\vec{\nabla}\vec{u})\,dx$$
$$- 2\int \Delta\vec{u}\cdot\vec{f}\,dx \tag{11.9}$$

The first one allows to control the L^2 norm:

$$\|\vec{u}(t,.)\|_2 \le \|\vec{u}_0\|_2 + \int_0^t \|\vec{f}(s,.)\|_2\,ds. \tag{11.10}$$

The second one aims to control the $\dot{H}^1$ norm.

Serrin [435] gave a very simple criterion to ensure the control of the $\dot{H}^1$ norm of $\vec{u}$ through Equation (11.9):

Serrin's criterion

Theorem 11.2.
Let $\vec{u}_0 \in H^1$ with div $\vec{u}_0 = 0$ and $\vec{f} \in L^2((0,+\infty), L^2)$. Let $\vec{u}$ be a solution of

$$\partial_t\vec{u} = \nu\Delta\vec{u} + \mathbb{P}(\vec{f} - \mathrm{div}(\vec{u}\otimes\vec{u}))$$

with $\vec{u} \in \mathcal{C}([0,T],H^1)\cap L^2((0,T),H^2)$ for all $T < T_{\mathrm{MAX}}$ and $\vec{u}(0,.) = \vec{u}_0$. Then, if $2/p + 3/q = 1$ with $2 \le p < +\infty$, we have

$$\|\vec{u}(T,.)\|_{\dot{H}^1}^2 \le (\|\vec{u}_0\|_{\dot{H}^1}^2 + \frac{1}{\nu}\|\vec{f}\|_{L^2 L^2}^2)e^{C_0\nu^{1-p}\int_0^T \|\vec{u}\|_q^p\,dt} \tag{11.11}$$

where the constant C_0 does not depend on T.

In particular, if the maximal existence time T_{MAX} satisfies $T_{\mathrm{MAX}} < +\infty$, then $\int_0^{T_{\mathrm{MAX}}} \|\vec{u}\|_q^p\,dt = +\infty$.

Proof. From (11.9), we get

$$\frac{d}{dt}\|\vec{u}\|_{\dot{H}^1}^2 \le -2\nu\|\Delta\vec{u}\|_2^2 + 2\,\|\Delta\vec{u}\|_2\|\vec{u}\cdot\vec{\nabla}\vec{u}\|_2 + 2\|\Delta\vec{u}\|_2\|\vec{f}\|_2$$

We then write for $\frac{1}{r} = \frac{1}{2} - \frac{1}{q}$ and $\sigma = 3(\frac{1}{2} - \frac{1}{r}) = 1 - \frac{2}{p}$:

$$\|\vec{u}\cdot\vec{\nabla}\vec{u}\|_2 \le \|\vec{u}\|_q\|\vec{\nabla}\otimes\vec{u}\|_r \le C\|\vec{u}\|_q\|\vec{\nabla}\otimes\vec{u}\|_{\dot{H}^\sigma} \le C\|\vec{u}\|_q\|\vec{u}\|_{\dot{H}^1}^{1-\sigma}\|\Delta\vec{u}\|_2^\sigma$$

and thus

$$\frac{d}{dt}\|\vec{u}\|_{\dot{H}^1}^2 \le \frac{1}{\nu}\|\vec{f}\|_2^2 + C_\sigma\nu^{-\frac{1+\sigma}{1-\sigma}}\|\vec{u}\|_q^{\frac{2}{1-\sigma}}\|\vec{u}\|_{\dot{H}^1}^2 = \frac{1}{\nu}\|\vec{f}\|_2^2 + C_\sigma\nu^{-\frac{1+\sigma}{1-\sigma}}\|\vec{u}\|_q^p\|\vec{u}\|_{\dot{H}^1}^2$$

and we conclude by Grönwall's lemma. □

As we shall see in Chapter 15, a theorem by Escauriaza, Seregin and Šverák [163] proves that the endpoint case $p = +\infty$, $q = 3$ of the Serrin criterion holds: if $T_{\text{MAX}} < +\infty$, then $\sup_{0<t<T_{\text{MAX}}} \|\vec{u}(t,.)\|_3 = +\infty$.

A former result of Kozono and Sohr [275] stated that, if $T_{\text{MAX}} < +\infty$ and $\vec{u}$ remained bounded in L^3 as $t \to T_{\text{MAX}}$, then there was a discontinuity of $\|\vec{u}\|_3$ at time T_{MAX}: there exists a positive constant γ such that

$$\limsup_{t \to T_{\text{MAX}}^-} \|\vec{u}(t,.) - \vec{u}(T_{\text{MAX}},.)\|_3 \geq \gamma\nu$$

Indeed, we split $\vec{u}(T_{\text{MAX}},.)$ into $\vec{v} + \vec{w}$, where $\vec{v} \in L^\infty$ and $\|\vec{w}\|_3$ is small. We get

$$\|\vec{u} \cdot \vec{\nabla}\vec{u}\|_2 \leq (\|\vec{u} - \vec{u}(T_{\text{MAX}},.)\|_3 + \|\vec{w}\|_3)\|\vec{\nabla} \otimes \vec{u}\|_6 + \|\vec{v}\|_\infty \|\vec{u}\|_{\dot{H}^1}$$

so that

$$2\|\Delta\vec{u}\|_2\|\vec{u} \cdot \vec{\nabla}\vec{u}\|_2 \leq C_0(\|\vec{u} - \vec{u}(T_{\text{MAX}},.)\|_3 + \|\vec{w}\|_3)\|\Delta\vec{u}\|_2^2 + \frac{\nu}{2}\|\Delta\vec{u}\|_2^2 + \frac{2}{\nu}\|\vec{v}\|_\infty^2\|\vec{u}\|_{\dot{H}^1}^2.$$

If we choose $\vec{w}$ such that $\|\vec{w}\|_3 < \frac{\nu}{4C_0}$ and if $\sup_{T_1<t<T_{\text{MAX}}} \|\vec{u} - \vec{u}(T_{\text{MAX}},.)\|_3 < \frac{\nu}{4C_0}$, we find that on (T_1, T_{MAX}) we have

$$\frac{d}{dt}\|\vec{u}\|_{\dot{H}^1}^2 \leq \frac{1}{\nu}\|\vec{f}\|_2^2 + \frac{2}{\nu}\|\vec{v}\|_\infty^2\|\vec{u}\|_{\dot{H}^1}^2.$$

Grönwall's lemma then gives the control on the $\dot{H}^1$ norm of $\vec{u}$, which is in contradiction with $T_{\text{MAX}} < +\infty$.

Theorem 11.2 has been generalized to the setting of Besov spaces, with the condition $\vec{u} \in L^p \dot{B}_{\infty,\infty}^\sigma$ with $1 \leq p \leq +\infty$, $-1 \leq \sigma \leq +1$ and $\frac{2}{p} = 1 + \sigma$. The case $2 < p < +\infty$ was treated by Kozono and Shimada [274]; the case $1 < p \leq 2$ may be found in the paper by Chen and Zhang [116]; the case $p = +\infty$ has been first discussed by May [354] as a generalization of the result of Kozono and Sohr (see also the more recent paper of Cheskidov and Shvydkoy [120]). The case $p = 1$ goes back to the criterion of Beale, Kato and Majda [27] which stated $T_{\text{MAX}} < +\infty \Rightarrow \int_0^{T_{/rmMAX}} \|\operatorname{curl}\vec{u}\|_\infty \, dt = +\infty$; the L^∞ norm was replaced by the weaker norm $\|\operatorname{curl}\vec{u}\|_{BMO}$ by Kozono and Taniuchi [278], then by the still weaker norm $\|\operatorname{curl}\vec{u}\|_{\dot{B}_{\infty,\infty}^0}$ by Kozono, Ogawa and Taniuchi [273].

Serrin's criterion and Besov spaces

Theorem 11.3.
Let $\vec{u}_0 \in H^1$ with $\operatorname{div}\vec{u}_0 = 0$ and $\vec{f} \in L^2((0, +\infty), L^2)$. Let $\vec{u}$ be a solution of

$$\partial_t\vec{u} = \nu\Delta\vec{u} + \mathbb{P}(\vec{f} - \operatorname{div}(\vec{u} \otimes \vec{u}))$$

with $\vec{u} \in \mathcal{C}([0,T], H^1) \cap L^2((0,T), H^2)$ for all $T < T_{\text{MAX}}$ and $\vec{u}(0,.) = \vec{u}_0$. Then

- *if $1 < p < +\infty$, $-1 < \sigma < 1$ and $\frac{2}{p} = 1 + \sigma$, we have*

$$\|\vec{u}(T,.)\|_{\dot{H}^1}^2 \leq (\|\vec{u}_0\|_{\dot{H}^1}^2 + \frac{1}{\nu}\|\vec{f}\|_{L^2L^2}^2)e^{C_0\nu^{1-p}\int_0^T \|\vec{u}\|_{\dot{B}_{\infty,\infty}^\sigma}^p \, dt} \tag{11.12}$$

where the constant C_0 does not depend on T.
In particular, if the maximal existence time T_{MAX} satisfies $T_{\text{MAX}} < +\infty$, then $\int_0^{T_{\text{MAX}}} \|\vec{u}\|_{\dot{B}_{\infty,\infty}^\sigma}^p \, dt = +\infty$.

- *case $p = +\infty$: there exists a positive constant γ such that, if $T_{\mathrm{MAX}} < +\infty$ and if $\sup_{0 < t < T_{\mathrm{MAX}}} \|\vec{u}(t,.)\|_{\dot{B}^{-1}_{\infty,\infty}} < +\infty$, then $\limsup_{t \to T^-_{\mathrm{MAX}}} \|\vec{u}(t,.) - \vec{u}(T_{\mathrm{MAX}},.)\|_{\dot{B}^{-1}_{\infty,\infty}} \geq \gamma\nu$.*

- *case $p = 1$: we have*

$$\|\vec{u}(T,.)\|^2_{\dot{H}^1} \leq (\|\vec{u}_0\|^2_{\dot{H}^1} + \frac{1}{\nu}\|\vec{f}\|^2_{L^2 L^2})e^{C_0 \int_0^T \|\vec{u}\|_{\dot{F}^1_{\infty,2}} dt} \tag{11.13}$$

where the constant C_0 does not depend on T.
In particular, if the maximal existence time T_{MAX} satisfies $T_{\mathrm{MAX}} < +\infty$, then $\int_0^{T_{\mathrm{MAX}}} \|\operatorname{curl} \vec{u}\|_{BMO}\, dt = +\infty$.

- *case $p = 1$ (continued): if $\vec{f} \in L^2 H^2$, then $\vec{u} \in \mathcal{C}((0, T_{\mathrm{MAX}}, H^3)$. Moreover we have, for any $\delta \in (0, T_{\mathrm{MAX}})$ and $\delta < T < T_{\max}$*

$$\|\vec{u}(T,.)\|^2_{\dot{H}^3} \leq C_0 e^{(\|\vec{u}(\delta,)\|^2_{H^3} + \frac{1}{\nu}\|\vec{f}\|^2_{L^2 H^2})e^{C_0 \int_\delta^T \|\vec{u}\|_{\dot{B}^1_{\infty,\infty}} dt}} \tag{11.14}$$

where the constant C_0 does not depend on T.
In particular, if the maximal existence time T_{MAX} satisfies $T_{\mathrm{MAX}} < +\infty$, then $\int_0^{T_{\mathrm{MAX}}} \|\operatorname{curl} \vec{u}\|_{\dot{B}^0_{\infty,\infty}}\, dt = +\infty$.

Proof. The proof is based on the Littlewood–Paley decomposition[1] and on the use of para-products. If u and v belong to L^2, we write

$$uv = \sum_{j \in \mathbb{Z}} S_{j-2} u \Delta_j v + \sum_{j \in \mathbb{Z}} S_{j-2} v \Delta_j u + \sum_{j \in \mathbb{Z}} \sum_{|k-j| \leq 2} \Delta_j u \Delta_k v$$

where $\Delta_j u$ is the j-th dyadic block of the Littlewood–Paley decomposition of $\vec{u}$: the Fourier transform $\mathcal{F}(\Delta_j u)$ is given by

$$\mathcal{F}(\Delta_j u)(\xi) = \psi(\frac{\xi}{2^j})\hat{u}(\xi)$$

where ψ is a smooth function supported in $\{\xi\ /\ \frac{1}{2} \leq |\xi| \leq 2\}$ and such that, for $\xi \neq 0$, $\sum_{j \in \mathbb{Z}} \psi(\frac{\xi}{2^j})) = 1$, while $S_j u = \sum_{k<j} \Delta_k u$. The term $\pi(u, v) = \sum_{j \in \mathbb{Z}} S_{j-2} u \Delta_j v$ is called the *paraproduct* of u and v; we shall write $R(u, v) = +\sum_{j \in \mathbb{Z}} \sum_{|k-j| \leq 2} \Delta_j u \Delta_k v$, so that

$$uv = \pi(u, v) + \pi(v, u) + R(u, v). \tag{11.15}$$

The important point is that the constituents of $\pi(u, v)$ and of $R(u, v)$ are localized in frequency variable: the support of $\mathcal{F}(S_{j-2} u \Delta_j v)$ is contained in $\{\xi\ /\ \frac{1}{4}2^j \leq |\xi| \leq \frac{9}{4}2^j\}$ while, for $|k - j| \leq 2$, the support of $\mathcal{F}(\Delta_j u \Delta_k v)$ is contained in $\{\xi\ /\ |\xi| \leq 10\,2^j\}$.

---

[1] See Lemarié-Rieusset [313] or Bahouri, Chemin and Danchin [15] for definitions and notations.

Case $1 < p < +\infty$:

We start from

$$\frac{d}{dt}\int |\vec{\nabla}\otimes\vec{u}(t,x)|^2\,dx = -2\nu\int |\Delta\vec{u}|^2\,dx - 2\int \Delta\vec{u}\cdot\vec{f}\,dx$$

$$-2\sum_{i=1}^{3}\int \partial_i\vec{u}\cdot((\partial_i\vec{u})\cdot\vec{\nabla})\vec{u}\,dx \tag{11.16}$$

and we estimate $\int \partial_i u\,\partial_j v\,\partial_k w\,dx$ for $u,v,w \in \dot{H}^1\cap\dot{H}^2\cap\dot{B}^\sigma_{\infty,\infty}$, with $-1 < \sigma < 1$. Let $2r = 3 - \sigma$, so that $1 < r < 2$. We then write

$$|\int \partial_i u\,\partial_j v\,\partial_k w\,dx| \leq C\|\partial_i u\|_{\dot{H}^{r-1}}(\|\pi(\partial_j v,\partial_k w)\|_{\dot{H}^{1-r}} + \|\pi(\partial_k w,\partial_j v)\|_{\dot{H}^{1-r}})$$

$$+ C\|\partial_i u\|_{\dot{B}^{\sigma-1}_{\infty,\infty}}\|R(\partial_j v,\partial_k w)\|_{\dot{B}^{1-\sigma}_{1,1}}.$$

We have $1 - r = r + \sigma - 2$ and $\sigma - 1 < 0$ so that

$$\|\pi(\partial_j v,\partial_k w)\|_{\dot{H}^{1-r}}^2 \leq C\sum_{j\in\mathbb{Z}} 2^{2j(r+\sigma-2)}\|S_{j-2}(\partial_j v)\|_\infty^2\|\Delta_j(\partial_k w)\|_2^2$$

$$\leq C\sup_{j\in\mathbb{Z}} 2^{2j(\sigma-1)}\|S_{j-2}(\partial_j v)\|_\infty^2 \sum_{j\in\mathbb{Z}} 2^{2j(r-1)}\|\Delta_j(\partial_k w)\|_2^2$$

$$\leq C'\|\partial_j v\|_{\dot{B}^{\sigma-1}_{\infty,\infty}}^2\|\partial_k w\|_{\dot{H}^{r-1}}^2$$

$$\leq C'\|v\|_{\dot{B}^\sigma_{\infty,\infty}}^2\|w\|_{\dot{H}^r}^2.$$

We have, of course, the similar estimate

$$\|\pi(\partial_k w,\partial_j v)\|_{\dot{H}^{1-r}}^2 \leq C\|w\|_{\dot{B}^\sigma_{\infty,\infty}}^2\|v\|_{\dot{H}^r}^2.$$

On the other hand, we have $1 - \sigma = 2r - 2$ and $1 - \sigma > 0$, so that

$$\|R(\partial_j v,\partial_k w)\|_{\dot{B}^{1-\sigma}_{1,1}} \leq C\sum_{j\in\mathbb{Z}} 2^{j(2r-2)}\|\sum_{|k-j|\leq 2}\Delta_j(\partial_j v)\Delta_k(\partial_k w)\|_1$$

$$\leq C'(\sum_{j\in\mathbb{Z}} 2^{2j(r-1)}\|\Delta_j(\partial_j v)\|_2^2)^{1/2}(\sum_{j\in\mathbb{Z}} 2^{2j(r-1)}\|\Delta_j(\partial_k w)\|_2^2)^{1/2}$$

$$\leq C''\|\partial_j v\|_{\dot{H}^{r-1}}\|\partial_k w\|_{\dot{H}^{r-1}}$$

$$\leq C''\|v\|_{\dot{H}^r}\|w\|_{\dot{H}^r}.$$

Thus, we find

$$|2\sum_{i=1}^{3}\int \partial_i\vec{u}.((\partial_i\vec{u})\cdot\vec{\nabla})\vec{u}\,dx| \leq C\|\vec{u}\|_{\dot{B}^\sigma_{\infty,\infty}}\|\vec{u}\|_{\dot{H}^r}^2$$

$$\leq C\|\vec{u}\|_{\dot{B}^\sigma_{\infty,\infty}}\|\vec{u}\|_{\dot{H}^1}^{4-2r}\|\vec{u}\|_{\dot{H}^2}^{2r-2} \tag{11.17}$$

$$= C\|\vec{u}\|_{\dot{B}^\sigma_{\infty,\infty}}\|\vec{u}\|_{\dot{H}^1}^{1+\sigma}\|\vec{u}\|_{\dot{H}^2}^{1-\sigma}$$

This gives

$$\frac{d}{dt}\|\vec{u}\|_{\dot{H}^1}^2 \leq \frac{1}{\nu}\|\vec{f}\|_2^2 + C_\sigma\nu^{-\frac{1+\sigma}{1-\sigma}}\|\vec{u}\|_{\dot{B}^\sigma_{\infty,\infty}}^{\frac{2}{1-\sigma}}\|\vec{u}\|_{\dot{H}^1}^2 = \frac{1}{\nu}\|\vec{f}\|_2^2 + C_\sigma\nu^{1-p}\|\vec{u}\|_{\dot{B}^\sigma_{\infty,\infty}}^p\|\vec{u}\|_{\dot{H}^1}^2$$

and we conclude by Grönwall's lemma.

Case $p = +\infty$:

For $\sigma = -1$, we get a similar estimate

$$|2\sum_{i=1}^{3}\int \partial_i \vec{u} \cdot ((\partial_i \vec{u}) \cdot \vec{\nabla})\vec{u}\, dx| \le C\|\vec{u}\|_{\dot{B}^{-1}_{\infty,\infty}}\|\vec{u}\|^2_{\dot{H}^2} \tag{11.18}$$

but the end of the proof would work only if $\vec{u}$ was small enough to grant that

$$(-\nu + C\|\vec{u}\|_{\dot{B}^{-1}_{\infty,\infty}})\|\Delta \vec{u}\|^2_2 < 0.$$

In Theorem 11.3, $\vec{u}$ is not assumed to be small, but only to have a small jump at time $t = T_{\text{MAX}}$. Let us make this statement more precise. First, we assume that

$$\sup_{0 < t < T_{\text{MAX}}} \|\vec{u}(t, .)\|_{\dot{B}^{-1}_{\infty,\infty}} < +\infty.$$

We shall see in Theorem 12.2 that the Navier–Stokes problem with initial value $\vec{u}_0$ and forcing term $\vec{f}$ admits global weak Leray solutions; moreover, from Theorem 12.3, those weak solutions will coincide with $\vec{u}$ on $(0, T_{\text{MAX}})$.

In particular, the map $t \mapsto \vec{u}(t, .)$ can be extended as a map from $[0, +\infty)$ to L^2 which is weakly continuous. Thus, $\vec{u}(t, .)$ has a limit $\vec{u}(T_{\text{MAX}}, .)$ when $t \to T^-_{\text{MAX}}$ (if $T_{\text{MAX}} < +\infty$). Moreover, as $\dot{B}^{-1}_{\infty,\infty}$ is a dual space, we find that $\vec{u}(T_{\text{MAX}}, .) \in \dot{B}^{-1}_{\infty,\infty}$ and

$$\|\vec{u}(T_{\text{MAX}}, .)\|_{\dot{B}^{-1}_{\infty,\infty}} \le \liminf_{t \to T^-_{\text{MAX}}} \|\vec{u}(t, .)\|_{\dot{B}^{-1}_{\infty,\infty}}.$$

Now, we want to prove that, if $T_{\text{MAX}} < +\infty$, then

$$\limsup_{t \to T^-_{\text{MAX}}} \|\vec{u}(t, .) - \vec{u}(T_{\text{MAX}}, .)\|_{\dot{B}^{-1}_{\infty,\infty}} \ge \gamma\nu.$$

Indeed, let $\epsilon = \limsup_{t \to T^-_{\text{MAX}}} \|\vec{u}(t, .) - \vec{u}(T_{\text{MAX}}, .)\|_{\dot{B}^{-1}_{\infty,\infty}}$, and let $\eta > \epsilon$. There is an interval $[T_0, T_{\text{MAX}})$ on which $\|\vec{u}(t, .) - \vec{u}(T_{\text{MAX}}, .)\|_{\dot{B}^{-1}_{\infty,\infty}} < \eta$. Moreover, chossing $T_0 > 0$, $\vec{u}(T_0, .) \in H^2 \subset L^\infty$, while $\|\vec{u}(t, .) - \vec{u}(T_0, .)\|_{\dot{B}^{-1}_{\infty,\infty}} < 2\eta$ on (T_0, T_{MAX}).

The next step is to estimate $\int \partial_i u \partial_j v \partial_k w\, dx$ for $u, v, w \in \dot{H}^1 \cap \dot{H}^2$, with $u = u_1 + u_2$, $u_1 \in L^\infty$ and $u_2 \in \dot{B}^{-1}_{\infty,\infty}$, and the same for $v = v_1 + v_2$ and $w = w_1 + w_2$. We write

$$\begin{aligned}
|\int \partial_i u \partial_j v \partial_k w\, dx| \le\; & C\|\partial_i u\|_2 (\|\pi(\partial_j v_1, \partial_k w)\|_2 + \|\pi(\partial_k w_1, \partial_j v)\|_2) \\
& + C\|\partial_i u\|_{\dot{H}^1} (\|\pi(\partial_j v_2, \partial_k w)\|_{\dot{H}^{-1}} + \|\pi(\partial_k w_2, \partial_j v)\|_{\dot{H}^{-1}}) \\
& + C\|u_1\|_\infty \|\partial_i R(\partial_j v, \partial_k w)\|_1 \\
& + C\|\partial_i u_2\|_{\dot{B}^{-2}_{\infty,\infty}} \|R(\partial_j v, \partial_k w)\|_{\dot{B}^2_{1,1}} \\
\le\; & C'\|u\|_{\dot{H}^1}(\|v_1\|_\infty \|w\|_{\dot{H}^2} + \|w_1\|_\infty \|v\|_{\dot{H}^2}) \\
& + C'\|u\|_{\dot{H}^2}(\|v_2\|_{\dot{B}^{-1}_{\infty,\infty}} \|w\|_{\dot{H}^2} + \|w_2\|_{\dot{B}^{-1}_{\infty,\infty}} \|v\|_{\dot{H}^2}) \\
& + C'\|u_1\|_\infty (\|v\|_{\dot{H}^1} \|w\|_{\dot{H}^2} + \|v\|_{\dot{H}^2} \|w\|_{\dot{H}^1}) \\
& + C'\|u_2\|_{\dot{B}^{-1}_{\infty,\infty}} \|v\|_{\dot{H}^2} \|w\|_{\dot{H}^2}.
\end{aligned}$$

Thus, we find on (T_0, T_{MAX}):

$$|2\sum_{i=1}^{3}\int \partial_i \vec{u} \cdot ((\partial_i \vec{u}) \cdot \vec{\nabla})\vec{u}\, dx| \le C_0 \eta \|\vec{u}\|^2_{\dot{H}^2} + C\|\vec{u}(T_0, .)\|_\infty \|\vec{u}\|_{\dot{H}^1} \|\vec{u}\|_{\dot{H}^2}$$

and thus

$$\frac{d}{dt}\|\vec{u}\|_{\dot{H}^1}^2 \le \frac{1}{\nu}\|\vec{f}\|_2^2 + C\frac{1}{\nu}\|\vec{u}(T_0,.)\|_\infty^2\|\vec{u}\|_{\dot{H}^1}^2 + (-\frac{\nu}{2} + C_0\eta)\|\vec{u}\|_{\dot{H}^2}^2.$$

If $\eta < \frac{\nu}{2C_0}$, we may apply Grönwall's lemma and get that $\vec{u}$ remains bounded in $\dot{H}^1$, which contradicts $T_{\mathrm{MAX}} < +\infty$.

Case $p = 1$:
For $\sigma = 1$, both estimates

$$\|\pi(\partial_j v, \partial_k w)\|_2^2 \le C\|\partial_j v\|_{\dot{B}_{\infty,\infty}^0}^2 \|\partial_k w\|_2^2$$

and

$$\|R(\partial_j v, \partial_k w)\|_{\dot{B}_{1,1}^0} \le C\|\partial_j v\|_2 \|\partial_k w\|_2.$$

fail. However, we may use the div-curl lemma of Coifman, Lions Meyer and Semmes [124] since div $\vec{u} = 0$ and write

$$\|\partial_j \vec{u} \cdot \vec{\nabla}\vec{u}\|_{\mathcal{H}^1} \le C\|\vec{u}\|_{\dot{H}^1}^2$$

where $\mathcal{H}^1$ is the Hardy space (whose dual is $BMO = \dot{F}_{\infty,2}^0$). Thus, we get

$$|2\sum_{i=1}^3 \int \partial_i \vec{u} \cdot ((\partial_i \vec{u}) \cdot \vec{\nabla})\vec{u} \, dx| \le C\|\vec{u}\|_{\dot{F}_{\infty,2}^1}\|\vec{u}\|_{\dot{H}^2}^2 \tag{11.19}$$

and we conclude by Grönwall's lemma.

Case $p = 1$ (continued):
Let $\vec{f} \in L^2 H^2$. For every $0 < \delta < T < T_{\mathrm{MAX}}$, we know that $\vec{u}$ will belong to $\mathcal{C}([T_0, T], H^3) \cap L^2((T_0, T), H^4)$.

We want to estimate

$$\|\vec{u}\|_{\dot{H}^3}^2 = \int |(-\Delta)^{3/2}\vec{u}|^2 \, dx.$$

We write

$$\begin{aligned}
\frac{d}{dt}(\|\vec{u}\|_{\dot{H}^3}^2) &= 2\langle(-\Delta)^3\vec{u}|\partial_t\vec{u}\rangle_{H^{-3}, H^3} \\
&= -2\nu\|\vec{u}\|_{\dot{H}^4}^2 - 2\langle(-\Delta)^3\vec{u}|\vec{u}\cdot\vec{\nabla}\vec{u}\rangle_{H^{-2}, H^2} \\
&\quad + 2\langle(-\Delta)^3\vec{u}|\vec{f}\rangle_{H^{-2}, H^2}
\end{aligned} \tag{11.20}$$

We have $(-\Delta)^3 = \sum_{|\alpha|=3} c_\alpha \partial^\alpha \partial^\alpha$. Integration by parts and Leibnitz rule give then

$$\int \partial^\alpha \partial^\alpha \vec{u} \cdot (\vec{u} \cdot \vec{\nabla}\vec{u}) \, dx = \sum_{\beta+\gamma=\alpha} c_{\beta,\gamma} \int \partial^\alpha \vec{u} \cdot (\partial^\beta \vec{u} \cdot \vec{\nabla}\partial^\gamma \vec{u}) \, dx.$$

As div $\vec{u} = 0$, we have

$$\int \partial^\alpha \vec{u} \cdot (\vec{u} \cdot \vec{\nabla}\partial^\alpha \vec{u}) \, dx = 0.$$

Thus, we need to estimate integrals

$$I_{\alpha,\beta,\gamma}(u,v,w) = \int \partial^\alpha u \, \partial^\beta v \, \partial^\gamma w \, dx$$

with $|\alpha| = 3$, $|\beta| + |\gamma| = 4$ and $|\beta| \ge 1$, $|\gamma \ge 1$, for $u, v, w \in \dot{H}^3 \cap \dot{B}_{\infty,\infty}^1$.

We have, for $|\beta| = 1$,

$$|I_{\alpha,\beta,\gamma}(u,v,w)| \leq \|u\|_{\dot{H}^3}\|\partial^\beta v\|_\infty \|w\|_{\dot{H}^3}.$$

For $|\beta| = 2$, $\partial^\beta = \partial_i \partial_j$, we use the Gagliardo–Nirenberg inequality $\|\partial^\beta v\|_4 \leq C\|\partial_i v\|_\infty^{1/2}\|\partial_i v\|_{\dot{H}^2}^{1/2}$. This inequality is easily established through Hedberg's inequality: write $\partial^\beta v = -\int_0^{+\infty} W_t * \Delta \partial^\beta v \, dt$; then write

$$|W_t * \Delta \partial^\beta v)(x)| \leq C \min(\frac{1}{\sqrt{t}}\mathcal{M}_{\Delta \partial_i v}(x), \frac{1}{t^{3/2}}\|\partial_i v\|_\infty)$$

to get

$$|\partial^\beta v(x)| \leq C\|\partial_i v\|_\infty^{1/2}\sqrt{\mathcal{M}_{\Delta \partial_i v}(x)}.$$

Thus, we get

$$|I_{\alpha,\beta,\gamma}(u,v,w)| \leq C\|u\|_{\dot{H}^3}\|\vec{\nabla} v\|_\infty^{1/2}\|v\|_{\dot{H}^3}\|\vec{\nabla} w\|_\infty^{1/2}\|w\|_{\dot{H}^3}.$$

Finally, we get

$$|\langle (-\Delta)^3 \vec{u} | \vec{u} \cdot \vec{\nabla} \vec{u} \rangle_{H^{-2}, H^2}| \leq C\|\vec{\nabla} \otimes \vec{u}\|_\infty \|\vec{u}\|_{\dot{H}^3}^2 \qquad (11.21)$$

We then use the logarithmic Sobolev inequality

$$\|\vec{\nabla} \otimes \vec{u}\|_\infty \leq C\left(\|\vec{u}\|_2 + 1 + \|\vec{u}\|_{\dot{B}^1_{\infty,\infty}} \ln(e + \|\vec{u}\|_{\dot{H}^3}^2)\right)$$

To establish this well-known inequality, just write $\partial_j v = -\int_0^{+\infty} W_t * \Delta \partial_j v \, dt$, with

$$|(W_t * \Delta \partial_j v)(x)| \leq C \min(\frac{1}{t^{3/4}}\|v\|_{\dot{H}^3}, \frac{1}{t}\|v\|_{\dot{B}^1_{\infty,\infty}}, \frac{1}{t^{9/4}}\|v\|_2);$$

if $\|v\|_{\dot{H}^3} \leq \|v\|_{\dot{B}^1_{\infty,\infty}}$, conclude by integrating $\int_0^1 \frac{1}{t^{3/4}}\|v\|_{\dot{H}^3}\, dt + \int_1^{+\infty} \frac{1}{t^{9/4}}\|v\|_2 \, dt$; if $\|v\|_{\dot{H}^3} > \|v\|_{\dot{B}^1_{\infty,\infty}}$, let $A = \left(\frac{\|v\|_{\dot{B}^1_{\infty,\infty}}}{\|v\|_{\dot{H}^3}}\right)^4$ and integrate

$$\int_0^A \frac{1}{t^{3/4}}\|v\|_{\dot{H}^3}\, dt + \int_A^1 \frac{1}{t}\|v\|_{\dot{B}^1_{\infty,\infty}}\, dt + \int_1^{+\infty} \frac{1}{t^{9/4}}\|v\|_2 \, dt.$$

We obtain $\|\partial_j v\|_\infty \leq C(\|v\|_2 + \|v\|_{\dot{B}^1_{\infty,\infty}}(1 + \ln^+\left(\frac{\|v\|_{\dot{H}^3}}{\|v\|_{\dot{B}^1_{\infty,\infty}}}\right))$; finally, if $\|v\|_{\dot{B}^1_{\infty,\infty}} \geq 1$, write $\ln^+\left(\frac{\|v\|_{\dot{H}^3}}{\|v\|_{\dot{B}^1_{\infty,\infty}}}\right) \leq \ln(e + \|v\|_{\dot{H}^3})$, if $\|v\|_{\dot{B}^1_{\infty,\infty}} \leq \min(1, \|v\|_{\dot{H}^3})$, write

$$\|v\|_{\dot{B}^1_{\infty,\infty}} \ln^+\left(\frac{\|v\|_{\dot{H}^3}}{\|v\|_{\dot{B}^1_{\infty,\infty}}}\right) = \|v\|_{\dot{B}^1_{\infty,\infty}}(\ln(\|v\|_{\dot{H}^3} - \ln(\|v\|_{\dot{B}^1_{\infty,\infty}}))$$

$$\leq \frac{1}{e} + \frac{1}{2}\|v\|_{\dot{B}^1_{\infty,\infty}} \ln(e + \|v\|_{\dot{H}^3}^2)$$

Thus far, we have obtained

$$\frac{d}{dt}(\|\vec{u}\|_{\dot{H}^3}^2) \leq \frac{1}{\nu}\|\vec{f}\|_{H^2}^2 + C(1 + \|\vec{u}\|_2)\|\vec{u}\|_{\dot{H}^3}^2 + C\|\vec{u}\|_{\dot{B}^1_{\infty,\infty}}\|\vec{u}\|_{\dot{H}^3}^2 \ln(e + \|\vec{u}\|_{\dot{H}^3}^2)$$

If $\Phi(t) = \ln(e + \|\vec{u}\|_{\dot{H}^3}^2)$, we obtain

$$\frac{d}{dt}\Phi \leq \frac{1}{\nu}\|\vec{f}\|_{H^2}^2 + C(1 + \|\vec{u}\|_2) + C\|\vec{u}\|_{\dot{B}^1_{\infty,\infty}}\Phi(t).$$

We then conclude by applying Grönwall's lemma. $\square$

In spite of the maximality of the Besov spaces $\dot{B}^\sigma_{\infty,\infty}$ (if E is a Banach space of distri-butions such that its norm if shift-invariant [$\|f(x-x_0)\|_E = \|f\|_E$] and homogeneous [$\|f(\lambda x)\|_E = \lambda^\sigma\|f\|_E$], then $E \subset \dot{B}^\sigma_{\infty,\infty}$), Theorem 11.3 may still be extended to some criteria based on weaker norms. For instance, Planchon [400] discussed how to replace the norm in $L^1\dot{B}^1_{\infty,\infty}$ ($\|\vec{u}\|_{L^1((0,T),\dot{B}^1_{\infty,\infty})} = \int_0^T \sup_{j\in\mathbb{Z}} 2^j\|\Delta_j\vec{u}\|_\infty\,dt$) by the norm in $\tilde{L}^1\dot{B}^1_{\infty,\infty}$ ($\|\vec{u}\|_{\tilde{L}^1((0,T),\dot{B}^1_{\infty,\infty})} = \sup_{j\in\mathbb{Z}} 2^j \int_0^T \|\Delta_j\vec{u}\|_\infty\,dt$). Spaces $\tilde{L}^p\dot{B}^s_{q,r}$ often occur in critical esti-mates for the Navier–Stokes equations, since the seminal paper of Chemin and Lerner [113] on quasi-Lipschitz flows.

Another way of extending Serrin's criterion is the remark by Montgomery-Smith [370] that the proof of Theorem 11.3 is based on the Grönwall lemma applied to (sub)linear estimates on the $\dot{H}^1$ norm of $\vec{u}$, while Grönwall's lemma applies to a slightly larger class of estimates:

Lemma 11.1 (Grönwall's lemma).
If $u \geq 0$ is defined on $[0,T)$ and satisfies

$$u(t) \leq a_0 + \int_0^t \Phi(u(s))\omega(s)\,ds$$

with $\Phi \geq 0$ a non-decreasing function such that

$$\int_1^{+\infty} \frac{dt}{\Phi(t)} = +\infty$$

and $\omega \geq 0$ with $\omega \in L^1((0,T))$, then $\sup_{0<t<T} u(t) \leq A^*$, where A^* is defined by $\int_{\max(1,a_0)}^{A^*} \frac{ds}{\Phi(s)} = \int_0^T \omega(s)\,ds$.

Proof. Let $a_1 = \max(1,a_0)$ and define $A(t) = a_1 + \int_0^t \Phi(u(s))\omega(s)\,ds$. We have $\frac{d}{dt}A = \omega\Phi(u) \leq \omega\Phi(A)$, so that $\int_0^t \frac{\frac{d}{dt}A}{\Phi(A)}\,ds \leq \int_0^t \omega(s)\,ds$. We write $\int_0^t \frac{\frac{d}{dt}A}{\Phi(A)}\,ds = \int_{a_1}^{A(t)} \frac{ds}{\Phi(s)}$. Thus, we find $\sup_{0<t<T} u(t) \leq \sup_{0<t<T} A(t) \leq A^*$. $\qquad\square$

We may now state Montgomery–Smith's result:

Proposition 11.3.
Let $1 < p < +\infty$ and $\sigma \in (-1,1)$ with $\frac{2}{p} = 1+\sigma$. Let $\Theta \geq 1$ be a non-decreasing function on $(0,+\infty)$ such that $\int_1^{+\infty} \frac{ds}{s\Theta(s)}\,ds = +\infty$. Let $k \in \{0,1,2\}$ with $k > \sigma + \frac{1}{2}$.

Let $\vec{u}_0 \in H^1$ with $\operatorname{div} \vec{u}_0 = 0$ and $\vec{f} \in L^2((0,+\infty), H^k)$. Let $\vec{u}$ be a solution of

$$\partial_t\vec{u} = \nu\Delta\vec{u} + \mathbb{P}(\vec{f} - \operatorname{div}(\vec{u}\otimes\vec{u}))$$

with $\vec{u} \in \mathcal{C}([0,T],H^1) \cap L^2((0,T),H^2)$ for all $T < T_{\mathrm{MAX}}$ and $\vec{u}(0,.) = \vec{u}_0$. If the maximal existence time T_{MAX} satisfies $T_{\mathrm{MAX}} < +\infty$, then $\int_0^{T_{\mathrm{MAX}}} \frac{\|\vec{u}\|_{\dot{B}^\sigma_{\infty,\infty}}^p}{\Theta(\|\vec{u}\|_{\dot{B}^\sigma_{\infty,\infty}})}\,dt = +\infty$.

Proof. For every $0 < T_0 < T < T_{\mathrm{MAX}}$, we know that $\vec{u}$ will belong to $\mathcal{C}([T_0,T],H^{k+1}) \cap L^2((T_0,T),H^{k+2})$.

We start from

$$\|\vec{u}\|_{\dot{H}^{k+1}}^2 = \int |(-\Delta)^{(k+1)/2}\vec{u}|^2\,dx$$

so that

$$
\begin{aligned}
\frac{d}{dt}(\|\vec{u}\|_{\dot{H}^{k+1}}^2) &= 2\langle(-\Delta)^{k+1}\vec{u}\,|\,\partial_t\vec{u}\rangle_{H^{-k},H^k}\\
&= -2\nu\|\vec{u}\|_{\dot{H}^{k+2}}^2 - 2\langle(-\Delta)^{k+1}\vec{u}\,|\,\vec{u}\cdot\vec{\nabla}\vec{u}\rangle_{H^{-k},H^k}\\
&\quad + 2\langle(-\Delta)^{k+1}\vec{u}\,|\,\vec{f}\rangle_{H^{-k},H^k}
\end{aligned}
\tag{11.22}
$$

Integration by parts and Leibnitz rule give then, for $|\alpha| = k+1$,

$$
\int \partial^\alpha \partial^\alpha \vec{u} \cdot (\vec{u}\cdot\vec{\nabla}\vec{u})\, dx = \sum_{\beta+\gamma=\alpha} c_{\beta,\gamma} \int \partial^\alpha \vec{u}\cdot(\partial^\beta\vec{u}\cdot\vec{\nabla}\partial^\gamma\vec{u})\, dx
$$

As $\operatorname{div}\vec{u} = 0$, we have

$$
\int \partial^\alpha \vec{u}\cdot(\vec{u}\cdot\vec{\nabla}\partial^\alpha\vec{u})\, dx = 0.
$$

Thus, we need to estimate integrals (in the sense of duality brackets)

$$
I_{\alpha,\beta,\gamma}(u,v,w) = \int \partial^\alpha u\,\partial^\beta v\,\partial^\gamma w\, dx
$$

with $|\alpha| = k+1$, $|\beta| + |\gamma| = k+2$ and $|\beta| \geq 1$, $|\gamma| \geq 1$, for $u,v,w \in \dot{H}^{k+1} \cap \dot{H}^{k+2} \cap \dot{B}_{\infty,\infty}^\sigma$, with $-1 < \sigma < 1$.

Let $2r = 2k+3-\sigma$, so that $k+1 < r < k+2$. We then write

$$
\begin{aligned}
|I_{\alpha,\beta,\gamma}(u,v,w)| &\leq C\|\partial^\alpha u\|_{\dot{H}^{r-k-1}}(\|\pi(\partial^\beta v,\partial^\gamma w)\|_{\dot{H}^{1+k-r}} + \|\pi(\partial^\gamma w,\partial_\beta v\|_{\dot{H}^{1+k-r}})\\
&\quad + C\|\partial^\alpha u\|_{\dot{B}_{\infty,\infty}^{\sigma-k-1}}\|R(\partial^\beta v,\partial^\gamma w)\|_{\dot{B}_{1,1}^{1+k-\sigma}}.
\end{aligned}
$$

We have $1+k-r = r+\sigma-k-2$ and $\sigma-|\beta| < 0$ so that

$$
\begin{aligned}
\|\pi(\partial^\beta v,\partial^\gamma w)\|_{\dot{H}^{1+k-r}}^2 &\leq C\sum_{j\in\mathbb{Z}} 2^{2j(r+\sigma-k-2)}\|S_{j-2}(\partial^\beta v)\|_\infty^2\|\Delta_j(\partial^\gamma w)\|_2^2\\
&\leq C\sup_{j\in\mathbb{Z}} 2^{2j(\sigma-|\beta|)}\|S_{j-2}(\partial^\beta v)\|_\infty^2 \sum_{j\in\mathbb{Z}} 2^{2j(r-|\gamma|)}\|\Delta_j(\partial^\gamma w)\|_2^2\\
&\leq C'\|\partial^\beta v\|_{\dot{B}_{\infty,\infty}^{\sigma-|\beta|}}^2\|\partial^\gamma w\|_{\dot{H}^{r-|\gamma|}}^2\\
&\leq C'\|v\|_{\dot{B}_{\infty,\infty}^\sigma}^2\|w\|_{\dot{H}^r}^2.
\end{aligned}
$$

We have, of course, the similar estimate

$$
\|\pi(\partial^\gamma w,\partial^\beta v)\|_{\dot{H}^{1+k-r}}^2 \leq C\|w\|_{\dot{B}_{\infty,\infty}^\sigma}^2\|v\|_{\dot{H}^r}^2.
$$

On the other hand, we have $1+k-\sigma = 2r-k-2$ and $1+k-\sigma > 0$, so that

$$
\begin{aligned}
\|R(\partial^\beta v,\partial^\gamma w)\|_{\dot{B}_{1,1}^{1+k-\sigma}} &\leq C\sum_{j\in\mathbb{Z}} 2^{j(2r-k-2)}\Big\|\sum_{|k-j|\leq 2}\Delta_j(\partial^\beta v)\Delta_k(\partial^\gamma w)\Big\|_1\\
&\leq C'\Big(\sum_{j\in\mathbb{Z}} 2^{2j(r-|\beta|)}\|\Delta_j(\partial^\beta v)\|_2^2\Big)^{1/2}\Big(\sum_{j\in\mathbb{Z}} 2^{2j(r-|\gamma|)}\|\Delta_j(\partial^\gamma w)\|_2^2\Big)^{1/2}\\
&\leq C''\|\partial^\beta v\|_{\dot{H}^{r-|\beta|}}\|\partial^\gamma w\|_{\dot{H}^{r-|\gamma|}}\\
&\leq C''\|v\|_{\dot{H}^r}\|w\|_{\dot{H}^r}.
\end{aligned}
$$

This gives

$$\frac{d}{dt}\|\vec{u}\|_{\dot{H}^{k+1}}^2 \le \frac{1}{\nu}\|\vec{f}\|_{\dot{H}^k}^2 + C_\sigma \nu^{1-p}\|\vec{u}\|_{\dot{B}^\sigma_{\infty,\infty}}^p \|\vec{u}\|_{\dot{H}^{k+1}}^2$$

Next, we write

$$\|\vec{u}\|_{\dot{B}^\sigma_{\infty,\infty}} \le D_\sigma \|\vec{u}\|_{H^{k+1}} \le D_\sigma (\|\vec{u}\|_2 + \|\vec{u}\|_{\dot{H}^{k+1}}).$$

If $T_{\mathrm{MAX}} < +\infty$ and $A_0 = \sup_{0<t<T_{\mathrm{MAX}}} \|\vec{u}(t,.)\|_2$, we find that

$$\|\vec{u}\|_{\dot{B}^\sigma_{\infty,\infty}} \le 2D_\sigma \frac{A_0^2 + \|\vec{u}\|_{\dot{H}^{k+1}}^2}{A_0}$$

so that

$$\Theta(\|\vec{u}\|_{\dot{B}^\sigma_{\infty,\infty}}) \le \Theta(2D_\sigma \frac{A_0^2 + \|\vec{u}\|_{\dot{H}^{k+1}}^2}{A_0}).$$

Let us write $B(t) = 2D_\sigma \frac{A_0^2 + \|\vec{u}\|_{\dot{H}^{k+1}}^2}{A_0}$, for $0 < T_0 < t < T_{\mathrm{MAX}}$, We have

$$B(t) \le B(T_0) + \frac{2D_\sigma}{A_0\nu} \int_0^{T_{\mathrm{MAX}}} \|\vec{f}\|_{\dot{H}^k}^2 \, ds + C_\sigma \frac{\nu^{1-p}}{A_0} \int_{T_0}^t \frac{\|\vec{u}\|_{\dot{B}^\sigma_{\infty,\infty}}^p}{\Theta(\|\vec{u}\|_{\dot{B}^\sigma_{\infty,\infty}})} B(s)\Theta(B(s)) \, ds$$

and we conclude by Grönwall's lemma. $\qquad\square$

Montgomery–Smith's result paved the way to numerous works on "logarithmic improvements" of Serrin's criterion. Many of them were quite uninspired, but some of them were very interesting. For instance, we shall describe Chan and Vasseur's result [103] (in a slightly more general statement):

Proposition 11.4.
Let $1 < p < +\infty$, $3 < q < +\infty$ with $\frac{2}{p} + \frac{3}{q} = 1$. Let $\Theta \ge 1$ be a non-decreasing function on $(0, +\infty)$ such that $\int_1^{+\infty} \frac{ds}{s\Theta(s)} \, ds = +\infty$.

Let $\vec{u}_0 \in H^1$ with $\operatorname{div} \vec{u}_0 = 0$ and $\vec{f} \in L^2((0, +\infty), H^1)$. Let $\vec{u}$ be a solution of

$$\partial_t \vec{u} = \nu\Delta\vec{u} + \mathbb{P}(\vec{f} - \operatorname{div}(\vec{u} \otimes \vec{u}))$$

with $\vec{u} \in \mathcal{C}([0,T], H^1) \cap L^2((0,T), H^2)$ for all $T < T_{\mathrm{MAX}}$ and $\vec{u}(0,.) = \vec{u}_0$. If the maximal existence time T_{MAX} satisfies $T_{\mathrm{MAX}} < +\infty$, then $\int_0^{T_{\mathrm{MAX}}} \|\frac{\vec{u}}{\Theta(|\vec{u}|^{1/p})}\|_{\dot{M}^{2,q}}^p \, dt = +\infty$.

Proof. For every $0 < T_0 < T < T_{\mathrm{MAX}}$, we know that $\vec{u}$ will belong to $\mathcal{C}([T_0, T], H^2) \cap L^2((T_0, T), H^3)$. Thus, $\vec{u}$ is bounded on $(T_0, T) \times \mathbb{R}^3$. We are going to prove that

$$\begin{aligned}
\|\vec{u}(t,.)\|_\infty \le &C_\nu(\|\vec{u}(T_0,.)\|_\infty + (1 + \sqrt{T - T_0})(\int_{T_0}^t \|\vec{f}(s,.)\|_{\dot{H}^1}^2 \, ds)^{1/2}) \\
&+ C_\nu \int_{T_0}^T \|\vec{u}\|_{\dot{M}^{2(p+1)/p,(p+1)q/p}}^{p+1} \, ds
\end{aligned} \tag{11.23}$$

This inequality may be proved in a very simple way: we write

$$\vec{u}(t,.) = W_{\nu(t-T_0)} * \vec{u}(T_0,.) + \int_{T_0}^t W_{\nu(t-s)} * \mathbb{P}\vec{f} \, ds - \int_{T_0}^t W_{\nu(t-s)} * \mathbb{P}\operatorname{div}(\vec{u} \otimes \vec{u}) \, ds.$$

We have, on (T_0, T)

$$\sup_{T_0 < t < T} \|W_{\nu(t-T_0)} * \vec{u}(T_0, .)\|_\infty \leq \|\vec{u}(T_0, .)\|_\infty$$

and

$$\sup_{T_0 < t < T} \left\| \int_{T_0}^t W_{\nu(t-s)} * \mathbb{P}\vec{f}\, ds \right\|_\infty \leq C_\nu(1 + \sqrt{T-T_0})\left(\int_0^T \|\vec{f}\|_{H^1}^2\, ds \right)^{1/2}$$

The key point is the estimation of the last term

$$I(t) = \int_{T_0}^t W_{\nu(t-s)} * \mathbb{P}\operatorname{div}(\vec{u} \otimes \vec{u})\, ds.$$

We have

$$\|W_{\nu(t-s)} * \mathbb{P}\operatorname{div}(\vec{u} \otimes \vec{u})\|_\infty \leq C \frac{1}{\sqrt{\nu(t-s)}} \|\vec{u}(s, .)\|_\infty^2$$

and

$$\|W_{\nu(t-s)} * \mathbb{P}\operatorname{div}(\vec{u} \otimes \vec{u})\|_\infty \leq \frac{C}{(\nu(t-s))^{\frac{1}{2} + \frac{3p}{q(p+1)}}} \|\vec{u}(s, .)\|_{\dot{M}^{\frac{2(p+1)}{p}, \frac{(p+1)q}{p}}}^2$$

so that, for every positive A, we have

$$|I(t)| \leq C \int_{t-A}^t 1_{s>T_0} \frac{1}{\sqrt{\nu(t-s)}} \|\vec{u}(s, .)\|_\infty^2\, ds$$

$$+ C \int_{-\infty}^{t-A} 1_{s>T_0} \frac{1}{(\nu(t-s))^{\frac{1}{2} + \frac{3p}{q(p+1)}}} \|\vec{u}(s, .)\|_{\dot{M}^{2(p+1)/p, (p+1)q/p}}^2\, ds$$

$$\leq C_\nu \sqrt{A} \left(\sup_{T_0 < s < t} \|\vec{u}(s, .)\|_\infty \right)^2$$

$$+ C_\nu \left(\int_A^{+\infty} \frac{1}{(t-s)^{(\frac{1}{2} + \frac{3p}{q(p+1)})\frac{p+1}{p-1}}}\, ds \right)^{1-\frac{2}{p+1}} \left(\int_{T_0}^t \|\vec{u}(s, .)\|_{\dot{M}^{2(p+1)/p, (p+1)q/p}}^{p+1}\, ds \right)^{\frac{2}{p+1}}$$

The first thing now is to check that

$$\left(\frac{1}{2} + \frac{3p}{q(p+1)} \right) \frac{p+1}{p-1} > 1.$$

But, recalling that $\frac{3}{q} = 1 - \frac{2}{p}$, we find that $\left(\frac{1}{2} + \frac{3p}{q(p+1)} \right)\frac{p+1}{p-1} = \frac{3}{2}$. Thus, we find that

$$\sup_{T_0 < t < T} |I(t)| \leq C_\nu \sqrt{A} \left(\sup_{T_0 < t < T} \|\vec{u}\|_\infty^2 + A^{-\frac{p}{p+1}} \|\vec{u}\|_{L^{p+1}\dot{M}^{2(p+1)/p, (p+1)q/p}}^2 \right).$$

Optimizing the choice of A, we get

$$\sup_{T_0 < t < T} |I(t)| \leq C_\nu \left(\sup_{T_0 < t < T} \|\vec{u}\|_\infty \right)^{\frac{p-1}{p}} \left(\|\vec{u}\|_{L^{p+1}\dot{M}^{2(p+1)/p, (p+1)q/p}} \right)^{\frac{p+1}{p}}$$

so that, by Young's inequality,

$$\sup_{T_0 < t < T} |I(t)| \leq \frac{1}{2} \sup_{T_0 < t < T} \|\vec{u}\|_\infty + C_\nu' \left(\|\vec{u}\|_{L^{p+1}\dot{M}^{2(p+1)/p, (p+1)q/p}} \right)^{p+1}$$

Thus, we have proved (11.23).

We now easily finish the proof. Recall that

$$\|u\|_{\dot{M}^{2(p+1)/p,(p+1)q/p}} \approx \sup_{x\in\mathbb{R}^3,\rho>0} \Big(\frac{1}{\rho^{3(1-\frac{2}{q})}}\int_{B(x,\rho)} |u(y)|^{2(p+1)/p}\,dy\Big)^{p/(2(p+1))}$$

and

$$\|v\|_{\dot{M}^{2,q}} \approx \sup_{x\in\mathbb{R}^3,\rho>0} \Big(\frac{1}{\rho^{3(1-\frac{2}{q})}}\int_{B(x,\rho)} |u(y)|^2\,dy\Big)^{1/2}$$

In particular, writing $v = \frac{|u|}{\Theta(|u|)^{1/p}}$, we find

$$\|u\|_{\dot{M}^{\frac{2(p+1)}{p},\frac{(p+1)q}{p}}} \approx \sup_{x\in\mathbb{R}^3,\rho>0} \Big(\frac{1}{\rho^{3(1-\frac{2}{q})}}\int_{B(x,\rho)} |u|^{2/p}\Theta(|u|)^{2/p}|v|^2\,dy\Big)^{p/(2(p+1))}$$

$$\leq \|u\|_\infty^{\frac{1}{p+1}}\|\Theta(|u|)\|_\infty^{\frac{1}{p+1}} \sup_{x\in\mathbb{R}^3,\rho>0} \Big(\frac{1}{\rho^{3(1-\frac{2}{q})}}\int_{B(x,\rho)} |v(y)|^2\,dy\Big)^{p/(2(p+1))}$$

and thus

$$\|\vec{u}\|^{p+1}_{\dot{M}^{\frac{2(p+1)}{p},\frac{(p+1)q}{p}}} \leq C\|\vec{u}\|_\infty\Theta(\|\vec{u}\|_\infty)\|\frac{\vec{u}}{\Theta(|\vec{u}|)^{1/p}}\|^p_{\dot{M}^{2,q}}. \tag{11.24}$$

From inequalities (11.23) and (11.24), and from Grönwall's lemma, we conclude that, if $\int_{T_0}^{T_{\mathrm{MAX}}} \|\frac{\vec{u}}{\Theta(|\vec{u}|)^{1/p}}\|^p_{\dot{M}^{2,q}} < +\infty$, then $\|\vec{u}(t,.)\|_\infty$ remains bounded on (T_0, T_{MAX}), so that $\vec{u} \in L^2((T_0, T_{\mathrm{MAX}}, L^\infty)$; but this contradicts $T_{\mathrm{MAX}} < +\infty$. $\qquad\square$

11.4 A Remark on Serrin's Criterion and Leray's Criterion

Serrin's criterion [435] is a very simple criterion to ensure the control of the $\dot{H}^1$ norm of $\vec{u}$ through Equation (11.9). However, if the force is slightly more regular, it is a simple consequence of a remark done by Leray in his 1934 paper [328]:

Leray's criterion

Theorem 11.4.
Let $\vec{u}_0 \in H^1$ with $\mathrm{div}\,\vec{u}_0 = 0$ and $\vec{f} \in L^\infty((0,+\infty), L^2)$. Let $\vec{u}$ be a solution of

$$\partial_t\vec{u} = \nu\Delta\vec{u} + \mathbb{P}(\vec{f} - \mathrm{div}(\vec{u}\otimes\vec{u}))$$

with $u \in \mathcal{C}([0,T], H^1)\cap L^2((0,T), H^2)$ for all $T < T_{\mathrm{MAX}}$ and $\vec{u}(0,.) = \vec{u}_0$. If $q > 3$, there exists a positive constant C_q such that, if the maximal existence time T_{MAX} satisfies $T_{\mathrm{MAX}} < +\infty$, then

$$\liminf_{T\to T_{\mathrm{MAX}}} (T^* - T)^{\frac{1}{2}(1-\frac{3}{q})}\|\vec{u}(T,.)\|_q > C_q.$$

Similarly, Kozono and Shimada's criterion [274] can be treated through Leray's criterion:

Leray's criterion and Besov spaces

Theorem 11.5.
Let $\vec{u}_0 \in H^1$ with $\operatorname{div} \vec{u}_0 = 0$ and $\vec{f} \in L^\infty((0,+\infty), L^2)$. Let $\vec{u}$ be a solution of

$$\partial_t \vec{u} = \nu \Delta \vec{u} + \mathbb{P}(\vec{f} - \operatorname{div}(\vec{u} \otimes \vec{u}))$$

with $\vec{u} \in \mathcal{C}([0,T], H^1) \cap L^2((0,T), H^2)$ for all $T < T_{\text{MAX}}$ and $\vec{u}(0,.) = \vec{u}_0$.
If $0 < \sigma < 1$, there exists a positive constant C_σ such that, if the maximal existence time T_{MAX} satisfies $T_{\text{MAX}} < +\infty$, then

$$\liminf_{T \to T_{\text{MAX}}} (T^* - T)^{\frac{1}{2}(1-\sigma)} \|\vec{u}(T,.)\|_{\dot{B}^{-\sigma}_{\infty,\infty}} > C_\sigma.$$

Proof. If $\vec{u}(T,.) \in \dot{B}^{-\sigma}_{\infty,\infty}$, then

$$\sup_{t>0} t^{\sigma/2} \|e^{t\Delta} \vec{u}(T,.)\|_\infty \leq C \|\vec{u}(T,.)\|_{\dot{B}^{-\sigma}_{\infty,\infty}}.$$

On the other hand,

$$\sup_{0<t<t_0} t^{\sigma/2} \|\int_0^t e^{(t-s)\Delta} \mathbb{P}\vec{f}(T+s,.)\, ds\|_\infty \leq C t_0^{\frac{\sigma}{2}+\frac{1}{4}} \|\vec{f}\|_{L^\infty L^2}$$

(where the constant C does not depend on t_0). We have

$$\sup_{0<t<t_0} t^{\sigma/2} \|\int_0^t e^{(t-s)\Delta} \mathbb{P}(\vec{u}(T+s,.) \otimes \vec{v}(T+s,.))\, ds\|_\infty$$

$$\leq C t_0^{\frac{1-\sigma}{2}} \sup_{0<t<t_0} t^{\sigma/2} \|\vec{u}(T+t,.)\|_\infty \sup_{0<t<t_0} t^{\sigma/2} \|\vec{v}(T+t,.)\|_\infty$$

This gives a solution $\vec{v}_T$ of the Cauchy problem for the Navier–Stokes problem on $[T, T+t_0]$ with initial value $\vec{u}(T,.)$, provided that

$$C_\sigma t_0^{\frac{1-\sigma}{2}} (\|\vec{u}(T,.)\|_{\dot{B}^{-\sigma}_{\infty,\infty}} + t_0^{\frac{\sigma}{2}+\frac{1}{4}} \|\vec{f}\|_{L^\infty L^2}) \leq 1.$$

By uniqueness of mild sollutions, this solution $\vec{v}_T$ will coincide with $\vec{u}$ on $[T, \min(T^*, T+t_0))$. This implies, if $T^* < +\infty$, that $t_0 < T^* - T$, as $\vec{u}(t,.)$ cannot remain bounded when t is approaching T^*. Thus,

$$1 \leq C_\sigma (T^* - T)^{\frac{1-\sigma}{2}} (\|\vec{u}(T,.)\|_{\dot{B}^{-\sigma}_{\infty,\infty}} + (T^* - T)^{\frac{\sigma}{2}+\frac{1}{4}} \|\vec{f}\|_{L^\infty L^2}). \qquad \square$$

11.5   Some Further Generalizations of Serrin's Criterion

Let us review what we have seen so far.

Let $\vec{u}_0 \in H^1$ with $\text{div}\,\vec{u}_0 = 0$ and $\vec{f} \in L^2((0,+\infty), L^2)$. Let $\vec{u}$ be a solution of

$$\partial_t \vec{u} = \nu \Delta \vec{u} + \mathbb{P}(\vec{f} - \text{div}(\vec{u} \otimes \vec{u}))$$

with $u \in \mathcal{C}([0,T], H^1) \cap L^2((0,T), H^2)$ for all $T < T_{\text{MAX}}$ and $\vec{u}(0,.) = \vec{u}_0$. Let us assume that the maximal existence time T_{MAX} satisfies $T_{\text{MAX}} < +\infty$. Then

- Serrin proved in 1963 [435] that, for $2/p + 3/q = 1$ with $2 \le p < \infty$, $\int_0^{T_{\text{MAX}}} \|\vec{u}\|_q^p \, dt = +\infty$.
  Escauriaza, Seregin and Šverák [163] proved in 2003 that the endpoint case $p = +\infty$, $q = 3$ of the Serrin criterion holds.

- Serrin's criterion was extended in 2004 by Kozono and Shimada [274] who proved that $\int_0^{T_{\text{MAX}}} \|\vec{u}\|_{\dot{B}_{\infty,\infty}^\sigma}^p \, dt = +\infty$ with $2/p = 1 + \sigma$, $2 < p < +\infty$ and $-1 < \sigma < 0$ (note that $L^q \subset \dot{B}_{\infty,\infty}^{-\frac{3}{q}}$).

- Beirão da Vega [29] proved in 1995 that, for $2/p + 3/r = 2$ with $1 < p < \infty$, $\int_0^{T_{\text{MAX}}} \|\vec{\nabla} \otimes \vec{u}\|_r^p \, dt = +\infty$.

- For $2 < p$, Beirão da Vega's criterion is a consequence of Serrin's criterion, as $\vec{\nabla} \otimes \vec{u} \in L^p L^r \Rightarrow \vec{u} \in L^p L^q$. This is no longer the case for $p = 2$, as we do not have the implication $\vec{\nabla} \otimes \vec{u} \in L^2 L^3 \Rightarrow \vec{u} \in L^2 L^\infty$, but only the implication $\vec{\nabla} \otimes \vec{u} \in L^2 L^3 \Rightarrow \vec{u} \in L^2 BMO$. However, this was generalized in 2000 by Kozono and Taniuchi [278] who proved that $\int_0^{T_{\text{MAX}}} \|\vec{u}\|_{BMO}^2 \, dt = +\infty$.

- Beirão da Vega's criterion was fully generalized in 2008 by Chen and Zhang [116] who proved that $\int_0^{T_{\text{MAX}}} \|\vec{u}\|_{\dot{B}_{\infty,\infty}^\sigma}^p \, dt = +\infty$ with $2/p = 1 + \sigma$, $1 < p < +\infty$ and $-1 < \sigma < 1$ (remark that $\vec{\nabla} \otimes \vec{u} \in L^p L^r \implies \vec{u} \in L^p \dot{B}_{\infty,\infty}^\sigma$ with $\sigma = 1 - \frac{3}{r}$).

- In 1984, Beale, Kato and Majda [27] proved that $\int_0^{T_{\text{MAX}}} \|\text{curl}\,\vec{u}\|_\infty \, dt = +\infty$; this was generalized in 2000 by Kozono and Taniuchi [278] to $\int_0^{T_{\text{MAX}}} \|\vec{\nabla} \otimes \vec{u}\|_{BMO} = +\infty$ and finally (if $\vec{f} \in L^2 H^1$) in 2002 by Kozono, Ogawa and Taniuchi [273] who proved that $\int_0^{T_{\text{MAX}}} \|\vec{u}\|_{\dot{B}_{\infty,\infty}^1} \, dt = +\infty$.

One could think that every possible criteria have been proposed (up to logarithmic improvements) in terms of the size of $\vec{u}$ or of $\vec{\nabla} \otimes \vec{u}$. However, several generalizations of Beirão da Vega's criterion were proposed:

- control of just one component: $\int_0^{T_{\text{MAX}}} \|\vec{\nabla} u_3\|_r^p \, dt = +\infty$ with $\frac{2}{p} + \frac{3}{r} = \frac{3}{2}$ and $2 \le p < +\infty$ (Neustupa, Novotný and Penel [374], He [231], Pokorný [404], Zhou [511])

- control of just one derivative: $\int_0^{T_{\text{MAX}}} \|\partial_3 \vec{u}\|_r^p \, dt = +\infty$ with $\frac{2}{p} + \frac{3}{r} = 2$ and $2 \le p \le 3$ (Penel and Pokorný [394], Kukavica and Ziane [289])

- control of the pressure: for $\varpi = -\frac{1}{\Delta}\text{div}(\vec{u} \cdot \vec{\nabla}\vec{u})$, $\int_0^{T_{\text{MAX}}} \|\vec{\nabla}\varpi\|_q^p \, dt = +\infty$ with $\frac{2}{p} + \frac{3}{q} = 3$ and $\frac{2}{3} < p < +\infty$ (Berselli and Galdi [39], Zhou [513, 512], Struwe [456]).

We are going to prove those three results. (Of course, they can be "logarithmically improved" in the spirit of Proposition 11.3.) Notice that recently Cao and Titi [88] have studied a global regularity criterion involving just one entry $\partial_i u_j$ of the velocity gradient tensor.

Proposition 11.5.
Let $\vec{u}_0 \in H^1$ with div $\vec{u}_0 = 0$ and $\vec{f} \in L^2((0, +\infty), H^1)$. Let $\vec{u}$ be a solution of

$$\partial_t \vec{u} = \nu \Delta \vec{u} + \mathbb{P}(\vec{f} - \text{div}(\vec{u} \otimes \vec{u}))$$

with $\vec{u} \in \mathcal{C}([0,T], H^1) \cap L^2((0,T), H^2)$ for all $T < T_{\text{MAX}}$ and $\vec{u}(0,.) = \vec{u}_0$. If the maximal existence time T_{MAX} satisfies $T_{\text{MAX}} < +\infty$, then $\int_0^{T_{\text{MAX}}} \|\vec{\nabla} u_3\|_r^p \, dt = +\infty$ with $\frac{2}{p} + \frac{3}{r} = \frac{3}{2}$ and $2 \le p < +\infty$.

Proof. As $\vec{f} \in L^2 L^2$, we know that, for every $0 < T_0 < T < T_{\text{MAX}}$, $\vec{u}$ will belong to $\mathcal{C}([T_0, T], H^1) \cap L^2((T_0, T), H^2)$. Thus, we may estimate at each positive time t the quantities

$$I(t) = \int \frac{|\vec{\nabla} \otimes \vec{u}|^2}{2} \, dx \text{ and } J(t) = 1 + \frac{(\int |\omega_3|^2 \, dx)^2}{4}$$

where $\vec{\omega} = \text{curl} \, \vec{u}$. We introduce as well two other quantities

$$A(t) = \|\Delta \vec{u}\|_2^2 \text{ and } B(t) = \|\omega_3\|_2^2 \|\vec{\nabla} \omega_3\|_2^2.$$

We shall first study J, in order to explain the choice of the exponent. Recall that we have $\vec{u} \in L^\infty L^2 \cap L^2 \dot{H}^1$:

$$\|\vec{u}(t,.)\|_2 \le \|\vec{u}_0\|_2 + \int_0^{T_{\text{MAX}}} \|\vec{f}\|_2 \, dx \tag{11.25}$$

and

$$\int_0^{T_{\text{MAX}}} \|\vec{\nabla} \otimes \vec{u}(s,.)\|_2^2 \, ds \le \frac{1}{\nu}(\|\vec{u}_0\|_2 + \int_0^{T_{\text{MAX}}} \|\vec{f}\|_2 \, dx)^2 \tag{11.26}$$

We shall write

$$N_0 = \|\vec{u}_0\|_2 + \int_0^{T_{\text{MAX}}} \|\vec{f}\|_2 \, dx.$$

In order to estimate J, we write and

$$\partial_t \vec{\omega} = \nu \Delta \vec{\omega} - \vec{u} \cdot \vec{\nabla} \vec{\omega} + \vec{\omega} \cdot \vec{\nabla} \vec{u} + \text{curl} \, \vec{f}$$

which gives

$$\begin{aligned}
\frac{d}{dt} J(t) = & \|\omega_3\|_2^2 \int \omega_3 \partial_t \omega_3 \, dx \\
= & -\nu \|\omega_3\|_2^2 \int |\vec{\nabla} \omega_3|^2 \, dx + \|\omega_3\|_2^2 \int f_1 \partial_2 \omega_3 - f_2 \partial_1 \omega_3 \, dx \\
& - \|\omega_3\|_2^2 \int \omega_3 \vec{u} \cdot \vec{\nabla} \omega_3 \, dx + \|\omega_3\|_2^2 \int \omega_3 \vec{\omega} \cdot \vec{\nabla} u_3 \, dx
\end{aligned} \tag{11.27}$$

As div $\vec{u} = 0$, we have $\int \omega_3 \vec{u} \cdot \vec{\nabla} \omega_3 \, dx = 0$. Moreover, we have $\|\omega\|_3 \le C\|\vec{u}\|_{\dot{H}^1}^{1/2} \|\Delta \vec{u}\|_2^{1/2}$. Thus, we find

$$\begin{aligned}
\frac{d}{dt} J(t) \le & -\nu \|\omega_3\|_2^2 \|\vec{\nabla} \omega_3\|_2^2 + \|\omega_3\|_2^2 \|\vec{\nabla} \omega_3\|_2 \|\vec{f}\|_2 \\
& + C \|\omega_3\|_2^2 \|\vec{\nabla} u_3\|_r \|\vec{u}\|_{\dot{H}^1}^{1/2} \|\Delta \vec{u}\|_2^{1/2} \|\omega_3\|_{\frac{3r}{2r-3}}
\end{aligned}$$

with $2 \le \frac{3r}{2r-3} < 6$. We have the interpolation inequality, for $\rho \in [2, 6]$,

$$\|\omega_3\|_\rho \le \|\omega_3\|_2^{\frac{3}{\rho} - \frac{1}{2}} \|\omega_3\|_6^{\frac{3}{2} - \frac{3}{\rho}}$$

so that

$$\|\omega_3\|_{\frac{3r}{2r-3}} \leq \|\omega_3\|_2^{\frac{3}{2}-\frac{3}{r}} \|\omega_3\|_6^{\frac{3}{r}-\frac{1}{2}}.$$

This gives (since $J \geq 1$, hence $J^{1/4} \leq J^{1/2}$)

$$\frac{d}{dt}J(t) \leq - \nu B(t) + \sqrt{2}B(t)^{\frac{1}{2}}J(t)^{\frac{1}{2}}\|\vec{f}\|_2 \qquad (11.28)$$
$$+ C\|\vec{\nabla}u_3\|_r\|\vec{u}\|_{\dot{H}^1}^{1/2}A(t)^{\frac{1}{4}}B(t)^{\frac{3}{2r}-\frac{1}{4}}J^{1-\frac{3}{2r}}$$

As $(\frac{1}{4}) + (\frac{3}{2r} - \frac{1}{4}) + (1 - \frac{3}{2r}) = 1$, we shall be able to use Grönwall's lemma.

Now, we rewrite the Navier–Stokes equations as

$$\partial_t \vec{u} = \nu \Delta \vec{u} - \vec{\omega} \wedge \vec{u} + \vec{f} - \vec{\nabla}(p + \frac{|\vec{u}|^2}{2})$$

(with div $\vec{u} = 0$) and obtain

$$\frac{d}{dt}I(t) = - \int \Delta \vec{u} \cdot \partial_t \vec{u}\, dx$$
$$= - \nu \int |\Delta \vec{u}|^2\, dx - \int \Delta \vec{u} \cdot \vec{f}\, dx + \int \Delta \vec{u} \cdot (\vec{\omega} \wedge \vec{u})\, dx \qquad (11.29)$$

which we expand into

$$\frac{d}{dt}I(t) = - \nu \int |\Delta \vec{u}|^2\, dx - \int \Delta \vec{u} \cdot \vec{f}\, dx$$
$$+ \int \Delta u_1 \omega_2 u_3 - \Delta u_1 \omega_3 u_2\, dx$$
$$+ \int \Delta u_2 \omega_3 u_1 - \Delta u_2 \omega_1 u_3\, dx$$
$$+ \int \Delta u_3 \omega_1 u_2 - \Delta u_3 \omega_2 u_1\, dx.$$

u_3 appears everywhere except in the term $\int \Delta u_2 \omega_3 u_1 - \Delta u_1 \omega_3 u_2\, dx$. For this last term, we may replace the control on u_3 by the control on ω_3 (given by (11.28)).

The natural scaling for $\vec{\nabla}u_3$ would have been $\frac{2}{p} + \frac{3}{r} = 2$ as for Beirão da Vega's criterion [29]. As a matter of fact, in our decomposition of $\frac{d}{dt}I(t)$, the scaling $\frac{2}{p} + \frac{3}{r} = 2$ would be enough to control the terms involving u_3; the scaling $\frac{2}{p} + \frac{3}{r} = \frac{3}{2}$ is necessary only for the terms involving ω_3.

We thus define $\frac{1}{p_0} = \frac{1}{2}(\frac{1}{p} + \frac{1}{2})$ and $\frac{1}{r_0} = \frac{1}{2}(\frac{1}{r} + \frac{1}{2})$, so that $\frac{2}{p_0} + \frac{3}{r_0} = 2$. We define X_{r_0} as L^{r_0} if $2 < r < 6$ (hence $2 < r_0 < 3$) and $L^{3,1}$ if $r = 6$ ($r_0 = 3$). Notice that:

- $2 < r \leq 6$, hence $2 < r_0 \leq 3$

- $X_{r_0} \subset L^{r_0}$

- we have $\int_0^{T_{MAX}} \|\vec{\nabla}u_3\|_{X_{r_0}}^{p_0}\, dt \leq C \int_0^{T_{MAX}} \|\vec{\nabla}u_3\|_2^{\frac{p_0}{2}} \|\vec{\nabla}u_3\|_{X_r}^{\frac{p_0}{2}}\, dt$, and thus $\int_0^{T_{MAX}} \|\vec{\nabla}u_3\|_{X_{r_0}}^{p_0}\, dt \leq C(\frac{N_0}{\sqrt{\nu}})^{\frac{p_0}{4}} (\int_0^{T_{MAX}} \|\vec{\nabla}u_3\|_r^p\, dt)^{\frac{p_0}{2p}}$

- the assumption $\vec{\nabla}u_3 \in L^{p_0}X_{r_0}$ gives $u_3 \in L^{p_0}L^{q_0}$ with $\frac{1}{q_0} = \frac{1}{r_0} - \frac{1}{3}$.

We then write

$$\int \Delta u_3 \omega_1 u_2 - \Delta u_3 \omega_2 u_1 \, dx = - \int \vec{\nabla} u_3 \cdot \vec{\nabla}(\omega_1 u_2 - \omega_2 u_1) \, dx$$

and

$$\frac{d}{dt} I(t) \leq - \nu \|\Delta \vec{u}\|_2^2 + \|\Delta \vec{u}\|_2 \|\vec{f}\|_2$$
$$+ C \|\Delta \vec{u}\|_2 \|\vec{\nabla} \otimes \vec{u}\|_{\frac{2q_0}{q_0-2}} \|u_3\|_{q_0} + C \|\Delta \vec{u}\|_2 \|\omega_3\|_3 \|\vec{u}\|_6$$
$$+ C \|\vec{\nabla} u_3\|_{r_0} \|\Delta \vec{u}\|_2 \|\vec{u}\|_{\frac{2r_0}{r_0-2}} + C \|\vec{\nabla} u_3\|_{r_0} \|\vec{\nabla} \otimes \vec{u}\|_{\frac{2r_0}{r_0-1}}^2$$

Notice that $2 \leq \frac{2q_0}{q_0-2} < 3$, $6 \leq \frac{2r_0}{r_0-2} < +\infty$ and $3 \leq \frac{2r_0}{r_0-1} < 4$. We then use the following interpolation inequalities:

- for $\rho \in [2,6]$, $\|\vec{\nabla} \otimes \vec{u}\|_\rho \leq C I(t)^{\frac{3}{2\rho}-\frac{1}{4}} A(t)^{\frac{3}{4}-\frac{3}{2\rho}}$

- for $\sigma \in [6, +\infty]$, $\|\vec{u}\|_\sigma \leq \|\vec{u}\|_6^{\frac{6}{\sigma}} \|\vec{u}\|_\infty^{1-\frac{6}{\sigma}} \leq C I(t)^{\frac{1}{4}+\frac{3}{2\sigma}} A(t)^{\frac{1}{4}-\frac{3}{2\sigma}}$

- $\|\omega_3\|_3 \leq \|\omega_3\|_2^{\frac{1}{2}} \|\omega_3\|_6^{\frac{1}{2}} \leq C B(t)^{\frac{1}{4}}$

- $\|\vec{u}\|_6 \leq C \|\vec{u}\|_{\dot{H}^1} \leq C \|\vec{u}\|_2^{\frac{1}{2}} \|\Delta \vec{u}\|_2^{\frac{1}{2}} \leq C \|\vec{u}\|_2^{\frac{1}{2}} A(t)^{\frac{1}{4}}$

This gives

$$\frac{d}{dt} I(t) \leq - \nu A(t) + A(t)^{\frac{1}{2}} \|\vec{f}\|_2$$
$$+ C \|\vec{\nabla} u_3\|_{X_{r_0}} A(t)^{\frac{3}{2r_0}} I(t)^{1-\frac{3}{2r_0}} \tag{11.30}$$
$$+ C \|u(t,.)\|_2^{\frac{1}{2}} A(t)^{\frac{3}{4}} B(t)^{\frac{1}{4}}$$

Thus, we obtain

$$\frac{d}{dt} I(t) \leq - \frac{\nu}{2} A(t) + C_1 \frac{1}{\nu} \|\vec{f}\|_2^2$$
$$+ C_1 \|\vec{\nabla} u_3\|_{X_{r_0}}^{\frac{2r_0}{2r_0-3}} \nu^{-\frac{3}{2r_0-3}} I(t)$$
$$+ C_1 N_0^2 \frac{1}{\nu^3} B(t)$$

(with $\frac{2r_0}{2r_0-3} = p_0$) and (for $\epsilon > 0$)

$$\frac{d}{dt} J(t) \leq - \frac{\nu}{2} B(t) + C \frac{1}{\nu} J(t) \|\vec{f}\|_2^2$$
$$+ \epsilon^4 \nu A(t) + C \epsilon^{-4/3} \nu^{-1/3} \|\vec{\nabla} u_3\|_r^{\frac{4}{3}} \|\vec{u}\|_{\dot{H}^1}^{\frac{2}{3}} B(t)^{\frac{2}{r}-\frac{1}{3}} J(t)^{\frac{4}{3}-\frac{2}{r}}$$
$$\leq - \frac{\nu}{4} B(t) + C_2 \frac{1}{\nu} J(t) \|\vec{f}\|_2^2$$
$$+ \epsilon^4 \nu A(t) + C_2 \epsilon^{-\frac{2r}{2r-3}} \nu^{-\frac{r}{2(2r-3)}} \|\vec{\nabla} u_3\|_r^{\frac{2r}{2r-3}} \|\vec{u}\|_{\dot{H}^1}^{\frac{r}{2r-3}} J(t)$$

Let

$$K(t) = I(t) + \lambda J(t).$$

We take $\lambda = \frac{4C_1 N_0^2}{\nu^4}$ and $\epsilon^4 = \frac{1}{2\lambda}$. We obtain

$$\frac{d}{dt}K(t) \leq C_1 \frac{1}{\nu}\|\vec{f}\|_2^2 + C_1 \|\vec{\nabla}u_3\|_{X_{r_0}}^{p_0} \nu^{1-p_0} K(t) + C_2 \frac{1}{\nu}\|\vec{f}\|_2^2 K(t)$$

$$+ C_2 \epsilon^{-\frac{2r}{2r-3}} \nu^{-\frac{r}{2(2r-3)}} \|\vec{\nabla}u_3\|_r^{\frac{2r}{2r-3}} \|\vec{u}\|_{\dot{H}^1}^{\frac{r}{2r-3}} K(t)$$

We then conclude by using Grönwall's lemma, as

- $\int_0^{T_{\text{MAX}}} \|\vec{f}\|_2 \, dt < +\infty$ by assumption on $\vec{f}$

- $\int_0^{T_{\text{MAX}}} \|\vec{\nabla}u_3\|_{X_{r_0}}^{p_0} \, dt \leq C(\frac{N_0}{\sqrt{\nu}})^{\frac{p_0}{4}} (\int_0^{T_{\text{MAX}}} \|\vec{\nabla}u_3\|_r^p \, dt)^{\frac{p_0}{2p}}$

- $\int_0^{T_{\text{MAX}}} \|\vec{\nabla}u_3\|_r^{\frac{2r}{2r-3}} \|\vec{u}\|_{\dot{H}^1}^{\frac{r}{2r-3}} \, dt \leq (\frac{N_0^2}{\nu})^{\frac{r}{2(2r-3)}} (\int_0^{T_{\text{MAX}}} \|\vec{\nabla}u_3\|_r^{\frac{4r}{3r-2}} \, dt)^{\frac{3(r-2)}{2(2r-3)}}$ with $p = \frac{4r}{3r-2}$.

Thus, if $\vec{u}$ blows up, we must have $\int_0^{T_{\text{MAX}}} \|\vec{\nabla}u_3\|_r^p \, dt = +\infty$. $\qquad\square$

Proposition 11.6.
Let $\vec{u}_0 \in H^1$ with div $\vec{u}_0 = 0$ and $\vec{f} \in L^2((0,+\infty), H^1)$. Let $\vec{u}$ be a solution of

$$\partial_t \vec{u} = \nu\Delta\vec{u} + \mathbb{P}(\vec{f} - \text{div}(\vec{u} \otimes \vec{u}))$$

with $\vec{u} \in \mathcal{C}([0,T], H^1) \cap L^2((0,T), H^2)$ for all $T < T_{\text{MAX}}$ and $\vec{u}(0,.) = \vec{u}_0$. If the maximal existence time T_{MAX} satisfies $T_{\text{MAX}} < +\infty$, then $\int_0^{T_{\text{MAX}}} \|\partial_3 \vec{u}\|_r^p \, dt = +\infty$ with $\frac{2}{p} + \frac{3}{r} = 2$ and $2 \leq p \leq 3$.

Proof. We follow the proof in Kukavica and Ziane [289]. As $\vec{f} \in L^2 L^2$, we know that, for every $0 < T_0 < T < T_{\text{MAX}}$, $\vec{u}$ will belong to $\mathcal{C}([T_0, T], H^1) \cap L^2((T_0, T), H^2)$. Thus, we may estimate at each positive time t the quantity

$$I(t) = 1 + \frac{(\|u_1\|_{\dot{H}^1}^2 + \|u_2\|_{\dot{H}^1}^2)^3}{2} + \frac{\|u_3\|_6^6}{6}.$$

We define

$$J(t) = \|u_1\|_{\dot{H}^1}^2 + \|u_2\|_{\dot{H}^1}^2, \quad K(t) = \|u_3\|_6^6$$

and

$$M(t) = \|\Delta u_1\|_2^2 + \|\Delta u_2\|_2^2, \quad N(t) = \|u_3^3\|_{\dot{H}^1}^2.$$

Using the equation

$$\partial_t \vec{u} = \nu\Delta\vec{u} - \vec{u} \cdot \vec{\nabla}\vec{u} + \vec{g} - \vec{\nabla}\varpi$$

with

$$\vec{g} = \mathbb{P}\vec{f} \quad \text{and} \quad \varpi = -\frac{1}{\Delta}\sum_{i=1}^{3}\sum_{j=1}^{3}\partial_i\partial_j(u_i u_j)$$

we get

$$\frac{d}{dt}I(t) = \int u_3^5 \partial_t u_3 \, dx - J(t)^2 (\int -\partial_t u_1 \Delta u_1 - \partial_t u_2 \Delta u_2 \, dx). \qquad (11.31)$$

Indeed, we have $\vec{u} \in L^{10}L^{10}((T_0, T) \times \mathbb{R}^3)$ (since $\|\vec{u}\|_{L^{10}L^{10}} \leq C\|\vec{u}\|_{L^\infty \dot{H}^1}^{4/5} \|\vec{u}\|_{L^2\dot{H}^2}^{1/5}$), so that we easily check that $\partial_t\vec{u}$, $\Delta\vec{u}$ and $|\vec{u}|^5$ belong to $L^2 L^2((T_0, T) \times \mathbb{R}^3)$.

This can be expanded into

$$\frac{d}{dt}I(t) = \nu \int u_3^5 \Delta u_3 \, dx - \nu J(t)^2 \int |\Delta u_1|^2 \, dx - \nu J(t)^2 \int |\Delta u_2|^2 \, dx$$
$$- \int u_3^5 \vec{u} \cdot \vec{\nabla} u_3 \, dx + J(t)^2 \int (\Delta u_1) \vec{u} \cdot \vec{\nabla} u_1 + (\Delta u_2) \vec{u} \cdot \vec{\nabla} u_2 \, dx$$
$$+ \int u_3^5 g_3 \, dx - J(t)^2 \int (g_1 \Delta u_1 + g_2 \Delta u_2) \, dx \qquad (11.32)$$
$$- \int u_3^5 \partial_3 \varpi \, dx + J(t)^2 \int (\Delta u_1) \partial_1 \varpi + (\Delta u_2) \partial_2 \varpi \, dx.$$

Next, we deal carefully with each term:

- Integration by parts gives

$$\nu \int u_3^5 \Delta u_3 \, dx = -\frac{5}{9} \nu \|\vec{\nabla}(u_3^3)\|_2^2 = -\frac{5}{9} \nu N(t) \qquad (11.33)$$

- We have obviously

$$-\nu \int |\Delta u_1|^2 \, dx - \nu \int |\Delta u_2|^2 \, dx = -\nu M(t) \qquad (11.34)$$

- Integration by parts gives (since div $\vec{u} = 0$)

$$A = -5 \int u_3^5 \vec{u} \cdot \vec{\nabla} u_3 \, dx = - \int u_3 \vec{u} \cdot \vec{\nabla}(u_3^5) \, dx = -5A = 0 \qquad (11.35)$$

- Writing $\Delta = \Delta_2 + \partial_3^2$, where $\Delta_2 = \partial_1^2 + \partial_2^2$, we get

$$\int (\Delta u_1) \vec{u} \cdot \vec{\nabla} u_1 + (\Delta u_2) \vec{u}.\vec{\nabla} u_2 \, dx =$$
$$- \int \partial_3 u_1 (\partial_3 \vec{u}) \vec{\nabla} u_1 \, dx - \int \partial_3 u_2 (\partial_3 \vec{u}) \cdot \vec{\nabla} u_2 \, dx$$
$$+ \int \Delta_2 u_1 u_3 \partial_3 u_1 \, dx + \int \Delta_2 u_2 u_3 \partial_3 u_2 \, dx$$
$$+ \int \Delta_2 u_1 (u_1 \partial_1 + u_2 \partial_2) u_1 \, dx$$
$$+ \int \Delta_2 u_2 (u_1 \partial_1 + u_2 \partial_2) u_2 \, dx$$

with

$$\int \Delta_2 u_1 (u_1 \partial_1 + u_2 \partial_2) u_1 \, dx$$
$$= - \int \partial_1 u_1 \partial_1 u_1 \partial_1 u_1 \, dx - \int \partial_1 u_1 \partial_1 u_2 \partial_2 u_1 \, dx$$
$$- \int \partial_2 u_1 \partial_2 u_1 \partial_1 u_1 \, dx - \int \partial_2 u_1 \partial_2 u_2 \partial_2 u_1 \, dx$$
$$- \int \partial_1 u_1 (u_1 \partial_1 + u_2 \partial_2) \partial_1 u_1 \, dx$$

$$-\int \partial_2 u_1 (u_1 \partial_1 + u_2 \partial_2) \partial_2 u_1 \, dx$$

$$= -\int \partial_1 u_1 \partial_1 u_1 \partial_1 u_1 \, dx - \int \partial_1 u_1 \partial_1 u_2 \partial_2 u_1 \, dx$$

$$-\int \partial_2 u_1 \partial_2 u_1 \partial_1 u_1 \, dx - \int \partial_2 u_1 \partial_2 u_2 \partial_2 u_1 \, dx$$

$$-\int \partial_3 u_3 \frac{|\partial_1 u_1|^2 + |\partial_2 u_1|^2}{2} \, dx$$

and similarly

$$\int \Delta_2 u_2 (u_1 \partial_1 + u_2 \partial_2) u_2 \, dx$$

$$= -\int \partial_1 u_2 \partial_1 u_1 \partial_1 u_2 \, dx - \int \partial_1 u_2 \partial_1 u_2 \partial_2 u_2 \, dx$$

$$-\int \partial_2 u_2 \partial_2 u_1 \partial_1 u_2 \, dx - \int \partial_2 u_2 \partial_2 u_2 \partial_2 u_2 \, dx$$

$$-\int \partial_3 u_3 \frac{|\partial_1 u_2|^2 + |\partial_2 u_2|^2}{2} \, dx$$

In particular, we shall have to deal with the term

$$A(t) = -\int \partial_1 u_1 \partial_1 u_1 \partial_1 u_1 \, dx - \int \partial_1 u_1 \partial_1 u_2 \partial_2 u_1 \, dx$$

$$-\int \partial_2 u_1 \partial_2 u_1 \partial_1 u_1 \, dx - \int \partial_2 u_1 \partial_2 u_2 \partial_2 u_1 \, dx$$

$$-\int \partial_1 u_2 \partial_1 u_1 \partial_1 u_2 \, dx - \int \partial_1 u_2 \partial_1 u_2 \partial_2 u_2 \, dx$$

$$-\int \partial_2 u_2 \partial_2 u_1 \partial_1 u_2 \, dx - \int \partial_2 u_2 \partial_2 u_2 \partial_2 u_2 \, dx$$

which we rewrite as

$$A(t) = -\int (\partial_1 u_1 + \partial_2 u_2)(|\partial_1 u_1|^2 + |\partial_2 u_2|^2) \, dx$$

$$+ \int (\partial_1 u_1 + \partial_2 u_2) \partial_1 u_1 \partial_2 u_2 \, dx$$

$$- \int (\partial_1 u_1 + \partial_2 u_2) \partial_1 u_2 \partial_2 u_1 \, dx$$

$$- \int (\partial_1 u_1 + \partial_2 u_2)(|\partial_2 u_1|^2 + |\partial_1 u_2|^2) \, dx$$

$$= \int \partial_3 u_3 (|\partial_1 u_1|^2 + |\partial_2 u_2|^2 + |\partial_2 u_1|^2 + |\partial_1 u_2|^2) \, dx$$

$$+ \int \partial_3 u_3 (\partial_1 u_2 \partial_2 u_1 - \partial_1 u_1 \partial_2 u_2) \, dx$$

This gives

$$\int (\Delta u_1)\vec{u} \cdot \vec{\nabla} u_1 + (\Delta u_2)\vec{u} \cdot \vec{\nabla} u_2 \, dx$$

$$= -\int \partial_3 u_1 (\partial_3 \vec{u}) \vec{\nabla} u_1 \, dx - \int \partial_3 u_2 (\partial_3 \vec{u}) \cdot \vec{\nabla} u_2 \, dx$$

$$+ \int \Delta_2 u_1 u_3 \partial_3 u_1 \, dx + \int \Delta_2 u_2 u_3 \partial_3 u_2 \, dx$$

$$+ \int \partial_3 u_3 \frac{|\partial_1 u_1|^2 + |\partial_2 u_1|^2 + |\partial_1 u_2|^2 + |\partial_2 u_2|^2}{2} \, dx$$

$$+ \int \partial_3 u_3 (\partial_1 u_2 \partial_2 u_1 - \partial_1 u_1 \partial_2 u_2) \, dx$$

$$\leq 2\|\partial_3 \vec{u}\|_r \|\vec{\nabla} u_1\|^2_{\frac{2r}{r-1}} + 2\|\partial_3 \vec{u}\|_r \|\vec{\nabla} u_2\|^2_{\frac{2r}{r-1}}$$

$$+ \|\Delta u_1\|_2 \|u_3\|_{\frac{2r}{r-2}} \|\partial_3 \vec{u}\|_r + \|\Delta u_2\|_2 \|u_3\|_{\frac{2r}{r-2}} \|\partial_3 \vec{u}\|_r$$

For $9/4 \leq r \leq 3$, we have $q_1 = \frac{2r}{r-1} \in [3, 18/5] \subset [2, 6]$. As we look for a control of $\vec{\nabla} u_1$ and $\vec{\nabla} u_2$ in $L^\infty L^2 \cap L^2 \dot{H}^1 \subset L^\infty L^2 \cap L^2 L^6$, we may use interpolation and write, for $i = 1, 2$

$$\|\vec{\nabla} u_i\|_{q_1} \leq \|\vec{\nabla} u_i\|_2^{\frac{3}{q_1} - \frac{1}{2}} \|\vec{\nabla} u_i\|_6^{\frac{3}{2} - \frac{3}{q_1}} \leq C \|u_i\|_{\dot{H}^1}^{1 - \frac{3}{2r}} \|\Delta u_i\|_2^{\frac{3}{2r}}$$

Similarly, we write that $q_2 = \frac{2r}{r-2} \in [6, 18]$ and we look for a control of u_3^3 in $L^\infty L^2 \cap L^2 \dot{H}^1 \subset L^\infty L^2 \cap L^2 L^6$, hence of u_3 in $L^\infty L^6 \cap L^6 L^{18}$. Thus we write

$$\|u_3\|_{q_2} \leq \|u_3\|_6^{\frac{9}{q_2} - \frac{1}{2}} \|u_3\|_{18}^{\frac{3}{2} - \frac{9}{q_2}} \leq C \|u_3\|_6^{4 - \frac{9}{r}} \|u_3^3\|_{\dot{H}^1}^{\frac{3}{r} - 1}$$

Thus, we get

$$\int (\Delta u_1)\vec{u} \cdot \vec{\nabla} u_1 + (\Delta u_2)\vec{u} \cdot \vec{\nabla} u_2 \, dx \leq$$

$$C \|\partial_3 \vec{u}\|_r J(t)^{1 - \frac{3}{2r}} M(t)^{\frac{3}{2r}} \tag{11.36}$$

$$+ C \|\partial_3 \vec{u}\|_r M(t)^{\frac{1}{2}} N(t)^{\frac{3}{2r} - \frac{1}{2}} K(t)^{\frac{2}{3} - \frac{3}{2r}}$$

- We have $\int u_3^5 g_3 \, dx \leq \|u_3\|_{10}^5 \|g_3\|_2$ with

$$\|u_3\|_{10} \leq \|u_3\|_6^{\frac{2}{5}} \|u_3\|_{18}^{\frac{3}{5}} \leq C \|u_3\|_6^{\frac{2}{5}} \|u_3^3\|_{\dot{H}^1}^{\frac{1}{5}}.$$

and thus

$$\int u_3^5 g_3 \, dx \leq C \|\vec{g}\|_2 K(t)^{\frac{1}{3}} N(t)^{\frac{1}{2}} \tag{11.37}$$

- We have obviously

$$-\int (g_1 \Delta u_1 + g_2 \Delta u_2) \, dx \leq \|\vec{g}\|_2 M(t)^{1/2} \tag{11.38}$$

- We have

$$-\int u_3^5 \partial_3 \varpi \, dx = 5 \int u_3^4 \partial_3 u_3 \varpi \, dx \leq 5 \|\partial_3 \vec{u}\|_r \|u_3\|_{\frac{6r}{r-1}}^4 \|\varpi\|_{\frac{3r}{r-1}}.$$

As the Riesz transforms are bounded on $L^{\frac{3r}{r-1}}$, we find that

$$\|\varpi\|_{\frac{3r}{r-1}} \le C \sum_{i=1}^{3}\sum_{j=1}^{3}\|u_i u_j\|_{\frac{3r}{r-1}} \cdot \le C' \sum_{i=1}^{3}\|u_i\|^2_{\frac{6r}{r-1}}.$$

Now, notice that $L^\infty \dot{H}^1 \cap L^2 \dot{H}^2 \subset L^\infty \dot{H}^1 \cap L^4 \dot{B}^{3/2}_{2,1} \subset L^\infty L^6 \cap L^4 L^\infty$ and that $q_3 = \frac{6r}{r-1} \in [9, \frac{54}{5}] \subset [6, 18]$, so that we may write, for $i = 1, 2$

$$\|u_i\|_{q_3} \le \|u_i\|_6^{\frac{6}{q_3}} \|u_i\|_\infty^{1-\frac{6}{q_3}} \le C\|u_i\|_{\dot{H}^1}^{\frac{1}{2}+\frac{3}{q_3}} \|\Delta u_i\|_2^{\frac{1}{2}-\frac{3}{q_3}} = C\|u_i\|_{\dot{H}^1}^{1-\frac{1}{2r}} \|\Delta u_i\|_2^{\frac{1}{2r}}$$

and

$$\|u_3\|_{q_3} \le \|u_3\|_6^{\frac{9}{q_3}-\frac{1}{2}} \|u_3\|_{18}^{\frac{3}{2}-\frac{9}{q_3}} \le C\|u_3\|_6^{1-\frac{3}{2r}} \|u_3\|_{\dot{H}^1}^{\frac{3}{2r}}$$

This gives

$$-\int u_3^5 \partial_3 \varpi \, dx \tag{11.39}$$
$$\le C\|\partial_3 \vec{u}\|_r (J(t)^{1-\frac{1}{2r}} M(t)^{\frac{1}{2r}} \left(K(t)^{1-\frac{3}{2r}} N(t)^{\frac{3}{2r}} \right)^{\frac{2}{3}} + K(t)^{1-\frac{3}{2r}} N(t)^{\frac{3}{2r}}).$$

• We write
  (as div $\vec{u} = 0$)

$$\Delta \varpi = -\sum_{i=1}^{3}\sum_{j=1}^{3} \partial_i \partial_j (u_i u_j) = -\sum_{i=1}^{3}\sum_{j=1}^{3} \partial_i u_j \partial_j u_i$$

so that

$$\int (\Delta u_1)\partial_1 \varpi + (\Delta u_2)\partial_2 \varpi \, dx = -\int (\partial_1 u_1 + \partial_2 u_2)\Delta \varpi \, dx$$
$$= -\int \partial_3 u_3 \sum_{i=1}^{3}\sum_{j=1}^{3} \partial_i u_j \partial_j u_i \, dx$$
$$= -\int \partial_3 u_3 \sum_{i=1}^{2}\sum_{j=1}^{2} \partial_i u_j \partial_j u_i \, dx$$
$$+ \int \partial_3 u_3 (\partial_1 u_1 + \partial_2 u_2)^2 \, dx$$
$$- 2\sum_{i=1}^{3} \int \partial_i (\partial_1 u_1 + \partial_2 u_2)u_3 \partial_3 u_i \, dx$$

and thus

$$\int (\Delta u_1)\partial_1 \varpi + (\Delta u_2)\partial_2 \varpi \, dx \le 4\|\partial_3 \vec{u}\|_r (\|\vec{\nabla} u_1\|^2_{\frac{2r}{r-1}} + \|\vec{\nabla} u_1\|^2_{\frac{2r}{r-1}})$$
$$+ 2\|\partial_3 \vec{u}\|_r \|u_3\|_{\frac{2r}{r-2}} (\|\Delta u_1\|_2 + \|\Delta u_2\|_2) \tag{11.40}$$
$$\le C\|\partial_3 \vec{u}\|_r J(t)^{1-\frac{3}{2r}} M(t)^{\frac{3}{2r}}$$
$$+ C\|\partial_3 \vec{u}\|_r M(t)^{\frac{1}{2}} N(t)^{\frac{3}{2r}-\frac{1}{2}} K(t)^{\frac{2}{3}-\frac{3}{2r}}$$

Collecting together all those inequalities, we find that

$$
\begin{aligned}
\frac{d}{dt}I(t) \leq &-\frac{5}{9}\nu N(t) - \nu J(t)^2 M(t) \\
&+ C\|\partial_3\vec{u}\|_r J(t)^2 J(t)^{1-\frac{3}{2r}} M(t)^{\frac{3}{2r}} \\
&+ C\|\partial_3\vec{u}\|_r J(t)^2 M(t)^{\frac{1}{2}} N(t)^{\frac{3}{2r}-\frac{1}{2}} K(t)^{\frac{2}{3}-\frac{3}{2r}} \\
&+ C\|\vec{g}\|_2 K(t)^{\frac{1}{3}} N(t)^{\frac{1}{2}} + J(t)^2 \|\vec{g}\|_2 M(t)^{1/2} \\
&+ C\|\partial_3\vec{u}\|_r J(t)^{1-\frac{1}{2r}} M(t)^{\frac{1}{2r}} \left(K(t)^{1-\frac{3}{2r}} N(t)^{\frac{3}{2r}} \right)^{\frac{2}{3}} \\
&+ C\|\partial_3\vec{u}\|_r K(t)^{1-\frac{3}{2r}} N(t)^{\frac{3}{2r}}
\end{aligned}
\tag{11.41}
$$

Using the fact that $K(t) \leq 6I(t)$, $J(t) \leq 2^{1/3}I(t)^{1/3}$ and $I(t) \geq 1$, we get

$$
\begin{aligned}
\frac{d}{dt}I(t) \leq &-\frac{5}{9}\nu N(t) - \nu J(t)^2 M(t) \\
&+ C\|\partial_3\vec{u}\|_r I(t)^{1-\frac{3}{2r}} \left(J^2(t)M(t) \right)^{\frac{3}{2r}} \\
&+ C\|\partial_3\vec{u}\|_r \left(J(t)^2 M(t) \right)^{\frac{1}{2}} N(t)^{\frac{3}{2r}-\frac{1}{2}} I(t)^{1-\frac{3}{2r}} \\
&+ C\|\vec{g}\|_2 I(t)^{\frac{1}{2}} N(t)^{\frac{1}{2}} + I(t)^{\frac{1}{2}} \|\vec{g}\|_2 \left(J^2(t)M(t) \right)^{1/2} \\
&+ C\|\partial_3\vec{u}\|_r I(t)^{1-\frac{3}{2r}} \left(J^2(t)M(t) \right)^{\frac{1}{2r}} N(t)^{\frac{2}{2r}} \\
&+ C\|\partial_3\vec{u}\|_r I(t)^{1-\frac{3}{2r}} N(t)^{\frac{3}{2r}}
\end{aligned}
\tag{11.42}
$$

Recalling that $\frac{1}{p} = 1 - \frac{3}{2r}$, we find that

$$
\frac{d}{dt}I(t) \leq C\frac{1}{\nu}\|\vec{g}\|_2^2 + C\nu^{1-p}\|\partial_3\vec{u}\|_3^p I(t)
\tag{11.43}
$$

If $T_{\text{MAX}} < +\infty$, $I(t)$ must blow up when $t \to T_{\text{MAX}}$ (as the L^6 norm of $\vec{u}$ blows up, according to Serrin's criterion), and thus, by Grönwall's lemma, $\int_0^{T_{\text{MAX}}} \|\partial_3\vec{u}\|_r^p \, dt = +\infty$. $\square$

Proposition 11.7.
Let $\vec{u}_0 \in H^1$ with $\operatorname{div} \vec{u}_0 = 0$ and $\vec{f} \in L^2((0,+\infty), H^1)$. Let $\vec{u}$ be a solution of

$$
\partial_t \vec{u} = \nu \Delta \vec{u} + \mathbb{P}(\vec{f} - \operatorname{div}(\vec{u} \otimes \vec{u}))
$$

with $\vec{u} \in \mathcal{C}([0,T], H^1) \cap L^2((0,T), H^2)$ for all $T < T_{\text{MAX}}$ and $\vec{u}(0,.) = \vec{u}_0$. If the maximal existence time T_{MAX} satisfies $T_{\text{MAX}} < +\infty$, then

- $\int_0^{T_{\text{MAX}}} \|\varpi\|_p^r \, dt = +\infty$ *for $\frac{2}{p} + \frac{3}{r} = 2$ and $1 < p < +\infty$ (and $3/2 < r < +\infty$).*

- *if $\vec{f} \in L^2 H^1$, $\int_0^{T_{\text{MAX}}} \|\vec{\nabla}\varpi\|_q^p \, dt = +\infty$ for $\frac{2}{p} + \frac{3}{q} = 3$ and $\frac{2}{3} < p < +\infty$ (and $1 < q < +\infty$).*

Proof. The first problem is, of course, to include the pressure ϖ in the estimates, and to exclude the term $\vec{u} \cdot \vec{\nabla}\vec{u}$ as we have no good control on this term (except for its divergence, as $\operatorname{div}(\vec{u} \cdot \vec{\nabla}u) = -\Delta\varpi$).

It means that the proof will no longer be based on a Grönwall lemma applied to some norm $\|\vec{u}(t,.)\|_{\dot{H}^\alpha}^2$ as the computations of $\frac{d}{dt}\|\vec{u}(t,.)\|_{\dot{H}^\alpha}^2$ would involve $\vec{u} \cdot \vec{\nabla}\vec{u}$ and would not depend on ϖ. Instead of that, we shall estimate $\frac{d}{dt}\|\vec{u}(t,.)\|_\theta^\theta$: it will not involve $\vec{u} \cdot \vec{\nabla}\vec{u}$ but, when $\theta \neq 2$, it will involve ϖ.

Under the assumption that $\vec{f} \in L^2 L^2$, we shall estimate $\|\vec{u}\|_4^4$. Under the assumption $\vec{f} \in L^2 H^1$, we shall estimate $\|\vec{u}\|_\theta^\theta$ with $\theta > 4$.

The quantity $\|\vec{u}\|_4^4$:
As $\vec{f} \in L^2 L^2$, we know that, for every $0 < T_0 < T < T_{\text{MAX}}$, $\vec{u}$ will belong to $\mathcal{C}([T_0, T], H^1) \cap L^2((T_0, T), H^2)$. Thus, $\vec{u}$ belongs to $\mathcal{C}([T_0, T], L^\theta(\mathbb{R}^3))$ for every $\theta \in (2, 6)$. Moreover, we have

$$\frac{d}{dt} \int |\vec{u}|^4 = 4 \int |\vec{u}|^2 \partial_t \vec{u} \cdot \vec{u} \, dx \tag{11.44}$$

This is obvious if $\vec{u}$ is regular enough; but if we write

$$\partial_t \vec{u} = \nu \Delta \vec{u} + \mathbb{P}\vec{f} - \vec{u} \cdot \vec{\nabla}\vec{u} - \vec{\nabla}\varpi$$

we find that $\partial_t \vec{u} \in L^2 L^2$ on (T_0, T_1), while we know that $\vec{u} \in \mathcal{C}L^4 \cap L^6 L^6$; we may then conclude by a density argument.

We have

$$\partial_j(|\vec{u}|^2 \vec{u}) = |\vec{u}|^2 \partial_j \vec{u} + 2 \sum_{k=1}^{3} \partial_j u_k u_k \vec{u}$$

and, as $\vec{u}$ is bounded in $L^\infty L^6$ and $\vec{\nabla} \otimes \vec{u}$ is bounded in $L^2 L^6$ on $[T_0, T] \times \mathbb{R}^3$, we find that $\partial_j(|\vec{u}|^{\theta-2}\vec{u}) \in L^\infty L^2$. Thus, we may write

$$\int \Delta\vec{u} \cdot |\vec{u}|^2 \vec{u} \, dx = -\sum_{j=1}^{3} \int \partial_j \vec{u} \cdot \partial_j(|\vec{u}|^2 \vec{u}) \, dx$$

$$= -\int |\vec{u}|^2 |\vec{\nabla} \otimes \vec{u}|^2 \, dx - 2\sum_{j=1}^{3} \int |\partial_j \vec{u} \cdot \vec{u}|^2 \, dx \tag{11.45}$$

$$= -\int |\vec{u}|^2 |\vec{\nabla} \otimes \vec{u}|^2 \, dx - \frac{1}{2} \int |\vec{\nabla}(|\vec{u}|^2)|^2 \, dx$$

The next term we study is $I = \int (\vec{u} \cdot \vec{\nabla}\vec{u}) \cdot |\vec{u}|^2 \vec{u} \, dx$. As $\text{div}\,\vec{u} = 0$, we have

$$I = -\int \vec{u} \cdot (\vec{u} \cdot \vec{\nabla}(|\vec{u}|^2 \vec{u})) \, dx$$

$$= -I - 2\sum_{j=1}^{3}\sum_{k=1}^{3} \int \partial_j u_k u_j u_k |\vec{u}|^2 \, dx$$

$$= -I - \int |\vec{u}|^2(\vec{u} \cdot \vec{\nabla}(|\vec{u}|^2)) \, dx$$

$$= -I$$

$$= 0$$

so that

$$\int (\vec{u} \cdot \vec{\nabla}\vec{u}) \cdot |\vec{u}|^2 \vec{u} \, dx = 0. \tag{11.46}$$

Thus far, we have proven that

$$\frac{d}{dt}(\|\vec{u}\|_4^4) \leq -4\nu \int |\vec{u}|^2 |\vec{\nabla} \otimes \vec{u}|^2 \, dx - 2\nu \| |\vec{u}|^2 \|_{\dot{H}^1}^2 + 4 \int |\vec{u}|^2 \vec{u} \cdot \mathbb{P}\vec{f} \, dx$$

$$- 4 \int |\vec{u}|^2 \vec{u} \cdot \vec{\nabla}\varpi \, dx.$$

We then write

$$
\begin{aligned}
\left| 4 \int |\vec{u}|^2 \vec{u} \cdot \mathbb{P}\vec{f}\, dx \right| &\leq 4 \||\vec{u}|^2\|_6 \|\vec{u}\|_3 \|\mathbb{P}\vec{f}\|_2 \\
&\leq C \||\vec{u}|^2\|_{\dot{H}^1} \|\vec{u}\|_3 \|\vec{f}\|_2 \\
&\leq \frac{\nu}{2} \||\vec{u}|^2\|_{\dot{H}^1}^2 + \frac{C^2}{2\nu} \|\vec{u}\|_3^2 \|\vec{f}\|_2^2 \\
&\leq \frac{\nu}{2} \||\vec{u}|^2\|_{\dot{H}^1}^2 + \frac{C^2}{2\nu} \frac{1}{2} \left(\frac{1}{3} \|\vec{u}\|_4^4 + \frac{2}{3} \|\vec{u}\|_2 \right) \|\vec{f}\|_2^2.
\end{aligned}
\tag{11.47}
$$

We are thus left with the task of estimating $J = 4 \int |\vec{u}|^2 \vec{u} \cdot \vec{\nabla} \varpi\, dx$. We first integrate by parts and find (since $\operatorname{div} \vec{u} = 0$)

$$
J = -4 \int \varpi \vec{u} \cdot \vec{\nabla} |\vec{u}|^2\, dx
$$

Thus, we get two different estimates for J:

$$
|J| \leq 4 \int |\vec{\nabla}\varpi| |\vec{u}|^3\, dx
\tag{11.48}
$$

and

$$
|J| \leq 4 \left(\int |\varpi|^2 |\vec{u}|^2\, dx \right)^{1/2} \left(\int |\vec{\nabla}(|\vec{u}|^2)|^2\, dx \right)^{1/2}
\tag{11.49}
$$

Case $\frac{2}{p} + \frac{3}{r} = 2$ and $1 < p < +\infty$:
Let $\frac{2}{\sigma} = \frac{1}{2}(1 - \frac{1}{r})$. We use estimate (11.49) to get

$$
\begin{aligned}
|J| &\leq 4 \|\varpi \vec{u}\|_2 \||\vec{u}|^2\|_{\dot{H}^1} \\
&\leq 4 \|\varpi\|_r^{1/2} \|\varpi\|_{\sigma/2}^{1/2} \|\vec{u}\|_\sigma \||\vec{u}|^2\|_{\dot{H}^1}
\end{aligned}
$$

As ϖ is given by Riesz transforms applied to $u_i u_j$ and as $\sigma/2 \in (1, +\infty)$, we have $\|\varpi\|_{\sigma/2}^{1/2} \leq C \|\vec{u}\|_\sigma$ and thus

$$
|J| \leq C \|\varpi\|_r^{1/2} \|\vec{u}\|_\sigma^2 \||\vec{u}|^2\|_{\dot{H}^1}
$$

As $3/2 < r < +\infty$, we find that $\sigma \in (\frac{1}{12}, \frac{1}{4})$; writing $\frac{1}{\sigma} = \lambda \frac{1}{4} + (1 - \lambda)\frac{1}{12}$, we get

$$
\|\vec{u}\|_\sigma \leq \|\vec{u}\|_4^\lambda \|\vec{u}\|_{12}^{1-\lambda} \leq C \|\vec{u}\|_4^\lambda \||\vec{u}|^2\|_2^{(1-\lambda)/2}
$$

so that

$$
\begin{aligned}
|J| &\leq C \|\varpi\|_r^{1/2} \|\vec{u}\|_4^{2\lambda} \||\vec{u}|^2\|_{\dot{H}^1}^{2-\lambda} \\
&\leq \frac{\nu}{2} \||\vec{u}|^2\|_{\dot{H}^1}^2 + C_\nu \|\varpi\|_r^{1/\lambda} \|\vec{u}\|_4^4
\end{aligned}
$$

Now, notice that we have $\lambda = \frac{6}{\sigma} - \frac{1}{2} = \frac{1}{1 - \frac{3}{2r}} = \frac{1}{p}$. It means that we finally get

$$
\frac{d}{dt}(\|\vec{u}\|_4^4) + \nu \||\vec{u}|^2\|_{\dot{H}^1}^2 \leq C_\nu (\|\vec{u}\|_4^4 + \|\vec{u}\|_2) \|\vec{f}\|_2^2 + C_\nu \|\varpi\|_r^p \|\vec{u}\|_4^4
\tag{11.50}
$$

and we may conclude by using Grönwall's lemma.

Case $\frac{2}{p} + \frac{3}{q} = 3$ and $1 < p < +\infty$:

We have $\frac{2}{p} + \frac{3}{q} = 3$, thus $1 < q < 3$. If $\frac{1}{r} = \frac{1}{q} - \frac{1}{3}$, then $\|\varpi\|_r \le C\|\vec{\nabla}\varpi\|_q$, $3/2 < r < +\infty$ and $\frac{2}{p} + \frac{3}{r} = 2$. This case has thus already been dealt with.

The quantity $\|\vec{u}\|_\theta^\theta$, $4 < \theta < +\infty$:

If $\vec{f} \in L^2 H^1$, we know that, for every $0 < T_0 < T < T_{\text{MAX}}$, $\vec{u}$ will belong to $\mathcal{C}([T_0, T], H^2) \cap L^2((T_0, T), H^3)$. Thus, $\vec{u}$ belongs to $\mathcal{C}([T_0, T], L^\theta(\mathbb{R}^3))$ for every $\theta \in (2, +\infty)$. Moreover, we have

$$\frac{d}{dt}\int |\vec{u}|^\theta = \theta \int |\vec{u}|^{\theta-2}\partial_t\vec{u} \cdot \vec{u}\,dx. \tag{11.51}$$

This is obvious if $\vec{u}$ is regular enough; but if we write

$$\partial_t\vec{u} = \nu\Delta\vec{u} + \mathbb{P}\vec{f} - \vec{u}\cdot\vec{\nabla}\vec{u} - \vec{\nabla}\varpi,$$

we find that $\partial_t\vec{u} \in L^2 L^2$ (and even in $L^2 H^1$) on (T_0, T_1), while we know that $\vec{u} \in CL^\theta \cap L^{2(\theta-1)}L^{2(\theta-1)}$; we may then conclude by a density argument.

We have

$$\partial_j(|\vec{u}|^{\theta-2}\vec{u}) = |\vec{u}|^{\theta-2}\partial_j\vec{u} + (\theta-2)\sum_{k=1}^{3}\partial_j u_k u_k |\vec{u}|^{\theta-4}\vec{u}$$

and, as $|\vec{u}|^{\theta-2}$ is bounded on $[T_0, T] \times \mathbb{R}^3$, we find that $\partial_j(|\vec{u}|^{\theta-2}\vec{u}) \in L^\infty L^2$. Thus, we may write

$$\int \Delta\vec{u} \cdot |\vec{u}|^{\theta-2}\vec{u}\,dx = -\sum_{j=1}^{3}\int \partial_j\vec{u} \cdot \partial_j(|\vec{u}|^{\theta-2}\vec{u})\,dx$$

$$= -\int |\vec{u}|^{\theta-2}|\vec{\nabla}\otimes\vec{u}|^2\,dx - \sum_{j=1}^{3}(\theta-2)\int |\partial_j\vec{u}\cdot\vec{u}|^2 |\vec{u}|^{\theta-4}\,dx \tag{11.52}$$

$$= -\int |\vec{u}|^{\theta-2}|\vec{\nabla}\otimes\vec{u}|^2\,dx - \frac{4}{\theta^2}(\theta-2)\int |\vec{\nabla}(|\vec{u}|^{\theta/2})|^2\,dx.$$

The next term we study is $I = \int(\vec{u}\cdot\vec{\nabla}\vec{u}).|\vec{u}|^{\theta-2}\vec{u}\,dx$. As div $\vec{u} = 0$, we have

$$I = -\int \vec{u}\cdot(\vec{u}\cdot\vec{\nabla}(|\vec{u}|^{\theta-2}\vec{u}))\,dx$$

$$= -I - (\theta-2)\sum_{j=1}^{3}\sum_{k=1}^{3}\int \partial_j u_k u_j u_k |\vec{u}|^{\theta-2}\,dx$$

$$= -I - \frac{2(\theta-2)}{\theta}\int |\vec{u}|^{\theta/2}(\vec{u}\cdot\vec{\nabla}(|\vec{u}|^{\theta/2})\,dx$$

$$= -I$$

$$= 0$$

so that

$$\int (\vec{u}\cdot\vec{\nabla}\vec{u})\cdot|\vec{u}|^{\theta-2}\vec{u}\,dx = 0. \tag{11.53}$$

Thus far, we have proven that

$$\frac{d}{dt}(\|\vec{u}\|_\theta^\theta) \le -\nu\frac{4(\theta-2)}{\theta}\||\vec{u}|^{\theta/2}\|_{\dot{H}^1}^2 + \theta\int |\vec{u}|^{\theta-2}\vec{u}\cdot\mathbb{P}\vec{f}\,dx - \theta\int |\vec{u}|^{\theta-2}\vec{u}\cdot\vec{\nabla}\varpi\,dx$$

We write, as $\theta \geq 2$, $|\vec{u}|^{\theta-1} \leq |\vec{u}|^{\theta/2} + |\vec{u}|^{\theta}$ and thus

$$|\theta \int |\vec{u}|^{\theta-2}\vec{u} \cdot \mathbb{P}\vec{f}\, dx| \leq \theta \|\, |\vec{u}|^{\theta/2}\|_2 \|\mathbb{P}\vec{f}\|_2 + \theta \|\, |\vec{u}|^{\theta/2}\|_2 \|\, |\vec{u}|^{\theta/2}\|_6 \|\mathbb{P}\vec{f}\|_3$$

$$\leq \theta \|\vec{u}\|_\theta^{\theta/2}\|\vec{f}\|_2 + C\theta \|\vec{u}\|_\theta^{\theta/2} \|\, |\vec{u}|^{\theta/2}\|_{\dot{H}^1}\|\vec{f}\|_{H^{1/2}} \qquad (11.54)$$

$$\leq \nu \frac{\theta-2}{\theta}\|\, |\vec{u}|^{\theta/2}\|_{\dot{H}^1}^2 + \frac{\theta^2}{4}\|\vec{f}\|_{H^1}^2 + \|\vec{u}\|_\theta^\theta (1 + C^2 \frac{\theta^3}{\theta-2}\frac{1}{\nu}\|\vec{f}\|_{\dot{H}^1}^2)$$

We are thus left with the task of estimating $J = \theta \int |\vec{u}|^{\theta-2}\vec{u} \cdot \vec{\nabla}\varpi\, dx$.
We first integrate by parts and find (since div $\vec{u} = 0$)

$$J = -\theta \int \varpi \vec{u} \cdot \vec{\nabla}|\vec{u}|^{\theta-2}\, dx$$

For $\theta \geq 4$, we may write $\vec{\nabla}|\vec{u}|^{\theta-2} = \frac{2(\theta-2)}{\theta}|\vec{u}|^{\theta/2-2}\vec{\nabla}(|\vec{u}|^{\theta/2})$. Thus, we get two different estimates for J:

$$|J| \leq \theta \int |\vec{\nabla}\varpi|\,|\vec{u}|^{\theta-1}\, dx \qquad (11.55)$$

and

$$|J| \leq 2(\theta-2)\left(\int |\varpi|^2 |\vec{u}|^{\theta-2}\, dx\right)^{1/2}\left(\int |\vec{\nabla}(|\vec{u}|^{\theta/2})|^2\, dx\right)^{1/2} \qquad (11.56)$$

Case $\frac{2}{p} + \frac{3}{q} = 3$ and $2/3 < p \leq 1$:
In this case, we have $q \in [3, +\infty)$. We use (11.55) and (11.56) to get

$$|J| \leq \theta \|\vec{\nabla}\varpi\|_q \|\vec{u}\|_{(\theta-1)\frac{q}{q-1}}^{\theta-1}$$

and (since the Riesz transforms are bounded on $L^{(\theta+2)/2}$)

$$|J| \leq C_\theta \|\vec{u}\|_{\theta+2}^{(\theta+2)/2} \|\, |\vec{u}|^{\theta/2}\|_{\dot{H}^1}.$$

Recall that we seek a control in terms of $\|\vec{u}\|_\theta$ and $\|\, |\vec{u}|^{\theta/2}\|_{\dot{H}^1}$ (hence in terms of $\|\vec{u}\|_{3\theta}$). Thus we need to choose $\theta \geq 4$ such that

$$\theta < (\theta-1)\frac{q}{q-1} < 3\theta \text{ and } \theta < \theta+2 < 3\theta.$$

The simplest choice would be

$$(\theta-1)\frac{q}{q-1} = \theta+2,$$

or equivalently $\theta = 3q - 2 \in [7, +\infty)$. For this choice of θ, writing

$$\|v\|_{\theta+2} \leq \|v\|_\theta^{\frac{\theta-1}{\theta+2}}\|v\|_{3\theta}^{\frac{3}{\theta+2}} \leq C_\theta \|v\|_\theta^{\frac{\theta-1}{\theta+2}}\|\, |v|^{\theta/2}\|_{\dot{H}^1}^{\frac{6}{\theta(\theta+2)}},$$

we find (since $\theta - 1 = 3(q-1) = \frac{2q}{p}$)

$$|J| \leq C_\theta \|\vec{\nabla}\varpi\|_q \|\vec{u}\|_{\theta+2}^{(\theta+2)(1-\frac{1}{q})} \leq C_\theta' \|\vec{\nabla}\varpi\|_q \|\vec{u}\|_\theta^{\frac{2(q-1)}{p}}\|\, |\vec{u}|^{\theta/2}\|_{\dot{H}^1}^{\frac{4}{\theta p}}$$

and

$$|J| \leq C_\theta \|\vec{u}\|_{\theta+2}^{(\theta+2)\frac{1}{2}}\|\, |\vec{u}|^{\theta/2}\|_{\dot{H}^1} \leq C_\theta' \|\vec{u}\|_\theta^{\frac{q}{p}}\|\, |\vec{u}|^{\theta/2}\|_{\dot{H}^1}^{1+\frac{3}{\theta}}$$

Finally, writing $|J| = |J|^{1/2}|J|^{1/2}$, we find

$$|J| \le C_\theta \|\vec{\nabla}\varpi\|_q^{1/2}\|\vec{u}\|_\theta^{\frac{\theta}{2p}} \||\vec{u}|^{\theta/2}\|_{\dot{H}^1}^{\frac{1+\frac{3}{\theta}}{2}+\frac{2}{\theta p}}$$

Now, we check[2] that the last exponent, $\frac{1+\frac{3}{\theta}}{2} + \frac{2}{\theta p}$, is equal to $\frac{2p-1}{p}$: equivalently, we must check that $\frac{3}{2} - \frac{1}{p} = \frac{1}{\theta}(\frac{3}{2} + \frac{2}{p})$, or that $\frac{3}{2q} = \frac{1}{\theta}(\frac{9}{2} - \frac{3}{q})$ and finally $\theta = 3q - 2$.

Thus, we have

$$|J| \le C_\theta \left(\|\vec{\nabla}\varpi\|_q^p \|\vec{u}\|_\theta^\theta \right)^{\frac{1}{2p}} \left(\||\vec{u}|^{\theta/2}\|_{\dot{H}^1} \right)^{1-\frac{1}{2p}} \tag{11.57}$$

From this and by Young's inequality, we get

$$\frac{d}{dt}(\|\vec{u}\|_\theta^\theta) \le \frac{\theta^2}{4}\|\vec{f}\|_{\dot{H}^1}^2 + \|\vec{u}\|_\theta^\theta(1 + C_\theta \frac{1}{\nu}\|\vec{f}\|_{\dot{H}^1}^2 + C_\theta \frac{1}{\nu^{2p-1}}\|\vec{\nabla}\varpi\|_q^p) \tag{11.58}$$

If $T_{\text{MAX}} < +\infty$, then, by Serrin's criterion, $\|\vec{u}\|_\theta^\theta$ must explode and thus, by Grönwall's lemma, one must have $\int_0^{T_{\text{MAX}}} \|\vec{\nabla}\varpi\|_q^p \, dx = +\infty$. □

11.6 Vorticity

Vorticity has always played a prominent role in the study of turbulent flows. In his book *Vorticity and Turbulence* [123], Chorin studied turbulence theory for incompressible flow described in terms of the vorticity field. Similarly, in their book *Vorticity and Incompressible Flow* [40], Bertozzi and Majda insist on *vortex dynamics*

> *which in lay terms refer to the interaction of local whirls or eddies in the fluid*

The celebrated Beale–Kato–Majda criterion [27] expresses the link between blow-up and high vorticity:

Beale–Kato–Majda criterion

Theorem 11.6.
Let $\vec{u}_0 \in H^3$ with div $\vec{u}_0 = 0$ *and $\vec{f} \in L^2((0, +\infty), H^2)$. Let $\vec{u}$ be a solution of*

$$\partial_t \vec{u} = \nu\Delta\vec{u} + \mathbb{P}(\vec{f} - \text{div}(\vec{u} \otimes \vec{u}))$$

with $\vec{u} \in \mathcal{C}([0,T], H^3) \cap L^2((0,T), H^4)$ for all $T < T_{\text{MAX}}$ and $\vec{u}(0, .) = \vec{u}_0$. Let $\vec{\omega} = \text{curl } \vec{u}$. Then we have, for any $T \in (0, T_{\text{MAX}})$

$$\|\vec{u}(T, .)\|_{\dot{H}^3}^2 \le C_0 e^{(\|\vec{u}_0\|_{H^3}^2 + \frac{1}{\nu}\|\vec{f}\|_{L^2 H^2}^2)e^{C_0 \int_0^T \|\vec{\omega}\|_\infty \, dt}} \tag{11.59}$$

where the constant C_0 does not depend on T.
In particular, if the maximal existence time T_{MAX} satisfies $T_{\text{MAX}} < +\infty$, then $\int_0^{T_{\text{MAX}}} \|\vec{\omega}\|_\infty \, dt = +\infty$.

---

[2]This could be done by a simple scaling argument.

As a matter of fact, the role of vorticity *per se* in this analytical criterion turned out not to be as significant as it seemed. Indeed, the criterion could be expressed in a norm on $\vec{\omega}$ that is equivalent on a norm on the whole gradient $\vec{\nabla} \otimes \vec{u}$: the L^∞ norm is not a good one as it is unstable under Riesz transforms while one needs Riesz transforms to express $\vec{\nabla} \otimes \vec{u}$ in terms of $\vec{\omega}$:

$$\vec{\nabla} \otimes \vec{u} = -\vec{\nabla} \otimes \frac{1}{\Delta}(\vec{\nabla} \wedge \vec{\omega}).$$

Recall that we have seen in Theorem 11.3 that the L^∞ norm may be replaced by the weaker norm $\|\vec{\omega}\|_{BMO}$ (Kozono and Taniuchi [278]), or by the still weaker norm $\|\vec{\omega}\|_{\dot{B}^0_{\infty,\infty}}$ (Kozono, Ogawa and Taniuchi [273]).

In order to highlight the role of vorticity, it is thus necessary to get into greater details into the geometrical aspects of this role. Taylor [466] insisted on the role of vortex stretching in the production (or dissipation) of vorticity. To explain this phenomenon, let us study the (local) *enstrophy* $\mathcal{E} = \frac{1}{2}|\vec{\omega}|^2$. If the flow is regular enough, we find

$$\partial_t \mathcal{E} = \vec{\omega} \cdot \partial_t \vec{\omega}$$
$$= \nu\vec{\omega} \cdot (\Delta\vec{\omega}) - \vec{\omega} \cdot (\vec{u} \cdot \vec{\nabla}\vec{\omega} - \vec{\omega} \cdot \vec{\nabla}\vec{u}) + \vec{\omega} \cdot \operatorname{curl} \vec{f}$$

As $\operatorname{div} \vec{u} = 0$, we have

$$\vec{\omega} \cdot (\vec{u} \cdot \vec{\nabla}\vec{\omega}) = \sum_{j=1}^3 u_j\vec{\omega}.\partial_j\vec{\omega} = \vec{u} \cdot \vec{\nabla}\mathcal{E}.$$

On the other hand, we have

$$\vec{\omega} \cdot (\vec{\omega} \cdot \vec{\nabla}\vec{u}) = \sum_{i=1}^3 \sum_{j=1}^3 \omega_i\omega_j\partial_j u_i = \sum_{i=1}^3 \sum_{j=1}^3 \epsilon_{i,j}\omega_i\omega_j$$

where ϵ is the strain tensor

$$\epsilon = \frac{1}{2}\left(Du + (Du)^T\right).$$

Finally, we write

$$\Delta\mathcal{E} = \vec{\omega} \cdot \Delta\vec{\omega} + |\vec{\nabla} \otimes \vec{\omega}|^2$$

and we get the equation that expresses the material derivative enstrophy:

$$\frac{D}{Dt}\mathcal{E} = \partial_t\mathcal{E} + \vec{u} \cdot \vec{\nabla}\mathcal{E} = \nu\Delta\mathcal{E} - \nu|\vec{\nabla} \otimes \vec{\omega}|^2 + \sum_{i=1}^3 \sum_{j=1}^3 \epsilon_{i,j}\omega_i\omega_j + \vec{\omega} \cdot \operatorname{curl} \vec{f} \qquad (11.60)$$

Thus, the inner production of enstrophy will be found in the regions where the quadratic form $\sum_{i=1}^3 \sum_{j=1}^3 \epsilon_{i,j}\omega_i\omega_j$ is positive, i.e., where the vorticity $\vec{\omega}$ aligns with the eigenvectors that correspond to positive eigenvalues of the tensor matrix (recall that the trace of ϵ is equal to the divergence of $\vec{u}$, hence is equal to 0, so that the eigenvalues cannot all be negative). One can find discussions on this production of enstrophy through the interaction between vorticity and strain and on its significance in the papers of Galanti, Gibbon and Heritage [193] and of Tsinober [483].

In this section, we will focus on the result of Constantin and Fefferman [128], which states that, whenever the direction of vorticity evolves regularly in the areas where the vorticity is large, the solution cannot blow up. We will more precisely prove the following generalization[3] by Beirão da Vega and Berselli [30, 31]:

---

[3]See also the survey by Berselli [38] and the recent paper by Giga and Miura [211].

Vorticity direction

Theorem 11.7.
Let $\vec{u}_0 \in H^3$ with $\operatorname{div} \vec{u}_0 = 0$ and $\vec{f} \in L^2((0,+\infty), H^2)$. Let $\vec{u}$ be a solution of

$$\partial_t \vec{u} = \nu \Delta \vec{u} + \mathbb{P}(\vec{f} - \operatorname{div}(\vec{u} \otimes \vec{u}))$$

with $\vec{u} \in \mathcal{C}([0,T], H^3) \cap L^2((0,T), H^4)$ for all $T < T_{\text{MAX}}$ and $\vec{u}(0,.) = \vec{u}_0$. Let $\vec{\omega} = \operatorname{curl} \vec{u}$ and, for $\vec{\omega}(t,x) \neq 0$, $\vec{\xi}(t,x) = \frac{\vec{\omega}(t,x)}{\|\vec{\omega}(t,x)\|}$.

Let $R > 0$. Then, if the maximal existence time T_{MAX} satisfies $T_{\text{MAX}} < +\infty$, we have

$$\limsup_{t \to T_{\text{MAX}}^-} \quad \sup_{|\vec{\omega}(t,x)|>R, |\vec{\omega}(t,y)|>R} \frac{|\vec{\xi}(t,x) \wedge \vec{\xi}(t,y)|}{|x-y|^{\frac{1}{2}}} = +\infty.$$

Proof. Let $M_R(t) = \sup_{|\vec{\omega}(t,x)|>R, |\vec{\omega}(t,y)|>R} \frac{|\vec{\xi}(t,x) \wedge \vec{\xi}(t,y)|}{|x-y|^{\frac{1}{2}}}$. We have

$$M_R(t) \leq C \frac{1}{R} \|\vec{\omega}\|_{\dot{B}^{1/2}_{\infty,\infty}} \leq C' \frac{1}{R} \|\Delta \vec{\omega}\|_2.$$

Thus, $M_R(t)$ is well defined as long as $\vec{u}$ remains controlled in the H^3 norm.

In order to show that $\vec{u}$ does not blow up in H^3 if $M_R(t)$ remains bounded, it is enough to show that $\vec{u}$ does not blow up in H^1; as $\vec{u}$ is controlled in $L^\infty L^2 \cap L^2 H^1$, we just have to show that $\vec{\omega}$ does not blow up in L^2. From (11.60), we see that

$$\frac{d}{dt}\left(\frac{\|\vec{\omega}\|_2^2}{2}\right) = -\nu\|\vec{\nabla} \otimes \vec{\omega}\|_2^2 + \int \vec{f} \cdot \operatorname{curl} \vec{\omega} \, dx + \int \sum_{i=1}^{3}\sum_{j=1}^{3} \epsilon_{i,j}\omega_i\omega_j \, dx$$

As $\epsilon_{i,j}$ is given by Riesz transforms of $\vec{\omega}$, a direct estimate would give

$$\frac{d}{dt}\left(\frac{\|\vec{\omega}\|_2^2}{2}\right) \leq -\frac{\nu}{2}\|\vec{\omega}\|_{\dot{H}^1}^2 + C\frac{1}{\nu}\|\vec{f}\|_{H^1}^2 + C\|\vec{\omega}\|_3^3$$

$$\leq -\frac{\nu}{2}\|\vec{\omega}\|_{\dot{H}^1}^2 + C\frac{1}{\nu}\|\vec{f}\|_{H^1}^2 + C'\|\vec{\omega}\|_2^{3/2}\|\vec{\omega}\|_{\dot{H}^1}^{3/2}$$

$$\leq C\frac{1}{\nu}\|\vec{f}\|_{H^1}^2 + C''\frac{1}{\nu^3}\|\vec{\omega}\|_2^6$$

If $\vec{\omega}$ would belong to $L^4 L^2$, we could control $\|\vec{\omega}\|_2$ by Grönwall's lemma; but we only know that $\vec{\omega}$ belongs to $L^2 L^2$. The information on $M_R(t)$ is then needed to lower the exponent of $\|\vec{\omega}\|_2$ in the last inequality from 6 to 4, to allow the use of Grönwall's lemma.

Write

$$\int \sum_{i=1}^{3}\sum_{j=1}^{3} \epsilon_{i,j}\omega_i\omega_j \, dx = \int \vec{\omega} \cdot (\vec{\omega} \cdot \vec{\nabla} \vec{u}) \, dx$$

with

$$\vec{u} = -\frac{1}{\Delta}(\vec{\nabla} \wedge \vec{\omega}).$$

If G is the Green function (the fundamental solution of $-\Delta G = \delta$), we find

$$\int \sum_{i=1}^{3}\sum_{j=1}^{3} \epsilon_{i,j}\omega_i\omega_j \, dx = -\iint \sum_{i=1}^{3}\sum_{j=1}^{3} \vec{\omega}_i(t,x)\vec{\omega}_j(t,x)\times$$

$$\times (\partial_j\partial_{\sigma(i)}G(y)\vec{\omega}_{\sigma^2(i)}(t,x-y) - \partial_j\partial_{\sigma^2(i)}G(y)\vec{\omega}_{\sigma(i)}(t,x-y))$$

$$= \iint \sum_{i=1}^{3}\sum_{j=1}^{3} |\vec{\omega}(t,x)|\,|\vec{\omega}(t,x)|\,|\vec{\omega}(t,x-y)|\mathcal{A}(t,x,y,x-y)\,dx\,dy$$

where σ is the permutation $1 \to 2 \to 3 \to 1$,

$$\mathcal{A}(t,x,y,z) =$$

$$-\sum_{i=1}^{3}\sum_{j=1}^{3} \xi_i(t,x)\xi_j(t,x)(\partial_j\partial_{\sigma(i)}G(y)\xi_{\sigma^2(i)}(t,z) - \partial_j\partial_{\sigma^2(i)}G(y)\xi_{\sigma(i)}(t,z))$$

and where the integrals are taken as principal values. The Fourier transform of $\mathcal{A}$ with respect to the y variable gives

$$\hat{\mathcal{A}}(t,x,\eta,z) = \sum_{i=1}^{3}\sum_{j=1}^{3} \xi_i(t,x)\xi_j(t,x)\eta_j\left(\eta_{\sigma(i)}\frac{1}{|\eta|^2}\xi_{\sigma^2(i)}(t,z) - \eta_{\sigma^2(i)}\frac{1}{|\eta|^2}\xi_{\sigma(i)}(t,z)\right)$$

which we may rewrite as

$$\hat{\mathcal{A}}(t,x,\eta,z) = \frac{1}{|\eta|^2}\left(\sum_{j=1}^{3} \eta_j\xi_j(t,x)\right) \text{Det}\,(\vec{\xi}(t,x),\vec{\eta},\vec{\xi}(t,z)).$$

The main point is then the identity

$$\hat{\mathcal{A}}(t,x,\eta,x) = 0$$

and thus

$$\mathcal{A}(t,x,y,x) = 0.$$

We then write $\vec{\omega} = \vec{\alpha} + \vec{\beta}$, where $\vec{\alpha} = 1_{|\vec{\omega}(t,x)|\leq R}\vec{\omega}$. We then have

$$\int \sum_{i=1}^{3}\sum_{j=1}^{3} \epsilon_{i,j}\omega_i\omega_j \, dx = -\int \vec{\omega}\cdot\left(\vec{\omega}\cdot\vec{\nabla}\left(\frac{1}{\Delta}(\vec{\nabla}\wedge\vec{\omega})\right)\right) dx$$

$$= -\int \vec{\alpha}\cdot\left(\vec{\alpha}\cdot\vec{\nabla}\left(\frac{1}{\Delta}(\vec{\nabla}\wedge\vec{\omega})\right)\right) dx$$

$$-\int \vec{\beta}\cdot\left(\vec{\beta}\cdot\vec{\nabla}\left(\frac{1}{\Delta}(\vec{\nabla}\wedge\vec{\alpha})\right)\right) dx$$

$$-\int \vec{\beta}\cdot\left(\vec{\beta}\cdot\vec{\nabla}\left(\frac{1}{\Delta}(\vec{\nabla}\wedge\vec{\beta})\right)\right) dx$$

with

$$-\int \vec{\alpha}\cdot\left(\vec{\alpha}\cdot\vec{\nabla}\left(\frac{1}{\Delta}\wedge(\vec{\nabla}\wedge\vec{\omega})\right)\right) dx \leq C\|\vec{\alpha}\|_3^2\|\vec{\omega}\|_3$$

$$\leq C\|\vec{\omega}\|_3^2\|\vec{\omega}\|_2^{2/3}\|\vec{\alpha}\|_\infty^{1/3}$$

$$\leq C'\|\vec{\omega}\|_2^{5/3}\|\vec{\omega}\|_{\dot{H}^1}R^{1/3}$$

$$-\int \vec{\beta} \cdot \left(\vec{\beta}\left(\frac{1}{\Delta} \wedge (\vec{\nabla} \wedge \vec{\alpha})\right)\right) dx \leq C\|\vec{\beta}\|_3^2 \|\vec{\alpha}\|_3$$

$$\leq C'\|\vec{\omega}\|_2^{5/3}\|\vec{\omega}\|_{\dot{H}^1} R^{1/3}$$

and

$$-\int \vec{\beta} \cdot \left(\vec{\beta} \cdot \vec{\nabla}\left(\frac{1}{\Delta} \wedge (\vec{\nabla} \wedge \vec{\beta})\right)\right) dx =$$

$$= \iint_{|\vec{\omega}(t,x)|>R,\, |\vec{\omega}(t,x)|>R} \sum_{i=1}^{3}\sum_{j=1}^{3} |\vec{\omega}(t,x)|\, |\vec{\omega}(t,x)|\, |\vec{\omega}(t,x-y)| \mathcal{A}(t,x,y,x-y)\, dx\, dy$$

In the last equality, as $\mathcal{A}(t,x,y,x) = 0$, we may replace $\mathcal{A}(t,x,y,x-y)$ with $\mathcal{A}(t,x,y,x-y) - \mathcal{A}(t,x,y,x)$ or with $\mathcal{A}(t,x,y,x-y) + \mathcal{A}(t,x,y,x)$. We have, for $|\vec{\omega}(t,x)| > R$ and $|\vec{\omega}(t,x)| > R$,

$$|\mathcal{A}(t,x,y,x-y) - \mathcal{A}(t,x,y,x)| \leq C\frac{1}{|y|^3}|\vec{\xi}(x-y) - \vec{\xi}(x)|$$

and

$$|\mathcal{A}(t,x,y,x-y) + \mathcal{A}(t,x,y,x)| \leq C\frac{1}{|y|^3}|\vec{\xi}(x-y) + \vec{\xi}(x)|$$

so that

$$|\mathcal{A}(t,x,y,x-y)| \leq C\frac{1}{|y|^3}\min(|\vec{\xi}(x-y) - \vec{\xi}(x)|, |\vec{\xi}(x-y) + \vec{\xi}(x)|)$$

$$\leq C'\frac{1}{|y|^{5/2}}M_R(t)$$

and thus

$$-\int \vec{\beta} \cdot \left(\vec{\beta} \cdot \vec{\nabla}\left(\frac{1}{\Delta} \wedge (\vec{\nabla} \wedge \vec{\beta})\right)\right) dx \leq C\|\vec{\omega}\|_3^2 \|\mathcal{I}_{1/2}\vec{\omega}\|_3 M_R(t)$$

$$\leq C'\|\vec{\omega}\|_3^2\|\vec{\omega}\|_2 M_R(t)$$

$$\leq C''\|\vec{\omega}\|_2^2\|\vec{\omega}\|_{\dot{H}^1} M_R(t).$$

Finally, we obtain

$$\frac{d}{dt}\left(\frac{\|\vec{\omega}\|_2^2}{2}\right) \leq -\frac{\nu}{2}\|\vec{\omega}\|_{\dot{H}^1}^2 + \frac{C}{\nu}\|\vec{f}\|_{H^1}^2 + C\|\vec{\omega}\|_2^{5/3}\|\vec{\omega}\|_{\dot{H}^1}R^{1/3} + C\|\vec{\omega}\|_2^2\|\vec{\omega}\|_{\dot{H}^1}M_R(t)$$

$$\leq C\frac{1}{\nu}\|\vec{f}\|_{H^1}^2 + C'\frac{1}{\nu}(R^{2/3}\|\vec{\omega}\|_2^{4/3} + M_R(t)^2\|\vec{\omega}\|_2^2)\|\vec{\omega}\|_2^2$$

and we conclude with Grönwall's lemma. $\qquad\qquad\square$

Remark: More generally, a similar proof gives that, if $2 \leq r \leq 3$ and $\int_0^{T_{\mathrm{MAX}}} \|\vec{\omega}\|_r^2\, dt < +\infty$, and if the maximal existence time T_{MAX} satisfies $T_{\mathrm{MAX}} < +\infty$, we have

$$sup_{0<t<T_{\mathrm{MAX}}} \sup_{|\vec{\omega}(t,x)|>R,\, |\vec{\omega}(t,y)|>R} \frac{|\vec{\xi}(t,x) \wedge \vec{\xi}(t,y)|}{|x-y|^{\frac{3}{r}-1}} = +\infty.$$

11.7   Squirts

Again, let $\vec{u}$ be a solution of

$$\partial_t \vec{u} = \nu \Delta \vec{u} + \mathbb{P}(\vec{f} - \operatorname{div}(\vec{u} \otimes \vec{u}))$$

with $\vec{u} \in \mathcal{C}([0,T], H^1) \cap L^2((0,T), H^2)$ for all $T < T_{\mathrm{MAX}}$ and $\vec{u}(0,.) = \vec{u}_0$, where $\vec{u}_0 \in H^1$ with $\operatorname{div} \vec{u}_0 = 0$ and $\vec{f} \in L^2((0,+\infty), L^2)$. Let us assume that the maximal existence time T_{MAX} satisfies $T_{\mathrm{MAX}} < +\infty$.

We are going to discuss the behavior of $\|\vec{u}\|_\infty$.

- As we have

$$\|\vec{u}\|_\infty \leq C \|\vec{u}\|_{\dot{B}^{3/2}_{2,1}} \leq C' \sqrt{\|\vec{u}\|_{\dot{H}^1} \|\vec{u}\|_{\dot{H}^2}}$$

we have that

$$\int_0^T \|\vec{u}\|_\infty^4 \, dt < +\infty \tag{11.61}$$

for all $T < T_{\mathrm{MAX}}$.

- As $T_{\mathrm{MAX}} < +\infty$, we have seen that

$$\int_0^{T_{\mathrm{MAX}}} \|\vec{u}\|_\infty^2 \, dx = +\infty. \tag{11.62}$$

- If $\vec{f} \in L^2 H^1$, we have a more precise estimate [210]:

$$\liminf \sqrt{T_{\mathrm{MAX}} - t} \|\vec{u}\|_\infty > 0. \tag{11.63}$$

Indeed, if $\vec{u}(t_0,.) \in L^\infty$, we may use Picard's algorithm to find a local solution of

$$\vec{v} = W_{\nu(t-t_0)} * \vec{u}(t_0,.) + \int_{t_0}^t W_{\nu(t-s)} * \mathbb{P}(\vec{f} - \operatorname{div}(\vec{v} \otimes \vec{v})) \, ds$$

in $L^\infty((t_0, t_0 + T), L^\infty)$. The existence time is estimated by the following inequality:

$$T \geq C_\nu \frac{1}{(\|\vec{u}(t_0,.)\|_\infty + \|\vec{f}\|_{L^2 H^1})^2}.$$

It is easy to check that $\vec{u} = \vec{v}$ on $(t_0, \in (T_{\mathrm{MAX}}, t_0 + T)$. In particular, we find that $T > T_{\mathrm{MAX}} - t_0$. This gives

$$\liminf_{t \to T_{\mathrm{MAX}}^-} \sqrt{T_{\mathrm{MAX}} - t} \|\vec{u}\|_\infty \geq \liminf_{t \to T_{\mathrm{MAX}}^-} \sqrt{C_\nu} - \sqrt{T_{\mathrm{MAX}} - t} \|\vec{f}\|_{L^2 H^1}$$

$$= \sqrt{C_\nu} > 0.$$

- However, $\|\vec{u}(t,.)\|_\infty$ cannot explode too fast, as we have

$$\int_0^{T_{\mathrm{MAX}}} \|\vec{u}\|_\infty \, dt < +\infty. \tag{11.64}$$

Indeed, we have seen that

$$\|\vec{u}(t,.)\|_2 \le \|\vec{u}_0\|_2^2 + \frac{1}{\nu} \int_0^t \|\vec{f}\|_2 \, ds$$

and

$$\nu \int_0^t \|\vec{\nabla} \otimes \vec{u}\|_2^2 \, dt \le (\|\vec{u}_0\|_2^2 + \frac{1}{\nu} \int_0^t \|\vec{f}\|_2 \, ds)^2.$$

Now, we write

$$\vec{u} = W_{\nu t} * \vec{u}(t_0,.) + \int_0^t W_{\nu(t-s)} * \mathbb{P}(\vec{f} - \text{div}(\vec{u} \otimes \vec{u})) \, ds$$

with

$$\int_0^{T_{\text{MAX}}} \|W_{\nu t} * \vec{u}(t_0,.)\|_\infty \, dt \le C \int_0^{T_{\text{MAX}}} \frac{1}{(\nu t)^{1/4}} \|\vec{u}_0\|_{H^1} \, dt < +\infty,$$

$$\int_0^{T_{\text{MAX}}} \| \int_0^t W_{\nu(t-s)} * \mathbb{P}\vec{f} \, ds\|_\infty \, dt \le \sqrt{T_{\text{MAX}}} \sqrt{\int_0^{T_{\text{MAX}}} \| \int_0^t W_{\nu(t-s)} * \mathbb{P}\vec{f} \, ds\|_\infty^2 \, dt)}$$

$$\le C\sqrt{T_{\text{MAX}}}(\int_0^{T_{\text{MAX}}} \| \int_0^t W_{\nu(t-s)} * \mathbb{P}\vec{f} \, ds\|_2^{1/4} \| \int_0^t W_{\nu(t-s)} * \mathbb{P}\vec{f} \, ds\|_{\dot{H}^2}^{3/4} \, dt)^{1/2}$$

$$\le C' \|\vec{f}\|_{L^2 L^2} (1 + T_{\text{MAX}}) < +\infty$$

and

$$\int_0^{T_{\text{MAX}}} \| \int_0^t W_{\nu(t-s)} * \mathbb{P} \, \text{div}(\vec{u} \otimes \vec{u}) \, ds\|_\infty \, dt$$

$$\le C \int_0^{T_{\text{MAX}}} \| \int_0^t W_{\nu(t-s)} * \mathbb{P} \, \text{div}(\vec{u} \otimes \vec{u}) \, ds\|_{\dot{B}_{2,1}^{3/2}} \, dt$$

$$\le C_\nu \int_0^{T_{\text{MAX}}} \| \text{div}(\vec{u} \otimes \vec{u})\|_{\dot{B}_{2,1}^{-1/2}} \, dt$$

$$\le C'_\nu \int_0^{T_{\text{MAX}}} \| \text{div}(\vec{u} \otimes \vec{u})\|_{L^{3/2,1}} \, dt$$

$$\le C''_\nu \int_0^{T_{\text{MAX}}} \|\vec{u}\|_{\dot{H}^1}^2 \, dt$$

$$\le C'''_\nu (\|\vec{u}_0\|_2^2 + \frac{1}{\nu} \int_0^{T_{\text{MAX}}} \|\vec{f}\|_2 \, ds)^2 < +\infty$$

where we have used the inequalities between Sobolev, Besov and Lorentz norms $uv\|_{L^{3/2,1}} \le C\|u\|_2\|v\|_{L^{6,2}}$, $\|v\|_{L^{6,2}} \le C\|v\|_{\dot{H}^1}$ and $\|v\|_{\dot{B}_{2,1}^{-1/2}} \le C\|v\|_{L^{3/2,1}}$ (which are easily deduced from the classical Sobolev inequalities through real interpolation) and the maximal regularity inequality for the heat kernel

$$\| \int_0^t W_{\nu(t-s)} * \Delta v \, ds\|_{L^1 \dot{B}_{2,1}^{-1/2}} \le C\|v\|_{L^1 \dot{B}_{2,1}^{-1/2}}$$

(see [313] for a proof).

In particular, inequality (11.64) precludes the possibility of *squirt singularities* at the blow-up time, as was noted by Córdoba, Fefferman and De la Llave [131]. They introduced the notion of squirt singularities to give a unified treatment for various singularities that had been studied for incompressible fluid mechanics, such as potato chip singularities, tube collapse singularities, and saddle point singularities. Roughly speaking, a squirt corresponds to a point x_0 from which fluid particles will be expelled at higher and higher speed: there exists a positive ϵ such that, for every $t < T_{\mathrm{MAX}}$, if a fluid particle lies in $B(x_0, \epsilon)$ at time t, then there will be a time $t' \in (t, T_{\mathrm{MAX}})$ where the particle will be expelled from the ball $B(x_0, 2\epsilon)$.

Of course, it means that we can follow the particle. The flow associated to the vector field $\vec{u}$ will have path lines given by the characteristic equation

$$\dot{X}(t) = \vec{u}(t, X(t)).$$

This equation will be solvable if $\vec{u} \in L_t^1 \mathrm{Lip}_x$. If we assume that the forcing term $\vec{f}$ belongs more precisely to $L^2 H^1$, then we know that, for $0 < T_0 < T_1 < T_{\mathrm{MAX}}$, we have $\vec{u} \in \mathcal{C}([T_0, T_1], H^2) \cap L^2((T_0, T_1), H^3)$, so that in particular

$$\int_{T_0}^{T_1} \|\vec{\nabla} \otimes \vec{u}\|_\infty^4 \, dt < +\infty.$$

Thus, we may follow the particles in the fluid.

Now, if we would have a squirt singularity at x_0, we would have for a particle lying in $B(x_0, \epsilon)$ at time t and outside $B(x_0, 2\epsilon)$ at time t'

$$\epsilon \leq |X(t') - X(t)| \leq \int_t^{t'} |\vec{u}(s, X(s))| \, ds \leq \int_t^{T_{\mathrm{MAX}}} \|\vec{u}\|_\infty \, ds$$

and thus

$$\liminf_{t \to T_{\mathrm{MAX}}^-} \int_t^{T_{\mathrm{MAX}}} \|\vec{u}\|_\infty \, ds \geq \epsilon > 0$$

But this is impossible due to inequality (11.64).

11.8   Eigenvalues of the Strain Matrix

Recall that the equation describing the evolution of the (local) enstrophy $\mathcal{E} = \frac{1}{2}|\vec{\omega}|^2$ is

$$\frac{D}{Dt}\mathcal{E} = \partial_t \mathcal{E} + \vec{u} \cdot \vec{\nabla}\mathcal{E} = \nu \Delta \mathcal{E} - \nu |\vec{\nabla} \otimes \vec{\omega}|^2 + \sum_{i=1}^{3}\sum_{j=1}^{3} \epsilon_{i,j}\omega_i\omega_j + \vec{\omega} \cdot \mathrm{curl}\, \vec{f} \qquad (11.65)$$

where ϵ is the strain tensor

$$\epsilon = \frac{1}{2}\left(Du + (Du)^T\right).$$

Enstrophy is increased in the regions where the vorticity $\vec{\omega}$ aligns with the eigenvectors that correspond to positive eigenvalues of the tensor matrix ϵ (Galanti, Gibbon and Heritage [193], Tsinober [483]). Recall that the eigenvalues $\lambda_i(t, x)$ of ϵ satisfy $\lambda_1 + \lambda_2 + \lambda_3 = \mathrm{div}\,\vec{u} = 0$, so that, if $\lambda_1 \leq \lambda_2 \leq \lambda_3$ we have $\lambda_1 \leq 0$ and $\lambda_3 \geq 0$. As a matter of fact, the sign of λ_2 plays an important role. In 1987, numerical simulations by Ashurst, Kerstein, Kerr, and Gibson [8] indicated that, when the fluid turns to turbulent, the vorticity aligns with the eigenvector associated to λ_2.

A criterion for blow up involving the positive part of λ_2 has recently been given by Miller [361] (see [97] for a related result of Chae):

Middle eigenvalue of the strain tensor

Theorem 11.8.
Let $\vec{u}_0 \in H^1$ with $\operatorname{div} \vec{u}_0 = 0$ and $\vec{f} \in L^2((0,+\infty), L^2)$. Let $\vec{u}$ be a solution of

$$\partial_t \vec{u} = \nu \Delta \vec{u} + \mathbb{P}(\vec{f} - \operatorname{div}(\vec{u} \otimes \vec{u}))$$

with $\vec{u} \in \mathcal{C}([0,T], H^1) \cap L^2((0,T), H^2)$ for all $T < T_{\text{MAX}}$ and $\vec{u}(0,.) = \vec{u}_0$. Let $\lambda_1(t,x) \leq \lambda_2(t,x) \leq \lambda_3(t,x)$ be the eigenvalues of the strain tensor $\epsilon(t,x)$. Then if and $\frac{2}{p} + \frac{3}{q} = 2$ with $3/2 < q \leq +\infty$, we have

$$\|\vec{u}(T,.)\|_{\dot{H}^1}^2 \leq (\|\vec{u}_0\|_{\dot{H}^1}^2 + \frac{1}{\nu}\|\vec{f}\|_{L^2 L^2}^2) e^{C_0 \nu^{1-\frac{p}{2}} \int_0^T \|\lambda_2^+(t,x)\|_{L^q(dx)}^p \, dt} \tag{11.66}$$

where $\lambda_2^+ = \max(0, \lambda_2)$ and where the constant C_0 does not depend on T.
In particular, if the maximal existence time T_{MAX} satisfies $T_{\text{MAX}} < +\infty$, then $\int_0^{T_{\text{MAX}}} \|\lambda_2^+(t,x)\|_{L^q(dx)}^p \, dt = +\infty$.

Proof. First, we remark that we have the identities, for $\vec{\omega} = \vec{\nabla} \wedge \vec{u}$ and $\epsilon = (\frac{\partial_j u_k + \partial_k u_j}{2})_{1 \leq j,k \leq 3}$

$$|\vec{\nabla} \otimes \vec{u}|^2 = \sum_{j=1}^{3} \sum_{k=1}^{3} |\partial_j u_k|^2 = \sum_{j=1}^{3} \sum_{k=1}^{3} \left| \frac{\partial_j u_k + \partial_k u_j}{2} + \frac{\partial_j u_k - \partial_k u_j}{2} \right|^2$$

$$= \sum_{j=1}^{3} \sum_{k=1}^{3} |\epsilon_{j,k}|^2 + \frac{1}{2} \sum_{l=1}^{3} |\omega_l|^2 = |\epsilon|^2 + \frac{1}{2}|\vec{\omega}|^2$$

and

$$-\Delta \vec{u} = \vec{\nabla} \wedge (\vec{\nabla} \wedge \vec{u}) - \vec{\nabla}(\operatorname{div} \vec{u})$$

so that, if $\vec{u} \in H^1$,

$$\int |\vec{\nabla} \otimes \vec{u}|^2 \, dx = \int |\vec{\omega}|^2 \, dx + \int |\operatorname{div} \vec{u}|^2 \, dx.$$

Thus, if $\operatorname{div} \vec{u} = 0$, we have

$$\|\vec{u}\|_{\dot{H}^1}^2 = \|\vec{\omega}\|_2^2 = 2\|\epsilon\|_2^2.$$

If $\vec{u} \in H^2$, we have

$$\int |\Delta \vec{u}|^2 \, dx = \int \sum_{1 \leq j \leq 3} \sum_{k=1}^{3} \partial_j^2 \vec{u} \cdot \partial_k^2 \vec{u} \, dx = \int \sum_{1 \leq j \leq 3} \sum_{k=1}^{3} |\partial_j \partial_k \vec{u}|^2 \, dx.$$

As $\partial_j \vec{\omega} = \vec{\nabla} \wedge (\partial_j \vec{u})$, we find as well, if $\vec{u} \in H^2$ and $\operatorname{div} \vec{u} = 0$,

$$\|\Delta \vec{u}\|_2^2 = \|\vec{\nabla} \otimes \vec{u}\|_{\dot{H}^1}^2 = \|\vec{\omega}\|_{\dot{H}^1}^2 = 2\|\epsilon\|_{\dot{H}^1}^2.$$

In order to show that $\|\vec{u}\|_{\dot{H}^1}$ remains bounded, we may equivalently show that $\|\epsilon\|_2$ remains bounded. We may now control the evolution of $\|\epsilon\|_2^2$ through the equalities:

- using $\|\vec{\omega}\|_2^2 = 2\|\epsilon\|_2^2$,

$$\frac{d}{dt}\|\epsilon\|_2^2 = \int \vec{\omega}\cdot\partial_t\vec{\omega}\,dx$$

$$= -\nu\int|\vec{\nabla}\otimes\vec{\omega}|^2\,dx - \int\vec{\omega}\cdot(\vec{u}\cdot\vec{\nabla}\vec{\omega} - \vec{\omega}\cdot\vec{\nabla}\vec{u})\,dx + \int\vec{\omega}\cdot\operatorname{curl}\vec{f}\,dx$$

$$= -\nu\|\vec{\omega}\|_{\dot{H}^1}^2 + \int(\vec{\omega}\otimes\vec{\omega})\cdot(\vec{\nabla}\otimes\vec{u})\,dx - \int\vec{f}\cdot\Delta\vec{u}\,dx$$

with

$$(\vec{\omega}\otimes\vec{\omega})\cdot(\vec{\nabla}\otimes\vec{u}) = \sum\sum_{1\le j,k\le 3}\omega_j\omega_k\partial_j u_k = (\vec{\omega}\otimes\vec{\omega})\cdot\epsilon,$$

so that

$$\frac{d}{dt}\|\epsilon\|_2^2 = -2\nu\|\epsilon\|_{\dot{H}^1}^2 + \int(\vec{\omega}\otimes\vec{\omega})\cdot\epsilon\,dx - \int\vec{f}\cdot\Delta\vec{u}\,dx \qquad (11.67)$$

- using $\|\vec{u}\|_{\dot{H}^1}^2 = 2\|\epsilon\|_2^2$,

$$\frac{d}{dt}\|\epsilon\|_2^2 = -\int\Delta\vec{u}\cdot\partial_t\vec{u}\,dx$$

$$= -\nu\int|\Delta\vec{u}|^2\,dx - \int\sum_{i=1}^{3}\partial_i\vec{u}\cdot\partial_i(\vec{u}\cdot\vec{\nabla}\vec{u})\,dx$$

$$+ \int\Delta\vec{u}\cdot\vec{\nabla}p\,dx - \int\Delta\vec{u}\cdot\vec{f}\,dx$$

$$= -\nu\|\Delta\vec{u}\|_2^2 - \int\sum_{i=1}^{3}\sum_{j=1}^{3}\sum_{k=1}^{3}\partial_i u_j\partial_i u_k\partial_k u_j\,dx - \int\vec{f}\cdot\Delta\vec{u}\,dx$$

with (writing $\Omega = \frac{1}{2}(\partial_i u_j - \partial_j u_i)_{1\le i,j\le 3}$)

$$\sum_{i=1}^{3}\sum_{j=1}^{3}\sum_{k=1}^{3}\partial_i u_j\partial_i u_k\partial_k u_j = \sum_{i=1}^{3}\sum_{j=1}^{3}\sum_{k=1}^{3}(\epsilon_{i,j}+\Omega_{i,j})(\epsilon_{i,k}+\Omega_{i,k})(\epsilon_{k,j}+\Omega_{k,j})$$

$$= \sum_{i=1}^{3}\sum_{j=1}^{3}\sum_{k=1}^{3}\epsilon_{i,j}\epsilon_{i,k}\epsilon_{k,j} + \sum_{i=1}^{3}\sum_{j=1}^{3}\sum_{k=1}^{3}\epsilon_{i,j}\epsilon_{i,k}\Omega_{k,j} + \epsilon_{i,j}\Omega_{i,k}\epsilon_{k,j} + \Omega_{i,j}\epsilon_{i,k}\epsilon_{k,j}$$

$$+ \sum_{i=1}^{3}\sum_{j=1}^{3}\sum_{k=1}^{3}\epsilon_{i,j}\Omega_{i,k}\Omega_{k,j} + \Omega_{i,j}\epsilon_{i,k}\Omega_{k,j} + \Omega_{i,j}\Omega_{i,k}\epsilon_{k,j} + \sum_{i=1}^{3}\sum_{j=1}^{3}\sum_{k=1}^{3}\Omega_{i,j}\Omega_{i,k}\Omega_{k,j}$$

$$= \sum_{i=1}^{3}\sum_{j=1}^{3}\sum_{k=1}^{3}\epsilon_{i,j}\epsilon_{i,k}\epsilon_{k,j} + \sum_{i=1}^{3}\sum_{j=1}^{3}\sum_{k=1}^{3}\epsilon_{i,j}\epsilon_{i,k}\Omega_{k,j} + \epsilon_{i,j}\Omega_{i,k}\epsilon_{k,j} + \Omega_{i,j}\epsilon_{i,k}\epsilon_{k,j}$$

$$+ \sum_{i=1}^{3}\sum_{j=1}^{3}\sum_{k=1}^{3}\epsilon_{i,j}\Omega_{i,k}\Omega_{k,j} + \Omega_{i,j}\epsilon_{i,k}\Omega_{k,j} + \Omega_{i,j}\Omega_{i,k}\epsilon_{k,j} + \sum_{i=1}^{3}\sum_{j=1}^{3}\sum_{k=1}^{3}\Omega_{i,j}\Omega_{i,k}\Omega_{k,j}$$

$$= \sum_{i=1}^{3}\sum_{j=1}^{3}\sum_{k=1}^{3}\epsilon_{j,i}\epsilon_{i,k}\epsilon_{k,j} + 3\sum_{i=1}^{3}\sum_{j=1}^{3}\sum_{k=1}^{3}\epsilon_{i,j}\epsilon_{i,k}\frac{(\Omega_{k,j}+\Omega_{j,k})}{2}$$

$$+\sum_{i=1}^{3}\sum_{j=1}^{3}\sum_{k=1}^{3}\epsilon_{i,j}(\Omega_{i,k}\Omega_{k,j}+\Omega_{i,k}\Omega_{j,k}+\Omega_{k,i}\Omega_{k,j})+\sum_{i=1}^{3}\sum_{j=1}^{3}\sum_{k=1}^{3}\frac{(\Omega_{i,j}+\Omega_{j,i})}{2}\Omega_{i,k}\Omega_{k,j}$$

$$=\sum_{i=1}^{3}\sum_{j=1}^{3}\sum_{k=1}^{3}\epsilon_{j,i}\epsilon_{i,k}\epsilon_{k,j}-\frac{1}{4}\sum_{i=1}^{3}\sum_{j=1}^{3}\epsilon_{i,j}\omega_i\omega_j$$

so that

$$\frac{d}{dt}\|\epsilon\|_2^2=-2\nu\|\epsilon\|_{\dot{H}^1}^2+\frac{1}{4}\int(\vec{\omega}\otimes\vec{\omega})\cdot\epsilon\,dx-\int\mathrm{tr}(\epsilon^3)\,dx-\int\vec{f}\cdot\Delta\vec{u}\,dx. \qquad (11.68)$$

Combining equations (11.67) and (11.68), we get rid of the term $\int(\vec{\omega}\otimes\vec{\omega})\cdot\epsilon\,dx$ and find

$$\frac{d}{dt}\|\epsilon\|_2^2=-2\nu\|\epsilon\|_{\dot{H}^1}^2-\frac{4}{3}\int\mathrm{tr}(\epsilon^3)\,dx-\int\vec{f}\cdot\Delta\vec{u}\,dx. \qquad (11.69)$$

Let $\lambda_1\le\lambda_2\le\lambda_3$ be the eigenvalues of ϵ. Then $\lambda_1+\lambda_2+\lambda_3=\mathrm{tr}(\epsilon)=\mathrm{div}\,\vec{u}=0$, $\lambda_1\lambda_2\lambda_3=\det(\epsilon)$, $|\epsilon|^2=\mathrm{tr}(\epsilon^2)=\lambda_1^2+\lambda_2^2+\lambda_3^2)$ and $\mathrm{tr}(\epsilon^3)=\lambda_1^3+\lambda_2^3+\lambda_3^3$. Moreover, we have

$$(\lambda_1+\lambda_2+\lambda_3)^3=-2(\lambda_1^3+\lambda_2^3+\lambda_3^3+3(\lambda_1+\lambda_2+\lambda_3)(\lambda_1^2+\lambda_2^2+\lambda_3^2)+6\lambda_1\lambda_2\lambda_3$$

so that

$$\mathrm{tr}(\epsilon^3)=3\det(\epsilon).$$

As $\lambda_1\lambda_3\le0$, we have

$$-\det(\epsilon)\le\lambda_2^+(-\lambda_1\lambda_3)\le\frac{1}{2}\lambda_2^+|\epsilon|^2.$$

Thus, we find:

$$\frac{d}{dt}\|\epsilon\|_2^2\le-2\nu\|\epsilon\|_{\dot{H}^1}^2+2\int\lambda_2^+|\epsilon|^2\,dx-\int\vec{f}\cdot\Delta\vec{u}\,dx. \qquad (11.70)$$

We. then write, for $3/2<q\le+\infty$, $r=\frac{q}{q-1}\in[1,3)$ and $\frac{1}{2r}=(1-\sigma)\frac{1}{2}+\sigma\frac{1}{6}$ ($\sigma\in[0,1)$)

$$2\int\lambda_2^+|\epsilon|^2\,dx\le 2\|\lambda_2^+\|_q\|\epsilon\|_{r/2}^2\le 2\|\lambda_2^+\|_q\|\epsilon\|_2^{2(1-\sigma)}\|\epsilon\|_6^{2\sigma}$$

$$\le\nu\|\epsilon\|_{\dot{H}^1}^2+C\nu^{-\frac{\sigma}{1-\sigma}}\|\lambda_2^+\|_q^{\frac{1}{1-\sigma}}\|\epsilon\|_2^2$$

and

$$-\int\vec{f}\cdot\Delta\vec{u}\,dx\le\frac{\nu}{2}\|\Delta\vec{u}\|_2^2+\frac{1}{2\nu}\|\vec{f}\|_2^2=\nu\|\epsilon\|_{\dot{H}^1}^2+\frac{1}{2\nu}\|\vec{f}\|_2^2$$

and we get

$$\frac{d}{dt}\|\epsilon\|_2^2\le C\nu^{-\frac{\sigma}{1-\sigma}}\|\lambda_2^+\|_q^{\frac{2}{1-\sigma}}\|\epsilon\|_2^2+\frac{1}{2\nu}\|\vec{f}\|_2^2. \qquad (11.71)$$

As $\|\vec{u}\|_{\dot{H}^1}^2=2\|\epsilon\|_2^2$ and $1-\sigma=\frac{3}{2r}-\frac{1}{2}=1-\frac{3}{2q}=\frac{2}{p}$, we get

$$\frac{d}{dt}\|\vec{u}\|_{\dot{H}^1}^2\le C\nu^{-\frac{\sigma}{1-\sigma}}\|\lambda_2^+\|_q^p\|\vec{u}\|_{\dot{H}^1}^2+\frac{1}{\nu}\|\vec{f}\|_2^2. \qquad (11.72)$$

and we conclude by Grönwall's lemma. $\qquad\square$

Remark: We can use the inequality of Lemarié-Rieusset in [316]:

$$\|fg\|_2\le C_r\|f\|_{\dot{M}^{2,\frac{3}{r}}}\|g\|_{\dot{B}_{2,1}^r}$$

for $0 < r < 1$, and by duality

$$\|fh\|_{\dot{B}_{2,\infty}^{-r}} \leq C_r \|f\|_{\dot{M}^{2,\frac{3}{r}}} \|h\|_2.$$

In particular, writing $k = \sqrt{|k|}\frac{k}{|k|}\sqrt{|k|}$, we get

$$\|kg\|_{\dot{B}_{2,\infty}^{-r}} \leq C_r^2 \|\sqrt{|k|}\|_{\dot{M}^{2,\frac{3}{r}}}^2 \|g\|_{\dot{B}_{2,1}^r} = C_r^2 \|k\|_{\dot{M}^{1,\frac{3}{2r}}}^2 \|g\|_{\dot{B}_{2,1}^r}$$

and thus, for $3/2 < q < +\infty$ and $\frac{2}{p} + \frac{3}{q} = 2$,

$$\int \lambda_2^+ |\epsilon|^2 \, dx \leq C \|\lambda_2^+\|_{\dot{M}^{1,q}} \|\epsilon\|_{\dot{B}_{2,1}^{\frac{3}{2q}}}^2 \leq C \|\lambda_2^+\|_{\dot{M}^{1,q}} \|\epsilon\|_2^{2-\frac{3}{q}} \|\epsilon\|_{\dot{H}^1}^{\frac{3}{q}} = C \|\lambda_2^+\|_{\dot{M}^{1,q}} \|\epsilon\|_2^{\frac{2}{p}} \|\epsilon\|_{\dot{H}^1}^{2(1-\frac{1}{p})}.$$

Thus, we find that a necessary condition for blow up in finite time is that $\int_0^{T_{\text{MAX}}} \|\lambda_2^+\|_{\dot{M}^{1,q}}^p \, dt = +\infty$.

Chapter 12

Leray's Weak Solutions

12.1   The Rellich Lemma

Existence of weak solutions relies on an energy estimate (Leray energy inequality) and on a compactness lemma in $(L_t^2 L_x^2)_{\text{loc}}$ which goes back to the Rellich lemma.

Rellich's lemma was published in 1930 [409]. In modern terms, it states that the set of functions $f \in W^{1,2}(\mathbb{R}^d)$ such that $\|f\|_{W^{1,2}} = \sqrt{\int |f|^2 + |\vec{\nabla} f|^2 \, dx} \leq 1$ and $f = 0$ for $|x| > 1$ is a compact subset of $L^2(\mathbb{R}^d)$. This can be easily generalized by replacing the $W^{1,2}$ norm with a H^s norm with positive s. (Another generalization was proved by Kondrashov in 1945, replacing L^2 norms by L^p norms [271]; Rellich's lemma is often quoted therefore as the Rellich–Kondrashov lemma).

When considering an evolution equation, it is often very useful to consider a variant of the (generalized) Rellich lemma that has been highlighted by Lions in 1961 [336]. (Again, this theorem has been extended from the context of Hilbert spaces to more general (reflexive) Banach spaces by Aubin [9] and Lions [337]; the extended theorem is known as the Aubin–Lions lemma).

Theorem 12.1 (Rellich–Lions theorem).
Let I be an open interval of $\mathbb{R}$ and Ω an open subset of $\mathbb{R}^d$. Let $(u_n)_{n \in \mathbb{N}}$ be a sequence of measurable functions on $I \times \Omega$ such that, for every $\varphi \in \mathcal{D}(I \times \Omega)$, we have:

- *for some positive $\alpha > 0$,*

$$\sup_{n \in \mathbb{N}} \|\varphi u_n\|_{L^2(I, H^\alpha(\mathbb{R}^d))} < +\infty$$

- *for some negative $\beta < 0$ and some $p \in (1, 2]$,*

$$\sup_{n \in \mathbb{N}} \|\varphi \partial_t u_n\|_{L^p(I, H^\beta(\mathbb{R}^d))} < +\infty$$

Then, there exists a subsequence $(u_{n_k})_{k \in \mathbb{N}}$ which converges strongly to a limit u in $(L_t^2 L_x^2)_{\text{loc}}$: for every $\varphi \in \mathcal{D}(I \times \Omega)$,

$$\lim_{k \to +\infty} \iint \varphi^2 |u_{n_k} - u|^2 \, dt \, dx = 0.$$

Proof. First, we consider a fixed φ. Let $v_n = \varphi u_n$. We have $\sup_{n \in \mathbb{N}} \|v_n\|_{L^2(I, H^\alpha(\mathbb{R}^d))} < +\infty$ and $\sup_{n \in \mathbb{N}} \|\partial_t v_n\|_{L^p(I, H^\beta(\mathbb{R}^d))} < +\infty$. In particular, if $\theta_r = c_r |t|^{r-1}$ is the inverse Fourier transform of $|\tau|^{-r}$ $(0 < r < 1)$ and if $r = \frac{1}{p} - \frac{1}{2}$, we find that $\sup_{n \in \mathbb{N}} \|\theta_r * \partial_t v_n\|_{L_t^2 H^\beta} < +\infty$. If V_n is the Fourier transform in t and x (i.e., on $\mathbb{R}^{d+1}$), we find by the Plancherel equality that

$$\sup_{n \in \mathbb{N}} \iint |V_n(\tau, \xi)|^2 (1 + |\xi|^2)^\alpha \, d\tau \, d\xi < +\infty$$

DOI: 10.1201/9781003042594-12

and

$$\sup_{n\in\mathbb{N}} \iint |\tau|^{2(1-r)} |V_n(\tau,\xi)|^2 (1+|\xi|^2)^\beta \, d\tau \, d\xi < +\infty$$

As $\beta < 0 < \alpha$, we may write as well

$$\sup_{n\in\mathbb{N}} \iint (1+|\tau|^2)^{1-r} |V_n(\tau,\xi)|^2 (1+|\xi|^2)^\beta \, d\tau \, d\xi < +\infty$$

(This inequality is valid as well in the case $p = 2$, $r = 0$). We now write for $0 < \gamma < 1 - r$, $\gamma = \gamma_1 + \gamma_2$,

$$\iint (1+|\tau|^2+|\xi|^2)^\gamma |V_n(\tau,\xi)|^2 \, d\tau \, d\xi \leq \iint (1+|\tau|^2)^\gamma (1+|\xi|^2)^\gamma |V_n(\tau,\xi)|^2 \, d\tau \, d\xi$$

$$\leq \left(\iint (1+|\tau|^2)^{1-r} |V_n(\tau,\xi)|^2 (1+|\xi|^2)^{\frac{(1-r)\gamma_1}{\gamma}} \, d\tau \, d\xi \right)^{\frac{\gamma}{1-r}} \times$$

$$\times \left(\iint |V_n(\tau,\xi)|^2 (1+|\xi|^2)^{\frac{(1-r)\gamma_2}{1-r-\gamma}} \, d\tau \, d\xi \right)^{1-\frac{\gamma}{1-r}}$$

The choice $\gamma = \frac{1-r}{1-r+\alpha-\beta}\alpha$, $\gamma_1 = \frac{\beta}{1-r+\alpha-\beta}\alpha$ and $\gamma_2 = \frac{1-r-\beta}{1-r+\alpha-\beta}\alpha$ gives then

$$\sup_{n\in\mathbb{N}} \iint (1+|\tau|^2+|\xi|^2)^\gamma |V_n(\tau,\xi)|^2 \, d\tau \, d\xi < +\infty$$

Thus, the sequence v_n is bounded in $H^\gamma(\mathbb{R} \times \mathbb{R}^d)$ (with $\gamma > 0$), and the support of v_n is contained in a fixed compact set (the support of φ); thus, we may apply Rellich's lemma and find a subsequence that is strongly convergent in $L_t^2 L_x^2$.

To finish the proof, it is then enough to consider an exhaustion of $I \times \Omega$ by compact sets $(K_l)_{l\in\mathbb{N}}$[1], test functions $\varphi_l \in \mathcal{D}(I \times \Omega)$ such that $\varphi_l = 1$ on K_l and to use the Cantor diagonal process. $\quad\square$

12.2   Leray's Weak Solutions

In 1934, Leray [328] exhibited global weak solutions for the Cauchy problem for the Navier–Stokes equations with a divergence-free initial value $\vec{u}_0 \in L^2$ and a forcing term $\vec{f} \in L^2((0,T), H^{-1}(\mathbb{R}^3))$. He first recalled Oseen's formula for regular solutions:

$$\vec{u} = W_{\nu t} * \vec{u}_0 + \int_0^t W_{\nu(t-s)} * \mathbb{P}(\vec{f} - \vec{u} \cdot \vec{\nabla} \vec{u}) \, ds \tag{12.1}$$

However, L^2 is not a good space for looking for solutions of this equation by Picard's iterative algorithm: if $\vec{u}_0 \in L^2$, then $W_{\nu t} * \vec{u}_0 \in L_t^\infty L_x^2 \cap L_t^2 \dot{H}_x^1$; conversely, if $\vec{g} \in L_t^2 H_x^{-1}$, then $\int_0^t W_{\nu(t-s)} * \mathbb{P}\vec{g} \, ds$ belongs to $L_t^\infty L_x^2 \cap L_t^2 \dot{H}_x^1$ on $[0,T]$ ($T < +\infty$) or on every bounded subinterval of $[0,+\infty)$ ($T = +\infty$). But the problem is that, for $\vec{u} \in L^\infty L^2 \cap L^2 \dot{H}^1$ with $\operatorname{div} \vec{u} = 0$, we do not have $\vec{u} \cdot \vec{\nabla} \vec{u} \in L^2 H^{-1}$ but only $\vec{u} \cdot \vec{\nabla} \vec{u} \in L^2 H^{-3/2}$.

Leray's idea was then to alleviate the non-linearity by replacing $\vec{u} \cdot \vec{\nabla} \vec{u}$ with $(\theta_\epsilon * \vec{u}) \cdot \vec{\nabla} \vec{u}$ in the equation, where θ is fixed in $\mathcal{D}(\mathbb{R}^3)$ with $\int \theta \, dx = 1$, $\theta \geq 0$, $\epsilon > 0$ and $\theta_\epsilon = \frac{1}{\epsilon^3}\theta(\frac{x}{\epsilon})$. In modern terms, θ_ϵ is called a **mollifier** (a term coined by Friedrichs in 1944 [183]).

---

[1] K_l is compact, K_l is contained in the interior of K_{l+1} and $\cup_{l\in\mathbb{N}} K_l = I \times \Omega$.

The strategy of proof then follows three steps:

- use the Picard algorithm to solve on a small interval of time the equation $\partial_t \vec{u} + \mathbb{P}((\vec{u} * \theta_\epsilon) \cdot \vec{\nabla}\vec{u}) = \nu\Delta\vec{u} + \mathbb{P}\vec{f}$, with a time of existence controlled by the size of $\|\vec{u}_0\|_2$

- establish an energy estimate on this solution to get a control on its size in L^2 and thus be able to extend it globally

- use the Rellich–Lions theorem to relax the mollification and get a solution to the Navier–Stokes equations $\partial_t\vec{u} + \mathbb{P}(\vec{u} \cdot \vec{\nabla}\vec{u}) = \nu\Delta\vec{u} + \mathbb{P}\vec{f}$

The solutions we will obtain by this method satisfy the Leray energy inequality (12.2) and will be called Leray weak solutions:

Definition 12.1.
Let $\vec{u}_0 \in L^2$ with $\operatorname{div}\vec{u}_0 = 0$ and $\vec{f} \in L^2_t H^{-1}_x$. A weak solution $\vec{u}$ of equations $\partial_t\vec{u} + \mathbb{P}(\vec{u}\cdot\vec{\nabla}\vec{u}) = \nu\Delta\vec{u} + \mathbb{P}\vec{f}$ on $(0,T) \times \mathbb{R}^3$ with initial value $\vec{u}_0$ is called a Leray weak solution if it satisfies

- $\vec{u} \in L^\infty_t L^2_x \cap L^2_t \dot{H}^1_x$

- *for every $t \in (0,T)$,*

$$\|\vec{u}(t,.)\|_2 \le \|\vec{u}_0\|_2^2 - 2\nu \int_0^t \|\vec{\nabla} \otimes \vec{u}\|_2^2 \, ds + 2\int_0^t \langle \vec{u}|\vec{f}\rangle_{H^1,H^{-1}} \, ds. \tag{12.2}$$

Leray's mollification

Theorem 12.2.
Let $\vec{u}_0 \in L^2$, with $\operatorname{div}\vec{u}_0 = 0$, and $\vec{f} \in L^2_t H^{-1}_x$ on $(0,T) \times \mathbb{R}^3$. Then

- *for $\epsilon > 0$, the problem associated to the mollifier θ_ϵ*

$$\partial_t\vec{u} + \mathbb{P}((\vec{u} * \theta_\epsilon) \cdot \vec{\nabla}\vec{u}) = \nu\Delta\vec{u} + \mathbb{P}\vec{f} \tag{12.3}$$

with initial value $\vec{u}(0,.) = \vec{u}_0$ has a unique solution $\vec{u}_\epsilon$ such that $\vec{u}_\epsilon \in L^\infty_t L^2_x \cap L^2_t H^1_x$ on every bounded subinterval of $[0,T]$.
Moreover we have the following inequality:

$$\|\vec{u}_\epsilon(t,.)\|_2^2 + \nu \int_0^t \|\vec{\nabla} \otimes \vec{u}_\epsilon(s,.)\|_2^2 \le (\|\vec{u}(0,.)\|_2^2 + \frac{1}{\nu} \int_0^T \|\vec{f}\|_{H^{-1}_x}^2 \, ds)e^{\nu t}. \tag{12.4}$$

- *there exists a sequence $\epsilon_k \to 0$ and a function $\vec{u} \in L^\infty_t L^2_x \cap L^2_t H^1_x$ (on every bounded subinterval of $[0,T]$) such that $\vec{u}_{(\epsilon_k)}$ is weakly convergent to $\vec{u}$.*
Moreover $\vec{u}$ is a Leray weak solution of the Navier–Stokes problem

$$\partial_t\vec{u} + \mathbb{P}(\vec{u} \cdot \vec{\nabla}\vec{u}) = \nu\Delta\vec{u} + \mathbb{P}\vec{f}, \quad \vec{u}(0,.) = \vec{u}_0. \tag{12.5}$$

Proof. • **First step: Local existence of $\vec{u}_\epsilon$.**
We start from the obvious inequality

$$\|\vec{u} * \theta_\epsilon\|_\infty \le \epsilon^{-3/2}\|\vec{u}\|_2\|\theta\|_2.$$

Thus, for $\vec{u}$ and $\vec{v}$ in $L^\infty L^2 \cap L^2 H^1$ with div $\vec{u} = 0$, we have for every $0 < T_0 < T$,

$$\|(\vec{u} * \theta_\epsilon) \cdot \vec{\nabla} \vec{v}\|_{L^2((0,T_0), \dot{H}^{-1})} = \| \operatorname{div}((\vec{u} * \theta_\epsilon) \otimes \vec{v}) \|_{L^2 \dot{H}^{-1}}$$
$$\leq C \|(\vec{u} * \theta_\epsilon) \otimes \vec{v}\|_{L^2 L^2}$$
$$\leq C' \sqrt{T_0}\, \epsilon^{-3/2} \|\vec{u}\|_{L^\infty L^2} \|\vec{v}\|_{L^\infty L^2}$$

Let $\|\vec{u}\|_{\nu, T_0} = \|\vec{u}\|_{L^\infty((0,T_0), L^2)} + \sqrt{\nu} \|\vec{u}\|_{L^2(0,T_0), \dot{H}^1)}$. We have

$$\|W_{\nu t} * \vec{u}_0 + \int_0^t W_{\nu(t-s)} * \mathbb{P}\vec{f}\, ds\|_{\nu, T_0} \leq C_0 (\|\vec{u}_0\|_2 + \frac{1}{\sqrt{\nu}} (1 + \sqrt{T_0 \nu}) \|\vec{f}\|_{L^2 \dot{H}^{-1}})$$

and

$$\| \int_0^t W_{\nu(t-s)} * \mathbb{P}((\vec{u} * \theta_\epsilon) \cdot \vec{\nabla} \vec{v})\, ds \|_{\nu, T_0} \leq C_0 \frac{1}{\sqrt{\nu}} \sqrt{T_0}\, \epsilon^{-3/2} \|\vec{u}\|_{L^\infty L^2} \|\vec{v}\|_{L^\infty L^2}.$$

Thus, we find existence (and uniqueness) of a solution $\vec{u} = \vec{u}_\epsilon$ of the equation

$$\vec{u} = W_{\nu t} * \vec{u}_0 + \int_0^t W_{\nu(t-s)} * \mathbb{P}(\vec{f} - (\vec{u} * \theta_\epsilon) \cdot \vec{\nabla} \vec{u})\, ds \tag{12.6}$$

for T_0 small enough to ensure

$$1 + \sqrt{T_0 \nu} \leq 2$$

and

$$T_0 \leq \epsilon^3 \nu \frac{1}{16\, C_0^4 (\|\vec{u}_0\|_2 + \frac{2}{\sqrt{\nu}} \|\vec{f}\|_{L^2 \dot{H}^{-1}})^2}. \tag{12.7}$$

- **Second step: Energy estimates and global existence of $\vec{u}_\epsilon$.**

To show the existence of a global solution to (12.3), it is then enough to show that the L^2 norm of $\vec{u}_\epsilon$ remains bounded (as the existence time T_0 is controlled by the L^2 norm of the Cauchy data by (12.7)).

Since div $(\vec{u}_\epsilon * \theta_\epsilon) = \theta_\epsilon * \operatorname{div} \vec{u}_\epsilon = 0$, we have

$$\int \vec{u}_\epsilon \cdot \left((\vec{u}_\epsilon * \theta_\epsilon) \cdot \vec{\nabla} \vec{u}_\epsilon \right) dx = 0$$

hence

$$\frac{d}{dt} \|\vec{u}_\epsilon\|_2^2 = 2 \int \partial_t \vec{u}_\epsilon \cdot \vec{u}_\epsilon\, dx$$
$$= -2\nu \|\vec{u}_\epsilon\|_{\dot{H}^1}^2 + 2\langle \vec{f} | \vec{u}_\epsilon \rangle_{H^{-1}, H^1} \tag{12.8}$$
$$\leq -\nu \|\vec{u}_\epsilon\|_{\dot{H}^1}^2 + \nu \|\vec{u}_\epsilon\|_2^2 + \frac{1}{\nu} \|\vec{f}\|_{\dot{H}^{-1}}^2$$

so that

$$\|\vec{u}_\epsilon(t,.)\|_2^2 + \nu \int_0^t \|\vec{u}_\epsilon\|_{\dot{H}^1}^2\, ds \leq \|\vec{u}_0\|_2^2 + \frac{1}{\nu} \|\vec{f}\|_{L^2 \dot{H}^{-1}}^2 + \nu \int_0^t \|\vec{u}_\epsilon(s,.)\|_2^2\, ds$$

We thus get (by Grönwall's lemma) the energy estimate (12.4) and, therefore, the global existence of $\vec{u}_\epsilon$.

• **Third step: Weak convergence.**

From the energy estimate (12.4), we know that $\vec{u}_\epsilon$ remains bounded in $L^\infty L^2 \cap L^2 \dot{H}^1$. Moreover, since $\|\theta\|_1 = 1$, we have $\|\vec{u}_\epsilon * \theta_\epsilon\|_2 \leq \|\vec{u}_\epsilon\|_2$. As we have

$$\partial_t \vec{u}_\epsilon = \nu \Delta \vec{u}_\epsilon + \mathbb{P}(\vec{f} - \text{div}((\vec{u}_\epsilon * \theta_\epsilon) \otimes \vec{u}_\epsilon))$$

we can see that $\partial_t \vec{u}_\epsilon$ remains bounded in $L^2 H^{-3/2}$ (on every bounded subinterval of $[0,T]$).

We may then use the Rellich–Lions theorem (Theorem 12.1) for the set of functions $\vec{v}_\epsilon$ defined on $(-T,T)$ by $\vec{v}_\epsilon(t,x) = \vec{u}_\epsilon(t,x)$ for $t > 0$ and $\vec{v}_\epsilon(t,x) = \vec{u}_\epsilon(-t,x)$ for $t < 0$[2] and find a sequence $\epsilon_n \to 0$ and a function $\vec{u}$ such that:

• on every bounded subinterval of $[0,T]$, $\vec{u}_{(\epsilon_n)}$ is *-weakly convergent to $\vec{u}$ in $L^\infty L^2$ and in $L^2 \dot{H}^1$

• $\vec{u}_{(\epsilon_n)}$ is strongly convergent to $\vec{u}$ in $L^2_{\text{loc}}([0,T) \times \mathbb{R}^3)$: for every compact subset K of $\mathbb{R}^3$ and every $T_0 < T$, $\lim_{n \to +\infty} \int_0^{T_0} \int_K |\vec{u}_{\epsilon_n} - \vec{u}|^2 \, dx \, dt = 0$.

Since $\vec{u}_\epsilon$ is bounded in $L^\infty L^2$, we get that $\vec{u}_{(\epsilon_n)} * \theta_{\epsilon_n}$ strongly converges $\vec{u}$ in $L^2_{\text{loc}}((0,T) \times \mathbb{R}^3)$ (and even in $(L^p_t L^2_x)_{\text{loc}}((0,T) \times \mathbb{R}^3)$ for every $p < +\infty$); as $\vec{\nabla} \otimes \vec{u}_{\epsilon_n}$ *-weakly converges to $\vec{\nabla} \otimes \vec{u}$ in L^2_{loc} we get that the sequence $(\vec{u}_{\epsilon_n} * \theta_{\epsilon_n}) \cdot \vec{\nabla} \vec{u}_{\epsilon_n}$ is *-weakly convergent to $\vec{u} \cdot \vec{\nabla} \vec{u}$ in $(L^q_t H^{-3/2}_x)_{\text{loc}}$ for every $1 < q < 2$; as the sequence is bounded in $L^2 H^{-3/2}$, we have *-weak convergence in $L^2 H^{-3/2}$ as well; as $\mathbb{P}$ is bounded on $H^{-3/2}$, we get the *-weak convergence of $\mathbb{P}\left((\vec{u}_{\epsilon_n} * \theta_{\epsilon_n}) \cdot \vec{\nabla} \vec{u}_{\epsilon_n}\right)$ to $\mathbb{P}(\vec{u} \cdot \vec{\nabla} \vec{u})$ in $L^2 H^{-3/2}$.

Thus, the weak limit $\vec{u}$ satisfies

$$\partial_t \vec{u} = \nu \Delta \vec{u} + \mathbb{P}(\vec{f} - \vec{u} \cdot \vec{\nabla} \vec{u}).$$

• **Fourth step: Global energy estimates for the weak limit.**

We remark that, for all $T < +\infty$, $\vec{u}_\epsilon$ belongs to $L^2((0,T), H^1)$ and $\partial_t \vec{u}_\epsilon$ belongs to $L^2((0,T), H^{-3/2})$, so that (from Lemma (6.1)), we can represent $\vec{u}_\epsilon$ as

$$\vec{u}_\epsilon(t,.) = \vec{u}_0 + \int_0^t \partial_t \vec{u}_\epsilon(s,.) \, ds$$

(so that $\vec{u}_\epsilon \in \mathcal{C}([0,T], H^{-3/2})$). Moreover, $\partial_t \vec{u}_\epsilon$ is bounded in $L^2((0,T), H^{-3/2})$, hence the weak convergence of $\vec{u}_{\epsilon_n}$ to $\vec{u}$ gives the weak convergence of $\partial_t \vec{u}_{\epsilon_n}$ to $\partial_t \vec{u}$ in $L^2((0,T), H^{-3/2})$, and then the weak convergence of $\vec{u}_{\epsilon_n}(t,.)$ to $\vec{u}(t,.)$ in $H^{-3/2}$, and in L^2 as $\vec{u}_\epsilon(t,.)$ is bounded in L^2.

For fixed t, we thus have the weak convergence of $(\vec{u}_{\epsilon_n}(t,.), 1_{0<s<t} \vec{u}_{\epsilon_n}(s,.))$ in $L^2 \times L^2((0,t), \dot{H}^1)$ and thus

$$\|\vec{u}(t,.)\|_2^2 + 2\nu \|\vec{u}\|_{L^2((0,t),\dot{H}^1)}^2 \leq \liminf_{n \to +\infty} \|\vec{u}_{\epsilon_n}(t,.)\|_2^2 + 2\nu \|\vec{u}_{\epsilon_n}\|_{L^2((0,t),\dot{H}^1)}^2$$

$$\leq \lim_{n \to +\infty} \|\vec{u}_0\|_2^2 + 2 \int_0^t \langle \vec{u}_{\epsilon_n} | \vec{f} \rangle_{H^1, H^{-1}} \, ds \qquad \square$$

$$= \|\vec{u}_0\|_2^2 + 2 \int_0^t \langle \vec{u} | \vec{f} \rangle_{H^1, H^{-1}} \, ds$$

---

[2] We check easily that $\partial_t \vec{v}_\epsilon$ is the distribution defined by $\partial_t \vec{u}_\epsilon(t,x)$ for $t < 0$ and $-\partial_t \vec{u}_\epsilon(-t,x)$ for $t < 0$.

As a matter of fact, we have a better convergence of $\vec{u}_{\epsilon_n}$ to $\vec{u}$ than just in L^2_{loc}:

Lemma 12.1.
If $T_0 \leq T$ and $T_0 < +\infty$, then $\lim_{\epsilon_n \to 0} \int_0^{T_0} \|\vec{u}_{\epsilon_n} - \vec{u}\|_2^2 \, dt = 0$.

Proof. From estimate (12.4), we know that, if $T_0 \leq T$ and $T_0 < +\infty$,

$$M_{T_0} = \sup_{\epsilon > 0} \|\vec{u}_\epsilon\|_{L^\infty((0,T_0),L^2)} + \sqrt{\nu}\|\vec{u}\|_{L^2((0,T_0),\dot{H}^1)} < +\infty.$$

We write

$$\mathbb{P}\left((\vec{u}_\epsilon * \theta_\epsilon) \cdot \vec{\nabla} \vec{u}_\epsilon \right) = (\vec{u}_\epsilon * \theta_\epsilon) \cdot \vec{\nabla} \vec{u}_\epsilon + \vec{\nabla} p_\epsilon.$$

Let $\phi \in \mathcal{D}(\mathbb{R}^3)$ be equal to 1 on the ball $B(0,1)$ and, for $R > 1$, let $\phi_R(x) = \phi(\frac{x}{R})$ and $\vec{u}_{R,\epsilon} = (1 - \phi_R(x))\vec{u}_\epsilon$. We have, for $t < T_0$,

$$\|\vec{u}_{\epsilon,R}(t,.)\|_2^2 - \|(1-\phi_R)\vec{u}_0\|_2^2 = 2\int_0^t \int (1-\phi_R)^2 \vec{u}_\epsilon \cdot \partial_t \vec{u}_\epsilon \, dx \, ds$$

$$= -2\sum_{k=1}^3 \int_0^t \int (1-\phi_R)^2 |\vec{\nabla} \otimes \vec{u}_\epsilon|^2 \, dx \, ds$$

$$-2\sum_{k=1}^3 \int_0^t \int (\partial_k(1-\phi_R)^2)\vec{u}_\epsilon \cdot \partial_k \vec{u}_\epsilon \, dx \, ds$$

$$+\int_0^t \int |\vec{u}_\epsilon|^2 (\vec{u}_\epsilon * \theta_\epsilon) \cdot \vec{\nabla}(1-\phi_R)^2 \, dx \, ds + 2\int_0^t \int p_\epsilon \vec{u}_\epsilon \cdot \vec{\nabla}(1-\phi_R)^2 \, dx \, ds$$

$$+2\int_0^t \int (1-\phi_R)^2 \vec{u}_\epsilon \cdot \mathbb{P}\vec{f} \, dx \, ds$$

$$\leq \frac{C}{R}\int_0^t \|\vec{u}_\epsilon\|_2 \|\vec{u}_\epsilon\|_{\dot{H}^{-1}} + \|\vec{u}_\epsilon\|_3^3 + \|\vec{u}_\epsilon\|_3 \|p_\epsilon\|_{3/2} \, dt$$

$$+2\int_0^t (\|\vec{u}_\epsilon\|_2 + \|\vec{u}_\epsilon\|_{\dot{H}^1})\|(1-\phi_R)^2 \mathbb{P}\vec{f}\|_{H^{-1}} \, ds$$

$$\leq \frac{C'}{R}\left(\sqrt{\frac{T_0}{\nu}}M_{T_0}^2 +\right) + C'(\sqrt{T_0} + \nu^{-1/2})M_{T_0}\|(1-\phi_R)^2\mathbb{P}\vec{f}\|_{L^2((0,T_0),H^{-1})}.$$

Thus, we have

$$\sup_{\epsilon > 0} \sup_{0 < t < T_0} \|(1-\phi_R)\vec{u}_\epsilon\|_2^2$$

$$\leq C_0(\frac{1}{R} + \|(1-\phi_R)\vec{u}_0\|_2^2 + \|(1-\phi_R)^2\mathbb{P}\vec{f}\|_{L^2((0,T_0),H^{-1})}) \tag{12.9}$$

where C_0 does not depend on R. As $(1-\phi_R)\vec{u}_{\epsilon_n}(t,.)$ is weakly convergent in L^2 to $(1-\phi_R)\vec{u}$, this estimate remains valid for $(1-\phi_R)\vec{u}$.

On the other hand, we have $\lim_{n \to +\infty} \|\phi_R(\vec{u}_{\epsilon_n} - \vec{u})\|_{L^2((0,T_0),L^2)} = 0$. Thus, we find that, for every $R > 1$, we have

$$\limsup_{n \to +\infty} \|\vec{u}_{\epsilon_n} - \vec{u}_\epsilon\|_{L^2((0,T_0),L^2)}$$

$$\leq 2C_0T_0(\frac{1}{R} + \|(1-\phi_R)\vec{u}_0\|_2^2 + \|(1-\phi_R)^2\mathbb{P}\vec{f}\|_{L^2((0,T_0),H^{-1})}).$$

To prove the Lemma, we just have to let R go to $+\infty$: if we write $\mathbb{P}\vec{f} = \vec{f}_0 + \sum_{k=1}^{3} \partial_k \vec{f}_k$ avec $\vec{f}_0, \dots, \vec{f}_3 \in L^2((0,T_0), L^2)$, we have

$$\|(1 - \phi_R)^2 \mathbb{P}\vec{f}\|_{L^2((0,T_0),H^{-1})} \leq \sum_{i=0}^{3} \|(1 - \phi_R)^2 \vec{f}_i\|_{L^2 L^2} + C\frac{1}{R} \sum_{i=1}^{3} \|\vec{f}_i\|_{L^2 L^2} = o(1). \qquad \square$$

Proposition 12.1 (Strong Leray energy inequality).
The solution $\vec{u}$ constructed in Theorem 12.2 satisfies the strong Leray energy inequality: for almost every t_0 in $(0,T)$ and for every $t \in (t_0, T)$, we have

$$\|\vec{u}(t, .)\|_2^2 + 2\nu \int_{t_0}^{t} \|\vec{u}\|_{\dot{H}^1}^2 \, ds \leq \|\vec{u}(t_0)\|_2^2 + 2 \int_{t_0}^{t} \langle \vec{f} | \vec{u} \rangle_{H^{-1}, H^1} \, ds \qquad (12.10)$$

Proof. [3]
Let $t_1 > 0$. For fixed $t_1 > t_0 \geq 0$, we have the weak convergence of $(\vec{u}_{\epsilon_n}(t_1, .), 1_{t_0 < s < t_1} \vec{u}_{\epsilon_n}(s, .))$ in $L^2 \times L^2((t_0, t_1), \dot{H}^1)$ and thus

$$\|\vec{u}(t_1, .)\|_2^2 + 2\nu \|\vec{u}\|_{L^2((t_0, t_1), \dot{H}^1)}^2 \leq \liminf_{n \to +\infty} \|\vec{u}_{\epsilon_n}(t_1, .)\|_2^2 + 2\nu \int_{t_0}^{t_1} \|\vec{u}_{\epsilon_n}\|_{\dot{H}^1}^2 \, ds$$

$$\leq \liminf_{n \to +\infty} \|\vec{u}_{\epsilon_n}(t_0, .)\|_2^2 + 2 \int_{t_0}^{t_1} \langle \vec{u} | \vec{f} \rangle_{H^1, H^{-1}} \, ds.$$

The problem is now to estimate $\liminf_{n \to +\infty} \|\vec{u}_{\epsilon_n}(t_0, .)\|_2^2$. By Lemma 12.1, we know that $\|\vec{u}_{\epsilon_n}(t, .) - \vec{u}(t, .)\|_2$ converges to 0 in L^2 norm on $(0,T_0)$ for every finite T_0, hence almost everywhere. Thus, for almost every t_0, we have $\lim_{n \to +\infty} \|\vec{u}_{\epsilon_n}(t_0, .)\|_2 = \|\vec{u}(t_0, .)\|_2$. $\qquad \square$

Proposition 12.2.
Let $\vec{u}$ be a Leray weak solution that satisfies the strong Leray energy inequality for almost every t_0 in $(0,T)$: for every $t \in (t_0, T)$, we have

$$\|\vec{u}(t, .)\|_2^2 + 2\nu \int_{t_0}^{t} \|\vec{u}\|_{\dot{H}^1}^2 \, ds \leq \|\vec{u}(t_0)\|_2^2 + 2 \int_{t_0}^{t} \langle \vec{f} | \vec{u} \rangle_{H^{-1}, H^1} \, ds. \qquad (12.11)$$

Then it satisfies inequality (12.11) for every Lebesgue point t_0 of the map $t \mapsto \|\vec{u}(t, .)\|_2^2$.

Proof. Let $t_0 < t$ and $\epsilon < t - t_0$. For almost every $t_1 \in (t_0, t_0 + \epsilon)$, we have

$$\|\vec{u}(t, .)\|_2^2 + 2\nu \int_{t_1}^{t} \|\vec{u}\|_{\dot{H}^1}^2 \, ds \leq \|\vec{u}(t_1)\|_2^2 + 2 \int_{t_1}^{t} \langle \vec{f} | \vec{u} \rangle_{H^{-1}, H^1} \, ds.$$

Integrating in t_1, we get

$$\|\vec{u}(t, .)\|_2^2 + 2\nu \frac{1}{\epsilon} \iint_{t_0 \leq t_1 \leq s \leq t} \|\vec{u}(s, .)\|_{\dot{H}^1}^2 \, ds \, dt_1$$

$$\leq \frac{1}{\epsilon} \int_{t_0}^{t_0 + \epsilon} \|\vec{u}(t_1)\|_2^2 \, dt_1 + 2\frac{1}{\epsilon} \iint_{t_0 \leq t_1 \leq s \leq t} \langle \vec{f} | \vec{u} \rangle_{H^{-1}, H^1} \, ds \, dt_1.$$

---

[3] Thanks to T. Tao's students who noticed that the proof given in the first edition was incorrect.

We let ϵ go to 0 and get

$$\|\vec{u}(t,.)\|_2^2 + 2\nu \int_{t_0 \le s \le t} \|\vec{u}(s,.)\|_{\dot{H}^1}^2 \, ds$$

$$\le \liminf_{\epsilon \to 0} \frac{1}{\epsilon} \int_{t_0}^{t_0+\epsilon} \|\vec{u}(t_1)\|_2^2 \, dt_1 + 2 \int_{t_0 \le s \le t} \langle \vec{f} | \vec{u} \rangle_{H^{-1}, H^1} \, ds.$$

If t_0 is a Lebesgue point of the map $t \mapsto \|\vec{u}(t,.)\|_2^2$, then

$$\lim_{\epsilon \to 0} \frac{1}{\epsilon} \int_{t_0}^{t_0+\epsilon} \|\vec{u}(t_1)\|_2^2 \, dt_1 = \|\vec{u}(t_0,.)\|_2^2 \qquad \square$$

12.3 Weak-Strong Uniqueness: The Prodi–Serrin Criterion

Theorem 12.2 shows global existence of weak Leray solutions (when $\vec{u}_0 \in L^2$ and $\vec{f} \in L_t^2 H_x^{-1}$) but gives no clue on whether those solutions are unique or not. If $\vec{u}_0$ belongs more precisely to H^1 and $\vec{f}$ to $L_t^2 L_x^2$, then Theorem 7.1 gives the local existence of a unique mild solution $\vec{u} \in L_t^\infty H^1 \cap L_t^2 H^2$. It is easy to check that, as long as this mild solution is defined, the Leray weak solutions coincide with this solution (and thus we have uniqueness in the class of Leray solutions). Such a result is called weak-strong uniqueness.

Weak-strong uniqueness

Theorem 12.3.
Let $\vec{u}_0 \in H^1$, with div $\vec{u}_0 = 0$, and $\vec{f} \in L_t^2 L_x^2$ on $(0,T) \times \mathbb{R}^3$. Assume that the Navier–Stokes problem

$$\partial_t \vec{u} + \mathbb{P}(\vec{u} \cdot \vec{\nabla} \vec{u}) = \nu \Delta \vec{u} + \mathbb{P}\vec{f}, \quad \vec{u}(0,.) = \vec{u}_0. \qquad (12.12)$$

has a solution $\vec{u}_1$ on $(0,T) \times \mathbb{R}^3$ such that $\vec{u}_1 \in L_t^\infty H^1 \cap L_t^2 H^2$. If $\vec{u}_2$ is a Leray weak solution of the same Navier–Stokes problem, then $\vec{u}_2 = \vec{u}_1$.

Proof. We have $\vec{u}_1 \in L^2 H^2$ with $\partial_t \vec{u}_1 \in L^2 H^{-1}$ while $\vec{u}_2 \in L^2 H^1$ with $\partial_t \vec{u}_2 \in L^2 H^{-2}$. This is enough to get that

$$\int \vec{u}_1(t,x) \cdot \vec{u}_2(t,x) \, dx = \|\vec{u}_0\|_2^2 + \int_0^t \langle \vec{u}_1 | \partial_t \vec{u}_2 \rangle_{H^2, H^{-2}} + \langle \partial_t \vec{u}_1 | \vec{u}_2 \rangle_{H^{-1}, H^1} \, ds.$$

If $\vec{w} = \vec{u}_2 - \vec{u}_1$, we write

$$\begin{aligned}
\|\vec{w}(t,.)\|_2^2 &= \|\vec{u}_2(t,.)\|_2^2 + \|\vec{u}_1(t,.)\|_2^2 - 2\langle \vec{u}_1(t,.) | \vec{u}_2(t,.) \rangle_{L^2, L^2} \\
&= \|\vec{u}_2(t,.)\|_2^2 - \|\vec{u}_1(t,.)\|_2^2 - 2\langle \vec{u}_1(t,.) | \vec{w}(t,.) \rangle_{L^2, L^2} \\
&= \|\vec{u}_2(t,.)\|_2^2 - \|\vec{u}_1(t,.)\|_2^2 - 2 \int_0^t \langle \vec{u}_1 | \partial_t \vec{w} \rangle_{H^2, H^{-2}} + \langle \partial_t \vec{u}_1 | \vec{w} \rangle_{H^{-1}, H^1} \, ds
\end{aligned}$$

where

- the Leray energy inequality gives

$$\|\vec{u}_2(t,.)\|_2^2 \le \|\vec{u}_0\|_2^2 + 2 \int_0^t \langle \vec{u}_2 | \vec{f} \rangle_{H^1, H^{-1}} \, ds - 2\nu \int_0^t \|\vec{\nabla} \otimes \vec{u}_2\|_2^2 \, ds$$

- the regularity of $\vec{u}_1$ gives $\int_0^t \int \vec{u}_1 \cdot (\vec{u}_1 \cdot \vec{\nabla})\vec{u}_1 \, dx \, ds = 0$ and thus

$$\|\vec{u}_1(t,.)\|_2^2 = \|\vec{u}_0\|_2^2 + 2 \int_0^t \langle \vec{u}_1|\vec{f}\rangle_{H^1,H^{-1}} \, ds - 2\nu \int_0^t \|\vec{\nabla} \otimes \vec{u}_1\|_2^2 \, ds$$

- using the Navier–Stokes equations on $\vec{u}_1$, we get

$$\int_0^t \langle \partial_t \vec{u}_1|\vec{w}\rangle_{H^{-1},H^1} \, ds = -\nu \int_0^t \langle \vec{\nabla} \otimes \vec{u}_1|\vec{\nabla} \otimes \vec{w}\rangle_{L^2,L^2} \, ds + \int_0^t \langle \vec{f}|\vec{w}\rangle_{H^{-1},H^1} \, ds$$
$$- \int_0^t \langle \vec{u}_1 \cdot \vec{\nabla}\vec{u}_1|\vec{w}\rangle_{H^{-1},H^1} \, ds$$

- using $\partial_t \vec{w} = \nu \Delta \vec{w} - \mathbb{P}(\vec{u}_2 \cdot \vec{\nabla}\vec{u}_2 - \vec{u}_1 \cdot \vec{\nabla}\vec{u}_1)$, we get

$$\int_0^t \langle \vec{u}_1|\partial_t \vec{w}\rangle_{H^2,H^{-2}}, ds = -\nu \int_0^t \langle \vec{\nabla} \otimes \vec{u}_1|\vec{\nabla} \otimes \vec{w}\rangle_{L^2,L^2} \, ds$$
$$- \int_0^t \langle \vec{u}_1|\vec{u}_2 \cdot \vec{\nabla}\vec{u}_2 - \vec{u}_1 \cdot \vec{\nabla}\vec{u}_1\rangle_{H^2,H^{-2}} \, ds$$

- Moreover, we have (since $\mathrm{div}\,\vec{u}_1 = \mathrm{div}\,\vec{u}_2 = 0$)

$$\langle \vec{u}_1|\vec{u}_1 \cdot \vec{\nabla}\vec{u}_1\rangle_{H^2,H^{-2}} = \langle \vec{u}_1|\vec{u}_2 \cdot \vec{\nabla}\vec{u}_1\rangle_{H^2,H^{-2}} = 0$$

and

$$\langle \vec{u}_1 \cdot \vec{\nabla}\vec{u}_1|\vec{w}\rangle_{H^{-1},H^1} = -\langle \vec{u}_1|\vec{u}_1 \cdot \vec{\nabla}\vec{w}\rangle_{H^2,H^{-2}}$$

so that

$$\langle \vec{u}_1 \cdot \vec{\nabla}\vec{u}_1|\vec{w}\rangle_{H^{-1},H^1} + \langle \vec{u}_1|\vec{u}_2 \cdot \vec{\nabla}\vec{u}_2 - \vec{u}_1 \cdot \vec{\nabla}\vec{u}_1\rangle_{H^2,H^{-2}} = \langle \vec{u}_1|\vec{w} \cdot \vec{\nabla}\vec{w}\rangle_{H^2,H^{-2}}$$

This gives finally that

$$\|\vec{w}(t,.)\|_2^2 \leq -2\nu \int_0^t \|\vec{w}\|_{\dot{H}^1}^2 \, ds + 2 \int_0^t \langle \vec{u}_1|\vec{w} \cdot \vec{\nabla}\vec{w}\rangle_{H^2,H^{-2}} \, ds \leq \frac{1}{2\nu} \int_0^t \|\vec{u}_1\|_\infty^2 \|\vec{w}\|_2^2 \, ds.$$

We then conclude by Grönwall's lemma, as $\int_0^T \|\vec{u}_1\|_\infty^2 \, ds \leq C\sqrt{\|\vec{u}_1\|_{L^2\dot{H}^1}\|\vec{u}_1\|_{L^2\dot{H}^2}}$. $\square$

Corollary 12.1.
Let $\vec{u}_0 \in L^2$ and $\vec{f} \in L^2((0,+\infty), L^2) \cap L^2((0,+\infty), \dot{H}^{-1})$. Then the Navier–Stokes problem

$$\partial_t \vec{u} + \mathbb{P}(\vec{u} \cdot \vec{\nabla}\vec{u}) = \nu \Delta \vec{u} + \mathbb{P}\vec{f}, \quad \vec{u}(0,.) = \vec{u}_0. \qquad (12.13)$$

has a weak solution $\vec{u}$ on $(0,+\infty) \times \mathbb{R}^3$ such that $\vec{u} \in L_t^\infty L^2 \cap L_t^2 \dot{H}^1$ and $\vec{u}$ satisfies the strong Leray inequality.
Moreover, we have

$$\lim_{t\to+\infty} \|\vec{u}(t,.)\|_2 = 0.$$

Proof. We construct $\vec{u}$ by Leray's mollification (see Theorem 12.2). The global control of $\vec{u}$ in $L_t^\infty L^2 \cap L_t^2 \dot{H}^1$ is provided by the inequality

$$\|\vec{u}(t,.)\|_2^2 + \nu \int_0^t \|\vec{\nabla} \otimes \vec{u}\|_2^2 \, ds \leq \|\vec{u}_0\|_2^2 + \frac{1}{\nu} \int_0^t \|\vec{f}\|_{\dot{H}^{-1}}^2 \, ds.$$

Recall that we have seen in Theorem 7.3 that if $\vec{u}(t_0, .) \in H^1$, $\vec{f} \in L^2((t_0, +\infty), L^2) \cap L^2((t_0, +\infty), \dot{H}^{-1/2})$ and if moreover $\|\vec{u}(t_0, .)\|_{\dot{H}^{1/2}} < \epsilon_0 \nu$ and $\int_{t_0}^{+\infty} \|\vec{f}(s, .)\|_{\dot{H}^{-\frac{1}{2}}}^2 \, ds < \epsilon_0^2 \nu^3$, then the Navier–Stokes problem with forcing term $\vec{f}$ and value $\vec{u}(t_0, .)$ at time $t = t_0$ has a global solution $\vec{v}$ on $(t_0, +\infty)$ which belongs to $\mathcal{C}([t_0, +\infty), H^1) \cap L^2(t_0, +\infty), \dot{H}^2)$.

As $\vec{u}$ belongs to $L_t^\infty L^2 \cap L_t^2 \dot{H}^1$, it belongs to $L_t^4 \dot{H}^{1/2}$. Thus, the set of times t such that $\|\vec{u}(t, .)\|_{\dot{H}^{1/2}} \geq \epsilon_0 \nu$ is of finite measure. As we have $\lim_{t \to +\infty} \int_t^{+\infty} \|\vec{f}(s, .)\|_{\dot{H}^{-\frac{1}{2}}}^2 \, ds = 0$ and as the set of Lebesgue points of $t \mapsto \|\vec{u}(t, .)\|_2^2$ has a complement of null measure, we may find a time t_0 such that

- $\|\vec{u}(t_0, .)\|_{\dot{H}^{1/2}} < \epsilon_0 \nu$

- $\int_{t_0}^{+\infty} \|\vec{f}(s, .)\|_{\dot{H}^{-\frac{1}{2}}}^2 \, ds < \epsilon_0^2 \nu^3$

- $\vec{u}$ is a weak Leray solution on $(t_0, +\infty)$: for every $t \in (t_0, +\infty)$, we have

$$\|\vec{u}(t, .)\|_2^2 + 2\nu \int_{t_0}^t \|\vec{u}\|_{\dot{H}^1}^2 \, ds \leq \|\vec{u}(t_0)\|_2^2 + 2\int_{t_0}^t \langle \vec{f} | \vec{u} \rangle_{H^{-1}, H^1} \, ds$$

Then, by the weak-strong uniqueness theorem of Serrin, we find that $\vec{u}$ coincides on $(t_0, +\infty)$ with the mild solution $\vec{v} \in \mathcal{C}([t_0, +\infty), H^1) \cap L^2((t_0, +\infty), \dot{H}^2)$. Thus, we shall prove the corollary if we prove that $\lim_{t \to +\infty} \|\vec{v}(t, .)\|_2 = 0$. If $t_0 < \tau < t$, we find that

$$\vec{v}(t, .) = W_{\nu(t-\tau)} * \vec{v}(\tau, .) + \int_\tau^t W_{\nu(t-s)} * \mathbb{P}(\vec{f} - \operatorname{div}(\vec{v} \otimes \vec{v})) \, ds$$

so that

$$\|\vec{v}(t, .)\|_2 \leq \|W_{\nu(t-\tau)} * \vec{v}(\tau, .)\|_2 + C\left(\int_\tau^t \|\mathbb{P}\vec{f}\|_{\dot{H}^{-1}}^2 + \|\mathbb{P}\operatorname{div}(\vec{v} \otimes \vec{v})\|_{\dot{H}^{-1}}^2 \, ds\right)^{1/2}$$

which gives

$$\limsup_{t \to +\infty} \|\vec{v}(t, .)\|_2 \leq$$
$$C\left(\left(\int_\tau^{+\infty} \|\vec{f}\|_{\dot{H}^{-1}}^2 \, ds\right)^{1/2} + \sup_{t > t_0} \|\vec{v}(t, .)\|_{\dot{H}^{1/2}} \left(\int_\tau^{+\infty} \|\vec{v}\|_{\dot{H}^1}^2 \, ds\right)^{1/2}\right).$$

Letting τ go to $+\infty$, we get

$$\lim_{t \to +\infty} \|\vec{v}(t, .)\|_2 = 0. \qquad \square$$

Weak-strong uniqueness has been proved under many various assumptions, in a generalization of the proof of Theorem 12.3. The idea is to consider two solutions $\vec{u}_1$ and $\vec{u}_2$ of the same Navier–Stokes problem associated to $\vec{u}_0 \in L^2$ and $\vec{f} \in L^2 H^{-1}$ such that $\vec{u}_1$ and $\vec{u}_2$ belong to $L^2 H^1 \cap L^\infty L^2$ (hence $\partial_t \vec{u}_1$ and $\partial_t \vec{u}_2$ belong to $L^2 H^{-3/2}$), and with assumptions that $\vec{u}_2$ is a Leray solution and that $\vec{u}_1$ satisfies $\vec{u}_1 \in \mathbb{X}$ for some well-chosen space $\mathbb{X}$, and to try to find a control on $\|\vec{w}(t, .)\|_2$ for $\vec{w} = \vec{u}_2 - \vec{u}_1$.

The first step is to write a convenient representation for $\|\vec{w}\|_2^2$. We write again

$$\|\vec{w}(t, .)\|_2^2 = \|\vec{u}_2(t, .)\|_2^2 - \langle \vec{u}_1 | \vec{u}_1 + 2\vec{w} \rangle_{L^2, L^2}.$$

We then use a mollifier θ_ϵ and write

$$\|\vec{w}(t, .)\|_2^2 = \|\vec{u}_2(t, .)\|_2^2 - \lim_{\epsilon \to 0^+} \langle \vec{u}_1 * \theta_\epsilon | \vec{u}_1 + 2\vec{w} \rangle_{L^2, L^2}.$$

We have $\vec{u}_1 * \theta_\epsilon \in L^2 H^{3/2}$ and $\partial_t(\vec{u}_1 * \theta_\epsilon) \in L^2 H^{-1}$, while $\vec{u}_1 + 2\vec{w} \in L^2 H^1$ and $\partial_t(\vec{u}_1 + 2\vec{w}) \in L^2 H^{-3/2}$. Thus, we may write

$$\langle \vec{u}_1 * \theta_\epsilon | \vec{u}_1 + 2\vec{w} \rangle_{L^2, L^2} = \langle \vec{u}_0 * \theta_\epsilon | \vec{u}_0 \rangle_{L^2, L^2}$$

$$+ \int_0^t \langle \partial_t(\vec{u}_1 * \theta_\epsilon) | \vec{u}_1 + 2\vec{w} \rangle_{H^{-1}, H^1} + \langle \vec{u}_1 * \theta_\epsilon | \partial_t(\vec{u}_1 + 2\vec{w}) \rangle_{H^{3/2}, H^{-3/2}} \, ds$$

and thus

$$\langle \vec{u}_1 | \vec{u}_1 + 2\vec{w} \rangle_{L^2, L^2} = \|\vec{u}_0\|_2^2 - 2\nu \int_0^t \langle \vec{\nabla} \otimes \vec{u}_1 | \vec{\nabla}(\vec{u}_1 + 2\vec{w}) \rangle_{L^2, L^2} \, ds$$

$$+ 2 \int_0^t \langle \vec{f} | \vec{u}_1 + \vec{w} \rangle_{H^{-1}, H^1} \, ds - \lim_{\epsilon \to 0^+} J_\epsilon$$

with

$$J_\epsilon = \int_0^t \langle (\vec{u}_1 \cdot \vec{\nabla} \vec{u}_1) * \theta_\epsilon | \vec{u}_1 \rangle_{H^{-1}, H^1} + \langle \vec{u}_1 * \theta_\epsilon | \vec{u}_1 \cdot \vec{\nabla} \vec{u}_1 \rangle_{H^{3/2}, H^{-3/2}} \, ds$$

$$+ 2 \int_0^t \langle (\vec{u}_1 \cdot \vec{\nabla} \vec{u}_1) * \theta_\epsilon | \vec{w} \rangle_{H^{-1}, H^1} + \langle \vec{u}_1 * \theta_\epsilon | - \vec{u}_1 \cdot \vec{\nabla} \vec{u}_1 + \vec{u}_2 \cdot \vec{\nabla} \vec{u}_2 \rangle_{H^{3/2}, H^{-3/2}} \, ds.$$

Recalling now that $\vec{u}_2$ satisfies the Leray energy inequality, we get

$$\|\vec{w}(t, .)\|_2^2 \leq - 2\nu \int_0^t \|\vec{\nabla} \otimes \vec{w}\|_2^2 \, ds + \lim_{\epsilon \to 0^+} J_\epsilon. \tag{12.14}$$

Inequality (12.14) has been established for any solution $\vec{u}_1$ in $L^2 H^1 \cap L^\infty L^2$. The problem is now to see for which spaces $\mathbb{X}$ the condition $\vec{u}_1 \in \mathbb{X}$ allows one to express the limit in (12.14) and to get $\vec{w} = 0$.

The Prodi–Serrin uniqueness criterion

Theorem 12.4.
Let $\vec{u}_0 \in L^2$, with div $\vec{u}_0 = 0$, and $\vec{f} \in L_t^2 H_x^{-1}$ on $(0, T) \times \mathbb{R}^3$. Assume that the Navier–Stokes problem

$$\partial_t \vec{u} + \mathbb{P}(\vec{u} \cdot \vec{\nabla} \vec{u}) = \nu \Delta \vec{u} + \mathbb{P} \vec{f}, \quad \vec{u}(0, .) = \vec{u}_0. \tag{12.15}$$

has a solution $\vec{u}_1$ on $(0, T) \times \mathbb{R}^3$ such that $\vec{u}_1 \in L_t^\infty L^2 \cap L_t^2 H^1 \cap \mathbb{X}_T^{(0)}$, where

- *$\mathbb{X}_T$ is the space of pointwise multipliers on $(0, T) \times \mathbb{R}^3$ from $L_t^\infty L^2 \cap L_t^2 \dot{H}^1$ to $L_t^2 L_x^2$, normed with $\|u\|_{\mathbb{X}_T} = \sup_{\|v\|_{L_t^\infty L^2} + \|v\|_{L_t^2 \dot{H}^1} \leq 1} \|uv\|_{L^2 L^2}$;*

- *$\mathbb{X}_T^{(0)}$ is the space of multipliers u in $\mathbb{X}_T$ such that, for every $t_0 \in [0, T)$, $\lim_{t_1 \to t_0^+} \|1_{(t_0, t_1)}(t) u(t, x)\|_{\mathbb{X}_T} = 0$.*

If $\vec{u}_2$ is a Leray weak solution of the same Navier–Stokes problem, then $\vec{u}_2 = \vec{u}_1$.

Proof. If $\vec{u}_1 \in L_t^\infty L^2 \cap L_t^2 H^1 \cap \mathbb{X}_T$ with div $\vec{u}_1 = 0$ and $\vec{v} \in L^2 H^1$, we write

$$\int_0^t \langle (\vec{u}_1 \cdot \vec{\nabla} \vec{u}_1) * \theta_\epsilon | \vec{v} \rangle_{H^{-1}, H^1} \, ds = - \int_0^t \langle (\vec{u}_1 \otimes \vec{u}_1) * \theta_\epsilon | \vec{\nabla} \otimes \vec{v} \rangle_{L^2, L^2};$$

since $\vec{u}_1 \otimes \vec{u}_1 \in L^2 L^2$, we have

$$\lim_{\epsilon \to 0^+} \int_0^t \langle (\vec{u}_1 \vec{\nabla} \vec{u}_1) * \theta_\epsilon | \vec{v} \rangle_{H^{-1}, H^1} \, ds = - \int_0^t \langle \vec{u}_1 \otimes \vec{u}_1 | \vec{\nabla} \otimes \vec{v} \rangle_{L^2, L^2}.$$

If $\vec{u}_1 \in L_t^\infty L^2 \cap L_t^2 H^1 \cap \mathbb{X}_T$ and $\vec{v} \in L_t^\infty L^2 \cap L^2 H^1$, we have that $\lim_{\epsilon \to 0^+} \| \vec{u}_1 * \theta_\epsilon - \vec{u}_1 \|_{L^2 L^6} = 0$, so that $\lim_{\epsilon \to 0^+} \| (\vec{u}_1 * \theta_\epsilon - \vec{u}_1) \otimes \vec{v} \|_{L^2 L^{3/2}} = 0$. Moreover

$$\| (\vec{u}_1 * \theta_\epsilon) \otimes \vec{v} \|_{L^2 L^2} \leq \int \| \vec{u}_1(t, x - y) \otimes \vec{v}(x) \|_{L^2 L^2} \, \theta_\epsilon(y) \, dy$$

$$= \int \| \vec{u}_1(t, x) \otimes \vec{v}(x + y) \|_{L^2 L^2} \, \theta_\epsilon(y) \, dy$$

$$\leq \| \vec{u}_1 \|_{\mathbb{X}_T} \int \| \vec{v}(t, x + y) \|_{L^\infty L^2 \cap L^2 \dot{H}^1} \, \theta_\epsilon(y) \, dy$$

$$= \| \vec{u}_1 \|_{\mathbb{X}_T} \| \vec{v} \|_{L^\infty L^2 \cap L^2 \dot{H}^1}.$$

This gives that $(\vec{u}_1 * \theta_\epsilon) \otimes \vec{v}$ is weakly convergent in $L^2 L^2$ to $\vec{u}_1 \otimes \vec{v}$ and thus

$$\lim_{\epsilon \to 0^+} \int_0^t \langle \vec{u}_1 * \theta_\epsilon | \vec{v} \cdot \vec{\nabla} \vec{v} \rangle_{H^{3/2}, H^{-3/2}} \, ds = \int_0^t \int \vec{u}_1 \cdot (\vec{v} \cdot \vec{\nabla}) \vec{v} \, dx \, ds.$$

Moreover, we have, for $\vec{v}_1$ and $\vec{v}_2$ in $\vec{v} \in L_t^\infty L^2 \cap L^2 \dot{H}^1$ (since div $\vec{u}_1 = 0$ and $(\vec{u}_1 * \theta_\epsilon) \otimes \vec{v}_i$ is weakly convergent in $L^2 L^2$ to $\vec{u}_1 \otimes \vec{v}_i$)

$$\int_0^t \int \vec{v}_1 \cdot (\vec{u}_1 \cdot \vec{\nabla}) \vec{v}_2 \, dx \, ds = \lim_{\epsilon \to 0^+} \int_0^t \int \vec{v}_1 \cdot ((\vec{u}_1 * \theta_\epsilon) \cdot \vec{\nabla}) \vec{v}_2 \, dx \, ds$$

$$= - \lim_{\epsilon \to 0^+} \int_0^t \int \vec{v}_2 \cdot ((\vec{u}_1 * \theta_\epsilon) \cdot \vec{\nabla}) \vec{v}_1 \, dx \, ds$$

$$= - \int_0^t \int \vec{v}_2 \cdot (\vec{u}_1 \cdot \vec{\nabla}) \vec{v}_1 \, dx \, ds$$

Thus, $\int_0^t \int \vec{u}_1 \cdot (\vec{u}_1 \cdot \vec{\nabla}) \vec{u}_1 \, dx \, ds = 0$ and we may transform inequality (12.14) into

$$\| \vec{w}(t, .) \|_2^2 \leq - 2\nu \int_0^t \| \vec{\nabla} \otimes \vec{w} \|_2^2 \, ds + 2 \int_0^t \int (\vec{u}_1 \cdot \vec{\nabla} \vec{u}_1) \cdot \vec{w} + \vec{u}_1 \cdot (\vec{u}_2 \cdot \vec{\nabla} \vec{w}) \, dx \, ds$$

$$= - 2\nu \int_0^t \| \vec{\nabla} \otimes \vec{w} \|_2^2 \, ds + 2 \int_0^t \int \vec{u}_1 \cdot (\vec{w} \cdot \vec{\nabla} \vec{w}) \, dx \, ds \tag{12.16}$$

If $0 \leq t_0 < t_1 < T$ are such that $\vec{w} = 0$ on $[0, t_0]$, we find that, on $[0, t_1]$,

$$\| \vec{w}(t, .) \|_2^2 + 2\nu \int_0^t \| \vec{\nabla} \otimes \vec{w} \|_2^2 \, ds \leq$$

$$2 (\| \vec{w} \|_{L^2((0, t_1), \dot{H}^1)} + \| \vec{w} \|_{L^\infty((0, t_1), L^2)})^2 \| \mathbb{1}_{(t_0, t_1)} \vec{u}_1 \|_{\mathbb{X}_T} \tag{12.17}$$

If $4(1 + \frac{1}{2\nu}) \| \mathbb{1}_{(t_0, t_1)} \vec{u}_1 \|_{\mathbb{X}_T} < 1$, we obtain $\vec{w} = 0$ on $[0, t_1]$. Thus, if $\vec{u}_1 \in \mathbb{X}_T^{(0)}$, we find that $\vec{w} = 0$ on $(0, T)$ and $\vec{u}_2 = \vec{u}_1$. $\square$

Remark: Of course, the assumption $\vec{u}_1 \in \mathbb{X}_T^{(0)}$ is very restrictive on $\vec{u}_0$: if $\vec{f} = 0$, we may apply the theory developed in Chapter 5. Indeed, we have obviously that $1_{0<t<T}\vec{u}_1 \in \mathcal{V}^{2,1}(\mathbb{R} \times \mathbb{R}^3)$. As the bilinear operator

$$B(\vec{u}, \vec{v}) = 1_{t>0} \int_0^t W_{\nu(t-s)} * \mathbb{P}\operatorname{div}(\vec{u} \otimes \vec{v}) \, ds$$

is bounded on $\mathcal{V}^{2,1}(\mathbb{R} \times \mathbb{R}^3)$, we find that $1_{0<t<T}W_{\nu t} * \vec{u}_0 \in \mathcal{V}^{2,1}(\mathbb{R} \times \mathbb{R}^3)$. Moreover, since $\vec{u}_1 \in \mathbb{X}_T^{(0)}$, we find that $\lim_{t_0 \to 0^+} \|1_{0<t<t_0}W_{\nu t} * \vec{u}_0\|_{\mathcal{V}^{2,1}(\mathbb{R}\times\mathbb{R}^3)} = 0$. This gives that $\vec{u}_1$ must be the mild solution associated to $\vec{u}_0$ through Picard's algorithm.

Proposition 12.3.
Let $\mathbb{X}_T^{(0)}$ be the space described in Theorem 12.4. Then

- *for $\frac{2}{p} + \frac{3}{q} = 1$ and $2 \leq p < +\infty$, we have $L_t^p L_x^q \subset \mathbb{X}_T^{(0)}$ (this is the original Prodi–Serrin criterion [406, 435])*

- *for $\frac{2}{p} + \frac{3}{q} = 1$ and $2 \leq p < +\infty$, we have $L_t^p \mathcal{M}(\dot{H}^{3/q} \mapsto L^2) \subset \mathbb{X}_T^{(0)}$ (Lemarié-Rieusset [313])*

- *for $\frac{2}{p} + \frac{3}{q} = 1$ and $2 \leq p < +\infty$, we have $L_t^p \dot{M}^{2,q} \subset \mathbb{X}_T^{(0)}$ (Lemarié–Rieusset [316])*

- *$\mathcal{C}([0,T], L^3) \subset \mathbb{X}_T^{(0)}$ (Von Wahl [494])*

- *if $\mathcal{V}_0^1$ is the closure of L^3 in $\mathcal{M}(\dot{H}^1 \mapsto L^2)$, then $\mathcal{C}([0,T], \mathcal{V}_0^1) \subset \mathbb{X}_T^{(0)}$ (Lemarié–Rieusset [313])*

Proof. For $q > 3/2$, we have $L^q \subset \mathcal{M}(\dot{H}^{3/q} \mapsto L^2) \subset \dot{M}^{2,q}$. Moreover, we have $\dot{M}^{2,q} = \mathcal{M}(\dot{B}_{2,1}^{3/q} \mapsto L^2)$. (A simple proof of this statement, using a decomposition on a wavelet basis, is given in Lemarié-Rieusset [316]).

When $2 < p < \infty$, we consider $L^p \dot{M}^{2,q}$. We have $\dot{B}_{2,1}^{3/q} = [L^2, \dot{H}^1]_{3/q,1}$, hence $\|v\|_{\dot{B}_{2,1}^{3/q}} \leq C\|v\|_2^{1-3/q}\|v\|_{\dot{H}^1}^{3/q}$ and thus $L^\infty L^2 \cap L^2 \dot{H}^1 \subset L^r \dot{B}_{2,1}^{3/q}$ with $\frac{1}{r} = \frac{3}{2q} = \frac{1}{2} - \frac{1}{p}$. Thus, $L^p \dot{M}^{2,q} \subset \mathbb{X}_T^{(0)}$.

When $p = 2$, we have obviously that $L^2 L^\infty$ is a pointwise multiplier from $L^\infty L^2$ to $L^2 L^2$ so that $L^2 L^\infty \subset \mathbb{X}_T^{(0)}$. (Remark: $L^\infty = \mathcal{M}(L^2 \mapsto L^2) = \dot{M}^{2,\infty}$.)

When $p = \infty$, we have obviously that $L^\infty \mathcal{V}^1$ is a pointwise multiplier from $L^2 \dot{H}^1$ to $L^2 L^2$. Moreover, if $\epsilon > 0$, we may split $u \in \mathcal{C}([0,T], \mathcal{V}_0^1)$ into $u = v + w$, where $v \in \mathcal{C}([0,T], \mathcal{V}_0^1) \cap L^\infty L^\infty$ and $\|w\|_{L^\infty \mathcal{V}^1} < \epsilon$. Thus, if $t_0 < t_1$, we find

$$\|1_{(t_0,t_1)}u\|_{\mathbb{X}_T} \leq \|1_{(t_0,t_1)}v\|_{\mathbb{X}_T} + \|1_{(t_0,t_1)}w\|_{\mathbb{X}_T} \leq \sqrt{t_1 - t_0}\|v\|_{L^\infty L^\infty} + \|w\|_{L^\infty \mathcal{V}^1}$$

so that

$$\limsup_{t_1 \to t_0^+} \|1_{(t_0,t_1)}u\|_{\mathbb{X}_T} \leq \epsilon$$

As ϵ is arbitrary, we find that $\mathcal{C}([0,T], \mathcal{V}_0^1) \subset \mathbb{X}_T^{(0)}$. $\square$

The endpoints of the Prodi–Serrin criterion have been slightly extended by Kozono and Taniuchi [277] when $p = 2$ and Kozono and Sohr [275] (extending a result of Masuda [350]) when $p = \infty$.

Proposition 12.4.
Let $\vec{u}_0 \in L^2$, with $\mathrm{div}\,\vec{u}_0 = 0$, and $\vec{f} \in L_t^2 H_x^{-1}$ on $(0,T) \times \mathbb{R}^3$. Assume that the Navier–Stokes problem

$$\partial_t \vec{u} + \mathbb{P}(\vec{u} \cdot \vec{\nabla} \vec{u}) = \nu \Delta \vec{u} + \mathbb{P}\vec{f}, \quad \vec{u}(0,.) = \vec{u}_0. \tag{12.18}$$

has two solutions $\vec{u}_1$ and $\vec{u}_2$ on $(0,T) \times \mathbb{R}^3$ such that $\vec{u}_1$ and $\vec{u}_2$ belong to $L_t^\infty L^2 \cap L_t^2 H^1$ and that $\vec{u}_2$ is a weak Leray solution. Then

- *If $\vec{u}_1$ belongs to $L_t^2 BMO$, then $\vec{u}_2 = \vec{u}_1$.*

- *If $\vec{u}_1$ belongs to $L^\infty L^3$ and $\vec{f}$ belongs to $L_t^\infty L^p$ for some $p \in (1,3)$, then $\vec{u}_2 = \vec{u}_1$.*

Proof.   (a) Case $\vec{u}_1 \in L_t^2 BMO$: we start from inequality (12.14):

$$\|\vec{w}(t,.\|_2^2 \leq -2\nu \int_0^t \|\vec{\nabla} \otimes \vec{w}\|_2^2 \, ds - \lim_{\epsilon \to 0^+} J_\epsilon.$$

The div-curl lemma of Coifman, Lions, Meyer, and Semmes [124, 313] gives that $\vec{u}_1 \cdot \vec{\nabla} \vec{u}_1$ and $\vec{u}_2 \cdot \vec{\nabla} \vec{u}_2$ belong to $L^2 \mathcal{H}^1$ (where $\mathcal{H}^1$ is the Hardy space, the pre-dual of BMO). Similarly, $\vec{u}_1 \cdot \vec{\nabla}(\theta_\epsilon * \vec{w})$ belongs to $L^2 \mathcal{H}^1$ and is controlled by $\|\vec{u}_1\|_{L^\infty L^2} \|\theta_\epsilon * \vec{w}\|_{L^2 \dot{H}^1}$. Thus, as we have the strong convergence of $\theta_\epsilon * \vec{w}$ to $\vec{w}$ in $L^2 \dot{H}^1$, we find that

$$\|\vec{w}(t,.)\|_2^2 \leq -2\nu \int_0^t \|\vec{\nabla} \otimes \vec{w}\|_2^2 \, ds$$

$$+ \int_0^t \langle \vec{u}_1 \cdot \vec{\nabla}\vec{u}_1 | \vec{u}_1 \rangle_{\mathcal{H}^1, BMO} + \langle \vec{u}_1 | \vec{u}_1 \cdot \vec{\nabla}\vec{u}_1 \rangle_{BMO, \mathcal{H}^1} \, ds$$

$$+ 2 \int_0^t - \langle \vec{u}_1 | \vec{u}_1 \cdot \vec{\nabla}\vec{w} \rangle_{BMO,,\mathcal{H}^1} + \langle \vec{u}_1 | -\vec{u}_1 \cdot \vec{\nabla}\vec{u}_1 + \vec{u}_2 \cdot \vec{\nabla}\vec{u}_2 \rangle_{BMO,\mathcal{H}^1} \, ds$$

Moreover, we have, for $j = 1, 2$,

$$\int_0^t \langle \vec{u}_1 * \theta_\epsilon | \vec{u}_j \cdot \vec{\nabla}\vec{u}_1 \rangle_{BMO, \mathcal{H}^1} \, ds = - \int_0^t \langle \vec{u}_1 | \vec{u}_j \cdot \vec{\nabla}(\theta_\epsilon * \vec{u}_1) \rangle_{BMO, \mathcal{H}^1} \, ds$$

so that

$$\int_0^t \langle \vec{u}_1 | \vec{u}_j \cdot \vec{\nabla}\vec{u}_1 \rangle_{BMO, \mathcal{H}^1} \, ds = - \int_0^t \langle \vec{u}_1 | \vec{u}_j \cdot \vec{\nabla}\vec{u}_1 \rangle_{BMO, \mathcal{H}^1} \, ds = 0.$$

This gives finally

$$\|\vec{w}(t,.)\|_2^2 \leq -2\nu \int_0^t \|\vec{\nabla} \otimes \vec{w}\|_2^2 \, ds + 2 \int_0^t \langle \vec{u}_1 | \vec{w} \cdot \vec{\nabla}\vec{w} \rangle_{BMO,,\mathcal{H}^1} \, ds$$

$$\leq -2\nu \int_0^t \|\vec{\nabla} \otimes \vec{w}\|_2^2 \, ds + 2C \int_0^t \|\vec{u}_1\|_{BMO} \|\vec{w}\|_2 \|\vec{w}\|_{\dot{H}^1} \, ds \tag{12.19}$$

$$\leq \frac{C^2}{2\nu} \int_0^t \|\vec{u}_1\|_{BMO}^2 \|\vec{w}\|_2^2 \, ds$$

and we conclude $\vec{w} = 0$ by Grönwall's lemma.

(b)   Case $\vec{u}_1 \in L_t^\infty L^3$: as $\partial_t \vec{u}_1$ and $\partial_t \vec{u}_2$ belong to $L^2 H^{-3/2}$, $\vec{u}_1$ and $\vec{u}_2$ are continuous from $[0,T)$ to $H^{-3/2}$, thus the set of times t such that $\vec{u}_1 = \vec{u}_2$ is closed. If $\vec{u}_1 \neq \vec{u}_2$, let T^* be the maximal time such that $\vec{u}_1 = \vec{u}_2$ on $[0, T^*]$. As $\vec{u}_1$ is bounded in L^3 and in L^2

and continuous in $H^{-3/2}$, we find that it is weakly continuous from $[0, T)$ to $L^3 \cap L^2$; in particular, $\vec{u}_1(T^*) \in L^3 \cap L^2$. It is easy to check that, following Theorem 7.5, we may construct a solution $\vec{u}_3$ on a small interval $[T^*, T^* + \epsilon]$ such that $\vec{u}_3$ belongs to $L_t^\infty L^2 \cap L_t^2 H^1 \cap C([T^*, T^* + \epsilon], L^3)$. But $\vec{u}_1$ belongs to X_T, so that $\partial_t \vec{u}_1 \in L^2 H^{-1}$ and $\vec{u}_1$ satisfies the Leray energy *equality* on $[0, T)$, while $\vec{u}_2$ satisfies the same Leray energy *equality* on $[0, T^*]$; thus, $\vec{u}_1$ and $\vec{u}_2$ are weak Leray solutions on $[T^*, T^* + \epsilon]$ and applying Proposition 12.3 to $\vec{u}_3$, we find that $\vec{u}_3 = \vec{u}_1$ and $\vec{u}_3 = \vec{u}_2$, so that $\vec{u}_1 = \vec{u}_2$ on $[0, T^* + \epsilon]$, which contradicts the definition of T^*.

$\square$

Of course, L^3 does not play a specific role in the Kozono–Sohr theorem. A general result is the following one:

Proposition 12.5.
Let $\mathcal{V}_0^1$ be the closure of $\mathcal{M}(\dot{H}^1 \mapsto L^2)$. Let E be a Banach space such that

- $E \subset \mathcal{V}_0^1$ *(continuous embedding)*

- E *is the dual of a Banach space E_0 such that $\mathcal{D}$ is dense in E_0.*

Let $\vec{u}_0 \in L^2$, with $\operatorname{div} \vec{u}_0 = 0$. Assume that the Navier–Stokes problem

$$\partial_t \vec{u} + \mathbb{P}(\vec{u} \cdot \vec{\nabla} \vec{u}) = \nu \Delta \vec{u}, \quad \vec{u}(0, .) = \vec{u}_0. \tag{12.20}$$

has two solutions $\vec{u}_1$ and $\vec{u}_2$ on $(0, T) \times \mathbb{R}^3$ such that $\vec{u}_1$ and $\vec{u}_2$ belong to $L_t^\infty L^2 \cap L_t^2 H^1$ and that $\vec{u}_2$ is a weak Leray solution. If $\vec{u}_1$ belongs to $L_t^\infty E$, then $\vec{u}_2 = \vec{u}_1$.

Proof. The proof is similar to the proof of Kozono and Sohr's theorem. As $\partial_t \vec{u}_1$ and $\partial_t \vec{u}_2$ belong to $L^2 H^{-3/2}$, $\vec{u}_1$ and $\vec{u}_2$ are continuous from $[0, T)$ to $H^{-3/2}$, thus the set of times t such that $\vec{u}_1 = \vec{u}_2$ is closed. If $\vec{u}_1 \neq \vec{u}_2$, let T^* be the maximal time such that $\vec{u}_1 = \vec{u}_2$ on $[0, T^*]$. As $\vec{u}_1$ is bounded in E and in L^2 and continuous in $H^{-3/2}$, we find that it is weakly continuous from $[0, T)$ to $E \cap L^2$; in particular, $\vec{u}_1(T^*) \in E \cap L^2$. It is easy to check that, following Theorem 8.2, we may construct a solution $\vec{u}_3$ on a small interval $[T^*, T^* + \epsilon]$ such that $\vec{u}_3$ belongs to $C([T^*, T^* + \epsilon], \mathcal{V}_0^1)$. Moreover, it is easy to check that this solution belongs to $L_t^\infty L^2 \cap L_t^2 H^1$.

$\vec{u}_1$ belongs to X_T, so that $\partial_t \vec{u}_1 \in L^2 H^{-1}$ and $\vec{u}_1$ satisfies the Leray energy *equality* on $[0, T)$, while $\vec{u}_2$ (which is equal to $\vec{u}_1$ on $[0, T^*]$) satisfies the same Leray energy *equality* on $[0, T^*]$; thus, $\vec{u}_1$ and $\vec{u}_2$ are weak Leray solutions on $[T^*, T^* + \epsilon]$ and applying Proposition 12.3 to $\vec{u}_3$, we find that $\vec{u}_3 = \vec{u}_1$ and $\vec{u}_3 = \vec{u}_2$, so that $\vec{u}_1 = \vec{u}_2$ on $[0, T^* + \epsilon]$, which contradicts the definition of T^*.

$\square$

Example: mixed-norm Lebesgue spaces.
Obvious examples of spaces E that fulfill the hypotheses of Proposition 12.5 are the Lorentz spaces $L^{3,q}$ with $1 \leq q < +\infty$. But we may find other examples, such as the case of mixed-norm Lebesgue spaces that has been recently considered by Phan and Robertson [396]:

Definition 12.2.
$L^{(p_1, p_2, p_3)} = L_{x_3}^{p_3} L_{x_2}^{p_2} L_{x_1}^{p_1}$ *is the space of measurable functions f on $\mathbb{R}^3$ such that*

$$\|f\|_{L^{(p_1, p_2, p_3)}} = \left(\int \left(\int \left(\int |f(x_1, x_2, x_3)|^{p_1} \, dx_1 \right)^{\frac{p_2}{p_1}} dx_2 \right)^{\frac{p_3}{p_2}} dx_3 \right)^{\frac{1}{p_3}} < +\infty.$$

Phan and Robertson's result states that the weak-strong uniqueness result stated in Proposition 12.5 holds for $E = L^{(p_1,p_2,p_3)}$, where $p_1, p_2, p_3 \in [2, +\infty)$, $\frac{1}{p_1} + \frac{1}{p_2} + \frac{1}{p_3} = 1$ and $p_3 > 2$. To prove this, we only need to check that $L^{(p_1,p_2,p_3)} \subset \mathcal{M}(\dot{H}^1 \mapsto L^2)$. As $L^{(p_1,p_2,p_3)} \subset \dot{\mathcal{M}}^{\min(p_1,p_2,p_3),3}$ for $\frac{1}{p_1} + \frac{1}{p_2} + \frac{1}{p_3} = 1$ and as $\dot{\mathcal{M}}^{p,3} \subset \mathcal{M}(\dot{H}^1 \mapsto L^2)$ for $2 < p \leq 3$, the result is obvious for $p_1 \neq 2$ and $p_2 \neq 2$. In order to prove the result for the general case (including the cases where p_1 or p_2 is equal to 2), Phan and Robertson use a Sobolev embedding theorem in mixed-norm spaces they found in the book by Besov, Il'in and Nikol'skiĭ [41]: for $2 \leq q_1, q_2, q_3 \leq +\infty$ with $\frac{1}{q_1} + \frac{1}{q_2} + \frac{1}{q_3} = \frac{1}{2}$ and $2 < q_3 < +\infty$, we have the continuous embedding $\dot{H}^1 \subset L^{q_1,q_2,q_3}$.

The proof of Kozono and Tanyuchi suggested to many authors a further extension of the Prodi-Serrin criterion for $1 \leq p < 2$, using paradifferential calculus (Ribaud [412], Gallagher and Planchon [201], Germain [204]). However, their results were generalized by Chen, Miao, and Zhang [115] in a very simple way that does not use para-differential calculus.

Proposition 12.6 (Chen, Miao, and Zhang).
Let $\vec{u}_0 \in L^2$, with div $\vec{u}_0 = 0$, and $\vec{f} \in L^2_t H^{-1}_x$ on $(0,T) \times \mathbb{R}^3$. Assume that the Navier–Stokes problem

$$\partial_t \vec{u} + \mathbb{P}(\vec{u} \cdot \vec{\nabla}\vec{u}) = \nu \Delta \vec{u} + \mathbb{P}\vec{f}, \quad \vec{u}(0,.) = \vec{u}_0 \tag{12.21}$$

has two solutions $\vec{u}_1$ and $\vec{u}_2$ on $(0,T) \times \mathbb{R}^3$ such that $\vec{u}_1$ and $\vec{u}_2$ belong to $L^\infty_t L^2 \cap L^2_t H^1$ and that $\vec{u}_2$ is a weak Leray solution. If $\vec{u}_1$ belongs to $(L^\infty_t L^2 \cap L^2_t H^1 \cap L^2 L^\infty) + (L^\infty_t L^2 \cap L^2_t H^1 \cap L^1 \dot{W}^{1,\infty})$, then $\vec{u}_2 = \vec{u}_1$.

In particular, if $\vec{u}_1 \in L^p \dot{B}^r_{\infty,\infty}$ with $1 < p < 2$ and $\frac{2}{p} = 1 + r$, then $\vec{u}_2 = \vec{u}_1$.

Proof. First, let us remark that $\dot{B}^r_{\infty,\infty}$ $(0 < r < 1)$ (the spaces of Hölderian functions of Hölder exponent r) and $\dot{W}^{1,\infty}$ (the space of Lipschitzian functions) are defined modulo the constants; however, on $L^2 \cap \dot{B}^r_{\infty,\infty}$ or $L^2 \cap \dot{W}^{1,\infty}$, the constants are fixed.

We now check that $L^\infty_t L^2 \cap L^2_t H^1 \cap L^{\frac{2}{1+r}} \dot{B}^r_{\infty,\infty} \subset (L^\infty_t L^2 \cap L^2_t H^1 \cap L^2 L^\infty) + (L^\infty_t L^2 \cap L^2_t H^1 \cap L^1 \dot{W}^{1,\infty})$. Indeed, we use a mollifier θ_ϵ and write

$$u(t,x) = u * \theta_{\epsilon(t)} + (u - u * \theta_{\epsilon(t)}) = U + V.$$

We have, of course, $U \in L^\infty_t L^2 \cap L^2_t H^1$ and $V \in L^\infty_t L^2 \cap L^2_t H^1$. Moreover, $\|\vec{\nabla}U(t,.)\|_\infty \leq C\|u(t,.)\|_{\dot{B}^r_{\infty,\infty}} \epsilon(t)^{-1+r}$ while $\|V(t,.)\|_\infty \leq C\|u(t,.)\|_{\dot{B}^r_{\infty,\infty}} \epsilon(t)^r$. The choice $\epsilon(t) = \|u(t,.)\|_{\dot{B}^r_{\infty,\infty}}^{-p/2}$ gives $U \in L^1 \dot{W}^{1,\infty}$ and $V \in L^2 L^\infty$.

We now prove the general case. We start again from inequality (12.14):

$$\|\vec{w}(t,.)\|_2^2 \leq -2\nu \int_0^t \|\vec{\nabla} \otimes \vec{w}\|_2^2 \, ds + \lim_{\epsilon \to 0^+} J_\epsilon$$

with

$$J_\epsilon = \int_0^t \langle (\vec{u}_1 \cdot \vec{\nabla}\vec{u}_1) * \theta_\epsilon | \vec{u}_1 \rangle_{H^{-1}, H^1} + \langle \vec{u}_1 * \theta_\epsilon | \vec{u}_1 \cdot \vec{\nabla}\vec{u}_1 \rangle_{H^{3/2}, H^{-3/2}} \, ds$$

$$+ 2 \int_0^t \langle (\vec{u}_1 \cdot \vec{\nabla}\vec{u}_1) * \theta_\epsilon | \vec{w} \rangle_{H^{-1}, H^1} + \langle \vec{u}_1 * \theta_\epsilon | - \vec{u}_1 \cdot \vec{\nabla}\vec{u}_1 + \vec{u}_2 \cdot \vec{\nabla}\vec{u}_2 \rangle_{H^{3/2}, H^{-3/2}} \, ds$$

We write $\vec{u}_1 = \vec{U} + \vec{V}$, where $\vec{U} \in (L_t^\infty L^2 \cap L_t^2 H^1 \cap L^1 \dot{W}^{1,\infty})$ and $\vec{V} \in (L_t^\infty L^2 \cap L_t^2 H^1 \cap L^2 L^\infty)$. Then, for $j = 1, 2$, we have

$$\int_0^t \langle (\vec{u}_1 \cdot \vec{\nabla} \vec{u}_1) * \theta_\epsilon | \vec{u}_j \rangle_{H^{-1}, H^1} \, ds = \int_0^t \langle (\vec{u}_1 \cdot \vec{\nabla} \vec{U}) * \theta_\epsilon | \vec{u}_j \rangle_{L^2, L^2} \, ds$$
$$- \int_0^t \langle (\vec{u}_1 \otimes \vec{V}) * \theta_\epsilon | \vec{\nabla} \otimes \vec{u}_j \rangle_{L^2, L^2}, \, ds$$

and

$$\int_0^t \langle \vec{u}_1 * \theta_\epsilon | \vec{u}_j \cdot \vec{\nabla} \vec{u}_j \rangle_{H^{3/2}, H^{-3/2}} \, ds = \int_0^t \langle \vec{V} * \theta_\epsilon | \vec{u}_j \cdot \vec{\nabla} \vec{u}_j \rangle_{L^\infty, L^1} \, ds$$
$$- \int_0^t \langle (\vec{\nabla} \otimes \vec{U}) * \theta_\epsilon | \vec{u}_j \otimes \vec{u}_j \rangle_{L^\infty, L^1} \, ds$$

By *-weak convergence in the space variable and then dominated convergence in the time variable, we find that

$$\|\vec{w}(t, .)\|_2^2 \leq - 2\nu \int_0^t \|\vec{\nabla} \otimes \vec{w}\|_2^2 \, ds$$
$$+ 2 \int_0^t \int (\vec{u}_1 \cdot \vec{\nabla} \vec{U}) \cdot \vec{w} \, dx \, ds - 2 \int_0^t \int \vec{V} \cdot (\vec{u}_1 \cdot \vec{\nabla} \vec{w}) \, dx \, ds$$
$$+ 2 \int_0^t \int \vec{V} \cdot (\vec{u}_2 \cdot \vec{\nabla} \vec{u}_2 - \vec{u}_1 \cdot \vec{\nabla} u_1) \, dx \, ds$$
$$- 2 \int_0^t \int \vec{u}_2 \cdot (\vec{u}_2 \cdot \vec{\nabla} \vec{U}) - \vec{u}_1 \cdot (\vec{u}_1 \cdot \vec{\nabla} \vec{U}) \, dx \, ds$$
$$= - 2\nu \int_0^t \|\vec{\nabla} \otimes \vec{w}\|_2^2 \, ds$$
$$- 2 \int_0^t \int \vec{w} \cdot (\vec{w} \cdot \vec{\nabla} \vec{U}) \, dx \, ds + 2 \int_0^t \int \vec{V} \cdot (\vec{w} \cdot \vec{\nabla} \vec{w}) \, dx \, ds$$
$$+ 2 \int_0^t \int \vec{V} \cdot (\vec{w} \cdot \vec{\nabla} \vec{u}_1) - \vec{u}_1 \cdot (\vec{w} \cdot \vec{\nabla} \vec{U}) \, dx \, ds.$$

A similar proof by mollification shows that

$$\int_0^t \int \vec{V} \cdot (\vec{w} \cdot \vec{\nabla} \vec{V}) \, dx \, ds = \int_0^t \int \vec{U} \cdot (\vec{w} \cdot \vec{\nabla} \vec{U}) \, dx \, ds = 0$$

so that finally we get

$$\|\vec{w}(t, .)\|_2^2 \leq - 2\nu \int_0^t \|\vec{\nabla} \otimes \vec{w}\|_2^2 \, ds$$
$$- 2 \int_0^t \int \vec{w} \cdot (\vec{w} \cdot \vec{\nabla} \vec{U}) \, dx \, ds + 2 \int_0^t \int \vec{V} \cdot (\vec{w} \cdot \vec{\nabla} \vec{w}) \, dx \, ds$$
$$\leq - 2\nu \int_0^t \|\vec{\nabla} \otimes \vec{w}\|_2^2 \, ds$$
$$+ 2 \int_0^t \|\vec{w}\|_2^2 \|\vec{\nabla} \otimes \vec{U}\|_\infty \, ds + 2 \int_0^t \|\vec{V}\|_\infty \|\vec{w}\|_2 \|\vec{\nabla} \otimes \vec{w}\|_2 \, ds$$
$$\leq 2 \int_0^t (\|\vec{\nabla} \otimes \vec{U}\|_\infty + \frac{1}{4\nu} \|\vec{V}\|_\infty^2) \|\vec{w}\|_2^2 \, ds$$

and we conclude by Grönwall's lemma. $\qquad \square$

The case of $\vec{u}_1 \in L^p \dot{B}^r_{\infty,\infty}$ with $\frac{2}{p} = r+1$ and $0 < r < 1$ could have been treated directly by elementary interpolation arguments:

Lemma 12.2.
If $\vec{u} \in \dot{B}^r_{\infty,\infty}$, $\vec{v} \in L^2$ with $\operatorname{div} \vec{v} = 0$ and $\vec{w} \in \dot{B}^{1-r}_{2,1}$, then

$$\left| \int \vec{u} . \operatorname{div}(\vec{v} \otimes \vec{w}) \, dx \right| \leq C \|\vec{u}\|_{\dot{B}^r_{\infty,\infty}} \|\vec{v}\|_2 \|\vec{w}\|_{\dot{B}^{1-r}_{2,1}}.$$

Proof. Let T be the operator $(\vec{v}, \vec{w}) \mapsto T(\vec{v}, \vec{w}) = \operatorname{div}((\mathbb{P}\vec{v}) \otimes \vec{w})$. If $\vec{v} \in L^2$ and $\vec{w} \in L^2$, then $\vec{v} \otimes \vec{w} \in L^1 \subset \dot{B}^0_{1,\infty}$, hence $T(\vec{v}, \vec{w}) \in \dot{B}^{-1}_{1,\infty}$. If $\vec{v} \in L^2$ and $\vec{w} \in \dot{H}^1$, then $T(\vec{v}, \vec{w}) = (\mathbb{P}\vec{v}) \cdot \vec{\nabla} \vec{w} \in L^1 \subset \dot{B}^0_{1,\infty}$. By interpolation, T is bounded from $L^2 \times [L^2, \dot{H}^1]_{1-r,1}$ to $[\dot{B}^{-1}_{1,\infty}, \dot{B}^0_{1,\infty}]_{1-r,1}$, hence from $L^2 \times \dot{B}^{1-r}_{2,1}$ to $\dot{B}^{-r}_{1,1}$. $\square$

Let us make a final remark. For $1 \leq p < +\infty$, the results of Propositions 12.3, 12.4, and 12.6, may (partially) be unified in a single statement: let $\mathcal{L}^{2,\lambda}$ be the Morrey–Campanato space of locally square integrable functions u such that

$$\|u\|_{\mathcal{L}^{2,\lambda}} = \sqrt{\sup_{x_0 \in \mathbb{R}^3, r>0} \frac{1}{r^{3+2\lambda}} \int_{B(x_0,r)} |u(x) - m_{B(x_0,r)}u|^2 \, dx} < +\infty$$

This space is defined modulo the constants.

- For $\lambda = 1$, we have $\mathcal{L}^{2,\lambda} = \dot{W}^{1,\infty}$: $u \in \dot{W}^{1,\infty} \Leftrightarrow \vec{\nabla} u \in L^\infty$.

- For $0 < \lambda < 1$, we have $\mathcal{L}^{2,\lambda} = \dot{B}^\lambda_{\infty,\infty}$ (where the homogeneous Besov space is defined by $u \in \dot{B}^\lambda_{\infty,\infty} \Leftrightarrow \vec{\nabla} u \in \dot{B}^{\lambda-1}_{\infty,\infty}$: as $\lambda - 1 < 0$, we have already defined $\dot{B}^{\lambda-1}_{\infty,\infty}$ unambiguously). This equality can be checked by using a decomposition on a wavelet basis, for instance.

- For $\lambda = 0$, we have $\mathcal{L}^{2,\lambda} = BMO$.

- For $-3/2 < \lambda < 0$, we can see that, for $u \in \mathcal{L}^{2,\lambda}$, $L(u) = \lim_{R\to+\infty} \frac{1}{|B(0,R)|} \int_{B(0,R)} u(x) \, dx$ exists; moreover, $u - L(u)$ belongs to $\dot{M}^{2,q}$, where $\frac{1}{q} + \frac{\lambda}{3} = 0$ and $\|u\|_{\mathcal{L}^{2,\lambda}} \approx \|u - L(u)\|_{\dot{M}^{2,q}}$. Of course, if $u \in L^2 \cap \mathcal{L}^{2,\lambda}$, then $L(u) = 0$.

Thus, we get:

The generalized Prodi–Serrin uniqueness criterion

Theorem 12.5.
Let $\vec{u}_0 \in L^2$, with $\operatorname{div} \vec{u}_0 = 0$, and $\vec{f} \in L^2_t H^{-1}_x$ on $(0,T) \times \mathbb{R}^3$. Assume that the Navier–Stokes problem

$$\partial_t \vec{u} + \mathbb{P}(\vec{u} \cdot \vec{\nabla} \vec{u}) = \nu \Delta \vec{u} + \mathbb{P}\vec{f}, \quad \vec{u}(0,.) = \vec{u}_0 \tag{12.22}$$

has a solution $\vec{u}_1$ on $(0,T) \times \mathbb{R}^3$ such that $\vec{u}_1 \in L^\infty_t L^2 \cap L^2_t H^1 \cap L^p \mathcal{L}^{2,\lambda}$, where $1 \leq p < +\infty$ and $\frac{2}{p} = 1 + \lambda$.
If $\vec{u}_2$ is a Leray weak solution of the same Navier–Stokes problem, then $\vec{u}_2 = \vec{u}_1$.

12.4   Weak-Strong Uniqueness and Morrey Spaces on the Product Space $\mathbb{R} \times \mathbb{R}^3$

In the preceding section, we have considered the inequality

$$\|\vec{w}(t,.)\|_2^2 \leq -2\nu \int_0^t \|\vec{w}\|_{\dot{H}^{-1}}^2 \, ds + 2 \int_0^t \int \vec{u}_1 \cdot (\vec{w} \cdot \vec{\nabla}\vec{w}) \, dx \, ds$$

and we have estimated the term $\int_0^t \int \vec{u}_1 \cdot (\vec{w} \cdot \vec{\nabla}\vec{w}) \, dx \, ds$ using only size estimates on $\vec{w}$ with respect to the time variable: $\vec{w} \in L_t^\infty L_x^2 \cap L_t^2 \dot{H}_x^1$. But we know that we have some time regularity on $\vec{w}$: $\partial_t \vec{w} \in L^2((0,T), H^{-3/2})$.

This suggests some generalizations of Theorem 12.4. We begin first with some lemmas on the Sobolev spaces $\dot{H}^r(\mathbb{R})$ with $0 < r < 1/2$.

Lemma 12.3.
If I is an interval of $\mathbb{R}$, then the pointwise multiplication by 1_I is bounded on $\dot{H}^r(\mathbb{R})$ with $0 < r < 1/2$:

$$\|1_I f\|_{\dot{H}^r} \leq C\|f\|_{\dot{H}^r}$$

where C does not depend on I.

Proof. It is enough to prove the theorem for $I = (0, +\infty)$ as

$$1_{(a,b)}(t) = 1_{(0,+\infty)}(t-a)(1 - 1_{(0,\infty)}(t-b))$$

(for $t \neq b$). But for $I = (0, +\infty)$, the boundedness of the multiplier 1_I on $\dot{H}^r$ is equivalent to the boundedness of the Hilbert transform on $L^2(|\tau|^{2r} \, d\tau)$ (by the Plancherel equality for the Fourier transform); as $|\tau|^{2r}$ is a Muckenhoupt weight in $\mathcal{A}_2$ for $0 < r < 1/2$ [448], the Hilbert transform is actually bounded on $L^2(|\tau|^{2r} \, d\tau)$. $\square$

From this lemma, we see that we can define $\dot{H}^r(I)$ as the space of functions f in $\dot{H}^r(\mathbb{R})$ which are equal to 0 outside I. We then have the following important lemma:

Lemma 12.4.
Let I be a bounded interval (a,b) of $\mathbb{R}$, w be a function defined on $I \times \mathbb{R}^3$ such that $w \in L_t^2(I, H_x^1)$ and $\partial_t w \in L_t^2(I, H_x^{-3/2})$ with $w(a,.) = 0$, then $w \in L_x^2 \dot{H}^{2/5}(I)$ and

$$\|w\|_{L_x^2 \dot{H}^{2/5}(I)} \leq C\|w\|_{L^2 H^1}^{3/5} \|\partial_t w\|_{L^2 H^{-3/2}}^{2/5} \tag{12.23}$$

where C does not depend on I.

Proof. If $I = (a,b)$, we define W on $(0,1) \times \mathbb{R}^3$ as

$$W(t,x) = \sqrt{b-a} \, w(a + t(b-a), x).$$

We have that $W \in L_t^2((0,1), H_x^1)$ and $\partial_t W \in L_t^2((0,1), H_x^{-3/2})$ and that

$$\|W\|_{L_t^2 H_x^1} = \|w\|_{L_t^2 H_x^1}, \|\partial_t W\|_{L_t^2 H_x^{-3/2}} = (b-a)\|\partial_t w\|_{L_t^2 H_x^{-3/2}},$$

while

$$\|W\|_{L_x^2 \dot{H}^{2/5}((0,1))} = (b-a)^{2/5}\|w\|_{L_x^2 \dot{H}^{2/5}(I)}.$$

Thus, we may only consider the case $I = (0,1)$.

If $I = (0, 1)$ we define ω on $\mathbb{R} \times \mathbb{R}^3$ as $\omega(t, x) = w(t, x)$ when $0 < t < 1$, $\omega(t, x) = w(2 - t, x)$ when $1 < t < 2$, and $\omega(t, x) = 0$ when $t < 0$ or $t > 2$. Then $\omega \in L_t^2(\mathbb{R}, H_x^1)$ and $\partial_t \omega \in L_t^2(I, H_x^{-3/2})$; in Fourier variables, we find $\hat{\omega}(1+|\xi|^2)^{1/2} \in L_\tau^2 L_\xi^2$ and $|\tau| \hat{\omega}(1+|\xi|)^{-3/2} \in L_\tau^2 L_\xi^2$ so that $|\tau|^{2/5} \hat{\omega} \in L_\tau^2 L_\xi^2$. Thus, $\omega \in L_x^2 \dot{H}_t^{2/5}$, and $w = 1_{(0,1)} \omega \in L_x^2 \dot{H}^{2/5}((0, 1))$. $\qquad\qquad\square$

We thus caan modify Theorem 12.4 by replacing multipliers from $L_t^\infty L_x^2 \cap L_t^2 \dot{H}_x^1$ to $L^2 L^2$ by multipliers from $L_t^\infty L_x^2 \cap L_t^2 \dot{H}_x^1 \cap L_x^2 \dot{H}_t^{2/5}$ to $L^2 L^2$:

Extension of the Prodi–Serrin uniqueness criterion

Theorem 12.6.
Let $\vec{u}_0 \in L^2$, with div $\vec{u}_0 = 0$, and $\vec{f} \in L_t^2 H_x^{-1}$ on $(0, T) \times \mathbb{R}^3$. Assume that the Navier–Stokes problem

$$\partial_t \vec{u} + \mathbb{P}(\vec{u} \cdot \vec{\nabla} \vec{u}) = \nu \Delta \vec{u} + \mathbb{P} \vec{f}, \quad \vec{u}(0, .) = \vec{u}_0. \tag{12.24}$$

has a solution $\vec{u}_1$ on $(0, T) \times \mathbb{R}^3$ such that $\vec{u}_1 \in L_t^\infty L^2 \cap L_t^2 H^1 \cap \mathbb{Y}_T^{(0)}$, where

- $\mathbb{Y}_T$ *is the space of pointwise multipliers on $(0, T) \times \mathbb{R}^3$ from $L_t^\infty L^2 \cap L_t^2 H^1 \cap L_x^2 \dot{H}_t^{2/5}$ to $L_t^2 L_x^2$, normed with*

$$\|u\|_{\mathbb{Y}_T} = \sup_{\|v\|_{L_t^\infty L^2} + \|v\|_{L_t^2 H^1} + \|v\|_{L_x^2 \dot{H}_t^{2/5}} \leq 1} \|uv\|_{L^2 L^2};$$

- $\mathbb{Y}_T^{(0)}$ *is the space of multipliers u in $\mathbb{Y}_T$ such that, for every $t_0 \in [0, T)$,*

$$\lim_{t_1 \to t_0^+} \|1_{(t_0, t_1)}(t) u(t, x)\|_{\mathbb{Y}_T} = 0.$$

If $\vec{u}_2$ is a Leray weak solution of the same Navier–Stokes problem, then $\vec{u}_2 = \vec{u}_1$.

Proof. We follow the proof of Theorem 12.4. If $\vec{u}_1 \in L_t^\infty L^2 \cap L_t^2 H^1 \cap \mathbb{Y}_T$ with div $\vec{u}_1 = 0$ and $\vec{v} \in L^2 H^1$, we write

$$\int_0^t \langle (\vec{u}_1 \cdot \vec{\nabla} \vec{u}_1) * \theta_\epsilon | \vec{v} \rangle_{H^{-1}, H^1} \, ds = -\int_0^t \langle (\vec{u}_1 \otimes \vec{u}_1) * \theta_\epsilon | \vec{\nabla} \otimes \vec{v} \rangle_{L^2, L^2}$$

and find

$$\lim_{\epsilon \to 0^+} \int_0^t \langle (\vec{u}_1 \cdot \vec{\nabla} \vec{u}_1) * \theta_\epsilon | \vec{v} \rangle_{H^{-1}, H^1} \, ds = -\int_0^t \langle \vec{u}_1 \otimes \vec{u}_1 | \vec{\nabla} \otimes \vec{v} \rangle_{L^2, L^2}.$$

If $\vec{u}_1 \in L_t^\infty L^2 \cap L_t^2 H^1 \cap \mathbb{Y}_T$ and $\vec{v} \in L_t^\infty L^2 \cap L^2 H^1$, we have that $\lim_{\epsilon \to 0^+} \|\vec{u}_1 * \theta_\epsilon - \vec{u}_1\|_{L^2 L^6} = 0$, so that $\lim_{\epsilon \to 0^+} \|(\vec{u}_1 * \theta_\epsilon - \vec{u}_1) \otimes \vec{v}\|_{L^2 L^{3/2}} = 0$. Moreover

$$\|(\vec{u}_1 * \theta_\epsilon) \otimes \vec{v}\|_{L^2 L^2} \leq \int \|\vec{u}_1(t, x - y) \otimes \vec{v}(x)\|_{L^2 L^2} \theta_\epsilon(y) \, dy$$

$$= \int \|\vec{u}_1(t, x) \otimes \vec{v}(x + y)\|_{L^2 L^2} \theta_\epsilon(y) \, dy$$

$$\leq \|\vec{u}_1\|_{\mathbb{Y}_T} \int \|\vec{v}(t, x + y)\|_{L^\infty L^2 \cap L^2 H^1 \cap L_x^2 \dot{H}_t^{2/5}} \theta_\epsilon(y) \, dy$$

$$= \|\vec{u}_1\|_{\mathbb{Y}_T} \|\vec{v}\|_{L^\infty L^2 \cap L^2 H^1 \cap L_x^2 \dot{H}_t^{2/5}}.$$

This gives that $(\vec{u}_1 * \theta_\epsilon) \otimes \vec{v}$ is weakly convergent in $L^2 L^2$ to $\vec{u}_1 \otimes \vec{v}$ and thus

$$\lim_{\epsilon \to 0^+} \int_0^t \langle \vec{u}_1 * \theta_\epsilon | \vec{v} \cdot \vec{\nabla} \vec{v} \rangle_{H^{3/2}, H^{-3/2}} \, ds = \int_0^t \int \vec{u}_1 \cdot (\vec{v} \cdot \vec{\nabla}) \vec{v} \, dx \, ds.$$

Moreover, we have, for $\vec{v}_1$ and $\vec{v}_2$ in $\vec{v} \in L_t^\infty L^2 \cap L^2 H^1$ (since $\operatorname{div} \vec{u}_1 = 0$ and $(\vec{u}_1 * \theta_\epsilon) \otimes \vec{v}_i$ is weakly convergent in $L^2 L^2$ to $\vec{u}_1 \otimes \vec{v}_i$)

$$\int_0^t \int \vec{v}_1 \cdot (\vec{u}_1 \cdot \vec{\nabla}) \vec{v}_2 \, dx \, ds = \lim_{\epsilon \to 0^+} \int_0^t \int \vec{v}_1 \cdot ((\vec{u}_1 * \theta_\epsilon) \cdot \vec{\nabla}) \vec{v}_2 \, dx \, ds$$

$$= - \lim_{\epsilon \to 0^+} \int_0^t \int \vec{v}_2 \cdot ((\vec{u}_1 * \theta_\epsilon) \cdot \vec{\nabla}) \vec{v}_1 \, dx \, ds$$

$$= - \int_0^t \int \vec{v}_2 \cdot (\vec{u}_1 \cdot \vec{\nabla}) \vec{v}_1 \, dx \, ds$$

Thus, $\int_0^t \int \vec{u}_1 \cdot (\vec{u}_1 \cdot \vec{\nabla}) \vec{u}_1 \, dx \, ds = 0$ and we may transform inequality (12.14) into

$$\|\vec{w}(t, .)\|_2^2 \leq - 2\nu \int_0^t \|\vec{\nabla} \otimes \vec{w}\|_2^2 \, ds + 2 \int_0^t \int (\vec{u}_1 \cdot \vec{\nabla} \vec{u}_1) \cdot \vec{w} + \vec{u}_1 \cdot (\vec{u}_2 \cdot \vec{\nabla} \vec{w}) \, dx \, ds$$

$$= - 2\nu \int_0^t \|\vec{\nabla} \otimes \vec{w}\|_2^2 \, ds + 2 \int_0^t \int \vec{u}_1 \cdot (\vec{w} \cdot \vec{\nabla} \vec{w}) \, dx \, ds \tag{12.25}$$

If $0 \leq t_0 < t_1 < T$ are such that $\vec{w} = 0$ on $[0, t_0]$, we find that, on $[0, t_1]$,

$$\|\vec{w}(t, .)\|_2^2 + 2\nu \int_0^t \|\vec{\nabla} \otimes \vec{w}\|_2^2 \, ds \leq \tag{12.26}$$

$$2(\|\vec{w}\|_{L^2((0,t_1), H^1)} + \|\vec{w}\|_{L^\infty((0,t_1), L^2)} + \|\vec{w}\|_{L_x^2 \dot{H}^{2/5}((0,t_1))})^2 \|1_{(t_0,t_1)} \vec{u}_1\|_{\mathbb{Y}_T}$$

We have $\partial_t \vec{w} = \Delta \vec{w} - \mathbb{P} \operatorname{div}(\vec{u}_2 \otimes \vec{w} + \vec{w} \otimes \vec{u}_1)$, so that

$$\|\vec{w}\|_{L_x^2 \dot{H}^{2/5}((0,t_1))} \leq C_0 (\|\vec{w}\|_{L^2((t_0,t_1), \dot{H}^1)} + \|\partial_t \vec{w}\|_{L^2((t_0,t_1), \dot{H}^{-3/2})})$$

$$\leq C_0 \|\vec{w}\|_{L^2((t_0,t_1), \dot{H}^1)} + C_1 (1 + \|\vec{u}_0\|_2) \|\vec{w}\|_{L^2((t_0,t_1), \dot{H}^1)}$$

If $4(1 + C_0 + \frac{1 + C_1(1 + \|\vec{u}_0\|_2)}{2\nu}) \|1_{(t_0,t_1)} \vec{u}_1\|_{\mathbb{Y}_T} < 1$, we obtain $\vec{w} = 0$ on $[0, t_1]$. Thus, if $\vec{u}_1 \in \mathbb{Y}_T^{(0)}$, we find that $\vec{w} = 0$ on $(0, T)$ and $\vec{u}_2 = \vec{u}_1$. $\qquad \square$

As the space $\mathbb{X}_T$ of Theorem 12.4 is obviously embedded into $\mathbb{Y}_T$, we obtain a larger class of weak-strong uniqueness. For instance, we have:

Proposition 12.7.
For $1 < p \leq 11/2$, let $\mathcal{M}_{5/2}^{p,11/2}$ be the Morrey space on $\mathbb{R} \times \mathbb{R}^3$ defined by $\|u\|_{\mathcal{M}_{5/2}^{p,11/2}} < +\infty$, where

$$\|u\|_{\mathcal{M}_{5/2}^{p,11/2}} = \sup_{x_0 \in \mathbb{R}^3, t_0 \in \mathbb{R}, R > 0} \left(\frac{1}{R^{11/2 - p}} \iint_{|t-t_0| < R^{5/2}, |x-x_0| < R} |u(t,x)|^p \, dt \, dx \right)^{1/p}.$$

Let $\mathbb{Z}_{p,T}$ be the space of the functions u defined on $(0,T) \times \mathbb{R}^3$ such that $1_{0<t<T} u \in \mathcal{M}_{5/2}^{p,11/2}$. Then, for $2 < p \leq 11/2$, we have:

- $\mathbb{Z}_{p,T} \subset \mathbb{Y}_T$

- *the closure* $\mathbb{Z}_{p,T}^{(0)}$ *of* $L_t^{11/2} L_x^{11/2}$ *in* $\mathbb{Z}_{p,T}$ *satisfies* $\mathbb{Z}_{p,T}^{(0)} \subset \mathbb{Y}_T^{(0)}$

Proof. Let X be the space of homogenous type $(\mathbb{R} \times \mathbb{R}^3, \delta_{5/2}, \mu)$, where $\delta_{5/2}$ is the parabolic (quasi)-distance

$$\delta_{5/2}((t,x),(s,y)) = |t-s|^{2/5} + |x-y| \tag{12.27}$$

and μ is the Lebesgue measure $d\mu = dt\,dx$.

Then the homogeneous dimension Q of X is equal to $11/2$ and $\mathcal{M}_{5/2}^{p,11/2}$ is the Morrey space $\dot{M}^{p,Q}(X)$. Then we may apply Theorem 5.3 and Proposition 5.4 on Riesz potentials to see that the elements of $\dot{M}^{p,Q}(X)$ with $2 < p \le Q$ are pointwise multipliers from $W^1(X)$ to $L_t^2 L_x^2$, where $u \in W^1(X)$ if and only if $u = \iint \frac{1}{\delta_{5/2}(t-s,x-y)^{Q-1}} v(s,y)\,ds\,dy$ with $v \in L_t^2 L_x^2$.

Moreover, by Proposition 5.6, we have $W^1(X) = W^{5/2,3/2}(\mathbb{R} \times \mathbb{R}^3) = L_t^2 \dot{H}_x^1 \cap L_x^2 \dot{H}_t^{2/5}$. Thus, we find that $\mathbb{Z}_{p,T} \subset \mathbb{Y}_T$. $\qquad\square$

We easily can check that, locally in space and time, $\mathcal{M}_{5/2}^{p,11/2}$ satisfies the parabolic scaling of mild solutions: recall that the space $\mathcal{M}_2^{p,5}$, $2 < p \le 5$, defined by $\|u\|_{\mathcal{M}_2^{p,5}} < +\infty$, where

$$\|u\|_{\mathcal{M}_2^{p,5}} = \sup_{x_0 \in \mathbb{R}^3, t_0 \in \mathbb{R}, R>0} \left(\frac{1}{R^{5-p}} \iint_{|t-t_0|<R^2, |x-x_0|<R} |u(t,x)|^p\, dt\,dx \right)^{1/p},$$

was discussed on page 98 in Chapter 5. If K is a compact subset of $\mathbb{R} \times \mathbb{R}^3$ and u belongs to $\mathcal{M}_{5/2}^{p,11/2}$, then $1_K(t,x)u$ belongs to $\mathcal{M}_2^{p,5}$. Indeed, if $R \ge 1$, we have $\iint_{|t-t_0|<R^2, |x-x_0|<R} |u(t,x)|^p\,dt\,dx \le R^{5-p} \iint_K |u(t,x)|^p\,dt\,dx$; if $R < 1$, we may split the interval $[t_0 - R^2, t_0 + R^2]$ into an union of $O(R^{-1/2})$ intervals $[t_{0,i} - R^{5/2}, t_{0,i} + R^{5/2}]$, so that

$$\iint_{|t-t_0|<R^2, |x-x_0|<R} |u(t,x)|^p\,dt\,dx = \sum_{i=1}^{O(R^{-1/2})} \iint_{|t-t_{0,i}|<R^{5/2}, |x-x_0|<R} |u(t,x)|^p\,dt\,dx$$

$$\le C\, R^{-1/2} \|u\|_{\mathcal{M}_{5/2}^{p,11/2}}^p R^{11/2-p}.$$

12.5   Almost Strong Solutions

Uniqueness remains an open problem in the class of Leray solutions. There have been many papers dealing with uniqueness in some subclasses of Leray solutions: one studies uniqueness in a class $L^\infty L^2 \cap L^2 \dot{H}^1 \cap \mathbb{X}$. Chemin proved uniqueness of solutions in $L^\infty L^2 \cap L^2 \dot{H}^1 \cap \mathcal{C}([0,T], \dot{B}_{\infty,\infty}^{-1})$ (which implies that $\vec{u}_0$ not only belongs to L^2 but belongs to $\dot{B}_{\infty,\infty}^{-1}$), under a further assumption: $\vec{u}_0$ belongs to the closure of test functions in some space $\dot{B}_{p,\infty}^{-1+\frac{3}{p}}$ with $p < +\infty$ [106]. Lemarié-Rieusset removed the assumption on $\vec{u}_0$ and proved uniqueness in $L^\infty L^2 \cap L^2 \dot{H}^1 \cap \mathcal{C}([0,T], \dot{B}_{\infty,\infty}^{-1})$ [317]. Then uniqueness was proved in $L^{\frac{2}{r+1}}((0,T), \dot{B}_{\infty,\infty}^r) \cap L^\infty L^2 \cap L^2 \dot{H}^1$ for $-1 < r \le 1$: the case $r < 0$ was proved by May in an extension of Lemarié-Rieusset's method [355], while the case $r > -1/2$ was proved by Chen, Miao and Zhang [115].

Definition 12.3.
Let $\vec{u}_0 \in L^2$ with div $\vec{u}_0 = 0$ and $\vec{f} \in L^2((0, +\infty), L^2)$. An almost strong *solution* $\vec{u}$ of the Navier–Stokes equations on $(0, T)$

$$\partial_t \vec{u} = \nu \Delta \vec{u} + \mathbb{P}(\vec{f} - \text{div}(\vec{u} \otimes \vec{u})), \quad \vec{u}(0, .) = \vec{u}_0$$

is a solution $\vec{u}$ *such that:*

- $\vec{u} \in L^\infty((0, T), L^2) \cap L^2((0, T), H^1)$

- $\lim_{t \to 0+} \|\vec{u}(t, .) - \vec{u}_0\|_2 = 0$

- $\vec{u} \in \mathcal{C}((0, T], H^1)$.

Note that such a solution satisfies Leray's energy equality: for all $0 \leq t_0 < t_1 \leq T$,

$$\|\vec{u}(t_1, .)\|_2 = \|\vec{u}(t_0, .)\|_2^2 - 2 \int_{t_0}^{t_1} \|\vec{\nabla} \otimes \vec{u}\|_2^2 \, ds + 2 \int_{t_0}^{t_1} \int \vec{u}(s, x) \cdot \vec{f}(s, x) \, dx \, ds.$$

Of course, if $\vec{u}_0 \in H^1$, then we may use Serrin's weak-strong uniqueness theorem to ensure that an almost strong solution $\vec{u}_{AS}$ and the classical strong solution $\vec{u}_{CL} \in \mathcal{C}([0, T], H^1)$ coincide: $\vec{u}_{AS} = \vec{u}_{CL}$.

Proposition 12.8.
Let $\vec{u}_0 \in L^2$ with div $\vec{u}_0 = 0$ and $\vec{f} \in L^2((0, +\infty), L^2)$. Assume that $\vec{u}$ is a solution of the Navier–Stokes equations on $(0, T)$

$$\partial_t \vec{u} = \nu \Delta \vec{u} + \mathbb{P}(\vec{f} - \text{div}(\vec{u} \otimes \vec{u})), \quad \vec{u}(0, .) = \vec{u}_0$$

such that $\vec{u} \in L^\infty((0, T), L^2) \cap L^2((0, T), H^1)$ *and assume moreover that, for some* $r \in (-1, 1)$, *we have*

$$\vec{u} \in L^{\frac{2}{r+1}}((0, T), \dot{B}^r_{\infty,\infty}).$$

Then $\vec{u}$ *is an almost strong solution.*

Proof. **First step: right-continuity of** $t \in [0, T) \mapsto \vec{u}(t, .) \in L^2$:

Case $r > 0$:
When $r > 0$, it is easy to check that, if $\vec{u} \in L^{\frac{2}{r+1}}((0, T), \dot{B}^r_{\infty,\infty}) \cap L^\infty L^2 \cap L^2 \dot{H}^1$ with div $\vec{u} = 0$, then $\vec{u} \cdot \vec{\nabla} \vec{u} \in L^2 \dot{H}^{-1} + L^1 L^2$. Just recall that we have seen that $L^{\frac{2}{1+r}} \dot{B}^r_{\infty,\infty} \subset L^1 \text{Lip} + L^2 L^\infty$ (see the proof of Proposition 12.6). Writing $\vec{u} = \vec{v} + \vec{w}$ with $\vec{v} \in L^1 \text{Lip}$ and $\vec{w} \in L^2 L^\infty$, we just write

$$\vec{u} \cdot \vec{\nabla} \vec{u} = \vec{u} \cdot \vec{\nabla} \vec{v} + \vec{\nabla} \cdot (\vec{u} \otimes \vec{w})$$

and get

$$\|\vec{u} \cdot \vec{\nabla} \vec{u}\|_{L^1 L^2 + L^2 \dot{H}^{-1}} \leq C \|\vec{u}\|_{L^\infty L^2} (\|\vec{w}\|_{L^1 \text{Lip}} + \|\vec{v}\|_{L^2 L^\infty}) \leq C' \|\vec{u}\|_{L^\infty L^2} \|\vec{u}\|_{L^{\frac{2}{1+r}} \dot{B}^r_{\infty,\infty}}.$$

From $\vec{u} \cdot \vec{\nabla} \vec{u} \in L^1 L^2 + \dot{H}^{-1}$, we get that $B(\vec{u}, \vec{u}) \in \mathcal{C}([0, T], L^2)$. As $\vec{u} = W_{\nu t} * \vec{u}_0 - B(\vec{u}, \vec{u})$, we see that $\vec{u} \in \mathcal{C}([0, T], L^2)$.

Case $r < 0$:

When $r < 0$, it is easy as well to check that, if $\vec{u} \in L^{\frac{2}{r+1}}((0,T), \dot{B}^r_{\infty,\infty}) \cap L^\infty L^2 \cap L^2 \dot{H}^1$ with div $\vec{u} = 0$, then $\vec{u} \cdot \vec{\nabla} \vec{u} \in L^2 \dot{H}^{-1} + L^1 L^2$. We write (using the Littlewood–Paley decomposition)

$$\|\Delta_j \vec{u} \otimes \Delta_k \vec{u}\|_2 \leq \|\Delta_{\min(j,k)} \vec{u}\|_\infty \|\Delta_{\max(j,k)} \vec{u}\|_2$$

which gives

$$\|(\vec{u} \otimes \vec{u})(t,.)\|_{\dot{H}^{1+r}} \leq \|\vec{u}(t,.)\|_{\dot{B}^r_{\infty,\infty}} \|\vec{u}(t,.)\|_{\dot{H}^1}$$

and

$$\|\vec{u} \cdot \vec{\nabla} u\|_{L^{\frac{2}{2+r}} \dot{H}^r} \leq \|\vec{u}(t,.)\|_{L^{\frac{2}{1+r}} \dot{B}^r_{\infty,\infty}} \|\vec{u}(t,.)\|_{L^2 \dot{H}^1}$$

and we end by using the embedding $L^{\frac{2}{2+r}} \dot{H}^r \subset L^1 L^2 + L^2 \dot{H}^{-1}$. Thus, again we get that $\vec{u} \in \mathcal{C}([0,T], L^2)$.

Case $r = 0$:

The case $r = 0$ is more delicate. We use a Littlewood–Paley decomposition and use the norm

$$\|\vec{u}\|_{LP} = (\sum_{j \in \mathbb{Z}} \|\Delta_j \vec{u}\|_2^2)^{1/2}$$

which is a Hilbertian norm equivalent to the L^2 norm. As $t \mapsto \vec{u}(t,.)$ is continuous from $[0,T]$ to $\mathcal{D}'$ and is bounded from $[0,T]$ to L^2, we find that it is *-weakly continuous from $[0,T]$ to L^2. In particular, we shall have that

$$\lim_{t \to t_0^+} \|\vec{u}(t,.) - \vec{u}(t_0,.)\|_2 = 0 \Leftrightarrow \lim_{t \to t_0^+} \|\vec{u}(t,.) - \vec{u}(t_0,.)\|_{LP} = 0$$

$$\Leftrightarrow \limsup_{t \to t_0^+} \|\vec{u}(t,.)\|_{LP}^2 - \|\vec{u}(t_0,.)\|_{LP}^2 \leq 0.$$

We have

$$\frac{d}{dt} \|\Delta_j \vec{u}(t,.)\|_2^2 = 2 \int \Delta_j \vec{u}(t,x) \cdot \Delta_j (\partial_t \vec{u})(t,x) \, dx$$

$$= 2 \int \Delta_j \vec{u} \cdot (\nu \Delta \Delta_j \vec{u} + \Delta_j \vec{f} - \Delta_j (\vec{u} \cdot \vec{\nabla} \vec{u}) \, dx$$

$$\leq -2\nu \|\vec{\nabla} \otimes \Delta_j \vec{u}\|_2^2 + \|\Delta_j \vec{u}\|_2^2 + \|\Delta_j \vec{f}\|_2^2 - 2 \int \Delta_j \vec{u} \cdot \Delta_j (\vec{u} \cdot \vec{\nabla} \vec{u}) \, dx$$

If $k \geq j + 5$ and $|l - k| \geq 4$, then $\int \Delta_j \vec{u} \cdot \Delta_j (\Delta_k \vec{u} \cdot \vec{\nabla} \Delta_l \vec{u}) \, dx = 0$, so that

$$\int \Delta_j \vec{u} \cdot \Delta_j (\vec{u} \cdot \vec{\nabla} \vec{u}) \, dx = \int \Delta_j \vec{u} \cdot \Delta_j (S_{j+5} \vec{u} \cdot \vec{\nabla} \vec{u}) \, dx$$

$$+ \sum_{k \geq j+5} \sum_{|k-l| \leq 3} \int \Delta_j \vec{u} \cdot \Delta_j (\Delta_k \vec{u} \cdot \vec{\nabla} \Delta_l \vec{u}) \, dx$$

We have, for $k \geq j + 5$ and $|l - k| \leq 3$,

$$A_{j,k,l} = -\int \Delta_j \vec{u} \cdot \Delta_j (\Delta_k \vec{u} \cdot \vec{\nabla} \Delta_l \vec{u}) \, dx$$

$$= \sum_{i=1}^{3} \int \Delta_k u_i (\Delta_l \vec{u} . \partial_i \Delta_j^* \Delta_j \vec{u}) \, dx$$

$$\leq C \|\Delta_k \vec{u}\|_\infty 2^j \|\Delta_j \vec{u}\|_2 \|\Delta_l \vec{u}\|_2$$

so that

$$\sum_{j\in\mathbb{Z}}\sum_{k\geq j+5}\sum_{|l-k|\leq 3} A_{j,k,l} \leq C\|\vec{u}\|_{\dot{B}^0_{\infty,\infty}}\sum_{j\in\mathbb{Z}}\|\Delta_j\vec{u}\|_2(\sum_{l\geq j+2}2^{j-l}2^l\|\Delta\vec{u}_l\|_2)$$

$$\leq C'\|\vec{u}\|_{\dot{B}^0_{\infty,\infty}}\|\vec{u}\|^2_{LP}\|\vec{\nabla}\otimes\vec{u}\|^2_{LP}$$

For the term $B_j = -\int \Delta_j\vec{u}\cdot\Delta_j(S_{j+5}\vec{u}\cdot\vec{\nabla}\vec{u})\,dx$, we note that

$$\int \Delta_j\vec{u}.(S_{j+5}\vec{u}\cdot\vec{\nabla}\Delta_j\vec{u})\,dx = 0$$

and we write

$$B_j = -\sum_{i=1}^3\int \Delta_j\vec{u}.[\Delta_j,S_{j+5}u_i]\partial_i\vec{u}\,dx \tag{12.28}$$

and

$$B_j = \sum_{i=1}^3 \partial_i\Delta_j\vec{u}.[\Delta_j,S_{j+5}u_i]\vec{u}\,dx. \tag{12.29}$$

We have (writing $\Delta_j h = \psi(\frac{D}{2^j})h$ and $\psi = \hat{\Psi}$)

$$[\Delta_j,S_{j+5}u_i]h = \int 2^{3j}\int (S_{j+5}u_i(y)-S_{j+5}u_i(x))\Psi(2^j(x-y))h(y)\,dy.$$

We have $\vec{\nabla}u_i\in\dot{B}^{-1}_{0,\infty}$, so that $\|\vec{\nabla}S_{j+5}u_i\|_\infty\leq C2^j$ and

$$|[\Delta_j,S_{j+5}u_i]h|\leq C\|\vec{u}\|_{\dot{B}^0_{\infty,\infty}}\int 2^{3j}\int 2^j|x-y||\Psi(2^j(x-y))||h(y)|\,dy$$
$$\leq C'\|\vec{u}\|_{\dot{B}^0_{\infty,\infty}}\mathcal{M}_h(x). \tag{12.30}$$

From (12.28) and (12.30), we get

$$B_j \leq C\|\vec{u}\|_{\dot{B}^0_{\infty,\infty}}\|\vec{\nabla}\otimes\vec{u}\|_2\|\Delta_j\vec{u}\|_2,$$

while we get, from (12.29) and (12.30),

$$B_j \leq C\|\vec{u}\|_{\dot{B}^0_{\infty,\infty}}\|\vec{u}\|_2 2^j\|\Delta_j\vec{u}\|_2.$$

and thus

$$\sum_{j\in\mathbb{Z}}B_j \leq C\|\vec{u}\|_{\dot{B}^0_{\infty,\infty}}\sqrt{\|\vec{u}\|_2\|\vec{\nabla}\otimes\vec{u}\|_2}\sum_{j\in\mathbb{Z}}2^{j/2}\|\Delta_j\vec{u}\|_2.$$

$$\leq C'\|\vec{u}\|_{\dot{B}^0_{\infty,\infty}}\sqrt{\|\vec{u}\|_2\|\vec{\nabla}\otimes\vec{u}\|_2}\|\vec{u}\|_{\dot{B}^{1/2}_{2,1}}$$

$$\leq C''\|\vec{u}\|_{\dot{B}^0_{\infty,\infty}}\|\vec{u}\|_2\|\vec{\nabla}\otimes\vec{u}\|_2.$$

Collecting all those estimates, we find

$$\frac{d}{dt}\|\vec{u}\|^2_{LP} \leq -2\nu\|\vec{\nabla}\otimes\vec{u}\|^2_{LP} + \|\vec{u}\|^2_{LP} + \|\vec{f}\|^2_{LP} + C\|\vec{u}\|_{\dot{B}^0_{\infty,\infty}}\|\vec{u}\|^2_{LP}\|\vec{\nabla}\otimes\vec{u}\|^2_{LP} \tag{12.31}$$

Thus, for $0 \leq t_0 < t_1$,

$$\|\vec{u}(t, .)\|_{LP}^2 \leq \|\vec{u}(t_0, .)\|_{LP}^2 + \int_{t_0}^t (\|\vec{u}\|_{LP}^2 + \|\vec{f}\|_{LP}^2) \, ds + C' \|\vec{u}\|_{L^\infty L^2}^2 \int_{t_0}^t \|\vec{u}\|_{\dot{B}_{\infty,\infty}^0}^2 \, ds$$

and we get

$$\limsup_{t \to t_0^+} \|\vec{u}(t, .)\|_{LP}^2 \leq \|\vec{u}(t_0, .)\|_{LP}^2.$$

Thus, $t \mapsto \vec{u}(t, .)$ is strongly right-continuous from $[0, T)$ to L^2.

Second step: energy equality.

By interpolation, we find that, for $\rho = \frac{1-r}{2} \in (0, 1)$,

$$\|\vec{u}\|_{L^3 \dot{B}_{3,3}^{1/3}} \leq C \|\vec{u}\|_{L^{\frac{2}{1+r}} \dot{B}_{\infty,\infty}^r}^{1/3} \qquad \|\vec{u}\|_{L^{\frac{2}{\rho}} \dot{B}_{2,2}^\rho}^{2/3} \leq C' \|\vec{u}\|_{L^{\frac{2}{1+r}} \dot{B}_{\infty,\infty}^r}^{1/3} \|\vec{u}\|_{L^\infty L^2}^{\frac{1-\rho}{3}} \|\vec{u}\|_{L^2 \dot{H}^1}^{\frac{\rho}{3}}.$$

In particular, $\vec{u}$ belongs to $L_t^3 b_{3,\infty}^{1/3}$, where $b_{3,\infty}^{1/3}$ is the closure of test functions in $\dot{B}_{3,\infty}^{1/3}$. Then, we may use Duchon and Robert's theorem [159] (see Theorem 13.7 below) and conclude that $\vec{u}$ satisfies the local energy equality:

$$\partial_t (\frac{|\vec{u}|^2}{2}) + \nu |\vec{\nabla} \otimes \vec{u}|^2 = \nu \Delta (\frac{|\vec{u}|^2}{2}) + 2\vec{u} \cdot \vec{f} - \mathrm{div}((\frac{|\vec{u}|^2}{2} + p)\vec{u}).$$

Using the right-continuity of $t \in [0, T) \mapsto \vec{u}(t, .) \in L^2$, we conclude that $\vec{u}$ satisfies the Leray energy equality: for all $0 \leq t_0 \leq t_1 \leq T$,

$$\|\vec{u}(t_1, .)\|_2^2 = \|\vec{u}(t_0, .)\|_2^2 - 2 \int_{t_0}^{t_1} \|\vec{\nabla} \otimes \vec{u}\|_2^2 \, ds + 2 \int_{t_0}^{t_1} \int \vec{u} \cdot \vec{f} \, dx \, ds$$

(see the discussion on page 119).

Third step: continuity in H^1 norm.

As $\vec{u} \in L^2 H^1$, we know that $\vec{u}(t, .)$ belongs to H^1 for almost every time. If $\vec{u}(t_0, .)$ belongs to H^1, then we know that there will be a solution $\vec{v}$ of the Navier-Stokes equations on some small interval $[t_0, t_0 + \delta]$ such that $\vec{v} \in \mathcal{C}([t_0, t_0 + \delta], H^1) \cap L^2 H^2$. By Serrin's weak-strong uniqueness theorem (Theorem 12.3), we know that $\vec{v} = \vec{u}$. Moreover, if we look at the maximal existence time of $\vec{u}$ as a solution in $\mathcal{C}([t_0, T^*), H^1)$, we know that $T^* = T$ as $\vec{u} \in L^{\frac{2}{r+1}} \dot{B}_{\infty,\infty}^r$ (see Theorem 11.3).

Thus, we find that $\vec{u} \in \mathcal{C}((0, T), H^1)$ so that $\vec{u}$ is an almost strong solution.

$\square$

We may now state the uniqueness theorem of May [355] and Chen, Miao and Zhang [115]:

Uniqueness for almost strong solutions

Theorem 12.7.
Let $\vec{u}_0 \in L^2$ with $\mathrm{div}\, \vec{u}_0 = 0$ and $\vec{f} \in L^2((0, +\infty), H^1)$. Assume that $\vec{u}$ and $\vec{v}$ are two solutions of the Navier-Stokes equations on $(0, T)$

$$\partial_t \vec{u} = \nu \Delta \vec{u} + \mathbb{P}(\vec{f} - \mathrm{div}(\vec{u} \otimes \vec{u})), \quad \vec{u}(0, .) = \vec{u}_0$$

such that $\vec{u}$ and $\vec{v}$ belong to $L^\infty((0,T),L^2)\cap L^2((0,T),H^1)$. *Assume moreover that, for some* $r_1, r_2 \in (-1,1)$, *we have*

$$\vec{u} \in L^{\frac{2}{r_1+1}}((0,T),\dot{B}^{r_1}_{\infty,\infty}) \ \text{and} \ \vec{v} \in L^{\frac{2}{r_2+1}}((0,T),\dot{B}^{r_2}_{\infty,\infty}).$$

Then $\vec{u} = \vec{v}$.

Proof. We may, of course, assume that $r_2 \le r_1$.

Case $r_1 > 0$:
In that case, we know that we have weak-strong uniqueness (Proposition 12.6). As $\vec{v}$ satisfies the Leray energy (in)equality and $\vec{u} \in L^{\frac{2}{r_1+1}}((0,T),\dot{B}^{r_1}_{\infty,\infty})$ with $r_1 > 0$, we know that $\vec{u} = \vec{v}$.

Case $r_1 < 0$:
As $\vec{f} \in L^2 H^1$, we find that we may enhance the regularity of $\vec{u}$ and $\vec{v}$ to $\mathcal{C}((0,T),H^2)$. Hence, they belong to $\mathcal{C}((0,T),L^\infty)$. In particular, we have that, for every $t_0 \in (0,T)$, the function $\eta_{t_0} : \tau \in [0,T-t_0) \mapsto \eta_{t_0}(\tau) = \sup_{0<\theta<\tau}\sqrt{\theta}\|\vec{u}(t_0+\theta,.)\|_\infty$ is continuous and satisfies $\eta_{t_0}(0) = 0$. Moreover, we have, for $\beta \in (\theta/4,\theta/2)$

$$\vec{u}(t_0+\theta,.) = W_{\nu(\theta-\beta)}*\vec{u}(t_0+\beta,.) - \int_\beta^\theta W_{\nu(\theta-s)}*\mathbb{P}\operatorname{div}(\vec{u}(t_0+s,.)\otimes\vec{u}(t_0+s))\,ds$$

so that

$$\sqrt{\theta}\|\vec{u}(t_0+\theta,.)\|_\infty \le C\theta^{\frac{1+r_1}{2}}\|\vec{u}(t_0+\beta,.)\|_{\dot{B}^{r_1}_{\infty,\infty}} + C\theta \sup_{\beta<s<\theta}\|\vec{u}(t_0+s,.)\|_\infty^2$$

Averaging over $(\theta/4,\theta/2)$, we find

$$\sqrt{\theta}\|\vec{u}(t_0+\theta,.)\|_\infty \le C\theta^{\frac{1+r_1}{2}}\frac{\int_{\theta/4}^{\theta/2}\|\vec{u}(t_0+\beta,.)\|_{\dot{B}^{r_1}_{\infty,\infty}}\,d\beta}{\theta} + C\eta_{t_0}(\theta)^2$$
$$\le C_0(\|\vec{u}\|_{L^{\frac{2}{1+r_1}}(t_0,t_0+\theta),\dot{B}^{r_1}_{\infty,\infty}} + \eta_{t_0}(\theta)^2)$$

and thus
$$\eta_{t_0}(\tau) \le C_0(\|\vec{u}\|_{L^{\frac{2}{1+r_1}}(t_0,t_0+\tau),\dot{B}^{r_1}_{\infty,\infty}} + \eta_{t_0}(\tau)^2). \tag{12.32}$$

If τ_0 is small enough to grant that

$$\sup_{0<t_0<T/2}\|\vec{u}\|_{L^{\frac{2}{1+r_1}}(t_0,t_0+\tau_0),\dot{B}^{r_1}_{\infty,\infty}} < \frac{1}{4C_0^2}$$

we get that, for $0 < t_0 < T/2$ and $0 \le \tau < \tau_0$

$$\eta_{t_0}(\tau) \le 2C_0\|\vec{u}\|_{L^{\frac{2}{1+r_1}}(t_0,t_0+\tau),\dot{B}^{r_1}_{\infty,\infty}}.$$

Letting t_0 go to 0, we find that, for $0 < t < T_0$,

$$\sqrt{t}\|\vec{u}(t,.)\|_\infty \le 2C_0\|\vec{u}\|_{L^{\frac{2}{1+r_1}}(0,t),\dot{B}^{r_1}_{\infty,\infty}}.$$

A similar estimate holds for $\vec{v}$. Now, if $\vec{w} = \vec{u} - \vec{v}$, we write

$$\vec{w} = -\int_0^t W_{\nu(t-s)}*\mathbb{P}\operatorname{div}(\vec{u}\otimes\vec{w}+\vec{w}\otimes\vec{v})\,ds$$

and

$$\|\vec{w}(t,.)\|_2 \le C \int_0^t \frac{1}{\sqrt{\nu(t-s)}} \|\vec{w}(s,.)\|_2 (\|\vec{u}(s,.)\|_\infty + \|\vec{v}(s,.)\|_\infty) \, ds$$

so that

$$\sup_{0<t<t_0} \|\vec{w}(t,.)\|_2 \le C \sup_{0<t<t_0} \|\vec{w}(t,.)\|_2 (\|\vec{u}\|_{L^{\frac{2}{1+r_1}}(0,t_0),\dot{B}^{r_1}_{\infty,\infty}} + \|\vec{v}\|_{L^{\frac{2}{1+r_2}}(0,t_0),\dot{B}^{r_2}_{\infty,\infty}}).$$

If t_0 is so small that

$$C(\|\vec{u}\|_{L^{\frac{2}{1+r_1}}(0,t_0),\dot{B}^{r_1}_{\infty,\infty}} + \|\vec{v}\|_{L^{\frac{2}{1+r_2}}(0,t_0),\dot{B}^{r_2}_{\infty,\infty}}) < 1,$$

we find $\vec{w} = 0$ on $(0, t_0]$, hence local uniqueness. This uniqueness propogates to $(0, T)$, as we have uniqueness in $\mathcal{C}([t_0, T], H^1)$.

Case $r_1 = 0$:
Let $\vec{w} = \vec{u} - \vec{v}$. One more time, we use a Littlewood–Paley decomposition but we do not use the norm

$$\|\vec{w}\|_{LP} = (\sum_{j\in\mathbb{Z}} \|\Delta_j \vec{w}\|_2^2)^{1/2}$$

which is a Hilbertian norm equivalent to the L^2 norm. Instead of it, we shall use the norm

$$\|\vec{w}\|_{LP,\sigma} = (\|S_0 \vec{w}\|_2^2 + \sum_{j\in\mathbb{N}} 2^{-2j\sigma} \|\Delta_j \vec{w}\|_2^2)^{1/2}$$

which is a Hilbertian norm equivalent to the $H^{-\sigma}$ norm. If $\sigma > 0$, we have the embedding $L^2 \subset H^{-\sigma}$, so that the map $t \in [0, T) \mapsto \vec{w}(t,.) \in H^{-\sigma}$ is (strongly) continuous. We write

$$\frac{d}{dt}\|\Delta_j \vec{w}(t,.)\|_2^2 = 2 \int \Delta_j \vec{w}(t,x) \cdot \Delta_j(\partial_t \vec{w})(t,x) \, dx$$

$$= 2 \int \Delta_j \vec{w} \cdot (\nu\Delta\Delta_j \vec{w} - \Delta_j(\vec{u}\cdot\vec{\nabla}\vec{u} - \vec{v}\vec{\nabla}\vec{v})) \, dx$$

$$= -2\nu\|\vec{\nabla}\otimes\Delta_j\vec{w}\|_2^2 - 2\int \Delta_j\vec{w}\cdot\Delta_j(\vec{u}\cdot\vec{\nabla}\vec{u} - \vec{v}\cdot\vec{\nabla}\vec{v})) \, dx$$

If $k \ge j+5$ and $|l-k| \ge 4$, then

$$\int \Delta_j\vec{w}\cdot\Delta_j(\Delta_k\vec{u}\cdot\vec{\nabla}\Delta_l\vec{u}) \, dx = \int \Delta_j\vec{w}\cdot\Delta_j(\Delta_k\vec{v}\cdot\vec{\nabla}\Delta_l\vec{v}) \, dx = 0,$$

so that

$$\int \Delta_j\vec{w}\cdot\Delta_j(\vec{u}\cdot\vec{\nabla}\vec{u} - \vec{v}\cdot\vec{\nabla}\vec{v}) \, dx = \int \Delta_j\vec{w}\cdot\Delta_j(S_{j+5}\vec{u}\cdot\vec{\nabla}\vec{u} - S_{j+5}\vec{v}\cdot\vec{\nabla}\vec{v}) \, dx$$

$$+ \sum_{k\ge j+5}\sum_{|k-l|\le 3} \int \Delta_j\vec{w}\cdot\Delta_j(\Delta_k\vec{u}\cdot\vec{\nabla}\Delta_l\vec{u} - \Delta_k\vec{v}\cdot\vec{\nabla}\Delta_l\vec{v}) \, dx$$

We have, for $k \ge j+5$ and $|l-k| \le 3$,

$$A_{j,k,l} = -\int \Delta_j\vec{w}\cdot\Delta_j(\Delta_k\vec{u}\cdot\vec{\nabla}\Delta_l\vec{u} - \Delta_k\vec{v}\cdot\vec{\nabla}\Delta_l\vec{v}) \, dx$$

$$= -\int \Delta_j\vec{w}\cdot\Delta_j(\Delta_k\vec{w}\cdot\vec{\nabla}\Delta_l\vec{u} + \Delta_k\vec{v}\cdot\vec{\nabla}\Delta_l\vec{w}) \, dx$$

$$= \int \Delta_l\vec{u}.(\Delta_k\vec{w}.\vec{\nabla}\Delta_j^*\Delta_j\vec{w}) \, dx + \int \Delta_l\vec{w}(\Delta_k\vec{v}.\vec{\nabla}\Delta_j^*\Delta_j\vec{w}) \, dx$$

$$< C2^j\|\Delta_j\vec{w}\|_2(2^{-kr_2}\|\vec{v}\|_{\dot{B}^{r_2}_{\infty,\infty}}\|\Delta_l\vec{w}\|_2 + \|\vec{u}\|_{\dot{B}^0_{\infty,\infty}}\|\Delta_k\vec{w}\|_2).$$

Thus, we get

$$\sum_{j\in\mathbb{N}} 2^{-2j\sigma} \sum_{k\geq j+5} \sum_{|k-l|\leq 3} A_{j,k,l}$$

$$\leq C \sum_{j\in\mathbb{N}} 2^{-2j\sigma} 2^{j} \|\Delta_j \vec{w}\|_2 \sum_{k\geq j+2} (2^{-kr_2} \|\vec{v}\|_{\dot{B}^{r_2}_{\infty,\infty}} + \|\vec{u}\|_{\dot{B}^{0}_{\infty,\infty}}) \|\Delta_k \vec{w}\|_2$$

$$= C \sum_{j\in\mathbb{N}} 2^{-j\sigma} 2^{j} \|\vec{v}\|_{\dot{B}^{r_2}_{\infty,\infty}} \|\Delta_j \vec{w}\|_2 \sum_{k\geq j+2} 2^{-(j-k)\sigma} 2^{-k(\sigma+r_2)} \|\Delta_k \vec{w}\|_2$$

$$+ C \sum_{j\in\mathbb{N}} 2^{-j\sigma} \|\vec{u}\|_{\dot{B}^{0}_{\infty,\infty}} 2^{j} \|\Delta_j \vec{w}\|_2 \sum_{k\geq j+2} 2^{-(j-k)-\sigma} 2^{-k\sigma} \|\Delta_k \vec{w}\|_2$$

As $\sigma > 0$, we get

$$\sum_{j\in\mathbb{N}} 2^{-2j\sigma} \sum_{k\geq j+5} \sum_{|k-l|\leq 3} A_{j,k,l} \leq$$

$$C \|\vec{v}\|_{\dot{B}^{r_2}_{\infty,\infty}} \left(\sum_{j\in\mathbb{N}} 2^{-2j(\sigma+r_2)} \|\Delta_j \vec{w}\|_2^2 \right)^{1/2} \left(\sum_{j\in\mathbb{N}} 2^{2j(1-\sigma)} \|\Delta_k \vec{w}\|_2^2 \right)^{1/2}$$

$$+ C \|\vec{u}\|_{\dot{B}^{0}_{\infty,\infty}} \left(\sum_{j\in\mathbb{N}} 2^{-2j\sigma} \|\Delta_j \vec{w}\|_2^2 \right)^{1/2} \left(\sum_{j\in\mathbb{N}} 2^{2j(1-\sigma)} \|\Delta_k \vec{w}\|_2^2 \right)^{1/2}$$

$$\leq C \|\vec{v}\|_{\dot{B}^{r_2}_{\infty,\infty}} \left(\sum_{j\in\mathbb{N}} 2^{-2j\sigma} \|\Delta_j \vec{w}\|_2^2 \right)^{\frac{1+r_2}{2}} \left(\sum_{j\in\mathbb{N}} 2^{2j(1-\sigma)} \|\Delta_k \vec{w}\|_2^2 \right)^{\frac{1-r_2}{2}}$$

$$+ C \|\vec{u}\|_{\dot{B}^{0}_{\infty,\infty}} \left(\sum_{j\in\mathbb{N}} 2^{-2j\sigma} \|\Delta_j \vec{w}\|_2^2 \right)^{1/2} \left(\sum_{j\in\mathbb{N}} 2^{2j(1-\sigma)} \|\Delta_k \vec{w}\|_2^2 \right)^{1/2}$$

(as $0 \leq -r_2 \leq 1$).

Now, we write (as $\int \Delta_j \vec{w}.(S_{j+5}\vec{v}\cdot\vec{\nabla}\Delta_j\vec{w}\, dx = 0)$

$$B_j = -\int \Delta_j \vec{w} \cdot \Delta_j (S_{j+5}\vec{u}\cdot\vec{\nabla}\vec{u} - S_{j+5}\vec{v}\cdot\vec{\nabla}\vec{v}))\, dx$$

$$= -\int \Delta_j \vec{w} \cdot \Delta_j (S_{j+5}\vec{w}\cdot\vec{\nabla}\vec{u} + S_{j+5}\vec{v}\cdot\vec{\nabla}\vec{w}))\, dx$$

$$= \int \Delta_j \vec{w} \cdot \Delta_j (S_{j+5}\vec{w}\cdot\vec{\nabla}\vec{u})\, dx - \int \Delta_j \vec{w} \cdot ([\Delta_j, S_{j+5}\vec{v}].\vec{\nabla}\vec{w})\, dx$$

$$= \int \Delta_j \vec{w} \cdot \Delta_j (S_{j+5}\vec{w}\cdot\vec{\nabla}S_{j+8}\vec{u})\, dx - \int \Delta_j \vec{w} \cdot ([\Delta_j, S_{j+5}\vec{v}].\vec{\nabla}S_{j+8}\vec{w})\, dx$$

$$\leq C \|\Delta_j \vec{w}\|_2 \|S_{j+5}\vec{w}\|_2 2^{j} \|\vec{u}\|_{\dot{B}^{0}_{\infty,\infty}} + C \|\Delta_j \vec{w}\|_2 2^{-jr_2} \|\vec{v}\|_{\dot{B}^{r_2}_{\infty,\infty}} \|\vec{\nabla}S_{j+8}\vec{w}\|_2.$$

Thus, we get

$$\sum_{j\in\mathbb{N}} 2^{-2j\sigma} B_j \leq C\|\vec{u}\|_{\dot{B}^0_{\infty,\infty}} \sum_{j\in\mathbb{N}} 2^{-2j\sigma} 2^j \|\Delta_j\vec{w}\|_2 (\|S_0\vec{w}\|_2 + \sum_{k\leq j+4} \|\Delta_k\vec{w}\|_2)$$

$$+C\|\vec{v}\|_{\dot{B}^{r_2}_{\infty,\infty}} \sum_{j\in\mathbb{N}} 2^{-2j\sigma} 2^{-jr_2} \|\Delta_j\vec{w}\|_2 (\|S_0\vec{w}\|_2 + \sum_{0\leq k\leq j+4} 2^k \|\Delta_k\vec{w}\|_2)$$

$$= C\|\vec{u}\|_{\dot{B}^0_{\infty,\infty}} \sum_{j\in\mathbb{N}} 2^{j(1-\sigma)} \|\Delta_j\vec{w}\|_2 (2^{-j\sigma}\|S_0\vec{w}\|_2 + \sum_{k\leq j+4} 2^{-(j-k)\sigma} 2^{-k\sigma} \|\Delta_k\vec{w}\|_2)$$

$$+C\|\vec{v}\|_{\dot{B}^{r_2}_{\infty,\infty}} \sum_{j\in\mathbb{N}} 2^{-j(\sigma+r_2)} \|\Delta_j\vec{w}\|_2 (2^{-j\sigma}\|S_0\vec{w}\|_2 + \sum_{k\leq j+4} 2^{-(j-k)\sigma} 2^{k(1-\sigma)} \|\Delta_k\vec{w}\|_2)$$

As $\sigma > 0$ and $0 \leq -r_2 \leq 1$, we find

$$\sum_{j\in\mathbb{N}} 2^{-2j\sigma} B_j$$

$$\leq C\|\vec{u}\|_{\dot{B}^0_{\infty,\infty}} \left(\sum_{j\in\mathbb{N}} 2^{2j(1-\sigma)} \|\Delta_j\vec{w}\|_2^2\right)^{1/2} \left(\|S_0\vec{w}\|_2^2 + \sum_{j\in\mathbb{N}} 2^{-2j\sigma} \|\Delta_j\vec{w}\|_2^2\right)^{1/2}$$

$$+C\|\vec{v}\|_{\dot{B}^{r_2}_{\infty,\infty}} \left(\sum_{j\in\mathbb{N}} 2^{-2j(\sigma+r_2)} \|\Delta_j\vec{w}\|_2^2\right)^{1/2} \left(\|S_0\vec{w}\|_2^2 + \sum_{j\in\mathbb{N}} 2^{2j(1-\sigma)} \|\Delta_j\vec{w}\|_2^2\right)^{1/2}$$

$$\leq C\|\vec{u}\|_{\dot{B}^0_{\infty,\infty}} \left(\sum_{j\in\mathbb{N}} 2^{2j(1-\sigma)} \|\Delta_j\vec{w}\|_2^2\right)^{1/2} \left(\|S_0\vec{w}\|_2^2 + \sum_{j\in\mathbb{N}} 2^{-2j\sigma} \|\Delta_j\vec{w}\|_2^2\right)^{1/2}$$

$$+C\|\vec{v}\|_{\dot{B}^{r_2}_{\infty,\infty}} \left(\sum_{j\in\mathbb{N}} 2^{-2j\sigma} \|\Delta_j\vec{w}\|_2^2\right)^{\frac{1+r_2}{2}} \left(\|S_0\vec{w}\|_2^2 + \sum_{j\in\mathbb{N}} 2^{2j(1-\sigma)} \|\Delta_j\vec{w}\|_2^2\right)^{\frac{1-r_2}{2}}$$

We can perform similar estimates when dealing with $\frac{d}{dt}\|S_0\vec{w}\|_2^2$, and we obtain finally

$$\frac{d}{dt}\|\vec{w}\|_{LP,\sigma}^2 \leq$$

$$-\nu\|\vec{\nabla}\otimes\vec{w}\|_{LP,\sigma}^2 + C\|\vec{u}\|_{\dot{B}^0_{\infty,\infty}} \left(\|S_0\vec{w}\|_2^2 + \sum_{j\in\mathbb{N}} 2^{2j(1-\sigma)} \|\Delta_j\vec{w}\|_2^2\right)^{1/2} \|\vec{w}\|_{LP,\sigma}$$

$$+C\|\vec{v}\|_{\dot{B}^{r_2}_{\infty,\infty}} \|\vec{w}\|_{LP,\sigma}^{\frac{1+r_2}{2}} \left(\|S_0\vec{w}\|_2^2 + \sum_{j\in\mathbb{N}} 2^{2j(1-\sigma)} \|\Delta_j\vec{w}\|_2^2\right)^{\frac{1-r_2}{2}}.$$

We then use Bernstein's inequality (for $j \geq 0$)

$$\|\Delta_j\vec{\nabla}\otimes\vec{w}\|_2^2 \geq \eta\|\Delta_j\vec{w}\|_2^2$$

(for a positive constant η) and Young's inequality

$$ca^\gamma b^{1-\gamma} \leq \gamma \left(\frac{c}{\epsilon}\right)^{\frac{1}{\gamma}} a + (1-\gamma)\epsilon^{\frac{1}{1-\gamma}} b$$

for $0 < \gamma < 1$ and positive a, b, c, ϵ, for the values $\gamma = \frac{1}{2}$ and $\gamma = \frac{1+r_2}{2}$, and for ϵ small enough to ensure that

$$\frac{1}{2}\epsilon^2 + \frac{1 - r_2}{2}\epsilon^{\frac{2}{1-r_2}} < \eta$$

and we get

$$\frac{d}{dt}\|\vec{w}\|^2_{LP,\sigma} \le C\|S_0\vec{w}\|^2_2 + C(\|\vec{u}\|^2_{\dot{B}^0_{\infty,\infty}} + \|\vec{v}\|^{\frac{2}{1+r_2}}_{\dot{B}^{r_2}_{\infty,\infty}})\|\vec{w}\|^2_{LP,\sigma}.$$

As $\|S_0\vec{w}\|_2 \le \|\vec{w}\|_{LP,\sigma}$, we may use the Grönwall lemma and get that $\vec{w} = 0$, hence $\vec{u} = \vec{v}$. $\qquad\square$

Theorem 12.7 may be generalized to the limit values $r_1 = 1$ or $r_2 = -1$ in the following way:

Theorem 12.8.
Let $\mathbb{X}^r_T$ be defined, for $-1 \le r \le 1$ as

- $\mathbb{X}^1_T = L^1((0,T), \mathrm{Lip})$

- *for $-1 < r < 1$, $\mathbb{X}^r_T = L^{\frac{2}{1+r}}((0,T), \dot{B}^r_{\infty,\infty})$*

- $\mathbb{X}^{-1}_T = \mathcal{C}([0,T], \dot{B}^{-1}_{\infty,\infty})$.

Let $\vec{u}_0 \in L^2$ with $\mathrm{div}\,\vec{u}_0 = 0$ and $\vec{f} \in L^2((0,+\infty), H^1)$. Assume that $\vec{u}$ and $\vec{v}$ are two solutions of the Navier–Stokes equations on $(0,T)$

$$\partial_t \vec{u} = \nu\Delta\vec{u} + \mathbb{P}(\vec{f} - \mathrm{div}(\vec{u} \otimes \vec{u})), \quad \vec{u}(0,.) = \vec{u}_0$$

such that $\vec{u}$ and $\vec{v}$ belong to $L^\infty((0,T), L^2) \cap L^2((0,T), H^1)$. Assume moreover that, for some $r_1, r_2 \in (-1, 1)$, we have

$$\vec{u} \in \mathbb{X}^{r_1}_T \text{ and } \vec{v} \in \mathbb{X}^{r_2}_T.$$

Then $\vec{u} = \vec{v}$.

Proof. **Step 1: almost strong solutions.**
The first step is to check that a solution $\vec{u} \in L^\infty L^2 \cap L^2 H^1 \cap \mathbb{X}^r_T$ (for some $r \in [-1,1]$) is indeed an almost strong solution. This has already been proved for $|r| < 1$. The proof for $r = 1$ is exactly the same as the one for $0 < r < 1$.

The proof for $r = -1$ is similar to the one for $-1 < r < 0$ but more delicate. We begin by the interpolation inequality

$$\|\vec{u}\|_{L^3 \dot{B}^{1/3}_{3,3}} \le \|\vec{u}\|^{1/3}_{L^\infty \dot{B}^{-1}_{\infty,\infty}} \|\vec{u}\|^{2/3}_{L^2 \dot{B}^1_{2,2}}.$$

In particular, $\vec{u}$ belongs to $L^3 b^{1/3}_{3,\infty}$ (where $b^{1/3}_{3,\infty}$ is the closure of test functions in $\dot{B}^{1/3}_{3,\infty}$). Then, we may again use Duchon and Robert's theorem [159] (see Theorem 13.7 below) and conclude that $\vec{u}$ satisfies the local energy equality:

$$\partial_t\left(\frac{|\vec{u}|^2}{2}\right) + \nu|\vec{\nabla} \otimes \vec{u}|^2 = \nu\Delta\left(\frac{|\vec{u}|^2}{2}\right) + 2\vec{u}\cdot\vec{f} - \mathrm{div}\left(\left(\frac{|\vec{u}|^2}{2} + p\right)\vec{u}\right).$$

Thus, for every Lebesgue point t_0 of the map $t \mapsto \|\vec{u}(t,.)\|_2$, we find that $\vec{u}$ satisfies the Leray energy equality: for all $t_0 \leq t_1 \leq T$,

$$\|\vec{u}(t_1,.)\|_2^2 = \|\vec{u}(t_0,.)\|_2^2 - 2 \int_{t_0}^{t_1} \|\vec{\nabla} \otimes \vec{u}\|_2^2 \, ds + 2 \int_{t_0}^{t_1} \int \vec{u} \cdot \vec{f} \, dx \, ds$$

(see the discussion on page 119).

As $\vec{u} \in L^2 H^1$, we know that $\vec{u}(t,.)$ belongs to H^1 for almost every time. If t_0 is a Lebesgue point of the map $t \mapsto \|\vec{u}(t,.)\|_2$ such that $\vec{u}(t_0,.)$ belongs to H^1, then we know that there will be a solution $\vec{v}$ of the Navier–Stokes equations on some small interval $[t_0, t_0 + \delta]$ such that $\vec{v} \in \mathcal{C}([t_0, t_0 + \delta], H^1) \cap L^2 H^2$. Moreover, by Serrin's weak-strong uniqueness theorem (Theorem 12.3), we know that $\vec{v} = \vec{u}$. Moreover, if we look at the maximal existence time of $\vec{u}$ as a solution in $\mathcal{C}([t_0, T^*), H^1)$, we know that $T^* = T$ as $\vec{u} \in \mathcal{C}([t_0, T], \dot{B}_{\infty,\infty}^{-1})$ (see Theorem 11.3). Hence, we get that $\vec{u} \in \mathcal{C}((0, T]), H^1)$.

As $\vec{f} \in L^2 H^1$, we find that we may enhance the regularity of $\vec{u}$ to $\mathcal{C}((0, T), H^2)$. Hence, it belongs to $\mathcal{C}((0, T), L^\infty)$. In particular, for every $t_0 \in (0, T)$, the function $\eta_{t_0} : \tau \in [0, T - t_0) \mapsto \eta_{t_0}(\tau) = \sup_{0 < \theta < \tau} \sqrt{\theta} \|\vec{u}(t_0 + \theta,.)\|_\infty$ is continuous and satisfies $\eta_{t_0}(0) = 0$. Moreover, we have

$$\vec{u}(t_0 + \theta,.) = W_{\nu(\theta/2)} * \vec{u}(t_0 + \theta/2,.) - \int_{\theta/2}^{\theta} W_{\nu(\theta-s)} * \mathbb{P} \operatorname{div}(\vec{u}(t_0 + s,.) \otimes \vec{u}(t_0 + s)) \, ds$$

so that

$$\sqrt{\theta} \|\vec{u}(t_0 + \theta,.)\|_\infty \leq \sqrt{\theta} \|W_{\nu(\theta/2)} * \vec{u}(t_0 + \theta/2,.)\|_\infty + C\theta \sup_{\theta/2 < s < \theta} \|\vec{u}(t_0 + s,.)\|_\infty^2$$

Let $b_{\infty,\infty}^{-1}$ be the closure of the Schwartz class $\mathcal{S}$ in $\dot{B}_{\infty,\infty}^{-1}$. As $\vec{u} \in L^4 \dot{H}^{1/2}$, we see that $\vec{u}(t,.) \in H^{1/2}$ for almost every t; as $\mathcal{S}$ is dense in $\dot{H}^{1/2}$ and $H^{1/2} \subset \dot{B}_{\infty,\infty}^{-1}$, we get that $\vec{u}(t,.) \in b_{\infty,\infty}^{-1}$ for almost every t; by continuity of $t \mapsto \vec{u}(t,.)$ in $\dot{B}_{\infty,\infty}^{-1}$ norm, we see that $\vec{u}(t,.) \in b_{\infty,\infty}^{-1}$ for every t. Thus, for $t \in [0, T]$ and $\epsilon > 0$, there exists $M(t, \epsilon)$, $\vec{\alpha}_t$ and $\vec{\beta}_t$ so that $\vec{u}(t,.) = \vec{\alpha}_t + \vec{\beta}_t$, $\|\vec{\alpha}_t\|_{\dot{B}_{\infty,\infty}^{-1}} < \epsilon$ and $\|\vec{\beta}_t\|_\infty < M(t, \epsilon)$. As $[0, T]$ is compact and thus $t \in [0, T] \mapsto \vec{u}(t,.) \in \dot{B}_{\infty,\infty}^{-1}$ is uniformly continuous, we can choose $M(t, \epsilon)$ independently from t. We thus obtain

$$\sqrt{\theta} \|\vec{u}(t_0 + \theta,.)\|_\infty \leq \sqrt{\theta} M(\epsilon) + C_0 \epsilon + C_0 \theta \sup_{\theta/2 < s < \theta} \|\vec{u}(t_0 + s,.)\|_\infty^2$$

and thus

$$\eta_{t_0}(\tau) \leq \sqrt{\tau} M(\epsilon) + C_0(\epsilon + \eta_{t_0}(\tau)^2). \tag{12.33}$$

For $\epsilon < \frac{1}{8C_0^2}$ and $T(\epsilon) = \frac{1}{(8C_0 M(\epsilon))^2}$, we get that, for $0 < t_0 < T$ and $0 \leq \tau < T(\epsilon)$

$$\eta_{t_0}(\tau) \leq 2(\sqrt{\tau} M(\epsilon) + C_0 \epsilon).$$

Letting t_0 go to 0, we find that, for $0 < t < T(\epsilon)$,

$$\sqrt{t} \|\vec{u}(t,.)\|_\infty \leq 2\sqrt{t} M(\epsilon) + C_0 \epsilon.$$

Thus, we get that $\sup_{0 < t < T} \sqrt{t} \|\vec{u}(t,.)\|_\infty < +\infty$ and

$$\lim_{t \to 0^+} \sqrt{t} \|\vec{u}(t,.)\|_\infty = 0. \tag{12.34}$$

This gives that $\vec{u}$ is an almost strong solution, as

$$\|\vec{u}(t,.) - \vec{u}_0\|_2 \leq \|W_{\nu t} * \vec{u}_0 - \vec{u}_0\|_2 + \|B(\vec{u}, \vec{u})(t,.)\|_2$$

$$\leq \|W_{\nu t} * \vec{u}_0 - \vec{u}_0\|_2 + C\|\vec{u}\|_{L^\infty L^2} \sup_{0 < s < t} \sqrt{s}\|\vec{u}(s,.)\|_\infty$$

so that

$$\lim_{t \to 0^+} \|\vec{u}(t,.) - \vec{u}_0\|_2 = 0.$$

Step 2: uniqueness.

Let $\vec{u} \in L^\infty L^2 \cap L^2 H^1 \cap \mathbb{X}_T^{r_1}$ and $\vec{v} \in L^\infty L^2 \cap L^2 H^1 \cap \mathbb{X}_T^{r_2}$, with $-1 \leq r_2 \leq r_1 \leq 1$, be two solutions of the same Navier–Stokes equations. We have already proved that $\vec{u} = \vec{v}$, in the case $-1 < r_2 \leq r_1 < 1$.

The proof for $r - 1 \leq r_2 \leq r_1$ and $0 < r_1 \leq 1$ is exactly the same as the one for $-1 < r_2 \leq r_1$ and $0 < r_1 < 1$.

The proof for $-1 \leq r_2 \leq r_1 < 0$ is exactly the same as the one for $-1 < r_2 \leq r_1 < 0$.

The proof for $-1 \leq r_2 \leq r_1 = 0$ is similar to the one for $-1 < r_2 \leq r_1 = 0$. We explain now how to modify the proof when $r_2 = -1$ and $r_1 = 0$. For $\epsilon > 0$, we may split $\vec{v}_2$ in $\vec{\alpha} + \vec{\beta}$ with $\sup_{0 \leq t \leq T} \|\vec{\alpha}(t,.)\|_{\dot{B}^{-1}_{\infty,\infty}} < \epsilon$ and $\sup_{0 \leq t \leq T} \|\vec{\beta}(t,.)\|_\infty \leq M(\epsilon) < +\infty$. As on page 382, we write $\vec{w} = \vec{u} - \vec{v}$ and compute $\frac{d}{dt}\|\vec{w}\|^2_{LP,\sigma}$.

We have

$$A_{j,k,l} = -\int \Delta_j \vec{w} \cdot \Delta_j(\Delta_k \vec{u} \cdot \vec{\nabla}\Delta_l \vec{u} - \Delta_k \vec{v} \cdot \vec{\nabla}\Delta_l \vec{v}) \, dx$$

$$\leq C 2^j \|\Delta_j \vec{w}\|_2 ((2^{-k r_2}\|\vec{\alpha}\|_{\dot{B}^{r_2}_{\infty,\infty}} + \|\vec{\beta}\|_\infty)\|\Delta_l \vec{w}\|_2 + \|\vec{u}\|_{\dot{B}^0_{\infty,\infty}}\|\Delta_k \vec{w}\|_2)$$

and

$$B_j = -\int \Delta_j \vec{w} \cdot \Delta_j(S_{j+5}\vec{u} \cdot \vec{\nabla}\vec{u} - S_{j+5}\vec{v} \cdot \vec{\nabla}\vec{v})) \, dx$$

$$\leq C\|\Delta_j \vec{w}\|_2 \|S_{j+5}\vec{w}\|_2 2^j \|\vec{u}\|_{\dot{B}^0_{\infty,\infty}} + C\|\Delta_j \vec{w}\|_2 (2^j \|\vec{\alpha}\|_{\dot{B}^{-1}_{\infty,\infty}} + \|\vec{\beta}\|_\infty)\|\vec{\nabla}S_{j+8}\vec{w}\|_2.$$

We obtain finally

$$\frac{d}{dt}\|\vec{w}\|^2_{LP,\sigma} \leq -\nu\|\vec{\nabla}\otimes\vec{w}\|^2_{LP,\sigma}$$

$$+ C(\|\vec{u}\|_{\dot{B}^0_{\infty,\infty}} + \|\vec{\beta}\|_\infty)\left(\|S_0\vec{w}\|^2_2 + \sum_{j\in\mathbb{N}} 2^{2j(1-\sigma)}\|\Delta_j\vec{w}\|^2_2\right)^{1/2}\|\vec{w}\|_{LP,\sigma}$$

$$+ C\|\vec{\alpha}\|_{\dot{B}^{-1}_{\infty,\infty}}\left(\|S_0\vec{w}\|^2_2 + \sum_{j\in\mathbb{N}} 2^{2j(1-\sigma)}\|\Delta_j\vec{w}\|^2_2\right).$$

We use again Bernstein's inequality and Young's inequality to get

$$\frac{d}{dt}\|\vec{w}\|^2_{LP,\sigma} \leq -\frac{1}{2}\nu\|\vec{\nabla}\otimes\vec{w}\|^2_{LP,\sigma} + C\|S_0\vec{w}\|^2_2 + C(\|\vec{u}\|^2_{\dot{B}^0_{\infty,\infty}} + \|\vec{\beta}\|^2_\infty)\|\vec{w}\|^2_{LP,\sigma}.$$

$$+ C\|\vec{\alpha}\|_{\dot{B}^{-1}_{\infty,\infty}}\left(\|S_0\vec{w}\|^2_2 + \sum_{j\in\mathbb{N}} 2^{2j(1-\sigma)}\|\Delta_j\vec{w}\|^2_2\right)$$

$$\leq -\frac{1}{2}\nu\|\vec{\nabla}\otimes\vec{w}\|^2_{LP,\sigma} + C\|S_0\vec{w}\|^2_2 + C(\|\vec{u}\|^2_{\dot{B}^0_{\infty,\infty}} + M(\epsilon)^2)\|\vec{w}\|^2_{LP,\sigma}.$$

$$+ C\epsilon(\|S_0\vec{w}\|^2_2 + \|\vec{\nabla}\otimes\vec{w}\|^2_{LP,\sigma}).$$

We choose ϵ such that $C\epsilon \leq \frac{\nu}{2}$, and then use the Grönwall lemma and get that $\vec{w} = 0$, hence $\vec{u} = \vec{v}$.

$\square$

Chen, Miao and Zhang [115] could further generalize Theorem 12.7 to the case $\vec{u} \in L^\infty L^2 \cap L^2 H^1 \cap L^1 \dot{B}^1_{\infty,\infty}$ and $\vec{v} \in L^\infty L^2 \cap L^2 H^1 \cap L^1 \dot{B}^1_{\infty,\infty}$. The proof was based on the losing regularity estimate for transportation through a Log-Lipschitz field (Chemin and Lerner [113], Danchin [142, 15])

Theorem 12.9.
Let $\vec{u}_0 \in L^2$ with $\operatorname{div} \vec{u}_0 = 0$ and $\vec{f} \in L^2((0, +\infty), H^1)$. Assume that $\vec{u}$ and $\vec{v}$ are two solutions of the Navier–Stokes equations on $(0, T)$

$$\partial_t \vec{u} = \nu \Delta \vec{u} + \mathbb{P}(\vec{f} - \operatorname{div}(\vec{u} \otimes \vec{u})), \quad \vec{u}(0, .) = \vec{u}_0$$

such that $\vec{u}$ and $\vec{v}$ belong to $L^\infty((0,T), L^2) \cap L^2((0,T), H^1) \cap L^1 \dot{B}^1_{\infty,\infty}$. Then $\vec{u} = \vec{v}$.

Proof. Remark that $\|S_0 \vec{u}\|_{L^\infty L^\infty} \leq C \|\vec{u}\|_{L^\infty L^2}$, so that $\vec{u}$ and $\vec{v}$ will belong to $L^1 \dot{B}^1_{\infty,\infty}$ (if $T < +\infty$). Let $\vec{w} = \vec{u} - \vec{v}$. One more time, we would like to compute the norm

$$\|\vec{w}\|_{LP,\sigma} = (\|S_0 \vec{w}\|_2^2 + \sum_{j \in \mathbb{N}} 2^{-2j\sigma} \|\Delta_j \vec{w}\|_2^2)^{1/2}$$

but we add some flexibility by allowing the regularity to worsen as time increases: we compute more precisely

$$\|\vec{w}\|_{LP,\sigma,\eta} = e^{-\eta_{-1}(t)} \|S_0 \vec{w}\|_2 + \sup_{j \in \mathbb{N}} e^{-\eta_j(t)} 2^{-j\sigma} \|\Delta_j \vec{w}\|_2$$

where η_j a time-dependent non-negative (increasing) function such that $\eta_j(0) = 0$. We write

$$\frac{d}{dt}(e^{-2\eta_j} \|\Delta_j \vec{w}(t,.)\|_2^2) = -2\frac{d\eta_j}{dt} e^{-2\eta_j} \|\Delta_j \vec{w}(t,.)\|_2^2$$

$$-2e^{-2\eta_j}(\nu \|\vec{\nabla} \otimes \Delta_j \vec{w}\|_2^2 + \int \Delta_j \vec{w} \cdot \Delta_j (\vec{u} \cdot \vec{\nabla} \vec{u} - \vec{v} \cdot \vec{\nabla} \vec{v}))\, dx)$$

We write again

$$\int \Delta_j \vec{w} \cdot \Delta_j (\vec{u} \cdot \vec{\nabla} \vec{u} - \vec{v} \cdot \vec{\nabla} \vec{v})\, dx = \int \Delta_j \vec{w} \cdot \Delta_j (S_{j+5} \vec{u} \cdot \vec{\nabla} \vec{u} - S_{j+5} \vec{v} \cdot \vec{\nabla} \vec{v}))\, dx$$

$$+ \sum_{k \geq j+5} \sum_{|k-l| \leq 3} \int \Delta_j \vec{w} \cdot \Delta_j (\Delta_k \vec{u} \cdot \vec{\nabla} \Delta_l \vec{u} - \Delta_k \vec{v} \cdot \vec{\nabla} \Delta_l \vec{v})\, dx$$

We have, for $k \geq j+5$ and $|l - k| \leq 3$,

$$A_{j,k,l} = -\int \Delta_j \vec{w} \cdot \Delta_j (\Delta_k \vec{u} \cdot \vec{\nabla} \Delta_l \vec{u} - \Delta_k \vec{v} \cdot \vec{\nabla} \Delta_l \vec{v})\, dx$$

$$\leq C 2^j \|\Delta_j \vec{w}\|_2 (2^{-l} \|\vec{v}\|_{\dot{B}^1_{\infty,\infty}} \|\Delta_l \vec{w}\|_2 + 2^{-k} \|\vec{u}\|_{\dot{B}^1_{\infty,\infty}} \|\Delta_k \vec{w}\|_2).$$

Now, we write (as $\int \Delta_j \vec{w}.(S_{j+5}\vec{v} \cdot \vec{\nabla}\Delta_j \vec{w}\, dx = 0$)

$$B_j = -\int \Delta_j \vec{w} \cdot \Delta_j(S_{j+5}\vec{u} \cdot \vec{\nabla}\vec{u} - S_{j+5}\vec{v} \cdot \vec{\nabla}\vec{v}))\, dx$$

$$= -\int \Delta_j \vec{w} \cdot \Delta_j(S_{j+5}\vec{w} \cdot \vec{\nabla} S_{j+8}\vec{u})\, dx - \int \Delta_j \vec{w} \cdot ([\Delta_j, S_{j+5}\vec{v}].\vec{\nabla} S_{j+8}\vec{w})\, dx$$

$$= \int S_{j+8}\vec{u} \cdot (\Delta_j(S_{j+5}\vec{w} \cdot \vec{\nabla}\Delta_j \vec{w})\, dx - \int \Delta_j \vec{w} \cdot ([\Delta_j, S_{j+5}\vec{v}].\vec{\nabla} S_{j+8}\vec{w})\, dx$$

$$\leq C\|\Delta_j \vec{w}\|_2 \|S_{j+5}\vec{w}\|_2 \|\vec{\nabla} \otimes S_{j+8}\vec{u}\|_\infty$$
$$+ C\|\Delta_j \vec{w}\|_2 2^{-j}\|\vec{\nabla} \otimes S_{j+5}\vec{v}\|_\infty \|\vec{\nabla} \otimes S_{j+8}\vec{w}\|_2$$

As

$$\frac{d}{dt}(e^{-2\eta_j}\|\Delta_j \vec{w}(t,.)\|_2^2) = 2e^{-\eta_j}\|\Delta_j \vec{w}(t,.)\|_2 \frac{d}{dt}(e^{-\eta_j}\|\Delta_j \vec{w}(t,.)\|_2)$$

we get (by dividing with $2e^{-\eta_j}\|\Delta_j \vec{w}\|_2$)

$$\frac{d}{dt}(e^{-\eta_j}\|\Delta_j \vec{w}(t,.)\|_2) \leq -\frac{d\eta_j}{dt}e^{-\eta_j}\|\Delta_j \vec{w}(t,.)\|_2$$
$$+ Ce^{-\eta_j}(\|\vec{u}\|_{B^1_{\infty,\infty}} + \|\vec{v}\|_{B^1_{\infty,\infty}}) \sum_{k\geq j+2} 2^{j-k}\|\Delta_k \vec{w}\|_2$$
$$+ Ce^{-\eta_j}(\|\vec{\nabla}\otimes S_{j+8}\vec{u}\|_\infty + \|\vec{\nabla} \otimes S_{j+8}\vec{v}\|_\infty)(\|S_0 \vec{w}\|_2 + \sum_{0\leq k\leq j+7}\|\Delta_k \vec{w}\|_2)$$

We take

$$\eta_j(t) = \lambda \int_0^t \|\vec{\nabla} \otimes S_0\vec{u}\|_\infty + \sum_{0\leq k\leq j+7}\|\vec{\nabla} \otimes \Delta_k\vec{u}\|_\infty + \|\vec{\nabla} \otimes S_0\vec{v}\|_\infty + \sum_{0\leq k\leq j+7}\|\vec{\nabla} \otimes \Delta_k\vec{v}\|_\infty\, ds$$

for some $\lambda \geq 0$ large enough (we shall fix the value of λ later). We have $\eta_j \leq \lambda(j + 9)(\|\vec{u}\|_{L^1 B^1_{\infty,\infty}} + \|\vec{v}\|_{L^1 B^1_{\infty,\infty}})$.

Let $\tau > 0$ (we shall fix the value of τ later). If $A(\tau) = \sup_{0<t<\tau}\|\vec{w}\|_{LP,\sigma,\eta}$, we find for $0 < t < \tau$

$$2^{-j\sigma}e^{-\eta_j(t)}\|\Delta_j \vec{w}(t,.)\|_2 + \int_0^t \frac{d\eta_j}{dt}2^{-j\sigma}e^{-\eta_j}\|\Delta_j \vec{w}\|_2\, ds$$

$$\leq CA(\tau)\int_0^t (\|\vec{u}\|_{B^1_{\infty,\infty}} + \|\vec{v}\|_{B^1_{\infty,\infty}}) \sum_{k\geq j+2} 2^{(j-k)(1-\sigma)}e^{\eta_k-\eta_j}\, ds$$

$$+ \frac{C}{\lambda}\int_0^t \frac{d\eta_j}{dt}2^{-j\sigma}e^{-\eta_j}(\|S_0 \vec{w}\|_2 + \sum_{0\leq k\leq j+7}\|\Delta_k \vec{w}\|_2)\, ds$$

$$= I + II.$$

As, for $k > j$, $\eta_k(t) - \eta_j(t) \leq \lambda(k - j)\int_0^t \|\vec{u}\|_{B^1_{\infty,\infty}} + \|\vec{v}\|_{B^1_{\infty,\infty}}\, ds$, we find that there exists $t_\lambda > 0$ (which does not depend on j nor k) such that

$$\text{for } t \in [0, t_\lambda], \quad \eta_k(t) - \eta_j(t) \leq \epsilon(k - j)$$

with $2^{-(1-\sigma)}e^\epsilon < 1$ and $2^{-\sigma}e^\epsilon < 1$, and thus, for $0 < t < \tau \leq t_\lambda$,

$$I \leq CA(\tau)\int_0^\tau \|\vec{u}\|_{B^1_{\infty,\infty}} + \|\vec{v}\|_{B^1_{\infty,\infty}}\, ds.$$

Similarly, we define

$$B(\tau) = \sup_{0<t<\tau} \left(\int_0^t \frac{d\eta_{-1}}{dt} e^{-\eta_{-1}} \|S_0 j \vec{w}\|_2 \, ds + \sup_{j\geq 0} \int_0^t \frac{d\eta_j}{dt} 2^{-j\sigma} e^{-\eta_j} \|\Delta_j \vec{w}\|_2 \, ds \right).$$

B is well defined and satisfies $B \leq C \|\vec{w}\|_2$. We have

$$II \leq III + IV + V$$

where

$$III = \frac{C}{\lambda} \int_0^t \frac{d\eta_{-1}}{dt} 2^{-j\sigma} e^{-\eta_j} \|S_0 \vec{w}\|_2 \, ds + \sum_{0\leq k\leq j} \frac{C}{\lambda} \int_0^t \frac{d\eta_k}{dt} 2^{-j\sigma} e^{-\eta_j} \|\Delta_k \vec{w}\|_2 \Big) \, ds$$

$$IV = \frac{C}{\lambda} \int_0^t \frac{d(\eta_j - \eta_{-1})}{dt} 2^{-j\sigma} e^{-\eta_j} \|S_0 \vec{w}\|_2 \, ds$$

$$+ \sum_{0\leq k\leq j} \frac{C}{\lambda} \int_0^t \frac{d(\eta_j - \eta_k)}{dt} 2^{-j\sigma} e^{-\eta_j} \|\Delta_k \vec{w}\|_2 \Big) \, ds$$

and

$$V = \frac{C}{\lambda} \int_0^t \frac{d\eta_j}{dt} 2^{-j\sigma} e^{-\eta_j} \Big(\sum_{j+1\leq k\leq j+7} \|\Delta_k \vec{w}\|_2 \Big) \, ds.$$

For $0 < t < \tau \leq t_\lambda$, we find

$$III \leq \frac{C}{\lambda} B(\tau) \sum_{-1\leq k\leq j} (2^{-\sigma} e^{\epsilon})^{j-k} \leq \frac{C'}{\lambda} B(\tau)$$

and, since $\frac{d\eta_j}{dt} \leq \frac{d\eta_j}{dt}$ when $j \leq k$,

$$V \leq \frac{C}{\lambda} B(\tau) \sum_{j+1\leq k\leq j+7} (2^{\sigma} e^{\epsilon})^{k-j} \leq \frac{C'}{\lambda} B(\tau).$$

Finally, we have

$$IV \leq \frac{C}{\lambda} A(\tau) \sum_{-1\leq k\leq j} 2^{-\sigma(j-k)} \int_0^t \frac{d(\eta_j - \eta_k)}{dt} e^{-\eta_j + \eta_k} \, ds \leq \frac{C'}{\lambda} A(\tau)$$

as

$$\int_0^t \frac{d(\eta_j - \eta_k)}{dt} e^{-\eta_j + \eta_k} \, ds = 1 - e^{-\eta_j(t) + \eta_k(t)} \leq 1$$

when $k \leq j$.

Similar estimates hold on $S_0 \vec{w}$. Finally, we find that, for $0 < t < \tau \leq t_\lambda$, we have

$$A(\tau) + B(\tau) \leq C_0 A(\tau) \int_0^\tau \|\vec{u}\|_{B^1_{\infty,\infty}} + \|\vec{v}\|_{B^1_{\infty,\infty}} \, ds + \frac{C_1}{\lambda} (A(\tau) + B(\tau))$$

where the constants C_0 and C_1 do not depend on λ nor τ. We then fix λ such that $\frac{C_1}{\lambda} < \frac{1}{2}$, then we fix $\tau \in (0, t_\lambda]$ such that $C_0 \int_0^\tau \|\vec{u}\|_{B^1_{\infty,\infty}} + \|\vec{v}\|_{B^1_{\infty,\infty}} \, ds \leq \frac{1}{4}$, and we obtain $A(\tau) = 0$. Thus, we have local uniqueness: $\vec{u} = \vec{v}$ on $[0, \tau]$. This uniqueness propagates to the whole interval $[0, T]$, as $t \mapsto \vec{w}(t, .)$ is weakly continuous in L^2: thus the maximal interval $[0, T^*]$ on which $\vec{w} = 0$ is closed, and as it must be open by local uniqueness, we conclude $T^* = T$. $\square$

12.6 Weak Perturbations of Mild Solutions

We have seen, up to now, essentially two classes of solutions: weak ones (obtained by mollification and then by the use of Rellich's theorem) and mild solutions (obtained through Picard's method). Sometimes, it is useful to combine the two approaches, i.e., to compute the solution $\vec{u}$ of

$$\partial_t \vec{u} = \nu \Delta \vec{u} + \mathbb{P}(\vec{f} - \operatorname{div}(\vec{u} \otimes \vec{u})), \quad \vec{u}(0,.) = \vec{u}_0$$

as the sum of a mild solution $\vec{w}$ of

$$\partial_t \vec{w} = \nu \Delta \vec{w} + \mathbb{P}(\vec{f} - \operatorname{div}(\vec{w} \otimes \vec{w})), \quad \vec{w}(0,.) = \vec{w}_0$$

and a weak solution $\vec{v}$ of

$$\partial_t \vec{v} = \nu \Delta \vec{v} - \mathbb{P}\operatorname{div}(\vec{v} \otimes \vec{v} + \vec{w} \otimes \vec{v} + \vec{v} \otimes \vec{w}), \quad \vec{v}(0,.) = \vec{v}_0$$

For instance, Calderón [77] considered the case of an initial value $\vec{u}_0 \in L^p$ with $2 < p < 3$ (and a forcing term $\vec{f} = 0$). Then he could show existence of a solution by splitting $\vec{u}_0$ into $\vec{v}_0 + \vec{w}_0$, with the norm of $\vec{w}_0$ small in L^3 (so that global existence of the mild solution $\vec{w}$ is granted) and $\vec{v}_0 \in L^2$. This kind of "mixed initial-values" which pave the way to a combination of weak and mild solutions was discussed by Lemarié-Rieusset (in the paper [310] and in the concluding chapter of [313]) and recently extended by Cui [134] who considered an initial value in $B_{\infty\infty}^{-1(\ln)} + B_{X_r}^{-1+r,\frac{2}{1-r}} + L^2$ (where X_r is the space of pointwise multipliers that map H^r to L^2).

In this section, we address the stability of mild solutions through some L^2 perturbation of the initial value. This issue has been recently considered by Karch, Pilarczyk, and Schonbek [251].

Existence of permanent solutions

Theorem 12.10.
Let $\mathbb{X}$ be a Banach space of Lebesque measurable functions such that

- *the pointwise product is bounded from $L^\infty \times \mathbb{X}$ to $\mathbb{X}$.*
- *the Hardy–Littlewood maximal function is a bounded operator on $\mathbb{X}$*
- *the Riesz transforms are bounded on $\mathbb{X}$*
- *the bilinear operator $(u,v) \mapsto \frac{1}{\sqrt{-\Delta}}(uv)$ is bounded on $\mathbb{X}$*
- *the bilinear operator $(u,v) \mapsto \frac{1}{(-\Delta)^{1/4}}(u\frac{1}{(-\Delta)^{1/4}}v)$ is bounded on $\mathbb{X}$*

Assume that the forcing term $\vec{f}$ corresponds to a permanent regime: $\vec{f}$ is steady (i.e., does not depend on time) or is time-periodic ($\vec{f}(t+T,x) = \vec{f}(t,x)$ for some positive T) and that $\vec{f}$ is small enough: for some $\epsilon_0 > 0$ (depending only on $\mathbb{X}$), we have

- *in the steady case: $\vec{f} = \Delta\vec{F}$ with $\vec{F} \in \mathbb{X}$ and $\|\vec{F}\|_{\mathbb{X}} < \epsilon_0 \nu^2$*
- *in the time periodic case:*

$\quad - \vec{f}$ belongs to $L^1_{\mathrm{per}}\mathbb{X}$ with $\int_0^T \|\vec{f}\|_{\mathbb{X}}\,dt < \epsilon_0 \nu$

$\quad -$ the mean value $\vec{f}_0 = \frac{1}{T}\int_0^T \vec{f}(s,.)\,ds$ can be written as $\vec{f}_0 = \Delta\vec{F}$ with $\vec{F} \in \mathbb{X}$
$\quad\quad$ with $\|\vec{F}\|_{\mathbb{X}} < \epsilon_0 \nu^2$.

Then, there exists a unique permanent solution $\vec{U}$ on $(0, +\infty) \times \mathbb{R}^3$ of the problem

$$\begin{cases} \partial_t \vec{U} - \mathrm{div}(\vec{U} \otimes \vec{U}) = \nu\Delta\vec{U} + \vec{f} - \vec{\nabla}p \\ \qquad\qquad\quad \mathrm{div}\,\vec{U} = 0 \end{cases} \qquad (12.35)$$

such that

- in the steady case: $\vec{U}$ is stationary $(\partial_t \vec{U} = 0)$, $\vec{U} \in \mathbb{X}$ and

$$\|\vec{U}\|_{\mathbb{X}} \le C_0 \frac{1}{\nu}\|\vec{F}\|_{\mathbb{X}}$$

- in the time-periodic case: $\vec{U}$ is time-periodic, $\vec{U} \in L^\infty_{\mathrm{per}}\mathcal{X}$ and more precisely

$$\| \sup_{t\in\mathbb{R}} |\vec{U}(t,.)| \|_{\mathbb{X}} \le C_0 \left(\int_0^T \|\vec{f}(s,.)\|_{\mathbb{X}} + \frac{1}{\nu}\|\vec{F}\|_{\mathbb{X}}\right).$$

where the constant $C_0 > 0$ depends only on $\mathbb{X}$.

Stability of permanent solutions

Theorem 12.11.
Let $\mathbb{X}$ satisfy the assumptions of Theorem 12.10. Assume that, for some $\epsilon_1 > 0$ (depending only on $\mathbb{X}$), we have the following assumptions on $\vec{f}$ and $\vec{u}_0$:

- *in the steady case: $\vec{f} = \Delta\vec{F}$ with $\vec{F} \in \mathbb{X}$ and $\|\vec{F}\|_{\mathbb{X}} < \epsilon_1 \nu^2$*

- *in the time-periodic case:*

$\quad - \vec{f}$ belongs to $L^1_{\mathrm{per}}\mathbb{X}$ with $\int_0^T \|\vec{f}\|_{\mathbb{X}}\,dt < \epsilon_1 \nu$

$\quad -$ the mean value $\vec{f}_0 = \frac{1}{T}\int_0^T \vec{f}(s,.)\,ds$ can be written as $\vec{f}_0 = \Delta\vec{F}$ with $\vec{F} \in \mathbb{X}$
$\quad\quad$ with $\|\vec{F}\|_{\mathbb{X}} < \epsilon_1 \nu^2$.

- *$\vec{u}_0$ can be written as the sum of two divergence-free vector fields $\vec{u}_0 = \vec{v}_0 + \vec{w}_0$, with $\vec{v}_0 \in L^2$, $\vec{w}_0 \in \mathbb{X}$ and $\|\vec{w}_0\|_{\mathbb{X}} < \epsilon_1 \nu$.*

We assume that $\epsilon_1 \le \epsilon_0$, so that existence of a permanent solution $\vec{U}$ is granted by Theorem 12.10.
$\quad$ *Then, there exists at least one solution $\vec{u}$ on $(0, +\infty) \times \mathbb{R}^3$ of the problem*

$$\begin{cases} \partial_t \vec{u} - \mathrm{div}(\vec{u} \otimes \vec{u}) = \nu\Delta\vec{u} + \vec{f} - \vec{\nabla}p \\ \qquad\qquad\qquad \mathrm{div}\,\vec{u} = 0 \\ \qquad\qquad\qquad \vec{u}(0,.) = \vec{u}_0 \end{cases} \qquad (12.36)$$

such that

- $\vec{u} = \vec{v} + \vec{w}$ with $\vec{w} \in L^\infty \mathbb{X}$ and $\vec{v} \in L^\infty L^2 \cap L^2 \dot{H}^1$

- $\|\sup_{t \in \mathbb{R}} |\vec{w}(t,.)| \|_{\mathbb{X}} \le C_1 \| \sup_{t \in \mathbb{R}} |\vec{U}(t,.)| \|_{\mathbb{X}}$ and

$$\sup_{t>0} t^{1/4} \|(-\Delta)^{1/4}(\vec{w}(t,.) - \vec{U}(t,.))\|_{\mathbb{X}} \le C_1 \nu^{-1/4} \| \sup_{t \in \mathbb{R}} |\vec{U}(t,.)| \|_{\mathbb{X}} \qquad (12.37)$$

  (where the constant $C_1 > 0$ depends only on $\mathbb{X}$)

- $\lim_{t \to +\infty} \|\vec{v}(t,.)\|_2 = 0$.

  In particular, $\vec{u}(t,.) - \vec{U}(t,.)$ converges to 0 in $\mathcal{S}'$ (and in L^2_{loc}) as t goes to $+\infty$.

Proof. Due to Theorems 10.11 and 10.15, we know that the existence result (Theorem 12.10) holds, as well as the stability result (Theorem 12.11) in the case $\vec{v}_0 = 0$.

When $\vec{v}_0 \ne 0$, we begin by solving the Navier–Stokes problem with initial value $\vec{w}_0$ and find a solution $\vec{w}$. Then, we study the problem

$$\partial_t \vec{v} = \nu \Delta \vec{v} - \mathbb{P} \operatorname{div}(\vec{v} \otimes \vec{v} + \vec{w} \otimes \vec{v} + \vec{v} \otimes \vec{w}), \quad \vec{v}(0,.) = \vec{v}_0.$$

The problem will be solved just as for the classical Navier–Stokes problem (i.e., when $\vec{w} = 0$).

Step 1: Leray's mollification.

As for Theorem 12.2, we study the problem associated to a mollifier θ_ϵ

$$\partial_t \vec{v} = \nu \Delta \vec{v} - \mathbb{P}((\theta_\epsilon * \vec{v}) \cdot \vec{\nabla} \vec{v}) - \mathbb{P} \operatorname{div}(\vec{w} \otimes \vec{v} + \vec{v} \otimes \vec{w}), \quad \vec{v}(0,.) = \vec{v}_0.$$

We have $\mathbb{X} \subset \mathcal{M}(\dot{H}^1 \mapsto L^2) = \mathcal{M}(L^2 \mapsto \dot{H}^{-1})$ and $\|\vec{w}\|_{L^\infty \mathbb{X}} \le 2C_1 C_0 \epsilon_1 \nu$. Thus, using again the norm

$$\|\vec{u}\|_{\nu, T_0} = \|\vec{u}\|_{L^\infty((0,T_0), L^2)} + \sqrt{\nu} \|\vec{u}\|_{L^2(0,T_0), \dot{H}^1)}$$

and the inequalities

$$\|W_{\nu t} * \vec{v}_0\|_{\nu, T_0} \le C_2 \|\vec{v}_0\|_2$$

and

$$\| \int_0^t W_{\nu(t-s)} * \vec{g} \, ds \|_{\nu, T_0} \le C_2 \frac{1}{\sqrt{\nu}} \|\vec{g}\|_{L^2 \dot{H}^{-1}}$$

we get local existence of the solution $\vec{v}$: as we have

$$\|(\vec{u} * \theta_\epsilon) \cdot \vec{\nabla} \vec{v}\|_{L^2((0,T_0), \dot{H}^{-1})} \le C_3 \sqrt{T_0}\, \epsilon^{-3/2} \|\vec{u}\|_{L^\infty L^2} \|\vec{v}\|_{L^\infty L^2}$$
$$\le C_3 \sqrt{T_0}\, \epsilon^{-3/2} \|\vec{u}\|_{\nu, T_0} \|\vec{v}\|_{\nu, T_0}$$

and

$$\| \operatorname{div}(\vec{w} \otimes \vec{v} + \vec{v} \otimes \vec{w})\|_{L^2((0,T_0), \dot{H}^{-1})} \le C_3 \|\vec{w}\|_{L^\infty \mathbb{X}} \|\vec{v}\|_{L^2 \dot{H}^1}$$
$$\le 2C_0 C_1 C_3 \epsilon_1 \sqrt{\nu} \|\vec{v}\|_{\nu, T_0},$$

we find that the Picard iterate shall converge to a solution $\vec{v}_\epsilon$ if ϵ_1 is small enough ($2C_0 C_1 C_3 \epsilon_1 < 1/4$) and T_0 is small enough ($2C_2^2 \frac{1}{\sqrt{\nu}} C_3 \sqrt{T_0}\, \epsilon^{-3/2} \|\vec{v}_0\|_2 < 1/4$).

We easily check that $\vec{v}_\epsilon$ is indeed a global solution: it is enough to show that the L^2 norm of $\vec{v}_\epsilon$ remains bounded (as the existence time T_0 is controlled by the L^2 norm of the Cauchy data). We have

$$
\begin{aligned}
\frac{d}{dt}\|\vec{v}_\epsilon\|_2^2 &= 2\int \partial_t \vec{v}_\epsilon \cdot \vec{v}_\epsilon \, dx \\
&= -2\nu\|\vec{v}_\epsilon\|_{\dot{H}^1}^2 - 2\langle \mathrm{div}(\vec{v}_\epsilon \otimes \vec{w}) | \vec{v}_\epsilon \rangle_{H^{-1},H^1} \\
&\leq -2(\nu - \|\vec{w}\|_{\mathcal{M}(\dot{H}^1 \mapsto L^2)})\|\vec{u}_\epsilon\|_{\dot{H}^1}^2 \\
&\leq -(2 - C_4\epsilon_1)\nu\|\vec{u}_\epsilon\|_{\dot{H}^1}^2
\end{aligned}
\tag{12.38}
$$

Thus, if ϵ_1 is small enough ($C_4\epsilon_1 < 1$), we find that

$$
\|\vec{v}_\epsilon(t,.)\|_2^2 + \nu\int_0^t \|\vec{v}_\epsilon\|_{\dot{H}^1}^2 \, ds \leq \|\vec{v}_0\|_2^2
\tag{12.39}
$$

Using this energy inequality, we find that we may then use the Rellich–Lions theorem (Theorem 12.1) and find a sequence $\epsilon_n \to 0$ and a function $\vec{v}$ such that:

- on every bounded subinterval of $[0, +\infty]$, $\vec{v}_{(\epsilon_n)}$ is *-weakly convergent to $\vec{v}$ in $L^\infty L^2$ and in $L^2 \dot{H}^1$

- $\vec{v}_{(\epsilon_n)}$ is strongly convergent to $\vec{v}$ in $L^2_{\mathrm{loc}}((0, +\infty) \times \mathbb{R}^3)$.

Moreover, the weak limit $\vec{v}$ satisfies

$$
\partial_t \vec{v} = \nu\Delta\vec{v} - \mathbb{P}\,\mathrm{div}(\vec{v} \otimes \vec{v} + \vec{w} \otimes \vec{v} + \vec{v} \otimes \vec{w})
$$

and the Leray energy inequality for every $t \in (0, +\infty)$, we have

$$
\|\vec{v}(t,.)\|_2^2 + 2\nu\int_0^t \|\vec{v}\|_{\dot{H}^1}^2 \, ds \leq \|\vec{v}_0\|_2^2 + 2\int_0^t \langle \vec{v} \otimes \vec{w} | \vec{\nabla} \otimes \vec{v} \rangle_{L^2,L^2} \, ds
\tag{12.40}
$$

It even fulfills the strong Leray energy inequality: for almost every t_0 in $(0, +\infty)$ and for every $t \in (t_0, +\infty)$, we have

$$
\|\vec{v}(t,.)\|_2^2 + 2\nu\int_{t_0}^t \|\vec{v}\|_{\dot{H}^1}^2 \, ds \leq \|\vec{v}(t_0)\|_2^2 + 2\int_{t_0}^t \langle \vec{v} \otimes \vec{w} | \vec{\nabla} \otimes \vec{v} \rangle_{L^2,L^2} \, ds
\tag{12.41}
$$

Step 2: Higher regularity estimates.

The proof of $\lim_{t\to+\infty}\|\vec{v}(t,.)\|_2 = 0$ then follows the proof of Corollary 12.1. However, we have a little difficulty to overcome: $\vec{w}$ is not regular enough to ensure that, when $\vec{v}_0$ is regular, then $\vec{v}$ is regular (we use the boundedness in H^1 in the proof of Corollary 12.1).

Thus, we shall study the behavior of $\vec{v}$ in a smaller space: the homogeneous Besov space $\dot{B}_{2,\infty}^{1/2}$. Let us remark that, as well as we have $X \subset \mathcal{M}(\dot{H}^1 \mapsto L^2) = \mathcal{V}^1$, we have $\dot{B}_{2,\infty}^{1/2} \subset \mathcal{V}^1$: interpolating the Sobolev embeddings $\dot{H}^0 = L^2 \subset L^2$ and $\dot{H}^1 \subset L^6$, we find

$$
\dot{B}_{2,\infty}^{1/2} = [\dot{H}^0, \dot{H}^1]_{1/2,\infty} \subset [L^2, L^6]_{1/2,\infty} \subset L^{3,\infty} \subset \mathcal{V}^1.
$$

Another useful remark is that $\mathcal{V}^1 = \mathcal{M}(\dot{H}^1 \mapsto L^2)$ coincides with $\mathcal{M}(L^2 \mapsto \dot{H}^{-1})$ (by duality, as pointwise multiplication is a self-adjoint operator) and thus (by interpolation) $\mathcal{V}^1 \subset \mathcal{M}(\dot{B}_{2,\infty}^{1/2} \mapsto \dot{B}_{2,\infty}^{-1/2})$.

A final remark is an inequality we already used (on page 147) when proving the uniqueness theorem for $\mathcal{C}([0,T], L^3)$ solutions:

$$\left\| \int_0^t W_{\nu(t-s)} * \vec{g}\, ds \right\|_{L^\infty \dot{B}_{2,\infty}^{-1/2}} \leq C_5 \frac{1}{\nu} \|\vec{g}\|_{L^\infty \dot{B}_{2,\infty}^{-3/2}}.$$

Thus, writing the inequalities

$$\|W_{\nu t} * \vec{v}_0\|_{L^\infty \dot{B}_{2,\infty}^{1/2}} \leq \|\vec{v}_0\|_{\dot{B}_{2,\infty}^{1/2}}$$

$$\| \operatorname{div}(\vec{u} \otimes \vec{v}) \|_{L^\infty \dot{B}_{2,\infty}^{-3/2}} \leq C_6 \|\vec{u}\|_{L^\infty \mathbb{X}} \|\vec{v}\|_{L^\infty \dot{B}_{2,\infty}^{1/2}}$$

$$\leq C_7 \|\vec{u}\|_{L^\infty \dot{B}_{2,\infty}^{1/2}} \|\vec{v}\|_{L^\infty \dot{B}_{2,\infty}^{1/2}}$$

and

$$\| \operatorname{div}(\vec{w} \otimes \vec{v} + \vec{v} \otimes \vec{w}) \|_{L^\infty \dot{B}_{2,\infty}^{-3/2}} \leq C_6 \|\vec{w}\|_{L^\infty \mathbb{X}} \|\vec{v}\|_{L^2 \dot{B}_{2,\infty}^{1/2}}$$

$$\leq C_7 \epsilon_1 \nu \|\vec{v}\|_{L^\infty \dot{B}_{2,\infty}^{1/2}}$$

Thus, if ϵ_1 is small enough ($C_5 C_7 \epsilon_1 < 1/4$) and $\vec{v}_0$ is small enough ($2C_5 C_7 \|\vec{v}_0\|_{\dot{B}_{2,\infty}^{1/2}} < \nu/4$), we have a global solution in $L^\infty \dot{B}_{2,\infty}^{1/2}$.

Of course, for ϵ_1 small enough and $\vec{v}_0$ small enough, this solution will still be in $L^\infty L^2 \cap L^2 \dot{H}^1$: just write

$$\| \operatorname{div}(\vec{u} \otimes \vec{v}) \|_{L^2 \dot{H}^{-1}} \leq C_8 \min(\|\vec{u}\|_{L^\infty \mathbb{X}} \|\vec{v}\|_{L^2 \dot{H}^1}, \|\vec{u}\|_{L^2 \dot{H}^1} \|\vec{v}\|_{L^\infty \mathbb{X}})$$

$$\leq C_9 \min(\|\vec{u}\|_{L^\infty \dot{B}_{2,\infty}^{1/2}} \|\vec{v}\|_{L^2 \dot{H}^1}, \|\vec{u}\|_{L^2 \dot{H}^1} \|\vec{v}\|_{L^\infty \dot{B}_{2,\infty}^{1/2}})$$

and

$$\| \operatorname{div}(\vec{w} \otimes \vec{v} + \vec{v} \otimes \vec{w}) \|_{L^2 \dot{H}^{-1}} \leq C_8 \|\vec{w}\|_{L^\infty \mathbb{X}} \|\vec{v}\|_{L^2 \dot{H}^1}$$

$$\leq C_9 \epsilon_1 \nu \|\vec{v}\|_{L^2 \dot{H}^1}$$

to check that the Picard iterates will converge in $L^\infty L^2 \cap L^2 \dot{H}^1$.

Step 3: Weak-strong uniqueness.

As for the classical Navier–Stokes problem (Theorem 12.3), we may prove weak-strong uniqueness for $\vec{v}$. More precisely, assume that we have two solutions $\vec{v}_1$ and $\vec{v}_2$ in $L^\infty L^2 \cap L^2 \dot{H}^1$ and that $\vec{v}_1$ is small enough in $L^\infty \dot{B}_{2,\infty}^{1/2}$ (and $\vec{w}$ small enough in $L^\infty \mathbb{X}$) while $\vec{v}_2$ satisfies the Leray energy inequality. Then we find that

$$\|\vec{v}_1 - \vec{v}_2(t,.)\|_2^2 \leq - 2\nu \int_0^t \|\vec{v}_1 - \vec{v}_2\|_{\dot{H}^1}^2\, ds$$

$$- 2 \int_0^t \langle (\vec{v}_1 - \vec{v}_2) \otimes \vec{w} | \vec{\nabla} \otimes (\vec{v}_1 - \vec{v}_2) \rangle_{L^2, L^2}\, ds$$

$$- 2 \int_0^t \langle (\vec{v}_1 - \vec{v}_2) \otimes \vec{v}_1 | \vec{\nabla} \otimes (\vec{v}_1 - \vec{v}_2) \rangle_{L^2, L^2}\, ds$$

$$\leq - 2(\nu - C_{10} \|\vec{w}\|_{L^\infty \mathbb{X}} - C_{10} \|\vec{v}_1\|_{L^\infty \dot{B}_{2,\infty}^{1/2}}) \int_0^t \|\vec{v}_1 - \vec{v}_2\|_{\dot{H}^1}^2\, ds.$$

Step 4: End of the proof.

The proof then follows the proof of Corollary 12.1. We have just seen that if ϵ_1 is small enough ($\epsilon_1 < C_{11}$) and $\vec{v}(t_0, .)$ is small enough ($\|\vec{v}(t_0, .)\|_{\dot{B}^{1/2}_{2,\infty}} < C_{11}\nu$), then the equations

$$\partial_t \vec{v} = \nu \Delta \vec{v} - \mathbb{P} \operatorname{div}(\vec{v} \otimes \vec{v} + \vec{w} \otimes \vec{v} + \vec{v} \otimes \vec{w}), \quad \vec{v}(t_0, .) = \vec{v}_0$$

has a solution $\vec{V}$ that belongs to $L^\infty((t_0, +\infty), L^2) \cap L^2((t_0, +\infty), \dot{H}^1) \cap L^\infty((t_0, +\infty), \dot{B}^{1/2}_{2,+\infty})$ and that we have weak-strong uniqueness for $\vec{V}$.

As $\vec{v}$ belongs to $L^\infty_t L^2 \cap L^2_t \dot{H}^1$, it belongs to $L^4_t \dot{B}^{1/2}_{2,+\infty}$. Thus, the set of times t such that $\|\vec{v}(t, .)\|_{\dot{B}^{1/2}_{2,+\infty}} \geq C_{11}\nu$ is of finite measure. As the set of Lebesgue points of $t \mapsto \|\vec{v}(t, .)\|_2$ has a complement of null measure, we may find a time t_0 such that

- $\|\vec{v}(t_0, .)\|_{\dot{H}^{1/2}} < C_{11}\nu$
- $\vec{v}$ is a weak Leray solution on $(t_0, +\infty)$: for every $t \in (t_0, +\infty)$, we have

$$\|\vec{v}(t, .)\|_2^2 + 2\nu \int_0^t \|\vec{v}\|_{\dot{H}^1}^2 \, ds \leq \|\vec{v}_0\|_2^2 + 2 \int_0^t \langle \vec{v} \otimes \vec{w} | \vec{\nabla} \otimes \vec{v} \rangle_{L^2, L^2} \, ds$$

Then, by weak-strong uniqueness, we find that $\vec{v}$ coincides on $(t_0, +\infty)$ with the mild solution $\vec{V}$.

We now write, for $t_0 < \tau < t$,

$$\|\vec{v}(t, .)\|_2 \leq \|W_{\nu(t-\tau)} * \vec{v}(\tau, .)\|_2 + C\left(\int_\tau^t \|\mathbb{P} \operatorname{div}(\vec{v} \otimes \vec{v} + \vec{w} \otimes \vec{v} + \vec{v} \otimes \vec{w})\|_{\dot{H}^{-1}}^2 \, ds\right)^{1/2}$$

which gives

$$\limsup_{t \to +\infty} \|\vec{v}(t, .)\|_2 \leq C\left(\sup_{t > t_0} \|\vec{v}(t, .)\|_{\dot{B}^{1/2}_{2,\infty}} + \sup_{t > t_0} \|\vec{w}(t, .\|_X)\right)\left(\int_\tau^{+\infty} \|\vec{v}\|_{\dot{H}^1}^2 \, ds\right)^{1/2}.$$

Letting τ go to $+\infty$, we get

$$\lim_{t \to +\infty} \|\vec{v}(t, .)\|_2 = 0. \qquad \square$$

12.7   Non-uniqueness of Weak Solutions

Very recently, some results have been published on the non-uniqueness of weak solutions. We have seen that weak-strong uniqueness holds for Leray weak solutions in presence of a mild solution, but other uniqueness issues could be considered:

- Q1) Does uniqueness hold for $L^\infty L^2 \cap L^2 H^1$ solutions in presence of a mild solution, but without assuming Leray's energy inequality?

- Q2) Does uniqueness hold for Leray weak solutions in absence of a mild solution?

- Q3) Does uniqueness hold for $L^\infty L^2$ solutions in presence of a mild solution?

While question Q1) is still open, answers to Q2) and Q3) have been proven to be negative by Buckmaster and Vicol [71] in 2019 and by Albritton, Brué and Colombo [5] in 2021.

Wild solutions on the torus

In this section, we will discuss solutions of the equations

$$\partial_t \vec{u} = \nu \Delta \vec{u} - \mathrm{div}(\vec{u} \otimes \vec{u}) - \vec{\nabla} p$$

on $(0, T) \times \mathbb{R}^3$ with the conditions

$$
\begin{cases}
\vec{u}(t, x + 2k\pi) = \vec{u}(t, x) \text{ for all } k \in \mathbb{Z}^3 \\
p(t, x + 2k\pi) = p(t, x) \text{ for all } k \in \mathbb{Z}^3 \\
\vec{u} \in L^\infty((0, T), L^2(\mathbb{R}^3/\mathbb{Z}^3)) \\
\mathrm{div}\,\vec{u} = 0 \\
\vec{u}(0, .) = \vec{u}_0 \\
\mathrm{div}\,\vec{u}_0 = 0 \text{ and } \displaystyle\int_{[-\pi,\pi]^3} \vec{u}_0(x)\,dx = 0
\end{cases}
$$

The periodicity of p will ensure that $\int_{[-\pi,\pi]^3} \vec{u}(t, x)\,dx = 0$ for all $t \in (0, T)$ and that $\vec{u} = -\frac{1}{\Delta}\vec{\nabla} \wedge (\vec{\nabla} \wedge \vec{u})$, where, for a Fourier series $f = \sum_{k \in \mathbb{Z}^3, k \neq 0} e^{ik \cdot x} a_k$,

$$\frac{1}{\Delta} f = -\sum_{k \in \mathbb{Z}^3, \neq 0} \frac{1}{|k|^2} e^{ik \cdot x} a_k.$$

Defining, for a periodic vector field $\vec{f}$ with zero mean,

$$\mathbb{P}\vec{f} = \vec{f} - \frac{1}{\Delta}\vec{\nabla}(\mathrm{div}\,\vec{f}) = -\frac{1}{\Delta}\vec{\nabla} \wedge (\vec{\nabla} \wedge \vec{f}),$$

we find that

$$\partial_t \vec{u} = \nu \Delta \vec{u} - \mathbb{P}\,\mathrm{div}(\vec{u} \otimes \vec{u}).$$

The analysis for mild or weak periodic solutions is then very similar to the analysis on the whole space. We start with the basic estimates:

- analysis of the heat kernel $e^{t\Delta} f = W_t * f$: if $f \in L^2(\mathbb{R}^3/\mathbb{Z}^3)$ with $\int_{[-\pi,\pi]^3} f(x)\,dx = 0$, then

$$\|e^{\nu t \Delta} f\|_{L^\infty((0,+\infty), L^2(\mathbb{R}^3/\mathbb{Z}^3))} = \|f\|_{L^2(\mathbb{R}^3/\mathbb{Z}^3)}$$

  and

$$\|\vec{\nabla} e^{\nu t \Delta} f\|_{L^2((0,+\infty), L^2(\mathbb{R}^3/\mathbb{Z}^3))} = \frac{1}{\sqrt{2\nu}} \|f\|_{L^2(\mathbb{R}^3/\mathbb{Z}^3)}$$

- Sobolev embedding: if $f \in H^1(\mathbb{R}^3/\mathbb{Z}^3)$ with $\int_{[-\pi,\pi]^3} f(x)\,dx = 0$, then

$$\|f\|_{L^6(\mathbb{R}^3/\mathbb{Z}^3)} \leq C\|\vec{\nabla} f\|_{L^2(\mathbb{R}^3/\mathbb{Z}^3)},$$

  and, if $f \in H^2(\mathbb{R}^3/\mathbb{Z}^3)$ with $\int_{[-\pi,\pi]^3} f(x)\,dx = 0$, then

$$\|f\|_{L^\infty(\mathbb{R}^3/\mathbb{Z}^3)} \leq C\|\vec{\nabla} f\|_{L^2(\mathbb{R}^3/\mathbb{Z}^3)}^{1/2} \|\Delta f\|_{L^2(\mathbb{R}^3/\mathbb{Z}^3)}^{1/2}.$$

- if $f \in L^2((0,+\infty), L^2(\mathbb{R}^3/\mathbb{Z}^3))$, then $F = \int_0^t e^{\nu(t-s)\Delta}\vec{\nabla} f\,ds$ belongs to $\mathcal{C}_b([0,+\infty), L^2(\mathbb{R}^3/\mathbb{Z}^3)) \cap L^2((0,+\infty), H^1(\mathbb{R}^3/\mathbb{Z}^3))$ and

$$\|F\|_{L^2((0,+\infty), L^2(\mathbb{R}^3/\mathbb{Z}^3))} \leq C\frac{1}{\sqrt{\nu}}\|f\|_{L^2((0,+\infty), L^2(\mathbb{R}^3/\mathbb{Z}^3))}$$

$$\|\vec{\nabla} \otimes F\|_{L^2((0,+\infty), L^2(\mathbb{R}^3/\mathbb{Z}^3))} \leq C\frac{1}{\nu}\|f\|_{L^2((0,+\infty), L^2(\mathbb{R}^3/\mathbb{Z}^3))}$$

We thus easily control the bilinear operator

$$B(\vec{u}, \vec{v}) = \int_0^t e^{(t-s)\Delta} \mathbb{P} \operatorname{div}(\vec{u} \otimes \vec{w}) \, ds.$$

We get:

- control in $\dot{H}^1$ norm:

$$\|B(\vec{u}, \vec{v})\|_{L^\infty((0,T),\dot{H}^1)} + \sqrt{\nu}\|B(\vec{u}, \vec{v})\|_{L^2((0,T),\dot{H}^2)}$$

$$\leq C \frac{1}{\sqrt{\nu}} \|\operatorname{div}(\vec{u} \otimes \vec{v})\|_{L^2((0,T),L^2)}$$

$$\leq C \frac{1}{\sqrt{\nu}} \|\vec{u}\|_{L^4((0,T),L^\infty)} \|\vec{v}\|_{L^4((0,T),\dot{H}^1)}$$

$$\leq C \frac{T^{1/4}}{\nu^{3/4}} \sqrt{\|\vec{u}\|_{L^\infty((0,T),\dot{H}^1)} \sqrt{\nu}\|\vec{u}\|_{L^2((0,T),\dot{H}^2)}} \|\vec{v}\|_{L^\infty((0,T),\dot{H}^1)}.$$

This control gives us existence of a solution of the Navier–Stokes equations in $\mathcal{C}_b([0,T), \dot{H}^1(\mathbb{R}^3/\mathbb{Z}^3)) \cap L^2((0,T), \dot{H}^2(\mathbb{R}^3/\mathbb{Z}^3))$ with $T = O(\frac{\nu^3}{\|\vec{u}_0\|_{\dot{H}^1}^4})$.

- control in L^2 norm:

$$\|B(\varphi_\epsilon * \vec{u}, \vec{v})\|_{L^\infty((0,T),L^2)} + \sqrt{\nu}\|B(\varphi_\epsilon * \vec{u}, \vec{v})\|_{L^2((0,T),\dot{H}^1)}$$

$$\leq C \frac{1}{\sqrt{\nu}} \|\vec{u} \otimes \vec{v}\|_{L^2((0,T),L^2)}$$

$$\leq C \frac{1}{\sqrt{\nu}} T^{1/2} \|\varphi_\epsilon * \vec{u}\|_{L^\infty((0,T),L^\infty)} \|\vec{v}\|_{L^\infty((0,T),L^2)}$$

$$\leq C' \frac{T^{1/2}}{\nu^{1/2}\epsilon^{3/2}} \|\vec{u}\|_{L^\infty((0,T),L^2)} \|\vec{v}\|_{L^\infty((0,T),L^2)}.$$

This control gives us existence of a solution $\vec{u}_\epsilon$ of the mollified Navier–Stokes equations in $\mathcal{C}_b([0,T_\epsilon), L^2(\mathbb{R}^3/\mathbb{Z}^3)) \cap L^2((0,T_\epsilon), \dot{H}^1(\mathbb{R}^3/\mathbb{Z}^3))$ with $T_\epsilon = O(\frac{\nu\epsilon^3}{\|\vec{u}_0\|_2^2})$.

- Then, we use the energy equality

$$\int_{[-\pi,\pi]^3} |\vec{u}_\epsilon(t,x)|^2 \, dx + 2\nu \int_0^t \int_{[-\pi,\pi]^3} |\vec{\nabla} \otimes \vec{u}_\epsilon(s,x)|^2 \, dx \, ds = \int_{[-\pi,\pi]^3} |\vec{u}_0(x)|^2 \, dx$$

to extend $\vec{u}_\epsilon$ as a global solution on $(0, +\infty) \times \mathbb{R}^3$. Then, applying the Rellich–Lions theorem, we get a weak solution $\vec{u}$ on $(0, +\infty) \times \mathbb{R}^3$ that satisfies Leray's energy inequality

$$\int_{[-\pi,\pi]^3} |\vec{u}(t,x)|^2 \, dx + 2\nu \int_0^t \int_{[-\pi,\pi]^3} |\vec{\nabla} \otimes \vec{u}(s,x)|^2 \, dx \, ds \leq \int_{[-\pi,\pi]^3} |\vec{u}_0(x)|^2 \, dx$$

Finally, we check easily that we have weak-strong uniqueness: if $\vec{u}_0 \in H^1$, and if a mild solution $\vec{u}$ is defined on $(0, T)$, then every weak Leray solution $\vec{v}$ coincides with $\vec{u}$ on $(0, T)$.

However, Buckmaster and Vicol [71] proved non-uniqueness in $\mathcal{C}([0,T], L^2)$:

Buckmaster and Vicol's theorem

Theorem 12.12.
There exists $\beta > 0$, such that for any nonnegative smooth function $e(t) : [0,T] \mapsto \mathbb{R}^+$, there exists $\vec{v} \in \mathcal{C}([0,T], H_x^\beta(\mathbb{R}^3/\mathbb{Z}^3))$ a very weak solution $\vec{v}$ of the Navier-Stokes equations, such that

$$\int_{[-\pi,\pi]^3} |\vec{v}(t,x)|^2 \, dx = e(t) \text{ for all } t \in [0,T].$$

To quote Buckmaster and Vicol,

> *In particular, the above theorem shows that $\vec{v} = 0$ is not the only weak solution which vanishes at a time slice, thereby implying the nonuniqueness of weak solutions.*

Buckmaster and Vicol's proof relies on convex integration tools, a technique developed by De Lellis and Székelyhidi Jr. [148, 149] for the study of Euler equations and the Onsager conjecture which was eventually fully proved by P. Isett in 2018 [241] (see [70, 72] for a survey).

Non-uniqueness results for Leray solutions.

Buckmaster and Vicol's solutions are less regular than Leray solutions. (Their solutions fulfill estimates in some Sobolev spaces $H^\beta(\mathbb{R}^3)$ with $\beta << 1$, but not in H^1, whereas a Leray solution should be controlled in $L^2 H^1$). Albritton, Brué and Colombo [5] published in 2022 the following result:

Albritton, Brué and Colombo's theorem

Theorem 12.13.
There exist $T > 0$, $\vec{f} \in L^1((0,T), L^2(\mathbb{R}^3))$, and two distinct suitable Leray solutions $\vec{u}_1$, $\vec{u}_2$ to the Navier–Stokes equations on $(0,T) \times \mathbb{R}^3$ with body force $\vec{f}$ and initial condition $\vec{u}_0 = 0$.

Non-uniqueness is to be seen at time $t = 0$: we have an asymptotic estimats that $\|\vec{u}_1(t,.) - \vec{u}_2(t,.)\| = \Omega(t^{a+\frac{1}{4}})$ for some positive a as $t \to 0$. Of course, $\vec{f}$ is not regular near $t = 0$: if we had $\vec{f} \in L^1((0,T), H^{1/2}(\mathbb{R}^3))$, then we would have, for a small time $T_0 > 0$, a mild solution $\vec{v}$ in $\mathcal{C}([0,T_0], H^{1/2})$ and, by weak-strong uniqueness, $\vec{u}_1 = \vec{v} = \vec{u}_2$; as a matter of fact, the force involved in the proof of Theorem 12.13 is such that $\|f(t,.)\|_{H^{1/2}} \sim Ct^{-1}$.

Albritton, Brué and Colombo's strategy of proof follows the idea developed by Guillod, Jia and Šverák [246, 223] to derive non-uniqueness from linear unstability of a self-similar profile for the underlying linearized problem. More precisely, if $\vec{U}(x)$ is a divergence-free vector field on $\mathbb{R}^3$, define $\vec{u}(t,x) = \frac{1}{\sqrt{t}}\vec{U}(\frac{x}{\sqrt{t}})$ and $\vec{f} = \partial_t \vec{u} - \Delta \vec{u} + \mathbb{P}(\vec{u} \cdot \vec{\nabla}\vec{u})$ (so that $\vec{f}(t,x) = \frac{1}{t^{3/2}}\vec{f}(1, \frac{x}{\sqrt{t}})$) [or equivalently define $\vec{f}(t,x) = \frac{1}{t^{3/2}}\vec{F}(\frac{x}{\sqrt{t}})$, where

$$\vec{F} = -\frac{1}{2}(\vec{U} + x \cdot \vec{\nabla}\vec{U}) - \Delta \vec{U} + \mathbb{P}(\vec{U} \cdot \vec{\nabla}\vec{U})].$$

If $\vec{U}$ is enough smooth and decaying (for instance, $\vec{U} \in H^2$ and $x^2\vec{U} \in L^2$), then $\vec{f}$ is in $L^1(]0,T[, L^2)$ for $T < +\infty$, and $\vec{u}$ is a Leray solution of the Cauchy problem with body force $\vec{f}$ and initial value $\vec{u}_0 = 0$. The idea for finding another solution $\vec{v}$ of the same Cauchy problem is to write the problem in the variables $\xi = \frac{x}{\sqrt{t}}$ and $\tau = \ln t$; writing $\vec{v}(t,x) = \frac{1}{\sqrt{t}}(\vec{U}(\xi) + \vec{V}(\tau,\xi))$, we find

$$\partial_\tau \vec{V} = \frac{1}{2}(\vec{V} + x \cdot \vec{\nabla}\vec{V}) + \Delta\vec{V} - \mathbb{P}(\vec{U} \cdot \vec{\nabla}\vec{V} + \vec{V} \cdot \vec{\nabla}\vec{U} + \vec{V} \cdot \vec{\nabla}\vec{V}) \qquad (12.42)$$

whose linearization gives

$$\partial_\tau \vec{W} = \frac{1}{2}(\vec{W} + x \cdot \vec{\nabla}\vec{W}) + \Delta\vec{W} - \mathbb{P}(\vec{U} \cdot \vec{\nabla}\vec{W} + \vec{W} \cdot \vec{\nabla}\vec{U}). \qquad (12.43)$$

If $\mathcal{L}(W) = \frac{1}{2}(\vec{W} + x \cdot \vec{\nabla}\vec{W}) + \Delta\vec{W} - \mathbb{P}(\vec{U} \cdot \vec{\nabla}\vec{W} + \vec{W} \cdot \vec{\nabla}\vec{U})$ and if $\mathcal{L}$ is linearly unstable, writing $\lambda = a + ib$ an eigenvalue of $\mathcal{L}$ with maximal real part $a > 0$, the authors pick up an eigenvector η of $\mathcal{L}$ such that $\mathcal{L}\eta = \lambda\eta$. Then $\vec{W} = \Re(e^{\lambda\tau}\eta)$ is a solution of (12.43) and we can find a solution $\vec{V}$ of (12.42) such that $\|\vec{V}(\tau,.) - \vec{W}(\tau,.)\|_2 \leq Ce^{2a\tau}$ as $\tau \to -\infty$ while $\|\vec{W}(\tau,.)\|_2 = \Omega(e^{a\tau})$. Thus, $\vec{V} \neq 0$, and we shall have two solutions of the same Cauchy problem.

The construction of $\vec{U}$ such that $\mathcal{L}$ is unstable is not easy [5]. It is derived from an unstable steady solution of the 2D Euler equations studied in two very recent papers of Višik [490, 491].

12.8   The Inviscid Limit

The Navier–Stokes equations for a viscous incompressible homogeneous Newtonian fluids (with no forcing term) read as

$$\begin{cases} \partial_t \vec{u} + \vec{u} \cdot \vec{\nabla}\vec{u} = \nu\Delta\vec{u} - \vec{\nabla}p, \\ \qquad\qquad\qquad \operatorname{div} \vec{u} = 0. \end{cases}$$

In the case of an inviscid fluid ($\nu = 0$), we have the Euler equations:

$$\begin{cases} \partial_t \vec{u} + \vec{u} \cdot \vec{\nabla}\vec{u} = -\vec{\nabla}p, \\ \qquad\qquad\quad \operatorname{div} \vec{u} = 0. \end{cases}$$

The inviscid limit problem is the study of the convergence of solutions $\vec{u}_\nu$ to the Cauchy problem for the Navier–Stokes equations with initial value $\vec{u}_0$ and viscosity ν when ν goes to 0; in particular do we have convergence to a solution of the Cauchy problem for the Euler equations with initial value $\vec{u}_0$?

This is a difficult problem when considering the problem in a domain Ω with a boundary $\partial\Omega$, as the natural boundary conditions for the Navier–Stokes equations and for the Euler equations are different: usually, the boundary condition for the Navier–Stokes equations is the no-slip conditions $\vec{u}_\nu = 0$ on $\partial\Omega$, whereas the boundary condition for the Euler equations is the no-flux condition $\vec{u} \cdot \vec{n} = 0$ on $\partial\Omega$ (where $\vec{n}$ is the normal vector to $\partial\Omega$). The curvature of the boundary may play a role in the convergence, as proved by Beirão da Veiga and Crispo in 2012 [33].

As we consider the problem on the whole space, we may ignore this discrepancy between the boundary conditions, and the solution is easy in case of regular flows. Page 72, we have already presented the results of Swann [458] on existence of a solution $\vec{u}_\nu$ on a time interval $(0, T)$ with T independent of ν when $\vec{u}_0$ is regular enough. In that case, the inviscid limit is easy to prove [458, 254]:

Theorem 12.14.
Let $\vec{u}_0 \in H^s(\mathbb{R}^3)$ with $s > 5/2$ and div $\vec{u}_0 = 0$ and $\mathbb{F} \in L^2((0, +\infty), H^{s+1})$. Then there exists $T > 0$ such that, for every $\nu > 0$, the Cauchy problem

$$\begin{cases} \partial_t \vec{u}_\nu + \vec{u}_\nu \cdot \vec{\nabla} \vec{u}_\nu = \nu \Delta \vec{u}_\nu - \vec{\nabla} p_\nu + \operatorname{div} \mathbb{F}, \\ \operatorname{div} \vec{u}_\nu = 0, \\ \vec{u}_\nu(0, .) = \vec{u}_0 \end{cases}$$

has a unique solution $\vec{u}_\nu$ in $\mathcal{C}([0, T], H^s) \cap L^2((0, T), H^{s+1})$.
Moreover, the Euler equations

$$\begin{cases} \partial_t \vec{u} + \vec{u} \cdot \vec{\nabla} \vec{u} = \vec{\nabla} p + \operatorname{div} \mathbb{F}, \\ \operatorname{div} \vec{u} = 0, \\ \vec{u}(0, .) = \vec{u}_0 \end{cases}$$

have a unique solution $\vec{u} \in \mathcal{C}([0, T], L^2) \cap L^\infty((0, T), H^s)$ and we have strong convergence of $\vec{u}_\nu$ to $\vec{u}$ in $\mathcal{C}([0, T], H^\sigma)$ for every $\sigma < s$.

Proof. Recall that we proved in Theorem 7.3 existence of a unique $\vec{u}_\nu$ on a time interval $(0, T_\nu)$. More precisely, we have the inequalities

$$\|W_{\nu t} * \vec{u}_0\|_{L^\infty H^s} \le \|\vec{u}_0\|_{H^s}, \quad \|\vec{\nabla} \otimes (W_{\nu t} * \vec{u}_0)\|_{L^2 H^s} \le \frac{1}{\sqrt{2\nu}} \|\vec{u}_0\|_{H^s}$$

$$\left\| \int_0^t W_{\nu(t-s)} * \mathbb{P} \operatorname{div} \mathbb{F} \, ds \right\|_{L^\infty H^s} \le \frac{1}{\sqrt{2\nu}} \|\mathbb{F}\|_{L^2 H^s},$$

$$\left\| \vec{\nabla} \otimes \left(\int_0^t W_{\nu(t-s)} * \mathbb{P} \operatorname{div} \mathbb{F} \, ds \right) \right\|_{L^2 H^{s+1}} \le \frac{1}{\nu} \|\mathbb{F}\|_{L^2 H^s},$$

and (since H^s is an algebra)

$$\|\vec{u} \otimes \vec{v}\|_{L^2((0, T_0), H^s)} \le C_0 T_0^{1/2} \|\vec{u}\|_{L^\infty((0, T_0), H^s)} \|\vec{v}\|_{L^\infty((0, T_0), H^s)}.$$

Thus, we find a solution $\vec{u}_\nu$ in $\mathcal{C}([0, T_\nu], H^s) \cap L^2((0, T_\nu), H^{s+1})$ with

$$T_\nu = \frac{2\nu}{\left(4C_0 (\|\vec{u}_0\|_{H^s} + \frac{1}{\sqrt{2\nu}} \|\mathbb{F}\|_{L^2 H^s}) \right)^2}.$$

From local-in-time existence and uniqueness of solutions, we find that we have a solution $\vec{u}_\nu$ on a maximal time interval $(0, T_\nu^*)$ (which belongs to $\mathcal{C}([0, T], H^s) \cap L^2((0, T), H^{s+1})$ for every $0 < T < T_\nu^*$). To prove that $T_\nu^* > T_0 > 0$, where T_0 does not depend on ν, we must prove that $\vec{u}$ remains bounded in H^s on $(0, \min(T_\nu^*, T_0))$. More precisely, we shall prove that $\|\vec{u}_\nu\|_{H^s} \le C_1$ on $(0, T_0)$, where neither T_0 nor C_1 depend on ν (but depend on $\vec{u}_0$ and $\mathbb{F}$).

We write the energy balance in H^s norm:

$$\frac{d}{dt}(\|\vec{u}_\nu\|_2^2 + \|(-\Delta)^{s/2}\vec{u}_\nu\|_2^2) = 2\int \partial_t\vec{u}_\nu \cdot (\vec{u}_\nu + (-\Delta)^s\vec{u}_\nu)\,dx$$

$$= -2\nu\int |\vec{\nabla}\otimes\vec{u}_\nu|^2 + |\vec{\nabla}\otimes(-\Delta)^{s/2}\vec{u}_\nu|^2\,dx$$

$$+ 2\int \vec{u}_\nu\cdot\operatorname{div}\mathbb{F}\,dx + 2\int (-\Delta)^{s/2}\vec{u}_\nu\cdot(-\Delta)^{s/2}\operatorname{div}\mathbb{F}\,dx$$

$$- 2\int \vec{u}_\nu\cdot(\vec{u}_\nu\cdot\vec{\nabla}\vec{u}_\nu)\,dx - 2\int (-\Delta)^{s/2}\vec{u}_\nu\cdot(\vec{u}_\nu\cdot\vec{\nabla}(-\Delta)^{s/2}\vec{u}_\nu)\,dx$$

$$+2\int (-\Delta)^{s/2}\vec{u}_\nu\cdot(\vec{u}_\nu\cdot\vec{\nabla}(-\Delta)^{s/2}\vec{u}_\nu)\,dx - 2\int (-\Delta)^{s/2}\vec{u}_\nu\cdot(-\Delta)^{s/2}(\vec{u}_\nu\cdot\vec{\nabla}\vec{u}_\nu)\,dx$$

As $\operatorname{div}\vec{u}_\nu = 0$, we have

$$\int \vec{u}_\nu\cdot(\vec{u}_\nu\cdot\vec{\nabla}\vec{u}_\nu)\,dx -= \int (-\Delta)^{s/2}\vec{u}_\nu\cdot(\vec{u}_\nu\cdot\vec{\nabla}(-\Delta)^{s/2}\vec{u}_\nu)\,dx = 0.$$

Moreover, we have

$$\|(-\Delta)^{\frac{s}{2}}(u\partial_k v) - u\partial_k(-\Delta)^{\frac{s}{2}}v\|_2 = (2\pi)^{-\frac{3}{2}}\|\int \eta_k\hat{u}(\xi-\eta)\hat{v}(\eta)(|\xi|^s - |\eta|^s)\,d\eta\|_2$$

$$\leq C\|\int |\eta_k||\hat{u}(\xi-\eta)||\hat{v}(\eta)||\xi-\eta|(|\xi-\eta|^{s-1} + |\eta|^{s-1})\,d\eta\|_2$$

$$\leq C(\||\xi|\hat{u}\|_1\||\xi|^s\hat{v}\|_2 + \||\xi|^s\hat{u}\|_2\||\xi|\hat{v}\|_1)$$

$$\leq C'\|u\|_{H^s}\|v\|_{H^s}.$$

Thus, we find that

$$\frac{d}{dt}(\|\vec{u}_\nu\|_2^2 + \|(-\Delta)^{s/2}\vec{u}_\nu\|_2^2)$$

$$\leq \|\vec{u}_\nu\|_2^2 + \|(-\Delta)^{s/2}\vec{u}_\nu\|_2^2 + \|\mathbb{F}\|_{H^{s+1}}^2 + C_0(\|\vec{u}_\nu\|_2^2 + \|(-\Delta)^{s/2}\vec{u}_\nu\|_2^2)^{3/2}$$

and we may conclude that

$$\|\vec{u}_\nu\|_2^2 + \|(-\Delta)^{s/2}\vec{u}_\nu\|_2^2 \leq 4(\|\vec{u}_0\|_2^2 + \|(-\Delta)^{s/2}\vec{u}_0\|_2^2 + 1)$$

on $(0, T_0)$, as long as T_0 is small enough to grant that

- $\int_0^{T_0}\|\mathbb{F}\|_{H^{s+1}}^2\,dt \leq 1$
- $T_0 \leq \frac{1}{4}$
- $T_0(\|\vec{u}_0\|_2^2 + \|(-\Delta)^{s/2}\vec{u}_0\|_2^2 + 1)^{1/2} \leq 1$.

Further, we may estimate $\vec{u}_\mu - \vec{u}_\nu$ for small μ and ν:

$$\frac{d}{dt}\|\vec{u}_\nu - \vec{u}_\mu\|_2^2 = 2\int (\vec{u}_\nu - \vec{u}_\mu)\cdot(\partial_t\vec{u}_\nu - \partial_t\vec{u}_\mu)\,dx$$

$$= 2\int (\vec{u}_\nu - \vec{u}_\mu)\cdot(\nu\Delta\vec{u}_\nu - \mu\Delta\vec{u}_\mu)\,dx$$

$$- 2\int (\vec{u}_\nu - \vec{u}_\mu)\cdot(\vec{u}_\nu\cdot\vec{\nabla}(\vec{u}_\nu - \vec{u}_\mu))\,dx$$

$$- 2\int (\vec{u}_\nu - \vec{u}_\mu)\cdot((\vec{u}_\nu - \vec{u}_\mu)\cdot\vec{\nabla}\vec{u}_\mu)\,dx$$

$$\leq 2\|\vec{u}_\nu - \vec{u}_\mu\|_2(\mu\|\Delta\vec{u}_\mu\|_2 + \nu\|\Delta\vec{u}_\nu\|_2)$$

$$+ 2\|\vec{\nabla}\otimes\vec{u}_\mu\|_\infty\|\vec{u}_\mu - \vec{u}_\nu\|_2^2$$

so that, on $(0, T_0)$, writing $M_0 = 2(\|\vec{u}_0\|_2^2 + \|(-\Delta)^{s/2}\vec{u}_0\|_2^2 + 1)^{1/2}$, we get

$$\frac{d}{dt}\|\vec{u}_\nu - \vec{u}_\mu\|_2^2 \leq 4M_0^2(\mu + \nu) + C_0 M_0 \|\vec{u}_\nu - \vec{u}_\mu\|_2^2$$

and

$$\|\vec{u}_\nu - \vec{u}_\mu\|_2^2 \leq T_0 4 M_0^2(\mu + \nu)e^{C_0 T_0}.$$

Thus, $\vec{u}_\mu$ is strongly convergent in $\mathcal{C}([0, T_0], L^2)$ as μ goes to 0, to a limit $\vec{u}$. As $\vec{u}_\mu$ is bounded in $L^\infty((0, T_0), H^s)$, it is strongly convergent in $\mathcal{C}([0, T_0]H^\sigma)$ for every $\sigma < s$. In particular, $\mu\Delta\vec{u}_\mu - \mathbb{P}\vec{u}_\mu$ converges in $\mathcal{C}([0, T_0], L^2)$ to $-\mathbb{P}\vec{u}$, and $\vec{u}$ is a solution to the Euler equations. $\quad\square$

Chapter 13

Partial Regularity Results for Weak Solutions

13.1 Interior Regularity

In this chapter, we shall work on the local behavior of the solution $\vec{u}$ of the Navier–Stokes equations

$$\partial_t \vec{u} = \nu \Delta \vec{u} - \vec{u} \cdot \vec{\nabla} \vec{u} + \vec{f} - \vec{\nabla} p, \quad \operatorname{div} \vec{u} = 0 \tag{13.1}$$

We assume that $\vec{f}$ is given on an open set $Q = I \times \Omega$, where $I = (a,b)$ is an interval of $\mathbb{R}$ and $\Omega = B(x_0, r)$ an open ball of $\mathbb{R}^3$.

The solution $\vec{u}$ is defined on Q and belongs to $L^\infty(I, L^2(\Omega)) \cap L^2(I, H^1(\Omega))$ and the equation (13.1) is fulfilled in a weak sense: for every $\vec{\varphi} \in \mathcal{D}(Q)$ with $\operatorname{div} \vec{\varphi} = 0$, we have

$$\langle \partial_t \vec{u} - \nu \Delta \vec{u} + \vec{u} \cdot \vec{\nabla} \vec{u} - \vec{f} | \vec{\varphi} \rangle_{\mathcal{D}', \mathcal{D}} = 0 \tag{13.2}$$

If $\vec{u}$ is a solution of (13.2), then it is easy to prove that there exists a distribution $p \in \mathcal{D}'(Q)$ such that (13.1) is fulfilled in $\mathcal{D}'$. Serrin [434] studied the local regularity of such solutions. His theory is based on the following theorem:

Local regularity theory

Theorem 13.1.
Let $Q = I \times \Omega$, where $I = (a,b)$ and $\Omega = B(x_0, r)$. Let $\vec{u} \in L^\infty(I, L^2(\Omega)) \cap L^2(I, H^1(\Omega))$, $\vec{f} \in L^2(I, L^2(\Omega))$ and $p \in \mathcal{D}'(Q)$, and assume that $\vec{u}$ is a weak solution on Q of the Navier–Stokes equations

$$\partial_t \vec{u} = \nu \Delta \vec{u} - \vec{u} \cdot \vec{\nabla} \vec{u} + \vec{f} - \vec{\nabla} p, \quad \operatorname{div} \vec{u} = 0.$$

Then, if moreover $\vec{u} \in L^\infty(Q)$, we have, for every $a < c < b$ and $0 < \rho < r$:

- *$\vec{u} \in L^\infty((c,b), H^1(B(x_0, \rho))) \cap L^2((c,b), H^2(B(x_0, \rho)))$*

- *if $k \in \mathbb{N}$ and $\vec{f} \in L^2(I, H^k(\Omega))$, then $\vec{u} \in L^\infty((c,b), H^{k+1}(B(x_0, \rho))) \cap L^2((c,b), H^{k+2}(B(x_0, \rho)))$*

In particular, if $\vec{f} \in \mathcal{C}^\infty(Q)$, then $\vec{u}$ is smooth on Q with respect to the space variable x.

Definition 13.1 (Serrin's regularity).
A solution $\vec{u}$ of the Navier–Stokes equations on $Q = I \times \Omega$ is regular in the sense of Serrin if $\vec{u}$ belongs to $L^\infty(I, L^2(\Omega)) \cap L^2(I, H^1(\Omega))$ and if moreover $\vec{u} \in L^\infty(Q)$.

Proof. **Step 1: Equation on the vorticity.**

DOI: 10.1201/9781003042594-13

To get rid of the unknown pressure p, we take the curl of the Equation (13.1) and get the following equation on $\vec{\omega} = \operatorname{curl} \vec{u}$:

$$\partial_t \vec{\omega} = \nu \Delta \vec{\omega} + \operatorname{curl} \vec{f} - \vec{\nabla} \wedge (\vec{u} \cdot \vec{\nabla} \vec{u})$$

As $\vec{u} \in L^2 H^1$ and $\operatorname{div} \vec{u} = 0$, we may develop

$$\vec{\nabla} \wedge (\vec{u} \cdot \vec{\nabla} \vec{u}) = \vec{\nabla} \wedge (\frac{1}{2} \vec{\nabla} |\vec{u}|^2 + \vec{\omega} \wedge \vec{u}) = \vec{u} \cdot \vec{\nabla} \vec{\omega} - \vec{\omega} \cdot \vec{\nabla} \vec{u} = \operatorname{div}(\vec{u} \otimes \vec{\omega} - \vec{\omega} \otimes \vec{u})$$

and we obtain

$$\partial_t \vec{\omega} = \nu \Delta \vec{\omega} + \operatorname{curl} \vec{f} - \operatorname{div}(\vec{u} \otimes \vec{\omega} - \vec{\omega} \otimes \vec{u})$$

Thus, $\vec{\omega}$ is solution of a linear heat equation

$$\partial_t \vec{\omega} = \nu \Delta \vec{\omega} + \vec{g} \tag{13.3}$$

with $\vec{g} \in L^2(I, H^{-1}(\Omega))$ and $\vec{\omega} \in L^2(Q)$.

Step 2: The heat equation.

We are going to prove that if $\omega \in L^2(Q)$ is solution of

$$\partial_t \omega = \nu \Delta \omega + g$$

with $g \in L^2(I, H^{k-1}(\Omega))$ for some $k \in \mathbb{N}$, then, for $a < c < b$ and $0 < \rho < r$, we have $\omega \in L^\infty((c, b), H^k(\Omega)) \cap L^2((c, b), H^{k+1}(\Omega))$.

As $\partial_t(\partial_j \omega) = \nu \Delta(\partial_j \omega) + \partial_j g$, this is done by induction on k, and we have just to consider the case $k = 0$.

We consider now a function $\phi \in \mathcal{D}(\mathbb{R} \times \mathbb{R}^3)$ which is equal to 1 on $[c, b] \times B(x_0, \rho)$ and is supported in $[\frac{a+c}{2}, b+1] \times B(x_0, \frac{r+\rho}{2})$. We define $\varpi = \phi \omega$. We have:

- $\varpi \in L^2((a, b) \times \mathbb{R}^3) \cap \mathcal{C}([a, b], H^{-2}(\mathbb{R}^3))$
- $\varpi(a.) = 0$
- $\partial_t \varpi = \nu \Delta \varpi + h$ with $h = \phi g - \nu \omega \Delta \phi - 2\nu \vec{\nabla} \omega \cdot \vec{\nabla} \phi + \omega \partial_t \phi \in L^2((a, b), H^{-1}(\mathbb{R}^3))$

Writing

$$\varpi = \int_a^t W_{\nu(t-s)} * h(s, .) \, ds$$

we see that $\varpi \in L^\infty((0, L^2)) \cap L^2 H^1$. Thus, ω is locally regular.

Step 3: regularity of $\vec{u}$.

We write

$$\vec{\nabla} \wedge \vec{\omega} = -\Delta \vec{u} + \vec{\nabla}(\operatorname{div} \vec{u}) = -\Delta \vec{u}.$$

If $\vec{f} \in L^2(I, H^k)$ and $\vec{u} \in L^\infty L^2 \cap L^2 H^1 \cap L^\infty_{t,x}$ on Q, then we shall see that for every $a' \in (a, b)$ and $r' \in (0, r)$, we have $\vec{\omega} \in L^\infty((a', b), H^k(B(x_0, r')) \cap L^2((a', b), H^{k+1}(B(x_0, r'))$. Thus $\Delta \vec{u} \in L^\infty((a', b), H^{k-1}(B(x_0, r')) \cap L^2((a', b), H^k (B(x_0, r'))$. We then pick up a function $\phi \in \mathcal{D}(\mathbb{R}^3)$ such that $\varphi = 1$ for $|x - x_0| < c$ and $= 0$ for $|x - x_0| > \frac{a'+c}{2}$. We have

$$\Delta(\varphi \vec{u}) = \varphi \Delta \vec{u} + (\Delta \varphi) \vec{u} + 2 \sum_{i=1}^{3} \partial_i \varphi \, \partial_i \vec{u}.$$

We know that $\vec{u} \in L^\infty L^2 \cap L^2 H^1$; if $\vec{u}$ is $L^\infty H^l \cap L^2 H^{l+1}$ for some $0 \leq l \leq k$ on $(a', b) \times B(x_0, r')$, then we find that $\Delta(\varphi \vec{u})$ is $L^\infty H^{l-1} \cap L^2 H^l$ and $\vec{u}$ is $L^\infty H^{l+1} \cap L^2 H^{l+2}$. Thus, $\vec{u}$ will be $L^\infty H^{k+1} \cap L^2 H^{k+2}$.

It remains to show that

$$\vec{\omega} \in L^\infty((a', b), H^k(B(x_0, r'))) \cap L^2((a', b), H^{k+1}(B(x_0, r'))).$$

We start from the equation

$$\partial_t \vec{\omega} = \nu \Delta \vec{\omega} + \vec{g}$$

with

$$\vec{g} = \operatorname{curl} \vec{f} - \operatorname{div}(\vec{u} \otimes \vec{\omega} - \vec{\omega} \otimes \vec{u}).$$

We know that $\vec{\omega} \in L^2 L^2$ and $\vec{u} \in L^\infty$, so that $\vec{g} \in L^2 H^{-1}$ and that, locally, $\vec{\omega} \in L^\infty L^2 \cap L^2 H^1$ and $\vec{u} \in L^\infty H^1 \cap L^2 H^2$. By induction, we assume that $\vec{u} \in L^\infty H^l \cap L^2 H^{l+1}$, for some $0 \leq l \leq k$. As

$$\vec{g} = \operatorname{curl} \vec{f} - \vec{\nabla} \wedge \operatorname{div}(\vec{u} \otimes \vec{u})$$

and that, for ψ supported in $(a', b] \times B(x_0, r')$,

$$\|\psi^2 \vec{u} \otimes \vec{u}\|_{L^2 H^{l+1}} \leq C \|\psi \vec{u}\|_{L^\infty(Q)} \|\psi \vec{u}\|_{L^2 H^{l+1}},$$

we can see that $\vec{g}$ is locally $L^2 H^{l-1}$, so that $\vec{\omega}$ is locally $L^\infty H^l \cap L^2 H^{l+1}$ and $\vec{u}$ is locally $L^\infty H^{l+1} \cap L^2 H^{l+2}$.

The theorem is thus proved.

□

It is important to notice that Theorem 13.1 does not convey any information on the time regularity of $\vec{u}$, because of the presence of the unknown pressure p: the control of p is equivalent to the control of $\partial_t \vec{u}$. Serrin gave the following example: if ψ is a harmonic function on $\mathbb{R}^3$ and α a bounded function on $\mathbb{R}$, define $\vec{u}$ on $(0, 1) \times B(0, 1)$ as $\vec{u} = \alpha(t) \vec{\nabla} \psi$. Then we have $\operatorname{div} \vec{u} = \alpha(t) \Delta \psi = 0$, $\operatorname{curl} \vec{u} = 0$, $\Delta \vec{u} = 0$ and

$$\vec{u} \cdot \vec{\nabla} \vec{u} = \vec{\nabla}(\frac{|\vec{u}|^2}{2}) + \operatorname{curl} \vec{u} \wedge \vec{u} = \vec{\nabla}(\frac{|\vec{u}|^2}{2}).$$

We get

$$\partial_t \vec{u} = \nu \Delta \vec{u} - \vec{u} \cdot \vec{\nabla} \vec{u} + \vec{f} - \vec{\nabla} p, \quad \operatorname{div} \vec{u} = 0$$

with $\vec{f} = 0$ and $p = -\frac{|\vec{u}|^2}{2} - \partial_t \alpha \, \psi$. Moreover, $\vec{u} \in L^\infty L^2 \cap L^2 H^1 \cap L^\infty_{t,x}$ on $(0, 1) \times B(0, 1)$; but, if α is not regular, $\vec{u}$ has no regularity with respect to time.

13.2   Serrin's Theorem on Interior Regularity

In view of Theorem 13.1, it is important to show that $\vec{u}$ is locally bounded in time and space variables. This may be done locally under the assumption $\vec{f} \in L^2(I, H^1(\Omega))$ and $\vec{u} \in L^p(I, L^q(\Omega))$ with $2/p + 3/q = 1$ and $q > 3$. The case $2/p + 3/q < 1$ was first proved by Serrin [434]; the case $2/p + 3/q = 1$ was then proved by Struwe [455] and Takahashi [459].

In order to state quite a general theorem, we use the space of multipliers introduced in Theorem 12.4:

- X is the space of pointwise multipliers on $\mathbb{R} \times \mathbb{R}^3$ from $L_t^\infty L^2 \cap L_t^2 H^1$ to $L_t^2 L_x^2$, normed with

$$\|u\|_{\mathbb{X}} = \sup_{\|v\|_{L_t^\infty L^2} + \|v\|_{L_t^2 H^1} \leq 1} \|uv\|_{L^2 L^2};$$

- $\mathbb{X}^{(0)}$ is the space of multipliers u in $\mathbb{X}$ such that, for every $t_0 \in \mathbb{R}$,

$$\lim_{\epsilon \to 0^+} \|1_{(t_0-\epsilon, t_0+\epsilon)}(t)u(t,x)\|_{\mathbb{X}} = 0.$$

Lemma 13.1.
If $u \in \mathbb{X}$ and $v \in L^2 L^2$, then $uv \in L_t^1 L^2 + L_t^2 H^{-1}$.

Proof. The dual of $L_t^1 L^2 + L_t^2 H^{-1}$ is $L_t^\infty L^2 \cap L_t^2 H^1$. Thus, for $w \in L_t^1 L^2 + L_t^2 H^{-1}$,

$$\|w\|_{L_t^1 L^2 + L_t^2 H^{-1}} \approx \sup\{|\langle w|z\rangle| \ / \ \|z\|_{L_t^\infty L^2} + \|z\|_{L_t^2 H^1} \leq 1\}$$

If $v \in \mathcal{D}(\mathbb{R} \times \mathbb{R}^3)$, then uv belongs to $L^2 L^2$ and has compact support, hence belong to $L^1 L^2$. Moreover, we have

$$\|uv\|_{L_t^1 L^2 + L_t^2 H^{-1}} \approx \sup\{|\langle v|uz\rangle| \ / \ \|z\|_{L_t^\infty L^2} + \|z\|_{L_t^2 H^1} \leq 1\} \leq \|v\|_{L^2 L^2}\|u\|_{\mathbb{X}}$$

We then conclude by the density of $\mathcal{D}$ in $L^2 L^2$ and the completeness of $L_t^1 L^2 + L_t^2 H^{-1}$. $\square$

Interior regularity

Theorem 13.2 (Serrin's theorem)**.**
Let $Q = I \times \Omega$, where $I = (a, b)$ and $\Omega = B(x_0, r)$. Let $\vec{u} \in L^\infty(I, L^2(\Omega)) \cap L^2(I, H^1(\Omega))$, $\vec{f} \in L^2(I, L^2(\Omega))$ and $p \in \mathcal{D}'(Q)$, and assume that $\vec{u}$ is a weak solution on Q of the Navier–Stokes equations

$$\partial_t \vec{u} = \nu \Delta \vec{u} - \vec{u} \cdot \vec{\nabla} \vec{u} + \vec{f} - \vec{\nabla} p, \quad \operatorname{div} \vec{u} = 0.$$

Then, if moreover $\vec{f} \in L^2(I, H^1(\Omega))$ and $1_Q \vec{u} \in \mathbb{X}^{(0)}$, we have, for every $a < c < b$ and $0 < \rho < r$, $\vec{u} \in L^\infty(((c, b), H^2(B(x_0, \rho)))$, so that $\vec{u}$ is locally bounded in time and space variables.

Proof. **First step: a linear heat equation.**
Let $\vec{\omega} = \operatorname{curl} \vec{u}$. We consider a function $\phi \in \mathcal{D}(\mathbb{R} \times \mathbb{R}^3)$ which is equal to 1 on $[c, b] \times B(x_0, \rho)$ and is supported in $[\frac{a+c}{2}, b+1] \times B(x_0, \frac{r+\rho}{2})$. We define $\vec{w} = \phi \vec{\omega}$. We have:

- $\vec{w} \in L^2((a, b) \times \mathbb{R}^3)$
- $\vec{w}(t, .) = 0$ for $a < t < \frac{b+c}{2}$
- $\partial_t \vec{w} = \nu \Delta \vec{w} - \operatorname{div}(1_Q \vec{u} \otimes \vec{w} - \vec{w} \otimes 1_Q \vec{u}) + \vec{g}$ with

$$\vec{g} = (\vec{\omega} \cdot \vec{\nabla}\phi)\vec{u} - (\vec{u} \cdot \vec{\nabla}\phi)\vec{\omega} - (\Delta\phi)\vec{\omega} - 2\sum_{i=1}^{3}(\partial_i\phi)\,\partial_i\vec{\omega} + \partial_t\phi\vec{\omega} + \phi\operatorname{curl}\vec{f}$$

From Lemma 13.1, we can see that $\vec{g} \in L^2((a, b), H^{-1}(\mathbb{R}^3)) + L^1((a, b), L^2(\mathbb{R}^3))$.

We now consider the solutions $\vec{z} \in L^2((a, b) \times \mathbb{R}^3)$ of the linear heat equation

$$\partial_t \vec{z} = \nu \Delta \vec{z} - \operatorname{div}(1_Q \vec{u} \otimes \vec{z} - \vec{z} \otimes 1_Q \vec{u}) + \vec{g} \tag{13.4}$$

with $\vec{g} \in L^2((a,b), H^{-1}(\mathbb{R}^3)) + L^1((a,b), L^2(\mathbb{R}^3))$. Using Lemma 13.1 again, we can see that $\partial_t \vec{z} \in L^1((a,b), H^{-2}(\mathbb{R}^3))$, so that $\vec{z} \in \mathcal{C}([a,b], H^{-2}(\mathbb{R}^3))$ and $\vec{z}(a,.)$ is well defined. Moreover, we have the following results:

- uniqueness: if $\vec{z}_1$ and $\vec{z}_2$ are two solutions in $L^2((a,b) \times \mathbb{R}^3)$ of Equation (13.4) and if $\vec{z}_1(a,.) = \vec{z}_2(a,.)$, then $\vec{z}_1 = \vec{z}_2$.

- regularity: if $\vec{z}$ is a solution in $L^2((a,b) \times \mathbb{R}^3)$ of Equation (13.4) and if $\vec{z}(a,.) \in L^2(\mathbb{R}^3)$, then $\vec{z} \in \mathcal{C}([a,b], L^2(\mathbb{R}^3)) \cap L^2((a,b), H^1(\mathbb{R}^3))$.

This can be easily checked. First, we see that there exists a constant C_0 such that, for every $t_0 \in \mathbb{R}$ and every $\delta \in (0,1)$, the map

$$\vec{h} \mapsto \int_{t_0}^t W_{\nu(t-s)} * \vec{h} \, ds = L_{t_0}(\vec{h})$$

satisfies:

- L_{t_0} is bounded from $L^2((t_0, t_0+\delta), H^{-2}(\mathbb{R}^3))$ to $L^2((t_0, t_0+\delta), L^2(\mathbb{R}^3))$ and

$$\|L_{t_0}(\vec{h})\|_{L^2 L^2} \leq C_0 \|\vec{h}\|_{L^2 H^{-2}}$$

- L_{t_0} is bounded from $L^1((t_0, t_0+\delta), H^{-1}(\mathbb{R}^3))$ to $L^2((t_0, t_0+\delta), L^2(\mathbb{R}^3))$ and

$$\|L_{t_0}(\vec{h})\|_{L^2 L^2} \leq C_0 \|\vec{h}\|_{L^2 H^{-1}}$$

- L_{t_0} is bounded from $L^2((t_0, t_0+\delta), H^{-1}(\mathbb{R}^3))$ to $\mathcal{C}([t_0, t_0+\delta], L^2(\mathbb{R}^3)) \cap L^2((t_0, t_0+\delta), H^1(\mathbb{R}^3))$ and

$$\|L_{t_0}(\vec{h})\|_{L^\infty L^2} + \|L_{t_0}(\vec{h})\|_{L^2 H^1} \leq C_0 \|\vec{h}\|_{L^2 H^{-1}}$$

- L_{t_0} is bounded from $L^1((t_0, t_0+\delta), L^2(\mathbb{R}^3))$ to $\mathcal{C}([t_0, t_0+\delta], L^2(\mathbb{R}^3)) \cap L^2((t_0, t_0+\delta), H^1(\mathbb{R}^3))$ and

$$\|L_{t_0}(\vec{h})\|_{L^\infty L^2} + \|L_{t_0}(\vec{h})\|_{L^2 H^1} \leq C_0 \|\vec{h}\|_{L^1 L^2}$$

Second, $1_Q \in \mathbb{X}^{(0)}$; hence, by compactness of $[a,b]$, for every $\epsilon > 0$, we may find a $\eta(\epsilon) \in (0,1)$ such that, for every $t_0 \in [a,b]$, $\|1_{[t_0, t_0+\eta(\epsilon)]}(t) 1_Q(t,x) \vec{u}\|_{\mathbb{X}} < \epsilon$.

We now prove our claims on the solutions of (13.4):

- uniqueness: if $\vec{z}_1 = \vec{z}_2$ on $[a, t_0]$ with $t_0 < b$, we write on $[t_0, t_0+\beta]$ with $t_0+\beta \leq b$:

$$\vec{z}_1 - \vec{z}_2 = \nu \Delta(\vec{z}_1 - \vec{z}_2) - \text{div}(1_Q \vec{u} \otimes (\vec{z}_1 - \vec{z}_2) - (\vec{z}_1 - \vec{z}_2) \otimes 1_Q \vec{u}).$$

Thus,

$$\vec{z}_1 - \vec{z}_2 = -L_{t_0}(\text{div}(1_Q \vec{u} \otimes (\vec{z}_1 - \vec{z}_2) - (\vec{z}_1 - \vec{z}_2) \otimes 1_Q \vec{u}))$$

If $\beta < 1$, we get

$$\|\vec{z}_1 - \vec{z}_2\|_{L^2((t_0,t_0+\beta),L^2)} \leq C_0 \|1_{[t_0,t_0+\beta]} 1_Q \vec{u}\|_{\mathbb{X}} \|\vec{z}_1 - \vec{z}_2\|_{L^2((t_0,t_0+\beta),L^2)}$$

Thus, if $\beta = \min(b - t_0, \eta(\frac{1}{2C_0}))$, we get $\vec{z}_1 = \vec{z}_2$ on $[t_0, t_0+\beta]$. Finally, we see that $\vec{z}_1 = \vec{z}_2$, by propagating the equality from $[a, a+k\eta]$ to $[a, a+(k+1)\eta]$ for $k \geq 0$.

- regularity: For $t_0 \in [a, b]$, $\vec{z}_0 \in L^2$ and $\eta = \eta(\frac{1}{4C_0})$ we consider the equation on $(t_0, t_0 + \eta) \times \mathbb{R}^3$

$$\vec{z} = \vec{Z} - L_{t_0}(\text{div}(1_Q \vec{u} \otimes \vec{z} - \vec{z} \otimes 1_Q \vec{u}))$$

with

$$\vec{Z} = W_{\nu(t-t_0)} * \vec{z}_0 + \int_{t_0}^t W_{\nu(t-s)} * 1_{[t_0, t_0 + \eta]}(s)\vec{g}(s, .) \, ds.$$

$\vec{Z}$ belongs to $\mathcal{C}([t_0, t_0 + \eta], L^2(\mathbb{R}^3)) \cap L^2((t_0, t_0 + \eta), H^1(\mathbb{R}^3))$. Moreover, $\vec{z} \mapsto L_{t_0}(\text{div}(1_Q \vec{u} \otimes \vec{z} - \vec{z} \otimes 1_Q \vec{u}))$ is bounded on $\mathcal{C}([t_0, t_0 + \eta], L^2(\mathbb{R}^3)) \cap L^2((t_0, t_0 + \eta), H^1(\mathbb{R}^3))$ and

$$\|L_{t_0}(\text{div}(1_Q \vec{u} \otimes \vec{z} - \vec{z} \otimes 1_Q \vec{u}))\|_{L^\infty L^2 + \cap L^2 H^1} \leq \frac{1}{2}(\|\vec{z}\|_{L^\infty L^2 \cap L^2 H^1})$$

By Banach's contraction principle, we can see that there exists one and only one solution $\vec{z}$ on $[t_0, t_0 + \eta]$.

Now, starting from $t_0 = 0$ and $\vec{z}_0 = \vec{z}(0, .)$, we construct our solution on $[0, \eta]$, then we reiterate the construction for $t_0 = \eta$ and $\vec{z}_0 = \vec{z}(\eta, .)$ and get a solution on $[\eta, 2\eta]$, and so on. Finally, we get a solution of (13.4) on the whole interval $[a, b]$ with $\vec{z} \in \mathcal{C}([a, b], L^2(\mathbb{R}^3)) \cap L^2((a, b), H^1(\mathbb{R}^3))$. By uniqueness of the solutions in $L^2((a, b), L^2)$, we see that a solution $\vec{z}$ of (13.4) that belongs to $L^2((a, b), L^2)$ and satisfies $\vec{z}(a, .) \in L^2$ must belong to $\mathcal{C}([a, b], L^2(\mathbb{R}^3)) \cap L^2((a, b), H^1(\mathbb{R}^3))$.

Second step: regularity estimates on the vorticity $\vec{\omega}$.

From our study of Equation (13.4), we have found that $\phi\vec{\omega}$ belongs to $\mathcal{C}([a, b], L^2(\mathbb{R}^3)) \cap L^2((a, b), H^1(\mathbb{R}^3))$. As $\Delta\vec{u} = -\operatorname{curl}\vec{\omega}$, we find that for every $a < c < b$ and $0 < \rho < r$, $\vec{u}$ belongs to $\mathcal{C}([c, b], H^1(B(x_0, \rho))) \cap L^2((c, b), H^2(B(x_0, \rho)))$.

We write again

$$\partial_t\vec{\omega} = \nu\Delta\vec{\omega} + \vec{g}$$

with

$$\vec{g} = \operatorname{curl}\vec{f} - \operatorname{div}(\vec{u} \otimes \vec{\omega} - \vec{\omega} \otimes \vec{u}).$$

We find that $\vec{g}$ is locally $L^2 H^{-1/2}$: for all $\phi \in \mathcal{D}((a, +\infty) \times B(x_0, r)))$, $1_{(a,b)}\phi\vec{g} \in L^2 H^{-1/2}$. This gives more local regularity on $\vec{\omega}$ and $\vec{u}$: for all $\phi \in \mathcal{D}((a, +\infty) \times B(x_0, r)))$, $1_{(a,b)}\phi\vec{\omega} \in \mathcal{C}([a, b], H^{1/2}(\mathbb{R}^3)) \cap L^2((a, b), H^{3/2})$ and $1_{(a,b)}\phi\vec{u} \in \mathcal{C}([a, b], H^{3/2}(\mathbb{R}^3)) \cap L^2((a, b), H^{5/2})$.

Those estimates on $\vec{\omega}$ and $\vec{u}$ give in turn more regularity on $\vec{g}$: $\vec{g}$ is locally $L^2 L^2$, so that $\vec{u}$ is locally $L^\infty H^2$: the theorem is proved.

$\square$

Of course, we find a proposition similar to Proposition 12.3:

Proposition 13.1.
Serrin's theorem on interior regularity holds in the following cases:

- $1_Q \vec{u} \in L_t^p L_x^q$ with $\frac{2}{p} + \frac{3}{q} = 1$ and $2 \leq p < +\infty$ *(this is the Struwe-Takahashi criterion [455, 459]*

- *more generally, $1_Q \vec{u} \in L_t^p \dot{M}_x^{2,q}$ with $\frac{2}{p} + \frac{3}{q} = 1$ and $2 \leq p < +\infty$*

- $1_Q \vec{u} \in \mathcal{C}([a, b], L^3)$

- $1_Q \vec{u} \in \mathcal{C}([a, b], , \mathcal{V}_0^1)$, *where $\mathcal{V}_0^1$ is the closure of L^3 in $\mathcal{M}(\dot{H}^1 \mapsto L^2)$*

13.3   O'Leary's Theorem on Interior Regularity

For $\alpha > 0$, let X_α be the space of homogenous type $(\mathbb{R} \times \mathbb{R}^3, \delta_\alpha, \mu)$, where δ_α is the parabolic (quasi)-distance

$$\delta_\alpha((t,x),(s,y)) = |t-s|^{1/\alpha} + |x-y| \tag{13.5}$$

and μ is the Lebesgue measure $d\mu = dt\,dx$. Then the homogeneous dimension Q of X_α is equal to $\alpha + 3$. Recall that we defined Morrey spaces $\mathcal{M}_\alpha^{p,q} = \dot{M}^{p,q}(X_\alpha)$ on X_α for $1 < p < q < +\infty$ by $u \in \mathcal{M}_\alpha^{p,q}$ if and only if u is locally $L_{t,x}^p$ and $\|u\|_{\mathcal{M}_\alpha^{p,q}} < +\infty$, where

$$\|u\|_{\mathcal{M}_\alpha^{p,q}} = \sup_{x_0 \in \mathbb{R}^3, t_0 \in \mathbb{R}, R>0} \left(\frac{1}{R^{(3+\alpha)(1-p/q)}} \iint_{|t-t_0|<R^\alpha, |x-x_0|<R} |u(t,x)|^p \, dt\,dx \right)^{1/p}.$$

In particular, the space $\mathcal{M}_2^{p,5}$, $2 < p \le 5$, was discussed on page 98 in Chapter 5.

O'Leary [379] gave the following variant of Serrin's theorem:

Theorem 13.3.
Let $Q = I \times \Omega$, where $I = (a,b)$ and $\Omega = B(x_0, r)$. Let $\vec{u} \in L^\infty(I, L^2(\Omega)) \cap L^2(I, H^1(\Omega))$, $\vec{f} \in L^2(I, H^1(\Omega))$ and $p \in \mathcal{D}'(Q)$, and assume that $\vec{u}$ is a weak solution on Q of the Navier–Stokes equations

$$\partial_t \vec{u} = \nu \Delta \vec{u} - \vec{u} \cdot \vec{\nabla} \vec{u} + \vec{f} - \vec{\nabla} p, \quad \operatorname{div} \vec{u} = 0.$$

Then, if moreover $1_Q \vec{u} \in \mathcal{M}_2^{p,q}(\mathbb{R} \times \mathbb{R}^3)$ for some $q > 5$ and some $2 < p \le q$, we have, for every $a < c < b$ and $0 < \rho < r$, $\vec{u} \in L^\infty((c,b) \times B(x_0, \rho))$.

Proof. With no loss of generality, we may assume, as Q is bounded, that $5 < q \le 6$. We write $\lambda = 1 - \frac{q-5}{5q} \in (0,1)$. We are going to show that, for $a < c < b$ and $0 < \rho < r$ and $Q_0 = (c,b) \times B(x_0, \rho)$, we have $1_{Q_0}\vec{u} \in \mathcal{M}_2^{p_0,q}$ with $p_0 = \min(\frac{p}{\lambda}, q)$.

One more time, we pick up functions $\phi, \psi \in \mathcal{D}(\mathbb{R} \times \mathbb{R}^3)$ supported in $(\frac{a+c}{2}, b) \times B(x_0, \frac{r+\rho}{2}))$ and such that $\psi\phi = \phi$ and such that $\phi(t,x) = 1$ on $(c,b) \times B(x_0, \rho))$, and we write, for $\vec{U} = \phi\vec{u}$,

$$\vec{U} = \psi(\frac{1}{\Delta}\Delta(\phi\vec{u})) = \psi\left(\frac{1}{\Delta}\left(\phi\Delta\vec{u} - (\Delta\phi)\vec{u} + 2\sum_{i=1}^3 \partial_i((\partial_i\phi)\vec{u}) \right) \right).$$

If $\vec{V} = \psi(\frac{1}{\Delta}(\phi\Delta\vec{u}))$, we find that

$$1_{a<t<b}(\vec{U} - \vec{V}) \in L_t^\infty L^6$$

and thus

$$1_Q(\vec{U} - \vec{V}) \in L^q = \mathcal{M}_2^{q,q} \subset \mathcal{M}_2^{p_0,q}.$$

Now, we write $\vec{V} = -\psi(\frac{1}{\Delta}(\phi\operatorname{curl}(\psi\vec{\omega}))$, where $\vec{\omega} = \operatorname{curl}\vec{u}$. Let $\vec{W} = \psi\vec{\omega}$. We have

$$\partial_t \vec{W} = (\partial_t\psi)\vec{\omega} + \nu\psi\Delta(\vec{\omega}) - \psi\operatorname{curl}(\sum_{i=1}^3 \partial_i(u_i\vec{u})) + \psi\operatorname{curl}\vec{f},$$

which we rewrite as

$$\partial_t \vec{W} = \nu\Delta\vec{W} + (\partial_t\psi)\vec{\omega} - \psi\operatorname{curl}(\sum_{i=1}^3 \partial_i(u_i\vec{u})) + \psi\operatorname{curl}\vec{f} + \nu(\Delta\psi)\vec{\omega} - 2\nu\sum_{i=1}^3 \partial_i((\partial_i\psi)\vec{\omega}).$$

This gives

$$\partial_t(\operatorname{curl}\vec{W}) = \nu\Delta(\operatorname{curl}\vec{W}) + \sum_{k=1}^{7}\vec{R}_k$$

with

$$\vec{R}_1 = \operatorname{curl}(\psi\operatorname{curl}\vec{f}), \quad \vec{R}_2 = \operatorname{curl}((\partial_t\psi + \nu\Delta\psi)\vec{\omega})$$

$$\vec{R}_3 = -2\nu\operatorname{curl}(\sum_{i=1}^{3}\partial_i((\partial_i\psi)\vec{\omega})), \quad \vec{R}_4 = \operatorname{curl}(\sum_{i=1}^{3}\partial_i(\vec{\nabla}\psi \wedge (u_i\vec{u})))$$

$$\vec{R}_5 = -\operatorname{curl}(\operatorname{curl}(\sum_{i=1}^{3}(\partial_i\psi)u_i\vec{u})). \quad \vec{R}_6 = \operatorname{curl}(\sum_{i=1}^{6}(\vec{\nabla}\partial_i\psi) \wedge (u_i\vec{u}))$$

$$\vec{R}_7 = -\operatorname{curl}(\operatorname{curl}(\sum_{i=1}^{3}\partial_i(\psi u_i\vec{u}))).$$

Thus, we have

$$\vec{W} = \sum_{i=1}^{7}\vec{S}_i = \sum_{i=1}^{7}\int_0^t W_{\nu(t-s)} * \vec{R}_i(s,.)\,ds.$$

As $\vec{R}_1$, $\vec{R}_2$ and $\vec{R}_3$ belong to L^2H^{-2}, we have that $\phi\vec{S}_1$, $\phi\vec{S}_2$, and $\phi\vec{S}_3$ belong to $L^\infty H^{-1}$, and even to $L^\infty\dot{H}^{-1}$ (as pointwise multiplication by ψ maps $L^\infty H^1$ to $L^\infty\dot{H}^{-1}$) and we get finally that $\psi\frac{1}{\Delta}(\phi\vec{S}_1)$, $\psi\frac{1}{\Delta}(\phi\vec{S}_2)$, and $\psi\frac{1}{\Delta}(\phi\vec{S}_3)$ belong to $L^\infty L^6$, and, being supported in a compact subset of $\mathbb{R} \times \mathbb{R}^3$, to $L^q_{t,x} = \mathcal{M}_2^{q,q} \subset \mathcal{M}_2^{p_0,q}$.

We must now estimate the non-linear terms $\psi\frac{1}{\Delta}(\phi\vec{S}_i)$, for $i = 4,\ldots,7$. For $i = 4,\ldots,6$, we write $\vec{S}_i = \operatorname{curl}\vec{T}_i$ and $\phi\vec{S}_i = \operatorname{curl}(\phi\vec{T}_i) - \vec{\nabla}\phi \wedge \vec{T}_i$. The estimations will be done with the Riesz potentials $\mathcal{I}_{\beta,\alpha}$ on X_α ($0 < \beta < 3 + \alpha$) defined by

$$\mathcal{I}_{\beta,\alpha}f(t,x) = \iint_{\mathbb{R}\times\mathbb{R}^3}\frac{1}{(|t-s|^{1/\alpha} + |x-y|)^{3+\alpha-\beta}}f(s,y)\,ds\,dy$$

We have $|\vec{T}_4| \leq C_Q\mathcal{I}_{1,2}(1_Q|\vec{u}|^2)$, $|\vec{T}_5| \leq C_Q\mathcal{I}_{1,2}(1_Q|\vec{u}|^2)$ and $|\vec{T}_6| \leq C_Q\mathcal{I}_{2,2}(1_Q|\vec{u}|^2)$.

Using Adams's inequality (see the Corollary 5.1 of Adams–Hedberg's inequality), we get:

- $\mathcal{I}_{1,2}$ maps $\mathcal{M}_2^{p/2,q/2}$ to $\mathcal{M}_2^{p_1,q_1}$ with $\frac{1}{q_1} = \frac{2}{q} - \frac{1}{5} = \frac{1}{q}\lambda$ and $\frac{1}{p_1} = \frac{1}{p}\lambda$

- Hence, as Q is bounded, $1_Q\mathcal{I}_{1,2}$ maps $\mathcal{M}_2^{p/2,q/2}$ to $\mathcal{M}_2^{p_0,q}$

- Let $r < 5/2$ close enough to $5/2$ to get that $\frac{\min(p/2,r)}{1-\frac{2r}{5}} > q$. $\mathcal{I}_{2,2}$ maps $\mathcal{M}_2^{\min(p/2,r),r}$ to $\mathcal{M}_2^{p_2,r_2}$ with $\frac{1}{r_2} = \frac{1}{r} - \frac{2}{5} = \frac{1}{r}\mu$ with $\mu = 1 - \frac{2r}{5} < 1$ and $\frac{1}{p_2} = \frac{1}{\min(p/2,r)}\mu < \frac{1}{q}$.

- Hence, as Q is bounded and $q > 5$, $f \mapsto \mathcal{I}_{2,2}(1_Qf)$ maps $\mathcal{M}_2^{p/2,q/2}$ to $\mathcal{M}_2^{p_2,r_2}$ and $f \mapsto 1_Q\mathcal{I}_{2,2}(1_Qf)$ maps $\mathcal{M}_2^{p/2,q/2}$ to $\mathcal{M}_2^{q,q} \subset \mathcal{M}_2^{p_0,q}$.

Moreover $L_1 : f \mapsto \psi\frac{\partial_k}{\Delta}(\phi f)$ and $L_2 : f \mapsto \psi\frac{1}{\Delta}((\partial_k\phi)f)$ are bounded on $\mathcal{M}_2^{p_0,q}$, as

$$|L_if(t,x)| \leq \int A(x-y)|f(t,y)|\,dy$$

with $A \in L^1(\mathbb{R}^3)$; as the norm of $\mathcal{M}_2^{p_0,q}$ is invariant by translation, we have $\|A * f\|_{\mathcal{M}_2^{p_0,q}} \leq \|A\|_1\|f\|_{\mathcal{M}_2^{p_0,q}}$. Thus, we find that $\psi\frac{1}{\Delta}(\phi\vec{S}_4)$, $\psi\frac{1}{\Delta}(\phi\vec{S}_5)$, and $\psi\frac{1}{\Delta}(\phi\vec{S}_6)$ belong to $\mathcal{M}_2^{p_0,q}$.

Finally, we write $\vec{S}_7 = \Delta \vec{T}_7$ with

$$\vec{T}_7 = -\sum_{i=1}^{3} \int_0^t \frac{1}{\Delta} \operatorname{curl}(\operatorname{curl}(\partial_i(W_{\nu(t-s)} * (\psi u_i \vec{u})))) \, ds)$$

We have $|\vec{T}_7| \leq C_Q \mathcal{I}_{1,2}(1_Q|\vec{u}|^2)$, so that $\vec{T}_7 \in \mathcal{M}_2^{p/\lambda,q/\lambda}$. Moreover, we have

$$\psi \frac{1}{\Delta}(\phi \Delta \vec{T}_7) = \phi \vec{T}_7 + \psi \frac{1}{\Delta}((\Delta\phi)\vec{T}_7) - 2\sum_{i=1}^{3} \psi \frac{\partial_i}{\Delta}((\partial_i \phi)\vec{T}_7)$$

and finally $\psi \frac{1}{\Delta}(\phi \vec{S}_7) \in \mathcal{M}_2^{p/\lambda,q/\lambda}$, and, as it is compactly supported, it belongs to $\mathcal{M}_2^{p_0,q}$.

Thus, we have seen that $1_{Q_0}\vec{u}$ belongs to $\mathcal{M}_2^{\min(p/\lambda,q),q}$. As $\lim_{n \to +\infty} p/\lambda^n = +\infty$, we see that we may reiterate the proof in finitely many steps on some smaller cylinders and get $1_{Q_0}\vec{u} \in \mathcal{M}_2^{q,q} = L_t^q L_x^q$. Thus, we may apply Serrin's theorem on interior regularity. $\quad\Box$

13.4   Further Results on Parabolic Morrey Spaces

While O'Leary's results deal with the condition $\vec{u} \in \mathcal{M}_2^{p,q}$, $2 < p \leq q$ and $q > 5$, which is subcritical with respect to the natural scaling of the Navier–Stokes equations, the case of critical scaling has been dealt with by Chen and Price [118], when $1_Q \vec{u}$ is small enough in $\mathcal{M}_2^{p,5}(\mathbb{R} \times \mathbb{R}^3)$ for some $\frac{7}{2} < p \leq 5$; as we shall see, the result is true when $2 < p \leq 5$, and even when the parabolic Morrey space $\mathcal{M}_2^{p,5}(\mathbb{R} \times \mathbb{R}^3)$ is replaced with the multiplier space $\mathcal{V}^{2,1} = \mathcal{M}(L_t^2\dot{H}_x^1 \cap L_t^2\dot{H}_t^{1/2} \mapsto L_t^2 L_x^2)$ described in Chapter 5. (Recall that for $2 < p \leq 5$, we have $L_{t,x}^5 \subset \mathcal{M}_2^{p,5} \subset \mathcal{V}^{1,2}(\mathbb{R} \times \mathbb{R}^3) \subset \mathcal{M}_2^{2,5}$).

Parabolic multipliers and interior regularity

Theorem 13.4.
Let $Q = I \times \Omega$, where $I = (a,b)$ and $\Omega = B(x_0, r)$. Let $\vec{u} \in L^\infty(I, L^2(\Omega)) \cap L^2(I, H^1(\Omega))$, $\vec{f} \in L^2(I, H^1(\Omega))$ and $p \in \mathcal{D}'(Q)$, and assume that $\vec{u}$ is a weak solution on Q of the Navier–Stokes equations

$$\partial_t \vec{u} = \nu \Delta \vec{u} - \vec{u} \cdot \vec{\nabla}\vec{u} + \vec{f} - \vec{\nabla}p, \quad \operatorname{div} \vec{u} = 0.$$

Then, if moreover $1_Q \vec{u} \in \mathcal{V}^{2,1}(\mathbb{R} \times \mathbb{R}^3) = \mathcal{M}(L_t^2\dot{H}_x^1 \cap L_t^2\dot{H}_t^{1/2} \mapsto L_t^2 L_x^2)$ and $\|1_Q \vec{u}\|_{\mathcal{V}^{2,1}}$ is small enough, we have, for every $a < c < b$ and $0 < \rho < r$, $\vec{u} \in L^\infty((c,b) \times B(x_0, \rho))$.

Proof. This theorem is a direct generalization of Theorem 13.2, and its proof is quite similar:

First step: uniqueness for a linear heat equation.
Let $\vec{\omega} = \operatorname{curl}\vec{u}$. We consider a function $\phi \in \mathcal{D}(\mathbb{R} \times \mathbb{R}^3)$ which is equal to 1 on $[c,b] \times B(x_0, \rho)$ and is supported in $[\frac{a+c}{2}, b+1] \times B(x_0, \frac{r+\rho}{2})$. We define $\vec{w} = \phi\vec{\omega}$. We have:

- $\vec{w} \in L^2((a,b) \times \mathbb{R}^3)$
- $\vec{w}(t,.) = 0$ for $a < t < \frac{b+c}{2}$

- $\partial_t \vec{w} = \nu \Delta \vec{w} - \operatorname{div}(1_Q \vec{u} \otimes \vec{w} - \vec{w} \otimes 1_Q \vec{u}) + \vec{g}$ with

$$\vec{g} = (\vec{\omega} \cdot \vec{\nabla}\phi)\vec{u} - (\vec{u} \cdot \vec{\nabla}\phi)\vec{\omega} - (\Delta\phi)\vec{\omega} - 2\sum_{i=1}^{3}(\partial_i\phi)\,\partial_i\vec{\omega} + \partial_t\phi\vec{\omega} + \phi\operatorname{curl}\vec{f}$$

We now prove uniqueness of the solutions $\vec{z} \in L^2((a,b) \times \mathbb{R}^3)$ of this linear heat equation

$$\partial_t\vec{z} = \nu\Delta\vec{z} - \operatorname{div}(1_Q\vec{u} \otimes \vec{z} - \vec{z} \otimes 1_Q\vec{u}) + \vec{g} \tag{13.6}$$

Let $\vec{z}_1$ and $\vec{z}_2$ be two solutions in $L^2((a,b) \times \mathbb{R}^3)$ of equation (13.6) with $\vec{z}_1(a,.) = \vec{z}_2(a,.)$. We write:

$$\vec{z}_1 - \vec{z}_2 = \nu\Delta(\vec{z}_1 - \vec{z}_2) - \operatorname{div}(1_Q\vec{u} \otimes (\vec{z}_1 - \vec{z}_2) - (\vec{z}_1 - \vec{z}_2) \otimes 1_Q\vec{u}).$$

Thus,

$$\vec{z}_1 - \vec{z}_2 = -L(\operatorname{div}(1_Q\vec{u} \otimes (\vec{z}_1 - \vec{z}_2) - (\vec{z}_1 - \vec{z}_2) \otimes 1_Q\vec{u}))$$

where L is defined by

$$L(\vec{h}) = \int_a^t W_{\nu(t-s)} * \vec{h}\,ds.$$

We have seen in Chapter 5 that

$$|L(\operatorname{div}\mathbb{H})| \le C_0 \mathcal{I}_1(|\mathbb{H}|)$$

where $\mathcal{I}_1$ is the parabolic Riesz potential

$$\mathcal{I}_1 f = \iint f(s,y)\frac{1}{(|t-s|^{1/2} + |x-y|)^4}\,dy\,ds.$$

Let $\frac{p}{p-1} < r \le 2$; we have $L^2_{t,x} = \mathcal{M}_2^{2,2} \subset \mathcal{M}_2^{r,2}$; pointwise multiplication with a function in $\mathcal{M}_2^{p,5}$ maps $\mathcal{M}_2^{r,2}$ into $\mathcal{M}_2^{\frac{rp}{p+r},\frac{10}{7}}$, while $\mathcal{I}_1$ maps $\mathcal{M}_2^{\frac{rp}{p+r},\frac{10}{7}}$ to $\mathcal{M}_2^{\frac{7rp}{5(p+r)},2}$; if $2p > 5r$ (such a choice of r is possible when $2p > 5\frac{p}{p-1}$, i.e. $p > 7/2$), we find

$$\|\vec{z}_1 - \vec{z}_2\|_{\mathcal{M}_2^{r,2}} \le C_1\|\vec{z}_1 - \vec{z}_2\|_{\mathcal{M}_2^{r,2}}\|1_Q\vec{u}\|_{\mathcal{M}_2^{p,5}}.$$

If $\|1_Q\vec{u}\|_{\mathcal{M}_2^{p,5}}$ is small enough ($\|1_Q\vec{u}\|_{\mathcal{M}_2^{p,5}} < \frac{1}{C_1}$), we have $\vec{z}_1 - \vec{z}_2 = 0$, hence uniqueness for the solutions of equation (13.6). Let $W = L^2_t\dot{H}^1_x \cap L^2_x\dot{H}^{1/2}_t$. We know that $\mathcal{I}_1$ maps $L^2_{t,x}$ to W, hence maps W' to L^2 by transposition; similarly, we know that for $V \in \mathcal{V}^{1,2}$, pointwise multiplication with V maps W to L^2, hence by transposition maps L^2 to W'. This gives:

$$\begin{aligned}
\|\vec{z}_1 - \vec{z}_2\|_{L^2((a,b),L^2(\mathbb{R}^3))} &\le 2C_0\|\mathcal{I}_1(1_Q|\vec{u}||\vec{z}_1 - \vec{z}_2|)\|_{L^2 L^2} \\
&\le C_1\|1_Q|\vec{u}||\vec{z}_1 - \vec{z}_2|\|_{W'} \\
&\le C_2\|1_Q\vec{u}\|_{\mathcal{V}^{1,2}}\|\vec{z}_1 - \vec{z}_2\|_{L^2((a,b),L^2(\mathbb{R}^3))}
\end{aligned}$$

Thus, when $\|1_Q\vec{u}\|_{\mathcal{V}^{1,2}}$ is small enough ($\|1_Q\vec{u}\|_{\mathcal{V}^{1,2}} < \frac{1}{C_2}$), we have $\vec{z}_1 - \vec{z}_2 = 0$, hence uniqueness for the solutions of Equation (13.6).

Second step: regular solutions for the linear heat equation.

We consider the equation on $(a,b) \times \mathbb{R}^3$

$$\vec{z} = \vec{Z} - L(\operatorname{div}(1_Q \vec{u} \otimes \vec{z} - \vec{z} \otimes 1_Q \vec{u}))$$

with

$$\vec{Z} = \int_a^t W_{\nu(t-s)} * \vec{g}(s,.)\, ds.$$

We are going to search solutions extended to $\mathbb{R} \times \mathbb{R}^3$ by extending $\vec{g}$ to 0 outside from (a,b) and defining L as

$$L(\vec{h}) = \int_{-\infty}^t W_{\nu(t-s)} * \vec{h}\, ds.$$

If $\vec{h}$ is equal to 0 on $(-\infty, a)$, we find that on (a,b) the new definition of $L(\vec{h})$ coincides with the old one.

We have $\vec{g} = \vec{g}_1 + \vec{g}_2$, with

$$\vec{g}_1 = 1_{(a,b)}(t)\left(-(\Delta\phi)\vec{\omega} - 2\sum_{i=1}^3 (\partial_i\phi)\,\partial_i\vec{\omega} + \partial_t\phi\vec{\omega} + \phi\operatorname{curl}\vec{f}\right) \in L^2(\mathbb{R}, H^{-1}(\mathbb{R}^3))$$

and $\vec{g}_2 = 1_Q(\vec{\omega}\cdot\vec{\nabla}\phi)\vec{u} - 1_Q(\vec{u}\cdot\vec{\nabla}\phi)\vec{\omega} \in W'$.

Moreover, we take a function $\psi \in \mathcal{D}(\mathbb{R})$ such that $\psi = 1$ on (a,b), and we study the solutions of

$$\vec{z} = \vec{Z} - L(\operatorname{div}(1_Q \vec{u} \otimes \vec{z} - \vec{z} \otimes 1_Q \vec{u})) \tag{13.7}$$

with

$$\vec{Z} = \psi(t)\int_{-\infty}^t W_{\nu(t-s)} * \vec{g}_1(s,.)\, ds + \int_{-\infty}^t W_{\nu(t-s)} * \vec{g}_2(s,.)\, ds = \vec{Z}_1 + \vec{Z}_2.$$

Then, we have:

- since $\vec{g}_1 \in L^2(H^{-1}(\mathbb{R}^3))$ and is equal to 0 for $t < a$, we know that $\vec{Z}_1 \in L_t^\infty L^2 \cap L^2 H^1$ moreover, we have

$$\partial_t \vec{Z}_1 = \nu\Delta\vec{Z}_1 + \vec{g}_1 + \partial_t\psi\int_{-\infty}^t W_{\nu(t-s)} * \vec{g}_i(s,.)\, ds \in L_t^2 H^{-1}.$$

  This gives $\vec{Z}_1 \in W$.

- Let us remark that, for a non-negative function f, we have

$$\mathcal{I}_2 f \le C\mathcal{I}_1(\mathcal{I}_1 f).$$

It is equivalent to prove that

$$\frac{1}{(|t|^{1/2}+|x|)^3} \le C \iint \frac{1}{(|t-s|^{1/2}+|x-y|)^4}\frac{dy\,ds}{(|s|^{1/2}+|y|)^4} = A(t,x).$$

If $|t|^{1/2} \le |x|$, we write $\frac{1}{(|t|^{1/2}+|x|)^3} \le \frac{1}{|x|^3}$ and

$$A(t,x) \ge \iint_{|y|<|s|^{1/2}<\frac{|x|}{2}} \frac{1}{(3|x|)^4}\frac{1}{(2|s|^{1/2})^4}\,dy\,ds = \frac{c}{|x|^3}.$$

If $|t|^{1/2} \ge |x|$, we write $\frac{1}{(|t|^{1/2}+|x|)^3} \le \frac{1}{|t|^{3/2}}$ and

$$A(t,x) \ge \iint_{|y|<|s|^{1/2}<\frac{|t|^{1/2}}{2}} \frac{1}{(3|t|^{1/2})^4}\frac{1}{(2|s|^{1/2})^4}\,dy\,ds = \frac{c}{|t|^{3/2}}.$$

- We have

$$|\vec{Z}_2| \leq C \int \frac{1}{(|t-s|^{1/2} + |x-y|)^3} |\vec{\omega}| 1_Q |\vec{u}| \, ds \, dy$$

so that

$$|\vec{Z}_2| \leq C \mathcal{I}_2(|\vec{\omega}| 1_Q |\vec{u}|) \leq C' \mathcal{I}_1(\mathcal{I}_1(|\vec{\omega}| 1_Q |\vec{u}|)).$$

We have $1_Q \vec{\omega} \in L^2_{t,x}$ and $1_Q \vec{u} \in \mathcal{V}^{1,2}$, so that $1_Q |\vec{u}||\vec{\omega}| \in W'$; as $\mathcal{I}_1$ maps W' to $L^2_{t,x}$ and $L^2_{t,x}$ to W, we find that $\vec{Z}_2 \in W$.

Now, if $\vec{z} \in W$, we have

$$\|L(\operatorname{div}(1_Q \vec{u} \otimes \vec{z} - \vec{z} \otimes 1_Q \vec{u}))\|_W \leq C \|1_Q \vec{u} \otimes \vec{z}\|_{L^2 L^2} \leq C_0 \|1_Q \vec{u}\|_{\mathcal{V}^{1,2}} \|\vec{z}\|_W.$$

If $\|1_Q \vec{u}\|_{\mathcal{V}^{1,2}}$ is small enough ($\|1_Q \vec{u}\|_{\mathcal{V}^{1,2}} < \frac{1}{C_0}$), the Banach contraction principle gives us the existence and uniqueness of solutions $\vec{z} \in W$ of Equation (13.7).

Third step: regularity estimates on the vorticity $\vec{\omega}$.

Recall that $\vec{w} = \phi \vec{\omega}$ is a solution on $(a,b) \times \mathbb{R}^3$ of Equation (13.6). We know that there exists a solution $\vec{z}$ in W of Equation (13.7). Let $1_{(a,b)}(t) \vec{z} = \vec{W}$. Then $\vec{W}$ is another solution on $(a,b) \times \mathbb{R}^3$ of Equation (13.6). Moreover, $\vec{W}$ belongs to $L^2((a,b), L^2)$:

- We have the Sobolev embedding $W \subset L^2_t \dot{H}^1_x \subset L^2_t L^6_x$, so that $1_{(a,b)}(t) 1_{B(x_0,3r)} \vec{z} \in L^2_t L^2_x$.
- As $\vec{z}$ is computed through a Picard iteration, we find that $\vec{z} = 0$ for $t < a$. We have

$$1_{|x-x_0| \geq 3r}(x)|L(\operatorname{div}(1_Q \vec{u} \otimes \vec{z} - \vec{z} \otimes 1_Q \vec{u}))| \leq$$
$$C \int_a^t \int_{|-y|>2r} \frac{1}{|x-y|^4} 1_Q(y)|\vec{u}(s,y)||\vec{z}(s,y)| \, dy \, ds$$

As $1_Q(y)|\vec{u}||\vec{z}| \in L^2 L^2$, we see that

$$1_{(a,b)}(t) 1_{|x-x_0| \geq 3r}(x) L(\operatorname{div}(1_Q \vec{u} \otimes \vec{z} - \vec{z} \otimes 1_Q \vec{u})) \in L^2_t L^2_x.$$

- We already know that $\vec{Z}_1 \in L^\infty L^2$, so that $1_{(a,b)}(t) 1_{|x-x_0| \geq 3r} \vec{Z}_1 \in L^2 L^2$.
- We have $\vec{u} \in L^\infty L^2$ and $\vec{\omega} \in L^2 L^2$, thus $\vec{g}_2 \in L^2 L^1$. As we have

$$1_{|x-x_0| \geq 3r}(x)|\vec{Z}_2| \leq C \int_a^t \int_{|-y|>2r} \frac{1}{|x-y|^3} |\vec{g}_2(s,y)| \, dy \, ds$$

we find that $1_{(a,b)}(t) 1_{|x-x_0| \geq 3r} \vec{Z}_2 \in L^2 L^2$.

Thus, $\vec{W} \in L^2 L^2$, and by uniqueness of solutions to Equation (13.6), we have $\vec{w} = \vec{W}$. In particular, $\phi \vec{u} \in L^2((a,b), H^2) \subset L^2 L^\infty$, and we may then finish the proof by applying Serrin's theorem on interior regularity.

$\square$

13.5   Hausdorff Measures

In the following sections, we shall recall the proofs that the set of singular points of a Leray solution is small, this smallness will be expressed in terms of Hausdorff dimensions.

Let (X, δ, μ) be a space of homogeneous type and Q its homogeneous dimension (see Definition 5.1). In particular:

- there is a positive constant κ such that:

$$\text{for all } x, y, z \in X, \delta(x, y) \leq \kappa(\delta(x, z) + \delta(z, y))$$

- there exists positive numbers $0 < A_0 \leq A_1$ which satisfy:

$$\text{for all } x \in X, \text{ for all } r > 0, A_0 r^Q \leq \int_{\delta(x,y)<r} d\mu(y) \leq A_1 r^Q$$

A basic useful property of spaces of homogeneous type is the Vitali covering lemma [215, 313].

Proposition 13.2 (The Vitali covering lemma).
Let $E \subset X$ be decomposed as a union of balls $E = \cup_{\alpha \in \mathcal{A}} B(x_\alpha, r_\alpha)$, where $(B(x_\alpha, r_\alpha))_{\alpha \in \mathcal{A}}$ is a family of balls so that $\sup_\alpha r_\alpha < \infty$. Then there exists a (countable) subfamily of balls $(B(x_\alpha, r_\alpha))_{\alpha \in \mathcal{B}}$ $(\mathcal{B} \subset \mathcal{A})$ so that $\alpha \neq \beta \Rightarrow B(x_\alpha, r_\alpha) \cap B(x_\beta, r_\beta) = \emptyset$ and so that $E \subset \cup_{\alpha \in \mathcal{B}} B(x_\alpha, 5\kappa^2 r_\alpha)$.

We may now introduce the Hausdorff measures on X.

Definition 13.2 (Hausdorff measure).
Let (X, δ, μ) be a separable space of homogeneous type (see Definition 5.1).

(i) *For a sequence of open balls $\mathcal{B} = (B(x_i, r_i))_{i \in \mathbb{N}}$ of X and for $\alpha > 0$, we define $r(\mathcal{B}) = \sup_{i \in \mathbb{N}} r_i$ and $\sigma_\alpha(\mathcal{B}) = \sum_{i \in \mathbb{N}} r_i^\alpha$.*

(ii) *The Hausdorff measure $\mathcal{H}^\alpha$ on X is defined for a Borel subset $B \subset X$ by*

$$\mathcal{H}^\alpha(B) = \lim_{\epsilon \to 0} \min\{\sigma_\alpha(\mathcal{B}) \ / \ \mathcal{B} = (B(x_i, r_i))_{i \in \mathbb{N}}, B \subset \cup_{i \in \mathbb{N}} B(x_i, r_i), r(\mathcal{B}) < \epsilon\}$$

We have obviously, if $\alpha < \beta$, $\sigma_\beta(\mathcal{B}) \leq r(\mathcal{B})^{\beta-\alpha} \sigma_\alpha(\mathcal{B})$; thus,

- $\mathcal{H}^\alpha(B) < +\infty \Rightarrow \mathcal{H}^\beta(B) = 0$

- $\mathcal{H}^\beta(B) > 0 \Rightarrow \mathcal{H}^\alpha(B) = +\infty$

Moreover, we have:

$$\alpha > Q \Rightarrow \mathcal{H}^\alpha(B) = 0.$$

Indeed, if $B = B(x_0, r)$, we use the Vitali lemma on the collection $(B(x, \epsilon))_{x \in B}$ to exhibit a family of disjoint balls $(B(x_i, \epsilon))$ such that $B \subset \cup_i B(x_i, 5\kappa^2 \epsilon)$. Let $\mathcal{B}_\epsilon = (B(x_i, 5\kappa^2 \epsilon))$. We have:

$$\sigma_Q(\mathcal{B}_\epsilon) \leq \frac{A_1}{A_0}(5\kappa^2)^Q \sum_i \mu(B(x_i, \epsilon)) \leq \frac{A_1}{A_0}(5\kappa^2)^Q A_1(\kappa(r+\epsilon))^Q$$

and thus

$$\mathcal{H}^Q(B(x_0, r)) \leq \frac{A_1^2}{A_0}(5r\kappa^3)^Q < +\infty.$$

If $\alpha > Q$ and $B \subset X$, we write, for a $x_0 \in X$,

$$\mathcal{H}^\alpha(B) \le \mathcal{H}^\alpha(X)) \le \sum_{N=1}^{+\infty} \mathcal{H}^\alpha(B(x_0, N)) = 0.$$

Definition 13.3 (Hausdorff dimension).
The Hausdorff dimension $d_{\mathcal{H}}(B)$ of a Borel subset of X is defined as

$$d_{\mathcal{H}}(B) = \inf\{\alpha > 0 \ / \ \mathcal{H}^\alpha(B) = 0\}.$$

If $d_{\mathcal{H}}(B) > 0$, it may be defined as well as

$$d_{\mathcal{H}}(B) = \sup\{\alpha > 0 \ / \ \mathcal{H}^\alpha(B) = +\infty\}.$$

13.6 Singular Times

A classical result (which goes back to the description of the structure of turbulent solutions by Leray [328]) states that the set of singular times for a Leray solution is very small.

We consider the Navier–Stokes problem

$$\partial_t \vec{u} = \nu \Delta \vec{u} + \mathbb{P} \operatorname{div}(\mathbb{F} \operatorname{div}(\vec{u} \otimes \vec{u})), \quad \vec{u}(0,.) = \vec{u}_0 \tag{13.8}$$

where $\vec{u}_0 \in L^2$ with $\operatorname{div} \vec{u} = 0$ and the tensor $\mathbb{F}$ is smooth on $(0, +\infty) \times \mathbb{R}^3$:

$$\mathbb{F} \in \cap_{k \in \mathbb{N}} H^k((0, +\infty) \times \mathbb{R}^3) = H^\infty((0, +\infty) \times \mathbb{R}^3) \tag{13.9}$$

We have seen in Proposition 12.1 that the solution $\vec{u}$ constructed in Theorem 12.2 satisfies the strong Leray energy inequality: for almost every t_0 in $(0, T)$ and for every $t \in (t_0, T)$, we have

$$\|\vec{u}(t,.)\|_2^2 + 2\nu \int_{t_0}^t \|\vec{u}\|_{\dot{H}^1}^2 \, ds \le \|\vec{u}(t_0)\|_2^2 + 2 \int_{t_0}^t \langle \operatorname{div} \mathbb{F} | \vec{u} \rangle_{H^{-1}, H^1} \, ds \tag{13.10}$$

Singular times

Theorem 13.5.
Let $\vec{u}_0 \in L^2$ with $\operatorname{div} \vec{u} = 0$ and $\mathbb{F} \in H^\infty((0, +\infty) \times \mathbb{R}^3)$. Let $\vec{u}$ be a weak Leray solution of the Navier–Stokes Equations (13.8) on $(0, \infty) \times \mathbb{R}^3$ which satisfies the strong energy inequality. Then there is compact set $\Sigma_t \subset [0, \infty)$ so that:

(i) $\vec{u}$ is smooth outside from $\Sigma_t \times \mathbb{R}^3$

(ii) $\mathcal{H}^{1/2}(\Sigma_t) = 0$ (where $\mathcal{H}^{1/2}$ is the Hausdorff measure on $\mathbb{R}$).

Proof. Let t_0 be a Lebesgue point of the map $t \mapsto \|\vec{u}(t,.)\|_2^2$ such that $\vec{u}(t_0,.) \in H^1$. From Theorem 7.1, we know that there exists a $t_1 > t_0$ a local solution $\vec{v}$ on (t_0, t_1) of the Navier–Stokes problem with initial value $\vec{u}(t_0,.)$ at $t = t_0$ such that $\vec{v} \in \mathcal{C}([t_0, t_1], (H^1)^3) \cap$

$L^2((t_0, t_1), (H^2)^3)$. If t_2 is the maximal existence time of this solution (for every $T < t_2$, $\vec{v} \in \mathcal{C}([t_0, T], (H^1)^3) \cap L^2((t_0, T), (H^2)^3)$) and $T < +\infty$, then $\int_{t_0}^{T} \|\vec{v}(s, .)\|_{\dot{H}^{3/2}}^2 \, ds = +\infty$. Moreover, by induction on k, we see that for every $k \in \mathbb{N}$ and $t_1 \in (t_0, t_2)$, $\vec{v}(t_1, .) \in H^k$ and thus (from Theorem 7.3) $\vec{v} \in \mathcal{C}([t_0, T], (H^k)^3) \cap L^2((t_0, T), (H^{k+1})^3)$ for every $T \in (t_1, t_2)$.

Moreover, by weak-strong uniqueness, we have $\vec{u} = \vec{v}$ on $[t_0, t_2]$. As $\vec{\nabla} p = \mathbb{P}(\operatorname{div}(\mathbb{F} - \vec{u} \otimes \vec{u}))$, we find by induction on k that for every $k \in \mathbb{N}$ and every $m \in \mathbb{N}$, and for every $t_0 < t_1 < T < t_2$, $\partial_t^k \vec{u} \in L^2((t_1, T), H^m)$; thus, for every $t_0 < t_1 < T < t_2$, $\vec{u} \in H^\infty((t_1, T) \times \mathbb{R}^3)$, and $\vec{u}$ is smooth on $(t_0, t_2) \times \mathbb{R}^3$.

Let $\mathcal{I}$ be the collection of open intervals $I \subset (0, +\infty)$ such that $\vec{u} \in H^\infty(J \times \mathbb{R}^3)$, $O = \cup_{I \in \mathcal{I}} I$ and $\Sigma_t = [0, +\infty) \setminus O$. By construction Σ_t is a closed subset of $[0, +\infty)$ and $\vec{u}$ is smooth outside $\Sigma_t \times \mathbb{R}^3$.

In order to check that Σ_t is compact, it is enough to show that it is bounded. Let us recall what we proved in Theorem 7.2 : there exists a positive constant ϵ_0 , such that, if $\|\vec{u}(t_0, .)\|_{\dot{H}^{1/2}} < \epsilon_0 \nu$ and $\int_{t_0}^{+\infty} \|\mathbb{F}(s, .)\|_{\dot{H}^{\frac{1}{2}}}^2 \, ds < \epsilon_0^2 \nu^3$, then there is a global solution $\vec{v}$ on $(t_0, +\infty)$ of the Navier–Stokes problem with initial value $\vec{u}(t_0, .)$ at $t = t_0$ such that $\vec{v} \in \mathcal{C}([t_0, +\infty], (H^1)^3) \cap L^2((t_0, +\infty), (\dot{H}^2)^3)$. If moreover t_0 is a Lebesgue point of the map $t \mapsto \|\vec{u}(t, .)\|_2^2$, then $\vec{u} = \vec{v}$ on $[t_{s_0}, +\infty)$ (by weak–strong uniqueness). As $\mathbb{F} \in L^2((0, +\infty), \dot{H}^{1/2})$, there exists a time T such that $\int_T^{+\infty} \|\mathbb{F}(s, .)\|_{\dot{H}^{\frac{1}{2}}}^2 \, ds < \epsilon_0^2 \nu^3$. As $\vec{u} \in L^\infty L^2 \cap L^2 \dot{H}^1 \subset L^4 \dot{H}^{1/2}$, we find that the measure of the set of points t such that $\|\vec{u}(t, .)\|_{\dot{H}^{1/2}} \geq \epsilon_0 \nu$ is finite; as almost every time is a Lebesgue point of the map $t \mapsto \|\vec{u}(t, .)\|_2^2$, we find that there exists a time $t_0 > T$ from which $\vec{u}$ will belong to $\mathcal{C}([t_0, +\infty], (H^1)^3) \cap L^2((t_0, +\infty), (\dot{H}^2)^3)$. We may conclude that $\Sigma_t \subset [0, t_0]$, and thus Σ_t is compact.

Now, we are going to estimate the Hausdorff dimension of Σ_t. Let $\tau \in \Sigma_t$ and let $s < \tau$ be a Lebesgue point of the map $t \mapsto \|\vec{u}(t, .)\|_2^2$ such that $\vec{u}(s, .) \in H^1$. By Theorem 7.1, we may find a local solution $\vec{v}$ on $(s, s + T)$ of the Navier–Stokes problem with initial value $\vec{u}(s, .)$ at $t = s$ such that $\vec{v} \in \mathcal{C}([s, s+T], (H^1)^3) \cap L^2((s, s+T), (H^2)^3)$, where the existence time is given by inequality 7.17:

$$T = \min(1, C_\nu \frac{1}{(\|\vec{u}(s, .)\|_{H^1} + \|\mathbb{F}\|_{L^2((s, s+1), H^1)})^4}).$$

Since $\vec{u}$ is a Leray solution on $(s, s + T)$, we find that $\vec{u} = \vec{v}$ on $(s, s + T)$ (due to weak-strong uniqueness). Thus, $\vec{u}$ is smooth on $(s, s + T) \times \mathbb{R}^3$ and $s + T < \tau$. As $T \geq C_{\nu, \mathbb{F}}(1 + \|\vec{u}(s, .)\|_{H^1})^{-4}$, we find that $1 + \|\vec{u}(s, .)\|_{H^1} \geq C_{\nu, \mathbb{F}}^{1/4}(\tau - s)^{-1/4}$.

Let $\epsilon > 0$ with $\epsilon < 16 C_{\nu, \mathbb{F}}$. We write

$$\Sigma_t \subset [0, \epsilon) \cup \cup_{\tau \in \Sigma_t, \tau \geq \epsilon}(\tau - \frac{1}{5}\epsilon, \tau + \frac{1}{5}\epsilon).$$

By the Vitali covering lemma, we may find $N \in \mathbb{N}$ and $\tau_1, \ldots, \tau_N \in \Sigma_t \cap [\epsilon, \infty)$ so that $\Sigma_t \subset [0, \epsilon) \cup_{i=1}^N (\tau_i - \epsilon, \tau_i + \epsilon)$ while $\min_{1 \leq i < j \leq N} |\tau_i - \tau_j| \geq 2\epsilon/5$. On $(\tau_i - \frac{2}{5}\epsilon, \tau_i)$, we have

$$\|\vec{u}(s, .)\|_{H^1} \geq \frac{1}{2} C_{\nu, \mathbb{F}}^{1/4}(\tau - s)^{-1/4}$$

and thus

$$\int_{(\tau_i - \frac{2}{5}\epsilon, \tau_i)} \|\vec{u}(s, .)\|_{H^1}^2 \, ds \geq \frac{1}{2} C_{\nu, \mathbb{F}}^{1/2}(\frac{2}{5}\epsilon)^{1/2}$$

Let $\mathcal{B} = ((\tau_i - \epsilon, \tau_i + \epsilon))_{0 \leq i \leq N}$ with $\tau_0 = 0$. We have

$$\sigma_{1/2}(\mathcal{B}) = (N+1)\epsilon^{1/2}$$

$$\leq \epsilon^{1/2} + \sum_{i=1}^{N} \sqrt{10} C_{\nu,\mathbb{F}}^{-1/2} \int_{\tau_i - 2\epsilon/5}^{\tau_i} \|\vec{u}\|_{H^1}^2 \, ds$$

$$\leq \epsilon^{1/2} + \sqrt{10} C_{\nu,\mathbb{F}}^{-1/2} \int_{d(s,\Sigma_t) \leq 2\epsilon/5} \|\vec{u}\|_{H^1}^2 \, ds.$$

Since we know that $\vec{u} \in (L^2 H^1)^3$, we find that $\mathcal{H}^{1/2}(\Sigma_t) \leq \sqrt{10} C_{\nu,\mathbb{F}}^{-1/2} \int_{\Sigma_t} \|\vec{u}\|_{H^1}^2 \, ds$. In particular, $\mathcal{H}^{1/2}(\Sigma_t) < \infty$; hence, the Lebesgue measure of Σ_t is equal to 0; this gives $\int_{\Sigma_t} \|\vec{u}\|_{H^1}^2 \, ds = 0$ and finally $\mathcal{H}^{1/2}(\Sigma_t) = 0$. □

13.7 The Local Energy Inequality

Scheffer studied the partial regularity of the Leray weak solutions. More precisely, he has been interested in the set Σ which is the complement in $(0, +\infty) \times \mathbb{R}^3$ of the set of regular points of the solution $\vec{u}$, i.e., of points (t, x) in the neighborhood of which $\vec{u}$ is a continuous function of time and space variables. We have, of course, $\Sigma \subset \Sigma_t \times \mathbb{R}^3$.

In Scheffer [425], he considered a maximal interval of regularity $I = (t_0, t_1)$ such that $t_0, t_1 \in \Sigma_t$ and $I \cap \Sigma_t = \emptyset$ and he showed that $\mathcal{H}^1(\Sigma \cap (\{t_1\} \times \mathbb{R}^3)) < \infty$. In [426], Scheffer then introduced the so-called *local energy inequality* and he was able to prove that $\mathcal{H}^2(\Sigma) < \infty$. This local energy inequality turned out to be a fundamental tool in the partial regularity theory of Caffarelli, Kohn and Nirenberg [74], in Lemarié-Rieusset's theory of uniformly locally square integrable solutions [313] and in Jia and Šverak's theory of self-similar solutions [245].

Local energy inequality

Theorem 13.6.
Let $\vec{u}_0 \in L^2$ with $\mathrm{div}\, \vec{u}_0 = 0$ and $\vec{f} \in L^2((0,T), H^{-1})$. The solution $\vec{u}$ of the Navier–Stokes problem with initial value $\vec{u}_0$ and forcing term $\vec{f}$ constructed in Theorem 12.2 satisfies the local Leray energy inequality: there exists a non-negative locally finite measure μ on $(0,T) \times \mathbb{R}^3$ such that

$$\partial_t |\vec{u}|^2 = \nu \Delta |\vec{u}|^2 - 2\nu |\vec{\nabla} \otimes \vec{u}|^2 - \mathrm{div}((|\vec{u}|^2 + 2p)\vec{u}) + 2\vec{u} \cdot \vec{f} - \mu \qquad (13.11)$$

Proof. Recall that $\vec{u}$ is constructed as the limit of $\vec{u}_{(\epsilon_n)}$, a sequence of solutions of the mollified equation, such that:

- on every bounded subinterval of $[0,T]$, $\vec{u}_{(\epsilon_n)}$ is *-weakly convergent to $\vec{u}$ in $L^\infty L^2$ and in $L^2 \dot{H}^1$

- $\vec{u}_{(\epsilon_n)}$ is strongly convergent to $\vec{u}$ in $L^2_{\mathrm{loc}}((0,T) \times \mathbb{R}^3)$.

We write

$$\partial_t \vec{u}_{(\epsilon)} = \nu \Delta \vec{u}_{(\epsilon)} + \vec{f} - (\vec{u}_{(\epsilon)} * \theta_\epsilon) \cdot \vec{\nabla} \vec{u}_{(\epsilon)} - \vec{\nabla} p_{(\epsilon)}$$

with

$$p_{(\epsilon)} = \frac{1}{\Delta} \operatorname{div}\left(\vec{f} - (\vec{u}_{(\epsilon)} * \theta_\epsilon) \cdot \vec{\nabla}\,\vec{u}_{(\epsilon)}\right)$$

and we write

$$\partial_t\left(\frac{|\vec{u}_{(\epsilon)}|^2}{2}\right) = \nu\Delta\left(\frac{|\vec{u}_{(\epsilon)}|^2}{2}\right) - \nu|\vec{\nabla} \otimes \vec{u}_{(\epsilon)}|^2 + \vec{f}\cdot\vec{u}_{(\epsilon)}$$

$$- \operatorname{div}\left(\frac{|\vec{u}_{(\epsilon)}|^2}{2}(\vec{u}_{(\epsilon)} * \theta_\epsilon)\right) - \operatorname{div}(p_{(\epsilon)}\vec{u}_{(\epsilon)})$$

We know that $\vec{u}_{(\epsilon_n)}$ converge strongly to $\vec{u}$ in $L^2_{\text{loc}}((0,T)\times\mathbb{R}^3)$; as the family is bounded in $L^{10/3}_t H^{3/5}_x \subset L^{10/3}_t L^{10/3}_x$, we find that we have strong convergence in $L^3_{\text{loc}}((0,T)\times\mathbb{R}^3)$ as well. Thus, we have the following convergence results in $\mathcal{D}'((0,T)\times\mathbb{R}^3)$: $\partial_t|\vec{u}_{(\epsilon_n)}|^2 \to \partial_t|\vec{u}|^2$, $\Delta|\vec{u}_{(\epsilon_n)}|^2 \to \Delta|\vec{u}|^2$, $\operatorname{div}(|\vec{u}_{(\epsilon_n)}|^2(\vec{u}_{(\epsilon_n)} * \theta_\epsilon)) \to \operatorname{div}(|\vec{u}|^2\vec{u})$ and $\vec{u}_{(\epsilon_n)}\cdot\vec{f} \to \vec{u}\cdot\vec{f}$. Similarly, we have that

$$\operatorname{div}\left(\left(\frac{1}{\Delta}\operatorname{div}\vec{f} + \sum_{j=1}^{3}\sum_{l=1}^{3}\frac{\partial_j}{\sqrt{-\Delta}}\frac{\partial_l}{\sqrt{-\Delta}}(u_{(\epsilon_n),j}(u_{(\epsilon_n,l)} * \theta_{\epsilon_n})))\right)\vec{u}_{(\epsilon_n)}\right)$$

converges in $\mathcal{D}'$ to $\operatorname{div}\left(\left(\frac{1}{\Delta}\operatorname{div} f + \sum_{j=1}^{3}\sum_{l=1}^{3}\frac{\partial_j}{\sqrt{-\Delta}}\frac{\partial_l}{\sqrt{-\Delta}}(u_j u_l))\vec{u}\right)\right.$.

Thus far, we have got that

$$\partial_t|\vec{u}|^2 = \nu\Delta|\vec{u}|^2 - \operatorname{div}((|\vec{u}|^2 + 2p)\vec{u}) + 2\vec{u}\cdot\vec{f} - \nu T$$

with

$$T = \lim_{\epsilon_n \to 0} 2|\vec{\nabla} \otimes \vec{u}_{(\epsilon_n)}|^2.$$

Let $\phi \in \mathcal{D}'((0,T)\times\mathbb{R}^3)$ be a non-negative function. As $\sqrt{\phi}\,\vec{\nabla} \otimes \vec{u}_{(\epsilon_n)}$ is weakly convergent to $\sqrt{\phi}\,\vec{\nabla} \otimes \vec{u}$ in $L^2_t L^2_x$, we find that $\|\sqrt{\phi}\,\vec{\nabla} \otimes \vec{u}\|^2_2 \le \liminf_{\epsilon_n \to 0}\|\sqrt{\phi}\,\vec{\nabla} \otimes \vec{u}_{(\epsilon_n)}\|^2_2$. Thus, we have

$$\langle T|\phi\rangle_{\mathcal{D}',\mathcal{D}} = 2\lim_{\epsilon_n \to 0}\iint|\vec{\nabla} \otimes \vec{u}_{(\epsilon_n)}|^2\,\phi(t,x)\,dt\,d$$

$$\ge 2\iint|\vec{\nabla} \otimes \vec{u}|^2\phi(t,x)\,dt\,dx.$$

Thus, $\nu T = 2\nu|\vec{\nabla} \otimes \vec{u}|^2 + \mu$, where μ is a non-negative locally finite measure. □

A natural question is to find a criterion when we have indeed *local energy equality*. An easy criterion is the following one:

Proposition 13.3.
Let $Q = I \times \Omega$, where $I = (a,b)$ and $\Omega = B(x_0, r)$. Let $\vec{u} \in L^\infty(I, L^2(\Omega)) \cap L^2(I, H^1(\Omega))$, $\vec{f} \in L^2(I, H^{-1}(\Omega))$ and $p \in \mathcal{D}'(Q)$, and assume that $\vec{u}$ is a weak solution on Q of the Navier–Stokes equations

$$\partial_t\vec{u} = \nu\Delta\vec{u} - \vec{u}\cdot\vec{\nabla}\vec{u} + \vec{f} - \vec{\nabla}p, \quad \operatorname{div}\vec{u} = 0.$$

If $p \subset L^1_{t,x}(Q)$, then

- *for every $0 < \rho < r$, $p \in L^1((a,b), L^2(B(x_0, \rho)))$*

- *the quantity*

$$\mu = -\partial_t|\vec{u}|^2 + \nu\Delta|\vec{u}|^2 - 2\nu|\vec{\nabla}\otimes\vec{u}|^2 - \operatorname{div}((|\vec{u}|^2 + 2p)\vec{u}) + 2\vec{u}\cdot\vec{f}$$

is well defined in $\mathcal{D}'(Q)$

- *if moreover* $\vec{u}\in L^4_{t,x}(Q)$, *then* $\mu = 0$

Proof. We first check the regularity of p. We take the divergence of the Navier–Stokes equations and get that, on Q, we have

$$\Delta p = -\sum_{i=1}^{3}\sum_{j=1}^{3}\partial_i\partial_j(u_iu_j) + \operatorname{div}\vec{f}$$

We then pick up a function $\phi\in\mathcal{D}(\mathbb{R}^3)$ supported in $B(x_0\frac{r+\rho}{2})$ such that $\phi(t,x) = 1$ on $B(x_0,\frac{r+3\rho}{4})$. We write

$$\Delta(\phi p) = -\phi\sum_{i=1}^{3}\sum_{j=1}^{3}\partial_i\partial_j(u_iu_j) + \phi\operatorname{div}\vec{f} - p\Delta\phi + 2\sum_{i=1}^{3}\partial_i(p\partial_i\phi).$$

We have:

- $R_1 = -\phi\sum_{i=1}^{3}\sum_{j=1}^{3}\partial_i\partial_j(u_iu_j)\in L^1((a,b),H^{-3/2}(\mathbb{R}^3))$ and is supported in the fixed compact set $\bar{B}(x_0,r)$, hence $R_1\in L^1((a,b),\dot{H}^{-2}+\dot{H}^{-1}(\mathbb{R}^3))$ and $\frac{1}{\Delta}R_1\in L^1((a,b),L^2+H^1)$ so that $1_Q\frac{1}{\Delta}R_1\in L^1L^2$

- similarly, $R_2 = \phi\operatorname{div}\vec{f}\in L^1((a,b),H^{-2}(\mathbb{R}^3))$ and is supported in the fixed compact set $\bar{B}(x_0,r)$, so that $R_2\in L^1((a,b),\dot{H}^{-2}+\dot{H}^{-1}(\mathbb{R}^3))$ and thus $1_Q\frac{1}{\Delta}R_2\in L^1L^2$

- $R_3 = -p\Delta\phi\in L^1L^1$, so that $\frac{1}{\Delta}R_3\in L^1((a,b),L^{3,\infty})$ so that $1_Q\frac{1}{\Delta}R_3\in L^1L^2$

- for estimating $R_4 = 2\sum_{i=1}^{3}\partial_i(p\partial_i\phi)$, we write $p\partial_i\phi\in L^1L^1$ so that $\frac{1}{\Delta}R_4\in L^1((a,b),L^{3/2,\infty})$ so that $1_Q\frac{1}{\Delta}R_4\in L^1L^{6/5}$

Thus far, we have just obtained that $\phi p\in L^1L^{6/5}$. But then reiterating the argument on a smaller ball, we find that, in estimating R_4, we may replace $p\partial_i\phi\in L^1L^1$ by $p\partial_i\phi\in L^1L^{6/5}$ and find that $1_Q\frac{1}{\Delta}R_4\in L^1L^2$.

Thus, we find that $\phi p\in L^1L^2$ and thus that μ is well defined. Moreover, if we consider a relatively compact open subset $O = (c,d)\times B(x,\rho)$ of Q and a mollifier θ_ϵ with $\epsilon < r-\rho$, we may define on O $\vec{u}_\epsilon = \vec{u}*\theta_\epsilon$; we have on O $\vec{u}_\epsilon\in L^\infty_t L^2_x$ and $\partial_t\vec{u}_\epsilon\in L^1_t L^2_x$, so that $\vec{u}_\epsilon\in\mathcal{C}([c,d],L^2)$; using the density of smooth functions in $\{\vec{v}\ /\ \vec{u}_\epsilon\in\mathcal{C}([c,d],L^2)$ and $\partial_t\vec{v}\in L^1L^2\}$, we find that

$$\partial_t|\vec{u}_\epsilon|^2 = 2\vec{u}_\epsilon\partial_t\vec{u}_\epsilon$$

and thus

$$\partial_t|\vec{u}_\epsilon|^2 = \nu\Delta|\vec{u}_\epsilon|^2 - 2\nu|\vec{\nabla}\otimes\vec{u}_\epsilon|^2 + 2\vec{u}_\epsilon*(\theta_\epsilon*\vec{f})$$
$$- 2\operatorname{div}((p*\theta_\epsilon)\vec{u}_\epsilon) - 2\vec{u}_\epsilon.\theta_\epsilon*\operatorname{div}(\vec{u}\otimes\vec{u})$$

We have the strong convergence of $\vec{u}_\epsilon$ to $\vec{u}$ in $L^2L^2(O)$, of $\vec{\nabla}\otimes\vec{u}_\epsilon$ to $\vec{\nabla}\otimes\vec{u}$ in $L^2L^2(O)$, of $\vec{u}_\epsilon$ to $\vec{u}$ in $L^2H^1(O)$, of $\theta_\epsilon*\vec{f}$ to $\vec{f}$ in $L^2H^{-1}(O)$, of $p*\theta_\epsilon$ to p in $L^1L^2(O)$ and the *-weak convergence of $\vec{u}_\epsilon$ to $\vec{u}$ in $L^\infty L^2$ so that we find

$$\mu = -\operatorname{div}(|\vec{u}|^2\vec{u}) + 2\lim_{\epsilon\to 0^+}\vec{u}_\epsilon.\theta_\epsilon*\operatorname{div}(\vec{u}\otimes\vec{u}) \tag{13.12}$$

Of course, when $\vec{u}\in L^4L^4$, we find that $\theta_\epsilon*\operatorname{div}(\vec{u}\otimes\vec{u})$ converges strongly to $\operatorname{div}(\vec{u}\otimes\vec{u})$ in $L^2H^{-1}(O)$ so that $\mu = 0$. $\qquad\square$

Of course, if $\vec{u}$ satisfies the hypotheses of Theorem 13.2 on interior regularity and if $p \in L^1 L^1$, then we find $\mu = 0$: we have $\vec{u} \in L^\infty L^2 \cap L^2 H^1$ and $\vec{u} \in \mathbb{X}$, hence $|\vec{u}|^2 \in L^2 L^2$, so that $\vec{u} \in L^4 L^4 \dots$

However, one may find a weaker assumption on $\vec{u}$ that grants that $\mu = 0$. This assumption has been described by Duchon and Robert [159], in a paper that generalizes the results of Constantin, E and Titi [127] on Onsager's conjecture [381]. This result of Duchon and Robert underlines the link between a minimal regularity of $\vec{u}$ and the fact that $\mu = 0$:

Energy equality

Theorem 13.7.
Let $Q = I \times \Omega$, where $I = (a, b)$ and $\Omega = B(x_0, r)$. Let $\vec{u} \in L^\infty(I, L^2(\Omega)) \cap L^2(I, H^1(\Omega))$, $\vec{f} \in L^2(I, H^{-1}(\Omega))$ and $p \in L^1_{t,x}(Q)$, and assume that $\vec{u}$ is a weak solution on Q of the Navier–Stokes equations

$$\partial_t \vec{u} = \nu \Delta \vec{u} - \vec{u} \cdot \vec{\nabla} \vec{u} + \vec{f} - \vec{\nabla} p, \quad \operatorname{div} \vec{u} = 0.$$

Let μ be the distribution

$$\mu = -\partial_t |\vec{u}|^2 + \nu \Delta |\vec{u}|^2 - 2\nu |\vec{\nabla} \otimes \vec{u}|^2 - \operatorname{div}((|\vec{u}|^2 + 2p)\vec{u}) + 2\vec{u} \cdot \vec{f}.$$

Let us define, for $0 < \rho < r$,

$$A(\rho) = \liminf_{\epsilon \to 0^+} \frac{1}{\epsilon^4} \int_a^b \int_{x \in B(x_0, \rho)} \int_{y \in B(0, \epsilon)} |\vec{u}(t, x) - \vec{u}(t, x+y)|^3 \, dt \, dx \, dy.$$

If $A(\rho) = 0$ for every $\rho \in (0, r)$, then $\mu = 0$.

Proof. We start from Equality (13.12) expresssing μ as a limit:

$$\mu = -\operatorname{div}(|\vec{u}|^2 \vec{u}) + 2 \lim_{\epsilon \to 0^+} \vec{u}_\epsilon . \theta_\epsilon * \operatorname{div}(\vec{u} \otimes \vec{u})$$

We introduce the distribution

$$T_\epsilon = \sum_{k=1}^{3} \int \partial_k \theta_\epsilon(y)(u_k(x-y) - u_k(x))|\vec{u}(x-y) - \vec{u}(x)|^2 \, dy$$

which is well defined on $(a, b) \times B(x_0, \rho)$ for $\epsilon < \frac{r-\rho}{2}$, as $\vec{u} \in L^\infty L^2 \cap L^2 H^1$ on Q, so that $\vec{u} \in L^4 L^3$ on $(a, b) \times B(x_0, (r + \rho)/2)$.

On $(a, b) \times B(x_0, \rho)$, we have $\sum_{k=1}^{3} \int \partial_k \theta_\epsilon(y)(u_k(x-y) - u_k(x)) \, dy = 0$ (as $\operatorname{div} \vec{u} = 0$), so that

$$T_\epsilon = \sum_{k=1}^{3} \int \partial_k \theta_\epsilon(y)(u_k(x-y) - u_k(x))(|\vec{u}(x-y)|^2 - 2\vec{u}(x-y) \cdot \vec{u}(x)) \, dy$$

Moreover,

$$\lim_{\epsilon \to 0} \theta_\epsilon * (|\vec{u}|^2 \vec{u}) - (\theta_\epsilon * |\vec{u}|^2)\vec{u} = 0 \text{ in } L^1_{t,x}((a, b) \times B(x_0, \rho))$$

so that

$$\sum_{k=1}^{3} \int \partial_k \theta_\epsilon(y)(u_k(x-y) - u_k(x))|\vec{u}(x-y)|^2 \, dy = \operatorname{div}(\theta_\epsilon * (|\vec{u}|^2 \vec{u}) - (\theta_\epsilon * |\vec{u}|^2)\vec{u}) \to 0$$

where the limit is taken in $\mathcal{D}'((a, b) \times B(x_0, \rho))$.

Similarly, we introduce the distribution

$$S_\epsilon = \sum_{k=1}^{3} \int \partial_k \theta_\epsilon(y)(u_k(x-y) - u_k(x))((\vec{u}(x-y) - \vec{u}(x)) . (\vec{u}_\epsilon(x) - \vec{u}(x)) \, dy$$

which is equal as well to

$$S_\epsilon = \sum_{k=1}^{3} \int \partial_k \theta_\epsilon(y)(u_k(x-y) - u_k(x))(\vec{u}(x-y) . (\vec{u}_\epsilon(x) - \vec{u}(x)) \, dy$$

We thus have $2S_\epsilon - T_\epsilon = A_\epsilon + B_\epsilon + C_\epsilon$ with

$$A_\epsilon = 2 \sum_{k=1}^{3} \int \partial_k \theta_\epsilon(y) u_k(x-y)\vec{u}(x-y) \cdot \vec{u}_\epsilon(x) \, dy = 2\vec{u}_\epsilon . \theta_\epsilon * \operatorname{div}(\vec{u} \otimes \vec{u})$$

$$B_\epsilon = -2 \sum_{k=1}^{3} \int \partial_k \theta_\epsilon(y) u_k(x)\vec{u}(x-y) \cdot \vec{u}_\epsilon(x) \, dy = -\, 2\vec{u}_\epsilon . (\vec{u} \cdot \vec{\nabla}\vec{u}_\epsilon)$$

$$= -\operatorname{div}(|\vec{u}_\epsilon|^2 \vec{u}) \to -\operatorname{div}(|\vec{u}|^2 \vec{u})$$

and

$$\lim_{\epsilon \to 0^+} C_\epsilon = 0 \text{ in } \mathcal{D}'.$$

Thus, we find that

$$\mu = \lim_{\epsilon \to 0} 2S_\epsilon - T_\epsilon.$$

This is the formula given by Duchon and Robert. We have

$$|T_\epsilon(t,x)| \le C\frac{1}{\epsilon^4} \int_{|y|<\epsilon} |\vec{u}(t, x+y) - \vec{u}(t,x)|^3 \, dy.$$

Similarly, writing $\vec{u}_\epsilon(t,x) - \vec{u}(t,x) = \int \theta_\epsilon(y)(\vec{u}(t, x-y) - \vec{u}(t,x)) \, dy$, we get

$$|S_\epsilon(t,x)| \le C\frac{1}{\epsilon^7} \left(\int_{|y|<\epsilon} |\vec{u}(t, x+y) - \vec{u}(t,x)|^2 \, dy \right) \left(\int_{|y|<\epsilon} |\vec{u}(t, x+y) - \vec{u}(t,x)| \, dy \right)$$

$$\le C'\frac{1}{\epsilon^4} \int_{|y|<\epsilon} |\vec{u}(t, x+y) - \vec{u}(t,x)|^3 \, dy.$$

Thus, if $A(\rho) = 0$, we find that $\lim_{\epsilon \to 0^+} \iint_{(a,b) \times B(x_0,\rho)} |2S_\epsilon - T_\epsilon| \, dt \, dx = 0$ and $\mu = 0$. $\quad\square$

Thus, we can see that the equality $\mu = 0$ is granted when locally $\vec{u}$ belongs to $L_t^3 b_{3,\infty}^{1/3}$, where $b_{3,\infty}^{1/3}$ is the closure of $\mathcal{D}$ in the Besov space $B_{3,\infty}^{1/3}$: if $\phi\vec{u} \in L_t^3 b_{3,\infty}^{1/3}$, then

- $\int |\phi(t,x)\vec{u}(t,x) - \phi(t, x+y)\vec{u}(t, x+y)|^3 \, dx \le C\|\phi\vec{u}\|_{B_{3,\infty}^{1/3}}^3 |y|$ and

$$\lim_{y \to 0} \frac{1}{|y|} \int |\phi(t,x)\vec{u}(t,x) - \phi(t, x+y)\vec{u}(t, x+y)|^3 \, dx = 0$$

- by dominated convergence, we get

$$\lim_{\epsilon \to 0^+} \frac{1}{\epsilon^4} \iiint_{|y|<\epsilon} |\phi(t,x)\vec{u}(t,x) - \phi(t, x+y)\vec{u}(t, x+y)|^3 \, dt \, dx \, dy = 0.$$

In particular, we may check that Duchon and Robert's criterion is based on a weaker assumption than $\vec{u} \in L^4_{t,x}(Q)$: if $v \in \dot{H}^1 \cap L^4$ then $v \in \dot{b}^{1/3}_{3,\infty}$ and

$$\|v\|_{\dot{B}^{1/3}_{3,\infty}} \leq C \|v\|^{1/3}_{\dot{H}^1} \|v\|^{2/3}_4. \tag{13.13}$$

Indeed let $I_p = \iint |v(t,x) - v(t, x+y)|^p \, dt \, dx$. We have $I_2 \leq \|v\|^2_{\dot{H}^1} |y|^2$ and $I_4 \leq 16\|v\|^4_4$ so that

$$I_3 \leq (I_2)^{1/2}(I_4)^{1/2} \leq 4\|v\|_{\dot{H}^1} \|v\|^2_4 \, |y|$$

Thus, (13.13) is proved.

13.8 The Caffarelli-Kohn-Nirenberg Theorem on Partial Regularity

The celebrated regularity criterion of Caffarelli, Kohn and Nirenberg [74] states that if $\vec{u}$ is a solution of the Navier–Stokes equations in a neighborhood of a point $(t_0, x_0) \in \mathbb{R} \times \mathbb{R}^3$ which satisfies "some conditions" on the velocity $\vec{u}$, the pressure p and the force $\vec{f}$ and if the number

$$\limsup_{r \to 0^+} \frac{1}{r} \iint_{(t_0 - r^2, t_0 + r^2) \times B(x_0, r)} |\vec{\nabla} \otimes \vec{u}|^2 \, ds \, dx$$

is "small enough," then (t_0, z_0) is a "regular point." The definitions of a regular point and the choice of the admissible conditions on $\vec{u}$, p and $\vec{f}$ have been discussed by many authors.

In the original paper of Caffarelli, Kohn and Nirenberg [74], assumptions on $\vec{u}$, p and $\vec{f}$ were:

1. $\vec{u}$, p and $\vec{f}$ are defined on a cylinder $Q = (T, T + R^2) \times B(X, R)$

2. on Q, $\vec{u}$ belongs to $L^\infty_t L^2_x \cap L^2_t H^1_x$

3. on Q, $\iint_Q |p(t,x)|^{5/4} \, dt \, dx < +\infty$

4. on Q, $\iint_Q |\vec{f}(t,x)|^q \, dt \, dx < +\infty$ for some $q > 5/2$

5. $\vec{u}$ is a solution of the Navier–Stokes equations on Q: div $\vec{u} = 0$ and

$$\partial_t \vec{u} = \nu \Delta \vec{u} - \vec{u} \cdot \vec{\nabla} \vec{u} + \vec{f} - \vec{\nabla} p \text{ in } \mathcal{D}'(Q) \tag{13.14}$$

6. $\vec{u}$ is a suitable solution, i.e., there exists a non-negative distribution μ such that

$$\partial_t |\vec{u}|^2 = \nu \Delta |\vec{u}|^2 - 2\nu |\vec{\nabla} \otimes \vec{u}|^2 - \text{div}((|\vec{u}|^2 + 2p)\vec{u}) + 2\vec{u} \cdot \vec{f} - \mu \tag{13.15}$$

7. regularity of $\vec{u}$ at (t_0, z_0) is meant in the sense of Definition 13.1: $\vec{u}$ is bounded in the neighborhood of (t_0, z_0)

The reason for the exponent $5/4$ in the assumption $p \in L^{5/4}_{t,x}(Q)$ was that $5/4$ was at that time the best exponent one could prove when exhibiting suitable solutions for the Navier–Stokes equations on a *bounded* domain associated with a square–integrable initial value $\vec{u}_0$. But Lin [335] proved the existence of suitable solutions with $p \in L^{3/2}_{loc}((0, T) \times \Omega)$ for bounded domains Ω, by using regularity estimates for the pressure obtained by Sohr and von Wahl [444]. The computations were much easier with the hypothesis $p \in L^{3/2}(Q)$, so Lin could give a simplified proof of the Caffarelli–Kohn–Nirenberg criterion.

While the pressure for Leray solutions on the whole space is entirely determined by the equation $\Delta p = \operatorname{div} \vec{f} - \sum_{i=1}^{3} \sum_{j=1}^{3} \partial_i \partial_j (u_i u_j)$, this is no longer the case when studying a local solution of the Navier–Stokes equations. Thus, pressure has to be dealt with very carefully. Some variants of the Caffarelli, Kohn and Nirenberg theorem involve different assumptions on the pressure: for instance, Vasseur [487] gave a proof (with $\vec{f} = 0$) under the assumption $p \in L^q_t L^1_x(Q)$ with a different method (instead of estimating quadratic means of $\vec{u}$ on small cylinders, as in the other references, he used an à la Di Giorgi method and estimated the measure of level sets $\{(t,x) \ / \ |\vec{u}| > \lambda\}$). Wolf [504] considered an extended version of suitable solutions in order to include in the pressure the harmonic term that is deleted when applying the divergence operator to the equation.

In this section, we are going to make the following assumptions on $\vec{u}$, $\vec{f}$ and p:

Hypotheses for the Caffarelli-Kohn-Nirenberg regularity criterion

Definition 13.4.
We call $(\mathcal{H}_{CKN})$ the following set of hypotheses:

1. *$\vec{u}$, p and $\vec{f}$ are defined on a domain $\Omega \subset \mathbb{R} \times \mathbb{R}^3$*

2. *on Ω, $\vec{u}$ belongs to $L^\infty_t L^2_x \cap L^2_t \dot{H}^1_x$:*

$$\sup_{t\in\mathbb{R}} \int_{(t,x)\in\Omega} |\vec{u}(t,x)|^2 \, dx < +\infty \quad and \quad \iint_\Omega |\vec{\nabla} \otimes \vec{u}|^2 \, dt \, dx < +\infty$$

3. *for some $q_0 > 1$, p belongs to $L^{q_0}_t L^1_x(\Omega)$:*

$$\int_\mathbb{R} (\int_{(t,x)\in\Omega} |p(t,x)| \, dx)^{q_0} \, dt < +\infty$$

4. *on Ω, $\vec{f}$ is a divergence free vector field in $L^{10/7}_{t,x}(\Omega)$:*

$$\operatorname{div} \vec{f} = 0 \quad and \quad \iint_\Omega |\vec{f}(t,x)|^{10/7} \, dt \, dx < +\infty$$

5. *$\vec{u}$ is a solution of the Navier–Stokes equations on Ω: $\operatorname{div} \vec{u} = 0$ and*

$$\partial_t \vec{u} = \nu \Delta \vec{u} - \vec{u} \cdot \vec{\nabla} \vec{u} + \vec{f} - \vec{\nabla} p \ in \ \mathcal{D}'(\Omega) \tag{13.16}$$

We have, of course, some further estimates on $\vec{u}$ and p that we can deduce from $(\mathcal{H}_{CKN})$. If I is a bounded interval of $\mathbb{R}$ and $B = B(x_B, r_B)$ a ball of $\mathbb{R}^3$ such that $I \times B(x_B, 2r_B) \subset \Omega$, then we have the following properties:

- $\iint_{I\times B} |\vec{u}||\vec{f}| \, dt \, dx < +\infty$: by Sobolev inequality, we have

$$\int_I (\int_B |\vec{u}(t,x)|^6 \, dx)^{1/3} \, dt \leq C \int_I \int_B \frac{|u|^2}{|B|^{2/3}} + |\vec{\nabla} \otimes \vec{u}|^2 \, dx \, dt$$
$$\leq C(\frac{|I|}{|B|^{2/3}} \|\vec{u}\|^2_{L^\infty_t L^2_x(\Omega)} + \|\vec{\nabla} \otimes \vec{u}\|^2_{L^2_t L^2_x(\Omega)}) \tag{13.17}$$

By interpolation between $L^\infty L^2$ and $L^2 L^6$, we find that

$$\vec{u} \in L_{t,x}^{10/3}(I \times B) \tag{13.18}$$

• $\iint_{I \times B} |\vec{u}||p|\, dt\, dx < +\infty$: taking the divergence of the Navier–Stokes equations and using div $\vec{u} = $ div $\vec{f} = 0$, we get:

$$\Delta p = -\operatorname{div}(\vec{u} \cdot \vec{\nabla} \vec{u}) = -\sum_{i=1}^{3} \sum_{j=1}^{3} \partial_i \partial_j (u_i u_j). \tag{13.19}$$

Now, we introduce a function $\omega \in \mathcal{D}(\mathbb{R}^3)$ with $\omega = 1$ on $B(0, 5/4)$ and with Supp $\omega \subset B(0, 7/4)$ and we define $\zeta_B(x) = \omega(\frac{x - x_B}{r_B})$. Let G be the fundamental solution of $-\Delta$ (so that $-\Delta G = \delta$):

$$G = \frac{1}{4\pi|x|}.$$

We have $\zeta_B p = G * (-\Delta(\zeta_B p))$, with

$$-\Delta(\zeta_B p) = p(-\Delta \zeta_B) - 2\sum_{i=1}^{3} \partial_i \zeta_B \partial_i p - \zeta_B \Delta p$$

$$= p(\Delta \zeta_B) - 2\sum_{i=1}^{3} \partial_i (p \partial_i \zeta_B) + \zeta_B \sum_{i=1}^{3}\sum_{j=1}^{3} \partial_i \partial_j (u_i u_j)$$

$$= p(\Delta \zeta_B) - 2\sum_{i=1}^{3} \partial_i (p \partial_i \zeta_B) + \sum_{i=1}^{3}\sum_{j=1}^{3} \partial_i \partial_j (\zeta_B u_i u_j)$$

$$+ \sum_{i=1}^{3}\sum_{j=1}^{3} u_i u_j \partial_i \partial_j \zeta_B - 2\sum_{i=1}^{3}\sum_{j=1}^{3} \partial_i (u_i u_j \partial_j \zeta_B)$$

We find:

$$\zeta_B p = \varpi_B + p_B + q_B \tag{13.20}$$

with

$$\begin{cases} \varpi_B = \displaystyle\sum_{j=1}^{3}\sum_{l=1}^{3} \partial_j \partial_l G * (\zeta_B u_j u_l) \\[2mm] q_B = \displaystyle -2\sum_{j=1}^{3}\sum_{l=1}^{3} \partial_j G * ((\partial_l \zeta_B) u_j u_l) + \sum_{j=1}^{3}\sum_{l=1}^{3} G * ((\partial_j \partial_l \zeta_B) u_j u_l) \\[2mm] p_B = \displaystyle -2\sum_{j=1}^{3} \partial_j G * ((\partial_j \zeta_B) p) + G * ((\Delta \zeta_B) p) \end{cases}$$

When $(t, x) \in I \times B$, we find that

$$\begin{cases} |q_B(t,x)| \leq C \displaystyle\sum_{j=1}^{3}\sum_{l=1}^{3} \frac{1}{r_B^3} \int_{B(x_B, 2r_B)} |u_j(t,y) u_l(t,y)|\, dy \\[2mm] |p_B(t,x)| \leq C \displaystyle\frac{1}{r_B^3} \int_{B(x_B, 2r_B)} |p(t,y)|\, dy \end{cases}$$

Thus, on $I \times B$, we have $p = \varpi_B + p_B + q_B$ with $\varpi_B \in L_{t,x}^{5/3}(I \times B)$, $p_B \in L_t^{q_0} L_x^{\infty}(I \times B)$ and $q_B \in L_{t,x}^{\infty}(I \times B)$ and we have, as $\vec{u} \in L_t^{\infty} L_x^2 \cap L_{t,x}^{10/3}(I \times B)$,

$$\iint_{I \times B} |p||\vec{u}|\, dx\, dt < +\infty \tag{13.21}$$

Thus, the distribution

$$\mu = -\partial_t |\vec{u}|^2 + \nu \Delta |\vec{u}|^2 - 2\nu |\vec{\nabla} \otimes \vec{u}|^2 + 2\vec{u} \cdot \vec{f} - \operatorname{div}((|\vec{u}|^2 + 2p)\vec{u}) \tag{13.22}$$

is well defined on Ω.

Suitable solutions

Definition 13.5.
The solution $\vec{u}$ is suitable if the distribution μ is a non-negative locally finite measure on Ω.

We are going to prove Caffarelli, Kohn and Nirenberg's result in the setting of parabolic Morrey spaces, following the papers by Ladyzhenskaya and Seregin [297] and by Kukavica [286] (a clear survey is given in the lecture notes of Robinson [415]).

Let ρ_2 be the parabolic "norm" given by $\rho_2(t, x) = |t|^{1/2} + |x|$. A function h on $\mathbb{R} \times \mathbb{R}^3$ is Hölderian of exponent $\alpha \in (0, 1)$ with respect to the parabolic distance if we have

$$|h(t, x) - h(s, y)| \le C_h(|t - s|^{1/2} + |x - y|)^{\alpha}.$$

A function h belongs to the parabolic Morrey space $\mathcal{M}_2^{q, \tau}$ $(1 < q \le \tau < +\infty)$ if and only if

$$\|h\|_{\mathcal{M}_2^{q, \tau}} < +\infty$$

with

$$\|h\|_{\mathcal{M}_2^{q, \tau}}^q = \sup_{(t,x) \in \mathbb{R} \times \mathbb{R}^3, r > 0} \frac{1}{r^{5(1 - \frac{q}{\tau})}} \iint_{\rho_2(t-s, x-y) < r} |h(s, y)|^q\, ds\, dy.$$

Of course, one may replace in this definition the balls $B((t, x), r)$ by the cylinders $Q_r(t, x) = (t - r^2, t + r^2) \times B(x, r)$, as we have $B((t, x), r) \subset Q_r(t, x) \subset B((t, x), 2r)$.

Moreover, when a function h is defined on a cylinder $Q_0 = Q_{r_0}(t_0, x_0)$, for estimating the parabolic Morrey of $1_{q_0} h$ (the function that is equal to h on Q_0 and to 0 elsewhere), i.e., to estimate

$$\sup_{(t,x) \in \mathbb{R} \times \mathbb{R}^3, r > 0} \frac{1}{r^{5(\frac{1}{q} - \frac{1}{\tau})}} I_r(t, x)^{1/q}$$

with

$$I_r(t, x) = \iint_{Q_r(t,x) \cap Q_0} |h(s, y)|^q\, ds\, dy,$$

there is no need to consider $r > r_0$: for $r \ge r_0$, we may write $I_r(t, x) \le I_{r_0}(t_0, x_0) \dots$ Moreover, if $r \le r_0$ and $Q_r(t, x) \cap Q_0 \ne \emptyset$, then there exists $(t_1, x_1) \in Q_0$ so that $I_r(t, x) \le I_{2r}(t_1, x_1)$; thus, there is no need as well to consider $(t, x) \notin Q_0$.

We may now state the theorem:

Caffarelli–Kohn–Nirenberg regularity criterion

Theorem 13.8.
Let Ω be a domain of $\mathbb{R} \times \mathbb{R}^3$. Let $(\vec{u}, p)$ a weak solution on Ω of the Navier–Stokes equations

$$\partial_t \vec{u} = \nu \Delta \vec{u} - \vec{u} \cdot \vec{\nabla} \vec{u} + \vec{f} - \vec{\nabla} p, \quad \operatorname{div} \vec{u} = 0.$$

Assume that

- *$(\vec{u}, p, \vec{f})$ satisfies the conditions $(\mathcal{H}_{CKN})$: $\vec{u} \in L^\infty L^2 \cap L^2 \dot{H}^1(\Omega)$, $p \in L^{q_0} L^1(\Omega)$ $(q_0 > 1)$, $\operatorname{div} \vec{f} = 0$ and $\vec{f} \in L^{10/7} L^{10/7}(\Omega)$*

- *$\vec{u}$ is suitable*

- *$1_\Omega(t,x)\vec{f} \in \mathcal{M}_2^{10/7, \tau_0}$ for some $\tau_0 > 5/2$.*

There exists a positive constant ϵ^* which depends only on ν and τ_0 such that, if for some $(t_0, x_0) \in \Omega$, we have

$$\limsup_{r \to 0} \frac{1}{r} \iint_{(t_0 - r^2, t_0 + r^2) \times B(x_0, r)} |\vec{\nabla} \otimes \vec{u}|^2 \, ds \, dx < \epsilon^*$$

then $\vec{u}$ is Hölderian (with respect to the quasi-norm $\delta(t,x) = |t|^{1/2} + |x|$) in a neighborhood of (t_0, x_0).

The proof relies on Campanato's lemma on Hölderian functions [79] applied to the regularity of solutions of parabolic equations:

Lemma 13.2 (Campanato's lemma).
Let $\rho_2(t,x) = |t|^{1/2} + |x|$, $Q_r(t,x) = \{(s,y) \in \mathbb{R} \times \mathbb{R}^3\} / \rho_2(t-s, x-y) < r\}$ and $M_{Q_r} f(t,x) = \frac{1}{|Q_r(t,x)|} |\iint_{Q_R(t,x)} f(s,y) \, ds \, dy$. Let $p \in [1, +\infty)$ and $f \in L^p_{loc}(dt \, dx)$. Let $0 < \alpha < 1$. Then f is Hölderian of exponent α with respect to the parabolic distance if and only if

$$\sup_{r>0} \sup_{t,x) \in \mathbb{R} \times \mathbb{R}^3} \frac{1}{r^\alpha} \Big(\frac{1}{|Q_r(t,x)|} \iint_{Q_R(t,x)} |f(s,y) - M_{Q_r(t,x)} f|^p \, ds \, dy\Big)^{1/p} < +\infty.$$

Proof. If f is Hölderian, just write

$$|f(s,y) - M_{Q_r(t,x)} f| = \frac{1}{|Q_r(t,x)|} |\iint_{Q_r(t,x)} f(s,y) - f(t,x) \, ds \, dy| \leq C_f r^\alpha.$$

Conversely, let

$$H_f = \sup_{r>0} \sup_{t,x) \in \mathbb{R} \times \mathbb{R}^3} \frac{1}{r^\alpha} \Big(\frac{1}{|Q_r(t,x)|} \iint_{Q_R(t,x)} |f(s,y) - M_{Q_r(t,x)} f|^p \, ds \, dy\Big)^{1/p}.$$

Let $\Phi \in \mathcal{D}(\mathbb{R} \times \mathbb{R}^3)$ supported in $Q_1(0,0)$ with $\iint \Phi \, dx \, dt = 1$ and $\Phi_\epsilon = \frac{1}{\epsilon^5} \Phi(\frac{t}{\epsilon^2}, \frac{x}{\epsilon})$. Let

$$F_\epsilon(l, x, s, y) = \Phi_\epsilon * f(t,x) - \Phi_\epsilon * f(s,y).$$

Then we have, for every t, x, y and ϵ,

$$\Phi_\epsilon * f(t,x) - \Phi_\epsilon * f(t,y)$$

$$= \iint (\Phi_\epsilon(t-\sigma, x-z) - \Phi_\epsilon(s-\sigma, y-z))f(\sigma,z)\,dz\,d\sigma$$

$$= \iint_{Q_{\epsilon+\rho_2(t-s,x-y)}(t,x)} (\Phi_\epsilon(t-\sigma, x-z) - \Phi_\epsilon(s-\sigma, y-z)) \times$$

$$\times (f(\sigma,z) - M_{Q_{\epsilon+\rho_2(t-s,x-y)}(t,x)}f)\,dz\,d\sigma$$

and

$$\Phi_\epsilon * f(t,x) - \Phi_{\epsilon/2} * f(t,x)$$

$$= \iint (\Phi_\epsilon(t-\sigma, x-z) - \Phi_{\epsilon/2}(t-\sigma, x-z))f(\sigma,z)\,dz\,d\sigma$$

$$= \iint_{Q_\epsilon(t,x)} (\Phi_\epsilon(t-\sigma, x-z) - \Phi_{\epsilon/2}(t-\sigma, x-z))(f(\sigma,z) - M_{Q_\epsilon(t,x)}f)\,dz\,d\sigma$$

We then write

$$|\Phi_\epsilon(t-\sigma, x-z)| \le \frac{1}{\epsilon^5}\|\Phi\|_\infty \le C\frac{1}{|Q_\epsilon(t,x)|}$$

so that

$$|\Phi_\epsilon * f(t,x) - \Phi_{\epsilon/2} * f(t,x)| \le CH_f\epsilon^\alpha$$

and

$$|\Phi_\epsilon(t-\sigma, x-z) - \Phi_\epsilon(s-\sigma, y-z)|$$

$$\le |\Phi_\epsilon(t-\sigma, x-z) - \Phi_\epsilon(t-\sigma, y-z)| + |\Phi_\epsilon(t-\sigma, y-z) - \Phi_\epsilon(s-\sigma, y-z)|$$

$$\le \frac{|t-s|^{1/2}}{\epsilon^6}\sqrt{2\|\Phi\|_\infty\|\partial_t\Phi\|_\infty} + \frac{|x-y|}{\epsilon^6}\|\vec{\nabla}\Phi\|_\infty$$

$$\le C\frac{1}{|Q_\epsilon(t,x)|}\frac{\rho_2(t-s, x-y)}{\epsilon}.$$

so that, for $\epsilon > \rho_2(t-s, x-y)$, we have

$$|\Phi_\epsilon * f(t,x) - \Phi_\epsilon * f(t,y)| \le CH_f\epsilon^\alpha\frac{\rho_2(t-s, x-y)}{\epsilon}.$$

Now, let (t_0, x_0) be a Lebesgue point of f. We have convergence in $\mathcal{D}'(\mathbb{R} \times \mathbb{R}^3)$ of

$$\sum_{j\in\mathbb{Z}} (\Phi_{2^j} * f(t,x) - \Phi_{2^j} * f(t_0, x_0)) - (\Phi_{2^{j+1}} * f(t,x) - \Phi_{2^{j+1}} * f(t_0, x_0))$$

to $f - f(t_0, x_0)$. Thus, as the series is uniformly convergent on every bounded subset of $\mathbb{R} \times \mathbb{R}^3$, the sum is a continuous function. Identifying f to the sum, we find finally that the function f is Hölderian of exponent α, as

$$\sum_{j\in\mathbb{Z}} 2^{\alpha j}\min(1, 2^{-j}\rho_2(t-s, x-y)) \le C\rho_2(t-s, x-y)^\alpha.$$

The lemma is proved. $\square$

We may now study the regularity of the heat equation. Regularity results on solutions of parabolic equations may be found in many references, such as the classical book by Ladyzhenskaya, Solonnikov and Uraltseva [298]. Here, we shall consider parabolic Hölderian regularity.

Proposition 13.4.

Let $f \in \mathcal{M}_2^{p,q_0}$ and $g \in \mathcal{M}_2^{p,q_1}$ with $1 \leq p \leq q_0 < q_1 < +\infty$, $\frac{1}{q_1} = \frac{1}{5} - \frac{\alpha}{5}$, $\frac{1}{q_0} = \frac{2}{5} - \frac{\alpha}{5}$, $0 < \alpha < 1$. Let σ be a smooth function on $\mathbb{R}^3 \setminus \{0\}$, homogeneous of exponent 1: $\sigma(\lambda\xi) = \lambda\sigma(\xi)$ for $\lambda > 0$, and let $\sigma(D)$ be the Fourier multiplier operator with symbol σ. Then the function h equal to 0 for $t \leq 0$ and to

$$h(t,x) = \int_0^t W_{\nu(t-s)} * (f(s,.) + \sigma(D)g(s,.)) \, ds$$

for $t > 0$ is Hölderian of exponent α with respect to the parabolic distance.

Proof. We may write h as a convolution in time and space variables

$$h = W_+ * (f_+ + \sigma(D)g_+)$$

with $W_+(t,x) = 1_{t>0}W_{\nu t}(x)$, $f_+ = 1_{t>0}f$ and $g_+ = 1_{t>0}g$. The size estimates on W_+ are easily established (see Ladyzhenskaya et al. [298] for instance, or our estimates in Chapter 5). In particular,

$$|W_+(t,x)| \leq C\rho_2(t,x)^{-3} \text{ and } |\sigma(D)W_+(t,x)| \leq C\rho_2(t,x)^{-4}$$

$$|\partial_t W_+(t,x)| \leq C\rho_2(t,x)^{-5} \text{ and } |\partial_t\sigma(D)W_+(t,x)| \leq C\rho_2(t,x)^{-6}$$

and

$$|\vec{\nabla}W_+(t,x)| \leq C\rho_2(t,x)^{-4} \text{ and } |\vec{\nabla}\sigma(D)W_+(t,x)| \leq C\rho_2(t,x)^{-5}.$$

We now estimate

$$\frac{1}{|Q_r(t,x)|}\iint_{Q_R(t,x)} |h(s,y) - M_{Q_r(t,x)}h|^p \, ds \, dy$$

$$\leq \frac{1}{|Q_r(t,x)|^2}\iiiint_{Q_R(t,x)\times Q_r(t,x)} |h(s,y) - h(\sigma,z)|^p \, ds \, dy \, d\sigma \, dz$$

Define $\Gamma_j = Q_{2^{j+1}r}(t,x)\setminus Q_{2^j r}(t,x)$, $f_j = 1_{t>0}1_{\Gamma_j}f$ and $g_j = 1_{t>0}1_{\Gamma_j}g$, so that $h = \sum_{j\in\mathbb{Z}} h_j$ with $h_j = W_+ * (f_j + \sigma(D)g_j)$. We have

- for $j \leq 5$, $|h_j(s,y)| \leq C(\iint 1_{\rho_2(\tau,\eta)<33r}\frac{1}{\rho_2(\tau,\eta)^3}|f_j(s - \tau, y - \eta)| \, d\tau \, d\eta + \iint 1_{\rho_2(\tau,\eta)<33r}\frac{1}{\rho_2(\tau,\eta)^4}|g_j(s - \tau, y - \eta)| \, d\tau \, d\eta)$, so that

$$\|h_j\|_p \leq C(\|1_{\rho_2(\tau,\eta)<33r}\frac{1}{\rho_2(\tau,\eta)^3}\|_1\|f_j\|_p + \|1_{\rho_2(\tau,\eta)<33r}\frac{1}{\rho_2(\tau,\eta)^4}\|_1\|g_j\|_p)$$

and thus

$$\left(\frac{1}{|Q_r(t,x)|^2}\iiiint_{Q_R(t,x)\times Q_r(t,x)} |h_j(s,y) - h_j(\sigma,z)|^p \, ds \, dy \, d\sigma \, dz\right)^{1/p}$$

$$\leq C(\|f\|_{\mathcal{M}_2^{p,q_0}} + \|g\|_{\mathcal{M}_2^{p,q_1}})r^\alpha 2^{5j(\frac{1}{p}-\frac{1}{q_1})}$$

- for $j \geq 6$, $(s,y) \in Q_r(t,x)$ and $(\sigma,z) \in Q_r(t,x)$, we have

$$|h_j(s,y) - h_j(\sigma,z)| \leq C(\frac{\rho_2(s-\sigma,y-z)}{(2^j r)^4}\iint |f_j(\tau,\eta)| \, d\tau \, d\eta$$

$$+ \frac{\rho_2(s-\sigma,y-z)}{(2^j r)^5}\iint |\vec{g}_j(\tau,\eta)| \, d\tau \, d\eta)$$

and thus

$$\left(\frac{1}{|Q_r(t,x)|^2} \iiiint_{Q_r(t,x) \times Q_r(t,x)} |h_j(s,y) - h_j(\sigma,z)|^p \, ds \, dy \, d\sigma \, dz\right)^{1/p}$$
$$\leq C(\|f\|_{\mathcal{M}_2^{p,q_0}} + \|\vec{g}\|_{\mathcal{M}_2^{p,q_1}}) r^\alpha 2^{j(\alpha-1)}.$$

Thus, we get

$$\sum_{j \in \mathbb{Z}} \left(\frac{1}{|Q_r(t,x)|^2} \iiiint_{Q_r(t,x) \times Q_r(t,x)} |h_j(s,y) - h_j(\sigma,z)|^p \, ds \, dy \, d\sigma \, dz\right)^{1/p}$$
$$\leq C(\|f\|_{\mathcal{M}_2^{p,q_0}} + \|\vec{g}\|_{\mathcal{M}_2^{p,q_1}}) r^\alpha$$

and the proposition is proved. $\qquad\qquad\qquad\qquad\qquad\qquad\qquad\qquad\qquad\qquad\square$

The strategy for the proof of the Caffarelli–Kohn–Nirenberg criterion is then clear. Let $r_1 > 0$ be fixed and $Q_1 = Q_{r_1}(t_0, x_0)$. We choose a non-negative function $\omega \in \mathcal{D}(\mathbb{R} \times \mathbb{R}^3)$ such that ω is supported in $(-1,1) \times B(0,1)$ and is equal to 1 on $(-1/4, 1/4) \times B(0,1/2)$, and we define

$$\psi(t,x) = \omega(\frac{t-t_0}{r_1^2}, \frac{x-x_0}{r_1}) \text{ and } \vec{v}(t,x) = \psi(t,x)\vec{u}(t,x)$$

$\vec{v}$ is defined on $\mathbb{R} \times \mathbb{R}^3$ with support in Q_1 and satisfies

$$\partial_t \vec{v} = \nu \Delta \vec{v} + \vec{g} + \sum_{i=1}^{3} \partial_i \vec{h}_i \qquad\qquad (13.23)$$

with

$$\begin{cases} \vec{g} = & (\partial_t \psi)\vec{u} + \nu(\Delta \psi)\vec{u} + (\vec{u} \cdot \vec{\nabla}\psi)\vec{u} + \psi\vec{\nabla}p + \psi\vec{f} \\ \vec{h}_i = & -2\nu(\partial_i \psi)\vec{u} - \psi \, u_i \, \vec{u} \end{cases}$$

As $\vec{v}$ coincides with $\vec{u}$ on $Q_{r_1/2}(t_0, x_0)$, we are going to estimate the Morrey norms $\mathcal{M}_2^{q_1, \tau_1/2}$ of $\vec{g}$ and $\mathcal{M}_2^{q_1, \tau_1}$ of $\vec{h}_i$ with $q_1 > 1$ and $\tau_1 > 5$, and conclude by using Proposition 13.4.

13.9 Proof of the Caffarelli–Kohn–Nirenberg Criterion

We list the quantities that we want to estimate for $(t,x) \in Q_{r_0}(t_0, x_0)$ and $r \leq r_0$ (we assume that r_0 is small enough to grant that $Q_{4r_0}(t_0, x_0) \subset \Omega$):

- $U_r(t,x) = \sup_{s \in (t-r^2, t+r^2)} \int_{B_r(t,x)} |\vec{u}(s,y)|^2 \, dx \, dy$

- $V_r(t,x) = \iint_{Q_r(t,x)} |\vec{\nabla} \otimes \vec{u}(s,y)|^2 \, ds \, dy$

- $W_r(t,x) = \iint_{Q_r(t,x)} |\vec{u}(s,y)|^3 \, ds \, dy$

- $\Omega_r(t,x) = \iint_{Q_r(t,x)} |\vec{u}(s,y)|^{10/3} \, ds \, dy$

- $P_r(t,x) = \iint_{Q_r(t,x)} |p(s,y)|^{q_0} \, ds \, dy$

- $\Pi_r(t,x) = \iint_{Q_r(t,x)} |\vec{\nabla} p(s,y)|^{q_1} \, ds \, dy$

- $F_r(t,x) = \iint_{Q_r(t,x)} |\vec{f}(s,y)|^{10/7} \, ds \, dy$

By assumptions ($\mathcal{H}_{CKN}$), we have $\vec{u} \in L^\infty L^2 \cap L^2 \dot{H}^1(\Omega)$, so that $U_r(t,x)$ and $V_r(t,x)$ are well defined on $Q_{r_0}(t_0, x_0)$; moreover, we have seen that, in that case, $\vec{u} \in L^{10/3} L^{10/3}(Q_{4r_0}(t_0, x_0))$, so that $W_r(t,x)$ and $\Omega_r(t,x)$ are well defined on $Q_{r_0}(t_0, x_0)$. We have $\vec{f} \in L_{t,x}^{10/7}(\Omega)$, so that $F_r(t,x)$ is well defined on $Q_{r_0}(t_0, x_0)$.

We have $p \in L_t^{q_0} L^1$ with $q_0 > 1$. With no loss of generality, we may assume $q_0 < 3/2$. In that case, using (13.20) for $B = B(x_0, 2r_0)$, we can see that we have $p \in L_{t,x}^{q_0}(Q_{2r_0}(t_0, x_0))$ so that $P_r(t,x)$ is well defined on $Q_{r_0}(t_0, x_0)$.

Finally, differentiating (13.20), we can see that $\vec{\nabla} p \in L_{t,x}^{q_1}(Q_{2r_0}(t_0, x_0))$ for every $1 < q_1 < \min(q_0, 6/5)$ (since $\vec{u}$ is locally $L^3 L^3$ and $L^2 H^1$). Thus, $\Pi_r(t,x)$ is well defined on $Q_{r_0}(t_0, x_0)$ for q_1 small enough.

We are going to estimate those quantities $U_r, V_r,...$ in terms of $U_\rho, V_\rho,...$, where $0 < r < \rho/2 < r_0/2$.

Step 1: The local energy inequality.

A consequence of the local energy inequality is that, for any smooth $\psi \in \mathcal{D}(Q_{4r_0}(t_0, x_0))$ with $\psi \geq 0$, we have, for $\tau \in (t_0 - 16r_0^2, t_0 + 16r_0^2)$

$$\int \psi(\tau, y)|\vec{u}(\tau, y)|^2 \, dy + 2\nu \int_{s<\tau} \int \psi(s,y)|\vec{\nabla} \otimes \vec{u}(s,y)|^2 \, dy \, ds$$
$$\leq \int_{s<\tau} \int (\partial_t \psi(s,y) + \nu \Delta \psi(s,y))|\vec{u}(s,y)|^2 \, dy \, ds$$
$$+ \int_{s<\tau} \int (|\vec{u}(s,y)|^2 + 2p(s,y))\vec{u}(s,y) \cdot \vec{\nabla}\psi(s,y) \, dy \, ds \tag{13.24}$$
$$+ 2 \int_{s<\tau} \int \psi(s,y)\vec{u}(s,y) \cdot \vec{f}(s,y) \, dy \, ds$$

Of course, the problem is to choose a good test function ψ. The choice of ψ has been given by Scheffer [426]: we choose a non-negative function $\omega \in \mathcal{D}(\mathbb{R} \times \mathbb{R}^3)$ such that ω is supported in $(-1,1) \times B(0, 3/4)$ and is equal to 1 on $(-1/4, 1/4) \times B(0, 1/2)$, a non-negative smooth function θ on $\mathbb{R}$ that is equal to 1 on $(-\infty, 1)$ and to 0 on $(2, +\infty)$ and we define

$$\psi(s,y) = r^3 \omega\left(\frac{s-t}{\rho^2}, \frac{y-x}{\rho}\right) \theta\left(\frac{s-t}{r^2}\right) H(4r^2 + t - s, x - y)$$

where $0 < r \leq \rho/2 \leq r_0/2$ and $H(t,x) = W_{\nu t}(x)$.

ψ enjoys many good properties (in the following estimates, C means some positive constant which depends on ν):

- ψ is smooth, non-negative and is supported in $Q_\rho(t,x)$
- $\psi(s,y) \leq C$ on $Q_\rho(t,x)$, and $\psi(s,y) \geq \frac{1}{C}$ on $Q_r(t,x)$
- $|\vec{\nabla}\psi(s,y)| \leq C\frac{1}{r}$ on $Q_\rho(t,x)$
- for $s < t + r^2$ and $(s,y) \notin Q_{\rho/2}(t,x)$, we have

$$H(4r^2 + t - s, x - y) \leq C\frac{1}{|4r^2 + t - s|^{3/2} + |x - y|^3} \leq C'\frac{1}{\rho^3}$$

- for $s < t + r^2$, we have

$$(\partial_s + \nu \Delta_y)(H(4r^2 + t - s, x - y)) = 0$$

while, for $(s, y) \in Q_{\rho/2}(t, x)$

$$(\partial_t + \nu \Delta)(\omega(\frac{s-t}{\rho^2}, \frac{y-x}{\rho})) = 0$$

so that for $(s, y) \in Q_\rho(t, x)$ with $s < t + r^2$, we have

$$|(\partial_s + \nu \Delta_y)\psi(s, y)| \le C\frac{r^3}{\rho^5}$$

Moreover, as div $\vec{u} = 0$, if $\Gamma_{\rho,\vec{u}}(s, t, x)$ is any function which does not depend on y, we have

$$\iint_{s<\tau} |\vec{u}(s, y)|^2 \vec{u}(s, y) \cdot \vec{\nabla}\psi(s, y) \, dy \, ds = \iint_{s<\tau} (|\vec{u}|^2 - \Gamma_{\rho,\vec{u}})\vec{u} \cdot \vec{\nabla}\psi \, dy \, ds.$$

We take

$$\Gamma_{\rho,\vec{u}}(s, t, x) = \frac{1}{|B(x, \rho)|} \int_{B(x,\rho)} |\vec{u}(s, y)|^2 \, dy.$$

(as a matter of fact, it does not depend on t).

We obtain

$$\max(U_r(t, x), 2\nu V_r(t, x)) \le C \iint_{Q_\rho(t,x)} \frac{r^3}{\rho^5} |\vec{u}(s, y)|^2 \, dy \, ds$$

$$+ C \iint_{Q_\rho(t,x)} \frac{1}{r} \left| |\vec{u}(s, y)|^2 - \Gamma_{\rho,\vec{u}}(s, t, s) \right| |\vec{u}(s, y)| \, dy \, ds$$

$$+ C \iint_{(t-\rho^2, t+\rho^2) \times B(x, \frac{3}{4}\rho)} \frac{1}{r} |p(s, y)| \, |\vec{u}(s, y)| \, dy \, ds$$

$$+ C \iint_{Q_\rho(t,x)} |\vec{u}(s, y)| \, |\vec{f}(s, y)| \, dy \, ds$$

The first term is easy to estimate:

$$\iint_{Q_\rho(t,x)} \frac{r^3}{\rho^5} |\vec{u}(s, y)|^2 \, dy \, ds \le 2\frac{r^3}{\rho^3} U_\rho(t, x) \tag{13.25}$$

For the second term, we write

$$\iint_{Q_\rho(t,x)} \frac{1}{r} \left| |\vec{u}(s, y)|^2 - \Gamma_{\rho,\vec{u}}(s, t, s) \right| |\vec{u}(s, y)| \, dy \, ds$$

$$\le C\frac{1}{r} (\iint_{Q_\rho(t,x)} \left| |\vec{u}(s, y)|^2 - \Gamma_{\rho,\vec{u}}(s, t, s) \right|^{3/2} dy \, ds)^{2/3} (\iint_{Q_\rho(t,x)} |\vec{u}(s, y)|^3 \, dy \, ds)^{1/3}$$

We write

$$(\iint_{Q_\rho(t,x)} |\vec{u}(s, y)|^{10/3} \, dy \, ds)^{3/10} \le \|\vec{u}\|_{L^\infty L^2(Q_\rho(t,x))}^{2/5} \|\vec{u}\|_{L^2 L^6(Q_\rho(t,x))}^{3/5}$$

$$\le C\|\vec{u}\|_{L^\infty L^2(Q_\rho(t,x))}^{2/5} ((\frac{\|\vec{u}\|_{L^2 L^2(Q_\rho(t,x))}}{\rho})^{3/5} + \|\vec{\nabla} \otimes \vec{u}\|_{L^2 L^2(Q_\rho(t,x))}^{3/5})$$

$$\le C'(U_\rho(t, x) + V_\rho(t, x))^{1/2}$$

In particular, we have

$$\left(\iint_{Q_\rho(t,x)} |\vec{u}(s,y)|^3 \, dy \, ds\right)^{1/3} \leq C\rho^{1/6} \left(\iint_{Q_\rho(t,x)} |\vec{u}(s,y)|^{10/3} \, dy \, ds\right)^{3/10}$$

$$\leq C'\rho^{1/6}(U_\rho(t,x) + V_\rho(t,x))^{1/2}$$

Moreover, by the Gagliardo–Nirenberg inequality, we have

$$\left(\int_{B(x,\rho)} \left||\vec{u}(s,y)|^2 - \Gamma_{\rho,\vec{u}}(s,t,s)\right|^{3/2} dy\right)^{2/3} \leq C \int_{B(x,\rho)} |\vec{\nabla}(|\vec{u}(s,y)|^2)| \, dy$$

so that

$$\left(\iint_{Q_\rho(t,x)} \left||\vec{u}(s,y)|^2 - \Gamma_{\rho,\vec{u}}(s,t,s)\right|^{3/2} dy \, ds\right)^{2/3}$$

$$\leq C\|\vec{u}\|_{L_t^6 L_x^2(Q_\rho(t,x))} \|\vec{\nabla} \otimes \vec{u}\|_{L_t^2 L_x^2(Q_\rho(t,x))}$$

$$\leq C\rho^{1/3} U_\rho(t,x)^{1/2} V_\rho(t,x)^{1/2}$$

and finally

$$\iint_{Q_\rho(t,x)} \frac{1}{r}\left||\vec{u}(s,y)|^2 - \Gamma_{\rho,\vec{u}}(s,t,s)\right| |\vec{u}(s,y)| \, dy \, ds \leq$$

$$C\frac{\rho^{1/2}}{r}(U_\rho(t,x) + V_\rho(t,x))V_\rho(t,x)^{1/2}. \tag{13.26}$$

The fourth term is then easy to control:

$$\iint_{Q_\rho(t,x)} |\vec{u}(s,y)|\,|\vec{f}(s,y)| \, dy \, ds \leq \|\vec{u}\|_{L^{10/3} L^{10/3}(Q_\rho(t,x))} \|\vec{f}\|_{L^{10/7} L^{10/7}(Q_\rho(t,x))}$$

and thus

$$\iint_{Q_\rho(t,x)} |\vec{u}(s,y)|\,|\vec{f}(s,y)| \, dy \, ds \leq C(U_\rho(t,x)+V_\rho(t,x))^{1/2} F_\rho(t,x)^{7/10} \tag{13.27}$$

The third term is more delicate to deal with. We introduce a function $\theta \in \mathcal{D}(\mathbb{R}^3)$ with $\theta = 1$ on $B(0, 13/16)$ and with Supp $\theta \subset B(0, 15/16)$ and we define $\zeta(y) = \theta(\frac{y-x}{\rho})$. On $(t - \rho^2, t + \rho^2) \times B(x, 3\rho/4)$, we have $\zeta p = p$. From $\Delta p = -\sum_{i=1}^3 \sum_{j=1}^3 \partial_i \partial_j (u_i u_j)$, we get

$$\Delta(\zeta p) = -\zeta \sum_{i=1}^3 \sum_{j=1}^3 \partial_i \partial_j (u_i u_j) + 2\sum_{j=1}^3 \partial_j(p\partial_j\zeta) - p\Delta\zeta$$

and we may write $\zeta(y)p(s,y) = p_{\rho,x}(s,y) + q_{\rho,x}(s,y)$ with

$$\begin{cases} q_{\rho,x} = \quad\quad \sum_{j=1}^3 \sum_{l=1}^3 G * (\zeta \partial_j \partial_l (u_j u_l)) \\ p_{\rho,x} = \quad -2\sum_{j=1}^3 \partial_j G * ((\partial_j \zeta)p) + G * ((\Delta\zeta)p) \end{cases}$$

We may, of course, replace, in the definition of $q_{\rho,x}$, the term $u_j(s,y)u_l(s,y)$ with $u_j u_l - \Gamma_{\rho,\vec{u},j,l}(s,x)$ with

$$\Gamma_{\rho,\vec{u},j,l}(s,x) = \frac{1}{|B(x,\rho)|} \int_{B(x,\rho)} u_j(s,y)u_l(s,y) \, dy.$$

For $(s, y) \in (t - \rho^2, t + \rho^2) \times B(x, 3\rho/4)$, we have

$$|p_{\rho,x}(s,y)| \leq C \frac{1}{\rho^3} \int_{B(x,\rho)} |p(s,z)| \, dz$$

so that

$$\|p_{\rho,x}\|_{L^{q_0} L^\infty((t-\rho^2,t+\rho^2) \times B(x,3\rho/4))} \leq C \rho^{-\frac{3}{q_0}} P_\rho(t,x)^{1/q_0}.$$

As we have

$$\|\vec{u}\|_{L^{\frac{q_0}{q_0-1}} L^1((t-\rho^2,t+\rho^2) \times B(x,3\rho/4))} \leq C \rho^{2(1-\frac{1}{q_0})} \rho^{3/2} U_\rho(t,x)^{1/2}$$

we get

$$\iint_{(t-\rho^2,t+\rho^2) \times B(x,\frac{3}{4}\rho)} \frac{1}{r} |p_{\rho,x}(s,y)| \, |\vec{u}(s,y)| \, dy \, ds \leq \tag{13.28}$$

$$C \frac{1}{r} \rho^{2+\frac{3}{2}-\frac{5}{q_0}} P_r(t,x)^{1/q_0} U_\rho(t,x)^{1/2}$$

For the term involving $q_{\rho,x}$, we write

$$q_{\rho,x} = \sum_{j=1}^{3} \sum_{l=1}^{3} \partial_j \partial_l G * (\zeta(u_j u_l - \Gamma_{\rho,\vec{u},j,l}))$$

$$- 2 \sum_{j=1}^{3} \sum_{l=1}^{3} \partial_j G * ((\partial_l \zeta)(u_j u_l - \Gamma_{\rho,\vec{u},j,l}))$$

$$+ \sum_{j=1}^{3} \sum_{l=1}^{3} G * ((\partial_j \partial_l \zeta)(u_j u_l - \Gamma_{\rho,\vec{u},j,l}))$$

Thus, again for $(s, y) \in (t - \rho^2, t + \rho^2) \times B(x, 3\rho/4)$, we have

$$|q_{\rho,x}(s,y)| \leq \sum_{j=1}^{3} \sum_{l=1}^{3} |\partial_j \partial_l G * (\zeta(u_j u_l - \Gamma_{\rho,\vec{u},j,l}))(s,y)|$$

$$+ C \sum_{j=1}^{3} \sum_{l=1}^{3} \mathcal{M}_{\mathbb{R}^3} (1_{Q_\rho(t,x)} (u_j u_l - \Gamma_{\rho,\vec{u},j,l}))(s,y)$$

where $\mathcal{M}_{\mathbb{R}^3}$ is the Hardy–Littlewood maximal function with respect to the space variable. Thus, we get

$$\|q_{\rho,x}\|_{L^{3/2} L^{3/2}((t-\rho^2,t+\rho^2) \times B(x,3\rho/4))} \leq C \|\vec{u}\|_{L^6_t L^2_x(Q_\rho(t,x))} \|\vec{\nabla} \otimes \vec{u}\|_{L^2_t L^2_x(Q_\rho(t,x))}$$

$$\leq C \rho^{1/3} U_\rho(t,x)^{1/2} V_\rho(t,x)^{1/2}$$

and finally

$$\iint_{(t-\rho^2,t+\rho^2) \times B(x,\frac{3}{4}\rho)} \frac{1}{r} |q_{\rho,x}(s,y)| \, |\vec{u}(s,y)| \, dy \, ds \leq \tag{13.29}$$

$$C \frac{\rho^{1/2}}{r} (U_\rho(t,x) + V_\rho(t,x)) V_\rho(t,x)^{1/2}.$$

Besides, let us notice that

$$P_r(t,x) \leq (\|p_{\rho,x}\|_{L^{q_0}L^{q_0}(Q_r(t,x))} + \|q_{\rho,x}\|_{L^{q_0}L^{q_0}(Q_r(t,x))})^{q_0}$$

$$\leq C(r^3\|p_{\rho,x}\|_{L^{q_0}L^\infty(Q_r(t,x))}^{q_0} + r^{5(1-\frac{2q_0}{3})}\|q_{\rho,x}\|_{L^{3/2}L^{3/2}(Q_r(t,x))}^{q_0})$$

$$\leq C(\frac{r^3}{\rho^3}P_\rho(t,x) + r^{5(1-\frac{2q_0}{3})}\rho^{q_0/3}U_\rho(t,x)^{q_0/2}V_\rho(t,x)^{q_0/2})$$

Summarizing all those estimates, we have shown:

Lemma 13.3.
Assume that

- $\vec{u} \in L^2_{t,x}(\Omega) \cap L^2_t \dot{H}^1_x(\Omega)$
- $p \in L^{q_0}_{t,x}(\Omega)$ *with* $1 < q_0 \leq 3/2)$
- $\vec{f} \in L^{10/7}_{t,x}(\Omega)$
- $\vec{u}$ *is suitable*

then, for $0 < r \leq \rho/2 \leq r_0/2$ *and* $(t,x) \in Q_{r_0}(t_0, x_0)$,

$$U_r(t,x) + V_r(t,x) \leq C\frac{r^3}{\rho^3}U_\rho(t,x)$$

$$+ C\frac{\rho^{1/2}}{r}(U_\rho(t,x) + V_\rho(t,x))V_\rho(t,x)^{1/2} \tag{13.30}$$

$$+ C\frac{1}{r}\rho^{2+\frac{3}{2}-\frac{5}{q_0}}P_\rho(t,x)^{1/q_0}U_\rho(t,x)^{1/2}$$

$$+ C(U_\rho(t,x) + V_\rho(t,x))^{1/2}F_\rho(t,x)^{7/10}$$

and

$$P_r(t,x) \leq C(\frac{r^3}{\rho^3}P_\rho(t,x) + r^{5(1-\frac{2q_0}{3})}\rho^{q_0/3}U_\rho(t,x)^{q_0/2}V_\rho(t,x)^{q_0/2}). \tag{13.31}$$

Step 2: Morrey estimates for the velocity and the pressure.
We are going to use the estimates of Lemma 13.3 to show the following result (inspired from Kukavica [286]):

Lemma 13.4.
Let $\vec{u}$ *be a suitable solution of the Navier–Stokes equations (with* $\vec{f} \in L^{10/7}_{t,x}(\Omega)$ *and* $p \in L^{q_0}_{t,x}(\Omega)$ *with* $1 < q_0 \leq 3/2$*). Assume moreover that* $1_\Omega \vec{f} \in \mathcal{M}^{10/7,\tau_0}_2$ *for some* $\tau_0 > 5/3$. *Let* τ_2 *be such that* $1 < \frac{\tau_2}{5} < \min(q_0, 2)$ *and* $2 - \frac{5}{\tau_0} + \frac{5}{\tau_2} > 0$.
There exists a positive constant ϵ^* *which depends only on* ν, q_0, τ_0 *and* τ_2 *such that, if* $(t_0, x_0) \in \Omega$ *and*

$$\limsup_{r\to 0} \frac{1}{r} \iint_{(t_0-r^2,t_0+r^2)\times B(x_0,r)} |\vec{\nabla} \otimes \vec{u}(s,y)|^2 \, ds \, dy < \epsilon^*$$

then there exists a neighborhood $Q_2 = Q_{r_2}(t_0, x_0)$ *of* (t_0, x_0) *such that* $1_{Q_2}\vec{u} \in \mathcal{M}^{3,\tau_2}_2$ *and* $1_{Q_2}p \in \mathcal{M}^{q_0,\tau_2/2}_2$.

Remark: Assumption $1_{Q_0}\vec{f} \in \mathcal{M}_2^{10/7,5/3+\epsilon}(\mathbb{R}\times\mathbb{R}^3))$ is borrowed from Kukavica [286]. The results that $1_{Q_0}\vec{u} \in \mathcal{M}_2^{3,5}$ and $1_{Q_0}p \in \mathcal{M}_2^{3/2,5/2}$ is underlined by Robinson [415] as providing a much easier proof for the following step than the sole control on $\vec{u} \in L^3(Q_0)$ and $p \in L^{3/2}(Q_0)$ provided by the original proof of Caffarelli, Kohn and Nirenberg [74].

The conclusion that $(1_{Q_2}\vec{u} \in \mathcal{M}_2^{3,5+\epsilon}(\mathbb{R} \times \mathbb{R}^3))$ is, of course, reminiscent of O'Leary's assumption in Theorem 13.3.

Proof. We want to prove, for $r < r_2$ and $(t,x) \in Q_2$,

$$W_r(t,x) \le Cr^{5(1-\frac{3}{\tau_2})} \text{ and } P_r(t,x) \le Cr^{5(1-\frac{2q_0}{\tau_2})} \tag{13.32}$$

If $0 < \kappa < 1$, it is enough to prove that, for every $n \in \mathbb{N}$, we have

$$\sup_{n\in\mathbb{N}} \sup_{(t,x)\in Q_{r_2}(t_0,x_0)} \frac{1}{(\kappa^n r_2)^{5(1-\frac{3}{\tau_2})}} \iint_{Q_{\kappa^n r_2}(t,x)\cap Q_2} |\vec{u}(s,y)|^3 \, dy \, ds < +\infty \tag{13.33}$$

and

$$\sup_{n\in\mathbb{N}} \sup_{(t,x)\in Q_{r_2}(t_0,x_0)} \frac{1}{(\kappa^n r_2)^{5(1-\frac{2q_0}{\tau_2})}} \iint_{Q_{\kappa^n r_2}(t,x)\cap Q_2} |p(s,y)|^{q_0} \, dy \, ds < +\infty. \tag{13.34}$$

We have seen that

$$W_r(t,x) \le Cr^{1/2}(U_r(t,x) + V_r(t,x))^{3/2}$$

hence it will be enough to prove that

$$U_r(t,x) + V_r(t,x) \le Cr^{3-\frac{10}{\tau_2}} \tag{13.35}$$

to get the control of $W_r(t,x)$.

We start from the cylinder $Q_0 = Q_{r_0}(x_0,t_0)$ discussed in the previous step. We have as assumptions that

$$F_r(t,x) \le C\|1_{Q_0}\vec{f}\|_{\mathcal{M}_2^{10/7,\tau_0}}^{10/7} r^{5(1-\frac{10}{7\tau_0})} \tag{13.36}$$

and

$$\limsup_{r\to 0} \frac{1}{r} V_r(t_0,x_0) < \epsilon^* \tag{13.37}$$

We introduce the reduced quantities

$$\alpha_r(t,x,\tau_2) = \frac{1}{r^{3-\frac{10}{\tau_2}}}(U_r(t,x) + V_r(t,x))$$

$$p_r(t,x,\tau_2) = \frac{1}{r^{5(1-\frac{2q_0}{\tau_2})}}P_r(t,x)$$

and

$$\beta_r(t,x) = \frac{1}{r}V_r(t,x).$$

We may rewrite the conclusions of Lemma 13.3 as

$$
\begin{aligned}
\alpha_r(t,x,\tau_2) \leq\; & C_0(\frac{r}{\rho})^{\frac{10}{\tau_2}}\alpha_\rho(t,x,\tau_2) \\
& + C_0(\frac{\rho}{r})^{4-\frac{10}{\tau_2}}\alpha_\rho(t,x,\tau_2)\beta_\rho(t,x)^{1/2} \\
& + C_0(\frac{\rho}{r})^{4-\frac{10}{\tau_2}}\rho^{1-\frac{5}{\tau_2}}p_\rho(t,x,\tau_2)^{1/q_0}\alpha_\rho(t,x,\tau_2)^{1/2} \\
& + C_0(\frac{\rho}{r})^{3-\frac{10}{\tau_2}}\rho^{2-\frac{5}{\tau_0}+\frac{5}{\tau_2}}\alpha_\rho(t,x,\tau_2)^{1/2}\|1_{Q_0}\vec{f}\|_{\mathcal{M}_2^{10/7,\tau_0}}
\end{aligned}
\tag{13.38}
$$

and

$$
\begin{aligned}
p_r(t,x,\tau_2) \leq\; & C_0(\frac{r}{\rho})^{\frac{10q_0}{\tau_2}-2}p_\rho(t,x,\tau_2) \\
& + C_0(\frac{\rho}{r})^{5q_0(\frac{2}{3}-\frac{2}{\tau_2})}\rho^{\frac{5q_0}{\tau_2}-q_0}\alpha_\rho(t,x,\tau_2)^{q_0/2}\beta_\rho(t,x)^{q_0/2}
\end{aligned}
\tag{13.39}
$$

where the constant C_0 does not depend on r, ρ, τ_0 nor τ_2. Inequality (13.39) cannot be used directly, as the exponent $\frac{5q_0}{\tau_2}-q_0$ is negative[1]. We therefore introduce the auxiliary quantity

$$
q_r(t,x,\tau_2) = r^{q_0(1-\frac{5}{\tau_2})-\eta}p_r(t,x,\tau_2)
$$

and rewrite our inequalities (since $\beta_\rho \leq \rho^{2-\frac{10}{\tau_2}}\alpha_\rho$) as

$$
\begin{aligned}
\alpha_r(t,x,\tau_2) \leq\; & C_0(\frac{r}{\rho})^{\frac{10}{\tau_2}}\alpha_\rho(t,x,\tau_2) \\
& + C_0(\frac{\rho}{r})^{4-\frac{10}{\tau_2}}\alpha_\rho(t,x,\tau_2)\beta_\rho(t,x)^{1/2} \\
& + C_0(\frac{\rho}{r})^{4-\frac{10}{\tau_2}}q_\rho(t,x,\tau_2)^{1/q_0}\alpha_\rho(t,x,\tau_2)^{1/2} \\
& + C_0(\frac{\rho}{r})^{3-\frac{10}{\tau_2}}\rho^{2-\frac{5}{\tau_0}+\frac{5}{\tau_2}}\alpha_\rho(t,x,\tau_2)^{1/2}\|1_{Q_0}\vec{f}\|_{\mathcal{M}_2^{10/7,\tau_0}},
\end{aligned}
\tag{13.40}
$$

$$
\begin{aligned}
q_r(t,x,\tau_2) \leq\; & C_0(\frac{r}{\rho})^{\frac{10q_0}{\tau_2}-2+q_0(1-\frac{5}{\tau_2})}q_\rho(t,x,\tau_2) \\
& + C_0(\frac{\rho}{r})^{5q_0(\frac{2}{3}-\frac{2}{\tau_2})+\frac{5q_0}{\tau_2}-q_0}\alpha_\rho(t,x,\tau_2)^{q_0/2}\beta_\rho(t,x)^{q_0/2}
\end{aligned}
\tag{13.41}
$$

and (since $\beta_\rho \leq \rho^{2-\frac{10}{\tau_2}}\alpha_\rho$) as

$$
\begin{aligned}
\alpha_r(t,x,\tau_2) \leq\; & C_0(\frac{r}{\rho})^{\frac{10}{\tau_2}}\alpha_\rho(t,x,\tau_2) \\
& + C_0(\frac{\rho}{r})^{4-\frac{10}{\tau_2}}\rho^{1-\frac{5}{\tau_2}}\alpha_\rho(t,x,\tau_2)^{3/2} \\
& + C_0(\frac{\rho}{r})^{4-\frac{10}{\tau_2}}q_\rho(t,x,\tau_2)^{1/q_0}\alpha_\rho(t,x,\tau_2)^{1/2} \\
& + C_0(\frac{\rho}{r})^{3-\frac{10}{\tau_2}}\rho^{2-\frac{5}{\tau_0}+\frac{5}{\tau_2}}\alpha_\rho(t,x,\tau_2)^{1/2}\|1_{Q_0}\vec{f}\|_{\mathcal{M}_2^{10/7,\tau_0}},
\end{aligned}
\tag{13.42}
$$

$$
\begin{aligned}
q_r(t,x,\tau_2) \leq\; & C_0(\frac{r}{\rho})^{\frac{10q_0}{\tau_2}-2+q_0(1-\frac{5}{\tau_2})}q_\rho(t,x,\tau_2) \\
& + C_0(\frac{\rho}{r})^{5q_0(\frac{2}{3}-\frac{2}{\tau_2})+\frac{5q_0}{\tau_2}-q_0}\rho^{\frac{q_0}{2}(1-\frac{5}{\tau_2})}\alpha_\rho(t,x,\tau_2)^{3q_0/2}
\end{aligned}
\tag{13.43}
$$

---

[1] Thanks to D. Chamorro and J. He for letting me know that the exponent in equation (13.39) was wrong in the first edition: it was estimated as the positive quantity $\frac{5q_0}{\tau_2}-\frac{q_0}{2}$ instead of the negative quantity $\frac{5q_0}{\tau_2}-q_0$.

$$p_r(t, x, \tau_2) \leq C_0 \left(\frac{r}{\rho}\right)^{\frac{10q_0}{\tau_2} - 2} p_\rho(t, x, \tau_2)$$

$$+ C_0 \left(\frac{\rho}{r}\right)^{5q_0\left(\frac{2}{3} - \frac{2}{\tau_2}\right)} \alpha_\rho(t, x, \tau_2)^{q_0}. \tag{13.44}$$

We first begin with inequalities (13.42), (13.43) and (13.44). Let λ be the positive exponent $\lambda = \min(\frac{10}{\tau_2}, \frac{10q_0}{\tau_2} - 2)$; we fix $\kappa \in (0, 1/2)$ such that $C_0 \kappa^\lambda \leq \frac{1}{4}$. For $r = \kappa\rho$ we obtain

$$\alpha_r(t, x, \tau_2) \leq \frac{1}{4}\alpha_\rho(t, x, \tau_2)$$

$$+ C_\kappa \rho^{1 - \frac{5}{\tau_2}} \alpha_\rho(t, x, \tau_2)^{3/2}$$

$$+ C_\kappa q_\rho(t, x, \tau_2)^{1/q_0} \alpha_\rho(t, x, \tau_2)^{1/2}$$

$$+ C_\kappa \rho^{2 - \frac{5}{\tau_0} + \frac{5}{\tau_2}} \alpha_\rho(t, x, \tau_2)^{1/2} \|1_{Q_0}\vec{f}\|_{\mathcal{M}_2^{10/7, \tau_0}}, \tag{13.45}$$

$$q_r(t, x, \tau_2) \leq \frac{1}{4}q_\rho(t, x, \tau_2) + C_\kappa \rho^{\frac{q_0}{2}\left(1 - \frac{5}{\tau_2}\right)} \alpha_\rho(t, x, \tau_2)^{3q_0/2}, \tag{13.46}$$

and

$$p_r(t, x, \tau_2) \leq \frac{1}{4}p_\rho(t, x, \tau_2) + C_\kappa \alpha_\rho(t, x, \tau_2)^{q_0}. \tag{13.47}$$

Now, since $5 < \tau_2$ and $2 - \frac{5}{\tau_0} + \frac{5}{\tau_2} > 0$, there exists a $\rho_0 = \rho_0(\kappa, \tau_0, \tau_2, \vec{f})$ such that

$$C_\kappa \rho_0^{1 - \frac{5}{\tau_2}} \leq \frac{1}{4}, \quad C_\kappa \rho_0^{2 - \frac{5}{\tau_0} + \frac{5}{\tau_2}} \|1_{Q_0}\vec{f}\|_{\mathcal{M}_2^{10/7, \tau}} \leq \frac{1}{4} \text{ and } C_\kappa \rho^{\frac{q_0}{2}\left(1 - \frac{5}{\tau_2}\right)} \leq \frac{3}{4}\left(\frac{1}{4c_\kappa}\right)^{q_0}.$$

Assume that for some $\rho = r_1 \leq \rho_0$, we have $\alpha_\rho(t, x, \tau_2) \leq 1$ and $q_\rho(t, x, \tau_2) \leq \left(\frac{1}{4c_\kappa}\right)^{q_0}$. Then the same will be true for $\kappa\rho$, and by induction for all $\kappa^n\rho$, $n \in \mathbb{N}$. Moreover, we will have $p_{\kappa^n r_1}(t, x, \tau_2) \leq p_{r_2}(t, x, \tau_2) + \frac{4}{3}C_\kappa$.

Thus, if $\alpha_{r_1}(t, x, \tau_2) \leq 1$ and $q_{r_1}(t, x, \tau_2) \leq \left(\frac{1}{4c_\kappa}\right)^{q_0}$ for every $(t, x) \in Q_{r_2}(t_0, x_0)$ for some $r_3 > 0$, we finally get that $1_{Q_{r_2}(t_0, x_0)}\vec{u} \in \mathcal{M}_2^{3, \tau_2}$ and $1_{Q_{r_2}(t_0, x_0)}p \in \mathcal{M}_2^{3/2, \tau_2/2}$. Thus, to finish the proof of Lemma 13.4, it is enough to prove that for r small enough, we have $\alpha_r(t_0, x_0, \tau_2) < 1$ and $q_r(t_0, x_0, \tau_2) < \left(\frac{1}{4c_\kappa}\right)^{q_0}$. This will be one by using inequalities (13.40) and (13.41) and the assumption that $\lim_{r \to 0} \beta_r(t_0, x_0) = 0$. We start from ρ_0 and define

$$\alpha_n = \alpha_{\kappa^n \rho_0}(t_0, x_0, \tau_2) \text{ and } q_n = q_{\kappa^n \rho_0}(t_0, x_0, \tau_2).$$

We rewrite inequalities (13.40) and (13.41) as

$$\alpha_{n+1} \leq \frac{1}{4}\alpha_n + u_n \alpha_n + C_\kappa q_n^{1/q_0} \alpha_n^{1/2} + v_n \alpha_n^{1/2} \tag{13.48}$$

and

$$q_{n+1} \leq \frac{1}{4}q_n + w_n \alpha_n^{\frac{q_0}{2}} \tag{13.49}$$

where u_n, v_n and w_n go to 0 when n goes to $+\infty$.

Let D be a large positive constant, and $\theta_n = \alpha_n + Dq_n^{\frac{2}{q_0}}$. We have

$$\theta_{n+1} \leq \frac{1}{4}\alpha_n + u_n \alpha_n + C_\kappa D^{-\frac{1}{2}}\theta_n + v_n \alpha_n + \frac{1}{4}v_n + D\left(\frac{1}{4}q_n + w_n \alpha_n^{\frac{q_0}{2}}\right)^{\frac{2}{q_0}}.$$

For $a, b, \epsilon > 0$, we have

$$(a+b)^{\frac{2}{q_0}} \leq (a^2 + b^2 + 2ab)^{\frac{1}{q_0}} \leq a^{\frac{2}{q_0}} + b^{\frac{2}{q_0}} + 2^{\frac{1}{q_0}} a^{\frac{1}{q_0}} b^{\frac{1}{q_0}} \leq (1+\epsilon) a^{\frac{2}{q_0}} + (1 + \frac{2^{\frac{2}{q_0}}}{4\epsilon}) b^{\frac{2}{q_0}}$$

and we get

$$\theta_{n+1} \leq \frac{1}{4}\alpha_n + u_n \theta_n + C_\kappa D^{-\frac{1}{2}} \theta_n + v_n \theta_n + \frac{1}{4} v_n$$

$$+ (1+\epsilon)(\frac{1}{4})^{\frac{2}{q_0}} D \; q_n^{\frac{2}{q_0}} + (1 + \frac{2^{\frac{2}{q_0}}}{4\epsilon}) D w_n^{\frac{2}{q_0}} \theta_n,$$

so that

$$\theta_{n+1} \leq \Gamma \theta_n + X_n \theta_n + Y_n$$

with

$$\Gamma = \max(\frac{1}{4} + C_\kappa D^{-\frac{1}{2}} \; (1+\epsilon)(\frac{1}{4})^{\frac{2}{q_0}}),$$

and

$$\lim_{n \to +\infty} X_n = \lim_{n \to +\infty} Y_n = 0.$$

If D is large enough and ϵ small enough, we have

$$\Gamma < 1$$

so that

$$\lim_{n \to +\infty} \theta_n = 0.$$

Thus, $\lim_{n \to +\infty} \alpha_n = \lim_{n \to +\infty} q_n = 0$. Lemma 13.4 is proved. $\qquad\square$

Step 3: Further estimates on the pressure and the velocity.

We shall now use a more precise representation for the pressure and the velocity. Let us first notice that the proof of Lemma 13.4 actually conveys more information on $\vec{u}$: we have indeed proved that

$$V_r(t, x) \leq Cr^{3 - \frac{10}{\tau_2}}$$

and thus that $1_{Q_2} \vec{\nabla} \otimes \vec{u}$ belongs to $\mathcal{M}_2^{2, \tau_3}$ with $\frac{1}{\tau_3} = \frac{1}{\tau_2} + \frac{1}{5}$, so that $\tau_3 > 5/2$.

Let $Q_3 = Q_{r_3}(t_0, x_0)$ of (t_0, x_0) with $r_3 < r_2$. We consider a function $\phi \in \mathcal{D}(\mathbb{R} \times \mathbb{R}^3)$ which is equal to 1 on Q_3 and is compactly supported in Q_2. We write

$$\vec{v} = \phi \vec{u}$$

and

$$\partial_t \vec{v} = \nu \Delta \vec{v} + \vec{g} + \sum_{i=1}^{3} \partial_i \vec{h}_i - \phi \vec{\nabla} p \tag{13.50}$$

with

$$\vec{g} = (\partial_t \phi)\vec{u} + (\Delta \phi)\vec{u} - \phi \vec{u} \cdot \vec{\nabla} \vec{u} + \phi \vec{f}$$

and

$$\vec{h}_i = -2(\partial_i \phi)\vec{u}$$

Now, we estimate $\phi\vec{\nabla}p$. We start from

$$\Delta p = -\sum_{j=1}^{3}\sum_{l=1}^{3}\partial_j\partial_l(u_j u_l)$$

and we consider a function $\zeta \in \mathcal{D}(\mathbb{R} \times \mathbb{R}^3)$ which is equal to 1 on a neighborhood of the support of ϕ and is compactly supported in Q_2. We have

$$\phi\vec{\nabla}p = \vec{\gamma} + \vec{\eta} + \sum_{j=1}^{3}\sum_{l=1}^{3}\vec{\nabla}\partial_j\partial_l G * (\phi u_j u_l) \tag{13.51}$$

with

$$\vec{\gamma} = \phi\vec{\nabla}G * ((\Delta\zeta)p) + \phi\sum_{j=1}^{3}\sum_{l=1}^{3}\vec{\nabla}G * ((\partial_j\partial_l\zeta)u_j u_l)$$

$$- 2\phi\sum_{j=1}^{3}\vec{\nabla}\partial_j * ((\partial_j\zeta)p) - \phi\sum_{j=1}^{3}\sum_{l=1}^{3}\vec{\nabla}\partial_j G * ((\partial_l\zeta)u_j u_l)$$

and

$$\vec{\eta} = -\sum_{j=1}^{3}\sum_{l=1}^{3}[\phi, \frac{\vec{\nabla}\partial_j\partial_l}{\Delta}](\zeta u_j u_l).$$

We finally find

$$|\vec{v}| \leq C1_{Q_2}(\mathcal{I}_2(|\vec{g}|) + \sum_{i=1}^{3}\mathcal{I}_1(|\vec{h}_i|) + \mathcal{I}_2(|\vec{\gamma}|) + \mathcal{I}_2(|\vec{\eta}|) + \sum_{j=1}^{3}\sum_{l=1}^{3}\mathcal{I}_1(\phi|u_j u_l|)) \tag{13.52}$$

(whre $\mathcal{I}_\alpha$ is the Riesz potential on the parabolic space $\mathbb{R} \times \mathbb{R}^3$ introduced in Theorem 5.3). This will allow us to prove:

Lemma 13.5.
Let $\vec{u}$ be a suitable solution of the Navier–Stokes equations (with $\vec{f} \in L_{t,x}^{10/7}(\Omega)$ and $p \in L_{t,x}^{q_0}(\Omega)$ with $1 < q_0 \leq 3/2$). Assume moreover that on some neighborhood $Q_2 = Q_{r_2}(t_0, x_0)$ of (t_0, x_0), we have

- $1_{Q_2}\vec{f} \in \mathcal{M}_2^{10/7,\tau_0}$ *for some $\tau_0 > 5/2$*
- $1_{Q_2}\vec{u}$ *belongs to $\mathcal{M}_2^{3,\tau_2}$ for some $\tau_2 > 5$*
- $1_{Q_2}\vec{\nabla}\otimes\vec{u}$ *belongs to $\mathcal{M}_2^{2,\tau_3}$ with $\frac{1}{\tau_3} = \frac{1}{\tau_2} + \frac{1}{5}$*

Then, for every $r_3 < r_2$, we have $1_{Q_3}\vec{u} \in \mathcal{M}_2^{3,\sigma}$ with $\frac{1}{\sigma} + \frac{1}{\tau_2} < \frac{1}{5}$.

Proof. We shall start from assumption $1_{Q_2}\vec{u} \in M^{3,\tau}$ with $\tau > 5$ and prove that $1_{Q_3}\vec{u} \in M_2^{3,\sigma}$ with $\sigma > \tau$. Some terms are easily controlled:

- $1_{Q_2}\mathcal{I}_2(|\vec{g}|) \leq 1_{Q_2}\mathcal{I}_2(A_1) + 1_{Q_2}\mathcal{I}_2(A_2)$ with $A_1 = |\partial_t\phi\,\vec{u}| + |\Delta\phi\,\vec{u}| + |\phi\vec{f}|$ and $A_2 = |\phi\vec{u}\cdot\vec{\nabla}\vec{u}|$. As A_1 belongs to $\mathcal{M}_2^{10/7,\rho}$ for every $\rho < 5/2$, we find that $1_{Q_2}\mathcal{I}_2 A_1$ belongs to $\mathcal{M}_2^{3,\sigma}$ for every $\sigma \geq 3$.

- as $\vec{h}_i$ belongs to $\mathcal{M}_2^{3,\rho}$ for every $\rho < 5$, we find that $1_{Q_2}\mathcal{I}_1(|\vec{h}_i|)$ belongs to $\mathcal{M}_2^{3,\sigma}$ for every $\sigma \geq 3$.

- as $1_{Q_2}p$ and $1_{Q_2}|\vec{u}|^2$ belong to $L_{t,x}^{q_0}$, we find that $\vec{\gamma}$ belongs to $L_t^{q_0}L_x^\infty \subset \mathcal{M}^{q_0,\frac{5q_0}{2}}$. As $\vec{\gamma}$ is compactly supported, we find that it belongs to $\mathcal{M}_2^{q_0,\rho}$ for every $\rho < 5/2$, so that $1_{Q_2}\mathcal{I}_2(|\vec{\gamma}|)$ belongs to $\mathcal{M}_2^{3,\sigma}$ for every $\sigma \geq 3$.

Thus, we are left with the estimation of $1_{Q_2}\mathcal{I}_2(A_2)$, $1_{Q_2}(|\vec{\eta}|)$ and $1_{Q_2}\mathcal{I}_1(\phi|u_ju_l|)$. First, let us notice that, when $1_{Q_2}|\vec{u}|^2$ belongs to $\mathcal{M}_2^{3/2,\gamma}$, then $\vec{\eta}$ belongs to $\mathcal{M}_2^{3/2,\delta}$ with $\frac{1}{\delta} = \frac{1}{\gamma} + \frac{1}{5}$: indeed, using the Calderón commutator theorem, we see that $w \mapsto [\phi, \frac{\partial_i\partial_j\partial_k}{\Delta}]w$ is bounded on $L_t^pL_x^q$ for $1 \leq p \leq \infty$, $1 < q < +\infty$. If we want to estimate $\iint_{Q_r(t,x)} |\vec{\eta}|^{3/2}\,ds\,dy$, it is enough to do it for $r < r_0$ (as $\vec{\eta}$ belongs to $L^{3/2}L^{3/2}$, hence the behavior on large cylinders is well controlled); then one writes $\vec{\eta} = \vec{\eta}_r + \vec{\eta}_{[r]}$ with

$$\vec{\eta}_r = -\sum_{j=1}^3\sum_{l=1}^3 [\phi, \frac{\vec{\nabla}\partial_j\partial_l}{\Delta}](1_{Q_{2r}(t,x)}\zeta u_j u_l).$$

and

$$\vec{\eta}_{[r]} = -\sum_{j=1}^3\sum_{l=1}^3 [\phi, \frac{\vec{\nabla}\partial_j\partial_l}{\Delta}]((1 - 1_{Q_{2r}(t,x)}\zeta u_j u_l).$$

Then, we have

$$\iint_{Q_r(t,x)} |\vec{\eta}_r|^{3/2}\,ds\,dy \leq C \iint_{Q_{2r}(t,x)\cap Q_2} |\vec{u}|^3\,ds\,dy \leq Cr^{5(1-\frac{3}{2\gamma})}\|1_{Q_2}|\vec{u}|^2\|_{\mathcal{M}_2^{3/2,\gamma}}^{3/2}$$

while, on $Q_r(t,x)$, we have

$$|\vec{\eta}_{[r]}(s,y)| \leq C \int_{|z|>r} \frac{1}{|z|^4}1_{Q_2}(y-z)|\vec{u}_2(s,y-z)|^2\,dz;$$

thus,

$$\|1_{Q_r(t,x)}\vec{\eta}_{[r]}\|_{\mathcal{M}_2^{3/2,\gamma}} \leq C\int_{|z|>r}\frac{dz}{|z|^4}\|1_{Q_2}|\vec{u}|^2\|_{\mathcal{M}_2^{3/2,\gamma}} = C'\frac{1}{r}\|1_{Q_2}|\vec{u}|^2\|_{\mathcal{M}_2^{3/2,\gamma}}$$

and

$$\iint_{Q_r(t,x)} |\vec{\eta}_{[r]}|^{3/2}\,ds\,dy \leq Cr^{-3/2}r^{5(1-\frac{3}{2\gamma})}\|1_{Q_2}|\vec{u}|^2\|_{\mathcal{M}_2^{3/2,\gamma}}^{3/2}$$
$$= Cr^{5(1-\frac{3}{2\delta})}\|1_{Q_2}|\vec{u}|^2\|_{\mathcal{M}_2^{3/2,\gamma}}^{3/2}$$

Let $\frac{1}{\tau_2} = \frac{1}{5} - \alpha$. Assume that $\frac{1}{\tau} > \frac{1}{\alpha}$. Then, we have:

- $\phi\vec{u}\cdot\vec{\nabla}\vec{u} \in \mathcal{M}_2^{\frac{6}{5},\rho}$ with $\frac{1}{\rho} = \frac{1}{\tau_3} + \frac{1}{\tau} = \frac{1}{\tau} + \frac{2}{5} - \alpha$. Hence, $\mathcal{I}_2(|\phi\vec{u}\cdot\vec{\nabla}\vec{u}|)$ belongs to $\mathcal{M}_2^{\frac{6}{5}\frac{\sigma}{\rho},\sigma}$ with $\frac{1}{\sigma} = \frac{1}{\rho} - \frac{2}{5} = \frac{1}{\tau} - \alpha$. Moreover, $\frac{1}{\rho} > \frac{3}{\tau} - \alpha > 3\frac{1}{\tau}$, so that $3 < \frac{6}{5}\frac{\sigma}{\rho}$, and $\mathcal{I}_2(|\phi\vec{u}\cdot\vec{\nabla}\vec{u}|)$ belongs to $\mathcal{M}_2^{3,\sigma}$

- $1_{Q_2}|\vec{u}|^2$ belongs to $\mathcal{M}_2^{3/2,\gamma}$ with $\frac{1}{\gamma} = \frac{1}{\tau_2} + \frac{1}{\tau}$, hence $\vec{\eta}$ belongs to $\mathcal{M}_2^{\frac{3}{2},\rho}$ with $\frac{1}{\rho} = \frac{1}{\tau} + \frac{2}{5} - \alpha$, and $\mathcal{I}_2(|\vec{\eta}|)$ belongs to $\mathcal{M}_2^{3,\sigma}$ with $\frac{1}{\sigma} = \frac{1}{\tau} - \alpha$.

- finally, we have that $\mathcal{I}_1(1_{Q_2}|\vec{u}|^2)$ belongs to $\mathcal{M}_2^{\frac{3\sigma}{2\gamma},\sigma}$ with $\frac{1}{\gamma} = \frac{1}{\tau_2} + \frac{1}{\tau}$ and $\frac{1}{\sigma} = \frac{1}{\tau} - \alpha < \frac{1}{2\gamma}$, and thus $\mathcal{I}_1(1_{Q_2}|\vec{u}|^2)$ belongs to $\mathcal{M}_2^{3,\sigma}$

We then iterate the estimate, changing $\frac{1}{\tau}$ into $\frac{1}{\tau} - \alpha$, until we obtain $\frac{1}{\sigma} \leq \alpha$ (and even $\frac{1}{\sigma} < \alpha$: if $\tau = \frac{1}{\alpha}$, we write $1_{Q_2}\vec{u} \in \mathcal{M}_2^{3,\tau'}$ with $\alpha < \frac{1}{\tau'} < 2\alpha\dots$) . $\quad\square$

Step 4: End of the proof.

We then end the proof with the lemma:

Lemma 13.6.
Let $\vec{u}$ be a suitable solution of the Navier–Stokes equations (with $\vec{f} \in L_{t,x}^{10/7}(\Omega)$ and $p \in L_{t,x}^{q_0}(\Omega)$ with $1 < q_0 \leq 3/2$). Assume moreover that on some neighborhood $Q_2 = Q_{r_2}(t_0, x_0)$ of (t_0, x_0), we have

- $1_{Q_2}\vec{f} \in \mathcal{M}_2^{10/7,\tau_0}$ *for some* $\tau_0 > 5/2$
- $1_{Q_2}\vec{u}$ *belongs to* $\mathcal{M}_2^{3,\tau_2}$ *for some* $\tau_2 > 5$
- $1_{Q_2}\vec{u}$ *belongs to* $\mathcal{M}_2^{3,\sigma}$ *for some σ with* $\frac{1}{\sigma} + \frac{1}{\tau_2} < \frac{1}{5}$
- $1_{Q_2}\vec{\nabla} \otimes \vec{u}$ *belongs to* $\mathcal{M}_2^{2,\tau_3}$ *with* $\frac{1}{\tau_3} = \frac{1}{\tau_2} + \frac{1}{5}$

Then, for every $r_3 < r_2$, $\vec{u}$ is Hölderian on $Q_{r_3}(t_0, x_0)$.

Proof. We write again

$$\partial_t \vec{v} = \nu \Delta \vec{v} + \vec{g} + \sum_{i=1}^{3} \partial_i \vec{h}_i - \vec{\gamma} - \vec{\eta} - \sum_{j=1}^{3}\sum_{l=1}^{3} \vec{\nabla}\partial_j\partial_l * (\phi u_j u_l)$$

with

- $\vec{g} = (\partial_t \phi)\vec{u} + (\Delta\phi)\vec{u} - \phi\vec{u}\cdot\vec{\nabla}\vec{u} + \phi\vec{f}$, where $(\partial_t\phi)\vec{u} + (\Delta\phi)\vec{u} + \phi\vec{f}$ belongs to $\mathcal{M}_2^{10/7,\min(\tau_0,\tau_2)}$ (with $\min(\tau_0,\tau_2) > 5/2$) and $\phi\vec{u}\cdot\vec{\nabla}\vec{u}$ belongs to $\mathcal{M}_2^{\frac{6}{5},\rho}$ with $\frac{1}{\rho} = \frac{1}{\tau_3} + \frac{1}{\sigma} = \frac{1}{\tau_2} + \frac{1}{5} + \frac{1}{\sigma} < \frac{2}{5}$ (so that $\rho > 5/2$)
- $\vec{h}_i = -2(\partial_i\phi)\vec{u}$ belongs to $\mathcal{M}_2^{3,\tau_2}$ with $\tau_2 > 5$
- $\vec{\gamma} = \phi\vec{\nabla}G * ((\Delta\zeta)p) + \phi\sum_{j=1}^{3}\sum_{l=1}^{3}\vec{\nabla}G*((\partial_j\partial_l\zeta)u_j u_l) - 2\phi\sum_{j=1}^{3}\vec{\nabla}\partial_j*((\partial_j\zeta)p) - \phi\sum_{j=1}^{3}\sum_{l=1}^{3}\vec{\nabla}\partial_j G*((\partial_l\zeta)u_j u_l)$ belongs to $\mathcal{M}^{q_0,\frac{5q_0}{2}}$ (with $\frac{5q_0}{2} > 5/2$)
- $\vec{\eta} = -\sum_{j=1}^{3}\sum_{l=1}^{3}[\phi, \frac{\vec{\nabla}\partial_j\partial_l}{\Delta}](\zeta u_j u_l)$: as $\zeta u_j u_l$ belongs to $\mathcal{M}_2^{3/2,\rho}$ with $\frac{1}{\rho} = \frac{1}{\tau_2} + \frac{1}{\tau}$, we find that $\vec{\eta}$ belongs to $\mathcal{M}_2^{3/2,\sigma}$ with $\frac{1}{\sigma} = \frac{1}{\tau_2} + \frac{1}{\sigma} + \frac{1}{5} < \frac{2}{5}$ (so that $\sigma > 5/2$)
- $\phi u_j u_l$ belongs to $\mathcal{M}_2^{3/2,\rho}$ with $\frac{1}{\rho} = \frac{1}{\tau_2} + \frac{1}{\tau} < \frac{1}{5}$ (so that $\rho > 5$).

and we apply Proposition 13.4. $\quad\square$

13.10 Parabolic Hausdorff Dimension of the Set of Singular Points

To more accurately describe the singularities in $\mathbb{R} \times \mathbb{R}^3$, Caffarelli, Kohn, and Nirenberg used a notion of parabolic Hausdorff dimension, adapted to the scaling properties of the Navier–Stokes equations:

Definition 13.6 (Parabolic Hausdorff measure).
(i) *For a sequence of open parabolic cylinders* $\mathcal{Q} = (Q((t_i, x_i), r_i))_{i \in \mathbb{N}}$ *of* $\mathbb{R} \times \mathbb{R}^d$ *(where* $Q((t_i, x_i), r_i) = \{(t, x) \ / \ |t - t_i| \le r_i^2 \text{ and } |x - x_i| \le r_i\})$ *and for* $\alpha > 0$, *we define* $\tilde{\sigma}_\alpha(\mathcal{Q}) = \sum_{i \in \mathbb{N}} r_i^\alpha$. *The parabolic Hausdorff measure* $\mathcal{P}^\alpha$ *on* $\mathbb{R} \times \mathbb{R}^d$ *is defined for a Borel subset* $B \subset \mathbb{R} \times \mathbb{R}^d$ *by*

$$\mathcal{P}^\alpha(B) = \lim_{\delta \to 0} \min\{\tilde{\sigma}_\alpha(\mathcal{Q}) \ / B \subset \cup_{i \in \mathbb{N}} Q((t_i, x_i), r_i), \sup_{i \in \mathbb{N}} r_i < \delta\}$$

(ii) *The parabolic Hausdorff dimension* $d_{\mathcal{H}}(B)$ *of a Borel subset of* $\mathbb{R} \times \mathbb{R}^d$ *is defined as*

$$d_{\mathcal{P}}(B) = \inf\{\alpha \ / \ \mathcal{P}^\alpha(B) = 0\} = \sup\{\alpha \ / \ \mathcal{P}^\alpha(B) = \infty\}.$$

We may now state the result of Caffarelli, Kohn, and Nirenberg:

Caffarelli–Kohn–Nirenberg regularity theorem

Theorem 13.9 (Dimension of the singular set).
Let $\vec{u}$ *be a weak solution for the Navier–Stokes equations on* $(0, T) \times \mathbb{R}^3$, *which is a suitable solution on the cylinder* $Q_0 = (a, b) \times B(x_0, r_0)$ *(with pressure* $p \in L_t^{q_0} L_x^1(Q)$ *with* $q_0 > 1$ *and* $1_{Q_0} \vec{f} \in \mathcal{M}_2^{10/7, \tau_0}$ *with* $\tau_0 > 5/2$). *Let* Σ *be the smallest closed set in* Q *so that* $\vec{u}$ *is locally bounded on* $Q_0 - \Sigma$. *Then* $\mathcal{P}^1(\Sigma) = 0$.

Proof. Let $\delta > 0$. Let $(t, x) \in \Sigma$. According to Theorem 13.8, we know that

$$\limsup_{r \to 0} \frac{1}{r} \iint_{Q(x,r)} |\vec{\nabla} \otimes \vec{u}|^2 \, dy \, ds \ge \epsilon^* > 0.$$

We fix ϵ_0 such that $0 < \epsilon_0 < \epsilon^*$. Then, we have $\Sigma \subset \cup_{Q \in \mathcal{Q}_\delta} Q$, where $\mathcal{Q}_\delta$ is the collection of open cylinders $Q((t, x), r) = (t - r^2, t + r^2) \times B(x, r)$ so that $Q \subset Q_0$, $r < \delta$ and $\iint_{Q} |\vec{\nabla} \otimes \vec{u}|^2 \, dy \, ds \ge \epsilon_0 r$. We use the parabolic distance $d_{\mathcal{P}}((t, x), (s, y)) = \max(|y - x|, \sqrt{2|t - s|})$; for this distance, an open cylinder $Q((t, x), r)$ is a ball $B(t, x), r)$, hence we may apply the Vitali covering lemma and find a countable subcollection $\mathcal{Q}_{[\delta]} = (Q((t_i, x_i), r_i))_{i \in \mathbb{N}}$ of $\mathcal{Q}_\delta$ so that $\Sigma \subset \cup_{i \in \mathbb{N}} Q((t_i, x_i), 5r_i)$ and $i \ne j \Rightarrow Q_i \cap Q_j = \emptyset$. We then have:

$$\tilde{\sigma}_1(\mathcal{Q}_{[\delta]}) = 5 \sum_{i \in \mathbb{N}} r_i \le \frac{5}{\epsilon_0} \int \int_{(t,x) \in Q_0, d_{\mathcal{P}}((t,x), \Sigma) \le \delta} |\vec{\nabla} \otimes \vec{u}|^2 \, dx \, dt.$$

This gives $\mathcal{P}^1(\Sigma) \le \frac{5}{\epsilon^*} \int\int_\Sigma |\vec{\nabla} \otimes \vec{u}|^2 \, dx \, dt$. In particular $\mathcal{P}^1(\Sigma) < \infty$; hence, the Lebesgue measure of Σ is equal to 0 and finally $\mathcal{P}^1(\Sigma) = 0$. $\square$

13.11   On the Role of the Pressure in the Caffarelli, Kohn, and Nirenberg Regularity Theorem

In June 2013, the 8th Japanese-German International Workshop on Mathematical Fluid Dynamics was held at Waseda University (Japan). Choe [122] announced that no assumption had to be made on the pressure to get the Caffarelli, Kohn and Nirenberg regularity theorem.

Of course, if no assumption is made on the pressure, one cannot speak of the local energy inequality and of suitable solutions, as the term $\operatorname{div}(p\vec{u})$ might be meaningless. Choe's talk explained that a new inequality, based on an idea of Jin about Cacciopolli inequalities for the Stokes problem, was satisfied by any weak solution $\vec{u} \in L^\infty L^2 \cap L^2 H^1$ and that this inequality allowed to prove that, under the assumption that the force was regular enough and that $\limsup_{r\to 0} \frac{1}{r} \iint_{(t_0-r^2,t_0+r^2)\times B(x_0,r)} |\vec{\nabla} \otimes \vec{u}(s,y)|^2 \, ds\, dy$ was small enough, $\vec{u}$ was Hölderian on a neighborhood of (t_0, x_0).

Chamorro, Lemarié-Rieusset and Mayoufi [102] studied Choe's ideas. As explained in Mayoufi's Ph.D. [356], the inequality may be not fulfilled by every weak solution (as Choe's proof implicitly used the equality, for a non-negative test function φ, $\iint \varphi \vec{u}.(\vec{u} \cdot \vec{\nabla}\vec{u}) \, dx\, dt = -\frac{1}{2}\iint |\vec{u}|^2 \vec{u}\cdot\vec{\nabla}\varphi \, dx\, dt$, while the term on the left-hand side of the equality is not well defined). Moreover, there is no hope of proving that $\vec{u}$ is locally Hölderian without any assumption on the pressure, as Serrin's counterexample (presented on page 406) has no regularity with respect to the time variable. Chamorro, Lemarié-Rieusset and Mayoufi extended the notion of suitable solutions by modifying Definitions 13.4 and 13.5 in the following way:

Definition 13.7.
We call $(\tilde{H}_{CKN})$ the following set of hypotheses:

1. *$\vec{u}$, p and $\vec{f}$ are defined on a domain $\Omega \subset \mathbb{R} \times \mathbb{R}^3$*

2. *on Ω, $\vec{u}$ belongs to $L_t^\infty L_x^2 \cap L_t^2 \dot{H}_x^1$:*

$$\sup_{t\in\mathbb{R}} \int_{(t,x)\in\Omega} |\vec{u}(t,x)|^2 \, dx < +\infty \quad \text{and} \quad \iint_\Omega |\vec{\nabla}\otimes\vec{u}|^2 \, dt\, dx < +\infty$$

3. *p belongs to $\mathcal{D}'(\Omega)$*

4. *on Ω, $\vec{f}$ is a divergence free vector field in $L_{t,x}^{10/7}(\Omega)$*

5. *$\vec{u}$ is a solution of the Navier–Stokes equations on Ω: $\operatorname{div}\vec{u} = 0$ and*

$$\partial_t \vec{u} = \nu\Delta\vec{u} - \vec{u}\cdot\vec{\nabla}\vec{u} + \vec{f} - \vec{\nabla}p \quad \text{in } \mathcal{D}'(\Omega) \tag{13.53}$$

This is the same definition as for $(\mathcal{H}_{CKN})$ (Definition 13.4), but for p: we no longer assume that p belongs to $L_t^{q_0} L_x^1(\Omega)$ for some $q_0 > 1$. It means that the distribution

$$\mu = -\partial_t|\vec{u}|^2 + \nu\Delta|\vec{u}|^2 - 2\nu|\vec{\nabla}\otimes\vec{u}|^2 + 2\vec{u}\cdot\vec{f} - \operatorname{div}((|\vec{u}|^2+2p)\vec{u})$$

is no longer well defined on Ω.

In order to overcome this difficulty, Chamorro, Lemarié-Rieusset and Mayoufi proved the following result:

Proposition 13.5.
Let $(\vec{u},p)$ satisfy $(\tilde{H}_{CKN})$. Let $\gamma \in \mathcal{D}(\mathbb{R})$ and let $\theta \in \mathcal{D}(\mathbb{R}^3)$, with $\int \gamma\, dt = \int \theta\, dx = 1$, and let $\varphi_{\epsilon,\alpha}(t,x) = \frac{1}{\epsilon^3\alpha}\gamma(\frac{t}{\alpha})\theta(\frac{x}{\epsilon})$ (where $\alpha > 0$ and $\epsilon > 0$). The distributions $\vec{u}*\varphi_{\epsilon,\alpha}$ and $p*\varphi_{\epsilon,\alpha}$ (with convolution in both time and space variables) are well defined on $O \subset \Omega$ as soon as we have $d(\bar{O}, \mathbb{R}\times\mathbb{R}^3 \setminus \Omega) > 2(\epsilon+\alpha)$.

Moreover, the limit $\lim_{\epsilon\to 0}\lim_{\alpha\to 0}\operatorname{div}((p*\varphi_{\epsilon,\alpha})(\vec{u}*\varphi_{\epsilon,\alpha}))$ is well defined in $\mathcal{D}'(\Omega)$ and does not depend on the choices of θ and γ. Thus, the distribution

$$\mu = -\partial_t|\vec{u}|^2 + \nu\Delta|\vec{u}|^2 - 2\nu|\vec{\nabla}\otimes\vec{u}|^2 + 2\vec{u}\cdot\vec{f} - \operatorname{div}(|\vec{u}|^2\vec{u})$$
$$- 2\lim_{\epsilon\to 0}\lim_{\alpha\to 0}\operatorname{div}((p*\varphi_{\epsilon,\alpha})(\vec{u}*\varphi_{\epsilon,\alpha})) \tag{13.54}$$

is well defined on Ω.

Thus, we may extend the notion of suitable solutions (Definition 13.5) to the notion of dissipative solutions (where the term "dissipative" is borrowed from Duchon and Robert [159]):

Dissipative solutions

Definition 13.8.
The solution $\vec{u}$ is dissipative *if the distribution μ is a non-negative locally finite measure on Ω.*

With this definition, we may extend Caffarelli, Kohn and Nirenberg's theorems [74]: Theorem 13.8 gave a criterion for local Hölderianity, Theorem 13.9 gave an estimate of the Hausdorff dimension of the singular set of a suitable solution. We have the following results for dissipative solutions:

Caffarelli–Kohn–Nirenberg regularity criterion

Theorem 13.10.
Let Ω be a bounded domain of $\mathbb{R} \times \mathbb{R}^3$. Let $(\vec{u}, p)$ a weak solution on Ω of the Navier–Stokes equations

$$\partial_t \vec{u} = \nu \Delta \vec{u} - \vec{u} \cdot \vec{\nabla} \vec{u} + \vec{f} - \vec{\nabla} p, \quad \operatorname{div} \vec{u} = 0.$$

Assume that

- *$(\vec{u}, p, \vec{f})$ satisfies the conditions $(\tilde{H}_{CKN})$: $\vec{u} \in L^\infty L^2 \cap L_t^2 H_x^1(\Omega)$, $p \in \mathcal{D}'(\Omega)$, $\operatorname{div} \vec{f} = 0$ and $\vec{f} \in L^{10/7} L^{10/7}(\Omega)$*

- *$\vec{u}$ is dissipative*

- *$\vec{f} \in L^2 H^1$.*

(A) There exists a positive constant ϵ^* which depends only on ν and τ_0 such that, if for some $(t_0, x_0) \in \Omega$, we have

$$\limsup_{r \to 0} \frac{1}{r} \iint_{(t_0 - r^2, t_0 + r^2) \times B(x_0, r)} |\vec{\nabla} \otimes \vec{u}|^2 \, ds \, dx < \epsilon^*$$

then $\vec{u}$ is bounded in a neighborhood of (t_0, x_0).

(B) Let Σ be the smallest closed set in Q so that $\vec{u}$ is locally bounded on $Q_0 - \Sigma$. Then $\mathcal{P}^1(\Sigma) = 0$.

Proof. We are now going to prove Proposition 13.5 and Theorem 13.10. With no loss of generality (due to the local character of the properties studied in Proposition 13.5 and Theorem 13.10), we may assume that $\Omega = (a, b) \times B$, where B is an open ball in $\mathbb{R}^3$.

Let $\vec{\omega} = \operatorname{curl} \vec{u}$. We may write the Navier–Stokes equations as $\operatorname{div} \vec{u} = 0$ and

$$\partial_t \vec{u} = \Delta \vec{u} - \vec{\omega} \wedge \vec{u} - \vec{\nabla}(p + \frac{|\vec{u}|^2}{2}) + \vec{f}.$$

Let O be compactly embedded into Ω: $\bar{O} \subset \Omega$, and let $\psi(t,x) = \alpha(t)\gamma(x)$ a function in $\mathcal{D}(\Omega)$ such that $\psi = 1$ on a neighborhood of $\bar{O}$. We define

$$\vec{v} = -\frac{1}{\Delta}\vec{\nabla} \wedge (\psi\vec{\omega}). \tag{13.55}$$

Note that $\vec{v}$ is defined on $\mathbb{R} \times \mathbb{R}^3$. It can be seen locally (i.e., on O) as a perturbation of $\vec{u}$: indeed, we have

$$\psi\vec{\omega} = \vec{\nabla} \wedge (\psi\vec{u}) - (\vec{\nabla}\psi) \wedge \vec{u}$$

and

$$\vec{\nabla} \wedge (\vec{\nabla} \wedge (\psi\vec{u})) = -\Delta(\psi\vec{u}) + \vec{\nabla}(\operatorname{div}(\psi\vec{u})) = -\Delta(\psi\vec{u}) + \vec{\nabla}(\vec{u} \cdot \vec{\nabla}\psi)$$

so that

$$\vec{v} = \psi\vec{u} - \frac{1}{\Delta}\left(\vec{\nabla}(\vec{u} \cdot \vec{\nabla}\psi) + \vec{\nabla} \wedge (\vec{u} \wedge \vec{\nabla}\psi)\right) = \psi\vec{u} + \vec{V}.$$

On O, $\psi\vec{u} = \vec{u}$; moreover, we write, for $(t,x) \in O$ and $\alpha \in \mathbb{N}^3$,

$$|\partial_x^\alpha \vec{V}(t,x)| \le C_\alpha \int \frac{|\vec{\nabla}\gamma(y)|}{|x-y|^{4+|\alpha|}}|\vec{u}(t,y)|\,dy \le C_{\alpha,O}\|1_B\vec{u}(t,.)\|_2$$

so that $\partial_x^\alpha \vec{V} \in L_{t,x}^\infty(O)$.

We have

$$\partial_t \vec{v} = -\frac{1}{\Delta}\vec{\nabla} \wedge ((\partial_t\psi)\vec{\omega}) - \frac{1}{\Delta}\vec{\nabla} \wedge (\psi\partial_t\vec{\omega})$$

with

$$\partial_t\vec{\omega} = \nu\Delta\vec{\omega} - \vec{\nabla} \wedge (\vec{\omega} \wedge \vec{u}) + \vec{\nabla} \wedge \vec{f}.$$

First, we notice that $\frac{1}{\Delta}\vec{\nabla} \wedge ((\partial_t\psi)\vec{\omega}) = 0$ on O.

Now, we write

$$\psi\Delta\vec{\omega} = \psi\Delta\vec{\nabla} \wedge \vec{u}$$

$$= \Delta(\psi\vec{\nabla} \wedge \vec{u}) + (\Delta\psi)\vec{\nabla} \wedge \vec{u} - 2\sum_i \partial_i((\partial_i\psi)\vec{\nabla} \wedge \vec{u})$$

$$= \Delta(\psi\vec{\omega}) + \vec{\nabla} \wedge ((\Delta\psi)\vec{u}) - (\vec{\nabla}\Delta\psi) \wedge \vec{u}$$
$$\quad - 2\sum_i \partial_i\vec{\nabla} \wedge ((\partial_i\psi)\vec{u}) + 2\sum_i \partial_i((\partial_i\vec{\nabla}\psi) \wedge \vec{u})$$

$$= \Delta(\psi\vec{\omega}) + \vec{W}$$

so that

$$-\frac{1}{\Delta}\vec{\nabla} \wedge (\psi\Delta\vec{\omega}) = -\vec{\nabla} \wedge (\psi\vec{\omega}) - \frac{1}{\Delta}\vec{\nabla} \wedge \vec{W}.$$

On O, we have $-\vec{\nabla} \wedge (\psi\vec{\omega}) = \vec{\nabla} \wedge \vec{\omega} = \Delta\vec{u} = \Delta\vec{v} - \Delta\vec{V} = \Delta\vec{v}$, as we have $\Delta\vec{V} = 0$ on O. Thus, we find that, in $\mathcal{D}'(O)$,

$$-\frac{1}{\Delta}\vec{\nabla} \wedge (\psi\Delta\vec{\omega}) = \Delta\vec{v} + \vec{G}_1$$

with $\operatorname{div}\vec{G}_1 = 0$ and $\vec{G}_1 = -\frac{1}{\Delta}\vec{\nabla} \wedge \vec{W} \in L_{t,x}^\infty(O)$.

Similarly, we write (using $\operatorname{div}\vec{f} = 0$)

$$-\frac{1}{\Delta}\vec{\nabla} \wedge (\psi\vec{\nabla} \wedge \vec{f}) = \psi\vec{f} - \frac{1}{\Delta}\vec{\nabla}(\vec{f} \cdot \vec{\nabla}\psi) - \frac{1}{\Delta}\vec{\nabla} \wedge (\vec{f} \wedge \vec{\nabla}\psi)$$

and we find that, in $\mathcal{D}'(O)$,

$$-\frac{1}{\Delta}\vec{\nabla}\wedge(\psi\vec{\nabla}\wedge\vec{f}) = \vec{f} + \vec{G}_2$$

with div $\vec{G}_2 = 0$ and $\vec{G}_2 \in L_t^2 L_x^\infty(O)$.

Finally, we write

$$\psi\vec{\nabla}\wedge(\vec{\omega}\wedge\vec{u}) = \vec{\nabla}\wedge(\vec{\omega}\wedge\psi\vec{u}) - (\vec{\nabla}\psi)\wedge(\vec{\omega}\wedge\vec{u})$$

so that

$$\frac{1}{\Delta}\vec{\nabla}\wedge(\psi\vec{\nabla}\wedge(\vec{\omega}\wedge\vec{u})) = -\vec{\omega}\wedge(\psi\vec{u}) - \vec{\nabla}P_3 + \vec{G}_3$$

with

$$P_3 = \frac{1}{\Delta}\,\text{div}\,(\vec{\omega}\wedge\psi\vec{u})$$

and

$$\vec{G}_3 = -\frac{1}{\Delta}\vec{\nabla}\wedge((\vec{\nabla}\psi)\wedge(\vec{\omega}\wedge\vec{u})).$$

On O, we have div $\vec{G}_3 = 0$ and $\vec{G}_3 \in L_t^2 L_x^\infty(O)$. Moreover, as $\psi\vec{u}$ belongs to $L^2 L^6 \cap L^\infty L^2 \subset L^6 L^{18/7}$, we have $\vec{\omega}\wedge\psi\vec{u} \in L^{3/2}L^{9/8}$ and thus $P_3 \in L^{3/2}L^{9/5}$. In particular, on O, we find $P_3 \in L^{3/2}L^{3/2}(O)$.

On O, we have $\vec{\omega} = \vec{\nabla}\wedge(\psi\vec{u})$. Recalling that $\psi\vec{u} = \vec{v} - \vec{V}$, we write that

$$-(\vec{\nabla}\wedge(\psi\vec{u}))\wedge(\psi\vec{u}) = -(\vec{\nabla}\wedge\vec{v})\wedge\vec{v} + \vec{Z}$$

with $\vec{Z} = (\vec{\nabla}\wedge\vec{V})\wedge(\psi\vec{u}) + (\vec{\nabla}\wedge(\psi\vec{u}))\wedge\vec{V} + (\vec{\nabla}\wedge\vec{V})\wedge\vec{V}$. On O, $\vec{Z}$ belongs to $L^2 L^2$, and we write

$$1_O\,\vec{Z} = \vec{G}_4 - \vec{\nabla}P_4$$

with $\vec{G}_4 = \mathbb{P}(1_O\,\vec{Z})$, so that div $\vec{G}_4 = 0$ and $\vec{G}_4 \in L^2 L^2$, while $P_4 = -\frac{1}{\Delta}\,\text{div}(1_O\,\vec{Z}) \in L^2 L^6$.

Thus far, we have obtained the following properties on $\vec{v} = -\frac{1}{\Delta}\vec{\nabla}\wedge(\psi\vec{\omega})$:

- on O, $\vec{v} = \vec{u} + \vec{V}$ with $\vec{V} \in L_t^\infty \text{Lip}_x(O)$

- $\vec{v} \in L_t^\infty L_x^2(O) \cap L_t^2 H_x^1(O)$ and div $\vec{v} = 0$

- $\vec{v}$ is a solution of the Navier–Stokes equations in $\mathcal{D}'(O)$:

$$\partial_t \vec{v} = \nu\Delta\vec{v} - \vec{v}\cdot\vec{\nabla}\vec{v} - \vec{\nabla}P + \vec{G}$$

with $\vec{G} = \nu\vec{G}_1 + \vec{f} + \vec{G}_2 + \vec{G}_3 + \vec{G}_4$, so that $\vec{G} \in L_t^2 L_x^2(O)$ and div $\vec{G} = 0$, and $P = P_3 + P_4 - \frac{|\vec{v}|^2}{2}$, so that $P \in L_t^{3/2}L_x^{3/2}(O)$.

The next step is a generalization of the formula of Duchon and Robert [159]. While in the proof of Duchon and Robert's theorem (Theorem 13.7), we used a mollifier $\theta_\epsilon(x) = \frac{1}{\epsilon^3}\theta(\frac{x}{\epsilon})$ in the space variable and computed $\partial_t(|\theta_\epsilon * \vec{u}|^2)$, our proof of Proposition 13.5 will use a mollifier $\varphi_{\epsilon,\alpha}$ in both time and space variables.

Let O' be a relatively compact open subset of O and let ϵ, α be small enough, so that $\vec{u}_{\epsilon,\alpha} = (\gamma_\alpha \otimes \theta_\epsilon) * \vec{u}$ and $\vec{v} = (\gamma_\alpha \otimes \theta_\epsilon) * \vec{v}$ are well defined on O' and involves only the values of $\vec{u}$ and $\vec{v}$ in O. As $\vec{u}_{\epsilon,\alpha}$ is now a smooth function, we find that

$$\partial_t |\vec{u}_{\epsilon,\alpha}|^2 = 2\vec{u}_{\epsilon,\alpha}.\partial_t \vec{u}_{\epsilon,\alpha}$$

and thus

$$\begin{aligned} \partial_t |\vec{u}_{\epsilon,\alpha}|^2 =& \nu\Delta|\vec{u}_{\epsilon,\alpha}|^2 - 2\nu|\vec{\nabla} \otimes \vec{u}_{\epsilon,\alpha}|^2 + 2\vec{u}_{\epsilon,\alpha}.(\varphi_{\epsilon,\alpha} * \vec{f}) \\ & - 2\operatorname{div}((p * \varphi_{\epsilon,\alpha})\vec{u}_{\epsilon,\alpha}) - 2\vec{u}_{\epsilon,\alpha}.\varphi_{\epsilon,\alpha} * \operatorname{div}(\vec{u} \otimes \vec{u}) \end{aligned}$$

We have similarly

$$\begin{aligned} \partial_t |\vec{v}_{\epsilon,\alpha}|^2 =& \nu\Delta|\vec{v}_{\epsilon,\alpha}|^2 - 2\nu|\vec{\nabla} \otimes \vec{v}_{\epsilon,\alpha}|^2 + 2\vec{v}_{\epsilon,\alpha}.(\varphi_{\epsilon,\alpha} * \vec{G}) \\ & - 2\operatorname{div}((P * \varphi_{\epsilon,\alpha})\vec{v}_{\epsilon,\alpha}) - 2\vec{v}_{\epsilon,\alpha}.\varphi_{\epsilon,\alpha} * \operatorname{div}(\vec{v} \otimes \vec{v}) \end{aligned}$$

Let $\vec{u}_\epsilon = \theta_\epsilon * \vec{u}$ (only convolution in the space variable is involved), so that $\vec{u}_{\epsilon,\alpha} = \gamma_\alpha * \vec{u}_\epsilon$. As $\alpha \to 0^+$, we have the strong convergence of $\vec{u}_{\epsilon,\alpha}$ to $\vec{u}_\epsilon$ in $L^2 L^2(O')$ and in $L^3 L^3(O')$, of $\vec{\nabla} \otimes \vec{u}_{\epsilon,\alpha}$ to $\vec{\nabla} \otimes \vec{u}_\epsilon$ in $L^2 L^2(O')$, of $\varphi_{\epsilon,\alpha} * (\vec{u} \otimes \vec{u})$ to $\theta_\epsilon * (\vec{u} \otimes \vec{u})$ in $L^{3/2} L^{3/2}(O')$, of $\varphi_{\epsilon,\alpha} * \vec{f}$ to $\theta_\epsilon \vec{f}$ in $L^2 L^2(O')$ and the *-weak convergence of $\vec{u}_{\epsilon,\alpha}$ to $\vec{u}_\epsilon$ in $L^\infty L^2$ so that we find

$$\begin{aligned} \partial_t |\vec{u}_\epsilon|^2 =& \lim_{\alpha \to 0^+} \partial_t |\vec{u}_{\epsilon,\alpha}|^2 \\ =& \nu\Delta|\vec{u}_\epsilon|^2 - 2\nu|\vec{\nabla} \otimes \vec{u}_\epsilon|^2 + 2\vec{u}_\epsilon.(\theta_\epsilon * \vec{f}) - 2\vec{u}_\epsilon.\theta_\epsilon * \operatorname{div}(|\vec{u} \otimes \vec{u}) \\ & - 2\lim_{\alpha \to 0^+} \operatorname{div}((p * \varphi_{\epsilon,\alpha})\vec{u}_{\epsilon,\alpha})). \end{aligned} \tag{13.56}$$

We now introduce

$$\mu_\epsilon = 2\vec{u}_\epsilon.\theta_\epsilon * \operatorname{div}(\vec{u} \otimes \vec{u}) - \operatorname{div}(|\vec{u}|^2\vec{u})$$

and

$$M_\epsilon = 2\vec{v}_\epsilon.\theta_\epsilon * \operatorname{div}(\vec{v} \otimes \vec{v}) - \operatorname{div}(|\vec{v}|^2\vec{v}).$$

We have the strong convergence of $\vec{u}_\epsilon$ to $\vec{u}$ in $L^2 L^2(O')$, of $\vec{\nabla} \otimes \vec{u}_\epsilon$ to $\vec{\nabla} \otimes \vec{u}$ in $L^2 L^2(O')$ and in $L^3 L^3(O')$, of $\vec{u}_\epsilon$ to $\vec{u}$ in $L^2 H^1(O')$, of $\theta_\epsilon * \vec{f}$ to $\vec{f}$ in $L^2 L^2(O')$ so that we find

$$\begin{aligned} \partial_t |\vec{u}|^2 =& \lim_{\epsilon \to 0^+} \partial_t |\vec{u}_\epsilon|^2 \\ =& \nu\Delta|\vec{u}|^2 - 2\nu|\vec{\nabla} \otimes \vec{u}|^2 + 2\vec{u} \cdot \vec{f} - \operatorname{div}(|\vec{u}|^2\vec{u}) \\ & - \lim_{\epsilon \to 0^+} (\mu_\epsilon + 2\lim_{\alpha \to 0^+} \operatorname{div}((p * \varphi_{\epsilon,\alpha})\vec{u}_{\epsilon,\alpha})). \end{aligned} \tag{13.57}$$

We have a similar result for $\vec{v}_\epsilon$, but with a better convergence: $\varphi_{\epsilon,\alpha} * P$ converges strongly to $\theta_\epsilon * P$ in $L^{3/2} L^{3/2}(O')$, and $\theta_\epsilon * P$ converges strongly to P in $L^{3/2} L^{3/2}(O')$, so that

$$\begin{aligned} \partial_t |\vec{v}|^2 =& \lim_{\epsilon \to 0^+} \partial_t |\vec{v}_\epsilon|^2 \\ =& \nu\Delta|\vec{v}|^2 - 2\nu|\vec{\nabla} \otimes \vec{v}|^2 + 2\vec{v} \cdot \vec{G} - \operatorname{div}(|\vec{v}|^2\vec{v}) \\ & - 2\operatorname{div}(P\vec{v}) - \lim_{\epsilon \to 0^+} M_\epsilon. \end{aligned} \tag{13.58}$$

We now rewrite μ_ϵ and M_ϵ as in the Theorem of Duchon and Robert. Let δ_y be defined by $\delta_y h(t,x) = h(t, x-y) - h(t,x)$ and let $\mathcal{T}_\epsilon$ be the trilinear operator

$$\mathcal{T}_\epsilon(\vec{u}, \vec{v}, \vec{w})(t,x) =$$

$$-\sum_{k=1}^{3} \int \partial_k \theta_\epsilon(y)\, \delta_y u_k(t,x)\, (\delta_y \vec{v}(t,x)\delta_y \vec{w}(t,x))\, dy$$

$$+2\sum_{k=1}^{3} \int \partial_k \theta_\epsilon(y)\, \delta_y u_k(t,x)\, (\delta_y \vec{v}(t,x).(\theta_\epsilon * \vec{w}(t,x) - \vec{w}(t,x)))\, dy$$

(notice that $\theta_\epsilon * \vec{w}(t,x) - \vec{w}(t,x) = \int \theta_\epsilon(z)\delta_z \vec{w}(t,x)\, dz$). Duchon and Robert proved that

$$\lim_{\epsilon \to 0} \mu_\epsilon - \mathcal{T}_\epsilon(\vec{u}, \vec{u}, \vec{u}) = 0$$

and we have similarly

$$\lim_{\epsilon \to 0} M_\epsilon - \mathcal{T}_\epsilon(\vec{v}, \vec{v}, \vec{v}) = 0$$

in $\mathcal{D}'(O')$.

Now, we write $\vec{v} = \vec{u} + \vec{V}$ on O, so that

$$\mathcal{T}_\epsilon(\vec{v}, \vec{v}, \vec{v}) - \mathcal{T}_\epsilon(\vec{u}, \vec{u}, \vec{u}) = \mathcal{T}_\epsilon(\vec{V}, \vec{v}, \vec{v}) + \mathcal{T}_\epsilon(\vec{u}, \vec{V}, \vec{v}) + \mathcal{T}_\epsilon(\vec{u}, \vec{u}, \vec{V})$$

This gives

$$\iint_{O'} |\mathcal{T}_\epsilon(\vec{v}, \vec{v}, \vec{v}) - \mathcal{T}_\epsilon(\vec{u}, \vec{u}, \vec{u})|\, dt\, dx \le$$

$$C \sup_{|y|<\epsilon} \sup_{(t,x)\in O} \frac{|\delta_y \vec{V}(t,x)|}{\epsilon} \iint_{O'} \int_{|y|<\epsilon} \frac{1}{\epsilon^3} (|\delta_y \vec{u}(t,x)|^2 + |\delta_y \vec{v}(t,x)|^2)\, dy\, dt\, dx$$

$$\le C\|\vec{V}\|_{L^\infty \mathrm{Lip}} \sup_{|y|<\epsilon} \iint_{O'} (|\delta_y \vec{u}(t,x)|^2 + |\delta_y \vec{v}(t,x)|^2)\, dt\, dx$$

$$\to_{\epsilon \to 0+} 0.$$

Thus, we find that

$$\lim_{\epsilon \to 0+} M_\epsilon - \mu_\epsilon = 0.$$

As $\mu = \lim_{\epsilon \to 0+} M_\epsilon$ exists (due to equality (13.58)), we find that $\lim_{\epsilon \to 0+} \mu_\epsilon$ exists (and is equal to the same limit μ). Equality (13.57) then gives the existence of $\lim_{\epsilon \to 0+} \lim_{\alpha \to 0+} \mathrm{div}((p * \varphi_{\epsilon,\alpha})\vec{u}_{\epsilon,\alpha})$: Proposition 13.5 is proved.

We now prove Theorem 13.10. Assumption that $\vec{u}$ is dissipative gives, using again equality (13.57), that $\mu = \lim_{\epsilon \to 0+} \mu_\epsilon$ is a non-negative locally finite measure. Using equality (13.58), we find that $\vec{v}$ is suitable.

Moreover, if $Q_r(t_0, x_0) \subset O$, we have

$$\frac{1}{r} \iint_{(t_0-r^2, t_0+r^2)\times B(x_0,r)} |\vec{\nabla} \otimes \vec{V}|^2\, ds\, dx \le C\|\vec{V}\|_{L^\infty \mathrm{Lip}}^2 r^4$$

so that, for $(t_0, x_0) \in O$,

$$\limsup_{r\to 0} \frac{1}{r} \iint_{(t_0-r^2, t_0+r^2)\times B(x_0,r)} |\vec{\nabla} \otimes \vec{v}|^2\, ds\, dx =$$

$$\limsup_{r\to 0} \frac{1}{r} \iint_{(t_0-r^2, t_0+r^2)\times B(x_0,r)} |\vec{\nabla} \otimes \vec{u}|^2\, ds\, dx.$$

Recall that the force $\vec{G}$ that appears in the Navier–Stokes equations whose $\vec{v}$ is a solution satisfies $\vec{G} \in L^2 L^2(O)$, so that $1_O \vec{G} \in \mathcal{M}_2^{10/7,2}$. Let us assume that

$$\limsup_{r \to 0} \frac{1}{r} \iint_{(t_0 - r^2, t_0 + r^2) \times B(x_0, r)} |\vec{\nabla} \otimes \vec{v}|^2 \, ds \, dx < \epsilon^*$$

for some small enough ϵ^*. As $2 < 5/2$, we cannot apply Theorem 13.8 (Caffarelli, Kohn and Nirenberg's theorem) to $\vec{v}$; but, as $2 > 5/3$, we may apply Lemma 13.4 (Kukavica's theorem) and find that, on a neighborhood of (t_0, x_0), $\vec{v}$ belongs to $\mathcal{M}_2^{3,\tau}$ for some $\tau > 5$. As $1_O \vec{V} \in L_{t,x}^\infty$, we have $1_O \vec{V} \in \mathcal{M}_2^{3,\tau}$, hence $\vec{u}$ belongs to $\mathcal{M}_2^{3,\tau}$ on a neighborhood of (t_0, x_0). Thus, as $\vec{f} \in L^2 H^1$, we can apply Theorem 13.3 (O'Leary's theorem) and find that $\vec{u}$ is bounded on a neighborhood of (t_0, x_0). Point (A) of Theorem 13.10 is proved.

The proof of point (B) is similar to the proof of Theorem 13.9. Theorem 13.10 is proved. $\qquad\square$

As a final remark, we may check that no regularity in the t variable is provided for the solution $\vec{u}$ in Theorem 13.10. Indeed, let us consider again Serrin's example on page 406. $\vec{u}$ is a solution of the Navier–Stokes equations on $(0,1) \times B(0,1)$ (with forcing term $\vec{f} = 0$) given by $\vec{u} = \zeta(t) \vec{\nabla} \psi$, where ψ is a harmonic function on $\mathbb{R}^3$ and ζ a bounded function on $\mathbb{R}$. The pressure p is given by $p = -\frac{|\vec{u}|^2}{2} - \partial_t \alpha \, \psi$. Let us compute the distribution μ given by Equation (13.54):

$$\mu = -\partial_t |\vec{u}|^2 + \nu \Delta |\vec{u}|^2 - 2\nu |\vec{\nabla} \otimes \vec{u}|^2 - \mathrm{div}(|\vec{u}|^2 \vec{u})$$
$$- 2 \lim_{\epsilon \to 0} \lim_{\alpha \to 0} \mathrm{div}((p * \varphi_{\epsilon,\alpha})(\vec{u} * \varphi_{\epsilon,\alpha}))$$

As $\Delta \vec{u} = 0$, we have $\nu \Delta |\vec{u}|^2 - 2\nu |\vec{\nabla} \otimes \vec{u}|^2 = 2\nu \vec{u}.\Delta \vec{u} = 0$, so that

$$\mu = -|\vec{\nabla}\psi|^2 \partial_t (\zeta(t)^2) - \mathrm{div}(|\vec{u}|^2 \vec{u})$$
$$- 2 \lim_{\epsilon \to 0} \lim_{\alpha \to 0} \mathrm{div}((p * \varphi_{\epsilon,\alpha})(\vec{u} * \varphi_{\epsilon,\alpha}))$$
$$= -|\vec{\nabla}\psi|^2 \partial_t (\zeta(t)^2) + \lim_{\epsilon \to 0} \lim_{\alpha \to 0} |\frac{1}{\epsilon^3}\theta(\frac{x}{\epsilon}) * \vec{\nabla}\psi|^2 \partial_t \left((\frac{1}{\alpha}\gamma(\frac{t}{\alpha}) * \zeta)^2 \right)$$
$$- \mathrm{div}(|\vec{u}|^2 \vec{u}) + \lim_{\epsilon \to 0} \lim_{\alpha \to 0} \mathrm{div}((|\vec{u}|^2 * \varphi_{\epsilon,\alpha})(\vec{u} * \varphi_{\epsilon,\alpha}))$$
$$= 0.$$

Thus, $\vec{u}$ is dissipative; however, $\vec{u}$ has no regularity at all with respect to t (if ζ is nowhere continuous).

Chapter 14

A Theory of Uniformly Locally L^2 Solutions

14.1   Uniformly Locally Square Integrable Solutions

We recall some basic results on uniformly locally square integrable solutions as they were described in Basson [24] and Lemarié-Rieusset [313], with a slight modification: we include forcing terms in the equations, as in Kikuchi and Seregin's paper [261].

We are thus considering the equations

$$\begin{cases} \partial_t \vec{u} + \mathrm{div}(\vec{u} \otimes \vec{u}) = \nu \Delta \vec{u} + \vec{f} - \vec{\nabla} p \\ \mathrm{div}\, \vec{u} = 0 \\ \vec{u}(0,.) = \vec{u}_0 \end{cases} \tag{14.1}$$

where $\vec{u}_0$ is a divergence-free uniformly locally square integrable vector field and $\vec{f} = \mathrm{div}\,\mathbb{F}$, where $\mathbb{F}$ is a uniformly square integrable tensor:

$$\sup_{x \in \mathbb{R}^d} \int_{|x-y|<1} |\vec{u}_0(y)|^2 \, dy < +\infty$$

and

$$\sup_{x \in \mathbb{R}^d} \int_0^1 \int_{|x-y|<1} |\mathbb{F}(s,y)|^2 \, dy\, ds < +\infty$$

and we are looking for a weak solution $\vec{u}$ on some $(0,T] \times \mathbb{R}^3$ (with $0 < T < 1$).

Recall that in our definition of weak solution (Definition 6.13), a weak solution $\vec{u}$ of Equations 14.1 on $(0,T) \times \mathbb{R}^3$ satisfies:

- $\vec{u} \in (L_t^\infty L_x^2)_{\mathrm{uloc}}$

- $\vec{\nabla} \otimes \vec{u} \in (L_t^2 L_x^2)_{\mathrm{uloc}}$

- p is locally $L^{3/2}L^{3/2}$ and $\vec{\nabla} p = (Id - \mathbb{P})\,\mathrm{div}(\mathbb{F} - \vec{u} \otimes \vec{u})$

A basic lemma on L^2_{uloc}, $(L_t^\infty L_x^2)_{\mathrm{uloc}}$ and $(L_t^2 L_x^2)_{\mathrm{uloc}}$ is the following one:

Lemma 14.1.
Let $f \in L^1(\mathbb{R}^3)$, $g \in L^2_{\mathrm{uloc}}$ and $h \in (L_t^2 L_x^2)_{\mathrm{uloc}}((0,T) \times \mathbb{R}^3)$. Then:

- $\|f * g\|_{L^2_{\mathrm{uloc}}} \leq C\|f\|_1 \|g\|_{L^2_{\mathrm{uloc}}}$

- $\|f * h\|_{(L_t^2 L_x^2)_{\mathrm{uloc}}} \leq C\|f\|_1 \|h\|_{(L_t^2 L_x^2)_{\mathrm{uloc}}}$

- $\|f * h\|_{(L_t^\infty L_x^2)_{\mathrm{uloc}}} \leq C\|f\|_1 \|h\|_{(L_t^\infty L_x^2)_{\mathrm{uloc}}}$

Moreover, if $\alpha > 0$, then

- $\|f * g\|_\infty \leq C_\alpha \|(1 + |x|)^{\frac{3+\alpha}{2}} f\|_2 \|g\|_{L^2_{\mathrm{uloc}}}$

DOI: 10.1201/9781003042594-14

- $\|f * h\|_{L^\infty_{t,x}} \le C_\alpha \|(1 + |x|)^{\frac{3+\alpha}{2}} f\|_2 \|h\|_{(L^\infty_t L^2_x)_{\mathrm{uloc}}}$

- $\|f * \int_0^T h \, dt\|_\infty \le C_\alpha \sqrt{T} \|(1 + |x|)^{\frac{3+\alpha}{2}} f\|_2 \|h\|_{(L^2_t L^2_x)_{\mathrm{uloc}}}$

Proof. Define $f_k = f(x) 1_{[0,1]^3}(x - k)$ and $Q_k = k + [0,1]^3$, $k \in \mathbb{Z}^3$. Then $\sum_{k \in \mathbb{Z}^3} \|f_k\|_1 = \|f\|_1$ and

$$\sum_{k \in \mathbb{Z}^3} \|f_k\|_2 \le C_\alpha \sum_{k \in \mathbb{Z}^3} (1 + |k|)^{-\frac{3+\alpha}{2}} \|1_{Q_k}(1 + |x|)^{\frac{3+\alpha}{2}} f\|_2 \le C'_\alpha \|(1 + |x|)^{\frac{3+\alpha}{2}} f\|_2.$$

We have

$$\int_{|x-y|<1} |f_k * g|^2 \, dy \le \|f_k\|_1^2 \int_{|x-k-z|<4} |g(z)|^2 \, dz$$

so that

$$\|f * g\|_{L^2_{\mathrm{uloc}}} \le \sum_{k \in \mathbb{Z}^3} \|f_k * g\|_{L^2_{\mathrm{uloc}}} \le C \sum_{k \in \mathbb{Z}^3} \|f_k\|_1 \|g\|_{L^2_{\mathrm{uloc}}} = C \|f\|_1 \|g\|_{L^2_{\mathrm{uloc}}}.$$

Inequalities $\|f * h\|_{(L^2_t L^2_x)_{\mathrm{uloc}}} \le C \|f\|_1 \|h\|_{(L^2_t L^2_x)_{\mathrm{uloc}}}$ and $\|f * h\|_{(L^\infty_t L^2_x)_{\mathrm{uloc}}} \le C \|f\|_1 \|h\|_{(L^\infty_t L^2_x)_{\mathrm{uloc}}}$ are proved in a similar way.

We have as well

$$\|f_k * g\|_\infty^2 \le \sup_{x \in \mathbb{R}^3} \|f_k\|_2^2 \int_{|x-k-z|<4} |g(z)|^2 \, dz$$

so that

$$\|f * g\|_\infty \le \sum_{k \in \mathbb{Z}^3} \|f_k * g\|_\infty \le C \sum_{k \in \mathbb{Z}^3} \|f_k\|_2 \|g\|_{L^2_{\mathrm{uloc}}} \le C' \|(1 + |x|)^{\frac{3+\alpha}{2}} f\|_2 \|g\|_{L^2_{\mathrm{uloc}}}.$$

The estimates on $f * h$ will be proved in the same way. $\square$

We have as well estimates on L^1_{uloc}: $\|f * g\|_{L^1_{\mathrm{uloc}}} \le C \|f\|_1 \|g\|_{L^1_{\mathrm{uloc}}}$. Thus, if $\mathbb{H} \in L^1_{\mathrm{uloc}}$, the equation

$$\vec{\nabla} p = (Id - \mathbb{P}) \operatorname{div} \mathbb{H}$$

is well defined, since we have

$$\vec{\nabla} p = \operatorname{div} \mathbb{H} - \vec{\nabla} \frac{1}{\Delta} \sum_{i=1}^3 \sum_{j=1}^3 \partial_i \partial_j H_{i,j}$$

where the operator $\frac{1}{\Delta} \partial_i \partial_j \partial_k$ is well defined on L^1_{uloc}: recall that $G = \frac{1}{4\pi|x|}$ is the Green function associated to $-\Delta$; then, if $\gamma \in \mathcal{D}$ satisfies $\gamma = 1$ on $|x| < 1$, if $f_0 = -\partial_i \partial_j \partial_k (\gamma G)$ and $f_1 = -\partial_i \partial_j \partial_k ((1 - \gamma) G)$, we have

$$\frac{1}{\Delta} \partial_i \partial_j \partial_k g = f_0 * g + f_1 * g$$

with $f_0 \in \mathcal{E}'$ is a compactly supported distribution (so that convolution with f_0 is well defined on $\mathcal{D}'$) and $f_1 \in L^1$ (so that convolution with f_1 is well defined on L^1_{uloc}).

Uniformly locally square integrable solutions

Theorem 14.1.
Let $\vec{u}_0 \in L^2_{\text{uloc}}$ with $\operatorname{div} \vec{u}_0 = 0$ and $\mathbb{F} \in (L^2_t L^2_x)_{\text{uloc}}((0,1) \times \mathbb{R}^3)$. Then there exists a solution $\vec{u}$ to the problem

$$\begin{cases} \partial_t \vec{u} = \nu \Delta \vec{u} + \mathbb{P}\operatorname{div}(\mathbb{F} - \vec{u} \otimes \vec{u}) \\[2mm] \vec{u}(0,.) = \vec{u}_0 \end{cases} \tag{14.2}$$

on $(0,T) \times \mathbb{R}^3$ with

$$T = \min\left(1, \frac{1}{C_0 \nu\left(1 + \frac{\|\vec{u}_0\|_{L^2_{\text{uloc}}}}{\nu} + \frac{\|\mathbb{F}\|_{(L^2 L^2)_{\text{uloc}}}}{\nu^{3/2}}\right)^4}\right)$$

and

$$\vec{u} \in (L^\infty_t L^2_x)_{\text{uloc}} \cap (L^2 H^1_x)_{\text{uloc}}$$

$$\sup_{0<t<T} \sup_{x\in\mathbb{R}^3} \left(\int_{|x-y|<1} |\vec{u}(t,y)|^2 \, dy\right)^{1/2} \leq 2(\|\vec{u}_0\|_{L^2_{\text{uloc}}} + C_0 \frac{\|\mathbb{F}\|_{(L^2 L^2)_{\text{uloc}}}}{\sqrt{\nu}})$$

$$\sup_{x\in\mathbb{R}^3} \left(\int_0^T \int_{|x-y|<1} |\vec{\nabla} \otimes \vec{u}(s,y)|^2 \, dy\, ds\right)^{1/2} \leq \frac{2}{\sqrt{\nu}}(\|\vec{u}_0\|_{L^2_{\text{uloc}}} + C_0 \frac{\|\mathbb{F}\|_{(L^2 L^2)_{\text{uloc}}}}{\sqrt{\nu}})$$

where the constant C_0 does not depend on ν.
   Moreover, this solution $\vec{u}$ is suitable: it satisfies in $\mathcal{D}'$ the local energy inequality

$$\partial_t\left(\frac{|\vec{u}|^2}{2}\right) \leq \nu\Delta\left(\frac{|\vec{u}|^2}{2}\right) - \nu|\vec{\nabla}\otimes\vec{u}|^2 - \operatorname{div}\left(\left(p + \frac{|\vec{u}|^2}{2}\right)\vec{u}\right) + \vec{u}\cdot\vec{f} \tag{14.3}$$

with $\vec{f} = \operatorname{div}\mathbb{F}$ and $\vec{\nabla}p = (Id - \mathbb{P})\operatorname{div}(\mathbb{F} - \vec{u}\otimes\vec{u})$.

Proof. **Step 1: local existence for the mollified problem.**
   As for the proof of the existence of Leray solutions, we start with a mollification of the non-linearity. We fix $\theta \in \mathcal{D}(\mathbb{R}^3)$ with $\int \theta \, dx = 1$ and we define, for $\epsilon > 0$, $\theta_\epsilon = \frac{1}{\epsilon^3}\theta(\frac{x}{\epsilon})$. We then shall look for a solution of

$$\begin{cases} \partial_t \vec{u} = \nu\Delta\vec{u} + \mathbb{P}(\vec{f} - (\theta_\epsilon * \vec{u}) \cdot \vec{\nabla}\vec{u}) \\[2mm] \vec{u}(0,.) = \vec{u}_0 \end{cases} \tag{14.4}$$

As usual, we use Picard's iterative scheme, and define inductively $\vec{U}_n$ by

$$\vec{U}_0 = W_{\nu t} * \vec{u}_0 + \int_0^t W_{\nu(t-s)} * \mathbb{P}\vec{f} \, ds \quad \text{and} \quad \vec{U}_{n+1} = \vec{U}_0 - B_\epsilon(\vec{U}_n, \vec{U}_n)$$

with

$$B_\epsilon(\vec{U}, \vec{V}) = \int_0^t W_{\nu(t-s)} * \mathbb{P}((\theta_\epsilon * \vec{U}) \cdot \vec{\nabla}\vec{V}) \, ds.$$

We shall show existence of a solution $\vec{u}$ on $(0, T_\epsilon) \times \mathbb{R}^3$ such that

$$\vec{u} \in E_\epsilon = \{\vec{u} \in \mathcal{D}'((0, T_\epsilon) \times \mathbb{R}^3) \, / \, \operatorname{div}\vec{u} = 0 \text{ and } \vec{u} \in (L^\infty_t L^2_x)_{\text{uloc}} \cap (L^2 H^1_x)_{\text{uloc}}\}.$$

E_ϵ is a Banach space for the norm

$$\|\vec{u}\|_{E_\epsilon} = \sup_{0<t<T_\epsilon} \sup_{x\in\mathbb{R}^3} \left(\int_{|x-y|<1} |\vec{u}(t,y)|^2 \, dy \right)^{\frac{1}{2}} + \sup_{x\in\mathbb{R}^3} \left(\int_0^{T_\epsilon} \int_{|x-y|<1} |\vec{\nabla}\otimes\vec{u}(s,y)|^2 \, dy \, ds \right)^{\frac{1}{2}}.$$

It is easy to check that $\vec{U}_0 \in E_\epsilon$: for $x \in \mathbb{R}^3$, we split $\vec{u}_0$ in $1_{|x-y|<3}\vec{u}_0(y) = \vec{u}_{1,x}(y)$ and $1_{|x-y|>3}\vec{u}_0(y) = \vec{u}_{2,x}(y)$, and similarly $\mathbb{F}$ in $1_{|x-y|<3}\mathbb{F}(t,y) = \mathbb{F}_{1,x}(t,y)$ and $1_{|x-y|>3}\mathbb{F}_{2,x}(t,y)(t,y)$; we then write:

- $\|W_{\nu t} * \vec{u}_{1,x}\|_{L^\infty L^2} = \|\vec{u}_{1,x}\|_2 \leq C\|\vec{u}_0\|_{L^2_{\text{uloc}}}$ and $\|W_{\nu t} * \vec{u}_{1,x}\|_{L^2\dot{H}^1} = \frac{1}{\sqrt{2\nu}}\|\vec{u}_{1,x}\|_2$

- on $|x-y|<1$, $|W_{\nu t}*\vec{u}_{2,x}(y)| \leq C\int_{|x-z|>3} \frac{\nu t}{|x-z|^5}|\vec{u}_0(z)|\,dz \leq C\nu T_\epsilon\|\vec{u}_0\|_{L^2_{\text{uloc}}}$ and $|\vec{\nabla}\otimes W_{\nu t}*\vec{u}_{2,x}(y)| \leq C\int_{|x-z|>3} \frac{1}{|x-z|^4}|\vec{u}_0(z)|\,dz \leq C\|\vec{u}_0\|_{L^2_{\text{uloc}}}$

- $\|\int_0^t W_{\nu(t-s)} * \mathbb{P}\operatorname{div}\mathbb{F}_{1,x}\,ds\|_{L^\infty((0,T_\epsilon),L^2)} \leq \frac{1}{\sqrt{2\nu}}\|\mathbb{F}_{1,x}\|_{L^2L^2} \leq \frac{C}{\sqrt{\nu}}\|\mathbb{F}\|_{(L^2L^2)_{\text{uloc}}}$ and $\|\int_0^t W_{\nu(t-s)} * \mathbb{P}\operatorname{div}\mathbb{F}_{1,x}\,ds\|_{L^2((0,T_\epsilon),\dot{H}^1)} \leq \frac{1}{\nu}\|\mathbb{F}_{1,x}\|_{L^2L^2}$

- on $|x-y|<1$, we have

$$\left| \int_0^t W_{\nu(t-s)} * \mathbb{P}\operatorname{div}\mathbb{F}_{2,x}(s,.)(y)\,ds \right| \leq C \int_0^t \int_{|x-z|>3} \frac{1}{|x-z|^4}|\mathbb{F}(s,z)|\,dz\,ds$$
$$\leq C\sqrt{T_\epsilon}\|\mathbb{F}\|_{(L^2L^2)_{\text{uloc}}}$$

and

$$\left| \vec{\nabla}\otimes \int_0^t W_{\nu(t-s)} * \mathbb{P}\operatorname{div}\mathbb{F}_{2,x}(s,.)(y)\,ds \right| \leq C \int_0^t \int_{|x-z|>3} \frac{1}{|x-z|^5}|\mathbb{F}(s,z)|\,dz\,ds$$
$$\leq C\sqrt{T_\epsilon}\|\mathbb{F}\|_{(L^2L^2)_{\text{uloc}}}$$

Thus, recalling that $T_\epsilon < 1$, we find that

$$\|W_{\nu t} * \vec{u}_0\|_{E_\epsilon} \leq C_\nu\|\vec{u}_0\|_{L^2_{\text{uloc}}} \tag{14.5}$$

and

$$\left\| \int_0^t W_{\nu(t-s)} * \mathbb{P}\operatorname{div}\mathbb{F}\,ds \right\|_{E_\epsilon} \leq C_\nu\|\mathbb{F}\|_{(L^2L^2)_{\text{uloc}}} \tag{14.6}$$

where the constant C_ν depends only on ν.

Inequality (14.6) gives as well:

$$\|B_\epsilon(\vec{U},\vec{V})\|_{E_\epsilon} \leq C_\nu\|(\theta_\epsilon * \vec{U})\otimes\vec{V}\|_{(L^2L^2)_{\text{uloc}}}$$
$$\leq CC_\nu T_\epsilon^{1/2}\|\theta_\epsilon * \vec{U}\|_{L^\infty L^\infty}\|\vec{V}\|_{L^\infty L^2_{\text{uloc}}} \tag{14.7}$$
$$\leq T_\epsilon^{1/2} C'\epsilon^{-3/4} C_\nu\|\vec{U}\|_{L^\infty L^2_{\text{uloc}}}\|\vec{V}\|_{L^\infty L^2_{\text{uloc}}}$$

Thus, Picard's algorithm will converge to a solution if T_ϵ is small enough:

$$T_\epsilon < \min(1, \frac{\epsilon^{3/2}}{C_\nu(\|\vec{u}_0\|_{L^2_{\text{uloc}}} + \|\mathbb{F}\|_{(L^2L^2)_{\text{uloc}}})^2}) \tag{14.8}$$

(where the constant C_ν depends only on ν).

Step 2: uniform existence time for the mollified problem.

The existence time T_ϵ we found in Equation (14.8) goes to 0 as ϵ goes to 0; however, if we want our approximation scheme to converge to a solution, we must find a time of existence which does not depend on ϵ. From (14.8), we can see that, as long as $\|\vec{u}\|_{L^2_{\text{uloc}}}$ remains bounded, we may extend the solution to a larger interval.

In order to control the size of $\vec{u}$, we introduce a non-negative compactly supported function $\varphi_0 \in \mathcal{D}(\mathbb{R}^3)$ such that

$$\sum_{k \in \mathbb{Z}^3} \varphi_0(x - k) = 1$$

and the set

$$\mathcal{B} = \{\varphi_{x_0} = \varphi_0(. - x_0) \ / \ x_0 \in \mathbb{R}^3\}$$

Then, we have

$$\|h\|_{L^2_{\text{uloc}}} \approx \sup_{\varphi \in \mathcal{B}} \|h\varphi\|_2 \text{ and } \|H\|_{(L^2 L^2)_{\text{uloc}}} \approx \sup_{\varphi \in \mathcal{B}} \|H\varphi\|_{L^2 L^2}$$

We have

$$\frac{d}{dt}\|\varphi\vec{u}\|_2^2 = 2\langle \nu\Delta\vec{u} + \vec{f} - \vec{\nabla}p - (\vec{u} * \theta_\epsilon) \cdot \vec{\nabla}\vec{u} \,|\, \varphi^2\vec{u}\rangle$$

We have seen that $\vec{\nabla}p$ is well defined, so p is defined up to a function $p(t)$ which does not depend on x. In order to estimate $\langle \vec{\nabla}p \,|\, \varphi^2\vec{u}\rangle$ with $\varphi = \varphi_{x_0}$, we shall fix the definition of $p(t, x)$ on the support of φ in the following way: let R_0 be such that the support of φ_0 is contained in the ball $B(0, R_0)$ and let $\mathbb{K}$ be the distribution kernel of $\frac{1}{\Delta}(\vec{\nabla} \otimes \vec{\nabla})$: $\frac{1}{\Delta}(\sum_{i=1}^3 \sum_{j=1}^3 \partial_i\partial_j H_{i,j}) = \mathbb{K} * \mathbb{H}$, then we define $p_{x_0}(t, x)$ as

$$p_{x_0}(t, x) = \frac{1}{\Delta}(\vec{\nabla} \otimes \vec{\nabla})(1_{B(x_0, 5R_0)}\mathbb{H}) + \int_{|y-x_0|>5R_0} (\mathbb{K}(x - y) - \mathbb{K}(x_0 - y))\mathbb{H}(t, y)\, dy$$

with

$$\mathbb{H} = \mathbb{F} - (\theta_\epsilon * \vec{u}) \otimes \vec{u}.$$

As $\mathbb{H}$ belongs to $(L^2 L^2)_{\text{uloc}}$, the singular integral operator $\frac{1}{\Delta}(\vec{\nabla} \otimes \vec{\nabla})$ is well defined on the compactly supported function $1_{B(x_0, 5R_0)}\mathbb{H}$.

We write

$$\varpi_{x_0} = \frac{1}{\Delta}(\vec{\nabla} \otimes \vec{\nabla})(1_{B(x_0, 5R_0)}\mathbb{F}),$$

$$\pi_{x_0} = \frac{1}{\Delta}(\vec{\nabla} \otimes \vec{\nabla})(1_{B(x_0, 5R_0)}(\theta_\epsilon * \vec{u}) \otimes \vec{u})$$

and, for $|x - x_0| < R_0$,

$$q_{x_0}(t, x) = \int_{|y-x_0|>5R_0} (\mathbb{K}(x - y) - \mathbb{K}(x_0 - y))\mathbb{F}(t, y)\, dy,$$

$$\rho_{x_0}(t, x) = \int_{|y-x_0|>5R_0} (\mathbb{K}(x - y) - \mathbb{K}(x_0 - y))(\theta_\epsilon * \vec{u}) \otimes \vec{u})(t, y)\, dy$$

As $\text{div}(\vec{u} * \theta_\epsilon) = 0$, we have (for $\text{Supp } \theta \subset B(0, R_1)$ and $\epsilon < \frac{R_0}{R_1}$)

$$2\int ((\vec{u} * \theta_\epsilon) \cdot \vec{\nabla}\vec{u}) \cdot (\varphi^2\vec{u})\, dx = -\int (\vec{u} * \theta_\epsilon) \otimes \vec{u} \cdot (\vec{\nabla}(\varphi^2) \otimes \vec{u})\, dx.$$

Thus, we have (for Supp $\theta \subset B(0, R_1)$ and $\epsilon < \frac{R_0}{R_1}$)

$$\frac{d}{dt}\|\varphi\vec{u}\|_2^2 = -2\nu\int(\vec{\nabla}\otimes\vec{u})\cdot\vec{\nabla}\otimes(\varphi^2\vec{u})\,dx - 2\int\mathbb{F}\cdot\vec{\nabla}\otimes(\varphi^2\vec{u})\,dx$$

$$+ 2\int p_{x_0}\,\mathrm{div}(\varphi^2\vec{u})\,dx + 2\int(\vec{u}*\theta_\epsilon)\otimes\vec{u}\cdot(\vec{\nabla}(\varphi^2)\otimes\vec{u})\,dx.$$

$$\leq -2\nu\int|\varphi\vec{\nabla}\otimes\vec{u}|^2\,dx + 4\nu\int|\varphi\vec{\nabla}\otimes\vec{u}|\,|\vec{\nabla}\varphi|\,|\vec{u}|\,dx$$

$$+ 2\int|\varphi\mathbb{F}|\,|\varphi\vec{\nabla}\otimes\vec{u}|\,dx + 4\int|\varphi\mathbb{F}|\,|\vec{\nabla}\varphi|\,|\vec{u}|\,dx$$

$$+ 2\int|p_{x_0}|\,|\varphi\vec{u}|\,|\vec{\nabla}\varphi|\,dx + 4\int|\vec{u}*\theta_\epsilon|\,|\vec{u}|^2\,|\varphi|\,|\vec{\nabla}\varphi|\,dx$$

$$\leq -\nu\int|\varphi\vec{\nabla}\otimes\vec{u}|^2\,dx + C_1\nu\int_{|x-x_0|<R_0}|\vec{u}|^2\,dx$$

$$+ C_1\frac{1}{\nu}\int_{|x-x_0|<R_0}|\mathbb{F}|^2\,dx + C_1\frac{1}{\nu}\int_{|x-x_0|<R_0}(|\varpi_{x_0}|^2 + (|q_{x_0}|^2)\,dx$$

$$+ C_1\int_{|x-x_0|<R_0}(|\pi_{x_0}|^{3/2} + |\rho_{x_0}|^{3/2})\,dx + C_1\int_{|x-x_0|<2R_0}|\vec{u}|^3\,ds$$

and finally we get

$$\frac{d}{dt}\|\varphi\vec{u}\|_2^2 \leq -\nu\int|\varphi\vec{\nabla}\otimes\vec{u}|^2\,dx + C_1\nu\int_{|x-x_0|<R_0}|\vec{u}|^2\,dx$$

$$+ C_2\frac{1}{\nu}\int_{|x-x_0|<5R_0}|\mathbb{F}|^2\,dx + C_2\frac{1}{\nu}\Big(\int_{|y-x_0|>5R_0}\frac{1}{|x_0-y|^4}|\mathbb{F}(t,y)|\,dy\Big)^2 \quad (14.9)$$

$$+ C_2\Big(\int_{|y-x_0|>5R_0}\frac{1}{|x_0-y|^4}|\vec{u}(t,y)|^2\,dy\Big)^{3/2} + C_2\int_{|x-x_0|<5R_0}|\vec{u}|^3\,dx$$

Let

$$\alpha(t) = \|\vec{u}\|_{L^2_{\mathrm{uloc}}} = \sup_{\varphi\in\mathcal{B}}\|\varphi\vec{u}\|_2$$

$$\beta(t) = \|\mathbb{F}\|_{(L^2L^2)_{\mathrm{uloc}}((0,t)\times\mathbb{R}^3)} = \sup_{\varphi\in\mathcal{B}}\Big(\int_0^t\int|\varphi(x)\mathbb{F}(s,x)|^2\,dx\,ds\Big)^{1/2}$$

$$\gamma(t) = \|\vec{\nabla}\otimes\vec{u}\|_{(L^2L^2)_{\mathrm{uloc}}((0,t)\times\mathbb{R}^3)} = \sup_{\varphi\in\mathcal{B}}\Big(\int_0^t\int|\varphi(x)\vec{\nabla}\otimes\vec{u}(s,x)|^2\,dx\,ds\Big)^{1/2}$$

$$\delta(t) = \|\vec{u}\|_{(L^3L^3)_{\mathrm{uloc}}((0,t)\times\mathbb{R}^3)} = \sup_{\varphi\in\mathcal{B}}\Big(\int_0^t\int|\varphi(x)\vec{u}(s,x)|^3\,dx\,ds\Big)^{1/3}$$

Integrating our inequality (14.9) and using Lemma 14.1, we get (for $t < \min(T^*, 1)$, where T^* is the maximal existence time)

$$\|\varphi\vec{u}\|_2^2 + \nu\int_0^t\int|\varphi\vec{\nabla}\otimes\vec{u}|^2\,dx$$

$$\leq\|\varphi\vec{u}_0\|_2^2 + C_3\nu\int_0^t\alpha(s)^2\,ds + C_3\frac{1}{\nu}\beta(t)^2 + C_3\delta(t)^3.$$

We have

$$\|\varphi\vec{u}\|_{L^3(dx)} \leq C\|\vec{\nabla}\otimes(\varphi\vec{u})\|_{L^2(dx)}^{1/2}\|\varphi\vec{u}\|_{L^2(dx)}^{1/2} \leq C\sqrt{\alpha(t)}\sqrt{\alpha(t) + \|\varphi\vec{\nabla}\otimes\vec{u}\|_2}$$

hence, for any $\eta > 0$,

$$\|\varphi \vec{u}\|_{L^3(dx)}^3 \leq C(\alpha(t)^3 + \eta^{-3}\alpha(t)^6 + \eta\|\varphi\vec{\nabla} \otimes \vec{u}\|_2^2).$$

Hence, we get

$$\|\varphi\vec{u}\|_2^2 + \nu\int_0^t \int |\varphi\vec{\nabla} \otimes \vec{u}|^2 \, dx$$

$$\leq \alpha(0)^2 + C_3\nu\int_0^t \alpha(s)^2 \, ds + C_4\eta\gamma(t)^2$$

$$+ C_3\frac{1}{\nu}\beta(t)^2 + C_4\left(\int_0^t \alpha(s)^3 \, ds + \eta^{-3}\int_0^t \alpha(s)^6 \, ds\right)$$

This gives in particular

$$\nu\gamma(t)^2 \leq \alpha(0)^2 + C_3\nu\int_0^t \alpha(s)^2 \, ds + C_4\eta\gamma(t)^2$$

$$+ C_3\frac{1}{\nu}\beta(t)^2 + C_4\left(\int_0^t \alpha(s)^3 \, ds + \eta^{-3}\int_0^t \alpha(s)^6 \, ds\right)$$

and, for $\eta = \frac{\nu}{2C_4}$,

$$\nu\gamma(t)^2 \leq 2\alpha(0)^2 + 2C_3\nu\int_0^t \alpha(s)^2 \, ds$$

$$+ 2C_3\frac{1}{\nu}\beta(t)^2 + 2C_4\left(\int_0^t \alpha(s)^3 \, ds + \eta^{-3}\int_0^t \alpha(s)^6 \, ds\right)$$

Now, we write

$$\|\varphi\vec{u}\|_2^2 \leq \alpha(0)^2 + C_3\nu\int_0^t \alpha(s)^2 \, ds + C_4\eta\gamma(t)^2$$

$$+ C_3\frac{1}{\nu}\beta(t)^2 + C_4\left(\int_0^t \alpha(s)^3 \, ds + \eta^{-3}\int_0^t \alpha(s)^6 \, ds\right)$$

$$\leq 2\alpha(0)^2 + 2C_3\nu\int_0^t \alpha(s)^2 \, ds$$

$$+ 2C_3\frac{1}{\nu}\beta(t)^2 + 2C_4\left(\int_0^t \alpha(s)^3 \, ds + 2(\frac{2C_4}{\nu})^{-3}\int_0^t \alpha(s)^6 \, ds\right)$$

and finally

$$\alpha(t)^2 \leq 2\alpha(0)^2 + C_5\nu\int_0^t \alpha(s)^2 \, ds$$

$$+ 2C_3\frac{1}{\nu}\beta(t)^2 + C_5(\frac{1}{\nu})^{-3}\int_0^t \alpha(s)^6 \, ds$$

Let

$$B_0 = \|\vec{u}_0\|_{L_{uloc}^2} + \sqrt{\frac{C_3}{\nu}}\|\mathbb{F}\|_{(L^2L^2)_{uloc}((0,1)\times\mathbb{R}^3))}$$

We have proved that (for $t < 1$)

$$\|\vec{u}(t,.)\|_{L_{uloc}^2}^2 = \alpha(t)^2 \leq 2B_0^2 + C_5\nu\int_0^t \alpha(s)^2 \, ds + C_5(\frac{1}{\nu})^{-3}\int_0^t \alpha(s)^6 \, ds$$

Thus $\|\vec{u}(t,.)\|_{L^2_{\text{uloc}}}$ will remain bounded by $2B_0$ as long as $t < 1$, $4C_5\nu t < 1$ and $64C_5 B_0^4 t < \nu^3$.

It means that the existence time of the mild solution $\vec{u}$ may be estimated independently from $\epsilon \in (0,1)$: $\vec{u}$ exists at least on $(0,T^*)$, where

$$T^* = \min(1, \frac{1}{4C_5\nu}, \frac{\nu^3}{64C_5(\|\vec{u}_0\|_{L^2_{\text{uloc}}} + \sqrt{\frac{C_3}{\nu}}\|\mathbb{F}\|_{(L^2L^2)_{\text{uloc}}((0,1)\times\mathbb{R}^3)})^4}). \qquad (14.10)$$

Step 3: Weak convergence.

Let $\vec{u}_\epsilon$ be the solution of the mollified problem (14.4). We have found a time T^* which is independent from $\epsilon \in (0,1)$ such that the solution $\vec{u}_\epsilon$ exists on $((0,T^*)\times\mathbb{R}^3)$ and satisfies

$$\sup_{0<t<T^*} \sup_{x\in\mathbb{R}^3} (\int_{|x-y|<1} |\vec{u}_\epsilon(t,y)|^2\, dy)^{1/2} \leq 2(\|\vec{u}_0\|_{L^2_{\text{uloc}}} + \sqrt{C_3}\frac{\|\mathbb{F}\|_{(L^2L^2)_{\text{uloc}}}}{\sqrt{\nu}})$$

$$\sup_{x\in\mathbb{R}^3} (\int_0^{T^*} \int_{|x-y|<1} |\vec{\nabla}\otimes\vec{u}_\epsilon(s,y)|^2\, dy\, ds)^{1/2} \leq \frac{2}{\sqrt{\nu}}(\|\vec{u}_0\|_{L^2_{\text{uloc}}} + \sqrt{C_3}\frac{\|\mathbb{F}\|_{(L^2L^2)_{\text{uloc}}}}{\sqrt{\nu}})$$

where C_3 does not depend on ϵ.

From those energy estimates, we can see that, for every test function $\phi \in \mathcal{D}'((0,T^*)\times\mathbb{R}^3)$, $\phi\vec{u}_\epsilon$ remains bounded in $L^\infty L^2 \cap L^2\dot{H}^1$. Moreover, we have

$$\partial_t\vec{u}_\epsilon = \nu\Delta\vec{u}_\epsilon + \mathbb{P}\operatorname{div}(\mathbb{F} - (\vec{u}_\epsilon * \theta_\epsilon)\otimes\vec{u}_\epsilon)$$

and we have seen that $\vec{u}_\epsilon$ remains bounded in $(L^3L^3)_{\text{uloc}}$; thus, we can see that $\phi\partial_t\vec{u}_\epsilon$ remains bounded in $L^{3/2}H^{-3/2}$.

We may then use the Rellich–Lions theorem (Theorem 12.1): we may find a sequence $\epsilon_n \to 0$ and a function $\vec{u} \in (L^\infty L^2)_{\text{uloc}} \cap (L^2\dot{H}^1)_{\text{uloc}}$ such that:

- $\vec{u}_{(\epsilon_n)}$ is *-weakly convergent to $\vec{u}$ in $(L^\infty L^2)_{\text{uloc}}$ and in $(L^2\dot{H}^1)_{\text{uloc}}$
- $\vec{u}_{(\epsilon_n)}$ is strongly convergent to $\vec{u}$ in $L^2_{\text{loc}}((0,T)\times\mathbb{R}^3)$.

In order to show that the weak limit $\vec{u}$ satisfies

$$\partial_t\vec{u} = \nu\Delta\vec{u} + \mathbb{P}(\vec{f} - \vec{u}\cdot\vec{\nabla}\vec{u}),$$

we have only to check that we have the convergence in $\mathcal{D}'$ of the non-linear term $\mathbb{P}\operatorname{div}((\vec{u}_\epsilon * \theta_\epsilon)\otimes\vec{u}_\epsilon)$ to $\mathbb{P}\operatorname{div}(\vec{u}\otimes\vec{u})$. As we know that $\vec{u}_\epsilon$ is bounded in $(L^3L^3)_{\text{uloc}}$ and is strongly convergent in $(L^2L^2)_{\text{loc}}$, we see that $(\vec{u}_\epsilon * \theta_\epsilon)\otimes\vec{u}_\epsilon$ is bounded in $(L^{6/5}L^{6/5})_{\text{uloc}}$ and strongly convergent to $\vec{u}\otimes\vec{u}$ in $(L^{6/5}L^{6/5})_{\text{loc}}$. This is enough to get the convergence of $\mathbb{P}\operatorname{div}((\vec{u}_\epsilon * \theta_\epsilon)\otimes\vec{u}_\epsilon)$.

Step 4: Local energy estimates for the weak limit.

We now check that $\vec{u}$ is more precisely a suitable weak solution (i.e., fulfills the local energy inequality). We work in the neighborhood $B(x_0, R_0)$ of a point x_0, and we write

$$\partial_t\vec{u}_\epsilon = \nu\Delta\vec{u}_\epsilon + \vec{f} - (\vec{u}_\epsilon * \theta_\epsilon)\cdot\vec{\nabla}\vec{u}_\epsilon - \vec{\nabla}p_{x_0,\epsilon}$$

with

$$p_{x_0,\epsilon}(t,x) = \frac{1}{\Delta}(\vec{\nabla}\otimes\vec{\nabla})(1_{B(x_0,5R_0)}\mathbb{H}_\epsilon) + \int_{|y-x_0|>5R_0} (\mathbb{K}(x-y) - \mathbb{K}(x_0-y))\mathbb{H}_\epsilon(t,y)\, dy$$

with

$$\mathbb{H}_\epsilon = \mathbb{F} - (\theta_\epsilon * \vec{u}_\epsilon) \otimes \vec{u}_\epsilon.$$

We then write

$$\partial_t(\frac{|\vec{u}_\epsilon|^2}{2}) = \nu\Delta(\frac{|\vec{u}_\epsilon|^2}{2}) - \nu|\vec{\nabla} \otimes \vec{u}_\epsilon|^2 + \vec{f} \cdot \vec{u}_\epsilon$$

$$- \operatorname{div}(\frac{|\vec{u}_\epsilon|^2}{2}(\vec{u}_\epsilon * \theta_\epsilon)) - \operatorname{div}(p_{x_0,\epsilon}\vec{u}_\epsilon).$$

We know that $\vec{u}_{\epsilon_n}$ converge strongly to $\vec{u}$ in $L^2_{\mathrm{loc}}((0,T)\times\mathbb{R}^3)$; as the family is bounded in $(L_t^{10/3}H_x^{3/5})_{\mathrm{uloc}} \subset (L_t^{10/3}L_x^{10/3})_{\mathrm{uloc}}$, we find that we have strong convergence in $L^3_{\mathrm{loc}}((0,T)\times\mathbb{R}^3)$ as well. Thus, we have the following convergence results in $\mathcal{D}'((0,T)\times\mathbb{R}^3)$: $\partial_t|\vec{u}_{\epsilon_n}|^2 \to \partial_t|\vec{u}|^2$, $\Delta|\vec{u}_{\epsilon_n}|^2 \to \Delta|\vec{u}|^2$, $\operatorname{div}(\frac{|\vec{u}_{\epsilon_n}|^2}{2}(\vec{u}_{\epsilon_n}*\theta_{\epsilon_n})) \to \operatorname{div}(\frac{|\vec{u}|^2}{2}\vec{u})$ and $\vec{u}_{(\epsilon_n)} \cdot \vec{f} \to \vec{u}\cdot\vec{f}$. Similarly, we find that p_{x_0,ϵ_n} converges weakly to p_{x_0} in $(L^{3/2}L^{3/2})(B(x_0,R_0))$ and the strong convergence of $\vec{u}_{\epsilon_n}$ in $(L^3L^3)_{\mathrm{loc}}$ gives the convergence of $\operatorname{div}(p\vec{u}_{\epsilon_n})$ to $\operatorname{div}(p\vec{u})$.

Thus far, we have got that

$$\partial_t|\vec{u}|^2 = \nu\Delta|\vec{u}|^2 - \operatorname{div}((|\vec{u}|^2 + 2p)\vec{u}) + 2\vec{u}\cdot\vec{f} - \nu T$$

with

$$T = \lim_{\epsilon_n\to 0} 2|\vec{\nabla}\otimes\vec{u}_{\epsilon_n}|^2.$$

Let $\phi \in \mathcal{D}'((0,T)\times\mathbb{R}^3)$ be a non-negative function. As $\sqrt{\phi}\,\vec{\nabla}\otimes\vec{u}_{\epsilon_n}$ is weakly convergent to $\sqrt{\phi}\,\vec{\nabla}\otimes\vec{u}$ in $L^2_t L^2_x$, we find that $\|\sqrt{\phi}\,\vec{\nabla}\otimes\vec{u}\|_2^2 \le \liminf_{\epsilon_n\to 0}\|\sqrt{\phi}\,\vec{\nabla}\otimes\vec{u}_{\epsilon_n}\|_2^2$. Thus, we have

$$\langle T|\phi\rangle_{\mathcal{D}',\mathcal{D}} = 2\lim_{\epsilon_n\to 0}\iint |\vec{\nabla}\otimes\vec{u}_{\epsilon_n}|^2\,\phi(t,x)\,dt\,d$$

$$\ge 2\iint |\vec{\nabla}\otimes\vec{u}|^2\phi(t,x)\,dt\,dx.$$

Thus, $T = 2|\vec{\nabla}\otimes\vec{u}|^2 + \mu$, where μ is a non-negative locally finite measure, and thus $\vec{u}$ is suitable.

$\square$

The solution $\vec{u}$ we have constructed is continuous in (local) L^2 norm at time $t = 0$:

Proposition 14.1.
The solution $\vec{u}$ constructed in the proof of Theorem 14.1 satisfies: for every compact subset K of $\mathbb{R}^3$, $\lim_{t\to 0+}\int_K |\vec{u}(t,x) - \vec{u}_0(t,x)|^2\,dx = 0$.

Proof. First, we remark that the solution $\vec{u}$ satisfies $\partial_t\vec{u} \in (L^1_t H^{-3/2})_{\mathrm{uloc}}$, so that $t \in [0,T) \mapsto \vec{u}(t,.)$ is continuous from $[0,T)$ to $\mathcal{D}'$. In particular, we have, for $\varphi = \varphi_{x_0} \in \mathcal{B}$,

$$\|\varphi\vec{u}_0\|_2 \le \liminf_{t\to 0+}\|\varphi\vec{u}(t,.)\|_2.$$

We must now estimate $\limsup_{t\to 0+}\|\varphi\vec{u}(t,.)\|_2$. Let $\gamma \in \mathcal{C}^\infty(\mathbb{R})$ be equal to 1 on $(-\infty,-1)$ and to 0 on $(-1/2,+\infty)$. We define for $t_0 \in (0,T^*)$ and $\eta < t_0$, $\gamma_{t_0,\eta}(t) = \gamma(\frac{t-t_0}{\eta})$. We have

$$\partial_t(\varphi^2\gamma_{t_0,\eta}\frac{|\vec{u}_\epsilon|^2}{2}) = \varphi^2\gamma_{t_0,\eta}(\nu\Delta(\frac{|\vec{u}_\epsilon|^2}{2}) - \nu|\vec{\nabla}\otimes\vec{u}_\epsilon|^2 + \vec{f}\cdot\vec{u}_\epsilon)$$

$$- \varphi^2\gamma_{t_0,\eta}(\operatorname{div}(\frac{|\vec{u}_\epsilon|^2}{2}(\vec{u}_\epsilon*\theta_\epsilon)) + \operatorname{div}(p_{x_0,\epsilon}\vec{u}_\epsilon)) + \varphi^2\frac{|\vec{u}_\epsilon|^2}{2}\partial_t\gamma_{t_0,\eta}$$

so that

$$-\int\int \varphi^2 \frac{|\vec{u}_\epsilon|^2}{2} \partial_t \gamma_{t_0,\eta} \, dx \, dt =$$

$$\int \varphi \frac{|\vec{u}_0|^2}{2} \, dx + \int_0^{t_0} \int \varphi^2 \gamma_{t_0,\eta} (\nu\Delta(\frac{|\vec{u}_\epsilon|^2}{2}) - \nu|\vec{\nabla}\otimes\vec{u}_\epsilon|^2 + \vec{f}\cdot\vec{u}_\epsilon) \, dx \, dt$$

$$-\int_0^{t_0} \int \varphi^2 \gamma_{t_0,\eta} (\mathrm{div}(\frac{|\vec{u}_\epsilon|^2}{2}(\vec{u}_\epsilon * \theta_\epsilon)) + \mathrm{div}(p_{x_0,\epsilon}\vec{u}_\epsilon)) \, dx \, dt.$$

If we let ϵ_n go to 0, we have proven that every integral will converge in this equality, except the integral $\int_0^{t_0} \int \varphi^2 \gamma_{t_0,\eta} |\vec{\nabla}\otimes\vec{u}_{\epsilon_n}|^2 \, dx \, dt$. But we have a control on this term:

$$\int_0^{t_0} \int \varphi^2 \gamma_{t_0,\eta} |\vec{\nabla}\otimes\vec{u}|^2 \, dx \, dt \leq \liminf_{\epsilon_n\to 0} \int_0^{t_0} \int \varphi^2 \gamma_{t_0,\eta} |\vec{\nabla}\otimes\vec{u}_{\epsilon_n}|^2 \, dx \, dt.$$

Thus, we have

$$-\int\int \varphi^2 \frac{|\vec{u}|^2}{2} \partial_t \gamma_{t_0,\eta} \, dx \, dt \leq \int \varphi^2 \frac{|\vec{u}_0|^2}{2} \, dx$$

$$+ \int_0^{t_0} \int \varphi^2 \gamma_{t_0,\eta} (\nu\Delta(\frac{|\vec{u}|^2}{2}) - \nu|\vec{\nabla}\otimes\vec{u}|^2 + \vec{f}\cdot\vec{u}) \, dx \, dt$$

$$- \int_0^{t_0} \int \varphi^2 \gamma_{t_0,\eta} (\mathrm{div}((\frac{|\vec{u}|^2}{2} + p_{x_0})\vec{u}) \, dx \, dt$$

If t_0 is a Lebesgue point of the map $t \mapsto \int \varphi^2(x)|\vec{u}(t,x)|^2 \, dx$, we have

$$\lim_{\eta\to 0} \left(-\int\int \varphi^2 \frac{|\vec{u}|^2}{2} \partial_t \gamma_{t_0,\eta} \, dx \, dt \right) = \int \varphi^2(x) \frac{|\vec{u}(t_0,x)|^2}{2} \, dx$$

and thus

$$\int \varphi(x)^2 \frac{|\vec{u}(t_0,x)|^2}{2} \, dx \leq \int \varphi^2 \frac{|\vec{u}_0|^2}{2} \, dx$$

$$+ \int_0^{t_0} \int \varphi^2 (\nu\Delta(\frac{|\vec{u}|^2}{2}) - \nu|\vec{\nabla}\otimes\vec{u}|^2 + \vec{f}\cdot\vec{u}) \, dx \, dt \qquad (14.11)$$

$$- \int_0^{t_0} \int \varphi^2 (\mathrm{div}((\frac{|\vec{u}|^2}{2} + p_{x_0})\vec{u}) \, dx \, dt$$

The right-hand side of inequality (14.11) is a continuous function of t_0, while the left-hand side is a lower semi-continuous function of t_0; thus, the inequality is fulfilled for every $t_0 \in (0,T)$. Letting t_0 go to 0 proves that

$$\limsup_{t\to 0^+} \|\varphi\vec{u}(t,.)\|_2 \leq \|\varphi\vec{u}_0\|_2.$$

Thus, we have $\lim_{t\to 0^+} \|\varphi\vec{u}(t,.)\|_2 \leq \|\varphi\vec{u}_0\|_2$. As we have weak convergence of $\varphi\vec{u}$ to $\varphi\vec{u}_0$ in L^2, we find that

$$\lim_{t\to 0^+} \|\varphi\vec{u}(t,.) - \varphi\vec{u}_0\|_2 = 0.$$

$\square$

The meaning of Proposition 14.1 is that we have constructed a local version of a Leray solution:

Definition 14.1 (Local Leray solution).
Let $\vec{u}_0 \in L^2_{\mathrm{uloc}}$ with $\mathrm{div}\,\vec{u}_0 = 0$ and $\mathbb{F} \in (L^2_t L^2_x)_{\mathrm{uloc}}((0,T) \times \mathbb{R}^3)$. A weak solution $\vec{u}$ on $(0,T) \times \mathbb{R}^3$ to the problem

$$\begin{cases} \partial_t \vec{u} = \nu \Delta \vec{u} + \mathbb{P}\,\mathrm{div}(\mathbb{F} - \vec{u} \otimes \vec{u}) \\[2mm] \vec{u}(0,.) = \vec{u}_0 \end{cases} \tag{14.12}$$

is a **local Leray solution** if it satisfies the following requirements:

- $\vec{u} \in (L^\infty_t L^2_x)_{\mathrm{uloc}} \cap (L^2 H^1_x)_{\mathrm{uloc}}$

- $\vec{u}$ is suitable

- for every compact subset K of $\mathbb{R}^3$, $\lim_{t \to 0^+} \int_K |\vec{u}(t,x) - \vec{u}_0(t,x)|^2\, dx = 0$.

14.2   Local Inequalities for Local Leray Solutions

The local energy inequalities we have derived in the previous section are valid for all local Leray solutions:

Local inequalities

Theorem 14.2.
Let $\vec{u}$ be a local Leray solution on $(0,T) \times \mathbb{R}^3$ to the problem

$$\begin{cases} \partial_t \vec{u} = \nu \Delta \vec{u} + \mathbb{P}\,\mathrm{div}(\mathbb{F} - \vec{u} \otimes \vec{u}) \\[2mm] \vec{u}(0,.) = \vec{u}_0 \end{cases} \tag{14.13}$$

where $\vec{u}_0 \in L^2_{\mathrm{uloc}}$ with $\mathrm{div}\,\vec{u}_0 = 0$ and $\mathbb{F} \in (L^2_t L^2_x)_{\mathrm{uloc}}((0,T) \times \mathbb{R}^3)$.
    Then, for a constant C_0 which does not depend on ν, for

$$T_0 = \min\!\left(T, 1, \frac{1}{C_0 \nu \left(1 + \frac{\|\vec{u}_0\|_{L^2_{\mathrm{uloc}}}}{\nu} + \frac{\|\mathbb{F}\|_{(L^2 L^2)_{\mathrm{uloc}}}}{\nu^{3/2}}\right)^4}\right)$$

we have

$$\sup_{0 < t < T_0}\ \sup_{x \in \mathbb{R}^3} \left(\int_{|x-y|<1} |\vec{u}(t,y)|^2\, dy\right)^{1/2} \leq 2\left(\|\vec{u}_0\|_{L^2_{\mathrm{uloc}}} + C_0 \frac{\|\mathbb{F}\|_{(L^2 L^2)_{\mathrm{uloc}}}}{\sqrt{\nu}}\right)$$

and

$$\sup_{x \in \mathbb{R}^3}\left(\int_0^{T_0} \int_{|x-y|<1} |\vec{\nabla} \otimes \vec{u}(s,y)|^2\, dy\, ds\right)^{1/2} \leq \frac{2}{\sqrt{\nu}}\left(\|\vec{u}_0\|_{L^2_{\mathrm{uloc}}} + C_0 \frac{\|\mathbb{F}\|_{(L^2 L^2)_{\mathrm{uloc}}}}{\sqrt{\nu}}\right).$$

Proof. We use the suitability of $\vec{u}$ and apply the local energy inequality to the test function $\phi(s, x) = \gamma_{t_0, \eta}(s)\varphi(x)$ (with $\varphi \in \mathcal{B}: \varphi(x) = \varphi_{x_0}(x) = \varphi_0(x - x_0)$) (where $\gamma_{t_0, \eta}$ is defined page 460). This gives

$$-\int\int \varphi \frac{|\vec{u}|^2}{2} \partial_t \gamma_{t_0, \eta} \, dx \, dt \leq \int \varphi \frac{|\vec{u}_0|^2}{2} \, dx$$

$$+ \int_0^{t_0} \int \varphi \gamma_{t_0, \eta} (\nu \Delta(\frac{|\vec{u}|^2}{2}) - \nu |\vec{\nabla} \otimes \vec{u}|^2 + \vec{f} \cdot \vec{u}) \, dx \, dt$$

$$- \int_0^{t_0} \int \varphi \gamma_{t_0, \eta} (\text{div}((\frac{|\vec{u}|^2}{2} + p_{x_0})\vec{u}) \, dx \, dt$$

Thus, we find again (letting η go to 0 for a Lebesgue point t_0 of $t \mapsto \int \varphi \chi_R^2 \frac{|\vec{u}|^2}{2} \, dx$, and then using the lower semi-continuity of the same map to get the control on other times t_0) that, for every $t \in (0, T)$:

$$\int \varphi \frac{|\vec{u}(t, x)|^2}{2} \, dx \, ds \leq \int \varphi \frac{|\vec{u}_0|^2}{2} \, dx$$

$$+ \int_0^t \int \varphi (\nu \Delta(\frac{|\vec{u}|^2}{2}) - \nu |\vec{\nabla} \otimes \vec{u}|^2 + \vec{f} \cdot \vec{u}) \, dx \, ds$$

$$- \int_0^t \int \varphi \, \text{div}((\frac{|\vec{u}|^2}{2} + p_{x_0})\vec{u}) \, dx \, ds$$

$$= \int \varphi \frac{|\vec{u}_0|^2}{2} \, dx - \nu \int_0^t \int \varphi |\vec{\nabla} \otimes \vec{u}|^2 \, dx \, ds$$

$$+ \nu \int_0^t \int (\Delta \varphi) \frac{|\vec{u}|^2}{2} \, dx \, ds$$

$$- \int_0^t \int \varphi \mathbb{F} \cdot \vec{\nabla} \otimes \vec{u} \, dx \, ds - \int_0^t \int \mathbb{F} \cdot \vec{\nabla} \varphi \otimes \vec{u} \, dx \, ds$$

$$+ \int_0^t \int (\vec{u} \cdot \vec{\nabla} \varphi)((\frac{|\vec{u}|^2}{2} + p_{x_0}) \, dx \, ds$$

Defining again

$$\alpha(t) = \|\vec{u}\|_{L^2_{\text{uloc}}} = \sup_{\varphi \in \mathcal{B}} \|\varphi \vec{u}\|_2$$

$$\beta(t) = \|\mathbb{F}\|_{(L^2 L^2)_{\text{uloc}}((0,t) \times \mathbb{R}^3)} = \sup_{\varphi \in \mathcal{B}} (\int_0^t \int |\varphi(x) \mathbb{F}(s, x)|^2 \, dx \, ds)^{1/2}$$

$$\gamma(t) = \|\vec{\nabla} \otimes \vec{u}\|_{(L^2 L^2)_{\text{uloc}}((0,t) \times \mathbb{R}^3)} = \sup_{\varphi \in \mathcal{B}} (\int_0^t \int |\varphi(x) \vec{\nabla} \otimes \vec{u}(s, x)|^2 \, dx \, ds)^{1/2}$$

$$\delta(t) = \|\vec{u}\|_{(L^3 L^3)_{\text{uloc}}((0,t) \times \mathbb{R}^3)} = \sup_{\varphi \in \mathcal{B}} (\int_0^t \int |\varphi(x) \vec{u}(s, x)|^3 \, dx \, ds)^{1/2}$$

we get (for $t < \min(T, 1)$

$$\|\varphi \vec{u}\|_2^2 + \nu \int_0^t \int |\varphi \vec{\nabla} \otimes \vec{u}|^2 \, dx \leq \|\varphi \vec{u}_0\|_2^2 + C_3 \nu \int_0^t \alpha(s)^2 \, ds$$

$$+ C_3 \frac{1}{\nu} \beta(t)^2 + C_3 \delta(t)^3$$

and we may conclude, following the lines on page 458. $\qquad \square$

The same computations show that, when letting $|x|$ go to $+\infty$, the behavior of our solution $\vec{u}$ depends only on the behavior of $\vec{u}_0$ and $\mathbb{F}$ near the infinity, and that the influence of the small values of x may be easily controlled:

Asymptotic behavior of local Leray solutions

Theorem 14.3.
Let $\vec{u}$ be a local Leray solution on $(0,T) \times \mathbb{R}^3$ to the problem

$$\begin{cases} \partial_t \vec{u} = \nu \Delta \vec{u} + \mathbb{P}\,\mathrm{div}(\mathbb{F} - \vec{u} \otimes \vec{u}) \\[2mm] \vec{u}(0,.) = \vec{u}_0 \end{cases} \qquad (14.14)$$

where $\vec{u}_0 \in L^2_{\mathrm{uloc}}$ with $\mathrm{div}\,\vec{u}_0 = 0$ and $\mathbb{F} \in (L^2_t L^2_x)_{\mathrm{uloc}}((0,T) \times \mathbb{R}^3)$.
Let $\omega \in \mathcal{D}(\mathbb{R}^3)$ be equal to 1 in a neighborhood of 0 and define $\chi_R(x) = 1 - \omega(x/R)$. Then there exists a positive constant C_T so that for all $0 < t < T$ and all $R > 1$, we have

$$\|\vec{u}(t,.)\chi_R\|_{L^2_{\mathrm{uloc}}} \leq C_T \left(\|\vec{u}_0 \chi_R\|_{L^2_{\mathrm{uloc}}} + \|\mathbb{F}\chi_R\|_{(L^2 L^2)_{\mathrm{uloc}}} + \sqrt{\frac{1 + \ln R}{R}}\right). \qquad (14.15)$$

The constant C_T depends only on ν, T, $\|\mathbb{F}\|_{(L^2 L^2)_{\mathrm{uloc}}}$, $\|\vec{\nabla} \otimes \vec{u}\|_{(L^2 L^2)_{\mathrm{uloc}}}$ and $\|\vec{u}\|_{(L^\infty L^2)_{\mathrm{uloc}}}$.

Proof. We use the suitability of $\vec{u}$ and apply the local energy inequality to the test function $\phi(s,x) = \gamma_{t_0,\eta}(s)\varphi(x)\chi_R^2(x)$ (with $\varphi \in \mathcal{B}$: $\varphi(x) = \varphi_{x_0}(x) = \varphi_0(x - x_0)$). This gives

$$-\iint \varphi\chi_R^2 \frac{|\vec{u}|^2}{2} \partial_t \gamma_{t_0,\eta}\,dx\,dt \leq$$

$$\int \varphi\chi_R^2 \frac{|\vec{u}_0|^2}{2}\,dx + \int_0^{t_0}\int \varphi\chi_R^2 \gamma_{t_0,\eta}(\nu\Delta(\frac{|\vec{u}|^2}{2}) - \nu|\vec{\nabla} \otimes \vec{u}|^2 + \vec{f}\cdot\vec{u})\,dx\,dt$$

$$-\int_0^{t_0}\int \varphi\chi_R^2 \gamma_{t_0,\eta}(\mathrm{div}((\frac{|\vec{u}|^2}{2} + p_{x_0})\vec{u})\,dx\,dt$$

Thus, we find again (letting η go to 0 for a Lebesgue point t_0 of $t \mapsto \int \varphi\chi_R^2 \frac{|\vec{u}|^2}{2}\,dx$, and then using the lower semi-continuity of the same map to get the control on other times t_0) that, for every $t \in (0,T)$:

$$\int \varphi\chi_R^2 \frac{|\vec{u}(t,x)|^2}{2}\,dx\,ds \leq \int \varphi\chi_R^2 \frac{|\vec{u}_0|^2}{2}\,dx$$

$$+ \int_0^t \int \varphi\chi_R^2(\nu\Delta(\frac{|\vec{u}|^2}{2}) - \nu|\vec{\nabla} \otimes \vec{u}|^2 + \vec{f}\cdot\vec{u})\,dx\,ds$$

$$-\int_0^t \int \varphi\chi_R^2(\mathrm{div}((\frac{|\vec{u}|^2}{2} + p_{x_0})\vec{u})\,dx\,ds$$

$$= \int \varphi \chi_R^2 \frac{|\vec{u}_0|^2}{2} \, dx - \nu \int_0^t \int \varphi \chi_R^2 |\vec{\nabla} \otimes \vec{u}|^2 \, dx \, ds$$

$$+ \nu \int_0^t \int (\Delta\varphi) \chi_R^2 \frac{|\vec{u}|^2}{2} \, dx \, ds$$

$$+ \nu \int_0^t \int \varphi \Delta(\chi_R^2) \frac{|\vec{u}|^2}{2} \, dx \, ds + \nu \int_0^t \int \chi_R (\vec{\nabla}\varphi \cdot \vec{\nabla}\chi_R) |\vec{u}|^2 \, dx \, ds$$

$$- \int_0^t \int \varphi \chi_R^2 \mathbb{F} \cdot \vec{\nabla} \otimes \vec{u} \, dx \, ds - \int_0^t \int \chi_R^2 \mathbb{F} \cdot \vec{\nabla}\varphi \otimes \vec{u} \, dx \, ds$$

$$- 2 \int_0^t \int \varphi \chi_R \mathbb{F} \cdot \vec{\nabla}\chi_R \otimes \vec{u} \, dx \, ds$$

$$+ \int_0^t \int (\vec{u} \cdot \vec{\nabla}\varphi) \chi_R^2 ((\frac{|\vec{u}|^2}{2} + p_{x_0}) \, dx \, ds + 2 \int_0^t \int (\vec{u} \cdot \vec{\nabla}\chi_R) \varphi \chi_R ((\frac{|\vec{u}|^2}{2} + p_{x_0}) \, dx \, ds$$

Let

$$\alpha(t) = \|\vec{u}\|_{L^2_{\mathrm{uloc}}}, \quad \beta(t) = \|\mathbb{F}\|_{(L^2 L^2)_{\mathrm{uloc}}((0,t) \times \mathbb{R}^3)}$$

$$\gamma(t) = \|\vec{\nabla} \otimes \vec{u}\|_{(L^2 L^2)_{\mathrm{uloc}}((0,t) \times \mathbb{R}^3)}, \quad \delta(t) = \|\vec{u}\|_{(L^3 L^3)_{\mathrm{uloc}}((0,t) \times \mathbb{R}^3)}$$

and similarly

$$\alpha_R(t) = \|\chi_R \vec{u}\|_{L^2_{\mathrm{uloc}}}, \quad \beta_R(t) = \|\chi_R \mathbb{F}\|_{(L^2 L^2)_{\mathrm{uloc}}((0,t) \times \mathbb{R}^3)}$$

$$\gamma_R(t) = \|\chi_R \vec{\nabla} \otimes \vec{u}\|_{(L^2 L^2)_{\mathrm{uloc}}((0,t) \times \mathbb{R}^3)}, \quad \delta_R(t) = \|\chi_R \vec{u}\|_{(L^3 L^3)_{\mathrm{uloc}}((0,t) \times \mathbb{R}^3)}$$

We have

$$\int \varphi \chi_R^2 |\vec{u}(t,x)|^2 \, dx \, ds + \nu \int_0^t \int \varphi \chi_R^2 |\vec{\nabla} \otimes \vec{u}|^2 \, dx \, ds$$

$$\leq \int \varphi \chi_R^2 |\vec{u}_0|^2 \, dx + C_1 \nu \int_0^t \alpha_R(s)^2 \, ds + C_1 \nu \frac{1}{R} \int_0^t \alpha(s)^2 \, ds$$

$$+ C_1 \frac{1}{\nu} \beta_R(t)^2 + C_1 \frac{1}{R} \beta(t)\gamma(t) + C_1 \nu \frac{1}{R} \beta(t)^2 + C_1 \frac{1}{R} \delta(t)^3$$

$$+ C_1 \delta(t) \delta_R(t)^2 + \int_0^t \int (\vec{u} \cdot \vec{\nabla}\varphi) \chi_R^2 p_{x_0} \, dx \, ds$$

The last term, which includes p_{x_0}, must be carefully dealt with, as p_{x_0} is given by a non-local operator which is linear with respect to $\mathbb{F}$ and quadratic with respect to $\vec{u}$. Recall that φ is supported by the ball $B(x_0, R_0)$ and that, on the ball $B(x_0, R_0)$, we have

$$p_{x_0} = \mathcal{T}_1(1_{|y-x_0|<5R_0}(\mathbb{F} - \vec{u} \otimes \vec{u})) + \mathcal{T}_2(1_{|y-x_0|>5R_0}(\mathbb{F} - \vec{u} \otimes \vec{u}))$$

where

$$\mathcal{T}_1 = \frac{1}{\Delta}(\vec{\nabla} \otimes \vec{\nabla})$$

and

$$\mathcal{T}_2 \mathbb{H} = \int (\mathbb{K}(x,y) - \mathbb{K}(x_0,y)) \mathbb{H}(y) \, dy.$$

Let M_{χ_R} be the pointwise multiplication by χ_R: $M_{\chi_R} g = \chi_R g$. We write

$$\int_0^t \int (\vec{u} \cdot \vec{\nabla}\varphi) \chi_R^2 p_{x_0} \, dx \, ds = \int_0^t \int (\vec{u} \cdot \vec{\nabla}\varphi) \chi_R \mathcal{T}_1(\chi_R 1_{|y-x_0|<5R_0} \mathbb{F}) \, dx \, ds$$

$$+ \int_0^t \int (\vec{u} \cdot \vec{\nabla}\varphi) \chi_R [M_{\chi_R}, \mathcal{T}_1](1_{|y-x_0|<5R_0} \mathbb{F}) \, dx \, ds$$

$$+ \int_0^t \int (\vec{u} \cdot \vec{\nabla}\varphi) \chi_R \mathcal{T}_2(\chi_R 1_{|y-x_0|>5R_0} \mathbb{F}) \, dx \, ds$$

$$+ \int_0^t \int (\vec{u} \cdot \vec{\nabla}\varphi) \chi_R [M_{\chi_R}, \mathcal{T}_2](1_{|y-x_0|>5R_0} \mathbb{F}) \, dx \, ds$$

$$- \int_0^t \int (\vec{u} \cdot \vec{\nabla}\varphi) \chi_R \mathcal{T}_1(\chi_R 1_{|y-x_0|<R_0} \vec{u} \otimes \vec{u}) \, dx \, ds$$

$$- \int_0^t \int (\vec{u} \cdot \vec{\nabla}\varphi) \chi_R [M_{\chi_R}, \mathcal{T}_1](1_{|y-x_0|<R_0} \vec{u} \otimes \vec{u}) \, dx \, ds$$

$$- \int_0^t \int (\vec{u} \cdot \vec{\nabla}\varphi) \chi_R \mathcal{T}_2(\chi_R 1_{|y-x_0|>5R_0} \vec{u} \otimes \vec{u}) \, dx \, ds$$

$$- \int_0^t \int (\vec{u} \cdot \vec{\nabla}\varphi) \chi_R [M_{\chi_R}, \mathcal{T}_2](1_{|y-x_0|>5R_0} \vec{u} \otimes \vec{u}) \, dx \, ds$$

$$= I_1 + \cdots + I_8$$

We already know how to control I_1, I_3, I_5 and I_7:

$$|I_1| + |I_3| \leq C_2 \nu \int_0^t \alpha_R(s)^2 \, ds + C_2 \frac{1}{\nu} \beta_R(t)^2$$

$$|I_5| + |I_7| \leq C_2 \delta(t) \delta_R(t)^2$$

We have

$$|I_4| \leq C_2 \int_0^t \int_{|x-x_0|<R_0} |\vec{u}(x)| \chi_R(x) \int_{|y-x|>4R_0} \frac{|\chi_R(x) - \chi_R(y)|}{|x-y|^4} |\mathbb{F}(s,y)| \, dy \, dx \, ds$$

$$\leq C_3 \int_0^t \int_{|x-x_0|<R_0} |\vec{u}(x)| \chi_R(x) \int_{|y-x|>R} \frac{1}{|x-y|^4} |\mathbb{F}(s,y)| \, dy \, dx \, ds$$

$$+ C_3 \frac{1}{R} \int_0^t \int_{|x-x_0|<R_0} |\vec{u}(x)| \chi_R(x) \int_{R>|y-x|>4R_0} \frac{1}{|x-y|^3} |\mathbb{F}(s,y)| \, dy \, dx \, ds$$

$$\leq C_4 \frac{1 + \ln^+(R/R_0)}{R} (\nu \int_0^t \alpha(s)^2 \, ds + \frac{1}{\nu} \beta(t)^2)$$

Similarly, we have

$$|I_8| \leq C_5 \frac{1 + \ln^+(R/R_0)}{R} \delta(t)^3.$$

The most difficult terms are I_2 and I_6. They will be dealt with the help of Calderón's lemma on the commutator between a pseudo-differential operator of order 1 and the pointwise multiplication with a Lipschitz function [313]: for $1 < p < +\infty$, we have for a Lipschitz and $\mathbb{H} \in L^p$,

$$\|[\vec{\nabla} \otimes \mathcal{T}_1, M_a] \mathbb{H}\|_p \leq C_p \|\vec{\nabla} a\|_\infty \|\mathbb{H}\|_p.$$

We thus write

$$\int_0^t \int (\vec{u} \cdot \vec{\nabla}\varphi)\chi_R[M_{\chi_R}, \mathcal{T}_1]\mathbb{H}\,dx\,ds = -2\int_0^t \int \varphi(\vec{u} \cdot \vec{\nabla}\chi_R)\chi_R\mathcal{T}_1\mathbb{H}\,dx\,ds$$
$$-\int_0^t \int \varphi\chi_R\vec{u}.[M_{\chi_R}, \vec{\nabla} \otimes \mathcal{T}_1]\mathbb{H}\,dx\,ds$$

and get

$$|I_2| \leq C_6\frac{1}{R}(\nu\int_0^t \alpha(s)^2\,ds + \frac{1}{\nu}\beta(t)^2)$$

and

$$|I_6| \leq C_6\frac{1}{R}\delta(t)^3.$$

Summing up all those estimates, we get:

$$\int \varphi\chi_R^2|\vec{u}(t,x)|^2\,dx\,ds + \nu\int_0^t \int \varphi\chi_R^2|\vec{\nabla} \otimes \vec{u}|^2\,dx\,ds$$
$$\leq \int \varphi\chi_R^2|\vec{u}_0|^2\,dx + C_7\frac{1}{\nu}\beta_R(t)^2$$
$$+ C_7\nu\frac{1 + \ln^+(R/R_0)}{R}(\nu\int_0^t \alpha(s)^2\,ds + \frac{1}{\nu}\beta(t)^2)$$
$$+ C_7\frac{1}{R}\beta(t)\gamma(t) + C_7\frac{1 + \ln^+(R/R_0)}{R}\delta(t)^3$$
$$+ C_7\nu\int_0^t \alpha_R(s)^2\,ds + C_7\delta(t)\delta_R(t)^2$$

so that, for a constant D_T (which depends only on ν, T, $\|\mathbb{F}\|_{(L^2L^2)_{\text{uloc}}}$, $\|\vec{\nabla} \otimes \vec{u}\|_{(L^2L^2)_{\text{uloc}}}$ and $\|\vec{u}\|_{(L^\infty L^2)_{\text{uloc}}}$)

$$\int \varphi\chi_R^2|\vec{u}(t,x)|^2\,dx\,ds + \nu\int_0^t \int \varphi\chi_R^2|\vec{\nabla} \otimes \vec{u}|^2\,dx\,ds$$
$$\leq D_T(\|\chi_R\vec{u}_0\|_{L^2_{\text{uloc}}}^2 + \|\chi_R\mathbb{F}\|_{(L^2L^2)_{\text{uloc}}}^2 + \frac{1 + \ln^+(R/R_0)}{R})$$
$$+ C_7\nu\int_0^t \alpha_R(s)^2\,ds + C_7\delta(T)\delta_R(t)^2$$

We then write

$$\delta_R(t)^2 \leq C_8(\int_0^t \alpha_R(s)^2\,ds + \gamma_R(t)(\int_0^T \alpha_R^6(s)\,ds)^{1/6})$$

and get

$$\int \varphi\chi_R^2|\vec{u}(t,x)|^2\,dx\,ds + \nu\int_0^t \int \varphi\chi_R^2|\vec{\nabla} \otimes \vec{u}|^2\,dx\,ds$$
$$\leq D_T(\|\chi_R\vec{u}_0\|_{L^2_{\text{uloc}}}^2 + \|\chi_R\mathbb{F}\|_{(L^2L^2)_{\text{uloc}}}^2 + \frac{1 + \ln^+(R/R_0)}{R})$$
$$+ (C_7(\nu + C_8)T^{2/3} + C_7^2C_8^2\delta(T)^2\frac{8}{\nu})(\int_0^t \alpha_R(s)^6\,ds)^{1/3}$$
$$+ \frac{\nu}{2}\gamma_R(t)^2$$

and thus, writing $\eta_R = \|\chi_R \vec{u}_0\|_{L^2_{\text{uloc}}} + \|\chi_R \mathbb{F}\|_{(L^2 L^2)_{\text{uloc}}} + \frac{1 + \ln^+(R/R_0)}{R}$,

$$\max(\alpha_R(t)^2, \nu \gamma_R(t)^2) \leq E_T (\eta_R^3 + \int_0^t \alpha_R(s)^6 \, ds)^{1/3} + \frac{\nu}{2} \gamma_R(t)^2$$

(for a constant E_T which depends only on ν, T, $\|\mathbb{F}\|_{(L^2 L^2)_{\text{uloc}}}$, $\|\vec{\nabla} \otimes \vec{u}\|_{(L^2 L^2)_{\text{uloc}}}$ and $\|\vec{u}\|_{(L^\infty L^2)_{\text{uloc}}}$). Finally, we can easily control $\nu \gamma_R(t)^2$ and get

$$\alpha_R(t)^2 \leq 2 E_T (\eta_R^3 + \int_0^t \alpha_R(s)^6 \, ds)^{1/3}$$

and thus

$$\frac{d}{dt} \left((\eta_R^3 + \int_0^t \alpha_R(s)^6 \, ds)^{1/3} \right) \leq (2 E_T)^3 (\eta_R^3 + \int_0^t \alpha_R(s)^6 \, ds)^{1/3}.$$

We then conclude by Grönwall's lemma.   □

14.3   The Caffarelli, Kohn and Nirenberg ϵ–Regularity Criterion

We give here a variant of the regularity criterion of Caffarelli, Kohn and Nirenberg [74] (Theorem 13.8).

Caffarelli–Kohn–Nirenberg ϵ–regularity criterion

Theorem 14.4.
Let $q > 5/2$. Let Ω be a domain of $\mathbb{R} \times \mathbb{R}^3$. Let $(\vec{u}, p)$ a weak solution on Ω of the Navier–Stokes equations

$$\partial_t \vec{u} = \nu \Delta \vec{u} - \vec{u} \cdot \vec{\nabla} \vec{u} + \vec{f} - \vec{\nabla} p, \quad \text{div } \vec{u} = 0.$$

Assume that

- $\vec{u} \in L^\infty L^2 \cap L^2 \dot{H}^1(\Omega)$

- $p \in L_t^{3/2} L_x^{3/2}(\Omega)$

- $\vec{f} \in L_t^q L_x^q(\Omega)$

- $\vec{u}$ *is suitable: it satisfies in $\mathcal{D}'$ the local energy inequality*

$$\partial_t (\frac{|\vec{u}|^2}{2}) \leq \nu \Delta (\frac{|\vec{u}|^2}{2}) - \nu |\vec{\nabla} \otimes \vec{u}|^2 - \text{div} \left((p + \frac{|\vec{u}|^2}{2}) \vec{u} \right) + \vec{u} \cdot \vec{f} \qquad (14.16)$$

Let $r_0 > 0$ and $Q_0 = Q_{r_0}(t_0, x_0) = (t_0 - r_0^2, t_0) \times B(x_0, r_0)$. There exists positive constants ϵ_0 and C_0 which depend only on ν and q (but not on x_0, t_0, r_0, $\vec{u}$ nor $\vec{f}$) such that, if

$$\begin{cases} 0 \leq \lambda \leq \epsilon_0 \\[2mm] \iint_{Q_0} |\vec{u}|^3 + |p|^{3/2} \, dx \, dt \leq \lambda^3 r_0^2 \\[2mm] \iint_{Q_0} |\vec{f}|^q \, dx \, dt \leq \lambda^{2q} r_0^{5 - 3q} \end{cases} \qquad (14.17)$$

> then $\vec{u}$ is bounded on $Q_1 = (t_0 - \frac{1}{4}r_0^2, t_0) \times B(x_0, r_0/2)$ and
>
> $$\sup_{(t,x)\in Q_1} |\vec{u}(t,x)| \leq C_0\lambda\frac{1}{r_0}.$$

Proof. The proof follows the lines of the proof of Theorem 13.8. For $(t, x) \in Q_{3r_0/4}(t_0, x_0)$ and $0 < r \leq r_0/8$, we define

- $U_r(t, x) = \sup_{s\in(t-r^2,t)} \int_{B_r(t,x)} |\vec{u}(s, y)|^2 \, dx \, dy$

- $V_r(t, x) = \int\int_{Q_r(t,x)} |\vec{\nabla} \otimes \vec{u}(s, y)|^2 \, ds \, dy$

- $W_r(t, x) = \int\int_{Q_r(t,x)} |\vec{u}(s, y)|^3 \, ds \, dy$

- $P_r(t, x) = \int\int_{Q_r(t,x)} |p(s, y)|^{3/2} \, ds \, dy$

- $F_r(t, x) = \int\int_{Q_r(t,x)} |\vec{f}(s, y)|^{3/2} \, ds \, dy$

We are going to estimate $U_r, \ldots$ with respect to $U_\rho, \ldots$ for $0 < r < \rho/2 < r_0/16$. We use the suitability of $\vec{u}$ and apply the local energy inequality to a variant of the test function ψ of Scheffer we defined on page 432 [426]. If $\psi \in \mathcal{D}(Q_{r_0}(t_0, x_0))$ with $\psi \geq 0$, we have, for $\tau \in (t_0 - r_0^2, t_0)$

$$\int \psi(\tau, y)|\vec{u}(\tau, y)|^2 \, dy + 2\nu \int_{s<\tau} \int \psi(s, y)|\vec{\nabla} \otimes \vec{u}(s, y)|^2 \, dy \, ds$$
$$\leq \int_{s<\tau} \int (\partial_t\psi(s, y) + \nu\Delta\psi(s, y))|\vec{u}(s, y)|^2 \, dy \, ds$$
$$+ \int_{s<\tau} \int (|\vec{u}(s, y)|^2 + p(s, y))\vec{u}(s, y) \cdot \vec{\nabla}\psi(s, y) \, dy \, ds \qquad (14.18)$$
$$+ 2 \int_{s<\tau} \int \psi(s, y)\vec{u}(s, y) \cdot \vec{f}(s, y) \, dy \, ds$$

The choice of ψ is then the following one: we choose a non-negative function $\omega \in \mathcal{D}(\mathbb{R} \times \mathbb{R}^3)$ such that ω is supported in $(-1, 1) \times B(0, 3/4)$ and is equal to 1 on $(-1/4, 1/4) \times B(0, 1/2)$, a non-negative smooth function θ on $\mathbb{R}$ that is equal to 1 on $(-\infty, \tau_1)$ and to 0 on $(\tau_2, +\infty)$ for some $\tau < \tau_1 < \tau_2 < t$ and we define

$$\psi(s, y) = r^3\omega\left(\frac{s-t}{\rho^2}, \frac{y-x}{\rho}\right)\theta\left(\frac{s-t}{r^2}\right)H(r^2 + t - s, x - y)$$

where $0 < r \leq \rho/2 \leq r_0/16$ and $H(t, x) = W_{\nu t}(x)$. We then obtain

$$\max(U_r(t, x), 2\nu V_r(t, x)) \leq C \int\int_{Q_\rho(t,x)} \frac{r^3}{\rho^5}|\vec{u}(s, y)|^2 \, dy \, ds$$
$$+ C \int\int_{Q_\rho(t,x)} \frac{1}{r}|\vec{u}(s, y)|^3 \, dy \, ds$$
$$+ C \int\int_{Q_\rho(t,x)} \frac{1}{r}|p(s, y)|^{3/2} \, dy \, ds$$
$$+ Cr^{1/2} \int\int_{Q_\rho(t,x)} |\vec{f}(s, y)|^{3/2} \, dy \, ds$$

As

$$\left(\iint_{Q_r(t,x)} |\vec{u}(s,y)|^3 \, dy \, ds\right)^{1/3} \leq C r^{1/6} (U_r(t,x) + V_r(t,x))^{1/2}$$

and

$$\iint_{Q_\rho(t,x)} |\vec{u}(s,y)|^2 \, dy \, ds \leq C \rho^{5/3} W_\rho^{2/3},$$

we obtain

$$W_r(t,x) \leq C(\frac{r}{\rho})^5 W_\rho(t,x) + C\frac{1}{r} W_\rho(t,x)^{3/2} + C\frac{1}{r} P_\rho(t,x)^{3/2} + C r^{5/4} F_\rho(t,x)^{3/2}$$

If $w_r = \frac{1}{r^2} W_r$, $p_r = \frac{1}{r^2} P_r$ and $f_r = \frac{1}{r^{1/2}} F_r$, we get

$$w_r(t,x) \leq C_1 ((\frac{r}{\rho})^3 w_\rho(t,x) + (\frac{\rho}{r})^3 w_\rho(t,x)^{3/2} + (\frac{\rho}{r})^3 p_\rho(t,x)^{3/2} + (\frac{\rho}{r})^{3/4} f_\rho^{3/2}) \qquad (14.19)$$

where the constant C_1 does not depend on r nor ρ.

We now turn our attention to the pressure. We introduce a function $\theta \in \mathcal{D}(\mathbb{R}^3)$ with $\theta = 1$ on $B(0, 13/16)$ and with $\operatorname{Supp} \theta \subset B(0, 15/16)$ and we define $\zeta_{\rho,t,x}(y) = \theta(\frac{y-x}{\rho})$. For the sake of simplicity, we write ζ for $\zeta_{\rho,t,x}$.

On $(t - \rho^2, t) \times B(x, 3\rho/4)$, we have $\zeta p = p$. From

$$\Delta p = \operatorname{div} \vec{f} - \sum_{i=1}^{3} \sum_{j=1}^{3} \partial_i \partial_j (u_i u_j),$$

we get

$$\Delta(\zeta p) = \zeta \operatorname{div} \vec{f} - \zeta \sum_{i=1}^{3} \sum_{j=1}^{3} u_i u_j + 2 \sum_{j=1}^{3} \partial_j (p \partial_j \zeta) - p \Delta \zeta$$

and we may write $\zeta(y) p(s,y) = p_{\rho,t,x}(s,y) + q_{\rho,t,x}(s,y) + \varpi_{\rho,t,x}(s,y)$ with

$$\begin{cases} q_{\rho,t,x} = & \sum_{j=1}^{3} \sum_{l=1}^{3} G * (\zeta \partial_j \partial_l (u_j u_l)) \\[2mm] p_{\rho,t,x} = & -2 \sum_{j=1}^{3} \partial_j G * ((\partial_j \zeta) p) + G * ((\Delta \zeta) p) \\[2mm] \varpi_{\rho,t,x} = & G * (\vec{f} \cdot \vec{\nabla} \zeta) - G \operatorname{div}(\zeta \vec{f}) \end{cases}$$

We have

$$q_{\rho,t,x} = \sum_{j=1}^{3} \sum_{l=1}^{3} G * (\partial_j \partial_l (\zeta u_j u_l)) - 2G * ((\partial_j \zeta) \partial_l (u_j u_l)) - G * ((\partial_j \partial_l \zeta) u_j u_l).$$

We have, on $Q_r(t,x)$,

$$|2G * ((\partial_j \zeta) \partial_l (u_j u_l)) + G * ((\partial_j \partial_l \zeta) u_j u_l)| \leq C M_{1_{Q_\rho(t,x)} u_j u_l}$$

so that

$$\|q_{\rho,t,x}\|_{L^{3/2}(Q_r(t,x))} \leq C \|\vec{u}\|^2_{L^3(Q_\rho(t,x))}.$$

On $Q_r(t,x)$, we have

$$|p_{\rho,t,x}(s,y)| \leq \frac{1}{\rho^3} \int_{B(x,\rho)} |p(s,z)| \, dz \leq C \frac{1}{\rho^2} \left(\int_{B(x,\rho)} |p(s,z)|^{3/2} \, dz\right)^{2/3}$$

so that

$$\|p_{\rho,t,x}\|_{L^{3/2}(Q_r(t,x))} \leq C \frac{r^2}{\rho^2} \|p\|_{L^{3/2}(Q_\rho(t,x))}.$$

Finally, we find

$$\|\varpi_{\rho,t,x}(s,.)\|_{L^{3/2}(B(x,r))} \leq Cr\|\varpi_{\rho,t,x}(s,.)\|_{L^3(B(x,r))} \leq C'r\|\vec{f}(s,.)\|_{L^{3/2}(B(x,\rho))}$$

so that

$$\|\varpi_{\rho,t,x}(s,.)\|_{L^{3/2}(Q_r(t,x))} \leq Cr\|\vec{f}(s,.)\|_{L^{3/2}(Q_\rho(t,x))}.$$

We thus obtain

$$P_r(t,x) \leq C(W_\rho(t,x) + (\frac{r}{\rho})^3 P_\rho(t,x) + r^{3/2} F_\rho(t,x))$$

Dividing by r^2, we obtain

$$p_r(t,x) \leq C_2(w_\rho(t,x) + (\frac{r}{\rho})^3 p_\rho(t,x) + f_\rho(t,x)) \tag{14.20}$$

where the constant C_2 does not depend on r nor ρ.

We shall now consider a sequence $\rho_n = \kappa^n r_0/8$, where $\kappa \in (0, 1/2)$ will be fixed below. Let

$$\chi_n(t,x) = w_{\rho_n}(t,x) + \eta p_{\rho_n}(t,x)$$

where $\eta > 0$ will be fixed below as well. From (14.19) and (14.20), we get

$$\begin{aligned}
\chi_{n+1}(t,x) \leq &\max(C_1, C_2)\kappa^3 \chi_n(t,x) \\
&+ \eta C_2 \chi_n(t,x) \\
&+ C_1 \kappa^{-3}(1 + \eta^{-3/2})\chi_n(t,x)^{3/2} \\
&+ C_1 \kappa^{-3/4} f_{\rho_n}^{3/2} + \eta C_2 f_{\rho_n}(t,x)
\end{aligned} \tag{14.21}$$

We then fix κ such that $\max(C_1, C_2)\kappa^3 \leq 1/5$ and η such that $\eta C_2 \leq 1/5$. If α is such that

$$C_1 \kappa^{-3}(1 + \eta^{-3/2})\alpha^{1/2} \leq \frac{1}{5} \text{ and } C_1 \kappa^{-3/4}\alpha^{1/2} \leq \frac{1}{5},$$

then we find that the inequalities

$$\chi_0(t,x) \leq \alpha$$

and

$$\sup_{\rho < r_0/8} f_\rho(t,x) \leq \alpha$$

imply that

$$\chi_n(t,x) \leq \alpha$$

for every $n \in \mathbb{N}$. Remark that we have

$$\chi_0(t,x) \leq C \frac{1}{r_0^2} \iint_{Q_0} |\vec{u}|^3 + \eta |p(t,x)|^{3/2} \, ds \, dy \leq C(1+\eta)\lambda^3$$

and

$$\sup_{\rho < r_0/8} f_\rho(t,x) \leq C\|\vec{f}\|_{L^{5/3}(Q_0)}^{3/2} \leq C'\|\vec{f}\|_{L^q(Q_0)}^{3/2} r_0^{\frac{3}{2}(3-\frac{5}{q})} \leq C'\lambda^3.$$

We have proved the following lemma:

Lemma 14.2.
Let $q \geq 5/3$. Let Ω be a domain of $\mathbb{R} \times \mathbb{R}^3$. Let $(\vec{u}, p)$ a weak solution on Ω of the Navier–Stokes equations

$$\partial_t \vec{u} = \nu \Delta \vec{u} - \vec{u} \cdot \vec{\nabla} \vec{u} + \vec{f} - \vec{\nabla} p, \quad \text{div } \vec{u} = 0.$$

Assume that

- $\vec{u} \in L^\infty L^2 \cap L^2 \dot{H}^1(\Omega)$

- $p \in L_t^{3/2} L_x^{3/2}(\Omega)$

- $\vec{f} \in L_t^q L_x^q(\Omega)$

- $\vec{u}$ *is suitable*

Let $r_0 > 0$ and $Q_0 = Q_{r_0}(t_0, x_0) = (t_0 - r_0^2, t_0) \times B(x_0, r_0)$. There exist positive constants ϵ_1 and C_3 which depend only on ν and q (but not on $x_0, t_0, r_0, \vec{u}$ nor $\vec{f}$) such that, if

$$
\begin{cases}
0 \leq \lambda \leq \epsilon_1 \\[2mm]
\iint_{Q_0} |\vec{u}|^3 + |p|^{3/2} \, dx \, dt \leq \lambda^3 r_0^2 \\[2mm]
\iint_{Q_0} |\vec{f}|^q \, dx \, dt \leq \lambda^{2q} r_0^{5-3q}
\end{cases}
$$

then, for every $(t, x) \in Q_{3r_0/4}(t_0, x_0)$ and $0 < r \leq r_0/8$, we have

$$\iint_{Q_r(t,x)} |\vec{u}|^3 + |p|^{3/2} \, dy \, ds \leq C_3 \lambda^3 r^2.$$

Then, we follow again the lines of Kukavica's paper [286] and we shall prove that, if $q > 5/3$, $1_{Q_{3r_0/4}(t_0, x_0)} \vec{u}$ belongs to a parabolic Morrey space $\mathcal{M}_2^{3,\tau}$ with $\tau > 5$. We introduce the reduced quantities

$$\alpha_r(t, x, \tau) = r^{5(-1+\frac{3}{\tau})} W_r(t, x) = r^{3(-1+\frac{5}{\tau})} w_r(t, x)$$

$$\beta_r(t, x, \tau) = r^{5(-1+\frac{3}{\tau})} P_r(t, x) = r^{3(-1+\frac{5}{\tau})} p_r(t, x)$$

and

$$\gamma_r(t, x, \tau) = r^{-\frac{7}{2}+\frac{15}{\tau}} F_r(t, x) = r^{3(-1+\frac{5}{\tau})} f_r(t, x)$$

Multiplying (14.19) and (14.20) by $r^{3(-1+\frac{5}{\tau})}$, we find

$$\alpha_r(t, x, \tau) \leq C_1\left(\left(\frac{r}{\rho}\right)^{\frac{15}{\tau}} \alpha_\rho(t, x, \tau) + \left(\frac{\rho}{r}\right)^{6-\frac{15}{\tau}} \alpha_\rho(t, x, \tau) w_\rho(t, x)^{1/2}\right)$$

$$+ C_1\left(\left(\frac{\rho}{r}\right)^{6-\frac{15}{\tau}} \beta_\rho(t, x, \tau) p_\rho(t, x)^{1/2} + \left(\frac{\rho}{r}\right)^{3(\frac{5}{4}-\frac{5}{\tau})} \gamma_\rho(t, x, \tau) f_\rho(t, x)^{1/2}\right) \tag{14.22}$$

and

$$\beta_r(t, x, \tau) \leq C_2\left(\left(\frac{\rho}{r}\right)^{3-\frac{15}{\tau}} \alpha_\rho(t, x, \tau) + \left(\frac{r}{\rho}\right)^{\frac{15}{\tau}} \beta_\rho(t, x, \tau) + \left(\frac{\rho}{r}\right)^{3-\frac{15}{\tau}} \gamma_\rho(t, x, \tau)\right) \tag{14.23}$$

Recalling that, due to Lemma 14.2, we have (if $\lambda \leq \epsilon_1$), $w_r \leq C_3 \lambda^3$ and $p_r \leq C_3 \lambda^3$, and that $f_r(t, x) \leq C_4 \lambda^3$ by assumption. We shall now consider a sequence $\rho_n = \kappa^n r_0/8$, where $\kappa \in (0, 1/2)$ will be fixed below. Let

$$\chi_n(t, x, \tau) = \alpha_{\rho_n}(t, x, \tau) + \eta \beta_{\rho_n}(t, x, \tau)$$

where $\eta > 0$ will be fixed below as well. From (14.22) and (14.23), we get

$$
\begin{aligned}
\chi_{n+1}(t,x,\tau) \leq\ & \max(C_1, C_2)\kappa^{\frac{15}{\tau}}\chi_n(t,x,\tau) \\
&+ \eta C_2 \kappa^{\frac{15}{\tau}-3}\chi_n(t,x,\tau) \\
&+ C_1\kappa^{\frac{15}{\tau}-6}(1+\eta^{-1})(C_3\lambda^3)^{1/2}\chi_n(t,x,\tau) \\
&+ C_1\kappa^{3(\frac{5}{\tau}-1)}(C_4\lambda^3)^{1/2}\gamma_{\rho_n}(t,x,\tau) + \eta C_2\kappa^{\frac{15}{\tau}-3}\gamma_{\rho_n}(t,x,\tau)
\end{aligned}
\tag{14.24}
$$

We then fix κ such that $\max(C_1, C_2)\kappa^{\frac{15}{\tau}} \leq 1/5$ and η such that $\eta C_2\kappa^{\frac{15}{\tau}-3} \leq 1/5$. If λ is small enough to grant that

$$
C_1\kappa^{\frac{15}{\tau}-6}(1+\eta^{-1})(C_3\lambda^3)^{1/2} \leq \frac{1}{5} \text{ and } C_1\kappa^{3(\frac{5}{\tau}-1)}(C_4\lambda^3)^{1/2} \leq \frac{1}{5}
$$

we find that

$$
\chi_n(t,x,\tau) \leq \max(\chi_0(t,x), \sup_{\rho < r_0/8} \gamma_\rho(t,x,\tau))
$$

for every $n \in \mathbb{N}$.

Remark that we have

$$
\chi_0(t,x,\tau) \leq C \frac{1}{r_0^{5(1-\frac{3}{\tau})}} \iint_{Q_0} |\vec{u}|^3 + \eta|p(t,x)|^{3/2}\, ds\, dy \leq C \frac{1}{r_0^{3(1-\frac{5}{\tau})}}(1+\eta)\lambda^3
$$

and for $\frac{1}{\sigma} = \frac{2}{\tau} + \frac{1}{5}$,

$$
\sup_{\rho < r_0/8} \gamma_\rho(t,x,\tau) \leq C\|\vec{f}\|_{L^\sigma(Q_0)}^{3/2};
$$

thus, if τ is chosen with $\frac{2}{\tau} + \frac{1}{5} \geq \frac{1}{q}$, we find

$$
\sup_{\rho < r_0/8} \gamma_\rho(t,x,\tau) \leq C\|\vec{f}\|_{L^q(Q_0)}^{3/2} r_0^{\frac{15}{2}(\frac{2}{\tau}+\frac{1}{5}-\frac{1}{q})} \leq C' \frac{1}{r_0^{3(1-\frac{5}{\tau})}}\lambda^3.
$$

We have proved the following lemma:

Lemma 14.3.
Let $q > 5/3$ and $\tau > 5$ with $\frac{2}{\tau} + \frac{1}{5} \geq \frac{1}{q}$. Let Ω be a domain of $\mathbb{R} \times \mathbb{R}^3$. Let $(\vec{u}, p)$ a weak solution on Ω of the Navier–Stokes equations

$$
\partial_t \vec{u} = \nu\Delta\vec{u} - \vec{u}\cdot\vec{\nabla}\vec{u} + \vec{f} - \vec{\nabla}p, \quad \text{div } \vec{u} = 0.
$$

Assume that

- $\vec{u} \in L^\infty L^2 \cap L^2\dot{H}^1(\Omega)$

- $p \in L_t^{3/2}L_x^{3/2}(\Omega)$

- $\vec{f} \in L_t^q L_x^q(\Omega)$

- $\vec{u}$ *is suitable*

Let $r_0 > 0$ and $Q_0 = Q_{r_0}(t_0, x_0) = (t_0 - r_0^2, t_0) \times B(x_0, r_0)$. There exist positive constants ϵ_2 and C_5 which depend only on ν, q and τ (but not on x_0, t_0, r_0, $\vec{u}$ nor $\vec{f}$) such that, if

$$
\begin{cases}
0 \leq \lambda \leq \epsilon_2 \\
\iint_{Q_0} |\vec{u}|^3 + |p|^{3/2}\, dx\, dt \leq \lambda^3 r_0^2 \\
\iint_{Q_0} |\vec{f}|^q\, dx\, dt \leq \lambda^{2q} r_0^{5-3q}
\end{cases}
$$

then, $1_{Q_{3r_0/4}(t_0,x_0)}\vec{u} \in \mathcal{M}_2^{3,\tau}$ and $1_{Q_{3r_0/4}(t_0,x_0)}p \in \mathcal{M}_2^{3/2,\tau/2}$ with

$$\|1_{Q_{3r_0/4}(t_0,x_0)}\vec{u}\|_{\mathcal{M}_2^{3,\tau}} \leq C_5 \lambda r_0^{-1+\frac{5}{\tau}}$$

and

$$\|1_{Q_{3r_0/4}(t_0,x_0)}p\|_{\mathcal{M}_2^{3/2,\tau/2}} \leq C_5 \lambda^2 r_0^{2(-1+\frac{5}{\tau})}$$

The next move is to use the subcritical estimates on $\vec{u}$ and $\vec{p}$ to bootstrap those regularity estimates to higher regularity estimates. Indeed, let ϕ be a smooth function on $(-\infty,0] \times \mathbb{R}^3$ such that ϕ is equal to 1 on $(-1,0) \times B(0,1)$ and to 0 outside of $(-(3/2)^2,0) \times B(0,3/2)$. We define

$$\psi(t,x) = \phi\left(\frac{4(t-t_0)}{r_0^2}, \frac{2(x-x_0)}{r_0}\right).$$

Assume now that $q > 5/2$. Thus, we may choose τ in Lemma 14.3 such that

$$\tau > 10.$$

Let $\vec{v} = \psi \vec{u}$. We have

$$\partial_t \vec{v} = \nu \Delta \vec{v} + \vec{g} - \sum_{j=1}^{3} \partial_j \vec{h}_j$$

with

$$\vec{g} = \partial_t \psi \, \vec{u} + \nu \Delta \psi \, \vec{u} + (\vec{u} \cdot \vec{\nabla}\psi)\vec{u} + p\vec{\nabla}\psi + \psi\vec{f}$$

and

$$h_{j,l} = 2\partial_j \psi \, u_l + \psi u_j u_l + p\psi \, \delta_{j,l}.$$

We find that $h_{j,l} \in \mathcal{M}_2^{3/2,\tau/2}$ with

$$\|h_{j,l}\|_{\mathcal{M}_2^{3/2,\tau/2}} \leq C_6 \lambda r_0^{-2+\frac{10}{\tau}}$$

and, since $\tau < 4q$, that $\vec{g} \in \mathcal{M}_2^{3/2,\tau/4}$ with

$$\|\vec{g}\|_{\mathcal{M}_2^{3/2,\tau/4}} \leq C_6 \lambda r_0^{-3+\frac{20}{\tau}}$$

Let $\mathcal{C}^\alpha$ be the (homogeneous) space of parabolic Hölderian functions of exponent $\alpha \in (0,1)$:

$$\|F\|_{\mathcal{C}^\alpha} = \sup_{(t,x)\neq(s,y)} \frac{|F(t,x)-F(s,y)|}{(|t-s|^{1/2}+|x-y|)^\alpha}.$$

Applying Proposition 13.4, we find that $1_{t>t_0-r_0^2}\int_{t_0-r_0^2}^t W_{\nu(t-s)} * \vec{g}\,ds$ belongs to $\mathcal{C}_{t,x}^\alpha$ with $\alpha = 2 - \frac{20}{\tau}$ and that $1_{t>t_0-r_0^2}\int_{t_0-r_0^2}^t W_{\nu(t-s)} * \partial_j \vec{h}_j\,ds$ belongs to $\mathcal{C}_{t,x}^\beta$ with $\beta = 1 - \frac{10}{\tau}$. As $\vec{v}$ is equal to 0 when $|x-x_0| > r_0^2$, we find that $\vec{v}$ is bounded on $Q_{R_0}(t_0,x_0)$ by $C\lambda r_0^{-3+\frac{20}{\tau}}r_0^\alpha + Cr_0^{-2+\frac{10}{\tau}}r_0^\beta = C_7 \lambda r_0^{-1}$. The theorem is proved. $\qquad\square$

Combining Theorems 14.2, 14.3 and 14.4, we get the following corollary:

Inequalities in the L^∞ norm for local Leray solutions

Theorem 14.5.
Let $\vec{u}$ be a local Leray solution on $(0,T) \times \mathbb{R}^3$ to the problem

$$\begin{cases} \partial_t \vec{u} = \nu \Delta \vec{u} + \mathbb{P} \operatorname{div}(\mathbb{F} - \vec{u} \otimes \vec{u}) \\[2mm] \vec{u}(0,.) = \vec{u}_0 \end{cases} \tag{14.25}$$

where $\vec{u}_0 \in L^2_{\text{uloc}}$ with $\operatorname{div} \vec{u}_0 = 0$ and $\mathbb{F} \in (L^2_t L^2_x)_{\text{uloc}}((0,T) \times \mathbb{R}^3)$. Assume moreover that

- $\lim_{x \to +\infty} \int_{|x-y|<1} |\vec{u}_0(y)|^2 \, dy = 0$

- $\lim_{x \to +\infty} \int_0^T \int_{|x-y|<1} |\mathbb{F}(s,y)|^2 \, dy \, ds = 0$

- $\vec{f} = \operatorname{div} \mathbb{F}$ *satisfies the following requirements:*

  - *for $|x| > 1$: $1_{|x|>1} \vec{f}$ belongs to $(L^q_t L^q_{t,x})_{\text{uloc}}$ for some $q > 5/2$ and $\lim_{x \to +\infty} \int_0^T \int_{|x-y|<1} |\vec{f}(s,y)|^q \, dy \, ds = 0$*

  - *for $|x| \leq 1$: for some $\beta > 0$, $t^\beta 1_{|x|<1} \vec{f} \in L^2((0,T), L^2)$*

Then, for a constant C_0 which does not depend on ν, for

$$T_0 = \min\left(T, 1, \frac{1}{C_0 \nu \left(1 + \frac{\|\vec{u}_0\|_{L^2_{\text{uloc}}}}{\nu} + \frac{\|\mathbb{F}\|_{(L^2 L^2)_{\text{uloc}}}}{\nu^{3/2}}\right)^4}\right)$$

we have

$$\int_{T_0/2}^{T_0} \|\vec{u}(t,.)\|_\infty \, dt < +\infty$$

Proof. Recall that, on the neighborhood of x_0, the pressure can be defined as

$$p_{x_0}(t,x) = \frac{1}{\Delta}(\vec{\nabla} \otimes \vec{\nabla})(1_{B(x_0, 5R_0)} \mathbb{H}) + \int_{|y-x_0|>5R_0} (\mathbb{K}(x-y) - \mathbb{K}(x_0 - y)) \mathbb{H}(t,y) \, dy$$

where

$$\mathbb{H} = \mathbb{F} - \vec{u} \otimes \vec{u}$$

and $\mathbb{K}$ is the distribution kenel of $\frac{1}{\Delta}(\vec{\nabla} \otimes \vec{\nabla})$. Due to Theorem 14.2, we have a uniform control of $\vec{u}$ on $(0, T_0)$, where T_0 is given by

$$T_0 = \min\left(T, 1, \frac{1}{C_0 \nu \left(1 + \frac{\|\vec{u}_0\|_{L^2_{\text{uloc}}}}{\nu} + \frac{\|\mathbb{F}\|_{(L^2 L^2)_{\text{uloc}}}}{\nu^{3/2}}\right)^4}\right)$$

(where the constant C_0 does not depend on ν). This control is the following one:

$$\sup_{0<t<T_0} \sup_{x \in \mathbb{R}^3} \left(\int_{|x-y|<1} |\vec{u}(t,y)|^2 \, dy\right)^{1/2} \leq 2\left(\|\vec{u}_0\|_{L^2_{\text{uloc}}} + C_0 \frac{\|\mathbb{F}\|_{(L^2 L^2)_{\text{uloc}}}}{\sqrt{\nu}}\right)$$

and

$$\sup_{x \in \mathbb{R}^3} (\int_0^{T_0} \int_{|x-y|<1} |\vec{\nabla} \otimes \vec{u}(s,y)|^2 \, dy \, ds)^{1/2} \leq \frac{2}{\sqrt{\nu}}(\|\vec{u}_0\|_{L^2_{uloc}} + C_0 \frac{\|\mathbb{F}\|_{(L^2 L^2)_{uloc}}}{\sqrt{\nu}}).$$

Then (the proof of) Theorem 14.3 gives that, for $R > 1$,

$$\sup_{0<t<T_0} \sup_{|x|>R} (\int_{|x-y|<1} |\vec{u}(t,y)|^2 \, dy)^{1/2} + \sup_{|x|>R} (\int_0^{T_0} \int_{|x-y|<1} |\vec{\nabla} \otimes \vec{u}(s,y)|^2 \, dy \, ds)^{1/2}$$

$$\leq C_T (\sup_{|x|>R/2} (\int_{|x-y|<1} |\vec{u}_0(y)|^2 \, dy)^{1/2} + \sqrt{\frac{1 + \ln R}{R}}$$

$$+ \sup_{|x|>R/2} (\int_0^{T_0} \int_{|x-y|<1} |\mathbb{F}(s,y)|^2 \, dy \, ds)^{1/2})$$

where the constant C_T depends on ν, T, $\|\vec{u}_0\|_{L^2_{uloc}}$ and $\|\mathbb{F}\|_{(L^2 L^2)_{uloc}}$, but not on R. As $T_0 \leq 1$, we find

$$\sup_{|x|>R} \int_0^{T_0} \int_{|x-y|<1} |\vec{u}(t,y)|^3 \, dy \, ds$$

$$\leq C C_T^3 (\sup_{|x|>R/2} (\int_{|x-y|<1} |\vec{u}_0(y)|^2 \, dy)^{3/2} + (\sqrt{\frac{1 + \ln R}{R}})^3$$

$$+ \sup_{|x|>R/2} (\int_0^{T_0} \int_{|x-y|<1} |\mathbb{F}(s,y)|^2 \, dy \, ds)^{3/2})$$

Now, we estimate $\sup_{|x|>R} \int_0^{T_0} \int_{|x-y|<1} |p_x(t,y)|^{3/2} \, dy \, ds$. We assume that $R > \max(2, 5R_0)$. For $q_x = \frac{1}{\Delta}(\vec{\nabla} \otimes \vec{\nabla})(1_{B(x,5R_0)} \mathbb{H})$, we find

$$\int_0^{T_0} \int_{|x-y|<1} |q_x(t,y)|^{3/2} \, dy \, ds \leq C \int_0^0 \int_{|x-y|<5R_0} |\vec{u}(s,y)|^3 + |\mathbb{F}(s,y)|^{3/2} \, ds \, dy$$

so that

$$\sup_{|x|>R} \int_0^{T_0} \int_{|x-y|<1} |q_x(t,y)|^{3/2} \, dy \, ds$$

$$\leq C_T (\sup_{|x|>R/2} (\int_{|x-y|<1} |\vec{u}_0(y)|^2 \, dy)^{3/2} + (\sqrt{\frac{1 + \ln R}{R}})^3$$

$$+ \sup_{|x|>R/2} (\int_0^{T_0} \int_{|x-y|<1} |\mathbb{F}(s,y)|^2 \, dy \, ds)^{3/2}).$$

For $\varpi_x(t,y) = \int_{|z-x|>5R_0} (\mathbb{K}(y-z) - \mathbb{K}(x-z))\mathbb{H}(t,z) \, dz$, we write (for $|x| > R$ and $|x-y| < 1$)

$$|\varpi_x(t,y)| \leq C \int_{|z-x|>5R_0} \frac{1}{|z-x|^4} |\mathbb{H}(t,z)| \, dz.$$

and thus, since $\int_{|z|>5R_0} \frac{1}{(|z| \ln |z|)^3} \, dz < +\infty$,

$$\int_0^{T_0} \int_{|x-y|<1} |\varpi_x(t,y)|^{3/2} \, dy \, dt \leq C \int_{|z-x|>5R_0} \frac{(\ln(|x-z|))^{3/2}}{|z-x|^{9/2}} \int_0^{T_0} |\mathbb{H}(t,z)|^{3/2} \, dz \, dt$$

$$\leq \sum_{k \in \mathbb{Z}^3, |k| \geq 5R_0 - \sqrt{3}} \frac{(\ln |k|)^{3/2}}{|k|^{9/2}} \int_0^{T_0} \int_{z \in x+k+[0,1]^3} |\mathbb{H}(t,z)|^{3/2} \, dz \, dt$$

and splitting the last sum between $|k| > R/4$ and $|k| < R/4$, we find

$$\int_0^{T_0} \int_{|x-y|<1} |\varpi_x(t,y)|^{3/2} \, dy \, dt$$

$$\leq C \left(\frac{\ln R}{R} \right)^{3/2} \left(\sup_{x \in \mathbb{R}^3} \left(\int_{|x-y|<1} |\vec{u}_0(y)|^2 \, dy \right)^{3/2} \right.$$

$$+ \sup_{x \in \mathbb{R}^3} \left(\int_0^{T_0} \int_{|x-y|<1} |\mathbb{F}(s,y)|^2 \, dy \, ds \right)^{3/2} \right)$$

$$+ C_T \left(\sup_{|x|>R/2} \left(\int_{|x-y|<1} |\vec{u}_0(y)|^2 \, dy \right)^{3/2} + \left(\sqrt{\frac{1 + \ln R}{R}} \right)^3 \right.$$

$$+ \sup_{|x|>R/2} \left(\int_0^{T_0} \int_{|x-y|<1} |\mathbb{F}(s,y)|^2 \, dy \, ds \right)^{3/2} \right).$$

Thus, we find that we have, writing $Q_{x_0} = (0,T_0) \times B(x_0, \sqrt{T_0})$,

$$\lim_{x_0 \to +\infty} \iint_{Q_{x_0}} |\vec{u}(s,y)|^3 \, ds \, dy = 0$$

$$\lim_{x_0 \to +\infty} \iint_{Q_{x_0}} |p_x(s,y)|^{3/2} \, ds \, dy = 0$$

$$\lim_{x_0 \to +\infty} \iint_{Q_{x_0}} |\vec{f}(s,y)|^q \, ds \, dy = 0$$

Then, by applying Theorem 14.4, we find that there exists some $R > 0$ such that for $T_0/2 < t < T_0$ and $|x| > R$, we have $|\vec{u}(t,x)| \leq C_\nu \frac{1}{\sqrt{T_0}}$.

It remains to evaluate $|\vec{u}(t,x)|$ when $|x| < R$. We fix $\psi \in \mathcal{D}$ which is equal to 1 on $B(0, 3R)$ and we write, for $T_0/2 < t < T_0$,

$$\vec{u} = W_{\nu(t-T_0/2)} * (\psi \vec{u}(T_0/2,.)) + W_{\nu t} * ((1-\psi)\vec{u}(T_0/2,.))$$

$$+ \int_{T_0/2}^t W_{\nu(t-s)} * \mathbb{P} \operatorname{div}(\psi \mathbb{F}) \, ds + \int_{T_0/2}^t W_{\nu(t-s)} * \mathbb{P} \operatorname{div}((1-\psi)\mathbb{F}) \, ds$$

$$- \int_{T_0/2}^t W_{\nu(t-s)} * \mathbb{P} \operatorname{div}(\psi(\vec{u} \otimes \vec{u})) \, ds - \int_{T_0/2}^t W_{\nu(t-s)} * \mathbb{P} \operatorname{div}((1-\psi)(\vec{u} \otimes \vec{u})) \, ds$$

For $x \in B(0, R)$, we find

$$|\vec{u}(t,x)| \leq$$

$$\|W_{\nu(t-T_0/2)} * (\psi \vec{u}(T_0/2,.))\|_{L^\infty(dx)} + C \int_{|x-y|>2R} \frac{\sqrt{\nu t}}{|x-y|^4} |\vec{u}(T_0/2,y)| \, dy$$

$$+ |\int_{T_0/2}^t W_{\nu(t-s)} * \mathbb{P} \operatorname{div}(\psi \mathbb{F}) \, ds\|_{L^\infty(dx)} + C \int_{T_0/2}^t \int_{|x-y|>2R} |\mathbb{F}(s,y)| \frac{dy \, ds}{|x-y|^4}$$

$$+ \| \int_{T_0/2}^t W_{\nu(t-s)} * \mathbb{P} \operatorname{div}(\psi(\vec{u} \otimes \vec{u})) \, ds\|_{L^\infty(dx)}$$

$$+ C \int_{T_0/2}^t \int_{|x-y|>2R} |\vec{u} \otimes \vec{u}(s,y)| \frac{dy \, ds}{|x-y|^4}$$

The integration at the large ($|x - y| > 2R$) is easy to control, and we have

$$\int_{|x-y|>2R} \frac{\sqrt{\nu t}}{|x - y|^4} |\vec{u}(T_0/2, y)| \, dy \leq C_R \|\vec{u}(T_0/2, .)\|_{L^2_{uloc}}$$

$$\int_{T_0/2}^t \int_{|x-y|>2R} \frac{1}{|x - y|^4} |\mathbb{F}(s, y)| \, dy \, ds \leq C_R \|\mathbb{F}\|_{(L^2_t L^2_x)_{uloc}}$$

$$\int_{T_0/2}^t \int_{|x-y|>2R} |\vec{u} \otimes \vec{u}(s, y)| \frac{dy \, ds}{|x - y|^4} \leq C_R \|\vec{u}\|_{L^\infty L^2_{uloc}}^2$$

Moreover, we have

$$\|W_{\nu(t-T_0/2)} * (\psi \vec{u}(T_0/2, .))\|_{L^\infty(dx)} \leq C(\nu t)^{-3/4} \|\psi \vec{u}(T_0/2, .)\|_2$$

and

$$\int_{T_0/2}^{T_0} \left\| \int_0^t W_{\nu(t-s)} * \mathbb{P} \operatorname{div}(\psi(\vec{u} \otimes \vec{u})) \, ds \right\|_{L^\infty(dx)} dt \leq C \int_{T_0/2}^{T_0} \| \operatorname{div}(\psi(\vec{u} \otimes \vec{u})) \|_{\dot{B}_{2,1}^{-1/2}} \, dt$$

where

$$\| \operatorname{div}(\psi(\vec{u} \otimes \vec{u})) \|_{\dot{B}_{2,1}^{-1/2}} \leq C \|\vec{u}\|_{H^1(B(0,5R))}^2.$$

Finally, we have $\operatorname{div}(\psi \mathbb{F}) = \vec{\nabla}\psi.\mathbb{F} + \psi \vec{f} \in L^2((T_0/2, T_0), L^2)$, so that

$$\int_{T_0/2}^t W_{\nu(t-s)} * \mathbb{P} \operatorname{div}(\psi \mathbb{F}) \, ds \in L^4((T_0/2, T_0), \dot{B}_{2,1}^{3/2}) \subset L^4((T_0/2, T_0), L^\infty).$$

Thus, we have $1_{B(0,5R)}\vec{u} \in L^1((T_0/2, T_0), L^\infty)$. $\qquad\square$

Using Theorem 14.4, we may give a more quantitative statement:

Quantitative inequalities for the L^∞ norm of local Leray solutions

Theorem 14.6.
Let $\vec{u}$ be a local Leray solution on $(0, T) \times \mathbb{R}^3$ to the problem

$$\begin{cases} \partial_t \vec{u} = \nu \Delta \vec{u} + \mathbb{P} \operatorname{div}(\mathbb{F} - \vec{u} \otimes \vec{u}) \\ \\ \vec{u}(0, .) = \vec{u}_0 \end{cases} \tag{14.26}$$

where $\vec{u}_0 \in L^2_{uloc}$ with $\operatorname{div} \vec{u}_0 = 0$ and $\mathbb{F} \in (L^2_t L^2_x)_{uloc}((0, T) \times \mathbb{R}^3)$. Assume moreover that

- $|\vec{u}_0(x)| \leq C_0 \frac{1}{|x|}$

- $|\mathbb{F}(t, x)| \leq C_0 \frac{1}{(\sqrt{\nu t} + |x|)^2}$

- $|\operatorname{div} \mathbb{F}(t, x)| \leq C_0 \frac{1}{(\sqrt{\nu t} + |x|)^3}$

> Then, for constants $T_0 > 0$, $R_1 > 1$ and C_1 which depend only on ν, T and C_0, we have
>
> $$\sup_{|x|>R_1} \sup_{T_0/2<t<T_0} \sqrt{\frac{|x|}{\ln|x|}}|\vec{u}(t,x)| \leq C_1$$
>
> and
>
> $$\int_{T_0/2}^{T_0} \|\vec{u}(t,.)\|_\infty dt \leq C_1.$$

Proof. Remark that $\frac{1}{|x|^2} \in L^1_{uloc}$. Thus, $\vec{u}_0 \in L^2_{uloc}$, and its norm is controlled by C_0. Similarly, we have

$$\int_0^{+\infty} \int_{|x-y|<1} \frac{dy\,dt}{(\sqrt{\nu t}+|y|)^4} = \frac{1}{\nu}\int_0^{+\infty} \frac{dt}{(\sqrt{t}+1)^4}\int_{|x-y|<1}\frac{dy}{|y|^2}$$

so that $\mathbb{F} \in (L^2_t L^2_x)_{uloc}$ and its norm is controlled by C_0. It means that we have a control of the norms $\|\vec{u}\|_{L^\infty L^2_{uloc}}$ and $\|\vec{\nabla}\otimes\vec{u}\|_{(L^2L^2)_{uloc}}$ on a band $(0,T_0)\times\mathbb{R}^3$, where T_0 depends only on ν, T and C_0 (Theorem 14.2).

Now, we check that $\vec{u}$ and $\mathbb{F}$ fulfill the assumptions of Theorem 14.5:

- for $|x|>2$,
$$\int_{|x-y|<1}|\vec{u}_0(y)|^2\,dy \leq CC_0^2\frac{1}{|x|^2} \to_{x\to\infty} 0$$

- for $|x|>2$,
$$\int_0^T\int_{|x-y|<1}|\mathbb{F}(s,y)|^2\,dy\,ds \leq CC_0^2\frac{1}{|x|^2}\to_{x\to\infty}0$$

- $|\sqrt{t}\operatorname{div}\mathbb{F}(t,x)|\leq C_0\frac{1}{(\sqrt{\nu t}+|x|)^2}$, so that $\sqrt{t}\operatorname{div}\mathbb{F}\in(L^2_tL^2_x)_{uloc}$ and its norm is controlled by C_0.

- for $|x|>1$: $|1_{|x|>1}\operatorname{div}\mathbb{F}(t,x)|\leq C_0|x|^{-3}$; the function $1_{|x|>1}|x|^{-3}$ belongs to L^q for evert $q>1$. Moreover, for $|x|>2$,
$$\int_0^T\int_{|x-y|<1}|\operatorname{div}\mathbb{F}(s,y)|^q\,dy\,ds\leq CC_0^q|x|^{-3q}\to_{x\to\infty}0$$

Moreover, the proof of Theorem 14.5 gives us the following estimates for $|x|>5R_0$ (for constants C_2, C_3 which depend only on ν, T, C_0 (and q):

$$\int_0^{T_0}\int_{|x-y|<1}|\vec{u}(t,y)|^3\,dy\,ds\leq C_2(|x|^{-3}+\left(\frac{\ln|x|}{|x|}\right)^{3/2})\leq T_0^5\left(C_3\sqrt{\frac{\ln|x|}{|x|}}\right)^3$$

$$\int_0^{T_0}\int_{|x-y|<1}|p_x(t,y)|^{3/2}\,dy\,ds\leq C_2(|x|^{-3}+\left(\frac{\ln|x|}{|x|}\right)^{3/2})\leq T_0^5\left(C_3\sqrt{\frac{\ln|x|}{|x|}}\right)^3$$

$$\int_0^T\int_{|x-y|<1}|\operatorname{div}\mathbb{F}(s,y)|^q\,dy\,ds\leq C_2|x|^{-3q}\leq T_0^5\left(C_3\sqrt{\frac{\ln|x|}{|x|}}\right)^{2q}$$

Then, if R_1 is large enough to ensure that $C_3\sqrt{\frac{\ln|x|}{|x|}} < \epsilon_0$, Theorem 14.4 gives that, for $|x| > R_1$, we have, for $T_0/2 < t < T_0$,

$$|\vec{u}(t,x)| \le C_4(C_3\sqrt{\frac{\ln|x|}{|x|}})\frac{1}{T_0}.$$

Thus, the theorem is proved.   □

14.4   A Weak-Strong Uniqueness Result

In this section, we generalize the von Wahl weak-strong uniqueness theorem [494] (see Proposition 12.3), replacing the L^2 Leray solutions by L^2_{uloc} suitable solutions:

Weak-strong uniqueness

Theorem 14.7.
Let $\vec{u}_1$, $\vec{u}_2$ be two local Leray solutions on $(0,T) \times \mathbb{R}^3$ to the same problem

$$\begin{cases} \partial_t \vec{u} = \nu\Delta\vec{u} + \mathbb{P}\operatorname{div}(\mathbb{F} - \vec{u} \otimes \vec{u}) \\ \\ \vec{u}(0,.) = \vec{u}_0 \end{cases} \tag{14.27}$$

where $\vec{u}_0 \in L^2_{\text{uloc}}$ with $\operatorname{div}\vec{u}_0 = 0$ and $\mathbb{F} \in (L^2_t L^2_x)_{\text{uloc}}((0,T) \times \mathbb{R}^3)$.
  Assume moreover that $\vec{u}_1$ can be written as $\vec{u}_1 = \vec{u}_3 + \vec{u}_4$ with $\vec{u}_3 \in L^2((0,T), L^\infty)$ and $\vec{u}_4 \in L^\infty((0,T), \bar{\mathcal{V}}^1)$ (where $\bar{\mathcal{V}}^1 = \mathcal{M}(H^1 \mapsto L^2)$) and that

$$\sup_{0<t<T} \|\vec{u}_4(t,.)\|_{\bar{\mathcal{V}}^1} < \epsilon_0 \nu$$

where ϵ_0 is a small positive constant which does not depend on ν, $\vec{u}_0$, $\mathbb{F}$, T, $\vec{u}_1$ nor $\vec{u}_2$. Then $\vec{u}_1 = \vec{u}_2$.

Proof. Let p_i be the pressure associated to $\vec{u}_i$. We have, due to the suitability of $\vec{u}_2$ and the regularity of $\vec{u}_1$,

$$\begin{cases} \partial_t(\frac{|\vec{u}_1|^2}{2}) = & \nu\Delta(\frac{|\vec{u}_1|^2}{2}) - \nu|\vec{\nabla}\otimes\vec{u}_1|^2 - \operatorname{div}((p_1 + \frac{|\vec{u}_1|^2}{2})\vec{u}_1) + \vec{u}_1 \cdot \operatorname{div}\mathbb{F} \\ \\ \partial_t(\vec{u}_1 \cdot \vec{u}_2) = & \nu\Delta(\vec{u}_1 \cdot \vec{u}_2) - 2\nu(\vec{\nabla}\otimes\vec{u}_1).(\vec{\nabla}\otimes\vec{u}_2) - \operatorname{div}(p_2\vec{u}_1 + p_1 \cdot \vec{u}_2) \\ & -\vec{u}_1.(\vec{u}_2 \cdot \vec{\nabla}\vec{u}_2) - \vec{u}_2.(\vec{u}_1 \cdot \vec{\nabla}u_1) + (\vec{u}_1 + \vec{u}_2) \cdot \operatorname{div}\mathbb{F} \\ \\ \partial_t(\frac{|\vec{u}_2|^2}{2}) = & \nu\Delta(\frac{|\vec{u}_2|^2}{2}) - \nu|\vec{\nabla}\otimes\vec{u}_2|^2 - \operatorname{div}((p_2 + \frac{|\vec{u}_2|^2}{2})\vec{u}_2) + \vec{u}_2 \cdot \operatorname{div}\mathbb{F} - \mu \end{cases}$$

where μ is some non-negative locally finite measure.
  Let $\vec{w} = \vec{u}_1 - \vec{u}_2$ and $q = p_1 - p_2$. We obtain

$$\partial_t(\frac{|\vec{w}|^2}{2}) = \nu\Delta(\frac{|\vec{w}|^2}{2}) - \nu|\vec{\nabla}\otimes\vec{w}|^2 - \operatorname{div}(q\vec{w}) - A - \mu$$

with

$$A = \text{div}(\frac{|\vec{u}_1|^2}{2}\vec{u}_1 + \frac{|\vec{u}_2|^2}{2}\vec{u}_2) - (\vec{u}_1.(\vec{u}_2 \cdot \vec{\nabla}\vec{u}_2) + \vec{u}_2.(\vec{u}_1 \cdot \vec{\nabla}u_1)$$

$$= \frac{1}{2}\text{div}\left(|\vec{u}_1|^2\vec{u}_1 + |\vec{u}_2|\vec{u}_2 - (\vec{u}_1 \cdot \vec{u}_2)(\vec{u}_1 + \vec{u}_2)\right)$$

$$+ \frac{1}{2}\left(\vec{u}_2.(\vec{u}_2 \cdot \vec{\nabla}\vec{u}_1) + \vec{u}_1.(\vec{u}_1 \cdot \vec{\nabla}u_2) - \vec{u}_1.(\vec{u}_2 \cdot \vec{\nabla}\vec{u}_2) - \vec{u}_2.(\vec{u}_1 \cdot \vec{\nabla}u_1)\right)$$

$$= \frac{1}{2}\text{div}\left(|\vec{w}|^2\vec{u}_1 + (\vec{u}_1 \cdot \vec{w})\vec{w} - |\vec{w}|^2\vec{w}\right) + \frac{1}{2}\left(\vec{w}.(\vec{w} \cdot \vec{\nabla}\vec{u}_1) - \vec{u}_1.(\vec{w} \cdot \vec{\nabla}\vec{w})\right)$$

$$= \frac{1}{2}\text{div}\left(|\vec{w}|^2\vec{u}_1 + 2(\vec{u}_1 \cdot \vec{w})\vec{w} - |\vec{w}|^2\vec{w}\right) - \vec{u}_1.(\vec{w} \cdot \vec{\nabla}\vec{w}).$$

We then follow the lines of the proof of Theorem 14.2. For $\varphi(x) = \varphi_{x_0}(x) = \varphi_0(x - x_0)$, we write

$$\int \varphi \frac{|\vec{w}(t,x)|^2}{2}\, dx\, ds \leq -\nu \int_0^t \int \varphi|\vec{\nabla} \otimes \vec{w}|^2\, dx\, ds + \nu \int_0^t \int (\Delta\varphi)\frac{|\vec{w}|^2}{2}\, dx\, ds$$

$$+ \int_0^t \int (\vec{w} \cdot \vec{\nabla}\varphi)(q_{x_0} + \vec{w} \cdot \vec{u}_1 - \frac{|\vec{w}|^2}{2})\, dx\, ds$$

$$+ \int_0^t \int (\vec{u}_1 \cdot \vec{\nabla}\varphi)\frac{|\vec{w}|^2}{2}\, dx\, ds - \int_0^t \int \varphi\vec{u}_1.(\vec{w} \cdot \vec{\nabla}\vec{w})\, dx\, ds$$

where q_{x_0} is defined as where

$$q_{x_0}(t,x) = \frac{1}{\Delta}(\vec{\nabla} \otimes \vec{\nabla})(1_{B(x_0,5R_0)}\mathbb{H}) + \int_{|y-x_0|>5R_0}(\mathbb{K}(x - y) - \mathbb{K}(x_0 - y))\mathbb{H}(t,y)\, dy$$

and

$$\mathbb{H} = \vec{u}_1 \otimes \vec{u}_1 - \vec{u}_2 \otimes \vec{u}_2 = \vec{w} \otimes \vec{u}_1 + \vec{u}_1 \otimes \vec{w} - \vec{w} \otimes \vec{w}.$$

We then write

$$q_{x_0} = R_{1,x_0} + R_{2,x_0} + S_{1,x_0} + S_{2,x_0}$$

with

$$R_{1,x_0} = -\frac{1}{\Delta}(\vec{\nabla} \otimes \vec{\nabla})(1_{B(x_0,5R_0)}\vec{w} \otimes \vec{w})$$

$$R_{2,x_0} = \int_{|y-x_0|>5R_0}(\mathbb{K}(x - y) - \mathbb{K}(x_0 - y))\vec{w} \otimes \vec{w}(t,y)\, dy$$

$$S_{1,x_0} = -\frac{1}{\Delta}(\vec{\nabla} \otimes \vec{\nabla})(1_{B(x_0,5R_0)}(\vec{u}_1 \otimes \vec{w} + \vec{w} \otimes \vec{u}_1))$$

$$S_{1,x_0} = \int_{|y-x_0|>5R_0}(\mathbb{K}(x - y) - \mathbb{K}(x_0 - y))(\vec{u}_1 \otimes \vec{w} + \vec{w} \otimes \vec{u}_1)(t,y)\, dy$$

Defining again

$$\alpha(t) = \|\vec{w}\|_{L^2_{\text{uloc}}} = \sup_{\varphi \in \mathcal{B}} \|\varphi\vec{w}\|_2$$

$$\gamma(t) = \|\vec{\nabla} \otimes \vec{w}\|_{(L^2L^2)_{\text{uloc}}((0,t)\times\mathbb{R}^3)} = \sup_{\varphi \in \mathcal{B}} (\int_0^t \int |\varphi(x)\vec{\nabla} \otimes \vec{w}(s,x)|^2\, dx\, ds)^{1/2}$$

$$\delta(t) = \|\vec{w}\|_{(L^3L^3)_{\text{uloc}}((0,t)\times\mathbb{R}^3)} = \sup_{\varphi \in \mathcal{B}} (\int_0^t \int |\varphi(x)\vec{w}(s,x)|^3\, dx\, ds)^{1/2}$$

we find

$$\nu \int_0^t \int (\Delta\varphi)\frac{|\vec{w}|^2}{2}\,dx\,ds \leq C\nu \int_0^t \alpha^2(s)\,ds$$

$$-\int_0^t \int (\vec{w}\cdot\vec{\nabla}\varphi)\frac{|\vec{w}|^2}{2}\,dx\,ds \leq C\delta(t)^3$$

$$\int_0^t \int (\vec{w}\cdot\vec{\nabla}\varphi)(R_{1,x_0}+R_{2,x_0})\,dx\,ds \leq C\delta(t)^3$$

while

$$\int_0^t \int (\vec{w}\cdot\vec{\nabla}\varphi)(\vec{u}_1\cdot\vec{w})\,dx\,ds \leq C\int_0^t \|\vec{u}_3(s,.)\|_\infty \alpha(s)^2\,ds$$

$$+ C\|\vec{u}_4\|_{L^\infty \tilde{\mathcal{V}}^1}\gamma(t)\sqrt{\gamma(t)^2 + \int_0^t \alpha(s)^2\,ds}$$

$$\int_0^t \int (\vec{w}\cdot\vec{\nabla}\varphi)(S_{1,x_0}+S_{2,x_0})\,dx\,ds \leq C\int_0^t \|\vec{u}_3(s,.)\|_\infty \alpha(s)^2\,ds$$

$$+ C\|\vec{u}_4\|_{L^\infty \tilde{\mathcal{V}}^1}\gamma(t)\sqrt{\gamma(t)^2 + \int_0^t \alpha(s)^2\,ds}$$

$$\int_0^t \int (\vec{u}_1\cdot\vec{\nabla}\varphi)\frac{|\vec{w}|^2}{2}\,dx\,ds \leq C\int_0^t \|\vec{u}_3(s,.)\|_\infty \alpha(s)^2\,ds$$

$$+ C\|\vec{u}_4\|_{L^\infty \tilde{\mathcal{V}}^1}(\gamma(t)^2 + \int_0^t \alpha(s)^2\,ds)$$

$$-\int_0^t \int \varphi\vec{u}_1.(\vec{w}\cdot\vec{\nabla}\vec{w})\,dx\,ds \leq C\gamma(t)\sqrt{\int_0^t \|\vec{u}_3(s,.)\|_\infty^2 \alpha(s)^2\,ds}$$

$$+ C\|\vec{u}_4\|_{L^\infty \tilde{\mathcal{V}}^1}\gamma(t)\sqrt{\gamma(t)^2 + \int_0^t \alpha(s)^2\,ds}$$

Writing

$$\delta(t)^2 \leq C\Big(\int_0^t \alpha(s)^2\,ds + \gamma(t)\big(\int_0^t \alpha^6(s)\,ds\big)^{1/6}\Big),$$

we obtain (for every $\eta > 0$, with constants that do not depend on η)

$$\int \varphi\frac{|\vec{w}(t,x)|^2}{2}\,dx\,ds + \nu\int_0^t \int \varphi|\vec{\nabla}\otimes\vec{w}|^2\,dx\,ds$$

$$\leq C_0\nu\int_0^t \alpha^2(s)\,ds + C_0\big(\int_0^t \alpha^2(s)\,ds\big)^{3/2} + C_0\gamma(t)^{3/2}\big(\int_0^t \alpha^6(s)\,ds\big)^{1/4}$$

$$+C_0\int_0^t \|\vec{u}_3(s,.)\|_\infty \alpha(s)^2\,ds + C_0\gamma(t)\sqrt{\int_0^t \|\vec{u}_3(s,.)\|_\infty^2 \alpha(s)^2\,ds}$$

$$+C_0\|\vec{u}_4\|_{L^\infty \tilde{\mathcal{V}}^1}(\gamma(t)^2 + \int_0^t \alpha(s)^2\,ds) + C_0\|\vec{u}_4\|_{L^\infty \tilde{\mathcal{V}}^1}\gamma(t)\sqrt{\gamma(t)^2 + \int_0^t \alpha(s)^2\,ds}$$

$$\leq C_1 \nu \int_0^t \alpha^2(s)\, ds + C_1 \Big(\int_0^t \alpha^2(s)\, ds \Big)^{3/2}$$

$$+ C_1 \eta \gamma(t)^2 + C_1 \eta^{-3} \int_0^t \alpha^6(s)\, ds$$

$$+ C_1 \int_0^t \|\vec{u}_3(s,.)\|_\infty \alpha(s)^2\, ds$$

$$+ C_1 \eta^{-1} \int_0^t \|\vec{u}_3(s,.)\|_\infty^2 \alpha(s)^2\, ds + C_1 \eta \gamma(t)^2$$

$$+ C_1 \|\vec{u}_4\|_{L^\infty \tilde{V}^1} \Big(\gamma(t)^2 + \int_0^t \alpha(s)^2\, ds \Big)$$

and thus

$$\max(\alpha(t)^2, \nu \gamma(t)^2) \leq C_1 \nu \int_0^t \alpha^2(s)\, ds + C_1 \Big(\int_0^t \alpha^2(s)\, ds \Big)^{3/2} + C_1 \eta^{-3} \int_0^t \alpha^6(s)\, ds$$

$$+ C_1 \int_0^t \|\vec{u}_3(s,.)\|_\infty \alpha(s)^2\, ds + C_1 \|\vec{u}_4\|_{L^\infty \tilde{V}^1} \int_0^t \alpha(s)^2\, ds$$

$$+ C_1 \eta^{-1} \int_0^t \|\vec{u}_3(s,.)\|_\infty^2 \alpha(s)^2\, ds$$

$$+ C_1 (2\eta + \|\vec{u}_4\|_{L^\infty \tilde{V}^1}) \gamma(t)^2$$

Thus, if η is chosen such that $C_1 \eta < \frac{1}{4}\nu$ and if $C_1 \epsilon_0 < \frac{1}{4}$, we find, for a constant C_ν which depends on ν,

$$\alpha(t)^6 \leq C_\nu t^2 \int_0^t \alpha(s)^6\, ds + C_\nu t^3 \Big(\int_0^t \alpha(s)^6\, ds \Big)^{3/2} + C_\nu \Big(\int_0^t \alpha(s)^6\, ds \Big)^6$$

$$+ C_\nu \Big(\int_0^t \|\vec{u}_3(s,.)\|_\infty\, ds \Big)^2 \int_0^t \|\vec{u}_3(s,.)\|_\infty \alpha(s)^6\, ds$$

$$+ C_\nu \|\vec{u}_4\|_{L^\infty \tilde{V}^1}^3 t^2 \int_0^t \alpha(s)^6\, ds$$

$$+ C_\nu \Big(\int_0^t \|\vec{u}_3(s,.)\|_\infty^2\, ds \Big)^2 \int_0^t \|\vec{u}_3(s,.)\|_\infty^2 \alpha(s)^6\, ds$$

As long as $\int_0^t \alpha(s)^6\, ds < 1$, we find that

$$\alpha(t)^6 \leq C_\nu \int_0^t A(s)\alpha(s)^6\, ds$$

with

$$A(s) = T^2 + T^3 + 1 + T\|\vec{u}_3\|_{L^2 L^\infty}^2 \|\vec{u}_3(s,.)\|_\infty + T^2\|\vec{u}_4\|_{L^\infty \tilde{V}^1}^3 + \|\vec{u}_3\|_{L^2 L^\infty}^2 \|\vec{u}_3(s,.)\|_\infty^2.$$

We then conclude by Grönwall's lemma that $\bar{\alpha} = 0$, i.e. $\vec{u}_1 = \vec{u}_2$. $\qquad\square$

14.5 Global Existence for Local Leray Solutions

In this section, we show how to turn the local existence result of Theorem 14.1 into a global existence result, assuming the initial data and the forcing term vanish at infinity.

Definition 14.2. *[Vanishing at infinity functions]*
Let $\varphi_0 \in \mathcal{D}(\mathbb{R}^3)$, $\varphi_0 \geq 0$, such that

$$\sum_{k \in \mathbb{Z}^3} \varphi_0(x - k) = 1$$

and let

$$\mathcal{B} = \{\varphi_{x_0} = \varphi_0(\,. - x_0) \ / \ x_0 \in \mathbb{R}^3\}$$

Then define $L^p_{\mathrm{uloc}}(\mathbb{R}^3)$ for $1 \leq p < +\infty$ by

$$\|h\|_{L^p_{\mathrm{uloc}}} = \sup_{\varphi \in \mathcal{B}} \|h\varphi\|_p.$$

We define E^p, the space of functions in L^p_{uloc} that vanish at infinity, by

$$f \in E^p \Leftrightarrow f \in L^p_{\mathrm{uloc}} \ \text{and} \ \lim_{x_0 \to \infty} \|f\varphi_{x_0}\|_p = 0.$$

Lemma 14.4.
$\mathcal{D}(\mathbb{R}^3)$ *is dense in E^p.*

Proof. Let E^p_{comp} be the space of compactly supported functions in E^p. Then $E^p_{\mathrm{comp}} \subset L^p \subset E^p$ continuous embeddings). As $\mathcal{D}(\mathbb{R}^3)$ is dense in L^p, we just have to check that E^p_{comp} is dense in E^p. But this is obvious, since $\lim_{N \to \infty} \|f \sum_{|k|>N} \varphi_0(x - k)\|_{L^p_{\mathrm{uloc}}} = 0$ for $f \in E^p$. $\qquad \square$

Lemma 14.5.
For $1 \leq p < +\infty$, let E^p_σ be the space of divergence-free vector fields in $(E^p)^3$ and let $E^p_{\mathrm{comp},\sigma}$ be the space of compactly supported functions divergence-free vector fields in $(E^p)^3$. Then $E^p_{\mathrm{comp},\sigma}$ is dense in E^p_σ.

Proof. This is proved via a decomposition on a divergence-free wavelet basis . Let us recall the results of [309, 313]: there exists compactly supported $\mathcal{C}^1$ vector fields $\vec{\phi}_i$ $(1 \leq i \leq 3)$, $\vec{\phi}_i^*$ $(1 \leq i \leq 3)$, $\vec{\psi}_l$ $(1 \leq l \leq 14)$ and $\vec{\psi}_l^*$ $(1 \leq l \leq 14)$ so that

- the scaling functions $\vec{\phi}_i$ generate a bi-orthogonal multi-resolution analysis $\vec{V}_j$ of $(L^2(\mathbb{R}^3))^3$ while $\vec{\phi}_i^*$ generate the dual multi-resolution analysis $\vec{\tilde{V}}_j$

- If $\vec{P}_j$ is the associated projection operator

$$\vec{P}_j(\vec{f}) = \sum_{k \in \mathbb{Z}^3} \sum_{i=1}^{3} \langle \vec{f} | \vec{\phi}_{i,j,k}^* \rangle \vec{\phi}_{i,j,k}$$

with $\vec{\phi}_{i,j,k}(x) = 2^{3j/2} \vec{\phi}_i(2^j x - k)$ and $\vec{\phi}_{i,j,k}^*(x) = 2^{3j/2} \vec{\phi}_i^*(2^j x - k)$ and if $\vec{f} \in L^1_{\mathrm{loc}}$ with div $\vec{f} = 0$, then div $\vec{P}_j(\vec{f}) = 0$. In particular, if $\vec{f} \in E^p_\sigma$, then $\vec{P}_j(\vec{f}) \in E^p_\sigma$, $\lim_{j \to +\infty} \|\vec{P}_j(\vec{f}) - \vec{f}\|_{L^p_{\mathrm{uloc}}} = 0$ and $\lim_{j \to -\infty} \|\vec{P}_j(\vec{f})\|_{L^p_{\mathrm{uloc}}} = 0$

- div $\vec{\psi}_l = 0$ $(1 \leq l \leq 14)$ and, when $\vec{f} \in E^p_\sigma$, we have

$$(\vec{P}_{j+1} - \vec{P}_j)(\vec{f}) = \sum_{k \in \mathbb{Z}^3} \sum_{i=1}^{14} \langle \vec{f} | \vec{\psi}_{i,j,k}^* \rangle \vec{\psi}_{i,j,k}$$

with $\vec{\psi}_{i,j,k}(x) = 2^{3j/2}\vec{\psi}_i(2^j x - k)$ and $\vec{\psi}^*_{i,j,k}(x) = 2^{3j/2}\vec{\psi}^*_i(2^j x - k)$. In particular, the operator $\vec{\Pi}$ defined by

$$\vec{\Pi}(\vec{f}) = \sum_{j\in\mathbb{Z}}\sum_{k\in\mathbb{Z}^3}\sum_{i=1}^{14}\langle \vec{f}|\vec{\psi}^*_{i,j,k}\rangle \vec{\psi}_{i,j,k}$$

is a Calderón–Zygmund operators and is a bounded projection operator from $(L^2)^3$ onto the space L^2_σ of divergence-free square integrable vector fields; but this not an orthogonal projection operator (div $\vec{\psi}^*_i \neq 0$) and $\vec{\Pi} \neq \mathbb{P}$.

- for $\vec{f} \in E^p_\sigma$ and $j \in \mathbb{Z}$, we have

$$\lim_{N\to+\infty}\|\sum_{|k|>N}\sum_{i=1}^{14}\langle\vec{f}|\vec{\psi}^*_{i,j,k}\rangle\vec{\psi}_{i,j,}\|_{L^p_{\text{uloc}}} = 0$$

Thus, we have

$$\lim_{J\to+\infty}\lim_{N\to+\infty}\left(\|\vec{f}-\vec{P}_{J+1}(\vec{f})\|_{L^p_{\text{uloc}}} + \sum_{j=-J}^{J}\|\sum_{|k|>N}\sum_{i=1}^{14}\langle\vec{f}|\vec{\psi}^*_{i,j,k}\rangle\vec{\psi}_{i,j,k}\|_{L^p_{\text{uloc}}}\right.$$
$$\left. +\|\vec{P}_{-J}(\vec{f})\|_{L^p_{\text{uloc}}}\right) = 0$$

and thus

$$\lim_{J\to+\infty}\lim_{N\to+\infty}\|\vec{f}-\sum_{j=-J}^{J}\sum_{|k|\leq N}\sum_{i=1}^{14}\langle\vec{f}|\vec{\psi}^*_{i,j,k}\rangle\vec{\psi}_{i,j,k}\|_{L^p_{\text{uloc}}} = 0.$$

The lemma is proved, since $\sum_{j=-J}^{J}\sum_{|k|\leq N}\sum_{i=1}^{14}\langle\vec{f}|\vec{\psi}^*_{i,j,k}\rangle\vec{\psi}_{i,j,k}$ belongs to $E^p_{\text{comp},\sigma}$. □

We give a definition forces that vanish at infinity similar to Definition 14.2:

Definition 14.3 (Vanishing at infinity forces).
Let $\varphi_0 \in \mathcal{D}(\mathbb{R}^3)$, $\varphi_0 \geq 0$, such that

$$\sum_{k\in\mathbb{Z}^3}\varphi_0(x-k) = 1$$

and let
$$\mathcal{B} = \{\varphi_{x_0} = \varphi_0(.-x_0) \ / \ x_0 \in \mathbb{R}^3\}$$
Then define $(L^p H^s)_{\text{uloc}}((0,T)\times\mathbb{R}^3)$, for $1 \leq p < +\infty$ and $s \in \mathbb{R}$, by

$$\|f\|_{(L^p H^s)_{\text{uloc}}} = \sup_{\varphi\in\mathcal{B}}\|f\varphi\|_{L^p H^s}.$$

We define $F^{s,p}$, the space of functions in $(L^p H^s)_{\text{uloc}}$ that vanish at infinity, by

$$f \in F^{s,p} \Leftrightarrow f \in (L^p H^s)_{\text{uloc}} \text{ and } \lim_{x_0\to\infty}\|f\varphi_{x_0}\|_{L^p H^s} = 0.$$

Lemma 14.6.
$L^p H^s$ is dense in $F^{s,p}$.

Proof. Just check that $\lim_{N\to\infty}\|f\sum_{|k|>N}\varphi_0(x-k)\|_{(L^p H^s)_{\text{uloc}}} = 0$ for $f \in F^{s,p}$. □

We may now discuss the existence of global solutions in L^2_{uloc}. Existence of global weak solutions, generalizing the result of Leray for $\vec{u}_0 \in (L^2)^3$ was first established for $\vec{u}_0 \in (L^p)^3$ ($2 \leq p < \infty$) by C. Calderón [77] and later by Lemarié-Rieusset [310]. The case $\vec{u}_0 \in L^2_{uloc}$ in absence of force was discussed by Lemarié-Rieusset in [313].

Global uniformly locally square integrable solutions

Theorem 14.8.
Let $\vec{u}_0 \in L^2_{uloc}$ with $\operatorname{div} \vec{u}_0 = 0$, vanishing at infinity: $\vec{u}_0 \in E^2_\sigma$. Let

$$\mathbb{F} \in \bigcap_{0 < T < +\infty} F^{2,3}((0,T) \times \mathbb{R}^3) \subset \bigcap_{0 < T < +\infty} (L^3 H^2)_{uloc}((0,T) \times \mathbb{R}^3).$$

Then there exists a global suitable weak solution $\vec{u}$ to the problem

$$\begin{cases} \partial_t \vec{u} = \nu \Delta \vec{u} + \mathbb{P} \operatorname{div}(\mathbb{F} - \vec{u} \otimes \vec{u}) \\ \\ \vec{u}(0,.) = \vec{u}_0 \end{cases} \tag{14.28}$$

on $(0, +\infty) \times \mathbb{R}^3$ such that

$$\vec{u} \in \bigcap_{0 < T < +\infty} (L^\infty_t L^2_x)_{uloc}((0,T) \times \mathbb{R}^3) \cap (L^2 \dot{H}^1_x)_{uloc}((0,T) \times \mathbb{R}^3).$$

Proof. **Step 1: the local solution with initial data in E^2_σ.**
  By Theorem 14.1, we know that there exists a local Leray solution $\vec{u}_1$ defined on an interval $(0, T_1)$. By Theorem 14.3, we even know that $\vec{u}_1 \in L^\infty((0, T_1), E^2)$. Moreover, we know by Theorem 14.5 that $\vec{u}_1 \in L^1(T_1/2, T_1), L^\infty)$. As $L^\infty \cap E_2 \subset E_3$ with $\|\vec{u}_1\|_{L^3_{uloc}} \leq \|\vec{u}_1\|_{L^2_{uloc}}^{2/3} \|\vec{u}_1\|_\infty^{1/3}$, we find that $\vec{u}_1 \in L^3((T_1/2, T_1), E^3)$.

  Moreover, for almost every $t \in (0, T_1)$, we have

$$\lim_{s \to t, s > t} \|\vec{u}_1(t,.) - \vec{u}_1(s,.)\|_{L^2_{uloc}} = 0. \tag{14.29}$$

  Indeed, as $\vec{u}_1$ is suitable, we use the local energy inequality o get that for all $\varphi \in \mathcal{B}$ and for all Lebesgue points t, τ of $\|\vec{u}\varphi\|_2$ with $t < \tau$, we have the inequality

$$\|\vec{u}(\tau,.)\varphi(x)\|_2^2 + 2\nu \int_t^\tau \|\vec{\nabla} \otimes \vec{u}(s,.))\varphi(x)\|_2^2 \, ds \leq$$

$$\|\vec{u}(t,.)\varphi(x)\|_2^2 + \nu \int\!\!\int_t^\tau |\vec{u}|^2 \, \Delta(\varphi^2(x)) \, dx \, ds + \int\!\!\int_t^\tau (|\vec{u}|^2 + 2(p - p_\varphi))(\vec{u}.\vec{\nabla})\varphi \, ds \, dx.$$

  Then, we use the weak continuity of $t \to \vec{u}(t,.)$ to conclude that this inequality is valid for almost all t and for all $\tau > t$. Thus, for all $\varphi \in \mathcal{B}$ and for all Lebesgue points t of $\|\vec{u}\varphi\|_2$, we have $\lim_{\tau > t, \tau \to t} \|(\vec{u}(\tau) - \vec{u}(t))\varphi\|_2 = 0$. Moreover, Theorem 14.3 implies a good uniform (in the time variable) control of the decay of $\|\vec{u}(\tau)\varphi_0(x - k)\|_2$ when k goes to ∞, whereas we have a good control of $\|(\vec{u}(\tau) - \vec{u}(t))\varphi_0(x - k)\|_2$ on the points t which are Lebesgue points for all $\|\vec{u}\varphi_0(x - k)\|_2, k \in \mathbb{Z}^3$; hence, for almost all t. Since $\sum_k \varphi_0(x - k) = 1$, this gives (14.29).

Step 2: the local solution with initial data in E_σ^3.

We now consider a time $t_1 \in (T_1/2, T_1)$ such that $\vec{u}_1(t_1, .) \in E^3$ and such that

$$\lim_{s \to t_1, s > t_1} \|\vec{u}_1(t_1, .) - \vec{u}_1(s, .)\|_{L_{uloc}^2} = 0.$$

In particular, $\vec{u}_1$ is a local Leray solution on (t_1, T_1) of the Cauchy problem for the Navier—Stokes equations with initial value $\vec{u}_1(t_1, .)$. We are going to associate to this Cauchy problem two other local Leray solutions.

First, we remark that $\tau \geq 0 \mapsto e^{\nu\tau\Delta}\vec{u}_1(t_1, .)$ is continuous from $[0, +\infty)$ to E^3 and

$$\sup_{0 < \tau < 1} \|e^{\nu\tau\Delta}\vec{u}_1(t_1, .)\|_{L_{uloc}^3} \leq \|\vec{u}_1(t_1, .)\|_{L_{uloc}^3}.$$

Moreover,

$$\sup_{0 < \tau < 1} \sqrt{\tau}\|e^{\nu\tau\Delta}\vec{u}_1(t_1, .)\|_\infty \leq C_\nu \|\vec{u}_1(t_1, .)\|_{L_{uloc}^3}$$

and

$$\lim_{\tau \to 0} \sqrt{\tau}\|e^{\nu\tau\Delta}\vec{u}_1(t_1, .)\|_\infty = 0.$$

In particular, $\tau^{1/4}e^{\nu\tau\Delta}\vec{u}(t_1, .) \in L^\infty((0,1), E^6)$ and

$$\lim_{\tau \to 0} \tau^{1/4}\|e^{\nu\tau\Delta}\vec{u}_1(t_1, .)\|_{L_{uloc}^6} = 0.$$

Finally, as $L_{uloc}^3 \subset L_{uloc}^2$, we have

$$\|e^{\nu\tau\Delta}\vec{u}_1(t_1, .)\|_{(L^2 H^1)_{uloc}((0,1) \times \mathbb{R}^3)} \leq C_\nu \|\vec{u}_1(t_1, .)\|_{L_{uloc}^3}.$$

Secondly, we write $\vec{W} = \int_{t_1}^t e^{\nu(t-s)\Delta}\mathbb{P}\operatorname{div}\mathbb{F}(s, .)\,ds$ for $t_1 < t < t_1 + 1$. We fix $\psi_0 \in \mathcal{D}$ such that ψ is non-negative and is identically equal to 1 on a neighborhood of the support of φ_0 ($\psi_0(x) = 1$ when the distance of x to the support of φ_0 is no more than 5). Writing

$$\varphi_0(x - k)|\vec{W}(t, .)|$$

$$\leq \left| \int_{t_1}^t e^{\nu(t-s)\Delta}\mathbb{P}\operatorname{div}(\psi_0(. - k)\mathbb{F}(s, .))\,ds \right| + C \int_{t_1}^t \int_{|x-y|>5} \frac{1}{|x-y|^4}|\mathbb{F}(s, y)|\,dy$$

$$\leq C' \int_{t_1}^t \frac{1}{\sqrt{\nu(t-s)}}\|\psi_0(-k)\mathbb{F}(s, .)\|_{H^2}\,ds$$

$$+ C' \sum_{j \in \mathbb{Z}^3, j \neq k} \frac{1}{|j-k|^4} \int_{t_1}^t \|\varphi_0(. - j)\mathbb{F}(s, .)\|_{H^2}\,ds$$

$$\leq C_\nu'' \|\mathbb{F}\|_{(L^3 H^2)_{uloc}((t_1, t_1+1) \times \mathbb{R}^3)}.$$

Thus, we find that $\vec{W}$ belongs to $L^\infty((t_1, t_1 + 1) \times \mathbb{R}^3)$ and that we have

$$\sup_{0 < \tau < 1} \|\vec{W}(t_1 + \tau, .)\|_{L_{uloc}^3} \leq C \sup_{0 < \tau < 1} \|\vec{W}(t_1 + \tau, .)\|_\infty \leq C_\nu \|\mathbb{F}\|_{(L^3 H^2)_{uloc}((t_1, t_1+1) \times \mathbb{R}^3)}$$

and

$$\lim_{\tau \to 0} \sqrt{\tau}\|\vec{W}(t_1 + \tau, .)\|_\infty = 0.$$

More precisely, $\vec{W}$ belongs to $\mathcal{C}([t_1, t_1 + 1], E^3)$ (due to the density of $L^3 H^2$ in $F^{2,3}$). Finally we check that

$$\varphi_0(x - k)|\vec{\nabla} \otimes \vec{W}|$$

$$\leq \left| \int_{t_1}^t e^{\nu(t-s)\Delta} \vec{\nabla} \otimes \mathbb{P} \operatorname{div}(\psi_0(. - k)F(s, .)) \, ds \right| + C \int_{t_1}^t \int_{|x-y|>5} \frac{1}{|x-y|^5} |\mathbb{F}(s, y)| \, dy$$

so that

$$\|\vec{W}\|_{(L^2 H^1)_{\text{uloc}}((t_1,t_1+1)\times\mathbb{R}^3)} \leq C_\nu \|\mathbb{F}\|_{L^2 L^2_{\text{uloc}}((t_1,t_1+1)\times\mathbb{R}^3)}.$$

Those estimates allow us to get a mild solution $\vec{v}_1$ in $\mathcal{C}([t_1, t_1 + \tau_1], E^3)$ of the Cauchy problem for the Navier—Stokes equations with initial value $\vec{u}_1(t_1, .)$, for τ_1 small enough. Indeed, we look for $\vec{v}_1$ as a solution of

$$\vec{v}_1(t, .) = e^{\nu(t-t_1)\Delta} \vec{u}_1(t_1, .) + \vec{W} - B(\vec{v}_1, \vec{v}_1)$$

where

$$B(\vec{v}, \vec{w}) = \int_{t_1}^t e^{\nu(t-s)\Delta} \mathbb{P} \operatorname{div}(\vec{v} \otimes \vec{w}) \, ds.$$

We have the estimates (uniformly in t for $t_1 < t < t_1 + \tau_1 \leq t_1 + 1$)

$$\|B(\vec{v}, \vec{w})(t, .)\|_\infty \leq C_\nu \int_{t_1}^t \frac{1}{(t-s)^{3/4}} \|\vec{v}(s, .)\|_{L^6_{\text{uloc}}} \|\vec{w}(s, .)\|_\infty \, ds$$

$$\leq C_{0,\nu} \frac{1}{\sqrt{t-t_1}} \sup_{t_1<s<t_1+\tau_1} |s - t_1|^{1/4} \|\vec{v}(s, .)\|_{L^6_{\text{uloc}}} \sup_{t_1<s<t_1+\tau_1} \sqrt{(s-t_1)} \|\vec{w}(s, .)\|_\infty$$

and

$$\|B(\vec{v}, \vec{w})(t, .)\|_{L^6_{\text{uloc}}} \leq C_\nu \int_{t_1}^t \frac{1}{(t-s)^{1/2}} \|\vec{v}(s, .)\|_{L^6_{\text{uloc}}} \|\vec{w}(s, .)\|_\infty \, ds$$

$$\leq C_{0,\nu} \frac{1}{|t-t_1|^{1/4}} \sup_{t_1<s<t_1+\tau_1} |s - t_1|^{1/4} \|\vec{v}(s, .)\|_{L^6_{\text{uloc}}} \sup_{t_1<s<t_1+\tau_1} \sqrt{(s-t_1)} \|\vec{w}(s, .)\|_\infty.$$

This gives the existence of a mild solution $\vec{v}_1$ on $[t_1, t_1 + \tau_1]$, provided that

$$\sup_{0<\tau<\tau_1} \sqrt{\tau} \|e^{\nu\tau\Delta}(t_1, .)\|_\infty + \sup_{0<\tau<\tau_1} |\tau|^{1/4} \|e^{\nu\tau\Delta}(t_1, .)\|_{L^6_{\text{uloc}}} < \frac{1}{8C_{0,\nu}}$$

and

$$\sup_{0<\tau<\tau_1} \sqrt{\tau} \|\vec{W}(t_1 + \tau, .)\|_\infty + \sup_{0<\tau<\tau_1} |\tau|^{1/4} \|\vec{W}(t_1 + \tau, .)\|_{L^6_{\text{uloc}}} < \frac{1}{8C_{0,\nu}}.$$

Moreover, this solution $\vec{v}_1$ belongs to $L^\infty([t_1, t_1 + \tau_1], E^3)$, since

$$\|B(\vec{v}, \vec{w})(t, .)\|_{E^3} \leq C_\nu \int_{t_1}^t \frac{1}{(t-s)^{1/2}} \|\vec{v}(s, .)\|_{E^3} \|\vec{w}(s, .)\|_{E^3} \, ds$$

$$\leq C'_\nu \sup_{t_1<s<t_1+\tau_1} |s - t_1|^{1/4} \|\vec{v}(s, .)\|_{E^6} \sup_{t_1<s<t_1+\tau_1} |s - t_1|^{1/4} \|\vec{w}(s, .)\|_{E^6}.$$

It belongs more precisely to $\mathcal{C}([t_1, t_1 + \tau_1], E^3)$: for the continuity of $B(\vec{v}_1, \vec{v}_1)$ in E^3, apply Theorem 8.1.

We remark finally that $\vec{v}_1$ belongs to $(L^2 H^1)_{\text{uloc}}((t_1, t_1 + \tau_1) \times \mathbb{R}^3)$, since

$$\|B(\vec{v}_1, \vec{v}_1)\|_{(L^2 H^1)_{\text{uloc}}((t_1, t_1+1) \times \mathbb{R}^3)} \leq C_\nu \|\vec{v}_1 \otimes \vec{v}_1\|_{L^2 L^2_{\text{uloc}}((t_1, t_1+1) \times \mathbb{R}^3)}$$

$$\leq C'_\nu \sqrt{\tau_1} \sup_{t_1 < s < t_1 + \tau_1} |s - t_1|^{1/4} \|\vec{v}_1(s, .)\|_{E^6} \sup_{t_1 < s < t_1 + \tau_1} \|\vec{v}_1(s, .)\|_{E^3}.$$

As $\vec{v}_1$ is regular enough, being locally $L^4 L^4$, it satisfies the energy equality, thus $\vec{v}_1$ is a local Leray solution as well.

We now construct a third local Leray solution $\vec{w}_1$ on $[t_1, t_1 + 1]$. We remark that we have the inequality

$$\|fg\|_2 \leq C \|f\|_{E^3} \|g\|_{H^1} \tag{14.30}$$

since

$$\|fg\|_2^2 \approx \sum_{k \in \mathbb{Z}^3} \|\varphi_0(x - k) fg\|_2^2, \quad \|g\|_{H^1}^2 \approx \sum_{k \in \mathbb{Z}^3} \|\varphi_0(x - k) g\|_{H^1}^2$$

and

$$\|\varphi_0(x - k) fg\|_2 \leq C \|\psi_0(x - k) f\|_3 \|\varphi_0(x - k) g\|_{H^1}.$$

We then decompose $\vec{u}_1(t_1, .)$ in $\vec{u}_1(t_1, .) = \vec{\alpha}_{1,t_1} + \vec{\beta}_{1,t_1}$ with $\vec{\alpha}_{1,t_1}$ small in E^3_σ and $\vec{\beta}_{1,t_1} \in E^3_{\text{comp}, \sigma}$, and we decompose $\mathbb{F}$ in $\mathbb{F} = \mathbb{G}_1 + \mathbb{H}_1$ with $\mathbb{G}_1$ small in $F^{2,3}((t_1, t_1 + 1) \times \mathbb{R}^3)$ and $\mathbb{H}_1 \in L^3((t_1, t_1 + 1), H^2)$. If $\vec{\alpha}_{1,t_1}$ and $\mathbb{G}_1$ are small enough, we have a (small) solution $\vec{\alpha}_1$ in $\mathcal{C}([t_1, t_1 + 1], E^3)$ of the equations

$$\begin{cases} \partial_t \vec{\alpha}_1 = \nu \Delta \vec{\alpha}_1 + \mathbb{P} \operatorname{div}(\mathbb{G}_1 - \vec{\alpha}_1 \otimes \vec{\alpha}_1), \\ \vec{\alpha}_1(t_1, .) = \vec{\alpha}_{1,t_1}. \end{cases}$$

In order to construct $\vec{w}_1$, we are going to construct $\vec{\beta}_1 = \vec{w}_1 - \vec{\alpha}_1$. Thus, we require $\vec{\beta}_1$ to be solution of the problem

$$\begin{cases} \partial_t \vec{\beta}_1 = \nu \Delta \vec{\beta}_1 + \mathbb{P} \operatorname{div}(\mathbb{H}_1 - \vec{\beta}_1 \otimes \vec{\beta}_1 - \vec{\alpha}_1 \otimes \vec{\beta}_1 - \vec{\beta}_1 \otimes \vec{\alpha}_1), \\ \vec{\beta}_1(t_1, .) = \vec{\beta}_{1,t_1}. \end{cases} \tag{14.31}$$

This is solved through the Leray mollification. We take $\theta \in \mathcal{D}(\mathbb{R}^3)$ with $\int \theta \, dx = 1$ and define, for $\epsilon > 0$, $\theta_\epsilon(x) = \frac{1}{\epsilon^3} \theta(\frac{x}{\epsilon})$. We replace the system (14.31) with

$$\begin{cases} \partial_t \vec{\beta}_{1,\epsilon} = \nu \Delta \vec{\beta}_{1,\epsilon} + \mathbb{P} \operatorname{div}(\mathbb{H}_1 - (\theta_\epsilon * \vec{\beta}_{1,\epsilon}) \otimes \vec{\beta}_{1,\epsilon} - \vec{\alpha}_1 \otimes \vec{\beta}_{1,\epsilon} - \vec{\beta}_{1,\epsilon} \otimes \vec{\alpha}_1), \\ \vec{\beta}_{1,\epsilon}(t_1, .) = \vec{\beta}_{1,t_1}. \end{cases} \tag{14.32}$$

For $\vec{U} \in L^2((t_1, t_1 + 1), H^1)$ and $\vec{V} = \int_{t_1}^t e^{\nu(t-s)\Delta} \mathbb{P} \operatorname{div}(\vec{\alpha}_1 \otimes \vec{U} - \vec{U} \otimes \vec{\alpha}_1) \, ds$, we have

$$\|\vec{V}\|_{L^\infty((t_1, t_1+1), L^2)} \leq \frac{2}{\sqrt{\nu}} \|\vec{U} \otimes \vec{\alpha}_1\|_{L^2 L^2} \leq \frac{C}{\sqrt{\nu}} \|\vec{\alpha}_1\|_{L^\infty E^3} \|\vec{U}\|_{L^2 H^1}$$

and

$$\|\vec{\nabla} \otimes \vec{V}\|_{L^2((t_1, t_1+1), L^2)} \leq \frac{2}{\nu} \|\vec{U} \otimes \vec{\alpha}_1\|_{L^2 L^2} \leq \frac{C}{\nu} \|\vec{\alpha}_1\|_{L^\infty E^3} \|\vec{U}\|_{L^2 H^1}.$$

As $\vec{\beta}_{1,t_1} \in L^2$ and $\mathbb{H}_1 \in L^2((t_1, t_1 + 1))$, Leray's formalism gives a solution $\vec{\beta}_{1,\epsilon} \in \mathcal{C}([t_1, t_1 + \tau_{1,\epsilon}], L^2) \cap L^2([t_1, t_1 + \tau_{1,\epsilon}], L^2)$ for a small interval $[t_1, t_1 + \tau_{1,\epsilon}]$ whose size depends on ϵ, on $\mathbb{H}$ and on $\|\vec{\beta}_{1,t_1}\|_2$, and this solution is then extended to $[t_1, t_1 + 1]$ as the L^2 norm of $\vec{\beta}_{1,\epsilon}$ can easily be controlled by the equality

$$\partial_t \|\vec{\beta}_{1,\epsilon}\|_2^2 + 2\nu \|\vec{\nabla} \otimes \vec{\beta}_{1,\epsilon}\|_2^2 = 2\int \vec{\alpha}_1 \cdot (\vec{\beta}_{1,\epsilon} \cdot \vec{\nabla} \vec{\beta}_{1,\epsilon}) \, dx - 2\int \mathbb{H}_1 \cdot (\vec{\nabla} \otimes \vec{\beta}_{1,\epsilon}) \, dx$$

and thus

$$\partial_t \|\vec{\beta}_{1,\epsilon}\|_2^2 + 2\nu \|\vec{\nabla} \otimes \vec{\beta}_{1,\epsilon}\|_2^2 \leq C\|\alpha_1\|_{E^3}\|\vec{\nabla} \otimes \vec{\beta}_{1,\epsilon}\|_2^2 + C\|\vec{\nabla} \otimes \vec{\beta}_{1,\epsilon}\|_2(\|\alpha_1\|_{E^3}\|\vec{\beta}_{1,\epsilon}\|_2 + \|\mathbb{H}_1\|_2).$$

if $\vec{\alpha}_1$ is small enough, we have $C\|\vec{\alpha}_1\|_{L^\infty E_3} < \nu$ and the L^2 norm of $\vec{\beta}_{1,\epsilon}$ is well controlled.

Thus, we have a solution $\vec{w}_{1,\epsilon} = \vec{\alpha}_1 + \vec{\beta}_{1,\epsilon}$ of the problem

$$\begin{cases} \partial_t \vec{w}_{1,\epsilon} = \nu \Delta \vec{w}_{1,\epsilon} + \mathbb{P} \operatorname{div}(\mathbb{F}_1 - \vec{w}_{1,\epsilon} \otimes \vec{w}_{1,\epsilon} - (\theta_\epsilon * \vec{\beta}_{1,\epsilon}) \otimes \vec{\beta}_{1,\epsilon} + \vec{\beta}_{1,\epsilon} \otimes \vec{\beta}_{1,\epsilon}), \\ \vec{w}_{1,\epsilon}(t_1, .) = \vec{u}_1(t_1, .). \end{cases}$$

As $\vec{w}_{1,\epsilon}$ is controlled in $(L^2 H^1)_{\text{uloc}}$ and $\partial_t \vec{w}_{1,\epsilon}$ is controlled in $(L^2 H^{-2})_{\text{uloc}}$, we can apply the Rellich–Lions theorem (Theorem 12.1): we may find a sequence $\epsilon_n \to 0$ and a function $\vec{w}_1 \in (L^\infty L^2)_{\text{uloc}} \cap (L^2 H^1)_{\text{uloc}}$ such that:

- $\vec{w}_{1,\epsilon_n}$ is *-weakly convergent to $\vec{w}_1$ in $(L^\infty L^2)_{\text{uloc}}$ and in $(L^2 H^1)_{\text{uloc}}$
- $\vec{w}_{1,\epsilon_n}$ is strongly convergent to $\vec{w}_1$ in $L^2_{\text{loc}}((0, T) \times \mathbb{R}^3)$.

It is then easy to check that $\vec{w}_1$ is a solution of

$$\begin{cases} \partial_t \vec{w}_1 = \nu \Delta \vec{w}_1 + \mathbb{P} \operatorname{div}(\mathbb{F} - \vec{w}_1 \otimes \vec{w}_1), \\ \vec{w}_1(t_1, .) = \vec{u}_1(t_1, .). \end{cases}$$

We then follow the same lines as for the proof of Theorem 14.1 and find that $\vec{w}_1$ is suitable:

$$\partial_t(\frac{|\vec{w}_1|^2}{2}) \leq \nu \Delta(\frac{|\vec{w}_1|^2}{2}) - \nu |\vec{\nabla} \otimes \vec{w}_1|^2 - \operatorname{div}\left((p_1 + \frac{|\vec{w}_1|^2}{2})\vec{w}_1\right) + \vec{w}_1 \cdot \operatorname{div} \mathbb{F} \quad (14.33)$$

with $\vec{\nabla} p_1 = (Id - \mathbb{P}) \operatorname{div}(\mathbb{F} - \vec{w}_1 \otimes \vec{w}_1)$. Another useful property of $\vec{w}$ is that $\vec{w} - \vec{\alpha}_1 \in L^\infty L^2 \cap L^2 HH^1 \subset L^4 L^3$, so that $\vec{w}_1 \in L^4 E^3$.

Thus, on for $t_1^* = \min(t_1 + \tau_1, T_1)$, we have three local Leray solutions $\vec{u}_1$, $\vec{v}_1$ and $\vec{w}_1$ on $(t_1, t_1^*) \times \mathbb{R}^3$, to the same problem

$$\begin{cases} \partial_t \vec{u} = \nu \Delta \vec{u} + \mathbb{P} \operatorname{div}(\mathbb{F} - \vec{u} \otimes \vec{u}) \\ \\ \vec{u}(t_1, .) = \vec{u}_1(t_1, .) \end{cases} \quad (14.34)$$

where $\vec{u}_1(t_1, .) \in L^2_{\text{uloc}}$ with $\operatorname{div} \vec{u}_1(t_1, .) = 0$ and $\mathbb{F} \in (L^2_t L^2_x)_{\text{uloc}}((0, T) \times \mathbb{R}^3)$. As $\vec{v}_1 \in \mathcal{C}([t_1, t_1^*], E^3)$, it can be written as $\vec{v}_1 = \vec{v}_{1,1} + \vec{v}_{1,2}$ with $\vec{v}_{1,1} \in L^2((t_1, t_1^*), L^\infty)$ and $\vec{v}_{1,2} \in L^\infty((0, T), E^3)$ with $\|\vec{v}_{1,2}(t, .)\|_{E^3}$ small enough. By the strong-weak uniqueness theorem (Theorem 14.7), we have $\vec{u}_1 = \vec{v}_1$ on (t_1, t_1^*), and similarly $\vec{w}_1 = \vec{v}_1$ on (t_1, t_1^*).

Let $T_2 = t_1 + 1$. We get a local Leray solution $\vec{u}_2$ for the Navier—Stokes problem on $(0, T_2) \times \mathbb{R}^3$ by defining $\vec{u}_2(t, x) = \vec{u}_1(t, x)$ for $0 < t < T_1$ and $\vec{u}_2(t, x) = \vec{w}_1(t, x)$ for $t_1 < t < t_1 + 1$.

Step 3: the global solution. Assume that, for some $N \geq 2$, we have a local Leray solution $\vec{u}_N$ for the Navier–Stokes problem on $(0, T_N) \times \mathbb{R}^3$ with $T_N > 1$, such that $\vec{u}_N \in L^3((0, T_N), E^3)$. We consider a time $t_N \in (T_N - \frac{1}{2}, T_N)$ such that $\vec{u}_N(t_N, .) \in E^3$ and such that

$$\lim_{s \to t_N, s > t_N} \|\vec{u}_N(t_N, .) - \vec{u}_N(s, .)\|_{L^2_{\mathrm{uloc}}} = 0.$$

so that $\vec{u}_N$ is a local Leray solution on (t_N, T_N) of the Cauchy problem for the Navier–Stokes equations with initial value $\vec{u}_N(t_N, .)$. We associate to this Cauchy problem two other local Leray solutions, a solution $\vec{v}_N$ such that $\vec{v}_N \in ([t_N, t_N + \tau_N), E^3)$ and a solution $\vec{w}_N$ defined on $[t_N, t_N + 1]$. For $t_N^* = \min(T_N, t_N + \tau_N)$, using the strong-weak uniqueness theorem, we find that $\vec{u}_N = \vec{v}_N = \vec{w}_N$ on (t_N, t_N^*). Thus, we have a solution $\vec{u}_{N+1}$ on $(0, T_{N+1}$, with $T_{N+1} = t_N + 1$, with $\vec{u}_{N+1}(t, x) = \vec{u}_N(t, x)$ for $0 < t < T_N$ and $\vec{u}_{N+1}(t, x) = \vec{w}_N(t, x)$ for $t_N < t < t_N + 1$.

As $T_{N+1} \geq T_N + \frac{1}{2}$, we have $\lim_{N \to +\infty} T_N = +\infty$. We then have a global solution by defining $\vec{u}(t, x) = \vec{u}_2(t, x)$ on $(0, T_2)$, and $\vec{u}(t, x) = \vec{u}_{N+1}(t, x)$ on $(T_N, T_{N+1}]$. $\qquad \square$

14.6 Weighted Estimates

Local Leray solutions to the Navier–Stokes equations allowed Jia and Šverák [245] to construct in 2014 self-similar solutions for large (homogeneous of degree -1) smooth data. Their result has been extended in 2016 by Lemarié-Rieusset [319] to solutions for rough locally square integrable data. We remark that an homogeneous (of degree -1) and locally square integrable data is automatically uniformly locally L^2.

Recently, Bradshaw and Tsai [57] and Chae and Wolf [100] considered the case of solutions which are self-similar according to a discrete subgroup of dilations. Those solutions are related to an initial data which is self-similar only for a discrete group of dilations; in contrast to the case of self-similar solutions for all dilations, such an initial data, when locally L^2, is not necessarily uniformly locally L^2, therefore their results are no consequence of the theory of local Leray solutions.

In this section, we follow Fernández-Dalgo and Lemarié-Rieusset [173] and construct an alternative theory to obtain infinite-energy global weak solutions for large initial data, which include the discretely self-similar locally square integrable data. More specifically, we consider the weight

$$\Phi(x) = \frac{1}{1 + |x|^2}$$

and the space

$$L^2_\Phi = L^2(\Phi \, dx).$$

(The construction by Bradshaw, Kukavica and Tsai is very similar [56] in the sightly more general condition $\lim_{R \to +\infty} \frac{1}{R^2} \int_{B(0,R)} |\vec{u}_0(x)|^2 \, dx = 0$. We prefer to work in the space L^2_Φ, as it is a Hilbert space, so that results are easier to state.) In this context, we adapt the definition of local Leray solution into the following one:

Definition 14.4 (Weighted Leray solution).
Let $\vec{u}_0 \in L^2_\Phi$ (where $\Phi(x) = \frac{1}{1+|x|^2}$) with div $\vec{u}_0 = 0$ *and* $\mathbb{F} \in L^2((0, T), L^2_\Phi(\mathbb{R}^3))$. *A weak*

solution $\vec{u}$ on $(0,T) \times \mathbb{R}^3$ to the problem

$$\begin{cases} \partial_t \vec{u} = \nu \Delta \vec{u} + \mathbb{P} \operatorname{div}(\mathbb{F} - \vec{u} \otimes \vec{u}) \\ \\ \vec{u}(0,.) = \vec{u}_0 \end{cases} \qquad (14.35)$$

is a **weighted Leray solution** if it satisfies the following requirements:

- $\vec{u} \in L^\infty((0,T), L^2_\Phi)$

- $\vec{\nabla} \otimes \vec{u} \in L^2((0,T), L^2_\Phi)$

- $\vec{u}$ is suitable

- $\lim_{t\to 0^+} \int |\vec{u}(t,x) - \vec{u}_0(t,x)|^2 \, \Phi(x) \, dx = 0.$

Theorem 14.1 becomes:

Weighted square integrable solutions

Theorem 14.9.
Let $\vec{u}_0 \in L^2_\Phi$ (where $\Phi(x) = \frac{1}{1+|x|^2}$) with $\operatorname{div} \vec{u}_0 = 0$ and $\mathbb{F} \in L^2((0,T), L^2_\Phi(\mathbb{R}^3))$. Then, if $T < +\infty$, there exists a weighted Leray solution $\vec{u}$ to the problem

$$\begin{cases} \partial_t \vec{u} = \nu \Delta \vec{u} + \mathbb{P} \operatorname{div}(\mathbb{F} - \vec{u} \otimes \vec{u}) \\ \\ \vec{u}(0,.) = \vec{u}_0 \end{cases} \qquad (14.36)$$

on $(0,T) \times \mathbb{R}^3$.
   If $T = +\infty$, there exists a global solution $\vec{u}$ which is a weighted Leray solution on every bounded interval $(0, T_0)$.

Proof. The proof of the theorem in the case of L^2_Φ wil be simpler than the proof in the case of L^2_{uloc}, as we may use more easily the Riesz transforms, which are bounded on $L^2(\Phi) = L^2(\frac{1}{1+|x|^2} \, dx)$ and on $L^{4/3}(\frac{1}{(1+|x|^2)^{4/3}} \, dx)$ (since the weight $\frac{1}{(1+|x|^2)^\gamma}$ belongs to the Muckenhoupt class $\mathcal{A}_p(\mathbb{R}^3)$ for every $\gamma \in [0, 3/2)$ and every $p \in (1, +\infty)$ [215]).

Step 1: local existence.
   In contrast with the proof for Leray solutions in the L^2 case or for local Leray solutions in the case of L^2_{uloc}, it is useless to start with a mollification of the non-linearity. The mollified equation (14.4), i.e.

$$\begin{cases} \partial_t \vec{u} = \nu \Delta \vec{u} + \mathbb{P}(\operatorname{div} \mathbb{F} - (\theta_\epsilon * \vec{u}) \cdot \vec{\nabla} \vec{u}) \\ \\ \vec{u}(0,.) = \vec{u}_0 \end{cases}$$

cannot be solved by Picard's iterative scheme as the operator

$$B_\epsilon(\vec{U}, \vec{V}) = \int_0^t W_{\nu(t-s)} * \mathbb{P} \operatorname{div}((\theta_\epsilon * \vec{U}) \otimes \vec{V}) \, ds$$

is not bounded on $L^\infty_t L^2_\Phi$: the mollified drift $\theta_\epsilon * \vec{U}$ will not belong to $L^\infty_{t,x}$.

The simplest way to get a solution is by approximating the problem with the problem in L^2; we take a function $\theta \in \mathcal{D}(\mathbb{R}^3)$ which is equal to 1 on $B(0,1)$ and, for $R > 1$, we define $\vec{u}_{0,R} = \mathbb{P}(\theta(\frac{x}{R})\vec{u}_0)$ and consider the Cauchy problem

$$\begin{cases} \partial_t \vec{u}_R = \nu\Delta\vec{u}_R + \mathbb{P}(\operatorname{div}(\theta(\frac{x}{R})\mathbb{F}) - \vec{u}_R \cdot \vec{\nabla}\vec{u}_R) \\ \\ \vec{u}_R(0,.) = \vec{u}_{0,R} \end{cases}$$

As $\vec{u}_{0,R}$ belongs to L^2, we know, by Theorems 12.2 and 13.6, that we may find a suitable weak Leray solution $\vec{u}_R$ defined on $(0,T)$. We are going to estimate $\vec{u}_R$ and $\vec{\nabla} \otimes \vec{u}_R$ in L^2_Φ independently from R and then let R go to $+\infty$.

In order to control $\vec{u}_R$, we shall use the local energy inequality (inequality (14.3)). Let $0 < t_0 < T$. We use the test function

$$\Omega(t,x) = \Phi(x)\theta_\epsilon(t)\varphi^2(x/S) \tag{14.37}$$

where $\varphi \in \mathcal{D}(\mathbb{R}^3)$ satisfies $\varphi(x) = 1$ on $B(0,1)$, and where

$$\theta_\epsilon(t) = \alpha(\frac{t-\epsilon}{\epsilon}) - \alpha(\frac{t-t_0+2\epsilon}{\epsilon})$$

with α a smooth non-decreasing function on $\mathbb{R}$ such that $\alpha(s) = 0$ when $s \le 1$ and $\alpha(s) = 1$ for $s \ge 2$, $S > 0$, $0 < \epsilon < t_0/3$.

Integrating inequality (14.3) against $\Omega(t,x)$, we find that:

$$-\frac{1}{2}\int_0^T \theta_\epsilon'(t)\|\vec{u}_R(t,.)\varphi(\frac{\cdot}{S})\|_{L^2_\Phi}^2 \, dt \le$$

$$\frac{\nu}{2S^2}\int_0^T \theta_\epsilon(t)(\int |\vec{u}_R(t,x)|^2\Delta(\varphi^2)(\frac{x}{S})\Phi(x)\, dx)\, dt$$

$$+\frac{\nu}{S}\int_0^T \theta_\epsilon(t)(\int \varphi(\frac{x}{S})\Phi(x)\vec{\nabla}\vec{u}_R(t,x)\cdot\vec{\nabla}\varphi(\frac{x}{S})\, dx)\, dt$$

$$+\frac{\nu}{2}\int_0^T \theta_\epsilon(t)(\int |\vec{u}_R(t,x)|^2\varphi^2(\frac{x}{S})\Delta\Phi(x)\, dx)\, dt$$

$$-\nu\int_0^T \theta_\epsilon(t)\|\varphi(\frac{x}{S})(\vec{\nabla}\otimes\vec{u}_R)\|_{L^2_\Phi}^2 \, dt$$

$$+\frac{2}{S}\int_0^T \theta_\epsilon(t)(\int \Phi(x)\varphi(\frac{x}{S})(p_R + \frac{1}{2}|\vec{u}_R|^2)\vec{u}_R\cdot\vec{\nabla}\varphi(\frac{x}{S})\, dx)\, dt$$

$$+\int_0^T \theta_\epsilon(t)(\int \varphi^2(\frac{x}{S})(p_R + \frac{1}{2}|\vec{u}_R|^2)\vec{u}_R\cdot\vec{\nabla}\Phi(x)\, dx)\, dt$$

$$-\int_0^T \theta_\epsilon(t)(\int \Phi(x)\varphi^2(\frac{x}{S})\,(\vec{\nabla}\otimes\vec{u}_R)\cdot\mathbb{F}\, dx)\, dt$$

$$-\frac{2}{S}\int_0^T \theta_\epsilon(t)(\int \Phi(x)\varphi(\frac{x}{S})\,(\vec{\nabla}\varphi(\frac{x}{S})\otimes\vec{u}_R)\cdot\mathbb{F}\, dx)\, dt$$

$$-\int_0^T \theta_\epsilon(t)(\int \varphi^2(\frac{x}{S})\,(\vec{\nabla}\Phi(x)\otimes\vec{u}_R)\cdot\mathbb{F}\, dx)\, dt$$

Noticing that $|\vec{\nabla}\Phi| \le 2\Phi^{3/2} \le 2\Phi$ and $|\Delta\Phi| \le 6\Phi^{3/2} \le 6\Phi$, and letting S go to ∞, we get

$$-\frac{1}{2}\int_0^T \theta_\epsilon'(t)\|\vec{u}_R(t,.)\|_{L_\Phi^2}^2\,dt \le \frac{\nu}{2}\int_0^T \theta_\epsilon(t)(\int |\vec{u}_R(t,x)|^2 \Delta\Phi(x)\,dx)\,dt$$

$$-\nu\int_0^T \theta_\epsilon(t)\|\vec{\nabla}\otimes\vec{u}_R\|_{L_\Phi^2}^2\,dt$$

$$+\int_0^T \theta_\epsilon(t)(\int (p_R + \frac{1}{2}|\vec{u}_R|^2)\vec{u}_R\cdot\vec{\nabla}\Phi(x)\,dx)\,dt$$

$$-\int_0^T \theta_\epsilon(t)(\int \Phi(x)\,(\vec{\nabla}\otimes\vec{u}_R)\cdot\mathbb{F}\,dx)\,dt$$

$$-\int_0^T \theta_\epsilon(t)(\int (\vec{\nabla}\Phi(x)\otimes\vec{u}_R)\cdot\mathbb{F}\,dx)\,dt$$

$$\le 3\nu\int_0^T \theta_\epsilon(t)\|\vec{u}_R(t,.)\|_{L_\Phi^2}^2\,dt$$

$$-\nu\int_0^T \theta_\epsilon(t)\|\vec{\nabla}\otimes\vec{u}_R\|_{L_\Phi^2}^2\,dt$$

$$+2\int_0^T \int \theta_\epsilon(t)(|p_R| + \frac{1}{2}|\vec{u}_R|^2)|\vec{u}_R|\Phi^{3/2}(x)\,dx\,dt$$

$$+\int_0^T \int \theta_\epsilon(t)\Phi(x)|\vec{\nabla}\otimes\vec{u}_R|\,|\mathbb{F}|\,dx\,dt$$

$$+2\int_0^T \int \theta_\epsilon(t)|\vec{u}_R|\Phi(x)\,dx\,dt$$

If t_0 is a Lebesgue point of $t \mapsto \|\vec{u}_R(t,.)\|_{L_\Phi^2}^2$ (and as 0 is a continuity point of $t \mapsto \|\vec{u}_R(t,.)\|_{L_\Phi^2}^2$), we find that

$$\|\vec{u}_R(t_0,.)\|_{L_\Phi^2}^2 \le \|\vec{u}_{0,R}\|_{L_\Phi^2}^2 - 2\nu\int_0^{t_0}\|\vec{\nabla}\otimes\vec{u}_R\|_{L_\Phi^2}^2\,dt + 6\nu\int_0^{t_0}\|\vec{u}_R\|_{L_\Phi^2}^2\,dt$$

$$+2\int_0^{t_0}\|\vec{\nabla}\otimes\vec{u}_R\|_{L_\Phi^2}\|\mathbb{F}\|_{L_\Phi^2}\,dt + 4\int_0^{t_0}\|\vec{u}_R\|_{L_\Phi^2}\|\mathbb{F}\|_{L_\Phi^2}\,dt \qquad (14.38)$$

$$+4\int_0^{t_0}\int (|p_R| + \frac{1}{2}|\vec{u}_R|^2)|\vec{u}_R|\Phi^{3/2}(x)\,dx\,dt$$

This inequality is thus satisfied for almost every t_0, and even for every t_0 as $t \mapsto \vec{u}_R(t,.)$ is weakly continuous from $[0,T)$ to L^2, hence from $[0,T)$ to L_Φ^2.

We then write

$$\Delta p_R + \sum_{i=1}^3\sum_{j=1}^3 \partial_i\partial_j(u_{i,R}u_{j,R}) = \sum_{i=1}^3\sum_{j=1}^3 \partial_i\partial_j F_{i,j}$$

so that

$$p_R = q_R - \varpi = \sum_{i=1}^3\sum_{j=1}^3 R_iR_j(u_{i,R}u_{j,R}) - \sum_{i=1}^3\sum_{j=1}^3 R_iR_j F_{i,j}.$$

As the Riesz transforms are bounded on L_Φ^2, we find that

$$\|\varpi\|_{L_\Phi^2} \le C\|\mathbb{F}\|_{L_\Phi^2}.$$

Moreover, $\sqrt{\Phi}\vec{u}_R$ is controlled in H^1, hence in L^3:

$$\begin{aligned}
\|\sqrt{\Phi}\vec{u}_R\|_3 &\leq C\|\sqrt{\Phi}\vec{u}_R\|_2^{1/2}\|\vec{\nabla}\otimes(\sqrt{\Phi}\vec{u}_R)\|_2^{1/2}\\
&\leq C\|\sqrt{\Phi}\vec{u}_R\|_2^{1/2}(\|\sqrt{\Phi}\vec{u}_R\|_2^{1/2}+\|\sqrt{\Phi}\vec{\nabla}\otimes\vec{u}_R\|_2^{1/2})\\
&= C\|\vec{u}_R\|_{L_\Phi^2}^{1/2}(\|\vec{u}_R\|_{L_\Phi^2}^{1/2}+\|\vec{\nabla}\otimes\vec{u}_R\|_{L_\Phi^2}^{1/2}).
\end{aligned}$$

In particular, $\Phi u_{i,R}u_{j,R}$ is controlled in $L^{3/2}$, hence $u_{i,R}u_{j,R}$ is controlled in $L^{3/2}(\Phi^{3/2})$. But this control cannot be transferred to a control on q_R as the Riesz transforms are not bounded on $L^{3/2}(\Phi^{3/2})$.

Instead, we shall use the control of $\sqrt{\Phi}\vec{u}_R$ in $L^{\frac{8}{3}}$ to control Φq_R in $L^{4/3}$ (as the Riesz transforms are bounded on $L^{4/3}(\Phi^{4/3}\,dx)$) and the control of $\sqrt{\Phi}\vec{u}_R$ in L^4 to control $\int_0^{t_0}\int|q_R||\vec{u}_R|\Phi^{3/2}\,dx\,dt$. We have

$$\begin{aligned}
\|\sqrt{\Phi}\vec{u}_R\|_{8/3} &\leq \|\sqrt{\Phi}\vec{u}_R\|_2^{5/8}\|\sqrt{\Phi}\vec{u}_R\|_6^{3/8}\\
&\leq C\|\sqrt{\Phi}\vec{u}_R\|_2^{5/8}\|\vec{\nabla}\otimes(\sqrt{\Phi}\vec{u}_R)\|_2^{3/8}\\
&\leq C\|\vec{u}_R\|_{L_\Phi^2}^{5/8}(\|\vec{u}_R\|_{L_\Phi^2}^{3/8}+\|\vec{\nabla}\otimes\vec{u}_R\|_{L_\Phi^2}^{3/8})
\end{aligned}$$

and

$$\begin{aligned}
\|\sqrt{\Phi}\vec{u}_R\|_4 &\leq \|\sqrt{\Phi}\vec{u}_R\|_2^{1/4}\|\sqrt{\Phi}\vec{u}_R\|_6^{3/4}\\
&\leq C\|\vec{u}_R\|_{L_\Phi^2}^{1/4}(\|\vec{u}_R\|_{L_\Phi^2}^{3/4}+\|\vec{\nabla}\otimes\vec{u}_R\|_{L_\Phi^2}^{3/4}).
\end{aligned}$$

Thus far, (14.38) becomes

$$\begin{aligned}
\|\vec{u}_R(t_0,.)\|_{L_\Phi^2}^2 \leq &\|\vec{u}_{0,R}\|_{L_\Phi^2}^2 - 2\nu\int_0^{t_0}\|\vec{\nabla}\otimes\vec{u}_R\|_{L_\Phi^2}^2\,dt + 6\nu\int_0^{t_0}\|\vec{u}_R\|_{L_\Phi^2}^2\,dt\\
&+ C_0\int_0^{t_0}\|\vec{\nabla}\otimes\vec{u}_R\|_{L_\Phi^2}\|\mathbb{F}\|_{L_\Phi^2}\,dt + C_1\int_0^{t_0}\|\vec{u}_R\|_{L_\Phi^2}\|\mathbb{F}\|_{L_\Phi^2}\,dt\\
&+ C_2\int_0^{t_0}\|\vec{u}_R\|_{L_\Phi^2}^3\,dt + C_3\int_0^{t_0}\|\vec{u}_R\|_{L_\Phi^2}^{3/2}\|\vec{\nabla}\otimes\vec{u}_R\|_{L_\Phi^2}^{3/2}\,dt
\end{aligned}$$

and finally

$$\begin{aligned}
\|\vec{u}_R(t_0,.)\|_{L_\Phi^2}^2 \leq &\|\vec{u}_{0,R}\|_{L_\Phi^2}^2 - \nu\int_0^{t_0}\|\vec{\nabla}\otimes\vec{u}_R\|_{L_\Phi^2}^2\,dt\\
&+ \frac{25}{4}\nu\int_0^{t_0}\|\vec{u}_R\|_{L_\Phi^2}^2\,dt + \frac{2C_0^2+C_1^2}{\nu}\int_0^{t_0}\|\mathbb{F}\|_{L_\Phi^2}^2\,dt \qquad (14.39)\\
&+ C_2\int_0^{t_0}\|\vec{u}_R\|_{L_\Phi^2}^3\,dt + \frac{C_3^4}{4\nu^3}\int_0^{t_0}\|\vec{u}_R\|_{L_\Phi^2}^6\,dt.
\end{aligned}$$

As the Riesz transforms are bounded on L_Φ^2, we know that $\|\vec{u}_{0,R}\|_{L_\Phi^2}\leq A_0\|\vec{u}_0\|_{L_\Phi^2}$. Let

$$A_1^2 = A_0^2\|\vec{u}_0\|_{L_\Phi^2}^2 + \frac{2C_0^2+C_1^2}{\nu}\int_0^T\|\mathbb{F}\|_{L_\Phi^2}^2\,dt. \qquad (14.40)$$

From (14.39), we see that we have

$$\|\vec{u}_R(t_0,.)\|_{L_\Phi^2}^2 + \nu\int_0^{t_0}\|\vec{\nabla}\otimes\vec{u}_R\|_{L_\Phi^2}^2\,dt \leq 4A_1^2 \qquad (14.41)$$

as long as

$$t_0\left(25\nu + 8C_2 A_1 + \frac{16C_3^4}{\nu^3} A_1^4\right) \leq 3.$$

As $8C_2 A_1 \leq 6\nu + 2C_2^4 \frac{A_1^4}{\nu^3}$, we see that, on $(0, T_0)$ with

$$T_0 = \frac{3}{31\nu + \frac{16C_3^4 + 2C_2^4}{\nu^3} A_1^4},$$

inequality (16.23) is fulfilled.

As the control (16.23) does not depend on R, we may apply the Rellich–Lions theorem (Theorem 12.1) and get a sequence R_k such that $\vec{u}_{R_k}$ is strongly convergent to a limit $\vec{u}$ in $L^2_{\text{loc}}([0, T_0) \times \mathbb{R}^3)$. Moreover, $\vec{u}_{R_k}$ is weakly-* convergent to $\vec{u}$ in $L^\infty((0, T_0), L^2_\Phi)$, $\vec{\nabla} \otimes \vec{u}_{R_k}$ is weakly convergent to $\vec{\nabla} \otimes \vec{u}$ in $L^2((0, T_0), L^2_\Phi)$ and $p_{R_k} = q_{R_k} - \varpi$ is convergent to $p = q - \varpi$ with weak convergence of q_{R_k} to q in $L^{8/3}((0, T_0), L^{4/3}(\Phi^{4/3}\, dx))$. In particular, we find that $\vec{u}$ is solution to

$$\partial_t \vec{u} + \vec{u} \cdot \vec{\nabla} \vec{u} = \nu \Delta \vec{u} - \vec{\nabla} p + \operatorname{div} \mathbb{F}$$

and

$$\operatorname{div} \vec{u} = 0$$

on $(0, T_0) \times \mathbb{R}^3$. Moreover, we know by Theorem 6.2, that the limit $\vec{u}$ (as a weak limit of uniformly controlled suitable solutions) satisfies the local Leray energy inequality (i.e. $\vec{u}$ is suitable).

As the Riesz transforms are bounded on L^2_Φ, we have as well the strong convergence of $\vec{u}_{0,R}$ to $\vec{u}_0$ in L^2. For every $\varphi \in \mathcal{D}(\mathbb{R}^3)$, $\varphi \vec{u}_{R_k}$ belongs to $L^2((0, T_0), H^1)$ and $\varphi \partial_t \vec{u}_{R_k}$ belongs to $L^2((0, T_0), H^{-3/2})$, so that (from Lemma (6.1)), we can represent $\vec{u}_{R_k}$ as

$$\vec{u}_{R_k}(t, .) = \vec{u}_{0,R_k} + \int_0^t \partial_t \vec{u}_{r_k}(s, .)\, ds$$

(so that $\varphi \vec{u}_{R_k} \in \mathcal{C}([0, T_0], H^{-3/2})$). Moreover, $\varphi \partial_t \vec{u}_{R_k}$ is bounded in $L^2((0, T_0), H^{-3/2})$, hence the weak convergence of $\vec{u}_{R_k}$ to $\vec{u}$ gives the weak convergence of $\varphi \partial_t \vec{u}_{R_k}$ to $\varphi \partial_t \vec{u}$ in $L^2((0, T_0), H^{-3/2})$, and then the weak convergence of $\varphi \vec{u}_{R_k}(t, .)$ to $\varphi \vec{u}(t, .)$ in $H^{-3/2}$; finally, we get the weak convergence of $\vec{u}_{R_k}(t, .)$ to $\vec{u}(t, .) = \vec{u}_0 + \int_0^t \partial_t \vec{u}(s, .)\, ds$ in L^2_Φ as $\vec{u}_{R_k}(t, .)$ is bounded in L^2_Φ.

For fixed t, we thus have the weak convergence of

$$(\vec{u}_{R_k}(t, .), 1_{0<s<t} \vec{\nabla} \otimes \vec{u}_{R_k}(s, .))$$

in $L^2_\Phi \times L^2((0, t), L^2_\Phi)$ and thus

$$\|\vec{u}(t, .)\|^2_{L^2_\Phi} + \nu \|\vec{\nabla} \otimes \vec{u}\|^2_{L^2((0,t), L^2_\Phi)}$$

$$\leq \liminf_{k \to +\infty} \|\vec{u}_{R_k}(t, .)\|^2_{L^2_\Phi} + 2\nu \|\vec{\nabla} \otimes \vec{u}_{R_k}\|^2_{L^2((0,t), L^2_\Phi)}$$

$$\leq \|\vec{u}_0\|^2_{L^2_\Phi} + t A_1^2 \left(25\nu + 8C_2 A_1 + \frac{16C_3^4}{\nu^3} A_1^4\right).$$

Thus, $\limsup_{t \to 0} \|\vec{u}(t, .)\|_{L^2_\Phi} \leq \|\vec{u}_0\|_{L^2_\Phi}$. As $t \mapsto \vec{u}(t, .) = \vec{u}_0 + \int_0^t \partial_t \vec{u}(s, .)\, ds$ is continuous from $[0, T_0]$ to $\mathcal{D}'$ and is bounded in L^2_Φ, it is weakly continuous to L^2_Φ, and $\vec{u}_0$ is the weak limit of $\vec{u}(t, .)$ ass t decreases to 0. Thus, $\|\vec{u}_0\|_{L^2_\Phi} \leq \liminf_{t \to 0} \|\vec{u}(t, .)\|_{L^2_\Phi}$.

Hence, we have $\|\vec{u}_0\|_{L^2_\Phi} = \lim_{t\to 0} \|\vec{u}(t,.)\|_{L^2_\Phi}$, and this turns the weak convergence in L^2_Φ into strong convergence:

$$\liminf_{t\to 0} \|\vec{u}(t,.) - \vec{u}_0\|_{L^2_\Phi} = 0.$$

We have proved the existence of a weighted Leray solution on $(0, T_0)$.

Step 2: size estimates for weighted Leray solutions.

We have shown the existence of a weighted Leray solution $\vec{u}$ on $(0, T_0)$, where

$$T_0 = \frac{3}{31\nu + \frac{16C_3^4 + 2C_2^4}{\nu^3} \left(A_0^2 \|\vec{u}_0\|_{L^2_\Phi}^2 + \frac{2C_0^2 + C_1^2}{\nu} \int_0^T \|\mathbb{F}\|_{L^2_\Phi}^2 \, dt \right)^2},$$

which is of the order of magnitude

$$T_0 \approx C_5 \frac{\nu^5}{\nu^6 + \nu^2 \|\vec{u}_0\|_{L^2_\Phi}^4 + (\int_0^T \|\mathbb{F}\|_{L^2_\Phi}^2 \, dt)^2}.$$

Now, if $\vec{v}$ is another weighted Leray solution on $(0, T_0)$ (with associated pressure q), we show that we have a control of $\vec{v}$ in $L^\infty((0, T_0), L^2_\Phi)$ and of $\vec{\nabla} \otimes \vec{v}$ in $L^2((0, T_0), L^2_\Phi)$

In order to control $\vec{v}$, we shall use again the local energy inequality (inequality (14.3)). Let $0 < t_0 < T_0$. We use the test function $\Omega(t, x) = \Phi(x)\theta_\epsilon(t)\varphi^2(x/S)$, where $\varphi \in \mathcal{D}(\mathbb{R}^3)$ satisfies $\varphi(x) = 1$ on $B(0, 1)$, and where

$$\theta_\epsilon(t) = \alpha(\frac{t - \epsilon}{\epsilon}) - \alpha(\frac{t - t_0 + 2\epsilon}{\epsilon})$$

with α a smooth non-decreasing function on $\mathbb{R}$ such that $\alpha(s) = 0$ when $s \leq 1$ and $\alpha(s) = 1$ for $s \geq 2$, $S > 0$, $0 < \epsilon < t_0/3$.

Integrating inequality (14.3) against $\Omega(t, x)$, we find that:

$$-\frac{1}{2} \int_0^T \theta'_\epsilon(t) \|\vec{v}(t,.)\varphi(\frac{\cdot}{S})\|_{L^2_\Phi}^2 \, dt \leq$$

$$\frac{\nu}{2S^2} \int_0^T \theta_\epsilon(t) \left(\int |\vec{v}(t,x)|^2 \Delta(\varphi^2)(\frac{x}{S})\Phi(x) \, dx \right) dt$$

$$+\frac{\nu}{S} \int_0^T \theta_\epsilon(t) \left(\int \varphi(\frac{x}{S})\Phi(x)\vec{\nabla}\vec{v}(t,x) \cdot \vec{\nabla}\varphi(\frac{x}{S}) \, dx \right) dt$$

$$+\frac{\nu}{2} \int_0^T \theta_\epsilon(t) \left(\int |\vec{v}(t,x)|^2 \varphi^2(\frac{x}{S})\Delta\Phi(x) \, dx \right) dt$$

$$-\nu \int_0^T \theta_\epsilon(t) \|\varphi(\frac{x}{S})(\vec{\nabla} \otimes \vec{v})\|_{L^2_\Phi}^2 \, dt$$

$$+\frac{2}{S} \int_0^T \theta_\epsilon(t) \left(\int \Phi(x)\varphi(\frac{x}{S})(q + \frac{1}{2}|\vec{v}|^2)\vec{v} \cdot \vec{\nabla}\varphi(\frac{x}{S}) \, dx \right) dt$$

$$+\int_0^T \theta_\epsilon(t) \left(\int \varphi^2(\frac{x}{S})(q + \frac{1}{2}|\vec{v}|^2)\vec{v} \cdot \vec{\nabla}\Phi(x) \, dx \right) dt$$

$$-\int_0^T \theta_\epsilon(t) \left(\int \Phi(x)\varphi^2(\frac{x}{S}) (\vec{\nabla} \otimes \vec{v}) \cdot \mathbb{F} \, dx \right) dt$$

$$-\frac{2}{S} \int_0^T \theta_\epsilon(t) \left(\int \Phi(x)\varphi(\frac{x}{S}) (\vec{\nabla}\varphi(\frac{x}{S}) \otimes \vec{v}) \cdot \mathbb{F} \, dx \right) dt$$

$$-\int_0^T \theta_\epsilon(t) \left(\int \varphi^2(\frac{x}{S}) (\vec{\nabla}\Phi(x) \otimes \vec{v}) \cdot \mathbb{F} \, dx \right) dt.$$

Letting S go to ∞, we get

$$-\frac{1}{2}\int_0^T \theta'_\epsilon(t)\|\vec{v}(t,.)\|_{L^2_\Phi}^2\,dt \leq \frac{\nu}{2}\int_0^T \theta_\epsilon(t)\Big(\int |\vec{v}(t,x)|^2\Delta\Phi(x)\,dx\Big)\,dt$$

$$-\nu\int_0^T \theta_\epsilon(t)\|\vec{\nabla}\otimes\vec{v}\|_{L^2_\Phi}^2\,dt$$

$$+\int_0^T \theta_\epsilon(t)\Big(\int (q+\frac{1}{2}|\vec{v}|^2)\vec{v}\cdot\vec{\nabla}\Phi(x)\,dx\Big)\,dt$$

$$-\int_0^T \theta_\epsilon(t)\Big(\int \Phi(x)\,(\vec{\nabla}\otimes\vec{v})\cdot\mathbb{F}\,dx\Big)\,dt$$

$$-\int_0^T \theta_\epsilon(t)\Big(\int (\vec{\nabla}\Phi(x)\otimes\vec{v})\cdot\mathbb{F}\,dx\Big)\,dt$$

$$\leq 3\nu\int_0^T \theta_\epsilon(t)\|\vec{v}(t,.)\|_{L^2_\Phi}^2\,dt$$

$$-\nu\int_0^T \theta_\epsilon(t)\|\vec{\nabla}\otimes\vec{v}\|_{L^2_\Phi}^2\,dt$$

$$+2\int_0^T\int \theta_\epsilon(t)(|q|+\frac{1}{2}|\vec{v}|^2)|\vec{v}|\Phi^{3/2}(x)\,dx\,dt$$

$$+\int_0^T\int \theta_\epsilon(t)\Phi(x)|\vec{\nabla}\otimes\vec{v}|\,|\mathbb{F}|\,dx\,dt$$

$$+2\int_0^T\int \theta_\epsilon(t)|\vec{v}|\Phi(x)\,|\mathbb{F}|\,dx\,dt.$$

If t_0 is a Lebesgue point of $t\mapsto\|\vec{v}(t,.)\|_{L^2_\Phi}^2$ (and as 0 is a continuity point of $t\mapsto \|\vec{v}(t,.)\|_{L^2_\Phi}^2$), we find that

$$\|\vec{v}(t_0,.)\|_{L^2_\Phi}^2 \leq \|\vec{u}_0\|_{L^2_\Phi}^2 - 2\nu\int_0^{t_0}\|\vec{\nabla}\otimes\vec{v}\|_{L^2_\Phi}^2\,dt + 6\nu\int_0^{t_0}\|\vec{v}\|_{L^2_\Phi}^2\,dt$$

$$+2\int_0^{t_0}\|\vec{\nabla}\otimes\vec{v}\|_{L^2_\Phi}\|\mathbb{F}\|_{L^2_\Phi}\,dt + 4\int_0^{t_0}\|\vec{v}\|_{L^2_\Phi}\|\mathbb{F}\|_{L^2_\Phi}\,dt \qquad (14.42)$$

$$+4\int_0^{t_0}\int (|q|+\frac{1}{2}|\vec{v}|^2)|\vec{v}|\Phi^{3/2}(x)\,dx\,dt$$

This inequality is thus satisfied for almost every t_0, and even for every t_0 as $t\mapsto\vec{v}(t,.)$ is weakly continuous from $[0,T_0)$ to L^2_Φ.

We then write

$$\Delta q + \sum_{i=1}^3\sum_{j=1}^3 \partial_i\partial_j(v_iv_j) = \sum_{i=1}^3\sum_{j=1}^3 \partial_i\partial_j F_{i,j}$$

so that

$$q = \varpi_1 - \varpi_2 = \sum_{i=1}^3\sum_{j=1}^3 R_iR_j(v_iv_j) - \sum_{i=1}^3\sum_{j=1}^3 R_iR_jF_{i,j}.$$

We have the inequalities

$$\|\varpi_2\|_{L^2_\Phi} \leq C\|\mathbb{F}\|_{L^2_\Phi}$$

and

$$\|\varpi_1\|_{L^{4/3}_{\Phi^{4/3}}} \leq C\|\vec{v}\|^2_{L^{8/3}_{\Phi^{4/3}}}.$$

Moreover, we have

$$\|\sqrt{\Phi}\vec{v}\|_{8/3} \leq C\|\vec{v}\|^{5/8}_{L^2_\Phi}(\|\vec{v}\|^{3/8}_{L^2_\Phi} + \|\vec{\nabla}\otimes\vec{v}\|^{3/8}_{L^2_\Phi})$$

and

$$\|\sqrt{\Phi}\vec{v}\|_4 \leq C\|\vec{v}\|^{1/4}_{L^2_\Phi}(\|\vec{v}\|^{3/4}_{L^2_\Phi} + \|\vec{\nabla}\otimes\vec{v}\|^{3/4}_{L^2_\Phi}).$$

Thus far, (14.42) becomes

$$
\begin{aligned}
\|\vec{v}(t_0,.)\|^2_{L^2_\Phi} \leq & \|\vec{u}_0\|^2_{L^2_\Phi} - \nu\int_0^{t_0} \|\vec{\nabla}\otimes\vec{v}\|^2_{L^2_\Phi} dt \\
& + \frac{25}{4}\nu\int_0^{t_0} \|\vec{v}\|^2_{L^2_\Phi} dt + \frac{2C_0^2 + C_1^2}{\nu}\int_0^{t_0} \|\mathbb{F}\|^2_{L^2_\Phi} dt \\
& + C_2\int_0^{t_0} \|\vec{v}\|^3_{L^2_\Phi} dt + \frac{C_3^4}{4\nu^3}\int_0^{t_0} \|\vec{v}\|^6_{L^2_\Phi} dt.
\end{aligned}
\tag{14.43}
$$

with the same constants C_0, C_1, C_2, C_3 as in inequality (16.23).

This gives that the solution $\vec{v}$ satisfies on $(0, T_0)$ the inequality

$$
\begin{aligned}
\|\vec{v}(t,.)\|^2_{L^2_\Phi} + \nu\int_0^t \|\vec{\nabla}\otimes\vec{v}\|^2_{L^2_\Phi} ds \\
\leq 4(A_0^2\|\vec{u}_0\|^2_{L^2_\Phi} + \frac{2C_0^2 + C_1^2}{\nu}\int_0^T \|\mathbb{F}\|^2_{L^2_\Phi} dt)^2.
\end{aligned}
\tag{14.44}
$$

Thus, the control of $\vec{v}$ does not depend on the specific solution $\vec{v}$ but is the same as the control on $\vec{u}$.

Step 3: global existence.
With no loss of generality, we may assume that $\mathbb{F} \in L^2((0,+\infty), L^2_\Phi(\mathbb{R}^3))$. (If $\mathbb{F}$ is defined only on $(0,T)\times\mathbb{R}^3$, we extend $\mathbb{F}$ by 0 for $t > T$.)

In order to prove global existence, we use the scaling properties of the Navier–Stokes equations: if $\vec{v}$ is a solution on $(0,T)$ (with associated pressure q) of the Cauchy problem with initial value $\vec{v}_0$ and forcing term $\vec{f} = \text{div }\mathbb{G}$, then for $\lambda > 0$, $\lambda\vec{v}(\lambda^2 t, \lambda x)$ is a solution on $(0, \lambda^{-2}T)$ (with associated pressure $\lambda^2 q(\lambda^2 t, \lambda x)$) of the Cauchy problem with initial value $\lambda\vec{v}_0(\lambda x)$ and forcing term $\text{div }\mathbb{G}_\lambda$, where $\mathbb{G}_\lambda = \lambda^2\mathbb{G}(\lambda^2 t, \lambda x)$.

Thus, we consider $\lambda > 1$ and for $n \in \mathbb{N}$ we consider the Cauchy problem with initial value $\vec{v}_{0,n} = \lambda^n\vec{u}_0(\lambda^n x)$ and forcing tensor $\mathbb{F}_n = \lambda^{2n}\mathbb{F}(\lambda^{2n}t, \lambda^n x)$. If we find a solution $\vec{v}_n$ on $(0, T_n)$, then we have a solution $\vec{u}_n = \lambda^{-n}\vec{v}_n(\lambda^{-2n}t, \lambda^{-n}x)$, defined on $(0, \lambda^{2n}T_n)$, of the Cauchy problem with initial value $\vec{u}_0$ and forcing tensor $\mathbb{F}$.

We know that we have a weighted Leray solution $\vec{v}_n$ on $(0, T_n)$ with

$$T_n \approx C_5 \frac{\nu^5}{\nu^6 + \nu^2\|\vec{v}_{0,n}\|^4_{L^2_\Phi} + (\int_0^{+\infty} \|\mathbb{F}_n\|^2_{L^2_\Phi} dt)^2}.$$

This gives a weighted Leray solution $\vec{u}_n$ on $(0, \lambda^{2n}T_n)$. We easily check that

$$\lim_{n\to+\infty} \lambda^{2n}T_n = +\infty.$$

Indeed, we have

$$\|\vec{v}_{0,n}\|_{L^2_\Phi}^2 = \int |\vec{u}_0(y)|^2 \lambda^{-n}\Phi(\lambda^{-n}y)\,dy = \int |\vec{u}_0(y)|^2 \Phi(y)\frac{\lambda^n(1+|y|^2)}{(\lambda^{2n}+|y|^2)}\,dy = o(\lambda^n)$$

and

$$\int_0^{+\infty}\|\mathbb{F}_n\|_{L^2_\Phi}^2\,dt = \int_0^{+\infty}\int |\mathbb{F}(t,y)|^2 \lambda^{-n}\Phi(\lambda^{-n}y)\,dy$$

so that

$$\int_0^{+\infty}\|\mathbb{F}_n\|_{L^2_\Phi}^2\,dt = \int_0^{+\infty}\int |\mathbb{F}(t,y)|^2 \Phi(y)\frac{\lambda^n(1+|y|^2)}{(\lambda^{2n}+|y|^2)}\,dy = o(\lambda^n).$$

Moreover, $\lambda^k \vec{u}_n(\lambda^{2k}t,\lambda^k x)$ is a weighted Leray solution, defined on $(0,\lambda^{2(n-k)}T_n)$, of the Cauchy problem with initial value $\vec{u}_{0,k}$ and forcing tensor $\mathbb{F}_k$. Thus, if $\lambda^{2(n-k)}T_n \geq T_k$, we have a control on $(0,T_k)$ given by inequality (14.44)

$$\|\lambda^k \vec{u}_n(\lambda^{2k}t,\lambda^k\cdot)\|_{L^2_\Phi}^2 + \nu\int_0^t \lambda^{4k}\|(\vec{\nabla}\otimes\vec{u}_n)(\lambda^{2k}s,\lambda^k\cdot)\|_{L^2_\Phi}^2\,ds$$
$$\leq 4(A_0^2\|\vec{u}_{0,k}\|_{L^2_\Phi}^2 + \frac{2C_0^2+C_1^2}{\nu}\int_0^{+\infty}\|\mathbb{F}_k\|_{L^2_\Phi}^2\,dt)^2. \tag{14.45}$$

This gives a control for $\vec{u}_n$ on $(0,\lambda^{2k}T_k)$, writing $\Phi_k(x)=\Phi(\lambda^{-k}x)$,

$$\|\vec{u}_n(t,\cdot)\|_{L^2_{\Phi_k}}^2 + \nu\int_0^t \|\vec{\nabla}\otimes\vec{u}_n(s,\cdot)\|_{L^2_{\Phi_k}}^2\,ds$$
$$\leq 4(A_0^2\|\vec{u}_0\|_{L^2_{\Phi_k}}^2 + \frac{2C_0^2+C_1^2}{\nu}\int_0^{+\infty}\|\mathbb{F}\|_{L^2_{\Phi_k}}^2\,dt)^2.$$

This can be rewritten in terms of the weight Φ, as $\Phi\leq\Phi_k\leq\lambda^{2k}\Phi$,

$$\|\vec{u}_n(t,\cdot)\|_{L^2_\Phi}^2 + \nu\int_0^t \|\vec{\nabla}\otimes\vec{u}_n(s,\cdot)\|_{L^2_\Phi}^2\,ds$$
$$\leq 4\lambda^{2k}(A_0^2\|\vec{u}_0\|_{L^2_\Phi}^2 + \frac{2C_0^2+C_1^2}{\nu}\int_0^{+\infty}\|\mathbb{F}\|_{L^2_\Phi}^2\,dt)^2. \tag{14.46}$$

As the control (14.46) does not depend on n, we may again apply the Rellich–Lions theorem (Theorem 12.1) and get a sequence n_j such that $\vec{u}_{n_j}$ is strongly convergent to a limit $\vec{u}$ in $L^2_{loc}([0,\lambda^{2k}T_k)\times\mathbb{R}^3)$. Using Cantor's diagonal process, we get a sequence n_q such that $\vec{u}_{n_q}$ is strongly convergent to a limit $\vec{u}$ in $L^2_{loc}([0,+\infty)\times\mathbb{R}^3)$. By Theorem 6.2, the limit $\vec{u}$ satisfies the local Leray energy inequality (i.e. $\vec{u}$ is suitable) and is a weighted Leray solution on every bounded interval $(0,T)$.

$\square$

14.7 A Stability Estimate

When comparing two weighted weak Leray solutions, we find a problem as the interaction of the solutions may grow too fast at infinity. Thus, we must assume that one of the solutions remains bounded. We will prove the following stability estimate:

Lemma 14.7.
Let $\vec{u}_1$, $\vec{u}_2$ be two weighted Leray solutions on $(0,T) \times \mathbb{R}^3$ to the problems

$$\begin{cases} \partial_t \vec{u}_i = \nu \Delta \vec{u}_i - \vec{u}_i \cdot \vec{\nabla} \vec{u}_i - \vec{\nabla} p_i \\ \\ \vec{u}_i(0,.) = \vec{u}_{0,i} \end{cases} \tag{14.47}$$

where $\vec{u}_{0,i} \in L^2(\frac{1}{1+|x|^2} dx)$ with $\operatorname{div} \vec{u}_{0,i} = 0$.
Assume moreover that $\vec{u}_1 \in L^2((0,T), L^\infty)$. Then, we have, for every $t_0 \in (0,T)$,

$$\|\vec{u}_1(t_0,.) - \vec{u}_2(t_0,.)\|_{L^2_\Phi}^2$$

$$\leq e^{Ct_0(\nu + \frac{\sup_{0<s<t_0}(\|\vec{u}_1(s,.)\|_{L^2_\Phi} + \|\vec{u}_2(s,.)\|_{L^2_\Phi})^4}{\nu^3})} e^{C \int_0^{t_0} \frac{\|\vec{u}_1\|_\infty^2}{\nu} dt} \|\vec{u}_1(0,.) - \vec{u}_2(0,.)\|_{L^2_\Phi}^2, \tag{14.48}$$

where the constant C does not depend on $\vec{u}_1$, $\vec{u}_2$, ν nor T.

Proof. Let p_i be the pressure associated to $\vec{u}_i$. We have, due to the suitability of $\vec{u}_2$ and the regularity of $\vec{u}_1$,

$$\begin{cases} \partial_t(\frac{|\vec{u}_1|^2}{2}) = \quad \nu\Delta(\frac{|\vec{u}_1|^2}{2}) - \nu|\vec{\nabla} \otimes \vec{u}_1|^2 - \operatorname{div}((p_1 + \frac{|\vec{u}_1|^2}{2})\vec{u}_1) \\ \\ \partial_t(\vec{u}_1 \cdot \vec{u}_2) = \quad \nu\Delta(\vec{u}_1 \cdot \vec{u}_2) - 2\nu(\vec{\nabla} \otimes \vec{u}_1).(\vec{\nabla} \otimes \vec{u}_2) - \operatorname{div}(p_2\vec{u}_1 + p_1 \cdot \vec{u}_2) \\ \qquad\qquad\qquad -\vec{u}_1.(\vec{u}_2 \cdot \vec{\nabla}\vec{u}_2) - \vec{u}_2.(\vec{u}_1 \cdot \vec{\nabla}u_1) \\ \\ \partial_t(\frac{|\vec{u}_2|^2}{2}) = \quad \nu\Delta(\frac{|\vec{u}_2|^2}{2}) - \nu|\vec{\nabla} \otimes \vec{u}_2|^2 - \operatorname{div}((p_2 + \frac{|\vec{u}_2|^2}{2})\vec{u}_2) - \mu \end{cases}$$

where μ is some non-negative locally finite measure.
Let $\vec{w} = \vec{u}_1 - \vec{u}_2$ and $q = p_1 - p_2$. We obtain

$$\partial_t(\frac{|\vec{w}|^2}{2}) = \nu\Delta(\frac{|\vec{w}|^2}{2}) - \nu|\vec{\nabla} \otimes \vec{w}|^2 - \operatorname{div}(q\vec{w}) - A - \mu$$

with

$$A = \operatorname{div}(\frac{|\vec{u}_1|^2}{2}\vec{u}_1 + \frac{|\vec{u}_2|^2}{2}\vec{u}_2) - (\vec{u}_1.(\vec{u}_2 \cdot \vec{\nabla}\vec{u}_2) + \vec{u}_2.(\vec{u}_1 \cdot \vec{\nabla}u_1)$$

$$= \frac{1}{2} \operatorname{div}\left(|\vec{u}_1|^2\vec{u}_1 + |\vec{u}_2|\vec{u}_2 - (\vec{u}_1 \cdot \vec{u}_2)(\vec{u}_1 + \vec{u}_2)\right)$$

$$\quad + \frac{1}{2}\left(\vec{u}_2.(\vec{u}_2 \cdot \vec{\nabla}\vec{u}_1) + \vec{u}_1.(\vec{u}_1 \cdot \vec{\nabla}u_2) - \vec{u}_1.(\vec{u}_2 \cdot \vec{\nabla}\vec{u}_2) - \vec{u}_2.(\vec{u}_1 \cdot \vec{\nabla}u_1)\right)$$

$$= \frac{1}{2} \operatorname{div}\left(|\vec{w}|^2\vec{u}_1 + (\vec{u}_1 \cdot \vec{w})\vec{w} - |\vec{w}|^2\vec{w}\right) + \frac{1}{2}\left(\vec{w}.(\vec{w} \cdot \vec{\nabla}\vec{u}_1) - \vec{u}_1.(\vec{w} \cdot \vec{\nabla}\vec{w})\right)$$

$$= \frac{1}{2} \operatorname{div}\left(|\vec{w}|^2\vec{u}_1 + 2(\vec{u}_1 \cdot \vec{w})\vec{w} - |\vec{w}|^2\vec{w}\right) - \vec{u}_1.(\vec{w} \cdot \vec{\nabla}\vec{w}).$$

We use again the test function $\Omega(t,x) = \Phi(x)\theta_\epsilon(t)\varphi^2(x/S)$ described in equation (14.37) and compute $\iint \Omega(t,x)\partial_t(\frac{|\vec{w}|^2}{2}) dt\, dx$.

Integrating inequality (14.3) against $\Omega(t,x)$, we find that:

$$-\frac{1}{2}\int_0^T \theta_\epsilon'(t)\|\vec{w}(t,.)\varphi(\frac{.}{S})\|_{L_\Phi^2}^2 \, dt \leq$$

$$\frac{\nu}{2S^2}\int_0^T \theta_\epsilon(t)(\int |\vec{w}(t,x)|^2 \Delta(\varphi^2)(\frac{x}{S})\Phi(x)\,dx)\,dt$$

$$+\frac{\nu}{S}\int_0^T \theta_\epsilon(t)(\int \varphi(\frac{x}{S})\Phi(x)\vec{\nabla}\vec{w}(t,x)\cdot\vec{\nabla}\varphi(\frac{x}{S})\,dx)\,dt$$

$$+\frac{\nu}{2}\int_0^T \theta_\epsilon(t)(\int |\vec{w}(t,x)|^2\varphi^2(\frac{x}{S})\Delta\Phi(x)\,dx)\,dt$$

$$-\nu\int_0^T \theta_\epsilon(t)\|\varphi(\frac{x}{S})(\vec{\nabla}\otimes\vec{w})\|_{L_\Phi^2}^2 \, dt$$

$$+\frac{2}{S}\int_0^T \theta_\epsilon(t)(\int \Phi(x)\varphi(\frac{x}{S})(q+\vec{u}_1\cdot\vec{w}-\frac{1}{2}|\vec{w}|^2)\vec{w}\cdot\vec{\nabla}\varphi(\frac{x}{S})\,dx)\,dt$$

$$+\int_0^T \theta_\epsilon(t)(\int \varphi^2(\frac{x}{S})(q+\vec{u}_1\cdot\vec{w}-\frac{1}{2}|\vec{w}|^2)\vec{w}\cdot\vec{\nabla}\Phi(x)\,dx)\,dt$$

$$+\frac{2}{S}\int_0^T \theta_\epsilon(t)(\int \Phi(x)\varphi(\frac{x}{S})(\frac{1}{2}|\vec{w}|^2)\vec{u}_1\cdot\vec{\nabla}\varphi(\frac{x}{S})\,dx)\,dt$$

$$+\int_0^T \theta_\epsilon(t)(\int \varphi^2(\frac{x}{S})(\frac{1}{2}|\vec{w}|^2)\vec{u}_1\cdot\vec{\nabla}\Phi(x)\,dx)\,dt$$

$$+\int_0^T \theta_\epsilon(t)(\int \Phi(x)\varphi^2(\frac{x}{S})\,\vec{u}_1.(\vec{w}\cdot\vec{\nabla}\vec{w})\,dx)\,dt$$

Noticing that $|\vec{\nabla}\Phi| \leq 2\Phi^{3/2} \leq 2\Phi$ and $|\Delta\Phi| \leq 6\Phi^2 \leq 6\Phi$, and letting S go to ∞, we get

$$-\frac{1}{2}\int_0^T \theta_\epsilon'(t)\|\vec{u}_R(t,.)\|_{L_\Phi^2}^2 \, dt \leq 3\nu\int_0^T \theta_\epsilon(t)\|\vec{w}(t,.)\|_{L_\Phi^2}^2 \, dt$$

$$-\nu\int_0^T \theta_\epsilon(t)\|\vec{\nabla}\otimes\vec{w}\|_{L_\Phi^2}^2 \, dt$$

$$+\int_0^T \int \theta_\epsilon(t)(|q|+|\vec{u}_1\cdot\vec{w}|+\frac{1}{2}|\vec{w}|^2)|\vec{w}|\Phi^{3/2}(x)\,dx\,dt$$

$$+\frac{1}{2}\int_0^T \int \theta_\epsilon(t)|\vec{w}|^2|\vec{u}_1|\Phi^{3/2}(x)\,dx\,dt$$

$$+\int_0^T \int \theta_\epsilon(t)\Phi(x)\,dx\,\vec{u}_1.(\vec{w}\cdot\vec{\nabla}\vec{w})\,dt.$$

If t_0 is a Lebesgue point of $t \mapsto \|\vec{w}(t,.)\|_{L^2_\Phi}^2$ (and as 0 is a continuity point of $t \mapsto \|\vec{w}(t,.)\|_{L^2_\Phi}^2$), we find that

$$
\begin{aligned}
\|\vec{w}(t_0,.)\|_{L^2_\Phi}^2 \leq &\|\vec{w}(0,.)\|_{L^2_\Phi}^2 - 2\nu \int_0^{t_0} \|\vec{\nabla} \otimes \vec{w}\|_{L^2_\Phi}^2 \, dt + 6\nu \int_0^{t_0} \|\vec{w}\|_{L^2_\Phi}^2 \, dt \\
&+ 2 \int_0^{t_0} \int (|q| + |\vec{u}_1 \cdot \vec{w}| + \frac{1}{2}|\vec{w}|^2)|\vec{w}|\Phi^{3/2}(x) \, dx \, dt \\
&+ \int_0^{t_0} \int |\vec{w}|^2 |\vec{u}_1|\Phi^{3/2}(x) \, dx \, dt \\
&+ 2 \int_0^{t_0} \int \Phi(x) \, dx \, \vec{u}_1.(\vec{w} \cdot \vec{\nabla}\vec{w}) \, dt.
\end{aligned}
$$
(14.49)

This inequality is thus satisfied for almost every t_0, and even for every t_0 as $t \mapsto \vec{w}(t,.)$ is weakly continuous from $[0,T)$ to L^2_Φ.

We then write

$$
\Delta q = \sum_{i=1}^3 \sum_{j=1}^3 \partial_i \partial_j (u_{2,i} u_{2,j} - u_{1,i} u_{1,j}) = \sum_{i=1}^3 \sum_{j=1}^3 \partial_i \partial_j j(w_i w_j - u_{1,i} w_j - w_i u_{1,j})
$$

so that

$$
q = \varpi_0 - \varpi_1 = \sum_{i=1}^3 \sum_{j=1}^3 R_i R_j (w_i w_j) - \sum_{i=1}^3 \sum_{j=1}^3 R_i R_j (u_{1,i} w_j + w_i u_{1,j}).
$$

As the Riesz transforms are bounded on L^2_Φ, we find

$$
\|\varpi_1\|_{L^2_\Phi} \leq C \|\vec{u}_1\|_\infty \|\vec{w}\|_{L^2_\Phi}.
$$

On the other hand we know that

$$
\|\Phi\varpi_0\|_{4/3} \leq C \|\sqrt{\Phi}\vec{w}\|_{4/3}^2 \leq C' \|\vec{w}\|_{L^2_\Phi}^{5/4} (\|\vec{w}\|_{L^2_\Phi}^{3/4} + \|\vec{\nabla} \otimes w\|_{L^2_\Phi}^{3/4}).
$$

Finally, we get the inequality

$$
\begin{aligned}
\|\vec{w}(t_0,.)\|_{L^2_\Phi}^2 \leq &\|\vec{w}(0,.)\|_{L^2_\Phi}^2 - 2\nu \int_0^{t_0} \|\vec{\nabla} \otimes \vec{w}\|_{L^2_\Phi}^2 \, dt + 6\nu \int_0^{t_0} \|\vec{w}\|_{L^2_\Phi}^2 \, dt \\
&+ C \int_0^{t_0} \|\vec{u}_1\|_\infty \|\vec{w}\|_{L^2_\Phi}^2 \, dt \\
&+ C \int_0^{t_0} \int \|\vec{u}_1\|_\infty \|\vec{w}\|_{L^2_\Phi} \|\vec{\nabla} \otimes \vec{w}\|_{L^2_\Phi} \, dt \\
&+ C \int_0^{t_0} \|\vec{w}\|_{L^2_\Phi}^3 \, dt + C \int_0^{t_0} \|\vec{w}\|_{L^2_\Phi}^{3/2} \|\vec{\nabla} \otimes \vec{w}\|_{L^2_\Phi}^{3/2} \, dt
\end{aligned}
$$
(14.50)

and thus

$$
\begin{aligned}
\|\vec{w}(t_0,.)\|_{L^2_\Phi}^2 \leq &\|\vec{w}(0,.)\|_{L^2_\Phi}^2 - \nu \int_0^{t_0} \|\vec{\nabla} \otimes \vec{w}\|_{L^2_\Phi}^2 \, dt \\
&+ C \int_0^{t_0} (\nu + \frac{\|\vec{u}_1\|_\infty^2}{\nu} + \frac{\|\vec{u}_1\|_{L^2_\Phi}^4}{\nu^3} + \frac{\|\vec{u}_2\|_{L^2_\Phi}^4}{\nu^3}) \|\vec{w}\|_{L^2_\Phi}^2 \, dt.
\end{aligned}
$$
(14.51)

This gives, for every $t_0 \in (0, T)$,

$$\|\vec{w}(t_0, .)\|_{L_\Phi^2}^2 \le e^{Ct_0\left(\nu + \frac{1}{\nu^3}\sup_{0<s<t_0}(\|\vec{u}_1(s,.)\|_{L_\Phi^2}^4 + \|\vec{u}_2(s,.)\|_{L_\Phi^2}^4)\right)} e^{C\int_0^{t_0} \frac{\|\vec{u}_1\|_\infty^2}{\nu} dt} \|\vec{w}(0, .)\|_{L_\Phi^2}^2.$$

$$(14.52)$$

□

Of course, this stability estimate is much more easy to get when we consider weak Leray solutions, as the terms involving the pressures disappear in the energy balance:

Lemma 14.8.

Let $\vec{u}_1$, $\vec{u}_2$ be two weak Leray solutions on $(0, T) \times \mathbb{R}^3$ to the problems

$$\begin{cases} \partial_t \vec{u}_i = \nu \Delta \vec{u}_i - \vec{u}_i \cdot \vec{\nabla} \vec{u}_i - \vec{\nabla} p_i \\ \\ \vec{u}_i(0, .) = \vec{u}_{0,i} \end{cases} \tag{14.53}$$

where $\vec{u}_{0,i} \in L^2$ with $\operatorname{div} \vec{u}_{0,i} = 0$.
Assume moreover that $\vec{u}_1 \in L^2((0, T), L^\infty)$. Then, we have, for every $t_0 \in (0, T)$,

$$\|\vec{u}_1(t_0, .) - \vec{u}_2(t_0, .)\|_2^2 + 2\nu \int_0^{t_0} \|\vec{\nabla} \otimes (\vec{u}_1 - \vec{u}_2)\|_2^2 \, ds$$

$$\le \|\vec{u}_1(0, .) - \vec{u}_2(0, .)\|_2^2 + 2 \iint_0^{t_0} \vec{u}_1 \cdot ((\vec{u}_1 - \vec{u}_2) \cdot \vec{\nabla}(\vec{u}_1 - \vec{u}_2)) \, dx \, ds. \tag{14.54}$$

14.8　Barker's Theorem on Weak-Strong Uniqueness

Let us recall the weak-strong uniqueness criterion (Theorem 12.4) given by Prodi and Serrin [406, 435] for Leray solutions of the Navier–Stokes problem

$$\begin{cases} \partial_t \vec{u} + \vec{u} \cdot \vec{\nabla} \vec{u} = \Delta \vec{u} - \vec{\nabla} p \\ \operatorname{div} \vec{u} = 0 \\ \vec{u}(0, .) = \vec{u}_0 \end{cases}$$

where $\vec{u}_0$ is a square-integrable divergence-free vector field on the space $\mathbb{R}^3$: if the Navier–Stokes equations have a solution $\vec{u}$ on $(0, T)$ such that

$$\vec{u} \in L_t^p L_x^q \text{ with } \frac{2}{p} + \frac{3}{q} \le 1 \text{ and } 2 < p < +\infty$$

then, if $\vec{v}$ is a Leray solution with the same initial value $\vec{u}_0$, we have $\vec{u} = \vec{v}$ on $(0, T)$. Let us remark that the existence of such a solution $\vec{u}$ restricts the range of the initial value $\vec{u}_0$: when $2 < p < +\infty$, existence of a time $T > 0$ and of a solution $\vec{u} \in L_t^p L_x^q$ is equivalent to the fact that $\vec{u}_0$ belongs to the Besov space $B_{q,p}^{-\frac{2}{p}}$

A natural endpoint case for this criterion is the assumption that $\vec{u}_0 \in L^2 \cap bmo^{-1}$, or more precisely to $L^2 \cap bmo_0^{-1}$, where the restriction to bmo_0^{-1} grants existence of a mild solution, due to the Koch and Tataru theorem [266].

Proposition 14.2.

For $0 < T < \infty$, define

$$\|\vec{u}\|_{X_T} = \sup_{0<t<T} \sqrt{t}\|\vec{u}(t, .)\|_\infty + \sup_{0<t<T, x_0 \in \mathbb{R}^3} (t^{-3/2} \int_0^t \int_{B(x_0, \sqrt{t})} |\vec{u}(s, y)|^2 \, dy \, ds)^{1/2}.$$

Then $\vec{u}_0 \in bmo^{-1}$ if and only if $(e^{t\Delta}\vec{u}_0)_{0<t<T} \in X_T$ *(with equivalence of the norms* $\|\vec{u}_0\|_{bmo^{-1}}$ *and* $\|e^{t\Delta}\vec{u}_0\|_{X_T}$*).*

Koch and Tataru's theorem is then the following one:

Theorem 14.10.
There exists C_0 (which does not depend on T) such that, if $\vec{u}$ and $\vec{v}$ are defined on $(0,T) \times \mathbb{R}^3$, then

$$\|B(\vec{u},\vec{v})\|_{X_T} \leq C_0 \|\vec{u}\|_{X_T} \|\vec{v}\|_{X_T},$$

where $B(\vec{u},\vec{v}) = \int_0^t e^{(t-s)\Delta} \mathbb{P} \operatorname{div}(\vec{u} \otimes \vec{v}) \, ds$.

Corollary 14.1.
Let $\vec{u}_0 \in bmo^{-1}$ with $\operatorname{div} \vec{u}_0 = 0$. If $\|e^{t\Delta}\vec{u}_0\|_{X_T} < \frac{1}{4C_0}$, then the integral Navier–Stokes equations have a solution on $(0,T)$ such that $\|\vec{u}\|_{X_T} \leq 2\|e^{t\Delta}\vec{u}_0\|_{X_T}$.
This is the unique solution such that $\|\vec{u}\|_{X_T} \leq \frac{1}{2C_0}$.

This Corollary grants local existence of a solution for the Navier–Stokes equations when the initial value belongs to the space bmo_0^{-1}:

Definition 14.5.
$\vec{u}_0 \in bmo_0^{-1}$ if $\vec{u} \in bmo^{-1}$ and $\lim_{T \to 0} \|e^{t\Delta}\vec{u}_0\|_{X_T} = 0$.

Let us remark that the initial values for the Prodi–Serrin criterion satisfy $\vec{u}_0 \in L^2 \cap B_{q,p}^{-1+\frac{3}{q}}$ with $1 < p < +\infty$ and $\frac{2}{p} + \frac{3}{q} \leq 1$, hence belong to $L^2 \cap bmo_0^{-1}$. However, there is no weak-strong uniqueness result of Leray weak solutions for initial values in $L^2 \cap bmo_0^{-1}$. Barker noticed that $L^2 \cap bmo_0^{-1} \subset B_{\infty,q}^{-1+\frac{2}{q}}$, while the Prodi-Serrin criterion requires a higher regularity ($\vec{u}_0 \in B_{\infty,q}^{-1+\frac{3}{q}}$). Barker's theorem [18] states that weak-strong uniqueness holds with only a slight improvement in regularity ($\vec{u}_0 \in L^2 \cap bmo_0^{-1} \cap B_{\infty,q}^s$ with $s > -1 + \frac{2}{q}$).

It is easy to check that, if $0 < s < 1 - \frac{2}{q}$, if $\vec{u}_0 \in bmo_0^{-1}$, and if $\vec{u}$ is the mild solution with $\|\vec{u}\|_{X_T} \leq \frac{1}{2C_0}$, then $\vec{u}_0 \in B_{q,\infty}^{-s}$ is equivalent to

$$\vec{u} \in L^{\frac{2}{s},\infty}((0,T), L^q)$$

or to

$$\sup_{0<t<T} t^{s/2} \|\vec{u}(t,.)\|_q < +\infty.$$

Lemarié-Rieusset [324] proved a generalization of Barker's result by relaxing the integrability requirement by a weighted integrability assumption (and restricting weak-strong uniqueness to suitable Leray solutions):

Theorem 14.11.
Let $\vec{u}_0$ be a divergence-free vector field with $\vec{u}_0 \in L^2 \cap bmo_0^{-1}$. Assume moreover that the mild solution $\vec{u}$ of the Navier–Stokes equations with initial value $\vec{u}_0$ such that $\|\vec{u}\|_{X_T} < \frac{1}{2C_0}$ is such that

$$\sup_{0<t<T} t^{s/2} \|\vec{u}\|_{L^q(\frac{1}{(1+|x|)^N} dx)} < +\infty$$

with

$$N \geq 0, 2 < q < +\infty \text{ and } 0 \leq s < 1 - \frac{2}{q}.$$

If $\vec{v}$ is a suitable weak Leray solution of the Navier–Stokes equations with the same initial value $\vec{u}_0$, then $\vec{u} = \vec{v}$ on $(0,T)$.

As in Barker's proof [18], Theorem 14.11 will be a consequence of another weak-strong uniquess theorem:

> **Theorem 14.12.**
> *Let $\vec{u}_0$ be a divergence-free vector field with $\vec{u}_0 \in L^2 \cap bmo_0^{-1}$. Let $\vec{u}$ be the mild solution of the Navier–Stokes equations with initial value $\vec{u}_0$ such that $\|\vec{u}\|_{X_T} < \frac{1}{2C_0}$ (with $T > 0$ such that $\|e^{t\Delta}\vec{u}_0\|_{X_T} < \frac{1}{4C_0}$). Assume moreover that, for some $\gamma < 1$ and some $\theta \in (0,1)$, $\vec{u}_0$ belongs to $[(L^2(\frac{1}{1+|x|^2}\,dx))_\sigma, (B_{\infty,\infty}^{-\gamma})_\sigma]_{\theta,\infty}$ (where σ stands for divergence-free). If $\vec{v}$ is a suitable weak Leray solution of the Navier–Stokes equations with the same initial value $\vec{u}_0$, then $\vec{u} = \vec{v}$ on $(0,T)$.*

Proof. **Step 1.**

We first check that the mild solution $\vec{u}$ in X_T of the Navier–Stokes equations with initial value $\vec{u}_0 \in bmo_0^{-1}$ (with $T > 0$ such that $\|e^{t\Delta}\vec{u}_0\|_{X_T} < \frac{1}{4C_0}$) is a suitable Leray solution if moreover $\vec{u}_0 \in L^2$ or a weighted Leray solution if $\vec{u}_0 \in L^2(\frac{1}{1+|x|^2}\,dx)$.

We write E for L^2 or $L^2(\frac{1}{1+|x|^2}\,dx)$. Let δ such that $\|e^{t\Delta}\vec{u}_0\|_{X_T} \leq \delta < \frac{1}{4C_0}$. We consider the Picard iterates $\vec{U}_n$ defined by $\vec{U}_0 = e^{t\Delta}\vec{u}_0$ and $\vec{U}_{n+1} = \vec{U}_0 - B(\vec{U}_n, \vec{U}_n)$, where $B(\vec{v}, \vec{w}) = \int_0^t e^{(t-s)\Delta}\mathbb{P}\operatorname{div}(\vec{v} \otimes \vec{w})\,ds$. We know that $\|\vec{U}_n\|_{X_T} \leq 2\delta$, that $\vec{U}_n$ converge to $\vec{u}$ and that $\|\vec{U}_{n+1} - \vec{U}_n\|_{X_T} \leq \frac{1}{4}(4\delta C_0)^{n+1}\delta$.

We have $|\vec{U}_0(t,x)| \leq \mathcal{M}_{\vec{u}_0}(x)$, so that $\|\vec{U}_0(t,.)\|_E \leq C_E\|\vec{u}_0\|_E$. Moreover,

$$|e^{(t-s)\Delta}\mathbb{P}\operatorname{div}(\vec{U}_n \otimes \vec{U}_n - \vec{U}_{n-1} \otimes \vec{U}_{n-1})|$$
$$\leq C\frac{1}{\sqrt{t-s}}\|\vec{U}_n(s,.) - \vec{U}_{n-1}(s,.)\|_\infty(\mathcal{M}_{\vec{U}_n(s,.)}(x) + (\mathcal{M}_{\vec{U}_{n-1}(s,.)}(x))$$

From this, we get

$$\|\vec{u}\|_{L^\infty((0,T),E)} \leq C_{E,\delta}\|\vec{u}_0\|_E.$$

Moreover, we have

$$\|\vec{u}(t,.) - \vec{u}_0\|_E \leq \|e^{t\Delta}\vec{u}_0 - \vec{u}_0\|_E + C \sup_{0<s<t}\|\vec{u}(s,.)\|_E \sup_{0<s<t}\sqrt{s}\|\vec{u}(s,.)\|_\infty = o(1).$$

Finally, we remark that the mild solution $\vec{u}$ is smooth on $(0,T) \times \mathbb{R}^3$, so that, for $0 < t_0 \leq t < T$,

$$\partial_t(|\vec{u}|^2) + 2|\vec{\nabla} \otimes \vec{u}|^2 = \Delta(|\vec{u}|^2) - \operatorname{div}((2p + |\vec{u}|^2)\vec{u})$$

and thus

$$\int \phi_R(x)|\vec{u}(t,x)|^2\,dx + 2\int_{t_0}^t \int \phi_R(x)|\vec{\nabla} \otimes \vec{u}(s,x)|^2\,dx\,ds$$
$$= \int \phi_R(x)|\vec{u}(t_0,x)|^2\,dx + \int_{t_0}^t \int \Delta(\phi_R(x))|\vec{u}(t,x)|^2\,dx\,ds$$
$$+ \int_{t_0}^t \int (2p + |\vec{u}|^2)\vec{u} \cdot \vec{\nabla}(\phi_R(x))\,dx\,ds,$$

where $\phi_R(x) = \theta(\frac{x}{R})w(x)$, θ is smooth and equal to 1 in a neighborhood of 0 and $w(x) = 1$ or $w(x) = \frac{1}{1+|x|^2}$ (so that $E = L^2(w\,dx)$). We have that $\vec{u} \in L^\infty(L^2(w\,dx))$, $\sqrt{t}u_iu_j \in L^\infty(L^2(w\,dx))$, and thus $\sqrt{t}(2p + |\vec{u}|^2) \in L^\infty(L^2(w\,dx))$ (as $w \in \mathcal{A}_2$ and $p = -\sum_{1\le i\le 3}\sum_{j=1}^{3}\frac{\partial_i\partial_j}{\Delta}(u_iu_j)$), so that

$$\int \phi_R(x)|\vec{u}(t,x)|^2\,dx + 2\int_{t_0}^{t}\int \phi_R(x)|\vec{\nabla}\otimes\vec{u}(s,x)|^2\,dx\,ds$$

$$\le C\sup_{0<s<T}\int |\vec{u}(s,x)|^2\,w(x)\,dx + C\int_0^T\int |\vec{u}(s,x)|^2\,w(x)\,dx\,ds$$

$$+ \int_0^T\int \sqrt{s}\big|2p + |\vec{u}|^2\big|\,|\vec{u}|\,w(x)\,dx\,\frac{ds}{\sqrt{s}} < +\infty.$$

We then let R go to $+\infty$ and t_0 go to 0.

A similar proof gives that, if $\vec{u}_0 \in bmo_0^{-1}$ (with $T > 0$ such that $\|e^{t\Delta}\vec{u}_0\|_{X_T} \le \delta < \frac{1}{4C_0}$) and if moreover $\vec{u}_0 \in B_{\infty,\infty}^{-\gamma}$ with $0 < \gamma < 1$, ten thhe Picard iterates $\vec{U}_n$ satisfy

$$|e^{(t-s)\Delta}\mathbb{P}\operatorname{div}(\vec{U}_n\otimes\vec{U}_n - \vec{U}_{n-1}\otimes\vec{U}_{n-1})|$$

$$\le C\frac{1}{\sqrt{t-s}}\|\vec{U}_n(s,.)-\vec{U}_{n-1}(s,.)\|_\infty\|\vec{U}_n(s,.)\|_\infty + \|\vec{U}_{n-1}\|_\infty)$$

so that, for $0 < t < \min(T,1)$,

$$\|\vec{u}(t,.)\|_\infty \le C_{\delta,\gamma}t^{-\gamma/2}\|\vec{u}_0\|_{B_{\infty,\infty}^{-\gamma}}.$$

Step 2.

We now check that Theorem 14.11 is a corollary of Theorem 14.12.

The first step is to diminish the value of N. We know that the mild solution $\vec{u}$ is a weak Leray solution as well. (As a matter of fact, the solutions of the Leray mollification will converge in $\mathcal{D}'$ to the mild solution and to a weak Leray solution [313]). In particular, we have $\sup_{0<t<T}\|\vec{u}(t,.)\|_2 < +\infty$, while $\sup_{0<t<T}t^{1/2}\|\vec{u}(t,.)\|_\infty \le \|\vec{u}\|_{X_T} < +\infty$. Thus,

$$\sup_{0<s<T} t^{\frac{1}{2}-\frac{1}{q}}\|\vec{u}\|_q < +\infty.$$

If $0 \le \alpha \le 1$, we find that

$$\sup_{0<t<T}(\sqrt{t})^{(1-\alpha)(1-\frac{2}{q})+\alpha s}\|\vec{u}\|_{L^q(\frac{1}{(1+|x|)^{\alpha N}}\,dx)} < +\infty.$$

For $0 < \alpha < \min(1, \frac{4}{Nq})$, we have $0 < s_\alpha = (1-\alpha)(1-\frac{2}{q}) + \alpha s < 1 - \frac{2}{q}$ and $\alpha N < \frac{4}{q}$. As $\frac{4}{q} < 3$, we find that the weight $\frac{1}{(1+|x|)^{\alpha N}}$ belongs to the Muckenhoupt class $\mathcal{A}_q$, so that the fact that the mild solution $\vec{u}_0$ satisfies $\vec{u} \in L^{\frac{2}{s_\alpha},\infty}((0,T), L^q(\frac{1}{(1+|x|)^{\alpha N}}\,dx)$ is equivalent to the fact that $\vec{u}_0$ belongs to the Besov space $B_{L^q(\frac{1}{(1+|x|)^{\alpha N}}\,dx),\infty}^{-s_\alpha}$ (i.e. that $\sup_{0<t<T}\sup_{0<t<T}(\sqrt{t})^{s_\alpha}\|e^{t\Delta}\vec{u}_0\|_{L^q(\frac{1}{(1+|x|)^{\alpha N}}\,dx)} < +\infty$). For

$$s_\alpha < \sigma < 1 - \frac{2}{q},$$

we have

$$B^{-s_\alpha}_{L^q(\frac{1}{(1+|x|)^{\alpha N}}\,dx),\infty} \subset H^{-\sigma}_{L^q(\frac{1}{(1+|x|)^{\alpha N}}\,dx)} = (Id-\Delta)^\sigma(L^q(\frac{1}{(1+|x|)^{\alpha N}}\,dx)).$$

We now recall the result proved in [324] on the complex interpolation of weighted Sobolev spaces. Let $\theta \in (0,1)$, s_0, s_1 be real numbers, $1 < p_0, p_1 < +\infty$ and $s = (1-\theta)s_0 + \theta s_1$ and $\frac{1}{p} = (1-\theta)\frac{1}{p_0} + \theta\frac{1}{p_1}$. Then, if w_0 is a weight in the Muckenhoupt class $\mathcal{A}_{p_0}$ and w_1 is a weight in the Muckenhoupt class $\mathcal{A}_{p_1}$,

$$(Id-\Delta)^s L^p(w_0^{1-\theta}w_1^\theta\,dx) = [(Id-\Delta)^{s_0}L^{p_0}(w_0\,dx), (Id-\Delta)^{s_1}L^{p_1}(w_1\,dx)]_\theta.$$

We are interested in $s = \sigma$, $w_0^{1-\theta}w_1^\theta = \frac{1}{(1+|x|)^{\alpha N}}$, $p_0 = 2$, $s_0 = 0$ and $w_1 = 1$. We pick $\theta \in (0,1)$ such that

$$\max(0, \frac{N\alpha}{2}, \frac{2}{q} - 2(1-\sigma-\frac{2}{q})) < 1-\theta < \frac{2}{q}.$$

We obtain, for

$$p = q, s = \sigma, w_0^{1-\theta}w_1^\theta = \frac{1}{(1+|x|)^{\alpha N}},$$

and

$$p_0 = 2, s_0 = 0, w_1 = 1,$$

the values

$$\begin{cases} \frac{1}{p_1} = \frac{1}{\theta}(\frac{1}{q} - \frac{1-\theta}{2}) & \text{with } q < p_1 < +\infty \\ s_1 = \frac{\sigma}{\theta} \\ w_0 = \frac{1}{(1+|x|)^{\frac{\alpha N}{1-\theta}}} & \text{with } \frac{\alpha N}{1-\theta} < 2 \end{cases}$$

so that $w_0 \in \mathcal{A}_{p_0}$ and $w_1 \in \mathcal{A}_{p_1}$. Moreover, $L^2(\frac{1}{(1+|x|)^{\frac{\alpha N}{1-\theta}}}\,dx) \subset L^2(\frac{1}{1+|x|^2}\,dx)$ and $H^{-s_1}_{p_1} \subset B^{-\gamma}_{\infty,\infty}$ with

$$\gamma = s_1 + \frac{3}{p_1} = \frac{1}{\theta}(\sigma + \frac{3}{q} - 3\frac{1-\theta}{2}) = 1 + \frac{1}{\theta}(\frac{1}{q} + \sigma + \frac{2}{q} - 1 - \frac{1-\theta}{2}) < 1.$$

Thus, under the assumptions of Theorem 14.11, we find that

$$\vec{u}_0 \in [L^2(w_0\,dx), H^{-s_1}_{p_1}]_\theta \subset [L^2(w_0\,dx), H^{-s_1}_{p_1}]_{\theta,\infty}.$$

As $\vec{u}_0$ is divergence free and as the Leray projection operator $\mathbb{P}$ is bounded on $L^2(w_0\,dx)$ and on $H^{-s_1}_{p_1}$, we find that

$$\vec{u}_0 \in [(L^2(w_0\,dx))_\sigma, (H^{-s_1}_{p_1})_\sigma]_{\theta,\infty} \subset [(L^2(\frac{1}{1+|x|^2}\,dx))_\sigma, (B^{-\gamma}_{\infty,\infty})_\sigma]_{\theta,\infty}$$

so that $\vec{u}_0$ fulfills the assumptions of Theorem 14.12.

Step 3.

Now, we assume that $\vec{u}_0 \in L^2 \cap bmo^{-1} \cap [(L^2(\frac{1}{1+|x|^2}\,dx))_\sigma, (B^{-\gamma}_{\infty,\infty})_\sigma]_{\theta,\infty}$ with $\gamma < 1$ and $0 < \theta < 1$, and we shall prove the following lemma of Barker:

There exists a constant C_1 (depending on $\vec{u}_0$) such that, for every $t \in (0, T)$, for every suitable weak Leray solution $\vec{v}$ of the Navier–Stokes equations with initial value $\vec{u}_0$, we have

$$\|\vec{v}(t,.) - \vec{u}(t,.)\|_{L^2(\frac{1}{1+|x|^2} \, dx)} \leq C_1 t^\eta \tag{14.55}$$

with $\eta = \frac{\theta(1-\gamma)}{2(1-\theta)}$, where $\|\vec{u}\|_{X_T} \leq \frac{1}{2C_0}$ and $\vec{u}$ is the mild solution on $(0, T)$.

Of course, we need to prove (14.55) only for $t < T_0$ for some T_0 depending on $\vec{u}_0$, since for $t > T_0$ we can write

$$\|\vec{v}(t,.) - \vec{u}(t,.)\|_{L^2(\frac{1}{1+|x|^2} \, dx)} \leq \|\vec{v}(t,.)\|_2 + \|\vec{u}(t,.)\|_2 \leq 2\|\vec{u}_0\|_2 \left(\frac{t}{T_0}\right)^\eta.$$

For every $\epsilon \in (0, 1)$ we can split $\vec{u}_0$ in

$$\vec{u}_0 = \vec{v}_{0,\epsilon} + \vec{w}_{0,\epsilon}$$

with

$$\operatorname{div} \vec{v}_{0,\epsilon} = \operatorname{div} \vec{w}_{0,\epsilon} = 0, \, \|\vec{v}_{0,\epsilon}\|_{\dot{B}_{\infty,\infty}^{-\gamma}} \leq C_2 \epsilon^{\theta-1} \text{ and } \|\vec{w}_{0,\epsilon}\|_{L^2(\frac{1}{1+|x|^2} \, dx)} \leq C_2 \epsilon^\theta,$$

where C_2 depends only on $\vec{u}_0$. For $0 < t \leq 1$, $\|e^{t\Delta} \vec{v}_{0,\epsilon}\|_\infty \leq C_3 t^{-\gamma/2} \epsilon^{\theta-1}$. If $0 < T_1 < 1$, we have

$$\sup_{0 < t < T_1} \sqrt{t} \|e^{t\Delta} \vec{v}_{0,\epsilon}\|_\infty \leq C_3 \epsilon^{\theta-1} T_1^{\frac{1-\gamma}{2}}$$

and

$$\sup_{0 < t < T_1, x \in \mathbb{R}^3} \sqrt{\frac{1}{t^{3/2}} \int_0^t \int_{B(x,\sqrt{t})} |e^{t\Delta} \vec{v}_{0,\epsilon}|^2 \, dx} \leq C_4 \epsilon^{\theta-1} T_1^{\frac{1-\gamma}{2}}$$

so that $\|e^{t\Delta} \vec{v}_{0,\epsilon}\|_{X_{T_1}} \leq (C_3 + C_4) \epsilon^{\theta-1} T_1^{\frac{1-\delta}{2}} < \frac{1}{8C_0}$ if $T_1 < C_5 \epsilon^{\frac{2}{1-\gamma}(1-\theta)})$ [with $C_5 < 1$ so that $T_1 < 1$].

According to Step 1, we know that the Navier–Stokes equations with initial value $\vec{v}_{0,\epsilon}$ will have a solution $\vec{v}_\epsilon$ on $(0, T_1)$ such that $\|\vec{v}_\epsilon(t,.)\|_\infty \leq C_6 t^{-\gamma/2} \epsilon^{\theta-1}$. Moreover, by Step 1, $\vec{v}_\epsilon$ is a weighted Leray weak solution (since $\vec{v}_{0,\epsilon} = \vec{u}_0 - \vec{w}_{0,\epsilon} \in L^2(\frac{1}{1+|x|^2} \, dx)$).

Now, if $\vec{v}$ is a suitable weak Leray solution of the Navier–Stokes equations with initial value $\vec{u}_0$, $\vec{v}$ is a weighted Leray weak solution as well and, by Lemma 14.7 (since $\vec{v}_\epsilon \in L^2((0, T_1), L^\infty)$), we know that for every $t_0 \in (0, T_1)$ we have (writing L_Φ^2 for $L^2(\frac{1}{1+|x|^2} \, dx)$)

$$\|\vec{v}(t_0,.) - \vec{v}_\epsilon(t_0,.)\|_{L_\Phi^2}^2$$
$$\leq e^{C_7 t_0 (\nu + \frac{\sup_{0 < s < t_0} (\|\vec{v}(s,.)\|_{L_\Phi^2} + \|\vec{v}_\epsilon(s,.)\|_{L_\Phi^2})^4}{\nu^3})} e^{C_7 \int_0^{t_0} \frac{\|\vec{v}_\epsilon\|_\infty^2}{\nu} \, dt} \|\vec{w}_{0,\epsilon}\|_{L_\Phi^2}^2. \tag{14.56}$$

Recall that $\|\vec{w}_{0,\epsilon}\|_{L^2(\frac{1}{1+|x|^2} \, dx)} \leq C_2 \epsilon^\theta$, $\|\vec{v}\|_{L^2(\frac{1}{1+|x|^2} \, dx)} \leq \|\vec{v}(s,.)\|_2 \leq \|\vec{u}_0\|_2$,

$$\int_0^{t_0} \|\vec{v}_\epsilon\|_\infty^2 \, ds \leq C_6^2 \epsilon^{2(\theta-1)} \frac{1}{1-\gamma} t_0^{1-\gamma} \leq C_6^2 \frac{1}{1-\gamma} C_4^{1-\gamma},$$

and, by Step 1,

$$\sup_{0 < s < T_1} \|\vec{v}_\epsilon\|_{L^2(\frac{1}{1+|x|^2} \, dx)} \leq C_8 \|\vec{v}_{0,\epsilon}\|_{L^2(\frac{1}{1+|x|^2} \, dx)} \leq C_8 (\|\vec{u}_0\|_2 + C_2 \epsilon^\theta).$$

Thus, for a constant C_9 depending only on $\vec{u}_0$, ν and T, we get

$$\|\vec{v}(t_0,.) - \vec{u}(t_0,.)\|_{L_\Phi^2}^2 \leq \|\vec{v}(t_0,.) - \vec{v}_\epsilon(t_0,.)\|_{L_\Phi^2}^2 + \|\vec{u}(t_0,.) - \vec{v}_\epsilon(t_0,.)\|_{L_\Phi^2}^2 \leq C_9 \epsilon^\theta \quad (14.57)$$

In particular, for $t = \frac{1}{2} C_5 \epsilon^{\frac{2}{1-\gamma}(1-\theta)}$, we find

$$\|\vec{v}(t,.) - \vec{u}(t,.)\|_{L_\Phi^2}^2 \leq C_{10} t^\eta. \quad (14.58)$$

This inequality has thus been proved for every $t \in (0, T_0)$ with $T_0 = \min(T, \frac{2}{C_5})$.

Step 4.

We now prove weak-strong uniqueness when $\vec{u}_0 \in L^2 \cap bmo^{-1} \cap [(L^2(\frac{1}{1+|x|^2} dx))_\sigma,$ $(B_{\infty,\infty}^{-\gamma})_\sigma]_{\theta,\infty}$ with $\gamma < 1$.

Let $\vec{u}$ be the mild solution of the Navier–Stokes equations with initial value $\vec{u}_0$ such that $\|\vec{u}\|_{X_T} < \frac{1}{2C_0}$ and let $\vec{v}$ be a suitable weak Leray solution of the Navier–Stokes equations with the same initial value $\vec{u}_0$. As $\vec{u} \in L^2((\epsilon, T), L^\infty)$ for every $\epsilon \in (0, T)$ and as $\vec{v}$ is a suitable Leray solution on (t_0, T) for almost every $t_0 \in (0, T)$, we may apply Lemma 14.8 and find for every $t \in (t_0, T)$

$$\|\vec{u}(t,.) - \vec{v}(t.)\|_2^2 + 2\nu \int_{t_0}^t \|\vec{\nabla} \otimes (\vec{u} - \vec{v})\|_2^2 \, ds$$
$$\leq \|\vec{u}(t_0,.) - \vec{v}(t_0,.)\|_2^2 + 2 \iint_{t_0}^t \vec{u} \cdot ((\vec{u} - \vec{v}) \cdot \vec{\nabla}(\vec{u} - \vec{v})) \, dx \, ds. \quad (14.59)$$

We get

$$\|\vec{u}(t,.) - \vec{v}(t.)\|_2^2 \leq \|\vec{u}(t_0,.) - \vec{v}(t_0,.)\|_2^2 + \frac{1}{\nu} \int_{t_0}^t \|\vec{u}\|_\infty^2 \|\vec{u} - \vec{v}\|_2^2 \, ds. \quad (14.60)$$

Letting t_0 go to 0, we find

$$\|\vec{u}(t,.) - \vec{v}(t.)\|_2^2 \leq \frac{1}{\nu} \int_0^t \|\vec{u}\|_\infty^2 \|\vec{u} - \vec{v}\|_2^2 \, ds. \quad (14.61)$$

We know by Step 3 that $\|\vec{u}(t,.) - \vec{v}(t.)\|_2^2 \leq C t^{2\eta}$, so that

$$t^{-2\eta} \|\vec{u}(t,.) - \vec{v}(t.)\|_2^2 \leq \frac{1}{\nu} (\sup_{0<s<t} \sqrt{s}\|\vec{u}(s,.)\|_\infty)^2 \frac{1}{2\eta} \sup_{0<s<t} s^{-2\eta}\|\vec{u}(s,.) - \vec{v}(s.)\|_2^2$$

For t_0 such that $\sup_{0<s<t_0} \sqrt{s}\|\vec{u}(s,.)\|_\infty \leq \sqrt{2\eta\nu}$, we find

$$\sup_{0<s<t_0} s^{-2\delta}\|\vec{u}(s,.) - \vec{v}(s.)\|_2^2 = 0$$

so that $\vec{u} = \vec{v}$ on $(0, t_0)$; on the other hand, we get

$$\|\vec{u}(t,.) - \vec{v}(t.)\|_2^2 \leq \frac{1}{\nu} \sup_{0<s<T} s\|\vec{u}(s,.)\|_\infty^2 \int_0^t \frac{1}{t_0} \|\vec{u} - \vec{v}\|_2^2 \, ds$$

and thus $\vec{u} = \vec{v}$ on $(0, T)$.

$$\square$$

14.9 Further Results on Global Existence of Suitable Weak Solutions

We have seen various cases of existence of global suitable weak solutions of the Cauchy problem for the Navier–Stokes equations

$$\begin{cases} \partial_t \vec{u} = \nu \Delta \vec{u} - \vec{u} \cdot \vec{\nabla} \vec{u} - \vec{\nabla} p \\ \operatorname{div} \vec{u} = 0 \\ \vec{u}(0,.) = \vec{u}_0 \end{cases} \tag{14.62}$$

where $\vec{u}_0$ is a locally square integrable divergence free vector field:

- when $\vec{u}_0 \in L^2$, Leray's mollification and Rellich's theorem give a solution $\vec{u} \in L^\infty((0,+\infty), L^2) \cap L^2((0,+\infty), \dot{H}^1)$ (Theorem 12.2) which is suitable (Theorem 13.6);

- when $\vec{u}_0 \in L^2(\mathbb{R}^3/2\pi\mathbb{Z}^3)$ with $\int_{(-\pi,\pi)^3} \vec{u}_0\,dx = 0$, Leray's mollification and Rellich's theorem give a periodical solution $\vec{u} \in L^\infty((0,+\infty), L^2(\mathbb{R}^3/2\pi\mathbb{Z}^3))$ with $\vec{\nabla} \otimes \vec{u} \in L^2((0,+\infty), L^2(\mathbb{R}^3/2\pi\mathbb{Z}^3))$ (see page 398); we can prove that this solution is suitable in exactly the same way as for the case $\vec{u}_0 \in L^2$;

- when $\vec{u}_0 \in E^2$, i.e. when $\vec{u}_0 \in L^2_{\text{uloc}}$ and $\lim_{x_0 \to \infty} \int_{B(x_0,1)} |\vec{u}_0|^2\,dx = 0$, Theorem 14.8 provides a suitable weak solution $\vec{u}$ such that

$$\vec{u} \in \bigcap_{0<T<+\infty} (L^\infty_t L^2_x)_{\text{uloc}}((0,T) \times \mathbb{R}^3) \cap (L^2 H^1_x)_{\text{uloc}}((0,T) \times \mathbb{R}^3);$$

- when $\vec{u}_0 \in L^2(\Phi\,dx)$ with $\Phi(x) = \frac{1}{1+|x|^2}$, Theorem 14.9 provides a suitable weak solution $\vec{u}$ such that

$$\vec{u} \in \bigcap_{0<T<+\infty} L^\infty((0,T), L^2(\Phi\,dx)) \text{ and } \vec{\nabla} \otimes \vec{u} \in \bigcap_{0<T<+\infty} L^2((0,T), L^2(\Phi\,dx)).$$

We remark that the control in $(L^\infty_t L^2_x)_{\text{uloc}}((0,T) \times \mathbb{R}^3)$ when $\vec{u}_0 \in E^2$ or in $L^\infty((0,T), L^2(\Phi\,dx))$ when $\vec{u}_0 \in L^2(\Phi\,dx)$ are not uniform with respect to T and may be less and less precise when T goes to $+\infty$. However, it is possible to recover uniform controls in time for special classes of weak solutions:

Proposition 14.3.
For $0 < \gamma < 1$, let $\Phi_\gamma(x) = \frac{1}{(1+|x|)^\gamma}$. When $\vec{u}_0 \in L^2(\Phi_\gamma\,dx)$, problem (14.62) has a suitable weak solution $\vec{u}$ such that

$$\vec{u} \in L^\infty((0,+\infty), L^2(\Phi_\gamma\,dx)) \text{ and } \vec{\nabla} \otimes \vec{u} \in L^2((0,+\infty), L^2(\Phi_\gamma\,dx)).$$

Proof. A proof similar to the proof of Theorem 14.9 provides, when $\vec{u}_0$ belongs to a weighted Lebesgue space $L^2(\Phi\,dx)$, a suitable weak solution $\vec{u}$ such that

$$\vec{u} \in \bigcap_{0<T<+\infty} L^\infty((0,T), L^2(\Phi\,dx)) \text{ and } \vec{\nabla} \otimes \vec{u} \in \bigcap_{0<T<+\infty} L^2((0,T), L^2(\Phi\,dx)),$$

under the following conditions on the weight Φ:

- **(H0)** Φ is a continuous Lipschitz function on $\mathbb{R}^3$

- **(H1)** $0 < \Phi \leq 1$.

- **(H2)** There exists $C_1 > 0$ such that $|\vec{\nabla}\Phi| \leq C_1 \Phi^{\frac{3}{2}}$

- **(H3)** $\Phi^{4/3} \in \mathcal{A}_{4/3}$ (where $\mathcal{A}_{4/3}$ is the Muckenhoupt class of weights).

- **(H4)** There exists $C_2 > 0$ such that $\Phi(x) \leq \Phi(\frac{x}{\lambda}) \leq C_2\lambda^2\Phi(x)$, for all $\lambda \geq 1$.

(For details, see Fernández-Dalgo and Lemarié-Rieusset [173, 174]). In particular, we have the inequality

$$\frac{d}{dt}\|\sqrt{\Phi}\vec{u}\|_2^2 + 2\nu\|\sqrt{\Phi}\vec{\nabla}\otimes\vec{u}\|_2^2 \leq -\nu\int \vec{\nabla}(|\vec{u}|^2)\cdot\vec{\nabla}\Phi\,dx + \int(|\vec{u}|^2 + 2p)\vec{u}\cdot\vec{\nabla}\Phi\,dx.$$

We have

$$\|\vec{\nabla}\otimes(\sqrt{\Phi}\vec{u})\|_2^2 \leq 2\|\sqrt{\Phi}\vec{\nabla}\otimes\vec{u}\|_2^2 + 2C_1^2\|\sqrt{\Phi}\vec{u}\|_2^2$$

so that $\sqrt{\Phi}\vec{u} \in \dot{H}^1$, hence

$$\|\sqrt{\Phi}\vec{u}\|_6 \leq C\|\vec{\nabla}\otimes(\sqrt{\Phi}\vec{u})\|_2$$

Writing (since $\Phi^{4/3} \in \mathcal{A}_{4/3}$)

$$\|\Phi(|\vec{u}|^2 + 2p)\|_{4/3} \leq C\|\sqrt{\Phi}\vec{u}\|_{8/3}^2 \leq C\|\sqrt{\Phi}\vec{u}\|_2^{5/4}\|\sqrt{\Phi}\vec{u}\|_6^{3/4},$$

while

$$\|\sqrt{\Phi}\vec{u}\|_4 \leq C\|\sqrt{\Phi}\vec{u}\|_2^{1/4}\|\sqrt{\Phi}\vec{u}\|_6^{3/4}.$$

If we follow the proof of Theorem 14.9, we then write the inequality

$$\frac{d}{dt}\|\sqrt{\Phi}\vec{u}\|_2^2 + 2\nu\|\sqrt{\Phi}\vec{\nabla}\otimes\vec{u}\|_2^2$$

$$\leq 2C_1\nu\|\sqrt{\Phi}\vec{u}\|_2\|\sqrt{\Phi}\vec{\nabla}\otimes\vec{u}\|_2$$

$$+ CC_1\|\sqrt{\Phi}\vec{u}\|_2^{3/2}(\|\sqrt{\Phi}\vec{\nabla}\otimes\vec{u}\|_2^{3/2} + C_1^{3/2}\|\sqrt{\Phi}\vec{u}\|_2^{3/2})$$

$$\leq \nu\|\sqrt{\Phi}\vec{\nabla}\otimes\vec{u}\|_2^2 + 2C_1^2\nu\|\sqrt{\Phi}\vec{u}\|_2^2$$

$$+ CC_1^{5/2}\|\sqrt{\Phi}\vec{u}\|_2^3 + (CC_1)^4\left(\frac{2}{\nu}\right)^3\|\sqrt{\Phi}\vec{u}\|_2^6$$

which provides a local-in-time control of $\vec{u}$: we have

$$\|\sqrt{\Phi}\vec{u}\|_2^2 + 2\nu\int_0^t\|\sqrt{\Phi}\vec{\nabla}\otimes\vec{u}\|_2^2\,ds \leq 2\|\sqrt{\Phi}\vec{u}_0\|_2^2$$

on $(0, T)$ with $C_1^2\nu T(4 + CC_1^2\nu^{-4}\|\sqrt{\Phi}\vec{u}_0\|_2^4) = 1/8$.

In the case of $\Phi = \Phi_\gamma = \frac{1}{(1+|x|)^\gamma}$ with $0 < \gamma < 1$, , we can modify the proof in the following way:

- We have

$$|\partial_i\partial_j\Phi_\gamma| \leq C_\gamma\frac{1}{|x|}\Phi_\gamma;$$

as

$$\left\|\frac{1}{|x|}\sqrt{\Phi_\gamma}\vec{u}\right\|_2 \leq C\|\vec{\nabla}\otimes(\sqrt{\Phi_\gamma}\vec{u})\|_2,$$

we can write

$$-\nu \int \vec{\nabla}(|\vec{u}|^2) \cdot \vec{\nabla}\Phi_\gamma \, dx = \nu \int |\vec{u}|^2 \Delta\Phi_\gamma \, dx$$

$$= \nu \int |\vec{u}|^2 \left(\frac{\gamma(1+\gamma)}{(1+|x|)^{\gamma+2}} - 2\gamma \frac{1}{|x|(1+|x|)^{\gamma+1}} \right) dx$$

$$= -\nu\gamma \int |\vec{u}|^2 \frac{(1-\gamma)|x| + 2}{|x|(1+|x|)^{\gamma+2}} \, dx$$

$$\leq -\nu\gamma(1-\gamma) \int |\vec{u}|^2 \frac{1}{(1+|x|)^{\gamma+1}} \, dx.$$

- We write (as $\Phi_\gamma^{3/2} \in \mathcal{A}_{3/2}$)

$$\int (|\vec{u}|^2 + 2p)\vec{u} \cdot \vec{\nabla}\Phi_\gamma \, dx \leq \|\Phi_\gamma(|\vec{u}|^2 + 2p)\|_{3/2} \frac{1}{\Phi_\gamma}\vec{u} \cdot \vec{\nabla}\Phi\|_3$$

$$\leq C\|\sqrt{\Phi_\gamma}\vec{u}\|_3^2 \|\frac{1}{\Phi_\gamma}\vec{u} \cdot \vec{\nabla}\Phi_\gamma\|_6^{1/2} \|\frac{1}{\Phi_\gamma}\vec{u} \cdot \vec{\nabla}\Phi_\gamma\|_2^{1/2}$$

$$\leq CC_1^{1/2}\|\sqrt{\Phi_\gamma}\vec{u}\|_2 \|\sqrt{\Phi_\gamma}\vec{u}\|_6^{3/2} \|\frac{\gamma}{1+|x|}\vec{u}\|_2^{1/2}$$

$$\leq \nu\gamma(1-\gamma)\|\frac{1}{1+|x|}\vec{u}\|_2^2$$

$$+ C'\left(\frac{\gamma}{\nu(1-\gamma)}\right)^{1/3} C_1^{2/3}\|\sqrt{\Phi_\gamma}\vec{u}\|_2^{4/3} \|\sqrt{\Phi_\gamma}\vec{u}\|_6^2.$$

- As $\int |\vec{u}|^2 \frac{1}{(1+|x|)^2} \, dx \leq \int |\vec{u}|^2 \frac{1}{(1+|x|)^{\gamma+1}} \, dx$, we get

$$\frac{d}{dt}\|\sqrt{\Phi_\gamma}\vec{u}\|_2^2 + 2\nu\|\sqrt{\Phi_\gamma}\vec{\nabla} \otimes \vec{u}\|_2^2 \leq C_{\nu,\gamma}\|\sqrt{\Phi_\gamma}\vec{u}\|_2^{4/3} \|\sqrt{\Phi_\gamma}\vec{u}\|_6^2.$$

We have

$$\|\sqrt{\Phi_\gamma}\vec{u}\|_6 \leq C\|\vec{\nabla} \otimes (\sqrt{\Phi_\gamma}\vec{u})\|_2 \leq C\|\sqrt{\Phi_\gamma}\vec{\nabla} \otimes \vec{u}\|_2 + C\|(\vec{\nabla}\sqrt{\Phi_\gamma}) \otimes \vec{u}\|_2.$$

Writing $v = \sqrt{\Phi_\gamma}|\vec{\nabla} \otimes \vec{u}|$, we have

$$|(\vec{\nabla}\sqrt{\Phi_\gamma}) \otimes \vec{u}| = |\sum_{k=1}^{3} (\vec{\nabla}\sqrt{\Phi_\gamma}) \otimes \frac{\partial_k}{\Delta}\partial_k\vec{u}|$$

$$\leq C\frac{1}{1+|x|}\sqrt{\Phi_\gamma}\frac{1}{\sqrt{-\Delta}}(\frac{1}{\sqrt{\Phi_\gamma}}v).$$

We have

$$(1+|y|)^{\gamma/2} \leq C((1+|x|)^{\gamma/2} + |x-y|^{\gamma/2})$$

so that

$$|(\vec{\nabla}\sqrt{\Phi_\gamma}) \otimes \vec{u}| \leq C\frac{1}{1+|x|}\left(\frac{1}{\sqrt{-\Delta}}v + \sqrt{\Phi_\gamma}\frac{1}{(\sqrt{-\Delta})^{1+\frac{\gamma}{2}}}v\right).$$

Thus,

$$\||(\vec{\nabla}\sqrt{\Phi_\gamma}) \otimes \vec{u}|\|_2 \leq C\|\frac{1}{1+|x|}\|_{L^{3,\infty}}\|\frac{1}{\sqrt{-\Delta}}v\|_{L^{6,2}}$$

$$+ C\|\frac{1}{1+|x|}\sqrt{\Phi_\gamma}\|_{L^{\frac{6}{2+\gamma},\infty}}\|\frac{1}{(\sqrt{-\Delta})^{1+\frac{\gamma}{2}}}v)\|_{L^{\frac{6}{1-\gamma},2}}.$$

$$\leq C'\|v\|_2 = C'\|\sqrt{\Phi_\gamma}\vec{\nabla} \otimes \vec{u}\|_2.$$

Summing up those estimates, we find

$$\frac{d}{dt}\|\sqrt{\Phi_\gamma}\,\vec{u}\|_2^2 + 2\nu\|\sqrt{\Phi_\gamma}\,\vec\nabla\otimes\vec{u}\|_2^2 \leq C_{\nu,\gamma}\|\sqrt{\Phi_\gamma}\,\vec{u}\|_2^{4/3}\|\sqrt{\Phi_\gamma}\,\vec\nabla\otimes\vec{u}\|_2^2. \tag{14.63}$$

Thus, if $C_{\nu,\gamma}\|\sqrt{\Phi_\gamma}\,\vec{u}_0\|_2^{4/3} < \nu$, we find that $\|\sqrt{\Phi_\gamma}\,\vec{u}\|_2^2$ is non-increasing and get a uniform control on $(0,+\infty)$:

$$\|\sqrt{\Phi_\gamma}\,\vec{u}(t,.)\|_2^2 + \nu\int_0^t\|\sqrt{\Phi_\gamma}\,\vec\nabla\otimes\vec{u}\|_2^2\,ds \leq \|\sqrt{\Phi_\gamma}\,\vec{u}_0(t,.)\|_2^2.$$

We then finish the proof by noticing that, for $\vec{u} \in L^2(\Phi_\gamma\,dx)$, we have $\lim_{\lambda\to+\infty}\|\lambda\vec{u}(\lambda\cdot)\|_{L^2(\Phi_\gamma\,dx}=0$; thus, for some $\lambda_0 > 1$, we have a control on $\vec{u}_{\lambda_0}(t,x) = \lambda_0\vec{u}(\lambda_0^2 t,\lambda_0 x)$:

$$\|\sqrt{\Phi_\gamma}\,\vec{u}_{\lambda_0}(t,.)\|_2^2 + \nu\int_0^t\|\sqrt{\Phi_\gamma}\,\vec\nabla\otimes\vec{u}_{\lambda_0}\|_2^2\,ds \leq \|\sqrt{\Phi_\gamma}\,\vec{u}_{\lambda_0,0}\|_2^2,$$

or equivalently:

$$\left\|\sqrt{\Phi_\gamma(\tfrac{x}{\lambda_0})}\,\vec{u}(t,.)\right\|_2^2 + \nu\int_0^t\left\|\sqrt{\Phi_\gamma(\tfrac{x}{\lambda_0})}\,\vec\nabla\otimes\vec{u}\right\|_2^2\,ds \leq \left\|\sqrt{\Phi_\gamma(\tfrac{x}{\lambda_0})}\,\vec{u}_0\right\|_2^2.$$

Finally, since $\Phi_\gamma(x) \leq \Phi_\gamma(\tfrac{x}{\lambda_0}) \leq \lambda_0^\gamma\Phi_\gamma(x)$, we get

$$\|\sqrt{\Phi_\gamma}\,\vec{u}(t,.)\|_2^2 + \nu\int_0^t\|\sqrt{\Phi_\gamma}\,\vec\nabla\otimes\vec{u}\|_2^2\,ds \leq \lambda_0^\gamma\|\sqrt{\Phi_\gamma}\,\vec{u}_0(t,.)\|_2^2. \qquad \square$$

We may as well discuss the control in L^2_{uloc} of a weak solution associated to a large initial value which does not vanish at infinity (so that Theorem 14.8 can not be used). A way to get such a control is to assume that the uniform control in L^2_{loc} (uniform with respect to spatial shifts of the argument) can be extended to a uniform control in $L^2(\frac{1}{(1+|x|)^\gamma}\,dx)$:

Proposition 14.4.
For $0 < \gamma < 2$, let $\Phi_\gamma(x) = \frac{1}{(1+|x|)^\gamma}$. When $\vec{u}_0 \in L^2(\Phi_\gamma\,dx)$ is such that

$$\sup_{x_0\in\mathbb{R}^3}\|\vec{u}_0(x-x_0)\|_{L^2(\Phi_\gamma\,dx)} < +\infty,$$

problem (14.62) has a suitable weak solution $\vec{u}$ such that

$$\vec{u}\in\bigcap_{0<T<+\infty}(L^\infty_t L^2_x)_{uloc}((0,T)\times\mathbb{R}^3)\cap(L^2 H^1_x)_{uloc}((0,T)\times\mathbb{R}^3).$$

Proof. In Theorem 14.2, we saw that every local Leray solution could be controlled on a small interval time whose size depends on the norm of $\vec{u}_0$ in L^2_{uloc}. Such a result is valid for a weighted local Leray solution [173]: if $\vec{u}$ is a suitable weak solution to (14.62) such that, for some $\gamma\in(0,2)$,

$$\vec{u}\in\bigcap_{0<T<+\infty}L^\infty((0,T),L^2(\Phi_\gamma\,dx))\quad\text{and}\quad\vec\nabla\otimes\vec{u}\in\bigcap_{0<T<+\infty}L^2((0,T),L^2(\Phi_\gamma\,dx))$$

(with $\lim_{t\to 0} \int_K |\vec{u}(t,x) - \vec{u}_0(x)|^2\, dx = 0$ for every compact subset K of $\mathbb{R}^2$), then for every $\lambda > 1$ and $T_\lambda \approx \frac{1}{C_{\gamma,\nu}(1+\|\lambda\vec{u}_0(\lambda x)\|_{L^2(\Phi_\gamma\, dx)})^4}$, we have a control for $\vec{u}_\lambda(t,x) = \lambda^2\vec{u}(\lambda^2 t, \lambda x)$, which gives a control of $\vec{u}$ on $(0, \lambda^2 T_\lambda)$:

$$\sup_{0 < t < \lambda^2 T_\lambda} \|\vec{u}(t,.)\|^2_{L^2(\Phi_\gamma\, dx)} + \int_0^{\lambda^2 T_\lambda} \|\vec{\nabla} \otimes u(t,.)\|^2_{L^2(\Phi_\gamma\, dx)}\, dt$$

$$\leq C_{\gamma,\nu}\lambda\|\lambda\vec{u}_0(\lambda x)\|^2_{L^2(\Phi_\gamma\, dx)}$$

$$\leq C_{\gamma,\nu}\lambda^\gamma\|\vec{u}_0(x)\|^2_{L^2(\Phi_\gamma\, dx)}. \tag{14.64}$$

Assume now that $\sup_{x_0 \in \mathbb{R}^3} \|\vec{u}_0(x - x_0)\|_{L^2(\Phi_\gamma\, dx)} < +\infty$ for somme $\gamma < 2$. With no loss of generality, we may asssume that $1 < \gamma < 2$ (as $\Phi_{\gamma_2} \leq \Phi_{\gamma_1}$ for $\gamma_1 < \gamma_2$). Applying the control (14.64) to $\vec{u}(x - x_0)$ instead of $\vec{u}_0$, we find that $\|\lambda\vec{u}_0(\lambda x - x_0)\|_{L^2(\Phi_\gamma\, dx)} \leq \lambda^{\frac{\gamma-1}{2}}\|\vec{u}_0(x - x_0)\|_{L^2(\Phi_\gamma\, dx)}$ which is uniformly small (with respect to x_0) when λ is great, so that, for $T \geq T_0$, T_0 large enough and independent from x_0, we have

$$\sup_{0 < t < T} \|\vec{u}(t, x - x_0)\|^2_{L^2(\Phi_\gamma\, dx)} + \int_0^T \|\vec{\nabla} \otimes u(t,.)\|^2_{L^2(\Phi_\gamma\, dx)}\, dt$$

$$\leq C_{\gamma,\nu}\left(\frac{T}{T_0}\right)^{\gamma/2}\|\vec{u}_0(x - x_0)\|^2_{L^2(\Phi_\gamma\, dx)}.$$

Thus, $\vec{u}$ is controlled in L^2_{uloc}. $\square$

Our next example deals with a sum of plane waves with large amplitudes:

Theorem 14.13.
Let $N \geq 1$, $\vec{\omega}_1, \ldots, \vec{\omega}_N$ N vectors in $\mathbb{R}^3 \setminus \{0\}$, $\vec{A}_1 \ldots, \vec{A}_N$ N vectors in $\mathbb{R}^3 \setminus \{0\}$, $\vec{B}_1 = \vec{A}_1 \wedge \vec{\omega}_1, \ldots, \vec{B}_N = \vec{A}_N \wedge \vec{\omega}_N$. For $\theta = (\theta_1, \ldots, \theta_N) \in \mathbb{R}^N/\mathbb{Z}^N = \mathbb{T}^N$, define

$$\vec{u}_0(x, \theta) = \sum_{k=1}^N \cos(\vec{\omega}_k \cdot x + 2\pi\theta_k)\vec{B}_k.$$

Let $3 < \gamma \leq 4$ and let $\Phi_\gamma(x) = \frac{1}{(1+|x|)^\gamma}$. Then, for almost very $\theta \in \mathbb{T}^N$, the problem

$$\begin{cases} \partial_t \vec{u} = \nu\Delta\vec{u} - \vec{u} \cdot \vec{\nabla}\vec{u} - \vec{\nabla}p \\ \text{div } \vec{u} = 0 \\ \vec{u}(0, \cdot) = \vec{u}_0(\cdot, \theta) \end{cases} \tag{14.65}$$

has a global weak solution $\vec{u}_\theta$ with $\vec{u}, \vec{\nabla} \otimes \vec{u}_\theta \in L^2((0,T), L^2(\Phi_\gamma\, dx))$ for every $T > 0$.

Proof. We have seen in Proposition 8.1 that, if the amplitudes $\vec{B}_k$ are small enough, then we have a global mild solution, as proved by Dinaburg and Sinai [153] or as a consequence of the Koch and Tataru theorem since $\vec{u}_{0,\theta} \in BMO^{-1}$. Thus, the problem we study deals with large amplitudes. We have $\vec{u}_0(\cdot, \theta) \in L^2_{\text{uloc}}$, but it does not vanish at infinity[1], so we cannot use Theorem 14.8. Similarly, we cannot use Theorem 14.9, as $\vec{u}_0(., \theta) \in L^2(\Phi_\gamma\, dx)$ implies $\gamma > 3$, whereas we can construct weak solutions in $L^2(\Phi_\gamma\, dx)$ only for $\gamma \leq 2$. Thus, we need new ideas. Theorem 19.2 is a special case of the theory of homogeneous statistical

---

[1] As a matter of fact, $\vec{u}_0$ vanishes at infinity in the sense that $\lim_{t\to+\infty} \|e^{t\Delta}\vec{u}_0\|_\infty = 0$ but it does not belong to E^2

solutions developed by Višik and Fursikov (see Višik and Fursikov [189, 190], Foias and Temam [180, 179] or Basson [23]).

The main property of the family of divergence-free vector fields $(\vec{u}_0(\cdot, \theta))_{\theta \in \mathbb{T}^N}$ is its stability under shifts of the argument: if $\tau_{x_0} f(x) = f(x - x_0)$, we have

$$\tau_{x_0}(\vec{u}_0(\cdot, \theta)) = \vec{u}_0(\cdot, \tau_{x_0}^* \theta)$$

with

$$\tau_{x_0}^* \theta = (\theta_1 - \frac{1}{2\pi} \vec{\omega}_1 \cdot x_0, \ldots, \theta_N - \frac{1}{2\pi} \vec{\omega}_N \cdot x_0).$$

The transform $\theta \mapsto \tau_{x_0}^* \theta$ preserves the Lebesgue measure on $\mathbb{T}^N$.

We follow Basson's ideas and approximate the (shift-invariant) family $(\vec{u}_0(\cdot, \theta))_{\theta \in \mathbb{T}^N}$ by another shift-invariant family $(\vec{u}_{0,\alpha}(\cdot, \theta))_{\theta \in \mathbb{T}^N}$ (indexed by $\alpha > 0$) defined by

$$\vec{u}_{0,\alpha}(x, \theta) = \sum_{k=1}^{N} \cos(\vec{\omega}_{k,\alpha} \cdot x + 2\pi \theta_k) \vec{B}_{k,\alpha}$$

with $\vec{\omega}_{k,\alpha} \in \mathbb{Q}^3 \setminus \{0\}$, $|\vec{\omega}_k - \vec{\omega}_{k,\alpha}| < \alpha$ and $\vec{B}_{k,\alpha} = \vec{A}_k \wedge \vec{\omega}_{k,\alpha}$. On any ball $B(0, R)$, we have

$$|\vec{u}_{0,\alpha}(x, \theta) - \vec{u}_0(x, \theta)| \le \alpha \sum_{k=1}^{N} |\vec{A}_k|(R + |\vec{\omega}_k|).$$

An important property of the family $(\vec{u}_{0,\alpha}(\cdot, \theta))_{\theta \in \mathbb{T}^N}$ is its periodicity: for some $L_\alpha > 0$, we have for every $x \in \mathbb{R}^3$, $\theta \in \mathbb{T}^N$ and $k \in \mathbb{Z}^3$

$$\vec{u}_{0,\alpha}(x + k L_\alpha, \theta) = \vec{u}_{0,\alpha}(x, \theta).$$

We then consider a mollified Cauchy problem for the Navier–Stokes equations with initial data $\vec{u}_{0,\alpha}(\cdot, \theta)$: we take $\varphi \in \mathcal{D}(\mathbb{R}^3)$ with $\int \varphi \, dx = 1$, we define, for $\epsilon > 0$, $\varphi_\epsilon(x) = \frac{1}{\epsilon^3}\varphi(\frac{x}{\epsilon})$; then we define $\vec{v}_{\epsilon,\alpha,\theta}(t, x)$ the solution in $L^\infty((0, +\infty), L^2(\mathbb{R}^3/L_\alpha \mathbb{Z}^3))$ of

$$\begin{cases} \partial_t \vec{v}_{\epsilon,\alpha,\theta} = \nu \Delta \vec{v}_{\epsilon,\alpha,\theta} - (\varphi_\epsilon * \vec{v}_{\epsilon,\alpha,\theta}) \cdot \vec{\nabla} \vec{v}_{\epsilon,\alpha,\theta} - \vec{\nabla} p_{\epsilon,\alpha,\theta} \\ \operatorname{div} \vec{v}_{\epsilon,\alpha,\theta} = 0 \\ \vec{v}_{\epsilon,\alpha,\theta}(0, \cdot) = \vec{u}_{0,\alpha}(\cdot, \theta) \end{cases} \qquad (14.66)$$

By usual arguments in the study of the mollified Cauchy problem, we see that we have a unique global solution such that $v_{\epsilon,\alpha,\theta} \in L^\infty((0, +\infty), L^2(\mathbb{R}^3/L_\alpha \mathbb{Z}^3))$, this solution is smooth: for every $T > 0$, $j \in \mathbb{N}$ and $k \in \mathbb{N}^3$,

$$\sup_{0 < t < T} \sup_{x_0 \in \mathbb{R}^3} \sup_{\theta \in \mathbb{T}^N} |\partial_t^j \partial_x^k \vec{v}_{\epsilon,\alpha,\theta}(t, x)| + |\partial_t^j \partial_x^k p_{\epsilon,\alpha,\theta}(t, x)| < +\infty.$$

We have continuity in θ as well: for every $T > 0$, $j \in \mathbb{N}$ and $k \in \mathbb{N}^3$, there exists a constant $C_{\alpha, T, j, k, \epsilon}$ which depends on $\alpha, T, j, k, \epsilon, \vec{\omega}_1, \ldots, \vec{\omega}_N, \vec{B}_1, \ldots, \vec{B}_N$, such that

$$|\partial_t^j \partial_x^k \vec{v}_{\epsilon,\alpha,\theta}(t, x) - \partial_t^j \partial_x^k \vec{v}_{\epsilon,\alpha,\eta}(t, x)| + |\partial_t^j \partial_x^k p_{\epsilon,\alpha,\theta}(t, x) - \partial_t^j \partial_x^k p_{\epsilon,\alpha,\eta}(t, x)| \le C_{\alpha, T, j, k, \epsilon} |\theta - \eta|.$$

Equations (14.66) preserve the stability of the family $(\vec{u}_{0,\alpha}(\cdot, \theta))_{\theta \in \mathbb{T}^N}$ under the shifts τ_{x_0}:

$$\tau_{x_0} \vec{v}_{\epsilon,\alpha,\theta}(t, \cdot) = \vec{v}_{\epsilon,\alpha,\tau_{x_0}^* \theta}(t, \cdot).$$

We have the local energy balance

$$\begin{aligned} \partial_t(|\vec{v}_{\epsilon,\alpha,\theta}|^2) = &\Delta(|\vec{v}_{\epsilon,\alpha,\theta}|^2) - 2|\vec{\nabla} \otimes \vec{v}_{\epsilon,\alpha,\theta}|^2 \\ &- 2\operatorname{div}(((\vec{v}_{\epsilon,\alpha,\theta} \cdot (\varphi_\epsilon * \vec{v}_{\epsilon,\alpha,\theta}))\vec{v}_{\epsilon,\alpha,\theta}) - 2\operatorname{div}(p_{\epsilon,\alpha,\theta}\vec{v}_{\epsilon,\alpha,\theta}) \end{aligned} \qquad (14.67)$$

We multiply by $e^{-t}\Phi_\gamma(x)$ (with $\gamma > 3$ so that $\Phi_\gamma \in L^1(\mathbb{R}^3)$) and integrate on $\Delta = (0, +\infty) \times \mathbb{R}^3 \times \mathbb{T}^N$ and get

$$\iiint_\Delta e^{-s} |\vec{v}_{\epsilon,\alpha,\theta}(s,x)|^2 \Phi_\gamma(x)\, dx\, d\theta\, ds$$

$$+ 2\iiint_\Delta e^{-s} |\vec{\nabla} \otimes \vec{v}_{\epsilon,\alpha,\theta}(s,x)|^2 \Phi_\gamma(x)\, dx\, d\theta\, ds$$

$$= \int\int_{\mathbb{T}^N} |\vec{u}_{0,\alpha}(x,\theta)|^2 \Phi_\gamma(x)\, dx\, d\theta + \iiint_\Delta e^{-s}\Delta(|\vec{v}_{\epsilon,\alpha,\theta}|^2)\Phi_\gamma(x)\, dx\, d\theta\, ds \qquad (14.68)$$

$$- 2\iiint_\Delta e^{-s}\operatorname{div}(((\vec{v}_{\epsilon,\alpha,\theta}\cdot(\varphi_\epsilon * \vec{v}_{\epsilon,\alpha,\theta}))\vec{v}_{\epsilon,\alpha,\theta})\Phi_\gamma(x)\, dx\, d\theta\, ds$$

$$- 2\iiint_\Delta e^{-s}\operatorname{div}(p_{\epsilon,\alpha,\theta}\vec{v}_{\epsilon,\alpha,\theta})\Phi_\gamma(x)\, dx\, d\theta\, ds$$

Now, we remark that if $F(x,\theta)$ is a bounded continuous function on $\mathbb{R}^3 \times \mathbb{T}^N$ such that $F(x - x_0, \theta) = F(x, \tau - x_0 * \theta)$, we have

$$\int\int_{\mathbb{T}^N} F(x,\theta)\Phi_\gamma(x)\, dx\, d\theta = \int\int_{\mathbb{T}^N} F(x, \tau^*_{x_0}\theta)\Phi_\gamma(x - x_0)\, dx\, d\theta$$

$$= \int\int_{\mathbb{T}^N} F(x,\theta)\Phi_\gamma(x - x_0)\, dx\, d\theta = \int\int_{\mathbb{T}^N} F(x + x_0, \theta)\Phi_\gamma(x)\, dx\, d\theta.$$

If F is periodic ($F(x + kL_\alpha, \theta) = F(x, \theta)$ for $k \in \mathbb{Z}^3$), we integrate this equality on $x_0 \in (0, L_\alpha)^3$ and obtain

$$\int\int_{\mathbb{T}^N} F(x,\theta)\Phi_\gamma(x)\, dx\, d\theta = \|\Phi_\gamma\|_1 \frac{1}{L_\alpha^3}\int_{(0,L^\alpha)^3}\int_{\mathbb{T}^N} F(x,\theta)\, dx\, d\theta.$$

We obtain

$$\int_{\mathbb{T}^N} \|e^{-t/2}\vec{v}_{\epsilon,\alpha,\theta}(t,x)\|^2_{L^2((0,+\infty), L^2(\Phi_\gamma(x)\, dx))}\, d\theta$$

$$+ 2\int_{\mathbb{T}^N} \|e^{-t/2}\vec{\nabla} \otimes \vec{v}_{\epsilon,\alpha,\theta}(t,x)\|^2_{L^2((0,+\infty), L^2(\Phi_\gamma(x)\, dx))}\, d\theta$$

$$= \int\int_{\mathbb{T}^N} |\vec{u}_{0,\alpha}(x,\theta)|^2 \Phi_\gamma(x)\, dx\, d\theta \qquad (14.69)$$

$$\leq \|\Phi_\gamma\|_1 (\sum_{k=1}^{N} |\vec{B}_k| + \alpha|\vec{A}_k|)^2.$$

We take $(\epsilon_n, \alpha_n) \to_{n\to+\infty} 0$ (with $\alpha_n \leq 1$) and we define

$$M_n(\theta) = \|e^{-t/2}\vec{v}_{\epsilon_n,\alpha_n,\theta}(t,x)\|^2_{L^2((0,+\infty), L^2(\Phi_\gamma(x)\, dx))}$$

$$+ 2\|e^{-t/2}\vec{\nabla} \otimes \vec{v}_{\epsilon_n,\alpha_n,\theta}(t,x)\|^2_{L^2((0,+\infty), L^2(\Phi_\gamma(x)\, dx))}.$$

and

$$\Sigma = \{\theta \in \mathbb{T}^N \ / \ \lim_{n\to+\infty} M_n(\theta) = +\infty\}.$$

We may write

$$\Sigma = \bigcap_{j\in\mathbb{N}} \bigcup_{k\in\mathbb{N}} \bigcap_{n\in\mathbb{N}, n\geq k} \{\theta \in \mathbb{T}^N \ / \ M_n(\theta) > 2^j\}.$$

We have (noting $|E|$ the Lebesgue measure of a subset E of $\mathbb{T}^N$)

$$\left|\bigcap_{n\in\mathbb{N},n\geq k}\{\theta\in\mathbb{T}^N\ /\ M_n(\theta)>2^j\}\right|\leq\left|\{\theta\in\mathbb{T}^N\ /\ M_k(\theta)>2^j\}\right|$$

$$\leq 2^{-j}\|\Phi_\gamma\|_1\left(\sum_{p=1}^N|\vec{B}_p|+|\vec{A}_p|\right)^2$$

and

$$|\Sigma|=\lim_{j\to+\infty}\lim_{k\to+\infty}\left|\bigcap_{n\in\mathbb{N},n\geq k}\{\theta\in\mathbb{T}^N\ /\ M_n(\theta)>2^j\}\right|=0.$$

Now, we consider $\theta\notin\Sigma$. We know that there exists a sequence $(\epsilon_{(n)},\alpha_{(n)})$ with $(\epsilon_{(n)},\alpha_{(n)})\to_{n\to+\infty}0$ and $\sup_{n\in\mathbb{N}}M_{(n)}(\theta)<+\infty$. In particular, the sequence $(\vec{v}_{\epsilon_{(n)},\alpha_{(n)},\theta})_{n\in\mathbb{N}}$ is locally (in time and space variables) bounded in L^2H^1. In order to apply the Aubin-Lions-Simon lemma (a generalization of the Rellich–Lions theorem (Theorem 12.1) in the case of a control of the time derivative in L^1H^β with $\beta<0$) [9, 337, 437] .

By Proposition 6.3 and Lemma 6.4, we know that $\vec\nabla p_{\epsilon,\alpha,\theta}$ can be computed as

$$\vec\nabla p_{\epsilon,\alpha,\theta}=\lim_{R\to+\infty}\vec\nabla(\chi_R G)*\mathrm{div}(\mathrm{div}((\varphi_\epsilon*\vec{v}_{\epsilon,\alpha,\theta})\otimes\vec{v}_{\epsilon,\alpha,\theta}))$$

where G is the Green function (fundamental solution of $-\Delta$), $\chi_R(x)=\chi(x/R)$ and $\chi\in\mathcal{D}$ is equal to 1 on a neighborhood of 0. Thus, if $\psi_1,\psi_2,\psi_3\in\mathcal{D}(\mathbb{R}^3)$ with $\psi_2=1$ on a neighborhood of the support of ψ_1 and $\psi_3=1$ on a neighborhood of the support of ψ_2, and $\psi\in\mathcal{C}_C^\infty([0,+\infty))$, we have

$$\psi(t)\psi_1(x)\vec\nabla p_{\epsilon,\alpha,\theta}=\vec{q_1}+\vec{q_2}$$

with

$$\vec{q_1}=\psi\psi_1\vec\nabla G*\mathrm{div}(\psi_3(\varphi_\epsilon*\vec{v}_{\epsilon,\alpha,\theta})\cdot\vec\nabla(\psi_2\vec{v}_{\epsilon,\alpha,\theta}))$$

and

$$\vec{q_2}=\psi\psi_1\vec\nabla G*\mathrm{div}(\mathrm{div}((1-\psi_2)(\varphi_\epsilon*\vec{v}_{\epsilon,\alpha,\theta})\otimes\vec{v}_{\epsilon,\alpha,\theta})).$$

We have (for $\epsilon\in(0,1)$)

$$\|\vec{q_1}\|_{3/2}\leq|\psi(t)|\|\psi_3(\varphi_\epsilon*\vec{v}_{\epsilon,\alpha,\theta})(t,.)\|_6\|\psi_2\vec{v}_{\epsilon,\alpha,\theta}\|_{H^1}$$

$$\leq C|\psi(t)|C_{\psi_1,\psi_2,\psi_3}\int(|\vec{v}_{\epsilon,\alpha,\theta}(t,y)|^2+|\vec\nabla\otimes\vec{v}_{\epsilon,\alpha,\theta}|^2)\frac{1}{(1+|y|)^4}\,dy$$

(so that $\vec{q_1}\in L^1([0,+\infty),L^{3/2})\subset L^1H^{-1}$) and

$$|\vec{q_2}|\leq|\psi(t)|C_{\psi_1,\psi_2}\int|(\varphi_\epsilon*\vec{v}_{\epsilon,\alpha,\theta})(t,y)\otimes\vec{v}_{\epsilon,\alpha,\theta}(t,y)|\frac{1}{(1+|y|)^4}\,dy$$

$$\leq C|\psi(t)|C_{\psi_1,\psi_2}\int|\vec{v}_{\epsilon,\alpha,\theta}(t,y)|^2\frac{1}{(1+|y|)^4}\,dy$$

(so that $\vec{q_2}\in L^1L^\infty$). Thus, we find

$$\sup_{n\in\mathbb{N}}\|\psi(t)\psi_1(x)\partial_t\vec{v}_{\epsilon_{(n)},\alpha_{(n)},\theta}\|_{L^1H^{-1}}<+\infty.$$

We then apply the Aubin-Lions-Simon lemma and find a sequence $(\vec{v}_{\epsilon_{[n]},\alpha_{[n]},\theta})$ which converges strongly in $(L^2L^2)_{\mathrm{loc}}$ to a limit $\vec{u}_\theta$. We have the convergence (as distributions) of $\partial_t \vec{v}_{\epsilon_{[n]},\alpha_{[n]},\theta}$ to $\partial_t \vec{u}_\theta$, of $\Delta \vec{v}_{\epsilon_{[n]},\alpha_{[n]},\theta}$ to $\Delta \vec{u}_\theta$, of $\mathrm{div}((\varphi_{\epsilon_{[n]}} * \vec{v}_{\epsilon_{[n]},\alpha_{[n]},\theta}) \otimes \vec{\nabla} \vec{v}_{\epsilon_{[n]},\alpha_{,[n]}\theta})$ to $\mathrm{div}(\vec{u}_\theta \otimes \vec{u}_\theta)$ and of $\vec{\nabla} p_{\epsilon_{[n]},\alpha_{[n]},\theta}$ to $\lim_{R \to +\infty} \vec{\nabla}(\chi_R G) * \mathrm{div}(\mathrm{div}(\vec{u}_\theta \otimes \vec{u}_\theta))$. Moreover, we find that $\vec{u}_{0,\alpha,\theta} + \int_0^t \partial_t \vec{v}_{\epsilon_{[n]},\alpha_{[n]},\theta}\, ds$ converges to $\vec{u}_{0,\theta} + \int_0^t \partial_t \vec{u}_\theta\, ds$, so that $\vec{u}_\theta$ is a solution of the Cauchy problem for the Navier-Stokes equations with initial value $\vec{u}_{0,\theta}$. $\qquad\square$

Using probabilistic tools of the theory of homogeneous statistical solutions developed by Višik and Fursikov [190], Basson [23] could prove a much stronger result: for $0 < \epsilon < 1$, $\vec{u}_\theta \in L^\infty(L^2(\Phi_{4+\epsilon}))$, $\vec{\nabla} \otimes \vec{u}_\theta \in L^2(L^2(\Phi_{4-\epsilon}))$, p_θ is locally $L^{3/2}L^{3/2}$ and $\vec{u}_\theta$ is suitable in the sense of Caffarelli, Kohn and Nirenberg.

Chapter 15

The L^3 Theory of Suitable Solutions

In this chapter, we use the theory of local Leray solutions to get two major recent results: the $L^\infty_t L^3_x$ regularity result of Escauriaza, Seregin and Šverák [163] for suitable solutions of the Navier–Stokes equations and the result of Jia and Šverák [244] on the (potential) existence of a minimal-norm initial value for a blowing-up mild solution to the Navier–Stokes Cauchy problem (first established by Rusin and Šverák [417] and by Gallagher, Koch and Planchon [199]).

15.1 Local Leray Solutions with an Initial Value in L^3

We first begin with a new construction of a local Leray solution associated to an initial value in L^3 that was initially studied by Calderón [77] (and further studied by Jia and Šverák [244]):

Proposition 15.1.
Let $M > 0$. Let $\vec{u}_0 \in L^3$ with $\operatorname{div} \vec{u}_0 = 0$ and $\|\vec{u}_0\|_3 \leq M$. Let $T_0 > 0$. Then there exists a local Leray solution on $(0, T_0) \times \mathbb{R}^3$ to the Navier–Stokes problem

$$\partial_t \vec{u} = \nu \Delta \vec{u} - \mathbb{P} \operatorname{div}(\vec{u} \otimes \vec{u}), \quad \vec{u}(0, .) = \vec{u}_0$$

that satisfies:

- $\vec{u} \in (L^\infty L^2)_{\text{uloc}} \cap (L^2 H^1)_{\text{uloc}}$

- $\vec{u} \in \mathcal{C}([0, T_0], H^{-3/2}_{\text{uloc}})$

- $\|\vec{u}(t, .) - W_{\nu t} * \vec{u}_0\|_{L^2_{\text{uloc}}} \leq \eta(t)$, *where the function η depends only on ν, T_0 and M and satisfies:*

$$\lim_{t \to 0^+} \eta(t) = 0.$$

We begin the proof with an easy lemma:

Lemma 15.1.
Let $\vec{v}_0 \in L^2_{\text{uloc}}$ and $\mathbb{G} \in (L^2 L^2)_{\text{uloc}}$ on $(0, T) \times \mathbb{R}^3$ with $T < +\infty$. Then

$$\vec{v} = W_{\nu t} * \vec{v}_0 + \int_0^t W_{\nu(t-s)} * \mathbb{P} \operatorname{div} \mathbb{G} \, ds$$

satisfies

$$\vec{v} \in (L^\infty_t L^2_x)_{\text{uloc}} \cap (L^2 H^1_x)_{\text{uloc}}$$

on $(0, T) \times \mathbb{R}^3$.

DOI: 10.1201/9781003042594-15

Proof. For a ball $K = B(x_0, 1)$ and, for $R > 2$, write $\vec{v}_0 = \vec{A}_{K,R} + \vec{B}_{K,R}$ and $\mathbb{G} = \mathbb{C}_{K,R} + \mathbb{D}_{K,R}$, where $\vec{A}_{K,R} = 1_{B(x_0,R)}\vec{v}_0$ and $\mathbb{C}_{K,R} = 1_{B(x_0,R)}\mathbb{G}$.

As $\vec{A}_{K,R} \in L^2$, we have $W_{\nu t} * \vec{A}_{K,R} \in L^\infty L^2 \cap L^2 \dot{H}^1$.

We have, on K,

$$|W_{\nu t} * \vec{B}_{K,R}(x)| \leq C \int_{|x-y|>R/2} \frac{\sqrt{\nu t}}{|x-y|^4} |\vec{v}_0(y)|\, dy \leq C\sqrt{\nu t}\, \frac{1}{R}\|\vec{v}_0\|_{L^2_{uloc}}$$

and, for $l = 1, \ldots, 3$,

$$|\partial_l(W_{\nu t} * \vec{B}_{K,R})(x)| \leq C \int_{|x-y|>R/2} \frac{\sqrt{\nu t}}{|x-y|^5} |\vec{v}_0(y)|\, dy \leq C\sqrt{\nu t}\, \frac{1}{R^2}\|\vec{v}_0\|_{L^2_{uloc}}.$$

Thus, on $(0,T)$ with $T < +\infty$, we have $1_K(W_{\nu t} * \vec{B}_{K,R}) \in L^\infty L^2 \cap L^2 \dot{H}^1$.

As $\mathbb{C}_{K,R} \in L^2((0,T), L^2)$, we have $\int_0^t W_{\nu(t-s)} * \mathbb{P} \operatorname{div} \mathbb{C}_{K,R}\, ds \in L^\infty L^2 \cap L^2 \dot{H}^1$ on $(0,T) \times \mathbb{R}^3$.

We have, on K,

$$\left|\int_0^t W_{\nu(t-s)} * \mathbb{P}\operatorname{div}\mathbb{D}_{K,R}\, ds(x)\right| \leq C \int_0^t \int_{|x-y|>R/2} |\mathbb{G}(s,y)| \frac{dy\, ds}{|x-y|^4}$$
$$\leq C \frac{\|\mathbb{G}\|_{(L^2 L^2)_{uloc}}}{R}$$

and, for $l = 1, \ldots, 3$,

$$\left|\partial_l \left(\int_0^t W_{\nu(t-s)} * \mathbb{P}\operatorname{div}\mathbb{D}_{K,R}\, ds\right)(x)\right| \leq C \int_0^t \int_{|x-y|>R/2} |\mathbb{G}(s,y)| \frac{dy\, ds}{|x-y|^5}$$
$$\leq C \frac{\|\mathbb{G}\|_{(L^2 L^2)_{uloc}}}{R^2}.$$

Thus, on $(0,T)$ with $T < +\infty$, we have $1_K(\int_0^t W_{\nu(t-s)} * \mathbb{P}\operatorname{div}\mathbb{D}_{K,R}\, ds) \in L^\infty L^2 \cap L^2 \dot{H}^1$. $\square$

Proof of Proposition 15.1:

If $\vec{u}_0 \in L^3$ with $\operatorname{div}\vec{u}_0 = 0$, then, for any $\eta > 0$, we may split it as $\vec{u}_0 = \vec{\alpha}_\eta + \vec{\beta}_\eta$, where $\vec{\alpha}_\eta \in L^2$, $\operatorname{div}\vec{\alpha}_\eta = 0$, $\vec{\beta}_\eta \in L^6$, $\operatorname{div}\vec{\beta}_\eta = 0$ and

$$\|\vec{\alpha}_\eta\|_2 \leq C_2\eta\|\vec{u}_0\|_3, \quad \|\vec{\beta}_\eta\|_6 \leq C_2\eta^{-1}\|\vec{u}_0\|_3$$

(just use the embedding $L^3 \subset [L^2, L^6]_{1/2,\infty}$ and the boundedness of the Leray projection operator $\mathbb{P}$ on L^2 and on L^6).

Let θ be a mollifier, as we already used it for Leray mollifications, and $\theta_\epsilon = \frac{1}{\epsilon^3}\theta(\frac{x}{\epsilon})$. Let

$$B(\vec{u}, \vec{v}) = \int_0^t W_{\nu(t-s)} * \mathbb{P}\left(\sum_{i=1}^3 \partial_i(u_i\vec{v})\right) ds.$$

Let

$$E_T = \{(\vec{u}, \vec{v}) \ / \ \vec{u} \in L^4((0,T), \dot{H}^{1/2}), \vec{v} \in L^\infty((0,T), L^6)\}$$

normed with

$$\|(\vec{u}, \vec{v})\|_{E_T} = \|\vec{u}\|_{L^4 \dot{H}^{1/2}} + \|\vec{v}\|_{L^\infty L^6}.$$

For $\vec{u}_0 \in L^2 + L^6 \subset L^2_{\text{uloc}}$, we looked for a solution $\vec{V} \in L^\infty L^2_{\text{uloc}}$ of the mollified Navier–Stokes equations

$$\begin{cases} \partial_t \vec{V} = \nu \Delta \vec{V} - \mathbb{P}(\sum_{i=1}^3 \partial_i((\theta_\epsilon * V_i)\vec{V})) \\ \\ \vec{V}(0,.) = \vec{u}_0 \end{cases} \tag{15.1}$$

or equivalently of

$$\vec{V} = W_{\nu t} * \vec{u}_0 - B(\theta_\epsilon * \vec{V}, \vec{V}).$$

Writing $\vec{u}_0 = \vec{\alpha} + \vec{\beta}$, with $\vec{\alpha} \in L^2$ and $\vec{\beta} \in L^6$, we now look more precisely for $\vec{V} = \vec{\gamma} + \vec{\delta}$ with $(\vec{\gamma}, \vec{\delta}) \in E_T$ and

$$\begin{cases} \partial_t \vec{\gamma} = \nu \Delta \vec{\gamma} - \mathbb{P}(\sum_{i=1}^3 \partial_i \left((\theta_\epsilon * \gamma_i)\vec{\gamma} + (\theta_\epsilon * \gamma_i)\vec{\delta} + (\theta_\epsilon * \delta_i)\vec{\gamma}\right)) \\ \\ \vec{\gamma}(0,.) = \vec{\alpha} \\ \\ \partial_t \vec{\delta} = \nu \Delta \vec{\delta} - \mathbb{P}(\sum_{i=1}^3 (\theta_\epsilon * \delta_i).\vec{\nabla}\vec{\delta}) \\ \\ \vec{\delta}(0,.) = \vec{\beta} \end{cases} \tag{15.2}$$

or equivalently of

$$\vec{\gamma} = W_{\nu t} * \vec{\alpha} - B(\theta_\epsilon * \vec{\gamma}, \vec{\gamma}) - B(\theta_\epsilon * \vec{\gamma}, \vec{\delta}) - B(\theta_\epsilon * \vec{\delta}, \vec{\gamma})$$

and

$$\vec{\delta} = W_{\nu t} * \vec{\beta} - B(\theta_\epsilon * \vec{\delta}, \vec{\delta}).$$

Let $\mathcal{B}_\epsilon$ be the bilinear operator on $E_T \times E_T$

$$\mathcal{B}_\epsilon((\vec{\gamma}_1, \vec{\delta}_1), (\vec{\gamma}_2, \vec{\delta}_2)) = (B(\theta_\epsilon * \vec{\gamma}_1, \vec{\gamma}_2) + B(\theta_\epsilon * \vec{\gamma}_1, \vec{\delta}_2) + B(\theta_\epsilon * \vec{\delta}_1, \vec{\gamma}_2), B(\theta_\epsilon * \vec{\delta}_1, \vec{\delta}_2)).$$

We have, for every $T > 0$,

$$\begin{aligned} \|B(\theta_\epsilon * \vec{\gamma}_1, \vec{\gamma}_2)\|_{L^4((0,T), \dot{H}^{1/2})} &\leq C_\nu T^{1/8} \|(\theta_\epsilon * \vec{\gamma}_1) \otimes \vec{\gamma}_2)\|_{L^2((0,T), \dot{H}^{1/4})} \\ &\leq C'_\nu T^{1/8} \|\vec{\gamma}_1\|_{L^4((0,T), \dot{H}^{7/4})} \|\vec{\gamma}_2)\|_{L^4((0,T), \dot{H}^{1/2})} \\ &\leq C_3 \frac{T^{1/8}}{\epsilon^{5/4}} \|\vec{\gamma}_1\|_{L^4((0,T), \dot{H}^{1/2})} \|\vec{\gamma}_2)\|_{L^4((0,T), \dot{H}^{1/2})} \\ \|B(\theta_\epsilon * \vec{\gamma}_1, \vec{\delta}_2)\|_{L^4((0,T), \dot{H}^{1/2})} &\leq C_\nu \|(\theta_\epsilon * \vec{\gamma}_1) \otimes \vec{\delta}_2)\|_{L^2((0,T), L^2)} \\ &\leq C_3 T^{1/4} \|\vec{\gamma}_1\|_{L^4((0,T), \dot{H}^{1/2})} \|\vec{\delta}_2\|_{L^\infty((0,T), L^6)} \\ \|B(\theta_\epsilon * \vec{\delta}_1, \vec{\gamma}_2)\|_{L^4((0,T), \dot{H}^{1/2})} &\leq C_3 T^{1/4} \|\vec{\gamma}_2\|_{L^4((0,T), \dot{H}^{1/2})} \|\vec{\delta}_1\|_{L^\infty((0,T), L^6)} \\ \|B(\theta_\epsilon * \vec{\delta}_1, \vec{\delta}_2)\|_{L^\infty((0,T), L^6)} &\leq C_3 T^{1/4} \|\vec{\delta}_1\|_{L^\infty((0,T), L^6)} \|\vec{\delta}_2\|_{L^\infty((0,T), L^6)} \end{aligned}$$

so that

$$\|\mathcal{B}_\epsilon((\vec{\gamma}_1, \vec{\delta}_1), (\vec{\gamma}_2, \vec{\delta}_2))\|_{E_T} \leq C_4 \min(T^{1/4}, \frac{T^{1/8}}{\epsilon^{5/4}}) \|(\vec{\gamma}_1, \vec{\delta}_1)\|_{E_T} \|(\vec{\gamma}_2, \vec{\delta}_2))\|_{E_T} \tag{15.3}$$

where C_4 does not depend on ϵ nor T.

We have

$$\|(W_{\nu t} * \vec{\alpha}, W_{\nu t} * \vec{\beta})\|_{E_T} \leq C_5(\|\vec{\alpha}\|_2 + \|\vec{\beta}\|_6).$$

If a number A is such that $C_5(\|\vec{\alpha}\|_2 + \|\vec{\beta}\|_6) < A$ and if T is such that $C_4 \min(T^{1/4}, \frac{T^{1/8}}{\epsilon^{5/4}}) < \frac{1}{8A}$, then the Picard iterates will converge to a solution $(\vec{\gamma}, \vec{\delta})$ on $(0, T) \times \mathbb{R}^3$.

Moreover, we have $\vec{\gamma} \in \mathcal{C}([0, T], L^2) \cap L^2((0, T), \dot{H}^1)$ with

$$\|\vec{\gamma}\|_{L^\infty((0,T),L^2)} + \|\vec{\gamma}\|_{L^2((0,T),\dot{H}^1)}$$
$$\leq C_\nu(\|\vec{\alpha}\|_2 + \|(\theta_\epsilon * \vec{\gamma})) \otimes \vec{\gamma} + (\theta_\epsilon * \vec{\gamma})) \otimes \vec{\delta} + (\theta_\epsilon * \vec{\delta})) \otimes \vec{\gamma}\|_{L^2((0,T),L^2)}$$
$$C'_\nu(\|\vec{\alpha}\|_2 + \epsilon^{-1/2}\|\vec{\gamma}\|^2_{L^4((0,T),\dot{H}^{1/2})} + T^{1/4}\|\vec{\gamma}\|_{L^4((0,T),\dot{H}^{1/2})}\|\vec{\delta}\|_{L^\infty((0,T),L^6)})$$

If moreover, we have $\|\vec{\gamma}(T, .)\|_2 + \|\vec{\delta}(T, .)\|_6 < \frac{2A}{C_5}$, then we can reiterate the construction on $(T, 2T) \times \mathbb{R}^3$, and so on, on $(kT, (k+1)T)$ as long as $\|\vec{\gamma}(kT, .)\|_2 + \|\vec{\delta}(kT, .)\|_6 < \frac{2A}{C_5}$.

Remark that the equation on $\vec{\delta}$ does not involve $\vec{\gamma}$. From

$$\|B(\theta_\epsilon * \vec{\delta}_1, \vec{\delta}_2)\|_{L^\infty((0,T_1),L^6)} \leq C_3 T_1^{1/4} \|\vec{\delta}_1\|_{L^\infty((0,T_1),L^6)} \|\vec{\delta}_2\|_{L^\infty((0,T_1),L^6)}$$

and

$$\|W_{\nu t} * \vec{\beta}\|_{L^\infty L^6} \leq \|\vec{\beta}\|_6,$$

we find that, if

$$T_1^{1/4} C_3 \|\vec{\beta}\|_6 < \frac{1}{4},$$

then we can define the solution $\vec{\delta}$ on $(0, T_1) \times \mathbb{R}^3$ and

$$\|\vec{\delta}\|_{L^\infty((0,T_1),L^6)} \leq 2\|\vec{\beta}\|_6.$$

Moreover, if we can define $\vec{\gamma}$ up to $t = kT \leq T_1$, we have $\vec{\gamma} \in L^2((0, kT), \dot{H}^1)$ and $\partial_t \vec{\gamma} \in L^2((0, kT), \dot{H}^{-1})$ so that

$$\partial_t \|\vec{\gamma}\|_2^2 = 2\langle \partial_t \vec{\gamma} | \vec{\gamma} \rangle_{\dot{H}^{-1}, \dot{H}^1}$$
$$= -2\nu\|\vec{\nabla} \otimes \vec{\gamma}\|_2^2 + 2 \int \vec{\delta}.((\theta_\epsilon * \vec{\gamma}).\vec{\nabla}\vec{\gamma}) \, dx$$
$$\leq -2\nu\|\vec{\gamma}\|_{\dot{H}^1}^2 + C\|\vec{\delta}\|_6 \|\vec{\gamma}\|_2^{1/2} \|\vec{\gamma}\|_{\dot{H}^1}^{3/2}$$
$$\leq -\nu\|\vec{\gamma}\|_{\dot{H}^1}^2 + C_6 \nu^{-4} \|\vec{\delta}\|_6^4 \|\vec{\gamma}\|_2^2$$

and thus

$$\|\vec{\gamma}(kT, .)\|_2^2 \leq \|\vec{\alpha}\|_2^2 e^{16 T_1 C_6 \nu^{-4} \|\vec{\beta}\|_6^4}$$

Thus, $\vec{\gamma}$ will be defined up to $t = T_1$ provided that

$$16 T_1 C_6 \nu^{-4} \|\vec{\beta}\|_6^4 < 2 \ln 2,$$

and we will have

$$\|\vec{\gamma}\|_{L^\infty((0,T_1),L^2)} \leq 2\|\vec{\alpha}\|_2$$

and

$$\sqrt{\nu}\|\vec{\gamma}\|_{L^2((0,T_1),\dot{H}^1)} \leq 2\|\vec{\alpha}\|_2.$$

Finally, by Lemma 15.1, we know that we control $\vec{\delta}$ in $L^\infty L^2_{uloc} \cap (L^2 H^1)_{uloc}$ independently from ϵ.

Thus, if $T_1\|\vec{\beta}\|_6^4 < \min(\frac{\nu^4}{8C_6} \ln 2, \frac{1}{256 C_3^4})$, we will have solutions $(\vec{\gamma}_\epsilon, \vec{\delta}_\epsilon)$ of the mollified equations on $(0, T_1) \times \mathbb{R}^3$ with controls in $L^\infty L^2_{uloc} \cap (L^2 L^2)_{uloc}$ that are uniform with respect to ϵ. This will allow to use the Rellich–Lions theorem (Theorem 12.1) and to find a local Leray solution $\vec{V} = \vec{\gamma} + \vec{\delta}$ with

$$
\begin{cases}
\partial_t \vec{\delta} = \nu \Delta \vec{\delta} - \mathbb{P}(\sum_{i=1}^3 \partial_i \left(\delta_i \vec{\delta} \right)) \\[2mm]
\vec{\delta}(0,.) = \vec{\beta} \\[2mm]
\|\vec{\delta}\|_{L^\infty((0,T_1),L^6)} \le 2\|\vec{\beta}\|_2
\end{cases}
$$

and

$$
\begin{cases}
\partial_t \vec{\gamma} = \nu \Delta \vec{\gamma} - \mathbb{P}(\sum_{i=1}^3 \partial_i \left(\gamma_i \vec{\gamma} + \gamma_i \vec{\delta} + \delta_i \vec{\gamma} \right)) \\[2mm]
\vec{\gamma}(0,.) = \vec{\alpha} \\[2mm]
\|\vec{\gamma}\|_{L^\infty((0,T_1),L^2)} \le 2\|\vec{\alpha}\|_2
\end{cases}
$$

Now, the splitting of $\vec{u}_0$ into $\vec{u}_0 = \vec{\alpha}_\eta + \vec{\beta}_\eta$ obviously depends on η. For each η, we may consider our splitting of the mollified Equation (15.1). By uniqueness of the solution of the mollified equation, all the $\vec{V}_{\eta,\epsilon}$, for fixed ϵ, will coincide as long as they are defined. By a Cantor diagonal process, considering a decreasing sequence $\eta_n \to 0$, we may ensure that $\vec{V}_{\eta_n,\epsilon_k}$ converges to $\vec{V}_{\eta_n}$ for every n (the convergence occurs on $(0,T_{\eta_n}) \times \mathbb{R}^3$), and that $\vec{V}_{\eta_n} = \vec{V}_{\eta_{n+1}}$ on $(0,T_{\eta_{n+1}})$, where $T_{\eta_n} = O(\min(\frac{\nu^4}{8C_6} \ln 2, \frac{1}{256 C_3^4})\|\beta_{\eta_n}\|_6^{-4}) = O(\eta_n^4 \|\vec{u}_0\|_3^{-4})$.

Of course, we begin with η_0 large enough to ensure that $T_0 < T_{\eta_0} \approx \eta_0^4 \|\vec{u}_0\|_3^{-4}$. We have our local Leray solution $\vec{u}$ on $(0,T_0) \times \mathbb{R}^3$, with $\vec{u} = \vec{V}_{\eta_0}$.

We check easily that $\vec{u} \in \mathcal{C}([0,T_0], H_{\mathrm{uloc}}^{-3/2})$. Indeed, we have $\vec{u} \in (L^\infty L^2)_{\mathrm{uloc}} \subset (L^1 H^{-3/2})_{\mathrm{uloc}}$ and $\partial_t \vec{u} \subset (L^1 H^{-3/2})_{\mathrm{uloc}}$. If g is a distribution on $(0,T_0) \times \mathbb{R}^3$ such that $g \in L_t^1 H^{-3/2}$ and $\partial_t g \in L^1 H^{-3/2}$, then $g \in \mathcal{C}([0,T_0], H^{-3/2})$ and, for $0 \le t \le \tau \le 1$,

$$
\|g(t,.) - g(\tau,.)\|_{H^{-3/2}} \le \int_t^\tau \|\partial_t g(s,.)\|_{H^{-3/2}} \, ds.
$$

To check it, it is enough to take ζ a smooth function on $\mathbb{R}$ which is equal to 1 on $(-\infty, 1/4)$ and to 0 on $(3/4, +\infty)$ and to define $\zeta_T(s) = \zeta(\frac{s}{T})$; we have

$$
g(t,.) = \int_0^t \partial_t((1 - \zeta_{T_0/3})g) \, ds \qquad \text{if } T_0/4 < t \le T_0
$$

and

$$
g(t,.) = -\int_t^{T_0} \partial_t(\zeta_{3T_0} g) \, ds \qquad \text{if } 0 \le t < 3T_0/4.
$$

It remains to estimate $\|\vec{u}(t,.) - W_{\nu t} * \vec{u}_0\|_{L^2_{\mathrm{uloc}}}$. We have, of course,

$$
\|\vec{u}(t,.) - W_{\nu t} * \vec{u}_0\|_{L^2_{\mathrm{uloc}}} \le 2\|\vec{u}\|_{(L^\infty L^2)_{\mathrm{uloc}}} \le 2(\|\vec{\gamma}_{\eta_0}\|_{(L^\infty L^2)_{\mathrm{uloc}}} + \|\vec{\delta}_{\eta_0}\|_{(L^\infty L^2)_{\mathrm{uloc}}})
$$

and thus

$$
\|\vec{u}(t,.) - W_{\nu t} * \vec{u}_0\|_{L^2_{\mathrm{uloc}}} \le C\|\vec{u}_0\|_3 \max(\eta_0, \eta_0^{-1})
$$

with $\eta_0 \approx T_0^{1/4} \|\vec{u}_0\|_3$.

Now, we go back to the splitting $\vec{u} = \vec{\gamma}_\eta + \vec{\delta}_\eta$, valid on $(0, T_\eta) \times \mathbb{R}^3$. We write, for $0 < t < T_\eta$,

$$\|\vec{u}(t,.) - W_{\nu t} * \vec{u}_0\|_{L^2_{\text{uloc}}} \leq \|\vec{\gamma}_\eta(t,.) - W_{\nu t} * \vec{\alpha}_\eta\|_{L^2_{\text{uloc}}} + \|\vec{\delta}_\eta(t,.) - W_{\nu t} * \vec{\beta}_\eta\|_{L^2_{\text{uloc}}}$$

$$\leq \|\vec{\gamma}_\eta(t,.)\|_2 + \|W_{\nu t} * \vec{\alpha}_\eta\|_2 + \|\int_0^t W_{\nu(t-s)} * \mathbb{P}\text{div}(\vec{\delta}_\eta \otimes \vec{\delta}_\eta)\,ds\|_{L^2_{\text{uloc}}}$$

with

$$\|\vec{\gamma}_\eta(t,.)\|_2 + \|W_{\nu t} * \vec{\alpha}_\eta\|_2 \leq 3\|\vec{\alpha}_\eta\|_2 \leq 3C_2\|\vec{u}_0\|_3 \, \eta$$

and (following Lemma 15.1)

$$\|\int_0^t W_{\nu(t-s)} * \mathbb{P}\text{div}(\vec{\delta}_\eta \otimes \vec{\delta}_\eta)\,ds\|_{L^2_{\text{uloc}}} \leq C_7\|\vec{\delta}_\eta \otimes \vec{\delta}_\eta\|_{(L^2 L^2)_{\text{uloc}}((0,t)\times\mathbb{R}^3)}$$

$$\leq C_8 t^{1/4}\|\vec{\delta}_\eta\|^2_{L^\infty L^6}$$

$$\leq 4C_8 C_2^2\|\vec{u}_0\|_3^2 \eta^{-2} t^{1/4}.$$

Thus, if η is small enough to get that $3C_2 M \, \eta < \epsilon/2$ and $0 < T_{[\epsilon]} < T_\eta$ is small enough to get that $4C_8 C_2^2 M^2 \eta^{-2} T_{[\epsilon]}^{1/4} < \epsilon/2$, we find that:

$$\text{for } 0 < t < T_{[\epsilon]}, \quad \|\vec{u}(t,.) - W_{\nu t} * \vec{u}_0\|_{L^2_{\text{uloc}}} < \epsilon.$$

As $T_{[\epsilon]}$ depends only on ϵ, ν and $\|\vec{u}_0\|_3$, the proposition is proved. $\qquad\square$

15.2 Blow up in Finite Time

We apply the theory of local Leray solutions developed in Chapter 14 to get a first criterion to check that a solution $\vec{u} \in \mathcal{C}([0,T), L^3)$ blows up at time $T^* = T$, or that a local Leray solution on $(0,T) \times \mathbb{R}^3$ with initial value in L^3 blows up at time $T^* \leq T$. Let us recall that we defined the cylinder $Q_r(t,x)$ as $Q_r(t,x) = (t - r^2, t) \times B(x, r)$.

Theorem 15.1.
Let $\vec{u}_0 \in L^3$ be a divergence free vector field on $\mathbb{R}^3$. Let $\vec{u}$ be the solution of the Navier–Stokes equations equations

$$\partial_t \vec{u} = \nu\Delta\vec{u} - \mathbb{P}\text{div}(\vec{u} \otimes \vec{u}), \quad \vec{u}(0,.) = \vec{u}_0$$

in $\mathcal{C}([0,T^*), L^3_x)$, where T^* is the maximal existence time of the solution. Let $0 < T < +\infty$ and let $\vec{v}$ be a local Leray solution on $(0,T) \times \mathbb{R}^3$ to the Navier–Stokes problem

$$\partial_t \vec{v} = \nu\Delta\vec{v} - \mathbb{P}\text{div}(\vec{v} \otimes \vec{v}), \quad \vec{u}(0,.) = \vec{u}_0$$

that satisfies $\vec{v} \in (L^\infty L^2)_{\text{uloc}} \cap (L^2 H^1)_{\text{uloc}}$. Then:

(A) $\vec{u}$ is a local Leray solution.

(B) $\vec{v} = \vec{u}$ on $(0, \min(T, T^*))$.

(C) $T < T^*$ if and only if, for every $(t,x) \in (0,T] \times \mathbb{R}^3$, there exists $r > 0$ such that $\vec{v}$ is bounded on $Q_r(t,x)$.

Proof. (A) $\vec{u}$ belongs to $\mathcal{C}([0, S], L^3)$ for every $0 < S < T^*$, hence to $L^\infty([0, S], L^3)$. Moreover, $\sqrt{t}\vec{u}$ is bounded on $(0, S) \times \mathbb{R}^3$. Hence, $t^{\frac{1}{6}}\vec{u} \in L^\infty((0, S), L^4)$. As $\vec{u}_0 \in L^3 \subset L^2_{\text{uloc}}$, we have $W_{\nu t} * \vec{u}_0 \in (L^\infty((0, S), L^2))_{\text{uloc}} \cap (L^2((0, S), H^1))_{\text{uloc}}$. As $\vec{u} \in L^4((0, S), L^4)$, we have

$$\int_0^t W_{\nu(t-s)} * \mathbb{P}\,\text{div}(\vec{u} \otimes \vec{u})\,ds \in \mathcal{C}([0, S], L^2) \cap L^2((0, S), H^1).$$

Moreover, $\vec{u}$ is smooth on $(0, T^*) \times \mathbb{R}^3$, hence suitable; Finally, for every compact subset K of $\mathbb{R}^3$, we have

$$\lim_{t \to 0} \int_K |\vec{u}(t, .) - \vec{u}_0|^2\,dx \leq \lim_{t \to 0} |K|^{1/3} \|\vec{u}(t, .) - \vec{u}_0\|_3^2 = 0.$$

(B) We have $\vec{v} = \vec{u}$ on $(0, \min(T, T^*))$ by the weak-strong uniqueness theorem (Theorem 14.7).

(C) If $T < T^*$, we have that $\vec{v} = \vec{u}$ on $[0, T]$, hence is continuous from $[0, T]$ to L^3. Moreover, $\sqrt{t}\vec{v}$ is bounded on $[0, T] \times \mathbb{R}^3$, thus, for $0 < t \leq T$ and $x \in \mathbb{R}^3$, $\vec{v}$ is bounded on $Q_r(t, x)$ for every $r \in (0, \sqrt{t}]$.

Conversely, assume that $T \geq T^*$, and $T < +\infty$. Thus, $T^* < +\infty$, and we know that

$$\sup_{T^*/2 < t < T^*} \|\vec{u}(t, .)\|_\infty = +\infty.$$

By Proposition 15.1, we know that $\vec{u}$ coincides on $(0, T^*)$ with a local Leray solution defined on $(0, \frac{3}{2}T^*)$. From the proof of Theorem 14.5, we see that there exists $R > 0$ and $M > 0$ such that

$$\sup_{\frac{3}{4}T^* < t < \frac{3}{2}T^*, |x| > R} |\vec{w}(t, x)| \leq M.$$

Let us assume that for every $x \in \mathbb{R}^3$ there exists $r_x > 0$ such that $\vec{w}$ is bounded on $Q_{r_x}(T^*, x)$. As $\overline{B(0, R)}$ is a compact set, we may find a finite covering

$$\overline{B(0, R)} \subset \cup_{i=1}^N B(x_i, r_{x_i})$$

so that $\vec{w}$ is bounded on $(T^* - \min_{1 \leq i \leq N} r_{x_i}^2, T^*) \times \overline{B(0, R)}$, and finally on $(T_0, T^*) \times \mathbb{R}^3$, with $T_0 = \max(T^* - \min_{1 \leq i \leq N} r_{x_i}^2, \frac{3}{4}T^*)$. As $\vec{u}$ is bounded on $(T^*/2, T_0)$, we get a contradiction.

$\square$

For $\vec{v}$ a local Leray solution on $(0, T)$ of

$$\begin{cases} \partial_t \vec{v} = \nu \Delta \vec{v} - \vec{v} \cdot \vec{\nabla} \vec{v} - \vec{\nabla} p = \nu \Delta \vec{v} - \mathbb{P}(\vec{v} \cdot \vec{\nabla} \vec{v}) \\ \text{div } \vec{v} = 0 \\ \vec{v}(0, .) = \vec{u}_0, \end{cases} \tag{15.4}$$

we say that a point $(t, x) \in (0, T] \times \mathbb{R}^3$ is regular if there exists $r > 0$ such that $\vec{v}$ is bounded on $Q_r(t, x)$, and singular otherwise. Thus, Theorem 15.1 states that if $T^* < +\infty$ there exists at least one singular point (T^*, x) for $\vec{u}$. Let ϵ_0 be the constant in Theorem 14.4. We write $m_{r,x}f$ for the average value of f on the ball $B(x, r)$: $m_{r,x}f = \frac{1}{|B(x,r)|} \int_{B_{x,r}} f(y)\,dy$. We have the following characterization of singular or regular points:

Theorem 15.2.
Let $\vec{u}_0 \in L^2_{\text{uloc}}$ be a divergence free vector field on $\mathbb{R}^3$ and let $\vec{v}$ a local Leray solution on $(0,T)$ of equations (15.4) which belongs to $(L^\infty L^2)_{\text{uloc}} \cap (L^2 H^1)_{\text{uloc}}$. Let $t \in (0,T)$ and $x \in \mathbb{R}^3$. Then:

(A) (t,x) is regular if and only if

$$\lim_{r \to 0} \frac{1}{r^2} \iint_{Q_r(t,x)} |\vec{v}(s,y)|^3 + |p(s,y) - m_{r,x}p(s,.)|^{3/2} \, dy \, ds = 0.$$

(B) (t,x) is singular if and only if

$$\inf_{0 < r < \sqrt{t}} \frac{1}{r^2} \iint_{Q_r(t,x)} |\vec{v}(s,y)|^3 + |p(s,y) - m_{r,x}p(s,.)|^{3/2} \, dy \, ds \geq \epsilon_0^3.$$

Proof. First, we remark that, by the Caffarelli–Kohn–Nirenberg ϵ–regularity criterion Theorem 14.4, if there exists $r \in (0, \sqrt{t})$ such that

$$\frac{1}{r^2} \iint_{Q_r(t,x)} |\vec{v}(s,y)|^3 + |p(s,y) - m_{r,x}p(s,.)|^{3/2} \, dy \, ds < \epsilon_0^3$$

then (t,x) is regular.

We now prove (A). If $\lim_{r \to 0} \frac{1}{r^2} \iint_{Q_r(t,x)} |\vec{v}(s,y)|^3 + |p(s,y) - m_{r,x}p(s,.)|^{3/2} \, dy \, ds = 0$, then $\frac{1}{r^2} \iint_{Q_r(t,x)} |\vec{v}(s,y)|^3 + |p(s,y) - m_{r,x}p(s,.)|^{3/2} \, dy \, ds < \epsilon_0$ for r small enough, and (t,x) is regular. Conversely, let us assume that (t,x) is regular and let $\rho > 0$ such that $\vec{v}$ is bounded on $Q_\rho(t,x)$. On $Q_{\rho/2}(t,x)$, we have

$$p(s,y) = \varpi(s,x) + \sum_{i=1}^{3} \sum_{j=1}^{3} R_i R_j (\mathbb{1}_{B(x,\rho)} v_i v_j)$$

$$+ \sum_{i=1}^{3} \sum_{j=1}^{3} \int_{|x-z|>\rho} (\partial_i \partial_j G(y-z) - \partial_i \partial_j G(x-z)) v_i(s,z) v_j(s,z) \, dz$$

$$= p_{0,x}(s) + p_{1,x}(s,y) + p_{2,x}(s,y).$$

As $\mathbb{1}_{B(x,\rho)} v_i v_j \in L^\infty((t-\rho^2, t), L^1 \cap L^\infty)$, we have $p_{1,x} \in L^\infty((t-\rho^2,t), L^3)$, so that, for $r < \rho/2$,

$$\frac{1}{r^2} \iint_{Q_r(t,x)} |p_{1,x}(s,y) - m_{r,x}p_{1,x}(s,.)|^{3/2} \, dy \, ds \leq C r^{3/2} \|p_{1,x}\|_{L^\infty L^3}^{3/2}.$$

On the other hand, on $Q_{\rho/2}(t,x)$,

$$|p_{2,x}(s,y)| \leq C \frac{1}{\rho^4} \sup_{z \in \mathbb{R}^3} \|\mathbb{1}_{B(z,\rho)} \vec{v}(s,.)\|_2^2$$

so that $p_{2,x}$ is bounded on $Q_{\rho/2}(t,x)$, and, for $r < \rho/2$,

$$\frac{1}{r^2} \iint_{Q_r(t,x)} |p_{2,x}(s,y) - m_{r,x}p_{2,x}(s,.)|^{3/2} \, dy \, ds \leq C r^3 \|\mathbb{1}_{Q_{\rho/2}(t,x)} p_{2,x}\|_{L^\infty L^\infty}^{3/2}.$$

Similarly, we have

$$\frac{1}{r^2} \iint_{Q_r(t,x)} |\vec{v}|^3 \, dy \, ds \leq C r^3 \|\mathbb{1}_{Q_{\rho/2}(t,x)} \vec{v}\|_{L^\infty L^\infty}^3.$$

Thus, $\lim_{r \to 0} \frac{1}{r^2} \iint_{Q_r(t,x)} |\vec{v}(s,y)|^3 + |p(s,y) - m_{r,x}p(s,.)|^{3/2} \, dy \, ds = 0.$ □

15.3 Backward Uniqueness for Local Leray Solutions

Let us recall Escauriaza, Seregin and Šverák's theorem [164] on backward uniqueness for parabolic systems in a half-space $\mathbb{R}_+^3 = \mathbb{R}^2 \times (0, +\infty)$:

Backward uniqueness in a half-space

Theorem 15.3.
Let $\vec{\omega}$ be a vector field on $Q_+ = (-1, 0) \times \mathbb{R}_+^3$ such that

- *for every bounded subdomain Ω of Q_+, $\vec{\omega}$ and its weak derivatives $\partial_t \vec{\omega}$, $\partial_i \vec{\omega}$ $(1 \leq i \leq 3)$ and $\partial_i \partial_j \vec{\omega}$ $(1 \leq i \leq 3, 1 \leq j \leq 3)$ are square-integrable on Ω*

- *for some positive constant C_0, we have*

$$|\partial_t \vec{\omega} - \Delta \vec{\omega}| \leq C_0 (|\vec{\omega}| + |\vec{\nabla} \otimes \vec{\omega}|) \text{ on } Q_+$$

- *for some positive constants C_1 and M, we have*

$$|\vec{\omega}(t,x)| \leq C_1 e^{M|x|^2} \text{ on } Q_+$$

- *$\vec{\omega}(0,.) = 0$ on $\mathbb{R}_+^3$.*

Then $\vec{\omega} = 0$ on Q_+.

The reader will find the proof of Theorem 15.3 in the papers of Escauriaza, Seregin and Šverák [164, 163] or in Seregin's book [429].

As we shall see later, Escauriaza, Seregin and Šverák applied their theorem to prove an endpoint version of Serrin's blow-up criterion [163] . In this section, we consider the case of local Leray solutions for the Navier–Stokes problem with no force ($\vec{f} = 0$) and initial value in L^3.

We first see the consequences of Theorem 15.3 for local Leray solutions for the Navier–Stokes problem. We consider a local Leray solution $\vec{u}$ on $(T_0, T_1) \times \mathbb{R}^3$ of

$$\partial_t \vec{u} = \nu \Delta \vec{u} - \mathbb{P} \operatorname{div}(\vec{u} \otimes \vec{u})$$

with $\vec{u}(T_0,.) \in L^3$. We have seen in the preceding section that $\partial_t \vec{u} \in (L^1 H^{-3/2})_{\text{uloc}}$ and $\partial_t \vec{u} \in (L^1 H^{-3/2})_{\text{uloc}}$, so that the map $t \mapsto \vec{u}(t,.)$ is continuous from $[T_0, T_1]$ to $H_{\text{uloc}}^{-3/2}$, and in particular $\vec{u}(T_1,.)$ is well defined. We then have the following theorem:

Backward uniqueness for local Leray solutions

Theorem 15.4.
Let $\vec{u}_0 \in L^3(\mathbb{R}^3)$ with $\operatorname{div} \vec{u}_0 = 0$. Let $\vec{u}$ be a local Leray solution on $(T_0, T_1) \times \mathbb{R}^3$ of

$$\partial_t \vec{u} = \nu \Delta \vec{u} - \mathbb{P} \operatorname{div}(\vec{u} \otimes \vec{u})$$

with $\vec{u}(T_0,.) = \vec{u}_0$. If $\vec{u}(T_1,.) = 0$, then $\vec{u} = 0$ on $(T_0, T_1) \times \mathbb{R}^3$.

Proof. **Step 1: behavior of $\vec{u}$ near $t = T_0$.**
We know that we find a local-in-time solution of the Navier–Stokes problem in $\mathcal{C}([T_0, T_2], L^3)$ for a small enough $T_2 > T_0$. By the weak-strong uniqueness theorem (Theorem 14.7), this solution coincides with $\vec{u}$ on $(T_0, T_2) \times \mathbb{R}^3$.

In particular, for every T in (T_0, T_2), $\vec{u}$ is bounded on $[T, T_2] \times \mathbb{R}^3$. By Theorem 9.12, $\vec{u}$ is analytic in space and time variables on $(T_0, T_2) \times \mathbb{R}^3$.

Step 2: behavior of $\vec{u}$ near $x = \infty$.
First, let us rematk that for almost every $T_3 \in (T_0, T_1)$, $\vec{u}$ is a local Leray solution on (T_3, T_2). The only thing we have to check is the strong convergence of $\vec{u}(t,.)$ to $\vec{u}(T_3,.)$ in L^2_{loc} when $t \to T_3^+$. But this is a consequence of the energy inequalities due to the suitability of $\vec{u}$: to get the convergence, it is enough to have that, for every $N \in \mathbb{N}$, T_3 is a Lebesgue point of $t \mapsto \|1_{B(0,N)}\vec{u}(t,.)\|_2$.

Let $T \in (T_0, T_1)$. Let

$$S = \frac{1}{C_0 \nu (1 + \frac{\|\vec{u}\|^2_{(L^\infty L^2)_{\text{uloc}}}}{\nu})^4}$$

where C_0 is the constant in Theorem 14.5. We pick up $0 < T_3 < T$ such that $T_3 - T < \min(1, S)$ and $\vec{u}$ is a local Leray solution on $(T_3, T) \times \mathbb{R}^3$. Then (the proof of) Theorem 14.5 shows that, for $|x|$ large enough ($|x| > R$, where R depends on $\vec{u}$ and T), we have $|\vec{u}(t,x)| < C\frac{1}{\sqrt{T-T_3}}$ on $(\frac{T+T_3}{2}, T)$.

Let $T_4 \in (T_0, T_1)$. We can reiterate the argument, descending from T_1 to $T_4/2$ by steps of size $\frac{1}{2}\min(1, S)$, and we find that there exists some $R > 0$ and $M > 0$ such that

$$|\vec{u}(t,x)| \leq M \text{ on } (T_4/2, T_1) \times (\mathbb{R}^3 \setminus B(0, R)).$$

Then we use the local regularity theory of Serrin [434] (see Theorem 13.1) and conclude that there exists some constant M_0 such that

$$\text{for } |\alpha| \leq 3, \quad |\partial^\alpha \vec{u}(t,x)| \leq M_0 \text{ on } (T_4, T_1) \times (\mathbb{R}^3 \setminus B(0, 2R)).$$

Step 3: Backward uniqueness for the vorticity.
Let $\vec{\omega} = \vec{\nabla} \wedge \vec{u}$. Let $Q_+ = (T_4, T_1) \times \mathbb{R}^2 \times (2R, +\infty)$. For $|\alpha| \leq 2$, we have $|\partial^\alpha \vec{\omega}(t,x)| \leq 2M_0$ on Q_+. Moreover, as

$$\partial_t \vec{\omega} = \nu \Delta \vec{\omega} - \vec{u}.\vec{\nabla}\vec{\omega} + \vec{\omega}.\vec{\nabla}\vec{u}$$

we find that $|\partial_t \vec{\omega}(t,x)| \leq 6M_0(\nu + M_0)$ on Q_+.

Similarly, we have

$$|\partial_t \vec{\omega} - \nu \Delta \vec{\omega}| \leq |\vec{u}.\vec{\nabla}\vec{\omega}| + |\vec{\omega}.\vec{\nabla}\vec{u}| \leq M_0|\vec{\nabla} \otimes \vec{\omega}| + 3M_0|\vec{\omega}| \text{ on } Q_+.$$

Finally, as $\vec{u}(T_1, .) = 0$, we find that $\vec{\omega}(T_1, x) = 0$ for $x_3 > 2R$. Applying Theorem 15.3, we find that $\vec{\omega} = 0$ on Q_+.

In particular, we have $\vec{\omega} = 0$ on $(T_4, T_2) \times \mathbb{R}^2 \times (2R, +\infty)$. As $\vec{\omega}$ is analytic in space and time variables on $(T_0, T_2) \times \mathbb{R}^3$, we find that $\vec{\omega} = 0$ on $(T_0, T_2) \times \mathbb{R}^3$.

Step 4: Backward uniqueness for the velocity.
As $-\Delta \vec{u} = \vec{\nabla} \wedge \vec{\omega}$, we find that $-\Delta \vec{u} = 0$ on $(T_0, T_2) \times \mathbb{R}^3$. But $\vec{u} \in \mathcal{C}([T_0, T_2], L^3)$; in L^3, $-\Delta \vec{u} = 0$ implies $\vec{u} = 0$. Thus, we find that $\vec{u} = 0$ on $[T_0, T_2] \times \mathbb{R}^3$. In particular, $\vec{u}_0 = 0$. By the weak-strong uniqueness theorem (Theorem 14.7), we find that the local Leray solution $\vec{u}$ must then coincide with the null solution of the Cauchy problem, and thus $\vec{u} = 0$ on $(T_0, T_1) \times \mathbb{R}^3$.

$\square$

15.4   Seregin's Theorem

In 2003 Escauriaza, Seregin and Šverák [163] extended the celebrated $L^p L^q$ criterion of Serrin (Theorem 11.2) to the limit case $L_t^\infty L^3$: in the case of a null force, they showed that a mild solution that remains bounded in the L^3 norm cannot blow up. Seregin [428] then gave a more precise statement: the L^3 norm goes to $+\infty$ near the blow-up time:

Seregin's theorem

Theorem 15.5.
Let $\vec{u}_0 \in L^3(\mathbb{R}^3))$ with $\operatorname{div} \vec{u}_0 = 0$.
   Let $\vec{u}$ be the solution of the Navier–Stokes equations equations

$$\partial_t \vec{u} = \nu \Delta \vec{u} - \mathbb{P} \operatorname{div}(\vec{u} \otimes \vec{u}), \quad \vec{u}(0, .) = \vec{u}_0$$

in $\mathcal{C}([0, T^*), L_x^3)$, where T^* is the maximal existence time of the solution. Then, if $T^* < +\infty$, we have

$$\lim_{t \to T^*} \|\vec{u}(t, .)\|_3 = +\infty.$$

Proof. We shall show that the assumption $\liminf_{t \to T^*} \|\vec{u}(t, .)\|_3 < +\infty$ leads to a contradiction. Thus, we assume that there exists some $M < +\infty$ and some sequence $T_n \uparrow T^*$ such that $\|\vec{u}(T_n, .)\|_3 \leq M$.

From Theorem 15.1, we know that there exists a point $x_0 \in \mathbb{R}^3$ such that

$$\text{for every } r \in (0, \sqrt{T^*}), \qquad \sup_{(s,y) \in Q_r(x_0)} |\vec{u}(s, y)| = +\infty,$$

where $Q_r(x_0) = (T^* - r^2, T^*) \times B(x_0, r)$. Changing $\vec{u}$ into $\vec{u}(t, x + x_0)$, we may assume with no loss of generality that $x_0 = 0$. Similarly, changing $\vec{u}$ into $\sqrt{T^*}\vec{u}(T^*t, \sqrt{T^*}x)$, we may assume that $T^* = 1$.

Now, let us assume that there exists some $M < +\infty$ and some sequence $T_n \to 1^-$ such that $\|\vec{u}(T_n, .)\|_3 \leq M$. We define

$$\vec{u}_n(t, x) = \sqrt{1 - T_n}\vec{u}(T_n + t(1 - T_n), \sqrt{1 - T_n}x).$$

We have $\vec{u}_n \in \mathcal{C}([0, 1), L^3)$, $\|\vec{u}_n(0, .)\|_3 \leq M$ and $\vec{u}_n$ blows up at $(1, 0)$.

By Proposition 15.1 (and the weak-strong uniqueness theorem Theorem 14.7), we know that $\vec{u}_n$ coincides on $(0,1)$ with a local Leray solution $\vec{v}_n$ defined on $(0,2)$ with

$$\sup_{0<t<2} \|\vec{v}_n(t,.)\|_{L^2_{\text{uloc}}} \le C_{\nu,M}$$

and

$$\|\vec{v}_n\|_{(L^2 H^1)_{\text{uloc}}} \le C_{\nu,M}.$$

Thus, for every test function $\phi \in \mathcal{D}'((0,T^*) \times \mathbb{R}^3)$, $\phi\vec{v}_n$ remains bounded in $L^\infty L^2 \cap L^2 \dot{H}^1$, while $\phi\partial_t\vec{v}_n$ remains bounded in $L^{3/2}H^{-3/2}$. We then use the Rellich–Lions theorem (Theorem 12.1): we may find a sequence $n_k \to +\infty$ and a function $\vec{v}_\infty \in (L^\infty L^2)_{\text{uloc}} \cap (L^2\dot{H}^1)_{\text{uloc}}$ such that:

- $\vec{v}_{n_k}$ is weak* convergent to $\vec{v}_\infty$ in $(L^\infty L^2)_{\text{uloc}}$ and in $(L^2\dot{H}^1)_{\text{uloc}}$

- $\vec{v}_{n_k}$ is strongly convergent to $\vec{v}_\infty$ in $L^2_{\text{loc}}([0,2] \times \mathbb{R}^3)$.

The limit $\vec{v}_\infty$ is a solution of the Navier–Stokes equations

$$\partial_t\vec{v}_\infty = \nu\Delta\vec{v}_\infty - \mathbb{P}\operatorname{div}(\vec{v}_\infty \otimes \vec{v}_\infty) = \nu\Delta\vec{v}_\infty - \operatorname{div}(\vec{v}_\infty \otimes \vec{v}_\infty) - \vec{\nabla}p_\infty$$

and satisfies the local energy inequality:

$$\partial_t\left(\frac{|\vec{v}_\infty|^2}{2}\right) \le \nu\Delta\left(\frac{|\vec{v}_\infty|^2}{2}\right) - \nu|\vec{\nabla}\otimes\vec{v}_\infty|^2 - \operatorname{div}\left((p_\infty + \frac{|\vec{v}_\infty|^2}{2})\vec{v}_\infty\right).$$

Moreover, $\vec{v}_{n_k}$ and $\partial_t\vec{v}_{n_k}$ are bounded in $(L^2 H^{-3/2})_{\text{uloc}}$, so that, writing for $0 \le t \le 1$ $\vec{v}_{n_k}(t,.) = -\int_t^2 \partial_t(\zeta\vec{v}_\infty)\,ds.$, where ζ is a smooth function on $\mathbb{R}$ which is equal to 1 on $(-\infty, 5/4)$ and to 0 on $(7/4, +\infty)$, we find that $\vec{v}_\infty \in \mathcal{C}([0,1], H^{-3/2}_{\text{uloc}})$ and that, for every $t \in [0,1]$, $\vec{v}_\infty(t,.)$ is the weak* limit in $H^{-3/2}_{\text{uloc}}$ of $\vec{v}_{n_k}(t,.)$.

As $\|\vec{v}_{n_k}(0,.)\|_3 = \|\vec{u}(T_{n_k},.)\|_3 \le M$, we find as well that $\vec{v}_\infty(0,.) \in L^3$. Moreover, by Proposition 15.1, we know that, for $0 < t < 1$,

$$\|\vec{u}_n(t,.) - W_{\nu t} * \vec{u}(T_n,.)\|_{L^2_{\text{uloc}}} \le \eta(t)$$

where the function η depends only on ν and M and satisfies:

$$\lim_{t\to 0^+} \eta(t) = 0.$$

Thus, $\|\vec{v}_\infty(t,.) - W_{\nu t} * \vec{v}_\infty(0,.)\|_{L^2_{\text{uloc}}} \le \eta(t)$, and since $\vec{v}_\infty(0,.) \in L^3$, $\lim_{t\to 0^+}\|\vec{v}_\infty(t,.) - \vec{v}_\infty(0,.)\|_{L^2_{\text{uloc}}} = 0$. Thus, $\vec{v}_\infty$ is a local Leray solution.

Now, we shall prove that $\vec{v}_\infty$ blows up at $(1,0)$: for every $r \in (0,1)$, we have $\|\vec{v}_\infty\|_{L^\infty(Q_r)} = +\infty$, where $Q_r = (1-r^2,1) \times B(0,r)$. Indeed, let us assume that, for some r_0, we have $\|\vec{v}_\infty\|_{L^\infty(Q_{r_0})} < +\infty$. On $Q_{r_0/2}$, we have, for $r_0 < R$,

$$p_\infty(s,y) = \varpi_\infty(s) + \sum_{i=1}^{3}\sum_{j=1}^{3} R_i R_j(\mathbb{1}_{B(0,r_0)})v_{\infty,i}v_{\infty,j})$$

$$+ \sum_{i=1}^{3}\sum_{j=1}^{3}\int_{r_0<|x-z|<R}(\partial_i\partial_j G(y-z) - \partial_i\partial_j G(-z))v_{\infty,i}(s,z)v_{\infty,j}(s,z)\,dz$$

$$+ \sum_{i=1}^{3}\sum_{j=1}^{3}\int_{|z|>R}(\partial_i\partial_j G(y-z) - \partial_i\partial_j G(-z))v_{\infty,i}(s,z)v_{\infty,j}(s,z)\,dz$$

$$= \varpi_\infty(s) + p_{\infty,1}(s,y) + p_{\infty,2,R}(s,y) + p_{\infty,3,R}(s,y).$$

For $0 < r < r_0/2$, we have the following estimates:

- As $\vec{v}_\infty \in L^\infty(Q_{r_0})$,

$$\frac{1}{r^2} \iint_{Q_r} |\vec{v}_\infty(s,y)|^3 \, dy \, ds \leq Cr^3 \|\vec{v}_\infty\|_{L^\infty(Q_{r_0})}^3.$$

- As $\mathbb{1}_{B(0,r_0)} v_{\infty,i} v_{\infty,j} \in L^\infty((t-r_0^2,t), L^1 \cap L^\infty)$, we find that $p_{\infty,1} \in L^\infty((1-r_0^2,1), L^3)$, so that, for $r < r_0/2$,

$$\frac{1}{r^2} \iint_{Q_r} |p_{\infty,1}(s,y) - m_{r,0} p_{\infty,1}(s,.)|^{3/2} \, dy \, ds \leq Cr^{3/2} \|p_{\infty,1}\|_{L^\infty L^3}^{3/2}.$$

- On $Q_{r_0/2}$, $|p_{\infty,2,R}(s,y)| \leq C \frac{1}{r_0^4} \|\mathbb{1}_{B(0,R)} \vec{v}_\infty(s,.)\|_2^2$ so that $p_{\infty,2,R}$ is bounded on $Q_{r_0/2}$ and, for $0 < r < r_0/2$,

$$\frac{1}{r^2} \iint_{Q_r} |p_{\infty,2,R}(s,y) - m_{r,0} p_{\infty,2,R}(s,.)|^{3/2} \, dy \, ds \leq Cr^3 \|\mathbb{1}_{Q_{r_0/2}} p_{\infty,2,R}\|_{L^\infty L^\infty}^{3/2}.$$

- On $Q_{r_0/2}$, $|p_{\infty,3,R}(s,y)| \leq C \frac{1}{r_0^3 R} \sup_{z \in \mathbb{R}^3} \|\mathbb{1}_{B(z,r_0)} \vec{v}_\infty(s,.)\|_2^2$ so that $p_{\infty,3,R}$ is bounded on $Q_{r_0/2}$ and, for $0 < r < r_0/2$,

$$\frac{1}{r^2} \iint_{Q_r} |p_{\infty,3,R}(s,y) - m_{r,0} p_{\infty,3,R}(s,.)|^{3/2} \, dy \, ds \leq Cr^3 \|\mathbb{1}_{Q_{r_0/2}} p_{\infty,3,R}\|_{L^\infty L^\infty}^{3/2}.$$

Thus, there exists $r_1 \in (0, r_0/2)$ such that

$$\frac{1}{r_1^2} \iint_{Q_{r_1}} |\vec{v}_\infty(s,y)|^3 + |p_\infty(s,y) - m_{r_1,0} p_\infty(s,.)|^{3/2} \, dy \, ds < \frac{\epsilon_0^3}{4} \tag{15.5}$$

We have, for a constant C_0 which does not depend on r_0, r_1, R, nor n,

$$\iint_{Q_{r_1}} |\vec{v}_n(s,y) - \vec{v}_\infty(s,y)|^3 + |p_n(s,y) - p_\infty(s,y) - m_{r_1,0}(p_n - p_\infty)(s,.)|^{3/2} dy \, ds$$

$$\leq \iint_{Q_{r_0}} |\vec{v}_n(s,y) - \vec{v}_\infty(s,y)|^3 \, dy \, ds$$

$$+ C_0 \iint_{Q_{r_0}} |\vec{v}_n \otimes \vec{v}_n(s,y) - \vec{v}_\infty \otimes \vec{v}_\infty(s,y)|^{3/2} \, dy \, ds$$

$$+ C_0 \frac{1}{r_0^{3/2}} \int_{1-r_1^2}^{1} \int_{B(0,R)} |\vec{v}_n \otimes \vec{v}_n(s,y) - \vec{v}_\infty \otimes \vec{v}_\infty(s,y)|^{3/2} \, dy \, ds$$

$$+ C_0 \frac{1}{R^{3/2}} \frac{r_1^5}{r_0^{9/2}} (1 + r_0)^{9/2} \|\vec{v}_n - \vec{v}_\infty\|_{L^\infty L^2_{uloc}}^3$$

$$= I + II + III + IV.$$

First, we notice that, as the $\vec{v}_n$ are bounded in $L^\infty((0,1), L^2_{uloc})$, the term IV is $O(R^{-3/2})$ uniformly in n. Then, we use the strong convergence of $\vec{v}_{n_k}$ to $\vec{v}_\infty$ in $L^3((1 - r_0^2, 1) \times B(0,R))$: $\vec{v}_{n_k}$ is bounded in $L^\infty((1 - r_0^2, 1), L^2(B(0,R)) \cap L^2((1 - r_0^2, 1), L^6(B(0,R))$ and strongly convergent in $L^2((1-r_0^2, 1), L^2(B(0,R)))$, so that it is strongly convergent in $L^2((1- r_0^2, 1), L^3(B(0,R)))$; as it is bounded in $L^4((1 - r_0^2, 1), L^3(B(0,R)))$, it is strongly convergent in $L^3((1-r_0^2,1), L^3(B(0,R)))$. Thus, the terms I and II are $o(1)$, and so is III when R is fixed. We conclude that, for k large enough, we have

$$\frac{1}{r_1^2} \iint_{Q_{r_1}} |\vec{v}_{n_k}(s,y)|^3 + |p_{n_k}(s,y) - m_{r_1,0}p_{n_k}(s,.)|^{3/2} \, dy \, ds < \frac{\epsilon_0^3}{2}. \qquad (15.6)$$

By Theorem 14.4, we conclude that $\vec{v}_{n_k}$ s bounded on $Q_{r/2}$; as $\vec{v}_{n_k}$ blows up at $(1,0)$, this is however not possible.

Thus, we have proved that $\vec{v}_\infty$ blows up at $(1,0)$. We will now use Theorem 15.4, to get $\vec{v}_\infty = 0$, which is a contradiction with the fact that $\vec{v}_\infty$ blows up at $(1,0)$. We already know that $\vec{v}_\infty$ is a local Leray solution on $(0,1) \times \mathbb{R}^3$, and that $\vec{v}_\infty(0,.) \in L^3$. We also know that $\vec{v}_\infty(1,.)$ is the weak* limit of $\vec{v}_{n_k}(1,.)$ in $H_{uloc}^{-3/2}$.

Moreover, $\vec{v}_{n_k}(1,.)$ is the (strong) limit of $\vec{v}_{n_k}(t,.)$ in $H_{uloc}^{-3/2}$ when $t \to 1$. For $t < 1$,

$$\vec{v}_{n_k}(t,x) = \vec{u}_{n_k}(t,x) = \sqrt{1-T_{n_k}}\vec{u}(T_{n_k} + t(1-T_{n_k}), \sqrt{1-T_{n_k}}x).$$

As $\vec{u} \in \mathcal{C}([0,1], H_{uloc}^{-3/2})$, we find that

$$\vec{v}_{n_k}(1,x) = \sqrt{1-T_{n_k}}\vec{u}(1, \sqrt{1-T_{n_k}}x).$$

Moreover, $\vec{u}(1,.)$ is the limit in $H_{uloc}^{-3/2}$ of $\vec{u}(T_{n_k},.)$, with $\|\vec{u}(T_{n_k},.)\|_3 \le M$; thus, $\vec{u}(1,.) \in L^3$. Thus, $\vec{v}_\infty(1,.)$ is the weak* limit of $\sqrt{1-T_{n_k}}\vec{u}(1, \sqrt{1-T_{n_k}}x)$ in $H_{uloc}^{-3/2}$, and this limit is equal to 0 since $\vec{u}(1,.) \in L^3$. By Theorem 15.4, $\vec{v}_\infty = 0$ $\qquad\square$

15.5 Further Comments on Seregin's Theorem

Phuc [397] proved in 2015 that a solution that blows up in finite time T^* must satisfy $\sup_{0<t<T^*} \|\vec{u}(t,.)\|_{L^{3,q}} = +\infty$ with $q < +\infty$. An analogous result holds in the setting of Besov spaces $\dot{B}_{p,q}^{-1+3/p}$ ($3 < p, q < +\infty$) (Gallagher, Koch and Planchon [199] or Albritton [4]). Seregin's theorem states that, more precisely, $\lim_{t\to T^*} \|\vec{u}(t,.)\|_3 = +\infty$. L^3 does not occupy a specific place in the theory we just described. It can be replaced with broader critical spaces such as the Lorentz spaces $L^{3,q}$ with $3 \le q < +\infty$:

Theorem 15.6.
Let $\vec{u}_0$ be a divergence free vector field on $\mathbb{R}^3$ and let $\vec{u}$ be a local Leray solution of the Navier–Stokes equations equations

$$\partial_t \vec{u} = \nu \Delta \vec{u} - \mathbb{P}\operatorname{div}(\vec{u} \otimes \vec{u}), \quad \vec{u}(0,.) = \vec{u}_0$$

such that $\vec{u} \in L_{loc}^\infty([0,T^*), L^{p,q})$ with $3 \le p < +\infty$, $1 \le q \le +\infty$, $(p,q) \ne (3,+\infty)$. Assume that T^* is the maximal existence time of such a solution. Then, if $T^* < +\infty$, we have

a) if $p > 3$, for every $t < T^*$,

$$\|\vec{u}(t,.)\|_{L^{p,q}} \ge C_{p,q}\nu \frac{1}{(\nu(T^*-t))^{\frac{1}{2}-\frac{3}{2p}}}$$

where the positive constant $C_{p,q}$ does not depend on $\vec{u}_0$ nor on ν.

b) If $p = 3$,

$$\lim_{t \to T^*} \|\vec{u}(t, .)\|_{L^{p,q}} = +\infty.$$

Proof. Let $p > 3$. The bilinear operator B

$$B(\vec{u}, \vec{v}) = \int_0^t W_{\nu(t-s)} * \mathbb{P} \operatorname{div}(\vec{u} \otimes \vec{v}) \, ds$$

is bounded on the space $E_{p,q,T} = L^\infty((0,T), L^{p,q})$ or on the space $F_{p,T} = \{\vec{u} \in L^\infty((0,T), L^{p,q}) \ / \ \sup_{0<t<T} t^{\frac{3}{2p}} \|\vec{u}(s,.)\|_\infty < +\infty\}$: we have

$$\|B(\vec{u}, \vec{v})\|_{E_{p,q,T}} \leq C_{p,q,0} \frac{1}{\nu}(\nu T)^{\frac{1}{2} - \frac{3}{2p}} \|\vec{u}\|_{E_{p,q,T}} \|\vec{v}\|_{E_{p,q,T}}$$

and

$$\|B(\vec{u}, \vec{v})\|_{F_{p,q,\nu,T}} \leq C_{p,q,1} \frac{1}{\nu}(\nu T)^{\frac{1}{2} - \frac{3}{2p}} \|\vec{u}\|_{F_{p,q,\nu,T}} \|\vec{v}\|_{E_{p,q,\nu,T}}$$

where

$$\|\vec{u}\|_{E_{p,q,T}} = \sup_{0<s<T} \|\vec{u}(s,.)\|_{L^{p,q}}$$

and

$$\|\vec{u}\|_{F_{p,q,\nu,T}} = \sup_{0<s<T} \|\vec{u}(s,.)\|_{L^{p,q}} + \sup_{0<s<T} (\nu s)^{\frac{3}{2p}} \|\vec{u}(s,.)\|_\infty.$$

As $\|W_{\nu t} * \vec{u}_0\|_{F_{p,q,\nu,T}} \leq C_{p,q,2} \|\vec{u}_0\|_{L^{p,q}}$, we find that the Navier–Stokes equations with initial value $\vec{u}_0 \in L^{p,q}$ has a solution $\vec{u} \in F_{p,q,\nu,T}$ with $\|\vec{u}\|_{F_{p,q,\nu,T}} \leq 2C_{p,q,2} \|\vec{u}_0\|_{L^{p,q}}$ if $C_{p,q,1} C_{p,q,2} \frac{1}{\nu}(\nu T)^{\frac{1}{2} - \frac{3}{2p}} \|\vec{u}_0\|_{L^{p,q}} \leq \frac{1}{8}$.

Assume now that $\vec{u} \in L^\infty_{\text{loc}}([0, T^*), L^{p,q})$ blows up at a finite time $t = T^*$. Let

$$A(t) = \frac{1}{\nu} \left(\frac{\nu}{8 C_{p,q,1} C_{p,q,2} \|\vec{u}(t,.)\|_{L^{p,q}}} \right)^{\frac{2p}{p-3}}.$$

$\vec{u}$ cannot blow up on $(t, t + A(t))$, so that $T^* - t \geq A(t)$, and thus

$$\|\vec{u}(t,.)\|_{L^{p,q}} \geq \frac{1}{8 C_{p,q,1} C_{p,q,2}} \nu \frac{1}{(\nu(T^* - t))^{\frac{1}{2} - \frac{3}{2p}}}.$$

Assume now $p = 3$ and $q < +\infty$. If $q \leq 3$, we have $L^{3,q} \subset L^3$ and we may conclude by Seregin's theorem. For $q > 3$, we just check that all the steps in Seregin's proof remain valid:

- Proposition 15.1 remains valid, as $L^{3,q} \subset L^{3,\infty} \subset [L^2, L^6]_{1/2,\infty}$.

- Theorem 15.1 remains valid: in (A), we use that a solution in $\mathcal{C}([0,T], L^{3,q})$ satisfies the inequalities $\sup_{0<s<T} \|\vec{u}(t,.)\|_{L^{3,q}} < +\infty$ and $\sup_{0<s<T} \sqrt{t} \|\vec{u}(t,.)\|_\infty < +\infty$, so that $\sup_{0<s<T} t^{1/6} \|\vec{u}(t,.)\|_4 < +\infty$; in (B), we used the weak-strong uniqueness theorem (Theorem 14.7), and in (C), we used the proof of Theorem 14.5, which are still valid for solutions in $L^{3,q}$, $q < +\infty$.

- Theorem 15.4 remains valid: step 1 of the proof is based on the weak-strong uniqueness theorem; step 2 is based on (the proof of) Theorem 14.5; step 4 used that harmonic functions that belong to L^3 are null, which remains valid wen L^3 is changed in $L^{3,q}$.

- The last final key points in Seregin's proof are that if $\vec{v}$ is a weak* limit of a sequence v_n which is bounded in L^3 then $v \in L^3$, and that if $v \in L^3$ and $\epsilon_n \to 0$, then the weak* limit of $\epsilon_n v(\epsilon_n x)$ is equal to 0. Both points remain valid in $L^{3,q}$, $q < +\infty$.

Theorem 15.6 does not preclude the existence of a blowing up solution in $L^\infty((0,T^*),L^{3,\infty})$, in particular since the weak* limit of $\epsilon_n v(\epsilon_n x)$ may be different from 0. $\qquad\square$

We don't know whether blow up occurs in finite time, but many studies were done on solutions that would blow up slowly:

Definition 15.1.
Let $\vec{u}_0$ be a divergence free vector field on $\mathbb{R}^3$ and let $\vec{u}$ be a local Leray solution of the Navier–Stokes equations equations

$$\partial_t \vec{u} = \nu\Delta\vec{u} - \mathbb{P}\operatorname{div}(\vec{u}\otimes\vec{u}), \quad \vec{u}(0,.) = \vec{u}_0$$

such that $\vec{u} \in L^\infty_{loc}((0,T^*),L^\infty)$. Let T^* be the maximal existence time of such a solution. Assume that $T^* < +\infty$.

 a) *$\vec{u}$ has a Type I blow up in L^∞ if $\sup_{T_0<t<T^*}(T^*-t)^{1/2}\|\vec{u}(t,.)\|_\infty < +\infty$ for some $0 < T_0 < T^*$.*

 b) *Let $3 < p < +\infty$, $1 \le q \le +\infty$. $\vec{u}$ has a Type I blow up in $L^{p,q}$ if $\sup_{T_0<t<T^*}(T^*-t)^{\frac{1}{2}-\frac{3}{2p}}\|\vec{u}(t,.)\|_{L^{p,q}} < +\infty$ for some $0 < T_0 < T^*$.*

 c) *$\vec{u}$ has a Type I blow up in $L^{3,\infty}$ if $\sup_{T_0<t<T^*}\|\vec{u}(t,.)\|_{L^{3,\infty}} < +\infty$ for some $0 < T_0 < T^*$.*

Proposition 15.2.
Let $3 < p < +\infty$, $1 \le q \le +\infty$. If a local Leray solution $\vec{u}$ of the Navier–Stokes equations equations has a Type I blow up in $L^{p,q}$, then $\vec{u}$ has a Type I blow up in L^∞.

Proof. . If $C_{p,q,1}C_{p,q,2}\frac{1}{\nu}(\nu\theta)^{\frac{1}{2}-\frac{3}{2p}}\|\vec{u}(t,.)\|_{L^{p,q}} \le \frac{1}{8}$, the solution $\vec{u}$ is controlled in L^∞ norm on $(t,t+\theta)$ by

$$(\nu(s-t))^{\frac{3}{2p}}\|\vec{u}(s,.)\|_\infty \le 2C_{p,q,2}\|\vec{u}(t,.)\|_{L^{p,q}}.$$

Thus, if $\|\vec{u}(t,.)\|_{L^{p,q}} \le C_0(T^*-t)^{-\frac{1}{2}+\frac{3}{2p}}$ on (T_0,T^*), defining $\gamma = \frac{1}{\nu}\left(\frac{\nu}{8C_0C_{p,q,1}C_{p,q,2}}\right)^{\frac{2p}{p-3}}$, we find that for $T_0 < t < T^*$, $\vec{u}$ cannot blow up on $(t,t+\gamma(T^*-t)]$ (so that $\gamma < 1$) and that

$$\|\vec{u}((1-\frac{\gamma}{2})t + \frac{\gamma}{2}T^*,.)\|_\infty \le 2C_{p,q,2}\|\vec{u}(t,.)\|_{L^{p,q}}\left(\frac{2}{\nu\gamma(T^*-t)}\right)^{\frac{3}{2p}}$$

$$\le 2C_0C_{p,q,2}\left(\frac{2}{\nu\gamma}\right)^{\frac{3}{2p}}(T^*-t)^{-\frac{1}{2}}.$$

Thus, for $T_1 = (1-\frac{\gamma}{2})T_0 + \frac{\gamma}{2}T^* < s < T^*$, we have

$$\|\vec{u}(s,.)\|_\infty \le 2C_0C_{p,q,2}\left(\frac{2}{\nu\gamma}\right)^{\frac{3}{2p}}(1-\frac{\gamma}{2})^{1/2}(T^*-s)^{-\frac{1}{2}}. \qquad\square$$

On the other hand, we have no control in L^∞ norm of a solution that would have a Type I blow up in $L^{3,\infty}$. Okhitani [378] discusses possible scale-invariant singularities of the form $|\vec{u}(t,x)| \le C_0\frac{\nu}{|x-x_0|+\sqrt{\nu(T^*-t)}} = C_0 f_\nu(T^*-t,x-x_0)$. We would have for such a singularity a blow up of Type I in $L^{3,\infty}$ and in L^∞.

While for blowing up solutions $\vec{u}$ in L^p on (T_0,T^*) for $p > 3$, we have a uniform control on $\|\vec{u}(t,.)\|_p$ of the form $\|\vec{u}(t,.)\|_p \ge C_{p,\nu}(T^*-t)^{-\frac{1}{2}+\frac{3}{2p}}$, where $C_{p,\nu} > 0$ does not depend on $\vec{u}$,

T_0 nor T^*, we only know, when $p = 3$, that $\|\vec{u}(t, .)\|_3 \geq C_\nu > 0$ and $\lim_{t \to T^*} \|\vec{u}(t, .)\|_3 = +\infty$. But there is no hope that we would have an estimate $\|\vec{u}(t, .)\|_3 \geq f(T^* - t)$, where f would be a universal function (that would depend on neither on $\vec{u}$, T_0 nor T^*) such that $\lim_{s \to 0} f(s) = +\infty$. This is commented in Barker's thesis, following an idea of Seregin [17]: if $T_0 < T_1 < T^*$, then $\vec{v}_A(t, , x) = \sqrt{A}\vec{u}(T_1 + A(t - T_1), \sqrt{A}x)$, defined on $(T_1 - \frac{T_1 - T_0}{A}, T_1 + \frac{T^* - T_1}{A})$ would blow up at $\frac{T^* - T_1}{A}$, so that one should have $\|\vec{u}(T_1, .)\|_3 = \|\vec{v}_A(0, .)\|_3 \geq f(\frac{T^* - T_1}{A})$; letting A go to $+\infty$, we find $\|\vec{u}(T_1, .)\|_3 = +\infty$.

However, if we allow f to depend on T_0, then Tao [463] proved such an a priori estimate:

Quantitative regularity for L^3 solutions.

Theorem 15.7. *[Tao's quanitative bound]*
Let $\vec{u} \in \mathcal{C}([0, T], L^3)$ be a solution to the Navier–Stokes equations (with null force and viscosity $\nu = 1$). If $\sup_{0 < t < T} \|\vec{u}(t, .)\|_3 \leq A$, then

$$\sup_{0 < t < T, x \in \mathbb{R}^3} \sqrt{t}|\vec{u}(t, x)| \leq e^{e^{e^{(A+2)^\beta}}}$$

where the constant β does not depend on A nor on $\vec{u}$.

This can be adapted without difficulty to the case $\nu \neq 1$ ot $T_0 \neq 0$. If $\vec{u} \in \mathcal{C}([T_0, T], L^3)$ is a solution to the Navier–Stokes equations with viscosity ν, then $\vec{v} = \vec{u}(T_0 + \nu t, \nu x)$ is a solution in $\mathcal{C}([0, \frac{T - T_0}{\nu}], L^3)$ with viscosity 1. Moreover, $\sup_{0 < t < \frac{T - T_0}{\nu}} \|\vec{v}(t, .)\|_3 = \frac{\sup_{T_0 < t < T} \|\vec{u}(t, .)\|_3}{\nu}$. Thus, we find

$$\|\vec{u}(T, .)\|_\infty \leq \frac{\nu}{\sqrt{\nu(T - T_0)}} e^{e^{e^{(2 + \frac{\sup_{T_0 < t < T} \|\vec{u}(t, .)\|_3}{\nu})^\beta}}}. \tag{15.7}$$

On the other hand, if $\vec{u} \in \mathcal{C}([T_0, T^*), L^3)$ is a solution to the Navier–Stokes equations which blows up at $t = T^*$, we have, for $T_0 < T < T^*$

$$\|\vec{u}(T, .)\|_\infty \geq C_0 \frac{\nu}{\sqrt{\nu(T^* - T)}}. \tag{15.8}$$

Thus, we deduce from Tao's theorem (Theorem 15.7) the following blow up estimate:

Corollary 15.1.
Let $\vec{u} \in \mathcal{C}([T_0, T^*), L^3)$ be a solution to the Navier–Stokes equations which blows up at $t = T^*$. If T is close enough to T^*:

$$\frac{T^* - T}{T - T_0} < \frac{1}{C_0^2} e^{-2e^{e^{4\beta}}}$$

then

$$\frac{\sup_{T_0 < t < T} \|\vec{u}(t, .)\|_3}{\nu} > \frac{1}{2}(\ln(\ln(\ln(C_0\sqrt{\frac{T - T_0}{T^* - T}}))))^{1/\beta}.$$

In particular,

$$\limsup_{T \to T^*} \frac{\sup_{T_0 < t < T} \|\vec{u}(t, .)\|_3}{(\ln(\ln(\ln(\frac{1}{T^* - T}))))^{1/\beta}} \geq \frac{\nu}{2}.$$

Tao's proof for Theorem 15.7 relies on Fourier analysis. Barker and Prange [20] proposed a different proof, developing arguments on locally defined quantities. They proved the following theorem:

Theorem 15.8.
Let $\vec{u} \in \mathcal{C}([T_0, T^*), L^3)$ be a solution to the Navier–Stokes equations which blows up at $t = T^*$. Assume that $\vec{u}$ has a Type I blow up in $L^{3,\infty}$: $\sup_{T_0 < t < T^*} \|\vec{u}(t, .)\|_{L^{3,\infty}} \leq M < +\infty$. Then

$$\liminf_{t \to T^*} \frac{\|\vec{u}(t, .)\|_3}{\ln(\frac{1}{T^* - t})} > 0.$$

Another result of Barker and Prange [19] answers a conjecture of Tao 15.7 on proving blow up in an Orlicz norm of the type $L^3(ln(ln(lnL)))^{-c}$:

Theorem 15.9.
Let $\vec{u} \in \mathcal{C}([T_0, T^*), L^3)$ be a solution to the Navier–Stokes equations which blows up at $t = T^*$. Assume that $\vec{u}_0 \in L^2 \cap L^4$. Then

$$\limsup_{t \to T^*} \int \frac{|\vec{u}(t, x)|^3}{\left(\ln \ln \ln \left((\ln(e^{e^{3e^e}} + |\vec{u}(t, x)|))^{1/3}\right)\right)^{\theta}} \, dx = +\infty$$

for some constant $\theta \in (0, 1)$ which does not depend on $\vec{u}$

See [21] for a survey on those recent developments.

15.6 Critical Elements for the Blow-up of the Cauchy Problem in L^3

In this section, we give a result of Jia and Šverák [244] on the (potential) existence of a minimal-norm initial value for a blowing-up mild solution to the Navier–Stokes Cauchy problem (first established by Rusin and Šverák [417] and by Gallagher, Koch and Planchon [199]).

We are interested in the properties of the set $\mathcal{B}_3$ of divergence-free vector fields $\vec{u}_0 \in L^3(\mathbb{R}^3)$ such that the L^3 solution of the Cauchy problem for the Navier–Stokes equations with null force ($\vec{f} = 0$) and initial value $\vec{u}_0$ blows up in finite time.

Jia and Šverák's theorem on blow-up

Theorem 15.10.
If $\mathcal{B}_3 \neq \emptyset$, then there exists an element of $\mathcal{B}_3$ with minimal norm in L^3:

$$\vec{u}_0 \in \mathcal{B}_3 \text{ and } \|\vec{u}_0\|_3 = \min_{\vec{w} \in \mathcal{B}_3} \|\vec{w}\|_3.$$

The proof will rely on a compactness property of $\mathcal{B}_3$. If we assume that $\mathcal{B}_3$ is not empty, then the set $\mathcal{B}_{3,M} = \{\vec{w} \in \mathcal{B}_3 / \|\vec{w}\|_3 \leq M\}$ is not empty for M large enough. It is easy to see that $\mathcal{B}_{3,M}$ is not a compact subset of L^3, neither for the strong topology, nor for the *-weak topology: if $\vec{w} \in \mathcal{B}_{3,M}$, then $\vec{w}_n(x) = \vec{w}(x - n\vec{e}_1)$ defines a sequence in $\mathcal{B}_{3,M}$ such that $\vec{w}_n \overset{w^*}{\to} 0 \notin \mathcal{B}_{3,M}$. This lack of compactness however can be overpassed:

Theorem 15.11.

If $\mathcal{B}_{3,M} \neq \emptyset$, then for every sequence $\vec{w}_n$ with values in $\mathcal{B}_{3,M}$, there exists an element $\vec{w}_\infty \in \mathcal{B}_{3,M}$, a subsequence $(\vec{w}_{n_k})_{k \in \mathbb{N}}$, and two sequences λ_{n_k} with values in $(0, +\infty)$ and x_{n_k} with values in $\mathbb{R}^3$ such that $\hat{w}_{n_k}$ defined by

$$\hat{w}_{n_k} = \lambda_{n_k} \vec{w}_{n_k}(\lambda_{n_k}(x + x_{n_k}))$$

satisfies $\hat{w}_{n_k} \in \mathbb{B}_{3,M}$, $\|\hat{w}_{n_k}\|_3 = \|\vec{w}_{n_k}\|_3$ and $\hat{w}_{n_k} \xrightarrow{w^*} \vec{w}_\infty$.
If moreover $\lim_{n \to +\infty} \|\vec{w}_n\|_3 = \inf_{\vec{w} \in \mathcal{B}_3} \|\vec{w}\|_3$, then

$$\lim_{n \to +\infty} \|\hat{w}_{n_k} - \vec{w}_\infty\|_3 = 0.$$

Proof. **Step 1: choice of λ_n.**

Let $\vec{v}_n \in \mathcal{C}([0, T_n^*), L^3)$ be the (unique) solution of the Cauchy problem

$$\partial_t \vec{v}_n = \nu \Delta \vec{v}_n - \mathbb{P} \operatorname{div}(\vec{v}_n \otimes \vec{v}_n), \quad \vec{v}_n(0, .) = \vec{w}_n$$

and T_n^* be its maximal time of existence, so that, by Theorem 15.5,

$$\lim_{t \to T_n^*} \|\vec{v}_n(t, .)\|_3 = +\infty.$$

We then choose

$$\lambda_n = \sqrt{T_n^*}.$$

We define $\tilde{w}_n(x) = \lambda_n \vec{w}_n(\lambda_n x)$ and $\tilde{v}_n(t, x) = \lambda_n \vec{v}_n(\lambda_n^2 t, \lambda_n x)$.

We have $\|\tilde{w}_n\|_3 = \|\vec{w}_n\|_3$, while $\tilde{v}_n$ is the solution of the Cauchy problem for the Navier–Stokes equations with $\tilde{w}_n$ as initial value. Thus, $\tilde{w}_n \in \mathcal{B}_{3,M}$ with a solution that blows up at time $T^* = 1$.

Step 2: $\tilde{v}_n$ as a local Leray solution.

As in the proof of Theorem 15.5, we get, from Proposition 15.1 (and the weak-strong uniqueness theorem Theorem 14.7), that $\tilde{v}_n$ coincides on $(0, 1)$ with a local Leray solution $\vec{v}_n$ defined on $(0, 2)$ and that

$$\sup_{0 < t < 2} \|\vec{v}_n(t, .)\|_{L^2_{\text{uloc}}} \leq C_{\nu, M}$$

and

$$\|\vec{v}_n\|_{(L^2 H^1)_{\text{uloc}}} \leq C_{\nu, M}.$$

Step 3: choice of x_n.

$\tilde{v}_n$ blows up at time $t = 1$. From Theorem 15.1, we know that there exists a point $x_n \in \mathbb{R}^3$ such that

$$\text{for every } r \in (0, 1), \quad \sup_{(s,y) \in Q_r(x_0)} |\tilde{v}_n(s, y)| = +\infty,$$

where $Q_r(x_0) = (1 - r^2, 1) \times B(x_0, r)$.

We then define $\hat{w}_n(x) = \lambda_n \vec{w}_n(\lambda_n(x + x_n))$ and $\hat{v}_n(t, x) = \lambda_n \vec{v}_n(\lambda_n^2 t, \lambda_n(x + x_n))$.

We have $\|\hat{w}_n\|_3 = \|\vec{w}_n\|_3$, while $\hat{v}_n$ is the solution of the Cauchy problem for the Navier–Stokes equations with $\hat{w}_n$ as initial value. Thus, $\hat{w}_n \in \mathcal{B}_{3,M}$ with a solution $\hat{v}_n$ that blows up at time $T^* = 1$ and point $X = 0$.

Step 4: choice of n_k.

As in the proof of Theorem 15.5, we have the estimates

$$\|\hat{v}_n\|_{L^\infty((0,1),L^2_{uloc})} \leq C_{\nu,M} \text{ and } \sqrt{\nu}\|\hat{v}_n\|_{(L^2\dot{H}^1)_{uloc}((0,1)\times\mathbb{R}^3))} \leq C_{\nu,M},$$

and may use the Rellich–Lions theorem (Theorem 12.1) to find a sequence $n_k \to +\infty$ and a function $\vec{v}_\infty \in (L^\infty L^2)_{uloc} \cap (L^2\dot{H}^1)_{uloc}$ such that:

- $\hat{v}_{n_k}$ is *-weakly convergent to $\vec{v}_\infty$ in $(L^\infty L^2)_{uloc}$ and in $(L^2\dot{H}^1)_{uloc}$
- $\hat{v}_{n_k}$ is strongly convergent to $\vec{v}_\infty$ in $L^2_{loc}((0,T)\times\mathbb{R}^3)$.

The limit $\vec{v}_\infty$ is a solution of the Navier–Stokes equations

$$\partial_t\vec{v}_\infty = \nu\Delta\vec{v}_\infty - \mathbb{P}\,\text{div}(\vec{v}_\infty \otimes \vec{v}_\infty) = \nu\Delta\vec{v}_\infty - \text{div}(\vec{v}_\infty \otimes \vec{v}_\infty) - \vec{\nabla}p_\infty$$

and satisfies the local energy inequality:

$$\partial_t(\frac{|\vec{v}_\infty|^2}{2}) \leq \nu\Delta(\frac{|\vec{v}_\infty|^2}{2}) - \nu|\vec{\nabla}\otimes\vec{v}_\infty|^2 - \text{div}((p_\infty + \frac{|\vec{v}_\infty|^2}{2})\vec{v}_\infty).$$

Still following the proof of Theorem 15.5, we find that $\vec{v}_\infty \in \mathcal{C}([0,1], H^{-3/2}_{uloc})$ and that, for every $t \in [0,1]$, $\vec{v}_\infty(t,.)$ is the weak* limit in $H^{-3/2}_{uloc}$ of $\hat{v}_{n_k}(t,.)$. As $\|\vec{v}_{n_k}(0,.)\|_3 \leq M$, we find as well that $\vec{v}_\infty(0,.) \in L^3$. Moreover, $\vec{v}_\infty$ is a local Leray solution. Finally, $\vec{v}_\infty$ blows up at $(1,0)$.

Step 5: determining $\vec{w}_\infty$.

We have $\hat{w}_{n_k} \overset{w^*}{\rightharpoonup} \vec{w}_\infty = \vec{v}_\infty(0,.)$. Let $\vec{v}$ be the mild solution associated to $\vec{w}_\infty$ in $\mathcal{C}([0,T^*), L^3)$. If $T^* > 1$, $\vec{v}$ and $\vec{v}_\infty$ would coincide on $[0,1]$, but $\vec{v}_\infty$ has a singular point at $(1,0)$. Thus, $T^* \leq 1$ and $\vec{w}_\infty \in \mathcal{B}_{3,M}$.

Step 6: end of the proof.

Finally, let us assume that $\lim_{n\to+\infty}\|\vec{w}_n\|_3 = \inf_{\vec{w}\in\mathcal{B}_3}\|\vec{w}\|_3$. Then as we have $\|\vec{w}_\infty\|_3 \leq \liminf_{k\to+\infty}\|\hat{w}_{n_k}\|_3$ and $\vec{w}_\infty \in \mathcal{B}_3$, we find that

$$\|\vec{w}_\infty\|_3 = \min_{\vec{w}\in\mathcal{B}_3}\|\vec{w}\|_3.$$

As L^3 is a uniformly convex space, the fact that $\hat{w}_{n_k} \overset{w^*}{\rightharpoonup} \vec{w}_\infty$ in L^3 and $\|\vec{w}_\infty\|_3 = \liminf_{k\to+\infty}\|\hat{w}_{n_k}\|_3$ implies the strong convergence of the sequence in L^3:

$$\lim_{n\to+\infty}\|\hat{w}_{n_k} - \vec{w}_\infty\|_3 = 0.$$

The theorem is proved.

$\square$

15.7 Known Results on the Cauchy Problem for the Navier–Stokes Equations in Presence of a Force

In this section, we recall some known results on mild solutions in $\mathcal{C}([0,T], L^3(\mathbb{R}^3))$ of the Cauchy problem for the Navier–Stokes equations

$$\begin{cases} \partial_t\vec{u} = \nu\Delta\vec{u} - \text{div}(\vec{u}\otimes\vec{u}) + \vec{f} - \vec{\nabla}p \\ \text{div}\,\vec{u} = 0 \\ \vec{u}(0,x) = \vec{u}_0(x) \end{cases} \qquad (15.9)$$

where $\vec{u}_0 \in L^3$ with div $\vec{u}_0 = 0$.

Existence of a solution is given by the following theorem of Cannone and Planchon [85]:

Theorem 15.12.
If the forcing term $\vec{f}$ satisfies

$$\operatorname{div} \vec{f} = 0$$

and, for every $T < +\infty$,

$$\vec{f} \in L_t^p((0,T), L^q(\mathbb{R}^3)) \text{ with } \frac{2}{p} + \frac{3}{q} = 3 \text{ and } 3/2 < q < 3$$

and if $\vec{u}_0 \in L^3(\mathbb{R}^3))$ with $\operatorname{div} \vec{u}_0 = 0$, then there exists a time $0 < T_0 < +\infty$ such that Equations (15.9) have a unique solution $\vec{u}$ such that

$$\vec{u} \in \mathcal{C}([0,T_0], L_x^3) \cap L_t^4((0,T_0), L_x^6).$$

Moreover, we have

- $\sup_{t >> 0} \sqrt{t} \|\vec{\nabla} \otimes W_{\nu t} * \vec{u}_0\|_3 < +\infty$ *and* $\vec{\nabla} \otimes (\vec{u} - W_{\nu t} * \vec{u}_0) \in L^2((0,T_0), L_x^3)$, *so that $\vec{u}$ is weakly regular*

- $p = -\sum_{i=1}^{3} \sum_{j=1}^{3} \frac{\partial_i \partial_j}{\Delta}(u_i u_j)$ *so that $p \in \mathcal{C}([0,T_0], L_x^{3/2}) \cap L_t^2((0,T_0), L_x^3)$*

Proof. If we look for a solution $\vec{u}$ of equations (15.9) such that $\vec{u} \in L^\infty([0,T_0], L_x^3)$, then we may remark that $\vec{u}$ belongs to the closure of $\mathcal{D}((0,T_0) \times \mathbb{R}^3)$ in $(L^2 L^2)_{\text{uloc}}$: it is enough to check that

$$\lim_{x \to \infty} \iint_{(0,T_0) \times B(x,1)} |\vec{u}(t,y)|^2 \, dt \, dy = 0.$$

This latter point is obvious: $\int_{B(x,1)} |\vec{u}(t,y)|^2 \, dy \leq |B(0,1)|^{1/3} \|\vec{u}\|_{L^\infty L^3}^2$ and $\lim_{x \to +\infty} \int_{B(x,1)} |\vec{u}(t,y)|^2 \, dy = 0$, hence we may conclude by the dominated convergence theorem.

Thus, we may apply Proposition 6.5 and get that

$$\vec{\nabla}p = \mathbb{P}(\operatorname{div}(\vec{u} \otimes \vec{u})) - \operatorname{div}(\vec{u} \otimes \vec{u}) = -\vec{\nabla}\left(\sum_{i=1}^{3} \sum_{j=1}^{3} \frac{\partial_i \partial_j}{\Delta}(u_i u_j) \right).$$

Now, we know that we are studying the equations

$$\begin{cases} \partial_t \vec{u} = \nu \Delta \vec{u} - \mathbb{P}\operatorname{div}(\vec{u} \otimes \vec{u}) + \vec{f} \\ \vec{u}(0,x) = \vec{u}_0(x) \end{cases} \tag{15.10}$$

By the uniqueness theorem for L^3 solutions (Theorem 7.7), we know that there exists at most one solution of Equations (15.10) in $\mathcal{C}([0,T_0], L^3)$.

We still have to prove existence of one solution. This will be done by using Picard's iterative scheme. First, we estimate $\vec{U}_0 = W_{\nu t} * \vec{u}_0 + \int_0^t W_{\nu(t-s)} * \vec{f}(s) \, ds$.

- We start from the embeddings $L^3 \subset \dot{B}_{3,3}^0 \subset \dot{B}_{6,3}^{-\frac{1}{2}} \subset \dot{B}_{6,4}^{-\frac{1}{2}}$. Using the thermic characterization of $\dot{B}_{r,\sigma}^{-2/\sigma}$ (i.e., $\|F\|_{\dot{B}_{r,\sigma}^{-2/\sigma}} \approx (\int_0^{+\infty} \|W_{\nu t} * F\|_r^\sigma \, dt)^{1/\sigma}$ for $1 \leq \sigma < +\infty$ and $1 \leq r \leq +\infty$), we find that $W_{\nu t} * \vec{u}_0$ belongs to $\mathcal{C}([0,T], L_x^3) \cap L_t^4((0,T), L_x^6)$ for every positive T.

- Now, we use the embeddings $L^{3/2} \subset \dot{B}^0_{3/2,2} \subset \dot{B}^{-2+\frac{3}{r}}_{r,2}$ for $3/2 \leq r \leq +\infty$. We obtain that, for $g_0 \in L^{3/2}$, we have

$$\text{for } 3/2 < r \leq 3, \ 1_{t>0} W_{\nu t} * g_0 \in L^\sigma_t L^r_x \text{ with } \frac{2}{\sigma} + \frac{3}{r} = 2.$$

By duality, we obtain that, if $\vec{f} \in L^p_t L^q_x$ on $(0,T) \times \mathbb{R}^3$ with $\frac{2}{p} + \frac{3}{q} = 3$ and $3/2 \leq q < 3$ then $\int_0^t W_{\nu(t-s)} * \vec{f} \, ds$ belongs to $L^\infty((0,T), L^3)$. Moreover, for $q > 3/2$, we have $p < +\infty$, so that compactly supported smooth functions are dense in $L^p_t L^q_x$, and we find more precisely that $\int_0^t W_{\nu(t-s)} * \vec{f} \, ds$ belongs to $\mathcal{C}([0,T], L^3)$.

- Finally, we write

$$\|W_{\nu(t-s)} * \vec{f}\|_6 \leq C\|\vec{f}\|_q (t-s)^{-\frac{3}{2}(\frac{1}{q} - \frac{1}{6})} = C\|\vec{f}\|_q (t-s)^{-1+\frac{1}{p}-\frac{1}{4}}$$

and the Hardy–Littlewood–Sobolev inequality gives then that $\int_0^t W_{\nu(t-s)} * \vec{f} \, ds$ belongs to $L^4_t L^6_x((0,T) \times \mathbb{R}^3)$.

We now define, as usual, $\vec{U}_{n+1} = \vec{U}_0 - B(\vec{U}_n, \vec{U}_n)$, with

$$B(\vec{v}, \vec{w}) = \int_0^t W_{\nu(t-s)} * \mathbb{P}\operatorname{div}(\vec{v} \otimes \vec{w}) \, ds.$$

We will perform the Picard iterations in the Banach space $\mathcal{X}_T = L^8_t((0,T), L^4_x) \cap L^4_t((0,T), L^6_x)$. Notice that this is a larger space than the one we are studying:

$$\mathcal{C}([0,T], L^3_x) \cap L^4_t((0,T), L^6_x) \subset \mathcal{X}_T.$$

Indeed, we have obviously

$$\|\vec{v}\|_{L^8 L^4} \leq \sqrt{\|\vec{v}\|_{L^\infty L^3} \|\vec{v}\|_{L^4 L^6}}.$$

We are going to check that, for $\vec{v}$ and $\vec{w}$ in $\mathcal{X}_T$, we have $B(\vec{v}, \vec{w}) \in \mathcal{C}([0,T], L^3_x) \cap \mathcal{X}_T$ and

$$\|B(\vec{v}, \vec{w})\|_{L^\infty L^3} + \|B(\vec{v}, \vec{w})\|_{\mathcal{X}_T} \leq C_0 \|\vec{v}\|_{\mathcal{X}_T} \|\vec{w}\|_{\mathcal{X}_T}$$

where C_0 does not depend on T.

- For $\vec{v}$ and $\vec{w}$ in $\mathcal{X}_T$, $\vec{v} \otimes \vec{w}$ belongs to $L^4 L^2$. Thus, we have $\mathbb{P}\operatorname{div}(\vec{v} \otimes \vec{w}) \in L^4 \dot{H}^{-1}$ and the maximal $L^2 L^2$ regularity for the heat kernel gives that $B(\vec{v}, \vec{w})$ belongs to $L^4 \dot{H}^1 \subset L^4 L^6$.

- From the embedding $\sqrt{-\Delta} L^{3/2} \subset \dot{B}^{-1}_{3/2,2}$, we find that, for $g_0 \in L^{3/2}$, we have $1_{t>0} \sqrt{-\Delta} W_{\nu t} * g_0 \in L^2_t L^{3/2}_x$. By duality, we find that, for $h \in L^2 L^3$, we have $\int_0^t W_{\nu(t-s)} * \sqrt{-\Delta} h \, ds \in L^\infty L^3$. Again, that is precised into $\int_0^t W_{\nu(t-s)} * \sqrt{-\Delta} h \, ds \in \mathcal{C}([0,T], L^3)$ by density of smooth compactly supported functions.

Thus far, we know that we will have a solution $\vec{u}$ in $\mathcal{X}_{T_0}$ as soon as $\|\vec{U}_0\|_{\mathcal{X}_{T_0}} \leq \frac{1}{4C_0}$. As we have $\vec{U}_0 \in \mathcal{X}_T$ and $\lim_{T_0 \to 0} \|\vec{U}_0\|_{\mathcal{X}_{T_0}} = 0$, such a T_0 exists. Now, we have moreover $\vec{U}_0 \in \mathcal{C}([0,T_0], L^3)$ and we have seen, since $\vec{u} \in \mathcal{X}_{T_0}$, that $B(\vec{u}, \vec{u})$ belongs to $\mathcal{C}([0,T_0], L^3)$. Thus, the solution $\vec{u}$ belongs to $\mathcal{C}([0,T_0], L^3) \cap L^4_t L^6_x$.

Moreover, we have $\sup_{t>>0} \sqrt{t} \|\vec{\nabla} \otimes W_{\nu t} * \vec{u}_0\|_3 < +\infty$. Thus, in order to prove that $\vec{\nabla} \otimes \vec{u}$ belongs locally to $L^2 L^2$, it will be enough to check that $\vec{\nabla} \otimes \int \int_0^t W_{\nu(t-s)} * \vec{f} \, ds$ and $\vec{\nabla} \otimes B(\vec{u}, \vec{u})$ belongs to $L^2_t L^3_x$.

- For $\vec{\nabla} \otimes B(\vec{u}, \vec{u})$, we just use the fact that $\vec{u} \otimes \vec{u}$ belongs to $L_t^2 L_x^3$ and the $L^2 L^3$ maximal regularity of the heat kernel[1].

- For $\vec{\nabla} \otimes \iint_0^t W_{\nu(t-s)} * \vec{f}\, ds$, we write

$$\|\vec{\nabla} \otimes (W_{\nu(t-s)} * \vec{f})\|_3 \leq C \|\vec{f}\|_q (t-s)^{-\frac{1}{2} - \frac{3}{2}(\frac{1}{q} - \frac{1}{3})} = C \|\vec{f}\|_q (t-s)^{-1 + \frac{1}{p} - \frac{1}{2}}$$

and, using the Hardy–Littlewood–Sobolev inequality, we finally get that $\int_0^t \vec{\nabla} \otimes (W_{\nu(t-s)} * \vec{f})\, ds$ belongs to $L_t^2 L_x^3((0, T_0) \times \mathbb{R}^3)$.

The theorem is proved.   □

In particular, we have that $\vec{u}$ is locally $L^\infty L^2 \cap L^2 H^1$, p is locally $L^1 L^1$, $\vec{f}$ is locally $L^2 H^{-1}$ and moreover $\vec{u}$ is locally $L_{t,x}^4$ so that, applying Proposition 13.3, we get:

Corollary 15.2.
The solution constructed in Theorem 15.12 satisfies the local energy equality in $\mathcal{D}'((0, T_0) \times \mathbb{R}^3$:

$$\partial_t |\vec{u}|^2 = \nu \Delta |\vec{u}|^2 - 2\nu |\vec{\nabla} \otimes \vec{u}|^2 - \mathrm{div}((|\vec{u}|^2 + 2p)\vec{u}) + 2\vec{u}.\vec{f}.$$

Of course, it means that $\vec{u}$ is suitable. We thus may apply the Caffarelli–Kohn–Nirenberg regularity criterion (Theorem 13.8); let us recall that it proves local regularity in a neighborhood $Q_{r_0}(t_0, x_0)$ of a suitable solution of the Navier–Stokes problem $\vec{u}$ under the assumption that on a larger neighborhood $Q = Q_R(t_0, x_0)$, we have $\vec{u} \in L_{t,x}^3(Q)$, $p \in L_{t,x}^{3/2}(Q)$ $1_Q(t, x)\vec{f} \in \mathcal{M}_2^{10/7, \tau_0}$ for some $\tau_0 > 5/2$ and

$$\limsup_{r \to 0} \frac{1}{r} \iint_{(t_0 - r^2, t_0 + r^2) \times B(x_0, r)} |\vec{\nabla} \otimes \vec{u}|^2\, ds\, dx < \epsilon^*.$$

The solution given in Theorem 15.12 satisfies that locally $\vec{\nabla} \otimes \vec{u} \in L_t^2 L_x^3$; as we have

$$\frac{1}{r} \iint_{(t_0 - r^2, t_0 + r^2) \times B(x_0, r)} |\vec{\nabla} \otimes \vec{u}|^2\, ds\, dx \leq \left(\iint_{(t_0 - r^2, t_0 + r^2) \times B(x_0, r)} |\vec{\nabla} \otimes \vec{u}|^3\, dx \right)^{2/3} ds$$

with

$$\lim_{r \to 0} \left(\iint_{(t_0 - r^2, t_0 + r^2) \times B(x_0, r)} |\vec{\nabla} \otimes \vec{u}|^3\, dx \right)^{2/3} ds = 0$$

and as $L^p L^q \subset \mathcal{M}_2^{3/2, \tau_0} \subset \mathcal{M}_2^{10/7, \tau_0}$ with $\frac{5}{\tau_0} = \frac{2}{p} + \frac{3}{q}$, we have the following result:

Corollary 15.3.
If moreover $\frac{2}{p} + \frac{3}{q} < 2$, then the solution constructed in Theorem 15.12 is locally Hölderian on $(0, T_0) \times \mathbb{R}^3$.

Moreover, we have, at fixed t_0 and r,

$$\lim_{x_0 \to +\infty} \left(\iint_{(t_0 - r^2, t_0 + r^2) \times B(x_0, r)} |\vec{\nabla} \otimes \vec{u}|^3\, dx \right)^{2/3} ds = 0$$

A consequence is that the size r_0 of the neighborhood where Hölderianity holds can be chosen independently from x_0 for large x_0; by compactness for bounded closed balls of $\mathbb{R}^3$, the same will then be true for every x_0. Thus, we find that $\vec{u}$ belongs locally on $(0, T_0)$ to $L_t^\infty(L^\infty(\mathbb{R}^3))$.

---

[1] For a proof of the $L^p L^q$ maximal regularity property with $1 < p < +\infty$ and $1 < q < +\infty$

$$\left\| \int_0^t W_{\nu(t-s)} * \Delta h\, ds \right\|_{L^p L^q} \leq C_{p,q} \|h\|_{L^p L^q},$$

see for instance Lemarié-Rieusset[313].

Definition 15.2.
Let $T_1 < T_2$ be two real numbers and $\vec{f} \in L_t^p L_x^q((T_1, T_2) \times \mathbb{R}^3)$ with div $\vec{f} = 0$, $\frac{3}{2} < q < 3$ and $\frac{2}{p} + \frac{3}{q} < 2$. $\mathcal{NS}_{T_1, T_2, \vec{f}}$ is defined as the set of divergence-free vector fields $\vec{u}$ on $(T_1, T_2) \times \mathbb{R}^3$ such that

- *$\vec{u} \in L_t^\infty L_x^3$*

- *there exists $p \in \mathcal{D}'((T_1, T_2) \times \mathbb{R}^3)$ such that*

$$\partial_t \vec{u} = \nu \Delta \vec{u} - \mathrm{div}(\vec{u} \otimes \vec{u}) + \vec{f} - \vec{\nabla} p.$$

Proposition 15.3.
Let $T_1 < T_2$ be two real numbers and $\vec{f} \in L_t^p L_x^q((T_1, T_2) \times \mathbb{R}^3)$ with div $\vec{f} = 0$, $\frac{3}{2} < q < 3$ and $\frac{2}{p} + \frac{3}{q} < 2$. Then, for $\vec{u} \in \mathcal{NS}_{T_0, T_1, \vec{f}}$, the following assertions are equivalent:

- *(A) $\vec{u} \in \mathcal{C}((T_1, T_2), L^3)$*

- *(B) for all compact subintervals $[T_3, T_4] \subset (T_1, T_2)$, $\vec{u} \in L_t^4 L_x^6$ on $[T_3, T_4] \times \mathbb{R}^3$*

- *(C) for all compact subintervals $[T_3, T_4] \subset (T_1, T_2)$, $\vec{u} \in L_{t,x}^\infty$ on $[T_3, T_4] \times \mathbb{R}^3$*

15.8 Local Estimates for Suitable Solutions

In this section, $\vec{f} \in L_t^p L_x^q((T_1, T_2) \times \mathbb{R}^3)$ with div $\vec{f} = 0$, $\frac{3}{2} < q < 3$ and $\frac{2}{p} + \frac{3}{q} < 2$, and a vector field $\vec{u} \in \mathcal{NS}_{T_1, T_2, \vec{f}}$ such that $\vec{u}$ is suitable. (Since $\vec{u}$ is locally $L^2 L^6$ and globally $L^\infty L^3$, it is locally $L^4 L^4$ and the local energy inequality is indeed an energy equality.) Thus, we have the following assumptions on $\vec{u}$:

- $\vec{u} \in L_t^\infty L_x^3$

- $\vec{\nabla} \otimes \vec{u} \in (L^2 L^2)_{\mathrm{loc}}((T_1, T_2) \times \mathbb{R}^3)$

- there exists $p \in L_t^\infty L^{3/2}((T_1, T_2) \times \mathbb{R}^3)$ such that

$$\partial_t \vec{u} = \nu \Delta \vec{u} - \mathrm{div}(\vec{u} \otimes \vec{u}) + \vec{f} - \vec{\nabla} p.$$

- $\vec{u}$ is suitable:

$$\partial_t(\frac{|\vec{u}|^2}{2}) = \nu \Delta(\frac{|\vec{u}|^2}{2}) - \nu|\vec{\nabla} \otimes \vec{u}|^2 - \mathrm{div}\left((p + \frac{|\vec{u}|^2}{2})\vec{u}\right) + \vec{u}.\vec{f} \qquad (15.11)$$

Let $\Omega = (T_0, T_1) \times \mathbb{R}^3$. We are going to estimate the behavior of $\vec{u}$ on cylinders $Q_r(t_0, x_0) \cap \Omega$, where $Q_r(x_0, t_0) = (t_0 - r^2, t_0 + r^2) \times B(x_0, r)$, $x_0 \in \mathbb{R}^3$, $T_1 + 2r^2 < t_0 \le T_2$.

We fix $\varphi \in \mathcal{D}(\mathbb{R}^3)$ which is non-negative, supported in $B(0, 2)$ and equal to 1 on $B(0, 1)$, and a function $\theta \in \mathcal{C}^\infty(\mathbb{R})$ which is equal to 0 on $(-\infty, 0]$, to 1 on $[1, +\infty)$ and is non-decreasing. For $\tau \in (t_0 - r^2, \min(t_0 + r^2, T_2))$ and $\epsilon \in (0, T_2 - \tau)$, we introduce the function

$$\varphi_{r, \tau, \epsilon} = \varphi(\frac{x - x_0}{r})\theta(\frac{t - t_0 - 2r^2}{r^2})(1 - \theta(\frac{\tau - t}{\epsilon}))$$

Applying Equality (15.11) to $\langle \partial_t(\frac{|\vec{u}|^2}{2})|\varphi_{r,\tau,\epsilon}\rangle_{\mathcal{D}',\mathcal{D}}$, we find

$$\iint \frac{|\vec{u}|^2}{2}\varphi(\frac{x-x_0}{r})\theta(\frac{t-t_0-2r^2}{r^2})\frac{1}{\epsilon}(\theta'(\frac{\tau-t}{\epsilon}) - \theta'(\frac{t-T_3}{\epsilon})) \, dt \, dx$$

$$+\nu \iint \varphi_{r,\tau,\epsilon}|\vec{\nabla} \otimes \vec{u}|^2 \, dt \, dx =$$

$$\nu \iint \frac{|\vec{u}|^2}{2}\frac{1}{r^2}(\Delta\varphi)(\frac{x-x_0}{r})\theta(\frac{t-t_0-2r^2}{r^2})(1 - \theta(\frac{\tau-t}{\epsilon})) \, dt \, dx$$

$$+ \iint (p + \frac{|\vec{u}|^2}{2})\frac{1}{r}(\vec{u}(t,x).(\vec{\nabla}\varphi)(\frac{x-x_0}{r}))\theta(\frac{t-t_0-2r^2}{r^2})(1 - \theta(\frac{\tau-t}{\epsilon})) \, dt \, dx$$

$$+ \iint \vec{u}.\vec{f}\varphi_{r,\tau,\epsilon} \, dt \, dx$$

$$+ \iint \frac{|\vec{u}|^2}{2}\varphi(\frac{x-x_0}{r})\frac{1}{r^2}\theta'(\frac{t-t_0-2r^2}{r^2})(1 - \theta(\frac{\tau-t}{\epsilon})) \, dt \, dx$$

As $\vec{u}$ is locally L^2H^1 and $\partial_t\vec{u} - \Delta\vec{u}$ is locally L^2H^{-1}, we find that $t \mapsto \vec{u}(t,.)$ is continuous from (T_1,T_2) to $L^2_{\text{loc}}(\mathbb{R}^3$. Thus, we find (by letting ϵ go to 0) that

$$\int_{B(x_0,r)} \frac{|\vec{u}(\tau,x)|^2}{2} \, dx$$

$$+\nu \iint_{(t,x)\in Q_r(t_0,x_0),t<\tau} |\vec{\nabla} \otimes \vec{u}|^2 \, dt \, dx \leq$$

$$C\nu\frac{1}{r^2}\iint_{Q_{2r}(t_0,x_0)\cap\Omega} |\vec{u}|^2 \, dt \, dx$$

$$+ C\frac{1}{r}\iint_{Q_{2r}(t_0,x_0)\cap\Omega} (|p| + |\vec{u}|^2)|\vec{u}| \, dt \, dx$$

$$+ C\iint_{Q_{2r}(t_0,x_0)\cap\Omega} |\vec{u}.\vec{f}| \, dt \, dx$$

$$+ C\frac{1}{r^2}\iint_{Q_{2r}(t_0,x_0)\cap\Omega} |\vec{u}|^2 \, dt \, dx$$

$$\leq C_\nu \left(\frac{\iint_{Q_{2r}(t_0,x_0)\cap\Omega}|\vec{u}|^3 \, dt \, dx}{r^2}\right)^{1/3} r \left(\|\vec{u}\|_{L^\infty L^3} + \|\vec{u}\|^2_{L^\infty L^3} + r^{3-(\frac{2}{p}+\frac{3}{q})}\|\vec{f}\|_{L^p L^q}\right)$$

Notice that we find again the same critical scaling $\frac{2}{p} + \frac{3}{q} = 3$ for $\vec{f}$ as the one we used in Lemma 13.4 in the proof of the Caffarelli–Kohn–Nirenberg regularity criterion and in Theorem 15.12, to get a control in $L^\infty L^3$. However, in the following sections, we shall consider another critical scaling is $\frac{2}{p} + \frac{3}{q} = 2$, as in the end of the proof of the Caffarelli–Kohn–Nirenberg regularity criterion.

We thus obtain the following result:

Theorem 15.13.
Let $T_1 < T_2$ be two real numbers and $p,q > 1$ with $\frac{3}{2} < q < 3$ and $\frac{2}{p} + \frac{3}{q} < 2$. Then:

(A) *Let $\vec{f} \in L^p_t L^q_x((T_1,T_2) \times \mathbb{R}^3)$ with div $\vec{f} = 0$ and $\vec{u} \in \mathcal{NS}_{T_1,T_2,\vec{f}}$ such that $\vec{u}$ is suitable. Then there exists a constant C_ν which does not depend on $\vec{f}$ nor on $\vec{u}$ nor on T_1 and*

T_2 such that for all $t_0 \in (T_1, T_2]$ and all $r \in (0, \sqrt{t_0 - T_1})$, we have

$$\sup_{x_0 \in \mathbb{R}^3} \iint_{Q_r(t_0,x_0) \cap \Omega} |\vec{\nabla} \otimes \vec{u}|^2 \, dt \, dx$$

$$\leq C_\nu r \|\vec{u}\|_{L^\infty L^3} \left(\|\vec{u}\|_{L^\infty L^3} + \|\vec{u}\|^2_{L^\infty L^3} + \sqrt{T_2 - T_1}^{3-(\frac{2}{p}+\frac{3}{q})} \|\vec{f}\|_{L^p L^q} \right)$$

(B) Let $\vec{f}_n$ be a sequence of vector fields such that $\vec{f}_n \in L^p_t L^q_x((T_1, T_2) \times \mathbb{R}^3)$ with $\text{div} \, \vec{f}_n = 0$ and $\sup_{n \in \mathbb{N}} \|\vec{f}_n\|_{L^p L^q} < +\infty$, and $\vec{u}_n$ be a sequence of vector fields such that $\vec{u}_n \in \mathcal{NS}_{T_1,T_2,\vec{f}_n}$ and $\sup_{n \in \mathbb{N}} \|\vec{u}_n\|_{L^\infty L^3} < +\infty$. If each $\vec{u}_n$ satisfies $\vec{\nabla} \otimes \vec{u}_n \in (L^2 L^2)_{\text{loc}}((T_1, T_2) \times \mathbb{R}^3)$ and is suitable, then

- there exists subsequences $\vec{f}_{n_k}$ and $\vec{u}_{n_k}$ that converge *-weakly (respectively in $L^p L^q$ and $L^\infty L^3$) to limits $\vec{f}_\infty$ and $\vec{u}_\infty$ and moreover such that $\vec{u}_{n_k}$ converges strongly in $(L^3_t L^3_x)_{\text{loc}}((T_1, T_2) \times \mathbb{R}^3)$
- in particular, $\vec{u}_\infty \in \mathcal{NS}_{T_1,T_2,\vec{f}_\infty}$ and is suitable

Proof. We have already proved Point (A). Now, for Point (B) let us notice that existence of limit points $\vec{u}_\infty$ and $\vec{f}_\infty$ is a straightforward consequence of the Banach–Alaoglu theorem, as both $L^\infty L^3$ and $L^p L^q$ are dual Banach spaces of separable spaces. Similarly, as $L^\infty L^{3/2}$ is a dual space, we may assume that p_{n_k} converges *-weakly to some $p_\infty \in L^\infty L^{3/2}$. The problem is to have a better convergence of $\vec{u}_{n_k}$ to allow to deal with the non-linearity $\vec{u} \otimes \vec{u}$ in the Navier–Stokes equations, as we try to get $\vec{u}_\infty \in \mathcal{NS}_{T_1,T_2,\vec{f}_\infty}$. This is done with the help of the Rellich–Lions theorem (Theorem 12.1): we control by the $L^\infty L^3$ norm of $\vec{u}$ and the $L^p L^q$ norm of $\vec{f}$ the local $L^2 H^1$ norm of $\vec{u}_n$ and the local $L^2 H^{-1}$ norm of $\partial_t \vec{u}_n$, thus we know that we have strong convergence in $(L^2 L^2)_{\text{loc}}$ of a subsequence $\vec{u}_{n_k}$. But we know as well that $\vec{u}_n$ is bounded in the local $L^2 L^6$ and in $L^\infty L^3$, hence in $(L^4 L^4)_{\text{loc}}$; by interpolation between $L^2 L^2$ and $L^4 L^4$, we find that we have strong convergence in $(L^3_t L^3_x)_{\text{loc}}((T_1, T_2) \times \mathbb{R}^3)$. Moreover, this strong convergence of $\vec{u}_{n_k}$ in $(L^2 L^2)_{\text{loc}}$ and the weak convergence of $\vec{\nabla} \otimes \vec{u}_{n_k}$ $(L^2 L^2)_{\text{loc}}$ allows us to conclude that $\vec{u}_\infty \in \mathcal{NS}_{T_1,T_2,\vec{f}_\infty}$ and is suitable. $\qquad \square$

15.9 Uniqueness for Suitable Solutions

We prove an intermediary result (as we shall see, Escauriaza, Seregin and Šverák's theorem (Theorem 15.21) actually implies that a $L^\infty_t L^3_x$ suitable solution belongs to $\mathcal{C}_t L^3_x$, so that uniqueness is, afterward, obvious).

Uniqueness for suitable solutions

Theorem 15.14.
Let $T_1 < T_2$ be two real numbers and $p, q > 1$ with $\frac{3}{2} < q < 3$ and $\frac{2}{p} + \frac{3}{q} < 2$. Let $\vec{f} \in L^p_t L^q_x((T_1, T_2) \times \mathbb{R}^3)$ with $\text{div} \, \vec{f} = 0$ and let $\vec{u}_1, \vec{u}_2 \in \mathcal{NS}_{T_1,T_2,\vec{f}}$ two suitable solutions of the Navier–Stokes equations.
If for some $T_3 \in (T_1, T_2)$, we have $\vec{u}_1(T_3, .) = \vec{u}_2(T_3, .)$, then $\vec{u}_1 = \vec{u}_2$ on $[T_3, T_2) \times \mathbb{R}^3$.

Proof. The proof follows the lines of the weak–strong uniqueness result of Kozono and Sohr [275] (Theorem 12.4). As $t \mapsto \vec{u}_i(t,.)$ is continuous from (T_1, T_2) to $L^2_{\mathrm{loc}}(\mathbb{R}^3)$, if $T_4 = \sup\{T \geq T_3 \ / \ \vec{u}_1 = \vec{u}_2 \text{ on } [T_3, T]\}$, then $\vec{u}_1 = \vec{u}_2$ on $[T_3, T_4]$. If $T_4 < T_2$, we must show that equality goes beyond T_4.

Let $\vec{u}_3$ be the solution in $\mathcal{C}([T_4, T_5), L^3)$ which originates with $\vec{u}_1(T_4, .)$ at time $t = T_4$. We are going to show that $\vec{u}_1$ (and $\vec{u}_2$) coincides with $\vec{u}_3$ on a small interval $[T_4, T_5]$. This will give the desired uniqueness result.

Let $\varphi \in \mathcal{D}(\mathbb{R}^3$ be fixed such that $0 \leq \varphi \leq 1$ and $\sum_{k \in \mathbb{Z}^3} \varphi(x - k) =$, and let $R > 0$. We are going to estimate

$$\alpha_R(t) = \sup_{x_0 \in \mathbb{R}^3} \sup_{T_4 \leq s \leq t} \int |\vec{u}_1(s,x) - \vec{u}_3(s,x)|^2 \varphi(\frac{x - x_0}{R}) \, dx$$

and

$$\beta_R(t) = \sup_{x_0 \in \mathbb{R}^3} \int\int_{T_4}^t |\vec{\nabla} \otimes (\vec{u}_1 - \vec{u}_3)(s,x)|^2 \varphi(\frac{x - x_0}{R}) \, dx$$

From Theorem 15.13, we already know that, for large R, $\alpha_R(t)$ is controlled by $R(\|\vec{u}_1\|^2_{L^\infty L^3} + \|\vec{u}_3\|^2_{L^\infty L^3})$ and $\beta_R(t)$ by

$$R^3 \|\vec{u}_1\|_{L^\infty L^3} \left(\|\vec{u}_1\|_{L^\infty L^3} + \|\vec{u}_1\|^2_{L^\infty L^3} + \sqrt{T_2 - T_1}^{3 - (\frac{2}{p} + \frac{3}{q})} \|\vec{f}\|_{L^p L^q} \right)$$

$$+ R^3 \|\vec{u}_3\|_{L^\infty L^3} \left(\|\vec{u}_3\|_{L^\infty L^3} + \|\vec{u}_3\|^2_{L^\infty L^3} + \sqrt{T_2 - T_1}^{3 - (\frac{2}{p} + \frac{3}{q})} \|\vec{f}\|_{L^p L^q} \right)$$

We use the identity

$$\partial_t |\vec{u}_1(t,x) - \vec{u}_3(t,x)|^2 = 2(\vec{u}_1(t,x) - \vec{u}_3(t,x)).\partial_t(\vec{u}_1(t,x) - \vec{u}_3(t,x))$$

and we find

$$\int \frac{|\vec{u}_1(t,x) - \vec{u}_3(t,x)|^2}{2} \varphi(\frac{x - x_0}{R}) \, dx$$

$$+ \nu \int\int_{T_4}^t \varphi(\frac{x - x_0}{R}) |\vec{\nabla} \otimes (\vec{u}_1 - \vec{u}_3)|^2 \, ds \, dx =$$

$$\nu \int\int_{T_4}^t \frac{|\vec{u}_1 - \vec{u}_3|^2}{2} \frac{1}{R^2} (\Delta\varphi)(\frac{x - x_0}{R}) \, ds \, dx$$

$$- \int\int_{T_4}^t \varphi(\frac{x - x_0}{R})(\vec{u}_1 - \vec{u}_3).(\vec{u}_1.\vec{\nabla}\vec{u}_1 - \vec{u}_3.\vec{\nabla}\vec{u}_3) \, ds \, dx$$

$$- \int\int_{T_4}^t \varphi(\frac{x - x_0}{R})(\vec{u}_1 - \vec{u}_3).\vec{\nabla}(p_1 - p_3) \, ds \, dx$$

$$= I_1 + I_2 + I_3$$

Let us notice that there exist C_0 and N such that $|\Delta\varphi(x)| \leq C_0 \sum_{k \in \mathbb{Z}^3, |k| < N} \varphi(x - k)$, so that

$$I_1 = \nu \int\int_{T_4}^t \frac{|\vec{u}_1 - \vec{u}_3|^2}{2} \frac{1}{R^2}(\Delta\varphi)(\frac{x - x_0}{R}) \, ds \, dx \leq C_0 N\nu \frac{1}{R^2} \int_{T_4}^t \alpha_R(s) \, ds.$$

In order to estimate I_2, we write

$$\vec{u}_1.\vec{\nabla}\vec{u}_1 - \vec{u}_3.\vec{\nabla}\vec{u}_3 = \vec{u}_1.\vec{\nabla}(\vec{u}_1 - \vec{u}_3) + (\vec{u}_1 - \vec{u}_3).\vec{\nabla}\vec{u}_3$$

and we find

$$
\begin{aligned}
I_2 = & \frac{1}{2R} \iint_{T_4}^{t} |\vec{u}_1 - \vec{u}_3|^2 \vec{u}_1(t,x).(\vec{\nabla}\varphi)(\frac{x-x_0}{R})\, ds\, dx \\
& + \frac{1}{R} \iint_{T_4}^{t} (\vec{u}_1 - \vec{u}_3.\vec{u}_3)(\vec{u}_1 - \vec{u}_3)(t,x).(\vec{\nabla}\varphi)(\frac{x-x_0}{R})\, ds\, dx \\
& + \iint_{T_4}^{t} \varphi(\frac{x-x_0}{R})\vec{u}_3.((\vec{u}_1 - \vec{u}_3).\vec{\nabla}(\vec{u}_1 - \vec{u}_3))\, ds\, dx \\
= & I_4 + I_5 + I_6.
\end{aligned}
$$

If B is a ball that contains the support of φ and if $\chi_{x_0,R}(x) = 1_B(\frac{x-x_0}{R})$, we have

$$
I_4 \leq C\frac{1}{R}\|\vec{u}_1\|_{L^\infty L^3}\int_{T_4}^{t} \|(\vec{u}_1 - \vec{u}_3)\chi_{x_0,R}\|_2\|(\vec{u}_1 - \vec{u}_3)\chi_{x_0,R}\|_6\, ds
$$

We have

$$
\|(\vec{u}_1 - \vec{u}_3)\chi_{x_0,R}\|_6 \leq C\sqrt{(\|(\vec{\nabla}\otimes(\vec{u}_1 - \vec{u}_3))\chi_{x_0,R}\|_2 \frac{1}{R}\|(\vec{u}_1 - \vec{u}_3)\chi_{x_0,R}\|_2}
$$

and

$$
1_B \leq C_0 \sum_{k\in\mathbb{Z}^3, |k|<N} \varphi(x-k)
$$

so that

$$
I_4 \leq C\|\vec{u}_1\|_{L^\infty L^3}(\frac{1}{R}\beta_R(t) + \frac{1}{R^2}\int_{T_4}^{t}\alpha_R(s)\, ds).
$$

Similarly, we have

$$
I_5 \leq C\|\vec{u}_3\|_{L^\infty L^3}(\frac{1}{R}\beta_R(t) + \frac{1}{R^2}\int_{T_4}^{t}\alpha_R(s)\, ds).
$$

In order to estimate I_6, let us assume that $\vec{u}_3 \in \mathcal{C}([T_4,T_5])$ Then, for every $\epsilon > 0$, we may split $\vec{u}_3$ on $[T_4,T_5]$ into $\vec{u}_3 = \vec{U}_3 + \vec{V}_3$ with $\|\vec{U}_3\|_{L^\infty L^3} < \epsilon$ and $\vec{V}_3 \in L^\infty_{t,x}$. Let

$$
M_\epsilon = \|\vec{V}_3\|_{L^\infty_{t,x}}.
$$

We have

$$
I_6 \leq C\int_{T_4}^{t} \|(\vec{\nabla}\otimes(\vec{u}_1 - \vec{u}_3))\chi_{x_0,R}\|_2 \times
$$
$$
\times (\epsilon\|(\vec{\nabla}\otimes(\vec{u}_1 - \vec{u}_3))\chi_{x_0,R}\|_2 + M_\epsilon\|(\vec{u}_1 - \vec{u}_3)\chi_{x_0,R}\|_2)\, ds
$$

and thus

$$
I_6 \leq C(\epsilon\beta_R(t) + \frac{M_\epsilon^2}{\epsilon}\int_0^{t}\alpha_R(s)\, ds).
$$

In order to estimate I_3 we write $p_1 - p_3 = q_1 + q_2$, where

$$
q_1 = -\sum_{i=1}^{3}\sum_{j=1}^{3}\frac{\partial_i\partial_j}{\Delta}(\omega_{x_0,R}(u_{1,i}u_{1,j} - u_{3,i}u_{3,j}))
$$

and ω is a smooth compactly supported function on $\mathbb{R}$, compactly supported on a ball $B(0, 3R_0)$, equal to 1 on $B(0, 2R_0)$, where R_0 satisfies that φ is supported in $B(0, R_0)$.

We write

$$I_3 = \frac{1}{R} \iint_{T_4}^t \varphi(\frac{x - x_0}{R}) q_1(s, x)(\vec{u}_1 - \vec{u}_3)(s, x).(\vec{\nabla}\varphi)(\frac{x - x_0}{R}) \, ds \, dx$$

$$- \iint_{T_4}^t \varphi(\frac{x - x_0}{R})(\vec{u}_1 - \vec{u}_3).\vec{\nabla}q_2 \, ds \, dx$$

$$= I_7 + I_8.$$

We have

$$I_7 \leq C \frac{1}{R} \int_{T_4}^t \|(\vec{u}_1 - \vec{u}_3)\chi_{x_0,R}\|_6 \|q_1\|_{6/5} \, ds$$

$$\leq C' \frac{1}{R}(\|\vec{u}_1\|_{L^\infty L^3} + \vec{u}_3\|_{L^\infty L^3}) \int_{T_4}^t \|(\vec{u}_1 - \vec{u}_3)\chi_{x_0,R}\|_6 \|(\vec{u}_1 - \vec{u}_3)\omega_{x_0,R}\|_2 \, ds$$

$$\leq C''(\|\vec{u}_1\|_{L^\infty L^3} + \|\vec{u}_3\|_{L^\infty L^3})(\frac{1}{R}\beta_R(t) + \frac{1}{R^2}\int_{T_4}^t \alpha_R(s) \, ds).$$

Finally, we estimate I_8:

$$I_8 \leq C \int_{T_4}^t \|(\vec{u}_1 - \vec{u}_3)\chi_{x_0,R}\|_6 \|\chi_{x_0,R}\vec{\nabla}q_2\|_{6/5} \, ds.$$

Let

$$N_R(g) = \sup_{x_0 \in \mathbb{R}^3} \|g(x) \, \varphi(\frac{x - x_0}{R})\|_{6/5}.$$

The norm N_R is shift invariant: $N_R(g(. - x_0)) = N_R(g)$. Thus, for $h \in L^1(\mathbb{R}^3)$, we have

$$N_R(g * h) \leq N_R(g)$$

(see Lemarié-Rieusset [313] for a theory of shift-invariant Banach spaces of distributions). As we have, on the support of $\chi_{x_0,R}$,

$$|\vec{\nabla}q_2(s, x)| \leq C \int_{|x-y|\geq R} \frac{1}{|x - y|^4} |u_{1,i}(s, y)u_{1,j}(s, y) - u_{3,i}(s, y)u_{3,j}(s, y)| \, dy$$

we find

$$\|\chi_{x_0,R}\vec{\nabla}q_2\|_{6/5} \leq C \frac{1}{R} N_R(u_{1,i}u_{1,j} - u_{3,i}u_{3,j}\|_{N_R} \leq C \frac{1}{R}(\|\vec{u}_1\|_{L^\infty L^3} + \|\vec{u}_3\|_{L^\infty L^3})\alpha_R(s)$$

so that

$$I_8 \leq C(\|\vec{u}_1\|_{L^\infty L^3} + \vec{u}_3\|_{L^\infty L^3})(\frac{1}{R}\beta_R(t) + \frac{1}{R^2}\int_{T_4}^t \alpha_R(s) \, ds).$$

Summing up all those estimates, we find that for a constant C_0 which does not depend on $\vec{u}_1$, $\vec{u}_3$, ϵ nor R, and for a constant $C_{R,\epsilon,\vec{u}_3}$ (which depends on R, ϵ and $\vec{u}_3$) and for

$T_4 \leq t \leq T_5$, we have

$$\int \frac{|\vec{u}_1(t,x) - \vec{u}_3(t,x)|^2}{2} \varphi(\frac{x - x_0}{R}) \, dx$$
$$+ \nu \int\int_{T_4}^{t} \varphi(\frac{x - x_0}{R}) |\vec{\nabla} \otimes (\vec{u}_1 - \vec{u}_3)|^2 \, ds \, dx \leq \tag{15.12}$$
$$C_{R,\epsilon,\vec{u}_3}(\nu + 1 + \|\vec{u}_1\|_{L^\infty L^3}) \int_{T_4}^{t} \alpha_R(s) \, ds$$
$$+ C_0(\frac{\|\vec{u}_1\|_{L^\infty L^3} + \|\vec{u}_3\|_{L^\infty L^3}}{R} + \epsilon)\beta_R(t)$$

Taking the supremum with respect to x_0, we get

$$\beta_R(t) \leq \frac{C_{R,\epsilon,\vec{u}_3}}{\nu}(\nu + 1 + \|\vec{u}_1\|_{L^\infty L^3}) \int_{T_4}^{t} \alpha_R(s) \, ds + C_0(\frac{\|\vec{u}_1\|_{L^\infty L^3} + \|\vec{u}_3\|_{L^\infty L^3}}{R\nu} + \epsilon)\beta_R(t)$$

Choosing R large enough and ϵ small enough to grant that

$$C_0(\frac{\|\vec{u}_1\|_{L^\infty L^3} + \|\vec{u}_3\|_{L^\infty L^3}}{R\nu} + \epsilon) < 1/2,$$

we obtain

$$\beta_R(t) \leq \frac{C_{R,\epsilon,\vec{u}_3}}{2\nu}(\nu + 1 + \|\vec{u}_1\|_{L^\infty L^3}) \int_{T_4}^{t} \alpha_R(s) \, ds$$

and thus

$$\int \frac{|\vec{u}_1(t,x) - \vec{u}_3(t,x)|^2}{2} \varphi(\frac{x - x_0}{R}) \, dx \leq \tag{15.13}$$
$$\frac{5}{4} C_{R,\epsilon,\vec{u}_3}(\nu + 1 + \|\vec{u}_1\|_{L^\infty L^3}) \int_{T_4}^{t} \alpha_R(s) \, ds$$

Taking one more time the supremum with respect to x_0, we get

$$\alpha_R(t) \leq \frac{5}{2} C_{R,\epsilon,\vec{u}_3}(\nu + 1 + \|\vec{u}_1\|_{L^\infty L^3}) \int_{T_4}^{t} \alpha_R(s) \, ds \tag{15.14}$$

and we may conclude by using Grönwall's lemma that $\alpha_R = 0$, hence $\vec{u}_1 = \vec{u}_3$. $\qquad\square$

15.10 A Quantitative One-scale Estimate for the Caffarelli–Kohn–Nirenberg Regularity Criterion

We now discuss again the regularity criterion of Caffarelli, Kohn and Nirenberg [74]. We study a suitable $\vec{u} \in \mathcal{NS}_{T_1,T_2,\vec{f}}$, where div $\vec{f} = 0$ and $\vec{f} \in L^p L^q$ with $\frac{3}{2} < q < 3$ and $\frac{2}{p} + \frac{3}{q} < 2$. We shall discuss regularity in the interior, as in Theorem 13.8 but also regularity at the top of the domain, i.e., at a point (t_0, x_0) with $t_0 = T_2$. A *regular point* of $\vec{u}$ will be a point (t_0, x_0) in $(T_1, T_2] \times \mathbb{R}^3$ such that, for a neighborhood $Q_r(t_0, x_0)$, we have that $\vec{u}$ is Hölderian on $Q_r(t_0, x_0) \cap \Omega$, where $\Omega = (T_1, T_2) \times \mathbb{R}^3$, and $\omega_{x_0,R}(x) = \omega(\frac{x - x_0}{R})$.

We recall Theorem 13.8 where we proved:

Theorem 15.15.
Let $\Omega = (T_1, T_2) \times \mathbb{R}^3$. Let p, q with $3/2 < q < 3$ and $\frac{2}{p} + \frac{3}{q} = \frac{5}{\tau_0} < 2$. Let $\vec{f} \in L^p_t L^q_x((T_1, T_2) \times \mathbb{R}^3$ with $\operatorname{div} f = 0$ and let $\vec{u}$ be a suitable solution of the Navier–Stokes equations in $\mathcal{NS}_{T_1, T_2, \vec{f}}$.
There exists a positive constant ϵ^* which depends only on ν and τ_0 such that, if for some $(t_0, x_0) \in \Omega$, we have

$$\limsup_{r \to 0} \frac{1}{r} \iint_{Q_r(t_0, x_0) \cap \Omega} |\vec{\nabla} \otimes \vec{u}|^2 \, ds \, dx < \epsilon^* \tag{15.15}$$

then $\vec{u}$ is Hölderian on $Q_{r_0}(t_0, x_0) \cap \Omega$ for a neighborhood $Q_{r_0}(t_0, x_0)$ of (t_0, x_0).

We have indeed proved the theorem only for interior points (t_0, x_0) with $t_0 < T_2$, but the proof is the same for $t_0 = T_2$. Theorem 15.15 seems to rely on the behavior of $\vec{u}$ at all (small) scales, but a closer look on the proof of the theorem shows that it is indeed enough to have a control on just *one* scale (this is the second criterion of Caffarelli, Kohn and Nirenberg).

The second Caffarelli, Kohn and Nirenberg criterion for $L^\infty_t L^3_x$ solutions

Theorem 15.16.
Let $T_1 < T_2$ be two real numbers and $p, q > 1$ with $\frac{3}{2} < q < 3$ and $\frac{2}{p} + \frac{3}{q} < 2$. Let $\vec{f} \in L^p_t L^q_x((T_1, T_2) \times \mathbb{R}^3)$ with $\operatorname{div} \vec{f} = 0$ and $\vec{u} \in \mathcal{NS}_{T_1, T_2, \vec{f}}$ such that $\vec{u}$ is suitable. Define

$$M_0 = \|\vec{u}\|_{L^\infty L^3} + \|\vec{u}\|^2_{L^\infty L^3} + \sqrt{T_2 - T_1}^{5 - \frac{2}{\tau_0}} \|\vec{f}\|_{L^p L^q}.$$

Then, there exists a constant ϵ_0 (which does not depend on $\vec{f}$, $\vec{u}$ nor on M_0) such that, if $t_0 \in (T_1, T_2]$ and $x_0 \in \mathbb{R}^3$ and, for some $r_1 \in (0, \sqrt{t_0 - T_1})$,

---

$$\frac{1}{r_1^2} \iint_{Q_{r_1}(t_0, x_0) \cap \Omega} |\vec{u}(t, x)|^3 \, dt \, dx \le \epsilon_0 \min\left(\frac{1}{M_0}, \frac{1}{M_0^3}\right) \tag{15.16}$$

then $\vec{u}$ is Hölderian on $Q_{r_0}(t_0, x_0) \cap \Omega$ for a neighborhood $Q_{r_0}(t_0, x_0)$ of (t_0, x_0).

Proof. From Theorem 15.15, it is enough to show that for $r < \frac{1}{2} r_1$, we have

$$\frac{1}{r} \iint_{Q_r(t_0, x_0) \cap \Omega} |\vec{\nabla} \otimes \vec{u}|^2 \, ds \, dx \le \frac{1}{2} \epsilon^*.$$

As in the proof of Theorem 13.8, we introduce the quantities

- $U_r(t, x) = \sup_{s \in (t - r^2, \min(t + r^2, T_2))} \int_{B_r(t, x)} |\vec{u}(s, y)|^2 \, dx \, dy$

- $V_r(t, x) = \iint_{Q_r(t, x) \cap \Omega} |\vec{\nabla} \otimes \vec{u}(s, y)|^2 \, ds \, dy$

- $W_r(t, x) = \iint_{Q_r(t, x) \cap \Omega} |\vec{u}(s, y)|^3 \, ds \, dy$

and we introduce the reduced quantities

$$\alpha_r = \frac{1}{r}(U_r(t_0,x_0) + V_r(t_0,x_0)), \ \gamma_r = \frac{1}{r^2}W_r(t_0,x_0) \text{ and } p_r = \frac{1}{r^2}P_r(t_0,x_0).$$

Recall that we have

$$W_r(t,x) \leq C_0 r^{1/2}(U_r(t,x) + V_r(t,x))^{3/2}$$

so that

$$\gamma_r \leq C_0 \alpha_r^{3/2}. \tag{15.17}$$

Moreover, we have seen just before stating Theorem 15.13 that

$$\sup_{\tau \in (t_0 - r^2, \min(t_0 + r^2, T_2))} \int_{B(x_0,r)} \frac{|\vec{u}(\tau,x)|^2}{2} \, dx$$

$$+ \nu \iint_{(t,x) \in Q_r(t_0,x_0) \cap \Omega} |\vec{\nabla} \otimes \vec{u}|^2 \, dt \, dx \leq$$

$$C_\nu \left(\frac{1}{r^2} \iint_{Q_{2r}(t_0,x_0) \cap \Omega} |\vec{u}|^3 \, dt \, dx\right)^{1/3} \times$$

$$\times \left(r\|\vec{u}\|_{L^\infty L^3} + r\|\vec{u}\|_{L^\infty L^3}^2 + r \times r^{3-(\frac{2}{p}+\frac{3}{q})}\|\vec{f}\|_{L^p L^q}\right)$$

so that

$$\alpha_r \leq C_\nu \gamma_{2r}(\|\vec{u}\|_{L^\infty L^3} + \|\vec{u}\|_{L^\infty L^3}^2 + \sqrt{T_2 - T_1}^{5-\frac{2}{\tau_0}}\|\vec{f}\|_{L^p L^q}) \tag{15.18}$$

We want to prove $\alpha_r \leq \frac{1}{2}\epsilon^*$, thus it is enough to check that $C_\nu \gamma_{2r} M_0 \leq \frac{1}{2}\epsilon^*$. Moreover, we have

$$\gamma_r \leq \gamma_{2r}\gamma_{2r}^{1/2}C_0(C_\nu M_0)^{3/2}$$

thus we shall have $\gamma_r \leq \gamma_{2r}$ if $\gamma_{2r}C_0^2(C_\nu M_0)^3 \leq 1$.

It means that we shall have for every $r < \frac{1}{2}r_1$

$$\frac{1}{r}\iint_{Q_r(t_0,x_0)\cap\Omega} |\vec{\nabla} \otimes \vec{u}|^2 \, ds \, dx \leq \frac{1}{2}\epsilon^*$$

provided that for every $\rho \in (r_1/2, r_1)$, we have $\gamma_r \leq \min(\frac{\epsilon^*}{2C_\nu M_0}, \frac{1}{C_0^2 C_\nu^3 M_0^3})$. As, for $r \in (r_1/2, r_1)$, we have

$$\gamma_\rho \leq 4\frac{1}{r_1^2}\iint_{Q_{r_1}(t_0,x_0)\cap\Omega} |\vec{u}(t,x)|^3 \, dx,$$

the Theorem is proved. $\qquad\qquad\square$

15.11 The Topological Structure of the Set of Suitable Solutions

We may now give some interesting corollaries of Theorems 15.13 and 15.16:

Corollary 15.4.
Let $T_1 < T_2$ be two real numbers and $p, q > 1$ with $\frac{3}{2} < q < 3$ and $\frac{2}{p} + \frac{3}{q} < 2$. Let $\vec{f} \in L_t^p L_x^q((T_1, T_2) \times \mathbb{R}^3)$ with $\text{div } \vec{f} = 0$ and $\vec{u} \in \mathcal{NS}_{T_1,T_2,\vec{f}}$ such that $\vec{u}$ is suitable. Define

$$M_0 = \|\vec{u}\|_{L^\infty L^3} + \|\vec{u}\|_{L^\infty L^3}^2 + \sqrt{T_2 - T_1}^{5-\frac{2}{\tau_0}}\|\vec{f}\|_{L^p L^q}.$$

Then, for $t_0 \in (T_1, T_2]$ and $x_0 \in \mathbb{R}^3$, we are in one of those two cases:

- *either (t_0, x_0) is regular and then $\lim_{r \to 0} \frac{1}{r^2} \iint_{Q_r(t_0,x_0) \cap \Omega} |\vec{u}(t,x)|^3 \, dt \, dx = 0$*

- *or (t_0, x_0) is singular and then for every $r \in (0\sqrt{t_0 - T_1})$, we have*

$$\frac{1}{r^2} \iint_{Q_r(t_0,x_0) \cap \Omega} |\vec{u}(t,x)|^3 \, dt \, dx \geq \epsilon_0 \min(\frac{1}{M_0}, \frac{1}{M_0^3})$$

where $\epsilon_0 > 0$ is the constant described in Theorem 15.16.

Proof. If there exists some $r \in (0\sqrt{t_0 - T_1})$ such that

$$\frac{1}{r^2} \iint_{Q_r(t_0,x_0) \cap \Omega} |\vec{u}(t,x)|^3 \, dt \, dx \geq \epsilon_0 \min(\frac{1}{M_0}, \frac{1}{M_0^3}),$$

we know that (t_0, x_0) is regular. Conversely, if (t_0, x_0) is regular, $\vec{u}$ is bounded on a cylinder $Q_{r_0}(t_0, x_0) \cap \Omega$ and we have

$$\frac{1}{r^2} \iint_{Q_r(t_0,x_0) \cap \Omega} |\vec{u}(t,x)|^3 \, dt \, dx = O(r^3) \to_{r \to 0} 0. \qquad \square$$

Theorem 15.17.
Let $T_1 < T_2$ be two real numbers and $p, q > 1$ with $\frac{3}{2} < q < 3$ and $\frac{2}{p} + \frac{3}{q} < 2$. Let $\vec{f}_n$ be a sequence of vector fields such that $\vec{f}_n \in L_t^p L_x^q((T_1, T_2) \times \mathbb{R}^3)$ with $\operatorname{div} \vec{f}_n = 0$ and $\sup_{n \in \mathbb{N}} \|\vec{f}_n\|_{L^p L^q} < +\infty$, and $\vec{u}_n$ be a sequence of vector fields such that $\vec{u}_n \in \mathcal{NS}_{T_1,T_2,\vec{f}_n}$ and $\sup_{n \in \mathbb{N}} \|\vec{u}_n\|_{L^\infty L^3} < +\infty$. If each $\vec{u}_n$ satisfies $\vec{\nabla} \otimes \vec{u}_n \in (L^2 L^2)_{\mathrm{loc}}((T_1, T_2) \times \mathbb{R}^3)$ and is suitable, then

- *there exists subsequences $\vec{f}_{n_k}$ and $\vec{u}_{n_k}$ that converge *-weakly (respectively in $L^p L^q$ and $L^\infty L^3$) to limits $\vec{f}_\infty$ and $\vec{u}_\infty$ and moreover such that $\vec{u}_{n_k}$ converges strongly in $(L_t^3 L_x^3)_{\mathrm{loc}}((T_1, T_2) \times \mathbb{R}^3)$*

- *in particular, $\vec{u}_\infty \in \mathcal{NS}_{T_1,T_2,\vec{f}_\infty}$ and is suitable*

- *if $T_1 < t_0 \leq T_2$ and $x_0 \in \mathbb{R}^3$ and if (t_0, x_0) is a regular point of $\vec{u}_\infty$, then for k large enough, (t_0, x_0) is a regular point of $\vec{u}_{n_k}$*

Proof. Let

$$M_0 = \sup_{n \in \mathbb{N}} \|\vec{u}_n\|_{L^\infty L^3} + \|\vec{u}_n\|_{L^\infty L^3}^2 + \sqrt{T_2 - T_1}^{5 - \frac{2}{r_0}} \|\vec{f}_n\|_{L^p L^q}.$$

If (t_0, x_0) is regular, then for r small enough, we have

$$\frac{1}{r^2} \iint_{Q_r(t_0,x_0) \cap \Omega} |\vec{u}_\infty(t,x)|^3 \, dt \, dx < \frac{1}{2} \epsilon_0 \min(\frac{1}{M_0}, \frac{1}{M_0^3})$$

By strong convergence, we find that, for k large enough (for fixed r),

$$\frac{1}{r^2} \iint_{Q_r(t_0,x_0) \cap \Omega} |\vec{u}_{n_k}(t,x)|^3 \, dt \, dx < \epsilon_0 \min(\frac{1}{M_0}, \frac{1}{M_0^3})$$

and (t_0, x_0) is a regular point of $\vec{u}_{n_k}$. $\qquad \square$

Theorem 15.18.
Let $T_1 < T_2$ be two real numbers and $p, q > 1$ with $\frac{3}{2} < q < 3$ and $\frac{2}{p} + \frac{3}{q} < 2$.

Let $\vec{f} \in L_t^p L_x^q ((T_1, T_2) \times \mathbb{R}^3)$ with div $\vec{f} = 0$ and $\vec{u} \in \mathcal{NS}_{T_1, T_2, \vec{f}}$ such that $\vec{u}$ is suitable. Then $\vec{u} \in \mathcal{C}((T_1, T_2), L^3)$ if and only if every point $(t_0, x_0) \in (T_1, T_2) \times \mathbb{R}^3$ is regular.

Proof. We have to prove the fact that regularity at each point implies that $\vec{u} \in \mathcal{C}((T_1, T_2), L^3)$.

Let us fix $t_0 \in (T_1, T_2)$ and $r \in (0, \sqrt{t - T_1})$. We have

$$\lim_{|x_0| \to +\infty} \frac{1}{r^2} \iint_{Q_r(t_0, x_0) \cap \Omega} |\vec{u}(t, x)|^3 \, dt \, dx = 0$$

so that, for x_0 large enough,

$$\frac{1}{r^2} \iint_{Q_r(t_0, x_0) \cap \Omega} |\vec{u}(t, x)|^3 \, dt \, dx \leq \epsilon_0 \min\left(\frac{1}{M_0}, \frac{1}{M_0^3}\right)$$

where

$$M_0 = \|\vec{u}\|_{L^\infty L^3} + \|\vec{u}\|_{L^\infty L^3}^2 + \sqrt{T_2 - T_1}^{\,5 - \frac{2}{r_0}} \|\vec{f}\|_{L^p L^q}.$$

As all the size estimates will then be uniform with respect to x_0, we find that there exist $R > 0$, $M > 0$ and $r_0 > 0$ such that, for $|x_0| > R$, we have $|\vec{u}(t, x)| \leq M$ on $Q_{r_0}(t_0, x_0) \cap \Omega$. Moreover, for each x_0 the compact set $\bar{B}(0, R)$, there exists a neighborhood $Q_{r(x_0)}(t_0, x_0)$ such that $\vec{u}$ is bounded on $Q_r(t_0, x_0) \cap \Omega$. By compactness, we may cover $\bar{B}(0, R)$ by a finite number N of balls $B(x_i, r(x_i))$, $1 \leq i \leq N$. Defining $\rho = \min(r, r(x_1), \ldots, r(x_N))$, we find that $\vec{u}$ is bounded on $(t_0 - \rho^2, \min(t_0 + \rho^2, T_2)) \times \mathbb{R}^3$. Thus, for all compact subintervals $[T_3, T_4] \subset (T_1, T_2)$, $\vec{u} \in L_{t,x}^\infty$ on $[T_3, T_4] \times \mathbb{R}^3$ and we may conclude. $\square$

A similar proof gives the following proposition[2]:

Theorem 15.19.
Let $T_1 < T_2 < T_3$ be three real numbers and $p, q > 1$ with $\frac{3}{2} < q < 3$ and $\frac{2}{p} + \frac{3}{q} < 2$. Let $\vec{f} \in L_t^p L_x^q ((T_1, T_3) \times \mathbb{R}^3)$ with div $\vec{f} = 0$ and $\vec{u} \in \mathcal{NS}_{T_1, T_2, \vec{f}} \cap \mathcal{C}((T_1, T_2), L^3)$. Then the following assertions are equivalent:

(A) $\vec{u}$ may be extended to (T_1, T_4) for some $T_2 < T_4 < T_3$ such that $\vec{u} \in \mathcal{NS}_{T_1, T_4, \vec{f}} \cap \mathcal{C}((T_1, T_4), L^3)$.

(B) Every point (T_2, x_0), $x_0 \in \mathbb{R}^3$, is regular.

---

[2]This proposition is in a way void, as we shall see that any such solution $\vec{u}$ has indeed no singular point at $t = T_2$ (Theorem 15.21).

Finally, we shall consider a last result on the *–weak convergence of $L^\infty L^3$ suitable solutions:

Strong local L^2 convergence

Theorem 15.20.
Let $T_1 < T_2$ be two real numbers and $p, q > 1$ with $\frac{3}{2} < q < 3$ and $\frac{2}{p} + \frac{3}{q} < 2$. Let $\vec{f}_n$ be a sequence of vector fields such that $\vec{f}_n \in L^p_t L^q_x((T_1, T_2) \times \mathbb{R}^3)$ with $\mathrm{div}\, \vec{f}_n = 0$ and $\sup_{n \in \mathbb{N}} \|\vec{f}_n\|_{L^p L^q} < +\infty$, and $\vec{u}_n$ be a sequence of vector fields such that $\vec{u}_n \in \mathcal{NS}_{T_1, T_2, \vec{f}_n}$ and $\sup_{n \in \mathbb{N}} \|\vec{u}_n\|_{L^\infty L^3} < +\infty$. If moreover

- *each $\vec{u}_n$ satisfies $\vec{\nabla} \otimes \vec{u}_n \in (L^2 L^2)_{\mathrm{loc}}((T_1, T_2) \times \mathbb{R}^3$ and is suitable*

- *$\vec{f}_n$ and $\vec{u}_n$ converge *-weakly (respectively in $L^p L^q$ and $L^\infty L^3$) to limits $\vec{f}_\infty$ and $\vec{u}_\infty$*

- *$\vec{u}_n$ converges strongly in $(L^3_t L^3_x)_{\mathrm{loc}}((T_1, T_2) \times \mathbb{R}^3)$*

then for every $t \in (T_1, T_2)$ such that each point (t, x) is regular for $\vec{u}_\infty$ and for every compact subset K of $\mathbb{R}^3$, we have

$$\lim_{n \to +\infty} \int_K |\vec{u}_n(t, x) - \vec{u}_\infty(t, x)|^2 \, dx = 0. \tag{15.19}$$

Proof. Let Ω be a relatively compact neighborhood of K. Let $\tau < t$ such that $\vec{u}_\infty$ is bounded on $[\tau, t] \times \Omega$. Let α a smooth function on $\mathbb{R}$ with $\alpha(s) = 0$ for $s < \tau$ and $\alpha(s) = 1$ for $s > t$, and $\varphi \in \mathcal{D}(\mathbb{R}^3)$ which is equal to 1 on K and is supported in Ω. Let

$$\Phi_n(s) = \alpha(s) \int \varphi(x) |\vec{u}_n(s, x) - \vec{u}_\infty(s, x)|^2 \, dx.$$

We have obviously

$$\int_K |\vec{u}_n(t, x) - \vec{u}_\infty(t, x)|^2 \, dx \le \Phi_n(t) - \Phi_n(\tau) = \int_\tau^t \frac{d}{ds} \Phi_n(s) \, ds.$$

Moreover,

$$\int_\tau^t \frac{d}{ds} \Phi_n(s) \, ds = 2 \iint_\tau^t (\vec{u}_n - \vec{u}_\infty) \cdot \partial_t(\vec{u}_n - \vec{u}_\infty) \, \alpha(s) \, \varphi(x) \, ds \, dx$$

$$+ \iint_\tau^t |\vec{u}_n(s, x) - \vec{u}_\infty(s, x)|^2 \alpha'(s) \, \varphi(x) \, ds \, dx$$

$$= \iint_\tau^t |\vec{u}_n(s, x) - \vec{u}_\infty(s, x)|^2 (\alpha'(s) \, \varphi(x) + \nu \alpha(s) \Delta \varphi(x)) \, ds \, dx$$

$$- 2\nu \iint_\tau^t |\vec{\nabla} \otimes (\vec{u}_n - \vec{u}_\infty)|^2 \, ds \, dx$$

$$+ 2 \iint_\tau^t (p_n - p_\infty)(\vec{u}_n - \vec{u}_\infty) \cdot \vec{\nabla} \varphi(x) \, \alpha(s) \, ds \, dx$$

$$+ \iint_\tau^t |\vec{u}_n - \vec{u}_\infty|^2 \, \vec{u}_n \cdot \vec{\nabla}\varphi \, \alpha(s) \, ds \, dx$$

$$+ 2 \iint_\tau^t \alpha(s)((\vec{u}_n - \vec{u}_\infty) \cdot \vec{u}_\infty)((\vec{u}_n - \vec{u}_\infty) \cdot \vec{\nabla}\varphi) \, ds \, dx$$

$$+ 2 \iint_\tau^t \alpha(s)\varphi(x)\vec{u}_\infty \cdot ((\vec{u}_n - \vec{u}_\infty) \cdot \vec{\nabla}(\vec{u}_n - \vec{u}_\infty)) \, ds \, dx$$

$$+ 2 \iint_\tau^t (\vec{u}_n - \vec{u}_\infty) \cdot (\vec{f}_n - \vec{f}_\infty)\alpha(s)\varphi(x) \, ds \, dx$$

$$\leq C_{\alpha,\phi,\nu} \iint_{[\tau,t]\times\Omega} |\vec{u}_n(s,x) - \vec{u}_\infty(s,x)|^2 \, ds \, dx$$

$$+ C_{\alpha,\phi,\nu}\|p_n - p_\infty\|_{L^\infty L^{3/2}} \Big(\iint_{[\tau,t]\times\Omega} |\vec{u}_n - \vec{u}_\infty|^3 \, ds \, dx \Big)^{1/3}$$

$$+ C_{\alpha,\phi,\nu}(\|\vec{u}_n\|_{L^\infty L^3} + \|\vec{u}_\infty\|_{L^\infty L^3}) \Big(\iint_{[\tau,t]\times\Omega} |\vec{u}_n - \vec{u}_\infty|^3 \, ds \, dx \Big)^{1/3}$$

$$+ C_{\alpha,\phi,\nu} \iint_{[\tau,t]\times\Omega} |\vec{u}_\infty(s,x)|^2 |\vec{u}_n(s,x) - \vec{u}_\infty(s,x)|^2 \, ds \, dx$$

$$+ C_{\alpha,\phi,\nu}\|\vec{f}_n - \vec{f}_\infty\|_{L^p L^q}(t-\tau)^{\frac{2}{3}-\frac{1}{p}}|\Omega|^{\frac{2}{3}-\frac{1}{q}} \Big(\iint_{[\tau,t]\times\Omega} |\vec{u}_n - \vec{u}_\infty|^3 \, ds \, dx \Big)^{1/3}$$

As $\vec{u}_\infty$ is bounded on $[\tau,t] \times \Omega$, we find that $\lim_{n\to+\infty} \Phi_n(t) = 0$. $\qquad\square$

15.12 Escauriaza, Seregin and Šverák's Theorem

We may now state and prove Escauriaza, Seregin and Šverák's theorem on $L_t^\infty L_x^3$ suitable solutions [163]:

Escauriaza, Seregin and Šverák's theorem

Theorem 15.21.
Let the forcing term $\vec{f}$ be defined on $(0,+\infty) \times \mathbb{R}^3$ and satisfy

$$\text{div } \vec{f} = 0$$

and, for every $T < +\infty$,

$$\vec{f} \in L_t^p((0,T), L^q(\mathbb{R}^3)) \text{ with } \frac{2}{p} + \frac{3}{q} = 3 \text{ and } 3/2 < q < 2$$

anf let $\vec{u}_0 \in L^3(\mathbb{R}^3))$ with div $\vec{u}_0 = 0$.
Let $\vec{u}$ be the solution of the Navier–Stokes Equations (15.9) in $\mathcal{C}([0,T^*), L_x^3)$, where T^* is the maximal existence time of the solution. Then, if $T^* < +\infty$, we have

$$\sup_{0<t<T^*} \|\vec{u}(t,.)\|_3 = +\infty.$$

Proof. Kenig and Koch [258], and Gallagher, Koch and Planchon [199] proposed a proof based on the method of critical elements and the theory of profile decomposition of Gallagher [196] and Koch [264]. However, Seregin's theorem (Theorem 15.5) will provide a much simpler proof, which we give here.

Assume that $T^* < +\infty$ and $\vec{u} \in L^\infty_t L^3_x((0, T^*) \times \mathbb{R}^3)$. By Theorem 15.19, we know that there exists a point (T^*, x_0) which is singular.

We define now $T_0 = -T^*$, $\vec{v}_0 = \vec{u}(t + T^*, x - x_0)$ and $\vec{f}_0 = \vec{f}(t + T^*, x - x_0)$. Clearly, $\vec{v}_0$ and $\vec{f}_0$ are defined on $(T_0, 0)$, with $\vec{f}_0 \in L^p L^q$, $\vec{v}_0 \in \mathcal{C}((T_0, 0), L^3) \cap L^\infty L^3$, $\vec{v}_0 \in \mathcal{NS}_{T_0, 0, \vec{f}_0}$ and $(0, 0)$ is a singular point of $\vec{v}_0$.

Let $T_n = 2^n T_0$, $\vec{v}_n = 2^{-n} \vec{v}_0(4^{-n} t, 2^{-n} x)$ and $\vec{f}_n = 2^{-3n} \vec{f}_0(4^{-n} t, 2^{-n} x)$. Then, due to the scaling invariance of the Navier–Stokes equations, we see that $\vec{v}_n$ and $\vec{f}_n$ are defined on $(T_n, 0)$, with $\vec{f}_n \in L^p L^q$, $\vec{v}_n \in \mathcal{C}((T_n, 0), L^3) \cap L^\infty L^3$, $\vec{v}_n \in \mathcal{NS}_{T_n, 0, \vec{f}_n}$ and $(0, 0)$ is a singular point of $\vec{v}_n$.

Moreover, we have:

$$\|\vec{v}_n\|_{L^\infty L^3} = \|\vec{v}_0\|_{L^\infty L^3}$$

and

$$\|\vec{f}_n\|_{L^p L^q} = 2^{-n(3 - \frac{2}{p} - \frac{3}{q})} \|\vec{f}_0\|_{L^p L^q} \to_{n \to +\infty} 0.$$

Thus, due to Theorems 15.17 and 15.18, we know that we can find a subsequence $\vec{v}_{n_k}$ and a vector field $\vec{v}$ so that

- on all intervals $(T, 0)$, $\vec{v}_{n_k}$ converges to $\vec{v}$ *-weakly in $L^\infty L^3$

- $\vec{v}$ is a suitable solution of the Navier–Stokes equations with a null force on $(-\infty, 0) \times \mathbb{R}^3$: $\vec{v} \in \mathcal{NS}_{-\infty, 0, \vec{0}}$

- $(0, 0)$ is a singular point of $\vec{v}$.

For almost every $T < 0$, $\vec{v}$ is a local Leray solution on $(T, 0) \times \mathbb{R}^3$. We may thus assume that $\vec{v}$ is a local Leray solution on $(-1, 0) \times \mathbb{R}^3$ (changing $\vec{v}$ into $\sqrt{|T|}\vec{u}(-Tt, \sqrt{|T|}x)$). Due to uniqueness of suitable solutions, we know that if $\vec{w}$ is the maximal solutions of the Navier–Stokes equations in $\mathcal{C}[-1, T^*(\vec{w}))$ evolving from $\vec{v}(-1, .)$ at time $t = -1$, then $\vec{w} = \vec{v}$ on $[-1, \min(0, T^*))$. Moreover, $T^* \leq 0$ as $(0, 0)$ is a singular point of $\vec{v}$. Thus, if a solution $\vec{u}$ exists on $[0, T^*)$ with forcing term $\vec{f}$ and $T^* < +\infty$, another solution $\vec{w}(t - 1, .)$ exists with null forcing term and with a finite existence time (i.e., with a singularity occuring in finite time). But this is impossible, as we have seen it with Theorem 15.5. $\qquad \square$

Chapter 16

Self-similarity and the Leray–Schauder Principle

In this chapter, we give examples of application of the Leray–Schauder principle: existence of large steady-state solutions existence of large self-similar solutions, existence of large discretely self-similar solutions, existence of large time-periodic solutions.

16.1 The Leray–Schauder Principle

The Leray–Schauder principle was introduced by Leray and Schauder [331] as a powerful tool for proving existence of fixed-points for continuous compact (non-linear) operators. In 1927, Schauder extended Brouwer's fixed-point theorem to the case of infinite-dimensional Banach spaces [424], while, in 1933, Leray [327] developed a *continuation method* to prove existence of solutions to integral equations in case where the usual contraction argument did not work (in particular, when uniqueness was not granted). They combined their theories in a new one, by defining what is now known as the Leray–Schauder degree for compact perturbations of the identity. In this chapter, we shall use a simple form of this principle, known as Schaefer's fixed–point theorem [423]:

Schaefer's fixed-point theorem

Theorem 16.1.
Assume that E is a Banach space and that $T : E \mapsto E$ is a continuous mapping. Assume moreover that

- *T is compact: the image $T(e_n)$, $n \in \mathbb{N}$, of any bounded sequence is a relatively compact sequence*

- *There exists a finite M such that, for every $\lambda \in [0,1]$, $e = \lambda T(e) \Rightarrow \|e\|_E \leq M$.*

Then there exists at least one $e \in E$ such that $T(e) = e$.

16.2 Steady-state Solutions

In order to illustrate the use of Schaefer's theorem, we shall first give a simpler example: the case of steady-state solutions. Since the first works of Finn [176] and Ladyzhenskaya

DOI: 10.1201/9781003042594-16

[292, 293], this is a well-documented topic (see the book of Galdi [194]). Here, we shall prove the following theorem:

Steady-state solutions

Theorem 16.2.
Let $\vec{f} \in \dot{H}^{-1}(\mathbb{R}^3)$ and $\nu > 0$. Then there exists at least one vector field $\vec{u} \in \dot{H}^1(\mathbb{R}^3)$ such that:

- $\vec{u}$ *is divergence-free:* div $\vec{u} = 0$

- $\vec{u}$ *is a solution of the steady-state Navier–Stokes equations*

$$\nu \Delta \vec{u} - \sum_{i=1}^{3} u_i \partial_i \vec{u} - \vec{\nabla} p + \vec{f} = 0.$$

Proof. First, we introduce a cut-off function $\theta \in \mathcal{D}(\mathbb{R}^3)$ such that $0 \le \theta \le 1$, $\theta(x) = 1$ for $|x| < 1$ and $\theta(x) = 0$ for $|x| > 2$. We solve the problem

$$\nu \Delta \vec{u}_R - \sum_{i=1}^{3} \theta(\frac{x}{R}) u_{R,i} \partial_i (\theta(\frac{x}{R}) \vec{u}_R) - \vec{\nabla} p_R + \vec{f} = 0 \qquad (16.1)$$

with $\vec{u}_R \in \dot{H}^1$ and div $\vec{u}_R = 0$. This is viewed as a fixed-point problem

$$\vec{u}_R = T_R(\vec{u}_R)$$

with

$$T_R(\vec{h}) = \frac{1}{\nu} \mathbb{P} \frac{1}{\Delta} \left(\sum_{i=1}^{3} \theta(\frac{x}{R}) h_i \partial_i (\theta(\frac{x}{R}) \vec{h}) - \vec{f} \right).$$

We prove the existence of a solution $\vec{u}_R$ by applying the Leray–Schauder principle:

- choice of the Banach space: E is the space of divergence-free vector fields $\vec{h}$ such that $\vec{h} \in \dot{H}^1$

- continuity of T_R: as $\vec{f} \in \dot{H}^{-1}$, we have obviously that $\mathbb{P}\frac{1}{\Delta}\vec{f} \in \dot{H}^1$; we focus on the variable part of T_R: for $\vec{h} \in E$, we have $\vec{h} \in L^6$, hence $\theta(\frac{x}{R})\vec{h} \in L^3$ and $\partial_i(\theta(\frac{x}{R})\vec{h}) \in L^2$, so that $\sum_{i=1}^{3} \theta(\frac{x}{R}) h_i \partial_i(\theta(\frac{x}{R})\vec{h}) \in L^{6/5} \subset \dot{H}^{-1}$ and we get finally that $\mathbb{P}\frac{1}{\Delta}\left(\sum_{i=1}^{3} \theta(\frac{x}{R}) h_i \partial_i(\theta(\frac{x}{R})\vec{h})\right) \in E$.

- compactness of T_R: we write

$$\mathbb{P}\frac{1}{\Delta}\left(\sum_{i=1}^{3} \theta(\frac{x}{R}) h_i \partial_i(\theta(\frac{x}{R})\vec{h}) \right) = \mathbb{P}\frac{1}{\Delta} \text{div } \mathbb{K}_1 + \mathbb{P}\frac{1}{\Delta}\vec{k}_2$$

with

$$\mathbb{K}_1 = \theta(\frac{x}{R})^2 \vec{h} \otimes \vec{h}$$

and

$$\vec{k}_2 = \theta(\frac{x}{R})(\vec{h} \cdot \vec{\nabla}[\theta(\frac{x}{R})])\vec{h}.$$

To control $\mathbb{P}\frac{1}{\Delta}\left(\sum_{i=1}^{3}\theta(\frac{x}{R})h_i\partial_i(\theta(\frac{x}{R})\vec{h})\right)$ in $\dot{H}^1$, we must control $\mathbb{K}_1$ in L^2 and $\vec{k}_2$ in $L^{6/5}$ (as $L^{6/5}\subset\dot{H}^{-1}$). Let us assume that $\sup_{n\in\mathbb{N}}\|\vec{h}_n\|_{\dot{H}^1}<+\infty$; by the Rellich theorem, as $\theta(\frac{x}{4R})\vec{h}_n$ is bounded in H^1 and its support contained in the fixed bounded set $B(0,8R)$, we know that we may find a subsequence that is strongly convergent in L^2, hence (as it is bounded in L^6 and has fixed bounded support) strongly in L^q for every $q\in[1,6)$; the case $q=4$ gives the control of $\mathbb{K}_1$, and the case $q=12/5$ gives the control of $\vec{k}_2$. Thus, T_R is a compact operator.

- a-priori estimate: if $\vec{v}=\lambda T_R(\vec{v})$, where $0\le\lambda\le1$, we have

$$\nu\Delta\vec{v}=\lambda\mathbb{P}(\sum_{i=1}^{3}\theta(\frac{x}{R})v_i\partial_i(\theta(\frac{x}{R})\vec{v})-\vec{f}).$$

Taking the product scalar with $\vec{v}$ (or, more precisely, using the duality bracket between $\dot{H}^1$ and $\dot{H}^{-1}$), we find

$$\nu\|\vec{v}\|_{\dot{H}^1}^2=-\nu\langle\vec{v}|\Delta\vec{v}\rangle$$
$$=\lambda\langle\vec{v}|\vec{f}\rangle-\lambda\int((\theta(\frac{x}{R})\vec{v}).(\vec{v}\cdot\vec{\nabla}(\theta(\frac{x}{R})\vec{v}))\,dx$$
$$=\lambda\langle\vec{v}|\vec{f}\rangle$$
$$\le\lambda\|\vec{v}\|_{\dot{H}^1}\|\vec{f}\|_{\dot{H}^{-1}}$$

so that
$$\|\vec{v}\|_{\dot{H}^1}\le\frac{1}{\nu}\|\vec{f}\|_{\dot{H}^{-1}}.$$

Applying Schaefer's theorem, we find that Equation (16.1) has at least one solution $\vec{u}_R$, with
$$\|\vec{u}_R\|_{\dot{H}^1}\le\frac{1}{\nu}\|\vec{f}\|_{\dot{H}^{-1}}.$$

As $\vec{u}_R$ is bounded in $\dot{H}^1$, hence in L^6, we find that for every $\varphi\in\mathcal{D}$, we have $\sup_{R>0}\|\varphi\vec{u}_R\|_{H^1}<+\infty$. Applying the Rellich theorem (and Cantor's diagonal process), we find that there is a sequence $R_k\to+\infty$ such that $\vec{u}_{R_k}$ converges to a limit $\vec{u}$ strongly in L^q_{loc} for every $q\in[1,6)$. By Banach-Steinhaus, we have that $\vec{u}\in\dot{H}^1$; moreover, we have the convergence in $\mathcal{D}'$ of $\nu\Delta\vec{u}_{R_k}+\vec{f}-\theta(\frac{x}{R_k})\text{div}(\theta(\frac{x}{R_k})\vec{u}_{R_k}\otimes\vec{u}_{R_k})$ to $\nu\Delta\vec{u}+\vec{f}-\text{div}(\vec{u}\otimes\vec{u})$; this gives that $\nu\Delta\vec{u}+\vec{f}-\text{div}(\vec{u}\otimes\vec{u})$ is irrotational, hence there exists a distribution p such that
$$\nu\Delta\vec{u}+\vec{f}-\text{div}(\vec{u}\otimes\vec{u})=\vec{\nabla}p.$$

The existence of $\vec{u}$ is proved. ☐

16.3 The Liouville Problem for Steady Solutions

Let us consider again the problem

$$\nu\Delta\vec{u}-\text{div}(\vec{u}\otimes\vec{u})-\vec{\nabla}p+\vec{f}=0,\text{div}\,\vec{u}=0 \tag{16.2}$$

with $\vec{f} \in \dot{H}^{-1}(\mathbb{R}^3)$. Theorem 16.2 states that there exists a solution $\vec{u} \in \dot{H}^1$ with $\|\vec{u}\|_{\dot{H}^1} \leq \frac{1}{\nu}\|\vec{f}\|_{\dot{H}^{-1}}$. However, the proof does not allow to get a uniqueness result for solutions in $\dot{H}^1$, even when $\vec{f} = 0$. Of course, if $\vec{f} = 0$, the only solution that satisfies $\|\vec{u}\|_{\dot{H}^1} \leq \frac{1}{\nu}\|\vec{f}\|_{\dot{H}^{-1}} = 0$ is $\vec{u} = 0$, but that does not preclude the potential existence of a solution that would violate the energy inequality.

If we assume that moreover $\vec{u} \in L^{9/2} \cap \dot{H}^1$, we find that the solution must obey the energy inequality, as noticed by Kozono, Sohr and Yamazaki [276]:

Proposition 16.1.
If $\vec{u} \in L^{9/2} \cap \dot{H}^1$ of equation (16.2), where $\vec{f} \in \dot{H}^{-1}(\mathbb{R}^3)$, then

$$\nu\|\vec{u}\|_{\dot{H}^1}^2 = \int \vec{u} \cdot \vec{f}\, dx \leq \|\vec{f}\|_{\dot{H}^{-1}}\|\vec{u}\|_{\dot{H}^1}$$

and thus $\|\vec{u}\|_{\dot{H}^1} \leq \frac{1}{\nu}\|\vec{f}\|_{\dot{H}^{-1}}$.

Proof. Let $\theta \in \mathcal{D}(\mathbb{R}^3)$ be a non-negative function which is equal to 1 on $B(0,1)$, and define, for $R > 1$, $\theta_R(x) = \theta(\frac{x}{R})$. We have

$$\nu \int \theta_R |\vec{\nabla} \otimes \vec{u}|^2\, dx = -\nu \int \vec{u} \cdot \sum_{k=1}^{3} \partial_k(\theta_R \partial_k \vec{u})\, dx$$

$$= -\nu \int \theta_R \vec{u} \cdot \Delta \vec{u} - \sum_{k=1}^{3}(\partial_k \theta_R)\partial_k(\frac{|\vec{u}|^2}{2})\, dx$$

$$= \int \theta_R \vec{u} \cdot \vec{f}\, dx + \int (p + \frac{|\vec{u}|^2}{2})\vec{u} \cdot \vec{\nabla}\theta_R\, dx + \nu \int \frac{|\vec{u}|^2}{2}\Delta\theta_R\, dx$$

with

$$\left| \int (p + \frac{|\vec{u}|^2}{2})\vec{u} \cdot \vec{\nabla}\theta_R\, dx \right| \leq C\|\vec{u}\|_{9/2}^2 \|\vec{\nabla}\theta\|_3 \|\mathbb{1}_{|x|>R}\vec{u}\|_{9/2}$$

and

$$\nu\left| \int \theta_R |\vec{\nabla} \otimes \vec{v} \int \frac{|\vec{u}|^2}{2}\Delta\theta_R\, dx \right| \leq \frac{\nu}{2}\|\vec{u}\|_{9/2}^2\|\Delta\theta\|_{9/5} R^{-\frac{1}{3}}.$$

Letting R go to infinity, we find that

$$\nu\|\vec{u}\|_{\dot{H}^1}^2 = \int \vec{u} \cdot \vec{f}\, dx \leq \|\vec{f}\|_{\dot{H}^{-1}}\|\vec{u}\|_{\dot{H}^1}$$

and thus $\|\vec{u}\|_{\dot{H}^1} \leq \frac{1}{\nu}\|\vec{f}\|_{\dot{H}^{-1}}$.   □

What is more interesting is that the energy inequality is a consequence of equations (16.2) without assuming a priori control on the energy norm $\|\vec{u}\|_{\dot{H}^1}$ but only local regularity together with some integrabilty and oscillation requirements. In order to pass from local regularity to global regularity, one uses a Caccioppoli inequality and Bogovskiĭ's formula:

Lemma 16.1 (Caccioppoli's inequality).
Let $R > 0$ and let $\vec{f} \in H^1(B(0,R))$. If, for every $R/2 \leq \rho < r \leq R$, $\vec{f}$ fulfills the inequality

$$\int_{|x|<\rho} |\vec{\nabla} \otimes \vec{f}|^2\, dx \leq \gamma \int_{|x|<r} |\vec{\nabla} \otimes \vec{f}|^2\, dx + \sum_{j=1}^{N} \frac{A_j}{(r-\rho)^{d_j}}$$

where $0 < \gamma < 1$ and $0 \le d_1 < \cdots < d_N$ then

$$\int_{|x|<R/2} |\vec{\nabla} \otimes \vec{f}|^2 \, dx \le C_\gamma \left(\sum_{j=1}^{N} \frac{A_j}{R^{d_j}} \right)$$

where the constant C_γ depends only on γ and on d_N.

Proof. With no loss of generality, we may assume that $N = 1$: we have

$$\int_{|x|<\rho} |\vec{\nabla} \otimes \vec{f}|^2 \, dx \le \gamma \int_{|x|<r} |\vec{\nabla} \otimes \vec{f}|^2 \, dx + \frac{\sum_{j=1}^{N} R^{d_N - d_j} A_j}{(r - \rho)^{d_N}}$$

$$= \gamma \int_{|x|<r} |\vec{\nabla} \otimes \vec{f}|^2 \, dx + \frac{B_N}{(r - \rho)^{d_N}}.$$

Let $\rho_k = 2^{-\frac{1}{k}} R$, for $k \ge 1$, and let $I_k = \int_{|x|<\rho_k} |\vec{\nabla} \otimes \vec{f}|^2 \, dx$. We have

$$2^{-\frac{1}{k+1}} - 2^{-\frac{1}{k}} = 2^{-\frac{1}{k}} (2^{\frac{1}{k} - \frac{1}{k+1}} - 1) \sim \frac{1}{k^2}$$

when k goes to $+\infty$. Thus, we have $\rho_{k+1} - \rho_k \ge \frac{R}{c_0} \frac{1}{k^2}$, where the constant c_0 is positive. This gives

$$I_k \le \gamma I_{k+1} + \frac{B_N (c_0 k^2)^{d_N}}{R^{d_N}}.$$

By induction, we find

$$I_1 \le \gamma^{n+1} I_{n+1} + \frac{B_N c_0^{d_N}}{R^{d_N}} \sum_{k=1}^{n} k^{2d_N} \gamma^{k-1}.$$

We have $\lim_{n \to +\infty} \gamma^{n+1} I_{n+1} = 0$, as $I_{n+1} \le \int_{|x|<R} |\vec{\nabla} \otimes \vec{f}|^2 \, dx < +\infty$ and $0 < \gamma < 1$. Thus, the lemma is proved with

$$C_0 = c_0^{d_N} \sum_{k=1}^{+\infty} k^{2d_N} \gamma^{k-1}. \qquad \square$$

We recall Bogovskiĭ's formula [48] and its properties (proved in Galdi's book [194]).

Lemma 16.2 (Bogovskiĭ's formula).
Let $\theta \in \mathcal{D}(\mathbb{R}^3)$ such that θ is supported in the ball $B(0, 1/2)$. For $f \in \mathcal{D}(B(0,1))$ such that $\int_{B(0,1)} f \, dx = 0$, define

$$\mathcal{B}(f)(x) = \int_{B(0,1)} f(y)(x - y) \int_1^{+\infty} \theta(y + r(x - y)) r^2 \, dr \, dy. \qquad (16.3)$$

Then, for $1 < p < +\infty$, $\mathcal{B}(f)$ belongs to $W_0^{1,p}((B(0,1))$ [the closure of $\mathcal{D}(B(0,1))$ for the norm $\|\vec{\nabla} f\|_{L^p(B(0,1))}$] and we have

$$\operatorname{div} \mathcal{B}(f) = f$$

and

$$\|\vec{\nabla} \mathcal{B}(f)\|_p \le C_p \|f\|_p.$$

Proof. If f is supported in $\overline{B(0,R)}$ with $R < 1$, then the integrand

$$F(y)\theta(y + r(x - y))$$

is equal to 0 unless $|y| < R$ and $|y + r(x - y)| \leq 1/2$, so that we have $|x| = |\frac{r-1}{r}y + \frac{1}{r}(y + r(x - y))| \leq \max(R, 1/2)$. Thus, $\mathcal{B}(f)$ is supported in the ball $\overline{B(0, \max(R, 1/2))}$.

Writing

$$\mathcal{B}(f)(x) = \int f(x - z)z \int_1^{+\infty} \theta(x + (r - 1)z)r^2 \, dr \, dz,$$

wee se that $\mathcal{B}(f)$ is C^∞, as, for $|x| < 1$, $\alpha, \beta \in \mathbb{N}^3$,

$$\int |\partial^\alpha f(x - z)||z| \int_1^{+\infty} |\partial^\beta \theta(x + (r - 1)z)|r^2 \, dr \, dz$$

$$\leq \frac{5}{2}\|\partial^\alpha f\|_\infty\|\partial^\beta \theta\|_\infty \int_{B(0,2)} \int_1^{\frac{5}{2|z|}} r \, dr \, dz < +\infty.$$

In particular, if $g_{i,j} = \partial_i(\mathcal{B}(f)_j)$, we have

$$g_{i,j} = \lim_{\epsilon \to 0} \int_{|z|>\epsilon} \partial_i f(x - z)z_j \int_1^{+\infty} \theta(x + (r - 1)z)r^2 \, dr \, dz$$

$$+ \int_{|z|>\epsilon} f(x - z)z_j \int_1^{+\infty} \partial_i\theta(x + (r - 1)z)r^2 \, dr \, dz$$

$$= \lim_{\epsilon \to 0} \epsilon^3 \int_{S^2} f(x - \epsilon\sigma)\sigma_i\sigma_j \int_1^{+\infty} \theta(x + (r - 1)\epsilon\sigma)r^2 \, dr \, d\sigma$$

$$+ \int_{|z|>\epsilon} f(x - z)\delta_{i,j} \int_1^{+\infty} \theta(x + (r - 1)z)r^2 \, dr \, dz$$

$$+ \int_{|z|>\epsilon} f(x - z)z_j \int_1^{+\infty} \partial_i\theta(x + (r - 1)z)(r - 1)r^2 \, dr \, dz$$

$$+ \int_{|z|>\epsilon} f(x - z)z_j \int_1^{+\infty} \partial_i\theta(x + (r - 1)z)r^2 \, dr \, dz$$

$$= f(x) \int \frac{z_i z_j}{|z|^2}\theta(x + z) \, dz + \int f(x - z)A_{i,j}(x, z) \, dz$$

$$+ \lim_{\epsilon \to 0} \int_{|z|>\epsilon} f(x - z)K_{i,j}(x, z) \, dz$$

with

$$A_{i,j}(x, z) = \delta_{i,j} \int_0^{+\infty} \theta(x + \rho z)(1 + 2\rho) \, d\rho + z_j \int_0^{+\infty} \partial_i\theta((x + \rho z)(3\rho^2 + 3\rho + 1) \, d\rho$$

and

$$K_{i,j}(x, z) = \delta_{i,j} \int_0^{+\infty} \theta(x + \rho z)\rho^2 \, d\rho + z_j \int_0^{+\infty} \partial_i\theta((x + \rho z)\rho^3 \, d\rho.$$

For $x \in B(0, 1)$, we have

$$\left| \int \frac{z_i z_j}{|z|^2}\theta(x + z) \, dz \right| \leq \|\theta\|_1, \quad |A_{i,j}(x, z)| \leq C\|\vec{\nabla}\theta\|_\infty \mathbb{1}_{|z|\leq 5/2}\frac{1}{|z|^2}.$$

Thus, the map

$$f \in L^p(B(0,1)) \mapsto f(x) \int \frac{z_i z_j}{|z|^2} \theta(x+z)\, dz + \int f(x-z) A_{i,j}(x,z)\, dz$$

is bounded from $L^p(B(0,1))$ to $L^p(B(0,1))$ for $1 \le p \le +\infty$. Moreover, the kernel $K_{i,j}$ fullfills the assumptions of Calderón and Zygmund's theorem on singular integrals (Theorem 2 in [76]):

a) for every $\lambda > 0$, $K_{i,j}(x, \lambda z) = \lambda^{-n} K_{i,j}(x,z)$

b) $\sup_{x \in B(0,1)} \sup_{|z|=1} |K_{i,j}(x,z)| < +\infty$

c) for every $x \in B(0,1)$, we have $\int_{S^2} K_{i,j}(x,\sigma)\, d\sigma = 0$: just write

$$\int_{S^2} K_{i,j}(x,\sigma)\, d\sigma = \delta_{i,j} \int \theta(x+z)\, dz + \int z_j \partial_i \theta(x+z)\, dz = 0.$$

Thus, the map

$$f \mapsto \lim_{\epsilon \to 0} \int_{|x-y|>\epsilon} f(y) K_{i,j}(x, x-y)\, dy$$

is bounded from $L^p(B(0,1))$ to $L^p(B(0,1))$ for $1 < p < +\infty$.
Finally, we have

$$\operatorname{div} \mathcal{B}(f)(x) = \sum_{i=1}^{3} \lim_{\epsilon \to 0} \Big(\epsilon^3 \int_{S^2} f(x - \epsilon\sigma) \sigma_i^2 \int_1^{+\infty} \theta(x + (r-1)\epsilon\sigma) r^2\, dr\, d\sigma$$

$$+ \int_{|z|>\epsilon} f(x-z) \int_1^{+\infty} \theta(x+(r-1)z) r^2\, dr\, dz$$

$$+ \int_{|z|>\epsilon} f(x-z) z_i \int_1^{+\infty} \partial_i \theta(x+(r-1)z) r^3\, dr\, dz \Big)$$

$$= f(x) \int \sum_{i=1}^{3} \frac{z_i^2}{|z|^2} \theta(x+z)\, dz$$

$$+ \lim_{\epsilon \to 0} \int_{|z|>\epsilon} f(x-z) \int_1^{+\infty} \partial_r (\theta(x+(r-1)z) r^3)\, dr\, dz$$

$$= f(x) - \int f(x-z)\, dz = f(x).$$

The lemma is proved. ∎

Corollary 16.1.
Let $R > 0$ and $1 < p < +\infty$. For $f \in L^p(B(0,1))$ such that $\int_{B(0,1)} f\, dx = 0$, there exists $\vec{F} \in W_0^{1,p}((B(0,R))$ such that

$$\operatorname{div} \vec{F} = f$$

and

$$\|\vec{\nabla} \vec{F}\|_p \le C_p \|f\|_p.$$

[The constant C_p does not depend on R.]

With those two lemmas, we are going to prove the following energy inequality:

Theorem 16.3.

Let $\vec{u} \in L^2_{\mathrm{loc}}(\mathbb{R}^3)$ be a solution to the problem

$$\nu \Delta \vec{u} - \operatorname{div}(\vec{u} \otimes \vec{u}) - \vec{\nabla} p + \vec{f} = 0, \ \operatorname{div} \vec{u} = 0 \qquad (16.4)$$

with $\vec{f} \in \dot{H}^{-1}(\mathbb{R}^3)$. Assume moreover that $\vec{u}$ fullfills one of the following assumptions:

   a) $\vec{u} \in L^p$ for some p with $3 \leq p \leq \frac{9}{2}$

   b) $\vec{u} \in L^p$ for some p with $3/2 < p < 3$ and $\vec{u} \in H^1_{\mathrm{loc}}(\mathbb{R}^3)$

   c) $\vec{u} \in L^p$ for some p with $\frac{9}{2} < p \leq 6$ and there exists $\vec{\Omega}$ such that $\vec{u} = \vec{\nabla} \wedge \vec{\Omega}$ with
      $\operatorname{div} \vec{\Omega} = 0$ and

$$\sup_{R>1} \frac{1}{R} \|\vec{\Omega} - m_{B(0,R)}\vec{\Omega}\|_{L^q(B(0,R))} < +\infty,$$

   where $\frac{1}{q} = \frac{1}{2} - \frac{1}{q}$ and where $m_{B(0,R)}\vec{\Omega} = \frac{1}{|B(0,R)|} \int_{B(0,R)} \vec{\Omega} \, dx$.

Then $\vec{u} \in \dot{H}^1(\mathbb{R}^3)$,

$$\nu \|\vec{u}\|^2_{\dot{H}^1} = \int \vec{u} \cdot \vec{f} \, dx$$

and thus $\|\vec{u}\|_{\dot{H}^1} \leq \frac{1}{\nu} \|\vec{f}\|_{\dot{H}^{-1}}$.

Proof.

Local regularity of $\vec{u}$.

Under the assumptions of Theorem 16.3, we can see that $H^1_{\mathrm{loc}}(\mathbb{R}^3)$. We may assume of course that $p \geq 3$ (in the case b), we assume that $\vec{u}$ is regular). If $\vec{u} \in L^p$ with $3 \leq p \leq 6$, we define $\vec{v}(t,x) = \vec{u}(x)$. We have $\vec{v} \in \mathcal{C}([0,T], L^p)$ and $\vec{v}$ is a solution of

$$\begin{cases} \partial_t \vec{v} = \nu \Delta \vec{v} - \operatorname{div}(\vec{v} \otimes \vec{v}) - \vec{\nabla} p + \vec{f} \\ \operatorname{div} \vec{v} = 0 \\ \vec{v}(0,.) = \vec{u} \end{cases} \qquad (16.5)$$

Using the embedding $L^p \subset L^3 + L^6$ and using the density of $L^3 \cap L^6$ in L^3, we decompose $\vec{u}$ in $\vec{u}_1 + \vec{u}_2$, hence $\vec{v}$ in $\vec{v}_1 + \vec{v}_2$ with $\vec{v}_1$ small in $L^\infty((0,T), L^{3,\infty})$ (controlled by $\|\vec{u}_1\|_3$) and, for T small enough, $t^{1/4}\vec{v}_2$ small in $L^\infty((0,T), L^6)$ (controlled by $T^{1/4}\|\vec{u}_2\|_6$). If $\vec{w} = \vec{w}_1 + \vec{w}_2$ is another solution of this problem with $\vec{w}_1$ small in $L^\infty((0,T), L^{3,\infty})$ and $t^{1/4}\vec{w}_2$ small in $L^\infty((0,T), L^6)$, we find that $\vec{z} = \vec{v} - \vec{w}$ is a solution of

$$\vec{z} = -\int_0^t W_{\nu(t-s)} * \mathbb{P} \operatorname{div}(\vec{z} \otimes \vec{v} + \vec{w} \otimes \vec{z}) \, ds.$$

But the linear mapping

$$\vec{z} \mapsto \mathcal{A}(\vec{z}) = \int_0^t W_{\nu(t-s)} * \mathbb{P} \operatorname{div}(\vec{z} \otimes \vec{v} + \vec{w} \otimes \vec{z}) \, ds$$

is a contraction on the space

$$E_T = L^\infty((0,T), L^{3,\infty}) + \{\vec{z}/\ t^{-1/4}\vec{z} \in L^\infty((0,T), L^6)\}$$

endowed with the norm

$$\|\vec{z}\|_{E_T} = \min_{\vec{z}=\vec{z}_1+\vec{z}_2} \left(\sup_{0<s<t} \|\vec{z}_1(t,.)\|_{L^{3,\infty}} + \sup_{0<s<t} (\nu t)^{1/4}\|\vec{z}_2(s,.)\|_6 \right):$$

if $\vec{z} = \vec{z}_1 + \vec{z}_2$ with $\vec{z}_1$ in $L^\infty((0,T), L^{3,\infty})$ and $t^{1/4}\vec{z}_2$ in $L^\infty((0,T), L^6)$, we find that $\mathcal{A}(\vec{z})$ in $L^\infty((0,T), L^{3,\infty})$ with

$$\|\mathcal{A}(\vec{z})(t,.)\|_{L^{3,\infty}} \leq \| \int_0^t W_{\nu(t-s)} * \mathbb{P}\operatorname{div}(\vec{z}_1 \otimes \vec{v}_1 + \vec{w}_1 \otimes \vec{z}_1)\,ds\|_{L^{3,\infty}}$$

$$+ \int_0^t \|W_{\nu(t-s)} * \mathbb{P}\operatorname{div}(\vec{z}_1 \otimes \vec{v}_2 + \vec{w}_2 \otimes \vec{z}_1)\|_{L^{3,\infty}}\,ds$$

$$+ \int_0^t \|W_{\nu(t-s)} * \mathbb{P}\operatorname{div}(\vec{z}_2 \otimes \vec{v}_1 + \vec{w}_1 \otimes \vec{z}_2)|_{L^{3,\infty}}\,ds$$

$$+ \int_0^t \|W_{\nu(t-s)} * \mathbb{P}\operatorname{div}(\vec{z}_2 \otimes \vec{v}_2 + \vec{w}_2 \otimes \vec{z}_2)\|_{L^{3,\infty}}\,ds$$

$$\leq \frac{C}{\nu} \sup_{0<s<t} \|\vec{z}_1(s,.)\|_{L^{3,\infty}} (\sup_{0<s<t} \|\vec{v}_1(s,.)\|_{L^{3,\infty}} + \sup_{0<s<t} \|\vec{w}_1(s,.)\|_{L^{3,\infty}}))$$

$$+ C \int_0^t \frac{1}{(\nu(t-s))^{3/4}} \frac{1}{(\nu s)^{1/4}} \|\vec{z}_1(s,.)\|_{L^{3,\infty}} \times$$
$$\times ((\nu s)^{1/4}\|\vec{v}_2(s,.)\|_6 + (\nu s)^{1/4}\|\vec{w}_2(s,.)\|_6)\,ds$$

$$+ C \int_0^t \frac{1}{(\nu(t-s))^{3/4}} \frac{1}{(\nu s)^{1/4}} (\nu s)^{1/4}\|\vec{z}_2(s,.)\|_6 \times$$
$$\times (\|\vec{v}_1(s,.)\|_{L^{3,\infty}} + \|\vec{w}_1(s,.)\|_{L^{3,\infty}})\,ds$$

$$+ C \int_0^t \frac{1}{(\nu(t-s))^{1/2}} \frac{1}{(\nu s)^{1/2}} (\nu s)^{1/4}\|\vec{z}_2(s,.)\|_6 \times$$
$$\times ((\nu s)^{1/4}\|\vec{v}_2(s,.)\|_6 + (\nu s)^{1/4}\|\vec{w}_2(s,.)\|_6)\,ds$$

so that

$$\|\mathcal{A}(\vec{z})\|_{E_T} \leq \frac{C}{\nu}\|\vec{z}\|_{E_T}(\|\vec{v}\|_{E_T} + \|\vec{w}\|_{E_T}).$$

Thus, for T small enough, $\vec{z} = 0$ on $(0,T)$, hence $\vec{v} = \vec{w}$.

On the other hand, the bilinear operator

$$B(\vec{z}, \vec{w}) = \int_0^t W_{\nu(t-s)} * \mathbb{P}\operatorname{div}(\vec{z} \otimes \vec{w})\,ds$$

is bounded on the space

$$F_T = \{\vec{z}/\ t^{-1/4}\vec{z} \in L^\infty((0,T), L^6)\}$$

endowed with the norm

$$\|\vec{z}\|_{F_T} = \sup_{0<s<t} (\nu t)^{1/4}\|\vec{z}(s,.)\|_6 :$$

$$\|B(\vec{z}, \vec{w})(t, .)\|_6 \leq \int_0^t \|W_{\nu(t-s)} * \mathbb{P}\operatorname{div}(\vec{z} \otimes \vec{w})\|_6 \, ds$$

$$\leq C \int_0^t \frac{1}{(\nu(t-s))^{3/4}} \frac{1}{(\nu s)^{1/2}} (\nu s)^{1/4} \|\vec{z}(s, .)\|_6 (\nu s)^{1/4} \|\vec{w}(s, .)\|_6 \, ds$$

so that

$$\|B(\vec{z}, \vec{w})\|_{F_T} \leq \frac{C_0}{\nu} \|\vec{z}\|_{F_T} \|\vec{w}\|_{F_T}.$$

The constant C_0 does not depend on T. Moreover,

$$W_{\nu t} * \vec{u} + \int_0^t W_{\nu(t-s)} * \mathbb{P}\vec{f} \, ds$$

is small in F_T (for small T): we have $(\nu t)^{1/4}\|W_{\nu t} * \vec{u}_1\|_6 \leq C\|\vec{u}_1\|_3$, $(\nu t)^{1/4}\|W_{\nu t} * \vec{u}_2\|_6 \leq (\nu T)^{1/4}\|\vec{u}_2\|_6$ and, since

$$\int_0^t W_{\nu(t-s)} * \mathbb{P}\vec{f} \, ds = (Id - e^{\nu t \Delta})\mathbb{P}\frac{1}{\nu \Delta}\vec{f},$$

$(\nu t)^{1/4}\| \int_0^t W_{\nu(t-s)} * \mathbb{P}\vec{f} \, ds\|_6 \leq C(\nu T)^{1/4}\|\frac{1}{\nu \Delta}\vec{f}\|_6 \leq C'(\nu T)^{1/4}\frac{\|\vec{f}\|_{\dot{H}^{-1}}}{\nu}$. Thus, the problem (16.5) has a small solution $\vec{w}$ in F_T, hence in E_T. By uniqueness of small solutions in E_T, we find that $\vec{w} = \vec{v}$. Thus, writing $\vec{u}(x) = \vec{w}(T/2, x)$, we find that $\vec{u} \in L^6$.

We then write $\partial_j \vec{u} = \mathbb{P}\frac{1}{\Delta}\partial_j \operatorname{div}(\vec{u} \otimes \vec{u}) - \mathbb{P}\frac{1}{\Delta}\partial_j \vec{f} \in L^3 + L^2$, so that $\vec{u} \in H^1_{loc}$.

The energy balance on $B(0, R)$

Let $0 < R/2 \leq \rho < r < R$. Let $\theta_{\rho,r} \in \mathcal{D}(\mathbb{R}^3)$ be a function which is equal to 1 on $B(0, \rho)$ and to 0 on $\mathbb{R}^3 \setminus B(0, r)$, such that $0 \leq \theta_{\rho,r} \leq 1$ and $|\vec{\nabla}\theta_{\rho,r}| \leq C_0 \frac{1}{r-\rho}$ and $|\partial_i \partial_j \theta_{\rho,r}| \leq C_0 \frac{1}{(r-\rho)^2}$ (where C_0 will not depend on ρ nor r). We define $< \vec{u} >_r = \frac{1}{|B(0,r)|} \int_{B(0,r)} \vec{u} \, dx$ the mean value of $\vec{u}$ on the ball $B(0, r)$, and

$$\vec{u}_r = \vec{u} - < \vec{u} >_r .$$

Finally, we define $\vec{w}_{\rho,r} \in W_0^{1,p}(B(0, r))$ with

$$\operatorname{div} \vec{w}_{\rho,r} = \operatorname{div}(\theta_{\rho,r}\vec{u}_r) = \vec{u}_r \cdot \vec{\nabla}\theta_{\rho,r},$$

as provided by Corollary 16.1.

We rewrite equation (16.4) in the following way

$$\nu\Delta\vec{u}_r - \operatorname{div}(\vec{u} \otimes \vec{u}_r) - \vec{\nabla}p + \vec{f} = 0$$

and we take the scalar product of the equation with $\theta_{\rho,r}\vec{u}_r - \vec{w}_{\rho,r}$. By hypoellipticity of the Laplacian, we find that p is locally L^2, while $\theta_{\rho,r}\vec{u}_r - \vec{w}_{\rho,r}$ belongs to $W_0^{1,2}(B(0, r))$. We find

$$\nu \int \theta_{\rho,r}|\vec{\nabla} \otimes \vec{u}|^2 \, dx = -\nu \sum_{k=1}^3 \int (\partial_k\theta_{\rho,r})\vec{u}_r \cdot \partial_k\vec{u} \, dx + \nu \sum_{k=1}^3 \int \partial_k\vec{w}_{\rho,r} \cdot \partial_k\vec{u} \, dx$$

$$+ \frac{1}{2} \int |\vec{u}_r|^2 \vec{u} \cdot \vec{\nabla}\theta_{\rho,r} \, dx - \int \vec{u}_r \cdot (\vec{u} \cdot \vec{\nabla}\vec{w}_{\rho,r}) \, dx$$

$$+ \int \theta_{\rho,r}\vec{f} \cdot \vec{u}_r \, dx - \int \vec{w}_{\rho,r} \cdot \vec{f} \, dx$$

$$= I + II + III + IV + V + VI.$$

We have

$$\nu \int_{|x|<\rho} |\vec{\nabla} \otimes \vec{u}|^2 \, dx \leq \nu \int \theta_{\rho,r} |\vec{\nabla} \otimes \vec{u}|^2 \, dx = I + II + III + IV + V + VI.$$

The control of I, II, V and VI is easy:

a) We have

$$I \leq C \frac{\nu}{r - \rho} \|\vec{\nabla} \otimes \vec{u}\|_{L^2(B(0,r))} \|\vec{u}_r\|_{L^2(B(0,r)\setminus B(0,\rho))}.$$

If $2 \leq p < 6$, we write

$$\|\vec{u}_r\|_{L^2(B(0,r)\setminus B(0,\rho))} \leq C \|\vec{u}\|_p (r - \rho)^{\frac{3}{2} - \frac{3}{p}}$$

so that

$$I \leq \frac{\nu}{8} \int_{|x|<r} |\vec{\nabla} \otimes \vec{u}|^2 \, dx + C\nu \|\vec{u}\|_p^2 \frac{1}{(r - \rho)^{\frac{6-p}{p}}}.$$

If $p = 6$, we write

$$I = \nu \int (\Delta \theta_{\rho,r}) \vec{u} \cdot \vec{u}_r \, dx + \nu \sum_{k=1}^{3} \int (\partial_k \theta_{\rho,r}) \vec{u} \cdot \vec{\nabla} u \, dx$$

$$\leq C\nu \|\vec{u}\|_6 \|\vec{u} \mathbb{1}_{|x|>R/2}\|_6 + C\nu \|\vec{\nabla} \otimes \vec{u}\|_{L^2(B(0,r))} \|\vec{u} \mathbb{1}_{|x|>R/2}\|_6$$

so that

$$I \leq \frac{\nu}{8} \int_{|x|<r} |\vec{\nabla} \otimes \vec{u}|^2 \, dx + C\nu \|\vec{u}\|_6 \|\vec{u} \mathbb{1}_{|x|>R/2}\|_6.$$

If $p < 2$, we write $\frac{1}{2} = \lambda \frac{1}{p} + (1 - \lambda)\frac{1}{6}$ with $0 < \lambda < 1$, so that $\|\vec{u}_r\|_{L^2(B(0,r)\setminus B(0,\rho))} \leq C\|\vec{u}\|_p^\lambda \|\vec{\nabla} \otimes \vec{u}\|_{L^2(B(0,r))}^{1-\lambda}$ and we get

$$I \leq \frac{\nu}{8} \int_{|x|<r} |\vec{\nabla} \otimes \vec{u}|^2 \, dx + C\nu \|\vec{u}\|_p^2 \frac{1}{(r - \rho)^{\frac{2}{\lambda}}}$$

$$= \frac{\nu}{8} \int_{|x|<r} |\vec{\nabla} \otimes \vec{u}|^2 \, dx + C\nu \|\vec{u}\|_p^2 \frac{1}{(r - \rho)^{\frac{6-p}{p}}}.$$

b) We have

$$II \leq \nu \|\|\vec{\nabla} \otimes \vec{u}\|_{L^2(B(0,r))} \|\vec{\nabla} \otimes \vec{w}_{\rho,r}\|_2$$

$$\leq C\nu \|\|\vec{\nabla} \otimes \vec{u}\|_{L^2(B(0,r))} \|\vec{u}_r\|_{L^2(B(0,r)\setminus B(0,\rho))} \frac{1}{r - \rho}$$

and thus II is controlled in the same way as I.

c) We have

$$V \leq \|\vec{f}\|_{\dot{H}^{-1}} \|\vec{\nabla} \otimes (\theta_{\rho,r} \vec{u}_r)\|_2 \leq \|\vec{f}\|_{\dot{H}^{-1}} (\|\vec{\nabla} \otimes \vec{u}\|_{L^2(B(0,r))} + C \frac{1}{r - \rho} \|\vec{u}_r\|_{L^2(B(0,r)\setminus B(0,\rho))})$$

and we get

$$V \leq \frac{\nu}{8} \int_{|x|<r} |\vec{\nabla} \otimes \vec{u}|^2 \, dx + \frac{C}{\nu} \|\vec{f}\|_{\dot{H}^{-1}}^2 + C_\nu \|\vec{u}\|_p^2 \frac{1}{(r - \rho)^{\frac{6-p}{p}}}.$$

d) We have

$$VI \leq \|\vec{f}\|_{\dot{H}^{-1}} \|\vec{\nabla} \otimes \vec{w}_{\rho,r})\|_2 \leq C\|\vec{f}\|_{\dot{H}^{-1}} \frac{1}{r-\rho} \|\vec{u}_r\|_{L^2(B(0,r)\backslash B(0,\rho))}$$

and we get

$$VI \leq \frac{\nu}{8} \int_{|x|<r} |\vec{\nabla} \otimes \vec{u}|^2 \, dx + \frac{C}{\nu} \|\vec{f}\|^2_{\dot{H}^{-1}} + C_\nu \|\vec{u}\|^2_p \frac{1}{(r-\rho)^{\frac{6-p}{p}}}.$$

The case $3 \leq p \leq 9/2$

If $3 \leq p \leq 9/2$, we write

$$III \leq C \frac{1}{r-\rho} \int_{B(0,r)} |\vec{u}_r|^2 |\vec{u}| \, dx \leq C \frac{1}{r-\rho} \|\vec{u}\|^3_p R^{3-\frac{9}{p}}$$

and similarly

$$IV \leq C \int_{B(0,r)} |\vec{u}_r| |\vec{u}| |\vec{\nabla} \otimes \vec{w}_{\rho,r}| \, dx \leq C \frac{1}{r-\rho} \|\vec{u}\|^3_p R^{3-\frac{9}{p}}.$$

The Caccioppoli inequality then gives us that

$$\nu \int_{|x|<R/2} |\vec{\nabla} \otimes \vec{u}|^2 \, dx \leq C_\nu (\|\vec{u}\|^2_p \frac{1}{R^{\frac{6-p}{p}}} + \|\vec{f}\|^2_{\dot{H}^{-1}} + \frac{1}{R^{\frac{9}{p}-2}} \|\vec{u}\|^3_p).$$

Letting R go to $+\infty$, we find that $\vec{u} \in \dot{H}^1$, so that $\vec{u} \in L^p \cap \dot{H}^1 \subset L^{9/2} \cap \dot{H}^1$ and we conclude by applying Proposition 16.1.

The case $3/2 < p < 3$

If $3/2 < p < 3$, we write

$$III \leq C \frac{1}{r-\rho} \int_{B(0,r)} |\vec{u}_r|^2 |\vec{u}| \, dx \leq C \frac{1}{r-\rho} \|\vec{u}\|_p \|\vec{u}_r\|^2_{\frac{2p}{p-1}}$$

and similarly

$$IV \leq C \int_{B(0,r)} |\vec{u}_r| |\vec{u}| |\vec{\nabla} \otimes \vec{w}_{\rho,r}| \, dx \leq C \frac{1}{r-\rho} \|\vec{u}\|_p \|\vec{u}_r\|^2_{\frac{2p}{p-1}}.$$

Writing $\frac{1}{2} - \frac{1}{2p} = \lambda \frac{1}{6} + (1-\lambda)\frac{1}{p}$ with $0 < \lambda < 1$, we find (since $\|\vec{u}_r\|_6 \leq C\|\vec{\nabla} \otimes \vec{u}\|_{L^2(B(0,r))}$)

$$III + IV \leq \frac{1}{r-\rho} \|\vec{u}\|^{3-2\lambda}_p \|\vec{\nabla} \otimes \vec{u}\|^{2\lambda}_{L^2(B(0,r))}$$

$$\leq \frac{\nu}{8} \int_{|x|<r} |\vec{\nabla} \otimes \vec{u}|^2 \, dx + C_\nu \frac{1}{(r-\rho)^{\frac{1}{1-\lambda}}} \|\vec{u}\|^{\frac{3-2\lambda}{1-\lambda}}_p$$

$$= \frac{\nu}{8} \int_{|x|<r} |\vec{\nabla} \otimes \vec{u}|^2 \, dx + C_\nu \frac{1}{(r-\rho)^{\frac{6-p}{2p-3}}} \|\vec{u}\|^{\frac{3p}{2p-3}}_p.$$

The Caccioppoli inequality then gives us that

$$\nu \int_{|x|<R/2} |\vec{\nabla} \otimes \vec{u}|^2 \, dx \leq C_\nu (\|\vec{u}\|^2_p \frac{1}{R^{\frac{6-p}{p}}} + \|\vec{f}\|^2_{\dot{H}^{-1}} + \frac{1}{R^{\frac{6-p}{2p-3}}} \|\vec{u}\|^{\frac{3p}{2p-3}}_p).$$

Letting R go to $+\infty$, we find that $\vec{u} \in \dot{H}^1$, so that $\vec{u} \in L^p \cap \dot{H}^1 \subset L^{9/2} \cap \dot{H}^1$ and we conclude by applying again Proposition 16.1.

The case $3 < p \leq 6$

We want to estimate, for g and h in $W_0^{1,2}(B(0,r))$ the integral

$$A_0(g,h) = \int g\vec{u} \cdot \vec{\nabla} h \, dx.$$

This is a continuous bilinear form on $W_0^{1,2}(B(0,r))$. If g and h are smooth, we may write $\vec{u} = \vec{\nabla} \wedge \vec{\Omega}_r$, where $\Omega_r = \vec{\Omega} - m_{B(0,r)}\vec{\Omega}$ and integrate by parts:

$$A_0(g,h) = \int \vec{\Omega}_r \cdot \vec{\nabla} \wedge (g\vec{\nabla} h) \, dx = \int \vec{\Omega}_r \cdot ((\vec{\nabla} g) \wedge \vec{\nabla} h) \, dx = A_1(g,h)$$

where A_1 is again a continuous bilinear form on $W_0^{1,2}(B(0,r))$. Thus, the identity is valid for every $f, g \in W_0^{1,2}(B(0,r))$. This gives

$$III = \int \vec{\Omega}_r \cdot \left(\vec{\nabla}(\frac{|\vec{u}_r|^2}{2}) \wedge \vec{\nabla}\theta_{\rho,r} \right) dx$$

and

$$IV = -\int \vec{\Omega}_r \cdot \left(\sum_{j=1}^{3}(\vec{\nabla} u_{r,j}) \wedge \vec{\nabla} w_{\rho,r,j} \right) dx$$

. Recall that $\frac{1}{q} = \frac{1}{2} - \frac{1}{q}$, so that

$$III + IV \leq C\|\vec{\Omega}_r\|_{L^q(B(0,r))} \frac{1}{r-\rho}\|\vec{u}_r\|_{L^p(B(0,r)\setminus B(0,\rho))}\|\vec{\nabla} \otimes \vec{u}\|_{L^2(B(0,r))}$$

$$\leq \frac{\nu}{8}\int_{|x|<r}|\vec{\nabla} \otimes \vec{u}|^2 \, dx + C'\frac{1}{\nu}\frac{R^2}{(r-\rho)^2}\|\vec{u}\|_p^2 \frac{\|\vec{\Omega}_r\|_{L^q(B(0,r))}^2}{r^2}.$$

The Caccioppoli inequality then gives us that

$$\nu \int_{|x|<R/2}|\vec{\nabla} \otimes \vec{u}|^2 \, dx \leq C_\nu (\|\vec{u}\|_p^2 \frac{1}{R^{\frac{6-p}{p}}} + \|\vec{f}\|_{\dot{H}^{-1}}^2 + \|\vec{u}\|_p^2 \sup_{r>1}\frac{\|\vec{\Omega}_r\|_{L^q(B(0,r))}^2}{r^2}).$$

Letting R go to $+\infty$, we find that $\vec{u} \in \dot{H}^1$.

We now define $\vec{W}_R$ by $\vec{W}_R \in W_0^{1,p}$, $\text{div } \vec{W}_R = \vec{u} \cdot \vec{\nabla}\theta_{R/2,R}$ and we take the scalar product between

$$\nu\Delta\vec{u} - \text{div}(\vec{u} \otimes \vec{u}) - \vec{\nabla} p + \vec{f} = 0$$

and $\theta_{R/2,R}\vec{u} - \vec{W}_R$. We obtain

$$\nu \int \theta_{R/2,R}|\vec{\nabla} \otimes \vec{u}|^2 \, dx = J + JJ + JJJ$$

with

$$J = -\nu\sum_{k=1}^{3}\int (\partial_k\theta_{R/2,R})\vec{u} \cdot \partial_k\vec{u} \, dx + \nu\sum_{k=1}^{3}\int \partial_k\vec{W}_R \cdot \partial_k\vec{u} \, dx$$

$$JJ = \int \vec{u} \cdot (\vec{u} \cdot \vec{\nabla}(\theta_{R/2,R}\vec{u})) \, dx - \int \vec{u} \cdot (\vec{u} \cdot \vec{\nabla}\vec{W}_R) \, dx$$

$$JJJ = \int \theta_{R/2,R}\vec{f} \cdot \vec{u} \, dx - \int \vec{W}_R \cdot \vec{f} \, dx.$$

As $\vec{u} \in L^6$, we have that

$$|J| \leq C\nu \|\vec{u}\|_{\dot{H}^1} \|\vec{u}1_{|x|>R/2}\|_6.$$

For JJ, we remark that the identity

$$A_0(g,h) = A_1(g,h)$$

is still valid for $g \in W^{1,2}(B(0,R))$ and $h \in W_0^{1,2}(B(0,R))$ (i.e. g is no longer assumed to vanish at the boundary). We obtain

$$|JJ| \leq C\|\vec{\Omega}_R\|_{L^q(B(0,R))} \frac{1}{R} \|\vec{u}1_{|x|>R/2}\|_p \|\vec{u}\|_{\dot{H}^1}$$

Writing $\vec{f} = \sum_{j=1}^3 \partial_j \vec{f}_j$ with $\vec{f}_j \in L^2$, we see that

$$JJJ = -\int \theta_{R/2,R} \sum_{j=1}^3 \vec{f}_j \cdot \partial_j \vec{u} \, dx - \int \sum_{j=1}^3 \vec{f}_j \cdot ((\partial_j \theta_{R/2,R})\vec{u} - \partial_j \vec{W}_R) \, dx$$

with

$$\left| -\int \sum_{j=1}^3 \vec{f}_j \cdot ((\partial_j \theta_{R/2,R})\vec{u} - \partial_j \vec{W}_R) \, dx \right| \leq C\|\vec{f}\|_{\dot{H}^{-1}} \|\vec{u}1_{|x|>R/2}\|_6.$$

Letting R go to infinity, we obtain $\nu\|\vec{u}\|_{\dot{H}^1}^2 = \int \vec{u} \cdot \vec{f} \, dx$. □

Let us remark that we have the following obvious uniqueness theorem:

Theorem 16.4.
There exists a positive constant ϵ_0 such that, if the problem

$$\nu\Delta\vec{u} - \operatorname{div}(\vec{u} \otimes \vec{u}) - \vec{\nabla}p + \vec{f} = 0, \ \operatorname{div}\vec{u} = 0 \qquad (16.6)$$

with $\vec{f} \in \dot{H}^{-1}(\mathbb{R}^3)$ has a solution $\vec{u} \in L^3$ with $\|\vec{u}\|_3 < \epsilon_0$, then $\vec{u}$ is the unique solution of (16.8) in $\dot{H}^1$ such that $\nu\|\vec{u}\|_{\dot{H}^1}^2 = \int \vec{u} \cdot \vec{f} \, dx$
In particular, if $\vec{v}$ is a solution of (16.8) such that it fullfills one of the assumptions of Theorem 16.3 then $\vec{v} = \vec{u}$.

Proof. Let $\vec{v}$ be a solution of (16.8) with $\nu\|\vec{v}\|_{\dot{H}^1}^2 = \int \vec{v} \cdot \vec{f} \, dx$ (with a pressure q). We compute $\nu\|\vec{\nabla} \otimes (\vec{u} - \vec{v})\|_2^2$. As $\vec{u} \in L^3 \cap \dot{H}^1$ et $\vec{v} \in \dot{H}^1$, the scalar products $\int \vec{u} \cdot (\vec{v} \cdot \vec{\nabla}\vec{v} + \vec{\nabla}q) \, dx$ et $\int \vec{v} \cdot (\vec{u} \cdot \vec{\nabla}\vec{u} + \vec{\nabla}p) \, dx$ are well defined and we find

$$\nu\|\vec{\nabla} \otimes (\vec{u} - \vec{v})\|_2^2 = \int \vec{u} \cdot (\vec{v} \cdot \vec{\nabla}\vec{v}) \, dx + \int \vec{v} \cdot (\vec{u} \cdot \vec{\nabla}\vec{u}) \, dx$$

$$= \int \vec{u} \cdot (\vec{v} \cdot \vec{\nabla}(\vec{v} - \vec{u})) \, dx + \int (\vec{v} - \vec{u}) \cdot (\vec{u} \cdot \vec{\nabla}\vec{u}) \, dx$$

$$= \int \vec{u} \cdot ((\vec{v} - \vec{u}) \cdot \vec{\nabla}(\vec{v} - \vec{u})) \, dx$$

$$\leq C\|\vec{u}\|_3 \|\vec{\nabla} \otimes (\vec{u} - \vec{v})\|_2^2.$$

Thus, if $\|\vec{u}\|_3 < \frac{1}{C\nu}$, we have $\vec{v} = \vec{u}$. □

The *Liouville problem for steady solutions* corresponds to the study of the solutions of the equation

$$\nu \Delta \vec{u} - \mathrm{div}(\vec{u} \otimes \vec{u}) - \vec{\nabla} p = 0, \mathrm{div}\, \vec{u} = 0. \tag{16.7}$$

If $\vec{u}$ is a solution of (16.7) such that $\vec{u} \in L^p$, where L^p fullfills one of the assumptions of Theorem 16.3 then $\vec{u} = 0$. The case $p = 9/2$ has been proved by Galdi [194], the case $p = 6$ has been discussed by Seregin [430], the case $3 < p < 6$ has been discussed by Chamorro, Jarrín and Lemarié-Rieusset [101]. The case $3/2 < p < 3$ is new. (Remark: recently, Yuan and Xiao[1] published a paper in *JMAA* on the case $2 \le p < 3$, but their use of Caccioppoli inequalities was erroneous.)

A very nice criterion for the triviality of a steady solution of (16.7) has been given by Chae [99]:

Theorem 16.5.
Let $\vec{u} \in \dot{H}^1$ of the equations

$$\nu \Delta \vec{u} - \mathrm{div}(\vec{u} \otimes \vec{u}) - \vec{\nabla} p = 0, \ \mathrm{div}\, \vec{u} = 0 \tag{16.8}$$

with $\vec{f} \in \dot{H}^{-1}(\mathbb{R}^3)$. If $\Delta \vec{u} \in L^{6/5}$, then $\vec{u} = 0$.

Proof. As a solution of the Navier–Stokes problem (with no force) in a subcritical space, we find that $\vec{u}$ is smooth. Indeed, we write

$$\vec{u} = \frac{1}{\nu} \sum_{i=1}^{3} \mathbb{P} \frac{\partial_i}{\Delta}(u_i \vec{u}).$$

We easily check that if $\vec{u} \in L^6 \cap \dot{W}^{s,6}$ with $s \ge 0$ then $\vec{u} \in \dot{W}^{s+\frac{1}{2},6}$. Thus $\vec{u} \in \cap_{s \ge 1} \dot{W}^{s,6}$ and $\vec{u}$ is C^∞ and goes to 0 at infinity:

$$\lim_{x \to \infty} |\vec{u}(x)| = 0.$$

We may write as well

$$p = -\sum_{i=1}^{3} \sum_{j=1}^{3} \frac{\partial_i \partial_j}{\Delta}(u_i u_j).$$

Thus, $p \in \cap_{s \ge 1/2} \dot{W}^{s,6}$ and p is C^∞ and goes to 0 at infinity.

We consider the total pressure $Q(x) = p + \frac{|\vec{u}|^2}{2}$. It is a function in C^∞ which goes to 0 à at infinity. Moreover, it satisfies

$$\nu \Delta \vec{u} = \vec{\omega} \wedge \vec{u} + \vec{\nabla} Q \tag{16.9}$$

where $\vec{\omega} = \vec{\nabla} \wedge \vec{u}$.

In particular, we have the equivalence:

$$\vec{u} = 0 \Leftrightarrow Q = 0.$$

Indeed, if $Q = 0$, we have $\vec{u} \cdot \Delta \vec{u} = 0 = \Delta(\frac{|\vec{u}|^2}{2}) - |\vec{\nabla} \otimes \vec{u}|^2$, and thus

$$\int |\vec{\nabla} \otimes \vec{u}|^2 \, dx = 0.$$

---

[1]Yuan, Baoquan and Xiao, Yamin, Liouville-type theorems for the 3D stationary Navier-Stokes, MHD and Hall-MHD equations, J. Math. Anal. Appl. 491 (2020), no. 2, 124343

[More precisely, we use a truncature $\sigma_R = \sigma(\frac{x}{R})$, where σ is radially non-increasing, $\sigma > 0$ for $|x| < 1$, $= 0$ elsewhere, and $\sigma = 1$ for $|x| < 1/2$; we then have

$$\int \Delta(\sigma_R)|\vec{u}|^2 \, dx = 2 \int \sigma_R |\vec{\nabla} \otimes \vec{u}|^2 \, dx$$

and we go to the limit.]

Taking the scalar product of equation (16.9) with $\vec{u}$, we find:

$$\vec{u} \cdot \vec{\nabla} Q = \nu \vec{u} \cdot \Delta \vec{u} = \nu \Delta(\frac{|\vec{u}|^2}{2}) - \nu \sum_{i=1}^{3} \sum_{j=1}^{3} |\partial_i u_j|^2.$$

We thus have

$$-\Delta Q + \frac{1}{\nu} \vec{u} \cdot \vec{\nabla} Q = \sum_{i=1}^{3} \sum_{j=1}^{3} \partial_i \partial_j (u_i u_j) - |\partial_i u_j|^2 = -|\vec{\omega}|^2 \leq 0.$$

We have an easy maximum principle: let F be a C^2 function on $\mathbb{R}^3$ and $\vec{b} \in C_b$; if F satisfies

$$\Delta F + \vec{b}.\vec{\nabla} F \geq 0$$

and

$$\lim_{x \to \infty} F(x) = 0$$

then $F = 0$ or $F < 0$. Chae invoked this maximum principle to get

$$Q = 0 \text{ or } Q < 0.$$

The next step is to use Sard's lemma for the smooth function Q to get that the set

$$A = \{y \in \mathbb{R} \ / \ \exists x \in \mathbb{R}^3 \quad Q(x) = y \text{ et } \vec{\nabla} Q(x) = 0\}$$

has a Lebesgue measure equal to 0.

In the following, we take $\epsilon > 0$ such that $-\epsilon \notin A$. We note $D_-^\epsilon = \{x \in \mathbb{R}^3 \ / \ Q(x) < -\epsilon\}$. It is a bounded open subset of $\mathbb{R}^3$ and its boundary is exactly the set $\partial D_-^\epsilon = \{x \in \mathbb{R}^3 \ / \ Q(x) = -\epsilon\}$ (since $-\epsilon \notin A$, we may apply the theorem of implicit functions in the neighborhood of x such that $Q(x) = -\epsilon$). Moreover, by the theorem of the constant rank, ∂D_-^ϵ is a differentiable submanifold.

Assume that $Q < 0$. We have

$$\text{div} \, ((Q + \epsilon)\vec{u}) = \nu \vec{u} \cdot \Delta \vec{u} = \nu \Delta(\frac{|\vec{u}|^2}{2}) - \nu |\vec{\nabla} \otimes \vec{u}|^2.$$

Integration on D_-^ϵ gives that

$$0 = \int_{\partial D_-^\epsilon} (Q + \epsilon)\vec{u} \cdot \vec{\nu} \, d\sigma = \nu(\int_{Q < -\epsilon} (\Delta(\frac{|\vec{u}|^2}{2}) - |\vec{\nabla} \otimes \vec{u}|^2) \, dx).$$

We have

$$\Delta(\frac{|\vec{u}|^2}{2}) = \vec{u} \cdot \Delta \vec{u} + |\vec{\nabla} \otimes \vec{u}|^2 \in L^1$$

Dominated convergence theorem gives

$$0 = \int_{Q < 0} (\Delta(\frac{|\vec{u}|^2}{2}) - |\vec{\nabla} \otimes \vec{u}|^2) \, dx.$$

We thus find

$$\int |\vec{\nabla} \otimes \vec{u}|^2 \, dx = \int \Delta(\frac{|\vec{u}|^2}{2}) \, dx = \lim_{R \to +\infty} \int \frac{|\vec{u}|^2}{2} \Delta(\sigma_R) \, dx = 0. \qquad \square$$

16.4 Self-similarity

The theory of self-similar solutions is linked to the scale invariance of the Navier–Stokes equations

$$\begin{cases} \partial_t \vec{u} = & \nu \Delta \vec{u} - \sum_{i=1}^{3} u_i \partial_i \vec{u} - \vec{\nabla} p + \vec{f} \\ \text{div } \vec{u} = & \sum_{i=1}^{3} \partial_i u_i = 0 \end{cases} \tag{16.10}$$

For sake of simplicity when dealing with the incompressibility constraint, we assume that $\vec{f}$ is given in a divergence form $\vec{f} = \text{div } \mathbb{F} = \sum_{i=1}^{3} \partial_i \vec{F_i}$. Indeed, the Leray projection operator $\mathbb{P}$ on divergence-free vector fields is a singular integral operator, so that $\mathbb{P}\vec{h}$ might be uneasy to define for a general vector field $\vec{h}$; but the operator $\mathbb{P}\partial_i$ is much more easy to deal with: this is a (matrix of) convolution operator(s) with a kernel $\mathbb{K}_i$ which can be split into a sum $A_i + B_i$, where A_i is compactly supported (so that convolution with A_i is well defined on $\mathcal{D}'$) and B_i is much less singular: $B_i \in L^1$ (see Lemarié-Rieusset [313] for instance).

We assume that the fluid fills the whole space (i.e., that $\vec{u}$ is defined on $(T_0, T_1) \times \mathbb{R}^3$) and that $\vec{u}$ and $\mathbb{F}$ are "null at infinity," which allows to get rid of the pressure p: indeed, we have (by taking the divergence of the Navier–Stokes equations)

$$\Delta p = \sum_{i=1}^{3} \sum_{j=1}^{3} \partial_i \partial_j (F_{i,j} - u_i u_j).$$

Thus, if p is defined up to a (time-dependent) harmonic term; the hypothesis on $\vec{u}$ and $\mathbb{F}$ allows one to define $\vec{\nabla} p$ unambiguously by

$$\vec{\nabla} p = \frac{1}{\Delta} \vec{\nabla} (\sum_{i=1}^{3} \sum_{j=1}^{3} \partial_i \partial_j (F_{i,j} - u_i u_j))$$

– see for instance Lemarié-Rieusset [313] where the case of a locally L^2 solution is discussed with

$$\lim_{x_0 \to \infty} \int_{T_0}^{T_1} \int_{B(x_0,1)} |\vec{u}(t,x)|^2 \, dx \, dt = 0.$$

We thus consider mild solutions of

$$\partial_t \vec{u} = \nu \Delta \vec{u} + \mathbb{P} \text{div}(\mathbb{F} - \vec{u} \otimes \vec{u}).$$

An interesting feature of the Navier–Stokes equations is their scale invariance: if $\vec{u}$ is a solution on $(T_0, T_1) \times \mathbb{R}^3$ and $\lambda > 0$ (for the forcing term $\vec{f} = \text{div } \mathbb{F}$), then

$$\lambda \vec{u}(T_0 + \lambda^2(t - T_0), \lambda x)$$

is a solution on $(T_0, T_0 + \lambda^{-2}(T_1 - T_0)) \times \mathbb{R}^3$ (for $\vec{f_\lambda} = \text{div } \mathbb{F_\lambda}$ with $\mathbb{F_\lambda} = \lambda^2 \mathbb{F}(T_0 + \lambda^2(t - T_0), \lambda x))$ and

$$\lambda \vec{u}(T_1 - \lambda^2(T_1 - t), \lambda x)$$

is a solution on $(T_1 - \lambda^{-2}(T_1 - T_0), T_1) \times \mathbb{R}^3$ (for $\vec{f_\lambda} = \text{div } \mathbb{F_\lambda}$ with $\mathbb{F_\lambda} = \lambda^2 \mathbb{F}(T_1 - \lambda^2(T_1 - t), \lambda x))$.

A *forward self-similar solution* is a solution on $(0, +\infty) \times \mathbb{R}^3$ such that $\lambda \vec{u}(\lambda^2 t, \lambda x) = \vec{u}(t, x)$ for every $\lambda > 0$. We then have

$$\vec{u}(t,x) = \frac{1}{\sqrt{t}} \vec{U}(\frac{x}{\sqrt{t}}) \text{ with } \vec{U}(x) = \vec{u}(1, x).$$

The initial value $\vec{u}_0$ (corresponding to $t = 0$) will then be homogeneous:

$$\vec{u}_0(x) = \lambda \vec{u}_0(\lambda x)$$

and the force $\vec{f} = \operatorname{div} \mathbb{F}$ will have a self-similar profile

$$\mathbb{F} = \frac{1}{t}\mathbb{F}_0(\frac{x}{\sqrt{t}}).$$

A *backward self-similar solution* is a solution on $(-\infty, 0) \times \mathbb{R}^3$ such that $\lambda \vec{u}(\lambda^2 t, \lambda x) = \vec{u}(t, x)$ for every $\lambda > 0$. We then have

$$\vec{u}(t, x) = \frac{1}{\sqrt{|t|}}\vec{U}(\frac{x}{\sqrt{|t|}}) \text{ with } \vec{U}(x) = \vec{u}(-1, x)$$

and

$$\mathbb{F} = \frac{1}{|t|}\mathbb{F}_0(\frac{x}{\sqrt{|t|}}).$$

Backward self-similar solutions (with $\mathbb{F} = 0$) were first considered by Leray [328], as a potential model for blow-up of weak solutions. Of course, the null solution is a backward self-similar solution. Existence of other solutions was ruled out by Nečas, Růžička, and Šverák [375] and by Tsai [482]. In the case of a profile $\vec{U} \in L^3$, the fact that $\vec{U} = 0$ is a direct consequence of Escauriaza, Seregin and Šverák's theorem (Theorem 15.21) [163].

In contrast with the case of backward self-similar solutions, forward self-similar solutions exist. Those solutions are associated to homogeneous initial values, so that one must rule out initial values in a Lebesgue space L^p or a Sobolev space H^s. Thus, the study of self-similar solutions was allowed by the theory of mild solutions in more general spaces as Morrey spaces (Giga and Miyakawa [212]), homogeneous Besov spaces (Cannone [81]) or Lorentz spaces (Barraza [22]).This theory allowed to prove existence of self-similar solutions for *small* data (see Theorem 10.8).

When considering arbitrarily large initial values, the Picard iterative scheme does not work any longer, and one must use other tools to get existence of a solution. Recently, Jia and Šverák [245] proved existence of such solutions by using another fixed-point theorem, namely Schaefer's fixed point theorem, a consequence of the Leray–Schauder index theory for non-linear equations.

Jia and Šverák considered homogeneous initial values that are locally Hölderian on $\mathbb{R}^3 \setminus \{0\}$, but their main tool was the energy inequalities described by Lemarié-Rieusset [313] for uniformly locally L^2 initial values. It is therefore natural to prove their result in the setting of a homogeneous initial value that is only locally square integrable; this is the main purpose of this chapter.

More precisely, Jia and Šverák proved their theorem first for homogeneous initial values that are smooth on $\mathbb{R}^3 \setminus \{0\}$, then extended their result to the case of homogeneous initial values that are locally Hölderian on $\mathbb{R}^3 \setminus \{0\}$ (they noted that "more general initial data can be considered"); we shall prove the result first for homogeneous initial values that are locally bounded on $\mathbb{R}^3 \setminus \{0\}$ [the proof of this point is very similar to Jia and Šverák's proof, in the use of the Leray–Schauder principle; for the estimates, we shall use the setting of the space $E^{-\delta, 0}$ introduced by Cannone, Meyer and Planchon in the study of self-similar solutions [83, 84] and the local energy inequalities of Lemarié-Rieusset [313], then extend this result to the case of homogeneous initial values that are locally L^2 on $\mathbb{R}^3 \setminus \{0\}$. Moreover, we shall allow the presence of a self-similar forcing term, while Jia and Šverák stated their result under the assumption of a null force.

16.5 Statement of Jia and Šverák's Theorem

Before stating our results, we begin by a straightforward remark on locally square integrable homogeneous functions. If u is a locally square integrable function on $\mathbb{R}^3$ we shall say that u belongs to the space of uniformly square integrable functions L^2_{uloc} if

$$\|u\|^2_{L^2_{\text{uloc}}} = \sup_{x_0 \in \mathbb{R}^3} \int_{B(x_0,1)} |u(x)|^2 \, dx < +\infty.$$

Similarly, we say that u belongs to the Morrey space $\dot{M}^{2,3}$ if

$$\|u\|^2_{\dot{M}^{2,3}} = \sup_{x_0 \in \mathbb{R}^3, R>0} \frac{1}{R} \int_{B(x_0,R)} |u(x)|^2 \, dx < +\infty.$$

In particular, we have

$$\|u\|_{\dot{M}^{2,3}} = \sup_{\lambda>0} \|\lambda u(\lambda x)\|_{L^2_{\text{uloc}}}.$$

Our remark on homogeneous functions is the following one:

Lemma 16.3.
Let u be a function on $\mathbb{R}^3$ such that:

- *u is locally square integrable on $\mathbb{R}^3 \setminus \{0\}$*

- *u is homogeneous: for $\lambda > 0$, $u(\lambda x) = \lambda^{-1} u(x)$.*

Then u belongs to L^2_{uloc} and can be written as

$$u(x) = \frac{1}{|x|} \gamma\left(\frac{x}{|x|}\right)$$

where $\gamma \in L^2(S^2)$ and we have

$$\|u\|_{L^2_{\text{uloc}}} \approx \|\gamma\|_{L^2(S^2)} \approx \|u\|_{\dot{M}^{2,3}}.$$

Proof. We have

$$\|\gamma\|^2_{L^2(S^2)} = \int_{1<|x|<2} |u(x)|^2 \, dx$$

and

$$\int_{B(x_0,1)} |u(x)|^2 \, dx \leq 2\|\gamma\|^2_{L^2(S^2)}$$

so that the lemma is obvious. $\qquad \square$

The main result in this chapter is then the following one:

Self-similar solutions for rough data

Theorem 16.6.
Let $\vec{u}_0$ be a vector field on $\mathbb{R}^3$ and $\mathbb{F}_0$ be a tensor function such that:

- $\vec{u}_0$ *is divergence-free:* div $\vec{u}_0 = 0$

- $\vec{u}_0$ *is homogeneous: for all* $\lambda > 0$, $\lambda \vec{u}_0(\lambda x) = \vec{u}_0(x)$

- $\vec{u}_0$ *is locally square integrable on* $\mathbb{R}^3 \setminus \{0\}$

- $\sup_{x \in \mathbb{R}^3}(1 + |x|)^2 |\mathbb{F}_0(x)| < +\infty$

Define $\mathbb{F} = \frac{1}{t}\mathbb{F}_0(\frac{x}{\sqrt{t}})$. *Then there exists at least one self-similar vector field* $\vec{u}(t,x) = \frac{1}{\sqrt{t}}\vec{U}(\frac{x}{\sqrt{t}})$ *on* $(0, +\infty) \times \mathbb{R}^3$ *such that:*

- $\vec{u}$ *is divergence-free:* div $\vec{u} = 0$

- $\vec{u}$ *is a solution of the Navier–Stokes equations*

$$\partial_t \vec{u} = \Delta \vec{u} - \sum_{i=1}^{3} u_i \partial_i \vec{u} - \vec{\nabla} p + \text{div } \mathbb{F}$$

- $\vec{U}$ *belongs to* H^1_{loc}

- $\lim_{t \to 0^+} \vec{u}(t,.) = \vec{u}_0$ *strongly in* L^2_{loc}

The proof of this theorem will rely on the preliminary study of the case of a homogeneous divergence-free initial value that is locally bounded on $\mathbb{R}^3 \setminus \{0\}$ (Theorem 16.7). The link between Theorems 16.6 and 16.7 is the fact that we may approximate a locally square integrable divergence-free homogeneous initial value by a locally bounded divergence-free homogeneous initial value.

First, we characterize locally bounded homogeneous functions:

Lemma 16.4.
Let u be a function on $\mathbb{R}^3$ of the form

$$u(x) = \frac{1}{|x|}\, \gamma\left(\frac{x}{|x|}\right)$$

where $\gamma \in L^2(S^2)$. Then the following assertions are equivalent:

- u *is locally (essentially) bounded on* $\mathbb{R}^3 \setminus \{0\}$

- $\gamma \in L^\infty(S^2)$

- $\sup_{x \in \mathbb{R}^3, t > 0}(\sqrt{t} + |x|)|e^{t\Delta}u(x)| < +\infty$

Proof. We have, of course, for $R > 1$,

$$\|\gamma\|_{L^\infty(S^2)} = (\text{ess.})\sup_{1/R < |x| < R}|x||u(x)|$$

If W is the heat kernel, so that $e^{t\Delta}u = \frac{1}{t^{3/2}}W(\frac{x}{\sqrt{t}}) * u$, we have

$$\|\sqrt{t}e^{t\Delta}u\|_\infty \le \|\gamma\|_{L^\infty(S^2)}\|W\|_{L^{3/2,1}}$$

and

$$\| |x|e^{t\Delta}u\|_\infty \le \|\gamma\|_{L^\infty(S^2)}(\|W\|_1 + \| |x| W\|_{L^{3/2,1}}).$$

Conversely,

$$\|\gamma\|_{L^{\infty}(S^2)} = \||x||u|\|_{\infty} \le \liminf_{t \to 0} \||x|e^{t\Delta}u\|_{\infty}.$$

Thus, the lemma is obvious. $\qquad\qquad\qquad\qquad\qquad\qquad\qquad\qquad\qquad\qquad\square$

Then, we show that we may approximate a locally square integrable divergence-free homogeneous initial value by a locally bounded divergence-free homogeneous initial value:

Lemma 16.5.
Let $\vec{u}_0$ be a vector field on $\mathbb{R}^3$ such that:

- *$\vec{u}_0$ is divergence-free: div $\vec{u}_0 = 0$*

- *$\vec{u}_0$ is homogeneous: for $\lambda > 0$, $\vec{u}_0(\lambda x) = \lambda^{-1}\vec{u}_0(x)$..*

- *$\vec{u}_0$ is locally square integrable: $\vec{u}_0(x) = \frac{1}{|x|}\vec{\gamma}(\frac{x}{|x|})$ with $\vec{\gamma} \in L^2(S^2)$*

Then, for every $\epsilon > 0$, there exists a vector field $\vec{v}_{0,\epsilon}$ such that

- *$\vec{v}_{0,\epsilon}$ is divergence-free: div $\vec{v}_{0,\epsilon} = 0$*

- *$\vec{v}_{0,\epsilon}$ is homogeneous.*

- *$\vec{v}_{0,\epsilon}$ is locally square integrable: $\vec{v}_{0,\epsilon} = \frac{1}{|x|}\vec{\delta}_{\epsilon}(\frac{x}{|x|})$ with $\vec{\delta}_{\epsilon} \in L^2(S^2)$*

- *$\vec{\delta}_{\epsilon} \in L^{\infty}$ and $\|\vec{\gamma} - \vec{\delta}_{\epsilon}\|_{L^2(S^2)} < \epsilon$.*

Proof. We approximate $\vec{\gamma}$ in $L^2(S^2)$ by a smooth vector field $\vec{\eta}_{\epsilon}$: $\vec{\eta}_{\epsilon}$ is $\mathcal{C}^{\infty}(S^2)$ and $\|\vec{\gamma} - \vec{\eta}_{\epsilon}\|_{L^2(S^2)}$ is small. Then we define $\vec{w}_{0,\epsilon}$ and $\vec{v}_{0,\epsilon}$ as $\vec{w}_{0,\epsilon} = \frac{1}{|x|}\vec{\eta}_{\epsilon}(\frac{x}{|x|})$ $\vec{v}_{0,\epsilon} = \mathbb{P}\vec{w}_{0,\epsilon}$, where $\mathbb{P}$ is the Leray projection operator:

$$\mathbb{P}\vec{h} = \vec{h} - \frac{1}{\Delta}\vec{\nabla}(\text{div } \vec{h}).$$

As $\mathbb{P}$ is a singular integral operator homogeneous of degree 0, $\vec{v}_{0,\epsilon}$ is homogeneous of the same degree as $\vec{w}_{0,\epsilon}$. Moreover, $\mathbb{P}$ is bounded on $\dot{M}^{2,3}$, so that we conclude that $\vec{v}_{0,\epsilon}$ may be written as $\vec{v}_{0,\epsilon} = \frac{1}{|x|}\vec{\delta}_{\epsilon}(\frac{x}{|x|})$ with $\vec{\delta}_{\epsilon} \in L^2(S^2)$ and

$$\|\vec{\gamma} - \vec{\delta}_{\epsilon}\|_{L^2(S^2)} \approx \|\vec{u}_0 - \vec{v}_{0,\epsilon}\|_{\dot{M}^{2,3}} \le C|\vec{u}_0 - \vec{w}_{0,\epsilon}\|_{\dot{M}^{2,3}} \approx C\|\vec{\gamma} - \vec{\eta}_{\epsilon}\|_{L^2(S^2)}.$$

Moreover, $\Delta\vec{v}_{0,\epsilon} = \Delta\vec{w}_{0,\epsilon} - \vec{\nabla}(\text{div } \vec{w}_{0,\epsilon})$, so that $\Delta\vec{v}_{0,\epsilon}$ is smooth on $\mathbb{R}^3 \setminus \{0\}$, and so is $\vec{v}_{0,\epsilon}$ (by (hypo)ellipticity of Δ); thus $\vec{\delta}_{\epsilon}$ belongs to $\mathcal{C}^{\infty}(S^2)$. $\qquad\qquad\square$

Theorem 16.6 now is turned into:

Theorem 16.7.
Let $\vec{u}_0$ be a vector field on $\mathbb{R}^3$ and $\mathbb{F}_0$ be a tensor function such that:

- *$\vec{u}_0$ is divergence-free: div $\vec{u}_0 = 0$*

- *$\vec{u}_0$ is homogeneous: for all $\lambda > 0$, $\lambda\vec{u}_0(\lambda x) = \vec{u}_0(x)$*

- $\vec{u}_0$ is locally bounded on $\mathbb{R}^3 \setminus \{0\}$

- $\sup_{x\in\mathbb{R}^3}(1+|x|)^2|\mathbb{F}_0(x)| < +\infty$

- $\sup_{x\in\mathbb{R}^3}(1+|x|)^3 \operatorname{div}|\mathbb{F}_0(x)| < +\infty$.

Define $\mathbb{F} = \frac{1}{t}\mathbb{F}_0(\frac{x}{\sqrt{t}})$. Then there exists at least one self-similar vector field $\vec{u}(t,x) = \frac{1}{\sqrt{t}}\vec{U}(\frac{x}{\sqrt{t}})$ on $(0,+\infty)\times\mathbb{R}^3$ such that:

- $\vec{u}$ is divergence-free: $\operatorname{div}\vec{u} = 0$

- $\vec{u}$ is a solution of the Navier–Stokes equations

$$\partial_t\vec{u} = \Delta\vec{u} - \sum_{i=1}^{3} u_i\partial_i\vec{u} - \vec{\nabla}p + \operatorname{div}\mathbb{F}$$

- the profile $\vec{U}$ is a continuous function with

$$\sup_{x\in\mathbb{R}^3}(1+|x|)|\vec{U}(x)| < +\infty$$

- $\vec{U}$ belongs to the Lorentz space $L^{3,\infty}(\mathbb{R}^3)$

- $\lim_{t\to 0^+}\vec{u}(t,.) = \vec{u}_0$ *-weakly in $L^{3,\infty}$.

Let us remark that $\vec{u}_0 \in L^{3,\infty}$ is a weaker statement than $\vec{u}_0$ is locally bounded far away from 0. Indeed, we have the equivalence for a homogeneous initial value $\vec{u}_0$, that $\vec{u}_0 \in L^{3,\infty} \Leftrightarrow \vec{\gamma} \in L^3(S^2)$ (see Barraza [22] or Lemarié-Rieusset [313]).

16.6   The Case of Locally Bounded Initial Data

We shall now proceed to the proof of Theorem 16.7, by using Schaefer's fixed point theorem. We start with a homogeneous initial value $\vec{u}_0$ and a self-similar tensor $\mathbb{F}$ and we are looking for a self-similar solution $\vec{u}$ to the Navier–Stokes equations

$$\partial_t\vec{u} = \Delta\vec{u} - \sum_{i=1}^{3} u_i\partial_i\vec{u} - \vec{\nabla}p + \operatorname{div}\mathbb{F}.$$

The pressure p is computed as

$$p = \frac{\vec{\nabla}\operatorname{div}}{\Delta}(\mathbb{F} - \vec{u}\otimes\vec{u}).$$

Thus, we focus on finding $\vec{u}$ as a self-similar solution to

$$\partial_t\vec{u} = \Delta\vec{u} - \mathbb{P}(\sum_{i=1}^{3} u_i\partial_i\vec{u} - \operatorname{div}\mathbb{F}).$$

This problem is equivalent to find a self-similar solution $\vec{u}$ of

$$\vec{u} = W_{\nu t} * \vec{u}_0 - \int_0^t W_{\nu(t-s)} * \mathbb{P} \operatorname{div} \mathbb{F} \, ds - \int_0^t W_{\nu(t-s)} * \mathbb{P} \operatorname{div}(\vec{u} \otimes \vec{u}) \, ds.$$

Writing $\operatorname{div} \mathbb{T} = \sum_{j=1}^3 \partial_j \vec{F}_j$ and using Oseen's tensor described in Theorem 4.6, we want to find a self-similar solution of

$$\vec{u}(t,x) = \int \frac{1}{(\nu t)^{3/2}} W\left(\frac{x-y}{\sqrt{\nu t}}\right) \vec{u}_0(y) \, dy$$

$$- \sum_{j=1}^3 \int_0^t \int \frac{1}{(\nu(t-s))^2} (\partial_j \mathcal{O})\left(\frac{x-y}{\sqrt{\nu(t-s)}}\right) :: \vec{F}_j(s,y) \, dy \, ds$$

$$- \sum_{j=1}^3 \int_0^t \int \frac{1}{(\nu(t-s))^2} (\partial_j \mathcal{O})\left(\frac{x-y}{\sqrt{\nu(t-s)}}\right) :: u_j(s,y)\vec{u}(s,y) \, dy \, ds$$

$$= \int W\left(\frac{x}{\sqrt{\nu t}} - z\right) \vec{u}_0(\sqrt{\nu t}z) \, dz$$

$$- \sum_{j=1}^3 \int_0^1 \int \frac{\sqrt{t}}{\sqrt{\nu}} \frac{1}{(1-\sigma)^2} (\partial_j \mathcal{O})\left(\frac{\frac{x}{\sqrt{\nu t}} - z}{\sqrt{1-\sigma}}\right) :: \vec{F}_j(t\sigma, \sqrt{\nu t}z) \, dz \, d\sigma$$

$$- \sum_{j=1}^3 \int_0^1 \int \frac{\sqrt{t}}{\sqrt{\nu}} \frac{1}{(1-\sigma)^2} (\partial_j \mathcal{O})\left(\frac{\frac{x}{\sqrt{\nu t}} - z}{\sqrt{1-\sigma}}\right) :: u_j(t\sigma, \sqrt{\nu t}z)\vec{u}(t\sigma, \sqrt{\nu t}z) \, dz \, d\sigma$$

Using the homogeneity of $\vec{u}_0$ and the self-similarity of $\vec{u}$ and $\mathbb{F}$, we find

$$\frac{1}{\sqrt{t}} \vec{U}\left(\frac{x}{\sqrt{t}}\right) = \frac{1}{\sqrt{t}} \int W\left(\frac{x}{\sqrt{\nu t}} - z\right) \vec{u}_0(\sqrt{\nu}z) \, dz$$

$$- \frac{1}{\sqrt{t}} \sum_{j=1}^3 \int_0^1 \int \frac{1}{\sqrt{\nu}} \frac{1}{(1-\sigma)^2} (\partial_j \mathcal{O})\left(\frac{\frac{x}{\sqrt{\nu t}} - z}{\sqrt{1-\sigma}}\right) :: \frac{1}{\sigma} \vec{F}_{0,j}\left(\frac{\sqrt{\nu}z}{\sqrt{\sigma}}\right) \, dz \, d\sigma$$

$$- \frac{1}{\sqrt{t}} \sum_{j=1}^3 \int_0^1 \int \frac{1}{\sqrt{\nu}} \frac{1}{(1-\sigma)^2} (\partial_j \mathcal{O})\left(\frac{\frac{x}{\sqrt{\nu t}} - z}{\sqrt{1-\sigma}}\right) :: \frac{1}{\sigma} U_j\left(\frac{\sqrt{\nu}z}{\sqrt{\sigma}}\right) \vec{U}\left(\frac{\sqrt{\nu}z}{\sqrt{\sigma}}\right) \, dz \, d\sigma$$

Thus, the problem is to find a solution $\vec{U}$ of

$$\vec{U} = \vec{V}_0 - \mathcal{T}(\vec{U}) \tag{16.11}$$

with

$$\vec{V}_0 = \int W\left(\frac{x}{\sqrt{\nu}} - z\right) \vec{u}_0(\sqrt{\nu}z) \, dz$$

$$- \sum_{j=1}^3 \int_0^1 \int \frac{1}{\sqrt{\nu}} \frac{1}{(1-\sigma)^2} (\partial_j \mathcal{O})\left(\frac{\frac{x}{\sqrt{\nu}} - z}{\sqrt{1-\sigma}}\right) :: \frac{1}{\sigma} \vec{F}_{0,j}\left(\frac{\sqrt{\nu}z}{\sqrt{\sigma}}\right) \, dz \, d\sigma \tag{16.12}$$

and

$$\mathcal{T}(\vec{U}) = \sum_{j=1}^3 \int_0^1 \int \frac{1}{\sqrt{\nu}} \frac{1}{(1-\sigma)^2} (\partial_j \mathcal{O})\left(\frac{\frac{x}{\sqrt{\nu}} - z}{\sqrt{1-\sigma}}\right) :: \frac{1}{\sigma} U_j\left(\frac{\sqrt{\nu}z}{\sqrt{\sigma}}\right) \vec{U}\left(\frac{\sqrt{\nu}z}{\sqrt{\sigma}}\right) \, dz \, d\sigma \tag{16.13}$$

This integral equation has been discussed for instance in Planchon's thesis [399], as an alternative to the usual differential formulation given by Leray for self-similar solutions [328].

As we said, we are going to solve the fixed-point problem $\vec{U} = \vec{V}_0 - \mathcal{T}(\vec{U})$ by mean of Schaefer's fixed-point theorem.

- **choice of the Banach space:**

E is the space of continuous divergence-free vector fields $\vec{U}$ on $\mathbb{R}^3$ such that $\vec{U}$ is a bounded function and

$$\|\vec{U}\|_E = \sup_{x \in \mathbb{R}^3} (\sqrt{\nu} + |x|)|\vec{U}(x)| < +\infty.$$

E has a companion space $\mathbb{E}$, which is the space of continuous divergence-free vector fields $\vec{u}$ on $(0, +\infty) \times \mathbb{R}^3$ such that

$$\|\vec{u}\|_\mathbb{E} = \sup_{t>0, x \in \mathbb{R}^3} (\sqrt{\nu t} + |x|)|\vec{u}(t, x)| < +\infty.$$

If $\vec{U}$ belongs to E, and $\vec{u}(t, x) = \frac{1}{\sqrt{t}}\vec{U}(\frac{x}{\sqrt{t}})$, then $\vec{u}$ belongs to $\mathbb{E}$. Moreover, the mapping $\vec{U} \mapsto \vec{u}$ is an isometry between E and the subspace $\mathbb{E}_s$ of $\mathbb{E}$ of self-similar vector fields in $\mathbb{E}$.

- **continuity of $\mathcal{T}$:**

We define the bilinear operator B as

$$B(\vec{u}, \vec{v}) = \int_0^t W_{\nu(t-s)} * \mathbb{P}\operatorname{div}(\vec{u} \otimes \vec{v})\, ds.$$

We have seen in Lemma 5.5 that there exists a constant ϵ_0 such that

$$\iint_{\mathbb{R} \times \mathbb{R}^3} \frac{1}{\nu^2(t-s)^2 + |x-y|^4} \frac{1}{(\sqrt{\nu|s|} + |y|)^2}\, dy\, ds \leq \epsilon_0 \frac{1}{\nu} \frac{1}{\sqrt{\nu|t|} + |x|} \tag{16.14}$$

so that

$$\sup_{t>0, x \in \mathbb{R}^3} (\sqrt{\nu t} + |x|)|B(\vec{u}, \vec{v})(t, x)| \leq C\epsilon_0 \frac{1}{\nu}\|\vec{u}\|_\mathbb{E}\|\vec{v}\|_\mathbb{E}.$$

We check moreover that $B(\vec{u}, \vec{v})$ is continuous in space and time variables. We have $\|\vec{u}(t, .)\|_{L^{3,\infty}} \leq C\|\vec{u}\|_\mathbb{E}$ and $\|\vec{u}(t, .)\|_\infty \leq \frac{1}{\sqrt{\nu t}}\|\vec{u}\|_\mathbb{E}$. Thus, $\|\vec{u} \otimes \vec{v}\|_{L^{q,\infty}} \leq C_q(\nu t)^{-1+\frac{3}{2q}}\|\vec{u}\|_\mathbb{E}\|\vec{v}\|_\mathbb{E}$ for $3/2 \leq q \leq +\infty$. If $0 < \gamma < 1$ and $\gamma + \frac{3}{q} < 1$, the operator $W_{\nu(t-s)} * \mathbb{P}\operatorname{div}$ maps $L^{q,\infty} \subset \dot{B}_{\infty,\infty}^{-\frac{3}{q}}$ to $\dot{B}_{\infty,\infty}^\gamma$ with an operator norm $O((\nu(t-s))^{-\frac{1+\gamma+3/q}{2}})$. Thus, we find that $B(\vec{u}, \vec{v})(t, .)$ is Hölderian of exponent γ, and

$$\|B(\vec{u}, \vec{v})(t, .)\|_{\dot{B}_{\infty,\infty}^\gamma} \leq C_\gamma(\nu t)^{-\gamma/2}\|\vec{u}\|_\mathbb{E}\|\vec{v}\|_\mathbb{E}.$$

For the control in time, we want to compare $B(\vec{u}, \vec{v})(t_1, x)$ and $B(\vec{u}, \vec{v})(t_2, x)$ for $0 < t_0 \leq t_1 \leq t_2$ and $t_1 - t_2 \to 0$. Let $\alpha = t_2 - t_1$ and $t_3 = t_1 - \sqrt{t_0 \alpha}$ ($t_3 > 0$ if $\alpha < t_0/2$). We then write

$$B(\vec{u}, \vec{v})(t_2, x) - B(\vec{u}, \vec{v})(t_1, x) = -\vec{w}_1 + \vec{w}_2 + \vec{w}_3$$

with

$$\vec{w}_1 = \int_{t_3}^{t_1} W_{\nu(t_1-s)} * \mathbb{P}\operatorname{div}(\vec{u} \otimes \vec{v})\, ds, \quad \vec{w}_2 = \int_{t_3}^{t_2} W_{\nu(t_2-s)} * \mathbb{P}\operatorname{div}(\vec{u} \otimes \vec{v})\, ds$$

and

$$\vec{w}_3 = (W_{\nu(t_2-t_3)} - W_{\nu(t_1-t_3)}) * \int_0^{t_3} W_{\nu(t_3-s)} * \mathbb{P}\operatorname{div}(\vec{u} \otimes \vec{v})\, ds$$

$$= (W_{\nu(t_2-t_3)} - W_{\nu(t_1-t_3)}) * B(\vec{u},\vec{v})(t_3,.).$$

The control of $\vec{w}_3$ is easy: we know that $\|B(\vec{u},\vec{v})(t_3,.)\|_\infty \leq C\frac{1}{\sqrt{\nu t_3}}\|\vec{u}\|_{\mathbb{E}}\|\vec{v}\|_{\mathbb{E}}$; on the other hand,

$$W_{\nu(t_2-t_3)} - W_{\nu(t_1-t_3)} = \int_{t_1-t_3}^{t_2-t_3} \nu\Delta W_{\nu s}\, ds$$

so that

$$\|W_{\nu(t_2-t_3)} - W_{\nu(t_1-t_3)}\|_{L^1(dx)} \leq C\int_{t_1-t_3}^{t_2-t_3} \frac{ds}{s} = C\ln\Big(1 + \sqrt{\frac{\alpha}{t_0}}\Big).$$

In order to control $\vec{w}_1$ and $\vec{w}_2$, we take $3 < q < +\infty$ and use the fact that $W_{\nu(t-s)} * \mathbb{P}\operatorname{div}$ maps $L^{q,\infty}$ to L^∞ with an operator norm $O((\nu(t-s))^{-\frac{1+3/q}{2}})$ and we find that, for $j = 1$ or $j = 2$,

$$\|\vec{w}_j\|_\infty \leq C\|\vec{u}\|_{\mathbb{E}}\|\vec{v}\|_{\mathbb{E}} \int_{t_3}^{t_j} \frac{1}{(\nu(t_j-s))^{\frac{1+3/q}{2}}} \frac{1}{(\nu s)^{1-\frac{3}{2q}}}\, ds$$

with

$$\int_{t_3}^{t_j} \frac{1}{(t_j-s)^{\frac{1+3/q}{2}}} \frac{1}{s^{1-\frac{3}{2q}}}\, ds = \frac{1}{\sqrt{t_j}} \int_{1-\frac{t_j-t_3}{t_j}}^{1} \frac{1}{(1-s)^{\frac{1+3/q}{2}}} \frac{1}{s^{1-\frac{3}{2q}}}\, ds.$$

When α is small enough with respect to t_0, we find

$$\|\vec{w}_j\|_\infty \leq C\frac{1}{\nu}\|\vec{u}\|_{\mathbb{E}}\|\vec{v}\|_{\mathbb{E}} \frac{1}{\sqrt{\nu t_0}}\Big(\frac{\alpha}{t_0}\Big)^{\frac{1-3/q}{4}}.$$

We thus have proved that B maps continuously $\mathbb{E} \times \mathbb{E}$ to $\mathbb{E}$. Moreover, it maps $\mathbb{E}_s \times \mathbb{E}_s$ to $\mathbb{E}_s$.

If $\vec{U} \in E$ and $\vec{V} = \mathcal{T}(\vec{U})$, and if $\vec{u}$ and $\vec{v}$ are the self-similar vector fields associated to $\vec{U}$ and $\vec{V}$, we have $\vec{v} = B(\vec{u},\vec{u})$. Hence, from the boundedness of the bilinear operator B on $\mathbb{E}_s \times \mathbb{E}_s$, we conclude that the quadratic operator $\mathcal{T}$ is continuous on E.

- **proof that $\vec{V}_0 \in E$:**

Let $\vec{v}_0(t,x) = \frac{1}{\sqrt{t}}\vec{V}_0(\frac{x}{\sqrt{t}})$. Proving $\vec{V}_0 \in E$ is equivalent to proving $\vec{v}_0 \in \mathbb{E}$. $\vec{v}_0$ is the sum of two terms $W_{\nu t} * \vec{u}_0$ and $\vec{w}_0 = \int_0^t W_{\nu(t-s)} * \mathbb{P}\operatorname{div}\mathbb{F}\, ds$.

The fact that $W_{\nu t} * \vec{u}_0 \in \mathbb{E}$ has already been seen in many places in this book. For instance, one may write that

$$\|W_{\nu t} * \vec{u}_0\|_\infty \leq C\frac{1}{\sqrt{\nu t}}\|\vec{u}_0\|_{\dot{B}^{-1}_{\infty,\infty}} \leq C'\frac{1}{\sqrt{\nu t}}\|\vec{u}_0\|_{L^{3,\infty}}$$

and

$$|W_{\nu t} * \vec{u}_0(x)| \leq \mathcal{M}_{\vec{u}_0}(x) \leq C\||x|\,\vec{u}_0\|_\infty \frac{1}{|x|}.$$

The proof that $\vec{w}_0 \in \mathbb{E}$ follows exactly the same lines as the proof that B maps $\mathbb{E} \times \mathbb{E}$ to $\mathbb{E}$.

- **compactness of $\mathcal{T}$:**

In order to prove that $\mathcal{T}$ is a compact mapping, we introduce a lemma that we shall use as well for getting a priori estimates for the fixed points:

Lemma 16.6.

Let $\alpha \in (0, 3/2)$ Then

$$\int_0^t \int \frac{1}{(\sqrt{\nu(t-s)} + |x-y|)^4} \frac{1}{s} \frac{1}{(1+\frac{|y|}{\sqrt{\nu s}})^{2\alpha}} \, dy \, ds \le C_\alpha \frac{1}{\sqrt{\nu t}} \frac{1}{(1+\frac{|x|}{\sqrt{\nu t}})^{2\alpha}}.$$

Proof. The change of variables $s = t\sigma$ and $y = \sqrt{\nu t} z$ gives

$$\int_0^t \int \frac{1}{(\sqrt{\nu(t-s)} + |x-y|)^4} \frac{1}{(1+\frac{|y|}{\sqrt{\nu s}})^{2\alpha}} \, dy \, \frac{ds}{s}$$

$$= \frac{1}{\sqrt{\nu t}} \int_0^1 \int \frac{1}{(\sqrt{1-\sigma} + |\frac{x}{\sqrt{\nu t}} - z|)^4} \frac{\sigma^\alpha}{(\sqrt{\sigma} + |z|)^{2\alpha}} \, dz \, \frac{d\sigma}{\sigma}.$$

We write, for $0 < \epsilon < \min(1/2, \alpha)$,

$$\int_0^1 \int \frac{1}{(\sqrt{1-\sigma} + |\frac{x}{\sqrt{\nu t}} - z|)^4} \frac{\sigma^\alpha}{(\sqrt{\sigma} + |z|)^{2\alpha}} \, dz \, \frac{d\sigma}{\sigma}$$

$$\le \int_0^1 \int \frac{1}{(1-\sigma)^{1/2+\epsilon}} \frac{1}{|\frac{x}{\sqrt{\nu t}} - z|^{3-2\epsilon}} \frac{1}{\sigma^{1-\epsilon}} \frac{1}{|z|^{2\epsilon}} \, dz \, d\sigma$$

$$= C_\epsilon.$$

Now, we consider the decay at infinity, measure as a power of $w = \frac{x}{\sqrt{\nu t}}$: we split the domain of integration in $\Delta_1 \cup \Delta_2$, where

$$z \in \Delta_1 \Leftrightarrow |z| > \frac{1}{2}|w| \quad \text{and} \quad z \in \Delta_2 \Leftrightarrow |z| \le \frac{1}{2}|w|.$$

We have

$$\int_0^1 \int_{\Delta_1} \frac{1}{(\sqrt{1-\sigma} + |w-z|)^4} \frac{\sigma^\alpha}{(\sqrt{\sigma} + |z|)^{2\alpha}} \, dz \, \frac{d\sigma}{\sigma}$$

$$\le (\frac{2}{|w|})^{2\alpha} \int_0^1 \int \frac{1}{(\sqrt{1-\sigma} + |w-z|)^4} \, dz \, \frac{d\sigma}{\sigma^{1-\alpha}}$$

$$= C(\frac{2}{|x|})^{2\alpha} \int_0^1 \frac{1}{\sqrt{1-\sigma}} \frac{1}{\sigma^{1-\alpha}} \, ds$$

$$= C_\alpha \frac{1}{|w|^\alpha}$$

and

$$\int_0^1 \int_{\Delta_2} \frac{1}{(\sqrt{1-\sigma} + |w-z|)^4} \frac{\sigma^\alpha}{(\sqrt{\sigma} + |z|)^{2\alpha}} \, dz \, \frac{d\sigma}{\sigma}$$

$$\le (\frac{2}{|w|})^4 \int_0^1 \int_{\Delta_2} \frac{1}{\sigma^{1-\alpha}} \frac{1}{|z|^{2\alpha}} \, dz \, d\sigma$$

$$= C_\alpha \frac{1}{|w|^{1+2\alpha}}.$$

Thus, the lemma is proved. $\qquad\qquad\qquad\qquad\qquad\qquad\qquad\qquad\qquad\quad \square$

Now, let us consider a bounded sequence $\vec{U}_n$ in E, and let $\vec{V}_n = \mathcal{T}(\vec{U}_n)$. We have:

- $\sup_{n \in \mathbb{N}} \sup_{x \in \mathbb{R}^3} (1 + |x|)^2 |\vec{V}_n(x)| < +\infty$

- $\sup_{n \in \mathbb{N}} \|\vec{V}_n\|_{\dot{B}^{1/2}_{\infty,\infty}} < +\infty$

Using Ascoli's theorem (and a diagonal process), we may find a subsequence $\vec{U}_{n_k}$ such that $\vec{V}_{n_k}$ converges uniformly on every compact subset of $\mathbb{R}^3$ to a limit $\vec{V}$. We have $\vec{V} \in E$, and moreover, for every $R > 1$,

$$\limsup_{k \to +\infty} \|\vec{V}_{n_k} - \vec{V}\|_E \leq \limsup_{k \to +\infty} \sup_{x \in \mathbb{R}^3, |x| > R} (1 + |x|)|\vec{V}_{n_k}(x) - \vec{V}(x)| = O(\frac{1}{R}).$$

Thus, $\vec{V}_{n_k}$ converge to $\vec{V}$, and we may conclude that $\mathcal{T}$ is compact.

- **a-priori estimates for the fixed points:**

Assume that we are given $\vec{u}_0$ and $\mathbb{F}_0$ as in the assumptions of Theorem 16.7 and that we have, for some $\lambda \in [0,1]$, a solution of $\vec{U} \in E$ of the fixed-point problem

$$\vec{U} = \lambda(\vec{V}_0 - \mathcal{T}(\vec{U})),$$

where $\vec{V}_0$ is given by Equation (16.12). We want to estimate $\|\vec{U}\|_E$ independently from λ.

Equivalently, we want to estimate the norm $\|\vec{u}\|_E$ for a self-similar solution $\vec{u} \in E_s$ of the equation

$$\vec{u} = \lambda(W_{\nu t} * \vec{u}_0 + \int_0^t W_{\nu(t-s)} * \mathbb{P} \operatorname{div} \mathbb{F} \, ds - B(\vec{u}, \vec{u}))$$

This is equivalent (writing $\vec{v} = \lambda \vec{u}$) to estimate independently from λ the quantity $\frac{1}{\lambda}\|\vec{v}\|_E$ of a solution $\vec{v} \in E_s$ of

$$\vec{v} = W_{\nu t} * (\lambda^2 \vec{u}_0) + \int_0^t W_{\nu(t-s)} * \mathbb{P} \operatorname{div}(\lambda^2 \mathbb{F}) \, ds - B(\vec{v}, \vec{v}).$$

This will be done by using the theory of energy inequalities for solutions that are uniformly square integrable.

Step 1: $\vec{v}$ is a local Leray solution.

We first check that $\vec{v}$ is a local Leray solution, as defined in Definition 14.1. As $\vec{v} \in E$, we have, of course, $\vec{v} \in L^\infty L^2_{\text{uloc}}$ (as $|\vec{v}(t,x)| \leq \|\vec{v}\|_E \frac{1}{|x|}$).

We have

$$\vec{v} = W_{\nu t} * \vec{v}_0 + \int_0^t W_{\nu(t-s)} * \mathbb{P} \operatorname{div} \mathbb{G} \, ds$$

with $\vec{v}_0 = \lambda^2 \vec{u}_0$ and $\mathbb{G} = \lambda^2 \mathbb{F} - \vec{v} \otimes \vec{v}$. The important point is that

$$\sup_{x \in \mathbb{R}^3} |x| |\vec{v}_0(x)| < +\infty$$

and

$$\sup_{t > 0, x \in \mathbb{R}^3} (\sqrt{\nu t} + |x|)^2 |\mathbb{G}(t,x)| < +\infty.$$

In particular, we have $\mathbb{G} \in (L^2 L^2)_{\text{uloc}}$ (as we have seen in the proof of Theorem 14.6).

We then use the following lemma:

Lemma 16.7.
Let $\vec{v}_0 \in L^2_{\text{uloc}}$ and $\mathbb{G} \in (L^2 L^2)_{\text{uloc}}$ on $(0, T) \times \mathbb{R}^3$ with $T < +\infty$. Then

$$\vec{v} = W_{\nu t} * \vec{v}_0 + \int_0^t W_{\nu(t-s)} * \mathbb{P} \operatorname{div} \mathbb{G} \, ds$$

satisfies

- $\vec{v} \in (L^\infty_t L^2_x)_{\text{uloc}} \cap (L^2 H^1_x)_{\text{uloc}}$ on $(0, T) \times \mathbb{R}^3$
- *for every compact subset K of $\mathbb{R}^3$,* $\lim_{t \to 0^+} \int_K |\vec{v}(t, x) - \vec{v}_0(t, x)|^2 \, dx = 0$.

Proof. For a ball $K = B(x_0, R_0)$ with $R_0 > 1$ and for $R > 0$ with $R > 2R_0$, write $\vec{v}_0 = \vec{A}_{K,R} + \vec{B}_{K,R}$ and $\mathbb{G} = \mathbb{C}_{K,R} + \mathbb{D}_{K,R}$, where $\vec{A}_{K,R} = 1_{B(x_0,R)} \vec{v}_0$ and $\mathbb{C}_{K,R} = 1_{B(x_0,R)} \mathbb{G}$.

As $\vec{A}_{K,R} \in L^2$, we have $W_{\nu t} * \vec{A}_{K,R} \in L^\infty L^2 \cap L^2 \dot{H}^1$ and

$$\lim_{t \to 0^+} \|W_{\nu t} * \vec{A}_{K,R} - \vec{A}_{K,R}\|_2 = 0.$$

We have, on K,

$$|W_{\nu t} * \vec{B}_{K,R}(x)| \leq C \int_{|x-y| > R/2} \frac{\sqrt{\nu t}}{|x-y|^4} |\vec{v}_0(y)| \, dy \leq C \sqrt{\nu t} \frac{1}{R} \|\vec{v}_0\|_{L^2_{\text{uloc}}}$$

and, for $l = 1, \ldots, 3$,

$$|\partial_l (W_{\nu t} * \vec{B}_{K,R})(x)| \leq C \int_{|x-y| > R/2} \frac{\sqrt{\nu t}}{|x-y|^5} |\vec{v}_0(y)| \, dy \leq C \sqrt{\nu t} \frac{1}{R^2} \|\vec{v}_0\|_{L^2_{\text{uloc}}}.$$

Thus, on $(0, T)$ with $T < +\infty$, we have $1_K (W_{\nu t} * \vec{B}_{K,R}) \in L^\infty L^2 \cap L^2 \dot{H}^1$ and $\lim_{t \to 0^+} \|1_K W_{\nu t} * \vec{B}_{K,R}\|_2 = 0$.

As $\mathbb{C}_{K,R} \in L^2((0, T), L^2)$, we have $\int_0^t W_{\nu(t-s)} * \mathbb{P} \operatorname{div} \mathbb{C}_{K,R} \, ds \in L^\infty L^2 \cap L^2 \dot{H}^1$ on $(0, T) \times \mathbb{R}^3$; as $\|\int_0^t W_{\nu(t-s)} * \mathbb{P} \operatorname{div} \mathbb{C}_{K,R} \, ds\|_2 \leq C \|\mathbb{C}_{K,R}\|_{L^2((0,t),L^2)}$, we have $\lim_{t \to 0^+} \|\int_0^t W_{\nu(t-s)} * \mathbb{P} \operatorname{div} \mathbb{C}_{K,R} \, ds\|_2 = 0$.
We have, on K,

$$|\int_0^t W_{\nu(t-s)} * \mathbb{P} \operatorname{div} \mathbb{D}_{K,R} \, ds(x)| \leq C \int_0^t \int_{|x-y| > R/2} |\mathbb{G}(s, y)| \frac{dy \, ds}{|x-y|^4}$$

$$\leq C \frac{\|\mathbb{G}\|_{(L^2 L^2)_{\text{uloc}}}}{R}$$

and, for $l = 1, \ldots, 3$,

$$|\partial_l (\int_0^t W_{\nu(t-s)} * \mathbb{P} \operatorname{div} \mathbb{D}_{K,R} \, ds)(x)| \leq C \int_0^t \int_{|x-y| > R/2} |\mathbb{G}(s, y)| \frac{dy \, ds}{|x-y|^5}$$

$$\leq C \frac{\|\mathbb{G}\|_{(L^2 L^2)_{\text{uloc}}}}{R^2}.$$

Thus, on $(0, T)$ with $T < +\infty$, we have $1_K (\int_0^t W_{\nu(t-s)} * \mathbb{P} \operatorname{div} \mathbb{D}_{K,R} \, ds) \in L^\infty L^2 \cap L^2 \dot{H}^1$. Finally, if $S > R$, we write, for $x \in K$,

$$\int_0^t \int_{|x-y|>R/2} |\mathbb{G}(s,y)| \frac{dy \, ds}{|x-y|^4}$$

$$\leq C \frac{\sqrt{t}}{R^{5/2}} \sqrt{\int_0^t \int_{B(x_0,S)} |\mathbb{G}(s,y)|^2 \, ds \, dy} + C \frac{\|\mathbb{G}\|_{(L^2 L^2)_{\mathrm{uloc}}}}{S}$$

and thus $\lim_{t \to 0+} \|1_K \int_0^t W_{\nu(t-s)} * \mathbb{P} \operatorname{div} \mathbb{D}_{K,R} \, ds\|_2 = 0$. $\qquad\square$

Finally, we check that $\vec{v}$ is suitable. Let us write $\operatorname{div} \mathbb{G} = \sum_{j=1}^3 \partial_j \vec{G}_j$. As the pressure p may be computed as $\sum_{j=1}^3 \frac{\operatorname{div} \partial_j}{\Delta}(\vec{G}_j - v_j \vec{v})$, we see that, for every $[t_0, t_1] \subset (0, +\infty)$ and every $q \in (3/2, +\infty)$, we have $p \in L^\infty([t_0, t_1], L^q)$; thus, p is locally $(L^{3/2} L^{3/2})$. Moreover, for every $[t_0, t_1] \subset (0, +\infty)$ and every $q \in (3, +\infty)$, $\vec{v} \in L^\infty([t_0, t_1], L^q)$, so that $\vec{v}$ is locally $L^4 L^4$. In that case, we know that we have the local energy equality (Proposition 13.3).

Step 2: Case of λ close to 0.

Recall that we have

$$\vec{v} = W_{\nu t} * (\lambda^2 \vec{u}_0) + \int_0^t W_{\nu(t-s)} * \mathbb{P} \operatorname{div}(\lambda^2 \mathbb{F} - \vec{v} \otimes \vec{v}) \, ds.$$

As $\lambda^2 \vec{u}_0 \in L^2_{\mathrm{uloc}}$ and $\lambda^2 \mathbb{F} \in (L^2 L^2)_{\mathrm{uloc}}$, we have seen in Theorem 14.1 a way to construct a (local-in-time) solution $\vec{w}$ of

$$\vec{w} = W_{\nu t} * (\lambda^2 \vec{u}_0) + \int_0^t W_{\nu(t-s)} * \mathbb{P} \operatorname{div}(\lambda^2 \mathbb{F} - \vec{w} \otimes \vec{w}) \, ds.$$

We began by solving a mollified problem

$$\vec{w}_\epsilon = W_{\nu t} * (\lambda^2 \vec{u}_0) + \int_0^t W_{\nu(t-s)} * \mathbb{P} \operatorname{div}(\lambda^2 \mathbb{F} - (\theta_\epsilon * \vec{w}_\epsilon) \otimes \vec{w}_\epsilon) \, ds$$

and then used Rellich's theorem to find a subsequence $\vec{w}_\epsilon$ that would converge to $\vec{w}$ on $(0, T) \times \mathbb{R}^3$. Note that in our case T may be chosen independent of λ, as the norms of $\lambda^2 \vec{u}_0$ in L^2_{uloc} and of $\lambda^2 \mathbb{F}$ in $(L^2 L^2)_{\mathrm{uloc}}$ are bounded independently from $\lambda \in [0, 1]$. Using again the inequality

$$\iint_{\mathbb{R} \times \mathbb{R}^3} \frac{1}{\nu^2(t-s)^2 + |x-y|^4} \frac{1}{(\sqrt{\nu|s|} + |y|)^2} \, dy \, ds \leq C_0 \frac{1}{\nu} \frac{1}{\sqrt{\nu|t|} + |x|} \qquad (16.15)$$

together with

$$|\frac{1}{\epsilon^3} \theta(\frac{x-y}{\epsilon}) \frac{1}{\sqrt{\nu t} + |y|} \, dy \leq C_0 \frac{1}{\sqrt{\nu t} + |x|},$$

we find that there exists a constant ϵ_0 independent from ϵ and ν, such that, if

$$\sup_{x \in \mathbb{R}^3} |x| |\lambda^2 \vec{u}_0(x)| < \epsilon_0 \nu$$

and

$$\sup_{t>0,x\in\mathbb{R}^3} (\sqrt{\nu t} + |x|)^2|\lambda^2\mathbb{F}(t,x)| < \epsilon_0\nu^2$$

then the Picard iterates will converge to $\vec{w}_\epsilon$ in the norm $\|\vec{w}\| = \sup_{t>0,x\in\mathbb{R}^3}(\sqrt{\nu t} + |x|)|\vec{w}|$ and that we will have

$$\sup_{t>0,x\in\mathbb{R}^3} (\sqrt{\nu t} + |x|)|\vec{w}_\epsilon| \leq C_\nu(\||x|\vec{u}_0\|_\infty + \sup_{t>0,x\in\mathbb{R}^3} (\sqrt{\nu t} + |x|)^2|\mathbb{F}(t,x)|)\lambda^2.$$

This inequality is transfered to the limit $\vec{w}$. Thus, $\vec{w}$ is a local Leray solution on $(0,T) \times \mathbb{R}^3$, and moreover it belongs to $L^\infty L^{3,\infty}$ and $\|\vec{w}\|_{L^\infty((0,T),L^{3,\infty})} \leq C_\nu(\||x|\vec{u}_0\|_\infty + \sup_{t>0,x\in\mathbb{R}^3}(\sqrt{\nu t} + |x|)^2|\mathbb{F}(t,x)|)\lambda^2$. As $L^{3,\infty} \subset \mathcal{V}^1$, we may use, if λ is small enough, the weak-strong uniqueness theorem for local Leray solutions (Theorem 14.7) and conclude that $\vec{v} = \vec{w}$.

As $\vec{v}$ is self-similar, we have

$$\|\vec{v}\|_E = \sup_{x\in\mathbb{R}^3}(\sqrt{T/2} + |x|)|\vec{v}(T/2,x)|$$

$$= \sup_{x\in\mathbb{R}^3}(\sqrt{T/2} + |x|)|\vec{w}(T/2,x)|$$

$$\leq C_\nu(\||x|\vec{u}_0\|_\infty + \sup_{t>0,x\in\mathbb{R}^3} (\sqrt{\nu t} + |x|)^2|\mathbb{F}(t,x)|)\lambda^2.$$

Thus, we have shown that, for some $\lambda_0 > 0$, we have $\|\vec{v}\|_E \leq C_0\lambda^2$ for $0 \leq \lambda \leq \lambda_0$ (where λ_0 and C_0 depend only on ν, $\mathbb{F}_0$ and $\vec{u}_0$).

We thus proved that any solution $\vec{U} \in E$ of

$$\vec{U} = \lambda(\vec{V}_0 - \mathcal{T}(\vec{U})),$$

with $0 \leq \lambda \leq \lambda_0$ must satisfy $\|\vec{U}\|_E \leq C_0\lambda_0$.

Step 3: Case of λ far from 0.

We now consider $\lambda_0 \leq \lambda \leq 1$. We are going to show more precisely that $\sup_{0\leq\lambda\leq1}\|\vec{v}\|_E \leq C_0$, where C_0 depends only on ν, $\mathbb{F}_0$ and $\vec{u}_0$.

Obviously, the quantities $\sup_{x\in\mathbb{R}^3}|x||\lambda^2\vec{u}_0(x)|$, $\sup_{t>0,x\in\mathbb{R}^3}(\sqrt{t} + |x|)^2|\lambda^2\mathbb{F}(x)|$ and $\sup_{t>0,x\in\mathbb{R}^3}(\sqrt{t} + |x|)^3|\lambda^2\operatorname{div}\mathbb{F}(x)|$ are bounded independently from $\lambda \in [0,1]$. We may then use the quantitative estimates for local Leray solutions given by Theorem 14.6, and find that for some $T_0 > 0$, $R_0 > 0$ and $C_1 > 0$ that depend only on ν, $\mathbb{F}_0$ and $\vec{u}_0$, we have

$$\sup_{T_0/2\leq t\leq T_0} \sup_{|x|>R_0} |x|^{1/4}|\vec{v}(t,x)| \leq C_1$$

and

$$\int_{T_0/2}^{T_0} \|\vec{v}(t,.)\|_\infty \, dt \leq C_1.$$

Thus, $\vec{V} \in L^\infty$:

$$\|\vec{V}\|_\infty = \frac{2}{T_0}\int_{T_0/2}^{T_0} \sqrt{t}\|\vec{v}(t,.)\|_\infty \, dt \leq \frac{2C_1}{\sqrt{T_0}}.$$

Moreover, $\vec{V}(x) = \sqrt{T_0}\vec{v}(T_0, \sqrt{T_0}x)$, and thus

$$\sup_{|x|>\frac{R_0}{\sqrt{T_0}}} |x|^{1/4}|\vec{V}(x)| \leq C_1 T_0^{3/8}.$$

This gives

$$|\vec{v}(t,x)| \le C_2 \frac{1}{\sqrt{t}} \frac{1}{(1+\frac{|x|}{\sqrt{t}})^{1/4}}.$$

Lemma 16.6 then gives that

$$|B(\vec{v},\vec{v})(t,x)| \le C_3 \frac{1}{\sqrt{t}} \frac{1}{(1+\frac{|x|}{\sqrt{t}})^{1/2}},$$

so that

$$|\vec{v}(t,x)| \le C_4 \frac{1}{\sqrt{t}} \frac{1}{(1+\frac{|x|}{\sqrt{t}})^{1/2}}.$$

Using again Lemma 16.6, we get

$$|B(\vec{v},\vec{v})(t,x)| \le C_5 \frac{1}{\sqrt{t}} \frac{1}{1+\frac{|x|}{\sqrt{t}}},$$

so that

$$|\vec{v}(t,x)| \le C_0 \frac{1}{\sqrt{\nu t}} \frac{1}{1+\frac{|x|}{\sqrt{\nu t}}}.$$

We thus proved that any solution $\vec{U} \in E$ of

$$\vec{U} = \lambda(\vec{V_0} - \mathcal{T}(\vec{U})),$$

with $\lambda_0 \le \lambda \le 1$ must satisfy $\|\vec{U}\|_E \le C_0 \frac{1}{\lambda_0}$.

Thus, Theorem 16.7 is proved.

16.7 The Case of Rough Data

We may now prove Theorem 16.6. We thus consider $\vec{u}_0$ a vector field on $\mathbb{R}^3$ and $\mathbb{F}_0$ a tensor function such that:

- $\vec{u}_0$ is divergence-free: div $\vec{u}_0 = 0$

- $\vec{u}_0$ is homogeneous: for all $\lambda > 0$, $\lambda \vec{u}_0(\lambda x) = \vec{u}_0(x)$

- $\vec{u}_0$ is locally square integrable on $\mathbb{R}^3 \setminus \{0\}$ (thus, $\vec{U}_0(x) = \frac{1}{|x|}\vec{\gamma}(\frac{x}{|x|})$ with $\vec{\gamma} \in L^2(S^2)$)

- $\sup_{x\in\mathbb{R}^3}(1+|x|)^2|\mathbb{F}_0(x)| < +\infty$

Using Lemma 16.5, we approximate $\vec{u}_0$ with $\vec{v}_{0,\epsilon}$ such that:

- $\vec{v}_{0,\epsilon}$ is divergence-free: div $\vec{v}_{0,\epsilon} = 0$

- $\vec{v}_{0,\epsilon}$ is homogeneous.

- $\vec{v}_{0,\epsilon}$ is locally square integrable: $\vec{v}_{0,\epsilon} = \frac{1}{|x|}\vec{\delta}_\epsilon(\frac{x}{|x|})$ with $\vec{\delta}_\epsilon \in L^2(S^2)$

- $\vec{\delta}_\epsilon \in L^\infty$ and $\|\vec{\gamma} - \vec{\delta}_\epsilon\|_{L^2(S^2)} < \epsilon$.

Moreover, using a function $\theta \in \mathcal{D}$ such that $\int \theta \, dx = 1$ and $\theta = 1$ on $B(0, 1/2)$, we truncate and regularize $\mathbb{F}_0$ into $\mathbb{F}_{0,\epsilon} = \frac{1}{\epsilon^3} \theta(\frac{\cdot}{\epsilon}) * (\theta(\epsilon.) \mathbb{F}_0(.))$.

We may then apply Theorem 16.7 to the data $\vec{v}_{0,\epsilon}$ and $\mathbb{F}_{0,\epsilon}$ and find a self-similar solution $\vec{v}_\epsilon$ such that

- $\vec{v}_\epsilon$ is divergence-free: div $\vec{v}_\epsilon = 0$

- $\vec{v}_\epsilon$ is a solution of the Navier–Stokes equations

$$\partial_t \vec{v}_\epsilon = \Delta \vec{v}_\epsilon - \sum_{i=1}^{3} v_{\epsilon,i} \partial_i \vec{v}_\epsilon - \vec{\nabla} p_\epsilon + \text{ div } \mathbb{F}_\epsilon$$

  with $\mathbb{F}_\epsilon(t, x) = \frac{1}{t} \mathbb{F}_{0,\epsilon}(\frac{x}{\sqrt{t}})$

- $\vec{v}_\epsilon \in \mathbb{E}_s$

In particular, we have seen that $\vec{v}_\epsilon$ is a local Leray solution. As the norms of $\vec{v}_{0,\epsilon}$ in L^2_{uloc} and of $\mathbb{F}_\epsilon$ in $(L^2 L^2)_{\text{uloc}}$ are controlled independently from ϵ, we see that we may apply Theorem 14.2 and find a positive time T_0 (which does not depend on ϵ) such that the family $(\vec{v}_\epsilon)_{\epsilon > 0}$ is bounded in $L^\infty((0, T_0), L^2_{\text{uloc}}) \cap (L^2(0, T_0), H^1)_{\text{uloc}}$.

Following the same lines as the proof of Theorem 14.1, we see that we may apply Rellich's theorem to get a subsequence $\epsilon_k \to 0$ such that $\vec{v}_{\epsilon,k}$ converges to a limit $\vec{u}$ strongly in $L^2_{\text{loc}}((0, T_0) \times \mathbb{R}^3)$ and *-weakly in $(L^\infty((0, T_0), L^2_{\text{uloc}}) \cap (L^2(0, T_0), H^1)_{\text{uloc}}$. Moreover, $\mathbb{F}_{\epsilon_k}$ converges strongly in $L^2_{\text{loc}}((0, T_0) \times \mathbb{R}^3)$ to $\mathbb{F}$, where $\mathbb{F}(t, x) = \frac{1}{t} \mathbb{F}_0(\frac{x}{\sqrt{t}})$. Further, we have seen that p_{ϵ_k} is locally convergent in $(L^{3/2} L^{3/2})$. Thus, we find that the limit $\vec{u}$ is a solution to the Navier–Stokes equations on $(0, T_0) \times \mathbb{R}^3$.

The self-similarity of $\vec{v}_{\epsilon,k}$ gives that, for $s, t < T_0$, we have $\vec{u}(t, x) = \sqrt{\frac{t}{s}} \vec{u}(s, \sqrt{\frac{t}{s}} x)$. This allows to extend $\vec{u}$ as a self-similar solution on $(0, +\infty)$.

It remains to show that, for every compact set of $\mathbb{R}^3$, we have $\lim_{t \to 0^+} \| \mathbb{1}_K(x)(\vec{u}(t, .) - \vec{u}_0) \|_2 = 0$. First, we notice that $\partial_t \vec{v}_\epsilon$ is bounded in $(L^2 H^{-3/2})_{\text{uloc}}$ on $(0, T_0) \times \mathbb{R}^3$, while $\vec{v}_{0,\epsilon}$ converges strongly to $\vec{u}_0$ in L^2_{uloc}. Thus, $\partial_t \vec{v}_{\epsilon_k}$ is weakly convergent to $\vec{u}$ in $L^2((0, T_0), H^{-3/2}(\Omega))$ for a bounded neighborhood Ω of K. We find that $\vec{v}_{\epsilon_k}(t, .) = \vec{v}_{0,\epsilon} + \int_0^t \partial_t \vec{v}_{\epsilon_k} \, ds$ is weakly convergent to $\vec{u}_0 + \int_0^t \partial_t \vec{u} \, ds$ in $H^{-3/2}(\Omega)$, so that

$$\vec{u}(t, .) = \vec{u}_0 + \int_0^t \partial_t \vec{u} \, ds$$

and $\vec{u}_0$ is the weak limit (in $H^{-3/2}(\Omega)$) of $\vec{u}(t, .)$ as t goes to 0. In particular, as $\mathbb{1}_K \vec{u}(t, .)$ is bounded in L^2 for $0 < t < T_0$, we find that $\vec{u}(t, .)$ converges weakly to $\vec{u}_0$ in $L^2(K)$. In particular, for every $\varphi \in \mathcal{D}$, we have

$$\| \varphi \vec{u}_0 \|_2 \leq \liminf_{t \to 0^+} \| \varphi \vec{u}(t, .) \|_2. \tag{16.16}$$

Reasoning again with the weak convergence of $H^{-3/2}$ norm $\vec{v}_{\epsilon_k}(t, .)$ and using the fact that the L^2 norm of $\vec{v}_{\epsilon_k}(t, .)$ is bounded on every compact subset of $\mathbb{R}^3$, we have the weak convergence of $\varphi \vec{v}_{\epsilon_k}(t, .)$ to $\varphi \vec{u}(t, .)$ in L^2; thus,

$$\| \varphi \vec{u}(t, .) \|_2 \leq \liminf_{k \to +\infty} \| \varphi \vec{v}_{\epsilon_k}(t, .) \|_2. \tag{16.17}$$

On the other hand, we saw in our computations (page 463) that, defining

$$\alpha_\epsilon(t) = \| \vec{v}_\epsilon \|_{L^2_{\text{uloc}}} = \sup_{\varphi \in \mathcal{B}} \| \varphi \vec{v}_\epsilon \|_2$$

$$\gamma_\epsilon(t) = \| \vec{\nabla} \otimes \vec{v}_\epsilon \|_{(L^2 L^2)_{\text{uloc}}((0,t) \times \mathbb{R}^3)} = \sup_{\varphi \in \mathcal{B}} \left(\int_0^t \int |\varphi(x) \vec{\nabla} \otimes \vec{v}_\epsilon(s, x)|^2 \, dx \, ds \right)^{1/2}$$

we had, for $0 < t < T_0$ and $K = \text{Supp } \varphi$,

$$\|\varphi\vec{v}_\epsilon\|_2^2 + \nu \int_0^t \int |\varphi\vec{\nabla} \otimes \vec{v}_\epsilon|^2 \, dx \leq \|\varphi\vec{v}_{0,\epsilon}\|_2^2$$

$$+ C_0\nu \int_0^t \alpha_\epsilon(s)^2 \, ds + C_0 \frac{1}{\nu} \int_0^t \int_K |\mathbb{F}_\epsilon(s,y)|^2 \, dy \, ds$$

$$+ C_0 \left(\int_0^t \alpha_\epsilon(s)^2 \, ds + \gamma_\epsilon(t) \left(\int_0^t \alpha_\epsilon^6(s) \, ds \right)^{1/6} \right)^{3/2}$$

As $A = \sup_{0<\epsilon<1} \sup_{0<t<T_0} \alpha_\epsilon < +\infty$ and $\Gamma = \sup_{0<\epsilon<1} \gamma_\epsilon(T_0) < +\infty$, we obtain

$$\|\varphi\vec{v}_\epsilon\|_2^2 \leq \|\varphi\vec{v}_{0,\epsilon}\|_2^2 + C_0 \frac{1}{\nu} \int_0^t \int_K |\mathbb{F}_\epsilon(s,y)|^2 \, dy \, ds + C_0\nu A^2 t + C_0(A^2 t + A\Gamma t^{1/6})^{2/3}.$$

Inequality (16.17) then gives

$$\|\varphi\vec{u}(t,.)\|_2^2 \leq \|\varphi\vec{u}_0\|_2^2 + C_0 \frac{1}{\nu} \int_0^t \int_K |\mathbb{F}(s,y)|^2 \, dy \, ds + C_0\nu A^2 t + C_0(A^2 t + A\Gamma t^{1/6})^{2/3}.$$

and

$$\limsup_{t \to 0^+} \|\varphi\vec{u}(t,.)\|_2 \leq \|\varphi\vec{u}_0\|_2. \tag{16.18}$$

Inequalities (16.16) and (16.18) show that we have the convergence of $\|\varphi\vec{u}(t,.)\|_2$ to $\|\varphi\vec{u}_0\|_2$, hence the weak convergence of $\varphi\vec{u}(t,.)$ to $\varphi\vec{u}_0$ in L^2 becomes a strong convergence.

Thus, Theorem 16.6 is proved.

16.8 Non-existence of Backward Self-similar Solutions

In this section, we prove the theorem of Nečas, Růžička, and Šverák [375] and Tsai [482], which states that no non-trivial backward self-similar solution can exist (in the case of a null force).

Tsai's theorem

Theorem 16.8.
Let $\vec{U}$ be a vector field on $\mathbb{R}^3$ and P a distribution, and $\vec{u}(t,x) = \frac{1}{\sqrt{-t}}\vec{U}(\frac{x}{\sqrt{-t}})$ the associated self-similar vector field on $(-\infty, 0) \times \mathbb{R}^3$ and $p(t,x) = \frac{1}{-t}p(\frac{x}{\sqrt{-t}})$ the associated self-similar distribution, such that:

- *$\vec{U}$ is divergence-free: $\text{div } \vec{U} = 0$*

- *$\vec{u}$ is a solution of the Navier–Stokes equations*

$$\partial_t \vec{u} = \nu\Delta\vec{u} - \sum_{i=1}^3 u_i\partial_i\vec{u} - \vec{\nabla}p$$

- *$\sup_{-1<t<0} \int_{B(0,1)} |\vec{u}(t,x)|^2 \, dx < +\infty$*

- $\int_{-1}^{0} \int_{B(0,1)} |\vec{\nabla} \otimes \vec{u}(t,x)|^2 \, dx \, dt < +\infty$

Then $\vec{U} = 0$.

Proof. This theorem was first proved in the case $\vec{U} \in L^3$ by Nečas, Růžička, and Šverák [375] then extended to the case $\vec{U} \in \mathcal{C}_0$ by Tsai [482] (remark that, due to the uniqueness of L^3 solutions of the Navier–Stokes equations, the assumption $\vec{U} \in L^3$ implies that $\vec{U} \in \mathcal{C}_0$); in both cases, one finds that $\vec{U}$ is $\mathcal{C}^\infty$ and thus satisfies the assumptions of Theorem 16.8.

The proof of Theorem 16.8 relies on a lemma in Tsai [482] that is a variation of Hopf's strong maximum principle [237]:

Lemma 16.8 (Strong maximum principle).
Let Π, $b_1, \ldots, b_3$ be real-valued continuous functions on $\mathbb{R}^3$ so that Π is of class $\mathcal{C}^2$ and, for $j = 1, \ldots, 3$, $\lim_{x \to \infty} \frac{|b_j(x)|}{|x|} = 0$. Let $\vec{b} = (b_1, b_2, b_3)$ and let $\vec{X}$ be the identical vector field on $\mathbb{R}^3$: $\vec{X}(x) = (x_1, x_2, x_3)$. If

$$\nu \Delta \Pi - \frac{1}{2}(\vec{b} + \vec{X}) \cdot \vec{\nabla} \Pi \geq 0 \quad \text{and} \quad \lim_{x \to \infty} \frac{|\Pi(x)|}{|x|^2} = 0,$$

then Π is constant.

Proof. Let us define $\mathcal{L}$ as the differential operator $\mathcal{L}f = \nu \Delta f - \frac{1}{2}(\vec{b} + \vec{X}) \cdot \vec{\nabla} f$. The hypothesis is then $\mathcal{L}\Pi \geq 0$.

Let us assume that Π is not constant. We select two points X_0 and X_1 so that $\Pi(X_0) < \Pi(X_1)$. We select two real numbers R_0 and R_1 so that

- $0 < R_0 < |X_0 - X_1| < R_1$

- R_1 is large enough, so that, for $|x - X_0| \geq R_1$, $|\vec{b}(x) + \vec{X}_0| \leq |x - X_0|/2$

- R_0 is small enough, so that for $|x - X_0| \leq R_0$, $\Pi(x) \leq (\Pi(X_0) + \Pi(X_1))/2$

- $R_1^2 > 12\nu$.

We then consider the function

$$V(x) = \Pi(x) + \alpha(e^{-\beta|x - X_0|^2} - \gamma|x - X_0|^2),$$

where the positive numbers α, β and γ are chosen in the following way:

- first, we choose β big enough to ensure that the function $\phi_\beta(x) = e^{-\beta|x - X_0|^2}$ satisfies $\mathcal{L}\phi_\beta > 0$ on $|x - X_0| \geq R_0$. This is always possible, since we have

$$\mathcal{L}\phi_\beta = e^{-\beta|x - X_0|^2} \beta \left((4\beta\nu + 1)|x - X_0|^2 - 6\nu + (\vec{b} + \vec{X}_0).(\vec{X} - \vec{X}_0) \right):$$

  since $|\vec{b}(x)|$ may be bounded by $C(R_0)|x - X_0|$ for $|x - X_0| \geq R_0$, it is enough to take $\beta > 0$ with $4\beta\nu + 1 > C(R_0) + 6\nu R_0^{-2} + |X_0| R_0^{-1}$.

- next, we choose γ small enough to ensure that the function

$$\psi(x) = |x - X_0|^2$$

satisfies

$$\gamma \sup_{R_0 \leq |x - X_0| \leq R_1} |\mathcal{L}\psi(x)| < \min_{R_0 \leq |x - X_0| \leq R_1} \mathcal{L}\phi_\beta(x).$$

- finally, we choose α small enough to ensure that

$$\alpha \sup_{|x-X_0|\leq R_1} |\phi_\beta(x) - \gamma\psi(x)| < (\Pi(X_1) - \Pi(X_0))/4.$$

The function V is $\mathcal{C}^2$; we have $V(x) \sim -\alpha\gamma|x|^2$ when $x \to \infty$, thus V has a maximum at some point X_2; at this point, we must have $\vec{\nabla}V(X_2) = 0$ and $\Delta V(X_2) \leq 0$. Thus, we have $\mathcal{L}V(X_2) \leq 0$.

For $|x - X_0| \leq R_0$, we have $V(x) \leq \Pi(X_0) + \frac{3}{4}(\Pi(X_1) - \Pi(X_0)) < V(X_1)$. Thus, we cannot have $|X_2 - X_0| \leq R_0$.

For $R_0 \leq |x - X_0| \leq R_1$, we have $\gamma|\mathcal{L}\psi(x)| < \mathcal{L}\phi_\beta(x)$ and $\mathcal{L}\Pi(x) \geq 0$, hence $\mathcal{L}V(x) > 0$. Thus, we cannot have $R_0 \leq |X_2 - X_0| \leq R_1$.

For $|x - X_0| \geq R_1$, we have $\mathcal{L}\phi_\beta(x) > 0$, $\mathcal{L}\Pi(x) \geq 0$, and

$$\mathcal{L}\psi(x) = 6\nu - (\vec{b} + \vec{X}_0).(\vec{X} - \vec{X}_0) - |x - X_0|^2 \leq 6\nu - \frac{1}{2}|x - X_0|^2 < 0,$$

hence $\mathcal{L}V(x) > 0$. Thus, we cannot have $|X_2 - X_0| \geq R_1$.

Therefore, Π must be constant. $\qquad\square$

Proof of Theorem 16.8:

Step 1: the Leray equations.

The study of backward self–similar solutions was initiated by Leray [328], as they were a potential model for blow-up of weak solutions. The self-similarity assumption transforms the (parabolic) Navier–Stokes equations into (elliptic) equations, *the Leray equations*:

$$\begin{cases} \nu\Delta\vec{U} = \frac{1}{2}(\vec{U} + \vec{X}\cdot\vec{\nabla}\vec{U}) + \vec{U}\cdot\vec{\nabla}\vec{U} + \vec{\nabla}P \\ \operatorname{div}\vec{U} = 0 \end{cases} \tag{16.19}$$

Taking the divergence of those equations, we get

$$\Delta P = -\sum_{i=1}^{3}\sum_{j=1}^{3}\partial_i\partial_j((U_i - \frac{x_i}{2})U_j) = -\sum_{i=1}^{3}\sum_{j=1}^{3}\partial_i\partial_j(U_iU_j) \tag{16.20}$$

The assumption $\int_{-1}^{0}\int_{B(0,1)}|\vec{\nabla}\otimes\vec{u}(t,x)|^2\,dx\,dt < +\infty$ implies that $\vec{U}$ belongs to H^1_{loc}. From the hypoellipticity of the Laplace operator Δ (which implies that, when $\Delta f \in H^s_{\mathrm{loc}}$, then $f \in H^{s+2}_{\mathrm{loc}}$), we get that $\vec{U} \in H^s_{\mathrm{loc}}$ with $s \geq 1$ implies

- $\vec{U}\otimes\vec{U} \in H^{s-\frac{1}{2}}_{\mathrm{loc}}$ so that $P \in H^{s-\frac{1}{2}}_{\mathrm{loc}}$ (due to Equation (16.20))
- $\vec{U} \in H^{s+\frac{1}{2}}_{\mathrm{loc}}$ (due to Equations (16.19)).

Thus, P and $\vec{U}$ are $\mathcal{C}^\infty$ functions.

Step 2: the energy balance.

Writing

$$\partial_t\vec{u} = \nu\Delta\vec{u} - (\vec{\nabla}\wedge\vec{u})\wedge\vec{u} - \vec{\nabla}(p + \frac{1}{2}|\vec{u}|^2),$$

we get

$$\partial_t(\frac{1}{2}|\vec{u}|^2) = \nu\vec{u}\cdot\Delta\vec{u} - \vec{u}\cdot\vec{\nabla}(p + \frac{1}{2}|\vec{u}|^2) = \nu\Delta(\frac{1}{2}|\vec{u}|^2) - \nu|\vec{\nabla}\otimes\vec{u}|^2 - \vec{u}\cdot\vec{\nabla}(p + \frac{1}{2}|\vec{u}|^2).$$

Using the self-similarity assumption, we get

$$\frac{1}{2}\vec{U}\cdot(\vec{U} + \vec{X}\cdot\vec{\nabla}\vec{U}) = \nu\Delta(\frac{1}{2}|\vec{U}|^2) - \nu|\vec{\nabla}\otimes\vec{U}|^2 - \vec{U}\cdot\vec{\nabla}(P + \frac{1}{2}|\vec{U}|^2).$$

Inserting Equation (16.20), we get

$$(\nu\Delta - \vec{U}\cdot\vec{\nabla})(P + \frac{1}{2}|\vec{U}|^2) = \frac{1}{2}\nu|\vec{\nabla}\wedge\vec{U}|^2 + \frac{1}{2}\vec{U}\cdot(\vec{U} + \vec{X}\cdot\vec{\nabla}\vec{U})$$

Let

$$\mathcal{L} = \nu\Delta - \vec{U}\cdot\vec{\nabla} - \frac{1}{2}\vec{X}\cdot\vec{\nabla}.$$

We have

$$\mathcal{L}(P + \frac{1}{2}|\vec{U}|^2) = \frac{1}{2}\nu|\vec{\nabla}\wedge\vec{U}|^2 + R$$

with

$$\begin{aligned}
R &= \frac{1}{2}\vec{U}\cdot(\vec{U} + \vec{X}\cdot\vec{\nabla}\vec{U}) - \frac{1}{2}\vec{X}\cdot\vec{\nabla}(P + \frac{1}{2}|\vec{U}|^2)\\
&= \frac{1}{2}|\vec{U}|^2 - \frac{1}{2}\vec{X}\cdot\vec{\nabla}P\\
&= \frac{1}{2}|\vec{U}|^2 - \frac{1}{2}\vec{X}.(\nu\Delta\vec{U} - \frac{1}{2}(\vec{U} + \vec{X}\cdot\vec{\nabla}\vec{U}) - \vec{U}\cdot\vec{\nabla}\vec{U})\\
&= -\frac{1}{2}(\nu\Delta - \vec{U}\cdot\vec{\nabla} - \frac{1}{2}\vec{X}\cdot\vec{\nabla})(\vec{X}\cdot\vec{U})
\end{aligned}$$

Thus, if we define

$$\Pi = P + \frac{1}{2}|\vec{U}|^2 + \frac{1}{2}\vec{X}\cdot\vec{U}$$

we find that

$$\mathcal{L}\Pi = \frac{1}{2}\nu|\vec{\nabla}\wedge\vec{U}|^2 \geq 0.$$

The strategy of the proof is now clear: we are going to prove that $\vec{U} = o(|x|)$ and that $P = o(|x|^2)$ for $x \to \infty$. Applying Lemma 16.8, we shall find $\Pi = 0$, hence $\vec{\nabla}\wedge\vec{U} = 0$. This gives now $\Delta\vec{U} = -\vec{\nabla}\wedge(\vec{\nabla}\wedge\vec{U}) = 0$; hence $\vec{U}$ is harmonic and $o(|x|)$, thus $\vec{U}$ must be constant. We find that the constant is null by evaluating $\int_{B(0,1)}|\vec{u}(t,x)|^2\,dx = |\vec{U}|^2|B(0,1)|\frac{1}{|t|}$: since this must be bounded as t goes to 0^-, we have $\vec{U} = 0$.

Step 3: a formula for the pressure.

We are going to check that, given a test function $\omega \in \mathcal{D}(\mathbb{R}^3)$ with $\omega(x) = 1$ on $B(0,1)$, then P may be computed as

$$P = \lim_{R\to+\infty}\frac{1}{\Delta}(\omega(\frac{x}{R})\Delta P).$$

First we check that P has a slow increase at infinity, we have

$$\int_{B(0,R)}|\vec{U}(x)|^2\,dx = R\int_{B(0,1)}|\vec{u}(-R^{-2},y)|^2\,dy = O(R);$$

as $\vec{\nabla} P = \nu \Delta \vec{U} - \mathrm{div}((\vec{U} + \frac{1}{2}\vec{X}) \otimes \vec{U})$, we find that $\vec{\nabla} P$ belongs to $\mathcal{S}'$, and so does P. Next, we show the existence of

$$P_0 = \lim_{R \to +\infty} \frac{1}{\Delta}(\omega(\frac{x}{R})\Delta P).$$

We have

$$\frac{1}{\Delta}(\omega(\frac{x}{R})\Delta P) = -\sum_{i=1}^{3}\sum_{j=1}^{3}\frac{1}{\Delta}(\omega(\frac{x}{R})\partial_i\partial_j(U_iU_j))$$

$$= -\sum_{i=1}^{3}\sum_{j=1}^{3}\frac{\partial_i\partial_j}{\Delta}(\omega(\frac{x}{R})U_iU_j)$$

$$+\sum_{i=1}^{3}\sum_{j=1}^{3}\frac{1}{\Delta}(\frac{1}{R^2}\partial_i\partial_j\omega(\frac{x}{R})U_iU_j) + 2\sum_{i=1}^{3}\sum_{j=1}^{3}\frac{1}{\Delta}(\frac{1}{R}\partial_i\omega(\frac{x}{R})U_i\partial_jU_j)$$

We have, for $R \geq 1$, and for $T \geq R$,

$$\int_{B(0,R)}|\vec{\nabla} \otimes \vec{U}(x)|^2 \, dx \leq \int_{B(0,T)}|\vec{\nabla} \otimes \vec{U}(x)|^2 \, dx = \frac{1}{T}\int_{B(0,1)}|\vec{\nabla} \otimes \vec{u}(-T^{-2},y)|^2 \, dy$$

so that

$$\int_{B(0,R)}|\vec{\nabla} \otimes \vec{U}(x)|^2 \, dx \leq R\int_{-R^2}^{0}\int_{B(0,1)}|\vec{\nabla} \otimes \vec{u}(-s,y)|^2 \, dy \, ds = o(R).$$

A first consequence, through the Sobolev embedding that gives us

$$\|\vec{U}\|_{L^6(B(0,R))} \leq C(\frac{1}{R}\|\vec{U}\|_{L^2(B(0,R))} + \|\vec{\nabla} \otimes \vec{U}\|_{L^2(B(0,R))}) = o(\sqrt{R}),$$

is that, for $2 \leq q \leq 6$ and $\alpha > q/2$, we have

$$\int |\vec{U}(x)|^q \frac{dx}{(1+|x|)^\alpha} < +\infty.$$

If we choose $2 < q < 6$ and $q/2 < \alpha < 3$, we obtain that $U_iU_j \in L^{q/2}(w\,dx)$ for $w = (1+|x|)^{-\alpha}$; as $\alpha < 3$, we have that w belongs to the Muckenhoupt class $\mathcal{A}_{q/2}$ (with $q/2 > 1$), so that the singular integral operator $\frac{\partial_i\partial_j}{\Delta}$ is bounded on $L^{q/2}(w\,dx)$. As we have the weak convergence of $\omega(\frac{x}{R})U_iU_j$ to U_iU_j in $L^{q/2}(w\,dx)$, we obtain that

$$\lim_{R \to +\infty} -\sum_{i=1}^{3}\sum_{j=1}^{3}\frac{\partial_i\partial_j}{\Delta}(\omega(\frac{x}{R})U_iU_j) = -\sum_{i=1}^{3}\sum_{j=1}^{3}\frac{\partial_i\partial_j}{\Delta}(U_iU_j).$$

On the other hand, we have

$$\|\frac{1}{\Delta}(\frac{1}{R^2}\partial_i\partial_j\omega(\frac{x}{R})U_iU_j)\|_{L^{3,\infty}} \leq C\frac{1}{R^2}\|\partial_i\partial_j\omega(\frac{x}{R})U_iU_j\|_1 = O(\frac{1}{R}) \to 0$$

and

$$\|\frac{1}{\Delta}(\frac{1}{R}\partial_i\omega(\frac{x}{R})U_i\partial_jU_j)\|_{L^{3,\infty}} \leq C\frac{1}{R}\|\partial_j\omega(\frac{x}{R})U_i\partial_jU_j\|_1 = o(1) \to 0.$$

Thus, P_0 is well defined.

The distribution $P - P_0$ is a tempered distribution such that $\Delta(P - P_0) = 0$: it is a harmonic polynomial. Let us write

$$P - P_0 = \sum_{|\alpha| \leq N} c_\alpha x^\alpha.$$

If $\varphi \in \mathcal{D}$ is such that $\int \varphi \, dx = 1$ and $\int x^\alpha \varphi \, dx = 0$ for $0 < |\alpha| \leq N$, we have

$$c_\alpha = \frac{(-1)^{|\alpha|}}{\alpha!} \int (P - P_0) \partial^\alpha \varphi \, dx.$$

If $|\alpha| > 0$, we write $\partial^\alpha = \partial_j \partial^\beta$, and get

$$
\begin{aligned}
c_\alpha &= \frac{(-1)^{|\beta|}}{\alpha!} \left(\int \partial_j P \partial^\beta \varphi \, dx + \int P_0 \partial^\alpha \varphi \, dx \right) \\
&= \frac{(-1)^{|\beta|}}{\alpha!} \left(-\nu \int \vec{\nabla} U_j \cdot \vec{\nabla} \partial^\beta \varphi \, dx + \int U_j (\vec{U} + \tfrac{1}{2}\vec{X}) \cdot \vec{\nabla} \partial^\beta \varphi \, dx + \int P_0 \partial^\alpha \varphi \, dx \right)
\end{aligned}
$$

The same identity holds for $\varphi_R(x) = \frac{1}{R^3} \varphi(\frac{x}{R})$, so that we get

$$
\begin{aligned}
|c_\alpha| \leq C(\nu \, o(\sqrt{R}) R^{-\frac{3}{2} - |\alpha|} &+ O(R) R^{-3 - |\alpha|} + O(\sqrt{R}) RR^{-\frac{3}{2} - |\alpha|}) \\
&+ C\|\partial^\alpha \varphi_R\|_{L^{\frac{q}{q-2}}(w^{-\frac{2}{q-2}} dx)}
\end{aligned}
$$

As

$$\|\partial^\alpha \varphi_R\|_{L^{\frac{q}{q-2}}(w^{-\frac{2}{q-2}} dx)} \leq C R^{-3 - |\alpha|} (R^{3 + |\alpha|\frac{2}{q-2}})^{1 - \frac{2}{q}} = C R^{-\frac{6}{q} - |\alpha|(1 - \frac{2}{q})}$$

we find that $c_\alpha = 0$ for $\alpha \neq 0$.

As P is defined up to a constant, we may choose $P = P_0$.

Step 4: estimates for $\vec{U}$.

We want to show that $\vec{U}$ is $o(|x|)$. We cannot use Leray's Equations 16.19, as the term $\vec{X}$ cannot provide a $o(|x|)$ estimate. Thus, we go back to the Navier–Stokes equations, and work with $\vec{u}$ and p.

The assumption on $\vec{u}$ and $\vec{\nabla} \otimes \vec{u}$ gives $\vec{u} \in L^2((-1,0), H^1(B(0,1)) \subset L^2((-1,0), L^{6,2}(B(0,1))$. By self-similarity, we find that, for $R \geq 1$,

$$\|\vec{u}\|_{L^2((-1,0), L^{6,2}(B(0,R))} = \sqrt{R} \, \|\vec{u}\|_{L^2((-R^{-2},0), L^{6,2}(B(0,1))} = o(\sqrt{R}).$$

Let ω be a function in $\mathcal{D}$ supported in $B(0,2)$ and equal to 1 on $B(0,1)$, let $\omega_0 = \omega(\frac{x}{4})$, $\theta_0 = \omega_0(\frac{x}{2}) - \omega_0(x)$ and $\theta_j(x) = \theta(2^{-j}x)$. We write

$$p = -\frac{\vec{\nabla} \operatorname{div}}{\Delta}(\vec{u} \otimes \vec{u}) = \Pi_0 + \sum_{j=0}^{+\infty} Q_j$$

with

$$\Pi_0 = -\frac{\vec{\nabla} \operatorname{div}}{\Delta}(\omega_0 \vec{u} \otimes \vec{u})$$

and

$$Q_j = -\frac{\vec{\nabla} \operatorname{div}}{\Delta}(\theta_j \vec{u} \otimes \vec{u}).$$

We write also

$$\partial_t(\omega\vec{u}) = \nu\Delta(\omega\vec{u}) + \vec{\beta} + \sum_{j=0}^{+\infty}\vec{\beta}_j - \sum_{i=1}^{3}\partial_i\vec{\gamma}_i - \sum_{j=0}^{+\infty}\vec{\nabla}\delta_j$$

with

$$\begin{cases} \vec{\beta} = & (\Delta\omega)\vec{u} + (\vec{u}\cdot\vec{\nabla}\omega)\vec{u} + \Pi_0\vec{\nabla}\omega \\ \vec{\beta}_j = & Q_j\vec{\nabla}\omega \\ \vec{\gamma}_i = & 2(\partial_i\omega)\vec{u} + \omega u_i\vec{u} + \omega\Pi_0 \\ \delta_j = & \omega Q_j \end{cases}$$

Defining $\mathcal{T}$ as

$$Tf(t,x) = \int_{-1}^{t} W_{\nu(t-s)}(x-y)f(s,y)\,ds$$

we find that, for $-1 < t < 0$,

$$\omega\vec{u} = W_{\nu(t+1)} * (\omega\vec{u}(-1,.)) + \mathcal{T}(\vec{\beta} + \sum_{j=0}^{+\infty}\vec{\beta}_j - \sum_{i=1}^{3}\partial_i\vec{\gamma}_i - \sum_{j=0}^{+\infty}\vec{\nabla}\delta_j).$$

We may then show that $\vec{u} \in L^1((-1,0), L^\infty(B(0,1)))$ (or more precisely that $\omega\vec{u} \in L^1((-1,0), L^\infty)$):

- $\int_{-1}^{0} \|W_{\nu(t+1)} * (\omega\vec{u}(-1,.))\|_\infty\,dt \leq C\|\omega\vec{U}\|_{\dot{H}^1}\int_{-1}^{0}(t+1)^{-1/4}\,dt < +\infty.$
- We have, for a constant C_0 (which does not depend on ν), the inequality[2]

$$\|\mathcal{T}(\sqrt{-\Delta}f)\|_{L^1((-1,0),L^{3,1})} \leq C_0\|f\|_{L^1((-1,0),L^{3/2,1})}$$

and thus, since $\mathcal{T}(f) = \frac{1}{\sqrt{-\Delta}}\mathcal{T}(\sqrt{-\Delta}f)$, we have the inequality

$$\|\mathcal{T}(f)\|_{L^1((-1,0),L^\infty)} \leq C_0\|f\|_{L^1((-1,0),L^{3/2,1})}$$

As $\vec{\beta} \in L^1((-1,0), L^{3/2,1})$, we find that $\mathcal{T}\vec{\beta} \in L^1((-1,0), L^\infty)$.

- We have $\vec{\gamma}_i \in L^1((-1,0), \dot{B}_{2,1}^{1/2})$, so that

$$\|\mathcal{T}(\partial_i\vec{\gamma}_i)\|_{L^1((-1,0),\dot{B}_{2,1}^{3/2})} \leq C\|\vec{\gamma}_i\|_{L^1((-1,0),\dot{B}_{2,1}^{1/2})} < +\infty$$

From the embedding $\dot{B}_{2,1}^{3/2} \subset L^\infty$, we find that $\mathcal{T}(\partial_i\vec{\gamma}_i) \in L^1((-1,0), L^\infty)$.

- We have

$$\|\vec{\beta}_j(t,.)\|_\infty \leq C\frac{1}{2^{3j}}\int_{B(0,2^{j+4})}|\vec{u}(t,x)|^2\,dx = C\frac{1}{2^{3j}t}\int_{B(0,2^{j+4})}|\vec{U}(\frac{x}{\sqrt{t}})|^2\,dx$$

and thus

$$\|\vec{\beta}_j(t,.)\|_\infty \leq C\frac{\sqrt{t}}{2^{3j}}\int_{B(0,\sqrt{t}2^{j+4})}|\vec{U}(x)|^2\,dx \leq \frac{C'\sqrt{t}}{2^{3j}}\max(1,\sqrt{t}2^{j+4}) \leq C''2^{-2j}$$

so that

$$\|\mathcal{T}\vec{\beta}_j(t,.)\|_\infty \leq \int_{-1}^{t}\|\vec{\beta}_j(s,.)\|_\infty\,ds \leq C2^{-2j}.$$

and $\mathcal{T}(\vec{\beta}_j) \in L^\infty((-1,0), L^\infty) \subset L^1((-1,0), L^\infty)$.

---

[2]This inequality may be seen as an (indirect) dual inequality to Meyer's inequality $\|\mathcal{T}(\sqrt{-\Delta}f)\|_{L^\infty((-1,0),L^{3,\infty})} \leq C_0\|f\|_{L^1((-1,0),L^{3/2,\infty})}$. See Meyer [359] or Lemarié-Rieusset [313].

- Similarly, we have

$$\|\delta_j(t,.)\|_\infty \le C\frac{1}{2^{3j}}\int_{B(0,2^{j+4})}|\vec{u}(t,x)|^2\,dx \le C'2^{-2j}.$$

so that

$$\|\mathcal{T}(\vec{\nabla}\delta_j)(t,.)\|_\infty \le C\int_{-1}^{t}\frac{1}{\sqrt{t-s}}\|\delta_j(\tau,.)\|_\infty\,ds \le C2^{-2j}$$

and $\mathcal{T}(\vec{\nabla}\delta_j) \in L^\infty((-1,0),L^\infty) \subset L^1((-1,0),L^\infty)$.

Thus, $\vec{u} \in L^1((-1,0),L^\infty(B(0,1)))$. In particular, we find that, for $|x| \ge 1$, and $R > |x|$,

$$|\vec{U}(x)| = |\vec{u}(-1,x)| = |\frac{1}{R}\vec{u}(-\frac{1}{R^2},\frac{x}{R})| \le \frac{1}{R}\|\vec{u}(-\frac{1}{R^2},.)\|_{L^\infty(B(0,1))}$$

so that

$$|\vec{U}(x)| \le 2|x|^2\int_{-\frac{1}{|x|^2}}^{-\frac{1}{2|x|^2}}\frac{1}{\sqrt{-t}}\|\vec{u}(t,.)\|_{L^\infty(B(0,1))}\,dt = o(|x|).$$

Step 5: estimates for $\vec{\nabla}\otimes\vec{U}$.

We now show that $\vec{\nabla}\otimes\vec{u} \in L^1((-1,0),L^{3,1}(B(0,1)))$ (or more precisely that $\vec{\nabla}\otimes(\omega\vec{u}) \in L^1((-1,0),L^{3,1})$).

For $l = 1,\dots,3$, we write

$$\partial_l(\omega\vec{u}) = W_{\nu(t+1)}*\partial_l(\omega\vec{U}) + \mathcal{T}(\partial_l\vec{\beta} + \sum_{j=0}^{+\infty}\partial_l\vec{\beta}_j - \sum_{i=1}^{3}\partial_l\partial_i\vec{\gamma}_i - \sum_{j=0}^{+\infty}\partial_l\vec{\nabla}\delta_j).$$

and:

- $\int_{-1}^{0}\|W_{\nu(t+1)}*(\omega\vec{u}(-1,.))\|_{L^{3,1}}\,dt \le C\|\omega\vec{U}\|_{\dot{H}^1}\int_{-1}^{0}(t+1)^{-1/4}\,dt < +\infty$.
- Meyer's inequality

$$\|\mathcal{T}(\partial_l f)\|_{L^1((-1,0),L^{3,1})} \le C_0\|f\|_{L^1((-1,0),L^{3/2,1})}$$

and the fact that $\vec{\beta} \in L^1((-1,0),L^{3/2,1})$ give us that $\mathcal{T}\partial_l\vec{\beta} \in L^1((-1,0),L^{3,1})$.
- we have $\vec{\gamma}_i \in L^1((-1,0),\dot{B}_{2,1}^{1/2})$, so that

$$\|\mathcal{T}(\partial_l\partial_i\vec{\gamma}_i)\|_{L^1((-1,0),\dot{B}_{2,1}^{1/2})} \le C\|\vec{\gamma}_i\|_{L^1((-1,0),\dot{B}_{2,1}^{1/2})} < +\infty$$

From the embedding $\dot{B}_{2,1}^{1/2} \subset L^{3,1}$, we find that $\mathcal{T}(\partial_i\vec{\gamma}_i) \in L^1((-1,0),L^{3,1})$.
- We have $\partial_l\vec{\beta}_j = (\partial_l Q_j)\vec{\nabla}\omega + Q_j\partial_l\vec{\nabla}\omega$ and we find

$$\|\partial_l\vec{\beta}_j(t,.)\|_\infty \le C\frac{1}{2^{3j}}\int_{B(0,2^{j+4})}|\vec{u}(t,x)|^2\,dx \le C'2^{-2j}.$$

As $\vec{\beta}_j$ is supported in $B(0,4)$, we control the $L^{3,1}$ norm of $\partial_l\vec{\beta}_j$ by its L^∞ norm, so that

$$\|\mathcal{T}\vec{\beta}_j(t,.)\|_{L^{3,1}} \le \int_{-1}^{t}\|\vec{\beta}_j(s,.)\|_{L^{3,1}}\,ds \le C2^{-2j}.$$

and $\mathcal{T}(\vec{\beta}_j) \in L^\infty((-1,0),L^{3,1}) \subset L^1((-1,0),L^{3,1})$.

- Similarly, we have

$$\|\partial_t \delta_j(t,.)\|_\infty \le C \frac{1}{2^{3j}} \int_{B(0,2^{j+4})} |\vec{u}(t,x)|^2 \, dx \le C' 2^{-2j}.$$

so that

$$\|\mathcal{T}(\vec{\nabla}\delta_j)(t,.)\|_{L^{3,1}} \le C \int_{-1}^t \frac{1}{\sqrt{t-s}} \|\delta_j(\tau,.)\|_{L^{3,1}} \, ds \le C 2^{-2j}$$

and $\mathcal{T}(\vec{\nabla}\delta_j) \in L^\infty((-1,0), L^{3,1}) \subset L^1((-1,0), L^{3,1})$.

Thus, $\vec{\nabla} \otimes u \in L^1((-1,0), L^\infty(B(0,1)))$.

Step 6: estimates for P.

We write again

$$p = -\frac{\vec{\nabla}\operatorname{div}}{\Delta}(\vec{u} \otimes \vec{u}) = \Pi_0 + \sum_{j=0}^{+\infty} Q_j$$

with

$$\Pi_0 = -\frac{\vec{\nabla}\operatorname{div}}{\Delta}(\omega_0 \vec{u} \otimes \vec{u})$$

and

$$Q_j = -\frac{\vec{\nabla}\operatorname{div}}{\Delta}(\theta_j \vec{u} \otimes \vec{u}).$$

We are interested in estimating ωP on $(-1,0)$. We already know that

$$\|\omega Q_j(t,.)\|_\infty = \|\delta_j(t,.)\|_\infty \le C 2^{-2j}.$$

On the other hand, we have

$$\|\operatorname{div}(\omega_0 \vec{u} \otimes \vec{u})\|_{L^{3,1}} \le C(\|\vec{\nabla} \otimes \vec{u}\|_{L^{3,1}(B(0,4))} \|\vec{u}\|_{L^\infty(B(0,4))} + \|\vec{u}\|_{L^{6,2}(B(0,4))}^2).$$

Thus, we find that $p \in L^{1/2}((-1,0), B(0,1))$.

In particular, we find that, for $|x| \ge 1$, and $R > |x|$,

$$|P(x)| = |p(-1,x)| = |\frac{1}{R^2} p(-\frac{1}{R^2}, \frac{x}{R})| \le \frac{1}{R^2} \|p(-\frac{1}{R^2}, .)\|_{L^\infty(B(0,1))}$$

so that

$$|P(x)| \le \left(2|x|^2 \int_{-\frac{1}{|x|^2}}^{-\frac{1}{2|x|^2}} \frac{1}{\sqrt{-t}} \|p(t,.)\|_{L^\infty(B(0,1))}^{1/2} \, dt \right)^2 = o(|x|^2).$$

The theorem is proved.

$\square$

16.9   Discretely Self-similar Solutions

In this section, we are going to give a new proof of the results of Chae and Wolf [100] and Bradshaw and Tsai [57] on the existence of λ-DSS solutions of the Navier–Stokes problem.

Definition 16.1.
Let $\vec{u}_0 \in L^2_{\mathrm{loc}}(\mathbb{R}^3)$. We say that $\vec{u}_0$ is a λ-discretely self-similar function (λ-DSS) if there exists $\lambda > 1$ such that $\lambda \vec{u}_0(\lambda x) = \vec{u}_0(x)$.

A vector field $\vec{u} \in L^2_{\mathrm{loc}}([0, +\infty) \times \mathbb{R}^3)$ is λ-DSS if there exists $\lambda > 1$ such that $\lambda \vec{u}(\lambda^2 t, \lambda x) = \vec{u}(t, x)$.

A forcing tensor $\mathbb{F} \in L^2_{\mathrm{loc}}([0, +\infty) \times \mathbb{R}^3)$ is λ-DSS if there exists $\lambda > 1$ such that $\lambda^2 \mathbb{F}(\lambda^2 t, \lambda x) = \mathbb{F}(t, x)$.

$\vec{u}_0 \in L^2_{\mathrm{loc}}$ is self-similar (i.e. homogeneous of degree -1) if and only if it is of the form $\vec{u}_0 = \dfrac{\vec{w}_0(\frac{x}{|x|})}{|x|}$ with $\vec{w}_0 \in L^2(S^2)$. In particular, $\vec{u}_0 \in L^2_{\mathrm{uloc}}$. In contrast with self-similar functions, if $\vec{u}_0 \in L^2_{\mathrm{loc}}$ is discretely self-similar, it may fail to belong to L^2_{uloc}. If we want to apply the theory of weak solutions, we must use weighted Lebesgue spaces: we have $\vec{u}_0 \in L^2(\Phi_\gamma \, dx)$ with $\Phi_\gamma = \frac{1}{(1+|x|)^\gamma}$ and $\gamma > 1$.

Theorem 16.9.
Let $4/3 < \gamma < 2$ and $\lambda > 1$. If $\vec{u}_0$ is a λ-DSS divergence-free vector field (such that $\vec{u}_0 \in L^2(\Phi_\gamma \, dx)$) and if $\mathbb{F}$ is a λ-DSS tensor $\mathbb{F}(t, x) = (F_{i,j}(t, x))_{1 \leq i,j \leq 3}$ such that $\mathbb{F} \in L^2_{\mathrm{loc}}([0, +\infty) \times \mathbb{R}^3)$, then the Navier–Stokes equations with initial value $\vec{u}_0$

$$(NS) \begin{cases} \partial_t \vec{u} = \nu \Delta \vec{u} - (\vec{u} \cdot \vec{\nabla})\vec{u} - \vec{\nabla}p + \vec{\nabla} \cdot \mathbb{F} \\[2mm] \vec{\nabla} \cdot \vec{u} = 0, \qquad\qquad \vec{u}(0, .) = \vec{u}_0 \end{cases}$$

has a global weak solution $\vec{u}$ such that:

- *$\vec{u}$ is a λ-DSS vector field*

- *for every $0 < T < +\infty$, $\vec{u}$ belongs to $L^\infty((0, T), L^2(\Phi_\gamma \, dx))$ and $\vec{\nabla} \otimes \vec{u}$ belongs to $L^2((0, T), L^2(\Phi_\gamma \, dx))$*

- *the map $t \in [0, +\infty) \mapsto \vec{u}(t, .)$ is weakly continuous from $[0, +\infty)$ to $L^2(\Phi_\gamma \, dx)$ and is strongly continuous at $t = 0$:*

$$\lim_{t \to 0} \|\vec{u}(t, .) - \vec{u}_0\|_{L^2(\Phi_\gamma \, dx)} = 0.$$

- *the solution $\vec{u}$ is suitable: there exists a non-negative locally finite measure μ on $(0, +\infty) \times \mathbb{R}^3$ such that*

$$\partial_t \left(\frac{|\vec{u}|^2}{2}\right) = \nu \Delta \left(\frac{|\vec{u}|^2}{2}\right) - \nu |\vec{\nabla}\vec{u}|^2 - \vec{\nabla} \cdot \left(\left(\frac{|\vec{u}|^2}{2} + p\right)\vec{u} \right) + \vec{u} \cdot (\vec{\nabla} \cdot \mathbb{F}) - \mu.$$

Proof. The main idea of the proof (due to Chae and Wolf [100]) is to mollify the equation with a mollification which is stable under dilations. We choose $\theta \in \mathcal{D}(\mathbb{R}^3)$ with $\int \theta \, dx = 1$ and define, for $\epsilon > 0$,

$$\theta_\epsilon(t, x) = \frac{1}{(\epsilon\sqrt{t})^3}\theta(\frac{x}{\epsilon\sqrt{t}}).$$

We then consider the mollified the Navier–Stokes equations

$$(NS_\epsilon) \begin{cases} \partial_t \vec{u} = \nu\Delta\vec{u} - ((\theta_\epsilon *_x \vec{u}) \cdot \vec{\nabla})\vec{u} - \vec{\nabla}p + \vec{\nabla} \cdot \mathbb{F} \\ \vec{\nabla} \cdot \vec{u} = 0, \qquad \vec{u}(0, .) = \vec{u}_0 \end{cases}$$

where

$$\theta_\epsilon *_x \vec{u}(t, x) = \int_{\mathbb{R}^3} \theta_\epsilon(t, x - y)\vec{u}(t, y) \, dy.$$

Remark that, for $\vec{u}_\lambda(t, x) = \lambda\vec{u}(\lambda^2 t, \lambda x)$, we have

$$\theta_\epsilon *_x \vec{u}_\lambda(t, x) = \lambda\int_{\mathbb{R}^3} \theta_\epsilon(t, x - y)\vec{u}(\lambda^2 t, \lambda y) \, dy$$
$$= \lambda\int_{\mathbb{R}^3} \frac{1}{\lambda^3}\theta_\epsilon(t, x - \frac{z}{\lambda})\vec{u}(\lambda^2 t, z) \, dz$$
$$= \lambda\int_{\mathbb{R}^3} \frac{1}{\lambda^3}\frac{1}{(\epsilon\sqrt{t})^3}\theta(\frac{x - \frac{z}{\lambda}}{\sqrt{t}})\vec{u}(\lambda^2 t, z) \, dz$$
$$= \lambda\int_{\mathbb{R}^3} \frac{1}{(\epsilon\sqrt{\lambda^2 t})^3}\theta(\frac{\lambda x - z}{\sqrt{\lambda^2 t}})\vec{u}(\lambda^2 t, z) \, dz$$
$$= \lambda(\theta_\epsilon *_x \vec{u})(\lambda^2 t, \lambda x).$$

Thus, if $\vec{u}$ is a solution to (NS_ϵ), $\vec{u}_\lambda$ is a solution to (NS_ϵ) as well (with initial value $\lambda\vec{u}_0(\lambda x) = \vec{u}_0(x)$).

The next step is to linearize the problem (NS_ϵ) into

$$(LNS_\epsilon) \begin{cases} \partial_t \vec{u} = \nu\Delta\vec{u} - ((\theta_\epsilon *_x \vec{v}) \cdot \vec{\nabla})\vec{u} - \vec{\nabla}p + \vec{\nabla} \cdot \mathbb{F} \\ \vec{\nabla} \cdot \vec{u} = 0, \qquad \vec{u}(0, .) = \vec{u}_0 \end{cases}$$

where $\vec{v}$ is a divergence-free λ-DSS vector field such that $\vec{v} \in L^3_{loc}([0, +\infty)\times\mathbb{R}^3)$. Again, if $\vec{u}$ is a solution to (LNS_ϵ), $\vec{u}_\lambda$ is a solution to (LNS_ϵ) as well (with initial value $\lambda\vec{u}_0(\lambda x) = \vec{u}_0(x)$).

The last step is to further mollify (LNS_ϵ). We choose a non-negative $\phi \in \mathcal{D}(\mathbb{R}^3)$, equal to 1 on a neighborhood of 0, with $\int \phi \, dx = 1$ and define, for $\alpha > 0$,

$$\phi_\alpha(x) = \frac{1}{\alpha^3}\phi(\frac{x}{\alpha}).$$

We then consider the problem

$$(LNS_{\epsilon,\alpha}) \begin{cases} \partial_t \vec{u} = \nu\Delta\vec{u} - ((\phi_\alpha * \mathbb{P}(\phi(\alpha x)(\theta_\epsilon *_x \vec{v}))) \cdot \vec{\nabla})\vec{u} - \vec{\nabla}p + \vec{\nabla} \cdot (\phi(\alpha x)\mathbb{F}) \\ \vec{\nabla} \cdot \vec{u} = 0, \qquad \vec{u}(0, .) = \mathbb{P}(\phi(\alpha x)\vec{u}_0). \end{cases}$$

a) Solving the linear problem with two mollifications.

As $\vec{v}$ is a divergence-free a λ-DSS vector field which is locally in $L^3 L^3$, we have

$$\int_0^T \int |\vec{v}(t,x)|^3 \frac{1}{(1+|x|)^{3\gamma/2}} \, dx$$

$$\leq \int_0^T \int_{B(0,1)} |\vec{v}(t,x)|^3 \, dx \, dt + \sum_{k=0}^{+\infty} \lambda^{-3k\gamma/2} \int_0^T \int_{B(0,\lambda^{k+1})} |\vec{v}(t,x)|^3 \, dx \, dt$$

$$= \int_0^T \int_{B(0,1)} |\vec{v}(t,x)|^3 \, dx \, dt + \sum_{k=0}^{+\infty} \lambda^{-3k\gamma/2} \lambda^{2(k+1)} \int_0^{\lambda^{-k-1}T} \int_{B(0,1)} |\vec{v}(t,x)|^3 \, dx \, dt$$

$$\leq (\int_0^T \int_{B(0,1)} |\vec{v}(t,x)|^3 \, dx \, dt)(1 + \lambda^2 \sum_{k=0}^{+\infty} \left(\frac{1}{\lambda^{3/2}}\right)^{k(\gamma-\frac{4}{3})}).$$

Thus, for $\gamma > 4/3$, we have $\vec{v} \in L^3((0,T), L^3(\frac{1}{(1+|x|)^{3\gamma/2}} \, dx))$.

Similarly, as $\mathbb{F}$ is a λ-DSS tensor such that $\mathbb{F} \in L^2_{\text{loc}}([0,+\infty) \times \mathbb{R}^3)$, we have

$$\int_0^T \int |\mathbb{F}(t,x)|^2 \frac{1}{(1+|x|)^\gamma} \, dx$$

$$\leq \int_0^T \int_{B(0,1)} |\mathbb{F}(t,x)|^2 \, dx \, dt + \sum_{k=0}^{+\infty} \lambda^{-k\gamma} \int_0^T \int_{B(0,\lambda^{k+1})} |\mathbb{F}(t,x)|^2 \, dx \, dt$$

$$= \int_0^T \int_{B(0,1)} |\mathbb{F}(t,x)|^2 \, dx \, dt + \sum_{k=0}^{+\infty} \lambda^{-k\gamma} \lambda^{k+1} \int_0^{\lambda^{-k-1}T} \int_{B(0,1)} |\mathbb{F}(t,x)|^2 \, dx \, dt$$

$$\leq (\int_0^T \int_{B(0,1)} |\mathbb{F}(t,x)|^2 \, dx \, dt)(1 + \lambda \sum_{k=0}^{+\infty} \left(\frac{1}{\lambda}\right)^{k(\gamma-1)}).$$

Thus, for $\gamma > 1$, we have $\mathbb{F} \in L^2((0,T), L^2(\frac{1}{(1+|x|)^\gamma} \, dx))$.

We have

$$|\theta_\epsilon *_x \vec{v}| \leq C M_{\vec{v}(t,\cdot)}.$$

For $\gamma < 2$, the weight $\frac{1}{(1+|x|)^{3\gamma/2}}$ belongs to the Muckenhoupt class $\mathcal{A}_3$, so that

$$\mathbb{P}(\phi(\alpha x)(\theta_\epsilon *_x \vec{v})) \in L^3((0,T), L^3))$$

and, for $4/3 < \gamma < 2$,

$$\sup_{\alpha>0,\epsilon>0} \|\mathbb{P}(\phi(\alpha x)(\theta_\epsilon *_x \vec{v}))\|_{L^3((0,T),L^3(\frac{1}{(1+|x|)^{3\gamma/2}} \, dx))} < +\infty.$$

We have as well

$$\phi(\alpha x)\mathbb{F} \in L^2((0,T), L^2))$$

and, for $1 < \gamma$,

$$\sup_{\alpha>0} \|\phi(\alpha x)\mathbb{F}\|_{L^2((0,T),L^2(\frac{1}{(1+|x|)^\gamma} \, dx))} < +\infty.$$

Finally, we have

$$\mathbb{P}(\phi(\alpha x)\vec{u}_0) \in L^2$$

and, for $1 < \gamma$,

$$\sup_{\alpha>0} \|\mathbb{P}(\phi(\alpha x)\vec{u}_0)\|_{L^2(\frac{1}{(1+|x|)^\gamma} \, dx)} < +\infty.$$

A last preliminary estimate is

$$\|\phi_\alpha * \mathbb{P}(\phi(\alpha x)(\theta_\epsilon *_x \vec{v}))\|_{L^3((0,T),L^\infty)} \leq \frac{\|\phi\|_{3/2}}{\alpha}\|\mathbb{P}(\phi(\alpha x)(\theta_\epsilon *_x \vec{v}))\|_{L^\infty((0,T),L^3)}.$$

Thus, the problem $(LNS_{\epsilon,\alpha})$ can be written as

$$\begin{cases} \partial_t \vec{u} = \nu\Delta\vec{u} - (\vec{w}\cdot\vec{\nabla})\vec{u} - \vec{\nabla}p + \text{div }\mathbb{G} \\ \\ \vec{\nabla}\cdot\vec{u} = 0, \qquad\qquad \vec{u}(0,.) = \vec{U}_0 \end{cases}$$

with $\vec{U}_0 \in L^2$, div $\vec{U}_0 = 0$, $\vec{w} \in \cap_{T>0}L^3((0,T),L^\infty)$, div $\vec{w} = 0$ and $\mathbb{G} \in \cap_{T>0}L^2((0,T),L^2)$. This linear problem is easily solved. We write the problem as

$$\vec{u} = \vec{V}_0 - \mathcal{L}(\vec{u})$$

with

$$\vec{V}_0 = W_{\nu t} * \vec{U}_0 + \int_0^t w_{\nu(t-s)} * \mathbb{P}\text{ div }\mathbb{G}\,ds$$

and

$$\mathcal{L}(\vec{u}) = \int_0^t w_{\nu(t-s)} * \mathbb{P}\text{ div}(\vec{w}\otimes\vec{u})\,ds.$$

We have $\vec{V}_0 \in \cap_{0<T}L^\infty((0,T)L^2) \cap L^2((0,T),\dot{H}^1)$, while

$$\|\mathcal{L}(\vec{u})\|_{L^\infty((0,T)L^2)\cap L^2((0,T),\dot{H}^1)} \leq C_\nu T^{1/6}\|\vec{w}\|_{L^3((0,T),L^\infty)}\|\vec{u}\|_{L^\infty((0,T),L^2)}.$$

We thus find a solution $\vec{u}_1 \in \mathcal{C}([0,T_1],L^2) \cap L^2((0,T_1),L^2)$, where T_1 is the largest time such that $T_1^{1/6}\|\vec{w}\|_{L^3((0,T_1),L^\infty)} \leq \frac{1}{2C_\nu}$. Then, we iterate the construction by considering the problem from initial time T_1 and with initial value $\vec{u}_1(T_1,.)$. We find a solution $\vec{u}_2 \in \mathcal{C}([T_1,T_2],L^2) \cap L^2((T_1,T_2),L^2)$, where T_2 is the largest time such that $(T_2 - T_1)^{1/6}\|\vec{w}\|_{L^3((T_1,T_2),L^\infty)} \leq \frac{1}{2C_\nu}$. By induction, we find solutions $\vec{u}_n$ defined on (T_{n-1},T_n), where T_n is the largest time such that $(T_n - T_{n-1})^{1/6}\|\vec{w}\|_{L^3((T_{n-1},T_n),L^\infty)} \leq \frac{1}{2C_\nu}$. We have $\lim_{n\to+\infty} T_n = +\infty$, so that we find a global solution $\vec{u}$ for $(LNS_{\epsilon,\alpha})$ by defining $\vec{u}$ piecewise as $\vec{u} = \vec{u}_n$ on $[T_{n-1},T_n]$.

b) On the size of solutions of $(LNS_{\epsilon,\alpha})$.

We have found, for every $\alpha > 0$ and every $\epsilon > 0$, a solution $\vec{u}_{\alpha,\epsilon}$ for $(LNS_{\epsilon,\alpha})$; the next step is to prove estimates on $\vec{u}_{\alpha,\epsilon}$ which don't depend on α nor on ϵ. We have the identity

$$\begin{aligned} \partial_t|\vec{u}_{\alpha,\epsilon}|^2 =& \nu\Delta|\vec{u}_{\alpha,\epsilon}|^2 - 2\nu|\vec{\nabla}\otimes\vec{u}_{\alpha,\epsilon}|^2 - 2\text{ div}(p_{\alpha,\epsilon}\vec{u}_{\alpha,\epsilon}) \\ & - \text{div}(|\vec{u}_{\alpha,\epsilon}|^2(\phi_\alpha * \mathbb{P}(\phi(\alpha x)(\theta_\epsilon *_x \vec{v})))) \\ & + 2\vec{u}_{\alpha,\epsilon}\cdot\vec{\nabla}\cdot(\phi(\alpha x)\mathbb{F}). \end{aligned} \tag{16.21}$$

Integrating against Φ_γ, we obtain

$$\begin{aligned} \frac{d}{dt}(\|\vec{u}_{\alpha,\epsilon}\|^2_{L^2(\Phi_\gamma\,dx)}) =& \nu\Delta|\vec{u}_{\alpha,\epsilon}|^2 - 2\nu|\vec{\nabla}\otimes\vec{u}_{\alpha,\epsilon}|^2 - \text{div}(p_{\alpha,\epsilon}\vec{u}_{\alpha,\epsilon}) \\ & - \text{div}(|\vec{u}_{\alpha,\epsilon}|^2(\phi_\alpha * \mathbb{P}(\phi(\alpha x)(\theta_\epsilon *_x \vec{v})))) \\ & + \vec{u}_{\alpha,\epsilon}\cdot\vec{\nabla}\cdot(\phi(\alpha x)\mathbb{F}). \end{aligned} \tag{16.22}$$

For $1 < \gamma < 2$, we have $|\vec{\nabla}\Phi_\gamma| \leq \gamma\Phi_{\gamma+1} \leq 2\Phi_\gamma^{3/2} \leq 2\Phi_\gamma$ and $|\Delta\Phi_\gamma| \leq \gamma(\gamma+1)\Phi_{\gamma+1} \leq 6\Phi_\gamma$. We thus have

$$\frac{d}{dt}(\|\vec{u}_{\alpha,\epsilon}\|_{L^2(\Phi_\gamma \, dx)}^2) \leq 6\nu\|\vec{u}_{\alpha,\epsilon}\|_{L^2(\Phi_\gamma \, dx)}^2 - 2\nu\|\vec{\nabla}\otimes\vec{u}_{\alpha,\epsilon}\|_{L^2(\Phi_\gamma \, dx)}^2$$
$$+ 4\int |p_{\alpha,\epsilon}|\,|\vec{u}_{\alpha,\epsilon}|\,\Phi_\gamma^{3/2}\,dx$$
$$+ 2\int |\vec{u}_{\alpha,\epsilon}|^2|\phi_\alpha * \mathbb{P}(\phi(\alpha x)(\theta_\epsilon *_x \vec{v}))|\,\Phi_\gamma^{3/2}\,dx$$
$$+ 4\int |\vec{u}_{\alpha,\epsilon}|\,\phi(\alpha x)\,|\mathbb{F}|\Phi_\gamma\,dx + 2\int |\vec{\nabla}\otimes\vec{u}_{\alpha,\epsilon}|\,\phi(\alpha x)\,|\mathbb{F}|\Phi_\gamma\,dx$$

We then write

$$\Delta p_{\alpha,\epsilon} + \sum_{i=1}^{3}\sum_{j=1}^{3}\partial_i\partial_j(u_{\alpha,\epsilon,i}(\phi_\alpha * \mathbb{P}(\phi(\alpha x)(\theta_\epsilon *_x \vec{v})))_j) = \sum_{i=1}^{3}\sum_{j=1}^{3}\partial_i\partial_j(\phi(\alpha x)F_{i,j})$$

so that

$$p_{\alpha,\epsilon} = q_{\alpha,\epsilon} - \varpi_\alpha$$
$$= \sum_{i=1}^{3}\sum_{j=1}^{3}R_iR_j(u_{\alpha,\epsilon,i}(\phi_\alpha * \mathbb{P}(\phi(\alpha x)(\theta_\epsilon *_x \vec{v})))_j) - \sum_{i=1}^{3}\sum_{j=1}^{3}R_iR_j(\phi(\alpha x)F_{i,j}).$$

As the Riesz transforms are bounded on $L^2(\Phi_\gamma \, dx)$ (since $\gamma < 3$), we find

$$\|\varpi_\alpha\|_{L^2(\Phi_\gamma \, dx)} \leq C_\gamma\|\mathbb{F}\|_{L^2(\Phi_\gamma \, dx)}.$$

We have seen that, for $4/3 < \gamma < 2$,

$$\sup_{\alpha>0,\epsilon>0}\|\mathbb{P}(\phi(\alpha x)(\theta_\epsilon *_x \vec{v}))\|_{L^3((0,T),L^3(\frac{1}{(1+|x|)^{3\gamma/2}}\,dx)} < +\infty.$$

Moreover, $\sqrt{\Phi_\gamma}\vec{u}_{\alpha,\epsilon}$ is controlled in H^1, hence in L^3:

$$\|\sqrt{\Phi_\gamma}\vec{u}_{\alpha,\epsilon}\|_3 \leq C\|\sqrt{\Phi_\gamma}\vec{u}_{\alpha,\epsilon}\|_2^{1/2}\|\vec{\nabla}\otimes(\sqrt{\Phi_\gamma}\vec{u}_{\alpha,\epsilon})\|_2^{1/2}$$
$$\leq C'\|\sqrt{\Phi_\gamma}\vec{u}_{\alpha,\epsilon}\|_2^{1/2}(\|\sqrt{\Phi_\gamma}\vec{u}_{\alpha,\epsilon}\|_2^{1/2} + \|\sqrt{\Phi_\gamma}\vec{\nabla}\otimes\vec{u}_{\alpha,\epsilon}\|_2^{1/2})$$
$$= C'\|\vec{u}_{\alpha,\epsilon}\|_{L^2(\Phi_\gamma \, dx)}^{1/2}(\|\vec{u}_{\alpha,\epsilon}\|_{L^2(\Phi_\gamma \, dx)}^{1/2} + \|\vec{\nabla}\otimes\vec{u}_{\alpha,\epsilon}\|_{L^2(\Phi_\gamma \, dx)}^{1/2}).$$

As $3\frac{\gamma}{2} < 3$, we find that

$$\|q_{\alpha,\epsilon}\|_{L^{3/2}(\Phi_{3\gamma/2}\,dx)} \leq C_\gamma\|\sqrt{\Phi_\gamma}\vec{u}_{\alpha,\epsilon}\|_3\|\vec{v}\|_{L^3(\Phi_{3\gamma/2}\,dx)}.$$

We obtain, for constants C_i depending only on γ,

$$\frac{d}{dt}(\|\vec{u}_{\alpha,\epsilon}\|_{L^2(\Phi_\gamma \, dx)}^2) \leq 6\nu\|\vec{u}_{\alpha,\epsilon}\|_{L^2(\Phi_\gamma \, dx)}^2 - 2\nu\|\vec{\nabla}\otimes\vec{u}_{\alpha,\epsilon}\|_{L^2(\Phi_\gamma \, dx)}^2$$
$$+ C_0\|\vec{\nabla}\otimes\vec{u}_{\alpha,\epsilon}\|_{L^2(\Phi_\gamma \, dx)}\|\mathbb{F}\|_{L^2(\Phi_\gamma \, dx)} + C_1\|\vec{u}_{\alpha,\epsilon}\|_{L^2(\Phi_\gamma \, dx)}\|\mathbb{F}\|_{L^2(\Phi_\gamma \, dx)}$$
$$+ C_2\|\vec{u}_{\alpha,\epsilon}\|_{L^2(\Phi_\gamma \, dx)}^2\|\vec{v}\|_{L^3(\Phi_{3\gamma/2}\,dx)}$$
$$+ C_3\|\vec{u}_{\alpha,\epsilon}\|_{L^2(\Phi_\gamma \, dx)}\|\vec{\nabla}\otimes\vec{u}_{\alpha,\epsilon}\|_{L^2(\Phi_\gamma \, dx)}\|\vec{v}\|_{L^3(\Phi_{3\gamma/2}\,dx)}$$
$$\leq -\nu\|\vec{\nabla}\otimes\vec{u}_{\alpha,\epsilon}\|_{L^2(\Phi_\gamma \, dx)}^2$$
$$+ \frac{4}{27}(C_2 + C_3)\|\vec{v}\|_{L^3(\Phi_{3\gamma/2}\,dx)}^3 + \frac{2C_0^2 + C_1^2}{\nu}\|\mathbb{F}\|_{L^2(\Phi_\gamma \, dx)}^2$$
$$+ \frac{25}{4}\nu\|\vec{u}_{\alpha,\epsilon}\|_{L^2(\Phi_\gamma \, dx)}^2 + C_2\|\vec{u}_{\alpha,\epsilon}\|_{L^2(\Phi_\gamma \, dx)}^3 + \frac{C_3^4}{4\nu^3}\|\vec{u}_{\alpha,\epsilon}\|_{L^2(\Phi_\gamma \, dx)}^6$$

As the Riesz transforms are bounded on L^2_Φ, we know that $\|\mathbb{P}(\phi(\alpha x)\vec{u}_0).\|_{L^2(\Phi_\gamma\, dx)} \leq C_5\|\vec{u}_0\|_{L^2(\Phi_\gamma\, dx)}$. Let

$$A_1^2 = C_5^2\|\vec{u}_0\|^2_{L^2(\Phi_\gamma\, dx)} + \frac{2C_0^2 + C_1^2}{\nu}\int_0^1 \|\mathbb{F}\|^2_{L^2_\Phi}\, dt + \frac{4}{27}(C_2 + C_3)\int_0^1 \|\vec{v}\|^3_{L^3(\Phi_{3\gamma/2}\, dx)}\, dt.$$

We see that we have

$$\|\vec{u}_{\alpha,\epsilon}\|^2_{L^2(\Phi_\gamma\, dx)} + \nu\int_0^{t_0} \|\vec{\nabla}\otimes\vec{u}_{\alpha,\epsilon}\|^2_{L^2(\Phi_\gamma\, dx)}\, dt \leq 4A_1^2 \tag{16.23}$$

as long as $t_0 \leq 1$ and

$$t_0\left(25\nu + 8C_2A_1 + \frac{16C_3^4}{\nu^3}A_1^4\right) \leq 3.$$

As $8C_2A_1 \leq 6\nu + 2C_2^4\frac{A_1^4}{\nu^3}$, we see that, on $(0, \min(T_0, 1))$ with

$$T_0 = \frac{3}{31\nu + \frac{16C_3^4 + 2C_2^4}{\nu^3}A_1^4},$$

inequality (16.23) is fulfilled.

c) Solving the linear problem with one mollification.

Let us fix ϵ. The solutions $\vec{u}_{\alpha,\epsilon}$ are bounded in $L^\infty((0, \min(T_0, 1)), L^2(\Phi_\gamma\, dx)) \cap L^2((0, \min(T_0, 1)), H^1(\Phi_\gamma\, dx))$, so that there exists a subsequence $(\vec{u}_{\alpha_n,\epsilon})_{n\in\mathbb{N}}$ (with $\alpha_n \to 0$) such that $\vec{u}_{\alpha_n,\epsilon}$ converge *-weakly in $L^\infty((0, \min(T_0, 1)), L^2(\Phi_\gamma\, dx))$ and weakly in $L^2((0, \min(T_0, 1)), H^1(\Phi_\gamma\, dx))$ to a limit $\vec{u}_\epsilon$.

On the other hand, we have the strong convergence of $\phi(\alpha x)\mathbb{F}$ to $\mathbb{F}$ in $L^2((0, \min(T_0, 1)), L^2(\Phi_\gamma\, dx))$, of $\phi_\alpha * \mathbb{P}(\phi(\alpha x)(\theta_\epsilon *_x \vec{v}))$ to $\theta_\epsilon *_x \vec{v}$ in $L^3((0, \min(T_0, 1)), L^3(\Phi_{3\gamma/2}\, dx))$, and of $\mathbb{P}(\phi(\alpha x)\vec{u}_0)$ to $\vec{u}_0$ in $L^2(\Phi_\gamma\, dx)$. This gives the weak convergence of $p_{\alpha_n,\epsilon}$ to $p_\epsilon = q_\epsilon - \varpi$, where $q_\epsilon = \sum_{i=1}^3\sum_{j=1}^3 R_iR_j(u_{\epsilon,i}(\theta_\epsilon *_x \vec{v}))_j)$ is the weak limit of $q_{\alpha_n,\epsilon}$ in $L^{3/2}((0, \min(T_0, 1)), L^{3/2}(\Phi_{3\gamma/2}\, dx))$ and $\varpi = \sum_{i=1}^3\sum_{j=1}^3 R_iR_j(F_{i,j})$ is the strong limit of ϖ_{α_n} in $L^2((0, \min(T_0, 1)), L^2(\Phi_\gamma\, dx))$.

Thus, we find a solution $\vec{u}_\epsilon$ of (LNS_ϵ) on $(0, \min(T_0, 1))$. Writing $\vec{u}_\epsilon(t, .) = \vec{u}_0 + \int_0^t \partial_t\vec{u}_\epsilon\, ds$ as the weak limit of $\vec{u}_{\alpha_n,\epsilon}(t, .)$, we see that $t \mapsto \vec{u}_\epsilon(t, .)$ is weakly continuous from $[0, \min(T_0, 1)]$ to $L^2(\Phi_\gamma\, dx)$. As $\|\vec{u}_{\alpha_n,\epsilon}(t, .)\|_{L^2(\Phi_\gamma\, dx)} \leq 2A_1$ on $[0, \min(T_0, 1)]$, we find that

$$\|\vec{u}_\epsilon\|^2_{L^2(\Phi_\gamma\, dx)} \leq \|\vec{u}_0\|^2_{L^2(\Phi_\gamma\, dx)} - \nu\int_0^t \|\vec{\nabla}\otimes\vec{u}_\epsilon\|^2_{L^2(\Phi_\gamma\, dx)}\, ds$$
$$+ \frac{4}{27}(C_2 + C_3)\int_0^t \|\vec{v}\|^3_{L^3(\Phi_{3\gamma/2}\, dx)}\, ds + \frac{2C_0^2 + C_1^2}{\nu}\int_0^t \|\mathbb{F}\|^2_{L^2(\Phi_\gamma\, dx)}\, ds$$
$$+ t\left(25\nu A_1^2 + 8C_2A_1^3 + 16\frac{C_3^4}{\nu^3}\right),$$

so that

$$\limsup_{t\to 0} \|\vec{u}_\epsilon(t, .)\|_{L^2(\Phi_\gamma\, dx)} \leq \|\vec{u}_0\|_{L^2(\Phi_\gamma\, dx)}.$$

By weak continuity,

$$\|\vec{u}_0\|_{L^2(\Phi_\gamma\, dx)} \leq \liminf_{t\to 0} \|\vec{u}_\epsilon(t, .)\|_{L^2(\Phi_\gamma\, dx)}$$

and we get finally

$$\lim_{t\to 0}\|\vec{u}_\epsilon(t,.)-\vec{u}_0\|_{L^2(\Phi_\gamma\,dx)}=0.$$

Assume that $(\vec{w}_\epsilon, q_\epsilon)$ is another solution of (LNS_ϵ) on $(0, \min(T_0, 1))$ such that $w_\epsilon \in L^\infty((0, \min(T_0, 1)), L^2(\Phi_\gamma\,dx)) \cap L^2((0, \min(T_0, 1)), H^1(\Phi_\gamma\,dx))$ and $\lim_{t\to 0}\|\vec{w}_\epsilon(t,.) - \vec{u}_0\|_{L^2(\Phi_\gamma\,dx)} = 0$. Let $\vec{z}_\epsilon = \vec{u}_\epsilon - \vec{w}_\epsilon$ and $r_\epsilon = p_\epsilon - q_\epsilon$. Then $\vec{z}_\epsilon$ is a solution of

$$\begin{cases} \partial_t \vec{z}_\epsilon = \nu\Delta\vec{z}_\epsilon - ((\theta_\epsilon *_x \vec{v})\cdot\vec{\nabla})\vec{z}_\epsilon - \vec{\nabla}r_\epsilon \\[2mm] \vec{\nabla}\cdot\vec{z}_\epsilon = 0, \qquad\qquad \lim_{t\to 0}\|\vec{z}_\epsilon(t,.)\|_{L^2(\Phi_\gamma\,dx)} = 0 \end{cases}$$

with $r_\epsilon = \sum_{i=1}^3\sum_{j=1}^3 R_i R_j(z_{\epsilon,i}(\theta_\epsilon *_x \vec{v}))_j)$. As $\theta_\epsilon *_x \vec{v}$ is locally $L_t^3 L_x^\infty$, we have enough regularity to write (in $\mathcal{D}'((0, \min(T_0, 1)) \times \mathbb{R}^3)$)

$$\partial_t(|\vec{z}_\epsilon|^2) = \nu\Delta(|\vec{z}_\epsilon|^2) - 2\nu|\vec{\nabla}\otimes\vec{z}_\epsilon|^2 - \operatorname{div}(|\vec{z}_\epsilon|^2(\theta_\epsilon *_x \vec{v})) - 2\operatorname{div}(r_\epsilon\vec{z}_\epsilon).$$

We integrate this equality against the test function φ_{η,R,t_0} defined (for $0 < t_0 < \min(T_0, 1)$) in the following way:

- We take a function $\alpha \in \mathcal{C}^\infty(\mathbb{R})$ which is non-decreasing, with $\alpha(t)$ equal to 0 for $t < 1/2$ and equal to 1 for $t > 1$. For $0 < \eta < \min(T_0, 1) - t_0$, we define

$$\alpha_{\eta,t_0}(t) = \alpha(\frac{t-\eta}{\eta}) - \alpha(\frac{t-t_0}{\eta}).$$

- We take as well a non-negative function $\psi \in \mathcal{D}(\mathbb{R}^3)$ which is equal to 1 for $|x| \le 1$ and to 0 for $|x| \ge 2$. For $R > 0$, we define $\psi_R(x) = \psi(\frac{x}{R})$.

- Finally, we define $\Phi_{\gamma,R}(x) = \left(1 + \sqrt{\frac{1}{R^2} + |x|^2}\right)^{-\gamma}$.

Then, we choose

$$\varphi_{\eta,R,t_0}(t,x) = \alpha_{\eta,t_0}(t)\psi_R(x)\Phi_{\gamma,R}(x).$$

Letting R go to $+\infty$, we find

$$-\iint \partial_t\alpha_{\eta,t_0}|\vec{z}_\epsilon|^2\,\Phi_\gamma\,dx\,dt = -\nu\iint \alpha_{\eta,t_0}\vec{\nabla}(|\vec{z}_\epsilon|^2)\cdot\vec{\nabla}\Phi_\gamma\,dx\,dt$$

$$-2\nu\iint \alpha_{\eta,t_0}|\vec{\nabla}\otimes\vec{z}_\epsilon|^2\,\Phi_\gamma\,dx\,dt$$

$$+\iint \alpha_{\eta,t_0}\left(|\vec{z}_\epsilon|^2(\theta_\epsilon *_x \vec{v})\right) + 2r_\epsilon\vec{z}_\epsilon)\cdot\vec{\nabla}\Phi_\gamma\,dx\,dt.$$

For t_0 a Lebesgue point of $t \mapsto \|\vec{z}_\epsilon\|_{L^2(\Phi_\gamma\,dx)}$ (hence, for almost every t_0), we find (by letting η go to 0)

$$\|\vec{z}_\epsilon(t_0,.)\|_{L^2(\Phi_\gamma\,dx)} = -\nu\int_0^{t_0}\int \vec{\nabla}(|\vec{z}_\epsilon|^2)\cdot\vec{\nabla}\Phi_\gamma\,dx\,dt - 2\nu\int_0^{t_0}\int |\vec{\nabla}\otimes\vec{z}_\epsilon|^2\,\Phi_\gamma\,dx\,dt$$

$$+\int_0^{t_0}\int \left(|\vec{z}_\epsilon|^2(\theta_\epsilon *_x \vec{v})\right) + 2r_\epsilon\vec{z}_\epsilon)\cdot\vec{\nabla}\Phi_\gamma\,dx\,dt.$$

As $t \mapsto \vec{z}_\epsilon(t, .)$ is weakly continuous from $[0, \min(T_0, 1)]$ to $L^2(\Phi_\gamma \, dx)$, we find that for every t we have

$$
\|\vec{z}_\epsilon(t, .)\|_{L^2(\Phi_\gamma \, dx)}^2 \leq -\nu \int_0^t \int \vec{\nabla}(|\vec{z}_\epsilon|^2) \cdot \vec{\nabla}\Phi_\gamma \, dx \, ds - 2\nu \int_0^t \int |\vec{\nabla} \otimes \vec{z}_\epsilon|^2 \, \Phi_\gamma \, dx \, ds
$$

$$
+ \int_0^t \int \left(|\vec{z}_\epsilon|^2 (\theta_\epsilon *_x \vec{v}) \right) + 2r_\epsilon \vec{z}_\epsilon) \cdot \vec{\nabla}\Phi_\gamma \, dx \, ds
$$

$$
\leq 6\nu \int_0^t \|\vec{z}_\epsilon\|_{L^2(\Phi_\gamma \, dx)}^2 \, ds - 2\nu \int_0^t \|\vec{\nabla} \otimes \vec{z}_\epsilon\|_{L^2(\Phi_\gamma \, dx)}^2 \, ds
$$

$$
+ C_2 \int_0^t \|\vec{z}_\epsilon\|_{L^2(\Phi_\gamma \, dx)}^2 \|\vec{v}\|_{L^3(\Phi_{3\gamma/2} \, dx)} \, ds
$$

$$
+ C_3 \int_0^t \|\vec{z}_\epsilon\|_{L^2(\Phi_\gamma \, dx)} \|\vec{\nabla} \otimes \vec{z}_\epsilon\|_{L^2(\Phi_\gamma \, dx)} \|\vec{v}\|_{L^3(\Phi_{3\gamma/2} \, dx)} \, ds
$$

$$
\leq \int \left(6\nu + C_2 \|\vec{v}\|_{L^3(\Phi_{3\gamma/2} \, dx)} + \frac{C_3^2}{8\nu} \|\vec{v}\|_{L^3(\Phi_{3\gamma/2} \, dx)}^2 \right) \|\vec{z}_\epsilon\|_{L^2(\Phi_\gamma \, dx)}^2 \, ds.
$$

Grönwall's lemma then gives that $\vec{z}_\epsilon = 0$, so that $\vec{w}_\epsilon = \vec{u}_\epsilon$: the solution for (LNS_ϵ) is unique.

Since $\vec{u}_0$ is a λ-DSS vector field, we find however a second solution for (LNS_ϵ) on $(0, \min(T_0, 1))$, the rescaled vector field $\frac{1}{\lambda}\vec{u}_\epsilon(\frac{t}{\lambda^2}, \frac{x}{\lambda})$. By uniqueness of the solution, we see that $\vec{u}_\epsilon$ is a λ-DSS vector field.

d) Solving the non-linear problem with one mollification.

Let E be the space of divergence-free λ-DSS vector fields $\vec{v}$ such that $\vec{v} \in L^3_{loc}([0, +\infty) \times \mathbb{R}^3)$. This is a Banach space for the norm

$$
\|\vec{v}\|_E = \|\vec{v}\|_{L^3((0,1) \times B(0,1))}.
$$

We have seen that for $\vec{v} \in E$ and $\epsilon > 0$, there is a unique solution $\vec{u}_\epsilon$ for the problem (LNS_ϵ), where $\vec{u}_0 \in L^2_{loc}(\mathbb{R}^3)$ is a λ-DSS divergence-free vector field and $\mathbb{F} \in L^2_{loc}([0, +\infty) \times \mathbb{R}^3)$ is a λ-DSS tensor. The solution $\vec{u}_\epsilon$ is a divergence-free λ-DSS vector field that belongs locally to $L^\infty L^2 \cap L^2 H^1$, hence to $L^4 L^3$. In particular, $\vec{u}_\epsilon \in E$. We define H_ϵ the map $\vec{v} \mapsto \vec{u}_\epsilon$.

A divergence-free λ-DSS vector field $\vec{u}$ is a solution (locally in $L^\infty L^2 \cap L^2 H^1$) for the non-linear problem (NS_ϵ) if and only if $\vec{u} \in E$ is a fixed-point of H_ϵ. Existence of such a fixed point will be proved by using Schaefer's fixed point theorem. We are going to prove that H_ϵ is continuous and compact from E to E, and that there exists a constant M such that $\vec{u} = \mu H_\epsilon(\vec{u})$ and $0 \leq \mu \leq 1$ implies $\|\vec{u}\|_E \leq M$.

H_ϵ is compact.

Let $(\vec{v}_n)_{n \in \mathbb{N}}$ be a bounded sequence in E. Then it is a bounded sequence in $L^3((0, 1), L^3(\Phi_{3\gamma/2} \, dx))$, where $4/3 < \gamma < 2$. Let $A_2 = \sup_{n \in \mathbb{N}} \|\vec{v}_n\|_{L^3((0,1), L^3(\Phi_{3\gamma/2} \, dx))}$. Let

$$
A_3^2 = C_5^2 \|\vec{u}_0\|_{L^2(\Phi_\gamma \, dx)}^2 + \frac{2C_0^2 + C_1^2}{\nu} \int_0^1 \|\mathbb{F}\|_{L_\Phi^2}^2 \, dt + \frac{4}{27}(C_2 + C_3)A_1^3.
$$

We have seen that we have

$$
\|H_\epsilon(\vec{v}_n)(t, .)\|_{L^2(\Phi_\gamma \, dx)}^2 + \nu \int_0^{t_0} \|\vec{\nabla} \otimes H_\epsilon(\vec{v}_n)\|_{L^2(\Phi_\gamma \, dx)}^2 \, dt \leq 4A_3^2 \qquad (16.24)
$$

as long as $t_0 \leq T_1$, where $T_1 = \min(1, \frac{3}{31\nu + \frac{16C_3^4 + 2C_2^4}{\nu^3} A_3^4})$. We find that, uniformly wit respect to n, $H_\epsilon(\vec{v}_n)$ is locally bounded in $L^2((0,1), H^1)$ and $\partial_t H_\epsilon(\vec{v}_n)$ is locally bounded
in $L^{3/2}((0,1), H^{-3/2})$. Thus, we may apply the Rellich–Lions theorem (Theorem 12.1)
and find a sub-sequence $(H_\epsilon(\vec{v}_{n_k}))_{k \in \mathbb{N}}$ that is locally strongly convergent in $L^2((0,1), L^2)$.
As it is bounded in $L^\infty((0,1), L^2(B(0,1)))$, we see that we have strong convergence in
$L^q((0,1), L^2(B(0,1)))$ for $2 \leq q < +\infty$; as it is bounded in $L^2((0,1), L^6(B(0,1)))$, we find
that we have strong convergence in $L^r((0,1), L^3(B(0,1)))$ for every $2 \leq r < 4$. In particular,
we have convergence in $L^3((0,1) \times B(0,1))$, hence in E.

H_ϵ is continuous.

Let $(\vec{v}_n)_{n \in \mathbb{N}}$ be a strongly convergent sequence in E and let $\vec{v} = \lim_{n \to +\infty} \vec{v}_n$. Then it is a
bounded sequence in $L^3((0,1), L^3(\Phi_{3\gamma/2} \, dx))$, where $4/3 < \gamma < 2$, so that $(H_\epsilon(\vec{v}_n))_{n \in \mathbb{N}}$ is a
relatively compact sequence in E. In order to prove that it converges to $H_\epsilon(\vec{v})$ in E, it is
enough to prove that $H_\epsilon(\vec{v})$ is the only limit point of the sequence $(H_\epsilon(\vec{v}_n))_{n \in \mathbb{N}}$.

Let $\vec{w}$ be the limit of a subsequence $(H_\epsilon(\vec{v}_{n_k}))_{k \in \mathbb{N}}$. This subsequence converges strongly
in $L^3((0, \min(T_0, 1)), L^3(\Phi_{3\gamma/2} \, dx))$. Thus, $(\theta_\epsilon *_x \vec{v}_{n_k}) \otimes H_\epsilon(\vec{v}_{n_k})$ converges strongly in
$L^{3/2}((0, \min(T_0, 1)), L^3(\Phi_{3\gamma/2} \, dx))$ to the limit $(\theta_\epsilon *_x \vec{v}) \otimes \vec{w}$. We find that $\vec{w}$ is a solution of
(LNS_ϵ), i.e. $\vec{w} = H_\epsilon(\vec{v})$.

Estimates for the solution of $\vec{u} = \mu H_\epsilon(\vec{u})$.

Assume that $\vec{u}_\mu$ is a solution of $\vec{u}_\mu = \mu H_\epsilon(\vec{u}_\mu)$, where $0 < \mu \leq 1$. It means that

$$\begin{cases} \partial_t \frac{\vec{u}_\mu}{\mu} = \nu \Delta \frac{\vec{u}_\mu}{\mu} - ((\theta_\epsilon *_x \vec{u}_\mu) \cdot \vec{\nabla}) \frac{\vec{u}_\mu}{\mu} - \vec{\nabla} p_\mu + \vec{\nabla} \cdot \mathbb{F} \\ \\ \vec{\nabla} \cdot \frac{\vec{u}_\mu}{\mu} = 0, \qquad\qquad \frac{\vec{u}_\mu(0,.)}{\mu} = \vec{u}_0 \end{cases}$$

or equivalently

$$\begin{cases} \partial_t \vec{u}_\mu = \nu \Delta \vec{u}_\mu - ((\theta_\epsilon *_x \vec{u}_\mu) \cdot \vec{\nabla}) \vec{u}_\mu - \vec{\nabla}(\mu p_\mu) + \vec{\nabla} \cdot (\mu \mathbb{F}) \\ \\ \vec{\nabla} \cdot \vec{u}_\mu = 0, \qquad\qquad \vec{u}_\mu(0,.) = \mu \vec{u}_0 \end{cases}$$

Thus, we have

$\|\vec{u}_\mu(t,.)\|_{L^2(\Phi_\gamma \, dx)}^2$

$$\leq \mu^2 \|\vec{u}_0\|_{L^2(\Phi_\gamma \, dx)}^2 + 6\nu \int_0^t \|\vec{u}_\mu\|_{L^2(\Phi_\gamma \, dx)}^2 \, ds - 2\nu \int_0^t \|\vec{\nabla} \otimes \vec{u}_\mu\|_{L^2(\Phi_\gamma \, dx)}^2 \, ds$$

$$+ C_0 \mu \int_0^t \|\vec{\nabla} \otimes \vec{u}_\mu\|_{L^2(\Phi_\gamma \, dx)} \|\mathbb{F}\|_{L^2(\Phi_\gamma \, dx)} \, ds + C_1 \mu \int_0^t \|\vec{u}_\mu\|_{L^2(\Phi_\gamma \, dx)} \|\mathbb{F}\|_{L^2(\Phi_\gamma \, dx)} \, ds$$

$$+ C_2 \|\vec{u}_\mu\|_{L^2(\Phi_\gamma \, dx)}^2 \|\vec{u}_\mu\|_{L^3(\Phi_{3\gamma/2} \, dx)}$$

$$+ C_3 \|\vec{u}_\mu\|_{L^2(\Phi_\gamma \, dx)} \|\vec{\nabla} \otimes \vec{u}_\mu\|_{L^2(\Phi_\gamma \, dx)} \|\vec{u}_\mu\|_{L^3(\Phi_{3\gamma/2} \, dx)}$$

$$\leq \mu^2 \|\vec{u}_0\|_{L^2(\Phi_\gamma \, dx)}^2 + 6\nu \int_0^t \|\vec{u}_\mu\|_{L^2(\Phi_\gamma \, dx)}^2 \, ds - 2\nu \int_0^t \|\vec{\nabla} \otimes \vec{u}_\mu\|_{L^2(\Phi_\gamma \, dx)}^2 \, ds$$

$$+ C_0 \mu \int_0^t \|\vec{\nabla} \otimes \vec{u}_\mu\|_{L^2(\Phi_\gamma \, dx)} \|\mathbb{F}\|_{L^2(\Phi_\gamma \, dx)} \, ds + C_1 \mu \int_0^t \|\vec{u}_\mu\|_{L^2(\Phi_\gamma \, dx)} \|\mathbb{F}\|_{L^2(\Phi_\gamma \, dx)} \, ds$$

$$+ C_4 \int_0^t \|\vec{u}_\mu\|_{L^2(\Phi_\gamma \, dx)}^3 \, ds + C_5 \int_0^t \|\vec{u}_\mu\|_{L^2(\Phi_\gamma \, dx)}^{5/2} \|\vec{\nabla} \otimes \vec{u}_\mu\|_{L^2(\Phi_\gamma \, dx)}^{1/2} \, ds$$

$$+ C_6 \int_0^t \|\vec{u}_\mu\|_{L^2(\Phi_\gamma \, dx)}^3 \, ds + C_7 \int_0^t \|\vec{u}_\mu\|_{L^2(\Phi_\gamma \, dx)}^{3/2} \|\vec{\nabla} \otimes \vec{u}_\mu\|_{L^2(\Phi_\gamma \, dx)}^{3/2} \, ds$$

and thus

$$\|\vec{u}_\mu(t,.)\|^2_{L^2(\Phi_\gamma\, dx)} + \nu \int_0^t \|\vec{\nabla} \otimes \vec{u}_\mu\|^2_{L^2(\Phi_\gamma\, dx)}\, ds$$

$$\leq \mu^2 \|\vec{u}_0\|^2_{L^2(\Phi_\gamma\, dx)} + \mu^2 \frac{2C_0^2 + C_1^2}{\nu} \int_0^t \|\mathbb{F}\|^2_{L^2(\Phi_\gamma\, dx)}\, ds$$

$$+ C_8\nu \int_0^t \|\vec{u}_\mu\|^2_{L^2(\Phi_\gamma\, dx)}\, ds + \frac{C_9}{\nu^3} \int_0^t \|\vec{u}_\mu\|^6_{L^2(\Phi_\gamma\, dx)}\, ds.$$

Let

$$A_1^2 = \|\vec{u}_0\|^2_{L^2(\Phi_\gamma\, dx)} + \frac{2C_0^2 + C_1^2}{\nu} \int_0^1 \|\mathbb{F}\|^2_{L^2_\Phi}\, dt.$$

We see that we have (for $\mu \leq 1$)

$$\|\vec{u}_\mu\|^2_{L^2(\Phi_\gamma\, dx)} + \nu \int_0^{t_0} \|\vec{\nabla} \otimes \vec{u}_\mu\|^2_{L^2(\Phi_\gamma\, dx)}\, dt \leq 4A_1^2\mu^2 \tag{16.25}$$

as long as $t_0 \leq 1$ and

$$t_0\left(4C_8\nu + \frac{64C_0}{\nu^3} A_1^4\right) \leq 3.$$

If

$$T_0 = \min\left(1, \frac{3}{4C_8\nu + \frac{64C_0}{\nu^3} A_1^4}\right),$$

we find that

$$\|\vec{u}_\mu\|_{L^3((0,T_0),\Phi_{3\gamma/2}\, dx)} \leq C_\nu A_1 \mu.$$

As the norm in E is equivalent to $\|\vec{u}_\mu\|_{L^3((0,T_0),\Phi_{3\gamma/2}\, dx)}$, we may apply Schaefer's theorem and get a solution to (NS_ϵ).

e) Solving the non-linear problem.

For $\epsilon > 0$, we have a solution $\vec{u}_\epsilon$ of (NS_ϵ) such that, for $0 < t < T_0$ with

$$T_0 = \min\left(1, \frac{3}{4C_8\nu + \frac{64C_0}{\nu^3} A_1^4}\right),$$

(where $A_1^2 = \|\vec{u}_0\|^2_{L^2(\Phi_\gamma\, dx)} + \frac{2C_0^2 + C_1^2}{\nu} \int_0^1 \|\mathbb{F}\|^2_{L^2_\Phi}\, dt$), we have

$$\|\vec{u}_\epsilon\|^2_{L^2(\Phi_\gamma\, dx)} + \nu \int_0^{t_0} \|\vec{\nabla} \otimes \vec{u}_\epsilon\|^2_{L^2(\Phi_\gamma\, dx)}\, dt \leq 4A_1^2. \tag{16.26}$$

We find that, uniformly wit respect to ϵ, $\vec{u}_\epsilon$ is locally bounded in $L^2((0,T_0), H^1)$ and $\partial_t \vec{u}_\epsilon$ is locally bounded in $L^{3/2}((0,T_0), H^{-3/2})$. Thus, we may apply the Rellich–Lions theorem (Theorem 12.1) and find a sub-sequence $\vec{u}_{\epsilon_k})_{k\in\mathbb{N}}$ that is locally strongly convergent in $L^2([0,T_0], L^2)$ to a limit $\vec{u}$. We then check that we have weak convergence of $(\theta_\epsilon *_x \vec{u}_{\epsilon_k}) \otimes \vec{u}_{\epsilon_k}$ to $\vec{u} \otimes \vec{u}$ in $L^{3/2}((0,T_0), L^{3/2}(\Phi_{3\gamma/2}\, dx))$ and we conclude that $\vec{u}$ is a solution for (NS). $\square$

16.10   Time-periodic Weak Solutions

In Theorem 10.14, we have seen the results of Kyed [290] on time-periodic mild solutions which belong to $L_t^\infty \dot{H}^{1/2} \cap L_{\mathrm{per}}^2 \dot{H}^{3/2}$. In this section, we explain the results of Kyed on time-periodic weak solutions which belong to $L_{\mathrm{per}}^2 \dot{H}^{-1}$.

Given a time-periodic force $\vec{f}(t, x)$ (with period T), we want to find a time-periodic velocity $\vec{u}(t, x)$ and a time-periodic pressure $p(t, x)$ such that

$$\partial_t \vec{u} = \nu \Delta \vec{u} - \vec{u} \cdot \vec{\nabla} \vec{u} - \vec{\nabla} p + \vec{f}, \quad \mathrm{div}\, \vec{u} = 0 \qquad (16.27)$$

We assume that $\vec{f}$ belongs to $L_{\mathrm{per}}^2 \dot{H}^{-1}$ and are looking for a solution $\vec{u}$ belonging to $L_{\mathrm{per}}^2 \dot{H}^1$. (This is analogous with the condition $\vec{f} \in \dot{H}^{-1}$ and $\vec{u} \in \dot{H}^1$ for steady solutions (Theorem 16.2).) Kyed's theorem is then:

Weak time-periodic Navier–Stokes equations

Theorem 16.10.
Leet $\vec{f}_{\mathrm{per}}$ be a time-periodic vector field on $\mathbb{R} \times \mathbb{R}^3$ (with period T) such that $\vec{f}_{\mathrm{per}}$ belongs to $L_{\mathrm{per}}^2 \dot{H}^{-1}$. Then there exists a time-periodic solution $\vec{u}_{\mathrm{per}}$ of the Navier–Stokes problem (16.27) such that $\vec{u}_{\mathrm{per}} \in L_{\mathrm{per}}^2 \dot{H}^1$ and $\vec{u}_{\mathrm{per}} - \vec{u}_0 \in L_{\mathrm{per}}^\infty L^2$, where $\vec{u}_0$ is the mean value of $\vec{u}$:

$$\vec{u}_0(x) = \frac{1}{T} \int_0^T \vec{u}(t, x)\, dt.$$

Proof. We consider the projection in the time variable on the low frequencies:

$$D_N \Big(\sum_{k \in \mathbb{Z}} \vec{g}_k(x) e^{\frac{2\pi}{T} ikt} \Big) = \sum_{-N \le k \le N} \vec{g}_k(x) e^{\frac{2\pi}{T} ikt}.$$

We shall look for solutions in the space E of divergence free vector fields $\vec{U}$ such that $\vec{U} \in L_{\mathrm{per}}^2 \dot{H}^1$ and $\vec{U} - D_0(\vec{U}) \in L^\infty L^2$, with norm

$$\|\vec{U}\|_E = \Big(\int_0^T \|\vec{U}(t, .)\|_{\dot{H}^1}^2\, dt \Big)^{1/2} + \frac{1}{\sqrt{\nu}} \big\| \, \|(\vec{U} - D_0(\vec{U}))(t, .)\|_2 \, \big\|_\infty.$$

As for the proof of Theorem 10.14, we first study $\vec{U}_0 = \int_{-\infty}^t e^{\nu(t-s)\Delta} \mathbb{P} \vec{f}_{\mathrm{per}}\, ds$. We expand $\mathbb{P} \vec{f}_{\mathrm{per}}$ as a time-Fourier series

$$\mathbb{P} \vec{f}_{\mathrm{per}} = \sum_{k \in \mathbb{Z}} \vec{g}_k(x) e^{\frac{2\pi}{T} ikt}.$$

We have

$$\int_0^T \|\mathbb{P} \vec{f}_{\mathrm{per}}\|_{\dot{H}^{-1}}^2\, dt = T \sum_{k \in \mathbb{Z}} \|\vec{g}_k\|_{\dot{H}^{-1}}^2.$$

The Fourier expansion of $\vec{U}_0$ is

$$\vec{U}_0 = \sum_{k \in \mathbb{Z}} \vec{W}_k(x) e^{\frac{2\pi}{T} ikt}, \quad \text{with} \quad \vec{W}_k = \frac{1}{ik\frac{2\pi}{T} - \nu\Delta} \mathbb{P} \vec{g}_k.$$

We have $\|\vec{W}_k\|_{\dot{H}^1} \leq \frac{1}{\nu}\|\vec{g}_k\|_{\dot{H}^{-1}}$, and thus $\vec{U}_0 \in L^2_{\text{per}}\dot{H}^1$. Let $\vec{\Omega}_k$ be the Fourier transform of $\vec{g}_k$. We have:

$$(2\pi)^3\|\vec{U}_0(t,.) - \vec{W}_0\|_2^2 = \int \left| \sum_{k\neq 0} \frac{1}{ik\frac{2\pi}{T} + \nu|\xi|^2}\vec{\Omega}_k(\xi)e^{\frac{2\pi}{T}ikt} \right|^2 d\xi$$

$$\leq \int \left(\sum_{k\neq 0} \frac{1}{k^2\frac{4\pi^2}{T^2} + \nu^2|\xi|^4}\right)\left(\sum_{k\neq 0}|\vec{\Omega}_k(\xi)|^2\right) d\xi$$

In the proof of Theorem 10.14, we have seen that

$$\sum_{k\neq 0} \frac{1}{k^2\frac{4\pi^2}{T^2} + \nu^2|\xi|^4} \leq (4 + \frac{1}{2\pi^2})\frac{T}{\nu|\xi|^2}.$$

Thus, we find that $\vec{U}_0 - D_0(\vec{U}_0) \in L_t^\infty L^2$. Thus, $\vec{U}_0 \in E$ and

$$\|\vec{U}_0\|_E \leq C\frac{1}{\nu}\left(\int_0^T \|\vec{f}_{\text{per}}(t,.)\|_{\dot{H}^{-1}}^2 dt\right)^{1/2}.$$

We now rewrite equation (16.27) as

$$\vec{u} = \vec{U}_0 - B(\vec{u}, \vec{u}) \tag{16.28}$$

where B is the bilinear operator

$$B(\vec{U}, \vec{V}) = \int_{-\infty}^t W_{\nu(t-s)} * \mathbb{P}\,\text{div}(\vec{U} \otimes \vec{V})\,ds.$$

As in the proof of Theorem 16.2, we introduce a cut-off function $\theta \in \mathcal{D}(\mathbb{R}^3)$ such that $0 \leq \theta \leq 1$, $\theta(x) = 1$ for $|x| < 1$ and $\theta(x) = 0$ for $|x| > 2$, and we solve the following approximation of equation (16.27): find $\vec{u}_R$ such that

$$\vec{u}_R = \vec{U}_0 - B_R(\vec{u}_R, \vec{u}_R) \tag{16.29}$$

where B_R is the bilinear operator

$$B_R(\vec{U}, \vec{V}) = \int_{-\infty}^t W_{\nu(t-s)} * \mathbb{P}\left(\theta(\frac{x}{R})\vec{U} \cdot \vec{\nabla}(\theta(\frac{x}{R})\vec{V})\right) ds.$$

We need as well to mollify the equation by introducing a non-negative $\phi \in \mathcal{D}(\mathbb{R}^3)$, equal to 1 on a neighborhood of 0, supported in $\overline{B(0,1)}$ with $\int \phi\,dx = 1$, and defining, for $\alpha > 0$,

$$\phi_\alpha(x) = \frac{1}{\alpha^3}\phi(\frac{x}{\alpha}).$$

We then consider the problem

$$\vec{u}_{R,\alpha} = \vec{U}_0 - B_{R,\alpha}(\vec{u}_{R,\alpha}, \vec{u}_{R,\alpha}) \tag{16.30}$$

where $B_{R,\alpha}$ is the bilinear operator

$$B_{R,\alpha}(\vec{U}, \vec{V}) = \int_{-\infty}^t W_{\nu(t-s)} * \mathbb{P}\left(\theta(\frac{x}{R})(\phi_\alpha * \vec{U}) \cdot \vec{\nabla}(\theta(\frac{x}{R})\vec{V})\right) ds.$$

Further, we consider the problem

$$\vec{u}_{R,N,\alpha} = D_N(\vec{U}_0) - D_N(B_{R,\alpha}(\vec{u}_{R,N,\alpha}, \vec{u}_{R,N,\alpha})). \tag{16.31}$$

Finally, we linearize the problem by taking $\vec{U} \in E$ as a data and solve

$$\vec{w}_{R,N,\alpha} = D_N(\vec{U}_0) - D_N(B_{R,\alpha}(\vec{U}, \vec{w}_{R,N,\alpha})). \tag{16.32}$$

a) Existence of $\vec{w}_{R,N,\alpha}$.

$\vec{w}_{R,N,\alpha}$ is fixed-point of the map $\vec{V} \in E \mapsto D_N(\vec{U}_0) - D_N(B_{R,\alpha}(\vec{U}, D_N(\vec{V}))) = D_N(\vec{U}_0) - T_{R,N,\alpha}(\vec{V})$. If $\vec{U} \in E$, we have $\phi_\alpha * \vec{U} \in L^\infty \dot{H}^1$ with

$$\|\phi_\alpha * \vec{U}(t,.)\|_{\dot{H}^1} \leq \|\phi\|_1 \|D_0(U)\|_{\dot{H}^1} + \frac{1}{\alpha}\|\vec{\nabla}\phi\|_1\|(\vec{U} - D_0(\vec{U}))(t,.)\|_2$$

and $\theta(\frac{x}{R})D_N(\vec{V}) \in L^2_{per}\dot{H}^1$ with

$$\|\theta(\frac{x}{R})D_N(\vec{V})(t,.)\|_{\dot{H}^1} \leq (\|\theta\|_\infty + C\|\vec{\nabla}\theta\|_3)\|D_N(\vec{V})(t,.)\|_{\dot{H}^1}$$

and

$$\|D_N(\vec{V})\|_{L^2_{per}\dot{H}^1} \leq \|\vec{V}\|_{L^2_{per}\dot{H}^1}.$$

In particular, $\theta(\frac{x}{R})(\phi_\alpha * \vec{U}) \cdot \vec{\nabla}(\theta(\frac{x}{R})D_N(\vec{V}) \in L^2_{per}\dot{H}^{-1}$, since

$$\|\theta(\frac{x}{R})(\phi_\alpha * \vec{U}(t,.)) \cdot \vec{\nabla}(\theta(\frac{x}{R})D_N(\vec{V})(t,.))\|_{6/5}$$
$$\leq \sqrt{R}\|\theta\|_6\|\phi_\alpha * \vec{U}(t,.)\|_6\|\vec{\nabla}(\theta(\frac{x}{R})D_N(\vec{V})(t,.))\|_2.$$

Thus, the linear map $T_{R,N,\alpha}$ is bounded on E:

$$\|T_{R,N,\alpha}(\vec{V})\|_E \leq C\frac{1}{\nu}\sqrt{R}(\sqrt{T} + \frac{\sqrt{\nu}}{\alpha})\|\vec{U}\|_E\|\vec{V}\|_E.$$

We now check that $T_{R,N,\alpha}$ is compact. Let $(\vec{V}_n)_{n\in\mathbb{N}}$ be a bounded sequence in E. We write

$$\vec{V}_n = \sum_{k\in\mathbb{Z}} \vec{v}_{k,n}(x)e^{\frac{2\pi}{T}ikt}.$$

and

$$\theta(\frac{x}{R})D_N(\vec{V}_n) = \sum_{-N\leq k\leq N} \theta(\frac{x}{R})\vec{v}_{k,n}(x)e^{\frac{2\pi}{T}ikt}.$$

As $\vec{v}_{k,n}$ are bounded in $\dot{H}^1$, we may find a subsequence $(\vec{V}_{n_j})_{j\in\mathbb{N}}$ such that $\vec{v}_{k,n_j}$ is weakly convergent in $\dot{H}^1$ to a limit $\vec{v}_{k,\infty}$ for $-N \leq k \leq N$. As $\theta(\frac{x}{R})\vec{v}_{k,n_j}$ is bounded in H^1, since

$$\|\theta(\frac{x}{R})\vec{v}_{k,n_j}\|_{H^1} \leq C(R\|\theta\|_3 + \|\theta\|_\infty + \|\vec{\nabla}\theta\|_3)\|\vec{v}_{k,n_j}\|_{\dot{H}^1},$$

and since they are supported in a fixed compact set, i.e. $\overline{B(0,2R)}$, we find (due to Rellich's theorem) that $\theta(\frac{x}{R})\vec{v}_{k,n_j}$ is strongly convergent to $\theta(\frac{x}{R})\vec{v}_{k,\infty}$ in $H^{1/2}$. Let

$$\vec{v}_\infty = \sum_{-N\leq k\leq N} \vec{v}_{k,\infty}(x)e^{\frac{2\pi}{T}ikt}.$$

Writing

$$\theta(\frac{x}{R})(\phi_\alpha * \vec{U}(t,.)) \cdot \vec{\nabla}(\theta(\frac{x}{R})D_N(\vec{V})(t,.))$$
$$= \text{div}\left(\theta(\frac{x}{R})(\phi_\alpha * \vec{U}(t,.)) \otimes (\theta(\frac{x}{R})D_N(\vec{V})(t,.))\right)$$
$$- \left((\phi_\alpha * \vec{U}(t,.)) \cdot \vec{\nabla}(\theta(\frac{x}{R}))\right)\theta(\frac{x}{R})D_N(\vec{V})(t,.)$$

with

$$\left\|\theta(\frac{x}{R})(\phi_\alpha * \vec{U}(t,\cdot)) \otimes (\theta(\frac{x}{R})D_N(\vec{V})(t,\cdot))\right\|_2$$

$$\leq C\|\theta\|_\infty \|\phi_\alpha * \vec{U}(t,\cdot)\|_{\dot{H}^1} \|\theta(\frac{x}{R})D_N(\vec{V})(t,\cdot)\|_{\dot{H}^{1/2}}$$

and

$$\left\|\left((\phi_\alpha * \vec{U}(t,\cdot)) \cdot \vec{\nabla}(\theta(\frac{x}{R}))\right)\theta(\frac{x}{R})D_N(\vec{V})(t,\cdot)\right\|_{6/5}$$

$$\leq C\|\vec{\nabla}\theta\|_3 \|\phi_\alpha * \vec{U}(t,\cdot)\|_{\dot{H}^1} \|\theta(\frac{x}{R})D_N(\vec{V})(t,\cdot)\|_{\dot{H}^{1/2}},$$

we see that $\theta(\frac{x}{R})(\phi_\alpha * \vec{U}) \cdot \vec{\nabla}(\theta(\frac{x}{R})D_N(\vec{v}_{n_j}))$ is strongly convergent in $L^2_{per}\dot{H}^{-1}$ to $\theta(\frac{x}{R})(\phi_\alpha * \vec{U}) \cdot \vec{\nabla}(\theta(\frac{x}{R})D_N(\vec{v}_\infty))$, so that $T_{R,N,\alpha}(\vec{v}_{n_j})$ is strongly convergent to $T_{R,N,\alpha}(\vec{v}_\infty)$ in E.

The next step is to estimate the norm in E of a solution $\vec{w}$ of

$$\vec{w} = D_N(\vec{U}_0) - T_{R,N,\alpha}(\vec{w}).$$

$\vec{w}$ is a solution $\vec{w} = D_N(\vec{w})$ in E of

$$\partial_t \vec{w} = \nu\Delta\vec{w} + \mathbb{P}(D_N(\vec{f}_{per}) - D_N(\theta(\frac{x}{R})(\phi_\alpha * \vec{U}) \cdot \vec{\nabla}(\theta(\frac{x}{R})D_N(\vec{w})))). \tag{16.33}$$

For $\epsilon > 0$, let P_ϵ be defined as the projection on spatial frequencies less than ϵ:

$$P_\epsilon(f) = \mathcal{F}^{-1}(\mathbb{1}_{|\xi| < \epsilon}\hat{f})$$

and $Q_\epsilon = \mathrm{Id} - P_\epsilon$. Let $\vec{w}_\epsilon = \vec{w} - P_\epsilon D_0(\vec{w})$. If $\vec{w} \in E$, then $\vec{w}_\epsilon \in L^\infty L^2 \cap L^2_{per}\dot{H}^1$. If moreover $\vec{w}$ is a solution to (16.33), then $\partial_t\vec{w}_\epsilon = \partial_t\vec{w} \in L^2_{per}\dot{H}^{-1}$ and $t \mapsto \vec{w}_\epsilon$ is continuous from $\mathbb{R}$ to L^2. Thus, we have enough regularity to write

$$2\int_0^T \int \vec{w}_\epsilon \cdot \partial_t\vec{w} \, dx \, dt = \int_0^T \frac{d}{dt}(\|\vec{w}_\epsilon(t,\cdot)\|_2^2) \, dt = \|\vec{w}_\epsilon(T,\cdot)\|_2^2 - \|\vec{w}_\epsilon(0,\cdot)\|_2^2 = 0.$$

We write

$$\partial_t\vec{w} = \nu\Delta\vec{w}_\epsilon + \nu\Delta P_\epsilon D_0(\vec{w}) + \mathbb{P}(D_N(\vec{f}_{per}) - D_N(\theta(\frac{x}{R})(\phi_\alpha * \vec{U}) \cdot \vec{\nabla}(\theta(\frac{x}{R})\vec{w}_\epsilon)))$$

$$+ \mathbb{P}(D_N(\theta(\frac{x}{R})(\phi_\alpha * \vec{U}) \cdot \vec{\nabla}(\theta(\frac{x}{R})P_\epsilon D_0(\vec{w})))).$$

We obtain, by integrating against $\vec{w}_\epsilon$,

$$0 = -\nu\int_0^T \|\vec{\nabla} \otimes \vec{w}_\epsilon\|_2^2 \, dt + \nu\int_0^T \int \vec{w}_\epsilon \cdot \Delta P_\epsilon D_0(\vec{w}) + \int_0^T \int \vec{w}_\epsilon \cdot \vec{f}_{per} \, dx \, dt$$

$$+ \int_0^T \int \vec{w}_\epsilon \cdot (\theta(\frac{x}{R})(\phi_\alpha * \vec{U}) \cdot \vec{\nabla}(\theta(\frac{x}{R})P_\epsilon D_0(\vec{w}))) \, dx \, dt.$$

When ϵ goes to 0, we have $\lim_{\epsilon\to 0}\|P_\epsilon D_0(\vec{w})\|_{\dot{H}^1} = 0$, $\lim_{\epsilon\to 0}\|\vec{w} - \vec{w}_\epsilon\|_{L^2_{per}\dot{H}^1} = 0$ and $\lim_{\epsilon\to 0}\|\theta(\frac{x}{R})(\phi_\alpha * \vec{U}) \cdot \vec{\nabla}(\theta(\frac{x}{R})P_\epsilon D_0(\vec{w}))\|_{L^2_{per}\dot{H}^{-1}} = 0$, so that

$$\nu\int_0^T \|\vec{\nabla} \otimes \vec{w}\|_2^2 \, dt = \int_0^T \int \vec{w} \cdot \vec{f}_{per} \, dx \, dt$$

and thus

$$\|\vec{w}\|_{L^2_{\text{per}}\dot{H}^1} \le \frac{1}{\nu}\|\vec{f}_{\text{per}}\|_{L^2_{\text{per}}\dot{H}^{-1}}. \tag{16.34}$$

In particular, there is at most one solution to (16.33): if $\vec{w}_1$ and $\vec{w}_2$ are solutions to (16.33) associated to the same force $\vec{f}_{\text{per}}$, then $\vec{w}_1 - \vec{w}_2$ is a solution to (16.33) associated to the null force, and by (16.34), $\|\vec{w}_1 - \vec{w}_2\|_{L^2_{\text{per}}\dot{H}^1} = 0$.

We now estimate $\|\vec{w} - D_0(\vec{w})\|_{L^\infty L^2}$. Let

$$\vec{w} = \sum_{-N \le k \le N} \vec{W}_k(x)e^{\frac{2\pi}{T}ikt} \quad \text{and} \quad \mathbb{P}\vec{f}_{\text{per}} = \sum_{k \in \mathbb{Z}} \vec{g}_k(x)e^{\frac{2\pi}{T}ikt}..$$

For $k \ne 0$, we know that $\vec{W}_k \in L^2$. Moreover,

$$T\|\vec{W}_k\|_2^2 = \int_0^T \int \overline{\vec{W}_k(x)} \cdot (\vec{w} - D_0(\vec{w}))e^{-\frac{2\pi}{T}ikt}\, dx\, dt$$

$$= -\frac{iT}{2k\pi}\int_0^T \int \overline{\vec{W}_k(x)} \cdot \partial_t(\vec{w} - D_0(\vec{w}))e^{-\frac{2\pi}{T}ikt}\, dx\, dt$$

$$\le \nu\frac{T^2}{2k\pi}\|\vec{W}_k\|_{\dot{H}^1}^2 + \frac{T^2}{2k\pi}\|\vec{W}_k\|_{\dot{H}^1}\|\vec{g}_k\|_{\dot{H}^{-1}}$$

$$+ C\frac{T}{2k\pi}\int_0^T \|\vec{W}_k\|_3\|\theta\|_\infty\|\phi\|_1\|\vec{U}\|_{\dot{H}^1}(\|\theta\|_\infty + C\|\vec{\nabla}\theta\|_3)\|\vec{w}\|_{\dot{H}^1}\, dt$$

$$\le \nu\frac{T^2}{2k\pi}\|\vec{W}_k\|_{\dot{H}^1}^2 + \frac{T^2}{2k\pi}\|\vec{W}_k\|_{\dot{H}^1}\|\vec{g}_k\|_{\dot{H}^{-1}}$$

$$+ C'\frac{T}{2k\pi}\|\vec{W}_k\|_2^{1/2}\|\vec{W}_k\|_{\dot{H}^1}^{1/2}\|\vec{U}\|_{L^2_{\text{per}}\dot{H}^1}\|\vec{w}\|_{L^2_{\text{per}}\dot{H}^1}$$

$$\le \frac{T}{2}\|\vec{W}_k\|_2^2 + \nu(\frac{T^2}{k\pi} + T^2)\|\vec{W}_k\|_{\dot{H}^1}^2 + \frac{T^2}{8k\pi\nu}\|\vec{g}_k\|_{\dot{H}^{-1}}^2$$

$$+ C'\frac{\sqrt{T}}{k^2\sqrt{\nu}}\|\vec{U}\|_{L^2_{\text{per}}\dot{H}^1}^2\|\vec{w}\|_{L^2_{\text{per}}\dot{H}^1}^2.$$

We obtain

$$\|\vec{w} - D_0(\vec{w})\|_{L^2_{\text{per}}L^2}^2 \le C\frac{T}{\nu}\|\vec{f}_{\text{per}}\|_{L^2_{\text{per}}\dot{H}^{-1}}^2 + C\frac{\sqrt{T}}{\nu^2\sqrt{\nu}}\|\vec{U}\|_{L^2_{\text{per}}\dot{H}^1}^2\|\vec{f}_{\text{per}}\|_{L^2_{\text{per}}\dot{H}^{-1}}^2. \tag{16.35}$$

We now write, for $-T \le t_0 \le 0 \le t \le T$,

$$\|\vec{w}(t,.) - D_0(\vec{w})(t,.)\|_2^2 - \|\vec{w}(t_0,.) - D_0(\vec{w})(t_0,.)\|_2^2$$

$$= 2\int_{t_0}^t \int (\vec{w}(s,x) - D_0(\vec{w})(s,x)) \cdot \partial_t\vec{w}(s,x)\, dx\, ds$$

$$= -2\nu\int_{t_0}^t \int (\vec{\nabla} \otimes (\vec{w} - D_0(\vec{w}))) \cdot (\vec{\nabla} \otimes \vec{w})\, dx\, ds$$

$$+ 2\int_{t_0}^t \int (\vec{w} - D_0(\vec{w})) \cdot \vec{f}\, dx\, ds$$

$$-2\int_{t_0}^t \int (\vec{w} - D_0(\vec{w})) \cdot (\theta(\frac{x}{R})(\phi_\alpha * \vec{U}) \cdot \vec{\nabla}(\theta(\frac{x}{R})D_0(\vec{w})))\, dx\, ds$$

$$\le 4\nu\|\vec{w}\|_{L^2_{\text{per}}\dot{H}^1}^2 + 4\|\vec{w}\|_{L^2_{\text{per}}\dot{H}^1}\|\vec{f}\|_{L^2_{\text{per}}\dot{H}^{-1}}$$

$$+ C\|\vec{w} - D_0(\vec{w})\|_{L^2_{\text{per}}L^2}^{1/2}\|\vec{w} - D_0(\vec{w})\|_{L^2_{\text{per}}\dot{H}^1}^{1/2}\|\vec{U}\|_{L^2_{\text{per}}\dot{H}^1}\|D_0(\vec{w})\|_{\dot{H}^1}$$

$$\leq 4\nu \|\vec{w}\|^2_{L^2_{\mathrm{per}}\dot{H}^1} + 4\|\vec{w}\|_{L^2_{\mathrm{per}}\dot{H}^1}\|\vec{f}\|_{L^2_{\mathrm{per}}\dot{H}^{-1}}$$
$$+\frac{1}{T}\|\vec{w}-D_0(\vec{w})\|^2_{L^2_{\mathrm{per}}L^2} + CT^{-1/3}\|\vec{w}\|^2_{L^2_{\mathrm{per}}\dot{H}^1}\|\vec{U}\|^{4/3}_{L^2_{\mathrm{per}}\dot{H}^1}.$$

Integrating on $t_0 \in (-T, 0)$, we get

$$\|\vec{w}-D_0(\vec{w})\|^2_{L^\infty L^2} \leq \frac{2}{T}\|\vec{w}-D_0(\vec{w})\|^2_{L^2_{\mathrm{per}}L^2} + 4\nu\|\vec{w}\|^2_{L^2_{\mathrm{per}}\dot{H}^1}$$
$$+ 4\|\vec{w}\|_{L^2_{\mathrm{per}}\dot{H}^1}\|\vec{f}\|_{L^2_{\mathrm{per}}\dot{H}^{-1}}$$
$$+ CT^{-1/3}\|\vec{w}\|^2_{L^2_{\mathrm{per}}\dot{H}^1}\|\vec{U}\|^{4/3}_{L^2_{\mathrm{per}}\dot{H}^1}.$$

and thus

$$\|\vec{w}-D_0(\vec{w})\|^2_{L^\infty L^2} \leq C\frac{1}{\nu}\|\vec{f}_{\mathrm{per}}\|^2_{L^2_{\mathrm{per}}\dot{H}^{-1}}$$
$$+ C\frac{1}{\nu}\|\vec{f}_{\mathrm{per}}\|^2_{L^2_{\mathrm{per}}\dot{H}^{-1}}\left(\frac{\|\vec{U}\|^4_{L^2_{\mathrm{per}}\dot{H}^1}}{T\nu^3}\right)^{1/3} \tag{16.36}$$
$$+ C\frac{1}{\nu}\|\vec{f}_{\mathrm{per}}\|^2_{L^2_{\mathrm{per}}\dot{H}^{-1}}\left(\frac{\|\vec{U}\|^4_{L^2_{\mathrm{per}}\dot{H}^1}}{T\nu^3}\right)^{1/2}.$$

Thus, a solution $\vec{w}$ of a solution (16.33) is bounded by

$$\|\vec{w}\|_E \leq C_0\frac{1}{\nu}\|\vec{f}_{\mathrm{per}}\|_{L^2_{\mathrm{per}}\dot{H}^{-1}}\left(1 + \frac{\|\vec{U}\|_{L^2_{\mathrm{per}}\dot{H}^1}}{T^{1/4}\nu^{3/4}}\right) \tag{16.37}$$

where C_0 does not depend on R, N, α, T, ν, $\vec{U}$ nor $\vec{f}$.

The last step is to estimate the norm in E of a solution $\vec{w}_\mu$ of

$$\vec{w}_\mu = \mu\left(D_N(\vec{U}_0) - T_{R,N,\alpha}(\vec{w}_\mu)\right) = \mu\vec{w}^*_\mu$$

with $0 \leq \mu \leq 1$. Writing

$$\partial_t\vec{w}_\mu = \mu\left(\nu\Delta\vec{w}^*_\mu + \mathbb{P}(D_N(\vec{f}) - D_N(\theta(\frac{x}{R})(\phi_\alpha * \vec{U})\cdot\vec{\nabla}(\theta(\frac{x}{R})D_N(\vec{w}_\mu))))\right)$$
$$= \nu\Delta\vec{w}_\mu + \mathbb{P}(D_N(\mu\vec{f}) - D_N(\theta(\frac{x}{R})(\phi_\alpha * \mu\vec{U})\cdot\vec{\nabla}(\theta(\frac{x}{R})D_N(\vec{w}_\mu)))),$$

we find, (applying (16.37)),

$$\|\vec{w}_\mu\|_E \leq C_0\frac{1}{\nu}\|\mu\vec{f}_{\mathrm{per}}\|_{L^2_{\mathrm{per}}\dot{H}^{-1}}\left(1 + \frac{\|\mu\vec{U}\|_{L^2_{\mathrm{per}}\dot{H}^1}}{T^{1/4}\nu^{3/4}}\right)$$
$$\leq C_0\frac{1}{\nu}\|\vec{f}_{\mathrm{per}}\|_{L^2_{\mathrm{per}}\dot{H}^{-1}}\left(1 + \frac{\|\vec{U}\|_{L^2_{\mathrm{per}}\dot{H}^1}}{T^{1/4}\nu^{3/4}}\right).$$

Using Schaefer's fixed point theorem, we get existence of a solution $\vec{w}_{R,N,\alpha}$ to equation (16.32).

b) Existence of $\vec{u}_{R,N,\alpha}$.

Given $\vec{f} \in L^2_{\mathrm{per}} \dot{H}^{-1}$, we found, for every $\vec{U} \in E$, a unique solution $\vec{w}_{R,N,\alpha}$ to equation (16.32). Thus, we have a map

$$\vec{U} \in E \mapsto A_{R,N,\alpha}(\vec{U}) = \vec{w}_{R,N,\alpha} \in E.$$

A vector field $\vec{u}_{R,N,\alpha}$ is then a solution to (16.31) if and only if it is a fixed point of $A_{R,N,\alpha}$: $\vec{u}_{R,N,\alpha} = A_{R,N,\alpha}(\vec{u}_{R,N,\alpha})$. As $\vec{w}_{R,N,\alpha} = D_N(\vec{w}_{R,N,\alpha})$, we may consider as well $\vec{u}_{R,N,\alpha}$ as a fixed point of $\vec{U} \mapsto A_{R,N,\alpha}(D_N(\vec{U}))$.

We first check that the map $A_{R,N,\alpha} \circ D_N$ is compact. Let $(\vec{U}_n)_{n\in\mathbb{N}}$ be a bounded sequence in E. By (16.37), the sequence

$$(\vec{W}_n)_{n\in\mathbb{N}} = (A_{R,N,\alpha} \circ D_N(\vec{U}_n))_{n\in\mathbb{N}}$$

is bounded in E. We write

$$\vec{U}_n = \sum_{k\in\mathbb{Z}} \vec{u}_{k,n}(x)e^{\frac{2\pi}{T}ikt}.$$

and

$$\vec{W}_n = \sum_{-N\le k\le N} \vec{w}_{k,n}(x)e^{\frac{2\pi}{T}ikt}.$$

We have

$$\theta(\frac{x}{R})\phi_\alpha * D_N(\vec{U}_n) = \sum_{-N\le k\le N} \theta(\frac{x}{R})\theta(\frac{x}{4R})\vec{\phi}_\alpha * \vec{u}_{k,n}(x)e^{\frac{2\pi}{T}ikt}.$$

and

$$\theta(\frac{x}{R})\vec{W}_n = \sum_{-N\le k\le N} \theta(\frac{x}{R})\vec{w}_{k,n}(x)e^{\frac{2\pi}{T}ikt}.$$

As $\vec{u}_{k,n}$ and $\vec{w}_{k,n}$ are bounded in $\dot{H}^1$, we may find subsequences $(\vec{U}_{n_j})_{j\in\mathbb{N}}$ and $(\vec{W}_{n_j})_{j\in\mathbb{N}}$ such that, for $-N \le k \le N$, $\vec{u}_{k,n_j}$ is weakly convergent in $\dot{H}^1$ to a limit $\vec{u}_{k,\infty}$ and $\vec{w}_{k,n_j}$ is weakly convergent in $\dot{H}^1$ to a limit $\vec{w}_{k,\infty}$. As $\theta(\frac{x}{4R})\phi_\alpha * \vec{u}_{k,n_j}$ is bounded in H^1, and since they are supported in a fixed compact set, i.e. $\overline{B(0,8R)}$, we find (due to Rellich's theorem) that $\theta(\frac{x}{8R})\phi_\alpha * \vec{u}_{k,n_j}$ is strongly convergent to $\theta(\frac{x}{4R})\phi_\alpha * \vec{u}_{k,\infty}$ in $\dot{H}^{3/4}$. Similarly, $\theta(\frac{x}{R})\vec{w}_{k,n_j}$ is strongly convergent to $\theta(\frac{x}{R})\vec{w}_{k,\infty}$ in $\dot{H}^{3/4}$.

Let

$$\vec{u}_\infty = \sum_{-N\le k\le N} \vec{u}_{k,\infty}(x)e^{\frac{2\pi}{T}ikt} \text{ and } \vec{w}_\infty = \sum_{-N\le k\le N} \vec{w}_{k,\infty}(x)e^{\frac{2\pi}{T}ikt}.$$

Writing

$$\theta(\frac{x}{R})(\phi_\alpha * D_N(\vec{U})(t,.)) \cdot \vec{\nabla}(\theta(\frac{x}{R})D_N(\vec{V})(t,.))$$
$$= \mathrm{div}\left(\theta(\frac{x}{R})\theta(\frac{x}{4R})(\phi_\alpha * D_N(\vec{U})(t,.)) \otimes (\theta(\frac{x}{R})D_N(\vec{V})(t,.))\right)$$
$$- \left(\theta(\frac{x}{4R})(\phi_\alpha * D_N(\vec{U})(t,.)) \cdot \vec{\nabla}(\theta(\frac{x}{R}))\right)\theta(\frac{x}{R})D_N(\vec{V})(t,.)$$

with

$$\left\|\theta(\frac{x}{R})\theta(\frac{x}{4R})(\phi_\alpha * D_N(\vec{U})(t,.)) \otimes (\theta(\frac{x}{R})D_N(\vec{V})(t,.))\right\|_2$$
$$\le C\|\theta\|_\infty \|\theta(\frac{x}{4R})\phi_\alpha * D_N(\vec{U})(t,.)\|_{\dot{H}^{3/4}}\|\theta(\frac{x}{R})D_N(\vec{V})(t,.)\|_{\dot{H}^{3/4}}$$

and

$$\left\|\left(\theta(\frac{x}{4R})(\phi_\alpha * D_N(\vec{U})(t,.)) \cdot \vec{\nabla}(\theta(\frac{x}{R}))\right)\theta(\frac{x}{R})D_N(\vec{V})(t,.)\right\|_{6/5}$$

$$\leq C\|\vec{\nabla}\theta\|_3\|\theta(\frac{x}{4R})(\phi_\alpha * D_N(\vec{U})(t,.))\|_{\dot{H}^{3/4}}\|\theta(\frac{x}{R})D_N(\vec{V})(t,.)\|_{\dot{H}^{3/4}},$$

we see that $\theta(\frac{x}{R})(\phi_\alpha * D_N(\vec{U}_{n_j})) \cdot \vec{\nabla}(\theta(\frac{x}{R})D_N(\vec{W}_{n_j}))$ is strongly convergent in $L^\infty \dot{H}^{-1}$, hence in $L^2_{per}\dot{H}^{-1}$, to $\theta(\frac{x}{R})(\phi_\alpha * \vec{u}_\infty) \cdot \vec{\nabla}(\theta(\frac{x}{R})D_N(\vec{w}_\infty))$. Writing

$$\vec{w}_{n_j} = D_N(\vec{U}_0) - D_N(B_{R,\alpha}(D_N(\vec{u}_{n_j}), D_N(\vec{w}_{n,j}))),,$$

we see that $\vec{w}_{n_j}$ is strongly convergent to $D_N(\vec{U}_0) - D_N(B_{R,\alpha}(D_N(\vec{u}_\infty), D_N(\vec{w}_\infty)))$ in E. As it is weakly convergent in $L^2_{per}\dot{H}^1$ to $\vec{w}_\infty = D_N(\vec{w}_\infty)$, Thus, $\vec{w}_\infty = A_{R,N,\alpha} \circ D_N(\vec{u}_\infty)..$ Hence, the map $A_{R,N,\alpha} \circ D_N$ is compact on E.

Moreover, if $(\vec{U}_n)_{n\in\mathbb{N}}$ is strongly convergent in E to a limit $\vec{u}_\infty$, we see that the sequence $(\vec{W}_n)_{n\in\mathbb{N}}$ is relatively compact in E with only one limit value $A_{R,N,\alpha} \circ D_N(\vec{u}_\infty) = A_{R,N,\alpha} \circ D_N(\vec{U}_\infty)..$ Thus, $(\vec{W}_n)_{n\in\mathbb{N}}$ is strongly convergent in E to to $A_{R,N,\alpha} \circ D_N(\vec{U}_\infty)$. Thus, we see that the map $A_{R,N,\alpha} \circ D_N$ is continuous on E. The last step is to estimate the norm in E of a solution $\vec{u}_\mu$ of

$$\vec{u}_\mu = \mu A_{R,N,\alpha}(D_N(\vec{u}_\mu)) = \mu\vec{u}_\mu^*$$

with $0 \leq \mu \leq 1$. Writing

$$\partial_t\vec{u}_\mu = \mu\left(\nu\Delta\vec{u}_\mu^* + \mathbb{P}(D_N(\vec{f}) - D_N(\theta(\frac{x}{R})(\phi_\alpha * (D_N(\vec{u}_\mu)) \cdot \vec{\nabla}(\theta(\frac{x}{R})D_N(\vec{u}_\mu)))))\right)$$

$$= \nu\Delta\vec{u}_\mu + \mathbb{P}(D_N(\mu\vec{f}) - D_N(\theta(\frac{x}{R})(\phi_\alpha * \mu D_N(\vec{u}_\mu)) \cdot \vec{\nabla}(\theta(\frac{x}{R})D_N(\vec{u}_\mu)))),$$

we find, (applying (16.37) and (16.34)),

$$\|\vec{w}_\mu\|_E \leq C_0\frac{1}{\nu}\|\mu\vec{f}_{per}\|_{L^2_{per}\dot{H}^{-1}}(1 + \frac{\|\mu\vec{u}_\mu\|_{L^2_{per}\dot{H}^1}}{T^{1/4}\nu^{3/4}})$$

$$\leq C_0\frac{1}{\nu}\|\vec{f}_{per}\|_{L^2_{per}\dot{H}^{-1}}(1 + \frac{\|\vec{f}\|_{L^2_{per}\dot{H}^{-1}}}{T^{1/4}\nu^{7/4}}).$$

Using Schaefer's fixed point theorem, we get existence of a solution $\vec{u}_{R,N,\alpha}$ to equation (16.31).

c) Existence of $\vec{u}_{R,\alpha}$.

By (16.37) and (16.34), we know that

$$\|\vec{u}_{R,N,\alpha}\|_E \leq C_0\frac{1}{\nu}\|\vec{f}_{per}\|_{L^2_{per}\dot{H}^{-1}}(1 + \frac{\|\vec{f}\|_{L^2_{per}\dot{H}^{-1}}}{T^{1/4}\nu^{7/4}}).$$

We write, since $D_N(\vec{u}_{R,N,\alpha}) = \vec{u}_{R,N,\alpha}$,

$$\partial_t\vec{u}_{R,N,\alpha} = \nu\Delta\vec{u}_{R,N,\alpha} + \mathbb{P}D_N(\vec{f})$$
$$- \mathbb{P}D_N(\theta(\frac{x}{R})(\phi_\alpha * D_0(\vec{u}_{R,N,\alpha})) \cdot \vec{\nabla}(\theta(\frac{x}{R})\vec{u}_{R,N,\alpha})))$$
$$- \mathbb{P}D_N(\theta(\frac{x}{R})(\phi_\alpha * (\vec{u}_{R,N,\alpha} - D_0(\vec{u}_{R,N,\alpha})) \cdot \vec{\nabla}(\theta(\frac{x}{R})\vec{u}_{R,N,\alpha})))$$

with

$$\|\nu\Delta\vec{u}_{R,N,\alpha}\|_{L^2_{per}\dot{H}^{-1}} \leq \nu\|\vec{u}_{R,N,\alpha}\|_{L^2_{per}\dot{H}^1},$$

$$\|\mathbb{P}D_N(\vec{f})\|_{L^2_{per}\dot{H}^{-1}} \leq \|\vec{f}\|_{L^2_{per}\dot{H}^{-1}},$$

$$\| \mathbb{P}D_N(\theta(\frac{x}{R})(\phi_\alpha * D_0(\vec{u}_{R,N,\alpha})) \cdot \vec{\nabla}(\theta(\frac{x}{R})\vec{u}_{R,N,\alpha})))\|_{L^2_{per}L^{3/2}}$$
$$\leq C_1\|D_0(\vec{u}_{R,N,\alpha})\|_{\dot{H}^1}\|\vec{u}_{R,N,\alpha}\|_{L^2_{per}\dot{H}^1}$$
$$\leq C_1 T^{-1/2}\|\vec{u}_{R,N,\alpha}\|^2_{L^2_{per}\dot{H}^1},$$

and (sincce $L^1 \subset H^{-2}$)

$$\| \mathbb{P}D_N(\theta(\frac{x}{R})(\phi_\alpha * (\vec{u}_{R,N,\alpha} - D_0(\vec{u}_{R,N,\alpha}))) \cdot \vec{\nabla}(\theta(\frac{x}{R})\vec{u}_{R,N,\alpha})))\|_{L^2_{per}H^{-2}}$$
$$\leq C_1\|\vec{u}_{R,N,\alpha} - D_0(\vec{u}_{R,N,\alpha})\|_{L^\infty L^2}\|\vec{u}_{R,N,\alpha}\|_{L^2_{per}\dot{H}^1},$$

where C_1 does not depend on R, N, α, T, ν, nor $\vec{f}$.

Thus, for every $\omega \in \mathcal{D}(\mathbb{R}^3)$, $\omega\vec{u}_{R,N,\alpha}$ is bounded in $L^2_{per}H^1$ and $\omega\partial_t\vec{u}_{R,N,\alpha}$ is bounded in $L^2_{per}H^{-2}$. Thus, we may apply the Rellich–Lions theorem (Theorem 12.1) and find a sub-sequence $(\vec{u}_{R,N_k,\alpha})_{k\in\mathbb{N}}$ that is locally strongly convergent in $L^2([0,T],L^2)$ to a limit $\vec{u}_{R,\alpha}$.

We then have the following convergences in $\mathcal{D}'(\mathbb{R} \times \mathbb{R}^3)$:

- $\partial_t\vec{u}_{R,N_k,\alpha} \overset{*}{\rightharpoonup} \partial_t\vec{u}_{R,\alpha}$

- $\nu\Delta\vec{u}_{R,N_k,\alpha} \overset{*}{\rightharpoonup} \nu\Delta\vec{u}_{R,\alpha}$

- $\mathbb{P}D_N(\vec{f}) = D_N(\mathbb{P}\vec{f}) \overset{*}{\rightharpoonup} \mathbb{P}\vec{f}$

- for every $\omega \in \mathcal{D}(\mathbb{R}^3)$, $\omega\theta(\frac{x}{R})(\phi_\alpha * \vec{u}_{R,N_k,\alpha})) \otimes (\theta(\frac{x}{R})\vec{u}_{R,N_k,\alpha}))$ is strongly convergent in $L^1_{per}([0,T],L^1)$ to $\omega\theta(\frac{x}{R})(\phi_\alpha * \vec{u}_{R,\alpha})) \otimes (\theta(\frac{x}{R})\vec{u}_{R,\alpha}))$; moreover,

$$\|\theta(\frac{x}{R})(\phi_\alpha * \vec{u}_{R,N_k,\alpha})\|_{L^\infty L^2} + \|\theta(\frac{x}{R})\vec{u}_{R,N_k,\alpha}\|_{L^\infty L^2}$$
$$\leq C(\sqrt{\nu}\|\theta\|_\infty + \frac{R}{\sqrt{T}}\|\theta\|_3)\|\vec{u}_{R,N_k,\alpha}\|_E,$$

so that $\omega\theta(\frac{x}{R})(\phi_\alpha * \vec{u}_{R,N_k,\alpha}) \otimes (\theta(\frac{x}{R})\vec{u}_{R,N_k,\alpha})$ is bounded in $L^\infty_{per}([0,T],L^1)$. By interpolation, $\omega\theta(\frac{x}{R})(\phi_\alpha * \vec{u}_{R,N_k,\alpha}) \otimes (\theta(\frac{x}{R})\vec{u}_{R,N_k,\alpha})$ is strongly convergent in $L^2_{per}([0,T],L^1)$, hence in $L^2_{per}([0,T],H^{-2})$.

- As the operators D_N are equicontinuous on $L^2_{per}([0,T],H^{-2})$ and since, for $u \in L^2_{per}([0,T],H^{-2})$, $D_N u$ is strongly convergent to u in $L^2_{per}([0,T],H^{-2})$, we find that $\omega D_{N_k}(\theta(\frac{x}{R})(\phi_\alpha * \vec{u}_{R,N_k,\alpha})) \otimes (\theta(\frac{x}{R})\vec{u}_{R,N_k,\alpha})))$ is strongly convergent to $\omega(\theta(\frac{x}{R})(\phi_\alpha * \vec{u}_{R,\alpha})) \otimes (\theta(\frac{x}{R})\vec{u}_{R,\alpha}))$. Thus, we have the convergence, in $\mathcal{D}'$,

$$D_{N_k}(\theta(\frac{x}{R})(\phi_\alpha * \vec{u}_{R,N_k,\alpha}) \otimes (\theta(\frac{x}{R})\vec{u}_{R,N_k,\alpha}))$$
$$\overset{*}{\rightharpoonup}\theta(\frac{x}{R})(\phi_\alpha * \vec{u}_{R,\alpha}) \otimes (\theta(\frac{x}{R})\vec{u}_{R,\alpha}).$$

Moreover, the weak convergence holds in $L^2_{per}H^{-2}$.

- As $\mathbb{P}$ is continuous on $L^2_{\mathrm{per}}H^{-3}$, we have the convergence, in $\mathcal{D}'$,

$$\mathbb{P}\operatorname{div}(D_{N_k}(\theta(\tfrac{x}{R})(\phi_\alpha * \vec{u}_{R,N_k,\alpha}) \otimes (\theta(\tfrac{x}{R})\vec{u}_{R,N_k,\alpha})))$$
$$\overset{*}{\rightharpoonup} \mathbb{P}\operatorname{div}(\theta(\tfrac{x}{R})(\phi_\alpha * \vec{u}_{R,\alpha}) \otimes (\theta(\tfrac{x}{R})\vec{u}_{R,\alpha})).$$

- Similarly, $\omega\left(\phi_\alpha * \vec{u}_{R,N_k,\alpha} \cdot \vec{\nabla}(\theta(\tfrac{x}{R}))\right)\theta(\tfrac{x}{R})\vec{u}_{R,N_k,\alpha}$ is strongly convergent in $L^1_{\mathrm{per}}([0,T],$ $L^1)$ and is bounded in $L^\infty_{\mathrm{per}}([0,T], L^1)$, hence is strongly convergent in $L^2_{\mathrm{per}}([0,T], H^{-2})$. We then find the convergence, in $\mathcal{D}'$,

$$D_{N_k}\left(\left(\phi_\alpha * \vec{u}_{R,N_k,\alpha} \cdot \vec{\nabla}(\theta(\tfrac{x}{R}))\right)\theta(\tfrac{x}{R})\vec{u}_{R,N_k,\alpha}\right)$$
$$\overset{*}{\rightharpoonup} \left(\phi_\alpha * \vec{u}_{R,\alpha} \cdot \vec{\nabla}(\theta(\tfrac{x}{R}))\right)\theta(\tfrac{x}{R})\vec{u}_{R,\alpha}.$$

Moreover, the weak convergence holds in $L^2_{\mathrm{per}}H^{-2}$. Thus, we have

$$\mathbb{P}D_{N_k}\left(\left(\phi_\alpha * \vec{u}_{R,N_k,\alpha} \cdot \vec{\nabla}(\theta(\tfrac{x}{R}))\right)\theta(\tfrac{x}{R})\vec{u}_{R,N_k,\alpha}\right)$$
$$\overset{*}{\rightharpoonup} \mathbb{P}\left(\left(\phi_\alpha * \vec{u}_{R,\alpha} \cdot \vec{\nabla}(\theta(\tfrac{x}{R}))\right)\theta(\tfrac{x}{R})\vec{u}_{R,\alpha}\right).$$

Thus, $\vec{u}_{R,\alpha}$ is a solution to (16.30).

d) Existence of $\vec{u}_R$.

$\vec{u}_{R,\alpha}$ inherits the good controls on $\vec{u}_{R,N,\alpha}$:

$$\|\vec{u}_{R,\alpha}\|_E \leq C_0 \frac{1}{\nu}\|\vec{f}_{\mathrm{per}}\|_{L^2_{\mathrm{per}}\dot{H}^{-1}}\left(1 + \frac{\|\vec{f}\|_{L^2_{\mathrm{per}}\dot{H}^{-1}}}{T^{1/4}\nu^{7/4}}\right)$$

and

$$\partial_t \vec{u}_{R,\alpha} = \nu\Delta\vec{u}_{R,\alpha} + \mathbb{P}(\vec{f})$$
$$- \mathbb{P}(\theta(\tfrac{x}{R})(\phi_\alpha * D_0(\vec{u}_{R,\alpha})) \cdot \vec{\nabla}(\theta(\tfrac{x}{R})\vec{u}_{R,\alpha})))$$
$$- \mathbb{P}(\theta(\tfrac{x}{R})(\phi_\alpha * (\vec{u}_{R,\alpha} - D_0(\vec{u}_{R,\alpha})) \cdot \vec{\nabla}(\theta(\tfrac{x}{R})\vec{u}_{R,\alpha})))$$

with

$$\|\nu\Delta\vec{u}_{R,\alpha}\|_{L^2_{\mathrm{per}}\dot{H}^{-1}} \leq \nu\|\vec{u}_{R,\alpha}\|_{L^2_{\mathrm{per}}\dot{H}^1},$$

$$\|\mathbb{P}(\vec{f})\|_{L^2_{\mathrm{per}}\dot{H}^{-1}} \leq \|\vec{f}\|_{L^2_{\mathrm{per}}\dot{H}^{-1}},$$

$$\|\mathbb{P}(\theta(\tfrac{x}{R})(\phi_\alpha * D_0(\vec{u}_{R,\alpha})) \cdot \vec{\nabla}(\theta(\tfrac{x}{R})\vec{u}_{R,\alpha})))\|_{L^2_{\mathrm{per}}L^{3/2}}$$
$$\leq C_1 T^{-1/2}\|\vec{u}_{R,\alpha}\|^2_{L^2_{\mathrm{per}}\dot{H}^1},$$

and

$$\|\mathbb{P}(\theta(\tfrac{x}{R})(\phi_\alpha * (\vec{u}_{R,\alpha} - D_0(\vec{u}_{R,\alpha}))) \cdot \vec{\nabla}(\theta(\tfrac{x}{R})\vec{u}_{R,\alpha})))\|_{L^2_{\mathrm{per}}H^{-2}}$$
$$\leq C_1\|\vec{u}_{R,\alpha} - D_0(\vec{u}_{R,\alpha})\|_{L^\infty L^2}\|\vec{u}_{R,\alpha}\|_{L^2_{\mathrm{per}}\dot{H}^1}.$$

Again, we may apply the Rellich–Lions theorem (Theorem 12.1) and find a sequence $(\alpha_k)_{k\in\mathbb{N}}$ converging to 0 such that $(\vec{u}_{R,\alpha_k})_{k\in\mathbb{N}}$ is locally strongly convergent in $L^2([0,T],L^2)$ to a limit $\vec{u}_R$. In particular, $\theta(\frac{x}{R})\vec{u}_{R,\alpha_k}$ is strongly convergent to $\theta(\frac{x}{R})\vec{u}_R$ in $L^2_{per}L^2$, and, since for $\alpha < 1$, we have

$$\theta(\frac{x}{R})(\phi_\alpha * \vec{u}_{R,\alpha}) = \theta(\frac{x}{R})(\phi_\alpha * (\theta(\frac{x}{8R})\vec{u}_{R,\alpha}))$$

and since the convolutions with ϕ_α are equicontinuous on $L^2_{per}L^2$ and $\lim_{\alpha\to 0}\|\phi_\alpha * \vec{u}_R - \vec{u}_R\|_{L^2_{per}L^2} = 0$, we have as well strong convergence of $\theta(\frac{x}{R})(\phi_\alpha*\vec{u}_{R,\alpha_k})$ to $\theta(\frac{x}{R})\vec{u}_R$ in $L^2_{per}L^2$.

As $\theta(\frac{x}{R})(\phi_\alpha * \vec{u}_{R,\alpha_k}) \cdot \vec{\nabla}(\theta(\frac{x}{R})\vec{u}_{R,\alpha_k})$ is bounded in $L^2_{per}H^{-2}$, we find that

$$\lim_{k\to+\infty} \theta(\frac{x}{R})(\phi_\alpha * \vec{u}_{R,\alpha_k}) \cdot \vec{\nabla}(\theta(\frac{x}{R})\vec{u}_{R,\alpha_k}) = \theta(\frac{x}{R})\vec{u}_R \cdot \vec{\nabla}(\theta(\frac{x}{R})\vec{u}_R)$$

and $\vec{u}_R$ is a solution to (16.29).

e) Existence of $\vec{u}$.

Again, $\vec{u}_R$ inherits the good controls on $\vec{u}_{R,\alpha}$:

$$\|\vec{u}_R\|_E \le C_0\frac{1}{\nu}\|\vec{f}_{per}\|_{L^2_{per}\dot{H}^{-1}}(1 + \frac{\|\vec{f}\|_{L^2_{per}\dot{H}^{-1}}}{T^{1/4}\nu^{7/4}})$$

and

$$\partial_t\vec{u}_R = \nu\Delta\vec{u}_R + \mathbb{P}(\vec{f})$$
$$- \mathbb{P}(\theta(\frac{x}{R})D_0(\vec{u}_R) \cdot \vec{\nabla}(\theta(\frac{x}{R})\vec{u}_R))$$
$$- \mathbb{P}(\theta(\frac{x}{R})(\vec{u}_R - D_0(\vec{u}_R)) \cdot \vec{\nabla}(\theta(\frac{x}{R})\vec{u}_R))$$

with

$$\|\nu\Delta\vec{u}_R\|_{L^2_{per}\dot{H}^{-1}} \le \nu\|\vec{u}_R\|_{L^2_{per}\dot{H}^1},$$
$$\|\mathbb{P}(\vec{f})\|_{L^2_{per}\dot{H}^{-1}} \le \|\vec{f}\|_{L^2_{per}\dot{H}^{-1}},$$
$$\|\mathbb{P}(\theta(\frac{x}{R})D_0(\vec{u}_R) \cdot \vec{\nabla}(\theta(\frac{x}{R})\vec{u}_R))\|_{L^2_{per}L^{3/2}}$$
$$\le C_1 T^{-1/2}\|\vec{u}_R\|^2_{L^2_{per}\dot{H}^1},$$

and

$$\|\mathbb{P}(\theta(\frac{x}{R})(\vec{u}_R - D_0(\vec{u}_R)) \cdot \vec{\nabla}(\theta(\frac{x}{R})\vec{u}_R))\|_{L^2_{per}H^{-2}}$$
$$\le C_1\|\vec{u}_R - D_0(\vec{u}_R)\|_{L^\infty L^2}\|\vec{u}_R\|_{L^2_{per}\dot{H}^1}.$$

Again, we may apply the Rellich–Lions theorem (Theorem 12.1) and find a sequence $(R_k)_{k\in\mathbb{N}}$ converging to $+\infty$ such that $(\vec{u}_{R_k})_{k\in\mathbb{N}}$ is locally strongly convergent in $L^2([0,T],L^2)$ to a limit $\vec{u}$. In particular,

$$\theta(\frac{x}{R})\vec{u}_R \cdot \vec{\nabla}(\theta(\frac{x}{R})\vec{u}_R) \overset{*}{\rightharpoonup} \vec{u} \cdot \vec{\nabla}\vec{u}$$

(convergence in $\mathcal{D}'$, as well as in $L^2_{per}H^{-2}$ since they are bounded in $L^2_{per}H^{-2}$). We then conclude that

$$\mathbb{P}(\theta(\frac{x}{R})\vec{u}_R \cdot \vec{\nabla}(\theta(\frac{x}{R})\vec{u}_R)) \overset{*}{\rightharpoonup} \mathbb{P}(\vec{u} \cdot \vec{\nabla}\vec{u})$$

in $L^2_{per}H^{-2}$, and finally that $\vec{u}$ is a solution to (16.27). □

Chapter 17

α-Models

In this chapter, we describe some ways to approximate the Navier–Stokes equations that have been developed in the mathematical literature since the seminal work of Leray [328]. We focus more precisely on the α-models that have been studied by Holm, Titi and others. More approximation models will be studied in the next chapter.

17.1 Global Existence, Uniqueness and Convergence Issues for Approximated Equations

In the preceding chapters, we have seen a large variety of solutions and functional spaces where to look for solutions. When dealing with an approximation of the equations, we must say which kind of solutions we shall try and approximate. In this chapter, we shall only consider Leray weak solutions: we consider the Navier–Stokes equations

$$\partial_t \vec{u} + \mathbb{P}(\vec{u} \cdot \vec{\nabla} \vec{u}) = \nu \Delta \vec{u} + \mathbb{P}\vec{f} \tag{17.1}$$

with initial value $\vec{u}_0$ and forcing term $\vec{f}$. We assume that:

- $\vec{u}_0 \in L^2(\mathbb{R}^3)$ with $\operatorname{div} \vec{u}_0 = 0$

- $\vec{f} \in L^2_t H^{-1}_x$ on $(0,T) \times \mathbb{R}^3$ (with $T < +\infty$)

and we are looking for a solution $\vec{u} \in L^\infty_t L^2_x \cap L^2_t H^1_x$ of (17.1) which satisfies *Leray's energy inequality*: for all $t \in (0,T)$,

$$\|\vec{u}(t,.)\|_2^2 + 2\nu \int_0^t \|\vec{u}(s,.)\|_{\dot{H}^1}^2 \, ds \leq \|\vec{u}_0\|_2^2 + 2 \int_0^t \langle \vec{f}(s,.) | \vec{u}(s,.) \rangle_{H^{-1},H^1} \, ds \tag{17.2}$$

The issue will be threefold:

- prove that the solution of the approximated equations is unique and has a life span of at least as long as the known Leray solutions

- prove that one has weak convergence to the weak solutions when the perturbation is shrinking to 0

- (when possible) prove that one obtains the good energy estimates for the recovered solution: the global Leray estimate, or the local energy estimate that characterizes suitable solutions.

For some approximations, we shall assume more regularity on the forcing term:

$$\vec{f} \in L^2_t L^2_x.$$

If moreover $\vec{u}_0 \in H^1(\mathbb{R}^3)$, and if we have a mild solution in $\mathcal{C}_t H^1 \cap L^2 H^2$, we may expect strong convergence, at least in the $L^\infty_t L^2_x$ norm.

DOI: 10.1201/9781003042594-17

17.2   Leray's Mollification and the Leray-α Model

Recall that we want to solve Equations (17.1) with initial value $\vec{u}_0 \in L^2$ and $\vec{f} \in L_t^2 H_x^1$. We rewrite (17.1) into

$$\vec{u} = W_{\nu t} * \vec{u}_0 + \int_0^t W_{\nu(t-s)} * \mathbb{P}(\vec{f} - \vec{u} \cdot \vec{\nabla}\,\vec{u})\,ds \qquad (17.3)$$

We have the following estimates:

$$\|W_{\nu t} * \vec{u}_0\|_{L^\infty L^2} = \|\vec{u}_0\|_2, \quad \|W_{\nu t} * \vec{u}_0\|_{L^2 \dot{H}^1} = \frac{1}{\sqrt{2\nu}}\|\vec{u}_0\|_2$$

$$\left\| \int_0^t W_{\nu(t-s)} * \mathbb{P}\vec{f}\,ds \right\|_{L^\infty((0,T),L^2)} \le (\sqrt{T} + \frac{1}{\sqrt{2\nu}})\|\vec{f}\|_{L^2 H^{-1}}$$

$$\left\| \int_0^t W_{\nu(t-s)} * \mathbb{P}\vec{f}\,ds \right\|_{L^2((0,T),\dot{H}^1)} \le (CT + \frac{1}{\nu})\|\vec{f}\|_{L^2 H^{-1}}$$

However, Picard's iterative method to get a fixed point for

$$\vec{u} \mapsto W_{\nu t} * \vec{u}_0 + \int_0^t W_{\nu(t-s)} * \mathbb{P}(\vec{f} - \vec{u} \cdot \vec{\nabla}\,\vec{u})\,ds$$

does not work, since, for $\vec{u} \in L^\infty L^2 \cap L^2 \dot{H}^1$ with div $\vec{u} = 0$, we do not have $\vec{u} \cdot \vec{\nabla}\vec{u} \in L^2 H^{-1}$ but only $\vec{u} \cdot \vec{\nabla}\vec{u} \in L^2 H^{-3/2}$.

Leray's mollification aims to soften the transportation term $\vec{u} \cdot \vec{\nabla}\vec{u}$ by replacing it with $\vec{u}_\alpha \cdot \vec{\nabla}\vec{u}$ où $\vec{u}_\alpha$ corresponds to a smoothing of $\vec{u}$:

$$\vec{u}_\alpha = \frac{1}{\alpha^3} K(\frac{x}{\alpha}) * \vec{u} \qquad (17.4)$$

where

$$K \in L^1 \cap L^2 \text{ and } \int K\,dx = 1. \qquad (17.5)$$

In the initial paper of Leray [328], the kernel K was a compactly supported smooth function. But other kernels have been used, as for instance the kernel of the Hemlhotz operator $(Id - \Delta)^{-1}$:

$$\vec{u}_\alpha = (Id - \alpha^2 \Delta)^{-1}\vec{u}$$

in the Leray–α model discussed by Cheskidov, Holm, Olson and Titi [119], where the authors introduce the Leray–α model in the periodical setting and prove that the dimension of the global attractor is much smaller than expected, suggesting that the Leray–α model could become a good model for the large eddy simulation of turbulence.

Leray's mollification

Theorem 17.1.
Let $\vec{u}_0 \in L^2$, with div $\vec{u}_0 = 0$, *and $\vec{f} \in L_t^2 H_x^{-1}$ on $(0,T) \times \mathbb{R}^3$ $(T < +\infty)$. Let $K \in L^1 \cap L^2$ with $\int K\,dx = 1$. Then*

- *for $\alpha > 0$, the problem*

$$\partial_t \vec{u} + \mathbb{P}(\vec{u}_\alpha \cdot \vec{\nabla} \vec{u}) = \nu \Delta \vec{u} + \mathbb{P}\vec{f} \tag{17.6}$$

with initial value $\vec{u}(0,.) = \vec{u}_0$ *and*

$$\vec{u}_\alpha = \frac{1}{\alpha^3} K(\frac{x}{\alpha}) * \vec{u}$$

has a unique solution $\vec{u}_{(\alpha)}$ *such that* $\vec{u}_{(\alpha)} \in L_t^\infty L_x^2 \cap L_t^2 H_x^1$.
Moreover, we have the following inequality:

$$\|\vec{u}_{(\alpha)}(t,.)\|_2^2 + \nu \int_0^t \|\vec{\nabla} \otimes \vec{u}_{(\alpha)}(s,.)\|_2^2$$

$$\leq (\|\vec{u}(0,.)\|_2^2 + \frac{1}{\nu}\int_0^T \|\vec{f}\|_{H_x^{-1}}^2 \, ds) e^{\nu t}. \tag{17.7}$$

- *there exists a sequence* $\alpha_k \to 0$ *and a function* $\vec{u} \in L_t^\infty L_x^2 \cap L_t^2 H_x^1$ *such that* $\vec{u}_{(\alpha_k)}$ *is weakly convergent to* $\vec{u}$. *Moreover* $\vec{u}$ *is a suitable Leray weak solution of the Navier–Stokes problem*

$$\partial_t \vec{u} + \mathbb{P}(\vec{u} \cdot \vec{\nabla} \vec{u}) = \nu \Delta \vec{u} + \mathbb{P}\vec{f}$$

with initial value $\vec{u}_0$.

- *If* $\vec{u}_0 \in H^1$ *and* $\vec{f} \in L_t^2 L_x^2$ *and if the Navier-Stokes problem has a solution* $\vec{u} \in \mathcal{C}([0,T], H^1) \cap L^2([0,T], H^2)$, *then we have*

$$\lim_{\alpha \to 0} \|\vec{u}_{(\alpha)} - \vec{u}\|_{L_t^\infty L_x^2} = 0.$$

Proof. • **First step: Local existence of** $\vec{u}_{(\alpha)}$.
We have
$$\|\vec{u}_\alpha\|_\infty \leq \alpha^{-3/2}\|\vec{u}\|_2\|K\|_2.$$
Thus, for $\vec{u}$ and $\vec{v}$ in $L^\infty L^2 \cap L^2 H^1$ with div $\vec{u} = 0$, we have for every $0 < T_0 < T$,

$$\|\vec{u}_\alpha \cdot \vec{\nabla}\vec{v}\|_{L^2((0,T_0),H^{-1})} \leq C\sqrt{T_0}\,\alpha^{-3/2}\|\vec{u}\|_{L^\infty L^2}\|\vec{v}\|_{L^\infty L^2}$$

Let $\|\vec{u}\|_{\nu,T_0} = \|\vec{u}\|_{L^\infty((0,T_0),L^2)} + \sqrt{\nu}\|\vec{u}\|_{L^2(0,T_0),\dot{H}^1)}$. We have

$$\|W_{\nu t} * \vec{u}_0 + \int_0^t W_{\nu(t-s)} * \mathbb{P}\vec{f}\,ds\|_{\nu,T_0} \leq C_0(\|\vec{u}_0\|_2 + \frac{1}{\sqrt{\nu}}(1+T_0\nu)\|\vec{f}\|_{L^2 H^{-1}})$$

and

$$\|\int_0^t W_{\nu(t-s)} * \mathbb{P}(\vec{u}_\alpha \cdot \vec{\nabla}\vec{v})\,ds\|_{\nu,T_0} \leq C_0\frac{1}{\sqrt{\nu}}(1+T_0\nu)\sqrt{T_0}\,\alpha^{-3/2}\|\vec{u}\|_{L^\infty L^2}\|\vec{v}\|_{L^\infty L^2}$$

Thus, we find existence (and uniqueness) of a solution $\vec{u} = \vec{u}_{(\alpha)}$ of the equation

$$\vec{u} = W_{\nu t} * \vec{u}_0 + \int_0^t W_{\nu(t-s)} * \mathbb{P}(\vec{f} - \vec{u}_\alpha \cdot \vec{\nabla}\vec{u})\,ds \tag{17.8}$$

for T_0 small enough to ensure

$$1 + T_0 \nu \leq 2$$

and

$$T_0 \leq \alpha^3 \nu \frac{1}{16\, C_0^4 (\|\vec{u}_0\|_2 + \frac{2}{\sqrt{\nu}} \|\vec{f}\|_{L^2 \dot{H}^{-1}})^2}. \tag{17.9}$$

- **Second step: Energy estimates and global existence of $\vec{u}_{(\alpha)}$.**

To show the existence of a global solution to (17.6), it is then enough to show that the L^2 norm of $\vec{u}_{(\alpha)}$ remains bounded (as the existence time T_0 is controlled by the L^2 norm of the Cauchy data by (17.9)).

Since div $\left(\vec{u}_{(\alpha)} * \frac{1}{\alpha^3} K(\frac{x}{\alpha})\right) = 0$, we have

$$\int \vec{u}_{(\alpha)} \cdot \left((\vec{u}_{(\alpha)} * \frac{1}{\alpha^3} K(\frac{x}{\alpha})) \cdot \vec{\nabla} \vec{u}_{(\alpha)} \right) \, dx = 0$$

hence

$$\begin{aligned}
\frac{d}{dt} \|\vec{u}_{(\alpha)}\|_2^2 &= 2 \int \partial_t \vec{u}_{(\alpha)} \cdot \vec{u}_{(\alpha)} \, dx \\
&= -2\nu \|\vec{u}_{(\alpha)}\|_{\dot{H}^1}^2 + 2\langle \vec{f}|\vec{u}_{(\alpha)}\rangle_{H^{-1},H^1} \\
&\leq -\nu \|\vec{u}_{(\alpha)}\|_{\dot{H}^1}^2 + \nu \|\vec{u}_{(\alpha)}\|_2^2 + \frac{1}{\nu} \|\vec{f}\|_{H^{-1}}^2
\end{aligned} \tag{17.10}$$

so that

$$\|\vec{u}_{(\alpha)}(t,.)\|_2^2 + \nu \int_0^t \|\vec{u}_{(\alpha)}\|_{\dot{H}^1}^2 \, ds \leq \|\vec{u}_0\|_2^2 + \frac{1}{\nu} \|\vec{f}\|_{L^2 H^{-1}}^2 + \nu \int_0^t \|\vec{u}_{(\alpha)}(s,.)\|_2^2 \, ds$$

We thus get (by Grönwall's lemma) the energy estimate (17.7) and, therefore, the global existence of $\vec{u}_{(\alpha)}$.

- **Third step: Weak convergence.**

From the energy estimate (17.7), we know that $\vec{u}_{(\alpha)}$ remains bounded in $L^\infty L^2 \cap L^2 \dot{H}^1$. Moreover, since $K \in L^1$, we have $\|\vec{u}_{(\alpha)} * \frac{1}{\alpha^3} K(\frac{x}{\alpha})\|_2 \leq \|\vec{u}_{(\alpha)}\|_2 \|K\|_1$. As we have

$$\partial_t \vec{u}_{(\alpha)} = \nu \Delta \vec{u}_{(\alpha)} + \mathbb{P}(\vec{f} - (\vec{u}_{(\alpha)} * \frac{1}{\alpha^3} K(\frac{x}{\alpha})) \cdot \vec{\nabla} \vec{u}_{(\alpha)})$$

we can see that $\partial_t \vec{u}_{(\alpha)}$ remains bounded in $L^2 H^{-3/2}$.

We may then use the Rellich–Lions theorem (Theorem 12.1): we may find a sequence $\alpha_n \to 0$ and a function $\vec{u} \in L^\infty L^2 \cap L^2 \dot{H}^1$ such that:

- $\vec{u}_{(\alpha_n)}$ is *-weakly convergent to $\vec{u}$ in $L^\infty L^2$ and in $L^2 \dot{H}^1$

- $\vec{u}_{(\alpha_n)}$ is strongly convergent to $\vec{u}$ in $L^2_{\text{loc}}([0,T) \times \mathbb{R}^3)$.

Since $K \in L^1$ may be approximated by compactly supported smooth kernels, and since $\vec{u}_{(\alpha)}$ is bounded in $L^\infty L^2$, we get that $\vec{u}_{(\alpha_n)} * \frac{1}{\alpha_n^3} K(\frac{x}{\alpha_n})$ strongly converges $\vec{u}$ in $L^2_{\text{loc}}([0,T) \times \mathbb{R}^3)$ (and even in $(L^p_t L^2_x)_{\text{loc}}((0,T) \times \mathbb{R}^3)$ for every $p < +\infty$); as $\vec{\nabla} \otimes \vec{u}_{(\alpha_n)}$ *-weakly converges to $\vec{\nabla} \otimes \vec{u}$ in L^2_{loc} we get that the sequence $(\vec{u}_{(\alpha_n)} * \frac{1}{\alpha_n^3} K(\frac{x}{\alpha_n})) \cdot \vec{\nabla} \vec{u}_{(\alpha_n)}$ is *-weakly convergent to $\vec{u} \cdot \vec{\nabla} \vec{u}$ in $(L^q_t H^{-3/2}_x)_{\text{loc}}$ for every $1 < q < 2$; as the sequence is bounded in $L^2 H^{-3/2}$, we have *-weak convergence in $L^2 H^{-3/2}$ as well; as $\mathbb{P}$ is bounded on $H^{-3/2}$, we get the *-weak convergence of $\mathbb{P}\left((\vec{u}_{(\alpha_n)} * \frac{1}{\alpha_n^3} K(\frac{x}{\alpha_n})) \cdot \vec{\nabla} \, \vec{u}_{(\alpha_n)} \right)$ to $\mathbb{P}(\vec{u} \cdot \vec{\nabla} \vec{u})$ in $L^2 H^{-3/2}$.

Thus, the weak limit $\vec{u}$ satisfies

$$\partial_t \vec{u} = \nu \Delta \vec{u} + \mathbb{P}(\vec{f} - \vec{u} \cdot \vec{\nabla} \, \vec{u}).$$

- **Fourth step: Global energy estimates for the weak limit.**

We may easily check that the weak limit $\vec{u}$ fulfills the Leray energy inequality. Indeed, let $w(t)$ be a non-negative compactly supported smooth function on $\mathbb{R}$, supported within $[-\epsilon, \epsilon]$, with $\|w\|_1 = 1$. On $(\epsilon, T - \epsilon)$, we have

$$\|w * \vec{u}_{(\alpha)}(t, .)\|_2^2 \le w * \|\vec{u}_{(\alpha)}(t, .)\|_2^2$$

and

$$w * \|\vec{u}_{(\alpha)}(t, .)\|_2^2 + 2\nu w * \left(\int_0^t \|\vec{u}_{(\alpha)}\|_{\dot{H}^1}^2 \, ds \right) = \|\vec{u}_0\|_2^2 + 2w * \left(\int_0^t \langle \vec{f} | \vec{u}_{(\alpha_n)} \rangle_{H^{-1}, H^1} \, ds \right)$$

Besides, we have the *-weak convergence of $w * \vec{u}_{(\alpha_n)}$ to $w * \vec{u}$ in $L^2 \dot{H}^1$, the strong convergence of $w * \vec{u}_{(\alpha)}(t, .)$ to $w * \vec{u}(t, .)$ in $L^2_{\text{loc}}(\mathbb{R}^3)$, and thus (due to the control we have on the L^2 norm) the *-weak convergence of $w * \vec{u}_{(\alpha)}(t, .)$ to $w * \vec{u}(t, .)$ in $L^2(\mathbb{R}^3)$. We thus get that

$$\|w * \vec{u}(t, .)\|_2^2 + 2\nu w * \left(\int_0^t \|\vec{u}\|_{\dot{H}^1}^2 \, ds \right) \le$$

$$\liminf_{n \to +\infty} \|w * \vec{u}_{(\alpha_n)}(t, .)\|_2^2 + 2\nu w * \left(\int_0^t \|\vec{u}_{(\alpha_n)}\|_{\dot{H}^1}^2 \, ds \right)$$

and thus

$$\|w * \vec{u}(t, .)\|_2^2 + 2\nu w * \left(\int_0^t \|\vec{u}\|_{\dot{H}^1}^2 \, ds \right) \le \|\vec{u}_0\|_2^2 + 2w * \left(\int_0^t \langle \vec{f} | \vec{u} \rangle_{H^{-1}, H^1} \, ds \right)$$

When t_0 is a Lebesgue point of $t \mapsto \|\vec{u}(t, .)\|_2$, we get that:

$$\|\vec{u}(t_0, .)\|_2^2 + 2\nu \int_0^{t_0} \|\vec{u}\|_{\dot{H}^1}^2 \, ds \le \|\vec{u}_0\|_2^2 + 2 \int_0^{t_0} \langle \vec{f} | \vec{u} \rangle_{H^{-1}, H^1} \, ds \qquad (17.11)$$

This inequality may be extended to every t_0, due to the weak boundedness of $t \in (0, T) \mapsto \vec{u}(t, .) \in L^2$.

- **Fifth step: Local energy estimates for the weak limit.**

We now check that $\vec{u}$ is more precisely a suitable weak solution (i.e., fulfills the local energy inequality). We write

$$\partial_t \vec{u}_{(\alpha)} = \nu \Delta \vec{u}_{(\alpha)} + \vec{f} - (\vec{u}_{(\alpha)} * \frac{1}{\alpha^3} K(\frac{x}{\alpha})) \cdot \vec{\nabla} \, \vec{u}_{(\alpha)} - \vec{\nabla} p_{(\alpha)}$$

with

$$p_{(\alpha)} = \frac{1}{\Delta} \operatorname{div} \left(\vec{f} - (\vec{u}_{(\alpha)} * \frac{1}{\alpha^3} K(\frac{x}{\alpha})) \cdot \vec{\nabla} \, \vec{u}_{(\alpha)} \right)$$

and we write

$$\partial_t(\frac{|\vec{u}_{(\alpha)}|^2}{2}) = \nu\Delta(\frac{|\vec{u}_{(\alpha)}|^2}{2}) - \nu|\vec{\nabla}\otimes\vec{u}_{(\alpha)}|^2 + \vec{f}\cdot\vec{u}_{(\alpha)}$$

$$- \operatorname{div}(\frac{|\vec{u}_{(\alpha)}|^2}{2}(\vec{u}_{(\alpha)} * \frac{1}{\alpha^3}K(\frac{x}{\alpha}))) - \operatorname{div}(p_{(\alpha)}\vec{u}_{(\alpha)})$$

We know that $\vec{u}_{(\alpha_n)}$ converge strongly to $\vec{u}$ in $L^2_{loc}((0,T)\times\mathbb{R}^3)$; as the family is bounded in $L^{10/3}_t H^{3/5}_x \subset L^{10/3}_t L^{10/3}_x$, we find that we have strong convergence in $L^3_{loc}((0,T)\times\mathbb{R}^3)$ as well. Thus, we have the following convergence results in $\mathcal{D}'((0,T)\times\mathbb{R}^3)$: $\partial_t|\vec{u}_{(\alpha_n)}|^2 \to \partial_t|\vec{u}|^2$, $\Delta|\vec{u}_{(\alpha_n)}|^2 \to \Delta|\vec{u}|^2$, $\operatorname{div}(|\vec{u}_{(\alpha_n)}|^2(\vec{u}_{(\alpha_n)} * \frac{1}{\alpha_n^3}K(\frac{x}{\alpha_n}))) \to \operatorname{div}(|\vec{u}|^2\vec{u})$ and $\vec{u}_{(\alpha_n)}\cdot\vec{f} \to \vec{u}\cdot\vec{f}$. Similarly, we have that

$$\operatorname{div}\left((\frac{1}{\Delta}\operatorname{div}\vec{f} + \sum_{j=1}^{3}\sum_{l=1}^{3}\frac{\partial_j}{\sqrt{-\Delta}}\frac{\partial_l}{\sqrt{-\Delta}}(u_{(\alpha_n),j}(u_{(\alpha_n),l} * \frac{1}{\alpha_n^3}K(\frac{x}{\alpha_n}))))\vec{u}_{(\alpha_n)}\right)$$

converges in $\mathcal{D}'$ to $\operatorname{div}\left((\frac{1}{\Delta}\operatorname{div}\vec{f} + \sum_{j=1}^{3}\sum_{l=1}^{3}\frac{\partial_j}{\sqrt{-\Delta}}\frac{\partial_l}{\sqrt{-\Delta}}(u_j u_l))\vec{u}\right)$.

Thus far, we have got that

$$\partial_t|\vec{u}|^2 = \nu\Delta|\vec{u}|^2 - \operatorname{div}((|\vec{u}|^2 + 2p)\vec{u}) + 2\vec{u}\cdot\vec{f} - \nu T$$

with

$$T = \lim_{\alpha_n\to 0} 2|\vec{\nabla}\otimes\vec{u}_{(\alpha_n)}|^2.$$

Let $\phi \in \mathcal{D}'((0,T)\times\mathbb{R}^3)$ be a non-negative function. As $\sqrt{\phi}\,\vec{\nabla}\otimes\vec{u}_{(\alpha_n)}$ is weakly convergent to $\sqrt{\phi}\,\vec{\nabla}\otimes\vec{u}$ in $L^2_t L^2_x$, we find that $\|\sqrt{\phi}\,\vec{\nabla}\otimes\vec{u}\|_2^2 \leq \liminf_{\alpha_n\to 0}\|\sqrt{\phi}\,\vec{\nabla}\otimes\vec{u}_{(\alpha_n)}\|_2^2$. Thus, we have

$$\langle T|\phi\rangle_{\mathcal{D}',\mathcal{D}} = 2\lim_{\alpha_n\to 0}\iint |\vec{\nabla}\otimes\vec{u}_{(\alpha_n)}|^2\,\phi(t,x)\,dt\,d$$

$$\geq 2\iint |\vec{\nabla}\otimes\vec{u}|^2\phi(t,x)\,dt\,dx.$$

Thus, $T = 2|\vec{\nabla}\otimes\vec{u}|^2 + \mu$, where μ is a non-negative locally finite measure, and thus $\vec{u}$ is suitable.

- **Sixth step: Strong convergence.**

If $\vec{u}_0 \in H^1$ and $\vec{f} \in L^2_t L^2_x$, and if $\vec{u}$ is a solution of the Navier-Stokes problem with $\vec{u} \in \mathcal{C}([0,T],H^1)\cap L^2([0,T],H^2)$, it is easy to check the strong convergence of $\vec{u}_{(\alpha)}$ to $\vec{u}$: as $\vec{u}$ and $\vec{u}_{(\alpha)}$ are regular enough, we may write

$$\frac{d}{dt}\|\vec{u} - \vec{u}_{(\alpha)}\|_2^2 = 2\int (\vec{u} - \vec{u}_{(\alpha)})\cdot\partial_t(\vec{u} - \vec{u}_{(\alpha)})\,dx$$

$$= -2\nu\|\vec{\nabla}\otimes(\vec{u} - \vec{u}_{(\alpha)})\|_2^2$$

$$- 2\int (\vec{u} - \vec{u}_{(\alpha)}).(\vec{u}\cdot\vec{\nabla}\vec{u} - (\vec{u}_{(\alpha)} * \frac{1}{\alpha^3}K(\frac{x}{\alpha}))\cdot\vec{\nabla}\vec{u}_{(\alpha)})\,dx$$

$$= -2\nu\|\vec{\nabla}\otimes(\vec{u} - \vec{u}_{(\alpha)})\|_2^2$$

$$- 2\int (\vec{u} - \vec{u}_{(\alpha)}).((\vec{u} - \vec{u}_{(\alpha)}) * \frac{1}{\alpha^3}K(\frac{x}{\alpha}))\cdot\vec{\nabla}\vec{u})\,dx$$

$$-2\int (\vec{u}-\vec{u}_{(\alpha)}).(\vec{u}-\vec{u}*\frac{1}{\alpha^3}K(\frac{x}{\alpha}))\cdot\vec{\nabla}\vec{u})\,dx$$
$$\leq -2\nu\|\vec{\nabla}\otimes(\vec{u}-\vec{u}_{(\alpha)})\|_2^2+2\|K\|_1\|\vec{u}-\vec{u}_{(\alpha)}\|_6\|\vec{u}-\vec{u}_{(\alpha)}\|_2\|\vec{\nabla}\otimes\vec{u}\|_3$$
$$+2\|\vec{u}-\vec{u}_{(\alpha)}\|_6\|\vec{u}-\vec{u}*\frac{1}{\alpha^3}K(\frac{x}{\alpha})\|_2\|\vec{\nabla}\otimes\vec{u}\|_3$$
$$\leq\frac{C}{\nu}\|\vec{u}-\vec{u}_{(\alpha)}\|_2^2\|\vec{\nabla}\otimes\vec{u}\|_3^2+\frac{C}{\nu}\|\vec{u}-\vec{u}*\frac{1}{\alpha^3}K(\frac{x}{\alpha})\|_2^2\|\vec{\nabla}\otimes\vec{u}\|_3^2$$

This gives

$$\|\vec{u}-\vec{u}_{(\alpha)}\|_{L^\infty L^2}^2\leq e^{\int_0^T\frac{C}{\nu}\|\vec{\nabla}\otimes\vec{u}\|_3^2\,ds}\int_0^T\frac{C}{\nu}\|\vec{u}-\vec{u}*\frac{1}{\alpha^3}K(\frac{x}{\alpha})\|_2^2\|\vec{\nabla}\otimes\vec{u}\|_3^2\,ds$$

and thus $\lim_{\alpha\to 0}\sup_{0<t<T}\|\vec{u}-\vec{u}_{(\alpha)}\|_2=0$. □

17.3 The Navier–Stokes α-Model

The Navier–Stokes α-model, also known as viscous Camassa–Holm equations, have been studied by Chen, Foias, Holm, Olson, Titi and Wynne [117] as a good mathematical model to yield accurate predictions of turbulent flow profiles at very large Reynolds numbers.

Recall that in the Leray-α model the Navier–Stokes equations

$$\partial_t\vec{u}+\vec{u}\cdot\vec{\nabla}\,\vec{u}=\nu\Delta\vec{u}+\vec{f}-\vec{\nabla}p,\quad \operatorname{div}\vec{u}=0$$

were approximated by

$$\partial_t\vec{u}+\vec{u}_\alpha\cdot\vec{\nabla}\,\vec{u}=\nu\Delta\vec{u}+\vec{f}-\vec{\nabla}p,\quad \operatorname{div}\vec{u}=0$$

with $\vec{u}_\alpha=(Id-\alpha^2\Delta)^{-1}\vec{u}$. For the Navier–Stokes-α model, one approximates the Navier–Stokes equations written with use of the vorticity $\vec{\omega}=\operatorname{curl}\vec{u}$ and the total pressure Q (Equation (2.24))

$$\partial_t\vec{u}+\vec{\omega}\wedge\vec{u}=\nu\Delta\vec{u}+\vec{f}-\vec{\nabla}Q,\quad \operatorname{div}\vec{u}=0$$

by the equations

$$\partial_t\vec{u}+\vec{\omega}\wedge\vec{u}_\alpha=\nu\Delta\vec{u}+\vec{f}-\vec{\nabla}Q,\quad \operatorname{div}\vec{u}=0$$

where again $\vec{u}_\alpha=(Id-\alpha^2\Delta)^{-1}\vec{u}$. As we have

$$\vec{\omega}\wedge\vec{u}_\alpha=\vec{u}_\alpha\cdot\vec{\nabla}\vec{u}+\sum_{k=1}^3 u_k\vec{\nabla}u_{\alpha,k}-\vec{\nabla}(\vec{u}\cdot\vec{u}_\alpha),$$

the equations of the Navier–Stokes-α model are written as

$$\partial_t\vec{u}+\vec{u}_\alpha\cdot\vec{\nabla}\,\vec{u}=\nu\Delta\vec{u}-\sum_{k=1}^3 u_k\vec{\nabla}u_{\alpha,k}+\vec{f}-\vec{\nabla}p \tag{17.12}$$

with $\operatorname{div}\vec{u}=0$ and $\vec{u}(0,.)=\vec{u}_0$.

Equations (17.12) were first derived by Holm, Marsden and Ratiu [236] through variational principles in the Lagrangian formalism. Thereafter, there were given two more derivations of the equations:

- Foias, Holm and Titi [177] have shown that the extra term $\sum_{k=1}^{3} u_k \vec{\nabla} u_{\alpha,k}$ aims to restore Kelvin's circulation theorem which states that in the inviscid limit ($\nu = 0$) and in the absence of external forces ($\vec{f} = 0$), the circulation of the velocity around a fluid loop that moves with the velocity field remains constant. The fact that Leray's mollification did not satisfy Kelvin's circulation theorem had been underlined by Gallavotti [202].

- Guermond, Oden and Prudhomme [221] show that the correction from the Leray-α model

$$\partial_t \vec{u} + \vec{u} \cdot \vec{\nabla} \vec{u} = \nu \Delta \vec{u} + \vec{f} - \vec{\nabla} p + (\vec{u} - \vec{u}_\alpha) \cdot \vec{\nabla} \vec{u}$$

  to the Navier–Stokes-α model

$$\partial_t \vec{u} + \vec{u} \cdot \vec{\nabla} \vec{u} = \nu \Delta \vec{u} + \vec{f} - \vec{\nabla} p + (\vec{u} - \vec{u}_\alpha) \cdot \vec{\nabla} \vec{u} - \sum_{k=1}^{3} u_k \vec{\nabla} u_{\alpha,k}$$

  restores the material indifference principle that is violated by the forcing term $-\vec{\nabla} p + (\vec{u} - \vec{u}_\alpha) \cdot \vec{\nabla} \vec{u}$ but respected by the modified forcing term $-\vec{\nabla} p + (\vec{u} - \vec{u}_\alpha) \cdot \vec{\nabla} \vec{u} - \sum_{k=1}^{3} u_k \vec{\nabla} u_{\alpha,k}$.

The mathematical study of the Navier–Stokes-α model has been done by Foias, Holm and Titi [178]. Following them, one can prove:

Navier–Stokes-α model

Theorem 17.2.
Let $\vec{u}_0 \in L^2$, with div $\vec{u}_0 = 0$, and $\vec{f} \in L_t^2 H_x^{-1}$ on $(0,T) \times \mathbb{R}^3$ ($T < +\infty$). Then

- *for $\alpha > 0$, the problem*

$$\partial_t \vec{u} + \mathbb{P}(\vec{u}_\alpha \cdot \vec{\nabla} \vec{u} + \sum_{k=1}^{3} u_k \vec{\nabla} u_{\alpha,k}) = \nu \Delta \vec{u} + \mathbb{P} \vec{f} \qquad (17.13)$$

  with initial value $\vec{u}(0,.) = \vec{u}_0$ and

$$\vec{u}_\alpha = (Id - \alpha^2 \Delta)^{-1} \vec{u}$$

  has a unique solution $\vec{u}_{(\alpha)}$ such that $\vec{u}_{(\alpha)} \in L_t^\infty L_x^2 \cap L_t^2 H_x^1$.

- *$\vec{U}_{(\alpha)} = (Id - \alpha^2 \Delta)^{-1} \vec{u}_{(\alpha)}$ satisfies, with uniform bounds with respect to α, that $\vec{U}_{(\alpha)} \in L_t^\infty L_x^2 \cap L_t^2 H_x^1$ and $\alpha \vec{U}_{(\alpha)} \in L_t^\infty H_x^1 \cap L_t^2 H_x^2$*

- *there exists a sequence $\alpha_k \to 0$ and a function $\vec{u} \in L_t^\infty L_x^2 \cap L_t^2 H_x^1$ such that $\vec{u}_{(\alpha_k)}$ converges to $\vec{u}$ in $\mathcal{D}'((0,T) \times \mathbb{R}^3)$. Moreover $\vec{u}$ is a suitable Leray weak solution of the Navier–Stokes problem*

$$\partial_t \vec{u} + \mathbb{P}(\vec{u} \cdot \vec{\nabla} \vec{u}) = \nu \Delta \vec{u} + \mathbb{P} \vec{f}$$

  with initial value $\vec{u}_0$.

- *If $\vec{u}_0 \in H^1$ and $\vec{f} \in L_t^2 L_x^2$ and if the Navier-Stokes problem has a solution $\vec{u} \in \mathcal{C}([0,T], H^1) \cap L^2([0,T], H^2)$, then we have*

$$\lim_{\alpha \to 0} \|\vec{u}_{(\alpha)} - \vec{u}\|_{L_t^\infty L_x^2} = 0.$$

Proof. • **First step: Local existence of $\vec{u}_{(\alpha)}$.**

This step is very close to the same step in the proof for the Leray-α model (Theorem 17.1).

We just have to estimate, for $\vec{u}$ and $\vec{v}$ in $L^{\infty}L^2 \cap L^2 H^1$ with div $\vec{u} = 0$ and for $0 < T_0 < T$, the norms in $L^2((0,T_0), H^{-1})$ of $\vec{u}_{\alpha} \cdot \vec{\nabla}\vec{v}$ and of $\sum_{k=1}^{3} u_k \vec{\nabla} v_{\alpha,k}$: the kernel K of $(Id - \Delta)^{-1}$ satisfies $K \in L^2$ and $\vec{\nabla}K \in L^{3/2,\infty}$ so that $\|\vec{u}_{\alpha}\|_{\infty} \leq \alpha^{-3/2}\|\vec{u}\|_2\|K\|_2$ and $\|\vec{\nabla} \otimes \vec{v}_{\alpha}\|_6 \leq \alpha^{-2}\|\vec{v}\|_2\|K\|_{L^{3/2,\infty}}$. Thus, we have

$$\|\vec{u}_{\alpha} \cdot \vec{\nabla}\vec{v}\|_{L^2((0,T_0),H^{-1})} = \|\operatorname{div}(\vec{v} \otimes \vec{u}_{\alpha})\|_{L^2((0,T_0),H^{-1})} \leq C\frac{\sqrt{T_0}}{\alpha^{3/2}}\|\vec{u}\|_{L^{\infty}L^2}\|\vec{v}\|_{L^{\infty}L^2}$$

and

$$\|\sum_{k=1}^{3} u_k \vec{\nabla} v_{\alpha,k}\|_{L^2((0,T_0),H^{-1})} \leq C\sqrt{T_0}\|\vec{u}\|_{L^{\infty}L^2}\|\vec{\nabla} \otimes \vec{v}_{\alpha}\|_{L^{\infty}L^3}$$

$$\leq C'\frac{\sqrt{T_0}}{\alpha^{3/2}}\|\vec{u}\|_{L^{\infty}L^2}\|\vec{v}\|_{L^{\infty}L^2}$$

Thus, we find existence (and uniqueness) of a solution $\vec{u} = \vec{u}_{(\alpha)}$ of the Equation (17.13) for T_0 small enough to ensure

$$1 + T_0\nu \leq 2$$

and

$$T_0 \leq C_0\alpha^3 \frac{\nu}{(\|\vec{u}_0\|_2 + \frac{2}{\sqrt{\nu}}\|\vec{f}\|_{L^2\dot{H}^{-1}})^2}. \tag{17.14}$$

• **Second step: Energy estimates and global existence of $\vec{u}_{(\alpha)}$.**

To show the existence of a global solution to (17.13), it is then enough to show that the L^2 norm of $\vec{u}_{(\alpha)}$ remains bounded (as the existence time T_0 is controlled by the L^2 norm of the Cauchy data).

To alleviate the notations, we still write $\vec{u}$ for the solution, instead of $\vec{u}_{(\alpha)}$. We consider a time T^* such that, for all $T_0 < T^*$, $\vec{u}$ belongs to $L^{\infty}((0,T_0), L^2) \cap L^2((0,T_0), H^1)$ and we want to prove that $\vec{u}$ belongs to $L^{\infty}((0,T^*), L^2)$. This will be done by proving energy estimates on $\vec{u}_{\alpha}$, in H^2 norms: for any $T_0 < T^*$, we have $\vec{u}_{\alpha} \in L^2((0,T_0), H^3)$ and $\partial_t \vec{u}_{\alpha} = (Id - \alpha^2\Delta)^{-1}\partial_t\vec{u} \in L^2((0,T_0), H^1)$; thus, we have

$$\frac{d}{dt}\frac{\|\vec{u}_{\alpha}\|_2^2}{2} = \int \partial_t\vec{u}_{\alpha} \cdot \vec{u}_{\alpha}\,dx, \quad \frac{d}{dt}\frac{\|\vec{\nabla} \otimes \vec{u}_{\alpha}\|_2^2}{2} = -\int \partial_t\vec{u}_{\alpha} \cdot \Delta\vec{u}_{\alpha}\,dx$$

and

$$\frac{d}{dt}(\frac{\|\vec{u}\|_2^2}{2}) = \langle \partial_t\vec{u}|\vec{u}\rangle_{H^{-1},H^1}.$$

As $(Id - \alpha^2\Delta)\partial_t\vec{u}_{\alpha} = \partial_t\vec{u}$, we may write

$$\frac{d}{dt}(\frac{\|\vec{u}_{\alpha}\|_2^2}{2} + \alpha^2\frac{\|\vec{\nabla} \otimes \vec{u}_{\alpha}\|_2^2}{2}) = \langle \partial_t\vec{u}|\vec{u}_{\alpha}\rangle_{H^{-1},H^1}.$$

and thus, since $(\vec{\omega} \wedge \vec{u}_{\alpha}) \cdot \vec{u}_{\alpha} = 0$, we find

$$\frac{d}{dt}(\frac{\|\vec{u}_{\alpha}\|_2^2}{2} + \alpha^2\frac{\|\vec{\nabla} \otimes \vec{u}_{\alpha}\|_2^2}{2}) = \langle \nu\Delta\vec{u} - \vec{\omega} \wedge \vec{u}_{\alpha} + \vec{f}|\vec{u}_{\alpha}\rangle_{H^{-1},H^1}$$

$$= \langle \nu\Delta\vec{u}_{\alpha} - \nu\alpha^2\Delta^2\vec{u}_{\alpha} + \vec{f}|\vec{u}_{\alpha}\rangle_{H^{-1},H^1}$$

$$= -\nu\|\vec{\nabla} \otimes \vec{u}_{\alpha}\|_2^2 - \nu\alpha^2\|\Delta\vec{u}_{\alpha}\|_2^2 + \langle \vec{f}|\vec{u}_{\alpha}\rangle_{H^{-1},H^1}$$

$$\leq -\frac{\nu}{2}\|\vec{\nabla} \otimes \vec{u}_{\alpha}\|_2^2 - \nu\alpha^2\|\Delta\vec{u}_{\alpha}\|_2^2 + \frac{1}{2\nu}\|\vec{f}\|_{\dot{H}^{-1}}^2 + \frac{\nu}{2}\|\vec{u}_{\alpha}\|_2^2$$

so that

$$\frac{\|\vec{u}_\alpha\|_2^2}{2} + \alpha^2 \frac{\|\vec{\nabla} \otimes \vec{u}_\alpha\|_2^2}{2} + \nu \int_0^t \frac{1}{2} \|\vec{\nabla} \otimes \vec{u}_\alpha\|_2^2 + \alpha^2 \|\Delta \vec{u}_\alpha\|_2^2 \, ds$$

$$\leq e^{\nu t} \left(\frac{\|(Id - \alpha^2 \Delta)^{-1} \vec{u}_0\|_2^2}{2} + \alpha^2 \frac{\|\vec{\nabla} \otimes (Id - \alpha^2 \Delta)^{-1} \vec{u}_0\|_2^2}{2} + \frac{1}{2\nu} \int_0^t \|\vec{f}\|_{H^{-1}}^2 \, ds \right) \quad (17.15)$$

$$\leq e^{\nu t} \left(\|\vec{u}_0\|_2^2 + \frac{1}{2\nu} \int_0^t \|\vec{f}\|_{H^{-1}}^2 \, ds \right)$$

On the other hand, we have

$$\frac{d}{dt} \left(\frac{\|\vec{u}\|_2^2}{2} \right) = \langle \nu \Delta \vec{u} - \vec{u}_\alpha \cdot \vec{\nabla} \vec{u} - \sum_{k=1}^3 u_k \vec{\nabla} u_{\alpha,k}) + \vec{f} | \vec{u} \rangle_{H^{-1}, H^1}$$

$$= -\nu \|\vec{\nabla} \otimes \vec{u}\|_2^2 + \langle \vec{f} | \vec{u} \rangle_{H^{-1}, H^1} - \sum_{k=1}^3 \int \vec{u} \cdot (u_{\alpha,k} \vec{\nabla} u_{\alpha,k}) \, dx$$

$$\leq -\frac{\nu}{2} \|\vec{\nabla} \otimes \vec{u}\|_2^2 + \frac{1}{2\nu} \|\vec{f}\|_{H^{-1}}^2 + \frac{\nu}{2} \|\vec{u}\|_2^2 + C \|\vec{u}\|_2 \|\vec{u}_\alpha\|_3 \|\vec{\nabla} \otimes \vec{u}_\alpha\|_6$$

$$\leq -\frac{\nu}{2} \|\vec{\nabla} \otimes \vec{u}\|_2^2 + \frac{1}{2\nu} \|\vec{f}\|_{H^{-1}}^2 + \frac{\nu}{2} \|\vec{u}\|_2^2 + C' \frac{1}{\alpha^2} \|\vec{\nabla} \otimes \vec{u}_\alpha\|_2^{1/2} \|\vec{u}_\alpha\|_2^{1/2} \|\vec{u}\|_2^2$$

so that

$$\|\vec{u}\|_2^2 + \nu \int_0^t \|\vec{\nabla} \otimes \vec{u}\|_2 \, ds \leq$$

$$e^{\nu t + \frac{2C'}{\alpha^2} \int_0^t \|\vec{\nabla} \otimes \vec{u}_\alpha\|_2^{1/2} \|\vec{u}_\alpha\|_2^{1/2} \, ds} \left(\|\vec{u}_0\|_2^2 + \frac{1}{\nu} \int_0^t \|\vec{f}\|_{H^{-1}}^2 \, ds \right) \quad (17.16)$$

From the energy estimates (17.15) and (17.16), we get that the L^2 norm of $\vec{u}$ remains bounded in time[1] (with a constant that depends on α) and we get global existence of $\vec{u}_{(\alpha)}$.

- **Third step: Weak convergence.**

We write $\vec{u}_{(\alpha)}$ for the solution of the problem (17.13), and we write $\vec{U}_{(\alpha)} = (Id - \alpha^2 \Delta)^{-1} \vec{u}_{(\alpha)}$. From the energy estimate (17.15), we know that $\vec{U}_{(\alpha)}$ remains bounded in $L^\infty L^2 \cap L^2 \dot{H}^1$ independently from α. Moreover, $\alpha \Delta \vec{U}_{(\alpha)}$ remains bounded in $L_t^2 L_x^2$ independently from α, so that $\vec{u}_{(\alpha)}$ is bounded in $L^2 L^2$ independently from $\alpha \in (0, 1)$. Finally, we know that $\alpha \vec{\nabla} \otimes \vec{U}_{(\alpha)}$ is bounded in $L^\infty \dot{H}^1$ independently from α.

We have

$$\partial_t \vec{U}_{(\alpha)} = \nu \Delta \vec{U}_{(\alpha)} + (Id - \alpha^2 \Delta)^{-1} \mathbb{P} (\vec{f} - \sum_{k=1}^3 \partial_k (U_{(\alpha),k} \vec{u}_{(\alpha)}) - \sum_{k=1}^3 u_{(\alpha),k} \vec{\nabla} U_{(\alpha),k})$$

and we write

$$\mathbb{P} (\sum_{k=1}^3 u_{(\alpha),k} \vec{\nabla} U_{(\alpha),k}) = \mathbb{P} \vec{\nabla} \left(\frac{|\vec{U}_{(\alpha)}|^2}{2} \right) - \alpha^2 \mathbb{P} (\sum_{k=1}^3 \Delta U_{(\alpha),k} \vec{\nabla} U_{(\alpha),k})$$

$$= -\sum_{j=1}^3 \alpha^2 \mathbb{P} (\partial_j (\sum_{k=1}^3 \partial_j U_{(\alpha),k} \vec{\nabla} U_{(\alpha),k})) + \sum_{j=1}^3 \alpha^2 \mathbb{P} (\vec{\nabla} \frac{\partial_j |\vec{U}_{(\alpha)}|^2}{2})$$

$$= -\sum_{j-1}^3 \alpha^2 \mathbb{P} (\partial_j (\sum_{k-1}^3 \partial_j U_{(\alpha),k} \vec{\nabla} U_{(\alpha),k}))$$

---

[1] Recall that we work on a finite time interval $(0, T)$.

Thus, we can see that $\partial_t \vec{U}_{(\alpha)}$ remains bounded in $L^2 H^{-s}$ for $s > 5/2$ (independently from $\alpha \in (0,1)$):

$$\|\partial_t \vec{U}_{(\alpha)}\|_{L^2 H^{-s}} \leq C(\|\vec{f}\|_{L^2 H^{-1}} + \|\vec{U}_{(\alpha)}\|_{L^\infty L^2} \|\vec{u}_{(\alpha)}\|_{L^2 L^2} +$$
$$+ \alpha \|\vec{U}_{(\alpha)}\|_{L^2 \dot{H}^1} \alpha \|\vec{U}_{(\alpha)}\|_{L^\infty \dot{H}^1}).$$

We may then use the Rellich–Lions theorem (Theorem 12.1): we may find a sequence $\alpha_n \to 0$ and a function $\vec{u} \in L^\infty L^2 \cap L^2 \dot{H}^1$ such that:

- $\vec{U}_{(\alpha_n)}$ is *-weakly convergent to $\vec{u}$ in $L^\infty L^2$ and in $L^2 \dot{H}^1$

- $\vec{U}_{(\alpha_n)}$ is strongly convergent to $\vec{u}$ in $L^2_{\text{loc}}((0,T) \times \mathbb{R}^3)$.

Since $\vec{u}_{(\alpha)} = \vec{U}_{(\alpha)} - \alpha^2 \Delta \vec{U}_{(\alpha)}$, we see that $\vec{u}_{(\alpha_n)}$ converges in $\mathcal{D}'((0,T) \times \mathbb{R}^3)$ (or weakly in $L^2 L^2$) to $\vec{u}$. We find that, in $\mathcal{D}'((0,T) \times \mathbb{R}^3)$, we have

$$\partial_t \vec{u} = \nu \Delta \vec{u} + \mathbb{P}\vec{f} - \lim_{n \to +\infty} \mathbb{P}(\vec{U}_{(\alpha_n)} \cdot \vec{\nabla}\vec{u}_{(\alpha_n)} + \sum_{k=1}^{3} u_{(\alpha_n),k} \vec{\nabla} U_{(\alpha_n),k})$$

We have seen that $\mathbb{P}(\sum_{k=1}^{3} u_{(\alpha_n),k} \vec{\nabla} U_{(\alpha_n),k})$ is bounded in $L^2 H^{-s}$ $(s > 5/2)$ in $O(\alpha)$, so that its limit in $\mathcal{D}'$ is equal to 0. Moreover, we write

$$\mathbb{P}(\vec{U}_{(\alpha_n)} \cdot \vec{\nabla}\vec{u}_{(\alpha_n)}) = \sum_{j=1}^{3} \mathbb{P}(\partial_j (U_{(\alpha_n),j} \vec{u}_{(\alpha_n)})) :$$

$U_{(\alpha_n),j} \vec{u}_{(\alpha_n)}$ converges in $\mathcal{D}'$ to $u_j \vec{u}$ and remains bounded in $L^2 H^{-s}$ for $s > 3/2$, thus $\mathbb{P}(\vec{U}_{(\alpha_n)} \cdot \vec{\nabla}\vec{u}_{(\alpha_n)})$ converges weakly to $\mathbb{P}(\vec{u} \cdot \vec{\nabla}\vec{u})$ in $L^2 H^{-s}$ for $s > 5/2$.

Thus, the weak limit $\vec{u}$ satisfies

$$\partial_t \vec{u} = \nu \Delta \vec{u} + \mathbb{P}(\vec{f} - \vec{u} \cdot \vec{\nabla}\vec{u}).$$

- **Fourth step: Global energy estimates for the weak limit.**

We now check that the weak limit $\vec{u}$ fulfills the Leray energy inequality. Let $\omega(t)$ be a non-negative compactly supported smooth function on $\mathbb{R}$, supported within $[-\epsilon, \epsilon]$, with $\|\omega\|_1 = 1$. On $(\epsilon, T - \epsilon)$, we have

$$\|\omega * \vec{U}_{(\alpha)}(t,.)\|_2^2 \leq \omega * \|\vec{U}_{(\alpha)}(t,.)\|_2^2$$

and

$$\frac{d}{dt}\left(\frac{\|\vec{U}_\alpha\|_2^2}{2} + \alpha^2 \frac{\|\vec{\nabla} \otimes \vec{U}_\alpha\|_2^2}{2}\right) = -\nu\|\vec{\nabla} \otimes \vec{U}_\alpha\|_2^2 - \nu\alpha^2\|\Delta\vec{U}_\alpha\|_2^2 + \langle \vec{f}|\vec{U}_\alpha \rangle_{H^{-1},H^1}$$

so that

$$\omega * \|\vec{U}_{(\alpha)}(t,.)\|_2^2 + 2\nu\omega * \left(\int_0^t \|\vec{U}_{(\alpha)}\|_{\dot{H}^1}^2 \, ds\right) \leq \frac{\|\vec{U}_\alpha(0,.)\|_2^2}{2} + \alpha^2 \frac{\|\vec{\nabla} \otimes \vec{U}_\alpha(0,.)\|_2^2}{2}$$

$$+ 2\omega * \left(\int_0^t \langle \vec{f}|\vec{u}_{(\alpha_n)} \rangle_{H^{-1},H^1} \, ds\right)$$

Besides, we have the *-weak convergence of $\omega * \vec{U}_{(\alpha_n)}$ to $\omega * \vec{u}$ in $L^2 \dot{H}^1$, the strong convergence of $\omega * \vec{U}_{(\alpha)}(t,.)$ to $\omega * \vec{u}(t,.)$ in $L^2_{\mathrm{loc}}(\mathbb{R}^3)$, and thus (due to the control we have on the L^2 norm) the *-weak convergence of $\omega * \vec{u}_{(\alpha)}(t,.)$ to $\omega * \vec{u}(t,.)$ in $L^2(\mathbb{R}^3)$. We thus get that

$$\|\omega * \vec{u}(t,.)\|_2^2 + 2\nu\omega * \left(\int_0^t \|\vec{u}\|_{\dot{H}^1}^2 \, ds \right)$$

$$\leq \liminf_{n \to +\infty} \|\omega * \vec{U}_{(\alpha_n)}(t,.)\|_2^2 + 2\nu\omega * \left(\int_0^t \|\vec{U}_{(\alpha_n)}\|_{\dot{H}^1}^2 \, ds \right)$$

and thus

$$\|\omega * \vec{u}(t,.)\|_2^2 + 2\nu\omega * \left(\int_0^t \|\vec{u}\|_{\dot{H}^1}^2 \, ds \right) \leq \|\vec{u}_0\|_2^2 + 2\omega * \left(\int_0^t \langle \vec{f}|\vec{u}\rangle_{H^{-1},H^1} \, ds \right)$$

When t_0 is a Lebesgue point of $t \mapsto \|\vec{u}(t,.)\|_2$, we get that:

$$\|\vec{u}(t_0,.)\|_2^2 + 2\nu \int_0^{t_0} \|\vec{u}\|_{\dot{H}^1}^2 \, ds \leq \|\vec{u}_0\|_2^2 + 2\int_0^{t_0} \langle \vec{f}|\vec{u}\rangle_{H^{-1},H^1} \, ds$$

This inequality may be extended to every t_0, due to the weak boundedness of $t \in (0,T) \mapsto \vec{u}(t,.) \in L^2$.

- **Fifth step: Local energy estimates for the weak limit.**

We now check that $\vec{u}$ is more precisely a suitable weak solution (i.e., fulfills the local energy inequality). We write

$$\partial_t \vec{u}_{(\alpha)} = \nu\Delta\vec{u}_{(\alpha)} + \vec{f} - \vec{U}_{(\alpha)} \cdot \vec{\nabla}\,\vec{u}_{(\alpha)} - \sum_{k=1}^3 u_{(\alpha),k} \cdot \vec{\nabla}U_{(\alpha),k} - \vec{\nabla}p_{(\alpha)}$$

with

$$p_{(\alpha)} = \frac{1}{\Delta}\operatorname{div}\left(\vec{f} - \vec{U}_{(\alpha)} \cdot \vec{\nabla}\,\vec{u}_{(\alpha)} - \sum_{k=1}^3 u_{(\alpha),k} \cdot \vec{\nabla}U_{(\alpha),k} \right).$$

Similarly, we have

$$\partial_t \vec{u} = \nu\Delta\vec{u} + \vec{f} - \vec{u} \cdot \vec{\nabla}\,\vec{u} - \vec{\nabla}p$$

with

$$p = \frac{1}{\Delta}\operatorname{div}\left(\vec{f} - \vec{u} \cdot \vec{\nabla}\,\vec{u} \right).$$

As $\vec{U}_{(\alpha_n)}$ converge strongly to $\vec{u}$ in $L^2_{\mathrm{loc}}((0,T) \times \mathbb{R}^3)$, we have

$$\partial_t \left(\frac{|\vec{u}|^2}{2} \right) = \lim_{n \to +\infty} \partial_t \left(\frac{|\vec{U}_{(\alpha_n)}|^2}{2} \right) = \lim_{n \to +\infty} \vec{U}_{(\alpha_n)} \cdot \partial_t \vec{U}_{(\alpha_n)}$$

in $\mathcal{D}'((0,T) \times \mathbb{R}^3)$. Moreover, we know that $\vec{U}_{(\alpha)}$ is bounded in $L^2 H^1$ independently from α; we estimate $\partial_t \vec{U}_{(\alpha)}$ in $L^2 H^{-1}$. We have seen that

$$\partial_t \vec{U}_{(\alpha)} = \nu\Delta\vec{U}_{(\alpha)} + (Id - \alpha^2\Delta)^{-1}\mathbb{P}(\vec{f} - \sum_{k=1}^3 \partial_k (U_{(\alpha),k}\vec{u}_{(\alpha)})) - \sum_{k=1}^3 u_{(\alpha),k}\vec{\nabla}U_{(\alpha),k})$$

and that $\sum_{k=1}^{3} \partial_k (U_{(\alpha),k} \vec{u}_{(\alpha)}) - \sum_{k=1}^{3} u_{(\alpha),k} \vec{\nabla} U_{(\alpha),k})$ is bounded in $L^2 H^{-11/4}$ independently from α in $(0,1)$; this gives that $\alpha^{7/4} \partial_t \vec{U}_{(\alpha)}$ is bounded in $L^2 H^{-1}$ independently from $\alpha \in (0,1)$. Thus, we have

$$\lim_{\alpha \to 0} \alpha^2 \partial_t \vec{U}_{(\alpha)} \cdot \partial_k \vec{U}_{(\alpha)} = 0$$

in $\mathcal{D}'((0,T) \times \mathbb{R}^3)$.

$$\vec{U}_{(\alpha)} \cdot \partial_t \vec{U}_{(\alpha)} = \vec{U}_{(\alpha)} \cdot \partial_t \vec{u}_{(\alpha)}$$
$$+ \alpha^2 \left(\Delta \partial_t (\frac{|\vec{U}_{(\alpha)}|^2}{2}) - \partial_t (\frac{|\vec{\nabla} \otimes \vec{U}_{(\alpha)}|^2}{2}) - \sum_{k=1}^{3} \partial_k (\partial_k \vec{U}_{(\alpha)} \cdot \partial_t \vec{U}_{(\alpha)}) \right)$$

we find that

$$\lim_{\alpha \to 0} \vec{U}_{(\alpha)} \cdot \partial_t \vec{U}_{(\alpha)} - \vec{U}_{(\alpha)} \cdot \partial_t \vec{u}_{(\alpha)} = 0$$

in $\mathcal{D}'((0,T) \times \mathbb{R}^3)$. We may thus conclude that

$$\partial_t (\frac{|\vec{u}|^2}{2}) = \lim_{n \to +\infty} \vec{U}_{(\alpha_n)} \cdot \partial_t \vec{u}_{(\alpha_n)} \tag{17.17}$$

in $\mathcal{D}'((0,T) \times \mathbb{R}^3)$.
We have

$$\vec{U}_{(\alpha_n)} \cdot \partial_t \vec{u}_{(\alpha_n)} = \nu \vec{U}_{(\alpha_n)} \cdot \Delta \vec{u}_{(\alpha_n)} + \vec{f} \cdot \vec{U}_{(\alpha_n)} - \vec{U}_{(\alpha_n)} \cdot (\vec{U}_{(\alpha_n)} \cdot \vec{\nabla} \vec{u}_{(\alpha_n)}))$$
$$- \sum_{k=1}^{3} \vec{U}_{(\alpha_n)} \cdot (u_{(\alpha_n),k} \cdot \vec{\nabla} \vec{U}_{(\alpha_n),k}) - \vec{U}_{(\alpha_n)} \cdot \vec{\nabla} p_{(\alpha_n)}.$$

Writing

$$\text{curl } \vec{u}_{(\alpha_n)} \wedge \vec{U}_{(\alpha_n)} = \vec{U}_{(\alpha_n)} \cdot \vec{\nabla} \vec{u}_{(\alpha_n)}) + \sum_{k=1}^{3} u_{(\alpha_n),k} \cdot \vec{\nabla} \vec{U}_{(\alpha_n),k} - \vec{\nabla}(\vec{u}_{(\alpha_n)} \cdot \vec{U}_{(\alpha_n)})$$

we find

$$\vec{U}_{(\alpha_n)} \cdot \partial_t \vec{u}_{(\alpha_n)} = \nu \vec{U}_{(\alpha_n)} \cdot \Delta \vec{u}_{(\alpha_n)} + \vec{f} \cdot \vec{U}_{(\alpha_n)} - \vec{U}_{(\alpha_n)} \cdot \vec{\nabla}(p_{(\alpha_n)} + \vec{u}_{(\alpha_n)} \cdot \vec{U}_{(\alpha_n)})$$
$$= \nu \vec{U}_{(\alpha_n)} \cdot \Delta \vec{u}_{(\alpha_n)} + \vec{f} \cdot \vec{U}_{(\alpha_n)} - \sum_{j=1}^{3} \partial_j \left(U_{(\alpha_n),j}(p_{(\alpha_n)} + \vec{u}_{(\alpha_n)} \cdot \vec{U}_{(\alpha_n)}) \right).$$

We consider each of the terms in the last sum. We have

- $\vec{U}_{(\alpha_n)} \cdot \Delta \vec{u}_{(\alpha_n)} = \vec{U}_{(\alpha_n)} \cdot \Delta \vec{U}_{(\alpha_n)} - \alpha_n^2 \vec{U}_{(\alpha_n)} \cdot \Delta^2 \vec{U}_{(\alpha_n)}$. We have $\vec{U}_{(\alpha_n)} \cdot \Delta \vec{U}_{(\alpha_n)} = \Delta(\frac{|\vec{U}_{(\alpha_n)}|^2}{2}) - |\vec{\nabla} \otimes \vec{U}_{(\alpha_n)}|^2$, with $\lim_{n \to +\infty} \Delta(\frac{|\vec{U}_{(\alpha_n)}|^2}{2}) = \Delta(\frac{|\vec{u}|^2}{2})$ in $\mathcal{D}'((0,T) \times \mathbb{R}^3)$. Moreover, we have

$$\vec{U}_{(\alpha_n)} \cdot \Delta^2 \vec{U}_{(\alpha_n)} = \Delta(\vec{U}_{(\alpha_n)} \cdot \Delta \vec{U}_{(\alpha_n)}) - 2 \sum_{j=1}^{3} \partial_j (\partial_j \vec{U}_{(\alpha_n)} \cdot \Delta \vec{U}_{(\alpha_n)}) + |\Delta \vec{U}_{(\alpha_n)}|^2$$

$\vec{U}_{(\alpha_n)}$ is bounded in $L^2 H^1$, so that $\Delta \vec{U}_{(\alpha_n)}$ is bounded and $L^2 H^{-1}$ and thus $\lim_{n \to +\infty} \alpha_n^2 \Delta \vec{U}_{(\alpha_n)} \vec{U}_{(\alpha_n)} = 0$ in $\mathcal{D}'((0,T) \times \mathbb{R}^3)$. Similarly, $\vec{U}_{(\alpha_n)}$ is bounded in $L^2 H^1$

and $\alpha_n \Delta \vec{U}_{(\alpha_n)}$ is bounded in $L^2 L^2$, so that $\lim_{n \to +\infty} \alpha_n^2 \partial_j \vec{U}_{(\alpha_n)} \cdot \Delta \vec{U}_{(\alpha_n)} = 0$ in $\mathcal{D}'((0,T) \times \mathbb{R}^3)$. Thus, we have

$$\lim_{n \to +\infty} \vec{U}_{(\alpha_n)} \cdot \Delta \vec{u}_{(\alpha_n)} + |\vec{\nabla} \otimes \vec{U}_{(\alpha_n)}|^2 + \alpha_n^2 |\Delta \vec{U}_{(\alpha_n)}|^2 = \Delta(\frac{|\vec{u}|^2}{2})$$

in $\mathcal{D}'((0,T) \times \mathbb{R}^3)$.

- $\vec{U}_{(\alpha_n)}$ converges weakly to $\vec{u}$ in $L^2 H^1$, so that $\lim_{n \to +\infty} \vec{f} \cdot \vec{U}_{(\alpha_n)} = \vec{f} \cdot \vec{u}$ in $\mathcal{D}'((0,T) \times \mathbb{R}^3)$

- $\vec{u}_{(\alpha_n)} \cdot \vec{U}_{(\alpha_n)} = \vec{U}_{(\alpha_n)} \cdot \vec{U}_{(\alpha_n)} - \alpha_n^2 \Delta \vec{U}_{(\alpha_n)} \cdot \vec{U}_{(\alpha_n)}$. We know that we have

$$\alpha_n^2 \|U_{(\alpha_n),j} \Delta \vec{U}_{(\alpha_n)} \vec{U}_{(\alpha_n)}\|_{L^1 L^1}$$
$$\leq C\alpha_n^2 \|\vec{U}_{(\alpha_n)}\|_{L^\infty L^2}^{1/2} \|\vec{U}_{(\alpha_n)}\|_{L^\infty H^1}^{1/2} \|\vec{U}_{(\alpha_n)}\|_{L^2 H^2} \|\vec{U}_{(\alpha_n)}\|_{L^2 H^1}$$
$$= O(\alpha_n^{1/2})$$

and thus $\lim_{n \to +\infty} \alpha_n^2 U_{(\alpha_n),j} \Delta \vec{U}_{(\alpha_n)} \vec{U}_{(\alpha_n)} = 0$ in $\mathcal{D}'((0,T) \times \mathbb{R}^3)$. On the other hand $\vec{U}_{(\alpha_n)}$ is bounded in $L^2 L^6$ and $L^\infty L^2$ and converges strongly in $(L^2 L^2)_{\text{loc}}$, hence in $(L^6 L^2)_{\text{loc}}$ and in $(L^3 L^3)_{\text{loc}}$, so that

$$\lim_{n \to +\infty} \sum_{j=1}^{3} \partial_j (U_{(\alpha_n),j} \vec{u}_{(\alpha_n)} \cdot \vec{U}_{(\alpha_n)})) = \text{div}(|\vec{u}|^2 \vec{u})$$

in $\mathcal{D}'((0,T) \times \mathbb{R}^3)$

- we decompose $p_{(\alpha)}$ into $\varpi_{(\alpha)} - q_{(\alpha)} + r_{(\alpha)}$ with $q_{(\alpha)} = \frac{1}{2}|\vec{U}_{(\alpha)}|^2$, $\varpi_{(\alpha)} = \frac{1}{\Delta} \text{div} \left(\vec{f} - \vec{U}_{(\alpha)} \cdot \vec{\nabla} \vec{u}_{(\alpha)} \right)$ and $r_{(\alpha)} = \alpha^2 \frac{1}{\Delta} \text{div} \left(\sum_{k=1}^{3} \Delta U_{(\alpha),k} \cdot \vec{\nabla} U_{(\alpha),k} \right)$. We have seen that

$$\lim_{n \to +\infty} \vec{U}_{(\alpha_n)} \cdot \vec{\nabla} q_{(\alpha_n)} = \frac{1}{2} \text{div}(|\vec{u}|^2 \vec{u}) \text{ in } \mathcal{D}'((0,T) \times \mathbb{R}^3).$$

Moreover, we have

$$\|\vec{U}_{(\alpha_n)} r_{(\alpha_n)}\|_{L^1 L^1} \leq C\alpha_n^2 \|\vec{U}_{(\alpha_n)}\|_{L^\infty L^{3,1}} \|\vec{U}_{(\alpha_n)}\|_{L^2 H^2} \|\vec{U}_{(\alpha_n)}\|_{L^2 H^1}$$
$$\leq C'\alpha_n^2 \|\vec{U}_{(\alpha_n)}\|_{L^\infty L^2}^{1/2} \|\vec{U}_{(\alpha_n)}\|_{L^\infty H^1}^{1/2} \|\vec{U}_{(\alpha_n)}\|_{L^2 H^2} \|\vec{U}_{(\alpha_n)}\|_{L^2 H^1}$$
$$= O(\alpha_n^{1/2})$$

and $\lim_{n \to +\infty} \text{div}(r_{(\alpha_n)} \vec{U}_{(\alpha_n)}) = 0$ in $\mathcal{D}'((0,T) \times \mathbb{R}^3)$. We now write

$$\varpi_{(\alpha)} = \frac{1}{\Delta} \text{div} \vec{f} - \sum_{j=1}^{3} \frac{1}{\Delta} \text{div} \partial_j \left(U_{(\alpha),j} \vec{U}_{(\alpha)} \right) + \alpha^2 \frac{1}{\Delta} \text{div} \left(\vec{U}_{(\alpha)} \cdot \vec{\nabla} \Delta \vec{U}_{(\alpha)} \right).$$

We have, of course, $\lim_{n \to +\infty} \vec{U}_{(\alpha_n)} \cdot \frac{1}{\Delta} \text{div} \vec{f} = \vec{u} \cdot \frac{1}{\Delta} \text{div} \vec{f}$ in $\mathcal{D}'((0,T) \times \mathbb{R}^3)$ (as $\vec{U}_{(\alpha_n)}$ converges weakly to $\vec{u}$ in $L^2 H^1$). We know that $U_{(\alpha_n),j} \vec{U}_{(\alpha_n)}$ is bounded in $L^{3/2} L^{3/2}$ and converges strongly in $(L^{3/2} L^{3/2})_{\text{loc}}$ to $u_j \vec{u}$, hence we have weak convergence of $\sum_{j=1}^{3} \frac{1}{\Delta} \text{div} \partial_j \left(U_{(\alpha_n),j} \vec{U}_{(\alpha_n)} \right)$ to $\sum_{j=1}^{3} \frac{1}{\Delta} \text{div} \partial_j (u_j \vec{u})$

in $(L^{3/2}L^{3/2})_{\text{loc}}$. As $\vec{U}_{(\alpha_n)}$ converges strongly in $(L^3L^3)_{\text{loc}}$, we find that $\lim_{n\to+\infty} \vec{U}_{(\alpha_n)} \frac{1}{\Delta} \text{div}(\sum_{j=1}^{3} \partial_j (U_{(\alpha_n),j}\vec{U}_{(\alpha_n)})) = \vec{u}(\sum_{j=1}^{3} \frac{1}{\Delta} \text{div} \, \partial_j (u_j\vec{u}))$ in $\mathcal{D}'((0,T)\times \mathbb{R}^3)$. Finally, we write

$$\alpha_n^2 \frac{1}{\Delta} \text{div}\left(\vec{U}_{(\alpha_n)} \cdot \vec{\nabla}\Delta\vec{U}_{(\alpha_n)}\right)) = \alpha_n^2 \sum_{j=1}^{3} \frac{1}{\Delta} \text{div} \, \partial_j \left(\vec{U}_{(\alpha_n)} \cdot \vec{\nabla}\partial_j\vec{U}_{(\alpha_n)}\right))$$

$$- \alpha_n^2 \frac{1}{\Delta} \text{div}\left(\partial_j\vec{U}_{(\alpha_n)} \cdot \vec{\nabla}\partial_j\vec{U}_{(\alpha_n)}\right))$$

with $\alpha_n^2 \|\vec{U}_{(\alpha_n)} \frac{1}{\Delta} \text{div}\left(\partial_j\vec{U}_{(\alpha_n)} \cdot \vec{\nabla}\partial_j\vec{U}_{(\alpha_n)}\right))\|_{L^1L^1} = O(\alpha_n^{1/2})$ and

$$\alpha_n^2 \|\vec{U}_{(\alpha_n)} \frac{1}{\Delta} \text{div} \, \partial_j \left(\vec{U}_{(\alpha_n)} \cdot \vec{\nabla}\partial_j\vec{U}_{(\alpha_n)}\right))\|_{L^1L^1} = O(\alpha_n^{1/2}).$$

Thus, we find that

$$\lim_{n\to+\infty} \text{div}(\varpi_{(\alpha_n)}\vec{U}_{(\alpha_n)}) = \text{div}(p\vec{u})$$

in $\mathcal{D}'((0,T)\times \mathbb{R}^3)$.

Thus far, we have got that

$$\partial_t|\vec{u}|^2 = \nu\Delta|\vec{u}|^2 - \text{div}((|\vec{u}|^2 + 2p)\vec{u}) + 2\vec{u}\cdot\vec{f} - \nu T$$

with

$$T = \lim_{n\to+\infty} 2|\vec{\nabla}\otimes\vec{U}_{(\alpha_n)}|^2 + 2\alpha_n^2|\Delta\vec{U}_{(\alpha_n)}|^2.$$

Let $\phi \in \mathcal{D}'((0,T)\times\mathbb{R}^3)$ be a non-negative function. As $\sqrt{\phi}\,\vec{\nabla}\otimes\vec{U}_{(\alpha_n)}$ is weakly convergent to $\sqrt{\phi}\,\vec{\nabla}\otimes\vec{u}$ in $L_t^2L_x^2$, we find that

$$\langle T|\phi\rangle_{\mathcal{D}',\mathcal{D}} \geq 2\liminf_{n\to+\infty} \iint |\vec{\nabla}\otimes\vec{U}_{(\alpha_n)}|^2 \,\phi(t,x)\,dt\,d$$

$$\geq 2\iint |\vec{\nabla}\otimes\vec{u}|^2\phi(t,x)\,dt\,dx.$$

Thus, $T = 2|\vec{\nabla}\otimes\vec{u}|^2 + \mu$, where μ is a non-negative locally finite measure, and thus $\vec{u}$ is suitable.

- **Sixth step: Strong convergence.**

We write

$$\partial_t\vec{u} = \nu\Delta\vec{u} - (\vec{\nabla}\wedge\vec{u})\wedge\vec{u} - \vec{\nabla}Q + \vec{f}$$

and

$$\partial_t\vec{u}_{(\alpha)} = \nu\Delta\vec{u}_{(\alpha)} - \left((Id - \alpha^2\Delta)^{-1}\vec{\nabla}\wedge\vec{u}_{(\alpha)}\right)\wedge\vec{u}_{(\alpha)} - \vec{\nabla}Q_{(\alpha)} + \vec{f}.$$

As $\vec{u}$ is regular enough, we may write

$$\begin{cases} \int \partial_t \left(\frac{|\vec{u}|^2}{2} \right) dx = -\nu \int |\vec{\nabla} \otimes \vec{u}|^2 \, dx + \int \vec{u} \cdot \vec{f} \, dx \\[2mm] \int \partial_t \left(\frac{|\vec{u}_{(\alpha)}|^2}{2} \right) dx = -\nu \int |\vec{\nabla} \otimes \vec{u}_{(\alpha)}|^2 \, dx + \int \vec{u}_{(\alpha)} \cdot \vec{f} \, dx \\[2mm] \int (\partial_t \vec{u}) \cdot \vec{u}_{(\alpha)} \, dx = -\nu \int (\vec{\nabla} \otimes \vec{u}) \cdot (\vec{\nabla} \otimes \vec{u}_{(\alpha)}) \, dx + \int \vec{u}_{(\alpha)} \cdot \vec{f} \, dx \\[2mm] \qquad\qquad\qquad\qquad - \int \vec{u}_{(\alpha)} \cdot ((\vec{\nabla} \wedge \vec{u}) \wedge \vec{u}) \, dx \\[2mm] \int \vec{u} \cdot (\partial_t \vec{u}_{(\alpha)}) \, dx = -\nu \int (\vec{\nabla} \otimes \vec{u}) \cdot (\vec{\nabla} \otimes \vec{u}_{(\alpha)}) \, dx + \int \vec{u} \cdot \vec{f} \, dx \\[2mm] \qquad\qquad\qquad\qquad - \int \vec{u} \cdot \left(\left((Id - \alpha^2 \Delta)^{-1} \vec{\nabla} \wedge \vec{u}_{(\alpha)} \right) \wedge \vec{u}_{(\alpha)} \right) dx \end{cases}$$

so that

$$\frac{d}{dt} \left(\frac{\|\vec{u} - \vec{u}_{(\alpha)}\|_2^2}{2} \right) = -\nu \|\vec{\nabla} \otimes (\vec{u} - \vec{u}_{(\alpha)})\|_2^2 + \int \vec{u}_{(\alpha)} \cdot ((\vec{\nabla} \wedge \vec{u}) \wedge \vec{u}) \, dx$$
$$+ \int \vec{u} \cdot \left(\left((Id - \alpha^2 \Delta)^{-1} \vec{\nabla} \wedge \vec{u}_{(\alpha)} \right) \wedge \vec{u}_{(\alpha)} \right) dx.$$

We have

$$\int \vec{u} \cdot \left(\left((Id - \alpha^2 \Delta)^{-1} \vec{\nabla} \wedge \vec{u}_{(\alpha)} \right) \wedge \vec{u}_{(\alpha)} \right) dx$$
$$= \int \vec{u} \cdot \left(\left((Id - \alpha^2 \Delta)^{-1} \vec{\nabla} \wedge \vec{u}_{(\alpha)} \right) \wedge (\vec{u}_{(\alpha)} - \vec{u}) \right) dx$$
$$= \int \vec{u} \cdot \left(\left((Id - \alpha^2 \Delta)^{-1} \vec{\nabla} \wedge (\vec{u}_{(\alpha)} - \vec{u}) \right) \wedge (\vec{u}_{(\alpha)} - \vec{u}) \right) dx$$
$$+ \int \vec{u} \cdot \left(\left((Id - \alpha^2 \Delta)^{-1} \vec{\nabla} \wedge \vec{u} \right) \wedge (\vec{u}_{(\alpha)} - \vec{u}) \right) dx$$
$$\leq \|\vec{u}\|_6 \|\vec{\nabla} \wedge (\vec{u}_{(\alpha)} - \vec{u})\|_2 \|\vec{u}_{(\alpha)} - \vec{u}\|_3$$
$$+ \int \det(\vec{u}, (Id - \alpha^2 \Delta)^{-1} \vec{\nabla} \wedge \vec{u}, \vec{u}_{(\alpha)} - \vec{u}) \, dx.$$

We have as well

$$\int \vec{u}_{(\alpha)} \cdot ((\vec{\nabla} \wedge \vec{u}) \wedge \vec{u}) \, dx = \int (\vec{u}_{(\alpha)} - \vec{u}) \cdot ((\vec{\nabla} \wedge \vec{u}) \wedge \vec{u}) \, dx$$
$$= \int \det(\vec{u}_{(\alpha)} - \vec{u}, \vec{\nabla} \wedge \vec{u}, \vec{u}) \, dx.$$

Moreover, we have

$$\int \det(\vec{u}, (Id - \alpha^2 \Delta)^{-1} \vec{\nabla} \wedge \vec{u}, \vec{u}_{(\alpha)} - \vec{u}) + \det(\vec{u}_{(\alpha)} - \vec{u}, \vec{\nabla} \wedge \vec{u}, \vec{u}) \, dx$$
$$= \int \det(\vec{u}, (Id - \alpha^2 \Delta)^{-1} \vec{\nabla} \wedge \vec{u} - \vec{\nabla} \wedge \vec{u}, \vec{u}_{(\alpha)} - \vec{u}) \, dx$$
$$= \int \det(\vec{u}, \alpha^2 (Id - \alpha^2 \Delta)^{-1} \vec{\nabla} \wedge \Delta \vec{u}, \vec{u}_{(\alpha)} - \vec{u}) \, dx$$
$$\leq \frac{\alpha}{2} \|\vec{u}\|_6 \|\Delta \vec{u}\|_2 \|\vec{u}_{(\alpha)} - \vec{u}\|_3$$

and

$$\|\vec{u}\|_6 \|\vec{\nabla} \wedge (\vec{u}_{(\alpha)} - \vec{u})\|_2 \|\vec{u}_{(\alpha)} - \vec{u}\|_3 + \frac{\alpha}{2} \|\vec{u}\|_6 \|\Delta \vec{u}\|_2 \|\vec{u}_{(\alpha)} - \vec{u}\|_3$$

$$\leq C \|\vec{u}\|_{H^1} \|\vec{\nabla} \otimes (\vec{u}_{(\alpha)} - \vec{u})\|_2^{3/2} \|\vec{u}_{(\alpha)} - \vec{u}\|_2^{1/2}$$
$$+ C\alpha \|\vec{u}\|_{H^1} \|\vec{u}\|_{H^2} \|\vec{\nabla} \otimes (\vec{u}_{(\alpha)} - \vec{u})\|_2^{1/2} \|\vec{u}_{(\alpha)} - \vec{u}\|_2^{1/2}$$

$$\leq \nu \|\vec{\nabla} \otimes (\vec{u}_{(\alpha)} - \vec{u})\|_2^2 + C' \frac{\|\vec{u}\|_{H^1}^4}{\nu^3} \|\vec{u}_{(\alpha)} - \vec{u}\|_2^2 + C'\alpha^2 \nu^2 \|\vec{u}\|_{H^2}^2.$$

Thus, we find

$$\|\vec{u}_{(\alpha)} - \vec{u}\|_{L^\infty((0,T),L^2)} \leq C\alpha e^{CT \frac{\|\vec{u}\|_{L^\infty((0,T),H^1)}^4}{\nu^3}} \|\vec{u}\|_{L^2((0,T),H^2)}. \qquad \square$$

17.4 The Clark-α Model

Another α-model has been studied by Cao, Holm and Titi [87] and was proven to be an interesting model for large eddy simulation.

We have seen the Leray-α model

$$\partial_t \vec{u} + \vec{u}_\alpha \cdot \vec{\nabla}\, \vec{u} = \nu \Delta \vec{u} + \vec{f} - \vec{\nabla} p$$

and the Navier–Stokes-α model

$$\partial_t \vec{u} + \vec{u}_\alpha \cdot \vec{\nabla}\, \vec{u} = \nu \Delta \vec{u} - \sum_{k=1}^3 u_k \vec{\nabla} u_{\alpha,k} + \vec{f} - \vec{\nabla} p$$

The Clark-α model is given by

$$\partial_t \vec{u} + \vec{u}_\alpha \cdot \vec{\nabla}\, \vec{u} = \nu \Delta \vec{u} + (\vec{u}_\alpha - \vec{u}) \cdot \vec{\nabla} \vec{u}_\alpha + \alpha^2 \sum_{k=1}^3 (\partial_k \vec{u}_\alpha).\vec{\nabla}(\partial_k \vec{u}_\alpha) + \vec{f} - \vec{\nabla} p$$

where we have again $\vec{u}_\alpha = (Id - \alpha^2 \Delta)^{-1} \vec{u}$ and div $\vec{u} = 0$. As $\vec{u}_\alpha - \vec{u} = \alpha^2 \Delta \vec{u}_\alpha$, we may write as well

$$\partial_t \vec{u} + \vec{u}_\alpha \cdot \vec{\nabla}\, \vec{u} = \nu \Delta \vec{u} + \alpha^2 \sum_{k=1}^3 \partial_k \left((\partial_k \vec{u}_\alpha).\vec{\nabla} \vec{u}_\alpha \right) + \vec{f} - \vec{\nabla} p$$

Let us remark as well that

$$\vec{u}_\alpha \cdot \vec{\nabla}\, \vec{u} - (\vec{u}_\alpha - \vec{u}) \cdot \vec{\nabla} \vec{u}_\alpha = \vec{u}_\alpha \cdot \vec{\nabla} \vec{u}_\alpha - \alpha^2 \vec{u}_\alpha \cdot \vec{\nabla}(\Delta \vec{u}_\alpha) - \alpha^2 (\Delta \vec{u}_\alpha) \cdot \vec{\nabla} \vec{u}_\alpha$$

$$= (Id - \alpha^2 \Delta)(\vec{u}_\alpha \cdot \vec{\nabla} \vec{u}_\alpha) + 2\alpha^2 \sum_{k=1}^3 (\partial_k \vec{u}_\alpha).\vec{\nabla}(\partial_k \vec{u}_\alpha)$$

and thus

$$\partial_t \vec{u}_\alpha + \vec{u}_\alpha \cdot \vec{\nabla} \vec{u}_\alpha = \nu \Delta \vec{u}_\alpha + (Id - \alpha^2 \Delta)^{-1}(\vec{f} - \alpha^2 \sum_{k=1}^3 (\partial_k \vec{u}_\alpha).\vec{\nabla}(\partial_k \vec{u}_\alpha) - \vec{\nabla} p).$$

The mathematical analysis of the Clark-α model is closed to the Navier–Stokes-α model, with the same emphasis on the energy $\frac{\|\vec{u}_\alpha\|_2^2}{2} + \alpha^2 \frac{\|\vec{\nabla} \otimes \vec{u}_\alpha\|_2^2}{2}$.

Clark-α model

Theorem 17.3.
Let $\vec{u}_0 \in L^2$, with div $\vec{u}_0 = 0$, and $\vec{f} \in L^2_t H^{-1}_x$ on $(0,T) \times \mathbb{R}^3$ ($T < +\infty$).

- *for $\alpha > 0$, the problem*

$$\partial_t \vec{u} + \mathbb{P}(\vec{u}_\alpha \cdot \vec{\nabla}\,\vec{u} - \alpha^2 \sum_{k=1}^{3} \partial_k\left((\partial_k \vec{u}_\alpha).\vec{\nabla}\vec{u}_\alpha\right)) = \nu\Delta\vec{u} + \mathbb{P}\vec{f} \qquad (17.18)$$

with initial value $\vec{u}(0,.) = \vec{u}_0$ and

$$\vec{u}_\alpha = (Id - \alpha^2\Delta)^{-1}\vec{u}$$

has a unique solution $\vec{u}_{(\alpha)}$ such that $\vec{u}_{(\alpha)} \in L^\infty_t L^2_x \cap L^2_t H^1_x$.

- *$\vec{U}_{(\alpha)} = (Id - \alpha^2\Delta)^{-1}\vec{u}_{(\alpha)}$ satisfies, with uniform bounds with respect to α, that $\vec{U}_{(\alpha)} \in L^\infty_t L^2_x \cap L^2_t H^1_x$ and $\alpha\vec{U}_{(\alpha)} \in L^\infty_t H^1_x \cap L^2_t H^2_x$*

- *there exists a sequence $\alpha_k \to 0$ and a function $\vec{u} \in L^\infty_t L^2_x \cap L^2_t H^1_x$ such that $\vec{u}_{(\alpha_k)}$ converges to $\vec{u}$ in $\mathcal{D}'((0,T) \times \mathbb{R}^3)$. Moreover $\vec{u}$ is a suitable Leray weak solution of the Navier–Stokes problem*

$$\partial_t \vec{u} + \mathbb{P}(\vec{u} \cdot \vec{\nabla}\vec{u}) = \nu\Delta\vec{u} + \mathbb{P}\vec{f}$$

with initial value $\vec{u}_0$.

- *If $\vec{u}_0 \in H^1$ and $\vec{f} \in L^2_t L^2_x$ and if the Navier–Stokes problem has a solution $\vec{u} \in \mathcal{C}([0,T], H^1) \cap L^2([0,T], H^2)$, then we have*

$$\lim_{\alpha \to 0} \|\vec{U}_{(\alpha)} - \vec{u}\|_{L^\infty_t L^2_x} = 0.$$

Proof. • **First step: Local existence of $\vec{u}_{(\alpha)}$.**
We just have to estimate, for $\vec{u}$ and $\vec{v}$ in $L^\infty L^2 \cap L^2 H^1$ with div $\vec{u} = 0$ and for $0 < T_0 < T$, the norms in $L^2((0,T_0), \dot{H}^{-1})$ of $\vec{u}_\alpha \cdot \vec{\nabla}\vec{v}$ and of $\alpha^2 \sum_{k=1}^3 \partial_k(\partial_k\vec{u}_\alpha \cdot \vec{\nabla}\vec{v}_\alpha)$, we have

$$\|\vec{u}_\alpha \cdot \vec{\nabla}\vec{v}\|_{L^2((0,T_0),\dot{H}^{-1})} = \|\,\mathrm{div}(\vec{v} \otimes \vec{u}_\alpha)\|_{L^2((0,T_0),\dot{H}^{-1})}$$
$$\leq C\sqrt{T_0}\,\alpha^{-3/2}\|\vec{u}\|_{L^\infty L^2}\|\vec{v}\|_{L^\infty L^2}$$

and

$$\alpha^2\|\sum_{k=1}^3 \partial_k(\partial_k\vec{u}_\alpha \cdot \vec{\nabla}\vec{v}_\alpha)\|_{L^2((0,T_0),\dot{H}^{-1})} \leq C\alpha^2\sqrt{T_0}\|\vec{\nabla} \otimes \vec{u}_\alpha\|_{L^\infty L^6}\|\vec{\nabla} \otimes \vec{v}_\alpha\|_{L^\infty L^3}$$
$$\leq C'\sqrt{T_0}\,\alpha^{-3/2}\|\vec{u}\|_{L^\infty L^2}\|\vec{v}\|_{L^\infty L^2}.$$

Thus, we find existence (and uniqueness) of a solution $\vec{u} = \vec{u}_{(\alpha)}$ of Equation (17.18) for T_0 small enough to ensure

$$1 + T_0\nu \leq 2$$

and

$$T_0 \leq C_0\alpha^3 \frac{\nu}{(\|\vec{u}_0\|_2 + \frac{2}{\sqrt{\nu}}\|\vec{f}\|_{L^2\dot{H}^{-1}})^2}. \qquad (17.19)$$

• **Second step: Energy estimates and global existence of $\vec{u}_{(\alpha)}$.**

To show the existence of a global solution to (17.18), it is then enough to show that the L^2 norm of $\vec{u}_{(\alpha)}$ remains bounded (as the existence time T_0 is controlled by the L^2 norm of the Cauchy data).

To alleviate the notations, we still write $\vec{u}$ for the solution, instead of $\vec{u}_{(\alpha)}$. We consider a time T^* such that, for all $T_0 < T^*$, $\vec{u}$ belongs to $L^\infty((0, T_0), L^2) \cap L^2((0, T_0), H^1)$ and we want to prove that $\vec{u}$ belongs to $L^\infty((0, T^*), L^2)$. As in the case of the Navier–Stokes-α model, we may write

$$\frac{d}{dt}\left(\frac{\|\vec{u}_\alpha\|_2^2}{2} + \alpha^2 \frac{\|\vec{\nabla} \otimes \vec{u}_\alpha\|_2^2}{2}\right) = \langle \partial_t \vec{u} | \vec{u}_\alpha \rangle_{H^{-1}, H^1}$$

and

$$\frac{d}{dt}\left(\frac{\|\vec{u}\|_2^2}{2}\right) = \langle \partial_t \vec{u} | \vec{u} \rangle_{H^{-1}, H^1}.$$

First, we write

$$\langle \vec{u}_\alpha \cdot \vec{\nabla} \vec{u} - \alpha^2 \sum_{k=1}^3 \partial_k \left((\partial_k \vec{u}_\alpha) . \vec{\nabla} \vec{u}_\alpha\right) | \vec{u}_\alpha \rangle_{H^{-1}, H^1}$$

$$= \int \vec{u}_\alpha . (\vec{u}_\alpha \cdot \vec{\nabla} \vec{u}) + \alpha^2 \sum_{k=1}^3 \partial_k \vec{u}_\alpha . (\partial_k \vec{u}_\alpha \cdot \vec{\nabla} \vec{u}_\alpha) \, dx$$

$$= \int \vec{u}_\alpha . (\vec{u}_\alpha \cdot \vec{\nabla} \vec{u}_\alpha) - \alpha^2 \sum_{k=1}^3 \partial_k \vec{u}_\alpha . (\vec{u}_\alpha \cdot \partial_k \vec{\nabla} \vec{u}_\alpha) \, dx = 0$$

and thus we get

$$\frac{d}{dt}\left(\frac{\|\vec{u}_\alpha\|_2^2}{2} + \alpha^2 \frac{\|\vec{\nabla} \otimes \vec{u}_\alpha\|_2^2}{2}\right)$$

$$= \langle \nu \Delta \vec{u} - \vec{u}_\alpha \cdot \vec{\nabla} \vec{u} + \alpha^2 \sum_{k=1}^3 \partial_k \left((\partial_k \vec{u}_\alpha) . \vec{\nabla} \vec{u}_\alpha\right) + \vec{f} | \vec{u}_\alpha \rangle_{H^{-1}, H^1}$$

$$= \langle \nu \Delta \vec{u}_\alpha - \nu \alpha^2 \Delta^2 \vec{u}_\alpha + \vec{f} | \vec{u}_\alpha \rangle_{H^{-1}, H^1}$$

$$= -\nu \|\vec{\nabla} \otimes \vec{u}_\alpha\|_2^2 - \nu \alpha^2 \|\Delta \vec{u}_\alpha\|_2^2 + \langle \vec{f} | \vec{u}_\alpha \rangle_{H^{-1}, H^1}$$

$$\leq -\frac{\nu}{2}\|\vec{\nabla} \otimes \vec{u}_\alpha\|_2^2 - \nu \alpha^2 \|\Delta \vec{u}_\alpha\|_2^2 + \frac{1}{2\nu}\|\vec{f}\|_{H^{-1}}^2 + \frac{\nu}{2}\|\vec{u}_\alpha\|_2^2$$

so that

$$\frac{\|\vec{u}_\alpha\|_2^2}{2} + \alpha^2 \frac{\|\vec{\nabla} \otimes \vec{u}_\alpha\|_2^2}{2} + \nu \int_0^t \frac{1}{2}\|\vec{\nabla} \otimes \vec{u}_\alpha\|_2^2 + \alpha^2 \|\Delta \vec{u}_\alpha\|_2^2 \, ds$$

$$\leq e^{\nu t}(\|\vec{u}_0\|_2^2 + \frac{1}{2\nu} \int_0^t \|\vec{f}\|_{H^{-1}}^2 \, ds) \tag{17.20}$$

On the other hand, we have

$$\frac{d}{dt}\left(\frac{\|\vec{u}\|_2^2}{2}\right) = \langle \nu\Delta\vec{u} - \vec{u}_\alpha \cdot \vec{\nabla}\vec{u} + \alpha^2 \sum_{k=1}^3 \partial_k\left((\partial_k\vec{u}_\alpha).\vec{\nabla}\vec{u}_\alpha\right) + \vec{f}|\vec{u}\rangle_{H^{-1},H^1}$$

$$= -\nu\|\vec{\nabla}\otimes\vec{u}\|_2^2 + \langle\vec{f}|\vec{u}\rangle_{H^{-1},H^1} - \alpha^2\sum_{k=1}^3\int \partial_k\vec{u}.(\partial_k\vec{u}_\alpha\cdot\vec{\nabla}\vec{u}_\alpha)\,dx$$

$$\leq -\frac{\nu}{2}\|\vec{\nabla}\otimes\vec{u}\|_2^2 + \frac{1}{\nu}\|\vec{f}\|_{H^{-1}}^2 + \frac{\nu}{4}\|\vec{u}\|_2^2 + \frac{\alpha^4}{\nu}\|\vec{\nabla}\otimes\vec{u}_\alpha\|_4^4$$

$$\leq -\frac{\nu}{2}\|\vec{\nabla}\otimes\vec{u}\|_2^2 + \frac{1}{\nu}\|\vec{f}\|_{H^{-1}}^2 + \frac{\nu}{4}\|\vec{u}\|_2^2 + C\frac{\alpha^4}{\nu}\|\vec{\nabla}\otimes\vec{u}_\alpha\|_2\|\Delta\vec{u}_\alpha\|_2^3$$

$$\leq -\frac{\nu}{2}\|\vec{\nabla}\otimes\vec{u}\|_2^2 + \frac{1}{\nu}\|\vec{f}\|_{H^{-1}}^2 + \frac{\nu}{4}\|\vec{u}\|_2^2 + C\frac{1}{\nu}\|\vec{\nabla}\otimes\vec{u}_\alpha\|_2\|\Delta\vec{u}_\alpha\|_2\|\vec{u}\|_2^2$$

so that

$$\|\vec{u}\|_2^2 + \nu\int_0^t \|\vec{\nabla}\otimes\vec{u}\|_2\,ds \leq e^{\frac{\nu}{2}t+\frac{C}{\nu}\int_0^t\|\vec{\nabla}\otimes\vec{u}_\alpha\|_2\|\Delta\vec{u}_\alpha\|_2\,ds}\left(\|\vec{u}_0\|_2^2 + \frac{2}{\nu}\int_0^t\|\vec{f}\|_{H^{-1}}^2\,ds\right) \quad (17.21)$$

From the energy estimates (17.20) and (17.21), we get that the L^2 norm of $\vec{u}$ remains bounded in time (with a constant that depends on α) and we get global existence of $\vec{u}_{(\alpha)}$.

- **Third step: Weak convergence.**

We write $\vec{u}_{(\alpha)}$ for the solution of the problem (17.18), and we write $\vec{U}_{(\alpha)} = (Id - \alpha^2\Delta)^{-1}\vec{u}_{(\alpha)}$. From the energy estimate (17.20), we know that $\vec{U}_{(\alpha)}$ remains bounded in $L^\infty L^2 \cap L^2\dot{H}^1$ independently from α. Moreover, $\alpha\Delta\vec{U}_{(\alpha)}$ remains bounded in $L_t^2 L_x^2$ independently from α, so that $\vec{u}_{(\alpha)}$ is bounded in $L^2 L^2$ independently from $\alpha \in (0,1)$. Finally, we know that $\alpha\vec{\nabla}\otimes\vec{U}_{(\alpha)}$ is bounded in $L^\infty\dot{H}^1$ independently from α.

We have

$$\partial_t\vec{U}_{(\alpha)} = \nu\Delta\vec{U}_{(\alpha)} + (Id - \alpha^2\Delta)^{-1}\mathbb{P}(\vec{f} - \sum_{k=1}^3 \partial_k(U_{(\alpha),k}\vec{u}_{(\alpha)})) - \sum_{k=1}^3 u_{(\alpha),k}\vec{\nabla}U_{(\alpha),k}$$

and we write

$$\mathbb{P}(\sum_{k=1}^3 u_{(\alpha),k}\vec{\nabla}U_{(\alpha),k}) = \mathbb{P}\vec{\nabla}(\frac{|\vec{U}_{(\alpha)}|^2}{2}) - \alpha^2\mathbb{P}(\sum_{k=1}^3 \Delta U_{(\alpha),k}\vec{\nabla}U_{(\alpha),k})$$

$$= -\sum_{j=1}^3 \alpha^2\mathbb{P}(\partial_j(\sum_{k=1}^3 \partial_j U_{(\alpha),k}\vec{\nabla}U_{(\alpha),k})) + \sum_{j=1}^3 \alpha^2\mathbb{P}(\vec{\nabla}\frac{\partial_j|\vec{U}_{(\alpha)}|^2}{2})$$

$$= -\sum_{j=1}^3 \alpha^2\mathbb{P}(\partial_j(\sum_{k=1}^3 \partial_j U_{(\alpha),k}\vec{\nabla}U_{(\alpha),k}))$$

Thus, we can see that $\partial_t\vec{U}_{(\alpha)}$ remains bounded in $L^2 H^{-s}$ for $s > 5/2$ (independently from $\alpha \in (0,1)$):

$$\|\partial_t\vec{U}_{(\alpha)}\|_{L^2 H^{-s}}$$
$$\leq C(\|\vec{f}\|_{L^2 H^{-1}} + \|\vec{U}_{(\alpha)}\|_{L^\infty L^2}\|\vec{u}_{(\alpha)}\|_{L^2 L^2} + \alpha\|\vec{U}_{(\alpha)}\|_{L^2\dot{H}^1}\alpha\|\vec{U}_{(\alpha)}\|_{L^\infty\dot{H}^1})$$

We may then use the Rellich–Lions theorem (Theorem 12.1) and find a sequence $\alpha_n \to 0$ and a function $\vec{u} \in L^\infty L^2 \cap L^2 \dot{H}^1$ such that:

- $\vec{U}_{(\alpha_n)}$ is *-weakly convergent to $\vec{u}$ in $L^\infty L^2$ and in $L^2 \dot{H}^1$

- $\vec{U}_{(\alpha_n)}$ is strongly convergent to $\vec{u}$ in $L^2_{\text{loc}}((0,T) \times \mathbb{R}^3)$.

Since $\vec{u}_{(\alpha)} = \vec{U}_{(\alpha)} - \alpha^2 \Delta \vec{U}_{(\alpha)}$, we see that $\vec{u}_{(\alpha_n)}$ converges in $\mathcal{D}'((0,T) \times \mathbb{R}^3)$ (or weakly in $L^2 L^2$) to $\vec{u}$. We find that, in $\mathcal{D}'((0,T) \times \mathbb{R}^3)$, we have

$$\partial_t \vec{u} = \nu \Delta \vec{u} + \mathbb{P}\vec{f} - \lim_{n \to +\infty} \mathbb{P}(\vec{U}_{(\alpha_n)} \cdot \vec{\nabla} \vec{u}_{(\alpha_n)} + \sum_{k=1}^{3} u_{(\alpha_n),k} \vec{\nabla} U_{(\alpha_n),k})$$

We have seen that $\mathbb{P}(\sum_{k=1}^{3} u_{(\alpha_n),k} \vec{\nabla} U_{(\alpha_n),k})$ is bounded in $L^2 H^{-s}$ ($s > 5/2$) in $O(\alpha)$, so that its limit in $\mathcal{D}'$ is equal to 0. Moreover, we write

$$\mathbb{P}(\vec{U}_{(\alpha_n)} \cdot \vec{\nabla} \vec{u}_{(\alpha_n)}) = \sum_{j=1}^{3} \mathbb{P}(\partial_j (U_{(\alpha_n),j} \vec{u}_{(\alpha_n)}));$$

since $U_{(\alpha_n),j} \vec{u}_{(\alpha_n)}$ converges in $\mathcal{D}'$ to $u_j \vec{u}$ and remains bounded in $L^2 H^{-s}$ for $s > 3/2$, we find that $\mathbb{P}(\vec{U}_{(\alpha_n)} \cdot \vec{\nabla} \vec{u}_{(\alpha_n)})$ converges weakly to $\mathbb{P}(\vec{u} \cdot \vec{\nabla} \vec{u})$ in $L^2 H^{-s}$ for $s > 5/2$.

Thus, the weak limit $\vec{u}$ satisfies

$$\partial_t \vec{u} = \nu \Delta \vec{u} + \mathbb{P}(\vec{f} - \vec{u} \cdot \vec{\nabla} \vec{u}).$$

- **Fourth step: Global energy estimates for the weak limit.**

We now check that the weak limit $\vec{u}$ fulfills the Leray energy inequality. Let $\omega(t)$ be a non-negative compactly supported smooth function on $\mathbb{R}$, supported within $[-\epsilon, \epsilon]$, with $\|\omega\|_1 = 1$. On $(\epsilon, T - \epsilon)$, we have

$$\|\omega * \vec{U}_{(\alpha)}(t,.)\|_2^2 \leq \omega * \|\vec{U}_{(\alpha)}(t,.)\|_2^2$$

and

$$\frac{d}{dt}\left(\frac{\|\vec{U}_\alpha\|_2^2}{2} + \alpha^2 \frac{\|\vec{\nabla} \otimes \vec{U}_\alpha\|_2^2}{2}\right) = -\nu\|\vec{\nabla} \otimes \vec{U}_\alpha\|_2^2 - \nu\alpha^2\|\Delta \vec{U}_\alpha\|_2^2 + \langle \vec{f} | \vec{U}_\alpha \rangle_{H^{-1},H^1}$$

so that

$$\omega * \|\vec{U}_{(\alpha)}(t,.)\|_2^2 + 2\nu\omega * \left(\int_0^t \|\vec{U}_{(\alpha)}\|_{\dot{H}^1}^2 \, ds\right) \leq \frac{\|\vec{U}_\alpha(0,.)\|_2^2}{2} + \alpha^2 \frac{\|\vec{\nabla} \otimes \vec{U}_\alpha(0,.)\|_2^2}{2}$$

$$+ 2\omega * \left(\int_0^t \langle \vec{f} | \vec{u}_{(\alpha_n)} \rangle_{H^{-1},H^1} \, ds\right)$$

Besides, we have the *-weak convergence of $\omega * \vec{U}_{(\alpha_n)}$ to $\omega * \vec{u}$ in $L^2 \dot{H}^1$, the strong convergence of $\omega * \vec{U}_{(\alpha)}(t,.)$ to $\omega * \vec{u}(t,.)$ in $L^2_{\text{loc}}(\mathbb{R}^3)$, and thus (due to the control we have on the L^2 norm) the *-weak convergence of $\omega * \vec{u}_{(\alpha)}(t,.)$ to $\omega * \vec{u}(t,.)$ in $L^2(\mathbb{R}^3)$. We thus get that

$$\|\omega * \vec{u}(t,.)\|_2^2 + 2\nu\omega * \left(\int_0^t \|\vec{u}\|_{\dot{H}^1}^2 \, ds\right)$$

$$\leq \liminf_{n \to +\infty} \|\omega * \vec{U}_{(\alpha_n)}(t,.)\|_2^2 + 2\nu\omega * \left(\int_0^t \|\vec{U}_{(\alpha_n)}\|_{\dot{H}^1}^2 \, ds\right)$$

and thus

$$\|\omega * \vec{u}(t, .)\|_2^2 + 2\nu\omega * \left(\int_0^t \|\vec{u}\|_{\dot{H}^1}^2 \, ds\right) \leq \|\vec{u}_0\|_2^2 + 2\omega * \left(\int_0^t \langle \vec{f} | \vec{u} \rangle_{H^{-1}, H^1} \, ds\right)$$

When t_0 is a Lebesgue point of $t \mapsto \|\vec{u}(t, .)\|_2$, we get that:

$$\|\vec{u}(t_0, .)\|_2^2 + 2\nu \int_0^{t_0} \|\vec{u}\|_{\dot{H}^1}^2 \, ds \leq \|\vec{u}_0\|_2^2 + 2 \int_0^{t_0} \langle \vec{f} | \vec{u} \rangle_{H^{-1}, H^1} \, ds$$

This inequality may be extended to every t_0, due to the weak boundedness of $t \in (0, T) \mapsto \vec{u}(t, .) \in L^2$.

- **Fifth step: Local energy estimates for the weak limit.**

We now check that $\vec{u}$ is more precisely a suitable weak solution (i.e., fulfills the local energy inequality). Recall that we have

$$\partial_t \vec{U}_{(\alpha)} = \nu\Delta\vec{U}_{(\alpha)} - \vec{U}_{(\alpha)} \cdot \vec{\nabla}\vec{U}_{(\alpha)} + (Id - \alpha^2\Delta)^{-1}\left(\vec{f} - \alpha^2\vec{\nabla}(\frac{1}{2}|\vec{\nabla} \otimes \vec{U}_{(\alpha)}|^2) - \vec{\nabla}p_{(\alpha)}\right)$$

or equivalently

$$\partial_t \vec{U}_{(\alpha)} = \nu\Delta\vec{U}_{(\alpha)} - \vec{U}_{(\alpha)} \cdot \vec{\nabla}\vec{U}_{(\alpha)} + (Id - \alpha^2\Delta)^{-1}\vec{f} - \vec{\nabla}q_{(\alpha)}$$

with

$$q_{(\alpha)} = \frac{1}{\Delta} \operatorname{div}\left((Id - \alpha^2\Delta)^{-1}\vec{f} - \vec{U}_{(\alpha)} \cdot \vec{\nabla}\vec{U}_{(\alpha)}\right).$$

Similarly, we have

$$\partial_t \vec{u} = \nu\Delta\vec{u} + \vec{f} - \vec{u} \cdot \vec{\nabla}\vec{u} - \vec{\nabla}p$$

with

$$p = \frac{1}{\Delta} \operatorname{div}\left(\vec{f} - \vec{u} \cdot \vec{\nabla}\vec{u}\right).$$

As $\vec{U}_{(\alpha_n)}$ converge strongly to $\vec{u}$ in $L^2_{\text{loc}}((0, T) \times \mathbb{R}^3)$, we have

$$\partial_t(\frac{|\vec{u}|^2}{2}) = \lim_{n \to +\infty} \partial_t(\frac{|\vec{U}_{(\alpha_n)}|^2}{2}) = \lim_{n \to +\infty} \vec{U}_{(\alpha_n)} \cdot \partial_t \vec{U}_{(\alpha_n)}$$

in $\mathcal{D}'((0, T) \times \mathbb{R}^3)$. We have

$$\vec{U}_{(\alpha_n)} \cdot \partial_t \vec{U}_{(\alpha_n)} = \nu\vec{U}_{(\alpha_n)} \cdot \Delta\vec{U}_{(\alpha_n)} - \vec{U}_{(\alpha_n)}.(\vec{U}_{(\alpha_n)} \cdot \vec{\nabla}\vec{U}_{(\alpha_n)})$$
$$+ (Id - \alpha_n^2\Delta)^{-1}\vec{f} \cdot \vec{U}_{(\alpha_n)} - \operatorname{div}(q_{(\alpha_n)}\vec{U}_{(\alpha_n)}).$$

We consider each of the terms in the last sum. We have

- $\vec{U}_{(\alpha_n)} \cdot \Delta\vec{U}_{(\alpha_n)} = \Delta(\frac{1}{2}|\vec{U}_{(\alpha_n)}|^2) - |\vec{\nabla} \otimes \vec{U}_{(\alpha_n)}|^2$ so that we have

$$\lim_{n \to +\infty} \vec{U}_{(\alpha_n)} \cdot \Delta\vec{U}_{(\alpha_n)} + |\vec{\nabla} \otimes \vec{U}_{(\alpha_n)}|^2 = \Delta(\frac{|\vec{u}|^2}{2})$$

in $\mathcal{D}'((0, T) \times \mathbb{R}^3)$.

- $\vec{U}_{(\alpha_n)}$ converges weakly to $\vec{u}$ in $L^2 H^1$ and $(Id - \alpha_n^2\Delta)^{-1}\vec{f}$ converges strongly to $\vec{f}$ in $L^2 H^{-1}$, so that $\lim_{n \to +\infty} \vec{U}_{(\alpha_n)}.(Id - \alpha_n^2\Delta)^{-1}\vec{f} = \vec{f} \cdot \vec{U}$ in $\mathcal{D}'((0, T) \times \mathbb{R}^3)$

- We write $\vec{U}_{(\alpha_n)}.(\vec{U}_{(\alpha_n)} \cdot \vec{\nabla}\vec{U}_{(\alpha_n)}) = \sum_{j=1}^{3} \partial_j (U_{(\alpha_n),j}\frac{|\vec{U}_{(\alpha_n)}|^2}{2})$. Just in the same way as for the case of the Navier–Stokes-α model,

$$\lim_{n\to+\infty} \sum_{j=1}^{3} \partial_j (U_{(\alpha_n),j}\frac{|\vec{U}_{(\alpha_n)}|^2}{2}) = \operatorname{div}(\frac{|\vec{u}|^2}{2}\vec{u})$$

in $\mathcal{D}'((0,T)\times\mathbb{R}^3)$.

- we now study

$$q_{(\alpha)} = \frac{1}{\Delta}\operatorname{div}(Id - \alpha_n^2\Delta)^{-1}\vec{f} - \frac{1}{\Delta}\operatorname{div}\left(\vec{U}_{(\alpha)} \cdot \vec{\nabla}\vec{U}_{(\alpha)}\right).$$

We have, of course, $\lim_{n\to+\infty} \vec{U}_{(\alpha_n)}.\frac{1}{\Delta}\operatorname{div}(Id-\alpha_n^2\Delta)^{-1}\vec{f} = \vec{u}.\frac{1}{\Delta}\operatorname{div}\vec{f}$ in $\mathcal{D}'((0,T)\times\mathbb{R}^3)$ (as $\vec{U}_{(\alpha_n)}$ converges weakly to $\vec{u}$ in L^2H^1 and $\frac{1}{\Delta}\operatorname{div}(Id-\alpha_n^2\Delta)^{-1}\vec{f}$ converges strongly to $\frac{1}{\Delta}\operatorname{div}\vec{f}$ in $(L^2L^2)_{\mathrm{loc}}$). We know that $U_{(\alpha_n),j}\vec{U}_{(\alpha_n)}$ is bounded in $L^{3/2}L^{3/2}$ and converges strongly in $(L^{3/2}L^{3/2})_{\mathrm{loc}}$ to $u_j\vec{u}$, hence we have weak convergence of $\sum_{j=1}^{3}\frac{1}{\Delta}\operatorname{div}\partial_j\left(U_{(\alpha_n),j}\vec{U}_{(\alpha_n)}\right)$ to $\sum_{j=1}^{3}\frac{1}{\Delta}\operatorname{div}\partial_j (u_j\vec{u})$ in $(L^{3/2}L^{3/2})_{\mathrm{loc}}$. As $\vec{U}_{(\alpha_n)}$ converges strongly in $(L^3L^3)_{\mathrm{loc}}$, we find that $\lim_{n\to+\infty} \vec{U}_{(\alpha_n)}\frac{1}{\Delta}\operatorname{div}\left(\vec{U}_{(\alpha_n)} \cdot \vec{\nabla}\vec{U}_{(\alpha_n)}\right) = \vec{u}(\sum_{j=1}^{3}\frac{1}{\Delta}\operatorname{div}\partial_j (u_j\vec{u}))$ in $\mathcal{D}'((0,T)\times\mathbb{R}^3)$. Thus, we find that

$$\lim_{n\to+\infty} \operatorname{div}(q_{(\alpha_n)}\vec{U}_{(\alpha_n)}) = \operatorname{div}(p\vec{u})$$

in $\mathcal{D}'((0,T)\times\mathbb{R}^3)$.

Thus far, we have got that

$$\partial_t|\vec{u}|^2 = \nu\Delta|\vec{u}|^2 - \operatorname{div}((|\vec{u}|^2 + 2p)\vec{u}) + 2\vec{u}\cdot\vec{f} - \nu T$$

with

$$T = \lim_{n\to+\infty} 2|\vec{\nabla}\otimes\vec{U}_{(\alpha_n)}|^2.$$

Let $\phi\in\mathcal{D}'((0,T)\times\mathbb{R}^3)$ be a non-negative function. As $\sqrt{\phi}\,\vec{\nabla}\otimes\vec{U}_{(\alpha_n)}$ is weakly convergent to $\sqrt{\phi}\,\vec{\nabla}\otimes\vec{u}$ in $L_t^2 L_x^2$, we find that

$$\langle T|\phi\rangle_{\mathcal{D}',\mathcal{D}} = 2\lim_{n\to+\infty}\iint|\vec{\nabla}\otimes\vec{U}_{(\alpha_n)}|^2\,\phi(t,x)\,dt\,d$$

$$\geq 2\iint|\vec{\nabla}\otimes\vec{u}|^2\phi(t,x)\,dt\,dx.$$

Thus, $T = 2|\vec{\nabla}\otimes\vec{u}|^2 + \mu$, where μ is a non-negative locally finite measure, and thus $\vec{u}$ is suitable.

- **Sixth step: Strong convergence.**

If $\vec{u}_0\in H^1$ and $\vec{f}\in L_t^2 L_x^2$, and if $\vec{u}$ is a solution of the Navier–Stokes problem with $\vec{u}\in\mathcal{C}([0,T],H^1)\cap L^2([0,T],H^2)$, it is easy to check the strong convergence of $\vec{U}_{(\alpha)}$ to $\vec{u}$: as $\vec{u}$ and $\vec{U}_{(\alpha)}$ are regular enough, we may write

$$\frac{d}{dt}\|\vec{u} - \vec{U}_{(\alpha)}\|_2^2 = 2\int (\vec{u} - \vec{U}_{(\alpha)}) \cdot \partial_t(\vec{u} - \vec{U}_{(\alpha)})\, dx$$

$$= -2\nu\|\vec{\nabla}\otimes(\vec{u} - \vec{U}_{(\alpha)})\|_2^2 - 2\int (\vec{u} - \vec{U}_{(\alpha)}).(\vec{u}\cdot\vec{\nabla}\vec{u} - \vec{U}_{(\alpha)}\cdot\vec{\nabla}\vec{U}_{(\alpha)})\, dx$$

$$+ 2\int (\vec{u} - \vec{U}_{(\alpha)}).(\vec{f} - (Id - \alpha^2\Delta)^{-1}\vec{f})\, dx$$

$$= -2\nu\|\vec{\nabla}\otimes(\vec{u} - \vec{U}_{(\alpha)})\|_2^2 - 2\int (\vec{u} - \vec{U}_{(\alpha)}).((\vec{u} - \vec{U}_{(\alpha)})\cdot\vec{\nabla}\vec{u})\, dx$$

$$+ 2\int (\vec{u} - \vec{U}_{(\alpha)}).(\vec{f} - (Id - \alpha^2\Delta)^{-1}\vec{f})\, dx$$

$$\leq -2\nu\|\vec{\nabla}\otimes(\vec{u} - \vec{U}_{(\alpha)})\|_2^2 + C\|\vec{u} - \vec{U}_{(\alpha)}\|_6\|\vec{u} - \vec{U}_{(\alpha)}\|_2\|\vec{\nabla}\otimes\vec{u}\|_3$$

$$+ 2\|\vec{u} - \vec{U}_{(\alpha)}\|_2\|\vec{f} - (Id - \alpha^2\Delta)^{-1}\vec{f}\|_2$$

$$\leq \frac{C}{\nu}\|\vec{u} - \vec{u}_{(\alpha)}\|_2^2\|\vec{\nabla}\otimes\vec{u}\|_3^2 + \|\vec{u} - \vec{U}_{(\alpha)}\|_2^2 + \|\vec{f} - (Id - \alpha^2\Delta)^{-1}\vec{f}\|_2^2$$

This gives

$$\|\vec{u} - \vec{U}_{(\alpha)}\|_{L^\infty L^2}^2 \leq e^{T + \int_0^T \frac{C}{\nu}\|\vec{\nabla}\otimes\vec{u}\|_3^2\, ds} \int_0^T \frac{C}{\nu}\|\vec{f} - (Id - \alpha^2\Delta)^{-1}\vec{f}\|_2^2\, ds$$

and thus $\lim_{\alpha\to 0}\sup_{0<t<T}\|\vec{u} - \vec{U}_{(\alpha)}\|_2 = 0$.   □

17.5   The Simplified Bardina Model

The simplified Bardina model is another α-model studied by Cao, Lunasin and Titi [89]. This model is given by

$$\partial_t\vec{u} + \vec{u}_\alpha \cdot \vec{\nabla}\,\vec{u}_\alpha = \nu\Delta\vec{u} + \vec{f} - \vec{\nabla}p$$

where we have again $\vec{u}_\alpha = (Id - \alpha^2\Delta)^{-1}\vec{u}$ and div $\vec{u} = 0$. This model provides, as the other α-models, an efficient modelization for turbulent fluids at high Reynold numbers, while the computations are made easier than for the other models.

The simplified Bardina model

Theorem 17.4.
Let $\vec{u}_0 \in L^2$, with div $\vec{u}_0 = 0$, and $\vec{f} \in L_t^2 H_x^{-1}$ on $(0,T) \times \mathbb{R}^3$ $(T < +\infty)$.

- *for $\alpha > 0$, the problem*

$$\partial_t\vec{u} + \mathbb{P}(\vec{u}_\alpha \cdot \vec{\nabla}\,\vec{u}_\alpha) = \nu\Delta\vec{u} + \mathbb{P}\vec{f} \qquad (17.22)$$

with initial value $\vec{u}(0,.) = \vec{u}_0$ and

$$\vec{u}_\alpha = (Id - \alpha^2\Delta)^{-1}\vec{u}$$

has a unique solution $\vec{u}_{(\alpha)}$ such that $\vec{u}_{(\alpha)} \in L_t^\infty L_x^2 \cap L_t^2 H_x^1$.

- $\vec{U}_{(\alpha)} = (Id - \alpha^2 \Delta)^{-1} \vec{u}_{(\alpha)}$ satisfies, with uniform bounds with respect to α, that $\vec{U}_{(\alpha)} \in L_t^\infty L_x^2 \cap L_t^2 H_x^1$ and $\alpha \vec{U}_{(\alpha)} \in L_t^\infty H_x^1 \cap L_t^2 H_x^2$

- there exists a sequence $\alpha_k \to 0$ and a function $\vec{u} \in L_t^\infty L_x^2 \cap L_t^2 H_x^1$ such that $\vec{u}_{(\alpha_k)}$ converges to $\vec{u}$ in $\mathcal{D}'((0,T) \times \mathbb{R}^3)$. Moreover $\vec{u}$ is a suitable Leray weak solution of the Navier–Stokes problem

$$\partial_t \vec{u} + \mathbb{P}(\vec{u} \cdot \vec{\nabla} \vec{u}) = \nu \Delta \vec{u} + \mathbb{P}\vec{f}$$

  with initial value $\vec{u}_0$.

- If $\vec{u}_0 \in H^1$ and $\vec{f} \in L_t^2 L_x^2$ and if the Navier–Stokes problem has a solution $\vec{u} \in \mathcal{C}([0,T], H^1) \cap L^2([0,T], H^2)$, then we have

$$\lim_{\alpha \to 0} \|\vec{U}_{(\alpha)} - \vec{u}\|_{L_t^\infty L_x^2} = 0.$$

Proof. • **First step: Local existence of $\vec{u}_{(\alpha)}$.**

We just have to estimate, for $\vec{u}$ and $\vec{v}$ in $L^\infty L^2 \cap L^2 H^1$ with div $\vec{u} = 0$ and for $0 < T_0 < T$, the norm in $L^2((0, T_0), H^{-1})$ of $\vec{u}_\alpha \cdot \vec{\nabla} \vec{v}_\alpha$: we have

$$\|\vec{u}_\alpha \cdot \vec{\nabla} \vec{v}_\alpha\|_{L^2((0,T_0),H^{-1})} = \|\operatorname{div}(\vec{v}_\alpha \otimes \vec{u}_\alpha)\|_{L^2((0,T_0),H^{-1})}$$
$$\leq C\sqrt{T_0}\, \alpha^{-3/2} \|\vec{u}_\alpha\|_{L^\infty L^2} \|\vec{v}_\alpha\|_{L^\infty L^2}.$$

Thus, we find existence (and uniqueness) of a solution $\vec{u} = \vec{u}_{(\alpha)}$ of Equation (17.22) for T_0 small enough to ensure

$$1 + T_0 \nu \leq 2$$

and

$$T_0 \leq C_0 \alpha^3 \frac{\nu}{(\|\vec{u}_0\|_2 + \frac{2}{\sqrt{\nu}} \|\vec{f}\|_{L^2 \dot{H}^{-1}})^2}. \tag{17.23}$$

• **Second step: Energy estimates and global existence of $\vec{u}_{(\alpha)}$.**

To show the existence of a global solution to (17.18), it is then enough to show that the L^2 norm of $\vec{u}_{(\alpha)}$ remains bounded (as the existence time T_0 is controlled by the L^2 norm of the Cauchy data).

To alleviate the notations, we still write $\vec{u}$ for the solution, instead of $\vec{u}_{(\alpha)}$. We consider a time T^* such that, for all $T_0 < T^*$, $\vec{u}$ belongs to $L^\infty((0, T_0), L^2) \cap L^2((0, T_0), H^1)$ and we want to prove that $\vec{u}$ belongs to $L^\infty((0, T^*), L^2)$. As in the case of the Navier–Stokes-α model, we may write

$$\frac{d}{dt}\left(\frac{\|\vec{u}_\alpha\|_2^2}{2} + \alpha^2 \frac{\|\vec{\nabla} \otimes \vec{u}_\alpha\|_2^2}{2}\right) = \langle \partial_t \vec{u} | \vec{u}_\alpha \rangle_{H^{-1}, H^1}.$$

and

$$\frac{d}{dt}\left(\frac{\|\vec{u}\|_2^2}{2}\right) = \langle \partial_t \vec{u} | \vec{u} \rangle_{H^{-1}, H^1}.$$

We have

$$\frac{d}{dt}\Big(\frac{\|\vec{u}_\alpha\|_2^2}{2} + \alpha^2 \frac{\|\vec{\nabla} \otimes \vec{u}_\alpha\|_2^2}{2}\Big) = \langle \nu \Delta \vec{u} - \vec{u}_\alpha \cdot \vec{\nabla}\, \vec{u}_\alpha + \vec{f} | \vec{u}_\alpha \rangle_{H^{-1},H^1}$$

$$= \langle \nu \Delta \vec{u}_\alpha - \nu \alpha^2 \Delta^2 \vec{u}_\alpha + \vec{f} | \vec{u}_\alpha \rangle_{H^{-1},H^1}$$

$$\leq -\frac{\nu}{2}\|\vec{\nabla} \otimes \vec{u}_\alpha\|_2^2 - \nu\alpha^2\|\Delta \vec{u}_\alpha\|_2^2 + \frac{1}{2\nu}\|\vec{f}\|_{H^{-1}}^2 + \frac{\nu}{2}\|\vec{u}_\alpha\|_2^2$$

so that

$$\frac{\|\vec{u}_\alpha\|_2^2}{2} + \alpha^2\frac{\|\vec{\nabla}\otimes\vec{u}_\alpha\|_2^2}{2} + \nu\int_0^t \frac{1}{2}\|\vec{\nabla}\otimes\vec{u}_\alpha\|_2^2 + \alpha^2\|\Delta\vec{u}_\alpha\|_2^2 \, ds$$

$$\leq e^{\nu t}(\|\vec{u}_0\|_2^2 + \frac{1}{2\nu}\int_0^t \|\vec{f}\|_{H^{-1}}^2 \, ds) \tag{17.24}$$

On the other hand, we have

$$\frac{d}{dt}\Big(\frac{\|\vec{u}\|_2^2}{2}\Big) = \langle \nu \Delta \vec{u} - \vec{u}_\alpha \cdot \vec{\nabla}\,\vec{u}_\alpha + \vec{f} | \vec{u} \rangle_{H^{-1},H^1}$$

$$= -\nu\|\vec{\nabla}\otimes\vec{u}\|_2^2 + \langle \vec{f}|\vec{u}\rangle_{H^{-1},H^1} - \alpha^2 \sum_{k=1}^3 \int \partial_k \vec{u}_\alpha.(\partial_k \vec{u}_\alpha \cdot \vec{\nabla}\vec{u}_\alpha) \, dx$$

$$\leq -\frac{\nu}{2}\|\vec{\nabla}\otimes\vec{u}\|_2^2 + \frac{1}{\nu}\|\vec{f}\|_{H^{-1}}^2 + \frac{\nu}{4}\|\vec{u}\|_2^2 + \frac{\alpha^4}{\nu}\|\vec{\nabla}\otimes\vec{u}_\alpha\|_4^4$$

$$\leq -\frac{\nu}{2}\|\vec{\nabla}\otimes\vec{u}\|_2^2 + \frac{1}{\nu}\|\vec{f}\|_{H^{-1}}^2 + \frac{\nu}{4}\|\vec{u}\|_2^2 + C\frac{\alpha^4}{\nu}\|\vec{\nabla}\otimes\vec{u}_\alpha\|_2\|\Delta\vec{u}_\alpha\|_2^3$$

$$\leq -\frac{\nu}{2}\|\vec{\nabla}\otimes\vec{u}\|_2^2 + \frac{1}{\nu}\|\vec{f}\|_{H^{-1}}^2 + \frac{\nu}{4}\|\vec{u}\|_2^2 + C\frac{1}{\nu}\|\vec{\nabla}\otimes\vec{u}_\alpha\|_2\|\Delta\vec{u}_\alpha\|_2\|\vec{u}\|_2^2$$

so that

$$\|\vec{u}\|_2^2 + \nu\int_0^t \|\vec{\nabla}\otimes\vec{u}\|_2 \, ds \leq e^{\frac{\nu}{2}t + \frac{C}{\nu}\int_0^t \|\vec{\nabla}\otimes\vec{u}_\alpha\|_2\|\Delta\vec{u}_\alpha\|_2 \, ds}(\|\vec{u}_0\|_2^2 + \frac{2}{\nu}\int_0^t \|\vec{f}\|_{H^{-1}}^2 \, ds) \tag{17.25}$$

From the energy estimates (17.24) and (17.25), we get that the L^2 norm of $\vec{u}$ remains bounded in time (with a constant that depends on α) and we get global existence of $\vec{u}_{(\alpha)}$.

• Third step: Weak convergence.

We write $\vec{u}_{(\alpha)}$ for the solution of the problem (17.18), and we write $\vec{U}_{(\alpha)} = (Id - \alpha^2\Delta)^{-1}\vec{u}_{(\alpha)}$. From the energy estimate (17.20), we know that $\vec{U}_{(\alpha)}$ remains bounded in $L^\infty L^2 \cap L^2 \dot{H}^1$ independently from α. Moreover, $\alpha\Delta\vec{U}_{(\alpha)}$ remains bounded in $L_t^2 L_x^2$ independently from α, so that $\vec{u}_{(\alpha)}$ is bounded in $L^2 L^2$ independently from $\alpha \in (0,1)$. Finally, we know that $\alpha\vec{\nabla}\otimes\vec{U}_{(\alpha)}$ is bounded in $L^\infty \dot{H}^1$ independently from α.

Moreover, we can see that $\partial_t \vec{U}_{(\alpha)}$ remains bounded in $L^2 H^{-3/2}$ for $s > 5/2$ (independently from α):

$$\|\partial_t \vec{U}_{(\alpha)}\|_{L^2 H^{-3/2}} \leq C(\|\vec{f}\|_{L^2 H^{-1}} + \|\vec{U}_{(\alpha)}\|_{L^\infty L^2}\|\vec{U}_{(\alpha)}\|_{L^2 H^1}).$$

We may then use the Rellich–Lions theorem (Theorem 12.1) and find a sequence $\alpha_n \to 0$ and a function $\vec{u} \subset L^\infty L^2 \cap L^2 \dot{H}^1$ such that:

• $\vec{U}_{(\alpha_n)}$ is *-weakly convergent to $\vec{u}$ in $L^\infty L^2$ and in $L^2 \dot{H}^1$

- $\vec{U}_{(\alpha_n)}$ is strongly convergent to $\vec{u}$ in $L^2_{\text{loc}}((0,T) \times \mathbb{R}^3)$.

Since $\vec{u}_{(\alpha)} = \vec{U}_{(\alpha)} - \alpha^2 \Delta \vec{U}_{(\alpha)}$, we see that $\vec{u}_{(\alpha_n)}$ converges in $\mathcal{D}'((0,T) \times \mathbb{R}^3)$ (or weakly in $L^2 L^2$) to $\vec{u}$. We find that, in $\mathcal{D}'((0,T) \times \mathbb{R}^3)$, we have

$$\partial_t \vec{u} = \nu \Delta \vec{u} + \mathbb{P}\vec{f} - \lim_{n \to +\infty} \mathbb{P}(\vec{U}_{(\alpha_n)} \cdot \vec{\nabla}\vec{U}_{(\alpha_n)})$$

Since $\mathbb{P}(\vec{U}_{(\alpha_n)} \cdot \vec{\nabla}\vec{U}_{(\alpha_n)}) = \sum_{j=1}^3 \mathbb{P}(\partial_j(U_{(\alpha_n),j}\vec{u}_{(\alpha_n)}))$ and since $U_{(\alpha_n),j}\vec{u}_{(\alpha_n)}$ converges weakly to $u_j\vec{u}$ in $L^2 H^{-1/2}$, we get that $\mathbb{P}(\vec{U}_{(\alpha_n)} \cdot \vec{\nabla}\vec{u}_{(\alpha_n)})$ converges weakly to $\mathbb{P}(\vec{u}\cdot\vec{\nabla}\vec{u})$ in $L^2 H^{-3/2}$. Thus, the weak limit $\vec{u}$ satisfies

$$\partial_t \vec{u} = \nu \Delta \vec{u} + \mathbb{P}(\vec{f} - \vec{u} \cdot \vec{\nabla}\vec{u}).$$

- **Fourth step: Global energy estimates for the weak limit.**

This step is conducted in exactly the same way as for the Navier–Stokes-α model or the Clark-α model.

- **Fifth step: Local energy estimates for the weak limit.**

We now check that $\vec{u}$ is more precisely a suitable weak solution (i.e., fulfills the local energy inequality). Recall that we have

$$\partial_t \vec{u}_{(\alpha)} = \nu \Delta \vec{u}_{(\alpha)} - \vec{U}_{(\alpha)} \cdot \vec{\nabla}\vec{U}_{(\alpha)} + \vec{f} - \vec{\nabla}p_{(\alpha)}$$

with

$$p_{(\alpha)} = \frac{1}{\Delta}\operatorname{div}\left(\vec{f} - \vec{U}_{(\alpha)} \cdot \vec{\nabla}\vec{U}_{(\alpha)}\right).$$

Similarly, we have

$$\partial_t \vec{u} = \nu \Delta \vec{u} + \vec{f} - \vec{u} \cdot \vec{\nabla}\vec{u} - \vec{\nabla}p$$

with

$$p = \frac{1}{\Delta}\operatorname{div}\left(\vec{f} - \vec{u} \cdot \vec{\nabla}\vec{u}\right).$$

As $\vec{U}_{(\alpha_n)}$ converge strongly to $\vec{u}$ in $L^2_{\text{loc}}((0,T) \times \mathbb{R}^3)$, we have

$$\partial_t\left(\frac{|\vec{u}|^2}{2}\right) = \lim_{n \to +\infty} \partial_t\left(\frac{|\vec{U}_{(\alpha_n)}|^2}{2}\right) = \lim_{n \to +\infty} \vec{U}_{(\alpha_n)} \cdot \partial_t\vec{U}_{(\alpha_n)}$$

in $\mathcal{D}'((0,T) \times \mathbb{R}^3)$. We have

$$\vec{U}_{(\alpha_n)} \cdot \partial_t\vec{U}_{(\alpha_n)} = \nu\vec{U}_{(\alpha_n)} \cdot \Delta\vec{U}_{(\alpha_n)} - \vec{U}_{(\alpha_n)}.(Id - \alpha_n^2\Delta)^{-1}(\vec{U}_{(\alpha_n)} \cdot \vec{\nabla}\vec{U}_{(\alpha_n)})$$
$$+ (Id - \alpha_n^2\Delta)^{-1}\vec{f} \cdot \vec{U}_{(\alpha_n)} - \operatorname{div}(((Id - \alpha_n^2\Delta)^{-1}p_{(\alpha_n)})\vec{U}_{(\alpha_n)}).$$

We consider each of the terms in the last sum. We have

- $\vec{U}_{(\alpha_n)} \cdot \Delta\vec{U}_{(\alpha_n)} = \Delta(\frac{1}{2}|\vec{U}_{(\alpha_n)}|^2) - |\vec{\nabla} \otimes \vec{U}_{(\alpha_n)}|^2$ so that we have

$$\lim_{n \to +\infty} \vec{U}_{(\alpha_n)} \cdot \Delta\vec{U}_{(\alpha_n)} + |\vec{\nabla} \otimes \vec{U}_{(\alpha_n)}|^2 = \Delta\left(\frac{|\vec{u}|^2}{2}\right)$$

in $\mathcal{D}'((0,T) \times \mathbb{R}^3)$.

- $\vec{U}_{(\alpha_n)}$ converges weakly to $\vec{u}$ in $L^2 H^1$ and $(Id - \alpha_n^2 \Delta)^{-1} \vec{f}$ converges strongly to $\vec{f}$ in $L^2 H^{-1}$, so that $\lim_{n \to +\infty} \vec{U}_{(\alpha_n)} \cdot (Id - \alpha_n^2 \Delta)^{-1} \vec{f} = \vec{f} \cdot \vec{U}$ in $\mathcal{D}'((0,T) \times \mathbb{R}^3)$

- We write $\vec{U}_{(\alpha_n)} \cdot (Id - \alpha_n^2 \Delta)^{-1}(\vec{U}_{(\alpha_n)} \cdot \vec{\nabla} \vec{U}_{(\alpha_n)}) = \sum_{j=1}^3 \partial_j (U_{(\alpha_n),j} \frac{|\vec{U}_{(\alpha_n)}|^2}{2}) + \vec{U}_{(\alpha_n)} \cdot \alpha_n^2 \Delta (Id - \alpha_n^2 \Delta)^{-1}(\vec{U}_{(\alpha_n)} \cdot \vec{\nabla} \vec{U}_{(\alpha_n)})$. Just in the same way as for the case of the Navier–Stokes-α model,

$$\lim_{n \to +\infty} \sum_{j=1}^3 \partial_j (U_{(\alpha_n),j} \frac{|\vec{U}_{(\alpha_n)}|^2}{2}) = \operatorname{div}(\frac{|\vec{u}|^2}{2} \vec{u})$$

in $\mathcal{D}'((0,T) \times \mathbb{R}^3)$.

On the other hand, we have that

$$\alpha_n^2 \Delta (Id - \alpha_n^2 \Delta)^{-1}(\vec{U}_{(\alpha_n)} \cdot \vec{\nabla} \vec{U}_{(\alpha_n)})$$

$$= \sum_{j=1}^3 \alpha_n \partial_j (Id - \alpha_n^2 \Delta)^{-1}(\alpha_n \partial_j \vec{U}_{(\alpha_n)} \cdot \vec{\nabla} \vec{U}_{(\alpha_n)}) + \vec{U}_{(\alpha_n)} \cdot \alpha_n \vec{\nabla} \partial_j \vec{U}_{(\alpha_n)})$$

$\alpha_n \partial_j \vec{U}_{(\alpha_n)} \cdot \vec{\nabla} \vec{U}_{(\alpha_n)}) + \vec{U}_{(\alpha_n)} \cdot \alpha_n \vec{\nabla} \partial_j \vec{U}_{(\alpha_n)}$ is bounded in $L^2 L^1$ hence in $L^2 H^{-7/4}$ (uniformly with respect to n). Thus we find that we have $\alpha_n^2 \| \Delta (Id - \alpha_n^2 \Delta)^{-1}(\vec{U}_{(\alpha_n)} \cdot \vec{\nabla} \vec{U}_{(\alpha_n)}) \|_{L^2 H^{-1}} = O(\alpha_n^{1/4})$. As $\vec{U}_{(\alpha_n)}$ is bounded in $L^2 J^1$, we find that

$$\lim_{n \to +\infty} \vec{U}_{(\alpha_n)} \cdot \alpha_n^2 \Delta (Id - \alpha_n^2 \Delta)^{-1}(\vec{U}_{(\alpha_n)} \cdot \vec{\nabla} \vec{U}_{(\alpha_n)}) = 0$$

in $\mathcal{D}'((0,T) \times \mathbb{R}^3)$.

- we now study

$$p_{(\alpha)} = \frac{1}{\Delta} \operatorname{div}(Id - \alpha_n^2 \Delta)^{-1} \vec{f} - \frac{1}{\Delta} \operatorname{div}(Id - \alpha_n^2 \Delta)^{-1} \left(\vec{U}_{(\alpha)} \cdot \vec{\nabla} \vec{U}_{(\alpha)} \right).$$

We have, of course, $\lim_{n \to +\infty} \vec{U}_{(\alpha_n)} \cdot \frac{1}{\Delta} \operatorname{div}(Id - \alpha_n^2 \Delta)^{-1} \vec{f} = \vec{u} \cdot \frac{1}{\Delta} \operatorname{div} \vec{f}$ in $\mathcal{D}'((0,T) \times \mathbb{R}^3)$ (as $\vec{U}_{(\alpha_n)}$ converges weakly to $\vec{u}$ in $L^2 H^1$ and $\frac{1}{\Delta} \operatorname{div}(Id - \alpha_n^2 \Delta)^{-1} \vec{f}$ converges strongly to $\frac{1}{\Delta} \operatorname{div} \vec{f}$ in $(L^2 L^2)_{\mathrm{loc}}$). We know that $U_{(\alpha_n),j} \vec{U}_{(\alpha_n)}$ is bounded in $L^{3/2} L^{3/2}$ and converges strongly in $(L^{3/2} L^{3/2})_{\mathrm{loc}}$ to $u_j \vec{u}$, hence we have weak convergence of $\sum_{j=1}^3 \frac{1}{\Delta} \operatorname{div} \partial_j \left(U_{(\alpha_n),j} \vec{U}_{(\alpha_n)} \right)$ to $\sum_{j=1}^3 \frac{1}{\Delta} \operatorname{div} \partial_j (u_j \vec{u})$ in $(L^{3/2} L^{3/2})$; as we have strong convergence of $(Id - \alpha_n^2 \Delta)^{-1} \varphi$ to φ in $L^3 L^3$ for all function $\varphi \in L^3 L^3$, we find that we have weak convergence of $(Id - \alpha_n^2 \Delta)^{-1}(\sum_{j=1}^3 \frac{1}{\Delta} \operatorname{div} \partial_j \left(U_{(\alpha_n),j} \vec{U}_{(\alpha_n)} \right))$ to $\sum_{j=1}^3 \frac{1}{\Delta} \operatorname{div} \partial_j (u_j \vec{u})$ in $(L^{3/2} L^{3/2})$. As $\vec{U}_{(\alpha_n)}$ converges strongly in $(L^3 L^3)_{\mathrm{loc}}$, we find that

$$\lim_{n \to +\infty} \vec{U}_{(\alpha_n)} (Id - \alpha_n^2 \Delta)^{-1} \frac{1}{\Delta} \operatorname{div} \left(\vec{U}_{(\alpha_n)} \cdot \vec{\nabla} \vec{U}_{(\alpha_n)} \right) = \vec{u} (\sum_{j=1}^3 \frac{1}{\Delta} \operatorname{div} \partial_j (u_j \vec{u}))$$

in $\mathcal{D}'((0,T) \times \mathbb{R}^3)$. Thus, we find that

$$\lim_{n \to +\infty} \operatorname{div}(p_{(\alpha_n)} \vec{U}_{(\alpha_n)}) = \operatorname{div}(p \vec{u})$$

in $\mathcal{D}'((0,T) \times \mathbb{R}^3)$.

Thus far, we have got that

$$\partial_t|\vec{u}|^2 = \nu\Delta|\vec{u}|^2 - \operatorname{div}((|\vec{u}|^2 + 2p)\vec{u}) + 2\vec{u}\cdot\vec{f} - \nu T$$

with

$$T = \lim_{n\to+\infty} 2|\vec{\nabla}\otimes\vec{U}_{(\alpha_n)}|^2.$$

Let $\phi \in \mathcal{D}'((0,T)\times\mathbb{R}^3)$ be a non-negative function. As $\sqrt{\phi}\,\vec{\nabla}\otimes\vec{U}_{(\alpha_n)}$ is weakly convergent to $\sqrt{\phi}\,\vec{\nabla}\otimes\vec{u}$ in $L^2_t L^2_x$, we find that

$$\langle T|\phi\rangle_{\mathcal{D}',\mathcal{D}} = 2 \lim_{n\to+\infty} \iint |\vec{\nabla}\otimes\vec{U}_{(\alpha_n)}|^2\,\phi(t,x)\,dt\,d$$

$$\geq 2 \iint |\vec{\nabla}\otimes\vec{u}|^2\phi(t,x)\,dt\,dx.$$

Thus, $T = 2|\vec{\nabla}\otimes\vec{u}|^2 + \mu$, where μ is a non-negative locally finite measure, and thus $\vec{u}$ is suitable.

- **Sixth step: Strong convergence.**

If $\vec{u}_0 \in H^1$ and $\vec{f} \in L^2_t L^2_x$, and if $\vec{u}$ is a solution of the Navier–Stokes problem with $\vec{u} \in \mathcal{C}([0,T], H^1) \cap L^2([0,T], H^2)$, it is easy to check the strong convergence of $\vec{U}_{(\alpha)}$ to $\vec{u}$: as $\vec{u}$ and $\vec{U}_{(\alpha)}$ are regular enough, we may write

$$
\begin{aligned}
\frac{d}{dt}\|\vec{u} - \vec{U}_{(\alpha)}\|_2^2 = {}& 2\int (\vec{u} - \vec{U}_{(\alpha)})\cdot\partial_t(\vec{u} - \vec{U}_{(\alpha)})\,dx \\
= {}& -2\nu\|\vec{\nabla}\otimes(\vec{u} - \vec{U}_{(\alpha)})\|_2^2 \\
& -2\int (\vec{u} - \vec{U}_{(\alpha)}).(\vec{u}\cdot\vec{\nabla}\vec{u} - (Id - \alpha^2\Delta)^{-1}(\vec{U}_{(\alpha)}\cdot\vec{\nabla}\vec{U}_{(\alpha)}))\,dx \\
& +2\int (\vec{u} - \vec{U}_{(\alpha)}).(\vec{f} - (Id - \alpha^2\Delta)^{-1}\vec{f})\,dx \\
= {}& -2\nu\|\vec{\nabla}\otimes(\vec{u} - \vec{U}_{(\alpha)})\|_2^2 - 2\int (\vec{u} - \vec{U}_{(\alpha)}).((\vec{u} - \vec{U}_{(\alpha)})\cdot\vec{\nabla}\vec{u})\,dx \\
& +2\int (\vec{u} - \vec{U}_{(\alpha)}).(\alpha^2\Delta(Id - \alpha^2\Delta)^{-1}(\vec{U}_{(\alpha)}\cdot\vec{\nabla}\vec{U}_{(\alpha)}))\,dx \\
& +2\int (\vec{u} - \vec{U}_{(\alpha)}).(\vec{f} - (Id - \alpha^2\Delta)^{-1}\vec{f})\,dx \\
\leq {}& -2\nu\|\vec{\nabla}\otimes(\vec{u} - \vec{U}_{(\alpha)})\|_2^2 + C\|\vec{u} - \vec{U}_{(\alpha)}\|_6\|\vec{u} - \vec{U}_{(\alpha)}\|_2\|\vec{\nabla}\otimes\vec{u}\|_3 \\
& +2\alpha^2\|\vec{u} - \vec{U}_{(\alpha)}\|_{H^1}\|\Delta(Id - \alpha^2\Delta)^{-1}(\vec{U}_{(\alpha)}\cdot\vec{\nabla}\vec{U}_{(\alpha)}))\|_{H^{-1}} \\
& +2\|\vec{u} - \vec{U}_{(\alpha)}\|_2\|\vec{f} - (Id - \alpha^2\Delta)^{-1}\vec{f}\|_2 \\
\leq {}& \frac{C}{\nu}\|\vec{u} - \vec{u}_{(\alpha)}\|_2^2\|\vec{\nabla}\otimes\vec{u}\|_3^2 + C(1 + \nu)\|\vec{u} - \vec{U}_{(\alpha)}\|_2^2 \\
& +\|\vec{f} - (Id - \alpha^2\Delta)^{-1}\vec{f}\|_2^2 + C\frac{1}{\nu}\|\Delta(Id - \alpha^2\Delta)^{-1}(\vec{U}_{(\alpha)}\cdot\vec{\nabla}\vec{U}_{(\alpha)}))\|_{H^{-1}}^2
\end{aligned}
$$

This gives

$$\|\vec{u} - \vec{U}_{(\alpha)}\|_{L^\infty L^2}^2$$
$$\leq e^{CT + \int_0^T \frac{C}{\nu}\|\vec{\nabla}\otimes\vec{u}\|_3^2\,ds} \times$$
$$\times\left(\int_0^T \frac{C}{\nu}\|\vec{f} - (Id - \alpha^2\Delta)^{-1}\vec{f}\|_2^2\,ds + \frac{C}{\nu}\int_0^T \|\Delta(Id - \alpha^2\Delta)^{-1}(\vec{U}_{(\alpha)}\cdot\vec{\nabla}\vec{U}_{(\alpha)}))\|_{H^{-1}}^2\,ds\right)$$

and thus (as $\|\Delta(Id - \alpha^2\Delta)^{-1}(\vec{U}_{(\alpha)} \cdot \vec{\nabla}\vec{U}_{(\alpha)}))\|_{H^{-1}} = O(\alpha^{1/4})$)

$$\lim_{\alpha \to 0} \sup_{0 < t < T} \|\vec{u} - \vec{U}_{(\alpha)}\|_2 = 0. \qquad \square$$

17.6   Reynolds Tensor

Reynolds [411] introduced the decomposition of the flow into mean and fluctuating parts. The mean velocity is obtained by averaging the velocity in a way that commutes with space or time translation; as we consider the evolution problem on the whole space, we may shift the space variable without any restriction, while we must keep the times being positive. Thus, we shall consider only space averaging (while it is customary in turbulence to consider time averaging; for large scale of times, an ergodicity assumption will grant the equivalence between time and space averaging).

We thus consider a linear filtering operation $\vec{u} \mapsto \vec{U} = \Phi * \vec{u}$, where Φ is the impulsional response of the filter. $\vec{U}$ is no longer solution of the Navier–Stokes equations, because of the advective term $\vec{u} \cdot \vec{\nabla}\vec{u}$ which is non-linear. We obtain the Reynolds equations:

$$\partial_t\vec{U} + \mathbb{P}\operatorname{div}(\Phi * (\vec{u} \otimes \vec{u})) = \nu\Delta\vec{U} + \mathbb{P}(\Phi * \vec{f})$$

or equivalently

$$\partial_t\vec{U} + \mathbb{P}(\vec{U} \cdot \vec{\nabla}U) = \nu\Delta\vec{U} + \mathbb{P}(\Phi * \vec{f}) + \mathbb{P}\operatorname{div}((\Phi * \vec{u}) \otimes (\Phi * \vec{u}) - \Phi * (\vec{u} \otimes \vec{u})) \qquad (17.26)$$

The tensor $\mathbb{R} = (\Phi * \vec{u}) \otimes (\Phi * \vec{u}) - \Phi * (\vec{u} \otimes \vec{u})$ is the Reynolds stress which is a correction to the Navier–Stokes equations corresponding to the interaction with the fluctuating part.

Generally, Equations (17.26) are not closed: if Φ is a cut-off filter that eliminates the high frequencies, we cannot recover $\vec{u}$ from $\vec{U}$, so that the Reynolds stress cannot be defined only by the values of $\vec{U}$. However, Germano [207] pointed that when filtering by Φ is invertible (i.e., if the Fourier transform of Φ has no zero), then $\vec{u}$ may be computed from $\vec{U}$. In this case, Equations (17.26) are closed, but the computation of $\vec{u}$ is unstable, as the inverse $\frac{1}{\hat{\Phi}(\xi)}$ of the Fourier transform of Φ is unbounded.

Filtering $\vec{u}$ with an invertible filter cannot provide a simpler equation (as the process is, unstably, reversible). In the *large eddy simulation* (LES) only the large scales are numerically resolved, the fine scales being parametrized (Deardorff [150]). This amounts to approximate the Reynolds stress by a more amenable correction to the Navier–Stokes equations.

Guermond, Oden and Prudhomme [222] have presented a mathematical analysis of the LES models. First of all, one needs to select a filter which preserves the key features of the turbulent solutions. When the forcing term is a constant force with only low frequencies, one observes in the Kolmogorov model three domains of frequencies: the low frequencies where the force supplies energy, the inertial range where the energy is supplied by a transfer from lower frequencies to higher frequencies by the non-linear advective term, and then the high frequencies where the diffusion is strong enough to drastically damp the energy faster than its increase through the transfer phenomenon; the filter should capture the inertial range of frequencies.

In the α-models we discussed, we have $\vec{U} = (Id - \alpha^2\Delta)^{-1}\vec{u}$ so that $\hat{\Phi}(\xi) = \frac{1}{1+\alpha^2|\xi|^2}$: the frequencies with $|\xi| << 1/\alpha$ are kept intact, while the frequencies with $|\xi| >> 1$ are damped. The approximations of the Reynolds stress were then the following ones:

- the Leray-α model:

$$\mathbb{P}\,\mathrm{div}\,(\Phi * (\vec{u} \otimes \vec{u})) \approx \mathbb{P}\,\mathrm{div}\,(\Phi * (\vec{u} \otimes (\Phi * \vec{u})))$$

- the Navier–Stokes-α model:

$$\mathbb{P}\,\mathrm{div}\,(\Phi * (\vec{u} \otimes \vec{u})) = \mathbb{P}(\Phi * (\vec{\omega} \wedge \vec{u})) \approx \mathbb{P}\,(\Phi * (\vec{\omega} \wedge (\Phi * \vec{u})))$$

- the Clark-α model:

$$\mathbb{P}\,\mathrm{div}\,(\Phi * (\vec{u} \otimes \vec{u})) \approx$$

$$\mathbb{P}\,\mathrm{div}\,((\Phi * \vec{u}) \otimes (\Phi * \vec{u}))) + \alpha^2 \mathbb{P}\,\mathrm{div}\left(\Phi * (\sum_{k=1}^{3} \partial_k(\Phi * \vec{u}) \otimes \partial_k \vec{u})\right)$$

and thus

$$\mathbb{P}\,\mathrm{div}\,\mathbb{R} \approx -\alpha^2 \mathbb{P}\,\mathrm{div}\left(\Phi * (\sum_{k=1}^{3} \partial_k(\Phi * \vec{u}) \otimes \partial_k \vec{u})\right)$$

- the simplified Bardinal model:

$$\mathbb{P}\,\mathrm{div}\,(\Phi * (\vec{u} \otimes \vec{u})) \approx \mathbb{P}\,\mathrm{div}\,(\Phi * ((\Phi * \vec{u}) \otimes (\Phi * \vec{u})))$$

and thus

$$\mathbb{P}\,\mathrm{div}\,\mathbb{R} \approx -\alpha^2 \mathbb{P}\,\mathrm{div}\,(\Delta\Phi * ((\Phi * \vec{u}) \otimes (\Phi * \vec{u})))\,.$$

Chapter 18

Other Approximations of the Navier–Stokes Equations

In this chapter, we describe some further ways to approximate the Navier–Stokes equations that have been developed in the mathematical literature since the seminal work of Leray [328]. Some of them have been developed not as approximations but as corrections of the Navier–Stokes equations; in that case, one should be careful to respect the basic rules of Newtonian mechanics such as Galilean invariance or the material indifference principle. Others have been developed as simple discretization schemes for the numerical resolution of the equations: we present them in the setting of the whole space while they were developed in the setting of a bounded domain; we will miss some compactness properties useful to prove convergence, and we shall not address the delicate problem of defining the boundary conditions for the approximated problem.

18.1   Faedo–Galerkin Approximations

The Faedo–Galerkin method was first used by Hopf [238] and Ladyzhenskaya [262] to study the Navier–Stokes equations on a bounded domain. The idea is to project the equation on a finite-dimension space of functions and to turn the Partial Differential Equation into an Ordinary Differential Equation.

We shall thus consider a family of functions $(\vec{w}_n)_{n \in \mathbb{N}}$ such that:

- $\vec{w}_n \in H^1$ and $\operatorname{div} \vec{w}_n = 0$

- $\int \vec{w}_n \cdot \vec{w}_p \, dx = \delta_{n,p}$.

and we consider the orthogonal projections on L^2

$$P_N : \vec{w} \mapsto \sum_{n=0}^{N} \langle \vec{w} | \vec{w}_n \rangle_{L^2, L^2} \vec{w}_n.$$

P_N may be extended to H^{-1}. Then we approximate the Navier–Stokes problem

$$\partial_t \vec{u} = \mathbb{P}(\nu \Delta \vec{u} - \vec{u} \cdot \vec{\nabla} \vec{u} + \vec{f}), \quad \vec{u}(0,.) = \vec{u}_0$$

through the equations

$$\partial_t \vec{u}_{(N)} = P_N(\nu \Delta \vec{u}_{(N)} - \vec{u}_{(N)} \cdot \vec{\nabla} \vec{u}_{(N)} + \vec{f}), \quad \vec{u}_{(N)}(0,.) = P_N \vec{u}_0$$

This is the EDO on $X = (x_0(t), \ldots, x_n(t))$ given by

$$\dot{x}_j(t) = \sum_{p=0}^{N} a_{j,p} x_p(t) + \sum_{p=0}^{N} \sum_{q=0}^{N} b_{j,p,q} x_p(t) x_q(t) + f_j(t)$$

DOI: 10.1201/9781003042594-18

with $a_{j,p} = \nu\langle\Delta\vec{\omega}_p|\vec{\omega}_j\rangle_{H^{-1},H^1}$, $b_{j,p,q} = \langle\vec{\omega}_p \cdot \vec{\nabla}\vec{\omega}_q|\vec{\omega}_j\rangle_{H^{-1},H^1}$ and $f_j(t) = \langle\vec{f}(t,.)|\vec{\omega}_j\rangle_{H^{-1},H^1}$. The initial values are $x_j(0) = \langle\vec{u}_0|\vec{\omega}_j\rangle_{H^{-1},H^1}$..

While it is easy to prove that the EDO has a unique global solution $X(t) \in \mathcal{C}([0,T],\mathbb{R}^{N+1})$, and that we have (under reasonable assumptions) weak convergence to a Leray weak solution, the suitability of the limit is not granted (see the discussion by Biryuk, Craig and Ibrahim [43] for the case of Fourier series in the periodical setting, and the discussion by Guermond [220] for wavelets or finite elements).

A Faedo–Galerkin approximation scheme

Theorem 18.1.
Let $\vec{u}_0 \in L^2$, with div $\vec{u}_0 = 0$, and $\vec{f} \in L_t^2 H_x^{-1}$ on $(0,T) \times \mathbb{R}^3$ ($T < +\infty$).

- *Let $(\vec{w}_n)_{n\in\mathbb{N}}$ such that:*

  - *(H1) $\vec{w}_n \in H^1(\mathbb{R}^3)$ and div $\vec{w}_n = 0$*
  - *(H2) $\int \vec{w}_n \cdot \vec{w}_p\, dx = \delta_{n,p}$.*

  and let $P_N : H^{-1} \mapsto H^1$ be defined by

  $$P_N(\vec{w}) = \sum_{n=0}^{N}\langle\vec{w}|\vec{w}_n\rangle_{H^{-1},H^1}\vec{w}_n.$$

  Then, for $N \in \mathbb{N}$, the problem

  $$\partial_t\vec{u} = P_N(\nu\Delta\vec{u} - \vec{u}\cdot\vec{\nabla}\vec{u} + \vec{f}) \qquad (18.1)$$

  with initial value $\vec{u}(0,.) = P_N\vec{u}_0$ has a unique solution $\vec{u}_{(N)}$.

- *This solution $\vec{u}_{(N)}$ satisfies, with uniform bounds with respect to N, that $\vec{u}_{(N)} \in L_t^\infty L_x^2 \cap L_t^2 H_x^1$*

- *Assume moreover that $\vec{w}_n$ satisfies*

  - *(H3) $(\vec{w}_n)$ is a Hilbertian basis of $\{\vec{w} \in L^2\ /\ \text{div}\,\vec{w} = 0\}$*
  - *(H4) there is a sequence of positive numbers ϵ_n and a constant C_0 such that, for any finite sequence (λ_n), we have*

  $$\left\|\sum_n \lambda_n\vec{w}_n\right\|_{H^1} \le C_0\sqrt{\sum_n \epsilon_n\lambda_n^2}$$

  and

  $$\left\|\sum_n \lambda_n\vec{w}_n\right\|_{H^{-1}} \le C_0\sqrt{\sum_n \epsilon_n^{-1}\lambda_n^2}$$

  Then there exists a sequence $N_k \to +\infty$ and a function $\vec{u} \in L_t^\infty L_x^2 \cap L_t^2 H_x^1$ such that $\vec{u}_{(N_k)}$ converges to $\vec{u}$ in $\mathcal{D}'((0,T) \times \mathbb{R}^3)$. Moreover $\vec{u}$ is a Leray weak solution of the Navier–Stokes problem

  $$\partial_t\vec{u} + \mathbb{P}(\vec{u}\cdot\vec{\nabla}\vec{u}) = \nu\Delta\vec{u} + \mathbb{P}\vec{f}$$

  with initial value $\vec{u}_0$.

- If $\vec{u}_0 \in H^1$ and $\vec{f} \in L_t^2 L_x^2$ and if the Navier–Stokes problem has a solution $\vec{u} \in \mathcal{C}([0,T], H^1) \cap L^2([0,T], H^2)$, then we have

$$\lim_{N \to +\infty} \|\vec{u}_{(N)} - \vec{u}\|_{L_t^\infty L_x^2} = 0.$$

Proof. • **First step: Global existence of $\vec{u}_{(N)}$.**

Let X_N be the linear space generated by $\vec{w}_0, \dots, \vec{w}_N$, endowed with the norm $\|\vec{w}\|_{X_N} = \|\vec{w}\|_2$. We have $P_N \vec{f} \in L^2 X_N$. Moreover the linear operator $\vec{w} \mapsto \nu P_N(\Delta \vec{w})$ and the bilinear operator $(\vec{v}, \vec{w}) \mapsto P_N(\vec{w} \cdot \vec{\nabla} \vec{w})$ are bounded on X_N. Thus, if F is defined on $\mathcal{C}([0, T_0], X_N)$ as

$$F(\vec{w}) = P_N(\vec{u}_0) + \int_0^t \nu P_N(\Delta \vec{w}) - P_N(\vec{w} \cdot \vec{\nabla} \vec{w}) + P_N \vec{f} \, ds$$

is a contraction on $\{\vec{w} \in \mathcal{C}([0, T_0], X_N) \,/\, \sup_{0 < t < T_0} \|\vec{w}_N(t, .)\|_{X_N} \leq 2(\|P_N \vec{u}_0\|_{X_N} + \int_0^T \|P_N \vec{f}\|_{X_N} \, ds)\}$ for T_0 small enough:

$$T_0 \leq C_N \min(1, \frac{1}{\|P_N \vec{u}_0\|_{X_N} + \int_0^T \|P_N \vec{f}\|_{X_N} \, ds}).$$

This ensures the existence of a (unique) solution $\vec{u}_{(N)}$ of (18.1) on $[0, T_0]$.

To show the existence of a global solution to (18.1), it is then enough to show that the L^2 norm of $\vec{u}_{(N)}$ remains bounded (as the existence time T_0 is controlled by the L^2 norm of the Cauchy data).

We write

$$\frac{d}{dt}\left(\frac{\|\vec{u}_{(N)}\|_2^2}{2}\right) = \langle \partial_t \vec{u}_{(N)} | \vec{u}_{(N)} \rangle_{H^{-1}, H^1}$$

$$= \langle \nu \Delta \vec{u}_{(N)} - \vec{u}_{(N)} \cdot \vec{\nabla} \vec{u}_{(N)} + \vec{f} | \vec{u}_{(N)} \rangle_{H^{-1}, H^1}$$

$$= -\nu \|\vec{\nabla} \otimes \vec{u}_{(N)}\|_2^2 + \langle \vec{f} | \vec{u}_{(N)} \rangle_{H^{-1}, H^1}$$

$$\leq -\frac{\nu}{2} \|\vec{\nabla} \otimes \vec{u}_{(N)}\|_2^2 + \frac{1}{2\nu} \|\vec{f}\|_{H^{-1}}^2 + \frac{\nu}{2} \|\vec{u}_{(N)}\|_2^2$$

so that

$$\|\vec{u}_{(N)}\|_2^2 + \nu \int_0^t \|\vec{\nabla} \otimes \vec{u}_{(N)}\|_2 \, ds \leq e^{\nu t}(\|\vec{u}_0\|_2^2 + \frac{1}{\nu} \int_0^t \|\vec{f}\|_{H^{-1}}^2 \, ds) \qquad (18.2)$$

Thus, we get global existence of $\vec{u}_{(N)}$.

• **Second step: Weak convergence.**

From the energy estimate (18.2), we know that $\vec{u}_{(N)}$ remains bounded in $L^\infty L^2 \cap L^2 \dot{H}^1$ independently from N. We have

$$\partial_t \vec{u}_{(N)} = P_N(\nu \Delta \vec{u}_{(N)} + \vec{f} - \operatorname{div}(\vec{u}_{(N)} \otimes \vec{u}_{(N)}))$$

where

$$\|\nu \Delta \vec{u}_{(N)} + \vec{f} - \operatorname{div}(\vec{u}_{(N)} \otimes \vec{u}_{(N)})\|_{L^{4/3} H^{-1}} \leq$$

$$\nu \sqrt{T} \|\vec{u}_{(N)}\|_{L^2 H^1} + \sqrt{T} \|\vec{f}\|_{L^2 H^{-1}} + C \|\vec{u}_{(N)}\|_{L^\infty L^2}^{1/2} \|\vec{u}_{(N)}\|_{L^2 H^1}^{3/2}).$$

By assumption, the P_N are equicontinuous on H^{-1}: if $\vec{w} \in H^{-1}$, we have

$$\|P_N\vec{w}\|_{H^{-1}}^2 \leq C_0^2 \sum_{n=0}^{N} \epsilon_n^{-1} |\langle\vec{w}|\vec{w}_n\rangle_{H^{-1},H^1}|^2$$

$$= C_0^2 \langle\vec{w}| \sum_{n=0}^{N} \epsilon_n^{-1}\langle\vec{w}|\vec{w}_n\rangle_{H^{-1},H^1}\vec{w}_n\rangle_{H^{-1},H^1}$$

$$\leq C_0^3 \|\vec{w}\|_{H^{-1}} \sqrt{\sum_{n=0}^{N} \epsilon_n^{-1}|\langle\vec{w}|\vec{w}_n\rangle_{H^{-1},H^1}|^2}$$

This shows that $\sqrt{\sum_{n=0}^{N} \epsilon_n^{-1}|\langle\vec{w}|\vec{w}_n\rangle_{H^{-1},H^1}|^2} \leq C_0\|\vec{w}\|_{H^{-1}}$ and that $\|P_N\vec{w}\|_{H^{-1}} \leq C_0^2\|\vec{w}\|_{H^{-1}}$.

Thus, we can see that $\partial_t\vec{u}_{(N)}$ remains bounded in $L^{4/3}H^{-1}$ (independently from N). We may then use the Rellich–Lions theorem (Theorem 12.1) and find a sequence $N_k \to +\infty$ and a function $\vec{u} \in L^\infty L^2 \cap L^2\dot{H}^1$ such that:

- $\vec{u}_{(N_k)}$ is *-weakly convergent to $\vec{u}$ in $L^\infty L^2$ and in $L^2\dot{H}^1$

- $\vec{u}_{(N_k)}$ is strongly convergent to $\vec{u}$ in $L^2_{\text{loc}}((0,T) \times \mathbb{R}^3)$.

The P_N are equicontinuous on H^1: if $\vec{w} \in H^1$, we have $\|P_N\vec{w}\|_{H^1} \leq C_0^2\|\vec{w}\|_{H^1}$. If $\vec{w} \in L^2$, we have $P_N\vec{w} = P_N\mathbb{P}\vec{w}$ and thus, as $(\vec{w}_n)_{n\in\mathbb{N}}$ is a basis for L^2_σ, $\lim_{N\to+\infty}\|P_N\vec{w} - \mathbb{P}\vec{w}\|_2 = 0$. By equicontinuity of the P_N, we find that if $\vec{w} \in H^1$, then $\lim_{N\to+\infty}\|P_N\vec{w} - \mathbb{P}\vec{w}\|_{H^1} = 0$. Similarly, if $\vec{w}(t,x) \in L^4H^1$, we have $\lim_{N\to+\infty}\|P_N\vec{w} - \mathbb{P}\vec{w}\|_{L^4H^1} = 0$.

This gives that the *-weak convergence of $\nu\Delta\vec{u}_{(N)} + \vec{f} - \text{div}(\vec{u}_{(N)} \otimes \vec{u}_{(N)})$ to $\nu\Delta\vec{u} + \vec{f} - \text{div}(\vec{u} \otimes \vec{u})$ in $L^{4/3}H^{-1}$ implies the *-weak convergence of $\partial_t\vec{u}_{(N)} = P_N(\nu\Delta\vec{u}_{(N)} + \vec{f} - \text{div}(\vec{u}_{(N)} \otimes \vec{u}_{(N)}))$ to $\nu\Delta\vec{u} + \mathbb{P}(\vec{f} - \text{div}(\vec{u} \otimes \vec{u}))$.

Thus, the weak limit $\vec{u}$ satisfies

$$\partial_t\vec{u} = \nu\Delta\vec{u} + \mathbb{P}(\vec{f} - \vec{u} \cdot \vec{\nabla}\vec{u}).$$

- **Third step: Global energy estimates for the weak limit.**

We now check that the weak limit $\vec{u}$ fulfills the Leray energy inequality. Let $\omega(t)$ be a non-negative compactly supported smooth function on $\mathbb{R}$, supported within $[-\epsilon,\epsilon]$, with $\|\omega\|_1 = 1$. On $(\epsilon, T - \epsilon)$, we have

$$\|\omega * \vec{u}_{(N)}(t,.)\|_2^2 \leq \omega * \|\vec{u}_{(N)}(t,.)\|_2^2$$

and

$$\frac{d}{dt}\frac{\|\vec{u}_{(N)}\|_2^2}{2} = -\nu\|\vec{\nabla} \otimes \vec{u}_{(n)}\|_2^2 + \langle\vec{f}|\vec{u}_{(N)}\rangle_{H^{-1},H^1}$$

so that

$$\omega * \|\vec{u}_{(N)}(t,.)\|_2^2 + 2\nu\omega * \left(\int_0^t \|\vec{u}_{(N)}\|_{\dot{H}^1}^2 \, ds\right) \leq$$

$$\frac{\|\vec{u}(0,.)\|_2^2}{2} + 2\omega * \left(\int_0^t \langle\vec{f}|\vec{u}_{(N)}\rangle_{H^{-1},H^1} \, ds\right)$$

We have

$$\|\omega * \vec{u}(t,.)\|_2^2 + 2\nu\omega * (\int_0^t \|\vec{u}\|_{\dot{H}^1}^2 \, ds) \leq$$

$$\liminf_{k\to+\infty} \|\omega * \vec{U}_{(N_k)}(t,.)\|_2^2 + 2\nu\omega * (\int_0^t \|\vec{U}_{(N_k)}\|_{\dot{H}^1}^2 \, ds)$$

and thus

$$\|\omega * \vec{u}(t,.)\|_2^2 + 2\nu\omega * (\int_0^t \|\vec{u}\|_{\dot{H}^1}^2 \, ds) \leq \|\vec{u}_0\|_2^2 + 2\omega * (\int_0^t \langle \vec{f} | \vec{u} \rangle_{H^{-1}, H^1} \, ds)$$

When t_0 is a Lebesgue point of $t \mapsto \|\vec{u}(t,.)\|_2$, we get that:

$$\|\vec{u}(t_0,.)\|_2^2 + 2\nu \int_0^{t_0} \|\vec{u}\|_{\dot{H}^1}^2 \, ds \leq \|\vec{u}_0\|_2^2 + 2\int_0^{t_0} \langle \vec{f} | \vec{u} \rangle_{H^{-1}, H^1} \, ds$$

This inequality may be extended to every t_0, due to the weak boundedness of $t \in (0, T) \mapsto \vec{u}(t,.) \in L^2$.

- **Fourth step: Strong convergence.**

If $\vec{u}_0 \in H^1$ and $\vec{f} \in L_t^2 L_x^2$, and if $\vec{u}$ is a solution of the Navier–Stokes problem with $\vec{u} \in \mathcal{C}([0,T], H^1) \cap L^2([0,T], H^2)$, it is easy to check the strong convergence of $\vec{u}_{(N)}$ to $\vec{u}$: as $\vec{u}$ and $\vec{u}_{(N)}$ are regular enough, we may write

$$\frac{d}{dt} \|\vec{u} - \vec{u}_{(N)}\|_2^2 = 2\int (\vec{u} - \vec{u}_{(N)}) \cdot \partial_t(\vec{u} - \vec{u}_{(N)}) \, dx$$

$$= -2\nu\|\vec{\nabla} \otimes (\vec{u} - \vec{u}_{(N)})\|_2^2 - 2\int (\vec{u} - \vec{u}_{(N)}).(\vec{u} \cdot \vec{\nabla}\vec{u} - \vec{u}_{(N)} \cdot \vec{\nabla}\vec{u}_{(N)}) \, dx$$

$$- 2\int (Id - P_N)\vec{u}.(\nu\Delta\vec{u}_{(N)} + \vec{f} - \vec{u}_{(N)} \cdot \vec{\nabla}\vec{u}_{(N)}) \, dx$$

$$= -2\nu\|\vec{\nabla} \otimes (\vec{u} - \vec{u}_{(N)})\|_2^2 - 2\int (\vec{u} - \vec{u}_{(N)}).((\vec{u} - \vec{u}_{(N)}) \cdot \vec{\nabla}\vec{u}) \, dx$$

$$- 2\int (Id - P_N)\vec{u}.(\nu\Delta\vec{u}_{(N)} + \vec{f} - \vec{u}_{(N)} \cdot \vec{\nabla}\vec{u}_{(N)}) \, dx$$

$$\leq \frac{C}{\nu}\|\vec{u} - \vec{u}_{(N)}\|_2^2\|\vec{\nabla} \otimes \vec{u}\|_3^2$$

$$+ \|(Id - P_N)\vec{u}\|_{H^1}\|\nu\Delta\vec{u}_{(N)} + \vec{f} - \vec{u}_{(N)} \cdot \vec{\nabla}\vec{u}_{(N)}\|_{H^{-1}}.$$

This gives

$$\|\vec{u} - \vec{u}_{(N)}\|_{L^\infty L^2}^2$$

$$\leq e^{T + \int_0^T \frac{C}{\nu}\|\vec{\nabla}\otimes\vec{u}\|_3^2 \, ds} \|(Id - P_N)\vec{u}\|_{L^4 H^1}\|\nu\Delta\vec{u}_{(N)} + \vec{f} - \vec{u}_{(N)} \cdot \vec{\nabla}\vec{u}_{(N)}\|_{L^{4/3} H^{-1}}$$

and thus $\lim_{\alpha\to 0} \sup_{0<t<T} \|\vec{u} - \vec{u}_{(N)}\|_2 = 0$.   □

We shall now discuss the local energy estimates for the weak limit $\vec{u}$. Recall that we have

$$\partial_t \vec{u}_{(N)} = P_N(\nu\Delta\vec{u}_{(N)} - \vec{u}_{(N)} \cdot \vec{\nabla}\vec{u}_{(N)} + \vec{f})$$

$$= \nu\Delta\vec{u}_{(N)} - \vec{u}_{(N)} \cdot \vec{\nabla}\vec{u}_{(N)} + \vec{f} - \vec{\nabla}p_{(N)}$$

$$+ (\mathbb{P} - P_N)(\vec{f} - \vec{u}_{(N)} \cdot \vec{\nabla}\vec{u}_{(N)})$$

with

$$p_{(N)} = \frac{1}{\Delta} \operatorname{div}\left(\vec{f} - \vec{u}_{(N)} \cdot \vec{\nabla}\, \vec{u}_{(N)}\right).$$

Similarly, we have

$$\partial_t \vec{u} = \nu \Delta \vec{u} + \vec{f} - \vec{u} \cdot \vec{\nabla}\, \vec{u} - \vec{\nabla} p$$

with

$$p = \frac{1}{\Delta} \operatorname{div}\left(\vec{f} - \vec{u} \cdot \vec{\nabla}\, \vec{u}\right).$$

As $\vec{u}_{(N_k)}$ converge strongly to $\vec{u}$ in $L^2_{\text{loc}}((0,T) \times \mathbb{R}^3)$, we have

$$\partial_t\left(\frac{|\vec{u}|^2}{2}\right) = \lim_{n \to +\infty} \partial_t\left(\frac{|\vec{u}_{(N_k)}|^2}{2}\right) = \lim_{n \to +\infty} \vec{u}_{(N_k)} \cdot \partial_t \vec{u}_{(N_k)}$$

in $\mathcal{D}'((0,T) \times \mathbb{R}^3)$. We have

$$\begin{aligned}
\vec{u}_{(N_k)} \cdot \partial_t \vec{u}_{(N_k)} =& \nu \vec{u}_{(N_k)} \cdot \Delta \vec{u}_{(N_k)} - \vec{u}_{(N_k)} \cdot (\vec{u}_{(N_k)} \cdot \vec{\nabla} \vec{u}_{(N_k)}) \\
&+ \vec{f} \cdot \vec{u}_{(N_k)} - \operatorname{div}(p_{(N_k)} \vec{u}_{(N_k)}) \\
&+ \vec{u}_{(N_k)} \cdot (\mathbb{P} - P_{N_k})(\vec{f} - \vec{u}_{(N_k)} \cdot \vec{\nabla} \vec{u}_{(N_k)}).
\end{aligned}$$

We consider each of the terms in the last sum.

- We have

$$\lim_{k \to +\infty} \vec{u}_{(N_k)} \cdot \Delta \vec{u}_{(N_k)} + |\vec{\nabla} \otimes \vec{u}_{(N_k)}|^2 = \Delta\left(\frac{|\vec{u}|^2}{2}\right)$$

in $\mathcal{D}'((0,T) \times \mathbb{R}^3)$.

- $\lim_{k \to +\infty} \vec{u}_{(N_k)} \cdot \vec{f} = \vec{f} \cdot \vec{u}$ in $\mathcal{D}'((0,T) \times \mathbb{R}^3)$

- We write $\vec{u}_{(N_k)} \cdot (\vec{u}_{(N_k)} \cdot \vec{\nabla} \vec{u}_{(N_k)}) = \sum_{j=1}^3 \partial_j (u_{(N_k),j} \frac{|\vec{u}_{(N_k)}|^2}{2})$ and get

$$\lim_{k \to +\infty} \sum_{j=1}^3 \partial_j (u_{(N_k),j} \frac{|\vec{u}_{(N_k)}|^2}{2}) = \operatorname{div}\left(\frac{|\vec{u}|^2}{2} \vec{u}\right)$$

in $\mathcal{D}'((0,T) \times \mathbb{R}^3)$.

- We now study

$$p_{(N_k)} = \frac{1}{\Delta} \operatorname{div} \vec{f} - \frac{1}{\Delta} \operatorname{div}\left(\vec{u}_{(N_k)} \cdot \vec{\nabla} u_{(N_k)}\right).$$

We have, of course, $\lim_{k \to +\infty} \vec{u}_{(N_k)} \cdot \frac{1}{\Delta} \operatorname{div} \vec{f} = \vec{u} \cdot \frac{1}{\Delta} \operatorname{div} \vec{f}$ in $\mathcal{D}'((0,T) \times \mathbb{R}^3)$. We know that $u_{(N_k),j} \vec{u}_{(N_k)}$ is bounded in $L^{3/2} L^{3/2}$ and converges strongly in $(L^{3/2} L^{3/2})_{\text{loc}}$ to $u_j \vec{u}$, hence we have weak convergence of $\sum_{j=1}^3 \frac{1}{\Delta} \operatorname{div} \partial_j (u_{(N_k),j} \vec{u}_{(N_k)})$ to $\sum_{j=1}^3 \frac{1}{\Delta} \operatorname{div} \partial_j (u_j \vec{u})$ in $(L^{3/2} L^{3/2})_{\text{loc}}$. As $\vec{u}_{(N_k)}$ converges strongly in $(L^3 L^3)_{\text{loc}}$, we find that $\lim_{k \to +\infty} \vec{u}_{(N_k)} \frac{1}{\Delta} \operatorname{div}\left(\vec{u}_{(N_k)} \cdot \vec{\nabla} \vec{u}_{(N_k)}\right) = \vec{u}(\sum_{j=1}^3 \frac{1}{\Delta} \operatorname{div} \partial_j (u_j \vec{u}))$ in $\mathcal{D}'((0,T) \times \mathbb{R}^3)$. Thus, we find that

$$\lim_{k \to +\infty} \operatorname{div}(p_{(N_k)} \vec{u}_{(N_k)}) = \operatorname{div}(p \vec{u})$$

in $\mathcal{D}'((0,T) \times \mathbb{R}^3)$

The trouble comes, however, from the very last term. While we have

$$\lim_{k \to +\infty} \|(\mathbb{P} - P_{N_k})\vec{f}\|_{L^2 H^{-1}} = 0$$

so that

$$\lim_{k \to +\infty} \vec{u}_{(N_k)} \cdot (\mathbb{P} - P_{N_k})\vec{f} = 0$$

in $\mathcal{D}'((0,T) \times \mathbb{R}^3)$, the behavior of $\vec{u}_{(N_k)} \cdot (\mathbb{P} - P_{N_k})(\vec{u}_{(N_k)} \cdot \vec{\nabla}\vec{u}_{(N_k)})$ is far from being clear. For some specific choices of the basis $(\vec{w}_n)$, it is however possible to show that

$$\lim_{k \to +\infty} \vec{u}_{(N_k)} \cdot (\mathbb{P} - P_{N_k})(\vec{u}_{(N_k)} \cdot \vec{\nabla}\vec{u}_{(N_k)}) = 0$$

in $\mathcal{D}'((0,T) \times \mathbb{R}^3)$ so that $\vec{u}$ is suitable. We are going to prove it in the case of a wavelet basis (see Guermond [220]).

A Faedo–Galerkin wavelet scheme

Theorem 18.2.
Let $\vec{u}_0 \in L^2$, with div $\vec{u}_0 = 0$, and $\vec{f} \in L^2_t H^{-1}_x$ on $(0,T) \times \mathbb{R}^3$ $(T < +\infty)$. Assume that the basis $(\vec{w}_n)_{n \in \mathbb{N}}$ in Theorem 18.1 is a Battle–Federbush Hilbertian basis of C^2 divergence-free vector wavelets. Then the solutions provided by the Faedo–Galerkin approximation scheme are suitable weak solutions of the Navier–Stokes problem.

Proof. Let us recall that the divergence-free vector wavelets described in Battle and Feder-bush [26] are a Hilbertian basis $(\vec{\psi}_{\alpha,j,k})_{\alpha \in \mathcal{A}, j \in \mathbb{Z}, k \in \mathbb{Z}^3}$ of $L^2_\sigma = \{\vec{w} \in L^2 \ / \ \text{div}\,\vec{w} = 0\}$ of the form:

$$\vec{\psi}_{\alpha,j,k}(x) = 2^{3j/2}\vec{\psi}_\alpha(2^j x - k)$$

where $\mathcal{A}$ is a finite set, the $\vec{\psi}_\alpha$ are regular (C^2 or more) and have exponential decay as well as their first and second derivatives, and have vanishing moments up to order 2 ($\int \vec{\psi}_\alpha x_1^{k_1} x_2^{k_2} x_3^{k_3}\, dx = 0$ for $k_1 + k_2 + k_3 \leq 2$).

Let $\vec{\phi} \in \mathcal{D}((0,T) \times \mathbb{R}^3)$; we are going to show that $\vec{w} \mapsto (\mathbb{P} - P_N)(\phi\vec{w})$ is bounded from $L^\infty X_N$ to $L^\infty H^{1/2}$ with an operator norm that is $o(1)$. As we have

$$\langle \phi | \vec{u}_{(N_k)} \cdot (\mathbb{P} - P_{N_k}))(\vec{u}_{(N_k)} \cdot \vec{\nabla}\vec{u}_{(N_k)}) \rangle_{\mathcal{D},\mathcal{D}'} =$$
$$\langle (\mathbb{P} - P_{N_k})(\phi\vec{u}_{(N_k)}) | \vec{u}_{(N_k)} \cdot \vec{\nabla}\vec{u}_{(N_k)} \rangle_{L^\infty H^{1/2}, L^1 H^{-1/2}},$$

we shall get

$$|\langle \phi | \vec{u}_{(N_k)} \cdot (\mathbb{P} - P_{N_k})(\vec{u}_{(N_k)} \cdot \vec{\nabla}\vec{u}_{(N_k)}) \rangle_{\mathcal{D},\mathcal{D}'}| \leq o(1)\|\vec{u}_{N_k}\|_{L^\infty L^2}\|\vec{u}_{(N_k)}\|^2_{L^2 H^1}$$

so that

$$\lim_{k \to +\infty} \vec{u}_{(N_k)} \cdot (\mathbb{P} - P_{N_k})(\vec{u}_{(N_k)} \cdot \vec{\nabla}\vec{u}_{(N_k)}) = 0$$

in $\mathcal{D}'((0,T) \times \mathbb{R}^3)$ and $\vec{u}$ is suitable.

Let $Y_N = \{\vec{\omega}_n \ / \ 0 \leq n \leq N\}$. We have, for $\vec{w} \in L^\infty L^2_\sigma$,

$$(\mathbb{P} - P_N)(\phi \vec{w}) = \sum_{\vec{\psi}_{\alpha,j,k} \notin Y_N} \left(\int \phi(t,y) \vec{w}(t,y) \cdot \vec{\psi}_{\alpha,j,k}(y) \, dy \right) \vec{\psi}_{\alpha,j,k}(x)$$

and

$$\|(\mathbb{P} - P_N)(\phi \vec{w})\|^2_{H^{1/2}} \leq C \sum_{\vec{\psi}_{\alpha,j,k} \notin Y_N} |\int \phi(t,y) \vec{w}(t,y) \cdot \vec{\psi}_{\alpha,j,k}(y) \, dy|^2 (1 + 2^j)$$

We split the last sum in blocks

$$S_i = \sum_{\vec{\psi}_{\alpha,j,k} \in Z_i} |\int \phi(t,y) \vec{w}(t,y) \cdot \vec{\psi}_{\alpha,j,k}(y) \, dy|^2 (1 + 2^j)$$

where the Z_i are defined in the following way:

- small frequencies: $\vec{\psi}_{\alpha,j,k} \in Z_1 \Leftrightarrow \vec{\psi}_{\alpha,j,k} \notin Y_N$ and $j < j_0 < 0$

- high frequencies: $\vec{\psi}_{\alpha,j,k} \in Z_2 \Leftrightarrow \vec{\psi}_{\alpha,j,k} \notin Y_N$ and $j > j_1 > 1$

- remote localization: $\vec{\psi}_{\alpha,j,k} \in Z_3 \Leftrightarrow \vec{\psi}_{\alpha,j,k} \notin Y_N, j_0 \leq j \leq j_1$ and $|2^{-j}k| > R$

- close localization: $\vec{\psi}_{\alpha,j,k} \in Z_4 \Leftrightarrow \vec{\psi}_{\alpha,j,k} \notin Y_N, j_0 \leq j \leq j_1$ and $|2^{-j}k| \leq R$

For S_1, we just write

$$S_1 \leq 2 \sum_{\alpha \in \mathcal{A}, j < j_0, k \in \mathbb{Z}^3} |\int \phi(t,y) \vec{w}(t,y) \cdot \vec{\psi}_{\alpha,j,k}(y) \, dy|^2$$

$$\leq 2\|\vec{w}(t,.)\|^2_2 \sum_{\alpha \in \mathcal{A}, j < j_0, k \in \mathbb{Z}^3} |\int |\phi(t,y)|^2 |\vec{\psi}_{\alpha,j,k}(y)|^2 dy$$

$$= 2\|\vec{w}(t,.)\|^2_2 \int |\phi(t,y)|^2 \left(\sum_{\alpha \in \mathcal{A}, j < j_0, k \in \mathbb{Z}^3} 2^{3j} |\vec{\psi}_\alpha(2^j y - k|^2 \right) dy$$

$$\leq C\|\phi(t,.)\|^2_2 \|\vec{w}(t,.)\|^2_2 2^{3j_0}$$

For S_2, we use the fact that, when $\vec{w}(t,.) \in X_N$ and $n > N$, we have $\int \vec{w}(t,y) \cdot \vec{w}_n(y) \, dy = 0$, so that (for $\vec{w} \in L^\infty X_N$)

$$S_2 \leq 2 \sum_{\alpha \in \mathcal{A}, j > j_1, k \in \mathbb{Z}^3} 2^j |\int (\phi(t,y) - \phi(t,k/2^j)) \vec{w}(t,y) \cdot \vec{\psi}_{\alpha,j,k}(y) \, dy|^2$$

$$\leq 2 \sum_{\alpha \in \mathcal{A}, j > j_1, k \in \mathbb{Z}^3} 2^j |\int |\phi(t,y) - \phi(t,k/2^j)|^2 |\vec{w}(t,y)|^2 |\vec{\psi}_{\alpha,j,k}(y)| \, dy 2^{-3j/2} \|\vec{\psi}_\alpha\|_1$$

$$\leq 2 \max_{\alpha \in \mathcal{A}} \|\vec{\psi}_\alpha\|_1 \|\vec{\nabla}\phi(t,.)\|^2_\infty \int |\vec{w}(t,y)|^2 \left(\sum_{\alpha \in \mathcal{A}, j > j_1, k \in \mathbb{Z}^3} 2^j |y - \frac{k}{2^j}|^2 |\vec{\psi}_\alpha(2^j y - k)| \right) dy$$

$$\leq C\|\vec{\nabla}\phi(t,.)\|^2_\infty \|\vec{w}(t,.)\|^2_2 2^{-j_1}$$

For S_3, if ϕ is supported by $B(0, R_0)$ and if $R > 2R_0$, we write

$$S_3 \leq 2 \sum_{\alpha \in \mathcal{A}, j_0 \leq j \leq j_1, |k| > R2^j} | \int \phi(t, y) \vec{w}(t, y) \cdot \vec{\psi}_{\alpha, j, k}(y) \, dy|^2$$

$$\leq 2\|\vec{w}(t, .)\|_2^2 \sum_{\alpha \in \mathcal{A}, j_0 \leq \leq j_1, |k| > R2^j} | \int |\phi(t, y)|^2 |\vec{\psi}_{\alpha, j, k}(y)|^2 dy$$

$$\leq C\|\vec{w}(t, .)\|_2^2 \int |\phi(t, y)|^2 \left(\sum_{j_0 \leq j \leq j_1, |k| > R2^j} 2^{3j} \frac{1}{|k|^6} \right) dy$$

$$\leq C\|\phi(t, .)\|_2^2 \|\vec{w}(t, .)\|_2^2 2^{3(j_1 - j_0)} \frac{1}{R^3}$$

Finally, the set $Z_5(j_0, j_1, R) = \{\vec{\psi}_{\alpha, j, k} \, / \, \alpha \in \mathcal{A}, j_0 \leq j \leq j_1 \text{ and } |2^{-j}k| \leq R\}$ is finite, so that $Z_5(j_0, j_1, R) \subset Y_N$ when N is large enough, so that $Z_4 = \emptyset$ and $S_4 = 0$.

Thus, we find that there is a constant such that, for every $j_0 < 0$, $j_1 > 1$, $R > 2R_0$, there exists a $N_0(j_0, j_1, R)$ such that for every $\vec{w} \in L^\infty X_N$ with $N > N_0(j_0, j_1, R)$, we have

$$\|(\mathbb{P} - P_N)(\phi\vec{w})\|_{L^\infty H^{1/2}} \leq$$
$$C\|\vec{w}\|_{L^\infty L^2} (2^{3j_0/2}\|\phi\|_{L^\infty L^2} + 2^{-j_1/2}\|\vec{\nabla}\phi\|_{L^\infty L^\infty} + 2^{3(j_1 - j_0)/2} R^{-3/2}\|\phi\|_{L^2 L^2}).$$

For every $\epsilon > 0$, taking j_0 close to $-\infty$, j_1 close to $+\infty$, and then R large enough, we can find $N_0(\epsilon)$ so that the operator norm of $\vec{w} \in L^\infty X_N \mapsto (\mathbb{P} - P_N)(\phi\vec{w}) \in L^\infty H^{1/2}$ is smaller than ϵ for every $N > N_0(\epsilon)$. Thus, the theorem is proved. $\square$

18.2 Frequency Cut-off

In the preceding section, we did not actually need that the spaces X_N were finite dimensional; we need in fact only good approximation and regularity properties for X_N:

A generalized Faedo–Galerkin approximation scheme

Theorem 18.3.
Let $\vec{u}_0 \in L^2$, with div $\vec{u}_0 = 0$, and $\vec{f} \in L^2_t H^{-1}_x$ on $(0, T) \times \mathbb{R}^3$ $(T < +\infty)$.

- Let (X_N) be a sequence of closed subspaces of L^2_σ, and P_N the orthonormal projection operator on X_N. Assume moreover that P_N satisfies

  - (H1) X_N is continuously embedded into H^1.
  - (H2) P_N can be extended as a bounded operator from H^{-1} to X_N.

  Then, for $N \in \mathbb{N}$, the problem

  $$\partial_t \vec{u} = P_N(\nu \Delta \vec{u} - \vec{u} \cdot \vec{\nabla}\vec{u} + \vec{f}) \qquad (18.3)$$

  with initial value $\vec{u}(0, .) = P_N \vec{u}_0$ has a unique solution $\vec{u}_{(N)}$.

- This solution $\vec{u}_{(N)}$ satisfies, with uniform bounds with respect to N, that $\vec{u}_{(N)} \in L^\infty_t L^2_x \cap L^2_t H^1_x$

- *Assume moreover that $\vec{w}_n$ satisfies*

  - *(H3) $X_N \subset X_{N+1}$ and the space $\cup_{N\in\mathbb{N}} X_N$ is dense in L^2_σ*
  - *(H4) the operators P_N are equicontinuous from H^{-1} to H^{-1} and from H^1 to H^1*

  Then there exists a sequence $N_k \to +\infty$ and a function $\vec{u} \in L^\infty_t L^2_x \cap L^2_t H^1_x$ such that $\vec{u}_{(N_k)}$ converges to $\vec{u}$ in $\mathcal{D}'((0,T) \times \mathbb{R}^3)$. Moreover $\vec{u}$ is a Leray weak solution of the Navier–Stokes problem

  $$\partial_t \vec{u} + \mathbb{P}(\vec{u} \cdot \vec{\nabla}\vec{u}) = \nu\Delta\vec{u} + \mathbb{P}\vec{f}$$

  with initial value $\vec{u}_0$.

- *If $\vec{u}_0 \in H^1$ and $\vec{f} \in L^2_t L^2_x$ and if the Navier–Stokes problem has a solution $\vec{u} \in \mathcal{C}([0,T], H^1) \cap L^2([0,T], H^2)$, then we have*

  $$\lim_{N\to+\infty} \|\vec{u}_{(N)} - \vec{u}\|_{L^\infty_t L^2_x} = 0.$$

Examples: Classical examples of such operators P_N are:

- The frequency cut-off: $P_N = Q_N \mathbb{P}$, where $\mathcal{F}(Q_N f)(\xi) = 1_{|\xi|<2^N}\hat{f}(\xi)$ and $\mathbb{P}$ is the Leray projection operator

- For a Battle–Federbush Hilbertian basis of $\mathcal{C}^2$ divergence-free vector wavelets $(\vec{\psi}_{\alpha,j,k})_{\alpha\in\mathcal{A},j\in\mathbb{Z},k\in\mathbb{Z}^3}$, X_N is the closed linear span of $(\vec{\psi}_{\alpha,j,k})_{\alpha\in\mathcal{A},j<N,k\in\mathbb{Z}^3}$

Proof. • **First step: Global existence of $\vec{u}_{(N)}$.**

We have $P_N\vec{f} \in L^2 X_N$. Moreover the linear operator $\vec{w} \mapsto \nu P_N(\Delta\vec{w})$ and the bilinear operator $(\vec{v}, \vec{w}) \mapsto P_N(\vec{v} \cdot \vec{\nabla}\vec{w})$ are bounded on X_N. Thus, if F is defined on $\mathcal{C}([0,T_0], X_N)$ as

$$F(\vec{w}) = P_N(\vec{u}_0) + \int_0^t \nu P_N(\Delta\vec{w}) - P_N(\vec{w} \cdot \vec{\nabla}\vec{w}) + P_N f\, ds$$

is a contraction on $\{\vec{w} \in \mathcal{C}([0,T_0], X_N) \ / \ \sup_{0<t<T_0}\|\vec{w}_N(t,.)\|_2 \leq 2(\|P_N\vec{u}_0\|_2 + \int_0^T \|P_N\vec{f}\|_2\, ds)\}$ for T_0 small enough:

$$T_0 \leq C_N \min(1, \frac{1}{\|P_N\vec{u}_0\|_2 + \int_0^T \|P_N\vec{f}\|_2\, ds}).$$

This ensures the existence of a (unique) solution $\vec{u}_{(N)}$ of (18.3) on $[0,T_0]$.

To show the existence of a global solution to (18.1), it is then enough to show that the L^2 norm of $\vec{u}_{(N)}$ remains bounded (as the existence time T_0 is controlled by the L^2 norm of the Cauchy data).

We write

$$\frac{d}{dt}\left(\frac{\|\vec{u}_{(N)}\|_2^2}{2}\right) = \langle \partial_t \vec{u}_{(N)} | \vec{u}_{(N)} \rangle_{H^{-1},H^1}$$

$$= \langle \nu \Delta \vec{u}_{(N)} - \vec{u}_{(N)} \cdot \vec{\nabla} \vec{u}_{(N)} + \vec{f} | \vec{u}_{(N)} \rangle_{H^{-1},H^1}$$

$$= -\nu \|\vec{\nabla} \otimes \vec{u}_{(N)}\|_2^2 + \langle \vec{f} | \vec{u}_{(N)} \rangle_{H^{-1},H^1}$$

$$\leq -\frac{\nu}{2} \|\vec{\nabla} \otimes \vec{u}_{(N)}\|_2^2 + \frac{1}{2\nu} \|\vec{f}\|_{H^{-1}}^2 + \frac{\nu}{2} \|\vec{u}_{(N)}\|_2^2$$

so that

$$\|\vec{u}_{(N)}\|_2^2 + \nu \int_0^t \|\vec{\nabla} \otimes \vec{u}_{(N)}\|_2 \, ds \leq e^{\nu t}\left(\|\vec{u}_0\|_2^2 + \frac{1}{\nu} \int_0^t \|\vec{f}\|_{H^{-1}}^2 \, ds\right) \qquad (18.4)$$

Thus, we get global existence of $\vec{u}_{(N)}$.

- **Second step: Weak convergence.**

From the energy estimate (18.4), we know that $\vec{u}_{(N)}$ remains bounded in $L^\infty L^2 \cap L^2 \dot{H}^1$ independently from N. We have

$$\partial_t \vec{u}_{(N)} = P_N(\nu \Delta \vec{u}_{(N)} + \vec{f} - \operatorname{div}(\vec{u}_{(N)} \otimes \vec{u}_{(N)}))$$

where

$$\|\nu \Delta \vec{u}_{(N)} + \vec{f} - \operatorname{div}(\vec{u}_{(N)} \otimes \vec{u}_{(N)})\|_{L^{4/3}H^{-1}} \leq$$

$$\nu \sqrt{T} \|\vec{u}_{(N)}\|_{L^2 H^1} + \sqrt{T} \|\vec{f}\|_{L^2 H^{-1}} + C\|\vec{u}_{(N)}\|_{L^\infty L^2}^{1/2} \|\vec{u}_{(N)}\|_{L^2 H^1}^{3/2}).$$

By assumption, the P_N are equicontinuous on H^{-1}. Thus, we can see that $\partial_t \vec{u}_{(N)}$ remains bounded in $L^{4/3}H^{-1}$ (independently from N). We may then use the Rellich–Lions theorem (Theorem 12.1) and find a sequence $N_k \to +\infty$ and a function $\vec{u} \in L^\infty L^2 \cap L^2 \dot{H}^1$ such that:

- $\vec{u}_{(N_k)}$ is *-weakly convergent to $\vec{u}$ in $L^\infty L^2$ and in $L^2 \dot{H}^1$

- $\vec{u}_{(N_k)}$ is strongly convergent to $\vec{u}$ in $L^2_{\text{loc}}((0,T) \times \mathbb{R}^3)$.

If $\vec{w} \in L^2$, we have $P_N \vec{w} = P_N \mathbb{P} \vec{w}$ and $\lim_{N \to +\infty} \|P_N \vec{w} - \mathbb{P}\vec{w}\|_2 = 0$. By equicontinuity of the P_N on H^1, we find that if $\vec{w} \in H^1$, then $\lim_{N \to +\infty} \|P_N \vec{w} - \mathbb{P}\vec{w}\|_{H^1} = 0$. Similarly, if $\vec{w}(t,x) \in L^4 H^1$, we have $\lim_{N \to +\infty} \|P_N \vec{w} - \mathbb{P}\vec{w}\|_{L^4 H^1} = 0$.

This gives that the *-weak convergence of $\nu \Delta \vec{u}_{(N)} + \vec{f} - \operatorname{div}(\vec{u}_{(N)} \otimes \vec{u}_{(N)})$ to $\nu \Delta \vec{u} + \vec{f} - \operatorname{div}(\vec{u} \otimes \vec{u})$ in $L^{4/3}H^{-1}$ implies the *-weak convergence of $\partial_t \vec{u}_{(N)} = P_N(\nu \Delta \vec{u}_{(N)} + \vec{f} - \operatorname{div}(\vec{u}_{(N)} \otimes \vec{u}_{(N)}))$ to $\nu \Delta \vec{u} + \mathbb{P}(\vec{f} - \operatorname{div}(\vec{u} \otimes \vec{u}))$.

Thus, the weak limit $\vec{u}$ satisfies

$$\partial_t \vec{u} = \nu \Delta \vec{u} + \mathbb{P}(\vec{f} - \vec{u} \cdot \vec{\nabla} \vec{u}).$$

- **Third step: Global energy estimates for the weak limit.**

The proof is similar to the proof of Theorem 18.1.

- **Fourth step: Strong convergence.**

The proof is similar to the proof of Theorem 18.1.   □

As for the case of the Faedo–Galerkin algorithm, wavelets provide a good frame to find suitable solutions:

Wavelet approximations

Theorem 18.4.
Let $\vec{u}_0 \in L^2$, with $\operatorname{div} \vec{u}_0 = 0$, and $\vec{f} \in L^2_t H^{-1}_x$ on $(0, T) \times \mathbb{R}^3$ ($T < +\infty$). Assume that the projection operators $(P_N)_{N \in \mathbb{N}}$ in Theorem 18.3 are the projection on the large scale wavelets of a Battle–Federbush Hilbertian basis:

$$P_N(\vec{w})(x) = \sum_{\alpha \in \mathcal{A}, j < N, k \in \mathbb{Z}^3} \left(\int \vec{w}(y) \cdot \vec{\psi}_{\alpha,j,k}(y) \, dy \right) \vec{\psi}_{\alpha,j,k}(x)$$

Then the weak solutions provided by Theorem 18.3 are suitable weak solutions of the Navier–Stokes problem.

Proof. As for Theorem 18.2, it is enough to prove that, for $\phi \in \mathcal{D}((0, T) \times \mathbb{R}^3)$, we have that $\vec{w} \mapsto (\mathbb{P} - P_N)(\phi \vec{w})$ is bounded from $L^\infty X_N$ to $L^\infty H^{1/2}$ with an operator norm that is $o(1)$.

We have

$$\|(\mathbb{P} - P_N)(\phi \vec{w})\|^2_{H^{1/2}} \leq C \sum_{\alpha \in \mathcal{A}, j \geq N, k \in \mathbb{Z}^3} 2^j \left| \int \phi(t, y) \vec{w}(t, y) \cdot \vec{\psi}_{\alpha,j,k}(y) \, dy \right|^2$$

We have only high frequencies in the sum (it is the term S_2 in the proof of Theorem 18.2), hence we get

$$\|(\mathbb{P} - P_N)(\phi \vec{w})\|_{L^\infty H^{1/2}} \leq C 2^{-N/2} \|\vec{\nabla}\phi\|_{L^\infty L^\infty} \|\vec{w}\|_{L^\infty L^2}.$$

Thus, the theorem is proved. $\qquad \square$

18.3 Hyperviscosity

Beirão da Vega [28] proved that one could construct suitable solutions by using hyperviscosity. The idea is to modify the Navier–Stokes problem into

$$\partial_t \vec{u} = -\alpha \Delta^2 \vec{u} + \nu \Delta \vec{u} + \mathbb{P}(\vec{f} - \vec{u} \cdot \vec{\nabla} \vec{u})$$

for which one can prove existence of a unique global solution, then to let α go to 0 to exhibit suitable weak solutions of the Navier–Stokes problem.

Hyperviscous Navier–Stokes equations

Theorem 18.5.
Let $\vec{u}_0 \in L^2$, with $\operatorname{div} \vec{u}_0 = 0$, and $\vec{f} \in L^2_t H^{-1}_x$ on $(0, T) \times \mathbb{R}^3$ ($T < +\infty$). Then

- *for $\alpha > 0$, the problem*

$$\partial_t \vec{u} = -\alpha \Delta^2 \vec{u} + \nu \Delta \vec{u} + \mathbb{P}(\vec{f} - \vec{u} \cdot \vec{\nabla} \vec{u}) \tag{18.5}$$

with initial value $\vec{u}(0,.) = \vec{u}_0$ has a unique solution $\vec{u}_{(\alpha)}$.

- This solution $\vec{u}_{(\alpha)}$ is bounded in $L_t^\infty L_x^2 \cap L_t^2 H_x^1$ independently from α.

- there exists a sequence $\alpha_k \to 0$ and a function $\vec{u} \in L_t^\infty L_x^2 \cap L_t^2 H_x^1$ such that $\vec{u}_{(\alpha_k)}$ is weakly convergent to $\vec{u}$. Moreover $\vec{u}$ is a suitable Leray weak solution of the Navier–Stokes problem

$$\partial_t \vec{u} + \mathbb{P}(\vec{u} \cdot \vec{\nabla} \vec{u}) = \nu \Delta \vec{u} + \mathbb{P}\vec{f}$$

with initial value $\vec{u}_0$.

- If $\vec{u}_0 \in H^1$ and $\vec{f} \in L_t^2 L_x^2$ and if the Navier–Stokes problem has a solution $\vec{u} \in \mathcal{C}([0,T], H^1) \cap L^2([0,T], H^2)$, then we have

$$\lim_{\alpha \to 0} \|\vec{u}_{(\alpha)} - \vec{u}\|_{L_t^\infty L_x^2} = 0.$$

Proof.   • **First step: Local existence of $\vec{u}_{(\alpha)}$.**

Let $e^{-t\Delta^2}$ be the operator defined by

$$\mathcal{F}(e^{-t\Delta^2} g)(\xi) = e^{-t|\xi|^4} \hat{g}(\xi).$$

We have the following estimates:

$$\|e^{-\alpha t \Delta^2} g\|_2 = \|g\|_2$$

$$\int_0^{+\infty} \|e^{-\alpha t \Delta^2} g\|_{\dot{H}^2}^2 \, dt = \frac{1}{2\alpha} \|g\|_2^2$$

$$\|e^{-\alpha t \Delta^2} g\|_{L^2} \le C(1 + \frac{1}{(\alpha t)^{1/4}}) \|\vec{g}\|_{H^{-1}}$$

$$\|e^{-\alpha t \Delta^2} g\|_{\dot{H}^2} \le (1 + \frac{1}{(\alpha t)^{3/4}}) \|\vec{g}\|_{H^{-1}}$$

$$\| \int_0^t e^{-\alpha(t-s)\Delta^2} G(s,.) \, ds \|_{L^\infty((0,T_0), L^2)} \le \frac{1}{\sqrt{2\alpha}} \|G\|_{L^2((0,T_0), \dot{H}^{-2})}$$

$$\| \int_0^t e^{-\alpha(t-s)\Delta^2} G(s,.) \, ds \|_{L^2((0,T_0), \dot{H}^2)} \le \frac{1}{\alpha} \|G\|_{L^2((0,T_0), \dot{H}^{-2})}$$

$$\| \int_0^t e^{-\alpha(t-s)\Delta^2} G(s,.) \, ds \|_{L^\infty((0,T_0), L^2)} \le C T_0^{1/4} (T_0^{1/4} + \frac{1}{\alpha^{1/4}}) \|G\|_{L^2((0,T_0), H^{-1})}$$

$$\| \int_0^t e^{-\alpha(t-s)\Delta^2} G(s,.) \, ds \|_{L^2((0,T_0), H^2)} \le C T_0^{1/4} (T_0^{3/4} + \frac{1}{\alpha^{3/4}}) \|G\|_{L^2((0,T_0), H^{-1})}$$

We want to prove existence (and uniqueness) of a solution $\vec{u} = \vec{u}_{(\alpha)}$ in $L^\infty((0,T_0), L^2) \cap L^2(0,T_0), H^2)$ of the equation

$$\vec{u} = e^{-\alpha t \Delta^2} \vec{u}_0 + \int_0^t e^{-\alpha(t-s)\Delta^2} (\mathbb{P}(\vec{f} + \nu \Delta \vec{u} - \vec{u} \cdot \vec{\nabla} \vec{u}) \, ds \tag{18.6}$$

for T_0 small enough.

For $\vec{u}$ and $\vec{v}$ in $L^\infty L^2 \cap L^2 H^2$ with div $\vec{u} = 0$, we have for every $0 < T_0 < T$,

$$\|\vec{u}\cdot\vec{\nabla}\vec{v}\|_{L^2((0,T_0),H^{-1})} \leq C(\|\vec{u}\|_{L^\infty L^2}\|\vec{v}\|_{L^2 H^2} + \|\vec{u}\|_{L^2 H^2}\|\vec{v}\|_{L^\infty L^2})$$

and

$$\|\Delta\vec{u}\|_{L^2((0,T_0),\dot{H}^{-2})} \leq C\sqrt{T_0}\|\vec{u}\|_{L^\infty L^2}$$

Let $\|\vec{u}\|_{\alpha,T_0} = \|\vec{u}\|_{L^\infty((0,T_0),L^2)} + \sqrt{\alpha}\|\vec{u}\|_{L^2(0,T_0),\dot{H}^2)}$. We have

$$\|e^{-\alpha t\Delta^2}\vec{u}_0 + \int_0^t e^{-\alpha(t-s)\Delta^2}\mathbb{P}\vec{f}\,ds\|_{\alpha,T_0} \leq C_0(\|\vec{u}_0\|_2 + \frac{T_0^{1/4}}{\alpha^{1/4}}(1+(\alpha T_0)^{3/4})\|\vec{f}\|_{L^2 H^{-1}})$$

$$\nu\|\int_0^t e^{-\alpha(t-s)\Delta^2}\Delta\vec{u}\,ds\|_{\alpha,T_0} \leq C_0\nu\frac{\sqrt{T_0}}{\sqrt{\alpha}}\|\vec{u}\|_{L^\infty L^2}$$

and

$$\|\int_0^t e^{-\alpha(t-s)\Delta^2}\mathbb{P}(\vec{u}\cdot\vec{\nabla}\vec{v})\,ds\|_{\alpha,T_0} \leq C_0\frac{T_0^{1/4}}{\alpha^{1/4}}(\|\vec{u}\|_{L^\infty L^2}\|\vec{v}\|_{L^2 H^2} + \|\vec{u}\|_{L^2 H^2}\|\vec{v}\|_{L^\infty L^2})$$

Thus, we find existence (and uniqueness) of a solution $\vec{u} = \vec{u}_{(\alpha)}$ for T_0 small enough to ensure

$$\alpha T_0 \leq 1$$
$$T_0 < \frac{\alpha}{16 C_0^2 \nu^2}$$

and

$$T_0 \leq \alpha^3 \frac{1}{256\, C_0^8(\|\vec{u}_0\|_2 + \frac{2}{\sqrt{\alpha}}\|\vec{f}\|_{L^2\dot{H}^{-1}})^4}. \tag{18.7}$$

- **Second step: Energy estimates and global existence of $\vec{u}_{(\alpha)}$.**

To show the existence of a global solution to (18.5), it is then enough to show that the L^2 norm of $\vec{u}_{(\alpha)}$ remains bounded (as the existence time T_0 is controlled by the L^2 norm of the Cauchy data by (18.7)). As $\vec{u} \in L^2 H^2$ and $\partial_t\vec{u} \in L^2 H^{-2}$, we have

$$\frac{d}{dt}\|\vec{u}_{(\alpha)}\|_2^2 = 2\int \partial_t\vec{u}_{(\alpha)}\cdot\vec{u}_{(\alpha)}\,dx$$
$$= -2\alpha\|\Delta\vec{u}_{(\alpha)}\|_2^2 - 2\nu\|\vec{u}_{(\alpha)}\|_{\dot{H}^1}^2 + 2\langle\vec{f}|\vec{u}_{(\alpha)}\rangle_{H^{-1},H^1} \tag{18.8}$$
$$\leq -2\alpha\|\Delta\vec{u}_{(\alpha)}\|_2^2 - \nu\|\vec{u}_{(\alpha)}\|_{\dot{H}^1}^2 + \nu\|\vec{u}_{(\alpha)}\|_2^2 + \frac{1}{\nu}\|\vec{f}\|_{H^{-1}}^2$$

so that (by Grönwall's lemma)

$$\|\vec{u}_{(\alpha)}(t,.)\|_2^2 + 2\alpha\int_0^t\|\Delta\vec{u}_{(\alpha)}\|_2^2\,ds + \nu\int_0^t\|\vec{u}_{(\alpha)}\|_{\dot{H}^1}^2\,ds \leq$$
$$e^{\nu t}(\|\vec{u}_0\|_2^2 + \frac{1}{\nu}\|\vec{f}\|_{L^2 H^{-1}}^2) \tag{18.9}$$

- **Third step: Weak convergence.**

From the energy estimate (18.9), we know that $\vec{u}_{(\alpha)}$ remains bounded in $L^\infty L^2 \cap L^2 \dot{H}^1$ independently from α. Moreover, $\partial_t \vec{u}_{(\alpha)}$ remains bounded in $L^2 H^{-2}$ when $\alpha \in (0,1)$.

We may then use the Rellich–Lions theorem (Theorem 12.1) and find a sequence $\alpha_n \to 0$ and a function $\vec{u} \in L^\infty L^2 \cap L^2 \dot{H}^1$ such that:

- $\vec{u}_{(\alpha_n)}$ is *-weakly convergent to $\vec{u}$ in $L^\infty L^2$ and in $L^2 \dot{H}^1$

- $\vec{u}_{(\alpha_n)}$ is strongly convergent to $\vec{u}$ in $L^2_{loc}((0,T) \times \mathbb{R}^3)$.

We thus get that $-\alpha_n \Delta^2 \vec{u}_{(\alpha_n)} + \nu \Delta \vec{u}_{(\alpha_n)} + \mathrm{div}(\vec{u}_{(\alpha_n)} \otimes \vec{u}_{(\alpha_n)})$ converges to $\nu \Delta \vec{u} + \mathrm{div}(\vec{u} \otimes \vec{u})$ in $\mathcal{D}'((0,T) \times \mathbb{R}^3)$, hence *-weakly in $L^2 H^{-2}$; as $\mathbb{P}$ is bounded on H^{-2}, we get the *-weak convergence of $\mathbb{P}(-\alpha_n \Delta^2 \vec{u}_{(\alpha_n)} + \nu \Delta \vec{u}_{(\alpha_n)} + \mathrm{div}(\vec{u}_{(\alpha_n)} \otimes \vec{u}_{(\alpha_n)}))$ to $\mathbb{P}(\vec{u} \cdot \vec{\nabla} \vec{u})$ in $L^2 H^{-2}$.

Thus, the weak limit $\vec{u}$ satisfies

$$\partial_t \vec{u} = \nu \Delta \vec{u} + \mathbb{P}(\vec{f} - \vec{u} \cdot \vec{\nabla} \vec{u}).$$

- **Fourth step: Global energy estimates for the weak limit.**

We may easily check that the weak limit $\vec{u}$ fulfills the Leray energy inequality. Indeed, let $\omega(t)$ be a non-negative compactly supported smooth function on $\mathbb{R}$, supported within $[-\epsilon, \epsilon]$, with $\|\omega\|_1 = 1$. On $(\epsilon, T - \epsilon)$, we have

$$\|\omega * \vec{u}_{(\alpha)}(t,.)\|_2^2 \le \omega * \|\vec{u}_{(\alpha)}(t,.)\|_2^2$$

and

$$\omega * \|\vec{u}_{(\alpha)}(t,.)\|_2^2 + 2\nu\omega * \left(\int_0^t \|\vec{u}_{(\alpha)}\|_{\dot{H}^1}^2 \, ds\right) \le \|\vec{u}_0\|_2^2 + 2\omega * \left(\int_0^t \langle \vec{f}|\vec{u}_{(\alpha_n)}\rangle_{H^{-1},H^1} \, ds\right)$$

Besides, we have the *-weak convergence of $\omega * \vec{u}_{(\alpha_n)}$ to $\omega * \vec{u}$ in $L^2 \dot{H}^1$, the strong convergence of $\omega * \vec{u}_{(\alpha)}(t,.)$ to $\omega * \vec{u}(t,.)$ in $L^2_{loc}(\mathbb{R}^3)$, and thus (due to the control we have on the L^2 norm) the *-weak convergence of $\omega * \vec{u}_{(\alpha)}(t,.)$ to $\omega * \vec{u}(t,.)$ in $L^2(\mathbb{R}^3)$. We thus get that

$$\|\omega * \vec{u}(t,.)\|_2^2 + 2\nu\omega * \left(\int_0^t \|\vec{u}\|_{\dot{H}^1}^2 \, ds\right) \le$$

$$\liminf_{n\to+\infty} \|\omega*\vec{u}_{(\alpha_n)}(t,.)\|_2^2 + 2\nu\omega * \left(\int_0^t \|\vec{u}_{(\alpha_n)}\|_{\dot{H}^1}^2 \, ds\right)$$

and thus

$$\|\omega * \vec{u}(t,.)\|_2^2 + 2\nu\omega * \left(\int_0^t \|\vec{u}\|_{\dot{H}^1}^2 \, ds\right) \le \|\vec{u}_0\|_2^2 + 2\omega * \left(\int_0^t \langle \vec{f}|\vec{u}\rangle_{H^{-1},H^1} \, ds\right)$$

When t_0 is a Lebesgue point of $t \mapsto \|\vec{u}(t,.)\|_2$, we get that:

$$\|\vec{u}(t_0,.)\|_2^2 + 2\nu \int_0^{t_0} \|\vec{u}\|_{\dot{H}^1}^2 \, ds \le \|\vec{u}_0\|_2^2 + 2\int_0^{t_0} \langle \vec{f}|\vec{u}\rangle_{H^{-1},H^1} \, ds \qquad (18.10)$$

This inequality may be extended to every t_0, due to the weak boundedness of $t \in (0,T) \mapsto \vec{u}(t,.) \in L^2$.

- **Fifth step: Local energy estimates for the weak limit.**

We now check that $\vec{u}$ is more precisely a suitable weak solution (i.e., fulfills the local energy inequality). We write

$$\partial_t \vec{u}_{(\alpha)} = \nu\Delta\vec{u}_{(\alpha)} + \vec{f} - \vec{u}_{(\alpha)} \cdot \vec\nabla\,\vec{u}_{(\alpha)} - \alpha\Delta^2\vec{u}_{(\alpha)} - \vec\nabla p_{(\alpha)}$$

with

$$p_{(\alpha)} = \frac{1}{\Delta}\operatorname{div}\left(\vec{f} - \vec{u}_{(\alpha)} \cdot \vec\nabla\,\vec{u}_{(\alpha)}\right)$$

and we write

$$\partial_t\left(\frac{|\vec{u}_{(\alpha)}|^2}{2}\right) = \nu\Delta\left(\frac{|\vec{u}_{(\alpha)}|^2}{2}\right) - \nu|\vec\nabla\otimes\vec{u}_{(\alpha)}|^2 + \vec{f}\cdot\vec{u}_{(\alpha)}$$

$$- \operatorname{div}\left(\frac{|\vec{u}_{(\alpha)}|^2}{2}\vec{u}_{(\alpha)}\right) - \operatorname{div}(p_{(\alpha)}\vec{u}_{(\alpha)})$$

$$- \alpha\Delta(\vec{u}_{(\alpha)}\cdot\Delta\vec{u}_{(\alpha)}) + 2\alpha\sum_{j=1}^{3}\partial_j(\partial_j\vec{u}_{(\alpha)}\cdot\Delta\vec{u}_{(\alpha)}) - \alpha|\Delta\vec{u}_{(\alpha)}|^2$$

In the same way as for the Leray model, we find that we have the following convergence results in $\mathcal{D}'((0,T)\times\mathbb{R}^3)$: $\partial_t|\vec{u}_{(\alpha_n)}|^2 \to \partial_t|\vec{u}|^2$, $\Delta|\vec{u}_{(\alpha_n)}|^2 \to \Delta|\vec{u}|^2$, $\operatorname{div}(|\vec{u}_{(\alpha_n)}|^2\vec{u}_{(\alpha_n)}) \to \operatorname{div}(|\vec{u}|^2\vec{u})$, $\vec{u}_{(\alpha_n)}\cdot\vec{f} \to \vec{u}\cdot\vec{f}$ and $\operatorname{div}(p_{(\alpha_n)}\vec{u}_{(\alpha_n)}) \to \operatorname{div}(p\vec{u})$. Moreover $\sqrt{\alpha_n}\Delta\vec{u}_{(\alpha_n)}$ is bounded in L^2L^2, while $\vec{u}_{(\alpha_n)}$ is bounded in L^2H^1; thus we have the convergence $-\alpha_n\Delta(\vec{u}_{(\alpha_n)}\cdot\Delta\vec{u}_{(\alpha_n)}) + 2\alpha_n\sum_{j=1}^{3}\partial_j(\partial_j\vec{u}_{(\alpha_n)}\cdot\Delta\vec{u}_{(\alpha_n)}) \to 0$.
Thus far, we have got that

$$\partial_t|\vec{u}|^2 = \nu\Delta|\vec{u}|^2 - \operatorname{div}((|\vec{u}|^2 + 2p)\vec{u}) + 2\vec{u}\cdot\vec{f} - T$$

with

$$T = \lim_{\alpha_n\to 0} 2\nu|\vec\nabla\otimes\vec{u}_{(\alpha_n)}|^2 + 2\alpha|\Delta\vec{u}_{(\alpha_n)}|^2.$$

Let $\phi\in\mathcal{D}'((0,T)\times\mathbb{R}^3)$ be a non-negative function. We have

$$\langle T|\phi\rangle_{\mathcal{D}',\mathcal{D}} \geq 2\liminf_{\alpha_n\to 0}\iint|\vec\nabla\otimes\vec{u}_{(\alpha_n)}|^2\,\phi(t,x)\,dt\,d$$

$$\geq 2\iint|\vec\nabla\otimes\vec{u}|^2\phi(t,x)\,dt\,dx.$$

Thus, $T = 2|\vec\nabla\otimes\vec{u}|^2 + \mu$, where μ is a non-negative locally finite measure, and thus $\vec{u}$ is suitable.

- **Sixth step: Strong convergence.**

If $\vec{u}_0\in H^1$ and $\vec{f}\in L^2_t L^2_x$, and if $\vec{u}$ is a solution of the Navier–Stokes problem with $\vec{u}\in\mathcal{C}([0,T],H^1)\cap L^2([0,T],H^2)$, it is easy to check the strong convergence of $\vec{u}_{(\alpha)}$ to $\vec{u}$: as $\vec{u}$ and $\vec{u}_{(\alpha)}$ are regular enough, we may write

$$\frac{d}{dt}\|\vec{u} - \vec{u}_{(\alpha)}\|_2^2 = 2 \int (\vec{u} - \vec{u}_{(\alpha)}) \cdot \partial_t(\vec{u} - \vec{u}_{(\alpha)}) \, dx$$

$$= -2\nu\|\vec{\nabla} \otimes (\vec{u} - \vec{u}_{(\alpha)})\|_2^2 + 2\alpha\langle \vec{u} - \vec{u}_{(\alpha)}) | \Delta^2 \vec{u}_{(\alpha)} \rangle_{H^2, H^{-2}}$$

$$- 2 \int (\vec{u} - \vec{u}_{(\alpha)}).(\vec{u} \cdot \vec{\nabla}\vec{u} - \vec{u}_{(\alpha)} \cdot \vec{\nabla}\vec{u}_{(\alpha)}) \, dx$$

$$= -2\nu\|\vec{\nabla} \otimes (\vec{u} - \vec{u}_{(\alpha)})\|_2^2 - 2\alpha\|\Delta(\vec{u} - \vec{u}_{(\alpha)})\|_2^2$$

$$+ 2\alpha\langle \vec{u} - \vec{u}_{(\alpha)}) | \Delta^2 \vec{u} \rangle_{H^2, H^{-2}} - 2 \int (\vec{u} - \vec{u}_{(\alpha)}).((\vec{u} - \vec{u}_{(\alpha)}) \cdot \vec{\nabla}\vec{u}) \, dx$$

$$\leq -2\nu\|\vec{\nabla} \otimes (\vec{u} - \vec{u}_{(\alpha)})\|_2^2 - \alpha\|\Delta(\vec{u} - \vec{u}_{(\alpha)})\|_2^2$$

$$+ \alpha\|\Delta\vec{u}\|_2^2 + 2\|\vec{u} - \vec{u}_{(\alpha)}\|_6\|\vec{u} - \vec{u}_{(\alpha)}\|_2\|\vec{\nabla} \otimes \vec{u}\|_3$$

$$\leq \frac{C}{\nu}\|\vec{u} - \vec{u}_{(\alpha)}\|_2^2\|\vec{\nabla} \otimes \vec{u}\|_3^2 + \alpha\|\Delta\vec{u}\|_2^2$$

This gives

$$\|\vec{u} - \vec{u}_{(\alpha)}\|_{L^\infty L^2}^2 \leq \alpha \, e^{\int_0^T \frac{C}{\nu}\|\vec{\nabla} \otimes \vec{u}\|_3^2 \, ds} \int_0^T \|\Delta\vec{u}\|_2^2 \, ds$$

and thus $\lim_{\alpha \to 0} \sup_{0 < t < T} \|\vec{u} - \vec{u}_{(\alpha)}\|_2 = 0$. $\qquad\qquad\square$

18.4 Ladyzhenskaya's Model

As we saw in Section 2.7, the Navier–Stokes equations for an incompressible Stokesian fluid read as

$$\partial_t \vec{u} + \vec{u} \cdot \vec{\nabla}\vec{u} = \vec{f}_{visc} + \vec{f}_{ext} - \vec{\nabla}p$$

where the viscosity forces are given by the divergence of the stress tensor

$$\vec{f}_{visc} = \operatorname{div} C$$

and where, according to Serrin [431, 432] the viscous stress tensor can be expressed as

$$C = \alpha \, \mathbb{I}_3 + \beta \, \epsilon + \gamma\epsilon^2 \tag{18.11}$$

where α, β and γ are symmetric functions of the eigenvalues of the strain tensor $\epsilon = (\frac{1}{2}(\partial_i u_j + \partial_j u_i))_{1 \leq i,j \leq 3}$.

A special class of Stokesian fluids is given by the power-law Ansatz:

$$C = (2\nu + \alpha|\epsilon|^r)\epsilon \tag{18.12}$$

(with $\nu > 0$ and $\alpha > 0$). Nečas, Málek and co-workers [348] considered the general case with $r > -\frac{4}{5}$. For small values of r, one does find only measure-valued solutions (according to Bellout, Bloom and Nečas [35]), then, increasing r, weak solutions, that turn out to be strong solutions when r is large enough. They emphasize the fact that the case $r < 0$ which corresponds to shear thinning phenomena has more physical meaning than the case $r > 0$, which corresponds to shear thickening phenomena.

However, the original model proposed by Ladyzhenskaya in the late sixties as a way to modelize fluids with uniqueness and global existence [294], was based on $r > 0$ (as it tends to minimize the influence of high gradients of velocity). The case $r = 1$ was considered as well by Smagorinsky [440]. A similar model, with $|\vec{\nabla} \otimes \vec{u}|$ instead of $|\epsilon|$, was considered by Lions [337], but this model is not Stokesian and does not respect the principle of material indifference.

In this section, we consider only the easy case $r = 2$, study uniqueness and global existence for the Cauchy initial-value problem (for an initial value $\vec{u}_0 \in L^2$, with div $\vec{u}_0 = 0$, and a forcing term $\vec{f} \in L_t^2 L_x^2$ on $(0, T) \times \mathbb{R}^3$), and then convergence to the Navier–Stokes equations when α goes to 0.

Ladyzhenskaya's model

Theorem 18.6.

- Let $\alpha > 0$. Let $\vec{u}_0 \in L^2$, with div $\vec{u}_0 = 0$, and $\vec{f} \in L_t^2 L_x^2$ on $(0, T) \times \mathbb{R}^3$ $(T < +\infty)$. Then the problem

$$\partial_t \vec{u} + \mathbb{P}(\vec{u} \cdot \vec{\nabla}\, \vec{u}) = \mathbb{P}(\mathrm{div}(2\nu + \alpha |\epsilon|^2)\epsilon)) + \mathbb{P}\vec{f} \qquad (18.13)$$

  with initial value $\vec{u}(0, .) = \vec{u}_0$ has a unique solution $\vec{u}_{(\alpha)}$ such that $\vec{u}_{(\alpha)} \in L_t^\infty L_x^2 \cap L_t^2 H_x^1$, $\sqrt{t}\,\vec{u}_{(\alpha)} \in L_t^\infty H_x^1 \cap L_t^2 H_x^2$ and $\vec{\nabla} \otimes \vec{u}_{(\alpha)} \in L_t^4 L_x^4$.
  Moreover, we have the following inequality:

$$\|\vec{u}_{(\alpha)}(t, .)\|_2^2 + 2\nu \int_0^t \|\vec{\nabla} \otimes \vec{u}_{(\alpha)}(s, .)\|_2^2 + 2\alpha \int_0^t \|\epsilon(\vec{u}_{(\alpha)})\|_4^4 \, ds$$
$$\leq (\|\vec{u}(0, .)\|_2^2 + \int_0^T \|\vec{f}\|_2^2 \, ds)e^t. \qquad (18.14)$$

- there exists a sequence $\alpha_k \to 0$ and a function $\vec{u} \in L_t^\infty L_x^2 \cap L_t^2 H_x^1$ such that $\vec{u}_{\alpha_k}$ is weakly convergent to $\vec{u}$. Moreover $\vec{u}$ is a suitable weak solution of the Navier–Stokes problem

$$\partial_t \vec{u} + \mathbb{P}(\vec{u} \cdot \vec{\nabla}\vec{u}) = \nu \Delta \vec{u} + \mathbb{P}\vec{f}$$

  with initial value $\vec{u}_0$.

- If $\vec{u}_0 \in H^1$ and if the Navier–Stokes problem has a solution $\vec{u} \in L^\infty([0, T], H^1) \cap L^2([0, T], H^2)$, then we have $\lim_{\alpha \to 0} \|\vec{u}_\alpha - \vec{u}\|_{L_t^\infty L_x^2} = 0$.

Proof.   • **First step: Low frequency approximation.**
We first take a Fourier projection operator $\mathcal{P}$ defined by $\mathcal{F}(\mathcal{P}\varphi)(\xi) = \chi(\xi)\hat{\varphi}(\xi)$, where χ is the characteristic function of a bounded set K: $\chi(\xi) = 1$ when $\xi \in K$ and $\chi(\xi) = 0$ when $\xi \notin K$. In the following, constants that depend on K will be written with K as an index.
We consider the problem

$$\partial_t \vec{u} + \mathbb{P}\mathcal{P}(\vec{u} \cdot \vec{\nabla}\, \vec{u}) = \mathbb{P}\mathcal{P}(\mathrm{div}(2\nu + \alpha |\epsilon|^2)\epsilon)) + \mathbb{P}\mathcal{P}\vec{f} \qquad (18.15)$$

with initial value $\vec{u}(0, .) = \mathcal{P}\vec{u}_0$.

The solution is searched as a fixed point of the transform

$$\vec{u} \in \mathcal{C}([0,T], X) \mapsto \mathbb{P}W_{\nu t} * \vec{u}_0 + \mathbb{P}\mathbb{P}\int_0^t W_{\nu(t-s)} * \vec{f}(s,.)\,ds$$
$$- \mathcal{P}B(\vec{u}, \vec{u}) + \alpha \mathcal{P}T(\vec{u}, \vec{u}, \vec{u})$$

with $X = \mathbb{P}\mathcal{P}((L^2)^3)$,

$$B(\vec{u}, \vec{v}) = \int_0^t W_{\nu(s-t)} * \mathbb{P}\operatorname{div}(\vec{u} \otimes \vec{v})\,ds$$

and

$$T(\vec{u}, \vec{v}, \vec{w}) = \int_0^t W_{\nu(t-s)} * \mathbb{P}\operatorname{div}(\sum_{i=1}^3 \sum_{j=1}^3 \epsilon_{i,j}(\vec{u})\epsilon_{i,j}(\vec{v})\epsilon(\vec{w}))\,ds.$$

The bilinear transform $\mathcal{P}B$ and the trilinear transform $\mathcal{P}T$ are bounded on $\mathcal{C}([0,T], X)$:

$$\|\mathcal{P}B(\vec{u}, \vec{v})(t,.)\|_2 \le C_{1,K}\, t \sup_{0<s<t} \|\vec{u}(s,.)\|_2 \sup_{0<s<t} \|\vec{v}(s,.)\|_2$$

and

$$\|\mathcal{P}T(\vec{u}, \vec{v}, \vec{w})(t,.)\|_2 \le C_{2,K}\, t \sup_{0<s<t} \|\vec{u}(s,.)\|_2 \sup_{0<s<t} \|\vec{v}(s,.)\|_2 \sup_{0<s<t} \|\vec{w}(s,.)\|_2.$$

Thus, we shall have a solution of Equation (18.15) on an interval $[0, T_0]$ for $T_0 < T$ small enough to ensure

$$4C_{1,K}T_0(\|\vec{u}_0\|_2 + \sqrt{T_0}\|\vec{f}\|_{L^2 L^2}) + 12C_{2,K}\alpha T_0(\|\vec{u}_0\|_2 + \sqrt{T_0}\|\vec{f}\|_{L^2 L^2})^2 \le \frac{1}{2}$$

– for instance, we may choose

$$T_0 = \min(T, \frac{1}{24(C_{1,K} + C_{2,K}\alpha)(1 + \|\vec{u}_0\|_2 + \sqrt{T}\|\vec{f}\|_{L^2 L^2})^2}.$$

- **Step 2: Energy estimate.**

If $\vec{u}$ is a solution of (18.15) on $[0, T_0]$, we find that

$$\frac{d}{dt}\|\vec{u}\|_2^2 = 2\int \vec{u} \cdot \partial_t \vec{u}\,dx = 2\int \vec{u}.(-\vec{u} \cdot \vec{\nabla}\vec{u} + \operatorname{div}((2\nu + \alpha|\epsilon|^2)\epsilon) + \vec{f})\,dx$$

with

$$\int (\vec{u} \cdot \vec{\nabla}\vec{u}) \cdot \vec{u}\,dx = 0$$

(since $\operatorname{div}\vec{u} = 0$),

$$2\int \vec{u} \cdot \operatorname{div}\epsilon\,dx = \int \vec{u} \cdot \Delta\vec{u}\,dx = -\int |\vec{\nabla} \otimes \vec{u}|^2\,dx$$

and

$$\int \vec{u} \cdot \operatorname{div}(|\epsilon|^2\epsilon)\,dx = -\int |\epsilon|^4\,dx$$

Thus we obtain, for $0 < t < T_0$,

$$\|\vec{u}(t,.)\|_2^2 + 2\nu \int_0^t \|\vec{\nabla} \otimes \vec{u}(s,.)\|_2^2 + 2\alpha \int_0^t \|\epsilon(\vec{u})\|_4^4\,ds =$$

$$\|\vec{u}(0,.)\|_2^2 + 2\int_0^t |\int \vec{u} \cdot \vec{f}\,dx|\,ds$$

This gives

$$\|\vec{u}(t,.)\|_2^2 \leq \|\vec{u}(0,.)\|_2^2 + \int_0^T \|\vec{f}\|_2^2 \, ds + \int_0^t \|\vec{u}(s,.)\|_2^2 \, ds$$

hence

$$\|\vec{u}(0,.)\|_2^2 + \int_0^T \|\vec{f}\|_2^2 \, ds + \int_0^t \|\vec{u}(s,.)\|_2^2 \, ds \leq (\|\vec{u}(0,.)\|_2^2 + \int_0^T \|\vec{f}\|_2^2 \, ds)e^t$$

and finally

$$\|\vec{u}(t,.)\|_2^2 + 2\nu \int_0^t \|\vec{\nabla} \otimes \vec{u}(s,.)\|_2^2 + 2\alpha \int_0^t \|\epsilon(\vec{u})\|_4^4 \, ds$$

$$\leq (\|\vec{u}(0,.)\|_2^2 + \int_0^T \|\vec{f}\|_2^2 \, ds)e^t. \tag{18.16}$$

Thus, the L^2 norm of $\vec{u}$ cannot blow up and the solution $\vec{u}$ may be extended on the whole interval $[0, T]$.

- **Step 3: Further energy estimates.**

For a vector field $\vec{u}$, we have

$$\Delta \vec{u} = 2 \operatorname{div} \epsilon(\vec{u}) - \vec{\nabla}(\operatorname{div} \vec{u}),$$

so that

$$\partial_j u_k = \frac{\partial_j}{\sqrt{-\Delta}} \frac{\partial_k}{\sqrt{-\Delta}} \operatorname{div} \vec{u} - 2 \sum_{l=1}^3 \frac{\partial_j}{\sqrt{-\Delta}} \frac{\partial_l}{\sqrt{-\Delta}} \epsilon_{l,k}.$$

The boundedness of the Riesz transforms on L^p for $1 < p < +\infty$ gives then Korn's inequality

$$\|\vec{\nabla} \otimes \vec{u}\|_p \leq C_p(\|\epsilon(\vec{u})\|_p + \|\operatorname{div} \vec{u}\|_p). \tag{18.17}$$

Thus, we find from inequality (18.16) and from $\operatorname{div} \vec{u} = 0$ that the solution $\vec{u}$ of (18.15) satisfies

$$\int_0^T \|\vec{\nabla} \otimes \vec{u}\|_4^4 \, dt \leq C \frac{1}{\alpha}(\|\vec{u}(0,.)\|_2^2 + \int_0^T \|\vec{f}\|_2^2 \, ds)e^T. \tag{18.18}$$

Moreover, we have

$$\frac{d}{dt}\|\vec{\nabla} \otimes \vec{u}\|_2^2 = -2 \int \Delta \vec{u} \cdot \partial_t \vec{u} \, dx$$

$$= -2 \int \Delta \vec{u}.(-\vec{u} \cdot \vec{\nabla} \vec{u} + \operatorname{div}((2\nu + \alpha|\epsilon|^2)\epsilon) + \vec{f}) \, dx$$

$$= -2\nu\|\Delta \vec{u}\|_2^2 - 2\alpha \sum_{j=1}^3 \int \partial_j \epsilon \cdot \partial_j(|\epsilon|^2 \epsilon) \, dx + 2 \int \Delta \vec{u}.(\vec{u} \cdot \vec{\nabla} \vec{u}) \, dx - 2 \int \Delta \vec{u} \cdot \vec{f} \, dx.$$

We have

$$\int \partial_j \epsilon \cdot \partial_j(|\epsilon|^2 \epsilon) \, dx = \int |\epsilon|^2 |\partial_j \epsilon|^2 + \frac{1}{2} \left(\partial_j(|\epsilon|^2)\right)^2 \, dx$$

so that

$$\sum_{j=1}^3 \int \partial_j \epsilon \cdot \partial_j(|\epsilon|^2 \epsilon) \, dx \geq \frac{1}{2}\||\epsilon|^2\|_{\dot{H}^1}^2 \geq \frac{\gamma_0}{2}\|\epsilon\|_{12}^4$$

(where γ_0 is a positive constant) so that

$$\frac{d}{dt}\|\vec{\nabla}\otimes\vec{u}\|_2^2 \leq -\nu\|\Delta\vec{u}\|_2^2 - \alpha\gamma_0\|\epsilon\|_{12}^4 + \frac{2}{\nu}\|\vec{f}\|_2^2 + \frac{2}{\nu}\|\vec{u}\cdot\vec{\nabla}\vec{u}\|_2^2.$$

Moreover, we have

$$\frac{2}{\nu}\|\vec{u}\cdot\vec{\nabla}\vec{u}\|_2^2 \leq \frac{2}{\nu}\|\vec{u}\|_3^2\|\vec{\nabla}\otimes\vec{u}\|_6^2 \leq C\frac{2}{\nu}\|\vec{u}\|_3^2\|\epsilon\|_6^2$$

$$\leq C\frac{2}{\nu}\|\vec{u}\|_6\|\vec{u}\|_{L^2}\|\epsilon\|_4\|\epsilon\|_{12}$$

$$\leq \frac{C'}{\alpha^{2/3}\nu^2}\|\vec{u}\|_6^2\|\vec{u}\|_2^2 + \frac{1}{\alpha^{2/3}}\alpha\|\epsilon\|_4^4 + \frac{1}{2}\gamma_0\alpha\|\epsilon\|_{12}^4$$

and thus, for $t_0 < t < T$,

$$\|\vec{\nabla}\otimes\vec{u}\|_2^2 + \nu\int_{t_0}^t \|\Delta\vec{u}\|_2^2\,ds + \frac{1}{2}\gamma_0\alpha\int_{t_0}^t \|\epsilon\|_{12}^4\,ds$$

$$\leq \|\vec{\nabla}\otimes\vec{u}(t_0,.)\|_2^2 + \frac{2}{\nu}\int_{t_0}^T \|\vec{f}\|_2^2\,ds \qquad (18.19)$$

$$+ C\frac{\|\vec{u}\|_{L^\infty L^2}^2 + \nu^3}{\nu^3\alpha^{2/3}}((\|\vec{u}(0,.)\|_2^2 + \int_0^T \|\vec{f}\|_2^2\,ds))e^t$$

Integrating with respect to t_0, we find that

$$t\|\vec{\nabla}\otimes\vec{u}\|_2^2 + \nu\int_0^t s\|\Delta\vec{u}(s,.)\|_2^2\,ds\frac{1}{2}\gamma_0\alpha\int_0^t s\|\epsilon\|_{12}^4\,ds \leq C_0(1 + \frac{1}{\alpha^{2/3}}) \qquad (18.20)$$

where the constant C_0 depends on $\vec{u}_0$, $\vec{f}$, T, and ν, but neither on K nor on α.

- **Step 4: Solutions for Ladyzhenskaya's model.**

Let us write $\vec{u}_{(N)}$ for the solution of the problem

$$\partial_t\vec{u} + \mathbb{PP}_{(N)}(\vec{u}\cdot\vec{\nabla}\vec{u}) = \mathbb{PP}_{(N)}(\text{div}(2\nu + \alpha|\epsilon|^2)\epsilon)) + \mathbb{PP}_{(N)}\vec{f}$$

with initial value $\vec{u}(0,.) = \mathcal{P}_{(N)}\vec{u}_0$, with $\mathcal{P}_{(N)}$ defined by $\mathcal{F}(\mathcal{P}_{(N)}\varphi)(\xi) = \chi_{(N)}(\xi)\hat{\varphi}(\xi)$, where $\chi_{(N)}$ is the characteristic function of the ball $B(0, 2^N)$.

The operators $\mathcal{P}_{(N)}$ are equicontinuous on the Sobolev spaces H^s ($s \in \mathbb{R}$). From the energy estimates (18.16), (18.18), and (18.20), we find that the sequence of functions $\vec{u}_{(N)}(t, x)$ satisfies for every $t_0 > 0$

$$\sup_{N\in\mathbb{N}}\sup_{t\in(t_0,T)}\|\vec{u}_{(N)}\|_{H_x^1} < +\infty$$

$$\sup_{N\in\mathbb{N}}\int_{t_0}^T \|\vec{\nabla}\otimes\vec{u}_{(N)}\|_4^4\,dt < +\infty$$

$$\sup_{N\in\mathbb{N}}\int_{t_0}^T \|\vec{u}_{(N)}\|_{H_x^2}^2 < +\infty$$

and

$$\sup_{N\in\mathbb{N}}\int_{t_0}^T \|\partial_t\vec{u}_{(N)}\|_{H_x^{-3}}^2 < +\infty$$

where we used the inequality

$$\||\epsilon(\vec{u}_{(N)})|^2 \, \epsilon(\vec{u}_{(N)})\|_{L_t^2 H_x^{-2}} \leq C\||\epsilon(\vec{u}_{(N)})|^2 \, \epsilon(\vec{u}_{(N)})\|_{L_t^2 L_x^1}$$

$$\leq C'\|\vec{\nabla} \otimes \vec{u}_{(N)}\|_{L_t^4 L_x^4}^2 \|\vec{u}_{(N)}\|_{L_t^\infty H_x^1}.$$

Thus, we may use the Rellich–Lions theorem (Theorem 12.1): we find that, for every $\phi \in \mathcal{D}((0,T) \times \mathbb{R}^3)$, the sequence of functions $\phi(t,x)\vec{u}_{(N)}(t,x)$ is bounded in the Sobolev space $H^{\frac{2}{7}}(\mathbb{R} \times \mathbb{R}^3)$, and we may find a subsequence $\vec{u}_{(N_k)}$ ($k \in \mathbb{N}$, $N_k \to +\infty$) that is strongly convergent in $L^2_{\text{loc}}((0,T) \times \mathbb{R}^3)$ to a limit $\vec{u}$.

Let $\vec{u}$ be the weak limit of a sequence $\vec{u}_{(N_k)}$, as described previously. We have the energy estimates (18.16), (18.18), and (18.20) on $\vec{u}_{(N_k)}$, which are uniform for all N_k and thus give that $\vec{u} \in L_t^\infty L_x^2 \cap L_t^2 H_x^1$, $\sqrt{t}\,\vec{u} \in L_t^\infty H_x^1 \cap L_t^2 H_x^2$ and $\vec{\nabla} \otimes \vec{u} \in L_t^4 L_x^4$.

Moreover, the sequence $\vec{u}_{(N_k)}$ ($k \in \mathbb{N}$, $N_k \to +\infty$) is strongly convergent in $L^2_{\text{loc}}((0,T) \times \mathbb{R}^3)$ to $\vec{u}$, while it is uniformly bounded in $L^2([t_0, T], H_x^2)$ for every positive t_0. Thus, we have, locally in time and space, strong convergence in $L_t^2 H_x^1$ as well. Since $\vec{\nabla} \otimes \vec{u}_{(N_k)}$ is bounded in $L^4 L^4$, we have strong convergence of $\epsilon(\vec{u}_{(N_k)})$ in $(L^3 L^3)_{\text{loc}}$ as well. Thus, we have the weak convergence of $\text{div}(2\nu + \alpha|\epsilon(\vec{u}_{(N_k)})|^2)\epsilon(\vec{u}_{(N_k)})) + \vec{f} - \vec{u}_{(N_k)} \cdot \vec{\nabla} u_{(N_k)}$ to $\text{div}(2\nu + \alpha|\epsilon(\vec{u})|^2)\epsilon(\vec{u})) + \vec{f} - \vec{u} \cdot \vec{\nabla} u$ in $L_t^2 H^{-3}$. As we have, for all $\vec{g} \in L^2 H^3$, the strong convergence of $\mathbb{PP}_{N_k}\vec{g}$ to $\mathbb{P}\vec{g}$ in $L^2 H^3$, we find that $\vec{u}$ is a solution of problem (18.13).

We now prove the uniqueness of $\vec{u}$. If $\vec{v}$ is another solution of problem (18.13) such that $\vec{v} \in L_t^\infty L_x^2 \cap L_t^2 H_x^1$, $\sqrt{t}\,\vec{v} \in L_t^\infty H_x^1 \cap L_t^2 H_x^2$ and $\vec{\nabla} \otimes \vec{v} \in L_t^4 L_x^4$, then we have:

$$
\begin{aligned}
\frac{d}{dt}\|\vec{u} - \vec{v}\|_2^2 &= 2\int (\vec{u} - \vec{v}) \cdot \partial_t(\vec{u} - \vec{v})\, dx \\
&= -2\nu\|\vec{\nabla} \otimes (\vec{u} - \vec{v})\|_2^2 - 2\alpha \int (\epsilon(\vec{u}) - \epsilon(\vec{v}).(|\epsilon(\vec{u})|^2\epsilon(\vec{u}) - |\epsilon(\vec{v})|^2\epsilon(\vec{v}))\, dx \\
&\quad - 2\int (\vec{u} - \vec{v}).(\vec{u} \cdot \vec{\nabla} u - \vec{v} \cdot \vec{\nabla} v)\, dx \\
&= -2\nu\|\vec{\nabla} \otimes (\vec{u} - \vec{v})\|_2^2 \\
&\quad -\alpha \int |\epsilon(\vec{u}) - \epsilon(\vec{v})|^2\, (|\epsilon(\vec{u})|^2 + |\epsilon(\vec{v})|^2)\, dx - \alpha \int \left||\epsilon(\vec{u})|^2 - |\epsilon(\vec{v})|^2\right|^2\, dx \\
&\quad - 2\int (\vec{u} - \vec{v}).((\vec{u} - \vec{v}) \cdot \vec{\nabla} v)\, dx \\
&\leq -2\nu\|\vec{\nabla} \otimes (\vec{u} - \vec{v})\|_2^2 + \|\vec{u} - \vec{v}\|_2\|\vec{u} - \vec{v}\|_6\|\vec{\nabla} \otimes \vec{v}\|_3 \\
&\leq \frac{C}{\nu}\|\vec{u} - \vec{v}\|_2^2\|\vec{\nabla} \otimes \vec{v}\|_3^2.
\end{aligned}
$$

Since $\int_0^T \|\vec{\nabla} \otimes \vec{v}\|_3^2\, dx < +\infty$ and $\vec{u}(0,.) - \vec{v}(0,.) = 0$, we find $\vec{u} = \vec{v}$.

- **Step 5: Weak convergence.**

For $\alpha > 0$, let $\vec{u}_{(\alpha)}$ be the solution of Equations (18.13). From inequality (18.14), we see that the family $(\vec{u}_{(\alpha)})_{\alpha>0}$ is bounded in $L_T^\infty L_x^2 \cap L_t^2 H_x^1$. Moreover, the family $(\alpha^{1/4}\epsilon(\vec{u}_{(\alpha)}))_{\alpha>0}$ is bounded in $L_t^4 L_x^4$, so that the family $(\alpha|\epsilon(\vec{u}_{(\alpha)})|^2\epsilon(\vec{u}_{(\alpha)}))_{0<\alpha<1}$ is bounded in $L_t^{4/3} L_x^{4/3}$, hence in $L_t^{4/3} H_x^{-3/4}$; since $\vec{u}_{(\alpha)} \cdot \vec{\nabla} u_{(\alpha)}$ is bounded in $L_t^{4/3} H_x^{-1}$, we find that the family $(\partial_t \vec{u}_{(\alpha)})_{0<\alpha<1}$ is bounded in $L_t^{4/3} H_x^{-7/4}$.

Thus, we may use the Rellich–Lions theorem (Theorem 12.1): we find that, for every $\phi \in \mathcal{D}((0,T) \times \mathbb{R}^3)$, the family of functions $\phi(t,x)\vec{u}_{(\alpha)}(t,x)$, $0 < \alpha < 1$, is bounded in the

Sobolev space $H^{\frac{3}{14}}(\mathbb{R} \times \mathbb{R}^3)$, and we may find a subsequence $\vec{u}_{(\alpha_k)}$ ($k \in \mathbb{N}$, $\alpha_k \to 0$) that is strongly convergent in $L^2_{\text{loc}}((0,T) \times \mathbb{R}^3)$ to a limit $\vec{u}$.

Then we find that $\vec{u}_{(\alpha_k)} \cdot \vec{\nabla} \vec{u}_{(\alpha_k)} = \text{div}(\vec{u}_{(\alpha_k)} \otimes \vec{u}_{(\alpha_k)})$ is weakly convergent in $L^{4/3}_t H^{-1}$ to $\vec{u} \cdot \vec{\nabla}\vec{u}$, while $\alpha_k |\epsilon(\vec{u}_{(\alpha_k)})|^2 \epsilon(\vec{u}_{(\alpha_k)})$ is strongly convergent to 0 (on $O(\alpha^{1/4})$) in $L^{4/3}_t L^{4/3}_x$. This gives that we have the weak convergence of

$$\mathbb{P}\left(\text{div}(2\nu + \alpha_k |\epsilon(\vec{u}_{(\alpha_k)})|^2)\epsilon(\vec{u}_{(\alpha_k)}))) + \vec{f} - \vec{u}_{(\alpha_k)} \cdot \vec{\nabla} u_{(\alpha_k)}\right)$$

to $\nu\Delta\vec{u} + \mathbb{P}(\vec{f} - \vec{u} \cdot \vec{\nabla} u)$ in $L^{4/3}_t H^{-7/4}_x$. Thus, $\vec{u}$ is a solution to the Navier–Stokes problem.

- **Step 6: Energy estimates.**

The pressure associated to $\vec{u}_{(\alpha)}$ is given by

$$p_{(\alpha)} = \frac{1}{\Delta} \text{div } f + \sum_{j=1}^{3}\sum_{k=1}^{3} \frac{\partial_j}{\sqrt{-\Delta}} \frac{\partial_k}{\sqrt{-\Delta}} (u_{(\alpha),j} u_{(\alpha),k} - \alpha|\epsilon(\vec{u}_{(\alpha)})|^2 \epsilon_{j,k}(\vec{u}_{(\alpha)}))$$

We have regularity enough on $\vec{u}_{(\alpha)}$ to be allowed to write:

$$\begin{aligned}
\partial_t |\vec{u}_{(\alpha)}|^2 =&\, 2\partial_t \vec{u}_{(\alpha)} \cdot \vec{u}_{(\alpha)}\\
=&\, 2\nu\Delta\vec{u}_{(\alpha)} \cdot \vec{u}_{(\alpha)} + 2\alpha\vec{u}_{(\alpha)} \cdot \text{div}(|\epsilon(u_{(\alpha)})|^2 \epsilon(\vec{u}_{(\alpha)}))\\
&- 2\vec{u}_{(\alpha)}.(\vec{u}_{(\alpha)} \cdot \vec{\nabla}\vec{u}_{(\alpha)}) - 2\vec{u}_{(\alpha)} \cdot \vec{\nabla}p_{(\alpha)} + 2\vec{u}_{(\alpha)} \cdot \vec{f}\\
=&\, \Delta|\vec{u}_{(\alpha)}|^2 - 2|\vec{\nabla} \otimes \vec{u}_{(\alpha)}|^2 + 2\alpha\,\text{div}(|\epsilon(\vec{u}_{(\alpha)})|^2 \vec{u}_{(\alpha)}.\epsilon(\vec{u}_{(\alpha)}))\\
&- 2\alpha|\epsilon(\vec{u}_{(\alpha)})|^4 - \text{div}(|\vec{u}_{(\alpha)}|^2 \vec{u}_{(\alpha)}) - 2\,\text{div}(p_{(\alpha)}\vec{u}_{(\alpha)}) + 2\vec{u}_{(\alpha)} \cdot \vec{f}
\end{aligned}$$

We know that $\vec{u}_{(\alpha_k)}$ converge strongly to $\vec{u}$ in $L^2_{\text{loc}}((0,T) \times \mathbb{R}^3)$; as the family is bounded in $L^{10/3}_t H^{3/5}_x \subset L^{10/3}_t L^{10/3}_x$, we find that we have strong convergence in $L^3_{\text{loc}}((0,T) \times \mathbb{R}^3)$ as well. Thus, we have the following convergence results in $\mathcal{D}'((0,T) \times \mathbb{R}^3)$: $\partial_t |\vec{u}_{(\alpha_k)}|^2 \to \partial_t |\vec{u}|^2$, $\Delta|\vec{u}_{(\alpha_k)}|^2 \to \Delta|\vec{u}|^2$, $\text{div}(|\vec{u}_{(\alpha_k)}|^2 \vec{u}_{(\alpha_k)}) \to \text{div}(|\vec{u}|^2 \vec{u})$ and $\vec{u}_{(\alpha_k)} \cdot \vec{f} \to \vec{u} \cdot \vec{f}$. Similarly, we have that

$$\text{div}\left((\frac{1}{\Delta}\text{div } f + \sum_{j=1}^{3}\sum_{l=1}^{3} \frac{\partial_j}{\sqrt{-\Delta}} \frac{\partial_l}{\sqrt{-\Delta}} (u_{(\alpha_k),j} u_{(\alpha_k),l}))\vec{u}_{(\alpha_k)}\right)$$

converges in $\mathcal{D}'$ to $\text{div}\left((\frac{1}{\Delta}\text{div } f + \sum_{j=1}^{3}\sum_{l=1}^{3} \frac{\partial_j}{\sqrt{-\Delta}} \frac{\partial_l}{\sqrt{-\Delta}} (u_j u_l))\vec{u}\right)$. Moreover, we have

$$\begin{aligned}
\|\vec{u}_{(\alpha_k)}\|_{L^4 L^4} &\leq \sqrt{\|\vec{u}_{(\alpha_k)}\|_{L^2 L^\infty}\|\vec{u}_{(\alpha_k)}\|_{L^\infty L^2}}\\
&\leq C\|\vec{u}_{(\alpha_k)}\|_{L^\infty L^2}^{1/2}\|\vec{\nabla} \otimes \vec{u}_{(\alpha_k)}\|_{L^2 L^2}^{1/6}\|\vec{\nabla} \otimes \vec{u}_{(\alpha_k)}\|_{L^4 L^4}^{1/3} T^{1/12}\\
&= O(\alpha_k^{-1/12})
\end{aligned}$$

while $\alpha_k \||\epsilon(\vec{u}_{(\alpha_k)})|^2 \epsilon(\vec{u}_{(\alpha_k)})\|_{L^{4/3} L^{4/3}} = O(\alpha_k^{1/4})$. Thus, we find that $\alpha_k \text{div}(|\epsilon(\vec{u}_{(\alpha_k)})|^2 \vec{u}_{(\alpha_k)}. \epsilon(\vec{u}_{(\alpha_k)}))$ converges to 0 in $\mathcal{D}'$. Similarly, we find that $\text{div}((\sum_{j=1}^{3}\sum_{l=1}^{3} \frac{\partial_j}{\sqrt{-\Delta}} \frac{\partial_l}{\sqrt{-\Delta}} (\alpha_k|\epsilon(\vec{u}_{(\alpha_k)})|^2 \epsilon(\vec{u}_{(\alpha_k)})))\vec{u}_{(\alpha_k)})$ converges to 0 in $\mathcal{D}'$.

Thus far, we have got that

$$\partial_t |\vec{u}|^2 = \Delta|\vec{u}|^2 - \text{div}((|\vec{u}|^2 + 2p)\vec{u}) + 2\vec{u} \cdot \vec{f} - T$$

with

$$T = \lim_{\alpha_k \to 0} 2\nu |\vec{\nabla} \otimes \vec{u}_{(\alpha_k)}|^2 + 2\alpha_k |\epsilon(\vec{u}_{(\alpha_k)})|^4.$$

Let $\phi \in \mathcal{D}'((0,T) \times \mathbb{R}^3)$ be a non-negative function. As $\sqrt{\phi}\,\vec{\nabla} \otimes \vec{u}_{(\alpha_k)}$ is weakly convergent to $\sqrt{\phi}\,\vec{\nabla} \otimes \vec{u}$ in $L_t^2 L_x^2$, we find that $\|\sqrt{\phi}\,\vec{\nabla} \otimes \vec{u}\|_2^2 \leq \liminf_{\alpha_k \to 0} \|\sqrt{\phi}\,\vec{\nabla} \otimes \vec{u}_{(\alpha_k)}\|_2^2$. Thus, we have

$$\langle T | \phi \rangle_{\mathcal{D}',\mathcal{D}} = \lim_{\alpha_k \to 0} \iint (2\nu |\vec{\nabla} \otimes \vec{u}_{(\alpha_k)}|^2 + 2\alpha_k |\epsilon(\vec{u}_{(\alpha_k)})|^4) \phi(t,x)\, dt\, dx$$

$$\geq 2\nu \liminf_{\alpha_k \to 0} \iint |\vec{\nabla} \otimes \vec{u}_{(\alpha_k)}|^2 \phi(t,x)\, dt\, dx$$

$$\geq 2\nu \iint |\vec{\nabla} \otimes \vec{u}|^2 \phi(t,x)\, dt\, dx.$$

Thus, $T = 2\nu |\vec{\nabla} \otimes \vec{u}_{(\alpha_k)}|^2 + \mu$, where μ is a non-negative locally bounded measure. The solution $\vec{u}$ is thus suitable.

- **Step 7: Strong convergence.**

Now, if $\vec{u}_0$ is in H^1 and $\vec{u}$ is a solution of the Navier–Stokes problem with $\vec{u} \in L^\infty([0,T], H^1) \cap L^2([0,T], H^2)$, it is easy to check the strong convergence of $\vec{u}_{(\alpha)}$ to $\vec{u}$: as $\vec{u}$ and $\vec{u}_{(\alpha)}$ are regular enough, we may write

$$\frac{d}{dt} \|\vec{u} - \vec{u}_{(\alpha)}\|_2^2 = 2 \int (\vec{u} - \vec{u}_{(\alpha)}) \cdot \partial_t(\vec{u} - \vec{u}_{(\alpha)})\, dx$$

Indeed, the only delicate point is the integrability of $\int \vec{u} \cdot \operatorname{div}((|\epsilon(\vec{u}_{(\alpha)})|^2 \epsilon(\vec{u}_{(\alpha)}))\, dx$: we have $\vec{\nabla} \otimes \vec{u} \in L^4 L^3$, while we know from (18.14) that $\epsilon(\vec{u}_{(\alpha)})$ is in $L^4 L^4$ and from (18.19) that $\epsilon(\vec{u}_{(\alpha)})$ is in $L^4 L^{12}$. Thus, we may write:

$$\frac{d}{dt} \|\vec{u} - \vec{u}_{(\alpha)}\|_2^2 = 2 \int (\vec{u} - \vec{u}_{(\alpha)}) \cdot \partial_t(\vec{u} - \vec{u}_{(\alpha)})\, dx$$

$$= -2\nu \|\vec{\nabla} \otimes (\vec{u} - \vec{u}_{(\alpha)})\|_2^2 - 2 \int (\vec{u} - \vec{u}_{(\alpha)}) \cdot (\vec{u} \cdot \vec{\nabla}\vec{u} - \vec{u}_{(\alpha)} \cdot \vec{\nabla}\vec{u}_{(\alpha)})\, dx$$

$$- 2\alpha \int (\vec{u} - \vec{u}_{(\alpha)}) \cdot \operatorname{div}(|\epsilon(\vec{u}_{(\alpha)})|^2 \epsilon(\vec{u}_{(\alpha)}))\, dx$$

$$\leq -2\nu \|\vec{\nabla} \otimes (\vec{u} - \vec{u}_{(\alpha)})\|_2^2 - 2 \int (\vec{u} - \vec{u}_{(\alpha)}) \cdot ((\vec{u} - \vec{u}_{(\alpha)}) \cdot \vec{\nabla}\vec{u})\, dx$$

$$- 2\alpha \int \epsilon(\vec{u}) \cdot (|\epsilon(\vec{u}_{(\alpha)})|^2 \epsilon(\vec{u}_{(\alpha)}))\, dx$$

$$\leq -2\nu \|\vec{\nabla} \otimes (\vec{u} - \vec{u}_{(\alpha)})\|_2^2 + 2\|\vec{u} - \vec{u}_{(\alpha)}\|_6 \|\vec{u} - \vec{u}_{(\alpha)}\|_2 \|\vec{\nabla} \otimes \vec{u}\|_3$$

$$+ 2\alpha \|\epsilon(\vec{u})\|_3 \|\epsilon(\vec{u}_{(\alpha)})\|_4^{5/2} \|\epsilon(\vec{u}_{(\alpha)}))\|_{12}^{1/2}$$

$$\leq \frac{C}{\nu} \|\vec{u} - \vec{u}_{(\alpha)}\|_2^2 \|\vec{\nabla} \otimes \vec{u}\|_3^2 + 2\alpha \|\epsilon(\vec{u})\|_3 \|\epsilon(\vec{u}_{(\alpha)})\|_4^{5/2} \|\epsilon(\vec{u}_{(\alpha)}))\|_{12}^{1/2}$$

This gives

$$\sup_{0 < t < T} \|\vec{u} - \vec{u}_{(\alpha)}\|_2^2 \leq 2\alpha e^{\frac{C}{\nu} \int_0^T \|\vec{\nabla} \otimes \vec{u}\|_3^2\, ds} \int_0^T \|\epsilon(\vec{u})\|_3 \|\epsilon(\vec{u}_{(\alpha)})\|_4^{5/2} \|\epsilon(\vec{u}_{(\alpha)}))\|_{12}^{1/2}\, ds$$

$$\leq 2\alpha e^{\frac{C}{\nu} \int_0^T \|\vec{\nabla} \otimes \vec{u}\|_3^2\, ds} \left(\int_0^T \|\vec{\nabla} \otimes \vec{u}\|_3^4\, dt \right)^{1/4} \left(\int_0^T \|\epsilon(\vec{u}_{(\alpha)})\|_4^4\, dt \right)^{5/8} \left(\int_0^T \|\epsilon(\vec{u}_{(\alpha)})\|_{12}^4\, dt \right)^{1/8}$$

From (18.19) and (18.14), we know that for a constant C_0 that depends on $\vec{u}_0$, $\vec{f}$, T, and ν but not on α, we have, for $0 < \alpha < 1$,

$$\alpha\left(\int_0^T \|\epsilon(\vec{u}_{(\alpha)})\|_4^4\, dt\right)^{5/8}\left(\int_0^T \|\epsilon(\vec{u}_{(\alpha)})\|_{12}^4\, dt\right)^{1/8} \leq C_0\alpha\left(\frac{1}{\alpha}\right)^{5/8}\left(\frac{1}{\alpha^{5/3}}\right)^{1/8} = C_0\alpha^{1/6}$$

and thus $\lim_{\alpha\to 0}\sup_{0<t<T}\|\vec{u} - \vec{u}_{(\alpha)}\|_2 = 0$. $\qquad\qquad\qquad\qquad\qquad\qquad\square$

18.5   Damped Navier–Stokes Equations

Another way of enforcing uniqueness and global existence of weak solutions is to introduce a damping term in the Navier–Stokes equations. Damped Navier–Stokes equations have been studied for instance by Cai and Jiu [75].

Damped Navier–Stokes equations

Theorem 18.7.

- Let $\alpha > 0$. Let $\vec{u}_0 \in L^2$, with $\operatorname{div}\vec{u}_0 = 0$, and $\vec{f} \in L_t^2 H_x^{-1}$ on $(0, T) \times \mathbb{R}^3$ $(T < +\infty)$. Then the problem

$$\partial_t\vec{u} + \mathbb{P}(\vec{u} \cdot \vec{\nabla}\,\vec{u}) = \nu\Delta\vec{u} + \mathbb{P}(\vec{f} - \alpha|\vec{u}|^4\vec{u}) \qquad (18.21)$$

  with initial value $\vec{u}(0,.) = \vec{u}_0$ has a unique solution $\vec{u}_{(\alpha)}$ such that $\vec{u}_{(\alpha)} \in L_t^\infty L_x^2 \cap L_t^2 H_x^1 \cap L_t^6 L_x^6$.
  Moreover, we have the following inequality:

$$\|\vec{u}_{(\alpha)}(t,.)\|_2^2 + \nu\int_0^t \|\vec{\nabla}\otimes\vec{u}_{(\alpha)}(s,.)\|_2^2 + 2\alpha\int_0^t \|\vec{u}_{(\alpha)}\|_6^6\, ds$$

$$\leq \left(\|\vec{u}(0,.)\|_2^2 + \frac{2}{\nu}\int_0^T \|\vec{f}\|_{H^{-1}}^2\, ds\right)e^{\nu t}. \qquad (18.22)$$

- There exists a sequence $\alpha_k \to 0$ and a function $\vec{u} \in L_t^\infty L_x^2 \cap L_t^2 H_x^1$ such that $\vec{u}_{\alpha_k}$ is weakly convergent to $\vec{u}$. Moreover $\vec{u}$ is a suitable weak solution of the Navier–Stokes problem

$$\partial_t\vec{u} + \mathbb{P}(\vec{u} \cdot \vec{\nabla}\vec{u}) = \nu\Delta\vec{u} + \mathbb{P}\vec{f}$$

  with initial value $\vec{u}_0$.

- If $\vec{u}_0 \in H^1$ and if the Navier–Stokes problem has a solution $\vec{u} \in L^\infty([0, T], H^1) \cap L^2([0, T], H^2)$, then we have $\lim_{\alpha\to 0}\|\vec{u}_\alpha - \vec{u}\|_{L_t^\infty L_x^2} = 0$.

Proof. ● **First step: Low-frequency approximation.**
Again, we take a Fourier projection operator $\mathcal{P}$ defined by $\mathcal{F}(\mathcal{P}\varphi)(\xi) = \chi(\xi)\hat{\varphi}(\xi)$, where χ is the characteristic function of a bounded set K: $\chi(\xi) = 1$ when $\xi \in K$ and $\chi(\xi) = 0$

when $\xi \notin K$. We consider the problem

$$\partial_t \vec{u} + \mathbb{P}\mathcal{P}(\vec{u} \cdot \vec{\nabla}\, \vec{u}) = \nu \Delta \vec{u} + \mathbb{P}\mathcal{P}(\vec{f} - \alpha |\vec{u}|^4 \vec{u}) \qquad (18.23)$$

with initial value $\vec{u}(0,.) = \mathcal{P}\vec{u}_0$.

The solution is searched as a fixed point of the transform

$$\vec{u} \in \mathcal{C}([0,T], X) \mapsto \mathcal{P}W_{\nu t} * \vec{u}_0 + \mathbb{P}\mathcal{P} \int_0^t W_{\nu(t-s)} * \vec{f}(s,.)\, ds$$
$$- \mathcal{P}B(\vec{u}, \vec{u}) - \alpha \mathcal{P}Q(\vec{u}, \vec{u}, \vec{u}, \vec{u}, \vec{u})$$

with $X = \mathbb{P}\mathcal{P}((L^2)^3)$,

$$B(\vec{u}, \vec{v}) = \int_0^t W_{\nu(s-t)} * \mathbb{P} \operatorname{div}(\vec{u} \otimes \vec{v})\, ds$$

and

$$Q(\vec{u}_1, \ldots, \vec{u}_5) = \int_0^t W_{\nu(t-s)} * \mathbb{P} \left((\sum_{i=1}^3 u_{1,i} u_{2,i})(\sum_{j=1}^3 u_{3,j} u_{4,j}) \vec{u}_5 \right) ds.$$

The bilinear transform $\mathcal{P}B$ and the quintilinear transform $\mathcal{P}Q$ are bounded on $\mathcal{C}([0,T], X)$:

$$\|\mathcal{P}B(\vec{u}, \vec{v})(t,.)\|_2 \leq C_{1,K}\, t \sup_{0<s<t} \|\vec{u}(s,.)\|_2 \sup_{0<s<t} \|\vec{v}(s,.)\|_2$$

and

$$\|\mathcal{P}Q((\vec{u}_1, \ldots, \vec{u}_5)(t,.)\|_2 \leq C_{2,K}\, t \prod_{i=1}^5 \sup_{0<s<t} \|\vec{u}_i(s,.)\|_2$$

Thus, we shall have a solution of Equation (18.23) on an interval $[0, T_0]$ for $T_0 < T$ small enough to ensure

$$4C_{1,K}T_0(\|\vec{u}_0\|_2 + \sqrt{T_0}\|\vec{f}\|_{L^2 H^{-1}}) + 80 C_{2,K}\alpha T_0(\|\vec{u}_0\|_2 + \sqrt{T_0}\|\vec{f}\|_{L^2 H^{-1}})^4 \leq \frac{1}{2}$$

- **Step 2: Energy estimate.**

If $\vec{u}$ is a solution of (18.23) on $[0, T_0]$, we find that

$$\frac{d}{dt}\|\vec{u}\|_2^2 = 2 \int \vec{u} \cdot \partial_t \vec{u}\, dx = 2 \int \vec{u}.(-\vec{u} \cdot \vec{\nabla}\vec{u} + \nu \Delta \vec{u} - \alpha |\vec{u}|^4 \vec{u} + \vec{f})\, dx$$

Thus we obtain, for $0 < t < T_0$,

$$\|\vec{u}(t,.)\|_2^2 + 2\nu \int_0^t \|\vec{\nabla} \otimes \vec{u}(s,.)\|_2^2 + 2\alpha \int_0^t \|\vec{u}\|_6^6\, ds =$$
$$\|\vec{u}(0,.)\|_2^2 + 2 \int_0^t \langle \vec{u} | \vec{f} \rangle_{H^1, H^{-1}}\, ds$$

This gives

$$\|\vec{u}(t,.)\|_2^2 + \nu \int_0^t \|\vec{\nabla} \otimes \vec{u}(s,.)\|_2^2 + 2\alpha \int_0^t \|\vec{u}\|_6^6\, ds$$
$$\leq (\|\vec{u}(0,.)\|_2^2 + \frac{2}{\nu} \int_0^T \|\vec{f}\|_{H^{-1}}^2\, ds)e^{\nu t}. \qquad (18.24)$$

Thus, the L^2 norm of $\vec{u}$ cannot blow up and the solution $\vec{u}$ may be extended on the whole interval $[0, T]$.

- **Step 3: Solutions for the damped equations.**

Let us write $\vec{u}_{(N)}$ for the solution of the problem

$$\partial_t \vec{u} + \mathbb{P}\mathcal{P}_{(N)}(\vec{u} \cdot \vec{\nabla}\, \vec{u}) = \nu\Delta\vec{u} + \mathbb{P}\mathcal{P}_{(N)}(\vec{f} - \alpha|\vec{u}|^4\vec{u})$$

with initial value $\vec{u}(0, .) = \mathcal{P}_{(N)}\vec{u}_0$, with $\mathcal{P}_{(N)}$ defined by $\mathcal{F}(\mathcal{P}_{(N)}\varphi)(\xi) = \chi_{(N)}(\xi)\hat{\varphi}(\xi)$, where $\chi_{(N)}$ is the characteristic function of the ball $B(0, 2^N)$.

From the energy estimate (18.22), we find that the sequence of functions $\vec{u}_{(N)}(t, x)$ satisfies

$$\sup_{N \in \mathbb{N}} \|\vec{u}_{(N)}\|_{L^2((0,T,H_x^1)} < +\infty$$

Moreover, from $L^{6/5} \subset H^{-1}$, we find that

$$\sup_{N \in \mathbb{N}} \|\partial_t \vec{u}_{(N)}\|_{L_t^{6/5} H^{-1}} < +\infty$$

Thus, we may use the Rellich–Lions theorem (Theorem 12.1) and find a subsequence $\vec{u}_{(N_k)}$ ($k \in \mathbb{N}$, $N_k \to +\infty$) that is strongly convergent in $L^2_{\mathrm{loc}}((0, T) \times \mathbb{R}^3)$ to a limit $\vec{u}$.

As the sequence $\vec{u}_{(N_k)}$ ($k \in \mathbb{N}$, $N_k \to +\infty$) is uniformly bounded in $L_t^6 L_x^6$, we have, locally in time and space, strong convergence in $L^{27/5} L^{27/5}$ as well. Thus, we have the weak convergence of $\nu\Delta\vec{u}_{(N_k)} - \vec{u}_{(N_k)} \cdot \vec{\nabla}\vec{u}_{(N_k)} + \vec{f} - \alpha|\vec{u}_{(N_k)}|^4\vec{u}_{(N_k)}$ to $\nu\Delta\vec{u} - \alpha|\vec{u}|^4\vec{u} + \vec{f} - \vec{u} \cdot \vec{\nabla}u$ in $L_t^{27/25} H^{-23/18}$. We find that $\vec{u}$ is a solution of problem (18.21).

We now prove the uniqueness of $\vec{u}$. If $\vec{v}$ is another solution of problem (18.21) such that $\vec{v} \in L_t^\infty L_x^2 \cap L_t^2 H_x^1 \cap L_t^6 L_x^6$, then we have:

$$
\begin{aligned}
\frac{d}{dt}\|\vec{u} - \vec{v}\|_2^2 =& \; 2\int (\vec{u} - \vec{v}) \cdot \partial_t(\vec{u} - \vec{v})\, dx \\
=& -2\nu\|\vec{\nabla} \otimes (\vec{u} - \vec{v})\|_2^2 - 2\alpha\int (\vec{u} - \vec{v}).(|\vec{u}|^4\vec{u} - |\vec{v}|^4\vec{v})\, dx \\
& -2\int (\vec{u} - \vec{v}).(\vec{u} \cdot \vec{\nabla}\vec{u} - \vec{v} \cdot \vec{\nabla}\vec{v})\, dx \\
=& -2\nu\|\vec{\nabla} \otimes (\vec{u} - \vec{v})\|_2^2 \\
& -\alpha\int (|\vec{u}|^2 - |\vec{v}|^2)^2(|\vec{u}|^2 + |\vec{v}|^2)\, dx - \alpha\int |\vec{u} - \vec{v}|^2(|\vec{u}|^4 + |\vec{v}|^4)\, dx \\
& +2\int \vec{v}.((\vec{u} - \vec{v}) \cdot \vec{\nabla}(\vec{u} - \vec{v}))\, dx \\
\leq& -2\nu\|\vec{\nabla} \otimes (\vec{u} - \vec{v})\|_2^2 + C\|\vec{\nabla} \otimes (\vec{u} - \vec{v})\|_2^{3/2}\|\vec{u} - \vec{v}\|_2^{1/2}\|\vec{v}\|_6 \\
\leq& \frac{C^4}{\nu^3}\|\vec{u} - \vec{v}\|_2^2\|\vec{v}\|_6^4.
\end{aligned}
$$

Since $\int_0^T \|\vec{v}\|_6^4\, dx < +\infty$ and $\vec{u}(0, .) - \vec{v}(0, .) = 0$, we find $\vec{u} = \vec{v}$.

- **Step 4: Weak convergence.**

For $\alpha > 0$, let $\vec{u}_{(\alpha)}$ be the solution of Equations (18.21). From inequality (18.22), we see that the family $(\vec{u}_{(\alpha)})_{\alpha>0}$ is bounded in $L_T^\infty L_x^2 \cap L_t^2 H_x^1$. Moreover, the family $(\alpha^{1/6}\vec{u}_{(\alpha)})_{\alpha>0}$ is bounded in $L_t^6 L_x^6$, so that the family $(\alpha|\vec{u}_{(\alpha)}|^4\vec{u}_{(\alpha)})_{0<\alpha<1}$ is bounded in $L_t^{6/5} L_x^{6/5}$, hence in $L_t^{6/5} H_x^{-1}$; since $\vec{u}_{(\alpha)} \cdot \vec{\nabla}u_{(\alpha)}$ is bounded in $L_t^{4/3} H_x^{-1}$, we find that the family $(\partial_t \vec{u}_{(\alpha)})_{0<\alpha<1}$ is bounded in $L_t^{6/5} H_x^{-1}$.

Thus, we may use the Rellich–Lions theorem (Theorem 12.1) and find a subsequence $\vec{u}_{(\alpha_k)}$ ($k \in \mathbb{N}$, $\alpha_k \to 0$) that is strongly convergent in $L^2_{\mathrm{loc}}((0,T) \times \mathbb{R}^3)$ to a limit $\vec{u}$.

Then we find that $\vec{u}_{(\alpha_k)} \cdot \vec{\nabla} \vec{u}_{(\alpha_k)} = \mathrm{div}(\vec{u}_{(\alpha_k)} \otimes \vec{u}_{(\alpha_k)})$ is weakly convergent in $L^{4/3}_t H^{-1}$ to $\vec{u} \cdot \vec{\nabla} \vec{u}$, while $\alpha_k |\vec{u}_{(\alpha_k)}|^4 \vec{u}_{(\alpha_k)}$ is strongly convergent to 0 (on $O(\alpha^{1/6})$) in $L^{6/5}_t L^{6/5}_x$. This gives that we have the weak convergence of $\mathbb{P}\left(\nu \Delta \vec{u}_{(\alpha_k)} - \alpha_k |\vec{u}_{(\alpha_k)}|^4 \vec{u}_{(\alpha_k)} + \vec{f} - \vec{u}_{(\alpha_k)} \cdot \vec{\nabla} u_{(\alpha_k)}\right)$ to $\nu \Delta \vec{u} + \mathbb{P}(\vec{f} - \vec{u} \cdot \vec{\nabla} u)$ in $L^{6/5}_t H^{-1}_x$. Thus, $\vec{u}$ is a solution to the Navier–Stokes problem.

- **Step 5: Energy estimates.**

The pressure associated to $\vec{u}_{(\alpha)}$ is given by

$$p_{(\alpha)} = \frac{1}{\Delta} \mathrm{div}(\vec{f} - \alpha |\vec{u}_{(\alpha)}|^4 \vec{u}_{(\alpha)}) + \sum_{j=1}^{3} \sum_{k=1}^{3} \frac{\partial_j}{\sqrt{-\Delta}} \frac{\partial_k}{\sqrt{-\Delta}} (u_{(\alpha),j} u_{(\alpha),k})$$

We have regularity enough on $\vec{u}_{(\alpha)}$ to be allowed to write:

$$
\begin{aligned}
\partial_t |\vec{u}_{(\alpha)}|^2 &= 2\partial_t \vec{u}_{(\alpha)} \cdot \vec{u}_{(\alpha)} \\
&= 2\nu \Delta \vec{u}_{(\alpha)} \cdot \vec{u}_{(\alpha)} - \alpha |\vec{u}_{(\alpha)}|^6 \\
&\quad - 2\vec{u}_{(\alpha)} \cdot (\vec{u}_{(\alpha)} \cdot \vec{\nabla} \vec{u}_{(\alpha)}) - 2\vec{u}_{(\alpha)} \cdot \vec{\nabla} p_{(\alpha)} + 2\vec{u}_{(\alpha)} \cdot \vec{f} \\
&= \nu \Delta |\vec{u}_{(\alpha)}|^2 - 2\nu |\vec{\nabla} \otimes \vec{u}_{(\alpha)}|^2 - \alpha |\vec{u}_{(\alpha)}|^6 \\
&\quad - \mathrm{div}(|\vec{u}_{(\alpha)}|^2 \vec{u}_{(\alpha)}) - 2\,\mathrm{div}(p_{(\alpha)} \vec{u}_{(\alpha)}) + 2\vec{u}_{(\alpha)} \cdot \vec{f}
\end{aligned}
$$

We know that $\vec{u}_{(\alpha_k)}$ converge strongly to $\vec{u}$ in $L^2_{\mathrm{loc}}((0,T) \times \mathbb{R}^3)$; as the family is bounded in $L^{10/3}_t H^{3/5}_x \subset L^{10/3}_t L^{10/3}_x$, we find that we have strong convergence in $L^3_{\mathrm{loc}}((0,T) \times \mathbb{R}^3)$ as well. Thus, we have the following convergence results in $\mathcal{D}'((0,T) \times \mathbb{R}^3)$: $\partial_t |\vec{u}_{(\alpha_k)}|^2 \to \partial_t |\vec{u}|^2$, $\Delta |\vec{u}_{(\alpha_k)}|^2 \to \Delta |\vec{u}|^2$, $\mathrm{div}(|\vec{u}_{(\alpha_k)}|^2 \vec{u}_{(\alpha_k)}) \to \mathrm{div}(|\vec{u}|^2 \vec{u})$ and $\vec{u}_{(\alpha_k)} \cdot \vec{f} \to \vec{u} \cdot \vec{f}$. Similarly, we have that

$$\mathrm{div}\left((\frac{1}{\Delta} \mathrm{div}\, f + \sum_{j=1}^{3} \sum_{l=1}^{3} \frac{\partial_j}{\sqrt{-\Delta}} \frac{\partial_l}{\sqrt{-\Delta}} (u_{(\alpha_k),j} u_{(\alpha_k),l})) \vec{u}_{(\alpha_k)} \right)$$

converges in $\mathcal{D}'$ to $\mathrm{div}\left((\frac{1}{\Delta} \mathrm{div}\, f + \sum_{j=1}^{3} \sum_{l=1}^{3} \frac{\partial_j}{\sqrt{-\Delta}} \frac{\partial_l}{\sqrt{-\Delta}} (u_j u_l)) \vec{u} \right)$. Moreover, we have that $\alpha_k^{5/6}(\frac{1}{\Delta} \mathrm{div}(|\vec{u}_{(\alpha_k)}|^4 \vec{u}_{(\alpha_k)})$ is bounded in $L^{6/5} L^2$ and $\vec{u}_{(\alpha_k)}$ is bounded in $L^\infty L^2$, so that $\alpha_k \,\mathrm{div}((\frac{1}{\Delta} \mathrm{div}(|\vec{u}_{(\alpha_k)}|^4 \vec{u}_{(\alpha_k)}))\vec{u}_{(\alpha_k)})$ converges to 0 in $\mathcal{D}'$.

Thus far, we have got that

$$\partial_t |\vec{u}|^2 = \nu \Delta |\vec{u}|^2 - \mathrm{div}((|\vec{u}|^2 + 2p)\vec{u}) + 2\vec{u} \cdot \vec{f} - T$$

with

$$T = \lim_{\alpha_k \to 0} 2\nu |\vec{\nabla} \otimes \vec{u}_{(\alpha_k)}|^2 + 2\alpha_k |\vec{u}_{(\alpha_k)}|^6.$$

Let $\phi \in \mathcal{D}'((0,T) \times \mathbb{R}^3)$ be a non-negative function. We have

$$
\begin{aligned}
\langle T | \phi \rangle_{\mathcal{D}',\mathcal{D}} &= \lim_{\alpha_k \to 0} \iint (2\nu |\vec{\nabla} \otimes \vec{u}_{(\alpha_k)}|^2 + 2\alpha_k |\vec{u}_{(\alpha_k)}|^6)\phi(t,x)\, dt\, dx \\
&\geq 2\nu \liminf_{\alpha_k \to 0} \iint |\vec{\nabla} \otimes \vec{u}_{(\alpha_k)}|^2 \phi(t,x)\, dt\, dx \\
&\geq 2\nu \iint |\vec{\nabla} \otimes \vec{u}|^2 \phi(t,x)\, dt\, dx.
\end{aligned}
$$

Thus, $T = 2\nu |\vec{\nabla} \otimes \vec{u}_{(\alpha_k)}|^2 + \mu$, where μ is a non-negative locally bounded measure. The solution $\vec{u}$ is thus suitable.

- **Step 6: Strong convergence.**

Now, if $\vec{u}_0$ is in H^1 and $\vec{u}$ is a solution of the Navier–Stokes problem with $\vec{u} \in L^\infty([0,T], H^1) \cap L^2([0,T], H^2)$, it is easy to check the strong convergence of $\vec{u}_{(\alpha)}$ to $\vec{u}$: as $\vec{u}$ and $\vec{u}_{(\alpha)}$ are regular enough, we may write:

$$\frac{d}{dt}\|\vec{u} - \vec{u}_{(\alpha)}\|_2^2 = 2\int (\vec{u} - \vec{u}_{(\alpha)}) \cdot \partial_t(\vec{u} - \vec{u}_{(\alpha)})\, dx$$

$$= -2\nu\|\vec{\nabla} \otimes (\vec{u} - \vec{u}_{(\alpha)})\|_2^2 - 2\int (\vec{u} - \vec{u}_{(\alpha)}).(\vec{u} \cdot \vec{\nabla}\vec{u} - \vec{u}_{(\alpha)} \cdot \vec{\nabla}\vec{u}_{(\alpha)})\, dx$$

$$+ 2\alpha\int (\vec{u} - \vec{u}_{(\alpha)}).(|\vec{u}_{(\alpha)}|^4\vec{u}_{(\alpha)})\, dx$$

$$\leq -2\nu\|\vec{\nabla} \otimes (\vec{u} - \vec{u}_{(\alpha)})\|_2^2 - 2\int (\vec{u} - \vec{u}_{(\alpha)}).((\vec{u} - \vec{u}_{(\alpha)}) \cdot \vec{\nabla}\vec{u})\, dx$$

$$+ 2\alpha\int \vec{u}.(|\vec{u}_{(\alpha)}|^4\vec{u}_{(\alpha)})\, dx$$

$$\leq -2\nu\|\vec{\nabla} \otimes (\vec{u} - \vec{u}_{(\alpha)})\|_2^2 + 2\|\vec{u} - \vec{u}_{(\alpha)}\|_6\|\vec{u} - \vec{u}_{(\alpha)}\|_2\|\vec{\nabla} \otimes \vec{u}\|_3$$

$$+ 2\alpha\|\vec{u}\|_6\|\vec{u}_{(\alpha)}\|_6^5$$

$$\leq \frac{C}{\nu}\|\vec{u} - \vec{u}_{(\alpha)}\|_2^2\|\vec{\nabla} \otimes \vec{u}\|_3^2 + 2\alpha^{1/6}\|\vec{u}\|_6(\alpha^{1/6}\|\vec{u}_{(\alpha)}\|_6)^5$$

This gives

$$\sup_{0<t<T}\|\vec{u} - \vec{u}_{(\alpha)}\|_2^2 \leq 2\alpha^{1/6}e^{\frac{C}{\nu}\int_0^T \|\vec{\nabla} \otimes \vec{u}\|_3^2\, ds}\|\vec{u}\|_{L^6 L^6}(\alpha^{1/6}\|\vec{u}_{(\alpha)}\|_{L^6 L^6})^5$$

From (18.22), we find that $\lim_{\alpha \to 0} \sup_{0<t<T}\|\vec{u} - \vec{u}_{(\alpha)}\|_2 = 0$.   □

Chapter 19

Artificial Compressibility

In order to simplify the estimation of the pressure in the Navier–Stokes equations, some papers have presented an approximation of the equations by introducing a small amount of compressibility on $\vec{u}$ in order to turn the Navier–Stokes equations, which contains a non-local term $\vec{\nabla}p$ (given by the Leray projection operator, thus by a singular integral), into a system of partial differential operators that contain no non-local terms.

We are going to consider two models with artificial compressibility. As we want to keep good energy balances, we transform the identity (for regular and integrable enough vector fields)

$$\int \vec{u} \cdot (\vec{u} \cdot \vec{\nabla}\vec{u}) \, dx = 0$$

when $\vec{u}$ is divergence free (i.e., in the case of incompressible fluids) into the identity

$$\int \vec{u} \cdot \left(\frac{1}{2}(\operatorname{div}\vec{u})\vec{u} + \vec{u} \cdot \vec{\nabla}\vec{u}\right) dx = \frac{1}{2}\int \operatorname{div}(|\vec{u}|^2\,\vec{u}) \, dx = 0$$

for general vector fields.

19.1 Temam's Model

The first model was introduced by Temam [469, 470] on bounded domains and reads as

$$\begin{cases} \partial_t\vec{u} + \vec{u} \cdot \vec{\nabla}\vec{u} = \nu\Delta\vec{u} + \mathbb{P}\vec{f} - \vec{\nabla}p - \frac{1}{2}(\operatorname{div}\vec{u})\vec{u} \\[2mm] \alpha\partial_t p + \operatorname{div}\vec{u} = 0 \end{cases} \tag{19.1}$$

A variant studied by Oskolkov [386] uses a parabolic equation on p:

$$\alpha(\partial_t p - \Delta p) + \operatorname{div}\vec{u} = 0.$$

The existence of solutions on the whole space has been established by Donatelli and Marcati [156], where a key point is that p satisfies a wave equation

$$\alpha\partial_t^2 p - \Delta p = \operatorname{div}(\vec{u} \cdot \vec{\nabla}\vec{u} + \frac{1}{2}(\operatorname{div}\vec{u})\vec{u} - \nu\Delta\vec{u}). \tag{19.2}$$

Donatelli and Marcati then used Strichartz estimates to control the behavior of p.

The convergence to suitable solutions (when α goes to 0) was studied recently by Donatelli and Spirito [157]. However, their proof seems incomplete, as it relies on the identity $\partial_t(|\vec{u}|^2) = 2\partial_t\vec{u} \cdot \vec{u}$ which does not seem to be established for the solutions of (19.1).

DOI: 10.1201/9781003042594-19

Artificial compressibility (Temam's model)

Theorem 19.1.

- Let $\alpha > 0$. Let $\vec{u}_0 \in L^2$, $p_0 \in L^2$ and $\vec{f} \in L^2_t H^{-1}_x$ on $(0, T) \times \mathbb{R}^3$ $(T < +\infty)$. Then the problem

$$\begin{cases} \partial_t \vec{u} + \vec{u} \cdot \vec{\nabla} \vec{u} = \nu \Delta \vec{u} + \mathbb{P}\vec{f} - \vec{\nabla}p - \frac{1}{2}(\operatorname{div} \vec{u})\vec{u} \\ \\ \alpha \partial_t p + \operatorname{div} \vec{u} = 0 \end{cases} \tag{19.3}$$

  with initial values $\vec{u}(0, .) = \vec{u}_0$ and $p(0, .) = p_0$ has a weak solution $(\vec{u}_{(\alpha)}, p_{(\alpha)})$ such that $\vec{u}_{(\alpha)} \in L^\infty_t L^2_x \cap L^2_t H^1_x$ and $\sqrt{\alpha}\, p_{(\alpha)} \in L^\infty L^2$ (uniformly with respect to α).

- if $\operatorname{div} \vec{u}_0 = p_0 = 0$, then there exists a sequence $\alpha_k \to 0$ and a function $\vec{u} \in L^\infty_t L^2_x \cap L^2_t H^1_x$ such that $\vec{u}_{(\alpha_k)}$ is weakly convergent to $\vec{u}$. Moreover $\vec{u}$ is a Leray weak solution of the Navier–Stokes problem

$$\partial_t \vec{u} + \mathbb{P}(\vec{u} \cdot \vec{\nabla} \vec{u}) = \nu \Delta \vec{u} + \mathbb{P}\vec{f}$$

  with initial value $\vec{u}_0$.

- Let $\operatorname{div} \vec{u}_0 = p_0 = 0$. If $\vec{u}_0 \in H^1$ and if the Navier–Stokes problem has a solution $\vec{u} \in L^\infty([0, T], H^1) \cap L^2([0, T], H^2)$, then we have the strong convergence $\lim_{\alpha \to 0} \|\vec{u}_\alpha - \vec{u}\|_{L^\infty_t L^2_x} = 0$.

Proof.   ● **First step: Low-frequency approximation.**

Again, we take a Fourier projection operator $\mathcal{P}$ on low frequencies ($\mathcal{P}$ is defined by $\mathcal{F}(\mathcal{P}\varphi)(\xi) = \chi(\xi)\hat{\varphi}(\xi)$, where χ is the characteristic function of a bounded set K). We consider the problem

$$\begin{cases} \partial_t \vec{u} = \nu \Delta \vec{u} - \vec{\nabla}p + \mathcal{P}\left(\mathbb{P}\vec{f} - \vec{u} \cdot \vec{\nabla}\vec{u} - \frac{1}{2}(\operatorname{div} \vec{u})\vec{u}\right) \\ \\ \alpha \partial_t p + \mathcal{P} \operatorname{div} \vec{u} = 0 \end{cases} \tag{19.4}$$

with initial values $\vec{u}(0, .) = \mathcal{P}\vec{u}_0$ and $p(0, .) = \mathcal{P}p_0$.

The solution is searched as a fixed point of the transform

$$(\vec{u}, p) \in \mathcal{C}([0, T], X^4) \mapsto$$

$$\left(\mathcal{P}W_{\nu t} * \vec{u}_0 + \mathbb{P}\mathcal{P}\int_0^t W_{\nu(t-s)} * \vec{f}(s, .)\, ds - \mathcal{L}(p) - \mathcal{P}B(\vec{u}, \vec{u}),\right.$$

$$\left.\mathcal{P}p_0 - \frac{1}{\alpha}\mathcal{P}\mathcal{M}(\vec{u})\right)$$

with $X = \mathcal{P}(L^2)$,

$$B(\vec{u}, \vec{v}) = \int_0^t W_{\nu(s-t)} * \left(\vec{u} \cdot \vec{\nabla}\vec{v} + \frac{1}{2}(\operatorname{div} \vec{u})\vec{v}\right) ds$$

$$\mathcal{L}(p) = \int_0^t W_{\nu(s-t)} * \vec{\nabla}p\, ds$$

and

$$\mathcal{M}(\vec{u}) = \int_0^t \operatorname{div} \vec{u}\, ds$$

The bilinear transform $\mathcal{P}B$ and the linear transforms $\mathcal{L}$ and $\mathcal{P}\mathcal{M}$ are bounded on $\mathcal{C}([0,T], X^4)$:

$$\|\mathcal{P}B(\vec{u}, \vec{v})(t,.)\|_2 \leq C_{1,K}\, t \sup_{0<s<t} \|\vec{u}(s,.)\|_2 \sup_{0<s<t} \|\vec{v}(s,.)\|_2$$

$$\|\mathcal{L}(p))(t,.)\|_2 \leq C_{2,K}\, t \sup_{0<s<t} \|p(s,.)\|_2$$

and

$$\|\mathcal{P}\mathcal{M}(\vec{u})(t,.)\|_2 \leq C_{3,K}\, t \sup_{0<s<t} \|\vec{u}(s,.)\|_2$$

Thus, we shall have a solution of Equation (19.4) on an interval $[0, T_0]$ for $T_0 < T$ small enough to ensure

$$(C_{2,K} + \frac{C_{3,K}}{\alpha})T_0 + 4T_0 C_{1,K}(\|\vec{u}_0\|_2 + C_{4,K}\sqrt{T_0}\|\vec{f}\|_{L^2 H^{-1}} + \|p_0\|_2) \leq \frac{1}{2}$$

- **Step 2: Energy estimate.**

If $\vec{u}$ is a solution of (19.4) on $[0, T_0]$, we find that

$$\frac{d}{dt}(\frac{\|\vec{u}\|_2^2 + \alpha\|p\|_2^2}{2}) = \int \vec{u} \cdot \partial_t \vec{u} + \alpha p \partial_t p\, dx = \int \vec{u} \cdot (\nu\Delta\vec{u} + \mathbb{P}\vec{f})\, dx$$

(where we used $\int \vec{u} \cdot (\frac{1}{2}(\operatorname{div}\vec{u})\vec{u} + \vec{u} \cdot \vec{\nabla}\vec{u})\, dx = 0$ and $\int \vec{u} \cdot \vec{\nabla}p + p\operatorname{div}\vec{u}\, dx = 0$). Thus we obtain, for $0 < t < T_0$,

$$\|\vec{u}(t,.)\|_2^2 + \alpha\|p(t,.)\|_2^2 + 2\nu \int_0^t \|\vec{\nabla} \otimes \vec{u}(s,.)\|_2^2 =$$
$$\|\vec{u}(0,.)\|_2^2 + \alpha\|p_0\|_2^2 + 2\int_0^t \langle \vec{u}|\mathbb{P}\vec{f}\rangle_{H^1, H^{-1}}\, ds \tag{19.5}$$

This gives

$$\|\vec{u}(t,.)\|_2^2 + \alpha\|p(t,.)\|_2^2 + \nu \int_0^t \|\vec{\nabla} \otimes \vec{u}(s,.)\|_2^2\, ds$$
$$\leq (\|\vec{u}(0,.)\|_2^2 + \|p_0\|_2^2 + \frac{2}{\nu}\int_0^T \|\vec{f}\|_{H^{-1}}^2\, ds)e^{\nu t}. \tag{19.6}$$

Thus, the L^2 norms of $\vec{u}$ and p cannot blow up and the solution $(\vec{u}, p)$ may be extended on the whole interval $[0, T]$.

- **Step 3: Solutions for Temam's model.**

Let us write $\vec{u}_{(N)}$ for the solution of the problem (19.4) with $\mathcal{P}_{(N)}$ defined by $\mathcal{F}(\mathcal{P}_{(N)}\varphi)(\xi) = \chi_{(N)}(\xi)\hat{\varphi}(\xi)$, where $\chi_{(N)}$ is the characteristic function of the ball $B(0, 2^N)$.

From the energy estimate (19.6), we find that the sequence of functions $\vec{u}_{(N)}(t,x)$ satisfies

$$\sup_{N \in \mathbb{N}} \|\vec{u}_{(N)}\|_{L^\infty((0,T), L_x^2)} + \|\vec{u}_{(N)}\|_{L^2((0,T), H_x^1)} < +\infty$$

This gives that

$$\sup_{N \in \mathbb{N}} \|p_{(N)}\|_{L^\infty((0,T),L^2_x)} + \|\partial_t p_{(N)}\|_{L^2((0,T),L^2_x)} < +\infty.$$

Moreover, we get that

$$\sup_{N \in \mathbb{N}} \|\partial_t \vec{u}_{(N)}\|_{L^{4/3}_t H^{-3/2}} < +\infty$$

Thus, we may use the Rellich–Lions theorem (Theorem 12.1) and find a subsequence $(\vec{u}_{(N_k)}, p_{(N_k)})$ ($k \in \mathbb{N}$, $N_k \to +\infty$) such that the sequence $\vec{u}_{(N_k)}$ is strongly convergent in $L^2_{\text{loc}}((0,T) \times \mathbb{R}^3)$ to a limit $\vec{u}$, while the sequence $p_{(N_k)}$ is *-weakly convergent in $L^\infty L^2$.

As the sequence $\vec{u}_{(N_k)}$ ($k \in \mathbb{N}$, $N_k \to +\infty$) is uniformly bounded in $L^\infty_t L^2_x$. we have, locally in time and space, strong convergence in $L^6_t L^2_x$ as well; similarly, as $\vec{u}_{(N_k)}$ is uniformly bounded in $L^2_t L^6_x$, we have, locally in time and space, strong convergence in $L^3_t L^3_x$ as well. Thus, we have the weak convergence of $\nu \Delta \vec{u}_{(N_k)} - \vec{u}_{(N_k)} \cdot \vec{\nabla} \vec{u}_{(N_k)} - \frac{1}{2}(\operatorname{div} \vec{u}_{(N_k)})\vec{u}_{(N_k)} + \mathbb{P}\vec{f} - \vec{\nabla} p_{(N_k)}$ to $\nu \Delta \vec{u} - \vec{\nabla} p + \mathbb{P}\vec{f} - \vec{u} \cdot \vec{\nabla} u - \frac{1}{2}(\operatorname{div} \vec{u})\vec{u}$ in $L^{4/3}_t H^{-3/2}$. As the $\mathcal{P}_{(N_k)}$ are equicontinuous on $H^{-3/2}$, we find that $(\vec{u}, p)$ is a solution of problem (19.3).

- **Step 4: Weak convergence.**

For $\alpha > 0$, let $\vec{u}_{(\alpha)}$ be a solution of Equations (19.3). From inequality (19.6), we see that we may assume the family $(\vec{u}_{(\alpha)})_{\alpha>0}$ is bounded in $L^\infty_T L^2_x \cap L^2_t H^1_t$. The problem is to find a uniform control on $\partial_t \vec{u}_{(\alpha)}$, as the energy inequality provides only a control on $\sqrt{\alpha}\|p_{(\alpha)}\|_2$.

We split $\vec{u}_{(\alpha)}$ into $\vec{v}_{(\alpha)} + \vec{w}_{(\alpha)}$, where $\vec{v}_{(\alpha)} = \mathbb{P}\vec{u}_{(\alpha)}$. Thus, we have

$$\begin{cases} \partial_t \vec{v}_{(\alpha)} = \nu \Delta \vec{v}_{(\alpha)} + \mathbb{P}\vec{f} - \mathbb{P}\left(\vec{u}_{(\alpha)} \cdot \vec{\nabla} \vec{u}_{(\alpha)} + \frac{1}{2}(\operatorname{div} \vec{u}_{(\alpha)})\vec{u}_{(\alpha)}\right) \\ \vec{v}_{(\alpha)}(0,.) = \vec{u}_0 \end{cases}$$

and

$$\begin{cases} \partial_t \vec{w}_{(\alpha)} = \nu \Delta \vec{w}_{(\alpha)} - \vec{\nabla} p_{(\alpha)} - (Id - \mathbb{P})\left(\vec{u}_{(\alpha)} \cdot \vec{\nabla} \vec{u}_{(\alpha)} + \frac{1}{2}(\operatorname{div} \vec{u}_{(\alpha)})\vec{u}_{(\alpha)}\right) \\ \alpha \partial_t p_{(\alpha)} + \operatorname{div} \vec{w}_{(\alpha)} = 0 \\ \vec{w}_{(\alpha)}(0,.) = 0 \text{ and } p_{(\alpha)}(0,.) = 0 \end{cases}$$

We have a good control on $\vec{v}_{(\alpha)}$ in $L^2 H^1$ and of $\partial_t \vec{v}_{(\alpha)}$ in $L^{4/3} H^{-1}$. Thus, we may use the Rellich–Lions theorem (Theorem 12.1) and find a subsequence $\vec{v}_{(\alpha_k)}$ ($k \in \mathbb{N}$, $\alpha_k \to 0$) that is strongly convergent in $L^2_{\text{loc}}((0,T) \times \mathbb{R}^3)$ to a limit $\vec{v}$.

We have no uniform control on $\partial_t \vec{w}_{(\alpha)}$. We shall however prove directly that $\vec{w}_{(\alpha)}$ is strongly convergent in $L^2_{\text{loc}}((0,T) \times \mathbb{R}^3)$ to 0.

From the control on $\vec{u}_{(\alpha)}$ in $L^\infty_T L^2_x \cap L^2_t H^1_x$ and the (non-uniform) control of $p_{(\alpha)}$ in $L^\infty L^2$, we see that $\partial_t \vec{w}_{(\alpha)}$ belongs (non-uniformly with respect to α) to $L^{4/3} \dot{H}^{-1}$ (as $\vec{u}_{(\alpha)}$ belongs to $L^4 L^3 \cap L^2 H^1$, so that $u_{(\alpha),i}\partial_j u_{(\alpha),k}$ belongs to $L^{4/3} L^{6/5} \subset L^{4/3} \dot{H}^{-1}$) and thus $\vec{w}_{(\alpha)} \in L^\infty((0,T), \dot{H}^{-1})$ and $\operatorname{div} \vec{w}_{(\alpha)} \in L^\infty \dot{H}^{-2}$ and $p_{(\alpha)} \in L^\infty \dot{H}^{-2}$. We thus may define $W_{(\alpha)} = \frac{1}{\Delta} \operatorname{div} \vec{w}_{(\alpha)}(= \frac{1}{\Delta} \operatorname{div} \vec{u}_{(\alpha)})$ and $P_{(\alpha)} = \frac{1}{\Delta} p_{(\alpha)}$.

We have

$$\begin{cases} \partial_t W_{(\alpha)} = \nu \Delta W_{(\alpha)} - \Delta P_{(\alpha)} - \operatorname{div}\left(\vec{u}_{(\alpha)} \cdot \vec{\nabla} \vec{u}_{(\alpha)} + \frac{1}{2}(\operatorname{div} \vec{u}_{(\alpha)})\vec{u}_{(\alpha)}\right) \\ \alpha \partial_t P_{(\alpha)} + W_{(\alpha)} = 0 \\ W_{(\alpha)}(0,.) = 0 \text{ and } P_{(\alpha)}(0,.) = 0 \end{cases}$$

and thus

$$\begin{cases} \alpha \partial_t^2 P_{(\alpha)} - \Delta P_{(\alpha)} = -\nu \Delta W_{(\alpha)} + \operatorname{div}\left(\vec{u}_{(\alpha)} \cdot \vec{\nabla} \vec{u}_{(\alpha)} + \frac{1}{2}(\operatorname{div} \vec{u}_{(\alpha)})\vec{u}_{(\alpha)} - \nu \vec{u}_{(\alpha)}\right) \\ P_{(\alpha)}(0,.) = \partial_t P_{(\alpha)}(0,.) = 0 \end{cases}$$

Thus, we may write $P_{(\alpha)} = Q_{(\alpha)} + R_{(\alpha)}$ with

$$\begin{cases} \alpha \partial_t^2 Q_{(\alpha)} - \Delta Q_{(\alpha)} = \operatorname{div}\left(\vec{u}_{(\alpha)} \cdot \vec{\nabla}\vec{u}_{(\alpha)} + \frac{1}{2}(\operatorname{div}\vec{u}_{(\alpha)})\vec{u}_{(\alpha)}\right) \\ Q_{(\alpha)}(0,.) = \partial_t Q_{(\alpha)}(0,.) = 0 \end{cases}$$

and

$$\begin{cases} \alpha \partial_t^2 R_{(\alpha)} - \Delta R_{(\alpha)} = -\nu \operatorname{div}\vec{u}_{(\alpha)} \\ R_{(\alpha)}(0,.) = \partial_t R_{(\alpha)}(0,.) = 0 \end{cases}$$

We now use the Strichartz estimates for the wave equations [454]: this estimate reads as

$$\left\| \int_0^t e^{i(t-s)\sqrt{-\Delta}} F(s,.)\, ds \right\|_{L^4(\mathbb{R}\times\mathbb{R}^3)} \leq C\|F\|_{L_t^1 \dot{H}_x^{1/2}}. \tag{19.7}$$

Thus, if $\alpha \partial_t^2 A - \Delta A = B$ with $A(0,.) = 0$, $\partial_t A(0,.) = 0$ and $B \in L^1\dot{H}^{1/2}$, we find that

$$\partial_t A = \frac{1}{\alpha}\int_0^t \cos(\frac{(t-s)}{\sqrt{\alpha}}\sqrt{-\Delta})\, B(s,.)\, ds$$

and

$$\alpha^{7/8}\|\partial_t A\|_{L^4 L^4} \leq C\|B\|_{L^1 \dot{H}^{1/2}}.$$

This gives

$$\alpha^{7/8}\|\frac{1}{\Delta}\partial_t Q_\alpha\|_{L^4 L^4} \leq C\|\vec{u}_{(\alpha)}\|_{L^2 H^1}^2$$

and

$$\alpha^{7/8}\|\frac{1}{(-\Delta)^{1/4}}\partial_t R_\alpha\|_{L^4 L^4} \leq C\sqrt{T}\|\vec{u}_{(\alpha)}\|_{L^2 H^1}.$$

We may now estimate $\vec{w}_{(\alpha)}$ in $L^2 L^4$ norm. We use a Littlewood–Paley decomposition $\vec{w}_{(\alpha)} = \sum_{j\in\mathbb{Z}} \Delta_j \vec{w}_{(\alpha)}$. We have

$$\|\Delta_j \vec{w}_{(\alpha)}\|_4 \leq C\|\Delta_j \vec{w}_{(\alpha)}\|_2 2^{3j/4} \leq C'\|\vec{\nabla}\otimes\vec{w}_{(\alpha)}\|_2 2^{-j/4}$$

so that, for any j_0,

$$\|\sum_{j\geq j_0} \Delta_j \vec{w}_{(\alpha)}\|_{L^2 L^4} \leq C 2^{-j_0/4}\|\vec{u}_{(\alpha)}\|_{L^2 H^1}.$$

On the other hand, we have $\vec{w}_{(\alpha)} = \vec{\nabla}\frac{1}{\Delta}\operatorname{div}\vec{w}_{(\alpha)} = -\alpha\vec{\nabla}\partial_t P_{(\alpha)}$ so that

$$\|\Delta_j \vec{w}_{(\alpha)}\|_4 \leq C\alpha 2^j\|\Delta_j \partial_t P_{(\alpha)}\|_4$$
$$\leq C'\alpha^{1/8}(2^{3j}\|\frac{1}{\Delta}\partial_t Q_\alpha\|_4 + 2^{3j/2}\|\frac{1}{(-\Delta)^{1/4}}\partial_t R_\alpha\|_4)$$

so that

$$\|\sum_{j< j_0} \Delta_j \vec{w}_{(\alpha)}\|_{L^4 L^4} \leq C\alpha^{1/8}(2^{3j_0}\|\vec{u}_{(\alpha)}\|_{L^2 H^1}^2 + 2^{3j_0/2}\sqrt{T}\|\vec{u}_{(\alpha)}\|_{L^2 H^1})$$

Choosing $j_0 = j_0(\alpha)$ such that $\lim_{\alpha\to 0} 2^{-j_0/4} = \lim_{\alpha\to 0} \alpha^{1/8}2^{3j_0} = 0$, we find that

$$\lim_{\alpha\to 0} \|\vec{w}_{(\alpha)}\|_{L^2((0,T),L^4)} = 0.$$

Writing $\vec{u}_{(\alpha_k)} = \vec{v}_{(\alpha_k)} + \vec{w}_{(\alpha_k)}$, we can see that $\vec{u}_{(\alpha_k)}$ is strongly convergent in $(L^2 L^2)_{\text{loc}}$ to $\vec{u} = \vec{v} + 0$, where $\vec{v}$ is the strong limit of $\vec{v}_{(\alpha_k)}$ obtained by the Rellich theorem, and 0 is the strong limit of $\vec{w}_{(\alpha_k)}$. Moreover, we have $\partial_t \vec{u}_{(\alpha_k)} = \partial_t \vec{v}_{(\alpha_k)} + \partial_t \vec{w}_{(\alpha_k)}$ with

$$\lim_{k \to +\infty} \partial_t \vec{w}_{(\alpha_k)} = \partial_t \lim_{k \to +\infty} \vec{w}_{(\alpha_k)} = 0$$

in $\mathcal{D}'((0,T) \times \mathbb{R}^3)$, while

$$\lim_{k \to +\infty} \partial_t \vec{v}_{(\alpha_k)} = \lim_{k \to +\infty} \mathbb{P}(\nu \Delta \vec{u}_{(\alpha_k)} - \text{div}(\vec{u}_{(\alpha_k)} \otimes \vec{u}_{(\alpha_k)}) + \vec{f} + (\frac{1}{2} \text{div } \vec{u}_{(\alpha_k)}) \vec{u}_{(\alpha_k)})$$

with

$$\lim_{k \to +\infty} \mathbb{P}(\nu \Delta \vec{u}_{(\alpha_k)} - \text{div}((\vec{u}_{(\alpha_k)} \otimes \vec{u}_{(\alpha_k)}) + \vec{f}) = \mathbb{P}(\nu \Delta \vec{u} - \text{div}(\vec{u} \otimes \vec{u}) + \vec{f})$$

and

$$\lim_{k \to +\infty} \mathbb{P}((\text{div } \vec{u}_{(\alpha_k)}) \vec{u}_{(\alpha_k)}) = \lim_{k \to +\infty} \mathbb{P}((\text{div } \vec{w}_{(\alpha_k)}) \vec{u}_{(\alpha_k)}) = 0$$

(since $\text{div } \vec{w}_{(\alpha_k)}$ is strongly convergent to 0 in $(L^2 H^{-1})_{\text{loc}}$, $\vec{u}_{(\alpha_k)}$ is weakly convergent to $\vec{u}$ in $L^2 H^1$, and $(\text{div } \vec{w}_{(\alpha_k)}) \vec{u}_{(\alpha_k)}$ is bounded in $L^{4/3} H^{-1}$). Thus, we find that $\vec{u}$ is a solution to the Navier–Stokes problem.

- **Step 5: Global energy estimates for the weak limit.**

We write $\vec{u} = \lim_{k \to +\infty} \vec{u}_{(\alpha_k)}$, and $\vec{u}_{(\alpha_k)} = \lim_{j \to +\infty} \vec{u}_{(\alpha_k, N_j)}$. We start from inequality (19.5):

$$\|\vec{u}_{(\alpha_k, N_j)}(t,.)\|_2^2 + \alpha_k \|p(t,.)\|_2^2 + 2\nu \int_0^t \|\vec{\nabla} \otimes \vec{u}_{(\alpha_k, N_j)}(s,.)\|_2^2 \, ds$$

$$= \|\vec{u}(0,.)\|_2^2 + \alpha_k \|p_0\|_2^2 + 2 \int_0^t \langle \mathbb{P}\vec{f} | \vec{u}_{(\alpha_k, N_j)} \rangle_{H^{-1}, H^1} \, ds$$

with $p_0 = 0$. This gives

$$\|\vec{u}_{(\alpha_k, N_j)}(t,.)\|_2^2 + 2\nu \int_0^t \|\vec{\nabla} \otimes \vec{u}_{(\alpha_k, N_j)}(s,.)\|_2^2 \, ds$$

$$\leq \|\vec{u}(0,.)\|_2^2 + 2 \int_0^t \langle \mathbb{P}\vec{f} | \vec{u}_{(\alpha_k, N_j)} \rangle_{H^{-1}, H^1} \, ds$$

Let $\omega(t)$ be a non-negative compactly supported smooth function on $\mathbb{R}$, supported within $[-\epsilon, \epsilon]$, with $\|\omega\|_1 = 1$. Let $t \in (\epsilon, T-\epsilon)$. We define the Hilbert space $\mathbb{H}_t = L^2(\mathbb{R}^3) \times L^2((0,t) \times (0,t), H^1)$. We consider, for a function $\vec{v} \in L^2((0,T), H^1)$,

$$A_t(\vec{v}) = (\int \omega(t-s) \vec{v}(s,.) \, ds, (\sqrt{\omega(t-\theta)} \mathbb{1}_{s<\theta} \vec{\nabla} \otimes \vec{v}(s,.))) \in \mathbb{H}_t.$$

Since $\vec{u}_{(\alpha_k, N_j)}$ is bounded in $L^\infty L^2 \cap L^2 H^1$ and converges strongly to $\vec{u}_{(\alpha_k)}$ in $(L^2 L^2)_{\text{loc}}$, we have the weak convergence of $\omega * \vec{u}_{(\alpha_k, N_j)}(t,.)$ to $\omega * \vec{u}_{(\alpha_k)}(t,.)$ in $L^2(\mathbb{R}^3)$ and the weak convergence of $\vec{u}_{(\alpha_k, N_j)}$ to $\vec{u}_{(\alpha_k)}$ in $L^2 \dot{H}^1$, so that $A_t(\vec{u}_{(\alpha_k, N_j)})$ is weakly convergent to

$A_t(\vec{u}_{(\alpha_k)})$ in the Hilbert space $\mathbb{H}_t$. Thus, we have

$$\|\omega * \vec{u}_{(\alpha_k)}(t,.)\|_2^2 + 2\nu \int \int_0^\theta \omega(t-\theta)\|\vec{\nabla} \otimes \vec{u}_{(\alpha_k)}(s,.)\|_2^2 \, ds \, d\theta$$

$$\leq \liminf_{j \to +\infty} \|\omega * \vec{u}_{(\alpha_k,N_j)}(t,.)\|_2^2 + 2\nu \int \int_0^\theta \omega(t-\theta)\|\vec{\nabla} \otimes \vec{u}_{(\alpha_k,N_j)}(s,.)\|_2^2 \, ds \, d\theta$$

$$\leq \liminf_{j \to +\infty} \omega * \|\vec{u}_{(\alpha_k,N_j)}\|_2^2 + 2\nu \int \int_0^\theta \omega(t-\theta)\|\vec{\nabla} \otimes \vec{u}_{(\alpha_k,N_j)}(s,.)\|_2^2 \, ds \, d\theta$$

$$= \liminf_{j \to +\infty} \int \omega(t-\theta) \left(\|\vec{u}_{(\alpha_k,N_j)}(\theta,.)\|_2^2 + 2\nu \int_0^\theta \|\vec{\nabla} \otimes \vec{u}_{(\alpha_k,N_j)}(s,.)\|_2^2 \, ds \right) d\theta$$

$$\leq \|\vec{u}(0,.)\|_2^2 + 2 \int_0^t \langle \mathbb{P}\vec{f}|\vec{u}_{(\alpha_k,N_j)}\rangle_{H^{-1},H^1} \, ds.$$

Now, we integrate on an interval $[t_1 - \delta, t]$ with $t_1 \in (2\delta, T - \delta)$ and $\delta > \epsilon$. We obtain

$$\int_{t_1-\delta}^{t_1} \int |\omega * \vec{u}_{(\alpha_k)}(t,x)|^2 \, dt \, dx + 2\nu \int_{t_1-\delta}^{t_1} \int \int_0^\theta \omega(t-\theta)\|\vec{\nabla} \otimes \vec{u}_{(\alpha_k)}(s,.)\|_2^2 \, ds \, d\theta \, dt$$

$$\leq \delta \left(\|\vec{u}(0,.)\|_2^2 + 2 \int_0^t \langle \mathbb{P}\vec{f}|\vec{u}_{(\alpha_k,N_j)}\rangle_{H^{-1},H^1} \, ds \right).$$

As $\vec{u}_{(\alpha_k)}$ belongs to $L^\infty((0,T),L^2) \subset L^2((0,T),L^2)$ and as smooth compactly supported functions are dense in $L^2 L^2$, we find that (for $\omega_\epsilon(\tau) = \frac{1}{\tau}\omega_0(\frac{x}{\tau})$)

$$\lim_{\epsilon \to 0} \int_{t_1-\delta}^{t_1} \int |\omega_\epsilon * \vec{u}_{(\alpha_k)}(t,x)|^2 \, dt \, dx = \int_{t_1-\delta}^{t_1} \int |\vec{u}_{(\alpha_k)}(t,x)|^2 \, dt \, dx.$$

Moreover, by continuity of $\theta \mapsto \int_0^\theta \|\vec{\nabla} \otimes \vec{u}_{(\alpha_k)}(s,.)\|_2^2 \, ds$, we have, for $t > t_1 - \delta$,

$$\lim_{\epsilon \to 0} \int \int_0^\theta \omega_\epsilon(t-\theta)\|\vec{\nabla} \otimes \vec{u}_{(\alpha_k)}(s,.)\|_2^2 \, ds \, d\theta = \int_0^t \|\vec{\nabla} \otimes \vec{u}_{(\alpha_k)}(s,.)\|_2^2 \, ds;$$

since

$$\int \int_0^\theta \omega_\epsilon(t-\theta)\|\vec{\nabla} \otimes \vec{u}_{(\alpha_k)}(s,.)\|_2^2 \, ds \, d\theta \leq \int_0^T \|\vec{\nabla} \otimes \vec{u}_{(\alpha_k)}(s,.)\|_2^2 \, ds,$$

we use dominated convergence to conclude that

$$\lim_{\epsilon \to 0} \int_{t_1-\delta}^{t_1} \int \int_0^\theta \omega_\epsilon(t-\theta)\|\vec{\nabla} \otimes \vec{u}_{(\alpha_k)}(s,.)\|_2^2 \, ds \, d\theta = \int_{t_1-\delta}^{t_1} \int_0^t \int \|\vec{\nabla} \otimes \vec{u}_{(\alpha_k)}(s,.)\|_2^2 \, ds.$$

Thus, we proved that

$$\int_{t_1-\delta}^{t_1} \|\vec{u}_{(\alpha_k)}\|_2^2 \, dt \, dx + 2\nu \int_{t_1-\delta}^{t_1} \int_0^t \|\vec{\nabla} \otimes \vec{u}_{(\alpha_k)}(s,.)\|_2^2 \, ds \, dt$$

$$\leq \delta \left(\|\vec{u}(0,.)\|_2^2 + 2 \int_0^t \langle \mathbb{P}\vec{f}|\vec{u}_{(\alpha_k,N_j)}\rangle_{H^{-1},H^1} \, ds \right).$$

When t_1 is a Lebesgue point of $t \mapsto \|\vec{u}_{(\alpha_k)}(t, .)\|_2$, we get that:

$$\|\vec{u}_{(\alpha_k)}(t_1, .)\|_2^2 + 2\nu \int_0^{t_1} \|\vec{u}_{(\alpha_k)}\|_{\dot{H}^1}^2 \, ds \leq \|\vec{u}_0\|_2^2 + 2 \int_0^{t_1} \langle \mathbb{P}\vec{f}|\vec{u}_{(\alpha_k)}\rangle_{H^{-1}, H^1} \, ds$$

This inequality may be extended to every t_1, due to the weak boundedness of $t \in (0, T) \mapsto \vec{u}_{(\alpha_k)}(t, .) \in L^2$.

The same proof as for the convergence of $\vec{u}_{(\alpha_k, N_j)}$ to $\vec{u}_{(\alpha_k)}$ then works for the convergence of $\vec{u}_{(\alpha_k)}$ to $\vec{u}$ and gives

$$\|\vec{u}(t, .)\|_2^2 + 2\nu \int_0^t \|\vec{u}\|_{\dot{H}^1}^2 \, ds \leq \|\vec{u}_0\|_2^2 + 2 \int_0^{t_0} \langle \mathbb{P}\vec{f}|\vec{u}\rangle_{H^{-1}, H^1} \, ds \qquad (19.8)$$

- **Step 6: Strong convergence.**

Now, if $\vec{u}_0$ is in H^1 and $\vec{u}$ is a solution of the Navier—Stokes problem with $\vec{u} \in L^\infty([0, T], H^1) \cap L^2([0, T], H^2)$, we check the strong convergence of $\vec{u}_{(\alpha)}$ to $\vec{u}$ in $L^\infty L^2$. However, $\vec{u}_{(\alpha)}$ is not regular enough to allow to write

$$\frac{d}{dt} \|\vec{u} - \vec{u}_{(\alpha)}\|_2^2 = 2 \int (\vec{u} - \vec{u}_{(\alpha)}) \cdot \partial_t(\vec{u} - \vec{u}_{(\alpha)}) \, dx.$$

We have enough regularity on $\vec{v}_{(\alpha)} = \mathbb{P}\vec{u}_{(\alpha)}$ to be allowed to write

$$\frac{d}{dt} \langle \vec{u}|\vec{u}_{(\alpha)}\rangle_{L^2, L^2} = \frac{d}{dt} \langle \vec{u}|\vec{v}_{(\alpha)}\rangle_{L^2, L^2}$$
$$= \langle \partial_t\vec{u}|\vec{v}_{(\alpha)}\rangle_{H^{-1}, H^1} + \langle \vec{u}|\partial_t\vec{v}_{(\alpha)}\rangle_{H^2, H^{-2}}$$

since $\vec{u} \in L^2 H^2$, $\partial_t\vec{u} \in L^2 H^{-1}$, $\vec{v}_{(\alpha)} \in L^2 H^1$ and $\partial_t\vec{v}_{(\alpha)} \in L^2 H^{-2}$. We then write

$$\|\vec{u} - \vec{u}_{(\alpha)}\|_2^2 = \|\vec{u}\|_2^2 + \|\vec{u}_{(\alpha)}\|^2 - 2\langle \vec{u}|\vec{u}_{(\alpha)}\rangle_{L^2, L^2}$$

with

$$\|\vec{u}\|_2^2 = \|\vec{u}_0\|_2^2 - 2\nu \int_0^t \|\vec{\nabla} \otimes \vec{u}\|_2^2 \, ds + 2 \int_0^t \langle \mathbb{P}\vec{f}|\vec{u}\rangle_{H^{-1}, H^1} \, ds$$

$$\|\vec{u}_{(\alpha)}\|_2^2 \leq \|\vec{u}_0\|_2^2 - 2\nu \int_0^t \|\vec{\nabla} \otimes \vec{u}_{(\alpha)}\|_2^2 \, ds + 2 \int_0^t \langle \mathbb{P}\vec{f}|\vec{u}_{(\alpha)}\rangle_{H^{-1}, H^1} \, ds$$

and

$$\langle \vec{u}|\vec{u}_{(\alpha)}\rangle_{L^2, L^2} = \|\vec{u}_0\|_2^2 + \langle \partial_t\vec{u}|\vec{v}_{(\alpha)}\rangle_{H^{-1}, H^1} + \langle \vec{u}|\partial_t\vec{v}_{(\alpha)}\rangle_{H^2, H^{-2}}$$
$$= \|\vec{u}_0\|_2^2 - 2\nu \int_0^t \int \sum_{j=1}^3 \partial_j\vec{u} \cdot \partial_j\vec{u}_{(\alpha)} \, dx \, ds$$
$$- \int_0^t \int (\vec{u} \cdot \vec{\nabla}\vec{u}) \cdot (\vec{u}_{(\alpha)} - \vec{w}_{(\alpha)}) \, dx \, ds + \int_0^t \langle \mathbb{P}\vec{f}|\vec{u}_{(\alpha)}\rangle_{H^{-1}, H^1} \, ds$$
$$- \int_0^t \int \vec{u} \cdot (\vec{u}_{(\alpha)} \cdot \vec{\nabla}\vec{u}_{(\alpha)}) + \frac{1}{2}(\operatorname{div}\vec{u}_{(\alpha)})\vec{u}_{(\alpha)}) + \int_0^t \langle \mathbb{P}\vec{f}|\vec{u}\rangle_{H^{-1}, H^1} \, ds$$

Thus far, we have

$$\|\vec{u} - \vec{u}_{(\alpha)}\|_2^2 \leq -2\nu \int_0^t \|\vec{\nabla} \otimes (\vec{u} - \vec{u}_{(\alpha)})\|_2^2 \, ds$$

$$+ 2 \int_0^t \int (\vec{u} \cdot \vec{\nabla}\vec{u}) \cdot (\vec{u}_{(\alpha)} - \vec{w}_{(\alpha)}) \, dx \, ds$$

$$+ 2 \int_0^t \int \vec{u} \cdot (\vec{u}_{(\alpha)} \cdot \vec{\nabla}\vec{u}_{(\alpha)} + \frac{1}{2}(\mathrm{div}\, \vec{u}_{(\alpha)})\vec{u}_{(\alpha)})$$

$$= -2\nu \int_0^t \|\vec{\nabla} \otimes (\vec{u} - \vec{u}_{(\alpha)})\|_2^2 \, ds - 2\int_0^t \int (\vec{u} \cdot \vec{\nabla}\vec{u}) \cdot \vec{w}_{(\alpha)} \, dx \, ds$$

$$- 2 \int_0^t \int ((\vec{u} - \vec{u}_{(\alpha)}) \cdot \vec{\nabla}\vec{u}) \cdot (\vec{u} - \vec{u}_{(\alpha)}) \, ds$$

$$-2 \int_0^t \int |\vec{u}|^2 \, \mathrm{div}\, \vec{u}_{(\alpha)} \, dx \, ds + 3 \int_0^t \int (\mathrm{div}(\vec{u} - \vec{u}_{(\alpha)})\vec{u} \cdot (\vec{u} - \vec{u}_{(\alpha)}) \, dx \, ds$$

with $\mathrm{div}\, \vec{u}_{(\alpha)} = \mathrm{div}\, \vec{w}_{(\alpha)}$ so that

$$\int_0^t \int |\vec{u}|^2 \, \mathrm{div}\, \vec{u}_{(\alpha)} \, dx \, ds = - \int_0^t \int \vec{w}_{(\alpha)} \cdot \vec{\nabla}(|\vec{u}|^2) \, dx \, ds$$

and

$$\|\vec{u} - \vec{u}_{(\alpha)}\|_2^2 \leq -2\nu \int_0^t \|\vec{\nabla} \otimes (\vec{u} - \vec{u}_{(\alpha)})\|_2^2 \, ds + C \int_0^t \|\vec{w}_{(\alpha)}\|_4 \|\vec{u}\|_{H^1} \|\vec{u}\|_4 \, ds$$

$$+ C \int_0^t \|\vec{u} - \vec{u}_{(\alpha)}\|_2 \|\vec{\nabla} \otimes (\vec{u} - \vec{u}_{(\alpha)})\|_2 (\|\vec{\nabla} \otimes \vec{u}\|_3 + \|\vec{u}\|_\infty) \, ds$$

$$\leq \frac{C}{8\nu} \int_0^t \|\vec{u} - \vec{u}_{(\alpha)}\|_2^2 (\|\vec{\nabla} \otimes \vec{u}\|_3 + \|\vec{u}\|_\infty)^2 \, ds + C \int_0^t \|\vec{w}_{(\alpha)}\|_4 \|\vec{u}\|_{H^1} \|\vec{u}\|_4 \, ds$$

Grönwall's lemma gives then

$$\|\vec{u} - \vec{u}_{(\alpha)}\|_{L^\infty L^2} \leq C e^{\int_0^T (\|\vec{\nabla} \otimes \vec{u}\|_3 + \|\vec{u}\|_\infty)^2 \, ds} \|\vec{w}_{(\alpha)}\|_{L^2 L^4} \|\vec{u}\|_{L^2 H^1} \|\vec{u}\|_{L^\infty L^4}$$

As

$$\lim_{\alpha \to 0} \|\vec{w}_{(\alpha)}\|_{L^2((0,T),L^4)} = 0,$$

we find that $\lim_{\alpha \to 0} \sup_{0<t<T} \|\vec{u} - \vec{u}_{(\alpha)}\|_2 = 0$. $\qquad \square$

19.2 Višik and Fursikov's Model

The second model was introduced by Fursikov and Višik [189] for the study of the Cauchy problem with infinite-energy initial values:

$$\begin{cases} \partial_t \vec{u} + \vec{u} \cdot \vec{\nabla}\vec{u} = \nu\Delta\vec{u} + \mathbb{P}\vec{f} - \vec{\nabla}p - \alpha|\vec{u}|^4\vec{u} - \frac{1}{2}(\mathrm{div}\, \vec{u})\vec{u} \\ \\ \alpha p + \mathrm{div}\, \vec{u} = 0 \end{cases} \qquad (19.9)$$

The extra damping $-\alpha|\vec{u}|^4\vec{u}$ will ensure uniqueness (which is not granted for Equations (19.1), see Temam [468]) and convergence to suitable solutions[1].

The convergence to suitable solutions was discussed by Basson [23] and Lelièvre and Lemarié-Rieusset [308]. A key point in [308] is that p satisfies a heat equation

$$\alpha\partial_t p - (1+\alpha\nu)\Delta p = \operatorname{div}(\vec{u}\cdot\vec{\nabla}\vec{u} + \frac{1}{2}(\operatorname{div}\vec{u})\vec{u} + \alpha|\vec{u}|^4\vec{u}) \qquad (19.10)$$

Artificial compressibility (Višik and Fursikov's model)

Theorem 19.2.

- Let $\alpha > 0$. Let $\vec{u}_0 \in L^2$, $\operatorname{div}\vec{u}_0 = 0$ and $\vec{f} \in L^2_t H^{-1}_x$ on $(0,T)\times\mathbb{R}^3$ $(T < +\infty)$. Then the problem

$$\begin{cases} \partial_t\vec{u} + \vec{u}\cdot\vec{\nabla}\vec{u} = \nu\Delta\vec{u} + \mathbb{P}\vec{f} - \vec{\nabla}p - \alpha|\vec{u}|^4\vec{u} - \frac{1}{2}(\operatorname{div}\vec{u})\vec{u} \\[2mm] \alpha p + \operatorname{div}\vec{u} = 0 \end{cases} \qquad (19.11)$$

  with initial value $\vec{u}(0,.) = \vec{u}_0$ has a unique solution $(\vec{u}_{(\alpha)}, p_{(\alpha)})$ such that $\vec{u}_{(\alpha)} \in L^\infty_t L^2_x \cap L^2_t H^1_x$ and $\alpha^{1/6}\vec{u}_{(\alpha)} \in L^6_t L^6_x$ (uniformly with respect to α).

- there exists a sequence $\alpha_k \to 0$ and a function $\vec{u} \in L^\infty_t L^2_x \cap L^2_t H^1_x$ such that $\vec{u}_{(\alpha_k)}$ is weakly convergent to $\vec{u}$. Moreover $\vec{u}$ is a suitable Leray weak solution of the Navier–Stokes problem

$$\partial_t\vec{u} + \mathbb{P}(\vec{u}\cdot\vec{\nabla}\vec{u}) = \nu\Delta\vec{u} + \mathbb{P}\vec{f}$$

  with initial value $\vec{u}_0$.

- If $\vec{u}_0 \in H^1$ and if the Navier–Stokes problem has a solution $\vec{u} \in L^\infty([0,T], H^1) \cap L^2([0,T], H^2)$, then we have the strong convergence $\lim_{\alpha\to 0}\|\vec{u}_\alpha - \vec{u}\|_{L^\infty_t L^2_x} = 0$.

Proof.   • **First step: Low-frequency approximation.**

One more time, we take a Fourier projection operator $\mathcal{P}$ on low frequencies $\mathcal{P}f = \mathcal{F}^{-1}(1_K(\xi)\hat{f}(\xi))$. We consider the problem

$$\begin{cases} \partial_t\vec{u} = \nu\Delta\vec{u} - \vec{\nabla}p + \mathcal{P}\left(\mathbb{P}\vec{f} - \vec{u}\cdot\vec{\nabla}\vec{u} - \frac{1}{2}(\operatorname{div}\vec{u})\vec{u} - \alpha|\vec{u}|^4\vec{u}\right) \\[2mm] \alpha p + \operatorname{div}\vec{u} = 0 \end{cases} \qquad (19.12)$$

with initial value $\vec{u}(0,.) = \mathcal{P}\vec{u}_0$.

The solution is searched as a fixed point of the transform

$$\vec{u} \in \mathcal{C}([0,T], X) \mapsto \mathcal{P}W_{\nu t} * \vec{u}_0 + \mathbb{P}\mathcal{P}\int_0^t W_{\nu(t-s)} * \vec{f}(s,.)\, ds$$

$$+ \alpha\mathcal{P}L(\vec{u}) - \mathcal{P}B(\vec{u},\vec{u}) - \alpha\mathcal{P}Q(\vec{u},\vec{u},\vec{u},\vec{u},\vec{u})$$

---

[1]The case with no damping term $-\alpha|\vec{u}|^4\vec{u}$ has been recently discussed by Rusin [416].

with $X = \mathbb{PP}((L^2)^3)$,

$$L(\vec{u}) = \int_0^t W_{\nu(s-t)} * \vec{\nabla} \operatorname{div} \vec{u} \, ds,$$

$$B(\vec{u}, \vec{v}) = \int_0^t W_{\nu(s-t)} * (\vec{u} \cdot \vec{\nabla} \vec{v} + \frac{1}{2}(\operatorname{div} \vec{u})\vec{v}) \, ds$$

and

$$Q(\vec{u}_1, \ldots, \vec{u}_5) = \int_0^t W_{\nu(t-s)} * \left((\sum_{i=1}^3 u_{1,i} u_{2,i})(\sum_{j=1}^3 u_{3,j} u_{4,j}) \vec{u}_5 \right) ds.$$

The linear transform $\mathcal{P}L$, the bilinear transform $\mathcal{P}B$ and the quintilinear transform $\mathcal{P}Q$ are bounded on $\mathcal{C}([0,T], X)$:

$$\|\mathcal{P}L(\vec{u})(t,.)\|_2 \leq C_{1,K} \, t \sup_{0<s<t} \|\vec{u}(s,.)\|_2$$

$$\|\mathcal{P}B(\vec{u}, \vec{v})(t,.)\|_2 \leq C_{2,K} \, t \sup_{0<s<t} \|\vec{u}(s,.)\|_2 \sup_{0<s<t} \|\vec{v}(s,.)\|_2$$

and

$$\|\mathcal{P}Q((\vec{u}_1, \ldots, \vec{u}_5)(t,.)\|_2 \leq C_{3,K} \, t \prod_{i=1}^5 \sup_{0<s<t} \|\vec{u}_i(s,.)\|_2$$

Thus, we shall have a solution of Equation (19.4) on an interval $[0, T_0]$ for $T_0 < T$ small enough to ensure

$$\frac{C_{1,K}}{\alpha} + 4C_{2,K}(\|\vec{u}_0\|_2 + C_{4,K}\sqrt{T_0}\|\vec{f}\|_{L^2 H^{-1}})$$

$$+ 80\alpha C_{3,K}(\|\vec{u}_0\|_2 + C_{4,K}\sqrt{T_0}\|\vec{f}\|_{L^2 H^{-1}})^4 \leq \frac{1}{2T_0}$$

- **Step 2: Energy estimate.**

If $\vec{u}$ is a solution of (19.12) on $[0, T_0]$, we find that

$$\frac{d}{dt}(\frac{\|\vec{u}\|_2^2}{2}) = \int \vec{u} \cdot \partial_t \vec{u} + \alpha p \partial_t p \, dx = \int \vec{u} \cdot (\nu \Delta \vec{u} + \mathbb{P}\vec{f} - \alpha |\vec{u}|^4 + \frac{1}{\alpha}\vec{\nabla} \operatorname{div} \vec{u}) \, dx$$

(where we used $\int \vec{u} \cdot (\frac{1}{2}(\operatorname{div} \vec{u})\vec{u} + \vec{u} \cdot \vec{\nabla} \vec{u}) \, dx = 0$). Thus we obtain, for $0 < t < T_0$,

$$\|\vec{u}(t,.)\|_2^2 + 2\nu \int_0^t \|\vec{\nabla} \otimes \vec{u}(s,.)\|_2^2 + \alpha \int_0^t \|\vec{u}\|_6^6 \, ds + \frac{1}{\alpha} \int_0^t \|\operatorname{div} \vec{u}\|_2^2 \, ds =$$
$$\|\vec{u}(0,.)\|_2^2 + 2 \int_0^t \langle \vec{u} | \mathbb{P}\vec{f} \rangle_{H^1, H^{-1}} \, ds \tag{19.13}$$

This gives

$$\|\vec{u}(t,.)\|_2^2 + \nu \int_0^t \|\vec{\nabla} \otimes \vec{u}(s,.)\|_2^2 \, ds + \alpha \int_0^t \|\vec{u}\|_6^6 \, ds + \frac{1}{\alpha} \int_0^t \|\operatorname{div} \vec{u}\|_2^2 \, ds$$
$$\leq (\|\vec{u}(0,.)\|_2^2 + \frac{2}{\nu} \int_0^T \|\vec{f}\|_{H^{-1}}^2 \, ds) e^{\nu t}. \tag{19.14}$$

Thus, the L^2 norm of $\vec{u}$ cannot blow up and the solution $\vec{u}$ may be extended on the whole interval $[0, T]$.

- **Step 3: Solutions for Višik and Fursikov's model.**

Let us write $\vec{u}_{(N)}$ for the solution of the problem (19.12) with $\mathcal{P}_{(N)}$ defined by $\mathcal{F}(\mathcal{P}_{(N)}\varphi)(\xi) = \chi_{(N)}(\xi)\hat{\varphi}(\xi)$, where $\chi_{(N)}$ is the characteristic function of the ball $B(0, 2^N)$.

From the energy estimate (19.14), we find that the sequence of functions $\vec{u}_{(N)}(t, x)$ satisfies

$$\sup_{N \in \mathbb{N}} \|\vec{u}_{(N)}\|_{L^\infty((0,T),L_x^2)} + \|\vec{u}_{(N)}\|_{L^2((0,T),H_x^1)} + \alpha^{1/6}\|\vec{u}_{(N)}\|_{L^6((0,T),L^6} < +\infty$$

As $\mathcal{P}$ is bounded on every Sobolev space H^s, $s \in \mathbb{R}$, we get that

$$\sup_{N \in \mathbb{N}} \|\partial_t \vec{u}_{(N)}\|_{L_t^{6/5}H^{-1}} < +\infty$$

Thus, we may use the Rellich–Lions theorem (Theorem 12.1): we may find a subsequence $(\vec{u}_{(N_k)}, p_{(N_k)})$ $(k \in \mathbb{N}, N_k \to +\infty)$ that the sequence $\vec{u}_{(N_k)}$ is strongly convergent in $L^2_{\text{loc}}((0,T) \times \mathbb{R}^3)$ to a limit $\vec{u}$.

As the sequence $\vec{u}_{(N_k)}$ $(k \in \mathbb{N}, N_k \to +\infty)$ is uniformly bounded in $L_t^\infty L_x^2$, we have, locally in time and space, strong convergence in $L_t^6 L_x^2$ as well; similarly, as $\vec{u}_{(N_k)}$ is uniformly bounded in $L_t^2 L_x^6$ and in $L_t^6 L_x^6$, we have, locally in time and space, strong convergence in $L_t^3 L_x^3$ and in $L_t^5 L_x^5$ as well. Thus, we have the weak convergence of $\nu \Delta \vec{u}_{(N_k)} - \vec{u}_{(N_k)} \cdot \vec{\nabla} \vec{u}_{(N_k)} - \frac{1}{2}(\operatorname{div} \vec{u}_{(N_k)})\vec{u}_{(N_k)} + \mathbb{P}\vec{f} + \frac{1}{\alpha}\vec{\nabla} \operatorname{div} \vec{u}_{(N_k)} - \alpha|\vec{u}_{(N_k)}|^4\vec{u}_{(N_k)}$ to $\nu \Delta \vec{u} + \frac{1}{\alpha}\vec{\nabla} \operatorname{div} \vec{u} - \alpha|\vec{u}|^4\vec{u} + \mathbb{P}\vec{f} - \vec{u} \cdot \vec{\nabla} u - \frac{1}{2}(\operatorname{div} \vec{u})\vec{u}$ in $L_t^{6/5}H^{-1}$. As the $\mathcal{P}_{(N_k)}$ are equicontinuous on H^{-1}, we find that $\vec{u}$ is a solution of problem (19.11).

We now prove the uniqueness of $\vec{u}$. If $\vec{v}$ is another solution of problem (19.11) such that $\vec{v} \in L_t^\infty L_x^2 \cap L_t^2 H_x^1 \cap L_t^6 L_x^6$, then we have:

$$\frac{d}{dt}\|\vec{u} - \vec{v}\|_2^2 = 2\int (\vec{u} - \vec{v}) \cdot \partial_t(\vec{u} - \vec{v})\, dx$$

$$= -2\nu\|\vec{\nabla} \otimes (\vec{u} - \vec{v})\|_2^2 - 2\frac{1}{\alpha}\|\operatorname{div}(\vec{u} - \vec{v})\|_2^2$$

$$- 2\alpha\int (\vec{u} - \vec{v}) \cdot (|\vec{u}|^4\vec{u} - |\vec{v}|^4\vec{v})\, dx$$

$$- 2\int (\vec{u} - \vec{v}) \cdot (\vec{u} \cdot \vec{\nabla}\vec{u} - \vec{v} \cdot \vec{\nabla}\vec{v})\, dx - \int (\vec{u} - \vec{v}) \cdot ((\operatorname{div} \vec{u})\vec{u} - (\operatorname{div} \vec{v})\vec{v})\, dx$$

$$= -2\nu\|\vec{\nabla} \otimes (\vec{u} - \vec{v})\|_2^2 - 2\frac{1}{\alpha}\|\operatorname{div}(\vec{u} - \vec{v})\|_2^2$$

$$- \alpha\int (|\vec{u}|^2 - |\vec{v}|^2)^2(|\vec{u}|^2 + |\vec{v}|^2)\, dx - \alpha\int |\vec{u} - \vec{v}|^2(|\vec{u}|^4 + |\vec{v}|^4)\, dx$$

$$+ 2\int \vec{v} \cdot ((\vec{u} - \vec{v}) \cdot \vec{\nabla}(\vec{u} - \vec{v}))\, dx + \int (\operatorname{div}(\vec{u} - \vec{v}))(\vec{u} - \vec{v}) \cdot \vec{v}\, dx$$

$$\leq -2\nu\|\vec{\nabla} \otimes (\vec{u} - \vec{v})\|_2^2 + C\|\vec{\nabla} \otimes (\vec{u} - \vec{v})\|_2^{3/2}\|\vec{u} - \vec{v}\|_2^{1/2}\|\vec{v}\|_6$$

$$\leq \frac{C^4}{\nu^3}\|\vec{u} - \vec{v}\|_2^2\|\vec{v}\|_6^4.$$

Since $\int_0^T \|\vec{v}\|_6^4\, dx < +\infty$ and $\vec{u}(0, .) - \vec{v}(0, .) = 0$, we find $\vec{u} = \vec{v}$.

- ## Step 4: Weak convergence.

For $\alpha > 0$, let $\vec{u}_{(\alpha)}$ be the solution of Equations (19.11). From inequality (19.14), we see that the family $(\vec{u}_{(\alpha)})_{\alpha>0}$ is bounded in $L_T^\infty L_x^2 \cap L_t^2 H_x^1$. The problem is to find a uniform control on $\partial_t \vec{u}_{(\alpha)}$, as the energy inequality provides only a control on $\sqrt{\alpha} \| \operatorname{div} u_{(\alpha)} \|_2$.

We split $\vec{u}_{(\alpha)}$ into $\vec{v}_{(\alpha)} + \vec{w}_{(\alpha)}$, where $\vec{v}_{(\alpha)} = \mathbb{P}\vec{u}_{(\alpha)}$. Thus, we have

$$\begin{cases} \partial_t \vec{v}_{(\alpha)} = \nu \Delta \vec{v}_{(\alpha)} + \mathbb{P}\vec{f} - \mathbb{P}\left(\vec{u}_{(\alpha)} \cdot \vec{\nabla} \vec{u}_{(\alpha)} + \frac{1}{2}(\operatorname{div} \vec{u}_{(\alpha)})\vec{u}_{(\alpha)} + \alpha |\vec{u}_{(\alpha)}|^4 \vec{u}_{(\alpha)} \right) \\ \vec{v}_{(\alpha)}(0,.) = \vec{u}_0 \end{cases}$$

and

$$\begin{cases} \partial_t \vec{w}_{(\alpha)} = (\nu + \frac{1}{\alpha})\Delta \vec{w}_{(\alpha)} - (Id - \mathbb{P})\left(\vec{u}_{(\alpha)} \cdot \vec{\nabla} \vec{u}_{(\alpha)} + \frac{1}{2}(\operatorname{div} \vec{u}_{(\alpha)})\vec{u}_{(\alpha)} + \alpha |\vec{u}_{(\alpha)}|^4 \vec{u}_{(\alpha)} \right) \\ \vec{w}_{(\alpha)}(0,.) = 0 \end{cases}$$

The family $(\alpha^{1/6}\vec{u}_{(\alpha)})_{\alpha>0}$ is bounded in $L_t^6 L_x^6$, so that the family $(\alpha |\vec{u}_{(\alpha)}|^4 \vec{u}_{(\alpha)})_{0<\alpha<1}$ is bounded in $L_t^{6/5} L_x^{6/5}$, hence in $L_t^{6/5} H_x^{-1}$; $\vec{u}_{(\alpha)} \cdot \vec{\nabla}\vec{u}_{(\alpha)}$ and $(\operatorname{div}\vec{u}_{(\alpha)})\vec{u}_{(\alpha)}$ are bounded in $L_t^{4/3} H_x^{-1}$. We thus find that the family $(\partial_t \vec{v}_{(\alpha)})_{0<\alpha<1}$ is bounded in $L_t^{6/5} H_x^{-1}$.

We have a good control on $\vec{v}_{(\alpha)}$ in $L^2 H^1$ and of $\partial_t \vec{v}_{(\alpha)}$ in $L^{4/3} H^{-1}$. Thus, we may use the Rellich–Lions theorem (Theorem 12.1) and find a subsequence $\vec{v}_{(\alpha_k)}$ ($k \in \mathbb{N}$, $\alpha_k \to 0$) that is strongly convergent in $L^2_{\text{loc}}((0,T) \times \mathbb{R}^3)$ to a limit $\vec{v}$.

We have no uniform control on $\partial_t \vec{w}_{(\alpha)}$. We shall, however, prove directly that $\vec{w}_{(\alpha)}$ is strongly convergent in $L^2_{\text{loc}}((0,T) \times \mathbb{R}^3)$ to 0.

We write

$$\vec{w}_{(\alpha)} = \int_0^t W_{(\nu+\frac{1}{\alpha})(t-s)} * \vec{U}_\alpha \, ds$$

with

$$\vec{U}_\alpha = -(Id - \mathbb{P})\left(\vec{u}_{(\alpha)} \cdot \vec{\nabla}\vec{u}_{(\alpha)} + \frac{1}{2}(\operatorname{div}\vec{u}_{(\alpha)})\vec{u}_{(\alpha)} + \alpha|\vec{u}_{(\alpha)}|^4 \vec{u}_{(\alpha)} \right).$$

U_α is bounded in $L_t^{6/5} L_x^{6/5}$. We find that

$$\|\vec{w}_{(\alpha)}(t,.)\|_2 \leq C \int_0^t \left(\frac{1}{(\nu+\frac{1}{\alpha})(t-s)} \right)^{\frac{3}{2}(\frac{5}{6}-\frac{1}{2})} \|\vec{U}_\alpha(s,.)\|_{6/5} \, ds$$

Now, we use the fact that $|t|^{-1/2} \in L^{2,\infty}$ and that $L^{2,\infty} * L^{6/5} \subset L^3$ and we find

$$\|\vec{w}_{(\alpha)}\|_{L^3 L^2} \leq C\sqrt{\alpha}\|\vec{U}_\alpha\|_{L^{6/5} L^{6/5}}.$$

Thus we find that $\vec{u}_{(\alpha_k)}$ is strongly convergent in $(L^2 L^2)_{\text{loc}}$ to a limit $\vec{u}$; we have as well weak convergence in $L^2 L^6$, hence strong convergence in $(L^3 L^3)_{\text{loc}}$; hence $\vec{u}_{(\alpha_k)} \cdot \vec{\nabla}\vec{u}_{(\alpha_k)}$ is weakly convergent in $(L^{6/5} L^{6/5})_{\text{loc}}$ (and in $L_t^{4/3} H^{-1}$) to $\vec{u} \cdot \vec{\nabla}\vec{u}$, while $\alpha_k |\vec{u}_{(\alpha_k)}|^4 \vec{u}_{(\alpha_k)}$ is strongly convergent to 0 (on $O(\alpha^{1/6})$) in $L_t^{6/5} L_x^{6/5}$ and $(\operatorname{div}\vec{u}_{(\alpha_k)})\vec{u}_{(\alpha_k)}$ is weakly convergent to $(\operatorname{div}\vec{u})\vec{u}$ in $L_t^{4/3} H^{-1}$.

Moreover, we have $\|\operatorname{div}\vec{u}_{(\alpha_k)}\|_{L^2 L^2} \leq \sqrt{\alpha_k}(\|\vec{u}(0,.)\|_2^2 + \frac{2}{\nu}\int_0^T \|\vec{f}\|_{H^{-1}}^2 \, ds)e^{\nu T}$. This gives $\operatorname{div}\vec{u} = 0$. Thus, we have the weak convergence of $\mathbb{P}(\nu\Delta\vec{u}_{(\alpha_k)} - \alpha_k|\vec{u}_{(\alpha_k)}|^4 \vec{u}_{(\alpha_k)} + \vec{f} - \vec{u}_{(\alpha_k)} \cdot \vec{\nabla}\vec{u}_{(\alpha_k)} - \frac{1}{2}(\operatorname{div}\vec{u}_{(\alpha_k)})\vec{u}_{(\alpha_k)})$ to $\nu\Delta\vec{u} + \mathbb{P}(\vec{f} - \vec{u} \cdot \vec{\nabla}u)$ in $L_t^{6/5} H_x^{-1}$. Thus, $\vec{u}$ is a solution to the Navier–Stokes problem.

- ### Step 5: Energy estimates.

Let us write

$$\partial_t \vec{u}_{(\alpha)} + \vec{u}_{(\alpha)} \cdot \vec{\nabla} \vec{u}_{(\alpha)} = \nu \Delta \vec{u}_{(\alpha)} + \vec{f} - \vec{\nabla} p_{(\alpha)} - \alpha |\vec{u}_{(\alpha)}|^4 \vec{u}_{(\alpha)} - \frac{1}{2}(\operatorname{div} \vec{u}_{(\alpha)})\vec{u}_{(\alpha)}$$

where the pressure associated to $\vec{u}_{(\alpha)}$ is given by

$$p_{(\alpha)} = \frac{1}{\Delta} \operatorname{div} \vec{f} - \frac{1}{\alpha} \operatorname{div} \vec{u}_{(\alpha)}$$

We have regularity enough on $\vec{u}_{(\alpha)}$ to be allowed to write:

$$\begin{aligned}
\partial_t |\vec{u}_{(\alpha)}|^2 =\ & 2\partial_t \vec{u}_{(\alpha)} \cdot \vec{u}_{(\alpha)} \\
=\ & 2\nu \Delta \vec{u}_{(\alpha)} \cdot \vec{u}_{(\alpha)} - 2\alpha |\vec{u}_{(\alpha)}|^6 - 2\vec{u}_{(\alpha)} \cdot \vec{\nabla} p_{(\alpha)} + 2\vec{u}_{(\alpha)} \cdot \vec{f} \\
& - 2\vec{u}_{(\alpha)} \cdot (\vec{u}_{(\alpha)} \cdot \vec{\nabla} \vec{u}_{(\alpha)}) - \frac{1}{2}(\operatorname{div} \vec{u}_{(\alpha)})|\vec{u}_{(\alpha)}|^2 \\
=\ & \nu \Delta |\vec{u}_{(\alpha)}|^2 - 2\nu |\vec{\nabla} \otimes \vec{u}_{(\alpha)}|^2 - 2\alpha |\vec{u}_{(\alpha)}|^6 + 2\vec{u}_{(\alpha)} \cdot \vec{f} \\
& - 2\operatorname{div}(p_{(\alpha)}\vec{u}_{(\alpha)}) + 2(\operatorname{div} \vec{u}_{(\alpha)})\frac{1}{\Delta} \operatorname{div} \vec{f} - \frac{2}{\alpha}|\operatorname{div} \vec{u}_{(\alpha)}|^2 \\
& - \operatorname{div}(|\vec{u}_{(\alpha)}|^2 \vec{u}_{(\alpha)}) + \frac{1}{2}(\operatorname{div} \vec{u}_{(\alpha)})|\vec{u}_{(\alpha)}|^2
\end{aligned}$$

We know that $\vec{u}_{(\alpha_k)}$ converge strongly to $\vec{u}$ in $L^2_{\text{loc}}((0,T) \times \mathbb{R}^3)$; as the family is bounded in $L_t^{10/3} H_x^{3/5} \subset L_t^{10/3} L_x^{10/3}$, we find that we have strong convergence in $L^3_{\text{loc}}((0,T) \times \mathbb{R}^3)$ as well. Thus, we have the following convergence results in $\mathcal{D}'((0,T) \times \mathbb{R}^3)$: $\partial_t |\vec{u}_{(\alpha_k)}|^2 \to \partial_t |\vec{u}|^2$, $\Delta |\vec{u}_{(\alpha_k)}|^2 \to \Delta |\vec{u}|^2$, $\operatorname{div}(|\vec{u}_{(\alpha_k)}|^2 \vec{u}_{(\alpha_k)}) \to \operatorname{div}(|\vec{u}|^2\vec{u})$ and $\vec{u}_{(\alpha_k)} \cdot \vec{f} \to \vec{u} \cdot \vec{f}$.

We write $q_{(\alpha)} = -\frac{1}{\alpha} \operatorname{div} \vec{u}_{(\alpha)}$. We have

$$\partial_t q_{(\alpha)} = (\nu + \frac{1}{\alpha})\Delta q_{(\alpha)} + \frac{1}{\alpha} \operatorname{div}(\vec{u}_{(\alpha)} \cdot \vec{\nabla} \vec{u}_{(\alpha)} + \frac{1}{2}(\operatorname{div} \vec{u}_{(\alpha)})\vec{u}_{(\alpha)} + \alpha |\vec{u}_{(\alpha)}|^4 \vec{u}_{(\alpha)})$$

and $q_{(\alpha)}(0,.) = 0$. Thus,

$$q_{(\alpha)} = \int_0^t W_{(\nu+\frac{1}{\alpha})(t-s)} * (\nu + \frac{1}{\alpha})\Delta Q_\alpha \, ds$$

with

$$Q_\alpha = \frac{1}{1+\alpha\nu} \frac{1}{\Delta} \operatorname{div}(\vec{u}_{(\alpha)} \cdot \vec{\nabla} \vec{u}_{(\alpha)} + \frac{1}{2}(\operatorname{div} \vec{u}_{(\alpha)})\vec{u}_{(\alpha)} + \alpha |\vec{u}_{(\alpha)}|^4 \vec{u}_{(\alpha)})$$

As $\vec{u}_{(\alpha)} \cdot \vec{\nabla} \vec{u}_{(\alpha)} + \frac{1}{2}(\operatorname{div} \vec{u}_{(\alpha)})\vec{u}_{(\alpha)} + \alpha |\vec{u}_{(\alpha)}|^4 \vec{u}_{(\alpha)}$ is bounded in $L_t^{6/5} L_x^{6/5}$, we have that Q_α is bounded in $L_t^{6/5} L_x^2$; the maximal regularity of the heat kernel (see Lemarié-Rieusset [313]) gives us that $q_{(\alpha)}$ is bounded in $L_t^{6/5} L_x^2$.

We have $\frac{1}{\Delta} \operatorname{div} \vec{f} \in L_t^2(L_x^2 + L_x^6)$. Thus, $p_{(\alpha)}$ is bounded in $(L_t^{6/5} L_x^2)_{\text{loc}}$. Moreover, the weak convergence of $\vec{u}_{(\alpha_k)} \cdot \vec{\nabla} \vec{u}_{(\alpha_k)} + \frac{1}{2}(\operatorname{div} \vec{u}_{(\alpha_k)})\vec{u}_{(\alpha_k)} + \alpha_k |\vec{u}_{(\alpha_k)}|^4 \vec{u}_{(\alpha_k)}$ to $\vec{u} \cdot \vec{\nabla} \vec{u}$ gives us that $p_{(\alpha_k)}$ converges weakly to p in $(L_t^{6/5} L_x^2)_{\text{loc}}$. As $\vec{u}_{(\alpha_k)}$ is bounded in $L^\infty L^2$ and converges strongly to $\vec{u}$ in $(L^2 L^2)_{\text{loc}}$, it converges strongly in $(L^6 L^2)_{\text{loc}}$ as well, so that, in $\mathcal{D}'((0,T) \times \mathbb{R}^3)$, we have the convergence of $p_{(\alpha_k)}\vec{u}_{(\alpha_k)}$ to $p\vec{u}$. On the other hand, $\|\operatorname{div} \vec{u}_{(\alpha)}\|_{L^2 L^2} = O(\sqrt{\alpha})$ while $\||\vec{u}_{(\alpha)}\|_{L^3 L^3} = O(\alpha^{-1/3})$ and $\frac{1}{\Delta} \operatorname{div} \vec{f}$ belongs to $L_t^2(L_x^2 + L_x^6)$; thus, $(\operatorname{div} \vec{u}_{(\alpha)})\frac{1}{\Delta} \operatorname{div} \vec{f}$ and $(\operatorname{div} \vec{u}_{(\alpha)})|\vec{u}_{(\alpha)}|^2$ converge to 0.

Thus far, we have got that

$$\partial_t |\vec{u}|^2 = \nu \Delta |\vec{u}|^2 - \text{div}((|\vec{u}|^2 + 2p)\vec{u}) + 2\vec{u} \cdot \vec{f} - T$$

with

$$T = \lim_{\alpha_k \to 0} 2\nu |\vec{\nabla} \otimes \vec{u}_{(\alpha_k)}|^2 + 2\alpha_k |\vec{u}_{(\alpha_k)}|^6 + \frac{2}{\alpha_k} |\text{div}\, \vec{u}_{(\alpha_k)}|^2.$$

Let $\phi \in \mathcal{D}'((0,T) \times \mathbb{R}^3)$ be a non-negative function. We have

$$\langle T|\phi \rangle_{\mathcal{D}',\mathcal{D}} = \lim_{\alpha_k \to 0} \iint (2\nu |\vec{\nabla} \otimes \vec{u}_{(\alpha_k)}|^2 + 2\alpha_k |\vec{u}_{(\alpha_k)}|^6 \frac{2}{\alpha_k} |\text{div}\, \vec{u}_{(\alpha_k)}|^2)\phi(t,x)\, dt\, dx$$

$$\geq 2\nu \liminf_{\alpha_k \to 0} \iint |\vec{\nabla} \otimes \vec{u}_{(\alpha_k)}|^2 \phi(t,x)\, dt\, dx$$

$$\geq 2\nu \iint |\vec{\nabla} \otimes \vec{u}|^2 \phi(t,x)\, dt\, dx.$$

Thus, $T = 2\nu |\vec{\nabla} \otimes \vec{u}_{(\alpha_k)}|^2 + \mu$, where μ is a non-negative locally bounded measure. The solution $\vec{u}$ is thus suitable.

- **Step 6: Strong convergence.**

Now, if $\vec{u}_0$ is in H^1 and $\vec{u}$ is a solution of the Navier–Stokes problem with $\vec{u} \in L^\infty([0,T], H^1) \cap L^2([0,T], H^2)$, it is easy to check the strong convergence of $\vec{u}_{(\alpha)}$ to $\vec{u}$: as $\vec{u}$ and $\vec{u}_{(\alpha)}$ are regular enough, we may write:

$$\frac{d}{dt} \|\vec{u} - \vec{u}_{(\alpha)}\|_2^2 = 2\int (\vec{u} - \vec{u}_{(\alpha)}) \cdot \partial_t(\vec{u} - \vec{u}_{(\alpha)})\, dx$$

$$= -2\nu \|\vec{\nabla} \otimes (\vec{u} - \vec{u}_{(\alpha)})\|_2^2 + 2\alpha \int (\vec{u} - \vec{u}_{(\alpha)}) \cdot (|\vec{u}_{(\alpha)}|^4 \vec{u}_{(\alpha)})\, dx$$

$$- 2\int (\vec{u} - \vec{u}_{(\alpha)}) \cdot (\vec{u} \cdot \vec{\nabla}\vec{u} - \vec{u}_{(\alpha)} \cdot \vec{\nabla}\vec{u}_{(\alpha)})\, dx + \int (\text{div}\, \vec{u}_{(\alpha)})\vec{u}_{(\alpha)} \cdot \vec{u}\, ds$$

$$- 2\int (\vec{u} - \vec{u}_{(\alpha)}).\vec{\nabla}(p - p_{(\alpha)})\, dx$$

$$\leq -2\nu \|\vec{\nabla} \otimes (\vec{u} - \vec{u}_{(\alpha)})\|_2^2 + 2\alpha \int \vec{u} \cdot (|\vec{u}_{(\alpha)}|^4 \vec{u}_{(\alpha)})\, dx$$

$$- 2\int (\vec{u} - \vec{u}_{(\alpha)}) \cdot ((\vec{u}. - \vec{u}_{(\alpha)}) \cdot \vec{\nabla}\vec{u})\, dx + \int (\text{div}\, \vec{u}_{(\alpha)})(\vec{u} - \vec{u}_{(\alpha)}) \cdot \vec{u}\, ds$$

$$- 2\int (\text{div}\, \vec{u}_{(\alpha)})\frac{1}{\Delta}\text{div}(\vec{u} \cdot \vec{\nabla}\vec{u})\, dx$$

$$\leq -2\nu \|\vec{\nabla} \otimes (\vec{u} - \vec{u}_{(\alpha)})\|_2^2 + 2\|\vec{u} - \vec{u}_{(\alpha)}\|_6 \|\vec{u} - \vec{u}_{(\alpha)}\|_2 \|\vec{\nabla} \otimes \vec{u}\|_3$$
$$+ 2\|\text{div}(\vec{u} - \vec{u}_{(\alpha)})\|_2 \|\vec{u} - \vec{u}_{(\alpha)}\|_2 \|\vec{u}\|_\infty$$
$$+ 2\alpha \|\vec{u}\|_6 \|\vec{u}_{(\alpha)}\|_6^5 + 2\|\text{div}\, \vec{u}_{(\alpha)}\|_2 \|\vec{u}\|_4^2$$

$$\leq \frac{C}{\nu} \|\vec{u} - \vec{u}_{(\alpha)}\|_2^2 (\|\vec{\nabla} \otimes \vec{u}\|_3^2 + \|\vec{u}\|_\infty^2)$$

$$+ 2\alpha^{1/6}\|\vec{u}\|_6(\alpha^{1/6}\|\vec{u}_{(\alpha)}\|_6)^5 + 2\sqrt{\alpha}\|\vec{u}\|_4^2(\frac{\|\text{div}\, \vec{u}_{(\alpha)}\|_2}{\sqrt{\alpha}})$$

From (19.14) and Grönwall's lemma, we find that $\sup_{0<t<T} \|\vec{u} - \vec{u}_{(\alpha)}\|_2 = O(\alpha^{1/6}) \to_{\alpha \to 0} 0$. $\square$

19.3   Hyperbolic Approximation

As the heat equation

$$\partial_t \theta = \kappa \Delta \theta$$

implies instantaneous diffusion to the whole space, hence transfer of information with an infinite speed, it has been criticized as contradictory with the theory of relativity. To overcome this contradiction Cattaneo [92, 93, 94] introduced a hyperbolic correction to the equation

$$\frac{1}{C^2}\partial_t^2 \theta + \frac{1}{\kappa}\partial_t \theta = \Delta \theta$$

to bound the speed of propagation by a limit C.

Brenier, Natalini and Puel [64] studied a similar hyperbolic correction to the Navier–Stokes equations:

$$\alpha^2 \partial_t^2 \vec{u} + \partial_t \vec{u} = \nu \Delta \vec{u} + \vec{f} - \vec{u} \cdot \vec{\nabla} \vec{u} - \vec{\nabla} p, \quad \operatorname{div} \vec{u} = 0$$

Under strong assumptions on the regularity of the initial values, they studied global existence and convergence to the Navier–Stokes equations in the 2D case, by means of energy inequalities.

The extension to the 3D setting was conducted by Paicu and Raugel [390], with help of Strichartz estimates for a damped wave equation. In 3D, those (global) approximation results hold only for small enough regular initial values and forcing terms. In particular, one assumes that the solution of the limit Navier–Stokes problem is global. Hachicha [226] then proved the same results with minimal regularity assumptions leading to global mild solutions and through energy methods in fractional Sobolev spaces.

Brenier, Natalini and Puel's model however does not follow the aim of Cattaneo's relaxation of finite speed of propagation for the solutions of the relaxed equation, as the incompressibility of the fluid still induces instantaneous spreading. Hachicha [225] then combined Cattaneo's relaxation and Višik and Fursikov's artificial compressibility to get a hyperbolic model with finite speed of propagation.

First, let us remark that, for $\vec{u} \in H^2$, one has

$$\|\vec{u} \cdot \vec{\nabla} \vec{u} + \frac{1}{2}(\operatorname{div} \vec{u})\vec{u}\|_2 \le C\|\vec{u}\|_3\|\vec{\nabla} \otimes \vec{u}\|_6 \le C'\|\vec{u}\|_2^{1/2}\|\vec{\nabla} \otimes \vec{u}\|_2^{1/2}\|\Delta \vec{u}\|_2.$$

Hachicha's result is then the following one:

Hyperbolic approximation with finite speed of propagation

Theorem 19.3.
Let C_* be a constant such that, for any vector field in H^2, one has

$$\|\vec{u} \cdot \vec{\nabla} \vec{u} + \frac{1}{2}(\operatorname{div} \vec{u})\vec{u}\|_2 \le C_*\|\vec{u}\|_2^{1/2}\|\vec{\nabla} \otimes \vec{u}\|_2^{1/2}\|\Delta \vec{u}\|_2. \qquad (19.15)$$

Let $\vec{u}_0 \in H^1$ with $\operatorname{div} \vec{u}_0 = 0$, and $\vec{f} \in L_t^2 L_x^2 \cap L_t^2 \dot{H}_x^{-1}$ (i.e., $\vec{f} = \operatorname{div} \mathbb{F}$ with $\mathbb{F} \in L_t^2 H^1$), such that

$$(\|\vec{u}_0\|_2^2 + \frac{1}{\nu}\|\vec{f}\|_{L^2 \dot{H}^{-1}}^2)(\|\vec{u}_0\|_{\dot{H}^1}^2 + \frac{1}{\nu}\|\vec{f}\|_{L^2 L^2}^2) < \left(\frac{\nu^2}{4C_*^2}\right)^2. \qquad (19.16)$$

- Let $\delta > 0$ be small enough. Let $\alpha > 0$ and $\beta > 0$. Let $\vec{u}_{(\alpha),0} = \mathcal{F}^{-1}(1_{\sqrt{\alpha}|\xi|<\sqrt{\delta}}\mathcal{F}\vec{u}_0)$ and $\vec{f}_{(\alpha)} = \mathcal{F}_x^{-1}(1_{\sqrt{\alpha}|\xi|<\sqrt{\delta}}\mathcal{F}_x\vec{f})$. Then, if α is small enough (independently from β), the problem

$$\begin{cases} \alpha\partial_t^2\vec{u} + \partial_t\vec{u} + \vec{u}\cdot\vec{\nabla}\vec{u} = \nu\Delta\vec{u} + \vec{f}_{(\alpha)} - \vec{\nabla}p - \frac{1}{2}(\text{div }\vec{u})\vec{u} \\ \\ \beta p + \text{div }\vec{u} = 0 \end{cases} \quad (19.17)$$

with initial values $\vec{u}(0,.) = \vec{u}_{(\alpha),0}$ and $\partial_t\vec{u}(0,.) = 0$, has a unique solution $\vec{u}_{(\alpha,\beta)}$ such that $\vec{u}_{(\alpha,\beta)} \in \mathcal{C}H^2$ with $\partial_t\vec{u}_{(\alpha,\beta)} \in L^\infty H^1$. Moreover, we have $\vec{u}_{(\alpha,\beta)} \in L_t^\infty H_x^1 \cap L_t^2 H_x^2$ (uniformly with respect to α and β).

- Finite speed: the value of $\vec{u}_{(\alpha,\beta)}$ at (t_0, x_0) depends only on the values of $\vec{u}_{(\alpha,0)}$ for $|x - x_0| \le \sqrt{\frac{\nu+\frac{1}{\beta}}{\alpha}}t_0$ and of $\vec{f}_{(\alpha)}$ for $0 < t < t_0$ and $|x - x_0| \le \sqrt{\frac{\nu+\frac{1}{\beta}}{\alpha}}|t - t_0|$

- Let $\vec{u}$ be the solution to the Cauchy problem for the Navier–Stokes equations with initial value $\vec{u}_0$ and forcing term $\vec{f}$. We have the strong convergence $\lim_{(\alpha,\beta)\to(0,0)} \|\vec{u}_{\alpha,\beta} - \vec{u}\|_{L_t^\infty L_x^2} = 0$.

Proof. • **First step: Mild solution for the Navier–Stokes equations.**

We first check that, under the smallness condition (19.16), the Navier–Stokes problem has a global solution $\vec{u} \in L^\infty H^1 \cap L^2 H^2$. Using the identity $\min_{A>0}(Aa + \frac{b}{A}) = 2\sqrt{ab}$ for $a, b > 0$, we may fix A such that

$$\frac{A}{2}(\|\vec{u}_0\|_2^2 + \frac{1}{\nu}\|\vec{f}\|_{L^2\dot{H}^{-1}}^2) + \frac{1}{2A}(\|\vec{u}_0\|_{\dot{H}^1}^2 + \frac{1}{\nu}\|\vec{f}\|_{L^2 L^2}^2) < \frac{\nu^2}{4C_*^2}.$$

We know (Theorem 7.1) that we may locally in time solve the Navier–Stokes problem in $L^\infty H^1 \cap L^2\dot{H}^2$, and that the existence time depends only on the size of $\|\vec{u}_0\|_{H^1}$. Thus, we must check that the norm of $\vec{u}$ cannot explode. In order to do that, we define

$$E(t) = \frac{A}{2}\|\vec{u}\|_2^2 + \frac{1}{2A}\|\vec{u}\|_{\dot{H}^1}^2.$$

We have

$$\frac{d}{dt}E(t) = \int \partial_t\vec{u}.(A\vec{u} - \frac{1}{A}\Delta\vec{u})\,dx$$

$$= -A\nu\|\vec{\nabla}\otimes\vec{u}\|_2^2 - \frac{\nu}{A}\|\Delta\vec{u}\|_2^2 + A\int\vec{f}\cdot\vec{u}\,dx - \frac{1}{A}\vec{f}.\Delta\vec{u}\,dx + \frac{1}{A}\int\Delta\vec{u}.(\vec{u}\cdot\vec{\nabla}\vec{u})\,dx$$

$$\le -\frac{A\nu}{2}\|\vec{\nabla}\otimes\vec{u}\|_2^2 - \frac{\nu}{2A}\|\Delta\vec{u}\|_2^2 + \frac{A}{2\nu}\|\vec{f}\|_{\dot{H}^{-1}}^2 + \frac{1}{2A\nu}\|\vec{f}\|_2^2 + \frac{1}{A}\|\Delta\vec{u}\|_2\|\vec{u}\cdot\vec{\nabla}\vec{u}\|_2$$

$$\le -\frac{A\nu}{2}\|\vec{\nabla}\otimes\vec{u}\|_2^2 - \frac{\nu}{2A}\|\Delta\vec{u}\|_2^2(1 - \frac{2C_*}{\nu}\|\vec{u}\|_2^{1/2}\|\vec{u}\|_{\dot{H}^1}^{1/2}) + \frac{A}{2\nu}\|\vec{f}\|_{\dot{H}^{-1}}^2 + \frac{1}{2A\nu}\|\vec{f}\|_2^2$$

Thus, as long as $\|\vec{u}\|_2^{1/2}\|\vec{u}\|_{\dot{H}^1}^{1/2} < \frac{\nu}{2C_*}$, we have

$$\frac{A}{2}\|\vec{u}\|_2^2 + \frac{1}{2A}\|\vec{u}\|_{\dot{H}^1}^2 \le \frac{A}{2}\|\vec{u}_0\|_2^2 + \frac{1}{2A}\|\vec{u}_0\|_{\dot{H}^1}^2 + \frac{A}{2\nu}\|\vec{f}\|_{L^2\dot{H}^{-1}}^2 + \frac{1}{2A\nu}\|\vec{f}\|_{L^2 L^2}^2.$$

Since $\|\vec{u}\|_2 \|\vec{u}\|_{\dot{H}^1} \le \frac{A}{2}\|\vec{u}\|_2^2 + \frac{1}{2A}\|\vec{u}\|_{\dot{H}^1}^2$, we find that we have a global solution on $(0,T)$ in $L^\infty H^1 \cap L^2 H^2$ as soon as[2]

$$\left(\|\vec{u}_0\|_2^2 + \frac{1}{\nu}\|\vec{f}\|_{L^2((0,T),\dot{H}^{-1})}^2\right)\left(\|\vec{u}_0\|_{\dot{H}^1}^2 + \frac{1}{\nu}\|\vec{f}\|_{L^2((0,T),L^2)}^2\right) < \frac{\nu^2}{4C_*^2}.$$

- **Step 2: Local existence of $\vec{u}_{(\alpha,\beta)}$.**

Proving local existence of a solution $\vec{u}$ to the problem (19.17) amounts to solve a classical problem in non-linear wave equation: we write $\vec{v} = \mathbb{P}\vec{u}$ and $\vec{w} = (Id-\mathbb{P})\vec{u}$. They are solutions of the system

$$\begin{cases} (\alpha\partial_t^2 - \nu\Delta)\vec{v} = \mathbb{P}\left(-\partial_t(\vec{v}+\vec{w}) + \vec{f}_{(\alpha)} - (\vec{v}+\vec{w})\vec{\nabla}(\vec{v}+\vec{w}) - \frac{1}{2}(\text{div }\vec{w})(\vec{v}+\vec{w})\right) \\ (\alpha\partial_t^2 - \gamma\Delta)\vec{w} = (Id - \mathbb{P})\left(-\partial_t(\vec{v}+\vec{w}) + \vec{f}_{(\alpha)} - (\vec{v}+\vec{w})\vec{\nabla}(\vec{v}+\vec{w}) - \frac{\text{div }\vec{w}}{2}(\vec{v}+\vec{w})\right) \end{cases}$$

with $\gamma = \nu + \frac{1}{\beta}$.

The solutions of this system (with general initial values $\vec{v}(0,.) = \vec{V}_0$, $\partial_t\vec{v}(0,.) = \vec{V}_1$, $\vec{w}(0,.) = \vec{W}_0$, and $\partial_t\vec{w}(0,.) = \vec{W}_1$) are fixed point of the transform $(\vec{v},\vec{w}) \mapsto (\vec{V},\vec{W})$ with

$$\begin{cases} \vec{V} = \cos(\sqrt{\frac{\nu}{\alpha}}t\sqrt{-\Delta})\vec{V}_0 + \frac{\sin(\sqrt{\frac{\nu}{\alpha}}t\sqrt{-\Delta})}{\sqrt{\frac{\nu}{\alpha}}\sqrt{-\Delta}}\vec{V}_1 \\ \qquad\qquad + \mathbb{P}\int_0^t \frac{\sin(\sqrt{\frac{\nu}{\alpha}}(t-s)\sqrt{-\Delta})}{\sqrt{\alpha\nu}\sqrt{-\Delta}}\left(\vec{f}_{(\alpha)} - B(\vec{v},\vec{w})\right) ds \\ \vec{W} = \cos(\sqrt{\frac{\gamma}{\alpha}}t\sqrt{-\Delta})\vec{W}_0 + \frac{\sin(\sqrt{\frac{\gamma}{\alpha}}t\sqrt{-\Delta})}{\sqrt{\frac{\gamma}{\alpha}}\sqrt{-\Delta}}\vec{W}_1 \\ \qquad\qquad + (Id - \mathbb{P})\int_0^t \frac{\sin(\sqrt{\frac{\gamma}{\alpha}}(t-s)\sqrt{-\Delta})}{\sqrt{\gamma\alpha}\sqrt{-\Delta}}\left(\vec{f}_{(\alpha)} - B(\vec{v},\vec{w})\right) ds \end{cases}$$

where

$$B(\vec{v},\vec{w}) = (\vec{v}+\vec{w})\vec{\nabla}(\vec{v}+\vec{w}) + \frac{1}{2}(\text{div }\vec{w})(\vec{v}+\vec{w}) + \partial_t(\vec{v}+\vec{w}).$$

Let us write

$$\vec{A} = \cos(\sqrt{\frac{\nu}{\alpha}}t\sqrt{-\Delta})\vec{V}_0 + \frac{\sin(\sqrt{\frac{\nu}{\alpha}}t\sqrt{-\Delta})}{\sqrt{\frac{\nu}{\alpha}}\sqrt{-\Delta}}\vec{V}_1 + \mathbb{P}\int_0^t \frac{\sin(\sqrt{\frac{\nu}{\alpha}}(t-s)\sqrt{-\Delta})}{\sqrt{\alpha\nu}\sqrt{-\Delta}}\vec{f}_{(\alpha)}\, ds$$

and

$$\vec{B} = \cos(\sqrt{\frac{\gamma}{\alpha}}t\sqrt{-\Delta})\vec{W}_0 + \frac{\sin(\sqrt{\frac{\gamma}{\alpha}}t\sqrt{-\Delta})}{\sqrt{\frac{\gamma}{\alpha}}\sqrt{-\Delta}}\vec{W}_1 \\ \qquad\qquad + (Id - \mathbb{P})\int_0^t \frac{\sin(\sqrt{\frac{\gamma}{\alpha}}(t-s)\sqrt{-\Delta})}{\sqrt{\gamma\alpha}\sqrt{-\Delta}}\vec{f}_{(\alpha)}\, ds.$$

For $0 < T_0 < T$, we have

$$\sup_{0<t<T_0} \|\vec{A}(t,.)\|_{H^2} + \|\partial_t\vec{A}(t,.)\|_{H^1} \le$$

$$(1 + T_0 + \sqrt{\frac{\alpha}{\nu}})(\sqrt{\frac{\nu}{\alpha}}\|\vec{V}_0\|_{H^2} + \|\vec{V}_1\|_{H^1} + \frac{\sqrt{T_0}}{\alpha}\|\vec{f}_{(\alpha)}\|_{L^2 H^1})$$

---

[2] As a matter of fact, we saw a more precise condition in Theorem 7.2: $\|\vec{u}_0\|_{\dot{H}^{1/2}} < \epsilon_0\nu$ and $\int_0^T \|\vec{f}(s,.)\|_{\dot{H}^{-\frac{1}{2}}}^2\, ds < \epsilon_0^2\nu^3$ ensure that we have a global solution on $(0,T)$.

and

$$\sup_{0<t<T_0} \|\vec{B}(t,.)\|_{H^2} + \|\partial_t \vec{B}(t,.)\|_{H^1} \le$$

$$(1 + T_0 + \sqrt{\frac{\alpha}{\gamma}})(\sqrt{\frac{\gamma}{\alpha}}\|\vec{W}_0\|_{H^2} + \|\vec{W}_1\|_{H^1} + \frac{\sqrt{T_0}}{\alpha}\|\vec{f}_{(\alpha)}\|_{L^2 H^1}).$$

Similarly, we have, for $\vec{C}(\vec{v}, \vec{w}) = \mathbb{P} \int_0^t \frac{\sin(\sqrt{\frac{\nu}{\alpha}}(t-s)\sqrt{-\Delta})}{\sqrt{\alpha \nu}\sqrt{-\Delta}} B(\vec{v}, \vec{w}) \, ds$ and $\vec{D}(\vec{v}, \vec{w}) = (Id - \mathbb{P}) \int_0^t \frac{\sin(\sqrt{\frac{\gamma}{\alpha}}(t-s)\sqrt{-\Delta})}{\sqrt{\gamma \alpha}\sqrt{-\Delta}} B(\vec{v}, \vec{w}) \, ds$,

$$\sup_{0<t<T_0} \|\vec{C}(\vec{v}_1, \vec{w}_1) - \vec{C}(\vec{v}_2, \vec{w}_2)\|_{H^2} + \|\partial_t \vec{C}(\vec{v}_1, \vec{w}_1) - \partial_t \vec{C}(\vec{v}_2, \vec{w}_1)\|_{H^2}$$

$$\le \frac{T_0}{\alpha}(1 + T_0 + \sqrt{\frac{\alpha}{\nu}}) \sup_{0<t<T_0} (\|\vec{v}_1\|_{H^2} + \|\vec{w}_1\|_{H^2} + \|\vec{v}_2\|_{H^2} + \|\vec{w}_2\|_{H^2}) \times$$

$$\times \sup_{0<t<T_0} (\|\vec{v}_1 - \vec{v}_2\|_{H^2} + \|\vec{w}_1 - \vec{w}_2\|_{H^2})$$

$$+ \frac{T_0}{\alpha}(1 + T_0 + \sqrt{\frac{\alpha}{\nu}}) \sup_{0<t<T_0} (\|\partial_t \vec{v}_1 - \partial_t \vec{v}_2\|_{H^1} + \|\partial_t \vec{w}_1 - \partial_t \vec{w}_2\|_{H^1})$$

and

$$\sup_{0<t<T_0} \|\vec{D}(\vec{v}_1, \vec{w}_1) - \vec{D}(\vec{v}_2, \vec{w}_2)\|_{H^2} + \|\partial_t \vec{D}(\vec{v}_1, \vec{w}_1) - \partial_t \vec{D}(\vec{v}_2, \vec{w}_1)\|_{H^2}$$

$$\le \frac{T_0}{\alpha}(1 + T_0 + \sqrt{\frac{\alpha}{\gamma}}) \sup_{0<t<T_0} (\|\vec{v}_1\|_{H^2} + \|\vec{w}_1\|_{H^2} + \|\vec{v}_2\|_{H^2} + \|\vec{w}_2\|_{H^2})\|\vec{v}_1 - \vec{v}_2\|_{H^2}$$

$$+ \frac{T_0}{\alpha}(1 + T_0 + \sqrt{\frac{\alpha}{\gamma}}) \sup_{0<t<T_0} (\|\vec{v}_1\|_{H^2} + \|\vec{w}_1\|_{H^2} + \|\vec{v}_2\|_{H^2} + \|\vec{w}_2\|_{H^2})\|\vec{w}_1 - \vec{w}_2\|_{H^2}$$

$$+ \frac{T_0}{\alpha}(1 + T_0 + \sqrt{\frac{\alpha}{\gamma}}) \sup_{0<t<T_0} (\|\partial_t \vec{v}_1 - \partial_t \vec{v}_2\|_{H^1} + \|\partial_t \vec{w}_1 - \partial_t \vec{w}_2\|_{H^1})$$

This ensures the existence of a solution $\vec{u} = \vec{v} + \vec{w}$ on $(0, T_0)$, when T_0 is small enough to ensure that

$$\frac{T_0}{\alpha}(1 + T_0 + \sqrt{\frac{\alpha}{\nu}}) + \frac{T_0}{\alpha}(1 + T_0 + \sqrt{\frac{\alpha}{\gamma}}) \le \frac{1}{2}$$

and

$$2T_0(\frac{1 + T_0 + \sqrt{\frac{\alpha}{\nu}}}{\alpha} + \frac{1 + T_0 + \sqrt{\frac{\alpha}{\gamma}}}{\alpha}) \times$$

$$\left((1 + T_0 + \sqrt{\frac{\alpha}{\nu}})(\sqrt{\frac{\nu}{\alpha}}\|\vec{V}_0\|_{H^2} + \|\vec{V}_1\|_{H^1} + \frac{\sqrt{T_0}}{\alpha}\|\vec{f}_{(\alpha)}\|_{L^2 H^1}) +\right.$$

$$\left.(1 + T_0 + \sqrt{\frac{\alpha}{\gamma}})(\sqrt{\frac{\gamma}{\alpha}}\|\vec{W}_0\|_{H^2} + \|\vec{W}_1\|_{H^1} + \frac{\sqrt{T_0}}{\alpha}\|\vec{f}_{(\alpha)}\|_{L^2 H^1}))\right)$$

$$\le \frac{1}{2}$$

- **Step 3: Speed of propagation.**

In order to check the finite speed of propagation, one cannot use the Leray operator $\mathbb{P}$, as it has no compactly supported convolution kernel.

We know that the local solution $\vec{u}_{(\alpha,\beta)}$ may be computed through an iterative process: we start from $\vec{U}_{(0)}$ defined by

$$\alpha \partial_t^2 \vec{U}_{(0)} = \nu \Delta \vec{U}_{(0)} + \vec{f}_{(\alpha)} + \frac{1}{\beta} \vec{\nabla} \operatorname{div} \vec{U}_{(0)}$$

(with $\vec{U}_{(0)}(0,.) = \vec{u}_{(\alpha)}(0,.)$ and $\partial_t \vec{U}_{(0)}(0,.) = \partial_t \vec{u}_{(\alpha)}(0,.)$) and

$$\alpha \partial_t^2 \vec{U}_{(n+1)} = \nu \Delta \vec{U}_{n+1} + \vec{f}_{(\alpha)} + \frac{1}{\beta} \vec{\nabla} \operatorname{div} \vec{U}_{(n+1)} - \partial_t \vec{U}_{(n)} - \vec{U}_{(n)} \cdot \vec{\nabla} \vec{U}_{(n)} - \frac{1}{2} (\operatorname{div} \vec{U}_{(n)}) \vec{U}_{(n)}$$

(with $\vec{U}_{(n+1)}(0,.) = \vec{u}_{(\alpha)}(0,.)$ and $\partial_t \vec{U}_{(n+1)}(0,.) = \partial_t \vec{u}_{(\alpha)}(0,.)$) We then take as new unknown $Z_{(n)} = \operatorname{div} \vec{U}_{(n)}$. We have, for $\gamma = \nu + \frac{1}{\beta}$,

$$\alpha \partial_t^2 Z_{(0)} = \gamma \Delta Z_{(0)} + \vec{f}_{(\alpha)}$$

(with $Z_{(0)}(0,.) = \operatorname{div} \vec{u}_{(\alpha)}(0,.)$ and $\partial_t Z_{(0)}(0,.) = \operatorname{div} \partial_t \vec{u}_{(\alpha)}(0,.)$) and

$$\alpha \partial_t^2 \vec{U}_{(0)} = \nu \Delta \vec{U}_0 + \vec{f}_{(\alpha)} + \frac{1}{\beta} \vec{\nabla} Z_{(0)}$$

(with $\vec{U}_{(0)}(0,.) = \vec{u}_{(\alpha)}(0,.)$ and $\partial_t \vec{U}_{(0)}(0,.) = \partial_t \vec{u}_{(\alpha)}(0,.)$) Similarly, we have

$$\alpha \partial_t^2 Z_{(n+1)} = \gamma \Delta Z_{(n+1)} + \vec{f}_{(\alpha)} - \operatorname{div} \left(\partial_t \vec{U}_{(n)} + \vec{U}_{(n)} \cdot \vec{\nabla} \vec{U}_{(n)} + \frac{1}{2} (\operatorname{div} \vec{U}_{(n)}) \vec{U}_{(n)} \right)$$

(with $Z_{(n+1)}(0,.) = \operatorname{div} \vec{u}_{(\alpha)}(0,.)$ and $\partial_t Z_{(n+1)}(0,.) = \operatorname{div} \partial_t \vec{u}_{(\alpha)}(0,.)$) and

$$\alpha \partial_t^2 \vec{U}_{(n+1)} = \nu \Delta \vec{U}_{n+1} + \vec{f}_{(\alpha)} + \frac{1}{\beta} \vec{\nabla} Z_{(n+1)} - \partial_t \vec{U}_{(n)} - \vec{U}_{(n)} \cdot \vec{\nabla} \vec{U}_{(n)} - \frac{1}{2} (\operatorname{div} \vec{U}_{(n)}) \vec{U}_{(n)}$$

(with $\vec{U}_{(n+1)}(0,.) = \vec{u}_{(\alpha)}(0,.)$ and $\partial_t \vec{U}_{(n+1)}(0,.) = \partial_t \vec{u}_{(\alpha)}(0,.)$)

Thus, the value of $Z_{(0)}(t_0, x_0)$ depends only on the values of $\operatorname{div} \vec{u}_{(\alpha)}(0, x)$ and $\operatorname{div} \partial_t \vec{u}_{(\alpha)}(0, x)$ with $|x - x_0| = \sqrt{\frac{\gamma}{\alpha}} t_0$ and on the values of $\vec{f}_{(\alpha)}(t, x)$ with $0 < t < t_0$ and $|x - x_0| = \sqrt{\frac{\gamma}{\alpha}}(t_0 - t)$; similarly, the value of $\vec{U}_{(0)}(t_0, x_0)$ depends only on the values of $\vec{u}_{(\alpha)}(0, x)$ and $\partial_t \vec{u}_{(\alpha)}(0, x)$ with $|x - x_0| = \sqrt{\frac{\nu}{\alpha}} t_0$ and on the values of $\vec{f}_{(\alpha)}(t, x)$ and $\vec{\nabla} Z_{(0)}(t, x)$ with $0 < t < t_0$ and $|x - x_0| = \sqrt{\frac{\nu}{\alpha}}(t_0 - t)$.

Similarly, the value of $Z_{(n+1)}(t_0, x_0)$ depends only on the values of $\operatorname{div} \vec{u}_{(\alpha)}(0, x)$ and $\operatorname{div} \partial_t \vec{u}_{(\alpha)}(0, x)$ with $|x - x_0| = \sqrt{\frac{\gamma}{\alpha}} t_0$ and on the values of $\vec{f}_{(\alpha)}(t, x)$ and $\operatorname{div} \left(\partial_t \vec{U}_{(n)} + \vec{U}_{(n)} \cdot \vec{\nabla} \vec{U}_{(n)} + \frac{1}{2} (\operatorname{div} \vec{U}_{(n)}) \vec{U}_{(n)} \right) (t, x)$ with $0 < t < t_0$ and $|x - x_0| = \sqrt{\frac{\gamma}{\alpha}}(t_0 - t)$; similarly, the value of $\vec{U}_{(n+1)}(t_0, x_0)$ depends only on the values of $\vec{u}_{(\alpha)}(0, x)$ and $\partial_t \vec{u}_{(\alpha)}(0, x)$ with $|x - x_0| = \sqrt{\frac{\nu}{\alpha}} t_0$ and on the values of $\vec{f}_{(\alpha)}(t, x)$, $\left(\partial_t \vec{U}_{(n)} + \vec{U}_{(n)} \cdot \vec{\nabla} \vec{U}_{(n)} + \frac{1}{2} (\operatorname{div} \vec{U}_{(n)}) \vec{U}_{(n)} \right) (t, x)$ and $\vec{\nabla} Z_{(n+1)}(t, x)$ with $0 < t < t_0$ and $|x - x_0| = \sqrt{\frac{\nu}{\alpha}}(t_0 - t)$.

By induction, we find that the value of $\vec{U}_{(n)}(t_0, x_0)$, and thus of $\vec{u}_{(\alpha)}(t_0, x_0)$, depends only on the values of $\vec{u}_{(\alpha)}(0, x)$ and $\partial_t \vec{u}_{(\alpha)}(0, x)$ with $|x - x_0| \leq \sqrt{\frac{\gamma}{\alpha}} t_0$ and on the values of $\vec{f}_{(\alpha)}(t, x)$ with $0 < t < t_0$ and $|x - x_0| \leq \sqrt{\frac{\gamma}{\alpha}}(t_0 - t)$.

- **Step 4: Energy estimates.**

In this section, we consider a solution $\vec{u}$ of the linear problem

$$\begin{cases} \alpha \partial_t^2 \vec{u} + \partial_t \vec{u} = \nu \Delta \vec{u} + \frac{1}{\beta} \vec{\nabla} \operatorname{div} \vec{u} + \vec{F} \\ \\ \vec{u}(0,.) = \vec{U}_0, \quad \partial_t \vec{U}(0,.) = \vec{U}_1 \end{cases} \tag{19.18}$$

with $\vec{U}_0 \in H^1$, $\vec{U}_1 \in L^2$ and $\vec{F} \in L^1((0,T), L^2)$. In that case, we have a solution $\vec{u} \in \mathcal{C}([0,T], H^1)$ with $\partial_t \vec{u} \in \mathcal{C}([0,T], L^2)$.

To get energy estimates, we first take the scalar product of the equation with $\vec{u}$. We write

$$\begin{cases} \alpha \int \partial_t^2 \vec{u}.\vec{u}\, dx = \frac{d}{dt}(\alpha \int \partial_t \vec{u}.\vec{u} dx) - \alpha \int |\partial_t \vec{u}|^2\, dx \\ \int \partial_t \vec{u}.\vec{u}\, dx = \frac{d}{dt}(\int \frac{|\vec{u}|^2}{2}\, dx) \\ \nu \langle \Delta \vec{u}|\vec{u}\rangle_{H^{-1},H^1} = -\nu \int |\vec{\nabla} \otimes \vec{u}|^2\, dx \\ \frac{1}{\beta}\langle \vec{\nabla} \operatorname{div} \vec{u}|\vec{u}\rangle_{H^{-1},H^1} = -\frac{1}{\beta}\int |\operatorname{div} \vec{u}|^2\, dx \end{cases}$$

We then define

$$E_1(\vec{u})(t) = \alpha \int \partial_t \vec{u}.\vec{u}\, dx + \int \frac{|\vec{u}|^2}{2}\, dx$$

and we find

$$E_1(\vec{u})(t) = \alpha \int \vec{U}_0 \cdot \vec{U}_1\, dx + \int \frac{|\vec{U}_0|^2}{2}\, dx$$
$$+ \int_0^t \int (\vec{F}.\vec{u} + \alpha|\partial_t \vec{u}|^2 - \nu|\vec{\nabla} \otimes \vec{u}|^2 - \frac{1}{\beta}|\operatorname{div} \vec{u}|^2)\, dx\, ds$$

We would like now to take the scalar product of the equation with $\partial_t \vec{u}$. However, we do not know that $\partial_t \vec{u}$ belongs to $L^\infty H^1$, so we are in trouble for computing the scalar product between $\Delta \vec{u}$ (which is in $L^\infty H^{-1}$) and $\partial_t \vec{u}$ (which is in $L^\infty L^2$). Thus, we first assume that $\vec{U}_0 \in H^2$, $\vec{U}_1 \in H^1$ and $\vec{F} \in L^1 H^1$, so that $\vec{u} \in \mathcal{C}([0,T], H^2)$ and $\partial_t \vec{u} \in \mathcal{C}([0,T], H^1)$. We thus may write:

$$\begin{cases} \alpha \int \partial_t^2 \vec{u}.\partial_t \vec{u}\, dx = \frac{d}{dt}(\alpha \int \frac{|\partial_t \vec{u}|^2}{2}\, dx) \\ \nu \int \Delta \vec{u}.\partial_t \vec{u}\, dx = -\frac{d}{dt}(\nu \int \frac{|\vec{\nabla} \otimes \vec{u}|^2}{2}\, dx) \\ \frac{1}{\beta}\int \vec{\nabla} \operatorname{div} \vec{u}.\partial_t \vec{u}\, dx = -\frac{d}{dt}(\frac{1}{\beta}\int \frac{|\operatorname{div} \vec{u}|^2}{2}\, dx) \end{cases}$$

We then define

$$E_2(\vec{u})(t) = \alpha \int \frac{|\partial_t \vec{u}|^2}{2}\, dx + \nu \int \frac{|\vec{\nabla} \otimes \vec{u}|^2}{2}\, dx + \frac{1}{\beta}\int \frac{|\operatorname{div} \vec{u}|^2}{2}\, dx$$

and we find

$$E_2(\vec{u})(t) = \alpha \int \frac{|\vec{U}_1|^2}{2}\, dx + \nu \int \frac{|\vec{\nabla} \otimes \vec{U}_0|^2}{2}\, dx + \frac{1}{\beta}\int \frac{|\operatorname{div} \vec{U}_0|^2}{2}\, dx$$
$$+ \int_0^t \int (\vec{F}.\partial_t \vec{u} - |\partial_t \vec{u}|^2)\, dx\, ds$$

Now, returning to the general case $\vec{U}_0 \in H^1$, $\vec{U}_1 \in L^2$ and $\vec{F} \in L^1((0,T), L^2)$, we approximate $\vec{u}$ by $\vec{v}$, a solution of the equation with initial values $\vec{v}(0,.) = \vec{V}_0 \in H^2$, $\partial_t \vec{v}(0,.) = \vec{V}_1 \in H^1$ and forcing term $\vec{G} \in L^1((0,T), H^1)$. We have (for $\gamma = \nu + \frac{1}{\beta}$)

$$\sup_{0<t<T} \|\vec{u}(t,.) - \vec{v}(t,.)\|_{H^1} + \|\partial_t \vec{u}(t,.) - \partial_t \vec{v}(t,.)\|_2 \le$$

$$(1+T+\sqrt{\frac{\alpha}{\nu}})(\sqrt{\frac{\nu}{\alpha}}\|\vec{U}_0 - \vec{V}_0\|_{H^1} + \|\vec{U}_1 - V_1\|_2 + \frac{1}{\alpha}\|\vec{F} - \vec{G}\|_{L^1 L^2})$$

$$+(1+T+\sqrt{\frac{\alpha}{\gamma}})(\sqrt{\frac{\gamma}{\alpha}}\|\vec{U}_0 - \vec{V}_0\|_{H^1} + \|\vec{U}_1 - V_1\|_2 + \frac{1}{\alpha}\|\vec{F} - \vec{G}\|_{L^1 L^2}).$$

Thus, if $\vec{V}_0$ converges to $\vec{U}_0$ in H^1, $\vec{V}_1$ to $\vec{U}_1$ in L^2 and $\vec{G}$ to $\vec{F}$ in $L^1 L^2$, we shall have the uniform convergence of $E_2(\vec{v})$ to $E_2(\vec{u})$ on $(0,T)$. We shall have as well the uniform convergence of $\int_0^t \int (\vec{G}.\partial_t \vec{v} - |\partial_t \vec{v}|^2) \, dx \, ds$ to $\int_0^t \int (\vec{F}.\partial_t \vec{u} - |\partial_t \vec{u}|^2) \, dx \, ds$. Thus, the energy equality holds as well for $E_2(\vec{u})$.

Now, if $\vec{U}_0 \in H^2$, $\vec{U}_1 \in H^1$ and $\vec{F} \in L^1 H^1$, we get energy equalities for $E_1(\vec{u})$, $E_2(\vec{u})$, and for $E_1(\partial_j \vec{u})$, $E_2(\partial_j \vec{u})$ as well. We then define, for some $\theta > 0$,

$$E(\vec{u}) = A(E_1(\vec{u}) + \alpha(1+\theta)E_2(\vec{u})) + \frac{1}{A}(\sum_{j=1}^3 E_1(\partial_j \vec{u}) + \alpha(1+\theta)\sum_{j=1}^3 E_2(\partial_j \vec{u})).$$

or, equivalently,

$$E(\vec{u}) = A(\int \frac{|\vec{u} + \alpha\partial_t \vec{u}|^2}{2} + \theta\alpha^2 \frac{|\partial_t \vec{u}|^2}{2})$$

$$+ A(\alpha\nu(1+\theta)\frac{|\vec{\nabla} \otimes \vec{u}|^2}{2} + \frac{\alpha(1+\theta)}{\beta}\frac{|\operatorname{div} \vec{u}|^2}{2} \, dx)$$

$$+ \frac{1}{A}(\int \frac{|\vec{\nabla} \otimes (\vec{u} + \alpha\partial_t \vec{u})|^2}{2} + \theta\alpha^2 \frac{|\vec{\nabla} \otimes \partial_t \vec{u}|^2}{2})$$

$$+ \frac{1}{A}(\alpha\nu(1+\theta)\frac{|\Delta \vec{u}|^2}{2} + \frac{\alpha(1+\theta)}{\beta}\frac{|\vec{\nabla} \operatorname{div} \vec{u}|^2}{2} \, dx).$$

The energy equality then reads as

$$E(\vec{u})(t) - E(\vec{u})(0) =$$

$$A \int_0^t \int (\vec{F}.(\vec{u} + (1+\theta)\alpha\partial_t \vec{u}) - \theta\alpha|\partial_t \vec{u}|^2 - \nu|\vec{\nabla} \otimes \vec{u}|^2 - \frac{1}{\beta}|\operatorname{div} \vec{u}|^2) \, dx \, ds$$

$$+ \frac{1}{A} \int_0^t \int (\vec{\nabla} \otimes \vec{F} \cdot \vec{\nabla} \otimes (\vec{u} + (1+\theta)\alpha\partial_t \vec{u}) - \theta\alpha|\vec{\nabla} \otimes \partial_t \vec{u}|^2 - \nu|\Delta \vec{u}|^2) \, dx \, ds \qquad (19.19)$$

$$- \frac{1}{A} \int_0^t \int \frac{1}{\beta}|\vec{\nabla} \operatorname{div} \vec{u}|^2 \, dx \, ds$$

• **Step 5: Global existence of $\vec{u}_{(\alpha,\beta)}$.**

In order to show global existence of $\vec{u}_{(\alpha,\beta)}$, we must show that $\|\vec{u}_{(\alpha)}(t,.)\|_{H^2}$ and $\|\partial_t \vec{u}_{(\alpha)}(t,.)\|_{H^1}$ cannot blow up.

We start from the energy equality (19.19) with

$$\vec{F} = \vec{f}_{(\alpha)} - (\vec{u} \cdot \vec{\nabla}\vec{u} + \frac{1}{2}(\operatorname{div} \vec{u})\vec{u}).$$

We have

$$E(\vec{u})(t) - E(\vec{u})(0) =$$

$$-A \int_0^t \int (\theta\alpha|\partial_t\vec{u}|^2 + \nu|\vec{\nabla}\otimes\vec{u}|^2 + \frac{1}{\beta}|\operatorname{div}\vec{u}|^2)\,dx\,ds$$

$$-\frac{1}{A} \int_0^t \int (\theta\alpha|\vec{\nabla}\otimes\partial_t\vec{u}|^2 + \nu|\Delta\vec{u}|^2 + \frac{1}{\beta}|\vec{\nabla}\operatorname{div}\vec{u}|^2)\,dx\,ds$$

$$+A \int_0^t \int (\vec{f}_{(\alpha)}.(\vec{u} + (1+\theta)\alpha\partial_t\vec{u}) - (1+\theta)\alpha(\vec{u}\cdot\vec{\nabla}\vec{u} + \frac{1}{2}(\operatorname{div}\vec{u})\vec{u}).\partial_t\vec{u})\,dx\,ds$$

$$+\frac{1}{A} \int_0^t \int \vec{\nabla}\otimes\vec{f}_{(\alpha)}\cdot\vec{\nabla}\otimes(\vec{u} + (1+\theta)\alpha\partial_t\vec{u})\,dx\,ds$$

$$-\frac{1}{A} \int_0^t \int \vec{\nabla}\otimes(\vec{u}\cdot\vec{\nabla}\vec{u} + \frac{1}{2}(\operatorname{div}\vec{u})\vec{u}).\vec{\nabla}\otimes(\vec{u} + (1+\theta)\alpha\partial_t\vec{u})\,dx\,ds$$

Recall that we have

$$\|\vec{u}\cdot\vec{\nabla}\vec{u} + \frac{1}{2}(\operatorname{div}\vec{u})\vec{u}\|_2 \leq C_*\|\vec{u}\|_2^{1/2}\|\vec{\nabla}\otimes\vec{u}\|_2^{1/2}\|\Delta\vec{u}\|_2$$

and

$$\|\vec{u}\cdot\vec{\nabla}\vec{u} + \frac{1}{2}(\operatorname{div}\vec{u})\vec{u}\|_{\dot{H}^1} \leq C_0\|\vec{u}\|_{\dot{H}^1}^{1/2}\|\Delta\vec{u}\|_2^{3/2}$$

for some positive constant C_0.

This gives, for $\eta > 0$,:

$$E(\vec{u})(t) - E(\vec{u})(0) \leq$$

$$-A \int_0^t \frac{\theta\alpha}{2}(1 - \alpha\frac{(1+\theta)^2}{2\eta\theta}C_*\|\vec{u}\|_2^{1/2}\|\vec{\nabla}\otimes\vec{u}\|_2^{1/2})\|\partial_t\vec{u}\|_2^2\,ds$$

$$-A \int_0^t (\frac{\nu}{2}\|\vec{\nabla}\otimes\vec{u}\|_2^2 + \frac{1}{\beta}\|\operatorname{div}\vec{u}\|_2^2)\,ds$$

$$+\frac{A}{2\nu} \int_0^t \|\vec{f}_{(\alpha)}\|_{\dot{H}^{-1}}^2\,ds + \frac{(1+\theta)^2}{2\theta}A\alpha \int_0^t \|\vec{f}_{(\alpha)}\|_2^2\,ds$$

$$-\frac{1}{A} \int_0^t (\frac{\theta\alpha}{4}\|\vec{\nabla}\otimes\partial_t\vec{u}\|_2^2 + \frac{1}{\beta}\|\vec{\nabla}\operatorname{div}\vec{u}\|_2^2)\,ds$$

$$-\frac{1}{A} \int_0^t \frac{\nu}{2}(1 - \frac{2C_*(1+\eta A^2)}{\nu}\|\vec{u}\|_2^{1/2}\|\vec{u}\|_{\dot{H}^1}^{1/2} - \alpha\frac{2C_0^2(1+\theta)^2}{\nu\theta}\|\vec{u}\|_{\dot{H}^1}\|\Delta\vec{u}\|_2)\|\Delta\vec{u}\|_2^2\,ds$$

$$+\frac{1}{2A\nu} \int_0^t \|\vec{f}_{(\alpha)}\|_2^2\,ds + \frac{(1+\theta)^2}{2\theta}\frac{\alpha}{A} \int_0^t \|\vec{f}_{(\alpha)}\|_{\dot{H}^1}^2\,ds$$

We then use the Hilbertian inequality

$$\|x\|_H^2 \leq (1+\epsilon)\|x+y\|_H^2 + (1+\frac{1}{\epsilon})\|y\|_H^2$$

For $\epsilon = \frac{1}{\theta}$, we get

$$\|\vec{u}\|_2\|\vec{u}\|_{\dot{H}^1} \leq A\frac{\|\vec{u}\|_2^2}{2} + \frac{1}{A}\frac{\|\vec{u}\|_{\dot{H}^1}^2}{2} \leq (1+\epsilon)E(\vec{u})$$

while

$$\sqrt{\alpha}\|\Delta\vec{u}\|_2 \leq \sqrt{A}\sqrt{\frac{2\epsilon}{\nu(1+\epsilon)}}\sqrt{E(\vec{u})}.$$

Moreover, we have:

$$\|\vec{u}_{(\alpha),0}\|_2 \leq \|\vec{u}_0\|_2, \|\vec{u}_{(\alpha),0}\|_{\dot{H}^1} \leq \min(\|\vec{u}_0\|_{\dot{H}^1}, \sqrt{\frac{\delta}{\alpha}}\|\vec{u}_0\|_2), \|\Delta\vec{u}_{(\alpha),0}\|_2 \leq \sqrt{\frac{\delta}{\alpha}}\|\vec{u}_0\|_{\dot{H}^1}$$

and

$$\|\vec{f}_{(\alpha)}\|_{\dot{H}^{-1}} \leq \|\vec{f}\|_{\dot{H}^{-1}}, \|\vec{f}_{(\alpha)}\|_2 \leq \min(\|\vec{f}\|_2, \sqrt{\frac{\delta}{\alpha}}\|\vec{f}\|_{\dot{H}^{-1}}), \|\vec{f}_{(\alpha)}\|_{\dot{H}^1} \leq \sqrt{\frac{\delta}{\alpha}}\|\vec{f}\|_2$$

Thus, we have the following inequality:

$$\begin{aligned}
E(\vec{u})(t) \leq &\frac{A}{2}\|\vec{u}_0\|_2^2 + \frac{1}{2A}\|\vec{u}_0\|_{\dot{H}^1}^2 + \frac{A}{2\nu}\int_0^t \|\vec{f}\|_{\dot{H}^{-1}}^2 \, ds + \frac{1}{2A\nu}\int_0^t \|\vec{f}\|_2^2 \, ds \\
&+ \alpha\frac{A}{2}\nu(1+\theta)\|\vec{u}_0\|_{\dot{H}^1}^2 + \delta\frac{1}{2A}\nu(1+\theta)\|\vec{u}_0\|_{\dot{H}^1}^2 \\
&+ \alpha\frac{(1+\theta)^2}{2\theta}A\int_0^t \|\vec{f}\|_2^2 \, ds + \delta\frac{(1+\theta)^2}{2\theta}\frac{1}{A}\int_0^t \|\vec{f}\|_2^2 \, ds \\
&- A\int_0^t (\frac{\nu}{2}\|\vec{\nabla}\otimes\vec{u}\|_2^2 + \frac{1}{\beta}\|\operatorname{div}\vec{u}\|_2^2) \, ds - \frac{1}{A}\int_0^t (\frac{\theta\alpha}{4}\|\vec{\nabla}\otimes\partial_t\vec{u}\|_2^2 + \frac{1}{\beta}\|\vec{\nabla}\operatorname{div}\vec{u}\|_2^2) \, ds \\
&- A\int_0^t \frac{\theta\alpha}{2}(1 - \alpha\frac{(1+\theta)^2}{2\eta\theta}C_*\sqrt{1+\epsilon}\sqrt{E(\vec{u})})\|\partial_t\vec{u}\|_2^2 \, ds \\
&- \frac{1}{A}\int_0^t \frac{\nu}{2}(1 - \frac{2C_*(1+\eta A^2)}{\nu}\sqrt{1+\epsilon}\sqrt{E(\vec{u})} - A\sqrt{\alpha}\frac{2C_0^2(1+\theta)^2}{\nu\theta}2E(\vec{u})\sqrt{\frac{\epsilon}{\nu}})\|\Delta\vec{u}\|_2^2 \, ds
\end{aligned}$$

We shall now fix the parameters A, ϵ, η and δ.

- The value of A has been fixed in the first step: from the smallness condition (19.16), we fixed A such that

$$\frac{A}{2}(\|\vec{u}_0\|_2^2 + \frac{1}{\nu}\|\vec{f}\|_{L^2\dot{H}^{-1}}^2) + \frac{1}{2A}(\|\vec{u}_0\|_{\dot{H}^1}^2 + \frac{1}{\nu}\|\vec{f}\|_{L^2L^2}^2) < \frac{\nu^2}{4C_*^2}.$$

- Let $K_1 = \frac{A}{2}(\|\vec{u}_0\|_2^2 + \frac{1}{2\nu}\|\vec{f}\|_{L^2\dot{H}^{-1}}^2) + \frac{1}{2A}(\|\vec{u}_0\|_{\dot{H}^1}^2 + \frac{1}{\nu}\|\vec{f}\|_{L^2L^2}^2)$, $K_5 = \frac{\nu^2}{4C_*^2}$, $K_2 = \frac{3K_1+K_5}{4}$ and $K_3\frac{K_1+K_5}{2}$ and $K_4 = \frac{K_1+3K_5}{2}$: $K_1 < K_2 < K_3 < K_4 < K_5$.

- we choose ϵ and η small enough to have:

$$K_3(1+\epsilon)(1+\eta A^2)^2 < K_4.$$

- we choose δ small enough to have:

$$K_1 + \delta(\frac{1}{2A}\nu(1+\theta)\|\vec{u}_0\|_{\dot{H}^1}^2 + \frac{(1+\theta)^2}{2\theta}\frac{1}{A}\|\vec{f}\|_{L^2L^2}^2) < K_2$$

We thus have:

$$E(\vec{u})(t) \leq K_2 + \alpha\left(\frac{A}{2}\nu(1+\theta)\|\vec{u}_0\|_{\dot{H}^1}^2 + \frac{(1+\theta)^2}{2\theta}A\|\vec{f}\|_{L^2L^2}^2\right)$$

$$-A\int_0^t \left(\frac{\nu}{2}\|\vec{\nabla}\otimes\vec{u}\|_2^2 + \frac{1}{\beta}\|\operatorname{div}\vec{u}\|_2^2\right)ds - \frac{1}{A}\int_0^t\left(\frac{\theta\alpha}{4}\|\vec{\nabla}\otimes\partial_t\vec{u}\|_2^2 + \frac{1}{\beta}\|\vec{\nabla}\operatorname{div}\vec{u}\|_2^2\right)ds$$

$$-A\int_0^t \frac{\theta\alpha}{2}\left(1 - \alpha\frac{(1+\theta)^2}{2\eta\theta}C_*\sqrt{K_4}\sqrt{\frac{E(\vec{u})}{K_3}}\right)\|\partial_t\vec{u}\|_2^2\,ds$$

$$-\frac{1}{A}\int_0^t \frac{\nu}{2}\left(1 - \sqrt{\frac{K_4}{K_5}}\sqrt{\frac{E(\vec{u})}{K_3}} - \sqrt{\alpha}AK_3\frac{2C_0^2(1+\theta)^2}{\nu\theta}2\sqrt{\frac{\epsilon}{\nu}}\frac{E(\vec{u})}{K_3}\right)\|\Delta\vec{u}\|_2^2\,ds$$

Thus we find:

Global existence for small α

If α is small enough to grant that:

- $\alpha\left(\frac{A}{2}\nu(1+\theta)\|\vec{u}_0\|_{\dot{H}^1}^2 + \frac{(1+\theta)^2}{2\theta}A\|\vec{f}\|_{L^2L^2}^2\right) < K_3 - K_2$

- $\alpha\frac{(1+\theta)^2}{2\eta\theta}C_*\sqrt{K_4} < \frac{1}{2}$

- $\alpha\left(AK_3\frac{2C_0^2(1+\theta)^2}{\nu\theta}2\sqrt{\frac{\epsilon}{\nu}}\right)^2 < \frac{1}{4}\left(1 - \sqrt{\frac{K_4}{K_5}}\right)^2$

then, we have

$$E(\vec{u})(t) + A\int_0^t\left(\frac{\nu}{2}\|\vec{\nabla}\otimes\vec{u}\|_2^2 + \frac{1}{\beta}\|\operatorname{div}\vec{u}\|_2^2 + \frac{\theta\alpha}{4}\|\partial_t\vec{u}\|_2^2\right)ds$$

$$+\frac{1}{A}\int_0^t\left(\frac{\theta\alpha}{4}\|\vec{\nabla}\otimes\partial_t\vec{u}\|_2^2 + \frac{1}{\beta}\|\vec{\nabla}\operatorname{div}\vec{u}\|_2^2 + \frac{\nu}{4}\left(1 - \sqrt{\frac{K_4}{K_5}}\right)\|\Delta u\|_2^2\right)ds \qquad (19.20)$$

$$< K_3$$

Thus, we have

$$A\frac{\|\vec{u}\|_2^2}{2} + \frac{1}{A}\alpha\nu(1+\theta)\frac{\|\Delta\vec{u}\|_2^2}{2} + A\theta\alpha^2\frac{\|\partial_t\vec{u}\|_2^2}{2} + \frac{1}{A}\theta\alpha^2\frac{\|\vec{\nabla}\otimes\partial_t\vec{u}\|_2^2}{2} < K_3(1+\epsilon)$$

Hence, $\|\vec{u}\|_H^2$ et $\|\partial_t\vec{u}\|_{H^1}$ cannot blow up, and we have global existence.

- **Step 6: Strong convergence.**

The last step of the proof is to check the strong convergence of $\vec{u}_{(\alpha,\beta)}$ to $\vec{u}$. To avoid notational burden, we write $\vec{u}_{(\alpha)}$ instead of $\vec{u}_{(\alpha,\beta)}$.

We write

$$\|\vec{u} - \vec{u}_{(\alpha)}\|_2^2 \leq (1+\epsilon)\left(\|\vec{u}_{(\alpha)} - \vec{u} + \alpha\partial_t\vec{u}_{(\alpha)}\|_2^2 + \theta\alpha^2\|\partial_t\vec{u}_{(\alpha)}\|_2^2\right)$$

so that

$$\frac{\|\vec{u} - \vec{u}_{(\alpha)}\|_2^2}{1+\epsilon} \leq \|\vec{u}_{(\alpha),0} - \vec{u}_0\|_2^2$$

$$+2\int_0^t\int (\alpha\partial_t^2\vec{u}_{(\alpha)} + \partial_t\vec{u}_{(\alpha)} - \partial_t\vec{u}).(\alpha\partial_t\vec{u}_{(\alpha)} + \vec{u}_{(\alpha)} - \vec{u}) + \alpha^2\theta\,\partial_t^2\vec{u}_{(\alpha)}.\partial_t\vec{u}_{(\alpha)}\,dx\,ds$$

$$= \|\vec{u}_{(\alpha),0} - \vec{u}_0\|_2^2$$

$$+ 2 \int_0^t \int (\nu \Delta(\vec{u}_{(\alpha)} - \vec{u}) + \vec{f}_{(\alpha)} - \vec{f}).(\alpha \partial_t \vec{u}_{(\alpha)} + \vec{u}_{(\alpha)} - \vec{u}) \, dx \, ds$$

$$+ 2 \int_0^t \int ((\vec{u} \cdot \vec{\nabla})\vec{u} - (\vec{u}_{(\alpha)} \cdot \vec{\nabla})\vec{u}_{(\alpha)} - \frac{1}{2}(\text{div } \vec{u}_{(\alpha)})\vec{u}_{(\alpha)}).(\alpha \partial_t \vec{u}_{(\alpha)} + \vec{u}_{(\alpha)} - \vec{u}) \, dx \, ds$$

$$+ 2 \int_0^t \int (\vec{\nabla}p + \frac{1}{\beta}\vec{\nabla} \text{div } \vec{u}_{(\alpha)}).(\alpha \partial_t \vec{u}_{(\alpha)} + \vec{u}_{(\alpha)} - \vec{u}) \, dx \, ds$$

$$+ 2\alpha\theta \int_0^t \int (-\partial_t \vec{u}_{(\alpha)} + \nu \Delta \vec{u}_{(\alpha)} + \vec{f}_{(\alpha)}).\partial_t \vec{u}_{(\alpha)} \, dx \, ds$$

$$+ 2\alpha\theta \int_0^t \int (-\vec{u}_{(\alpha)} \cdot \vec{\nabla}\vec{u}_{(\alpha)} - \frac{1}{2}(\text{div } \vec{u}_{(\alpha)})\vec{u}_{(\alpha)} + \frac{1}{\beta}\vec{\nabla} \text{div } \vec{u}_{(\alpha)}).\partial_t \vec{u}_{(\alpha)} \, dx \, ds$$

We then write

- $\nu \int_0^t \int \Delta(\vec{u}_{(\alpha)} - \vec{u}).(\vec{u}_{(\alpha)} - \vec{u}) \, dx \, ds = -\nu \int_0^t \|\vec{u}_{(\alpha)} - \vec{u}\|_{\dot{H}^1}^2 \, ds$

- $|\int_0^t \int (\vec{f}_{(\alpha)} - \vec{f}).((\vec{u}_{(\alpha)} - \vec{u}) \, dx \, ds| \leq 2 \int_0^t \|\vec{f}_{(\alpha)} - \vec{f}\|_2^2 \, ds + 2 \int_0^t \|\vec{u}_{(\alpha)} - \vec{u}\|_2^2 \, ds$

- $|\int_0^t \int (-\nu \Delta \vec{u} + \vec{f}_{(\alpha)} - \vec{f}).\alpha \partial_t \vec{u}_{(\alpha)} \, dx \, ds| \leq \alpha \int_0^t (\nu\|\Delta \vec{u}\|_2 + \|\vec{f}\|_2)\|\partial_t \vec{u}_{(\alpha)}\|_2 \, ds$. Using (19.20), we get:

$$|\int_0^t \int (-\nu \Delta \vec{u} + \vec{f}_{(\alpha)} - \vec{f}).\alpha \partial_t \vec{u}_{(\alpha)} \, dx \, ds| \leq \sqrt{\alpha}\sqrt{\frac{4K_3}{\theta}}(\nu\|\Delta \vec{u}\|_{L^2 L^2} + \|\vec{f}\|_{L^2 L^2})$$

- $\int_0^t \int \nu \Delta \vec{u}_{(\alpha)}.\alpha \partial_t \vec{u}_{(\alpha)} \, dx \, ds = \frac{\nu\alpha}{2}(\|\vec{u}_{(\alpha),0}\|_{\dot{H}^1}^2 - \|\vec{u}_{(\alpha)}(t,.)\|_{\dot{H}^1}^2) \leq \frac{\nu\alpha}{2}\|\vec{u}_0\|_{\dot{H}^1}^2$

-

$$\alpha|\int_0^t \int (\vec{u}_{(\alpha)} \cdot \vec{\nabla}\vec{u}_{(\alpha)} + \frac{1}{2}(\text{div } \vec{u}_{(\alpha)})\vec{u}_{(\alpha)}).\partial_t \vec{u}_{(\alpha)} \, dx \, ds|$$

$$\leq \alpha\|\partial_t \vec{u}_{(\alpha)}\|_{L^2 \dot{H}^1}\|\vec{u}_{(\alpha)} \cdot \vec{\nabla}\vec{u}_{(\alpha)} + \frac{1}{2}(\text{div } \vec{u}_{(\alpha)})\vec{u}_{(\alpha)}\|_{L^2 \dot{H}^{-1}}$$

$$\leq C\alpha\|\partial_t \vec{u}_{(\alpha)}\|_{L^2 \dot{H}^1}\|\vec{u}_{(\alpha)}\|_{L^2 \dot{H}^1} \sup_{0<t<T} \|\vec{u}_{(\alpha)}\|_{\dot{H}^1}^{1/2}\|\vec{u}_{(\alpha)}\|_{L^2}^{1/2}.$$

Using again (19.20), we get:

- $\|\partial_t \vec{u}_{(\alpha)}\|_{L^2 \dot{H}^1} \leq \sqrt{\frac{4AK_3}{\alpha\theta}}$
- $\sup_{0<t<T} \|\vec{u}_{(\alpha)}\|_{\dot{H}^1}^{1/2}\|\vec{u}_{(\alpha)}\|_{L^2}^{1/2} \leq \sqrt{(1+\epsilon)K_3}$
- $\|\vec{u}_{(\alpha)}\|_{L^2 \dot{H}^1} \leq \sqrt{\frac{2K_3}{\nu A}}$

-

$$\alpha|\int_0^t \int (\vec{u} \cdot \vec{\nabla}\vec{u}).\partial_t \vec{u}_{(\alpha)} \, dx \, ds| \leq \alpha\|\partial_t \vec{u}_{(\alpha)}\|_{L^2 L^2}\|\vec{u} \cdot \vec{\nabla}\vec{u}\|_{L^2 \dot{L}^2}$$

$$\leq C\alpha\|\partial_t \vec{u}_{(\alpha)}\|_{L^2 L^2}\|\vec{u}\|_{L^2 H^2}\|\vec{u}\|_{L^\infty H^1}.$$

- since we have

$$\int (\vec{u}_{(\alpha)} \cdot \vec{\nabla}(\vec{u} - \vec{u}_{(\alpha)}) + \frac{1}{2}(\text{div } \vec{u}_{(\alpha)})(\vec{u} - \vec{u}_{(\alpha)})).(\vec{u} - \vec{u}_{(\alpha)}) \, dx = 0,$$

we have

$$\int_0^t \int \left((\vec{u} \cdot \vec{\nabla})\vec{u} - (\vec{u}_{(\alpha)} \cdot \vec{\nabla})\vec{u}_{(\alpha)} - \frac{1}{2}(\operatorname{div} \vec{u}_{(\alpha)})\vec{u}_{(\alpha)} \right) . (\vec{u}_{(\alpha)} - \vec{u}) \, dx \, ds$$

$$= \int_0^t \int \left((\vec{u} - \vec{u}_{(\alpha)}) \cdot \vec{\nabla} \right) \vec{u}.(\vec{u}_{(\alpha)} - \vec{u}) \, dx \, ds - \int_0^t \int \frac{1}{2}(\operatorname{div} \vec{u}_{(\alpha)}) \, \vec{u}.(\vec{u}_{(\alpha)} - \vec{u}) \, dx \, ds$$

$$\leq \int_0^t \|\vec{u}_{(\alpha)} - \vec{u}\|_6 (\|\vec{\nabla} \otimes \vec{u}\|_3 \|\vec{u}_{(\alpha)} - \vec{u}\|_2 + \frac{1}{2}\|\operatorname{div}(\vec{u}_{(\alpha)}\|_2 \|\vec{u}\|_3) \, ds$$

$$\leq \nu \int_0^t \|\vec{u}_{(\alpha)} - \vec{u}\|_{\dot{H}^1}^2 \, ds + C\frac{1}{\nu} \int_0^t \|\vec{\nabla} \otimes \vec{u}\|_3^2 \|\vec{u} - \vec{u}_{(\alpha)}\|_2^2 \, ds$$

$$+ C\frac{1}{\nu}\|\vec{u}\|_{L^\infty L^3}^2 \|\operatorname{div} \vec{u}_{(\alpha)}\|_2^2$$

with $\|\operatorname{div} \vec{u}_{(\alpha)}\|_2^2 \leq \frac{K_3}{A}\beta$.

- $|\int_0^t \int (\vec{\nabla}p).\alpha\partial_t\vec{u}_{(\alpha)} \, dx \, ds| \leq \alpha\|\vec{\nabla}p\|_{L^2 L^2}\|\partial_t\vec{u}_{(\alpha)}\|_{L^2 L^2}$, so that we have $|\int_0^t \int (\vec{\nabla}p).\alpha\partial_t\vec{u}_{(\alpha)} \, dx \, ds| \leq C\sqrt{\alpha}\sqrt{\frac{4K_3}{\theta}}(\|\vec{u}\|_{L^\infty H^1}\|\vec{u}\|_{L^2 H^2} + \|\vec{f}\|_{L^2 L^2})$

- $\int_0^t \int (\frac{1}{\beta}\vec{\nabla} \operatorname{div} \vec{u}_{(\alpha)}).\alpha\partial_t\vec{u}_{(\alpha)} \, dx \, ds = -\frac{\alpha}{\beta}\|\operatorname{div} \vec{u}_{(\alpha)}(t,.)\|_2^2 \leq 0$

- $\int_0^t \int (\vec{\nabla}p + \frac{1}{\beta}\vec{\nabla} \operatorname{div} \vec{u}_{(\alpha)}).(\vec{u}_{(\alpha)} - \vec{u}) \, dx \, ds = -\int_0^t \int p \operatorname{div} \vec{u}_{(\alpha)} \, dx \, ds - \frac{1}{\beta}\int_0^t \|\operatorname{div} \vec{u}_{(\alpha)}\|_2^2 \, ds \leq \frac{\beta}{2}\|p\|_{L^2 L^2} - \frac{1}{2\beta}\int_0^t \|\operatorname{div} \vec{u}_{(\alpha)}\|_2^2 \, ds$ with $\|p\|_{L^2 L^2} \leq C(\|\vec{u}\|_{L^\infty L^2}\|\vec{u}\|_{L^2 H^2} + \|\vec{f}\|_{L^2 \dot{H}^{-1}})$

and thus we get

$$\frac{\|\vec{u} - \vec{u}_{(\alpha)}\|_2^2}{1 + \epsilon} \leq 2C\frac{1}{\nu} \int_0^t \|\vec{\nabla} \otimes \vec{u}\|_3^2 \|\vec{u} - \vec{u}_{(\alpha)}\|_2^2 \, ds + 4 \int_0^t \|\vec{u}_{(\alpha)} - \vec{u}\|_2^2 \, ds$$

$$+ \|\vec{u}_{(\alpha),0} - \vec{u}_0\|_2^2 + 4 \int_0^t \|\vec{f}_{(\alpha)} - \vec{f}\|_2^2 \, ds + 2C\frac{1}{\nu}\frac{K_3}{A}\beta\|\vec{u}\|_{L^\infty L^3}^2$$

$$+ \sqrt{\alpha}\sqrt{\frac{4K_3}{\theta}}(\nu\|\Delta\vec{u}\|_{L^2 L^2} + (1 + \theta)\|\vec{f}\|_{L^2 L^2}) + \nu\alpha(1 + \theta)\|\vec{u}_0\|_{\dot{H}^1}^2$$

$$+ 2C\alpha\sqrt{\alpha}\sqrt{\frac{4K_3}{\theta}}\|\vec{u}\|_{L^2 H^2}\|\vec{u}\|_{L^\infty H^1} + 2C(1 + \theta)\sqrt{\alpha}\sqrt{\frac{8(1 + \epsilon)K_3^3}{\nu\theta}}$$

$$+ \beta C(\|\vec{u}\|_{L^\infty L^2}\|\vec{u}\|_{L^2 H^2} + \|\vec{f}\|_{L^2 \dot{H}^{-1}}) + 2C\sqrt{\alpha}\sqrt{\frac{4K_3}{\theta}}(\|\vec{u}\|_{L^\infty H^1}\|\vec{u}\|_{L^2 H^2} + \|\vec{f}\|_{L^2 L^2})$$

$$= 2C\frac{1}{\nu} \int_0^t \|\vec{\nabla} \otimes \vec{u}\|_3^2 \|\vec{u} - \vec{u}_{(\alpha)}\|_2^2 \, ds + 4 \int_0^t \|\vec{u}_{(\alpha)} - \vec{u}\|_2^2 \, ds + o(1)$$

We may then conclude by applying Grönwall's lemma. $\qquad\square$

Chapter 20

Conclusion

Twentyfour years after Wiegner's [501] survey on the "neverending challenge" of the Navier–Stokes equations, one may feel that very little advances have been gained in the core problem of global existence of large regular solutions. However, an impressive work has been done during those twentyfour years and a huge literature has been devoted to the study of Navier–Stokes equations on the whole space (we included in the references of this book around 235 significant papers published after 2000) [i.e. more than 50 papers more than for the first edition...].

Clearly, the gap between what we can control (essentially, the L^2 norm) and what we should control (the L^p norm for $p \geq 3$ or the H^s norm for $s \geq 1/2$) could not be bridged, as there are structural reasons to this gap (mainly, the scaling of the equations).

Leray's inequality gave a control in $L^\infty L^2 \cap L^2 \dot{H}^1$, so that we have a control on weak solutions in some $L_t^p L_x^q$ norms with $\frac{2}{p} + \frac{3}{q} = \frac{3}{2}$ (in the case of 3D equations; when dealing with the Navier–Stokes equations in 2D, the control works with $\frac{2}{p} + \frac{2}{q} = 1$). Serrin's criteria then showed that we had uniqueness and regularity of the $d - D$ weak solutions provided that we had a control in some $L_t^p L_x^q$ norms with $\frac{2}{p} + \frac{d}{q} \leq 1$; thus, in dimension 2, Navier–Stokes equations are well-posed in L^2, and in dimension 3 we lack the necessary controls on weak solutions.

20.1 Energy Inequalities

Let us consider the Navier–Stokes equations

$$\partial_t \vec{u} + \vec{u} \cdot \vec{\nabla} \vec{u} = \nu \Delta \vec{u} + \vec{f} - \vec{\nabla} p$$

with

$$\text{div } \vec{u} = 0$$

and

$$\vec{u}(0, .) = \vec{u}_0.$$

Up to now, except for solutions with special symmetries (such as axisymmetry), we still have only one energy inequality: Leray's inequality

$$\|\vec{u}(t, .)\|_2^2 + 2\nu \int_0^t \|\vec{u}(s, .)\|_{\dot{H}^1}^2 \, ds \leq \|\vec{u}_0(t, .)\|_2^2 + 2 \int_0^t \int \vec{u}(s, y) \cdot \vec{f}(s, y) \, dy \, ds$$

which was established in 1934 [328].

This inequality allowed Leray to prove the existence of global weak solutions to the Navier–Stokes equations when $\vec{u}_0 \in L^2$ (and $\vec{f} \in L^2 H^{-1}$).

In 1990, global existence of weak solutions was extended by Calderón [77] to initial values $\vec{u}_0 = \vec{v}_0 + \vec{w}_0$, where $\vec{w}_0$ was small in L^3 and $\vec{v}_0$ belonged to L^2. $\vec{w}_0$ generated a global mild solution $\vec{w}$ to the Navier–Stokes equations (with null force), due to Kato's [255] result in 1983 on solutions in L^p, and $\vec{v}_0$ generated a global weak solution to

$$\partial_t \vec{v} = \nu \Delta \vec{v} + \mathbb{P}(\vec{f} - \text{div}(\vec{v} \otimes \vec{v} + \vec{v} \otimes \vec{w} + \vec{w} \otimes \vec{v}));$$

the existence of the solution was established through the energy inequality

$$\|\vec{v}(t,.)\|_2^2 + 2\nu \int_0^t \|\vec{v}(s,.)\|_{\dot{H}^1}^2 \, ds \leq \|\vec{v}_0(t,.)\|_2^2 + 2 \int_0^t \int \vec{v}(s,y) \cdot \vec{f}(s,y) \, dy \, ds$$

$$+ 2 \int_0^t \int \vec{w}(s,y).(\vec{v}(s,y) \cdot \vec{\nabla} \vec{v}(s,y)) \, dy \, ds$$

(Calderón's splitting provides indeed global weak solutions for initial values $\vec{u}_0 \in L^2 + L^p$ with $3 \leq p < +\infty$: one writes $\vec{u}_0 = \vec{v}_{0,R} + \vec{w}_{0,R}$ with $\vec{v}_{0,R} \in L^2$ and $\vec{w}_{0,R}$ in L^p small enough to ensure existence and smallness of the associated mild solution $\vec{w}_R$ in $M^\infty((0,R), L^p)$, and then one proceeds as usual to an extraction to get a sequence of solutions $\vec{u}_{R_k}$ that converge to the solution $\vec{u}$).

Calderón's splitting was extended and systematized by Lemarié-Rieusset in 1998 [310] and 2002 [313] and in 2013 by Cui [134] and Karch, Pilarczyk, and Schonbek [251].

Another class of global weak solutions could be constructed by mean of energy inequality. In 1977, Scheffer [426] studied the local version of Leray's inequality

$$\partial_t \left(\frac{|\vec{u}|^2}{2}\right) = \nu \Delta \left(\frac{|\vec{u}|^2}{2}\right) - \nu |\vec{\nabla} \otimes \vec{u}|^2 + \vec{u} \cdot \vec{f} + \text{div}\left((p + \frac{|\vec{u}|^2}{2})\vec{u}\right) - \mu$$

where μ is a non-negative locally finite measure. In 1999, Lemarié-Rieusset [311, 313] used this inequality to prove local existence of weak solutions for initial value $\vec{u}_0$ in L^2_{uloc} (in the case of a null force; exposition of the results in the presence of an external force was given in 2007 by Kikuchi and Seregin [261]). In 2007, Lemarié-Rieusset [316] turned his local weak solutions into global ones in the case of a data $\vec{u}_0$ in the Morrey space $\dot{M}^{2,3}$.

Recently, Bradshaw, Kukavica and Tsai [56, 57] and Fernández-Dalgo and Lemarié-Rieusset [173] constructed infinite-energy global weak solutions for large initial data in weighted Lebesgue spaces.

20.2 Critical Spaces for Mild Solutions

While the search for weak solutions was based on compactness arguments (by using Rellich's lemma after establishing a control on $\vec{u}$ and $\vec{\nabla} \otimes \vec{u}$ through energy inequalities), the search for mild solutions is based on Banach's contraction principle. One looks for estimates on the bilinear operator

$$B(\vec{u}, \vec{v}) = \int_0^t W_{\nu(t-s)} * \mathbb{P} \, \text{div}(\vec{u} \otimes \vec{v}) \, ds$$

in order to establish the existence of a solution of

$$\vec{u} = \vec{U}_0 - B(\vec{u}, \vec{u})$$

where

$$\vec{U}_0 = W_{\nu t} * \vec{u}_0 + \int_0^t W_{\nu(t-s)} \mathbb{P}\vec{f}(s,.)\, ds.$$

Due to the scaling properties of the Navier–Stokes equations, the search has been focused on solutions $\vec{u}$ defined on $(0,T) \times \mathbb{R}^3$ such that $\vec{u}$ belongs to path spaces $\mathbb{X}_T$ whose norms are homogeneous with respect to parabolic dilations:

$$\text{for } \lambda > 0, \quad \|\lambda^\gamma \vec{u}(\lambda^2 t, \lambda x)\|_{\mathbb{X}_{\lambda^{-2} T}} = \|\vec{u}\|_{\mathbb{X}_T}$$

The initial value $\vec{u}_0$ belongs to a homogeneous Banach space $\mathbb{Y}$:

$$\lambda^\gamma \|\vec{u}_0(\lambda x)\|_{\mathbb{Y}} = \|\vec{u}_0\|_{\mathbb{Y}}.$$

and the force to a Banach space $\mathbb{Z}_T$ such that

$$\|\lambda^{\gamma+2} \vec{f}(\lambda^2 t, \lambda x)\|_{\mathbb{Z}_{\lambda^{-2} T}} = \|\vec{f}\|_{\mathbb{Z}_T}.$$

We assume as well

- monotonicity in T: if $\vec{u}$ is defined on $(0, T_1) \times \mathbb{R}^3$ and if $0 < T_0 < T_1$, $\vec{u}$ is obviously defined on the smaller set $(0, T_0) \times \mathbb{R}^3$ and we require that

$$\|\vec{u}\|_{\mathbb{X}_{T_0}} \le \|\vec{u}\|_{\mathbb{X}_{T_1}}$$

- compatibility between the given spaces:

$$\|\vec{U}_0\|_{\mathbb{X}_\infty} \le C_0 (\|\vec{u}_0\|_{\mathbb{Y}} + \|\vec{f}\|_{\mathbb{Z}_\infty}).$$

For $\lambda > 0$, let us define

$$\vec{u}_\lambda(t,x) = \lambda \vec{u}(\lambda^2 t, \lambda x).$$

Assume that we have

$$\|B(\vec{u}, \vec{v})\|_{\mathbb{X}_1} \le C_1 \|\vec{u}\|_{\mathbb{X}_1} \|\vec{v}\|_{\mathbb{X}_1}. \tag{20.1}$$

As we have $B(\vec{u}, \vec{v}) = B(\vec{u}_\lambda, \vec{v}_\lambda)_{1/\lambda}$, we find, for $\lambda > 0$,

$$\|B(\vec{u}, \vec{v})\|_{\mathbb{X}_{\lambda^2}} = \|B(\vec{u}_\lambda, \vec{v}_\lambda)_{1/\lambda}\|_{\mathbb{X}_{\lambda^2}} = \|B(\vec{u}_\lambda, \vec{v}_\lambda)\|_{\mathbb{X}_1} \lambda^{\gamma-1}$$

so that

$$\|B(\vec{u}, \vec{v})\|_{\mathbb{X}_{\lambda^2}} \le C_1 \lambda^{\gamma-1} \|\vec{u}_\lambda\|_{\mathbb{X}_1} \|\vec{v}_\lambda\|_{\mathbb{X}_1} = C_1 \lambda^{1-\gamma} \|\vec{u}\|_{\mathbb{X}_{\lambda^2}} \|\vec{v}\|_{\mathbb{X}_{\lambda^2}}.$$

Hence, we may draw the following conclusions from inequality (20.1) depending on the value of γ:

- if $\gamma > 1$ (case of a *super-critical* space):
  Using inequality (20.1), one would obtain, for $\vec{u}$ and $\vec{v} \in \mathbb{X}_\infty$ and $\lambda > 1$,

$$\|B(\vec{u}, \vec{v})\|_{\mathbb{X}_1} \le \|B(\vec{u}, \vec{v})\|_{\mathbb{X}_{\lambda^2}} \le C_1 \lambda^{1-\gamma} \|\vec{u}\|_{\mathbb{X}_\infty} \vec{v}\|_{\mathbb{X}_\infty}$$

and thus $B = 0$ on $\mathbb{X}_\infty \times \mathbb{X}_\infty$. It means that we could only consider spaces where the non-linear equations would become linear, which is quite never the case[1].

---

[1] See however the section on Trkalian flows, page 306.

- if $\gamma < 1$ (case of a *sub-critical* space):

  Using (20.1), one obtains local existence of mild solutions on $(0, T) \times \mathbb{R}^3$, provided that

  $$C_1 C_0 T^{\frac{1-\gamma}{2}} (\|\vec{u}_0\|_{\mathbb{Y}} + \|\vec{f}\|_{\mathbb{Z}_\infty}) < \frac{1}{4}.$$

- if $\gamma = 1$ (case of a *critical* space):

  Using (20.1), one obtains global existence of mild solutions on $(0, +\infty) \times \mathbb{R}^3$, provided that

  $$C_1 C_0 (\|\vec{u}_0\|_{\mathbb{Y}} + \|\vec{f}\|_{\mathbb{Z}_\infty}) < \frac{1}{4}.$$

Thus, the search for global mild solutions has been turned in many cases to the determination of critical spaces.

Critical spaces

Definition 20.1.

A Banach space $\mathbb{Y}$ will be called a critical Banach space for the Navier–Stokes equations if

- *we have homogeneity with respect to dilations:*

  $$\text{for } \lambda > 0, \quad \|\lambda^\gamma \vec{u}_0(, \lambda x)\|_{\mathbb{Y}} = \|\vec{u}_0\|_{\mathbb{Y}}$$

- *we have a Banach space $\mathbb{X}$ of functions on $(0, +\infty) \times \mathbb{R}^3$ such that*

  $$\|\vec{u}_0\|_{\mathbb{Y}} \approx \|W_{\nu t} * \vec{u}_0\|_{\mathbb{X}}$$

  and

  $$\text{for } \lambda > 0, \quad \|\lambda^\gamma \vec{u}(\lambda^2 t, \lambda x)\|_{\mathbb{X}} = \|\vec{u}\|_{\mathbb{X}}$$

- *the bilinear operator*

  $$B(\vec{u}, \vec{v}) = \int_0^t W_{\nu(t-s)} * \mathbb{P} \operatorname{div}(\vec{u} \otimes \vec{v}) \, ds$$

  is bounded on $\mathbb{X} \times \mathbb{X}$:

  $$\|B(\vec{u}, \vec{v})\|_{\mathbb{X}} \leq C_1 \|\vec{u}\|_{\mathbb{X}} \|\vec{v}\|_{\mathbb{X}}. \tag{20.2}$$

The study of critical spaces began with the Sobolev space $\mathbb{Y} = \dot{H}^{1/2}$ (and $\mathbb{X} = L^\infty \dot{H}^{1/2} \cap L^2 \dot{H}^{3/2}$) (Fujita and Kato in 1964 [185]) and the Lebesgue space $\mathbb{Y} = L^3$ (and $\mathbb{X} = L^\infty L^3 \cap \{\vec{u} \ / \ \sup_{t>0} \sqrt{t} \|\vec{u}(t, .)\|_\infty < +\infty\}$) (Kato in 1984 [255]), followed by other examples such as the Lorentz space $\mathbb{Y} = L^{3,\infty}$ (and $\mathbb{X} = L^\infty L^{3,\infty}$) (Kozono and Nakao in 1996 [272]) or the Morrey space $\mathbb{Y} = \dot{M}^{2,3}$ (and $\mathbb{X} = L^\infty \dot{M}^{2,3} \cap \{\vec{u} \ / \ \sup_{t>0} \sqrt{t} \|\vec{u}(t, .)\|_\infty < +\infty\}$) (Kato in 1992 [256]).

A systematic study of critical spaces was then made by Cannone in 1995 [81] and Meyer in 1999 [359] (see also Auscher and Tchamitchian [12] and Lemarié-Rieusset [313]).

Following Meyer's formalism, one requires as well that the norms of $\mathbb{Y}$ and $\mathbb{X}$ be shift-invariant (due to the fact that the Navier–Stpokes equations have constant coefficients):

$$\|\vec{u}_0(x - x_0)\|_{\mathbb{Y}} = \|\vec{u}_0(x)\|_{\mathbb{Y}} \text{ and } \|\vec{u}(t, x - x_0)\|_{\mathbb{X}} = \|\vec{u}(t, x)\|_{\mathbb{X}}.$$

Then one finds that $\mathbb{Y} \subset \dot{B}_{\infty,\infty}^{-1}$. This maximality of the Besov space confers to $\dot{B}_{\infty,\infty}^{-1}$ a special role in the study of the Navier–Stokes equations: see for instance the result of May [354] in 2003 on Serrin's criterion for blow-up of strong solutions or the results of Chemin [106] in 1999 and Lemarié-Rieusset [317] in 2007 on uniqueness of weak solutions in the class $L^\infty L^2 \cap L^2 \dot{H}^1 \cap \mathcal{C}([0,T], \dot{B}_{\infty,\infty}^{-1})$

In order to be able to define B on $\mathbb{X} \times \mathbb{X}$, one is led to require moreover that $\mathbb{X}$ is included in $(L^2 L^2)_{\text{uloc}}((0,T) \times \mathbb{R}^3)$ for every $T > 0$. Then, in 2001, Koch and Tataru [266] proved that we have $\mathbb{Y} \subset BMO^{-1}$. Moreover, they proved that the Navier–Stokes problem is well posed for a small initial value in BMO^{-1}. Thus, BMO^{-1} is the largest (shift-invariant) critical space for the Navier–Stokes equations.

We have the embeddings

$$\dot{B}_{\infty,2}^{-1} \subset BMO^{-1} \subset \dot{B}_{\infty,\infty}^{-1}$$

and, for $1 \leq p < +\infty$,

$$\dot{B}_{p,\infty}^{-1+\frac{3}{p}} \subset BMO^{-1}.$$

While the case of what happens for the problem in $\dot{B}_{p,\infty}^{-1+\frac{3}{p}}$ $(p < +\infty)$ is well understood since the work of Cannone [81], a lot of attention is still paid to the case of the Besov spaces with regularity exponent -1.

In 2008, Bourgain and Pavlović [52] proved that the Cauchy problem for the Navier–Stokes equations was ill-posed in $\dot{B}_{\infty,\infty}^{-1}$ (more precisely, they proved a phenomenon of *norm inflation*). Ill-posedness in the smaller spaces $\dot{B}_{\infty,q}^{-1}$ (for $2 < q < +\infty$) has been also discussed by Germain [205] in 2008 and Yoneda [510] in 2010. A common assumption was that the problem was well-posed in $\dot{B}_{\infty,q}^{-1}$ when $1 \leq q \leq 2$, as in that case $\dot{B}_{\infty,q}^{-1} \subset BMO^{-1}$. However, in 2015, Wang [495] proved that the Cauchy problem for the Navier–Stokes equations is ill-posed on every Besov space $\dot{B}_{\infty,q}^{-1}$, $1 \leq q \leq +\infty$. The norm inflation described by Bourgain and Pavlović occurs in all those spaces, while existence of mild solutions is granted for small data in $\dot{B}_{\infty,q}^{-1}$ with $1 \leq q \leq 2$: for those solutions, the control in BMO^{-1} norms will not retroact on the control of the norms in the Besov space.

Finally, let us state that, while the maximality of BMO^{-1} in the class of spaces $\mathbb{Y}$ such that the Navier–Stokes problem with small initial value in $\mathbb{Y}$ is well posed, the maximal space $\mathbb{X}$ where to search for a mild solution in presence of a forcing term is still unclear (Lemarié-Rieusset [323]).

20.3 Models for the (Potential) Blow-up

Up to now, we have no clear idea whether a regular solution to the Navier–Stokes equations may blow up in finite time or not.

As soon as the Millennium problem was presented, it aroused the interest of confirmed searchers and of amateurs who announced that they solved the conjecture (and sometimes some other unrelated Millenium problems as well). Some attempts remained confidential: the authors submitted their result as a manuscript sent to a peer-reviewed journal, and the reviewers were able to find the gap in the proofs. Other attempts, however, were publicized, through the posting of preprints on the Internet.

Ten (or more...) of those uncorrect proofs succeeded in being published in peer-reviewed scientific journals between 2008 and 2015. Of course, we may recall that some of those "academic" journals are published by editors of open-access journals that are labeled as "predatory" in the website *Scholarly Open Access* and are therefore included in Beall's list[2]. But others were published in serious academic journals, whose reviewers seem to have been extremely absent-minded when reviewing the papers.

Most of those erroneous proofs can, of course, be easily discarded, by merely checking the computations (especially the parts where miraculous and unsuspected cancellations are unveiled by the authors) or comparing the (optimistic) conclusions with what we already know on the actual behavior of generic solutions.

Basically, all those "proofs" do not really use the actual structure of the Navier–Stokes equations, but only some general features of the equations. For instance, the very long paper published in 2013 by Otelbaev [389] presented a proof that was based on an abstract approach of the equations (considered as a non-linear equation $\frac{d}{dt}u = Au + B(u,u) + f$ in a Hilbert H with some basic spectral estimates on the operators A and B that had been known for more than fifty years). In 2014, a couple of months later, Tao [462] posted on arXiv a paper that aimed to prove that such an abstract approach was doomed to fail, and developed the discussion on his blog[3]. His paper

> *roughly speaking asserts that it is not possible to establish global regularity by any "abstract" approach which only uses upper bound function space estimates on the nonlinear part of the equation, combined with the energy identity.*

Tao's example came after other toy models that imitated some features of the Navier–Stokes equations and yet allowed blow-up in finite time. If we just look at scaling properties, an obvious model is the cubic non-linear heat equation:

$$\partial_t u = \nu \Delta u + u^3.$$

For this equation, it has been known for a long time that, if $\nu \|\vec\nabla u_0\|_2^2 < 2\|u_0\|_4^4$, then blow-up occurs in fnite time (see Levine's 1973 paper [333]).

In order to imitate in a closer way the Navier–Stokes equations by introducing spatial "derivatives" in the right-hand side of the equation, Montgomery–Smith [369] studied what he called *cheap Navier–Stokes equation*:

$$\partial_t u = \nu \Delta u + \sqrt{-\Delta}(u^2)$$

and proved blow-up in finite time for large enough initial values with nonnegative Fourier transforms. Remarkably enough, this toy models preserves important features of the Navier–Stokes equations: it is a quadratic equation and for u_0 small enough in BMO^{-1} one has global solutions.

Montgomery-Smith's cheap Navier–Stokes equation is a scalar equation (u is real-valued). This equation has been adapted in 2009 by Gallagher and Paicu [200] into a vector equation ($\vec u$ is a vector field) which preserves the divergence-free condition.

A more striking example of blow-up has been given in 2010 by Li and Sinai [334] who considered the usual Navier–Stokes equations

$$\partial_t \vec u = \nu \Delta \vec u - \mathbb{P}\,\mathrm{div}(\vec u \otimes \vec u),$$

but allowed $\vec u$ to be complex-valued. In that case, they could prove that blow-up may occur in finite time.

Tao's [462] example introduces another important feature in the toy model. He considers the problem

$$\partial_t \vec u = \nu \Delta \vec u - \mathcal{B}(\vec u, \vec u)$$

---

[2] http://scholarlyoa.com/publishers/

[3] http://terrytao.wordpress.com/2014/02/04/finite-time-blowup-for-an-averaged-three-dimensional-navier-stokes-equation/

where $\mathcal{B}$ mimicks the operator $B(\vec{u}, \vec{v}) = \mathbb{P}\,\mathrm{div}(\vec{u} \otimes \vec{v})$ on three points:

- (A) $\mathrm{div}\,\mathcal{B}(\vec{u}, \vec{v}) = 0$

- (B) $\|\mathcal{B}(\vec{u}, \vec{v})\|_{L^2(\mathbb{R}^3)} \leq C(\|\vec{u}\|_4 \|\vec{\nabla} \otimes \vec{v}\|_4 + \|\vec{v}\|_4 \|\vec{\nabla} \otimes \vec{u}\|_4)$

- (C) $\int \mathcal{B}(\vec{u}, \vec{u}) \cdot \vec{u}\,dx = 0$ for $\vec{u} \in H^2$ with $\mathrm{div}\,\vec{u} = 0$

(Montgomery-Smith's model satisfies (B) and Gallagher and Paicu's model satisfies (A) and (B)). Tao could build such an operator $\mathcal{B}$ for which blow-up occurs in finite time. He indicates on his blog[4] the important perspectives his example offers:

> to my knowledge this is the first blowup result for a Navier-Stokes type equation in three dimensions that also obeys the energy identity. Intriguingly, the method of proof in fact hints at a possible route to establishing blowup for the true Navier-Stokes equations, which I am now increasingly inclined to believe is the case (albeit for a very small set of initial data).

Another active research field is the search for mechanisms that would prevent or cause blow-up. Besides the variations on Serrin's 1963 criterion [435] that control in $L^p L^q$ norm prevents blow-up (with $2/p + 3/q = 1$ and $2 \leq p < +\infty$), a criterion that has been extended to the case of a control in $L^p \dot{B}^\sigma_{\infty,\infty}$ with $\frac{2}{p} = 1 + \sigma$ and $1 \leq p \leq +\infty$ (Kozono, Ogawa and Taniuchi in 2002 [273] for $p = 1$; May in 2003 [354] for $p = +\infty$; Kozono and Shimada in 2004 [274] for $2 < p < +\infty$; Chen and Zhang in 2006 [116] for $1 < p \leq 2$) or even to the case of a control of only one derivative $\partial_i u_j$ of one component of the velocity vector field (Cao and Titi in 2011 [88]), a lot of attention has been paid to the behavior of *vorticity*.

Vortices were labeled in 1965 by Küchemann as "the sinews and muscles of fluid motions" [284], a simile that was reused in 1994 by Kida, Moffatt, and Ohkitani [260] for whom stretched vortices are "the sinews of turbulence". In 1993, Constantin and Fefferman [128] proved that, whenever the direction of vorticity evolves regularly in the areas where the vorticity is large, the solution cannot blow up. The regularity requirement on this evolution was further lowered from Lipschitz regularity to Hölder regularity by Beirão da Vega and Berselli in 2002 [30, 31].

In contrast with this line of results which aimed at establishing regularity, Pelz [393] insisted in 2002 on a program that could lead to the unveiling of a blow-up phenomenon. The idea is that, when constrained by symmetries, a fluid could be led to blow up by following unstable branches of its dynamics whereas unconstrained fluids would relax to more stable states:

> A group with a centre of symmetry indicates that in an associated flow a critical point exists around which a collapse may occur.

Such a scenario of blowing up centered at one single point has already been discussed in various ways. For instance, as soon as 1934, Leray [328] raised the issue of finding a backward self-similar solution

$$\vec{u}(t, x) = \frac{1}{\sqrt{T^* - t}} \vec{U}\left(\frac{x}{\sqrt{T^* - t}}\right)$$

as a model of blow-up. This scenario was ruled out in 1996 by Nečas, Růžička, and Šverák [375] for a profile $\vec{U} \in L^3$ and in 1998 by Tsai [482] in case of a local control of $\vec{u}$ in

---

[4]https://terrytao.wordpress.com/2014/02/04/finite-time-blowup-for-an-averaged-three-dimensional-navier-stokes-equation/

$L^\infty L^2 \cap L^2 H^1$. Even the case of asymptotic self-similarity was ruled out by Chae [98] in 2007.

More complex models were developped for describing potential blow-up. Many of them, suggested by numerical simulations, were described as *squirt singularities* (and ruled out) in 2004 by Cordóba, Fefferman and de la Llave [131]. A squirt corresponds to a point x_0 from which fluid particles will be expelled at higher and higher speed when approaching the blow-up time: there exists a positive ϵ such that, for every $t < T_{\text{MAX}}$, if a fluid particle lies in $B(x_0, \epsilon)$ at time t, then there will be a time $t' \in (t, T_{\text{MAX}})$ where the particle will be expelled from the ball $B(x_0, 2\epsilon)$.

Thus, we still do not know whether blow-up might occur or not. The only thing we know, due to the 1982 theory of Caffarelli, Kohn and Nirenberg [74], is that, when (and if) blow-up occurs at a time T^*, then it occurs on a very small set: the one-dimensional Hausdorff measure of the set of singular points x such that we get no control on the size of $\vec{u}(t, y)$ for y close enough to x and t close enough to T^* is equal to 0.

20.4   The Method of Critical Elements

As we are so unsuccessful to understand the behavior of turbulent solutions to the Navier–Stokes equations, a natural idea is to add constraints on the solution in order to have a more predictible behavior. We already evoked the approach of Pelz [393] who focused on symmetries of the Navier–Stokes solutions.

Another class of constrained solutions that naturally emerges in the study of the Navier–Stokes equations is the class of self-similar solutions. In the case of small data, they were first studied in 1989 by Giga and Miyakawa [212] and in 1995 by Cannone [81]. A priori estimates for large data were given in 2006 by Grujić [217], in 2009 by Brandolese [61] and in 2011 by Lelièvre and Lemarié-Rieusset [308]. In 2014, Jia and Šverák [245] proved existence of self-similar solutions for every large homogeneous data; however, we still have no clue on uniqueness of those solutions.

Another way to add constraints on the solution is the method of critical elements: in case of blow-up, one looks at the set of data or solutions that lead to blow-up and have a minimal norm. This method was successfully introduced by Kenig and Merle [259] for the study of the non-linear Schrödinger equation or the non-linear wave equation.

For the Navier–Stokes problem, blowing up solutions with minimal norms were considered in 2003 by Escauriaza, Seregin and Šverák [163] in the class $L_t^\infty L^3$; they could then prove that no blowing up solution can exist in this class $L_t^\infty L^3$, a very subtle limit case of the celebrated $L^p L^q$ criterion of Serrin. This was extended to other classes of critical spaces in 2013 by Gallagher, Koch and Planchon [199], using the recent theory of profile decomposition of Navier–Stokes solutions (Gallagher [196], Koch [264]).

We end with a recent and very promising result for the investigation of (potentially) blowing up solutions: if blow-up may occur, then one can find initial values with minimal norms in critical spaces that lead to blow-up. This has been proved in 2011 by Rusin and Šverák [417] for the $\dot{H}^{1/2}$ norm, and in 2013 by Jia and Šverák [244] for the L^3 norm. Moreover, the set of data with minimal norms enjoys a kind of compactness that should be useful in further studies.

20.5   Some Open Questions

Even in the "simple" case of incompressible homogeneous Newtonian fluids, many open questions are raised about solutions to the Navier–Stokes equations

$$\partial_t \vec{u} + \vec{u} \cdot \vec{\nabla} \vec{u} = \Delta \vec{u} - \vec{\nabla} p + \operatorname{div} \mathbb{F}, \quad \operatorname{div} \vec{u} = 0, \quad \vec{u}(0,.) = \vec{u}_0.$$

We list here some questions we stumbled on throughout the book:

Question 1 (the Clay problem).   *[Chapter 4]*
If $(1 + |x|^4)\partial^\alpha \vec{u}(0,.) \in L^\infty$ *for* $|\alpha| \leq 2$ *and* $\mathbb{F} = 0$, *do we have a global smooth solution?*

Question 2.   *[Chapter 8]*
Let $\tilde{M}^{2,3}$ *be the closure of* $\mathcal{D}$ *in* $\dot{M}^{2,3}$. *Do we have uniqueness of solutions in* $\mathcal{C}([0,T], \tilde{M}^{2,3})$ *?*

Question 3.   *[Chapter 9]*
Do we have existence of a mild solution for $\vec{u}_0 = 0$ *and* $\mathcal{L}(\mathbb{F})$ *small in* $\mathcal{M}(\dot{H}^{1/2,1} \mapsto L^2 L^2) + L^2 \mathcal{F} L^1$ *?*

Question 4.   *[Chapter 14]*
Do we have existence of mild solutions in weighted Lebesgue spaces or in weighted Sobolev spaces?

Question 5 (Barker).   *[Chapter 14]*
Do we have weak–strong uniqueness for Leray solutions for an initial value in $L^2 \cap BMO_0^{-1}$ *?*

Question 6.   *[Chapter 15]*
(For $\mathbb{F} = 0$*) If blowing up solutions exist wit* $\vec{u}_0 \in L^3$, *do we have existence of a solution with a type I blow up in* $L^{3,\infty}$ *?*

Question 7 (Seregin).   *[Chapter 16]*
Do we have existence of a non-trivial steady solution in $\dot{H}^1$ *(for* $\mathbb{F} = 0$*)? in* L^6 *?*

Question 8.   *[Chapter 16]*
Do we have existence of a non-trivial steady solution in L^p *with* $p < 3$ *(for* $\mathbb{F} = 0$*)?*

Question 9.   *[Chapter 16]*
Do we have existence of a non-trivial steady solution in $\dot{M}^{2,3}$ *(for* $\mathbb{F} = 0$*)? in* $\tilde{M}^{2,3}$ *?*

Notations and Glossary

Glossary

Cross-product: $\vec{u} \wedge \vec{v} = (u_2 v_3 - u_3 v_2, u_3 v_1 - u_3 v_1, u_1 v_2 - u_2 v_1)$.

Curl: $\operatorname{curl} \vec{u} = \vec{\nabla} \wedge \vec{u} = (\partial_2 u_3 - \partial_3 u_2, \partial_3 u_1 - \partial_1 u_3, \partial_1 u_2 - \partial_2 u_3)$.

Divergence: $\operatorname{div} \vec{u} = \vec{\nabla} \cdot \vec{u} = \partial_1 u_1 + \partial_2 u_2 + \partial_3 u_3$.

Euclidean norm: $|x| = \sqrt{x_1^2 + x_2^2 + x_3^2}$, $|\vec{u}| = \sqrt{u_1^2 + u_2^2 + u_3^2}$.

Green function: $G = \frac{1}{4\pi|x|}$, fundamental solution of the Poisson equation.

Heat kernel: $W_t = \frac{1}{(4\pi t)^{3/2}} e^{-\frac{x^2}{4t}}$, fundamental solution of the heat equation.

Leray projection operator: $\mathbb{P}$, the projection onto solenoidal vector fields.

Oseen tensor: $O = (O_{j,k})_{1 \leq j,k \leq 3}$, the fundamental solution of the Stokes equation $\partial_t \vec{u} = \Delta u$, $\vec{u}(0,.) = \mathbb{P}\vec{u}_0$.

Scalar product (in $\mathbb{R}^3$): $\vec{u} \cdot \vec{v} = u_1 v_1 + u_2 v_2 + u_3 v_3$.

Vorticity: The curl of the velocity.

Notations

Symbol Description

$\delta_{j,k}$	The Kronecker symbol
$\mathbb{N}$	The set of non-negative integers

Vector analysis.

$|\ |$	The Euclidean norm
$\cdot$	The scalar product in $\mathbb{R}^3$
$\wedge$	The cross-product

Differential calculus.

∂_j	Partial derivative with respect to the j-th space variable
∂_t	Partial derivative with respect to the time variable
$\vec{\nabla}$	The gradient operator
div	The divergence operator
$\vec{\nabla}\cdot$	The divergence operator
curl	The curl operator
$\vec{\nabla}\wedge$	The curl operator
Δ	The Laplace operator

The Navier–Stokes equations.

$$\partial\vec{u} = \nu\Delta\vec{u} + \vec{f} - (\vec{u}.\vec{\nabla})\vec{u} - \vec{\nabla}p$$

$\vec{u}$	The velocity
p	The kinematic pressure
$\vec{f}$	The (reduced) force density
ν	The kinematic viscosity

$$\partial\vec{u} = \nu\Delta\vec{u} + \vec{f} - \vec{\omega}\wedge\vec{u} - \vec{\nabla}Q$$

$\vec{\omega}$	The vorticity
Q	The (reduced) total pressure

Operators.

G	The Green function
$O_{j,k}$	The Oseen tensor
$\mathbb{P}$	Leray projection operator
W_t	The heat kernel

Banach spaces.

L^p	Lebesgue space
$L^p_t L^q_x$	Lebesgue space
$L^{p,q}$	Lorentz space
$L^{p,*}$	weak Lebesgue space $(= L^{p,\infty}$ if $1 < p < +\infty)$
L^p_{loc}	local Lebesgue spaces
L^p_{uloc}	uniform local Lebesgue spaces
$(L^p_t L^q_x)_{\mathrm{uloc}}$	uniform local Lebesgue spaces
L^p_Φ	weighted Lebesgue space $L^p(\Phi(x)\,dx)$
$\dot{M}^{p,q}$	(homogeneous) Morrey space
$\mathcal{M}^{p,q}_2$	parabolic Morrey space
H^s	Sobolev space
$\dot{H}^s$	homogeneous Sobolev space
$\dot{B}^s_{p,q}$	homogeneous Besov space
$\dot{F}^s_{p,q}$	homogeneous Triebel-Lizorkin space
$\dot{B}^s_{E,q}$	generalized Besov space
$\mathcal{H}^1$	Hardy space
BMO	the BMO space of John and Nirenberg
$\mathcal{V}^r$	multiplier space (from $\dot{H}^r$ to L^2)
$\mathcal{V}^{2,r}$	parabolic multiplier space

Bibliography

[1] Hamadi Abidi. Résultats de régularité de solutions axisymétriques pour le système de Navier–Stokes. *Bull. Sci. Math.*, 132:592–624, 2008.

[2] David Adams. A note on Riesz potentials. *Duke Math. J.*, 42:765–778, 1975.

[3] David Adams and Lars Hedberg. *Function Spaces and Potential Theory*. Springer, 1996.

[4] Dallas Albritton. Blow-up criteria for the Navier–Stokes equations in non-endpoint critical besov spaces. *Anal. PDE*, 11:1415–1456, 2018.

[5] Dallas Albritton, Elia Brué, and Maria Colombo. Non-uniqueness of leray solutions of the forced Navier–Stokes equations. *Annals of Math.*, 196:415–455, 2022.

[6] Rutherford Aris. *Vectors, Tensors, and the Basic Equations of Fluid Mechanics*. Dover Publ., 1989.

[7] Vladimir Arnol'd. *Huygens and Barrow, Newton and Hooke: Pioneers in Mathematical Analysis and Catastrophe Theory from Evolvents to Quasicrystals*. Birkhaüser Verlag, 1990.

[8] William Ashurst, Alan Kerstein, Robert Kerr, and Carl Gibson. Alignment of vorticity and scalar gradient with strain rate in simulated Navier–Stokes turbulence (doi: 10.1063/1.866513). *The Physics of Fluids*, 30, 1987.

[9] Jean-Pierre Aubin. Un théorème de compacité. *C. R. Acad. Sci. Paris*, 256:5042–5044, 1963.

[10] Pascal Auscher, Sandrine Dubois, and Philippe Tchamitchian. On the stability of global solutions of the Navier–Stokes equations in the space. *J. Math. Pures Appl.*, 83:673–697, 2004.

[11] Pascal Auscher and Dorothee Frey. On well-posedness of parabolic equations of Navier-Stokes type with $BMO^{-1}(\mathbb{R}^n)$ data. *J. Inst. Math. Jussieu*, 16:947–985, 2017.

[12] Pascal Auscher and Philippe Tchamitchian. Espaces critiques pour le système des équations deNavier–Stokes incompressibles. *ArXiv e-prints*, arXiv 0812.1158, 2008.

[13] Hantaek Bae, Animikh Biswas, and Eitan Tadmor. Analyticity and decay estimates of the Navier–Stokes equations in critical Besov spaces. *Arch. Ration. Mech. Anal.*, 205:963–991, 2012.

[14] Hyeong-Ohk Bae and Lorenzo Brandolese. On the effect of external forces on the motion of incompressible flows at large distances. *Ann. Univ. Ferrara, VII Sci.*, 55:225–238, 2009.

[15] Hajer Bahouri, Jean-Yves Chemin, and Raphaël Danchin. *Fourier Analysis and Nonlinear Partial Differential Equations*. Springer Berlin Heidelberg, 2011.

[16] Stefan Banach. Sur les opérations dans les ensembles abstraits et leur application aux équations intégrales. *Fund. Math.*, 3:133–181, 1922.

[17] Tobias Barker. *Uniqueness Results for Viscous Incompressible Fluids*. Ph.D. Thesis, University of Oxford, 2017.

[18] Tobias Barker. Uniqueness results for weak Leray–Hopf solutions of the Navier–Stokes system with initial values in critical spaces. *J. Math. Fluid Mech.*, 20:133–160, 2018.

[19] Tobias Barker and Christophe Prange. Mild criticality breaking for the Navier–Stokes equations. *J. Math. Fluid Mech.*, 23:66, 12p., 2021.

[20] Tobias Barker and Christophe Prange. Quantitative regularity for the Navier–Stokes equations via spatial concentration. *Comm. Math. Phys.*, 385:717–792, 2021.

[21] Tobias Barker and Christophe Prange. From concentration to quantitative regularity: a short survey of recent developments for the Navier–Stokes equations. *ArXiv e-prints*, arXiv 2211.16215, 2022.

[22] Oscar Barraza. Self-similar solutions in weak L^p-spaces of the Navier–Stokes equations. *Rev. Mat. Iberoamericana*, 12:411–439, 1996.

[23] Arnaud Basson. Homogeneous statistical solutions and local energy inequality for 3D Navier–Stokes equations. *Comm. Math. Phys.*, 266:17–35, 2006.

[24] Arnaud Basson. *Solutions spatialement homogènes adaptées des équations de Navier–Stokes*. Thesis. University of Evry, 2006.

[25] George K. Batchelor. *An Introduction to Fluid Dynamics*. Cambridge University Press, 2000.

[26] Guy Battle and Paul Federbush. Divergence–free vector wavelets. *Michigan Math. J.*, 40:181–195, 1993.

[27] J. Thomas Beale, Tosio Kato, and Andrei Majda. Remarks on the breakdown of smooth solutions for the 3-D Euler equations. *Comm. Math. Phys.*, 94:61–66, 1984.

[28] Hugo Beirão da Vega. On the suitable weak solutions to the Navier–Stokes equations in the whole space. *J. Math. Pures et Appl.*, 64:77–86, 1985.

[29] Hugo Beirão da Vega. A new regularity class for the Navier–Stokes equations in $\mathbb{R}^n$. *Chinese Ann. Math. Ser. B*, 16:407–412, 1995.

[30] Hugo Beirão da Vega. Vorticity and smoothness in viscous flows. In M.S. Birman, S. Hildebrandt, V.A. Solonnikov, and N.N. Uraltseva, editors, *Nonlinear Problems in Mathematical Physics and Related Topics, II*, pages 61–67. Kluwer, New York, 2002.

[31] Hugo Beirão da Vega and Luigi Berselli. On the regularizing effect of the vorticity direction in incompressible viscous flows. *Differential and Integral Equations*, 15:345–356, 2002.

[32] Hugo Beirão da Veiga and Francesca Crispo. Sharp inviscid limit results under Navier type boundary conditions. An L^p theory. *J. Math. Fluid Mech.*, 12:397–411, 2010.

[33] Hugo Beirão da Veiga and Francesca Crispo. The 3-D inviscid limit result under slip boundary conditions. A negative answer. *J. Math. Fluid Mech.*, 14:55–59, 2012.

[34] Ioan Bejenaru and Terence Tao. Sharp well-posedness and ill-posedness results for a quadratic non-linear schrödinger equation. *J. Funct. Anal.*, 233:228–256, 2006.

[35] Hamid Bellout, Frederick Bloom, and Jindrich Nečas. Young measure-valued solutions for non-Newtonian incompressible fluids. *Comm. Partial Differential Equations*, 19:1763–1803, 1994.

[36] Jöran Bergh and Jörgen Löfström. *Interpolation Spaces*. Springer-Verlag, 1976.

[37] Daniel Bernoulli. *Hydrodynamica, sive De viribus et motibus fluidorum commentarii*. Strasbourg, 1738.

[38] Luigi Berselli. Some criteria concerning the vorticity and the problem of global regularity for the 3D Navier–Stokes equations. *Ann. Univ. Ferrara Sez. VII Sci. Mat.*, 55:209–224, 2009.

[39] Luigi Berselli and Giovanni Galdi. Regularity criteria involving the pressure for the weak solutions of the Navier–Stokes equations. *Proc. Amer. Math. Soc.*, 130:3585–3595, 2002.

[40] Andrea Bertozzi and Andrew Majda. *Vorticity and Incompressible Flow*. Cambridge University Press, 2002.

[41] Oleg Besov, Valentin Il'in, and Sergey Nikol'skiĭ. *Integral Representations of Functions and Imbedding Theorems. Vol. I*. Wiley, 1978.

[42] Robert Betchov. On the non-Gaussian aspects of turbulence. *Archives of Mechanics*, 28:837–845, 1976.

[43] Andrei Biryuk, Walter Craig, and Slim Ibrahim. Construction of suitable weak solutions of the Navier–Stokes equations. In Gui-Qiang Chen, editor, *Stochastic Analysis and Partial Differential Equations. Emphasis Year 2004–2005 on Stochastic Analysis and Partial Differential Equations*, pages 1–18. American Mathematical Society, 2007.

[44] Clayton Bjorland, Lorenzo Brandolese, Dragoş Iftimie, and Maria Schonbek. L^p solutions of the steady-state Navier–Stokes equations with rough external forces. *Comm. Partial Differential Equations*, 36:216–246, 2011.

[45] Michel Blay. *La science du mouvement des eaux. De Torricelli à Lagrange*. Belin, Paris, 2007.

[46] Robert Blumenthal and Ronald Getoor. Some theorems on stable processes. *Trans. Amer. Math. Soc.*, 95:263–273, 1960.

[47] Salomon Bochner. The role of mathematics in the rise of mechanics. *Am. Sci.*, 50:294–311, 1962.

[48] Mikhail E. Bogovskiĭ. Solutions of some problems of vector analysis, associated with the operators div and grad [in russian]. In *Theory of cubature formulas and the application of functional analysis to problems of mathematical physics*, volume 1980 of *Trudy Sem. S. L. Soboleva, No. 1*, pages 5–40, 149. Akad. Nauk SSSR Sibirsk. Otdel., Inst. Mat., Novosibirsk, 1980.

[49] Robert Boisvert. *Group Analysis of the Navier–Stokes Equations*. Ph.D. Dissertation. Georgia Institute of Technology, 1982.

[50] Enrico Bombieri. The Riemann hypothesis. In J.A. Carlson, A. Jaffe, A. Wiles, Clay Mathematics Institute, and American Mathematical Society, editors, *The Millennium Prize Problems*, pages 107–124. American Mathematical Society, 2006.

[51] Jean-Michel Bony. Calcul symbolique et propagation des singularités pour les équations aux dérivées partielles non linéaires. *Ann. Sci. École Norm. Sup.*, 14:209–246, 1981.

[52] Jean Bourgain and Nataša Pavlović. Ill-posedness of the Navier–Stokes equations in a critical space in 3D. *J. Funct. Anal.*, 255:2233–2247, 2008.

[53] Joseph Boussinesq. Essai théorique sur les lois trouvées expérimentalement par M. Bazin sur l'écoulement uniforme de l'eau dans les canaux découverts. *C. R. Acad. Sci. Paris*, 71:389–393, 1870.

[54] Ruggero Giuseppe Bošković. *De Viribus Vivis*. Venice, 1745.

[55] Ruggero Giuseppe Bošković. *Theoria philosophiae naturalis*. (Translated as *A Theory of Natural Philosophy* by J. M. Child. English ed. Cambridge, Mass.: M. I. T. Press, 1966). Venice, 1758.

[56] Zachary Bradshaw, Igor Kukavica, and Tai-Peng Tsai. Existence of global weak solutions to the Navier–Stokes equations in weighted space. *Indiana University Math. J.*, 71:191–212, 2022.

[57] Zachary Bradshaw and Tai-Peng Tsai. Discretely self-similar solutions to the Navier–Stokes equations with data in L^2_{loc}. *Analysis and PDE*, 12:1943–1962, 2019.

[58] Zachary Bradshaw and Tai-Peng Tsai. On the local pressure expansion for the Navier–Stokes equations. *ArXiv e-prints*, arXiv 2001.11526, 2020.

[59] Lorenzo Brandolese. *Localisation, oscillations et comportement asymptotique pour les équations de Navier–Stokes*. Thèse, ENS Cachan, 2001.

[60] Lorenzo Brandolese. Space-time decay of Navier–Stokes flows invariant under rotations. *Math. Ann.*, 329:685–706, 2004.

[61] Lorenzo Brandolese. Fine properties of self–similar solutions of the Navier–Stokes equations. *Arch. Rational Mech. Anal.*, 192:375–401, 2009.

[62] Lorenzo Brandolese and Yves Meyer. On the instantaneous spreading for the Navier–Stokes system in the whole space. *ESAIM Contr. Optim. Calc. Var.*, 8:273–285, 2002.

[63] Lorenzo Brandolese and François Vigneron. New asymptotic profiles of nonstationary solutions of the Navier–Stokes system. *J. Math. Pures Appl.*, 88:64–86, 2007.

[64] Yann Brenier, Roberto Natalini, and Marjolaine Puel. On a relaxation of the incompressible Navier–Stokes equations. *Proc. Amer. Math. Soc.*, 132:1021–1028, 2004.

[65] Howard Brenner. Navier–Stokes revisited. *Phys. A*, 349:60–132, 2005.

[66] Haïm Brezis. Remarks on the preceding paper by M. Ben–Artzi "Global solutions of two–dimensional Navier–Stokes and Euler equations". *Arch. Ration. Mech. Anal.*, 128:359–360, 1994.

[67] Jean Bricmont and Alan Sokal. *Fashionable Nonsense.* Picador, 1999.

[68] Felix Browder. Nonlinear equations of evolution. *Ann. of Math.*, 80:485–523, 1964.

[69] Felix Browder. Mathematical challenges of the 21st century. *Notices A.M.S.*, 43:324, 2000.

[70] Tristan Buckmaster and Vlad Vicol. Convex integration and phenomenologies in turbulence. *EMS Surv. Math. Sci.*, 6:173–263, 2019.

[71] Tristan Buckmaster and Vlad Vicol. Nonuniqueness of weak solutions to the Navier-Stokes equation. *Ann. of Math.*, 101-144:189, 2019.

[72] Tristan Buckmaster and Vlad Vicol. Convex integration constructions in hydrodynamics. *Bull. Amer. Math. Soc. (N.S.)*, 58:1–44, 2021.

[73] V.O. Bytev. Group properties of the Navier–Stokes equations [in russian]. *Čisl. Metody Meh. Splošnoi Sredy*, 3:13–17, 1972.

[74] Luis Caffarelli, Robert Kohn, and Louis Nirenberg. Partial regularity of suitable weak solutions of the Navier–Stokes equations. *Comm. Pure Appl. Math.*, 35:771–831, 1982.

[75] Xiaojing Cai and Quansen Jiu. Weak and strong solutions for the incompressible Navier-Stokes equations with damping. *J. Math. Anal. Appl.*, 343:799–809, 2008.

[76] Alberto Calderón and Antony Zygmund. On singular integrals. *Amer. J. Math.*, 78:289–309, 1956.

[77] Calixto Calderón. Existence of weak solutions for the Navier–Stokes equations with initial data in L^p. *Trans. Amer. Math. Soc.*, 318:179–207, 1990.

[78] Calixto Calderón. Initial values of Navier–Stokes equations. *Proc. Amer. Math. Soc.*, 117:761–766, 1993.

[79] Sergio Campanato. Proprietà di hölderianità di alcune classi di funzioni. *Ann. Scuola Norm. Sup. Pisa*, 17:175–188, 1963.

[80] J. R. Cannon and George Knightly. A note on the Cauchy problem for the Navier–Stokes equations. *SIAM J. Appl. Math.*, 18:641–644, 1970.

[81] Marco Cannone. *Ondelettes, paraproduits et Navier–Stokes.* Diderot Editeur, Paris, 1995.

[82] Marco Cannone and Grzegorz Karch. Smooth or singular solutions to the Navier–Stokes system? *J. Differential Equations*, 197:247–274, 2004.

[83] Marco Cannone, Yves Meyer, and Fabrice Planchon. Solutions auto–similaires des équations de Navier–Stokes in $\mathbb{R}^3$, exposé n. VIII. In *Séminaire X-EDP*. École Polytechnique, 1994.

[84] Marco Cannone and Fabrice Planchon. Self–similar solutions of the Navier–Stokes equations in $\mathbb{R}^3$. *Comm. Partial Differential Equations*, 21:179–193, 1996.

[85] Marco Cannone and Fabrice Planchon. On the non stationary Navier–Stokes equations with an external force. *Adv. Differential Equations*, 4:697–730, 1999.

[86] Marco Cannone and Gang Wu. Global well–posedness for Navier–Stokes equations in critical Fourier–Herz spaces. *Nonlinear Anal.*, 75:3754–3760, 2012.

[87] Chongsheng Cao, Darryl D. Holm, and Edriss S. Titi. On the Clark−α model of turbulence: global regularity and long-time dynamics. *J. Turbul.*, 6:Paper 20, 11 pp. (electronic), 2005.

[88] Chongsheng Cao and Edriss S. Titi. Global regularity criterion for the 3D Navier–Stokes equations involving one entry of the velocity gradient tensor. *Arch. Ration. Mech. Anal.*, 202:919–932, 2011.

[89] Yanping Cao, Evelyn M. Lunasin, and Edriss S. Titi. Global well-posedness of the three-dimensional viscous and inviscid simplified Bardina turbulence models. *Commun. Math. Sci.*, 4:823–848, 2006.

[90] James Carlson. First Clay Mathematics Institute Millennium Prize announced today, press release, 18 March 2010 [http://www.claymath.org/poincare/millenniumprizefull.pdf].

[91] James Carlson, Arthur Jaffe, Andrew Wiles, Clay Mathematics Institute, and American Mathematical Society, editors. *The Millennium Prize Problems.* American Mathematical Society, 2006.

[92] Carlo Cattaneo. Sulla conduzione del calore. *Atti Semin. Mat. Fis. Univ. Modeno*, 3:83–101, 1948.

[93] Carlo Cattaneo. Sur une forme de l'équation de la chaleur éliminant le paradoxe d'une propagation instantanée. *C. R. Acad. Sci. Paris*, 247:431–433, 1958.

[94] Carlo Cattaneo. Sulla conduzione del calore. In A. Pignedoli, editor, *Some Aspects of Diffusion Theory. Lectures given at a Summer School of the Centro Internazionale Matematico Estivo (C.I.M.E.) held in Varenna (Como), Italy, September 9-27, 1966*, page 485. Springer, 2011.

[95] Augustin Cauchy. Sur la pression ou la tension des corps solides. *Exercices de mathématiques*, 2:42–57, 1827.

[96] Augustin Cauchy. Sur l'équilibre ou le mouvement d'un système de points matériels sollicités par des forces d'attraction ou de répulsion naturelle. *Exercices de mathématiques*, 4:129–139, 1829.

[97] Dongho Chae. On the spectral dynamics of the deformation tensor and new a priori estimates for the 3D Euler equations. *Comm. Math. Phys.*, 263:789–801, 2006.

[98] Dongho Chae. Nonexistence of asymptotically self-similar singularities in the Euler and the Navier–Stokes equations. *Math. Ann.*, 338:435–449, 2007.

[99] Dongho Chae. Liouville-type theorems for the forced Euler equations and the Navier-Stokes equations. *Comm. Math. Phys.*, 326:37–48, 2014.

[100] Dongho Chae and Jörg Wolf. Existence of discretely self-similar solutions to the Navier-Stokes equations for initial value in $L^2_{\mathrm{loc}}(\mathbb{R}^3)$. *Ann. Inst. H. Poincaré Anal. Non Linéaire*, 35:1019–1039, 2018.

[101] Diego Chamorro, Oscar Jarrín, and Pierre-Gilles Lemarié-Rieusset. Some Liouville theorems for stationary Navier-Stokes equations in Lebesgue and Morrey spaces. *Ann. Inst. H. Poincaré C Anal. Non Linéaire*, 38:689–710, 2021.

[102] Diego Chamorro, Pierre Gilles Lemarié-Rieusset, and Kawther Mayoufi. The role of the pressure in the partial regularity theory for weak solutions of the Navier–Stokes equations. *Arch. Rat. Mech. Anal.*, 228:237–277, 2018.

[103] Chi Hin Chan and Alexis Vasseur. Log improvement of the Prodi–Serrin criteria for Navier–Stokes equations. *Methods Appl. Anal.*, 14:197–212, 2007.

[104] Émilie du Châtelet. *Institutions de physique*. Paris, 1740.

[105] Jean-Yves Chemin. Remarques sur l'existence globale pour le système de Navier–Stokes incompressible. *SIAM J. Math. Anal.*, 23:20–28, 1992.

[106] Jean-Yves Chemin. Théorèmes d'unicité pour le système de Navier–Stokes tridimensionnel. *J. Anal. Math.*, 77:27–50, 1999.

[107] Jean-Yves Chemin and Isabelle Gallagher. On the global wellposedness of the 3-D incompressible Navier–Stokes equations. *Ann. Sci. École Norm. Sup.*, 39:679–698, 2006.

[108] Jean-Yves Chemin and Isabelle Gallagher. Wellposedness and stability results for the Navier–Stokes equations in $\mathbb{R}^3$. *Ann. Inst. H. Poincaré Anal. Non Linéaire*, 26:599–624, 2009.

[109] Jean-Yves Chemin and Isabelle Gallagher. Large, global solutions to the Navier–Stokes equations, slowly varying in one direction. *Trans. Amer. Math. Soc.*, 362:2859–2873, 2010.

[110] Jean-Yves Chemin, Isabelle Gallagher, and Chloé Mullaert. The role of spectral anisotropy in the resolution of the three–dimensional Navier–Stokes equations. In *Studies in Phase Space Analysis with Applications to PDEs, PNLDE vol. 84*, pages 53–79. Birkhäuser, 2013.

[111] Jean-Yves Chemin, Isabelle Gallagher, and Marius Paicu. Global regularity for some classes of large solutions to the Navier-Stokes equations. *Ann. of Math.*, 173:983–1012, 2011.

[112] Jean-Yves Chemin, Isabelle Gallagher, and Ping Zhang. Sums of large global solutions to the incompressible Navier–Stokes equations. *J. Reine Angew. Math.*, 681:65–82, 2013.

[113] Jean-Yves Chemin and Nicolas Lerner. Flot de champs de vecteurs non-lipschitziens et équations de Navier–Stokes. *J. Differential Equations*, 121:314–328, 1995.

[114] Hui Chen, Daoyuan Fang, and Ting Zhang. Regularity of 3D axisymmetric Navier–Stokes equations. *Discrete Contin. Dyn. Syst.*, 461:629–649, 2017.

[115] Qionglei Chen, Changxing Miao, and Zhifei Zhang. On the uniqueness of weak solutions for the 3D Navier–Stokes equations. *Ann. Inst. H. Poincaré Anal. Non Linéaire*, 26:2165–2180, 2009.

[116] Qionglei Chen and Zhifei Zhang. Space-time estimates in the Besov spaces and the 3D Navier–Stokes equations. *Methods Appl. Anal.*, 13:107–122, 2006.

[117] Shiyi Chen, Ciprian Foias, Darryl D. Holm, Eric Olson, Edriss S. Titi, and Shannon Wynne. The Camassa–Holm equations and turbulence. *Phys. D*, 133:49–65, 1999.

[118] Zhi-Min Chen and W. Geraint Price. Morrey space techniques applied to the interior regularity problem of the Navier–Stokes equations. *Nonlinearity*, 14:1453–1472, 2001.

[119] Alexey Cheskidov, Darryl D. Holm, Eric Olson, and Edriss S. Titi. On a Leray–α model of turbulence. *Proc. R. Soc. London Ser. A*, 37:1923–1939, 2005.

[120] Alexey Cheskidov and Roman Shvydkoy. The regularity of weak solutions of the 3D Navier–Stokes equations in $B_{\infty,\infty}^{-1}$. *Arch. Ration. Mech. Anal*, 195:159–169, 2010.

[121] Stephen Childress. *An Introduction to Theoretical Fluid Mechanics*. Courant Institute of Mathematical Sciences, 2009.

[122] Hi Jun Choe. Regularity question of the incompressible Navier–Stokes equations I. Presentation slides. The 8th Japanese-German International Workshop on Mathematical Fluid Dynamics, Waseda University, 2013.

[123] Alexandre Chorin. *Vorticity and Turbulence*. Springer, 1994.

[124] Ronald Coifman, Pierre-Louis Lions, Yves Meyer, and Stephen Semmes. Compensated compactness and Hardy spaces. *J. Math. Pures et Appl.*, 72:247–286, 1992.

[125] Ronald Coifman and Guido Weiss. *Analyse harmonique non-commutative sur certains espaces homogènes*. Lecture notes in mathematics. Springer-Verlag, 1971.

[126] Peter Constantin. Euler equations, Navier–Stokes equations and turbulence. In Marco Cannone and Tetsuro Miyakawa, editors, *Mathematical Foundations of Turbulent Viscous Flows*, Lecture notes in mathematics, pages 1–43. Springer-Verlag, 2006.

[127] Peter Constantin, Weinan E., and Edriss S. Titi. Onsager's conjecture on the energy conservation for solutions of Euler's equation. *Comm. Math. Phys.*, 165:207–209, 1994.

[128] Peter Constantin and Charles Fefferman. Direction of vorticity and the problem of global regularity for the Navier–Stokes equations. *Indiana Univ. Math. J.*, 42:775–789, 1993.

[129] Peter Constantin and Andrew Majda. The Beltrami spectrum for incompressible fluid flows. *Comm. Math. Phys.*, 115:435–456, 1988.

[130] Stephen Cook. The P versus NP problem. In J.A. Carlson, A. Jaffe, A. Wiles, Clay Mathematics Institute, and American Mathematical Society, editors, *The Millennium Prize Problems*, pages 87–104. American Mathematical Society, 2006.

[131] Diego Cordóba, Charles Fefferman, and Rafael de la Llave. On squirt singularities in hydrodynamics. *SIAM J. Math. Anal.*, 36:204–213, 2004.

[132] Maurice Couette. *Études sur le frottement des liquides*. Gauthiers-Villars, Paris, 1890.

[133] David Cruz-Uribe and Alberto Fiorenza. *Variable Lebesgue Spaces*. Birkhäuser, Basel, 2013.

[134] Shangbin Cui. Weak solutions for the Navier–Stokes equations with $B_{\infty\infty}^{-1(\ln)} + B_{X_r}^{-1+r,\frac{2}{1-r}} + L^2$ initial data. *J. Math. Phys.*, 54:051503, 18 pp., 2013.

[135] Shangbin Cui. Sharp well-posedness and ill-posedness of the Navier–Stokes initial value problem in Besov-type spaces. *ArXiv e-prints*, arXiv 1505.00865v2, 2018.

[136] Jean Le Rond D'Alembert. *Traité de dynamique*. David, Paris, 1743 (2nd edition: 1758).

[137] Jean Le Rond D'Alembert. *Traité de l'Équilibre et du Mouvement des fluides*. Paris, 1744.

[138] Jean Le Rond D'Alembert. *Réflexions sur la cause générale des Vents*. David, Paris, 1747.

[139] Jean Le Rond D'Alembert. *Essai d'une nouvelle théorie de la résistance des fluides*. David, Paris, 1752.

[140] Jean Le Rond D'Alembert. Conservation des forces vives. In *L'Encyclopédie, vol. 7*, pages 114–116. Paris, 1757.

[141] Jean Le Rond D'Alembert. Mémoire xxxiv. In *Opuscules Mathématiques, vol. 5*, pages 132–138. Paris, 1768.

[142] Raphaël Danchin. Estimate in Besov spaces for transport and transport–diffusion with almost Lipschitz coefficients. *Rev. Mat. Iberoamericana*, 21:861–886, 2005.

[143] Nguyen Anh Dao and Quoc Hung Nguyen. Nonstationary Navier–Stokes equations with singular time-dependent external forces. *C. R. Acad. Sci. Paris, Ser. I*, 355:966–972, 2017.

[144] Olivier Darrigol. Between hydrodynamics and elasticity theory: the five first births of the Navier–Stokes equation. *Arch. Hist. Exact Sci.*, 56:95–150, 2002.

[145] Olivier Darrigol. *Worlds of Flow*. Oxford University Press, 2005.

[146] Peter A. Davidson, Yukio Kaneda, Keith Moffat, and Katepalli R. Sreenivasan, editors. *A Voyage through Turbulence*. Cambridge University Press, 2011.

[147] Richard Dawkins. Postmodernism disrobed. *Nature*, 394:141–143, 1998.

[148] Camillo De Lellis and Jr. Székelyhidi, László. The euler equations as a differential inclusion. *Ann. of Math.*, 170:1417–1436, 2009.

[149] Camillo De Lellis and Jr. Székelyhidi, László. Dissipative continuous Euler flows. *Invent. Math.*, 193:377–407, 2013.

[150] James Deardorff. A numerical study of three-dimensional channel flow at large Reynolds numbers. *J. Fluid Mech.*, 41:453–480, 1970.

[151] Pierre Deligne. The Hodge conjecture. In J.A. Carlson, A. Jaffe, A. Wiles, Clay Mathematics Institute, and American Mathematical Society, editors, *The Millennium Prize Problems*, pages 45–53. American Mathematical Society, 2006.

[152] Keith Devlin. *The Millennium Problems: The Seven Greatest Unsolved Mathematical Puzzles of Our Time*. Granta Books, 2005.

[153] Efim Dinaburg and Yakov Sinai. Existence theorems for the 3d–Navier–Stokes system having as initial conditions sums of plane waves. In Claude Bardos and Andrei Fursikov, editors, *Instability in Models Connected with Fluid Flows I*, pages 289–300. Springer New York, 2008.

[154] Serguei Dobrokhotov and Andrei Shafarevich. Some integral identities and remarks on the decay at infinity of the solutions to the Navier–Stokes equations in the entire space. *Russ. J. Math. Phys.*, 2:133–135, 1994.

[155] Thierry Dombre, Uriel Frisch, John Greene, Michel Hénon, A. Mehr, and Andrew Soward. Chaotic streamlines in the ABC flows. *J. Fluid Mech*, 167:353–391, 1986.

[156] Donatella Donatelli and Pierangelo Marcati. A dispersive approach to the artificial compressibility approximations of the Navier–Stokes equations in 3-D. *J. Hyperbolic. Diff. Equ.*, 3:575–588, 2006.

[157] Donatella Donatelli and Stefano Spirito. Weak solutions of Navier–Stokes equations constructed by artificial compressibility method are suitable. *J. Hyperbolic. Diff. Equ.*, 8:101–113, 2011.

[158] Sandrine Dubois. What is a solution to the Navier–Stokes equations? *C. R. Acad. Sci. Paris*, 335:27–32, 2002.

[159] Jean Duchon and Raoul Robert. Inertial energy dissipation for weak solutions of incompressible Euler and Navier–Stokes equations. *Nonlinearity*, 13:249–255, 2000.

[160] Pierre Duhem. *Recherches sur l'hydrodynamique. Deuxième série: Les conditions aux limites. Le théorème de Lagrange et la viscosité. Les coefficients de viscosité et la viscosité au voisinage de l'état critique.* Gauthier-Villars, 1904.

[161] Pierre Duhem. *L'évolution de la mécanique.* Hermann, Paris, 1905.

[162] Bruno Eckhardt. Turbulence transition in pipe flow: some open problems. *Nonlinearity*, 21:T1–T11, 2008.

[163] Luis Escauriaza, Gregori Seregin, and Vladimir Šverák. $L^{3;\infty}$-solutions to the Navier–Stokes equations and backward uniqueness. *Russian Math. Surveys*, 58:211–250, 2003.

[164] Luis Escauriaza, Gregori Seregin, and Vladimir Šverák. On backward uniqueness for parabolic equations. *Arch. Ration. Mech. Anal.*, 169:147–157, 2003.

[165] Leonhard Euler. Recherches sur le mouvement des corps célestes en général. *Mémoires de l'Acad. des Sciences de Berlin*, 3:93–143, 1749.

[166] Leonhard Euler. Découverte d'un nouveau principe de mécanique. *Mémoires de l'Acad. des Sciences de Berlin*, 6:185–217, 1752.

[167] Leonhard Euler. Principes généraux du mouvement des fluides. *Mémoires de l'Acad. des Sciences de Berlin*, 11:274–315, 1757.

[168] Eugene Fabes, B. Frank Jones, and Nestor Rivière. The initial value problem for theNavier—Stokes equations with data in L^p. *Arch. Ration. Mech. Anal.*, 45:222–240, 1972.

[169] Paul Federbush. Navier and Stokes meet the wavelet. *Comm. Math. Phys.*, 155:219–248, 1993.

[170] Charles L. Fefferman. The uncertainty principle. *Bull. Amer. Math. Soc.*, 9:129–206, 1983.

[171] Charles L. Fefferman. Existence and smoothness of the Navier-Stokes equation. In J.A. Carlson, A. Jaffe, A. Wiles, Clay Mathematics Institute, and American Mathematical Society, editors, *The Millennium Prize Problems*, pages 57–67. American Mathematical Society, 2006.

[172] Eduard Feireisl and Alexis Vasseur. New perspectives in fluid dynamics: Mathematical analysis of a model proposed by Howard Brenner. In Andrei V. Fursikov, Giovanni Galdi, and Vladislav V. Pukhnachev, editors, *New Directions in Mathematical Fluid Mechanics. The Alexander V. Kazhikhov Memorial Volume*, pages 153–179. Birkhäuser. Basel, 2010.

[173] Pedro Fernández-Dalgo and Pierre Gilles Lemarié-Rieusset. Weak solutions for Navier–Stokes equations with initial data in weighted L^2 spaces. *Arch. Ration. Mech. Anal.*, 237:347–382, 2020.

[174] Pedro Fernández-Dalgo and Pierre Gilles Lemarié-Rieusset. Characterisation of the pressure term in the incompressible Navier–Stokes equations on the whole space. *Discrete & Continuous Dynamical Systems - S*, 14:2917–2931, 2021.

[175] Richard B. Feynman, Robert B. Leighton, and Matthew Sands. *The Feynman Lectures on Physics, vol. 2*. Addison–Wesley, 1964.

[176] Robert Finn. On the steady-state solutions of the Navier–Stokes equations, III. *Acta Math.*, 105:197–244, 1961.

[177] Ciprian Foias, Darryl D. Holm, and Edriss S. Titi. The Navier–Stokes$-\alpha$ model of fluid turbulence. *Phys. D*, 152–153:49–65, 2001.

[178] Ciprian Foias, Darryl D. Holm, and Edriss S. Titi. The three dimensional viscous Camassa-Holm equations, and their relation to the Navier–Stokes equations and turbulence theory. *J. Dynam. Differential Equations*, 14:1–35, 2002.

[179] Ciprian Foias, Oscar Manley, Ricardo Rosa, and Roger Temam. *Navier–Stokes Equations and Turbulence*, volume 83 of *Encyclopedia of Mathematics and its Applications*. Cambridge University Press, Cambridge, 2001.

[180] Ciprian Foias and Roger Temam. Homogeneous statistical solutions of Navier–Stokes equations. *Indiana Univ. Math. J.*, 29:913–957, 1980.

[181] Ciprian Foias and Roger Temam. Gevrey class regularity for the solutions of the Navier–Stokes equations. *J. Funct. Anal.*, 87:359–369, 1989.

[182] Robert Fox. The rise and fall of Laplacian physics. *Historical Studies in the Physical Sciences*, 4:89–136, 1972.

[183] Kurt Otto Friedrichs. The identity of weak and strong extensions of differential operators. *Trans. Amer. Math. Soc.*, 55:132–151, 1944.

[184] Uriel Frisch. *Turbulence. The Legacy of A.N. Kolmogorov*. Cambridge University Press, 1995.

[185] Hiroshi Fujita and Tosio Kato. On the Navier–Stokes initial value problem, I. *Arch. Ration. Mech. Anal.*, 16:269–315, 1964.

[186] Giulia Furioli, Pierre Gilles Lemarié-Rieusset, and Elide Terraneo. Sur l'unicité dans $L^3(\mathbb{R}^3)$ des solutions "mild" de l'équation de Navier–Stokes. *C. R. Acad. Sci. Paris, Série I*, 325:1253–1256, 1997.

[187] Giulia Furioli, Pierre Gilles Lemarié-Rieusset, and Elide Terraneo. Unicité dans $L^3(\mathbb{R}^3)$ et d'autres espaces limites pour Navier–Stokes. *Revista Mat. Iberoamericana*, 16:605–667, 2000.

[188] Giulia Furioli, Pierre Gilles Lemarié-Rieusset, Ezzedine Zahrouni, and Ali Zhioua. Un théorème de persistance de la régularité en norme d'espaces de Besov pour les solutions de Koch et Tataru des équations de Navier–Stokes dans $\mathbb{R}^3$. *C. R. Acad. Sci. Paris, Série I*, 330:339–342, 2000.

[189] Andrei Fursikov and Mark Višik. Solutions statistiques homogènes des systèmes différentiels paraboliques et du système de Navier–Stokes. *Ann. Scuola Norm. Sup. Pisa, série IV*, IV:531–576, 1977.

[190] Andrei Fursikov and Mark Višik. *Mathematical Problems of Statistical Hydromechanics*. Dordrecht: Kluwer Academic Publishers, 1988.

[191] Sadek Gala and Pierre Gilles Lemarié-Rieusset. Multipliers between Sobolev spaces and fractional differentiation. *J. Math. Anal. Appl.*, 322:1030–1054, 2006.

[192] Victor Galaktionov. On blow-up "twistors" for the Navier–Stokes equations in $\mathbb{R}^3$: a view from reaction-diffusion theory. *ArXiv e-prints*, arXiv 0901.4286, 2009.

[193] Barak Galanti, John Gibbon, and M. Heritage. Vorticity alignment results for the three-dimensional Euler and Navier–Stokes equations. *Nonlinearity*, 10:1675–1694, 1997.

[194] Giovanni Galdi. *An Introduction to the Mathematical Theory of the Navier-Stokes Equations Steady-State Problems*. Springer Verlag, 2011.

[195] Isabelle Gallagher. The tridimensional Navier–Stokes equations with almost bidimensional data: stability, uniqueness and life span. *Intern. Math. Res. Notices*, 18:919–935, 1997.

[196] Isabelle Gallagher. Profile decomposition for solutions of the Navier–Stokes equations. *Bull. Soc. Math. France*, 129:285–316, 2001.

[197] Isabelle Gallagher, Slim Ibrahim, and Mohamed Majdoub. Existence et unicité de solutions pour le système de Navier–Stokes axisymétrique. *Comm. Partial Differential Equations*, 26:883–907, 2001.

[198] Isabelle Gallagher, Dragoş Iftimie, and Fabrice Planchon. Asymptotics and stability for global solutions to the Navier–Stokes equations. *Ann. Inst. Fourier*, 53:1387–1424, 2003.

[199] Isabelle Gallagher, Gabriel Koch, and Fabrice Planchon. A profile decomposition approach to the $L_t^\infty(L_x^3)$ Navier–Stokes regularity criterion. *Math. Ann.*, 355:1527–1559, 2013.

[200] Isabelle Gallagher and Marius Paicu. Remarks on the blow-up of solutions to a toy model for the Navier–Stokes equations. *Proc. Amer. Math. Soc.*, 137:2075–2083, 2009.

[201] Isabelle Gallagher and Fabrice Planchon. On global infinite energy solutions to the Navier–Stokes equations. *Arch. Ration. Mech. Anal.*, 161:307–337, 2002.

[202] Giovanni Gallavotti. Some rigorous results about 3D Navier–Stokes. In R.. Benzi, C. Basdevant, and S. Ciliberto, editors, *Les Houches 1992 NATO-ASI Meeting on Turbulence in Spatially Extended Systems*, pages 45–81. Nova Science Publishers, New York, 1993.

[203] Thierry Gallay and Vladimir Šverák. Remarks on the Cauchy problem for the axisymmetric Navier–Stokes equations. *Confluentes Mathematici,*, 7:67–92, 2015.

[204] Pierre Germain. Multipliers, paramultipliers, and weak-strong uniqueness for the Navier–Stokes equations. *J. Differential Equations*, 226:373–428, 2006.

[205] Pierre Germain. The second iterate for the Navier–Stokes equations. *J. Funct. Anal.*, 255:2248–2264, 2008.

[206] Pierre Germain, Naraša Pavlović, and Gigliola Staffilani. Regularity of solutions to the Navier-Stokes equations evolving from small data in bmo^{-1}. *Int. Math. Res. Not.*, 21, Art. ID rnm087, 35 pp., 2007.

[207] Massimo Germano. Turbulence: the filtering approach. *J. Fluid Mech.*, 238:325–336, 1992.

[208] Yoshikazu Giga. Regularity criteria for weak solutions of the Navier-Stokes system. In Felix Browder, editor, *Nonlinear Functional Analysis and Its Applications, Part 1 (Proceedings of Symposia in Pure Mathematics 45)*, pages 449–453. American Mathematical Society, 1986.

[209] Yoshikazu Giga. Solutions of semilinear parabolic equations in l^p and regularity of weak solutions of the Navier–Stokes system. *J. Differential Equations*, 62:186–212, 1986.

[210] Yoshikazu Giga, Katsuya Inui, and Shin'ya Matsui. On the Cauchy problem for the Navier–Stokes equations with nondecaying initial data. Advances in fluid dynamics. *Quad. Mat.*, 4:27–68, 1999.

[211] Yoshikazu Giga and Hideyuki Miura. On vorticity directions near singularities for the Navier–Stokes flows with infinite energy. *Comm. Math. Phys.*, 303:289–300, 2011.

[212] Yoshikazu Giga and Tetsuro Miyakawa. Navier–Stokes flow in $\mathbb{R}^3$ with measures as initial vorticity and Morrey spaces. *Comm. Partial Differential Equations*, 14:577–618, 1989.

[213] Pierre-Simon Girard. Mémoire sur le mouvement des fluides dans les tubes capillaires et l'influence de la température sur ce mouvement. *Institut National des Sciences et des Arts. Mémoires de sciences mathématiques et physiques (1813-1815)*, pages 249–380, 1816.

[214] David Goldberg. A local version of real hardy spaces. *Duke Math. J.*, pages 27–42, 1979.

[215] Loukas Grafakos. *Classical Harmonic Analysis (2nd ed.)*. Springer, 2008.

[216] Jeremy Gray. A history of prizes in mathematics. In J.A. Carlson, A. Jaffe, A. Wiles, Clay Mathematics Institute, and American Mathematical Society, editors, *The Millennium Prize Problems*, pages 3–27. American Mathematical Society, 2006.

[217] Zoran Grujić. Regularity of forward-in-time self-similar solutions to the 3D Navier–Stokes equations. *Discrete Contin. Dyn. Syst.*, 14:837–843, 2006.

[218] Zoran Grujić and Igor Kukavica. Space analyticity for the Navier–Stokes and related equations with initial data in L^p. *J. Funct. Anal.*, 152:447–466, 1998.

[219] Rafaela Guberović. Smoothness of Koch–Tataru solutions to the Navier–Stokes equations revisited. *Discrete Contin. Dyn. Syst.*, 27:231–236, 2010.

[220] Jean-Luc Guermond. Finite-element-based Faedo–Galerkin weak solutions to the Navier–Stokes equations in the three–dimensional torus are suitable. *J. Math. Pures Appl.*, 85:451–464, 2006.

[221] Jean-Luc Guermond, John Tinsley Oden, and Serge Prudhomme. An interpretation of the Navier–Stokes-α as a frame-indifferent Leray regularization. *Phys. D*, 177:23–30, 2003.

[222] Jean-Luc Guermond, John Tinsley Oden, and Serge Prudhomme. Mathematical perspectives on large eddy simulation models for turbulent flows. *J. Math. Fluid Mech.*, 6:194–248, 2004.

[223] Julien Guillod and Vladimír Šverák. Numerical investigations of non-uniqueness for the Navier–Stokes initial value problem in borderline spaces. *ArXiv e-prints*, arXiv 1704.00560, 2017.

[224] Boling Guo, Baoxiang Wang, and Lifeng Zhao. Isometric decomposition operators, function spaces $E_{p,q}^\lambda$ and applications to nonlinear evolution equations. *J. Funct. Anal.*, 233:1–39, 2006.

[225] Imène Hachicha. A finite speed of propagation approximation of the Navier–Stokes equations. *ArXiv e-prints*, arXiv:1308.0542, 2013.

[226] Imène Hachicha. Global existence for a damped wave equation and convergence towards a solution of the Navier–Stokes problem. *Nonlinear Anal.*, 96:68–86, 2014.

[227] Gotthilf Hagen. Ueber die Bewegung des Wassers in engen zylindrischen Röhren. *Pogg. Ann.*, 46:423–442, 1839.

[228] Gotthilf Hagen. Ueber den Einfluss der Temperatur auf die Bewegung des Wassers in Röhren. *Math. Abh. Akad. Wiss. Berlin*, pages 17–98, 1854.

[229] G.J. Hancock. The self-propulsion of microscopic organisms through liquids. *Proc. Royal Soc. London*, A 217:96–121, 1953.

[230] Katherine Hayles. Gender encoding in fluid mechanics: Masculine channels and feminine flows. *Differences: a Journal of Feminist Cultural Studies*, 4:16–44, 1992.

[231] Cheng He. Regularity for solutions to the Navier–Stokes equations with one velocity component regular. *Electron. J. Differential Equations*, 29, 13pp, 2002.

[232] Lars Hedberg. On certain convolution inequalities. *Proc. Amer. Math. Soc.*, 10:505–510, 1972.

[233] Hermann von Helmholtz. Über Integrale der hydrodynamischen Gleichungen, welche den Wirbelbewegungen entsprechen. *J. Reine Angew. Math.*, 55:25–55, 1858.

[234] Carl Herz. Lipschitz spaces and Bernstein's theorem on absolutely convergent Fourier transforms. *J. Math. Mech.*, 18:283–323, 1968/69.

[235] Jenny Hogan. Has famous math problem been solved, and in only a month? (doi:10.1038/news061002-14). *Nature*, 2006.

[236] Darryl D Holm, Jerrold Marsden, and Tudor Ratiu. The Euler–Poincaré equations and semidirect products with applications to continuum theories. *Adv. Math.*, 137:1–81, 1998.

[237] Eberhard Hopf. Elementare Bemerkungen über die lösungen partieller Differential gleichungen zweiter Ordnung vom Elliptischen Typus. *Sitzungsberichte der Preussischen Akademie der Wissenschaften*, 19:147–152, 1927.

[238] Eberhard Hopf. Ueber die Anfangswertaufgabe für die hydrodynamischen Grundgleichungen. *Math. Nachr.*, 4:213–231, 1951.

[239] Dragoş Iftimie. The 3D Navier–Stokes equations seen as a perturbation of the 2D Navier–Stokes equations. *Bull. Soc. Math. France*, 127:473–517, 1999.

[240] Luce Irigaray. La "mécanique" des fluides. *L'arc*, 58:49–55, 1974.

[241] Philip Isett. A proof of Onsager's conjecture. *Ann. of Math.*, 188:871–963, 2018.

[242] Tsukasa Iwabuchi. Global well-posedness for Keller–Segel system in Besov type spaces. *J. Math. Anal. Appl.*, 379:930–948, 2011.

[243] Arthur Jaffe and Edward Witten. Quantum Yang–Mills theory. In J.A. Carlson, A. Jaffe, A. Wiles, Clay Mathematics Institute, and American Mathematical Society, editors, *The Millennium Prize Problems*, pages 129–152. American Mathematical Society, 2006.

[244] Hao Jia and Vladimir Šverák. Minimal L^3-initial data for potential Navier–Stokes singularities. *SIAM J. Math. Anal.*, 45:1448–1459, 2013.

[245] Hao Jia and Vladimir Šverák. Local-in-space estimates near initial time for weak solutions of the Navier–Stokes equations and forward self-similar solutions. *Invent. Math.*, 196:233–265, 2014.

[246] Hao Jia and Vladimir Šverák. Are the incompressible 3d Navier-Stokes equations locally ill-posed in the natural energy space? *J. Funct. Anal.*, 268:3734–3766, 2015.

[247] Fritz John and Louis Nirenberg. On functions of bounded mean oscillation. *Comm. Pure Appl. Math.*, 14:415–426, 1961.

[248] Jean-Lin Journé. *Calderón–Zygmund Operators, Pseudodifferential Operators and the Cauchy Integral of Calderón*. Lecture Notes in Mathematics, vol. 994. Springer, 1983.

[249] Charles Kahane. On the spatial analyticity of solutions of the Navier–Stokes equations. *Arch. Ration. Mech. Anal.*, 33:386–405, 1969.

[250] Nigel Kalton and Igor Verbitsky. Nonlinear equations and weighted norm inequalities. *Trans. Amer. Math. Soc.*, 351:3441–3497, 1999.

[251] Grzegorz Karch, Dominika Pilarczyk, and Maria E. Schonbek. L^2-asymptotic stability of singular solutions to the Navier–Stokes system of equations in $\mathbb{R}^3$. *J. Math. Pures Appl.*, 108:14–40, 2017.

[252] Theodore von Kármán. Mechanische Aehnlichkeit und Turbulenz. *Nach. Ges. Wiss. Göttingen, Math.-Phys.*, K1:58–76, 1930.

[253] Tosio Kato. Nonlinear evolution equations in Banach spaces. In *Proceedings of the Symposium on Applied Mathematics*, volume 17, pages 50–67. American Mathematical Society, 1965.

[254] Tosio Kato. Nonstationary flows of viscous and ideal fluids in $\mathbb{R}^3$. *J. Functional Anal.*, 9:296–305, 1972.

[255] Tosio Kato. Strong L^p solutions of the Navier–Stokes equations in $\mathbb{R}^m$ with applications to weak solutions. *Math. Z.*, 187:471–480, 1984.

[256] Tosio Kato. Strong solutions of the Navier–Stokes equations in Morrey spaces. *Bol. Soc. Brasil. Math.*, 22:127–155, 1992.

[257] Tadashi Kawanago. Stability estimate of strong solutions for the Navier–Stokes system and its application. *J. Differential Equations*, 15, 1998.

[258] Carlos Kenig and Gabriel Koch. An alternative approach to the Navier–Stokes equations in critical spaces. *Ann. Inst. H. Poincaré Anal. Non Linéaire*, 28:159–187, 2011.

[259] Carlos Kenig and Franck Merle. Global well-posedness, scattering and blow-up for the energy-critical focusing non-linear wave equation. *Acta Math.*, 201:147–212, 2008.

[260] Shigeo Kida, H. Keith Moffatt, and Koji Ohkitani. Stretched vortices - the sinews of turbulence; large-Reynolds-number asymptotics. *J. Fluid Mech.*, 259:241–264, 1994.

[261] Norio Kikuchi and Gregori Seregin. Weak solutions to the Cauchy problem for the Navier–Stokes equations satisfying the local energy inequality. In M.S. Birman and N.N. Uraltseva, editors, *Nonlinear Equations and Spectral Theory. Amer. Math. Soc. Transl. Ser. 2, 220*, pages 141–164. American Mathematical Society, 2007.

[262] Andrei Kiselev and Olga Ladyzhenskaya. On the existence and uniqueness of the solutions of the non-stationary problem for a viscous incompressible fluid. *Amer. Math. Soc. Transl.*, 24:79–106, 1963.

[263] George Knightly. On a class of global solutions of theNavier–Stokes equations. *Arch. Ration. Mech. Anal.*, 21:211–245, 1966.

[264] Gabriel Koch. Profile decompositions for critical Lebesgue and Besov space embeddings. *Indiana Univ. Math. J.*, 59:1801–1830, 2010.

[265] Gabriel Koch, Nikolai Nadirashvili, Gregori Seregin, and Vladimir Šverák. Liouville theorems for the Navier–Stokes equations and applications. *Acta Math.*, 203:83–105, 2009.

[266] Herbert Koch and Daniel Tataru. Well-posedness for the Navier–Stokes equations. *Adv. Math.*, 157:22–35, 2001.

[267] Andrei N. Kolmogorov. *Grundbegriffe der Wahrscheinlichkeitsrechnung*. Springer, 1933.

[268] Andrei N. Kolmogorov. Energy dissipation in locally isotropic turbulence. *Dokl. Akad. Nauk SSSR*, 32:19–21, 1941.

[269] Andrei N. Kolmogorov. The local structure of turbulence in incompressible viscous fluid for very large reynolds numbers [in Russian]. *Dokl. Akad. Nauk SSSR*, 30:9–13, 1941.

[270] Andrei N. Kolmogorov. A refinement of previous hypotheses concerning the local structure of turbulence in an incompressible viscous fluid at high reynolds number. *J. Fluid Mech.*, 13:82–85, 1962.

[271] Vladimir Kondrashov. On certain properties of functions from spaces L_p^ν. *Dokl. Akad. Nauk SSSR*, 48:563–565, 1945.

[272] Hideo Kozono and Mitsuhiro Nakao. Periodic solutions of the Navier–Stokes equations in unbounded domains. *Tohoku Math. J.*, 48:33–50, 1996.

[273] Hideo Kozono, Takayoshi Ogawa, and Yashushi Taniuchi. The critical Sobolev inequalities in Besov spaces and regularity criterion to some semi-linear evolution equations. *Math. Z.*, 242:251–278, 2002.

[274] Hideo Kozono and Yukihiro Shimada. Bilinear estimates in homogeneous Triebel–Lizorkin spaces and the Navier–Stokes equations. *Math. Nachr.*, 276:63–74, 2004.

[275] Hideo Kozono and Hermann Sohr. Regularity of weak solutions to the Navier–Stokes equation. *Adv. Differential Equations*, 2:535–554, 1997.

[276] Hideo Kozono, Hermann Sohr, and Maseo Yamazaki. Representation formula, net force and energy relation to the stationary Navier—Stokes equations in 3-dimensional exterior domains. *Kyushu J. Math.*, 51:239–260, 1997.

[277] Hideo Kozono and Yashushi Taniuchi. Bilinear estimates in BMO and Navier–Stokes equations. *Math.Z.*, 235:173–194, 2000.

[278] Hideo Kozono and Yashushi Taniuchi. Limiting case of the Sobolev inequality in BMO, with application to the Euler equations. *Comm. Math. Phys.*, 214:191–200, 2000.

[279] Hideo Kozono and Maseo Yamazaki. Semilinear heat equations and the Navier–Stokes equations with distributions in new function spaces as initial data. *Comm. Partial Differential Equations*, 19:959–1014, 1994.

[280] Hideo Kozono and Maseo Yamazaki. Small stable stationary solutions in Morrey spaces of the Navier–Stokes equation. *Proc. Japan Acad. Ser. A Math. Sci.*, 71:199–201, 1995.

[281] Nicola Krylov. On parabolic equations in Morrey spaces with *vmo a* and Morrey *b, c*. *ArXiv e-prints*, arXiv:2304.03736, 2023.

[282] Nicolai Krylov. On parabolic Adams's, the Chiarenza-Frasca theorems, and some other results related to parabolic Morrey spaces. *Math. Eng.*, 5:1–20, 2023.

[283] Henk Kubbinga. *L'histoire du concept de 'molécule'*. Springer, 2001.

[284] Dietrich Küchemann. Report on the IUTAM symposium on concentrated vortex motions in fluids. *Fluid Mech.*, 21:1–20, 1965.

[285] H.K. Kuiken and H.K. Lorentz. Sketches of his work on slow viscous flow and some other areas in fluid mechanics and the background against which it arose. *J. Engineering Math.*, 30:1–18, 1996.

[286] Igor Kukavica. The partial regularity results for the Navier–Stokes equations. In J.C. Robinson and J.L. Rodrigo, editors, *Partial Differential Equations and Fluid Mechanics*, pages 121–145. Warwick, UK, 2008.

[287] Igor Kukavica. Partial regularity for the Navier–Stokes equations with a force in a Morrey space. *J. Math. Anal. Appl.*, 374:573–584, 2011.

[288] Igor Kukavica, Walter Rusin, and Mohammed Ziane. A class of large BMO^{-1} non-oscillatory data for the Navier–Stokes equations. *J. Math. Fluid Mech.*, 16:293–305, 2014.

[289] Igor Kukavica and Mohammed Ziane. Navier–Stokes equations with regularity in one direction. *J. Math. Phys.*, 48(6), 2007.

[290] Mads Kyed. *Time-Periodic Solutions to the Navier-Stokes Equations. Habilitationsschrift.* Technische Universität Darmsdadt, 2012.

[291] Olga Ladyzhenskaya. *The Mixed Problem for a Hyperbolic Equation.* Gosudarstv. Izdat., Moscow (in Russian), 1953.

[292] Olga Ladyzhenskaya. The study of Navier–Stokes equations for stationary motion of an incompressible liquid. *Usp. Mat. Nauk*, 15:75–97, 1959.

[293] Olga Ladyzhenskaya. *The Mathematical Theory of Viscous Incompressible Flow.* Gordon and Breach, New York - London, 1963.

[294] Olga Ladyzhenskaya. On modifications of Navier–Stokes equations for large gradients of velocities. *Zap. Nauchn. Sem. LOMI*, 7:126–154, 1968.

[295] Olga Ladyzhenskaya. Unique solvability in the large of a three–dimensional Cauchy problem for the Navier–Stokes equations in the presence of axial symmetry. *Zap. Nauchn. Sem. LOMI*, 7:155–177, 1968.

[296] Olga Ladyzhenskaya. Sixth problem of the millennium: Navier–Stokes equations, existence and smoothness. *Russian Math. Surveys*, 58:251–286, 2003.

[297] Olga Ladyzhenskaya and Gregori Seregin. On partial regularity of suitable weak solutions to the three-dimensional Navier–Stokes equations. *J. Math. Fluid Mech.*, 1:356–387, 1999.

[298] Olga Ladyzhenskaya, Vsevolod Solonnikov, and Nina Uraltseva. *Linear and Quasilinear Equations of Parabolic Type.* English translation: American Mathematical Society, 1968.

[299] Joseph-Louis Lagrange. *Traité de méchanique analitique.* Paris, 1788.

[300] Akhlesh Lakhtakia. Viktor Trkal, Beltrami fields, and Trkalian flows. *Czechoslovak Journal of Physics*, 44:89–96, 1994.

[301] Lev D. Landau. A new exact solution of Navier–Stokes equations. *C. R. (Doklady) Acad. Sci. URSS (N.S.)*, 43:286–288, 1944.

[302] Lev D. Landau and Evgeni M. Lifshitz. *Fluid Mechanics.* Number vol. 6 in Course of Theoretical Physics. Butterworth-Heinemann, 1987.

[303] Pierre-Simon de Laplace. *Exposition du système du monde (3rd edition).* Paris, 1808.

[304] Eric Lauga, Michael Brenner, and Howard Stone. Microfluidics: the no-slip boundary condition. In J. Foss, C. Tropea, and A. Tarin, editors, *Handbook of Experimental Fluid Dynamics*, chapter 15. Springer NewYork, 2005.

[305] Yves Le Jan and Alain-Sol Sznitman. Cascades aléatoires et équations de Navier–Stokes. *C. R. Acad. Sci. Paris*, 324 Série I:823–826, 1997.

[306] Zhen Lei and Fanghua Lin. Global mild solutions of Navier–Stokes equations. *Comm. Pure Appl. Math*, 64:1297–1304, 2011.

[307] Zhen Lei and Qi S. Zhang. Criticality of the Axially Symmetric Navier–Stokes equations. *Pacific J. Math.*, 289:169–187, 2017.

[308] Frédéric Lelièvre and Pierre Gilles Lemarié-Rieusset. Suitable solutions for the Navier–Stokes problem with an homogeneous initial value. *Comm. Math. Phys.*, 307:133–156, 2011.

[309] Pierre Gilles Lemarié-Rieusset. Analyses multi-résolutions non orthogonales, commutation entre projecteurs et dérivation et ondelettes vecteurs à divergence nulle. *Rev. Mat. Iberoamericana*, 8:221–237, 1992.

[310] Pierre Gilles Lemarié-Rieusset. Quelques remarques sur les équations de Navier–Stokes dans $\mathbb{R}^3$, exposé n. IX. In *Séminaire X-EDP*. École Polytechnique, 1997-1998.

[311] Pierre Gilles Lemarié-Rieusset. Solutions faibles d'énergie infinie pour les équations de Navier–Stokes dans $\mathbb{R}^3$. *C. R. Acad. Sci. Paris, Serie I.*, 328:1133–1138, 1999.

[312] Pierre Gilles Lemarié-Rieusset. Une remarque sur l'analyticité des solutions milds des équations de Navier–Stokes dans $\mathbb{R}^3$. *C. R. Acad. Sci. Paris, Serie I.*, 330:183–186, 2000.

[313] Pierre Gilles Lemarié-Rieusset. *Recent Developments in the Navier–Stokes Problem*. CRC Press, 2002.

[314] Pierre Gilles Lemarié-Rieusset. Nouvelles remarques sur l'analyticité des solutions milds des equations de Navier–Stokes dans $\mathbb{R}^3$. *C. R. Acad. Sci. Paris, Serie I.*, 338:443–446, 2004.

[315] Pierre Gilles Lemarié-Rieusset. Point fixe d'une application non contractante. *Rev. Mat. Iberoamericana*, 22:339–356, 2006.

[316] Pierre Gilles Lemarié-Rieusset. The Navier–Stokes equations in the critical Morrey–Campanato space. *Rev. Mat. Iberoamericana*, 23:897–930, 2007.

[317] Pierre Gilles Lemarié-Rieusset. Uniqueness for the Navier–Stokes problem. Remarks on a theorem of Jean–Yves Chemin. *Nonlinearity*, 20:1475–1490, 2007.

[318] Pierre Gilles Lemarié-Rieusset. Multipliers and Morrey spaces. *Potential Anal.*, 38:741–752, 2013.

[319] Pierre Gilles Lemarié-Rieusset. *The Navier–Stokes Problem in the 21st Century*. CRC Press, 2016.

[320] Pierre Gilles Lemarié-Rieusset. Parabolic Morrey spaces and mild solutions to Navier–Stokes equations. An interesting answer to a stupid question through a silly method. In J. Robinson, J.L. Rodrigo, W. Sadowski, and Alejandro Vidal-López, editors, *Recent Progress in the Theory of the Euler and Navier–Stokes Equations*, pages 126–136. Cambridge University Press, 2016.

[321] Pierre Gilles Lemarié-Rieusset. Sobolev multipliers, maximal functions and parabolic equations with a quadratic nonlinearity. *J. Funct. Anal.*, 274:659–694, 2018.

[322] Pierre Gilles Lemarié-Rieusset. The Navier–Stokes equations in mixed-norm time-space parabolic Morrey spaces. *ArXiv e-prints*, 2023.

[323] Pierre Gilles Lemarié-Rieusset. Forces for the Navier–Stokes equations and the Koch and Tataru theorem. *J. Math. Fluid Mech.*, to appear.

[324] Pierre Gilles Lemarié-Rieusset. A remark on weak-strong uniqueness for suitable weak solutions of the Navier–Stokes equations. *Rev. Mat. Iberoamer.*, to appear.

[325] Pierre Gilles Lemarié-Rieusset and Ramzi May. Uniqueness for the Navier–Stokes equations and multiplier between Sobolev spaces. *Nonlinear Anal.*, 66:813–838, 2007.

[326] Salvatore Leonardi, Josef Málek, Jindrich Nečas, and Milan Pokorný. On axially symmetric flows in $\mathbb{R}^3$. *Z. Anal. Anwendungen*, 18:639–649, 1999.

[327] Jean Leray. études de diverses équations intégrales non linéaires et de quelques problèmes que pose l'hydrodynamique. *J. Math. Pures Appl.*, 12:1–82, 1933.

[328] Jean Leray. Essai sur le mouvement d'un fluide visqueux emplissant l'espace. *Acta Math.*, 63:193–248, 1934.

[329] Jean Leray. Essai sur les mouvements plans d'un liquide visqueux que limitent des parois. *J. Math. Pures Appl.*, 13:331–418, 1934.

[330] Jean Leray. Les problèmes non-linéaires. *Enseign. Math.*, 35:139–151, 1936.

[331] Jean Leray and Julius Schauder. Topologie et analyse fonctionnelle. *Ann. Sci. École Norm. Sup.*, 51:45–78, 1934.

[332] Nicolas Lerner. A note on the Oseen kernels. In A. Bove, D. Del Santo, and M.K. Venkatesha Murthy, editors, *Advances in Phase Space Analysis of Partial Differential Equations, PNLDE vol. 78*, pages 161–170. Birkhäuser, 2009.

[333] Howard Levine. Some nonexistence and instability theorems for solutions of formally parabolic equations of the form $Pu_t = -Au + \mathcal{F}u$. *Arch. Ration. Mech. Anal.*, 51:371–386, 1973.

[334] Dong Li and Yakov Sinai. Blowups of complex-valued solutions for some hydrodynamic models. *Regul. Chaotic Dyn.*, 15:521–531, 2010.

[335] Fanghua Lin. A new proof of the Caffarelli–Kohn–Nirenberg theorem. *Comm. Pure Applied Math.*, 51:240–257, 1998.

[336] Jacques-Louis Lions. *Équations différentielles opérationnelles et problèmes aux limites*. Etudes Mathématiques. Springer-Verlag, 1961.

[337] Jacques-Louis Lions. *Quelques méthodes de résolution des problèmes aux limites non linéaires*. Etudes Mathématiques. Gauthier-Villars, 1969.

[338] Jacques-Louis Lions and Jaak Peetre. Sur une classe d'espaces d'interpolation. *Inst. Hautes Études Sci. Publ. Math.*, 19:5–68, 1964.

[339] Jacques-Louis Lions and Giovanni Prodi. Un théorème d'existence et unicité dans les équations de Navier–Stokes en dimension 2. *C. R. Acad. Sci. Paris*, 248:3519–3521, 1959.

[340] Pierre-Louis Lions and Nader Masmoudi. Unicité des solutions faibles de Navier–Stokes dans $L^N(\Omega)$. *C. R. Acad. Sci. Paris, Série I*, 327:491–496, 1998.

[341] Stuart P. Lloyd. The infinitesimal group of the Navier–Stokes equations. *Acta Mech.*, 38:85–98, 1981.

[342] Hendrik Lorentz. Eene allgemeene stelling omtrent de beweging eener vloeistof met wrijving en eenige daaruit afgeleide gevolgen. *Zittingsverslag vand de Koninklijke Akademie van Wetenschappen te Amsterdam*, 5:168–175, 1896.

[343] John Lumley and Akiva Yaglom. A century of turbulence. *Flow, Turbulence and Combustion*, 66:241–286, 2001.

[344] Jesper Lützen. *The Prehistory of the Theory of Distributions*. Springer, 1982.

[345] Dana Mackenzie. The Poincaré conjecture – proved. *Science*, 314:1848–1849, 2006.

[346] Colin MacLaurin. *Treatise of Fluxions*. Edinburgh, 1742.

[347] A. Mahalov, Edriss Titi, and S. Leibovich. Invariant helical subspaces for the Navier–Stokes equations. *Arch. Ration. Mech. Anal.*, 112:193–222, 1990.

[348] Josef Málek, Jindrich Nečas, Mirko Rokyta, and Michael Růžička. *Weak and Measure-Valued Solutions to Evolutionary PDEs*. Taylor & Francis, 1996.

[349] Paolo Maremonti. Existence and stability of time-periodic solutions to the Navier–Stokes equations in the whole space. *Nonlinearity*, 4:503–529, 1991.

[350] Kyûya Masuda. On the analyticity and the unique continuation theorem for solutions of the Navier–Stokes equation. *Proc. Japan Acad.*, 43:827–832, 1967.

[351] Kyûya Masuda. Weak solutions of Navier–Stokes equations. *Tôhoku Math. Journ.*, 36:623–§'§, 1984.

[352] Jean Mawhin. Leray-Schauder degree: a half century of extensions and applications. *Topological Meth. Nonlinear Anal.*, 14:195–228, 1999.

[353] Ramzi May. *Régularité et unicité des solutions milds des équations de Navier–Stokes*. Ph.D. Thesis, Université d'Évry, 2002.

[354] Ramzi May. Rôle de l'espace de Besov $B_\infty^{-1,\infty}$ dans le contrôle de l'explosion éventuelle en temps fini des solutions régulières des équations de Navier–Stokes. *C. R. Acad. Sci. Paris*, 336:731–734, 2003.

[355] Ramzi May. Extension d'une classe d'unicité pour les équations de Navier–Stokes. *Ann. Inst. H. Poincaré Anal. Non Linéaire*, 27:705–718, 2010.

[356] Kawther Mayoufi. *Inégalités d'énergie locales en mécanique des fluides*. Ph.D. Thesis, Université d'Évry, 2016.

[357] Vladimir Maz'ya. On the theory of the n-dimensional Schrödinger operator [in Russian]. *Izv. Akad. Nauk SSSR (ser. Mat.,)*, 28:1145–1172, 1964.

[358] Vladimir Maz'ya and Igor Verbitsky. Capacitary inequalities for fractional integrals, with applications to partial differential equations and Sobolev multipliers. *Ark. Mat.*, 33:81–115, 1995.

[359] Yves Meyer. *Wavelets, Paraproducts and Navier–Stokes Equations*. Current developments in mathematics 1996, International Press, PO Box 38-2872, Cambridge, MA 02238-2872, 1999.

[360] Gleb K. Mikhailov. Early studies on the outflow of water from vessels and Daniel Bernoulli's in *Exercitationes quaedam mathematicae*. In *Die Werke von Daniel Bernoulli, Band 1: Medizin und Physiologie*, pages 199–255. Birkhäuser, 1996.

[361] Evan Miller. A regularity criterion for the Navier–Stokes equation involving only the middle eigenvalue of the strain tensor. *Arch. Ration. Mech. Anal.*, 235:99–139, 2020.

[362] John Milnor. The Poincaré conjecture. In J.A. Carlson, A. Jaffe, A. Wiles, Clay Mathematics Institute, and American Mathematical Society, editors, *The Millennium Prize Problems*, pages 71–84. American Mathematical Society, 2006.

[363] Marius Mitrea and Sylvie Monniaux. The nonlinear Hodge–Navier–Stokes equaqtions in Lipschitz domains. *Differential and Integral Equations*, 22:329–356, 2009.

[364] Hideyuki Miura. Remark on uniqueness of mild solutions to the Navier–Stokes equations. *J. Funct. Anal*, 218:110–129, 2005.

[365] Hideyuki Miura and Okihiro Sawada. Regularizing rate estimates for the Navier–Stokes equations. In F. Asakura, H. Aiso, S. Kawashima, A. Matsumura, S. Nishibata, and K. Nishihara, editors, *Hyperbolic Problems: Theory, Numerics and Applications. II.*, pages 173–180. Yokohama Publ., Yokohama, 2006.

[366] Henry Keith Moffat. Singularities in fluid dynamics and their resolution. In R.L. Ricca, editor, *Lectures on Topological Fluid Mechanics (Lecture Notes in Mathematics 1973)*, pages 157–166. Springer-Verlag Berlin Heidelberg, 2009.

[367] Andrei S. Monin and A.M. Yaglom. *Statistical Fluid Mechanics*. MIT Press, Cambridge: Massachusetts, 1971.

[368] Sylvie Monniaux. Uniqueness of mild solutions of the Navier–Stokes equation and maximal L^p-regularity. *C. R. Acad. Sci. Paris, Série I*, 328:663–668, 1999.

[369] Stephen Montgomery-Smith. Finite time blow up for a Navier–Stokes like equation. *Proc. Amer. Math. Soc.*, 129:3017–3023, 2001.

[370] Stephen Montgomery-Smith. Conditions implying regularity of the three dimensional Navier–Stokes equations. *Appl. Math.*, 50:451–464, 2005.

[371] Thomas Mullin. Experimental studies of transition to turbulence in a pipe. *Ann. Rev. Fluid Mech.*, 43:1–24, 2011.

[372] Claude-Louis Navier. Mémoire sur les lois de l'équilibre et du mouvement des corps élastiques [read in May 1821]. *Mémoires de l'Acad. des Sciences de l'Institut de France*, 7:375–394, 1827.

[373] Claude-Louis Navier. Mémoire sur les lois du mouvement des fluides [read in March and December 1822]. *Mémoires de l'Acad. des Sciences de l'Institut de France*, 6:389–440, 1827.

[374] Jiří Neustupa, Antonin Novotný, and Patrick Penel. An interior regularity of a weak solution to the Navier–Stokes equations in dependence on one component of velocity. In *Topics in Mathematical Fluid Mechanics, Quad. Mat. Vol. 10*, pages 163–183. Seconda Universita di Napoli, Caserta, 2002.

[375] Jindrich Nečas, Michael Růžička, and Vladimir Šverák. On Leray's self-similar solutions of the Navier–Stokes equations. *Acta Math.*, 176:283–294, 1996.

[376] Friedrich Nietzsche. *Jenseits von Gut und Böse*. Naumann: Leipzig, 1886.

[377] Walter Noll and Clifford Truesdell. *The Non-Linear Field Theories of Mechanics, 3rd edition*. Springer, 2004.

[378] Koji Okhitani. Characterization of blowup for the Navier–Stokes equations using vector potentials. *AIP Advances*, 7(1):015211, 2017.

[379] Mike O'Leary. Conditions for the local boundedness of solutions of the Navier–Stokes system in three dimensions. *Comm. Partial Differential Equations*, 28:617–636, 2003.

[380] Peder Olsen. Fractional integration, Morrey spaces and a Schrödinger equation. *Comm. Partial Differential Equations*, 20:2005–2055, 1995.

[381] Lars Onsager. Statistical hydrodynamics. *Nuovo Cimento (Supplemento)*, 6:279–287, 1949.

[382] Steven Orszag and G. Stuart Patterson. Numerical simulation of three–dimensional homogeneous turbulence. *Phys. Rev. Letter*, 28:76–79, 1972.

[383] Frédéric Oru. *Rôle des oscillations dans quelques problèmes d'analyse non linéaire*. Ph.D. Thesis, École Normale Supérieure de Cachan., 1998.

[384] Carl Wilhelm Oseen. Sur les formules de Green généralisées qui se présentent dans l'hydrodynamique et sur quelques unes de leurs applications. *Acta Math.*, 34:205–284, 1911.

[385] Carl Wilhelm Oseen. *Methoden und Ergebnisse in der Hydrodynamik*. Akademische Verlags–gesellschaft, Leipzig, 1927.

[386] Anatolii Oskolkov. A small-parameter quasi-linear parabolic system approximating the Navier–Stokes system. *J. Math. Sci.*, 1:452–470, 1973.

[387] Konrad Osterwalder and Robert Schrader. Axioms for Euclidean Green's functions. *Comm. Math. Phys.*, 31:83–112, 1973.

[388] Konrad Osterwalder and Robert Schrader. Axioms for Euclidean Green's functions. II. *Comm. Math. Phys.*, 42:281–305, 1975.

[389] Mukhtarbay Otelbaev. Existence of a strong solution of the Navier–Stokes equations. *Mathematical Journal (Kazakhstan)*, 13, 2013.

[390] Marius Paicu and Geneviève Raugel. A hyperbolic perturbation of the Navier–Stokes equations. (une perturbation hyperbolique des équations de Navier–Stokes). *ESAIM Proc.*, 21:65–87, 2007.

[391] Marius Paicu and Zhifei Zhang. Global well-posedness for 3D Navier–Stokes equations with ill-prepared initial data. *J. Inst. Math. Jussieu*, 13:395–411, 2014.

[392] Blaise Pascal. *De l'équilibre des liqueurs et de la masse de l'air*. Paris, 1663.

[393] Richard Pelz. Discrete groups, symmetric flows and hydrodynamic blowup. In K. Bajer and H. Moffat, editors, *Tubes, Sheets and Singularities in Fluid Dynamics (Zakopane, 2001)*, pages 269–283. Kluwer Acad. Publ., Dordrecht, 2002.

[394] Patrick Penel and Milan Pokorný. Some new regularity criteria for the Navier–Stokes equations containing gradient of the velocity. *Appl. Math.*, 48:483–493, 2004.

[395] Tuoc Van Phan and Nguyen Cong Phuc. Stationary Navier–Stokes equations with critically singular external forces: existence and stability results. *Adv. Math.*, 241:137–161, 2013.

[396] Tuoc Van Phan and Timothy Robertson. On Masuda uniqueness theorem for Leray–Hopf weak solutions in mixed-norm spaces. *Eur. J. Mech. B Fluids*, 90:18–28, 2021.

[397] Nguyen Cong Phuc. The Navier–Stokes equations in nonendpoint borderline Lorentz spaces. *J. Math. Fluid Mech.*, 17:741–760, 2015.

[398] Émile Picard. Mémoire sur la théorie des équations aux dérivées partielles et la méthode des approximations successives. *J. Math. Pures et Appl.*, 6:145–210, 1890.

[399] Fabrice Planchon. *Solutions globales et comportement asymptotique pour les équations de Navier–Stokes*. Ph.D. Thesis, École Polytechnique, 1996.

[400] Fabrice Planchon. An extension of the Beale–Kato–Majda criterion for the Euler equations. *Comm. Math. Phys.*, 232:319–326, 2003.

[401] Jean-Louis Poiseuille. Recherches expérimentales sur le mouvement des liquides dans les tubes de petit diamètre. *Mémoires des savants étrangers*, IX:433–543, 1846.

[402] Siméon Denis Poisson. Mémoire sur l'équilibre et le mouvement des corps élastiques. *Annales de Chimie et de Physique*, 37:337–355, 1828.

[403] Siméon Denis Poisson. Mémoire sur l'équilibre et le mouvement des corps élastiques. *Mémoires de l'Académie des Sciences*, 8:357–570, 1829.

[404] Milan Pokorný. On the result of He concerning the smoothness of solutions to the Navier–Stokes equations. *Electron. J. Differential Equations*, 10:1–8, 2003.

[405] Ludwig Prandtl. Ueber Flüssigkeitsbewegnung bei sehr kleiner Reihung. In A. Krazer, editor, *Verhandlungen des deitten Internationalen Mathermatiker Kongresses in Heidelberg, 1904*, pages 574–585. Teubner, Leipzig, 1905.

[406] Giovanni Prodi. Un teorema di unicitá per le equazioni di Navier–Stokes. *Ann. Mat. Pura Appl.*, 48:173–182, 1959.

[407] MH Lakshminarayana Reddy, S Kokou Dadzie, Raffaella Ocone, Matthew K Borg, and Jason M Reese. Recasting Navier–Stokes equations. *J. Phys. Commun.*, 3(10):105009, 2019.

[408] Markus Reiner. A mathematical theory of dilatancy. *Amer. J. Math.*, 67:350–362, 1945.

[409] Franz Rellich. Ein Satz über mittlere Konvergenz. *Nachr. Akad. Wiss. Göttingen*, 1930:30–35, 1930.

[410] Osborne Reynolds. An experimental investigation of the circumstances which determine whether the motion of water shall be direct or sinuous, and of the law of resistance in parallel channels. *Philos. Trans. Royal Soc. London*, 174:935–982, 1883.

[411] Osborne Reynolds. On the dynamical theory of incompressible viscous fluids and the determination of the criterion. *Philos. Trans. Royal Soc. London*, 86:123 164, 1895.

[412] Francis Ribaud. A remark on the uniqueness problem for the weak solutions of Navier–Stokes equations. *Ann. Fac. Sci. Toulouse Math.*, XI:225–238, 2002.

[413] Lewis Fry Richardson. *Weather Prediction by Numerical Process.* Cambridge University Press, 1922.

[414] Ronald S. Rivlin. The hydrodynamics of non-Newtonian fluids. *Philos. Trans. Royal Soc. London. Ser. A*, 193:260–281, 1948.

[415] James Robinson. An introduction to the classical theory of the Navier–Stokes equations. Lecture notes, IMECC-Unicamp., 2010.

[416] Walter Rusin. Incompressible 3D Navier–Stokes equations as a limit of a nonlinear parabolic system. *J. Math. Fluid Mech.*, 14:383–405, 2012.

[417] Walter Rusin and Vladimir Šverák. Minimal initial data for blow-up of energy. *J. Funct. Anal.*, 260:879–891, 2011.

[418] Adhémar Barré de Saint-Venant. Mémoire sur la dynamique des fluides, séance du 14 Avril 1834, Académie des Sciences, 1834.

[419] Adhémar Barré de Saint-Venant. Note à joindre au mémoire sur la dynamique des fluides, présenté le 14 Avril 1834. *C. R. Acad. Sci. Paris*, 17:1240–1243, 1843.

[420] Adhémar Barré de Saint-Venant. Mémoire sur des formules nouvelles pour la solution des problèmes relatifs aux eaux courantes. *C. R. Acad. Sci. Paris*, 31:283–286, 1850.

[421] Gennady Samorodnotsky and Murad Taqqu. *Stable Non-Gaussian Random Processes.* Chapman & Hall, 1994.

[422] Okihiro Sawada. A description of Bourgain–Pavlović's ill–posedness theorem of the Navier–Stokes equations in the critical Besov space. In Mitsuru Sugimoto and Tohru Ozawa, editors, *Harmonic Analysis and Nonlinear Partial Differential Equations*, pages 59–85. Research Institute for Mathematical Sciences (RIMS), Kyoto, 2012.

[423] Helmut Schaefer. über die Methode der a-priori Schranken. *Math. Ann.*, 129:415–416, 1955.

[424] Julius Schauder. Zur Theorie stetiger Abbildungen in Funktionalräumen. *Math. Z.*, 26:47–65, 1927.

[425] Vladimir Scheffer. Partial regularity of solutions to the Navier–Stokes equations. *Pacific J. Math.*, 66:535–552, 1976.

[426] Vladimir Scheffer. Hausdorff measure and the Navier–Stokes equations. *Comm. Math. Phys.*, 55:97–112, 1977.

[427] Leonid Sedov. *Similarity and Dimensional Methods in Mechanics (10th edition).* CRC Press, 1993.

[428] Gregory Seregin. A certain necessary condition of potential blow up for Navier–Stokes equations. *Comm. Math. Phys.*, 312:833–845, 2012.

[429] Gregory Seregin. *Lecture Notes on Regularity Theory for the Navier-Stokes Equations.* World Scientific Publishing Co. Pte. Ltd., Hackensack, NJ, 2015.

[430] Gregory Seregin. Liouville type theorem for stationary Navier-Stokes equations. *Nonlinearity*, 29:2191–2195, 2016.

[431] James Serrin. The derivation of stress-deformation relations for a Stokesian fluid. *J. Math. Mech.*, 8:459–469, 1959.

[432] James Serrin. Mathematical principles of classical fluid mechanics. In Flugge and Truesdell, editors, *Encyclopedia of Physics Volume VIII/1 FLUID DYNAMICS I*, pages 125–263. Springer, 1959.

[433] James Serrin. A note on the existence of periodic solutions of the Navier–Stokes equations. *Arch. Ration. Mech. Anal.*, 3:120–122, 1959.

[434] James Serrin. On the interior regularity of weak solutions of the Navier–Stokes equations. *Arch. Ration. Mech. Anal.*, 9:187–195, 1962.

[435] James Serrin. The initial value problem for the Navier–Stokes equations. In *Nonlinear Problems (Proc. Sympos., Madison, Wis., 1962)*, pages 69–98. Univ. of Wisconsin Press, Madison, Wis., 1963.

[436] Winfried Sickel, Dachun Yang, and Wen Yuan. *Morrey and Campanato meet Besov, Lizorkin and Triebel.* Lecture Notes in Math. 2005. Springer, 2010.

[437] Jacques Simon. Compact sets in the space $L^p(0, T; B)$. *Ann. Mat. Pura Appl. (4)*, 146:65–96, 1987.

[438] Yakov Sinai. Power series for solutions of the 3d Navier–Stokes system on $\mathbb{R}^3$. *J. Stat. Phys.*, 121:779–803, 2005.

[439] Nikolaï Slezkin. On an integrability case of full differential equations of the motion of a viscous fluid (in Russian). *Uchenye Zapinski MGU*, 2:89–90, 1934.

[440] Joseph Smagorinsky. General circulation experiments with the primitive equations. *Mon. Wea. Rev.*, 91:99–164, 1963.

[441] Joseph Smagorinsky and Syukuro Manabe. Numerical model for study of global general circulation. *Bull. Amer. Meteorolog. Soc.*, 43:673, 1962.

[442] Steve Smale. Mathematical problems for the next century. *Mathematical Intelligencer*, 20:7–15, 1998.

[443] Sergei Sobolev. Méthode nouvelle à résoudre le problème de Cauchy pour les équations linéaires hyperboliques normales. *Mat. Sbornik*, 43:39–72, 1936.

[444] Hermann Sohr and Wolf von Wahl. On the regularity of the pressure of weak solutions of Navier–Stokes equations. *Arch. Math.*, 46:428–439, 1986.

[445] Vsevolod A. Solonnikov. Solvability of the problem of evolution of a viscous incompressible fluid bounded by a free surface on a finite time interval. *St. Petersburg Math. J.*, 3:189–220, 1992.

[446] Vsevolod A. Solonnikov and V.E. Ščadilov. A certain boundary value problem for the stationary system of navier-stokes equations. [russian]. *Trudy Mat. Inst. Steklov*, 125:196–210, 1973.

[447] H.B. Squire. The round laminar jet. *Quart. J. Mech. Appl. Math.*, 4:321–329, 1951.

[448] Elias Stein. *Harmonic Analysis.* Princeton University Press, 1993.

[449] Simon Stevin. *De Beghinselen des Waterwichts.* Leiden, 1586.

[450] George Gabriel Stokes. On the steady motion of incompressible fluids. *Trans. Cambridge Phl. Soc.*, 1842.

[451] George Gabriel Stokes. On the theory of the internal friction of fluids in motion, and of the equilibrium and motion of elastic solids. *Trans. Cambridge Phl. Soc.*, 8:287–319, 1849.

[452] George Gabriel Stokes. On the effect of the internal friction of fluids on the motion of pendulum. *Trans. Cambridge Phl. Soc.*, 1850.

[453] Raymond Streater and Arthur Wightman. *PCT, Spin and Statistics, and All That.* W.A. Benjamin, New York, 1927.

[454] Robert Strichartz. Restriction of Fourier transforms to quadratic surfaces and decay of solutions of wave equations. *Duke Math. J.*, 44:705–714, 1977.

[455] Michael Struwe. On partial regularity results for the Navier–Stokes equations. *Comm. Pure Appl. Math.*, 41:437–458, 1988.

[456] Michael Struwe. On a Serrin-type regularity criterion for the Navier–Stokes equations in terms of the pressure. *J. Math. Fluid Mech.*, 9:235–242, 2007.

[457] Magnus Svärd. A new Eulerian model for viscous and heat conducting compressible flows. *Physica A: Statistical Mechanics and its Applications*, 506:350–375, 2018.

[458] Howard Swann. The convergence with vanishing viscosity of the nonstationary Navier–Stokes flow to ideal flow in $\mathbb{R}_3$. *Trans. Amer. Math. Soc.*, 157:373–397, 1971.

[459] Shuji Takahashi. On interior regularity criteria for weak solutions of the Navier–Stokes equations. *Manuscripta Math.*, 69:237–254, 1990.

[460] Terence Tao. *Structure and Randomness: Pages from Year One of a Mathematical Blog.* Amer. Math. Soc., 2009.

[461] Terence Tao. Localisation and compactness properties of the Navier–Stokes global regularity problem. *Anal. PDE*, 6:25–107, 2013.

[462] Terence Tao. Finite time blowup for an averaged three-dimensional Navier–Stokes equation. *ArXiv e-prints*, arXiv:1402.0290, 2014.

[463] Terence Tao. Quantitative bounds for critically bounded solutions to the Navier–Stokes equations. In A. Kechris, N. Makarov, D. Ramakrishnan, and X. Zhu, editors, *Nine Mathematical Challenges-an Elucidation. Proceedings of the Symposium on Applied Mathematics*, volume 104, pages 149–193. American Mathematical Society, 2021.

[464] Luc Tartar. *An Introduction to Navier–Stokes Equation and Oceanography.* Springer, 2006.

[465] George I. Taylor. Statistical theory of turbulence $I-III$. *Proc. Royal Soc.*, A151:421–464, 1935.

[466] George I. Taylor. Production and dissipation of vorticity in a turbulent fluid. *Proc. Royal Soc.*, A164:15–23, 1938.

[467] Michael Taylor. Analysis on Morrey spaces and applications to Navier–Stokes equations and other evolution equations. *Comm. Partial Differential Equations*, 17:1407–1456, 1992.

[468] Roger Temam. Une méthode d'approximation de la solution des équations de Navier–Stokes. *Bull. Soc. Math. France*, 96:115–152, 1968.

[469] Roger Temam. Sur l'approximation de la solution des équations de Navier–Stokes par la méthode des pas fractionnaires (i). *Arch. Ration. Mech. Anal.*, 32:135–153, 1969.

[470] Roger Temam. Sur l'approximation de la solution des équations de Navier–Stokes par la méthode des pas fractionnaires (ii). *Arch. Ration. Mech. Anal.*, 33:377–385, 1969.

[471] Roger Temam. *Navier–Stokes Equations. Theory and Numerical Analysis.* North Holland, 1977.

[472] William Thomson. On the propagation of laminar motion through a turbulently moving inviscid liquid. *Phil. Mag.*, 24:332–353, 1887.

[473] Gang Tian and Zhouping Xin. One-point singular solutions to the Navier–Stokes equations. *Topol. Methods Nonlinear Anal.*, 11:135–145, 1998.

[474] Evangelista Torricelli. De motu aquarum. In *Opera Geometrica.* Florence, 1644.

[475] Hans Triebel. *Theory of Functions Spaces.* Monographs in Mathematics 78, Birkhäuser Verlag, Basel, 1983.

[476] Hans Triebel. *Local Function Spaces, Heat and Navier–Stokes Equations.* EMS Tracts in Mathematics Vol. 20, EMS Publishing House, 2013.

[477] Viktor Trkal. Poznámka k hydrodynamice vazkých tekutin. *Časopis pro péstován mathematiky a fysiky*, 48:302–311, 1919.

[478] Clifford Truesdell. The mechanical foundations of elasticity and fluid dynamics. *J. Rational Mech. Anal.*, 1:125–300, 1952.

[479] Clifford Truesdell. Notes on the history of the general equations of hydrodynamics. *Amer. Math. Monthly*, 60:445–458, 1953.

[480] Clifford Truesdell. Rational fluid mechanics. In *L. Euler, Opera Omnia, vol. 12*, pages IX–CXXV. Orell Füssli Turici, Lausanne, 1954.

[481] Clifford Truesdell. A program toward rediscovering the rational mechanics of the Age of Reason. *Arch. Hist. Exact Sci.*, 1:3–36, 1960.

[482] Tai-Peng Tsai. On Leray's self–similar solutions of the Navier–Stokes equations satisfying local energy estimates. *Arch. Ration. Mech. Anal.*, 143:29–51, 1998.

[483] Arkady Tsinober. Is concentrated vorticity that important? *Eur. J. Mech., B/Fluids.*, 17:421–449, 1998.

[484] Arkady Tsinober. *An Informal Conceptual Introduction to Turbulence.* Springer, 2009.

[485] Andrey N. Tychonov. Théorèmes d'unicité pour l'équation de la chaleur. *Jour. Mat. Sb.*, 42:199–216, 1935.

[486] M. Uchovskii and Victor Yudovich. Axially symmetric flows of ideal and viscous fluids filling the whole space. *J. Appl. Math. Mech*, 32:52–69, 1968.

[487] Alexis Vasseur. A new proof of partial regularity of solutions to Navier–Stokes equations. *NoDEA Nonlinear Differential Equations Appl.*, 14:753–785, 2007.

[488] Giovanni Battista Venturi. *Recherches expérimentales sur le principe de la communication latérale dans les fluides.* Paris, 1797.

[489] Anatoly Vershik. What is good for mathematics? Thoughts on the Clay Millennium Prizes. *Notices A.M.S.*, 54:45–47, 2007.

[490] Misha Višik. Instability and non-uniqueness in the Cauchy problem for the Euler equations of an ideal incompressible fluid. part I. *ArXiv e-prints*, arXiv:1805.09426, 2018.

[491] Misha Višik. Instability and non-uniqueness in the Cauchy problem for the Euler equations of an ideal incompressible fluid. part II. *ArXiv e-prints*, arXiv:1805.09440, 2018.

[492] Vladimir Šverák. On Landau's solutions of the Navier–Stokes equations. *J. Math. Sci. (N. Y.)*, 179:208–228, 2011.

[493] V.G. Vyskrebtsov. New exact solutions of Navier–Stokes equations for axisymmetric self–similar fluid flows (in Russian). *Mat. Metodi Fiz.-Mekh. Polya*, 41:44–51, 1998.

[494] Wolf von Wahl. *The Equations of Navier–Stokes and Abstract Parabolic Equations*. Vieweg and Sohn, Wiesbaden, 1985.

[495] Baoxiang Wang. Ill-posedness for the Navier–Stokes equations in critical Besov spaces $\dot{B}_{\infty,q}^{-1}$. *Adv. Math.*, 268:350–372, 2015.

[496] Keyan Wang. Global regularity for a model of three-dimensional Navier–Stokes equation. *J. Differential Equations*, 258:2969–2982, 2015.

[497] Emil Warburg. Ueber den Ausfluss des Quecksilbers aus gläsernen Capillarröhren. *Annalen der Physik*, 216:367–379, 1870.

[498] Frederic Weissler. The Navier–Stokes initial value problem in L^p. *Arch. Ration. Mech. Anal.*, 74:219–230, 1981.

[499] Greg Whitlock. Roger J. Boscovich and Friedrich Nietzsche: a re-examination. In B.E. Babich and R.S Cohen, editors, *Nietzsche, Epistemology, and Philosophy of Science: Nietzsche and the Sciences II*, pages 187–202. Springer, 1999.

[500] Lancelot Law Whyte. *Essays on Atomism: From Democritus to 1960*. Harper & Row, 1963.

[501] Michael Wiegner. The Navier–Stokes equations - a neverending challenge? *Jahresber. Deutsch. Math.-Verein.*, 101:1–25, 1999.

[502] Ernest Wilczynski. An application of group theory to hydrodynamics. *Trans. Amer. Math. Soc.*, 1:339–352, 1900.

[503] Andrew Wiles. The Birch and Swinnerton-Dyer conjecture. In J.A. Carlson, A. Jaffe, A. Wiles, Clay Mathematics Institute, and American Mathematical Society, editors, *The Millennium Prize Problems*, pages 31–41. American Mathematical Society, 2006.

[504] Jörg Wolf. A direct proof of the Caffarelli–Kohn–Nirenberg theorem. In *Parabolic and Navier-Stokes Equations. Part 2*, pages 533–552. Polish Acad. Sci. Inst. Math., Warsaw, 2008.

[505] Percy Wong. Global wellposedness for a certain class of large initial data for the 3D Navier–Stokes equations. *Ann. Henri Poincaré*, 15:633–643, 2014.

[506] Jie Xiao. *Holomorphic Q-Classes.* Lecture notes in mathematics. Springer-Verlag, 2001.

[507] Jie Xiao. Homothetic variant of fractional Sobolev space with application to Navier–Stokes system. *Dyn. Partial Differ. Equ.*, 4:227–245, 2007.

[508] Yuelong Xiao and Zhouping Xin. On the vanishing viscosity limit for the $3D$ navier–stokes equations with a slip boundary condition. *Comm. Pure Appl. Math.*, 60:1027–1055, 2007.

[509] Masao Yamazaki. The Navier–Stokes equations in the weak-L^n space with time-dependent external force. *Math. Ann.*, 317:635–675, 2000.

[510] Tsuyoshi Yoneda. Ill-posedness for the 3D Navier–Stokes equations in a generalized Besov space near BMO^{-1}. *J. Funct. Anal.*, 258:3376–3387, 2010.

[511] Yong Zhou. A new regularity criterion for the Navier–Stokes equations in terms of the gradient of one velocity component. *Methods Appl. Anal.*, 9:563–578, 2002.

[512] Yong Zhou. On a regularity criterion in terms of the gradient of pressure for the Navier–Stokes equations in $\mathbb{R}^n$. *Z. Angew. Math. Phys.*, 57:384–392, 2006.

[513] Yong Zhou. On regularity criteria in terms of pressure for the Navier–Stokes equations in $\mathbb{R}^3$. *Proc. Amer. Math. Soc.*, 134:149–156, 2006.

Index

Printed in the United States
by Baker & Taylor Publisher Services